The Biomarker Guide
Second Edition
Volume 2

The second edition of *The Biomarker Guide* is a fully updated and expanded version of this essential reference. Now in two volumes, it provides a comprehensive account of the role that biomarker technology plays both in petroleum exploration and in understanding Earth history and processes.

Biomarkers and Isotopes in Petroleum Exploration and Earth History itemizes parameters used to genetically correlate petroleum and interpret thermal maturity and extent of biodegradation. It documents most known petroleum systems by geologic age throughout Earth history.

The Biomarker Guide is an invaluable resource for geologists, petroleum geochemists, biogeochemists, and environmental scientists.

KENNETH E. PETERS is currently Senior Research Geologist at the US Geological Survey in Menlo Park, California, where he is involved in three-dimensional petroleum system modeling for the North Slope of Alaska, the San Joaquin Basin, and elsewhere. He has 25 years of research experience with Chevron, Mobil, and ExxonMobil. Ken taught formal courses in petroleum geochemistry and thermal modeling both in industry and at various universities. He was Chair of the Organic Geochemistry Division of the Geochemical Society (2001–2004).

CLIFFORD C. WALTERS is Senior Research Associate with the ExxonMobil Research and Engineering Company, where he models oil generation and reservoir transformations, geomicrobiology, and processes of solids formation. He has over 20 years of industrial experience, including research at Sun Exploration and Production Company, and Mobil.

J. MICHAEL MOLDOWAN is Professor (Research) in the Department of Geological and Environmental Sciences at Stanford University. He joined Chevron's Biomarker Group in 1974, which, under the leadership of the late Dr Wolfgang K. Seifert, is largely credited with pioneering the application of biological marker technology to petroleum exploration.

The Biomarker Guide

Second Edition
II. Biomarkers and Isotopes in Petroleum Systems and Earth History

K. E. Peters
US Geological Survey, Menlo Park, CA, USA

C. C. Walters
*ExxonMobil Research and Engineering Co.,
Corporate Strategic Research, Annandale, NJ, USA*

J. M. Moldowan
*Department of Geological and Environmental Sciences,
School of Earth Sciences, Stanford, CA, USA*

CAMBRIDGE UNIVERSITY PRESS
Cambridge, New York, Melbourne, Madrid, Cape Town, Singapore, São Paulo

Cambridge University Press
The Edinburgh Building, Cambridge CB2 8RU, UK

Published in the United States of America by Cambridge University Press, New York

www.cambridge.org
Information on this title: www.cambridge.org/9780521837620

First edition © Chevron Texaco Exploration and Production Company 1993
Second edition © Cambridge University Press 2005

This publication is in copyright. Subject to statutory exception
and to the provisions of relevant collective licensing agreements,
no reproduction of any part may take place without the written
permission of Cambridge University Press.

First published 1993 by Prentice Hall, Inc.
Second edition published 2005 by Cambridge University Press
Reprinted with corrections 2005
This digitally printed version 2007

A catalogue record for this publication is available from the British Library

Library of Congress Cataloguing in Publication data

Peters, Kenneth E.
The biomarker guide. – 2nd ed. / K. E. Peters, C. C. Walters, and J. M. Moldowan.
  p. cm.
Includes bibliographical references and index.
Contents: 1. Biomarkers in the environment and human history – 2. Biomarkers in petroleum systems and Earth history.
ISBN 0 521 78158 2
1. Petroleum – Prospecting. 2. Biogeochemical prospecting. 3. Biochemical markers. I. Walters, C. C. (Clifford C.)
II. Moldowan, J. M. (J. Michael), 1946– III. Title.
TN271.P4P463 2004
622'.1828–dc22   2003065416

ISBN 978-0-521-78158-9 hardback vol. 1
ISBN 978-0-521-78697-3 paperback vol. 1

ISBN 978-0-521-83762-0 hardback vol. 2
ISBN 978-0-521-03998-7 paperback vol. 2

Dedicated to
Vanessa, Brent, and Miwok
Johnet
Mary

Contents

About the authors	page ix
Preface	xi
Purpose	xvi
Acknowledgments	xvii

PART II BIOMARKERS AND ISOTOPES IN PETROLEUM SYSTEMS AND EARTH HISTORY

12	**Geochemical correlation and chemometrics**	**475**
	Oil–source rock correlation	475
	Oil–oil correlation	477
	Chemometric analysis of large data sets	479
13	**Source- and age-related biomarker parameters**	**483**
	Source-related parameters	483
	Prediction of source-rock character from oil composition	487
	Age-related parameters	490
	Alkanes and acyclic isoprenoids	493
	Steranes and diasteranes	524
	Terpanes and similar compounds	538
	Aromatic biomarkers	580
	Porphyrins	602
	Exercise	606
14	**Maturity-related biomarker parameters**	**608**
	Criteria for biomarker maturity parameters	609
	Maturity assessment	610
	Terpanes	613
	Polycadinenes and related products	619
	Steranes	625
	Aromatic steroids	631
	Aromatic hopanoids	634
	Porphyrins	635
	Exercise	636
	Appendix A: gas chromatograms of crude oils from Moravia, Czech Republic	638
	Appendix B: terpane mass chromatograms of crude oils from Moravia, Czech Republic	639
	Appendix C: sterane mass chromatograms of crude oils from Moravia, Czech Republic	640
15	**Non-biomarker maturity parameters**	**641**
	Alkanes and isoprenoids	641
	Aromatic hydrocarbons	642
16	**Biodegradation parameters**	**645**
	Controls on petroleum biodegradation	646
	Rates of aerobic and anaerobic biodegradation	650
	Prediction of the occurrence and extent of biodegradation	650
	Aerobic hydrocarbon degradation pathways	650
	Anaerobic hydrocarbon degradation pathways	653
	Biodegradation of methane	656
	Biodegradation of subsurface crude oil: aerobic or anaerobic?	658
	Effects of biodegradation on petroleum composition	658
	Biodegradation parameters	664
	Biodegradation of coals and kerogen	702
	Biodegradation of oil seeps	704
	Prediction of physical properties	705
	Exercise	705
17	**Tectonic and biotic history of the Earth**	**709**
	Birth of the solar system (~4.7–4.6 Ga)	709

Origins of the oceans, atmosphere, and life: the Hadean Eon (~4.6–3.8 Ga)	709	
Age of the prokaryotes: the Archean Eon (~3.8–2.5 Ga)	711	
The oxygen holocaust: the Proterozoic Eon (~2.5–1.0 Ga)	714	
Rise of the eukaryotes: the Neoproterozoic Era (1000–544 Ma)	715	
Age of the metazoans: the Phanerozoic Eon (544 Ma–present)	719	
Mass extinctions	743	
Biomarkers and mass extinctions	749	

18 Petroleum systems through time — 751

Petroleum system nomenclature	751
Petroleum systems through time	755
Archean petroleum systems	755
Early Proterozoic petroleum systems	764
Middle Proterozoic petroleum systems	764
Late Proterozoic petroleum systems	768
Phanerozoic petroleum systems	774
Paleozoic Era: Cambrian source rocks	774
Ordovician source rocks	777
Silurian source rocks	790
Devonian source rocks	794
Carboniferous source rocks	807
Permian source rocks	812
Mesozoic Era: Triassic source rocks	820
Jurassic source rocks	828
Cretaceous source rocks	834
Tertiary source rocks	918
Exercise	961

19 Problem areas and further work — 964

Migration	964
Biomarker kinetics	966
Oil–oil and oil–source rock correlation	976
Source-rock age from biomarkers	979
Extraterrestrial biomarkers	980
Source input and depositional environment	982

Appendix: geologic time charts	984
Glossary	986
References	1030
Index	1132

About the authors

Kenneth E. Peters is currently Senior Research Geologist at the US Geological Survey in Menlo Park, California, where he is involved in one-dimensional (1D), two-dimensional (2D), and three-dimensional (3D) petroleum system modeling of the North Slope of Alaska, the San Joaquin Basin, and elsewhere. He attained B.A. and M.A. degrees in geology from University of California, Santa Barbara, and a Ph.D. in geochemistry from University of California, Los Angeles, (UCLA) in 1978. His experience includes 15 years with Chevron, 6 years with Mobil, and 2 years as Senior Research Associate with ExxonMobil. Ken taught formal courses in petroleum geochemistry and thermal modeling for Chevron, Mobil, ExxonMobil, Oil and Gas Consultants International, and at various universities, including University of California, Berkeley, and Stanford University. He served as Associate Editor for *Organic Geochemistry* and the *American Association of Petroleum Geologists Bulletin*. Ken and co-authors received the Organic Geochemistry Division of the Geochemical Society Best Paper Awards for publications in 1981 and 1989. He served as Chair of the Gordon Research Conference on Organic Geochemistry (1998) and Chair of the Organic Geochemistry Division of the Geochemical Society (2001–2004).

Clifford C. Walters received Bachelor degrees in chemistry and biology from Boston University in 1976. He attended the University of Maryland, where he worked on the chemistry of Martian soil and conducted field and laboratory research on metasediments from Isua, Greenland, the oldest sedimentary rocks on Earth. After receiving a Ph.D. in geochemistry in 1981, Cliff continued with postdoctoral research on the organic geochemistry of Precambrian sediments and meteorites. He joined Gulf Research and Development in 1982, where he implemented a program in biological marker compounds. In 1984, he moved to Sun Exploration & Production Company, where he was responsible for technical service and establishing biomarker geochemistry and thermal modeling as routine exploration tools. Mobil's Dallas Research Lab hired Cliff in 1988, where he became Supervisor of the Geochemical Laboratories in 1991. He is now Senior Research Associate with ExxonMobil Research and Engineering Company, where he conducts work on the modeling of oil generation and reservoir transformations, geomicrobiology, and processes of solids formation. Cliff published numerous papers, served as Editor of the ACS Geochemistry Division from 1990 to 1992, and is a current Associate Editor of *Organic Geochemistry*.

J. Michael Moldowan attained a B.S. in chemistry from Wayne State University, Detroit, Michigan, and a Ph.D. in chemistry from the University of Michigan in 1972. Following a postdoctoral fellowship in marine natural products with Professor Carl Djerassi at Stanford University, he joined Chevron's Biomarker Group in 1974. The Chevron biomarker team, led by the late Dr. Wolfgang K. Seifert in the mid 1970s to early 1980s, is largely credited with pioneering the application of biological marker technology to petroleum exploration. Mike joined the Department of Geological and Environmental Sciences of Stanford University as Professor (Research) in 1993. In 1986, he served as Chair of the Division of Geochemistry of the American Chemical Society, and he has twice been awarded the Organic Geochemistry Division of the Geochemical Society Best Paper Award for publications he co-authored in 1978 and 1989.

Preface

Biological markers (biomarkers) are complex molecular fossils derived from biochemicals, particularly lipids, in once-living organisms. Because biological markers can be measured in both crude oils and extracts of petroleum source rocks, they provide a method to relate the two (correlation) and can be used by geologists to interpret the characteristics of petroleum source rocks when only oil samples are available. Biomarkers are also useful because they can provide information on the organic matter in the source rock (source), environmental conditions during its deposition and burial (diagenesis), the thermal maturity experienced by rock or oil (catagenesis), the degree of biodegradation, some aspects of source rock mineralogy (lithology), and age. Because of their general resistance to weathering, biodegradation, evaporation, and other processes, biomarkers are commonly retained as indicators of petroleum contamination in the environment. They also occur with certain human artifacts, such as bitumen sealant for ancient boats, hafting material on spears and arrows, burial preservatives, and as coatings for medieval paintings.

Biomarker and non-biomarker geochemical parameters are best used together to provide the most reliable geologic interpretations to help solve exploration, development, production, and environmental or archeological problems. Prior to biomarker work, oil and rock samples are typically screened using non-biomarker analyses. The strength of biomarker parameters is that they provide more detailed information needed to answer questions about the source-rock depositional environment, thermal maturity, and the biodegradation of oils than non-biomarker analyses alone.

Distributions of biomarkers can be used to correlate oils and extracts. For example, C_{27}-C_{28}-C_{29} steranes or monoaromatic steroids distinguish oil-source families with high precision. Cutting-edge analytical techniques, such as linked-scan gas chromatography/mass spectrometry/mass spectrometry (GCMS/MS) provide sensitive measurements for correlation of light oils and condensates, where biomarkers are typically in low concentrations. Because biomarkers typically contain more than ~20 carbon atoms, they are useful for interpreting the origin of the liquid fraction of crude oil, but they do not necessarily indicate the origin of associated gases or condensates.

Different depositional environments are characterized by different assemblages of organisms and biomarkers. Commonly recognized classes of organisms include bacteria, algae, and higher plants. For example, some rocks and related oils contain botryococcane, a biomarker produced by the lacustrine, colonial alga *Botryococcus braunii*. *Botryococcus* is an organism that thrives only in lacustrine environments. Marine, terrigenous, deltaic, and hypersaline environments also show characteristic differences in biomarker composition.

The distribution, quantity, and quality of organic matter (organic facies) are factors that help to determine the hydrocarbon potential of a petroleum source rock. Optimal preservation of organic matter during and after sedimentation occurs in oxygen-depleted (anoxic) depositional environments, which commonly lead to organic-rich, oil-prone petroleum source rocks. Various biomarker parameters, such as the C_{35} homohopane index, can indicate the degree of oxicity under which marine sediments were deposited.

Biomarker parameters are an effective means to rank the relative maturity of petroleum throughout the entire oil-generative window. The rank of petroleum can be correlated with regions within the oil window (e.g. early, peak, or late generation). This information can provide a clue to the quantity and quality of the oil that may have been generated and, coupled with quantitative petroleum conversion measurements (e.g. thermal modeling programs), can help evaluate the timing of petroleum expulsion.

Biomarkers can be used to determine source and maturity, even for biodegraded oils. Ranking systems are based on the relative loss of n-alkanes, acyclic isoprenoids, steranes, terpanes, and aromatic steroids during biodegradation.

Biomarkers in oils provide information on the lithology of the source rock. For example, the absence of rearranged steranes can be used to indicate petroleum derived from clay-poor (usually carbonate) source rocks. Abundant gammacerane in some petroleum appears to be linked to a stratified water column (e.g. salinity stratification) during deposition of the source rock.

Biomarkers provide information on the age of the source rock for petroleum. Oleanane is a biomarker characteristic of angiosperms (flowering plants) found only in Tertiary and Upper Cretaceous rocks and oils. C_{26} norcholestanes originate from diatoms and can be used to distinguish Tertiary from Cretaceous and Cretaceous from older oils. Dinosterane is a marker for marine dinoflagellates, possibly distinguishing Mesozoic and Tertiary from Paleozoic source input. Unusual distributions of n-alkanes and cyclohexylalkanes are characteristic of *Gloeocapsomorpha prisca* found in early Paleozoic samples. 24-n-Propylcholestane is a marker for marine algae extending from at least the Devonian to the present.

Continued growth in the geologic, environmental, and archeological applications of biomarker technology is anticipated, particularly in the areas of age-specific biomarkers, the use of biomarkers to indicate source organic matter input and sedimentologic conditions, correlation of oils and rocks, and understanding the global cycle of carbon. New developments in analytical methods and instrumentation and the use of biomarkers to understand petroleum migration and kinetics are likely. Finally, early work suggests that biomarkers will continue to grow as tools to understand production, environmental, and archeological problems.

HOW TO USE THIS GUIDE

The Biomarker Guide is divided into two volumes. The first volume introduces some basic chemical principles and analytical techniques, concentrating on the study of biomarkers and isotopes in the environment and human history. The second volume expands on the uses of biomarkers and isotopes in the petroluem industry, and investigates their occurrence throughout Earth history.

The Biomarker Guide was written for a diverse audience, which might include the following:

- students of geology, environmental science, and archeology who wish to gain general knowledge of what biomarkers can do;
- practicing geologists and geochemical coordinators in the petroleum industry with both specific and general questions about which biomarker and/or non-biomarker parameters might best answer regional exploration, development, or production problems;
- experienced geochemists who require detailed information on specific parameters or methodology;
- managers or research directors who require a concise explanation for terms and methodology;
- refinery process chemists requiring a more detailed knowledge of petroleum; and
- archeologists and environmental scientists interested in a technology useful for characterizing petroleum in the environment.

The text in each chapter is supplemented by many references to related sections in the book and to the literature. Various parts of the guide, such as notes, highlight detailed discussions that supplement the text.

The following is a brief overview of each chapter in the two volumes.

PART I BIOMARKERS AND ISOTOPES IN THE ENVIRONMENT AND HUMAN HISTORY

1 Origin and preservation of organic matter

This chapter introduces biomarkers, the domains of life, primary productivity, and the carbon cycle on Earth. Morphological and biochemical differences among different life forms help to determine their environmental habitats and the character of the biomarkers that they contribute to sediments, source rocks, and petroleum. The discussion summarizes processes affecting the distribution, preservation, and alteration of biomarkers in sedimentary rocks. Various factors, such as type of organic matter input, redox potential during sedimentation, bioturbation, sediment grain size, and sedimentation rate, influence the quantity and quality of organic matter preserved in rocks during Earth's history.

2 Organic chemistry

A brief overview of organic chemistry includes explanations of structural nomenclature and stereochemistry necessary to understand biomarker parameters. The discussion includes an overview of compound classes in petroleum and concludes with examples of the structures and nomenclature for several biomarkers, their precursors in living organisms, and their geologic alteration products.

3 Biochemistry of biomarkers

This chapter provides an overview of the biochemical origins of the major biomarkers, including discussions of the function, biosynthesis, and occurrence of their precursors in living organisms. Some topics include lipid membranes and their chemical compositions, the biosynthesis of isoprenoids and cyclization of squalene, and examples of hopanoids, sterols, and porphyrins in the biosphere and geosphere.

4 Geochemical screening

This chapter describes how to select sediment, rock, and crude oil samples for advanced geochemical analyses by using rapid, inexpensive geochemical tools, such as Rock–Eval pyrolysis, total organic carbon, vitrinite reflectance, scanning fluorescence, gas chromatography, and stable isotope analyses. The discussion covers sample quality, selection, storage, and geochemical rock and oil standards. Other topics include how to test rock samples for indigenous bitumen, surface geochemical exploration using piston cores, geochemical logs and their interpretation, chromatographic fingerprinting for reservoir continuity, and how to deconvolute mixtures of oils from different production zones. Mass balance equations show how to calculate the extent of fractional conversion of kerogen to petroleum, source-rock expulsion efficiency, and the original richness of highly mature source rocks.

5 Refinery oil assays

Many refinery oil assays differ substantially from geochemical analyses conducted by petroleum or environmental geochemists, although interdisciplinary use of these tools is becoming more common. Some basic oil assays include API (American Petroleum Institute) gravity, pour point, cloud point, viscosity, trace metals, total acid number, refractive index, and wax content. More advanced oil assays include chemical group-type fractionation and field ionization mass spectrometry. A brief overview of refinery processes includes the fate of biomarkers in straight-run and processed refinery products with tips on how to distinguish refined from natural petroleum products in environmental or geological samples.

6 Stable isotope ratios

This chapter describes stable isotopes and their use to characterize petroleum, including gases, crude oils, sediment and source-rock extracts, and kerogen, with emphasis on stable carbon isotope ratios. The discussion includes isotopic standards and notation, principles of isotopic fractionation, and the use of various isotopic tools, such as stable carbon isotope-type curves, for correlation or quantification of petroleum mixtures. The chapter concludes with new developments in compound-specific isotope analysis, including its application to better understand the origin of carboxylic acids and the process of thermochemical sulfate reduction in petroleum reservoirs.

7 Ancillary geochemical methods

Ancillary geochemical tools (e.g. diamondoids, C_7 hydrocarbons, compound-specific isotopes, and fluid inclusions) can be used to evaluate the origin, thermal maturity, and extent of biodegradation or mixing of petroleum, even when the geological samples lack or have few biomarkers. Molecular modeling can be used to rationalize or predict the geochemical behavior of biomarkers and other compounds in the geosphere.

8 Biomarker separation and analysis

This chapter describes the organization of a biomarker laboratory and the methods used to prepare and separate crude oils and sediment or source-rock extracts into fractions prior to mass spectrometric analysis. The concept of mass spectrometry is explained. Many of these fundamentals, such as the difference between a mass chromatogram and a mass spectrum, or between selected ion and linked-scan modes of analysis, are critical to understanding later discussions of biomarker parameters. Several key topics, including analytical procedures, internal standards, and examples of gas chromatography/mass spectrometry (GCMS) data problems, help the reader to evaluate the quality of biomarker data and interpretations.

9 Origin of petroleum

This chapter describes evidence against the deep-earth gas hypothesis, which invokes an abiogenic origin for petroleum by polymerization of methane deep in the Earth's mantle. The deep-earth gas hypothesis has little scientific support but, if correct, could have major implications for petroleum exploration and the application of biomarkers to environmental science and archeology. The discussion covers experimental, geological, and geochemical evidence supporting the thermogenic origin of petroleum.

10 Biomarkers in the environment

This chapter explains how analyses of biomarkers and other environmental markers, such as polycyclic aromatic hydrocarbons, are used to characterize, identify, and assess the environmental impact of oil spills. The discussion covers processes affecting the composition of spilled oil, such as emulsification, oxidation, and biodegradation, as well as oil-spill mitigation and modeling. Field and laboratory procedures for sampling and analyzing spills are discussed, including program design, chemical fingerprinting, and data quality control. The chapter includes sections on smoke, natural gas, and gasoline and other light fuels as pollutants, and a detailed discussion of the controversial *Exxon Valdez* oil spill.

11 Biomarkers in archeology

This chapter provides examples of the growing use of biomarker and isotopic analyses to evaluate organic materials in archeology. Some of the topics include bitumens in Egyptian mummies, such as Cleopatra, archeological gums and resins, and biomarkers in art and ancient shipwrecks. The discussion covers the use of biomarkers and isotopes in studies of paleodiet and agricultural practices, including studies of ancient wine and beeswax. Other topics include archeological DNA, proteins, and evidence for ancient narcotics.

PART II BIOMARKERS AND ISOTOPES IN PETROLEUM AND EARTH HISTORY

12 Geochemical correlations and chemometrics

Geochemical correlation can be used to establish petroleum systems to improve exploration success, define reservoir compartments to enhance production, or identify the origin of petroleum contaminating the environment. This chapter explains how chemometrics simplifies genetic oil-oil and oil-source rock correlations and other interpretations of complex multivariate data sets.

13 Source- and age-related biomarker parameters

This chapter explains how biomarker analyses are used for oil-oil and oil-source rock correlation and how they help to identify characteristics of the source rock (e.g. lithology, geologic age, type or organic matter, redox conditions), even when samples of rock are not available. Biomarker parameters are arranged by groups of related compounds in the order: (1) alkanes and acyclic isoprenoids, (2) steranes and diasteranes, (3) terpanes and similar compounds, (4) aromatic steroids, hopanoids, and similar compounds, and (5) porphyrins. Critical information on specificity and the means for measurement are highlighted above the discussion for each parameter.

14 Maturity-related biomarker parameters

This chapter explains how biomarker analyses are used to assess thermal maturity. The parameters are arranged by groups of related compounds in the order (1) terpanes, (2) polycadinenes and related products, (3) steranes, (4) aromatic steroids, (5) aromatic hopanoids, and (6) porphyrins. Critical information on specificity and the means for measurement are highlighted in bold print above the discussion for each parameter.

15 Non-biomarker maturity parameters

This chapter explains how certain non-biomarker parameters, such as ratios involving n-alkanes and aromatic hydrocarbons, are used to assess thermal maturity. Critical information on specificity and the means for measurement are highlighted above the discussion for each parameter.

16 Biodegradation parameters

This chapter explains how biomarker and non-biomarker analyses are used to monitor the extent of biodegradation. Compound classes and parameters are discussed in the approximate order of increasing resistance to biodegradation. The discussion covers recent advances in our understanding of the controls and mechanisms of petroleum biodegradation and the relative significance of aerobic versus anaerobic degradation in both surface and subsurface environments. Examples show how to predict the original physical properties of crude oils prior to biodegradation.

17 Tectonic and biotic history of the Earth

The evolution of life is closely tied to biomarkers in petroleum. This chapter provides a brief tectonic history of the Earth in relation to the evolution of major life forms. Mass extinctions and their possible causes are discussed. The end of the section for each time period includes examples of source rocks and related crude oils with emphasis on the geochemistry of the oils. These examples are linked to more detailed discussion of petroleum systems in Chapter 18.

18 Petroleum systems through time

This chapter defines petroleum systems and provides examples of the geology, stratigraphy, and geochemistry of source rocks and crude oils through geologic time. Gas chromatograms, sterane and terpane mass chromatograms, stable isotope compositions, and other geochemical data are provided for representative crude oils generated from many worldwide source rocks.

19 Problem areas and further work

This chapter describes areas requiring further research, including the application of biomarkers to migration, the kinetics of petroleum generation, geochemical correlation and age assessment, and the search for extraterrestrial life.

Purpose

The Biomarker Guide provides a comprehensive discussion of the basic principles of biomarkers, their relationships with other parameters, and their applications to studies of maturation, correlation, source input, depositional environment, and biodegradation of the organic matter in petroleum source rocks, reservoirs, and the environment. It builds upon previous books by Tissot and Welte (1984), Waples and Machihara (1991), Bordenave (1993), Peters and Moldowan (1993), Hunt (1996), and Welte *et al.* (1997). The volumes were prepared for a broad audience, including students, company exploration geologists, geochemists, and environmental scientists for several reasons:

(1) Biomarker geochemistry is a rapidly growing discipline with important worldwide applications to petroleum exploration and production and environmental monitoring.
(2) Biomarker parameters are becoming increasingly prominent in exploration, production, and environmental reports.
(3) Different parameters are used within the industry, academia, service laboratories, and the literature.
(4) The quality of biomarker data and interpretation can vary considerably, depending on their source.

The objective of this guide is to provide a single, concise source of information on the various biomarker parameters and to create general guidelines for the use of selected parameters. An important aim is to clarify the relationships between biomarker and other geochemical parameters and to show how they can be used together to solve problems. It is not intended to teach interpretation of raw biomarker data. This is a job for a biomarker specialist with years of training in instrumentation and organic chemistry. A crash-course or cookbook approach cannot provide such training without the consequence of serious interpretive errors and a tarnished view of the applicability of biomarkers in general.

A final objective of the guide is to impart in each reader a feeling for the excitement and vigor of the new field of biomarker geochemistry. Expanding research efforts at geochemical laboratories worldwide have increased the rate of change and growth in our geochemical concepts. Applications of many of the biomarker parameters presented here will undoubtedly improve with time and further research. We anticipate that more than a few readers will be directly involved in making these improvements possible.

Acknowledgments

The authors gratefully acknowledge the support of Chevron management and technical personnel (now ChevronTexaco) during preparation of the precursor to this book, which was called *The Biomarker Guide*. In particular, the authors thank G. J. Demaison, C. Y. Lee, F. Fago, R. M. K. Carlson, P. Sundararaman, J. E. Dahl, M. Schoell, E. J. Gallegos, P. C. Henshaw, S. R. Jacobson, R. J. Hwang, D. K. Baskin, and M. A. McCaffrey for discussions, technical assistance, and helpful review comments.

We acknowledge the support of Mobil and ExxonMobil management and technical personnel during preparation of much of *The Biomarker Guide* Second Edition. In particular, the authors thank Ted Bence, Paul Mankiewicz, John Guthrie, Jim Gormly, and Roger Prince for their input. We also thank Bill Clendenen, Larry Baker, Gary Isaksen, Jim Stinnett (Mobil, retired), Al Young (Exxon, retired), and Steve Koch.

We acknowledge the support of management at the US Geological Survey during preparation of the book. Special thanks are due to Les Magoon, Bob Eganhouse, Mike Lewan, Keith Kvenvolden, Fran Hostettler, Tom Lorenson, and Ron Hill.

Special thanks are due to Steve Brown and John Zumberge of GeoMark Research, Inc. for allowing us to use various oil Information sheets from their Oil Information Library System (OILS) and several cross-plots of biomarker ratios. We also thank David Zinniker of Stanford University for input on terrigenous biomarkers.

Finally, we thank the many reviewers of various drafts of the Second Edition, who are listed in the following table. Their dedication to the sometimes thankless job of peer review is to be commended.

Chapter	Title	Reviewer	Affiliation
1	Origin and preservation of organic matter	Kirsten Laarkamp	ExxonMobil Upstream Research
		Phil Meyers	University of Michigan
2	Organic chemistry	Kirsten Laarkamp	ExxonMobil Upstream Research
3	Biochemistry of biomarkers	Robert Carlson	ChevronTexaco
4	Geochemical screening	Dave Baskin	OilTracers, L. L. C.
		George Claypool	Mobil (retired)
		Jim Gormly	ExxonMobil Upstream Research
		Tom Lorenson	US Geological Survey
5	Refinery oil assays	Owen BeMent	Shell Oil Company
		Paul Mankiewicz	ExxonMobil Upstream Research
		Robert McNeil	Shell Oil Company
6	Stable isotope ratios	Mike Engel	University of Oklahoma
		Martin Schoell	ChevronTexaco (retired)
		Zhengzheng Chen	Stanford University
		John Guthrie	ExxonMobil Upstream Research
		Jeffrey Sewald	Woods Hole Oceanographic Institution

(cont.)

Chapter	Title	Reviewer	Affiliation
7	Ancillary geochemical methods	Ron Hill	US Geological Survey
		Dan Jarvie	Humble Geochemical Services, Inc.
		Yitian Xiao	ExxonMobil Upstream Research
8	Biomarker separation and analysis	John Guthrie	ExxonMobil Upstream Research
		Robert Carlson	ChevronTexaco
9	Origin of petroleum	Kevin Bohacs	ExxonMobil Upstream Research
		Barbara Sherwood Lollar	University of Toronto
10	Biomarkers in the environment	Ted Bence, Rochelle Jozwiak, Mike Smith, Bill Burns (retired)	ExxonMobil Upstream Research
		Roger Prince	ExxonMobil Strategic Research
		Bob Eganhouse, Keith Kvenvolden, Fran Hostettler	US Geological Survey
		Ian Kaplan	UCLA (retired)
11	Biomarkers in archeology	Roger Prince	ExxonMobil Strategic Research
		Max Vityk	ExxonMobil Upstream Research
12	Geochemical correlations and chemometrics	Jaap Sinninghe Damsté	Netherlands Institute for Sea Research
		Paul Mankiewicz	ExxonMobil Upstream Research
		Scott Ramos, Brian Rohrback	Infometrix, Inc.
13	Source- and age-related biomarker parameters	Jaap Sinninghe Damsté	Netherlands Institute for Sea Research
		Leroy Ellis	Terra Nova Technologies
		Kliti Grice	University of Western Australia
		Paul Mankiewicz	ExxonMobil Upstream Research
		Roger Summons	Massachusetts Institute of Technology
		David Zinniker	Stanford University
14	Maturity-related biomarker parameters	Gary Isaksen	ExxonMobil Upstream Research
		Ron Noble	BHP Billiton
15	Non-biomarker maturity parameters	Gary Isaksen	ExxonMobil Upstream Research
		Ron Noble	BHP Billiton
16	Biodegradation parameters	Dave Converse	ExxonMobil Upstream Research
		Roger Prince	ExxonMobil Strategic Research
17	Tectonic and biotic history of the Earth	Kevin Bohacs	ExxonMobil Upstream Research
		Keith Kvenvolden	US Geological Survey
18	Petroleum systems through time	Steve Creaney	ExxonMobil Exploration Company
		Les Magoon	US Geological Survey
19	Problem areas and further work	John Guthrie	ExxonMobil Upstream Research
		Mike Lewan	U.S. Geological Survey
—	References	Jan Heagy, Marsha Harris	ExxonMobil Upstream Research
		Susie Bravos, Page Mosier, Emily Shen-Torbik	US Geological Survey

Part II
Biomarkers and isotopes in petroleum exploration and Earth history

12 · Geochemical correlation and chemometrics

> Geochemical correlation can be used to establish petroleum systems to improve exploration success, define reservoir compartments to enhance production, and identify the origin of petroleum contaminating the environment. This chapter explains how chemometrics simplifies genetic oil–oil and oil–source rock correlations and other interpretations of complex multivariate data sets.

The origin of spilled crude oil or refined product is commonly difficult to determine because they can be displaced from their point of origin by groundwater or ocean currents and many potential sources may be nearby, e.g. natural seeps, pipelines, tankers, and production platforms. Likewise, exploration is complicated by the fact that oil tends to migrate from fine-grained, organic-rich source rock to coarser-grained reservoir rocks. Because both short- and long-distance migration of petroleum can occur and several potential source rocks are generally available in a given basin, the source for many oils remains problematic.

> Note: Oil from the Monterey Formation in California is commonly found within or stratigraphically near the source rock. Oil in the Devonian pinnacle reefs of western Canada appears to have migrated from nearby sources, while oil at Athabasca probably migrated up to 100 km or more from the source rock.

Areas of concentrated petroleum occurrence can be predicted using mapping methods that integrate geology, geophysics, and geochemistry (Demaison, 1984). Biomarkers are a powerful component in this petroleum exploration approach. Combined with other exploration tools, biomarker geochemistry significantly reduces the risk associated with exploration for petroleum. The single most significant contribution of geochemistry to exploration efficiency is that it can be used to show genetic relationships among crude oils and source rocks. Geochemistry also increases exploration efficiency by accounting for many of the variables that control the volumes of petroleum available for entrapment (charge), including source-rock quality and richness, thermal maturity, and the timing of generation–migration–accumulation relative to trap formation (e.g. Murris, 1984; Hunt, 1996, pp. 604–614). Geochemistry is most powerful when used with other disciplines, such as seismic sequence stratigraphy and reservoir characterization (e.g. Isaksen and Bohacs, 1995; Kaufman *et al.*, 1990). Including geochemistry in prospect appraisal improves exploration efficiency (Figure 12.1). Figure 12.1 shows that forecasting efficiency based only on structural and reservoir data (e.g. geophysics) approximately doubles when geochemical charge and retention parameters are included in prospect evaluation. Costly exploration failures, such as the Mukluk OCS Y-0334 No. 1 well in Alaska (~$140 million in 1983 dollars) (Weimer, 1987), are painful reminders that large structures indicated by seismic data may lack oil and gas due to geochemical charge factors.

OIL–SOURCE ROCK CORRELATION

Correlations are geochemical comparisons among crude oils, refined products, and/or extracts from prospective source rocks to determine whether a genetic relationship exists (Peters and Moldowan, 1993; Waples and Curiale, 1999). Oil–source rock correlations are based on the concept that certain compositional parameters of migrated oil do not differ significantly from those of bitumen remaining in the source rock. This similarity through heritage can range from bulk properties, such as stable carbon isotope composition, to individual compound ratios, such as pristane/phytane. Detailed oil–source rock correlations provide important information on the origin and possible paths of

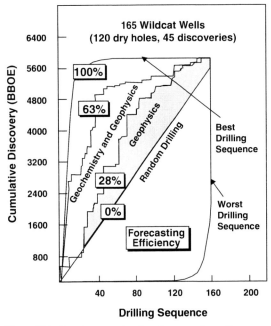

Figure 12.1. Petroleum geochemistry improves forecasting efficiency (FE) (modified from Murris, 1984). The figure shows results for 165 petroleum prospects evaluated before drilling. Vertical axis shows cumulative discovered volumes in place (billions of barrels); horizontal axis shows the well sequence number. The upper curve is the theoretical best possible ranking of prospects (100% FE) compared with actual outcomes. Random drilling would lead to a discovery sequence along the diagonal line (0% FE). Ranking based on trap size from structural and reservoir data gave the lower stepped curve (28% FE). Actual ranking based on complete prospect evaluation, including geochemistry, gave the upper stepped curve (63% FE). Reprinted by permission of the AAPG, whose permission is required for further use.

migration of oils that can lead to additional exploration plays.

Geochemical correlation commonly relies heavily on gas chromatography/mass spectrometry (GCMS) of biomarkers (see also Chapter 13) (Seifert and Moldowan, 1978; Seifert and Moldowan, 1981; Mackenzie, 1984). Biomarker correlations of crude oils with each other, and with their source rocks, improve understanding of reservoir relationships, petroleum migration pathways, and possible new exploration plays. Biomarkers can be used to identify sources of petroliferous contamination in the environment and the progress of remediation (Eganhouse, 1997). They can be used to evaluate thermal maturity (see Chapter 14) and/or biodegradation (see Chapter 16), thus, providing information needed to evaluate the distribution and producibility of petroleum in basins. Biomarkers provide information on regional variations in the character of oils and source rocks as controlled by organic matter input and characteristics of the depositional environment. Biomarkers in seep or oil samples can be used to indirectly predict source-rock quality, even when source-rock samples are not available for direct comparison (e.g. Dahl et al., 1994). Earlier works that review biomarkers, their origins, and applications to correlation include those by Connan (1981a), Philp (1982), Mackenzie (1984), Petrov (1987), Tissot and Welte (1984), Johns (1986), and Waples and Machihara (1991).

Ratios of adjacent homologs or compounds with similar structures, such as the source-dependent biomarker ratios (see Chapter 13), do not change from bitumen in the source rock to the migrated oil. For example, the ratio $C_{27}/(C_{27}–C_{29})$ steranes used in C_{27}–C_{28}–C_{29} ternary diagrams does not differ significantly between extracts from source rocks and genetically related oils throughout the oil-generative window (e.g. see Figure 13.38).

Correlations between samples generally become more reliable when more parameters are compared. In this multiparameter approach, independent measurements of biomarkers, stable carbon isotopes, and other genetic parameters support the inferred correlations. This approach was initiated early to answer correlation and other geochemical problems (Seifert and Moldowan, 1978). Several key geochemical advances that facilitate the successful correlation of oils and source rocks include innovations in GCMS, metastable reaction monitoring (MRM)/GCMS, GCMS/MS (e.g. Gallegos, 1976; Warburton and Zumberge, 1982) and compound-specific isotope analysis (CSIA) (e.g. Hayes et al., 1990). Future geochemical correlations are likely to be improved by better integration of geochemistry with source-rock sequence stratigraphy (e.g. Isaksen and Bohacs, 1995) and increased research on age-related biomarkers and isotopes (e.g. Chung et al., 1992; Moldowan et al., 1994a; Holba et al., 1998; Holba et al., 2000; Holba et al., 2001; Andrusevich et al., 2000). As discussed later, certain geochemical parameters based on ratios of homologous biomarkers are excellent source parameters (see Chapter 13), while some of the more useful maturity parameters are based on ratios of stereoisomers (see Chapter 14).

A positive correlation is not necessarily proof that samples are related. For example, different source rocks can show similar geochemical characteristics. Bitumen extracted from a potential source may be similar to an oil sample until further analyses show that other bitumens from different strata are even more similar to the oil. On the other hand, a negative correlation is strong evidence for lack of a relationship between samples.

The reader should be aware that compositional relationships between source-rock extracts and related, migrated oils *might* be obscured for several reasons. These potential problems are listed below in decreasing order of importance. Items (5) and (6) are rarely problems in correlation studies based on available data:

1. Bitumen extracted from proposed source rocks might contain undetected migrated oils or contaminants that are not representative of the indigenous hydrocarbons. Tests for indigenous bitumen are given in Chapter 5.
2. Most oil–source rock correlations require equivalent or at least similar levels of thermal maturity for the bitumen and oil samples to be compared. In many cases, however, the available source rock candidates are either more or less mature than the oil sample. For potential source rocks that have not been buried deeply, hydrous or closed-tube pyrolysis can be used to increase maturity artificially (see Figure 19.16).
3. Most oil–source rock studies are limited to a few selected source rocks that may or may not be representative of the composite section that generated the petroleum. Petroleum is a composite of migrated fluid and gases generated from thick source intevals.
4. Petroleum accumulations commonly consist of contributions from more than one source rock. For example, at least three source rocks on the North Slope of Alaska contribute to oil accumulations in Prudhoe Bay and vicinity (Seifert *et al.*, 1980; Masterson, 2001) and two source rocks contribute to the Beatrice Field in the Inner Moray Firth (Peters *et al.*, 1989).
5. Expulsion/migration affects the distribution of compounds showing radically different molecular weight, polarity, or adsorptivity. For example, migrated oils are typically enriched in saturated and aromatic hydrocarbons and depleted in NSO (nitrogen, sulfur, oxygen) compounds and asphaltenes compared with related bitumens (Hunt, 1996; Tissot and Welte, 1984; Peters *et al.*, 1990). Biomarkers showing only minor differences in polarity or stereochemistry can show changes in relative amounts due to migration through clay, as demonstrated in laboratory studies (Carlson and Chamberlain, 1986). These effects may reflect expulsion from the fine-grained source rock rather than migration through coarse-grained carrier beds to the reservoir.
6. The composition of petroleum in reservoirs may not coincide with that being generated at a given time from the source rock because migration continues throughout petroleum generation and different compounds form at different times during this process. For example, Mackenzie *et al.* (1985a) suggested that oil accumulations are mixtures of organic fluids representing a range of maturity. According to these authors, catagenesis results in decreases in concentrations of biomarkers that could exaggerate the contributions of less mature sources to a given reservoir.

Despite the potential problems, when adequate source-rock samples can be obtained successful oil–source rock correlations are the rule rather than the exception. Nonetheless, the origins of some important accumulations remain unclear, such as the 1.8 trillion barrels of heavy oil in Mannville-Grosmont reservoirs in the Western Canada Basin. This oil could be a mixture that originated from Jurassic (Nordegg or younger) and Mississippian (Exshaw/Bakken) source rocks, or it could represent a single family that originated from a unique source rock (Creaney *et al.*, 1994a).

OIL–OIL CORRELATION

Reliable biomarker interpretations are usually based on multiple biomarker parameters. Individual biomarker parameters become meaningful only when they agree with:

- other biomarker parameters;
- supporting geochemical parameters;
- reasonable geologic scenarios.

Only rarely does the initial biomarker assessment in a large study answer all correlation questions. In many cases, additional questions are raised by the data, requiring more samples and/or analyses. Anomalous results may be encountered in the initial examination of biomarker data from a large sample suite. Re-examination of the data may reveal interfering peaks in the GCMS analysis, gas chromatography resolution problems resulting from lack of optimization of instrumental operating parameters, poor signal-to-noise ratios due to low biomarker concentrations, instrument drift, or other problems. The potential effects of analytical problems on correlation are discussed in Chapter 8, where several examples are given in the text. Recognition of these problems generally requires a trained geochemist, mass spectroscopist, or technician with extensive experience.

Oil–oil correlation requires parameters that (1) distinguish oils from different sources and (2) are resistant to secondary processes, such as biodegradation and thermal maturation. In many cases, oil–oil correlations can be accomplished using only a few simple bulk parameters, such as gas chromatographic fingerprints, carbon or sulfur stable isotope ratios, or the relative contents of vanadium and nickel (V/Ni). For example, Clayton et al. (1987) distinguished three types of Paleozoic oils in the Northern Denver Basin based mainly on pristane/phytane and stable carbon isotope ratios. Additional data from gas chromatograms and biomarker analyses supported the classification. In cases where the principal correlation questions are answered using rapid and inexpensive analyses, the interpreter may choose to omit additional analyses, thereby saving time and money. However, complete biomarker analyses invariably add additional information to geochemical studies of crude oils and source rocks.

A critical problem in geochemistry is to distinguish the effects of source, including organic matter input and depositional environment, from those of thermal maturity on petroleum composition. As might be expected, geochemical parameters show a range of sensitivities to source and maturity effects. Variations in some parameters are clearly dominated by maturity (e.g. 20S/(20S + 20R) steranes), others are dominated by source input (e.g. C_{27}, C_{28}, or C_{29} steranes versus total C_{27}–C_{29} steranes), and many are affected by both source and maturity (e.g. Ts/(Ts + Tm) or diasteranes/steranes), as discussed in Chapters 13 and 14. One approach to understanding these controls is to apply principal component or eigenvector analysis to a large geochemical database that includes all types of parameters (e.g. Hughes et al., 1985). Parameters that are controlled mainly by thermal maturity tend to be associated with each other in one principal component.

Correlations that include biomarker ratios are superior to those using only bulk parameters because biomarkers are generally more resistant to secondary processes and provide more specific genetic information on source, depositional environment, and thermal maturity. For example, bulk parameters might be used to delineate several families of oils in a basin. Supplementary biomarker analyses on selected oils might then be used to determine organic facies variations in the source rock for each family, resulting in the recognition of regional oil subgroups. Another example is the supplementary use of biomarkers to correlate and describe the level of thermal maturity of biodegraded oils.

Light oils and condensates

Light oils and condensates present a special correlation problem because their generally high maturity results in mainly gasoline-range components. Higher-molecular-weight components, including the biomarkers, occur in low concentrations or are absent in many condensates. Low concentrations of biomarkers may introduce analytical difficulties that obscure their genetic relationships with normal oils derived from the same source rock. In addition, condensates may solubilize biomarkers from less mature rocks during migration (e.g. Curiale et al., 2000). Because indigenous biomarkers are scarce in most condensates, contaminating biomarkers may adversely affect various interpretations, including correlation, source organic matter input, and thermal maturity.

Solid bitumens

Solid bitumens complicate correlation studies because of their highly variable, sometimes refractory character. They consist of a wide variety of organic materials with common names like grahamite, anthraxolite, gilsonite, and albertite. These names are based on an antiquated classification scheme that requires measurements of solubility, fusibility, and atomic hydrogen/carbon (e.g. Hunt, 1996). This generic classification is of little use in correlating or describing the origins of these materials.

Curiale (1986) recommends that the above generic scheme be discarded in favor of a classification that divides solid bitumens into pre-oil and post-oil products using biomarkers and supporting data. Most pre-oil bitumens are associated intimately with their source rocks, while post-oil bitumens have undergone extensive migration before alteration. By accounting carefully for differences caused by maturation and biodegradation, genetic relationships between solid bitumens and oils or source-rock extracts can be established.

Gilsonite is solid bitumen that occurs as dike complexes in the Uinta Basin, Utah (see Figure 18.157). It is composed mainly of NSO compounds and asphaltenes with subordinate saturated and aromatic hydrocarbons. Gilsonite is a common drilling mud additive, but it also is used in asphalt paving, roofing and construction paper, paints, inks, explosives, carbon electrodes, and various fuels. At the surface, gilsonite dikes range from several millimeters to >5 m thick; some are exposed for up to 39 km, but most are <5 km in length. The dikes cut across various Lower Eocene to Oligocene rocks and are thought to originate as hydraulic extension fractures caused by local overpressure in oil shale of the Mahogany Ledge Member of the Eocene Green River Formation. Most dikes extend upward from roots in the Green River into the overlying Uinta and Duchesne River formations, but some extend downward into the lowermost Green River and uppermost Wasatch formations. The distribution of the gilsonite dikes correlates with net thickness of high quality (>15 gallons/ton (62.2 l/metric ton) retort yield) Green River oil shale (Monson and Parnell, 1992). Total ion chromatograms from pyrolysis/GCMS of Green River oil shale and gilsonite show similar distributions of pristene, steroids, and carotene (Hatcher *et al.*, 1992). Stable carbon isotope ratios of saturated, aromatic, NSO compounds and asphaltenes for gilsonites are similar to those for shale extracts from the Mahogany Ledge (Schoell *et al.*, 1994). They also found that steranes in the gilsonites are isotopically uniform throughout the area and resemble those in the Mahogany Zone in the Piceance Creek Basin >50 km to the east. Dike emplacement probably occurred when the Green River Formation achieved maximum burial depth between 10 and 35 Ma. Water flowed through the fracture system, precipitating limonite, calcite, and chlorite in sandstones along the dike walls, followed by emplacement of viscous bitumen that solidified to gilsonite (Monson and Parnell, 1992; Hatcher *et al.*, 1992).

Gilsonite is a low-maturity product based on biomarker analyses, which indicate equivalent vitrinite reflectance of <0.5% (Anders *et al.*, 1992). Use of the term "immature" to describe gilsonite is misleading because generation and expulsion can occur at different points on the vitrinite reflectance scale, depending on organic richness and kerogen type. As discussed above, many workers believe that the gilsonite dikes originated locally. However, Verbeek and Grout (1993) suggest that the gilsonite originated from more mature Green River Formation oil shale to the northwest of the dike complexes near the Altamont-Blue Bell Field.

Some hydrocarbons occur as minerals. For example, curtisite is a greenish crystalline solid that is commonly associated with mercury ore at New Idria and elsewhere in the San Joaquin Basin, California (Peabody, 1993). It is a mixture of PAHs, including picene, dibenzo[a,h]fluorene, 11H-indeno[2,1-*a*]phenanthrene, benzo[b]phenanthro[2,1-*d*]thiophene, indenofluorenes, and chrysene (Wise *et al.*, 1986).

CHEMOMETRIC ANALYSIS OF LARGE DATA SETS

Evaluation of multivariate data is commonly a weak link in oil–oil and oil–source rock correlation studies because many interpreters ignore or improperly apply computerized data analysis. Reliable genetic correlations require multiple parameters that may include chromatographic and mass spectral homolog peak ratios, isotopic compositions of saturated or aromatic hydrocarbon fractions, metal ratios (e.g. V/Ni), and other source-related parameters. Visual inspection of data or simple binary or ternary diagrams of compositions assist interpretation, but these traditional methods are poorly suited to the extraction of meaningful information from large data sets. This certainly does not imply that mass chromatograms or other raw data can be ignored. On the contrary, detailed comparison of mass chromatograms, for example, helps to evaluate data quality and identify important differences between samples.

Chemometrics is the use of multivariate statistics to recognize patterns and extract useful information from measured data (e.g. Kramer, 1998). The method can be used to identify and remove noise from the data, show affinities among samples or variables, and make accurate predictions about unknown samples. Chemometrics is well suited for regional geochemical correlations, which commonly involve large numbers of samples and data

(Christie et al., 1984; Christie, 1992). Multivariate techniques treat all of the data simultaneously using matrix algebra, thus allowing the unique description of a sample by a point in space that has as many axes as variables. In contrast, univariate or bivariate techniques treat only one or two variables at a time. For example, a plot of x versus y might yield two clusters of samples. This plot is a unique representation of each sample in two-dimensional space if only two variables were measured. However, for three measured variables, a unique point representation of the samples would require a three-dimensional plot of x versus y versus z. In this case, a simple plot of x versus y might result in overlap of genetically distinct samples that differ in z.

In chemometric exploratory data analysis (EDA), the data are processed and graphically displayed. These displays accentuate clusters and/or trends in the data that help the analyst to distinguish similar data patterns. Two of the most common techniques for EDA are hierarchical cluster analysis (HCA) and principal component analysis (PCA). In HCA, distances between samples (or variables) in the data set are calculated and compared, typically by means of a dendrogram (Figure 12.2). The HCA dendrogram provides a simple view of groups of samples in the data set, where cluster distance is a relative measure of the degree of similarity among samples. Replicate analyses of samples are needed for quality control and to assess the relative significance of cluster distances. Furthermore, HCA is an iterative process. Preliminary HCA generally reveals certain parameters that show high variance and contribute little to reasonably distinguish samples. These parameters can be removed to improve results. Comparison of the HCA results with known geology or chemistry is also useful. For example, oils from the same field or well commonly show genetic affinities to each other. Likewise, spilled oils collected from adjacent areas of the same beach are likely to be genetically related. Multiple HCA analyses high-grade those parameters with genetic significance and reduce unexplained variance, allowing one to surmise the most geologically or environmentally reasonable results.

Various cluster options are available for HCA, including nearest neighbor, farthest neighbor, and centroid. In data sets where the group clusters are distinct, all three options give similar clusters and outliers. However, where the groups are less distinct, different cluster options can lead to slightly different groupings. Selection of the optimal clustering option requires user experience.

The statistical analysis of multivariate geochemical data in Figure 12.2 was completed using a commercial chemometrics program (Pirouette, Infometrix Inc., Woodinville, WA, USA). The hierarchical cluster analysis was completed using autoscale preprocessing, Euclidean metric distance, and incremental linkage. The analysis included 15 source-related biomarker and isotopic parameters, not significantly affected by migration, biodegradation, or thermal maturation (Peters et al., 2000). These parameters included $\delta^{13}C_{saturates}$, $\delta^{13}C_{aromatics}$, $\%C_{27}$, $\%C_{28}$, and $\%C_{29}$ steranes, $\%C_{27}$, $\%C_{28}$, and $\%C_{29}$ diasteranes, $24/(24 + 26)$ tricyclic terpanes, $C_{29}/(C_{29} + C_{30})$ hopanes, oleanane/(oleanane + hopane), C_{24} tetracyclic/(C_{24} tetracyclic + C_{23} tricyclic) terpanes, and %rimuane, %isorimuane, and %isopimarane diterpane ratios. Each of these parameters is discussed in detail in Chapter 13.

Because the measurement scales of the biomarker and isotopic parameters vary considerably, it is customary to use autoscale preprocessing. In autoscale preprocessing, the data are scaled so that each parameter provides the same relative contribution to the distinction between the samples as measured by cluster distance. To autoscale the original geochemical data, the mean value for each variable is subtracted and the remainder is divided by the standard deviation. This results in equal weight or significance for different parameters, such as stable carbon isotope ratios (e.g. -27%) compared with $\%C_{27}/(C_{27}$ to $C_{29})$ steranes.

PCA or eigenvector analysis simplifies complex n-dimensional data, where n is the number of variables, by transforming the original variables into a new set of variables. These new variables concentrate relevant information into a few dimensions, while noise or irrelevant information composes the remaining variables. Thus, the objective of PCA is to reduce the effective dimensionality of the geochemical data to a few components that best explain the variation in the data. Before PCA, the data are autoscaled as in cluster analysis. The new variables are called eigenvectors, which can be thought of as a new set of orthogonal plot axes. These variables, also called principal components, are linear combinations of the original geochemical variables or parameters. The relationships between samples are not changed by this transformation, but because the new axes are

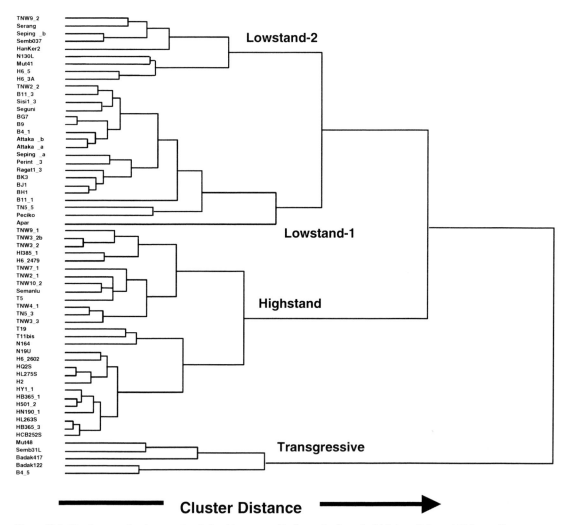

Figure 12.2. Dendrogram showing genetic relationships among 61 oil samples from the Mahakam Delta and Makassar Slope, Kalimantan, Indonesia, based on statistical analysis of selected source-related geochemical data (Peters et al., 2000). Cluster distance is a measure of genetic similarity indicated by the horizontal distance from any two samples on the left to their branch point on the right. Some samples lack reliable data for certain parameters but were included by replacing the missing parameter with the mean value for that parameter determined from all of the other oils. Abbreviated sample names include TNW3_2 and TNW3_2b, which are repeats of the same sample from DST 2 in the Tunu-TNW3 well. Reprinted by permission of the AAPG, whose permission is required for further use.

ordered by how much information they contain, we can graphically see the most significant differences between the samples using a two- or three-dimensional rather than n-dimensional plot. For PCA, the largest point scatter in multidimensional variable space coincides with the direction of the first principal component (PC1). The direction of the second vector (PC2) is the largest point scatter that is at right angles to the PC1 direction. Thus, a plot of PC1 versus PC2, also called a scores plot, represents the best two-dimensional separation of the samples into genetic groups from n-dimensional space (Figure 12.3). PCA can also be used to relate samples based on thermal maturity (e.g. Kruge, 2000). The plane for the first and second principal component vectors has higher information content than any other two-dimensional projection of the data. The sum

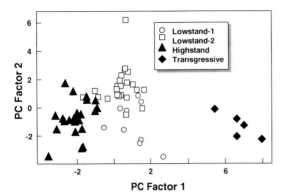

Figure 12.3. Principal component (PC) scores plot for 61 oil samples from the Mahakam Delta area of Kalimantan, Indonesia (Peters et al., 2000). The principal component analysis utilized autoscale preprocessing. The name for each oil group (inset) is based on correlation of the oils with extracts from source rocks deposited in specific sequence stratigraphic settings.

of the variances along the first and second principal component vectors quantifies that fraction of the total information captured by the PC1 versus PC2 plane projection. The geochemical variables responsible for the principal component axes can be viewed on a corresponding loadings plot.

Several early papers applied chemometrics to petroleum geochemistry (Øygard et al., 1984; Kvalheim et al., 1985; Peters et al., 1986; Zumberge, 1987a; Zumberge and Ramos, 1996). These early studies show that genetic classifications are enhanced by preselecting parameters that are known to be source-related. As described in this book, petroleum composition depends on source input, but many secondary processes are also important, including thermal maturation, biodegradation, water washing, and migration. Statistical purists commonly recommend input of all available data, where the principal components are used to separate source from secondary effects (e.g. Hughes et al., 1985). In practice, geochemists can identify the principal factors controlling many parameters. For example, sterane homolog and isomerization ratios are good source and maturity parameters, respectively. Oil–oil correlation might thus be initiated by preselecting sterane homolog ratios to be included in the chemometric analysis. New parameters can be added to this preselected data set to determine whether they co-vary with the known source parameters. A similar approach can be used to evaluate thermal maturity, where known maturity parameters, such as sterane isomerization ratios, are preselected and suspected maturity parameters are added to the data set before chemometric analysis.

Once a geologically or environmentally reasonable classification of genetic groups is established, the data can be used as a training set to generate a predictive K-nearest neighbor (KNN) and/or soft independent modeling of class analogy (SIMCA) model. Classification using KNN compares the n-dimensional distance between all samples, where n is the total number of geochemical variables. The three (or more) closest training set samples to each of the source-rock extracts are determined in order to predict the group to which each extract belongs. In SIMCA, separate PCA models for each genetic group are created using the selected geochemical variables. To predict the groups to which new extracts belong, their profiles are projected against each category and the SIMCA algorithm then uses statistical reliability to determine membership in the respective genetic groups.

Peters et al. (2000) used source-related geochemical data for the 61 oil samples from the Mahakam Delta area to generate a KNN model that was used to relate the oil groups to source-rock extracts and thus establish petroleum systems. The KNN model was used also to predict the genetic affinities for newly discovered oils that became available after completion of the geochemical study. The oil groups were named using sequence stratigraphic terms lowstand, highstand, and transgressive (Figures 12.2 and 12.3) based on detailed palynology and geology of the genetically related rock samples. This work resulted in a new geochemical-stratigraphic model for the Mahakam Delta and Makassar Slope that upgrades the potential of the outer shelf and justified the drilling of recent oil and gas-condensate discoveries.

13 · Source- and age-related biomarker parameters

> This chapter explains how biomarker analyses are used for oil–oil and oil–source rock correlation and how they help to identify characteristics of the source rock (e.g. lithology, geologic age, type of organic matter, redox conditions) even when samples of rock are not available. Biomarker parameters are arranged by groups of related compounds in the order (1) alkanes and acyclic isoprenoids, (2) steranes and diasteranes, (3) terpanes and similar compounds, (4) aromatic steroids, hopanoids, and similar compounds, and (5) porphyrins. Critical information on specificity and the means for measurement is highlighted before the discussion of each parameter.

Organic matter input and depositional conditions of the sediments that become source rocks exert primary control on the biomarker fingerprints of source-rock extracts and crude oils. An example of organic matter input is the distribution of C_{27}, C_{28}, and C_{29} sterols from eukaryotic organisms that reaches the sediment from an overlying water column. This initial distribution of sterols might be altered by many diagenetic factors during and after sedimentation, but ternary plots of the relative amounts of C_{27}, C_{28}, and C_{29} steranes largely reflect original source input. However, the extent of conversion of C_{27} sterols to C_{27} diasterenes reflects the acid catalytic or oxidative potential of minerals in the water column and sediment, which is an effect of the depositional environment. Thus, the diasterane/sterane ratio of crude oils can be used to indicate relative amounts of clays in the related source rock.

SOURCE-RELATED PARAMETERS

Under certain depositional conditions, large populations of one or a few organisms can produce abundant supplies of one or a few diagnostic precursors that give rise to biomarkers. For example, blooms of purple halophilic bacteria may account for the abundance of β-carotene in underlying lake-bottom sediments (Jiang Zhusheng and Fowler, 1986). The presence of unusually high concentrations of the saturated analogs of these biomarkers in petroleum, such as β-carotane, suggests source input dominated by one or a few species of halophilic bacteria and can also be used to infer something about the depositional conditions of the source rock, e.g. arid climate. As another example, unusually high concentrations of dinosterol in sediments appear to result from dinoflagellate blooms in the overlying nutrient-rich, upwelling marine waters (Withers, 1983). This situation was more likely to occur in sediments deposited after the radiation of dinoflagellates in the Triassic Period, which gives rise to the use of dinosteranes in crude oils as age-related biomarkers (see below).

Compared with other parameters, source-related biomarker ratios are particularly useful in order to describe the source rock even when only crude oil samples are available for analysis. For various reasons, samples of mature basinal source rocks are generally more difficult to obtain than oils. Nonetheless, biomarkers in the oils can be used to infer the depositional environment, organic input, thermal maturity, and even the age of the source rock. These data can be used to suggest likely source rocks that might be sampled later for detailed oil–source rock correlations.

Biomarker distributions in source rock and crude oil samples also show systematic variations related to sequence stratigraphy. For example, biomarkers in Lower-Middle Triassic mudrocks from the Barents Sea, Norway, co-vary with sea-level variations, proximity to the ancient shoreline, and petrographic estimates of the amount of terrigenous or marine algal organic matter (Isaksen and Bohacs, 1995). The mudrocks with the best petroleum-generative potential were deposited under reducing conditions in distal open-marine shelf

Table 13.1. *Acyclic biomarkers as indicators of biological input or depositional environment (assumes high concentration of component)*

Compound	Biological origin	Environment	Sample references
nC_{15}, nC_{17}, nC_{19}	Algae	Lacustrine, marine	Gelpi et al. (1970), Tissot and Welte (1984)
nC_{15}, nC_{17}, nC_{19}	~Ordovician, *G. prisca*	Tropical marine	Reed et al. (1986), Longman and Palmer (1987), Hoffmann et al. (1987), Jacobson et al. (1988)
nC_{27}, nC_{29}, nC_{31}	Higher plants	Terrigenous	Eglinton and Hamilton (1967), Tissot and Welte (1984)
nC_{23}–nC_{31} (odd)	Non-marine algae	Lacustrine	Gelpi et al. (1970), Moldowan et al. (1985)
2-Methyldocosane	Bacteria?	Hypersaline	Connan et al. (1986)
Mid-chain monomethylalkanes	Cyanobacteria	Hot springs, marine	Shiea et al. (1990), Thiel et al. (1999a)
Pristane/phytane (low)	Phototrophs, archaea	Anoxic, high salinity	Didyk et al. (1978), ten Haven et al. (1987), Fu Jiamo et al. (1986; 1990)
PMI (PME)	Archaea, methanogens, methanotrophs	Hypersaline, anoxic	Brassell et al. (1981), Risatti et al. (1984), Schouten et al. (1997b)
Crocetane	Archaea, methanotrophs?	Methane seeps?	Thiel et al. (2001)
C_{20} HBI	Diatoms	Marine, lacustrine	Yon et al. (1982), Kenig et al. (1990)
C_{25} HBI	Diatoms	Marine, lacustrine	Nichols et al. (1988), Volkman et al. (1994)
Squalane	Archaea	Hypersaline?	Ten Haven et al. (1986)
C_{31}–C_{40} head-to-head isoprenoids	Archaea	Unspecified	Seifert and Moldowan (1981), Risatti et al. (1984)
Botryococcane	Green algae, *Botryococcus*	Lacustrine-brackish-saline	Moldowan and Seifert (1980), McKirdy et al. (1986)
16-Desmethyl-botryococcane	Green algae, *Botryococcus*	Lacustrine-brackish-saline	Seifert and Moldowan (1981), Brassell et al. (1985)
Polymethylsqualanes	Green algae, *Botryococcus*	Lacustrine-brackish-saline	Summons et al. (2002b)

C_{20} HBI, 2,6,10-trimethyl-7-(3-methylbutyl)-dodecane; C_{25} HBI, 2,6,10,14-tetramethyl-7-(3-methylpentyl)-pentadecane; PMI, 2,6,10,15,19-pentamethylicosane (current IUPAC nomenclature), previously spelled pentamethyleicosane (PME).

environments during late transgressive and early highstand system tracts (see Figure 4.10).

Tables 13.1 and 13.2 summarize some of the more useful acyclic and cyclic biomarkers, respectively, which provide information on source-rock organic matter input or depositional environment. These tables must be used with caution. Although a compound or group of compounds may support a particular biological

Table 13.2. *Cyclic biomarkers as indicators of biological input or depositional environment (assumes high concentration of component)*

Compound	Biological origin	Environment	Sample references
Saturates			
C_{25}–C_{34} macrocyclic alkanes	Green algae, *Botryococcus*	Lacustrine-brackish	Audino *et al.* (2001b)
C_{15}–C_{23} cyclohexyl alkanes (odd)	~Ordovician, *G. prisca*	Marine	Reed *et al.* (1986), Rullkötter *et al.* (1986)
β-Carotane	Cyanobacteria, algae	Arid, hypersaline	Jiang Zhusheng and Fowler (1986), Koopmans *et al.* (1997)
Phyllocladanes	Conifers	Terrigenous	Noble *et al.* (1985a; 1985b; 1986)
4β-Eudesmane	Higher plants	Terrigenous	Alexander *et al.* (1983a)
C_{19}–C_{30} tricyclic terpanes	*Tasmanites*?	Marine, high latitude	Aquino Neto *et al.* (1983), Volkman *et al.* (1989), De Grande *et al.* (1993)
C_{24} tetracyclic terpane	Unknown	Hypersaline	Connan *et al.* (1986), Grice *et al.* (2001)
C_{27}–C_{29} steranes	Algae and higher plants	Various	Moldowan *et al.* (1985), Volkman (1986)
23,24-Dimethylcholestanes	Dinoflagellates?, haptophytes	Marine	Withers (1983), Volkman (1986)
C_{30} 24-*n*-propylcholestanes (4-desmethyl)	Chrysophyte algae	Marine	Moldowan *et al.* (1985), Peters *et al.* (1986), Moldowan *et al.* (1990)
4-Methylsteranes	Some bacteria, dinoflagellates	Lacustrine or marine	Brassell *et al.* (1985), Wolff *et al.* (1986)
Pregnane, homopregnane	Unknown	Hypersaline	Ten Haven *et al.* (1986)
Diasteranes	Algae/higher plants	Clay-rich rocks	Rubinstein *et al.* (1975), Van Kaam-Peters *et al.* (1998)
Dinosteranes	Dinoflagellates	Marine, Triassic or younger	Summons *et al.* (1987), Goodwin *et al.* (1988)
25,28,30-trisnorhopane	Bacteria	Anoxic marine, upwelling?	Grantham *et al.* (1980), Schouten *et al.* (2001)
28,30-Bisnorhopane			Seifert *et al.* (1978), Grantham *et al.* (1980), Katz and Elrod (1983), Schoell *et al.* (1992)
C_{35} 17α,21β(H)-hopane	Bacteria	Reducing to anoxic	Peters and Moldowan (1991), Köster *et al.* (1997)
Norhopane (C_{29} hopane)	Various	Carbonate/evaporite	Clark and Philp (1989)
2-Methylhopanes	Cyanobacteria	Enclosed basin	Summons *et al.* (1999)
3β-Methylhopanes	Methanotrophic bacteria	Lacustrine?	Summons and Jahnke (1992), Collister *et al.* (1992)
Bicadinanes	Higher plants	Terrigenous	Van Aarssen *et al.* (1990b; 1992)

(*cont.*)

Table 13.2. (cont.)

Compound	Biological origin	Environment	Sample references
23,28-Bisnorlupanes	Higher plants	Terrigenous	Rullkötter et al. (1982)
Gammacerane	Tetrahymanol in ciliates feeding on bacteria	Stratified water, sulfate-reducing, hypersaline (low sterols)	Hills et al. (1966), Moldowan et al. (1985), Fu Jiamo et al. (1986), ten Haven et al. (1988), Sinninghe Damsté et al. (1995), Grice et al. (1998a)
18α-Oleanane	Cretaceous or younger, higher plants	Paralic	Ekweozor et al. (1979a), Riva et al. (1988), Ekweozor and Udo (1988), Moldowan et al. (1994a)
Hexahydrobenzohopanes	Bacteria	Anoxic carbonate-anhydrite	Connan and Dessort (1987)
Aromatics			
Benzothiophenes, alkyldibenzothiophenes	Unknown	Carbonate/evaporite	Hughes (1984)
^{13}C-rich 2,3,6-trimethyl-substituted aryl isoprenoids, isorenieratene	*Chlorobiaceae*, anaerobic green sulfur bacteria	Photic zone anoxia	Summons and Powell (1987), Clark and Philp (1989), Requejo et al. (1992), Hartgers et al. (1994a), Koopmans et al. (1996a), Grice et al. (1997)
Methyl n-pristanyl and methy i-butyl maleimides	*Chlorobiaceae*, anaerobic green sulfur bacteria	Photic zone anoxia	Grice et al. (1997)
Isorenieratane	*Chlorobiaceae*, anaerobic green sulfur bacteria	Photic zone anoxia	Sinninghe Damsté et al. (1993b), Grice et al. (1996b; 1997), Koopmans et al. (1996a)
Trimethyl chromans*	Phytoplankton	Saline photic zone?	Sinninghe Damsté et al. (1987b), Schwark et al. (1998), Grice et al. (1998a)

* Trimethyl chromans = 2-methyl-2-(4,8,12-trimethyl-tridecyl)-chromans.

origin or paleoenvironment, exceptions are common. Volkman (1988) summarizes the use of biomarkers in crude oils for reconstructing source-rock depositional environment and organic matter input. Conclusions on correlations, source, and depositional environment should always be based on a thorough evaluation of all of the available geochemical information, including other biomarker, isotope, and supporting data. A few of the many examples of integrated geochemical correlations include Shi Ji-Yang et al. (1982), Palmer (1984a), Grantham et al. (1988), Peters et al. (1989), and Moldowan et al. (1992).

Statistical multivariate discriminate analysis combining selected biomarker and isotope data can effectively discriminate different organic facies (Peters et al., 1986, Chapter 12).

Caution should be applied when comparing biomarker parameters from different laboratories. The mass chromatograms and peaks used to determine each biomarker parameter described in the following sections might differ between laboratories. For this reason, publications that provide details on measurement procedures for each parameter are very useful (e.g. Fu Jiamo et al., 1990). Interlaboratory standards are discussed in Chapter 8.

Lakes generally show greater spatial and temporal variations in salinity, redox, depth, temperature, and organic matter input than marine settings (Katz, 1995c). For this reason, it is difficult to identify biomarkers in petroleum that are specific for lacustrine source rock. The following biomarkers or biomarker ratios indicate lacustrine settings, but their absence or low values of the ratios do not preclude lacustrine source rock:

- Elevated 4-methylsteranes from dinoflagellate algae are common in lacustrine source-rock extracts and crude oils (Summons et al., 1987; Goodwin et al., 1988). However, dinoflagellates are absent in Paleozoic rock samples. 4-Methylsteranes may also be low or absent in Paleozoic lacustrine source rocks.
- C_{26}/C_{25} tricyclic terpane ratios >1 are typical of lacustrine oils (see Figure 13.77) (Zumberge, 1987a), but exceptions occur.
- β-Carotane from halophilic bacteria is abundant in some crude oils from arid, lacustrine source-rock settings (see Figure 13.33) (Jiang Zhusheng and Fowler, 1986; Fu Jiamo et al., 1990), but it is rare and generally low in marine oils (Clark and Philp, 1989; Peters and Moldowan 1993).
- Botryococcane (see Figure 13.25) and polymethylsqualanes (see Figure 13.27) from the fresh- or brackish-water alga *Botryoccocus braunii* are highly specific for lacustrine oils (Seifert and Moldowan, 1981), and are common in many oils from Southeast Asia.
- Elevated tetracyclic polyprenoid ratios are typical of lacustrine oils (see Figure 13.81) (Holba et al., 2000).

PREDICTION OF SOURCE-ROCK CHARACTER FROM OIL COMPOSITION

Carbonate versus shale

In sparsely explored or frontier basins, petroleum source rocks are usually unknown, and only a few crude oils or seep samples may be available for study. Various geochemical methods can be applied to such oils in attempts to provide information on their source rocks. This information is useful to geologists who plan further exploration within a basin. For example, structural, mineralogical, and organic geochemical studies can help to delineate probable migration pathways or the types of petroleum that will be found. The reader is referred to several statistical studies of geochemical data obtained from oils as a means to determine source-rock organic matter type and/or mineralogy (Peters et al., 1986; Zumberge, 1987a).

Combinations of the above parameters can be used to describe the organic matter type, depositional environment, and mineralogy of the source rock from oil composition. Table 13.3 is an example showing characteristics of oils derived from carbonate versus shale source rocks. In the text, the term "carbonate rocks" refers to fine-grained sedimentary rocks containing 50% or more of carbonate minerals, typically associated with evaporitic, siliceous, and argillaceous components. The data in the table apply only to oils of comparable maturity up to peak oil generation.

Other environments can be differentiated using biomarkers. For example, Mello et al. (1988b) describe a series of biomarker and non-biomarker parameters that they used to separate offshore Brazilian oils and bitumens from the following depositional environments: lacustrine freshwater, lacustrine saline, marine evaporitic, marine carbonate, marine deltaic, marine calcareous, and marine siliceous lithology. Connan et al. (1986) show that detailed biomarker analysis can be used to distinguish anhydrites from carbonates in a core from a paleo-sabhka in Guatemala.

Table 13.3. *Some characteristics of petroleum from carbonate versus shale source rocks*

Characteristics	Shales	Carbonates	References
Non-biomarker parameters			
API, gravity	Medium–high	Low–medium	1, 2, 3
Sulfur, wt.%	Variable	High (marine)	1, 2, 3, 6, 9
Thiophenic sulfur	Low	High	1
Saturate/aromatic	Medium–high	Low–medium	1, 2, 3
Naphthenes/alkanes	Medium–low	Medium–high	1, 3
Carbon preference index (C_{22}–C_{32})	≥ 1	≤ 1	1, 2, 6, 9
Biomarker parameters			
Pristane/phytane	High (≥ 1)	Low (≤ 1)	1, 2, 6, 9, 10
Phytane/nC_{18}	Low (≤ 0.3)	High (≥ 0.3)	2, 6
Steranes/17α-hopanes	High	Low	7, 9
Diasteranes/steranes	High	Low	1
C_{24} tetra-/C_{26} tricyclic diterpanes	Low–medium	Medium–high	2, 7
C_{29}/C_{30} hopane	Low	High (>1)	10, 11
C_{35} homohopane index	Low	High	4, 10
Hexahydrobenzohopanes and benzohopanes	Low	High	5
Dia/(Reg + Dia) monoaromatic steroids	High	Low	8
Ts/(Ts + Tm)	High	Low	4
C_{29} monoaromatic steroids	Low	High	9

References: (1) Hughes (1984), (2) Palacas (1984), (3) Tissot and Welte (1984), (4) McKirdy *et al.* (1983), (5) Connan and Dessort (1987), (6) Connan (1981), (7) Connan *et al.* (1986), (8) Riolo *et al.* (1986), (9) Moldowan *et al.* (1985), (10) ten Haven *et al.* (1988), (11) Fan Pu *et al.* (1987).

Oil-prone source rocks, whether carbonates or shales, commonly show similar characteristics, including lamination, high total organic carbon (TOC), and hydrogen-rich organic matter. Despite these similarities, oils from carbonate rocks are typically richer in cyclic hydrocarbons and sulfur compared with those from shales. Jones (1984) describes differences in the rock matrix of carbonates and shales that result in different primary migration characteristics. He believes that carbonate source rocks do not have a lower minimum TOC compared with other source rocks.

No individual parameter in Table 13.3 should be considered as proof of the origin for oil. Exceptions to the general ranges of values used as a guide in the table are common. For example, oils from lacustrine carbonates, such as the Green River Formation, may be low in sulfur. High abundance of oleanane commonly relates to high terrigenous input to Tertiary source rocks. Most oils from lacustrine source rocks contain little sulfur, but some oils from source rocks deposited in hypersaline lacustrine systems have high sulfur. Many oils generated from source rocks deposited under anoxic marine conditions do not have high 28,30-bisnorhopane. Also, biomarkers and other components in highly mature oils are altered severely, making interpretations of source rock type difficult. The parameters in Table 13.3 should be used together to indicate the origin of oils.

Marine versus terrigenous organic matter

Few systematic studies address the effects of variations in the salinity of the source-rock depositional environment on geochemical parameters. Philp *et al.* (1989) examined oils thought to originate from source rocks deposited in brackish, saline, and freshwater lacustrine environments in China. They concluded that oils from source rocks deposited under brackish water showed high tricyclic terpanes compared with hopanes and a predominance of 24-methyl- and 24-ethylcholestanes (C_{28} and C_{29}) with few cholestanes (C_{27}). The oils from saline source rocks showed high gammacerane, while those from freshwater source rocks showed low tricyclic

Table 13.4. *Generalized geochemical properties* differ between non-biodegraded crude oils from marine, terrigenous, and lacustrine source-rock organic matter (modified from Peters and Moldowan, 1993)*

Property	Marine	Terrigenous	Lacustrine
Sulfur (wt.%)	High (anoxic)	Low	Low
C_{21}–C_{35} n-alkanes	Low	High	High
Pristane/phytane	<2	>3	~1–3
Pristane/nC_{17}	Low (<0.5)	High (>0.6)	–
4-Methylsteranes	Moderate	Low	High
C_{27}–C_{29} steranes	High C_{28}	High C_{29}	High C_{27}
C_{30} 24-n-propylcholestane	Low	Low or absent	Absent
Steranes/hopanes	High	Low	Low
Bicyclic sesquiterpanes	Low	High	Low
Tricyclic diterpanes	Low	High	High
Tetracyclic diterpanes	Low	High	Low
Lupanes, bisnorlupanes	Low	High	Low
28,30-Bisnorhopane	High (anoxic)	Low	Low
Oleananes	Low or absent	High	Low
β-Carotane	Absent	Absent	High (arid)
Botryococcane	Absent	Absent	High (brackish)
V/(V + Ni)	High (anoxic)	Low or absent	Low or absent

* Quoted properties encompass most samples, but exceptions occur. For example, many nearshore oxic marine environments resulted in source rocks that generated oils with low sulfur, and some very high-sulfur oils originated from source rocks deposited in hypersaline lacustrine settings. The terms "marine," "terrigenous," and "lacustrine" can be misleading. "Marine" oil might refer to (1) oil produced from marine reservoir rock, (2) oil generated from source rock deposited under marine conditions, or (3) oil derived from marine organic matter in the source rock. The table refers to provenance of the organic matter (3).

terpanes compared with hopanes, an unknown C_{30} pentacyclic terpane (probably C_{30}*), small amounts of C_{31}–C_{35} homohopanes compared with C_{30} hopane, and large amounts of 24-ethylcholestanes (C_{29}) compared with other steranes.

Fu Jiamo *et al.* (1990) conclude that the most useful parameters for distinguishing Chinese rocks deposited in freshwater, brackish, and hypersaline lacustrine environments include the relative abundance of n-alkanes, acyclic isoprenoids, 4-methylsteranes, the hopane/sterane ratio, and the gammacerane and homohopane indices. Samples from each of these groups were distinguished using principal component analyses of these data.

Some parameters are useful in order to indicate marine versus terrigenous organic matter in the source rocks for oils. Many of these parameters are described in the above discussion, although few are diagnostic in every case. For example, oleanane indicates higher plants, but its absence does not prove lack of that input. Sea grasses, such as *Zostera*, are basically vascular plants that are found in marine rather than terrigenous environments. For these reasons, we recommend against heavy reliance on one or a few of the above parameters. The most reliable statements about organic matter input are made based on multiple parameters. For example, Talukdar *et al.* (1986) used vanadium, sulfur, pristane/phytane, pristane/nC_{17}, sterane distributions, C_{19} and C_{20} diterpanes, oleanane, and hopane/sterane ratios to distinguish marine, terrigenous, and mixed oils in the Maracaibo Basin, Venezuela.

Some of the principal differences between non-biodegraded petroleum derived from marine, terrigenous, and lacustrine algal input are listed in Table 13.4. This table is greatly simplified because of the enormous diversity of environments and types of contributing organisms included within these categories, and because more or less terrigenous organic matter is commonly mixed with either marine or lacustrine organic matter. The reader is urged to consult the detailed discussions of various parameters in the text.

Terrigenous markers are commonly low in marine and lacustrine oils. However, source rocks deposited in marine deltaic environments contain mixtures of both marine and terrigenous organic matter. In the extreme case of overwhelming terrigenous input, marine deltaic source rocks generate terrigenous oils. Terrigenous oils also originate from lacustrine sediments that are dominated by higher-plant input. However, many lacustrine and non-marine oils originate from non-marine algal and bacterial organic matter in lacustrine source rocks. Note that hypersaline lacustrine environments are not included in Table 13.4. Concentrations of biomarkers also provide important information. For example, high concentrations of markers for vascular plants might be expected in lacustrine or estuarine sediments. Deep-sea sediments might receive smaller amounts of these vascular plant markers due to losses during transport by wind or turbidity currents.

AGE-RELATED PARAMETERS

Various compounds in source rocks show distributions through geologic time suggesting their use as age-related biomarkers in crude oils (Figure 13.1). Many age-related biomarkers can be related to specific taxa through natural-product chemistry, and the occurrence or relative abundance of these biomarkers parallels the

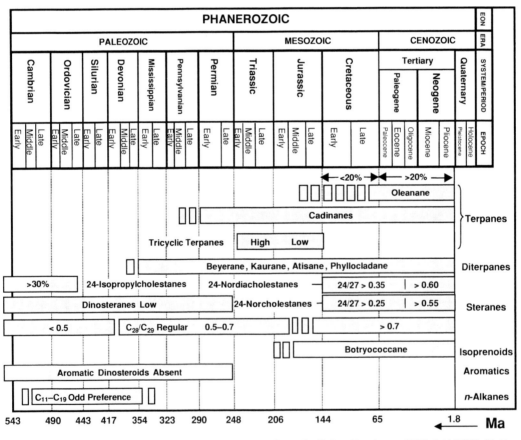

Figure 13.1. Age-related biomarkers help to infer the source rocks for crude oils (e.g. Grantham and Wakefield (1988); Moldowan (2000); Moldowan et al. (1994a); Moldowan et al. (1996); Moldowan et al. (2001a); Moldowan and Jacobson (2000); Holba et al. (1998); Holba et al. (2001)). Oleanane ratio = %Ol/(Ol + hopane). See text for further details on each parameter.

Table 13.5. *Potential age-related biomarkers*

Biomarker or biosignature	Related (main) organisms	Age range for high abundance	Other limitations
nC_{15}, nC_{17}, nC_{19}	*G. prisca*	Cambrian-Devonian	–
Botryococcane	*B. braunii*	Only known in Tertiary	Lacustrine, rare
24-Isopropylcholestane	Porifera (class Demospongiae)	Abundant in some Vendian-Ordovician	Marine
Oleananes, lupanes	Angiosperms	Cretaceous or younger, higher in some Tertiary	Rare in pre-Cretaceous
Dinosterane, triaromatic dinosteroids	Dinoflagellates	Abundant in some Triassic or younger	Abundant in some Paleozoic-Precambrian
Triaromatic 23,24-dimethylcholesteroids	Haptophytes, dinoflagellates	Common in Triassic or younger	Rare in Paleozoic
24-Norcholestane, 24-nordiacholestanes	Diatoms?	Abundant in some Cretaceous or younger, higher in some Tertiary	Maximum effect at high latitudes, verified only for oils
C_{20}, C_{25}, C_{30} highly branched isoprenoids	Diatoms	Cretaceous or younger	Few data
C_{28}–C_{29} tricyclic terpane/18α-22,29,30-trisnorneohopane (Ts)	Green algae?, bacteria	Decreases from Upper Triassic to Upper Jurassic	Few Phanerozoic data
Baccharane	Unknown	Triassic rocks (Adriatic)	Rare
C_{28}/C_{29} steranes	Algae	Increases from Precambrian-Tertiary	Marine settings without terrigenous input
Beyerane, kaurane, phyllocladane	Terrigenous plants	Devonian and younger	Few data
Cadinanes	Terrigenous plants	Jurassic and younger, present to at least Permian	Few data

taxonomic record. Some examples include oleanane from angiosperms (flowering plants, see Figure 13.94), dinosterane from dinoflagellates (Table 13.5), nordiacholestanes from diatoms (see Figure 13.55), and 24-isopropylcholestane from porifera (sponges, see Figure 13.44). Brocks *et al.* (1999) and Summons *et al.* (1999) show that most prokaryotic biomarkers have little practical potential to constrain geologic age, because they occur even in Archean rocks.

Note: Taxon-specific biomarkers commonly have a record of occurrence that parallels the radiation of the related organism, but they can also occur earlier in geologic time (Moldowan, 2000). For example, dinosteroids, which are believed to originate from dinoflagellates, occur in pre-Triassic and even Precambrian rocks, although dinoflagellate cysts are first recognized in Triassic rocks.
24-Norcholestanes occur throughout the Phanerozoic Eon, whereas their presumed precursor organisms, diatoms, first appeared in the Jurassic Period. Oleanane occurs in some Paleozoic rock extracts, although the first angiosperms that manufacture its precursor are known only from the earliest Cretaceous Period. The occurrence of biomarkers that predate the first macrofossil evidence for their precursor organisms suggests either that (1) older macrofossils remain to be

found or (2) the biosynthetic pathways for the precursor compounds existed in other organisms before complete development of the recognized morphology of the precursor organism.

Other relationships between biomarkers and specific taxa are empirical, such as the predominance of nC_{15}, nC_{17}, and nC_{19} in the gas chromatograms of extracts from Ordovician rocks rich in *Gloeocapsomorpha prisca* (see Figure 4.21, bottom chromatogram) (Jacobson *et al.*, 1988, Reed *et al.*, 1986). The correlation presumes a relationship between this biosignature and the extinct organism *G. prisca*. This *n*-alkane signature is most common in Ordovician oil and rock samples, but it also occurs in earlier (Cambrian) and later (Silurian-Devonian) samples. For example, Moldowan and Jacobson (2000) reported typical *G. prisca n*-alkane fingerprints for oils from Middle Cambrian reservoirs in Kentucky, which are unlikely to have originated from Ordovician source rocks (D. Silberman, 2001, personal communication).

The relationship between 24-norcholestane or 24-nordiacholestane and possible diatom biomarkers (Holba *et al.*, 1998) is supported by empirical evidence linking the relative abundance of 24-norcholestane or 24-nordiacholestane in oil to the appearance of diatom skeletal remains (frustules) in rocks of the same ages and paleolatitudes. The radiation of diatoms in the Cretaceous and Tertiary periods corresponds to increased 24-norcholestane abundance. The 24-norcholestane record from oil samples supports the affinity of diatoms for cold, high-latitude or upwelling waters. Siliceous sediments are particularly rich in the 24-norcholestanes (Suzuki *et al.*, 1993). However, occurrence of 24-norcholesterols in diatoms or other algal taxa has yet to be documented completely. They occur in marine filter feeders, where they are assumed to be of dietary origin, and in some diatom-rich marine algal assemblages, but they have not been found in laboratory-cultured diatoms (Holba *et al.*, 1998).

Environment of deposition and diagenesis are additional factors that determine age-related biomarker distributions. For example, oleanane is absent in some Cretaceous and Tertiary marine evaporitic or carbonate source rocks and related oils because terrigenous plant material is absent in those depositional environments. However, Murray *et al.* (1994) reported that oleanane preservation is enhanced by salty water and that freshwater environments are poor in this regard. Thus, some non-clastic marine source rocks with very little terrigenous input contain oleanane due to an excellent diagenetic environment for oleanane preservation, while some Cretaceous or Tertiary lacustrine rocks or coal formed in freshwater swamps lack oleanane despite abundant angiosperm input.

Large amounts of data were used to generate the age-related biomarker profiles in Figure 13.2. Table 13.5

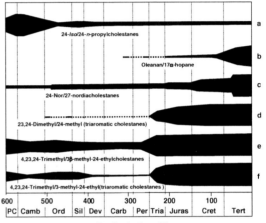

Figure 13.2. Average relative concentrations of taxon-specific biomarkers during geologic time (from Moldowan and Jacobson, 2000). (a) The ratio of $\alpha\alpha\alpha 20R + 20S$, 24-isopropylcholestane/24-*n*-propylcholestane in marine rock extracts reported by McCaffrey *et al.* (1994a). 24-Isopropylcholestane originates from sponges (Porifera) and the ratio maximizes in the Early Cambrian and Late Vendian epochs. (b) The ratio of $18\beta + 18\alpha$-oleananes/17α-hopane in marine rock extracts reported by Moldowan *et al.* (1994a). The ratio continues to increase from the Cretaceous Period through the Tertiary Period. Oleanane rarely occurs in Jurassic and older rocks. (c) 24-Norcholestanes are probably related to diatoms, and their ratio to 27-norcholestanes (algal, non-specific) in oil samples follows the known diatom record with some increase in the Cretaceous Period and high ratios in the Tertiary Period (Holba *et al.*, 1998). (d) In marine rock extracts, triaromatic 23,24-dimethylcholestanes occur almost exclusively in the Triassic and younger ages (Barbanti *et al.*, 1999). Precursors are known in dinoflagellates and haptophytes (coccolithophores). (e) 4α,23,24-Trimethylcholestanes (dinosteranes) and (f) triaromatic dinosteroids show highest relative concentrations (relative to 3β-methylstigmastane and triaromatic 3-methylstigmasteroid, respectively) in Triassic and younger marine rock extracts, in agreement with the record of their precursor dinoflagellates (Moldowan *et al.*, 1996; Moldowan *et al.*, 2001a). Pre-Triassic dinosteroids could be from unrecognized dinoflagellates or dinoflagellate ancestors (Moldowan and Talyzina, 1998).

also summarizes the biomarkers with age-related potential. Various age ranges have been attributed to other biomarkers in the literature, but most have not been documented sufficiently to allow their inclusion in Figure 13.2.

ALKANES AND ACYCLIC ISOPRENOIDS

n-Alkane ratios

> The terrigenous/aquatic ratio (TAR) is used as a crude indicator of relative terrigenous versus aquatic organic matter input but is sensitive to secondary processes. Measured using gas chromatographic peak heights or areas.

Ratios of certain n-alkanes can be used to identify changes in the relative amounts of terrigenous versus aquatic hydrocarbons in sediment or rock extracts. For example, higher terrigenous/aquatic ratios (TARs) in various recent sediment extracts indicate more terrigenous input from the surrounding watershed relative to aquatic sources (Bourbonniere and Meyers, 1996):

$$\text{TAR} = (nC_{27} + nC_{29} + nC_{31})/(nC_{15} + nC_{17} + nC_{19})$$

TAR must be used with caution because it is sensitive to thermal maturation and biodegradation. Furthermore, land-plant organic matter typically contains more n-alkanes than aquatic organic matter, resulting in disproportionate weight assigned to land-plant input. Certain non-marine algae (e.g. *Botryococcus braunii*) may contribute to the C_{27}–C_{31} n-alkanes (Moldowan *et al.*, 1985; Derenne *et al.*, 1988). Nonetheless, vertical distributions of TAR measurements are useful in order to determine relative changes in the contributions of land versus aquatic flora through time, particularly in young sediments (e.g. Meyers, 1997).

Other common n-alkane ratios are the carbon preference index (CPI) (Bray and Evans, 1961) and the improved odd-to-even preference (OEP) (Scalan and Smith, 1970) described in Chapter 15. These indices are similar to TAR because they are influenced by both source input and maturity, except that they compare the nC_{25}–nC_{35} odd-carbon-numbered n-alkanes with nC_{24}–nC_{34} even-numbered n-alkanes. Immature source rocks with significant input of land-plant organic matter are dominated by the odd-carbon-numbered n-alkanes, particularly nC_{27}, nC_{29}, and nC_{31}. These n-alkanes originate from epicuticular waxes and either are synthesized directly by higher plants or are defunctionalized even-numbered acids, alcohols, or esters. The nC_{24}–nC_{35} alkanes derived from marine organic matter tend to have little or no carbon-number preference or, in the case of some hypersaline carbonate and evaporite source rocks, a have slight preference for even-numbered n-alkanes. In all cases, carbon-number preferences decrease with increasing maturity. This is attributed to a combination of the generation of n-alkanes from kerogen derived from different biological precursors and from the thermal cracking of the early diagenetic hydrocarbons. Hence, while high CPI indicates low maturity and land-plant input, oils and source rocks with CPI $\sim$1 may arise from a predominance of marine input and/or thermal maturation.

Alkyl-substituted alkanes

> Some isomers are associated with cyanobacteria and with Precambrian rocks and oils. Various ions are used for mass spectrometric detection.

The monomethyl-, dimethyl-, and T-branched alkanes are common hydrocarbons in biological lipids, oils, and bitumens. Iso- (2-methyl) and anteiso- (3-methyl) alkanes are common in crude oils and source-rock bitumens. Although there are many examples of 2- and 3-methylalkanes and functionalized precursors as natural products, they are not strictly biomarkers because they lack biological specificity and can form by inorganic processes, such as isomeric equilibrium (Hoering, 1980; Klomp, 1986) or acid-catalysis of alkenes formed during thermal cracking (Kissin, 1993).

Dimethylalkanes are examined infrequently. Fowler and Douglas (1987) disputed the report by Klomp (1986) of dimethylalkanes in Precambrian oils. Modern cyanobacterial mats contain dimethylalkanes (de Leeuw *et al.*, 1985; Shiea *et al.*, 1990) and trimethylalkanes (Kenig *et al.*, 1995a) whose origins are largely unknown. The association of these compounds with mid-chain monomethylalkanes (MMAs) of similar length in bacterial mats suggests that cyanobacteria may be a common source. Functionalized dimethylalkanes are possible biological precursors because dimethyl carboxylic acids are known to occur in bacteria. As with the mid-chain MMAs, the known microbial dimethyl carboxylic acids have shorter chain lengths than dimethylalkanes in the geosphere. Chappe *et al.* (1980)

reported abundant 13,16-dimethyloctacosane liberated by ether cleavage of Messel oil shale kerogen, indicating a functionalized precursor. Sinninghe Damsté et al. (2000) identified the precursor for this compound as an archaeal dialkyl glycerol tetraether lipid in peats, coastal marine, and lake sediments.

T-branched alkanes are hydrocarbons consisting of linear carbon chains with a single n-alkyl substitution (Gough and Rowland, 1990). Thus, the simplest T-branched hydrocarbons are the MMAs. Higher-order substitutions up to heptane are known (Gough et al., 1992; Warton et al., 1997). Detection of these compounds is routine, although it is critical to remove interfering normal and cyclic alkanes during sample preparation. One preparative method isolates the normal and monomethylalkanes by urea adduction and then removes the n-alkanes using 5-Å molecular sieves (Fowler and Douglas, 1987). An alternative method is to first remove n-alkanes from the saturated hydrocarbon fraction and then isolate the branched alkanes using ZSM-5 (also called silicalite) (Hoering and Freeman, 1984; West et al., 1990). GCMS is required in order to identify the substitution position, which is accomplished by comparing the mass fragments at the secondary or tertiary carbon positions (Figure 13.3). The assignment of the position(s) of substitution can be ambiguous without using pure standards, particularly for the dimethylalkanes and T-branched ($>C_1$) alkanes.

The origins of mid-chain MMAs may be more specific than those of isoalkanes or anteisoalkanes. Cyanobacteria contain relatively high concentrations of 4-methyl- through 8-methylalkanes in the range of C_{16}–C_{21}, with the 7- and 8-methyl isomers predominant (Gelpi et al., 1970; Han and Calvin, 1970; Han et al., 1968; Shiea et al., 1990). This biological signature is preserved in the most recent sediments associated with microbial mats (Robinson and Eglinton, 1990; Shiea et al., 1991; Kenig et al., 1995a). Many oils and source rocks, particularly of Precambrian age, contain abundant MMAs (Bazhenova and Arefiev, 1996; Fowler and Douglas, 1987; Jackson et al., 1986; Makushina et al., 1978; Summons, 1987; Warton et al., 1997; Höld et al., 1999). Enrichment of mid-chain MMAs in these ancient samples may indicate cyanobacteria in the source-rock setting. Summons et al. (1999) found 2-methylhopanes in ~2500-million-year-old rocks, suggesting the presence of cyanobacteria during their deposition. The correlation of mid-chain MMAs to cyanobacteria, however,

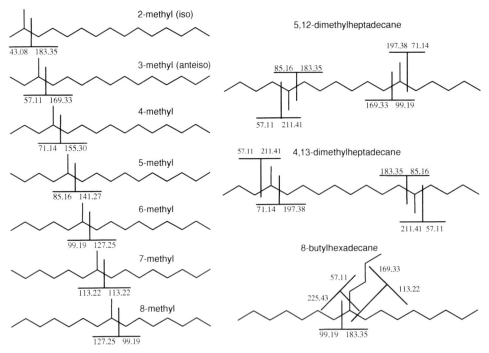

Figure 13.3. Examples of fragment ions that characterize monomethyl-, dimethyl-, and T-branched alkanes.

is not certain. In late Precambrian oil from Eastern Siberia, Makushina *et al.* (1978) and Fowler and Douglas (1987) identified a series of MMAs from C_{20}–C_{30}, dominated by 12- and 13-methylalkanes. Because no known modern precursors can give rise to these compounds, Fowler and Douglas (1987) argued for an extinct biological source. The proposed inorganic mechanisms of formation do not favor one isomer over another and cannot explain the predominance of the 12- and 13-methylalkanes. Potential bacterial precursors include diols, cyclopropyl acids, and other acids that occur in living organisms, but with shorter chain lengths (<C_{20}). Sulfur-bound C_{18}–C_{32} 9-methylalkanes in bitumen from an organic sulfur-rich limestone of the Upper Jurassic Calcaires en Plaquettes Formation (southern Jura, France) suggest the existence of such biological precursors (Van Kaam-Peters and Sinninghe Damsté, 1997). The predominance of higher mid-chain MMAs in Precambrian oils and source rocks and their declining significance in younger samples also suggest an origin from cyanobacteria. Thiel *et al.* (1999a) found abundant mid-chain branched carboxylic acids, dominated by methylhexadecanoic and methyloctadecanoic acids, in the lipid assemblages of the living fossil stromatoporoid *Astrosclera willeyana* and the desmosponge *Agelas oroides*. These complex mixtures of isomeric acids apparently originate from symbiotic bacteria living exclusively in the desmosponges. Comparison with hydrocarbon assemblages in ancient carbonates suggests that the mid-chain branched monocarboxylic acids are precursors of the analogous MMAs in these rocks and related crude oils. Mid-chain MMAs should not be used as markers of specific biological input or age without supporting evidence.

Extracts and hydrous pyrolyzates of Carboniferous-Permian torbanites from southern Africa and eastern Australia contain four unique ^{13}C-rich homologous series of MMAs ranging from C_{23} to C_{31+}, which occur in greater abundance than the more commonly occurring MMAs (Audino *et al.*, 2001a). Each unique series begins with the 2-methylalkane, while each member of a particular homologous series has a common alkyl group but differs in the number of carbon atoms in the second alkyl group. Each series differs from the next series by two carbon atoms (Figure 13.4).

Audino *et al.* (2001a) proposed that the four unique series of MMAs in torbanites may originate from botryals during intense bacterial reworking of algal

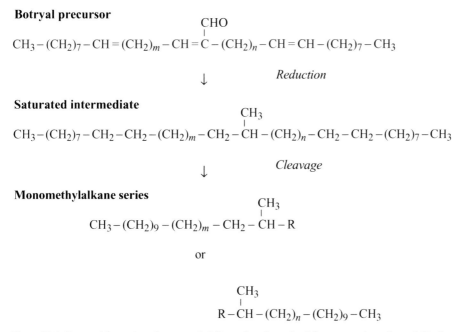

Figure 13.4. Proposed formation of monomethylalkanes from botryals of *Botryococcus braunii* race A (Audino *et al.*, 2001a). m = odd 15–21, n = even 14–20, R = CH_3, CH_2CH_3, $CH_2CH_2CH_3$, etc.

biomass. Botryals are C_{52}–C_{64} even-carbon-numbered, α-branched, α-unsaturated aldehyldes that represent up to 45% of the lipids in some strains of extant *B. braunii* race A (Metzger *et al.*, 1989). Reduction of these botryals would yield long-chain C_{52}–C_{64} even-carbon-numbered mid-branched MMAs. Cleavage of one chain at any position could yield the various series of MMAs, each differing by two carbon atoms from the adjacent homolog, as indicated in Figure 13.4. If botryals are the precursors for the four unique MMA series in torbanites, then the fossil organism must have produced them with somewhat different carbon-number distributions compared with those in living *B. braunii*. The carbon-number range for the first member of each series in the proposed reaction (C_{27}–C_{35}) differs from that observed in the torbanite samples (C_{23}–C_{31+}), implying that the carbon-number distribution in the torbanites requires *n* and *m* to begin at 10 and 11 carbon atoms, respectively (Figure 13.4), rather than at 14 and 15 carbon atoms as in the living organism. Intense bacterial reworking of torbanites that contain abundant microscopic remains of *B. braunii* may explain why they generally lack *B. braunii* biomarkers, such as botryoccane (Derenne *et al.*, 1988; Audino *et al.*, 2001a). Alternatively, some races of *B. braunii* do not produce these biomarkers (Metzger *et al.*, 1991).

Ethyl- and higher *n*-alkyl substituted T-branched alkanes appear to be mainly products of thermal maturation. Warton *et al.* (1997) found all possible isomers of C_{18} T-branched alkanes in proportions approximately equal to their relative thermal stability. Although some unusual ethyl-branched fatty acids are known to occur in goats and sheep, formation of ethyl- and higher *n*-alkyl T-branched alkanes likely occurs by isomeric equilibrium, as suggested by Hoering (1980) to explain the distribution of MMAs.

Schouten *et al.* (1998c) identified C_{27} and C_{29} 3-isopropylalkanes in the Kimmeridge Clay Formation. The biological origin and significance of these compounds is unknown, although they are enriched in ^{13}C ($\delta^{13}C \sim -17‰$), suggesting an origin from *Chlorobiaceae*.

Ether-bound methylated alkanes

> Isolated from low-maturity sediments via selective cleavage, these biomarkers are of uncertain affinity. They are possibly derived from tetraether lipids hypothesized to be present in archaea or eubacteria that have evolved from hyperthermophilic microorganisms as evolutionary adaptations to low-temperature terrigenous and/or marine environments.

Several isoalkanes occur as ether-cleaved products that have no known biological precursors (Figure 13.5). Chappe *et al.* (1980) found C_{30}

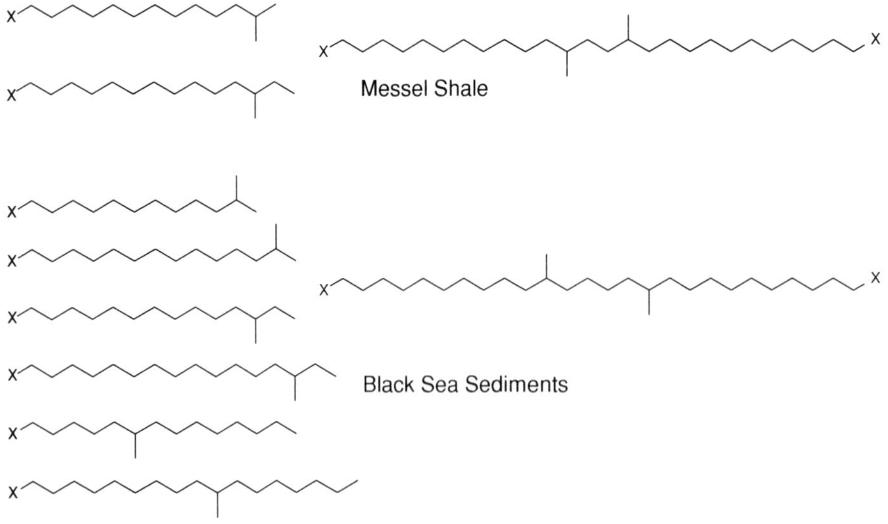

Figure 13.5. Ether-bound lipids release from Messel oil shale (Chappe *et al.*, 1980) and Black Sea sediments (Vella and Holzer, 1990). Ether-cleavage site indicated by x.

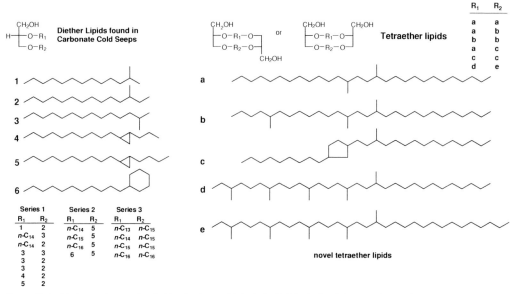

Figure 13.6. Diether and tetraether lipids of unknown biological affinities isolated from modern and ancient lacustrine and coastal marine sediments (see Pancost et al. (2001b) and Schouten et al. (2000c)).

13,16-dimethyloctacosane as the major peak in the degradation of Messel Shale that also liberated large quantities of C_{40} biphytanes. 13,16-Dimethyloctacosane is the dimer of isopentadecane (2-methylpentadecane), which, along with 3-methylpentadecane, was also released by ether cleavage. The linkage of isopentadecanes to form 13,16-dimethyloctacosane may be analogous to that between phytanes to form biphytanes, but it may be fortuitous. Vella and Holzer (1990) reported a similar compound, 12,17-dimethyloctacosane, and various MMAs in a Black Sea core. The biological origins of these compounds are not known, and they could be formed by condensation reactions during diagenesis.

Pancost et al. (2001b) identified several series of non-isoprenoid diether lipids in carbonate crusts precipitated from methane-rich bottom and pore waters associated with Mediterranean mud volcanoes. The diether lipids contained a variety of ether-bound C_{14}–C_{17} alkyl units (e.g. n-alkyl, branched alkyl, alkylcyclohexyl, alkylcyclopropyl, methylenealkyl, and 11,12-methylenehexadecyl). The sedimentary diether lipids are similar to those found in thermophilic eubacteria (e.g. *Thermodesulfobacterium commune*, *Ammonifex degensii*, and *Aquifex pyrophilus*). Schouten et al. (2000c) identified several novel tetraether lipids that are likely precursors for the larger dimethyl compounds (Figure 13.6). The lipids were present in both ancient and modern sediments that were deposited in lacustrine or coastal marine settings that received significant amounts of terrigenous organic matter. Only archaea are known to synthesize tetraether lipids; however, as noted above, diether lipids containing two isopentadecyl moieties are known to occur in several thermophilic eubacteria. Schouten et al. (2000c) speculated that these novel tetraether lipids might be synthesized by eubacteria that evolved from thermophilic species via lateral transfer of archaeal genes. The adaptation of tetraether lipid synthesis may allow these unknown eubacteria to thrive in low-temperature, terrigenous environments.

Regular (head-to-tail) acyclic isoprenoids

> Multiple sources. Measured using gas chromatography/flame ionization detection (GC/FID) or GCMS reconstructed ion chromatograms (e.g. m/z 183).

The homologous series of regular head-to-tail isoprenoid hydrocarbons extends from C_8 (2,6-dimethylheptane) to as high as C_{250+}. The high-molecular-weight isoprenoids (C_{80+}) originate from polyprenoids or dolichols (similar to polyprenoids, except that the OH-terminal isoprene residue is saturated). Lower-molecular-weight isoprenoids may originate from a host of biomolecules, including algal and bacterial chlorophylls, tocopherols, archaeal lipids, and proteins, as well as by thermal degradation of larger isoprenoids.

Regular head-to-tail $\leq C_{20}$ isoprenoid hydrocarbons are ubiquitous biomarkers that are usually the most prevalent class of hydrocarbons in crude oils and rock extracts after normal alkanes (Volkman and Maxwell, 1986). In contrast, regular C_{21+} isoprenoid hydrocarbons are usually trace compounds that are specific to the archaea.

Acyclic isoprenoids (>C_{20})

> Highly specific for certain archaea and for correlation. Measured using m/z 183, m/z 253, or other fragments (Figure 8.20).

The higher-molecular-weight acyclic isoprenoids include the regular isoprenoids (head-to-tail-linked) ranging up to C_{45} (Albaigés, 1980) and various irregular isoprenoids. Some examples of irregular isoprenoids include the head-to-head isoprenoids with a parent C_{40} compound (biphytane) and lower pseudohomologs (Moldowan and Seifert, 1979; Chappe et al., 1980; Albaigés, 1980), and various compounds like squalane (C_{30}, tail-to-tail), lycopane (C_{40}, tail-to-tail) (see Figure 8.20), and polymethylsqualanes (see Figure 13.27).

Head-to-head isoprenoids are specific markers of archaeal input to sediments. A series of C_{28}–C_{39} head-to-head-linked isoprenoids in mildly biodegraded oil from northwest Siberia are believed to originate from archaeal cell walls (Petrov et al., 1990). Most of these compounds consist of pristane coupled to a second isoprenoid unit in the range C_{10}–C_{20}. The C_{12} and C_{17} members of the second unit are absent because this would require the breaking of two carbon–carbon bonds at a branch position during maturation. Petrov et al. (1990) also found a range of components consisting of phytane coupled to lower isoprenoids (i.e. C_{20}–C_{19}, C_{20}–C_{18}, C_{20}–C_{16}, C_{20}–C_{13}, C_{20}–C_{11}). These results extend the carbon-number range and structural variety of these compounds previously reported by Moldowan and Seifert (1979).

The acyclic isoprenoids are commonly ignored in correlation studies because their analysis by GCMS requires removal of the n-alkanes, which have the same fragment ions and often dominate in intensity. Special gas chromatographic conditions may be required to separate the parent compounds (C_{40} homologs) of each series. This may be one reason why lycopane (tail-to-tail, C_{40}) (Figure 8.20) has been identified only rarely in oils (Albaigés et al., 1985). Another reason could be that lycopane originates from an unsaturated to polyunsaturated C_{40}-isoprenoid. These alkenes can be preserved as sulfur-bound species under anoxic marine conditions (de Leeuw and Sinninghe Damsté, 1990). For example, large concentrations of lycopane occur in hydrocarbons liberated from the sulfur-rich "red band" fraction isolated from immature, high-sulfur oils (P. Albrecht, 1991, personal communication).

Acyclic isoprenoids are important for correlation of crude oils from lacustrine source rocks where sterane concentrations are commonly very low. For example, distributions and relative concentrations of C_{33}–C_{40} head-to-head isoprenoids, C_{19}–C_{31} regular isoprenoids, and botryococcanes were used to genetically classify lacustrine oils from Sumatra (Figure 13.26) (Seifert and Moldowan, 1981).

Certain acyclic isoprenoids can be used as biological markers for methanogenic bacteria (e.g. C_{31}–C_{40} head-to-head isoprenoids) (Risatti et al., 1984; Rowland, 1990). Methanogens are found only under highly reducing conditions. They use the anaerobic process of fermentation to derive energy from various oxidized forms of organic matter. For example, Brassell et al. (1981) showed that 2,6,10,15,19-pentamethylicosane and squalane (Figure 2.15) in source rocks and petroleum indicate past biological methanogenesis. Squalane is common in petroleum but has not been used for correlation (Gardner and Whitehead, 1972).

Squalane has been used as a biomarker for archaea and hypersaline depositional environments (ten Haven et al., 1988). The precursor for squalane, squalene (Figure 3.15), is ubiquitous in sediments because it is the starting material for all polycyclic triterpenes and sterols. Squalene and its saturated analog squalane are acyclic and symmetrical, containing two farnesyl residues linked tail-to-tail (Figures 13.20 and 13.21). Squalene has been isolated from asphaltenes in petroleum (Samman et al., 1981). High concentrations of squalenes occur in living halophilic bacteria (Tornabene, 1978).

Pristane/phytane

> Low specificity for redox conditions in the source rock due to interference by thermal maturity and source input. Measured from FID/gas chromatography or reconstructed ion chromatogram (e.g. m/z 183).

The most abundant source of pristane (C_{19}) and phytane (C_{20}) is the phytyl side chain of chlorophyll a in phototrophic organisms and bacteriochlorophyll a and b in purple sulfur bacteria (e.g. Brooks et al., 1969; Powell and McKirdy, 1973). Reducing or anoxic conditions in sediments promote cleavage of the phytyl side chain to yield phytol, which undergoes reduction to dihydrophytol and then phytane. Oxic conditions promote the competing conversion of phytol to pristane by oxidation of phytol to phytenic acid, decarboxylation to pristene, and then reduction to pristane (Figure 13.7). A common precursor for pristane and phytane in crude oils is inferred by the similarity of their $\delta^{13}C$ values, which commonly differ by no more than $\pm\ 0.3$‰. Furthermore, these isoprenoids have $\delta^{13}C$ ~4–5‰ more negative than the associated tetrapyrroles, consistent with their derivation from chlorophyll (Hayes et al., 1990).

Chappe et al. (1982), Illich (1983), Goosens et al. (1984) and Rowland (1990) describe other likely sources for pristane and phytane, such as archaea. For example, dihydrophytol is a component in archaeal cell membranes and a building block for kerogen (Chappe et al., 1982).

Didyk et al. (1978) interpreted the redox conditions of the source-rock depositional environment for crude oil based on a model for the origin of pristane (Pr) and phytane (Ph) like that in Figure 13.7. According to these authors, Pr/Ph <1 in crude oil indicates anoxic source-rock deposition, particularly when accompanied by high porphyrin and sulfur contents, while Pr/Ph >1 indicates oxic deposition. Pr/Ph is commonly applied because Pr and Ph are measured easily using gas chromatography. Figure 13.8 shows how Pr/Ph and the canonical variable (Sofer, 1984) readily distinguish crude oils from different source rocks. Figure 13.8 is based on >500 worldwide crude oil samples used to predict source-rock depositional environments. Crude oils from coal/resin and paralic shale source rocks have high values for Pr/Ph and the canonical variable, as expected for oxic source-rock depositional environments.

For rock and oil samples within the oil-generative window, pristane/phytane correlates weakly with the depositional redox conditions. High Pr/Ph (>3.0) indicates terrigenous organic matter input under oxic conditions, while low values (<0.8) typify anoxic, commonly hypersaline or carbonate environments. However, most crude oils have Pr/Ph that fall within a fairly narrow range (0.8–3), which, while still influenced by depositional redox conditions, is influenced by other factors. Inferences from Pr/Ph on the redox potential of the source sediments should always be supported by

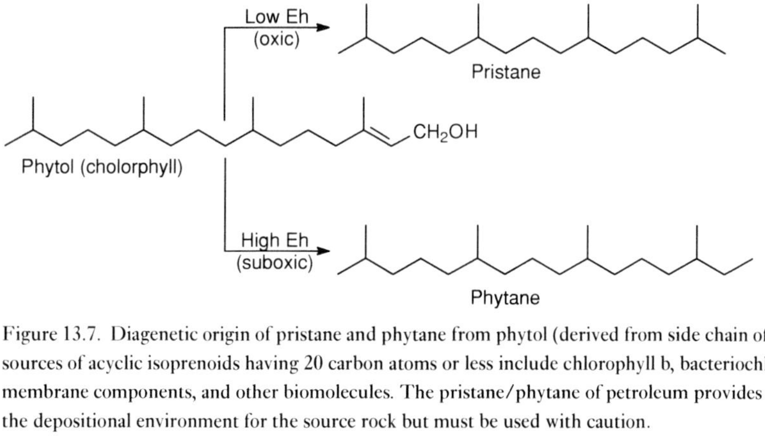

Figure 13.7. Diagenetic origin of pristane and phytane from phytol (derived from side chain of chlorophyll a) (see Figure 3.24). Other sources of acyclic isoprenoids having 20 carbon atoms or less include chlorophyll b, bacteriochlorophyll a, tocopherols, archaeal membrane components, and other biomolecules. The pristane/phytane of petroleum provides information on the redox potential of the depositional environment for the source rock but must be used with caution.

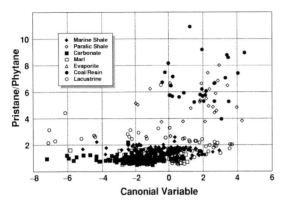

Figure 13.8. Crude oils from many coal/resin and paralic marine shale source rocks have high values for the canonical variable (CV) and pristane/phytane, consistent with oxic, terrigenous source-rock depositional settings. $CV = -2.53\,\delta^{13}C_{sat} + 2.22\,\delta^{13}C_{aro} - 11.65$ for the C_{15+} saturate and aromatic hydrocarbon fractions (Sofer, 1984). CV >0.47 indicates mainly waxy terrigenous oils, while CV <0.47 indicates mainly non-waxy marine oils. Coal/resin samples are from the Gippsland, Ardjuna, and Taranaki basins, Sumatra, Java, and the Gulf of Thailand, respectively, while paralic marine shale samples are from Nigeria, the Western Desert (Egypt), the Maturin Basin (Venezuela), and the Mahakam Delta (Indonesia). Figure courtesy of GeoMark Research, Inc. (J. E. Zumberge, 2000 personal communication).

The origins of pristane and phytane, however, are much more complex than simply the reduction or oxidation of the phytol side chain in chlorophylls. Although phytane and some pristane may evolve by this mechanism, most of the pristane in crude oils originates by thermal cleavage of isoprenoid moieties bound by non-hydrolyzable C–C and/or C–O bonds within the kerogen matrix. This was demonstrated by the generation of prist-1-ene using flash pyrolysis of low-maturity kerogens (0.4–0.7% R_o) (Larter et al., 1979; 1983). Goosens et al. (1984) suggested that tocopherols could be a major source of pristane. These compounds are minor constituents in plants and algal lipid membranes, where they serve primarily as antioxidants. Because these compounds form along the same biochemical pathways as chlorophyll, pristane derived from tocopherols has the same $\delta^{13}C$ as the phytol side chain. Tocopherols, however, are of substantially lower abundance than chlorophyll.

Methyltrimethyltridecylchromans (MTTCs) have been proposed as an alternative source of pristane and phytane. The chromans are similar but structurally unrelated to tocopherols. A biological origin was proposed for MTTCs because they occur in sediments and crude

other geochemical and geologic data. Typically, conditions of source-rock deposition inferred from Pr/Ph of oils agree with other indicators, such as sulfur content or the C_{35} homohopane index. For example, the Baghewala-1 oil from India has low Pr/Ph (0.9), high sulfur (1.2 wt.%), and high C_{35} homohopane index (12) (see Figure 13.86), consistent with anoxia during deposition of the source rock (Peters et al., 1995). As another example, many high-sulfur Gulf Coast oils have low Pr/Ph, and vice versa (Walters and Cassa, 1985). Pr/Ph for extracts from source rocks also commonly agree with other redox indicators. For example, Pr/(Pr + Ph) correlates with increasing relative amounts of C_{27} diasteranes in a sequence of Toarcian source rocks from southwest Germany (Figure 13.9). Oxicity and clay activity (acidity) in the water column and sediments during source-rock deposition apparently control the observed trend. These source rocks had a rather consistent organic matter input and are marginally mature throughout the 16-m depth interval in the study (Moldowan et al., 1986). Interpretation of depositional environment for more mature samples on this type of plot is not recommended.

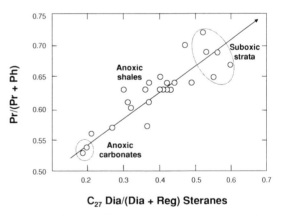

Figure 13.9. Extracts of Toarcian marine source rocks from the Dotternhausen Quarry in southwestern Germany, showing a positive correlation between pristane/(pristane + phytane) and C_{27} diasteranes/(diasteranes + regular steranes) that was controlled by the depositional environment (Moldowan et al., 1994b). Pr/(Pr + Ph) increases with clay content, as measured by increasing diasteranes, which parallels oxidative strength (Eh) of the water column during deposition of these rocks (Moldowan et al., 1986). The arrow shows the expected direction of a thermal maturity effect, which is not a factor for these low-maturity rocks.

oils only with specific alkylation patterns (Sinninghe Damsté et al., 1987b), possibly originating from eubacteria or archaea (de Leeuw and Sinninghe Damsté, 1990; Kenig et al., 1995b). Li et al. (1995a) proposed that MTTCs formed during diagenesis from chlorophyll and alkylphenols (Figure 13.10). The chromans are thus the result of condensation reactions that give rise to a "pseudo-biomarker," a compound that originated abiotically from different biochemicals giving the appearance of a molecule that could be biosynthesized. Whether the chromans are true biomarkers continues to be debated (Li and Larter, 1995; Sinninghe Damsté and de Leeuw, 1995). Nevertheless, they appear to correlate with paleosalinity (Schwark et al., 1998; Grice et al., 1998a).

Ten Haven et al. (1987) recommend against drawing conclusions on the oxicity of the environment of deposition from Pr/Ph alone. They show that low Pr/Ph (<1) is typical of hypersaline environments, where the ratio appears to be controlled by the effects of variable salinity on halophilic bacteria. For example, a relationship was found between Pr/Ph and the gammacerane index for oils generated from Angolan source rocks (see Figure 13.97). Considered in isolation, neither of these parameters could be used with confidence to assess the salinity of the depositional environment. However, the supporting relationship between Pr/Ph and gammacerane index reinforces the inferred salinity relationship. Likewise, Schwark et al. (1998) note a general decrease in Pr/Ph during deposition of the Permian Kupferschiefer sequence in Germany, which they interpret to indicate increasing paleosalinity. This interpretation gains support by a parallel increase in trimethylated 2-methyl-2-(4,8,12-trimethyltridecyl)chromans, aromatic compounds be-

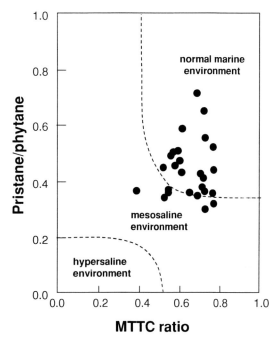

Figure 13.11. Relationship between pristane/phytane and the methyltrimethyltridecylchroman (MTTC) ratio in Jurassic Malm ζ stage carbonates. The MTTC ratio is defined as the 5,7,8-trimethylchroman/total MTTCs (Sinninghe Damsté et al., 1993c). Reprinted from Schwark et al. (1998). © Copyright 1998, with permission from Elsevier.

lieved to be markers of salinity (Sinninghe Damsté et al., 1987b; Grice et al., 1998a). Similar interpretations followed a study of the Jurassic Malm ζ stage carbonates in Eastern Bavaria (Figure 13.11) (Schwark et al., 1998).

We now know that the utility of Pr/Ph to accurately describe the redox state of paleoenvironments is limited by several factors:

- *Variable source input:* In addition to chlorophyll, many other biomolecules may give rise to pristane and/or phytane. These include unsaturated isoprenoids in zooplankton (Blumer et al., 1963; Blumer and Snyder, 1965) and higher animals (Blumer and Thomas, 1965), tocopherols (Goosens et al., 1984), and archaeal ether lipids (Risatti et al., 1984; Rowland, 1990; Navale, 1994).
- *Different rates of early generation:* Phytane is frequently abundant compared with pristane in low-maturity oils and source-rock extracts. The resulting low Pr/Ph ratios are inconsistent with higher

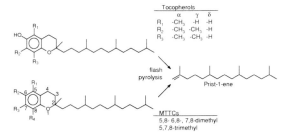

Figure 13.10. Tocopherols and methyltrimethyltridecylchromans as precursors for pristane (Li et al., 1995a).

Pr/Ph ratios observed for equivalent samples of higher maturity (Volkman and Maxwell, 1986). This discrepancy may be due to preferential release of sulfur-bound phytols from certain source rocks during early maturation (Kohnen, 1991; de Graaf et al., 1992).

- *Variations at higher maturity:* Petroleum Pr/Ph generally increases with increasing thermal maturity (Connan, 1974), while Ph/n-C_{18} decreases (ten Haven et al., 1987). However, Pr/Ph maturation trends are not systematic. For example, in a relatively uniform sequence of argillites from the Douala Basin in Cameroon, Pr/Ph first increases with depth to 4.9 in the principal zone of oil formation but then decreases to 1.5 at higher levels of maturity (Albrecht et al., 1976). Similarly, others found an increase in Pr/Ph to maxima at 0.7% (Connan, 1984), 0.9% (Radke et al., 1980), and 1.0% R_o (Brooks et al., 1969) for different coals, beyond which the ratio decreased. Conversely, Pr/Ph decreased for oils of increasing maturity generated by pyrolysis of Green River oil shale (Burnham et al., 1982). This may be due to preferential release of pristane compared with phytane precursors from this kerogen during early catagenesis. However, Koopmans et al. (1999) concluded that the increase in Pr/Ph with maturity is controlled mainly by more pristane precursors in source kerogens than different timing of generation and expulsion of pristane and phytane.
- *Analytical uncertainty:* Pr/Ph is usually determined from gas chromatographic analysis of whole-oil or C_{15+} saturate fractions. Co-elution with other isoprenoid hydrocarbons can perturb Pr/Ph. A highly branched, irregular isoprenoid in petroleum [2,6,10-trimethyl-7(3-methylbutyl)-dodecane] (see Figure 13.22) co-elutes with pristane on most capillary columns (Volkman and Maxwell, 1986). Mass chromatograms of m/z 168 can be used to show this compound, even in the presence of large amounts of pristane. Crocetane (see Figure 13.17) co-elutes with phytane.

In summary, we do not recommend that Pr/Ph ratios in the range of 0.8–3.0 be interpreted to indicate specific paleoenvironmental conditions without corroborating data. Pr/Ph >3.0 indicates terrigenous plant input deposited under oxic to suboxic conditions, while Pr/Ph <0.8 indicates saline to hypersaline conditions associated with evaporite and carbonate deposition.

C_{13}–C_{20} regular isoprenoids

> Low biological specificity. Used in oil–oil and oil–source rock correlations. Measured from FID/gas chromatogram or reconstructed ion chromatogram (e.g. m/z 183).

Although the entire regular isoprenoid series could be used for correlation, typically only the C_{13}–C_{20} isoprenoids are measured from whole-oil gas chromatograms. The C_9–C_{11} homologs are difficult to resolve from other light hydrocarbons, and C_{21+} homologs usually require additional sample fractionation and mass spectrometric detection.

Much like pristane and phytane, the C_{13}–C_{18} regular isoprenoids originate from multiple sources in addition to the phytol side chain of chlorophyll (Illich, 1983). Didyk et al. (1978) suggested that C_{13}–C_{20} regular isoprenoids could be used as markers for terrigenous input. However, isoprenoid side chains are common in many biomolecules without phylogenetic specificity. For example, bacteriochlorophylls c, d, e, and g have esterified farnesyl side chains, C_{15} isoprenoids are common constituents of archaeal lipids, and many protein chains terminate with farnesol. Diagenetic precursors of the C_{13}–C_{18} isoprenoid hydrocarbons, such as aldehydes and acids, are common constituents in sediments. For example, C_{13}–C_{18} ketones are found in various recent marine sediments (e.g. Leif and Simoneit, 1995) and in immature source-rock extracts (Azevedo et al., 2001).

While the general lack of precursor specificity limits the C_{13}–C_{18} regular isoprenoids as source indicators, their distributions are useful for correlation. Comparisons among crude oils can be made by line plots of the normalized or absolute abundance of the C_{13}–C_{20} isoprenoids or by ratios of individual compounds (e.g. Anders et al., 1985; Sulistyo, 1994). The C_{17} isoprenoid is typically low or absent and is not included in the distribution plots. Thermal maturation, phase separation, evaporation, and other non-source-related processes can alter the distribution of the C_{13}–C_{20} isoprenoids.

C_{21+} regular isoprenoids

> Probably originate from halophilic archaeal lipids. Specific for saline to hypersaline environments. Measured using FID/gas chromatogram or reconstructed ion chromatogram (e.g. m/z 183).

Regular head-to-tail acyclic isoprenoids with more than 20 carbons are diagnostic of saline to hypersaline sourcerock depositional environments. The C_{21}–C_{25} homologs are usually the most abundant within the series. The C_{25} compound was first proposed as a marker for saline lagoonal settings (Waples et al., 1974). Since then, these compounds have been observed in other hypersaline environments (Fu Jiamo et al., 1988; ten Haven et al., 1988; Keely and Maxwell, 1993; Sinninghe Damsté et al., 1993c; Grice et al., 1998a) and in crude oils generated from hypersaline source rocks (Albaigés, 1980). The sources of C_{21+} regular isoprenoids are not known with certainty, but they likely originate from halophilic archaea. The C_{21+} isoprenoids in extracts from the haliterich Sdom Formation in the Dead Sea Basin of Israel are ~7‰ enriched in ^{13}C compared with algal biomarkers, indicating a different source (Grice et al., 1998b).

Archaeal diether and tetraether lipids

> Highly specific for archaea. Usually measured indirectly by cleavage of the ether bonds and subsequent reduction to a saturated hydrocarbon.

Archaea use di- and tetraether-bound isoprenoids as main structural components of their membranes (Figure 13.12) (de Rosa et al., 1991). Typical diether archaeal lipids contain C_{20}, C_{25}, and C_{40} isoprenoids and can be considered counterparts of the diglycerides in the bacteria and eukaryotes. The tetraether lipids are common to many methanogens and most thermophilic *Crenarchaeota*. The C_{40} isoprenoids are bound by ether bonds and commonly consist of biphytane and/or isoprenoids with one to four pentacyclic rings. Highgrowth temperatures cause thermophiles to increase the degree of cyclization. Addition of pentacyclic rings into the polar lipids shortens the tetraethers and provides enhanced rigidity and more efficient packing. Psychrophilic marine *Crenarchaeota* add a hexacyclic ring.

The archaeal polar lipids are ubiquitous in sediments, but they are reported rarely because they require special analysis. Historically, sedimentary ether

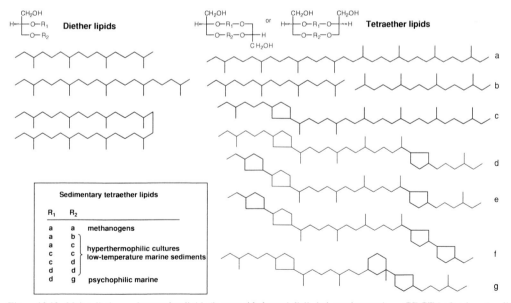

Figure 13.12. Major diether and tetraether lipids (isoprenoid glycerol dialkyl glycerol tetraethers, GDGTs) of archaea (modified from de Rosa et al. (1991) and Schouten et al. (2000c)). Extreme thermophilic *Crenarchaeota* can increase the number of pentacyclic rings as a response to increasing temperature. Psychrophilic marine *Crenarchaeota* add a hexyl ring to maintain membrane fluidity at low temperatures. Intact tetraethers that have been isolated in marine sediments are listed.

lipids were determined by an indirect chemical degradation method. Ether bonds are cleaved using hydrogen iodide (the preferred reagent), BCl_3, or BBr_3, producing alkyl halides. The alkyl halides are then reduced to saturated or deuterium-labeled hydrocarbons using either $HI/NaSCH_3$ or $LiAlD_4$ (Hoering, 1972; Michaelis and Albrecht, 1979; Chappe et al., 1982). Alternatively, the aklyl halides may be converted to methylthioethers using $NaSCH_3$ (Hoefs et al., 1997). These reactions usually are performed on the separated polar fraction, but they may be used on total rock extracts or even isolated kerogens and asphaltenes. Isolated alcohol fractions are derivatized to iodides or acetates before chromatography (e.g. Pauly and Van Vleet, 1986a; 1986b). The evolved saturated or derivatized hydrocarbons can be separated and detected using standard GCMS procedures. Diagnostic mass fragments can be used, although positive identification requires co-injection and/or full-scan mass spectra for comparison to standards that can be prepared by chemical degradation of lipids extracted from archaeal cultures. Chromatographic separation of the numerous C_{40} isoprenoids, such as biphytane and lycopane, may be difficult.

High-performance liquid chromatographic (HPLC) separation and mass spectrometric techniques for direct detection are now possible (Nichols et al., 1987; Hopmans et al., 2000). These advances allow the isolation of the major and minor GDGT lipids from microbial cultures, recent sediments, and source rocks. The individual, pure lipids can then be characterized fully by chemical degradation with $HI/LiAlH_4$ or $HI/NaSCH_3$ and by nuclear magnetic resonance (NMR).

Ether-bound isoprenoids occur in various marine and lacustrine rocks. Michaelis and Albrecht (1979) and Chappe et al. (1979; 1980) found ether-bound C_{15}, C_{20}, C_{30}, and C_{40} (biphytane and a monocyclic biphytane) isoprenoids in kerogen from the lacustrine Messel oil shale. Ether-bound acyclic and cyclic isoprenoids occur in nearly all source-rock depositional settings, including normal marine shales and marls, hypersaline evaporites (Chappe et al., 1982; Schouten et al., 1998a; Schouten et al., 2000c), and in many recent sediments, hydrothermal springs and vents, surface waters (Hoefs et al., 1997), and peat bogs (Pancost et al., 2000b). The ubiquitous occurrence of acyclic biphytanes and related compounds with one or more cyclic rings indicates that these cyclized C_{40} isoprenoids may originate from planktonic archaea rather than solely from hyperthermophiles (Hoefs et al., 1997; Schouten et al., 2000c).

While the archaeal polar lipids appear to be ubiquitous and probably occur in most low-maturity bitumens and oils, few studies report their presence. Chappe et al. (1982) reported biphytane released from the polar fractions of oils from the North Sea and Aquitaine basins and the dipentyl C_{40} isoprenoid from Mahakam Delta oil. While free biphytane has been reported in many oils, only one study of oil from offshore Spain identifies the free monopentacyclic isoprenoid (Albaigés et al., 1985).

Because the tetraether polar lipids are unique to archaea, many studies of recent sediments attempt to use them as measures of archaeal activity and input. Martz et al. (1983) used polar lipid concentrations to estimate methanogenic biomass and methane production rates. Sewer sludge and freshwater sediments were inferred to have average methanogen biomass slightly <1%, while in estuarine sediments the methanogenic biomass was estimated to be <0.01%. Pauly and Van Vleet (1986b) conducted a similar study of Florida swamp sediments and concluded that methanogenic lipids constitute ~0.1% of the total organic matter and that there are correlations between methanogenic lipids, methane production rates, and $\delta^{13}C$ distributions. Using the Florida swamp data and analyses of oceanic sediments, Pauly and Van Vleet (1986a) proposed that depth profiles of the archaeal lipids differentiate environments and that the zones with high lipid concentrations correspond to sulfate-reducing zones, suggesting a coupling of methanogenesis with methane oxidation. Using HPLC for the direct measurement of diether and tetraether polar lipids, Nichols et al. (1987) equated their concentrations to biomass in soil samples where methanogens were grown under controlled conditions, in sewer sludge, and in a waste treatment digester. Smith and Floodgate (1992) claim that lipid concentrations can be used to estimate methanogen biomass in fresh waters and polluted estuarine sediments, but their claims are unsubstantiated. Hopmans et al. (2000) coupled HPLC with atmospheric pressure mass spectrometry to analyze tetraether lipids in archaeal cells.

The correlation between microbial activity, biomass, and abundance of specific archaeal lipids in sediments is tenuous. In various open marine recent sediments, Schouten et al. (2000c) found that the acyclic biphytanyl GDGT of non-thermophilic methangens constituted a fairly consistent 30–45% of the total, even though a number of sites have little or no methane production. These samples also contained 29–42% of the GDGT associated with pyschrophilic marine-pelagic *Crenarchaeota*. Although significant, the amount of GDGT is low considering that these archaea probably have the greatest biomass (Karner et al., 2001). The remaining GDGTs (5–30%) were those with pentacyclic rings usually associated with hyperthermophilic archaea. The presence of these compounds in a low-temperature environment indicates that there must be alternative sources.

Considering the difficulties in using polar lipids to determine archaeal biomass in recent sediments, confusion over their significance in older sedimentary organic matter is understandable. The presence of ether-bound isoprenoids in kerogens (Chappe et al., 1980), surface sediments (Hoefs et al., 1997), and head-to-head isoprenoids in petroleum (Moldowan and Seifert, 1979) led these and others to conclude that the archaea may contribute significant amounts of organic matter to source rocks. Sinninghe Damsté and Schouten (1997) challenged this conclusion, pointing out that even in sediments richest in ether-bound C_{40} isoprenoids, these compounds still account for only <0.3% of the total organic carbon. Furthermore, Schouten et al. (1997a) found no correlation between biphytane concentration and the $\delta^{13}C$ of organic matter in the Monterey Formation. A more direct comparison between the $\delta^{13}C$ of 2,6,10,15,19-pentamethylicosane (PMI, a C_{25} isoprenoid from methanogens) and $\delta^{13}C$ of the total organic matter showed differences of 10‰. These results prompted Sinninghe Damsté and Schouten (1997) to conclude that there is little evidence for significant contributions of archaeal biomass to sedimentary organic matter. The same researchers reconsidered their conclusions upon examining samples from the Monterey Formation, noting that the concentrations of the acyclic and cyclic biphytanes vary drastically (∼0–1500 mg/kg bitumen) compared with sulfur-bound cholestane or C_{35} hopane (∼10–150 mg/kg bitumen) (Schouten et al., 1998a). The reasons for these large variations in the sedimentary column were not defined, which is understandable because the factors that control modern planktonic archaeal populations are not known. Nevertheless, because ether-bound isoprenoids can be ten times more abundant than the sulfur-bound algal and bacterial lipids, the archaea probably contribute significant amounts of organic matter to the geosphere.

Based on analysis of soluble and insoluble organic matter in black shales, Kuypers et al. (2001) concluded that up to 80 wt.% of the sedimentary carbon deposited during an Albian oceanic anoxic event (∼112 Ma) originated from marine, non-thermophilic archaea. Archaeal lipids are abundant in the organic-rich black shales of the anoxic event but are largely absent in adjacent rocks. Extractable archaeal markers included the acyclic isoprenoids 2,6,15,19-tetramethylicosane (TMI) and 2,6,10,15,19-pentamethylicosane (PMI). Treatment of the polar fraction from the black shale interval with HI/LiAlH$_4$ released ether-bound archaeal markers, including acyclic, monocyclic, bicyclic, and tricyclic biphytanes (C_{40}). HPLC/mass spectrometry revealed four isoprenoid GDGTs, which are the main constituents of archaeal membranes. Two of these GDGTs were characteristic of the *Crenarchaeota*, hyperthermophilic archaea that thrive at temperatures >60°C. Another compound containing a hexyl ring accounts for 60% of the total GDGTs and represents the first evidence for non-thermophilic *Crenarachaeota*. Components of unambiguous archaeal origin, such as PMI, TMI, and the biphytanes, have $\delta^{13}C$ >10‰ more positive than algal biomarkers, such as cholestane and 24-ethylcholestane. The weighted average of the $\delta^{13}C$ for the chemically released isoprenoids ($-14‰$) agrees with that of the insoluble organic matter in black shale interval that accounts for more than 95% of the bulk organic matter. Therefore, the sharp increase in $\delta^{13}C$ for the bulk organic carbon during the anoxic event is attributed to increased contribution of ^{13}C-rich archaeal-derived organic carbon.

Archaeal alcohols and acids

> Highly specific for archaea. Measured using GCMS or CSIA as either trimethylsilyl (Hinrichs et al., 2000) or methyl ester (Pearson et al., 2001) derivatives.

Dicarboxylic acids with acyclic and cyclic biphytane skeletons tentatively occur in sediment extracts from Morocco (Meunier-Christmann, 1988). Ahmed et al. (2001) and de Lemos Scofield (1990) reported similar compounds in extracts from the Monterey Formation. Biphytanyl diols also occur in modern sediments. Hoering (1972) first identified the head-to-head isoprenoid structure of biphytanol in sediments from the

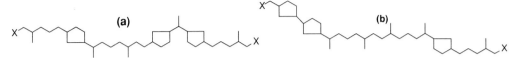

Figure 13.13. Proposed structure (a) for a tricyclic biphytane (ether-bound and as free diols) found in various sediments in contrast to (b) the tricyclic biphytane in the archaeal polar lipids.

Cariaco Trench. This work predates the first report of C_{40} ether-bound isoprenoids in a thermophilic organism (de Rosa et al., 1974) and their structural elucidation (de Rosa et al., 1977a; 1977b; 1977c). The detection of biphytanol is one of the earliest applications of the HI/LiAlH$_4$ or LiAlD$_4$ chemical degradation reaction to geologic samples (Hoering, 1971).

Schouten et al. (1998a) found diols with both acyclic and cyclic biphytane skeletons in various sediments, including the structure later identified as having a hexyl ring that originated from psychrophilic marine archaea (Sinninghe Damsté et al., 2002a). Schouten et al. (1998a) found both diols with acyclic and cyclic biphytane skeletons in various sediments. Biphytane in sediments is believed to have a different structure than that in the membranes of archaea (Figure 13.13). Schouten et al. (1998a) suggest that the sedimentary biphytane has the skeleton of tricyclic biphytane dicarboxylic acids described in the previously listed references.

The origins of the biphytane alcohols and acids are not known with certainty. Schouten et al. (1998a) noted several problems, with the conclusion that these compounds are diagenetic products of the archaeal ether-bound membrane lipids. Foremost is that the free diols and the ether-bound biphytanes have different distributions, suggesting independent sources. Schouten et al. (1998a) speculated that the poorly studied planktonic archaea may biosynthesize such diols. This could explain the origins of sulfur-bound biphytanes released from kerogens (Höld et al., 1998; Richnow et al., 1992) as free biosynthesized diols, which would be available for sulfur incorporation during early diagenesis.

Head-to-head acyclic isoprenoids

> Highly specific for archaea and as a correlation tool. Measured using m/z 183, m/z 253, or a variety of other fragments (e.g. see Figure 8.20).

The head-to-head acyclic isoprenoid hydrocarbons are highly specific for archaea. These biomarkers originate from $\omega\omega'$-biphytanediol, a key component in the di- and tetraether lipids that are major structural components of cell membranes in some archaea. These lipids are unique and distinguish archaea from eukarya and bacteria (Figure 13.14). Bacteria and eukaryotes use fatty acids and various derivatives of cyclized squalene as major structural components in their lipid membranes.

Moldowan and Seifert (1979) first reported head-to-head isoprenoids ranging from C_{32} to C_{40} in crude oils and source rock extracts and related their source to archaeal lipids. Petrov et al. (1990) extended the series down to C_{29}. The head-to-head isoprenoids consist of a pristane unit (iC_{19}) joined to one unit in the range iC_{10}–iC_{19} or a phytane unit (iC_{20}) joined to one unit in the range iC_{11}–iC_{20}. Linkages to iC_{12} or iC_{17} units are absent because this would require the breaking of two carbon–carbon bonds at a branch position during maturation. Petrov et al. (1990) also identified an unusual C_{28} head-to-head isoprenoid composed of linked iC_{14} units. The origins of this isoprenoid and the lower homologs of the 3,7,11,... polymethylalkane series ranging from C_{17} to C_{24} are uncertain, but they probably also originate from archaeal lipids. These head-to-head isoprenoids occur in oils from northwest Siberia (Petrov et al., 1973; 1990), Costa Rica (Haug and Curry, 1974), offshore Spain (Albaigés, 1980), the San Joaquin Valley, California (Moldowan and Seifert, 1979), and Sumatra, Indonesia (Seifert and Moldowan, 1981) (see Figure 13.26). Haug and Curry (1974) were the first to speculate that the irregular isoprenoids may originate from a C_{40} biphytane precursor, although an

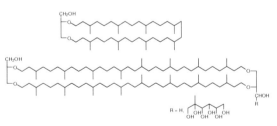

Figure 13.14. Diether (top) and tetraether (bottom) archaeal lipids containing biphytane.

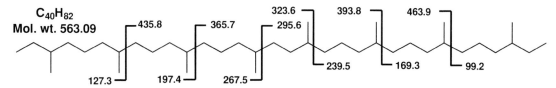

Figure 13.15. Biphytane ($C_{40}H_{82}$ head-to-head isoprenoid) yields diagnostic mass spectral fragment ions at m/z 197, 259, 267, 323, 383, 393, and 463.

archaeal source was not yet known. Albaigés et al. (1985) detected archaeal C_{40} biphytanes with head-to-head, tail-to-tail, and head-to-tail linkages in western Mediterranean oils (Amposta and Tarraco fields) and related the occurrence of C_{21}–C_{24} isoprenoids and quasi-isoprenoids to the thermal cracking of the C_{40} precursors.

The head-to-head acyclic isoprenoids are not used commonly in correlation studies because their analysis by GCMS requires removal of the n-alkanes, which have many of the same fragment ions and generally dominate in intensity. Biphytane (3,7,11,15,18,22,26,30-octamethyldotriacontane, also called bisphytane) (Figure 13.15) typically elutes during gas chromatography around nC_{34}–nC_{35} and can be measured using various mass fragments characteristic of the head-to-head isoprenoid linkage. The mass fragments typically deprotonate in the mass spectrometer, resulting in diagnostic doublets at m/z 196–197, 266–267, 280–281, 322–323, 382–383, and 462–463. The molecular ion at 562.64 amu is weak and not detected easily. Particular care must be taken to separate biphytane from other C_{40} isoprenoids, such as lycopane, and positive identification requires a full-scan mass spectrum.

Lower homologs show mass spectral fragments similar to those of biphytane. Doublets at m/z 252–253 and 308–309 are particularly diagnostic. All head-to-head isoprenoids yield m/z 183 mass fragments upon ionization, although it is not necessarily as abundant as for the regular head-to-tail isoprenoids. Consequently, single-ion monitoring of m/z 183 is diagnostic for both series and can be used for correlation studies. Albaigés and Albrecht (1979) used regular and irregular isoprenoid distributions to identify marine oil pollution, recognizing their resistance to biodegradation and weathering. The distributions and relative concentrations of C_{33}–C_{40} head-to-head isoprenoids compared with the C_{19}–C_{31} regular isoprenoids and botryococcanes differentiate various lacustrine Sumatran oils (Seifert and Moldowan, 1981). Because Sumatran oils contain very low steranes and have similar triterpane distributions, the acyclic isoprenoids prove useful for correlation of these oils (see Figure 13.26).

Tail-to-tail acyclic isoprenoids

Acyclic isoprenoid hydrocarbons with a tail-to-tail linkage have a range of source specificity. For example, squalane has little source specificity because its precursor, squalene, is the prescursor for all polycyclic triterpanoids and sterols. On the other hand, 2,6,10,15,19-pentamethylicosane (PMI) is highly specific for methanogenic and methanotrophic archaea.

2,6,10,15,19-Pentamethylicosane and crocetane

> Specific for methanogenic and methanotrophic archaea and useful as correlation tools. PMI measured using m/z 183, m/z 239, or various other fragments (Figure 13.16). Crocetane co-elutes with phytane and is measured using m/z 169 and m/z 197.

Crocetane (2,6,11,15-tetramethylhexadecane, $C_{20}H_{42}$) and PMI (2,6,10,15,19-pentamethylicosane, $C_{25}H_{52}$; sometimes spelled pentamethyleicosane or PME in older literature) are tail-to-tail-linked isoprenoid hydrocarbons that are diagnostic for methanogenic archaea and/or anaerobic methane oxidizing consortia comprised of methanotrophic archaea (reverse methanogenesis) and sulfate-reducing bacteria. Although an archaeal source for these compounds was long suspected, proof of their origins required compound-specific isotopic analysis, which reveals extreme depletion of ^{13}C that can arise only by oxidation of biogenic methane.

Crocetane is not observed easily because it typically co-elutes with phytane (Robson and Rowland, 1993). Crocetane can be separated partially from phytane

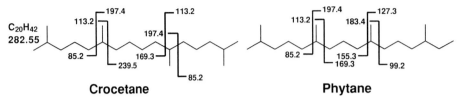

Figure 13.16. Comparison of major mass fragments (electron impact ionization) of crocetane and phytane. Crocetane has diagnostic mass fragments at m/z 113, 169, and 197. Phytane shares these ion fragments, but also yields a diagnostic ion at m/z 183.

(R = 0.6) using a 25-m squalane (CpSil 2CB) capillary column run isothermally at 170°C with hydrogen as a carrier gas (Thiel et al., 1999b). Barber et al. (2001a) reported that better resolution (R = 0.8) can be achieved using a 50-m capillary column with a permethyl-β-cyclodextrin (β-CYDEX from SGE) stationary phase. This separation allows for routine GCMS (selected ion monitoring, SIM). The mass spectra of crocetane and phytane are almost identical, but there are diagnostic differences. Unlike the regular isoprenoids, crocetane does not yield significant m/z 183 ions. Instead, mass fragments of m/z 113, 169, and 197 are diagnostic (Figure 13.16). Improved measurements can be obtained using GCMS/MS analysis with the mass transitions of 197→127/126 and 169→126 being diagnostic for crocetane and 183→127 being diagnostic for phytane (Greenwood and Summons, 2003).

PMI is measured easily by conventional GCMS, although its identification can be difficult because of co-elution with the regular C_{25} isoprenoid as well as some unidentified compounds. The mass spectrum of PMI contains the usual ion fragments associated with isoprenoid hydrocarbons (e.g. m/z 183) with diagnostic fragments at m/z 239 and 267 (Figure 13.17). The ratio of m/z 239 to 253 is >>1 for PMI, which distinguishes it from other C_{25} isoprenoids (2,6,10,14,18- and 2,6,10,14,19-pentamethylicosanes) that yield m/z 239/253 <1. GCMS/MS offers improved selectivity using the mass transitions of 352→267 and 352→239 (Greenwood and Summons, 2003). Positive identification of PMI and crocetane requires a full-scan mass spectrum and/or co-injected standard.

Brassell et al. (1981) found PMI in recent sediments and ancient rocks soon after it was identified as a free lipid in methanogenic archaea (Holzer et al., 1979; Tornabene et al., 1979). PMI is particularly abundant in Methanosarcina barkeri, where it is the principal free lipid and pyrolyzate hydrocarbon (Risatti et al., 1984; Rowland, 1990). Unsaturated 2,6,10,15,19-pentamethylicosenes containing three to five double bonds occur in various methanogenic archaea (Schouten et al., 1997b). Summons et al. (1996) showed large isotopic fractionations for PMI and other lipids in methanogens grown in the laboratory.

PMI and lycopane dominated the hydrocarbon fraction of particulates collected from water below the oxic/anoxic boundary in the Cariaco Trench (Wakeham and Ertel, 1988). In a similar study, Freeman et al. (1994) found PMI in both oxic and anoxic waters from the Black Sea and the Cariaco Trench. PMI in the water column seemed inconsistent with the occurrence of methanogenesis below the layer of bacterial sulfate reduction. Freeman et al. (1994) suggested a planktonic, non-archaeal source for PMI in the water column. These studies predate the general recognition of planktonic archaea that can occur in oxic and anoxic waters (De Long, 1992; De Long et al., 1994), and PMI is now considered

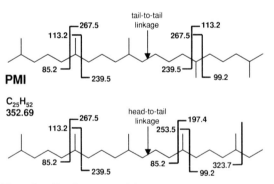

Figure 13.17. Comparison of major mass fragments (electron impact ionization) for 2,6,10,15,19-pentamethylicosane (PMI) with the regular C_{25} isoprenoid. PMI yields diagnostic mass spectral fragment ions at m/z 239 and 267. The regular C_{25} isoprenoid yields these ions plus diagnostic ions at m/z 253 and 323.

a reliable biomarker for methanogenic and methanotrophic archaea. Noble and Henk (1998) used the concentration of PMI as an indicator for methanogens and determined that the mid-outer neritic facies are responsible for the microbial gas in the Terang-Sirasun Field in the East Java Sea, Indonesia. This study showed that large microbial methane accumulations do not necessarily require early entrapment and that traps may fill with methane long after the deposition of source beds, provided that the conditions for active methanogenesis are maintained. Extracts of immature sediments associated with cold seeps and carbonates derived from the anaerobic oxidation of methane contain both saturated and unsaturated forms of crocetane and PMI (Figure 13.18) (Sinninghe Damsté et al., 1997; Schouten et al., 1997b; Elvert et al., 1999; Elvert et al., 2000). Cold seeps are sites of methane- or sulfide-rich fluid release from the sea floor commonly associated with chemosynthetic communities.

Syntrophic consortia of archaea and sulfate-reducing bacteria represent a major link in the global carbon cycle. In anoxic marine sediments, methane generation is confined to sulfate-depleted zones, but methane oxidation occurs within a transition zone where both methane and sulfate decline (Hoehler et al., 1994). They suggested that a consortium of methanogens and sulfate-reducing bacteria conduct anaerobic methane oxidation or "reverse methanogenesis" at cold seeps throughout the world. Ancient cold-seep deposits occur in Maastrichtian-Danian shale of the Moreno Formation, California (Schwartz et al., 2003), which generated crude oil in the nearby Oil City Field (Peters et al., 1994a).

Crocetane, PMI, and associated archaeal lipids may be major hydrocarbons in extracts from modern and ancient sediments associated with anaerobic methane oxidation. Bian (1994) first reported crocetane in modern marine sediments from Kattegat (Denmark/Sweden) with $\delta^{13}C$ similar to that of PMI, a marker for methanogens. The isotopic and structural similarities between these compounds suggested that crocetane may also be a marker for archaea. Crocetane and PMI and their unsaturated equivalents have now been studied at several sites of anaerobic methane oxidation. Their abundance indicates a pronounced role of archaea in the biogeochemical cycling of carbon at

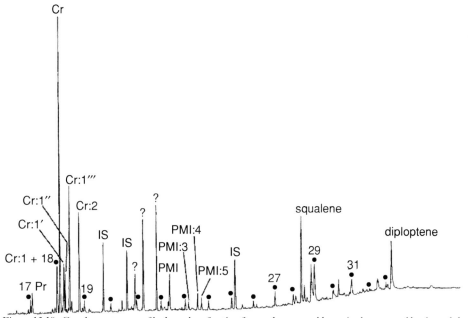

Figure 13.18. Gas chromatogram of hydrocarbon fraction from a deep-sea cold seep in the eastern Aleutian subduction zone. Normal alkanes indicated by • and carbon number. Reprinted from Elvert et al. (2000). © Copyright 2000, with permission from Elsevier. Cr, crocetane; Cr:1 and Cr:2, crocetenes with one or two double bonds, respectively; IS, internal standard; Ph, phytane; PMI, 2,6,10,15,19-pentamethylicosane; PMI:3, PMI:4, PMI:5, PMI with multiple double bonds; Pr, pristane; ?, unknown cyclic compounds.

Table 13.6. *Summary of the occurrence of crocetane (Cr), 2,6,10,15,19-pentamethylicosane (PMI), the isoprenoid gylcerol diether archaeols (Ar) and hydroxyarchaeols (hyAr) (adapted from Elvert* et al., *2000)*

Cold seep environment	Cr	PMI	Ar	hyAr	References
Recent, gas hydrate, Eel River Basin	None	None	Moderate	Moderate	Hinrichs et al. (1999)
Recent, Santa Barbara Basin	Low	Low	High	High	Hinrichs et al. (2000)
Recent, gas hydrate, carbonates, Hydrate Ridge, Cascadia continental margin	High	High	High	Moderate	Elvert et al. (1999)
Fossil carbonates ("Calcari a Lucina")	High	High	None[a]	None	Thiel et al. (1999b)
Recent mud volcanoes (Mediterranean Sea)	High	Moderate	High	High	Pancost et al. (2000a; 2001a; 2001b), Aloisi et al. (2002)
Recent carbonates, Aleutian Trench	High	Minor	High[b]	None[c]	Elvert et al. (2000)
Recent sediments, Kattegat (Denmark/Sweden).	Moderate	Moderate	n.d.	n.d.	Bian et al. (2001)
Recent carbonates, Ukrainian shelf in the northwestern Black Sea	High	High	n.d.	n.d.	Thiel et al. (2001)
CH_4 seeps Aleutian accretionary margin	High	n.d.	High	n.d.	Greinert et al. (2002)
Black Sea waters	n.d.	High	n.d.	n.d.	Wakeham et al. (2003)
Gulf of Mexico gas hydrates	Low	Low	High	High	Zhang et al. (2003)
Modern sediments, small meromictic basin, Ellis Fjord, Antarctica	Low	Moderate	n.d.	n.d	Greenwood and Summons (2003)
Microbial mat, Lee Stocking Island, Bahamas	Low	n.d.	n.d	n.d	Greenwood and Summons (2003)
Barney Creek Formation, McArthur Basin Northern Territory, Australia	Low	n.d.	n.d	n.d	Greenwood and Summons (2003)

[a] No archaeol, but detected a ^{13}C-depleted ether lipid containing a phytanyl moiety.
[b] Archaeol not found at all stations studied (e.g. TV-GKG 40).
[c] No free hydroxyarchaeol, but detected monounsaturated archaeols.
n.d., not detected.

methane seeps (Table 13.6). These cold seeps typically have massive authigenic carbonates that are isotopically depleted in ^{13}C (Elvert et al., 1999). The isoprenoid hydrocarbons are also depleted in ^{13}C, indicating that they originated from methanotrophic archaea (reverse methanogenesis) that were consuming, and further fractionating, isotopically light, biogenic methane (see Figure 16.8). Elvert et al. (1999) reported irregular isoprenoids with $\delta^{13}C$ as low as −130.3‰. This contrasts with co-occurring modern algal marine lipids and hydrocarbons that typically are ∼−30‰ (e.g. Bian et al., 2001; Thiel et al., 2001). Because crocetane commonly co-elutes with phytane, direct measurement of its carbon isotope ratio is not possible. The values can be inferred by measuring the $\delta^{13}C$ of pristane and the peak containing co-eluting phytane and crocetane, assuming that the $\delta^{13}C$ of pristane and phytane are equal, and calculating the $\delta^{13}C$ of crocetane by difference. Biogenic carbonates and the microbial mats also contain authigenic, framboidal pyrite and isotopically depleted fatty acids, including *iso-* and *anteiso-* branched compounds, most likely derived from sulfate-reducing bacteria (e.g. Hinrichs *et al.*, 1999; 2000). The close association of these organisms supports a syntrophic relationship between sulfate reducers and archaea responsible for the anaerobic oxidation of methane.

These worldwide studies established the association of crocetane, PMI, and the isoprenoid glycerol diethers archaeol and 2-hydroarchaeol with methanogenesis and anaerobic methane oxidation (Table 13.6). Differing biomarker distributions at various cold seep sites indicate that there are various syntrophic consortia that perform anaerobic methane oxidation. It is likely that different consortia are viable depending on environmental conditions (e.g. water depth, temperature, degree of anoxia, and supply of sufficient free methane) (Elvert *et al.*, 2000; Pancost *et al.*, 2001b). Once these criteria are established and the biochemical pathways of the consortia are known, then the distribution of these diagnostic biomarkers may serve as a sensitive proxy for depositional conditions.

Until recently, crocetane and PMI were rarely reported in petroleum, but its scarcity may be due to problems with analytical separation rather than lack of occurrence. Barber *et al.* (2001b) reported crocetane, PMI, and a series of C_{22}–C_{24} irregular isoprenoids in oils from the Canning and Perth basins of Western Australia. Unlike the measurements made by Thiel *et al.* (1999b), the $\delta^{13}C$ of the combined crocetane/phytane peak was not unusually depleted, although crocetane abundance was very low. Greenwood and Summons (2003) confirmed the occurrence of crocetane in several of the Canning and Perth basin oils studied by Barber *et al.* (2001b) using similar chromatographic procedures and GCMS/MS detection. Several samples where crocetane appeared to be present by GCMS could not be verified, and PMI could not be detected in any of the samples.

2,6,15,19-Tetramethylicosane

> Origins unknown. Likely related to PMI with similar measurement and usage.

Vink *et al.* (1998) found C_{24} 2,6,15,19-tetramethylicosane and tentatively identified C_{26} 10-ethyl-2,6,15,19-tetramethylicosane in the saturated hydrocarbon fractions and kerogen pyrolyzates of samples from Lower Albian black shale from southeastern France (Figure 13.19). These compounds appear to be structurally related to PMI, and their co-occurrence and similar $\delta^{13}C$ support a common origin. Vink *et al.* (1998) suggested that these compounds provide evidence for an alternative biosynthetic route for PMI via the methylation of 2,6,15,19-tetramethylicosane rather than the tail-to-tail coupling of farnesyl and geranyl groups.

Squalene and squalane

> Ubiquitous, associated with certain archaea, and generally reliable as a correlation tool. Squalene is measured using m/z 410; squalane is measured using m/z 183, m/z 239, or a variety of other fragments (see Figure 8.20).

Squalene is common to all of life's domains, where it serves as a precursor to polycyclic terpenoids, steroids, and carotenoids (Figure 13.20). Squalene

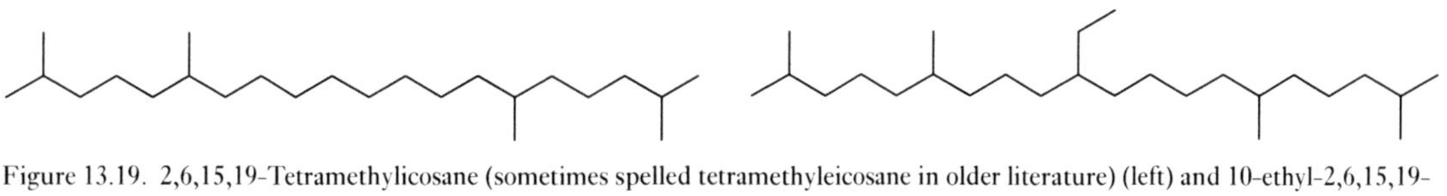

Figure 13.19. 2,6,15,19-Tetramethylicosane (sometimes spelled tetramethyleicosane in older literature) (left) and 10-ethyl-2,6,15,19-tetramethylicosane (right).

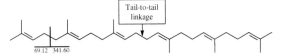

Figure 13.20. Squalene (2,6,10,15,19,23-hexamethyl-2,6,10,14,18,22-tetracosahexene).

(2,6,10,15,19,23-hexamethyl-2,6,10,14,18,22-tetracosahexene, $C_{30}H_{50}$) is a major lipid produced by methanogenic, halophilic, and thermoacidophilic archaea (Tornabene et al., 1978; 1979). Thus, abundant squalene in recent sediments may indicate archaeal input (Matsumoto and Watanuki, 1990). However, Colombo et al. (1996; 1997) showed that high squalene correlates with the selective preservation of terrigenous organic matter in sediments and sinking particulates from the Laurentian Trough. Squalene is common in diatom blooms (Matsueda et al., 1986), where it may be used to control buoyancy (Bieger et al., 1997).

Squalane (2,6,10,15,19,23-hexamethyltetracosane, $C_{30}H_{62}$) was one of the first biomarkers identified in crude oil (Gardner and Whitehead, 1972). Given the ubiquitous distribution of its unsaturated precursor (squalene), squalane probably occurs in most source rocks and oils. Squalane can be liberated by pyrolysis of asphaltenes (Samman et al., 1981; Strausz et al., 1982), or by desulfurization reactions with Raney nickel (Adam et al., 1993; 2000), and has been recovered from coal macerals by treatment with NaH (Stefanova, 2000). Squalane is easily separated and identified by standard GCMS methods. However, its mass fragmentation is common to most isoprenoids (Figure 13.21), and positive identification requires full-scan mass spectra and/or co-injection with a pure standard.

Because squalane has no unique biological origin, it is used infrequently as a biomarker in geochemical studies. High concentrations of squalane, however, are associated with input from archaea. Brassell et al. (1981) reported squalane as a major component of extracts from sediments with a correspondingly high concentration of 2,6,10,15,19-pentamethylicosane (PMI), a biomarker for methanogens. Freeman et al. (1994), however, found divergent $\delta^{13}C$ for PMI (-25.6‰) and two hydrogenated squalenes (hexahydrosqualene = -19.2‰, octahydrosqualene = -33.4‰) in sediments from the Cariaco Trench, indicating different biological sources for the three compounds.

High abundance of squalane in oils and rocks from hypersaline environments led ten Haven et al. (1988) to suggest it as a marker for hypersalinity. Mello et al. (1993; 1994a) exploited this association using squalane abundance to differentiate oils and bitumens from lacustrine sources deposited under varying water salinities. Grice et al. (1998b) provided evidence that squalane originates from halophilic bacteria in extracts from hypersaline deposits of the Sdom Formation near the Dead Sea by showing that squalane and regular isoprenoids (C_{21}–C_{25}) are isotopically heavier by >7‰ than the phytoplankton-derived steranes and hopanes. However, while hypersaline environments may promote the generation and preservation of squalane, the presence of this biomarker in oil or bitumen is not necessarily evidence of high salinity.

Highly branched isoprenoids

> Probable diatom markers in Jurassic(?) to Tertiary source rocks and related oils. Measured using a variety of mass fragments and specific metastable reaction monitoring (MRM) reactions (e.g. m/z 239 → 238).

The highly branched isoprenoids (HBIs) are a series of C_{20}, C_{25}, C_{30}, and C_{35} hydrocarbons first described in recent sediments from the Gulf of Mexico (Gearing et al., 1976) and in heavy oil from Rozel Point, Utah (Yon et al., 1982). They occur in many source rocks and crude oils, sometimes as the predominant compounds in the C_{15+} saturated hydrocarbon fraction.

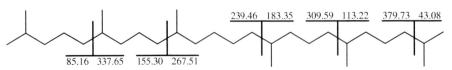

Figure 13.21. Squalane (2,6,10,15,19,23-hexamethyltetracosane).

Although HBIs were reported frequently, their structures remained controversial for a long time (Rowland and Robson, 1990). For example, Requejo and Quinn (1983) and Venkatesan (1988) reported that upon hydrogenation of a $C_{25}H_{48}$ HBI, a $C_{25}H_{50}$ hydrocarbon formed, leading to the conclusion that the parent compound was a monocyclic, monoene HBI. Rowland et al. (1990) showed later that the $C_{25}H_{48}$ HBI is actually a diene, where hydrogenation of one double bond is hindered.

C_{20} HBI was first identified as 2,6,10-trimethyl-7-(3-methylbutyl)-dodecane after isolating the compound from sulfur-rich Rozel Point crude oil, where it is the most abundant saturated hydrocarbon after phytane (Yon et al., 1982). C_{20} HBI co-elutes with pristane on most capillary columns (Volkman and Maxwell, 1986). Robson and Rowland (1986) later confirmed the structure of the sedimentary C_{25} HBI as 2,6,10,14-tetramethyl-7-(3-methylpentyl)-pentadecane. In addition to isolation, various NMR spectrometric analyses, and oxidative degradation, these structural identifications required the synthesis of reference alkanes for comparison of retention times and mass spectra of the naturally occurring and synthetic HBIs. Similar identification of the C_{30} HBI, 2,6,10,14-tetramethyl-7-(3-methylpentyl)-pentadecane (Robson and Rowland, 1988), and the tentative identification of the C_{35} HBI (Hoefs et al., 1995) followed.

HBIs are major compounds in the saturated hydrocarbon fractions of many modern coastal sediments. Their alkene precursors are prone to sulfurization during early diagenesis (Kohnen et al., 1990a; Werne et al., 2000). C_{25} HBI thiophenes, dibenzothiophene, C_1- and C_2-alkyldibenzothiophenes, and benzo[b]naphto[1,2-d]thiophene occur in Miocene to Pliocene sedimentary rocks in the Shinjo Basin, Japan (Katsumata and Shimoyama, 2001b). The C_{25} HBI thiophenes and C_{25} HBI alkane apparently originated by reaction of C_{25} HBI alkadiene and C_{25} HBI polyene, respectively, with reduced sulfur during early diagenesis.

Detection of the saturated HBIs by gas chromatography is now routine, but care must be taken to avoid misidentification. In crude oils and rock extracts, accurate assessment of HBIs requires enrichment using molecular sieves to remove the n-alkanes. The mass spectra of HBIs are similar to those of other isoprenoids, except for prominent fragment ions resulting from cleavage at the major side-chain branch point, including a predominant and unusual even-numbered ion fragment (Rowland and Robson, 1990). This fragment is m/z 238 for the C_{25} HBI (Figure 13.22). Because the mass fragments of all isoprenoid hydrocarbons are similar, scanning a single ion is insufficient for positive identification. Molecular ions of HBIs are weak. HBIs are characterized by fragmentation about the tertiary carbon followed by deprotonation, which results is a series of doublet ions. The deprotonated fragments are more abundant than the parent fragment, and the ratio of the doublet ions is diagnostic. For example, in the mass spectrum of the C_{25} HBI, the major mass fragments above m/z 85 are 238 and 239, where m/z 238 response is approximately twice that of m/z 239 (Figure 13.23). Doublets at m/z 210–211 and 266–267 are minor compared with the 238–239 ions, but they still show preference for the deprotonated ion. Signal-to-noise ratio can be improved using MRM techniques that focus on secondary fragmentation of the major doublet ions.

Determination of the biological origins of HBIs took longer than their molecular structures. Early studies inferred various marine planktonic and bacterial sources by comparing the abundance and isotopic composition of HBIs relative to their sedimentary distribution or to other biomarkers of known origins (see review by Rowland and Robson, 1990). Rowland et al. (1985) reported the first biological occurrence of HBIs (the C_{20} saturated HBI and related monoene and diene and a possible C_{25} diene) in two field samples of the green alga *Enteromorpha prolifera*. While this green alga could be the source, the authors cautioned that their field samples contained various other hydrocarbons, including an alkene typical of diatoms ($nC_{21:6}$) and that the HBI could be produced by epiphytic microalgae or bacteria associated with *Enteromorpha prolifera*. Summons et al. (1993) identified an isomer of the C_{25} HBI in a Western Australian benthic diatom community.

Suitable precursors for HBIs have been reported only in diatoms (Nichols et al., 1988; Volkman et al., 1994). The C_{25} HBIs, also called 2,6,10,14-tetramethyl-7-(3-methylpentyl)-pentadecane, occurs in crude oils and rock extracts as old as Cretaceous (Sinninghe

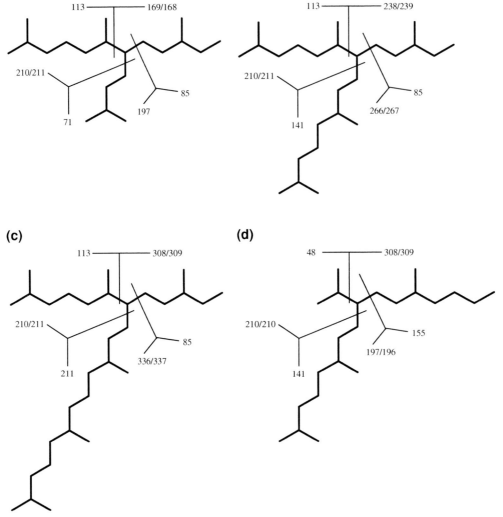

Figure 13.22. (a) C_{20} highly branched isoprenoid (HBI), [2,6,10-trimethyl-7-(3-methylbutyl)-dodecane]; (b) C_{25} HBI, [2,6,10,14-tetramethyl-7-(3-methylpentyl)-pentadecane]; (c) C_{30} HBI, [2,6,10,14,18-pentamethyl-7-(3-methylpentyl)-eicosane]; and (d) a C_{30} HBI pseudohomolog, [2,6,12,16-tetramethyl-9-isopropyl-octadecane].

Damsté et al., 1989a), suggesting that it may be a useful age marker. Diatoms originated in the Jurassic Period. The biological origins of HBIs remained uncertain until Nichols et al. (1988) identified high concentrations of the C_{25} diene in selected diatom blooms from Antarctica. In support, Summons et al. (1993) found a C_{25} HBI alkene in benthic diatomaceous microbial communities, and Volkman et al. (1994) discovered various C_{25} and C_{30} HBI alkenes in laboratory cultures of the marine diatoms *Halsea ostrearia* and *Rhizolenia setigera*. Biological precursors for C_{20} and C_{35} HBIs have not yet been found, but there are >15 000 species of diatoms to survey.

With diatoms established as the likely source of HBIs, research focused on environmental factors influencing the number and position of the unsaturated bonds. Seasonal variations were noted in various natural (Cooke et al., 1998; Hird and Rowland, 1995) and

Alkanes and acyclic isoprenoids   515

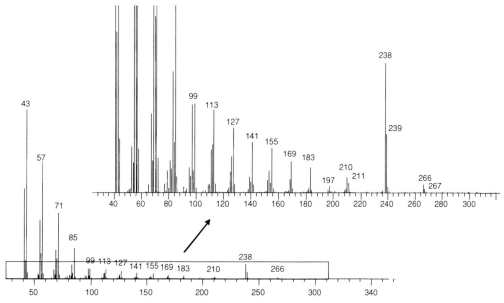

Figure 13.23. Mass spectrum of C_{25} HBI [2,6,10,14-tetramethyl-7-(3-methylpentyl)-pentadecane]. In the expanded view (top), spectral peaks below $\sim m/z$ 90 are truncated.

controlled environments (Wraige et al., 1997; 1998). Metabolic variations between different strains of diatom species also may be significant. Volkman et al. (1994) reported only C_{30} HBIs from a Pacific strain of *Rhizolenia setigera*, while Schouten et al. (1998b) and Sinninghe Damsté et al. (1999) reported only C_{25} HBIs in a North Atlantic strain of the same diatom species. Clearly, the genetic and environmental factors that determine the biological variability of HBI must be determined before these compounds can be used as markers for specific depositional conditions.

Tri- and tetra-unsaturated HBI alkenes were isolated from the diatom *Pleurosigma intermedium*, purified by column chromatography, and characterized by NMR spectroscopy and mass spectrometry (Belt et al., 2000). The compounds were used to identify previously unknown but common and abundant HBIs in many studies of sediments and biota. These HBIs are structurally different from those reported for other diatoms. The positions of the double bonds in the HBIs of *P. intermedium* are consistent with the positions of sulfur incorporation in some HBI thiolanes and thiophenes that were reported previously in sediments and oils. Diatoms that produce HBI alkenes are now known to include benthic, planktonic, marine, or freshwater species (Belt et al., 2001).

Diagenesis of HBIs in the geosphere is not understood fully. Diatoms appear to preferentially produce HBIs trienes, tetraenes, and pentaenes (Belt et al., 1996), while the HBIs in modern sediments consist of saturated compounds, monoenes, and dienes. Thus, hydrogenation of some of the double bonds is rapid, or the polyenes are degraded more rapidly, but the exact reaction pathways are not yet defined. The reactivity of the double bonds allows HBIs to readily incorporate inorganic sulfur during diagenesis. Sinninghe Damsté et al. (1989a) identified eight C_{20} and two C_{25} HBI thiophenes (HBITs) in a variety of sulfur-rich oil shales and oils, including Rozel Point oil. Since only a few of all possible isomers were detected, they concluded that sulfur incorporation into the unsaturated HBIs fixed the position of the double bond(s), limiting isomerization. However, the ever-increasing number of HBITs discovered with novel structures (de las Heras et al., 1997; Kohnen et al., 1990a; Rospondek et al., 1997; Sinninghe Damsté and Rijpstra, 1993) indicate that the reactions involving sulfur incorporation are complex.

Applications

Geochemical application of HBIs has changed with awareness of their origins. Initial observations suggested that HBIs were markers for marine plankton or bacteria. For example, Requejo and Quinn (1983) inferred marine input in Narrangansett Bay estuary sediment using HBIs, and Dunlop and Jefferies (1985) used HBIs to distinguish between deposition in oceanic and hypersaline environments in Western Australia. HBIs occur in some modern (Rowland and Robson, 1990) and ancient (Brassell et al., 1985) lacustrine sedimentary rocks, although typically in lower abundance than in recent marine sediments. HBIs are now considered to be ubiquitous in modern coastal sediments and are common in many lake sediments.

Recognition of diatoms as the likely biological source of HBIs promoted other geochemical applications. Diatoms evolved in the Jurassic Period and, thus, HBIs can be used as an age-related marker. These compounds are readily applicable to chemostratigraphic studies and may provide facies markers for source rocks deposited in upwelling zones when seasonal diatom blooms occur. For example, C_{25} HBI is the main saturated hydrocarbon in several facies of the Loma Chimico Shale, a low-maturity organic-rich source rock in Costa Rica (Figure 13.24). Abundant nutrients coupled with coastal upwelling resulted in high biological productivity and an expanded oxygen minimum zone that promoted preservation and the widespread occurrence of source facies dominated by diatom input (Erlich et al., 1996). The C_{25} HBI provides an ideal marker for mapping the distribution of this facies.

Botryococcanes

> Highly specific for *Botryococcus braunii* in brackish or lacustrine environments. Measured using m/z 183 chromatogram (or m/z 238, 239, 294, 295) (see Figure 8.20) of the saturated branched-cyclic fraction (isoprenoids separated from n-alkanes).

This irregular C_{34}-isoprenoid originates from botryococcene known to occur in only one living organism, *Botryococcus braunii*, a fresh-brackish water alga (Figure 13.25) (Moldowan and Seifert, 1980; McKirdy et al., 1986). Thus, botryococcane represents one of the most specific biomarkers used in paleoreconstruction. The occurrence of botryococcane and closely related 16-desmethylbotryococcane in petroleum is limited to a few regions, including Sumatra (Figure 13.26) (Seifert and Moldowan, 1981), Australia (McKirdy et al., 1986), and the Maoming oil shale, China (Brassell et al., 1985). Some extracts of *B. braunii* consist of a mixture of C_{30}–C_{37} branched alkenes and botryococcenes (Metzger et al., 1985a), which are the probable precursors for the more complex mixtures of botryococcane pseudo-homologs in a Sumatran oil (Seifert and Moldowan, 1981). Metzger et al. (1991) describe 16 C_{30}–C_{37} botryococcene structures that were determined from among

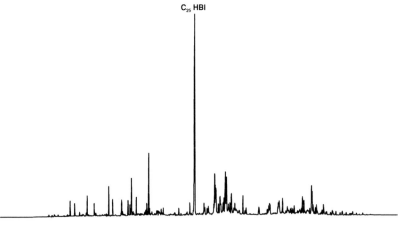

Figure 13.24. Gas chromatogram of the C_{15+} saturated hydrocarbon fraction extracted from the Upper Cretaceous Loma Chimico Shale (Morote #1 well), Costa Rica (see Erlich et al. (1996) for additional chromatograms of extracts from different organofacies of this source rock).

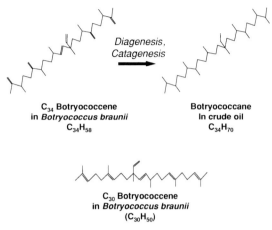

Figure 13.25. C_{34} botryococcane is a marker for the lacustrine alga, *Botryococcus braunii*. Botryococcane is identified on m/z 183 mass chromatograms of oils from Sumatra in Figure 13.27. The C_{30} botryococcene and other related botryococcene structures were identified in extracts from *Botryococcus braunii* by Metzger et al. (1991).

~50 chromatographic peaks isolated from *B. braunii*. The C_{30} botryococcene has a similar isoprenoid structure to that of the higher series members that are alkylated and cyclized in various ways. *B. braunii* is the major organism contributing to boghead coals, such as torbanite (Audino et al., 2001a; Grice et al., 2001), some Tertiary oil shales, and rubbery deposits of coorongite from Australia (Maxwell et al., 1968; Cane, 1969).

Three modern races of *B. braunii* have different hydrocarbon compositions (Metzger et al., 1991). Race A (formerly race PRBA, where PRB is polymeric resistant biopolymer) (Metzger et al., 1985b) yields odd-carbon-numbered straight-chain alkadienes and alkatrienes and, rarely, n-alkenes from C_{23} to C_{31}. Race L is characterized by lycopadiene, a C_{40} tetraterpene. Only race B (formerly race PRBB) synthesizes the C_{30}–C_{37} botryococcenes. Hence, the occurrence of botryococcane proves the presence of *B. braunii* in the depositional environment, but lack of botryococcane does not prove that *B. braunii* was absent (Derenne et al., 1988).

B. braunii is rich in hydrocarbons. In a major *B. braunii* bloom at Darwin River Reservoir, Australia, ~27–40% of the dry algal biomass was oil (Wake and Hillen, 1981). Some reports indicate 70–90 wt.% of botryococcenes could be harvested during the scenescent stage of the *B. braunii* life cycle (Maxwell et al.,

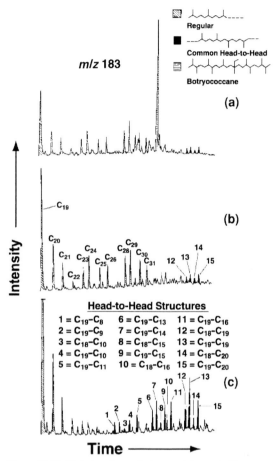

Figure 13.26. Variations in acyclic isoprenoids revealed by selected ion monitoring/gas chromatography/mass spectrometry (SIM/GCMS) m/z 183 of saturate fractions proved useful for correlation of Sumatran oils (Seifert and Moldowan, 1981). Like many other oils generated from lacustrine source rocks, these oils contain few steranes, which are commonly used for correlation. The isoprenoids were concentrated by urea adduction of the saturated fractions of the oils, thus removing the otherwise abundant n-alkanes, which interfere with the analysis. (a) Minas oil has botryococcane and low amounts of head-to-head isoprenoids compared with the other samples. (b) Petapahan oil has a regular and head-to-head isoprenoid distribution similar to Minas (a) but lacks botryococcane. (c) Damar oil has higher concentrations and more diverse head-to-head isoprenoids than either Minas (a) or Petapahan (b). Reprinted with permission by ChevronTexaco Exploration and Production Technology Company, a division of Chevron USA Inc.

1968; Knights et al., 1970). However, it is now clear that the three groups of characteristic hydrocarbons described above are race-related and are synthesized throughout all growth stages of race B (Metzger et al., 1991).

The absence of botryococcane in torbanites and some other deposits could be caused by heterotrophic microbial reworking and degradation of the precursors (i.e. botryococcenes), or transformation of the precursor by sequestration of reduced inorganic sulfur species during diagenesis (Audino et al., 2001a). However, botryococcenes undergo an aging process to form a substance called "*Botryococcus* rubber," presumably by oxidation and polymerization (Metzger et al., 1991). Little free botrycoccene remains after this process, which may be important in torbanite formation.

Polymethylsqualanes

> Highly specific for *B. braunii* and lacustrine environments. Measured by GCMS using m/z 350.

Summons et al. (2002b) determined the structure of 3,7,18,22-tetramethyl squalane (Figure 13.27, top) by synthesis of the standard, comparison of mass spectra, and co-elution with a component in the Maoming oil shale, China (Brassell et al., 1985). The structure of an isomer of this compound was inferred from the mass spectrum to be 3,7,11,14-tetramethyl squalane

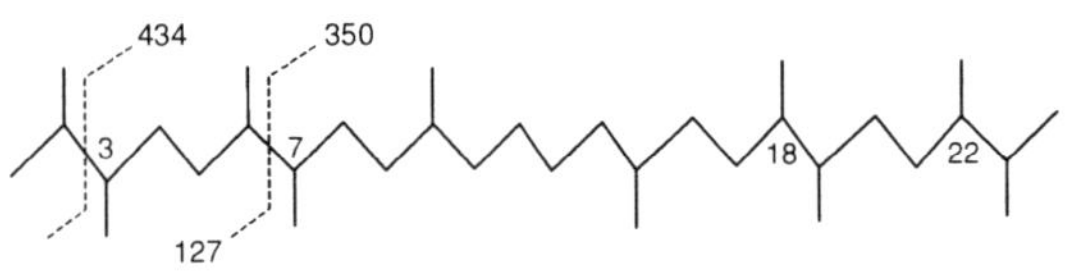

3,7,18,22-Tetramethylsqualane

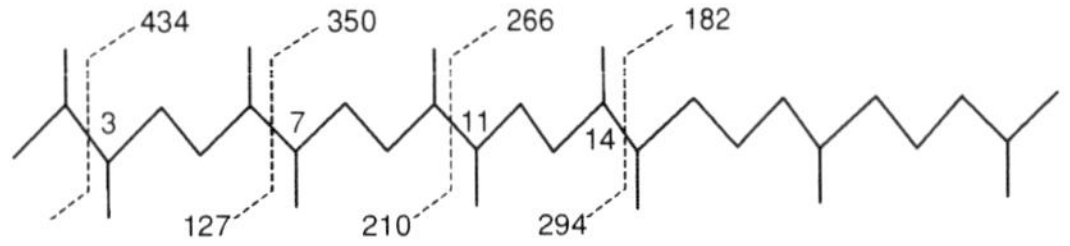

3,7,11,14-Tetramethylsqualane

Figure 13.27. Two polymethylsqualanes identified in extracts of Duri and Minas crude oil, Sumatra, Indonesia, by Summons et al. (2002b).

(Figure 13.27, bottom). They found that these polymethylsqualanes and botryococcane co-occur in extracts of Sumatran oil shale (Pematang Brown Shale) that contains abundant remains of the freshwater alga *B. braunii* and in the Duri and Minas crude oils that originated from this source rock (Figure 13.28). Polymethylsqualanes and botryoccanes also occur in coal and mudstone from a Maniguin Island core. The botryococcanes in these samples are enriched in ^{13}C, with $\delta^{13}C$ in the range of ∼−9 to −15‰. These values are similar to the $\delta^{13}C$ of the polymethylsqualanes of ∼−11 to −12‰, and differ substantially from other compounds in the extracts and oils, which range from −20‰ for C_{16}–C_{20} acyclic isoprenoids to −52‰ for C_{29} hopanes (Dowling et al., 1995). Co-occurrence with botryococcanes and similar ^{13}C-rich isotope compositions suggest strongly that polymethylsqualanes originate from *B. braunii*.

Lycopane

> Widespread in recent sediments and attributed to various biological precursors. Measured using various mass fragments or, preferably, full-scan mass spectrum.

Lycopane (2,6,10,14,19,23,27,31-octamethyl dotriacontane) (Figure 13.29) is a common biomarker in suspended particulate organic matter, recent sediments, and some Tertiary lacustrine source rocks, but it is identified infrequently in crude oils. The compound typically elutes near nC_{34}–nC_{35} and can be measured using various mass fragments characteristic of the tail-to-tail isoprenoid linkage. Particular care must be taken to separate lycopane from other C_{40} isoprenoids derived from archaeal lipids, and positive identification requires a full-scan mass spectrum.

Lycopane is widespread and is particularly abundant in lacustrine sediments that were deposited under anoxic conditions, such as the Messel (Freeman et al., 1990; Kimble et al., 1974a) and Condor shales (Freeman, 1991). Abundant lycopane has also been reported in hypersaline euxinic settings, such as the halite-rich dolostones of the Sdom Formation (Grice et al., 1998c) and in mesosaline carbonates, such as the Jurassic Malm ζ stage carbonates (Schwark et al., 1998). Lycopane is fairly common in modern marine sediments, such as the Cariaco Trench (Dastillung and Corbert, 1975;

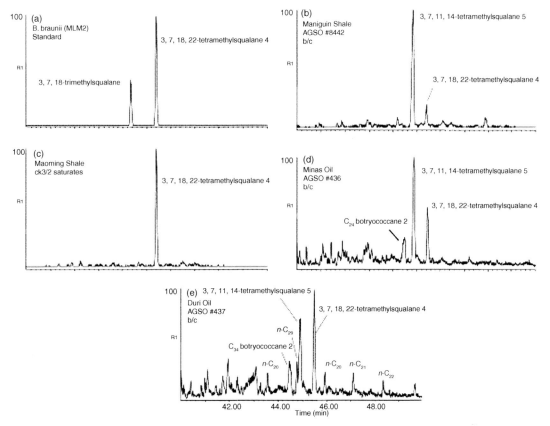

Figure 13.28. Reconstructed ion chromatograms (m/z 350) show tetramethylsqualanes in a standard and various geological samples. The trace for the standards (a) shows tri- and tetramethylsqualane prepared from precursor epoxides isolated from the MLM2 strain of *Botryococcus braunii*. 3,7,11,14-Tetramethylsqualane is the major polymethylsqualane in the Maoming oil shale (c) and the Duri oil (e). 3,7,11,14-Tetramethylsqualane is identified tentatively as the main isomer in samples of Maniguin Shale (b) and the Minas oil (d). Reprinted from Summons *et al.* (2002). © Copyright 2002, with permission from Elsevier.

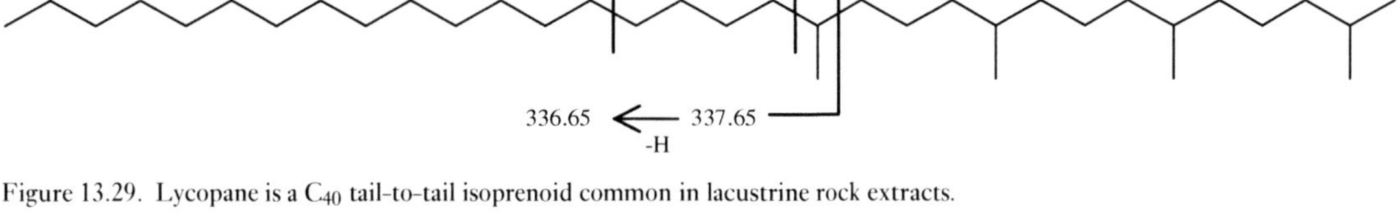

Figure 13.29. Lycopane is a C_{40} tail-to-tail isoprenoid common in lacustrine rock extracts.

Wakeham, 1990), Japan Trench (Brassell *et al.*, 1980), Walvis Bay (Wardroper, 1979), Peru upwelling zone (Farrington *et al.*, 1988a), Black Sea (Wakeham and Beier, 1991), and North Pacific gyre (Wakeham *et al.*, 1993), as well as marine source rocks, such as the Monterey Formation (Schouten *et al.*, 1997a).

Among many possible sources of lycopane, the most obvious is the C_{40} carotenoid lycopene (Figure 13.30), which occurs in numerous microorganisms, such as purple phototrophic sulfur bacteria, and in higher plants (Schmidt, 1978). An alternative source is *Botryococcus braunii* (L-strain only), which produces abundant

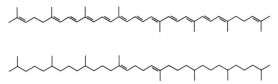

Figure 13.30. Lycopene (top) and lycopadiene (bottom).

lycopa-14(E),18(E)-diene (Metzger and Casadevall, 1987; Metzger et al., 1991). Brassell et al. (1981) suggested that methanogenic archaea might produce a precursor for lycopane, although this is now considered unlikely.

Although it can be a major component in the saturated hydrocarbon fraction of sediments, lycopane is identified rarely in crude oils. Albaigés et al. (1985) found lycopane in oils from the Amposta and Tarraco fields, offshore Spain. Lycopane, however, nearly co-eluted with two C_{40} head-to-head isoprenoids and required special peak smoothing and enhancement processing to resolve. Unsaturated lycopane precursors may also be altered through incorporation of sulfur during early diagenesis (de Leeuw and Sinninghe Damsté, 1990). These sulfurized species occur in low-maturity crude oils. For example, treatment of the sulfide fraction of Rozel Point oil with Raney nickel liberates abundant lycopane (Adam et al., 1993). Grice et al. (1998c) found a series of sulfurized biomarkers traced to lycopadiene in B. braunii. These organosulfur compounds are not formed directly from lycopadiene but originate from the early thermal release of polysulfide-bound lycopane precursors (Figure 13.31).

Macrocyclic alkanes

> Macrocyclic alkanes appear to originate from algaenan in the lacustrine alga B. braunii and consist of a homologous series from $\sim C_{15}$ to C_{34}. Measured using the homologous molecular ions in GCMS, or by MRM/GCMS using M+ $\to$ M $-$ 28 or the parent to m/z 111 transition.

Until a recent study by Audino et al. (2001b), non-isoprenoid macrocyclic alkanes were not known in the geosphere. They used co-elution with authentic standards to identify a series of macrocyclic alkanes (C_{15}–C_{34}, maximum at C_{21}) and their methylated homologs (C_{17}–C_{26}) that were particularly abundant in a Carboniferous torbanite that was rich in B. braunii. These compounds can be characterized by their strong molecular ions and by fragment ions characteristic of cycloalkanes (m/z 69, 83, 97, and 111) (Figure 13.32). Their distribution was similar to that of the torbanite n-alkanes, and they were liberated from B. braunii algaenan (Audino et al., 2001c).

The only previous identification was for the C_{10} and C_{16} series in semi-coking oil derived from an Estonian oil shale (Müürisepp et al., 1994). All torbanites contain abundant n-alkanes and novel macrocyclic alkanes with similar $\delta^{13}C$, thus apparently originating from algaenan in B. braunii (Grice et al., 2001). These macrocyclic alkanes were found with botryococcane in Tertiary crude oils from Minas and Duri in Sumatra (Audino et al.,

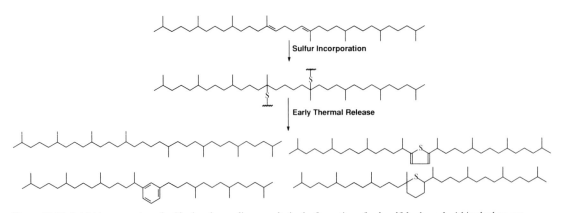

Figure 13.31. Initial incorporation of sulfur into lycopadiene results in the formation of polysulfides bound within the kerogen matrix. Early thermal release results in the saturated, aromatized, and organosulfur products (from Grice et al., 1998c).

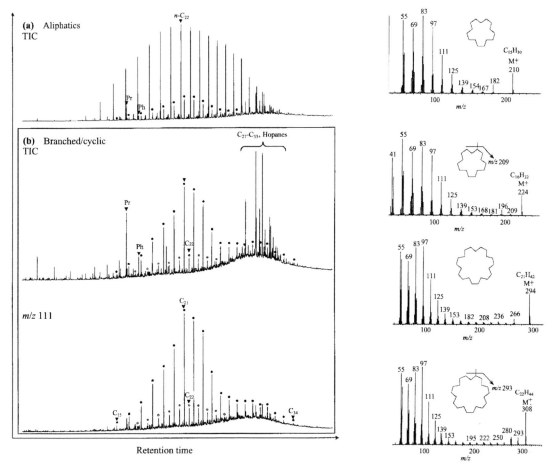

Figure 13.32. The macrocyclic alkane series occurs in the saturated hydrocarbon fraction extracted from torbanite. Filled circles indicate macrocycloalkanes, open circles indicate their methyl-substituted equivalents. Mass spectra of the macrocycloalkanes show a strong molecular ion and ($C_nH_{2n}-1$) fragments. Structures from top to bottom right are cyclopentadecane, 1-methylcyclopentadecane, C_{21} macrocyclic alkane, and C_{22} 1-methyl-macrocyclic alkane. Reprinted from Audino et al. (2001b). © Copyright 2001, with permission from Elsevier.

2001b) and probably originated from polyunsaturated precursors in the algaenan of B. braunii (Audino et al., 2001b). The macrocyclic alkanes in these samples are accompanied by a series of methyl macrocyclic alkanes in much lower abundance.

Mass spectra of the macrocyclic alkanes have a prominent molecular ion and a characteristic M − 28 fragment for loss of ethylene, while the methyl macrocyclic alkane mass spectra also include the M − 15 fragment for loss of methyl.

β-Carotane and related carotenoids

> Highly specific for lacustrine deposition. Measured using m/z 558 (M+) and/or m/z 125 fragmentograms (see Figure 8.20) or by capillary gas chromatography.

Carotenoids comprise a wide range of highly unsaturated C_{40} compounds produced mainly by photosynthetic organisms. Carotenoids in petroleum are generally saturated analogs. Because unsaturated carotenoids are easily oxidized, they are rarely found in sediments. Oxygen-containing carotenoids, the xanthophylls, are degraded rapidly in the water column (Repeta and Gagosian, 1987; Repeta, 1989). The preferential removal of xanthophylls relative to the carotenoid hydrocarbons results in dramatic changes in the relative proportion of these compounds in the water column and in recent sediments (Villanueva et al., 1994).

Under highly reducing conditions, the carotenoid carbon skeleton may be preserved in sediments. The most prominent of these compounds is β-carotane (perhydro-β-carotene), the fully saturated form of β-carotene (see Figure 2.15). One of the first biomarkers to be identified (Murphy et al., 1967), β-carotane is associated primarily with anoxic, saline lacustrine, or highly restricted marine settings where organisms such as the unicellular algae *Dunaliella* bloom and thrive as the dominant biota (Hall and Douglas, 1983; Jiang Zhusheng and Fowler, 1986; Irwin and Meyer, 1990; Fu Jiamo et al., 1990). These halotolerant algae synthesize massive amounts of β-carotene (up to 10% of their cellular mass) to prevent photoinhibition (Ben-Amotz et al., 1989; Avron and Ben-Amotz, 1992).

Many source rocks contain high concentrations of β-carotane, such as Green River Formation (Murphy et al., 1967; Gallegos, 1971), Carboniferous source rock for the Kelamayi oils (Jiang Zhusheng and Fowler, 1986), Upper Permian lacustrine oil shales of the Junggar Basin (Carroll et al., 1992; Carroll, 1998), the Eocene portion of the Shahejie Formation, Fulin Basin, Eastern China (Jianyu Chen et al., 1996), and strata from saline to hypersaline restricted basins in Brazil (Mello et al., 1993). Crude oils from these source facies also contain high concentrations of β-carotane. In the Green River Formation, β-carotane is found in gilsonite (Schoell et al., 1994) and related oils. In some cases, β-carotane may be the dominant saturated hydrocarbon. For example, in oils from Zhungeer Basin, China (Fu Jiamo et al., 1988) and Albania (Sinninghe Damsté and Koopmans, 1997), β-carotane is more abundant than n-alkanes, isoprenoids, or any other saturated compound.

In most oils where β-carotane is present, it is as a trace component similar in abundance to the typical cyclic biomarkers. Its presence suggests a saline

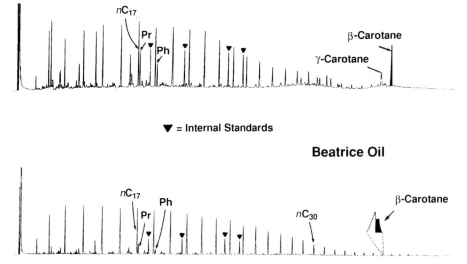

Figure 13.33. Gas chromatograms showing β-carotane in bitumen from lacustrine Devonian flagstone (top) and Beatrice oil (bottom) from the Moray Firth, UK. The compound was identified by co-injection of the authentic standard and by the similarity of the mass spectrum of the peak in the oil with that of the standard. γ-Carotane is also a prominent peak in the gas chromatogram of the Devonian sample. The small concentrations of β-carotane and other evidence helped to show that the Devonian flagstones are a co-source for the Beatrice oil (Peters et al., 1989).

Alkanes and acyclic isoprenoids 523

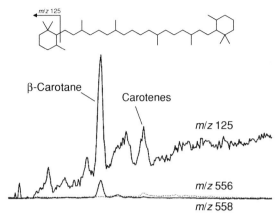

Figure 13.34. Oil-cemented sandstone breccia collected from a beach near Brora on the Inner Moray Firth coast of northeastern Scotland contains β-carotane (structure at top), indicative of low Eh, arid lacustrine source-rock deposition. The molecular ions for β-carotane and monounsaturated carotenes are m/z 558 and m/z 556, respectively. Reprinted from Peters *et al.* (1996b). © Copyright 1996, with permission from Elsevier.

lacustrine or highly restricted marine depositional setting. For example, β-carotane helped to identify lacustrine Devonian source rock as the co-source for the Beatrice oil in the Inner Moray Firth (Peters *et al.*, 1989) and Devonian input to mixed oil from Brora Beach, North Sea (Figures 13.33 and 13.34) (Peters *et al.*, 1999b).

Gamma-carotane (γ-carotane) commonly occurs with β-carotane (Jiang Zhusheng and Fowler, 1986). It has only one tetra-alkyl substituted cyclohexane ring compared with two in β-carotane. Consequently, γ-carotane has a molecular ion at m/z 560 but retains the prominent m/z 125 fragment typical of β-carotane (see Figure 8.20). Jiang Zhusheng and Fowler (1986) note that the ratio of γ- to β-carotane increases with thermal maturity and decreases with biodegradation, but in our experience the ratio also is affected by source organic matter input. The C_{33} carotenoid, lexane, is isotopically identical to carotane in gilsonite from the Uinta Basin, Utah, supporting a common C_{40} carotenoid precursor (Schoell *et al.*, 1994).

β-Carotane also may be accompanied by related aromatic hydrocarbons that form during early diagenesis. These were identified in extracts of Green River Shale and appear to be formed by cyclization and aromatization of the polyene isoprenoid chain and aromatization of the terminal β-cyclohexenyl groups to form 1,2-methylbenzyl groups (Figure 13.35). Isorenieratane is a product of β-carotene diagenesis (Koopmans *et al.*, 1996b). In the Green River Shale extracts, β-carotane accounts for 63% of the products attributed to β-carotene diagenesis, suggesting that hydrogenation of the double bonds is the preferred reaction.

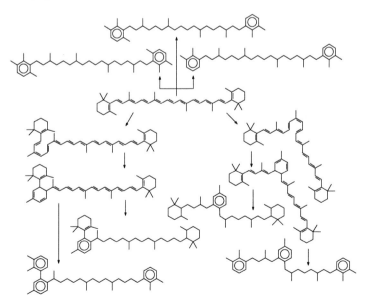

Figure 13.35. Diagenesis of β-carotene by cyclization and aromatization of the polyene chain and aromatization of the terminal cyclohexenyl groups (modified from Koopmans *et al.*, 1997).

β-Carotane and other perhydro-carotenoids can be liberated from sulfur complexes in high-sulfur oils and from kerogens (Schouten et al., 1995; Sinninghe Damsté and Koopmans, 1997). Apparently, the double bonds in β-carotene react with sulfur species in hypersaline anoxic marine environments, becoming part of a sulfur cross-linked system, while in low-sulfur anoxic lacustrine systems β-carotene is reduced to β-carotane. Lycopane (Figure 2.15) was also liberated from the sulfur complexes. Like β-carotane, lycopane also originates from a highly unsaturated precursor, in this case lycopene.

STERANES AND DIASTERANES

Regular steranes/17α-hopanes

> Moderate specificity for relative input from eukaryotes versus prokaryotes. Measured using GCMS/MS for steranes (M+ → m/z 217) or hopanes (M+ → m/z 191) or GCMS (m/z 217 or m/z 191, respectively). Also expressed as steranes/hopanes (St/H).

In steranes/17α-hopanes, the regular steranes consist of the C_{27}, C_{28}, and C_{29} $\alpha\alpha\alpha$(20S + 20R) and $\alpha\beta\beta$(20S + 20R) compounds and the 17α-hopanes consist of the C_{29}–C_{33} pseudohomologs, including 22S and 22R epimers for C_{31}–C_{33} homologs (Moldowan et al., 1985). The C_{29} and C_{30} compounds do not have 22S or 22R epimers. Improved accuracy and resolution of modern GCMS for measuring higher-molecular-weight compounds could allow extension of the 17α-hopanes in the denominator of the ratio to include the C_{34} and C_{35} homologs. Because of the differences in mass-spectral response for the various compounds, amounts of each are commonly measured in parts per million of the saturate fraction and then combined to give the ratio.

Regular steranes/17α-hopanes reflects input of eukaryotic (mainly algae and higher plants) versus prokaryotic (bacteria) organisms to the source rock. Thus, related oils of differing thermal maturity typically fall along a line on plots of sterane versus hopane concentration. Unrelated oils may or may not fall on this line. Because organisms vary widely in their steroid and hopanoid contents, differences in this ratio allow only qualitative assessment of eukaryote versus prokaryote input. Maturity may increase this ratio (Seifert and Moldowan, 1978).

In general, high concentrations of steranes and high steranes/hopanes (≥1) typify marine organic matter with major contributions from planktonic and/or benthic algae (e.g. Moldowan et al., 1985). Conversely, low steranes and low steranes/hopanes are more indicative of terrigenous and/or microbially reworked organic matter (e.g. Tissot and Welte, 1984). Regular steranes/17α-hopanes was generally lower (near zero) in non-marine compared with marine oil samples in a study of ~40 oils generated from different source rocks (Moldowan et al., 1985).

Andrusevich et al. (2000) demonstrated a paleolatitude effect on steranes/hopanes for Upper Jurassic crude oils (Figure 13.36). Figure 13.36 shows that crude oils generated from Upper Jurassic carbonate-rich source rocks deposited in equatorial marine settings contain relatively more bacterial input than oils from marine siliciclastic source rocks from higher latitudes. A similar paleolatitude effect was observed for the isotopic compositions of the oils.

Some workers use steranes/triterpanes as an indicator of organic matter input, assuming that steranes

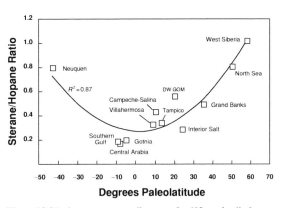

Figure 13.36. Average steranes/hopanes for 482 crude oils from Upper Jurassic source rocks vary systematically with paleolatitude (data from Andrusevich et al., 2000). Steranes/hopanes are controlled mainly by the relative abundance of algal and bacterial input to the precursor kerogen. The figure shows that crude oils generated from carbonate-rich source rocks deposited in equatorial marine settings contain relatively more bacterial input than oils from marine siliciclastic source rocks from higher latitudes. Solid curve is second-order polynomial best fit, where R^2 is the correlation coefficient. DW GOM, deep water, Gulf of Mexico.

originate from algae and higher plants while triterpanes come mainly from bacteria. For example, Connan et al. (1986) used low steranes/triterpanes (<0.05) and other molecular data for bitumen from anoxic, carbonate-anhydrite facies as evidence of high microbial input. We prefer to use steranes/hopanes rather than steranes/triterpanes. Both ratios are of limited use because of the variety of organisms that contribute to steranes, and especially triterpanes. For example, extended tricyclic terpanes (>C_{20}) could originate from terrigenous plants in addition to bacterial or algal sources.

C_{27}–C_{28}–C_{29} steranes

> Highly specific for correlation. Measured using GCMS/MS (M+ → 217) (see Figure 8.20). Attempts to measure these parameters using m/z 217 from routine SIM/GCMS can result in interference.

Based on a study of recent marine and terrigenous sediments, Huang and Meinschein (1979) showed that the ratio of cholest-5-en-3β-ol to 24-ethylcholest-5-en-3β-ol is a source parameter that can be used to differentiate depositional settings. They proposed that the distributions of C_{27}-, C_{28}-, and C_{29}-sterol homologs on a ternary diagram might be used to differentiate ecosystems. Attempts to apply this concept to steranes in source rocks and crude oils (e.g. C_{27}, C_{28}, and C_{29} steranes) have met with only limited success (Mackenzie et al., 1983a; Moldowan et al., 1985). Figure 13.37 shows a C_{27}–C_{28}–C_{29} sterane ternary diagram that represents a composite of data for oils from various source-rock depositional environments (Moldowan et al., 1985). There is so much overlap on this figure that the analysis is seldom used to differentiate depositional environments of the source rocks for crude oils, with the possible exception of certain samples containing predominantly higher-plant organic matter (e.g. non-marine shales in area B, Figure 13.37). Because of less overlap, monoaromatic steroid ternary diagrams (see Figure 13.105) are more useful in distinguishing petroleum based on the depositional setting of the source rock (Moldowan et al., 1985).

However, sterane ternary diagrams are used extensively to show relationships between oils and/or source-rock bitumens (e.g. Peters et al., 2000). Based on our experience (but see Curiale, 1986), plot locations on these

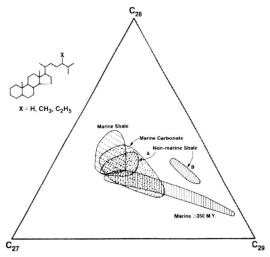

Figure 13.37. Ternary diagram showing the relative abundances of C_{27}, C_{28}, and C_{29} regular steranes [5α,14α,17α(H) 20S + 20R and 5α,14β,17β(H) 20S + 20R] in the saturate fractions of crude oils determined by gas chromatography/mass spectrometry (GCMS/MS) (M+ → 217). Labeled areas represent a composite of data for oils from known source rocks (Moldowan et al., 1985). Overlap between the different oil source types limits the use of sterane distributions to describe the source-rock depositional environment. However, the diagram can be used to infer genetic relationships among oil and bitumen samples if they plot in close proximity. Non-marine shale areas A and B refer to petroleum generated mainly from non-marine algal organic matter or terrigenous (higher plant) organic matter, respectively. Reprinted with permission by ChevronTexaco Exploration and Production Technology Company, a division of Chevron USA Inc.

diagrams do not change significantly throughout the oil-generative window. For example, Figure 13.38 suggests a close genetic relationship between the Piper oil and bitumen extracted from Upper Jurassic Kimmeridge Clay, as supported by other data (Peters et al., 1989). The plot location for the Beatrice oil suggests that it could represent a mixture of Devonian lacustrine and Middle Jurassic marine input. This inference is supported by independent geochemical evidence (e.g. Figure 13.106).

The principal use of C_{27}–C_{28}–C_{29} sterane ternary diagrams is to distinguish groups of crude oils from different source rocks or different organic facies of the same source rock. For example, Grantham et al. (1988) used a sterane ternary diagram, stable carbon isotope ratios, and other supporting data to classify five groups of Oman oils and relate most of them to source rock

526 Source- and age-related biomarker parameters

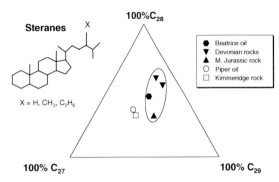

Figure 13.38. Sterane ternary diagram for oils and source rock extracts from the North Sea (Peters *et al.*, 1989). See also Figure 13.115. Reprinted by permission of the AAPG, whose permission is required for further use.

extracts. Palmer (1984a) used a sterane ternary diagram to differentiate two major lacustrine organic facies in the Eocene-Oligocene Elko Formation from northeastern Nevada: a lignitic siltstone and an oil shale. Bitumens from the siltstones showed a predominance of C_{29} steranes, high pristane/phytane (>1), abundant C_{19} and C_{20} tricyclic diterpanes, and other characteristics indicating higher-plant input. The oil shale bitumens showed more C_{27} and C_{28} steranes than those from the siltstones, and other characteristics of oil-prone organic matter, including low pristane/phytane (<0.5), high C_{28}, C_{29}, and C_{30} 4-methylsteranes, and gammacerane.

> **Note:** Because the identity of their source rock was questionable, one oil group found mainly in central Oman and the Ghaba Salt Basin was called Q oils (Grantham *et al.*, 1988). Biomarkers show that the source rock for the Q oils contained mainly type I/II kerogen deposited in a strongly evaporitic, but not hypersaline, carbonate environment. The Q oils show higher %C_{27} steranes (>45%) than other Oman oils in addition to other unusual characteristics, including stable carbon isotope ratios near −30‰, abundant tricyclic terpanes, and mid-chain monomethylalkanes. Source-rock data and generation modeling suggest that the Q oils originated from top-salt Dhahaban Formation source rocks in the Precambrian-Cambrian Ara Group (Terken and Frewin, 2000). Other Oman oils generated from the Ara Group show dominantly C_{29} steranes and more negative stable carbon isotope ratios near −35‰.

C_{28}/C_{29} steranes

> Age-related parameter for oils lacking terrigenous input. Measured using GCMS/MS of $m/z\ 414 \rightarrow 217$.

Data indicate a general increase in the relative content of C_{28} steranes and a decrease in C_{29} steranes in marine petroleum through geologic time (Moldowan *et al.*, 1985; Grantham and Wakefield, 1988). The increase in the C_{28} steranes may be related to increased diversification of phytoplankton assemblages, including diatoms, coccolithophores, and dinoflagellates in the Jurassic and Cretaceous periods. Although this approach is not sufficiently accurate to determine the age of the source rock for oil, it is possible to distinguish Upper Cretaceous and Tertiary oils from Paleozoic or older oils (Grantham and Wakefield, 1988). These authors observed that C_{28}/C_{29} steranes is <0.5 for Lower Paleozoic and older oils, 0.4–0.7 for Upper Paleozoic to Lower Jurassic oils, and greater than ∼0.7 for Upper Jurassic to Miocene oils (Figure 13.39). We do not recommend such an age differentiation without additional supporting age data (see Figure 13.1). The age relationship between the relative contents of C_{28} and C_{29} steranes described by Grantham and Wakefield (1988) applies only to samples from marine source rocks. Furthermore, we have observed many

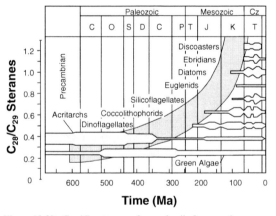

Figure 13.39. C_{28}/C_{29} steranes for crude oils from marine source rocks with little or no terrigenous organic matter input plot within the area of the shaded curve. The ratio is <0.5 for Lower Paleozoic and older oils, 0.4–0.7 for Upper Paleozoic to Lower Jurassic oils, and greater than ∼0.7 for Upper Jurassic to Miocene oils. Reprinted from Grantham and Wakefield (1988). © Copyright 1988, with permission from Elsevier.

exceptions to this relationship, even for crude oils that are clearly of marine origin.

$C_{30}/(C_{27}-C_{30})$ steranes (C_{30} sterane index)

> Highly specific for marine organic matter input. Measured using GCMS/MS or MRM/GCMS of M+ (414) → 217.

The presence of C_{30} 4-desmethylsteranes in crude oil is the most powerful means in order to identify input of marine organic matter to the source rock (Moldowan et al., 1985; Peters et al., 1986). Additional parameters that are useful to distinguish marine from terrigenous and lacustrine oils are listed in Table 13.4. These C_{30} steranes are identified as 24-n-propylcholestanes (Figure 2.15), which originate from 24-n-propylcholesterols (Moldowan et al., 1990). The latter are biosynthesized by marine Chrysophyte algae of the order *Sarcinochrysidales* and are common in marine invertebrates, presumably due to ingestion of algae (Raederstorff and Rohmer, 1984).

Positive identification of the 24-n-propylcholestanes by GCMS/MS or MRM/GCMS (m/z 414 → 217) can be difficult due to low concentrations and co-elution with the C_{30} 4α-methylsteranes. The 4α-methyl-24-ethylcholestanes have a minor m/z 217 fragment, yielding peaks from the m/z 414 → 217 transition. The retention times of the $\alpha\alpha\alpha$20R and $\alpha\beta\beta$20R stereoisomers are nearly identical to those of the same stereoisomers for 24-n-propylcholestane. For samples with high levels of 4α-methyl-24-ethylcholestanes, a "cross-talk" pattern appears on the m/z 414 → 217 chromatogram that resembles that of the 24-n-propylcholestanes. In such cases, small amounts of 24-n-propylcholestanes may be difficult to verify. The expected intensity of m/z 414 → 217 peaks for the 4α-methyl-24-ethylcholestanes is ~10% of their m/z 414 → 231 intensities. Therefore, careful calibration of cross-talk response factors with 4α-methyl-24-ethylcholestane standards can facilitate accurate measurement of 24-n-propylcholestanes.

Figure 13.40 shows GCMS/MS mass chromatograms of the C_{29} and C_{30} steranes for several crude oils and rock extracts from the North Sea (Peters et al., 1989). The C_{30} steranes in the samples with input from marine source rock (Beatrice oil, Piper oil, Kimmeridge extract, Middle Jurassic extract) elute at predicted retention times after their C_{29} homologs. The two Devonian extracts lack C_{30} steranes, consistent with geologic, paleontologic, and geochemical evidence for a lacustrine origin. Similar application of C_{30} sterane data supported conclusions on the origins of various Brazilian crude oils (Mello et al., 1988a; 1988b).

An unusual oil-cemented sandstone breccia collected from Brora Beach, Scotland, not far from the Beatrice Field (see Figure 18.118) contains C_{30} n-propylcholestanes (Figure 13.41) and β-carotane (see Figure 13.81), indicting a mixed marine and lacustrine origin (Peters et al., 1999b). This conclusion is supported by independent data (Holba et al., 2000).

When C_{30} steranes are low in petroleum, questions may arise regarding their correct identification. For example, Bailey et al. (1990) analyzed two Beatrice oil samples and concluded that C_{30} steranes were too minor to identify with confidence. The striking difference in signal-to-noise ratio between mass chromatograms for C_{30} steranes obtained by Peters et al. (1989) and Bailey et al. (1990) for Beatrice oil may be part of the explanation. Both groups of workers used MRM/GCMS, although the specific instruments may differ in sensitivity. Peters et al. (1989) analyzed a saturate fraction where n-alkanes had been removed, which concentrates components relative to the whole oil. Bailey et al. (1990) used whole oil for their published mass chromatogram. However, they stated that analyses of the saturate fraction of Beatrice oil also failed to detect C_{30} steranes.

Ratios of $C_{30}/(C_{27}-C_{30})$ steranes plotted against oleanane/hopane give a better assessment of marine versus terrigenous input to petroleum than either parameter alone (Figures 13.42 and 13.43). Moldowan et al. (1992) plotted $C_{30}/(C_{27}-C_{30})$ steranes versus C_{34} or $C_{35}/(C_{31}-C_{35})$ 17α-homohopanes for selected crude oils. They found that many oils derived from source rocks deposited under restricted saline to hypersaline lagoonal conditions show lower $C_{30}/(C_{27}-C_{30})$ steranes than those from open marine systems. They showed that the C_{30}-sterane ratio varies inversely with C_{34} 17α-homohopanes in Cretaceous and Liassic-Triassic oils from the Adriatic Basin. Low C_{30}-sterane ratios and high C_{34} 17α-homohopanes appear to indicate a restricted, evaporitic source-rock depositional environment.

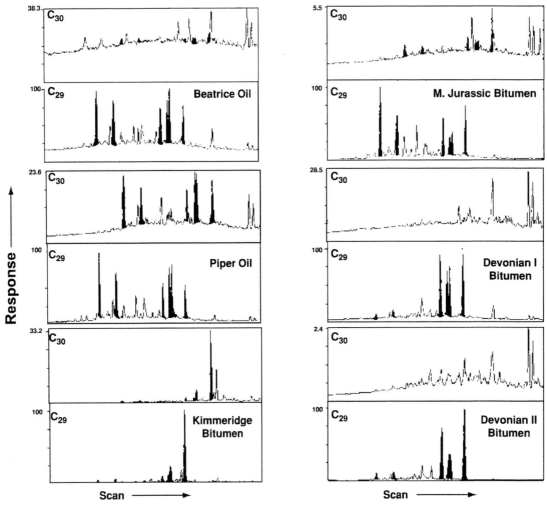

Figure 13.40. C_{29}- and C_{30}-sterane distributions for saturate fractions of oils and bitumens obtained by metastable reaction monitoring/gas chromatography/mass spectrometry (MRM/GCMS). The distributions were obtained by monitoring parent-to-daughter ion transitions, M+ → m/z 217, where M+ corresponds to molecular ions at m/z 400 and m/z 414 for C_{29} and C_{30} steranes, respectively. Note the interference of terpanes on the m/z 414 → 217 (C_{30}) chromatograms. The higher resolution of multisector mass spectrometers (monitoring collision-activated decomposition (CAD) ions) compared with the magnetic-sector instrument (monitoring metastable transitions) may reduce this interference. Reprinted with permission by ChevronTexaco Exploration and Production Technology Company, a division of Chevron USA Inc.

Paleontological and geochemical evidence suggest marine transgressions into the gigantic freshwater Lake Songliao (Dujie Hou et al., 2000). The transgressive shales contain abundant C_{30} 4-desmethylsteranes (24-n-propylcholestanes) (Moldowan et al., 1985) and various C_{28}–C_{31} 4-methylsteranes. The dinosteranes (4,23,24-trimethylcholestanes) are more abundant than their 24-ethyl counterparts in samples with a clear marine influence, whereas the latter compounds dominate the C_{30} 4-methylsteranes in samples typical of freshwater lacustrine sediments. C_{30} dinosteranes can be differentiated from their 24-ethyl counterparts and from C_{30} 4-desmethylsteranes by monitoring m/z 414 → 98 (e.g. see Figure 13.48).

Values of zero for the C_{30}-sterane ratio generally correspond to non-marine oils (e.g. Holba et al., 2000).

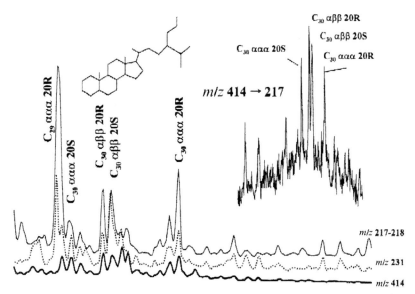

Figure 13.41. The presence of C_{30} 24-n-propylcholestanes (structure at top) provides evidence that oil generated from a marine source rock has contributed to the bitumen cementing a sandstone conglomerate collected near Brora on the Inner Moray Firth coast of northeastern Scotland. C_{30} 24-n-propylcholestanes were detected by gas chromatography/mass spectrometry (GCMS) (lower left) and confirmed by gas chromatography/ mass spectrometry/mass spectrometry (GCMS/MS) of parent and daughter ions (upper right). GCMS ions include m/z 217–218 (steranes), m/z 231 (methylsteranes), and m/z 414 (C_{30} steranes). Reprinted from Peters et al. (1996b). © Copyright 1996, with permission from Elsevier.

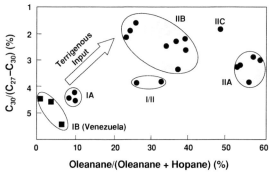

Figure 13.42. Oils and seep oils from Colombia (solid circles) can be compared with oils from the Maracaibo Basin in Venezuela (squares) using relative abundances of marine (24-n-propylcholestanes) and terrigenous (oleanane) markers. Oil groups IIA, IIB, and IIC have high oleanane and low C_{30} steranes, suggesting a deltaic source rock with strong terrigenous input. Group IA has higher C_{30} steranes and lower oleanane than the other oils, suggesting a marine source rock with less terrigenous input. Group IA is similar to the Venezuelan oils (Group IB), which probably originated from the La Luna Formation. Reprinted with permission by ChevronTexaco Exploration and Production Technology Company, a division of Chevron USA Inc.

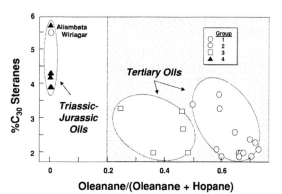

Figure 13.43. Oleanane versus C_{30} sterane (24-n-propylcholestane) ratios separate oil groups in eastern Indonesia (Peters et al., 1999a). Oleanane ratios for oils greater than 0.20 indicate Tertiary source rocks (groups 2 and 3), while lack of oleanane is consistent with a Jurassic or older source (group 4). The C_{30}-sterane ratio generally increases with marine versus terrigenous organic-matter input to the source rock. Wiriagar oil is problematic because biomarkers are low. This oil lacks oleanane but has high C_{30} steranes. Reprinted by permission of the AAPG, whose permission is required for further use.

Most of the small number of Cambrian and Precambrian rock extracts and oils that were examined initially lacked detectable C_{30} steranes (Moldowan *et al.*, 1985; Peters *et al.*, 1986). The absence of C_{30} steranes in petroleum older than ~500 Ma was interpreted as an evolutionary lag in the appearance of C_{30} sterols in marine organisms or domination of the marine biota by a few species that did not contain C_{30} sterols (Moldowan *et al.*, 1985). However, re-examination of some very old extracts (e.g. Precambrian Chuar Group in Arizona) and crude oils using new instrumentation with better sensitivity confirms low concentrations of 24-*n*-propylcholestanes in many of the samples (J. M. Moldowan, 2002, unpublished data). C_{30} steranes occur in 2700 Ma extracts from the Pilbara Craton, Australia, although no structures were specified (Brocks *et al.*, 1999). The pattern and relative retention times of the peaks for these compounds suggest that they are 24-*n*-propylcholestanes (Moldowan, 2000; Moldowan and Jacobson, 2000).

C_{30} 24-isopropylcholestanes

> Age-specific. Measured using GCMS/MS of m/z 414 → 217 in saturate fraction.

High 24-isopropylcholestane ratios [24-isopropyl-/(24-iso- + 24 *n*-propylcholestanes)] occur in Late Proterozoic and Early Cambrian oils and bitumens (McCaffrey *et al.*, 1994a). These high ratios appear to be due to input from stromatoporids, the dominant reef-building organisms during these times. Stromatoporids may be related genetically to modern Porifera

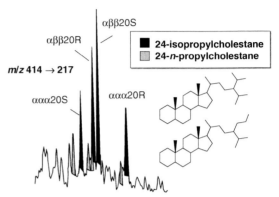

Figure 13.45. 24-Isopropylcholestanes predominate over 24-*n*-propylcholestanes on metastable reaction monitoring/gas chromatography/mass spectrometry (MRM/GCMS) fragmentogram of Infracambrian Baghewala-1 oil from Rajasthan in northwestern India (Peters *et al.*, 1995). Reprinted by permission of the AAPG, whose permission is required for further use.

(sponges), which contain abundant sterols having the 24-isopropylcholestane skeleton. Late Proterozoic (Vendian) and Early Cambrian oils and bitumens from Siberia, the Urals, Oman, Australia, and India have high 24-isopropylcholestane ratios compared with younger and older samples (Figure 13.44).

The 24-isopropylcholestane ratio and other data establish an Infracambrian age for the source rock of oil in the Baghewala-1 well in northwestern India (Figure 13.45) (Peters *et al.*, 1995). The oil has a 24-isopropylcholestane ratio near 70%, consistent with other data suggesting that it originated from Infracambrian (~600 Ma) source rocks like those in the Huqf Formation in Oman. Based on oil–oil and oil–source rock correlation and paleogeographic plate reconstruction, the Baghewala-1 oil correlates with Huqf source rocks in Oman, ~2000 km from the well.

4-Methylsteranes

> Potentially highly specific for marine or non-marine dinoflagellates or bacteria (e.g. 4-methylsteranes are abundant in certain lacustrine rocks, such as in China). Monitored using m/z 231 and 232 fragmentograms or preferably by GCMS/MS.

The 4-methylsteranes consist of two major classes: (1) C_{28}–C_{30} analogs of the steranes substituted at

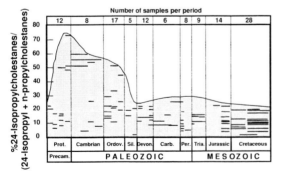

Figure 13.44. Late Proterozoic (Vendian) and Early Cambrian oils and bitumens from Siberia, the Urals, Oman, Australia, and India have high 24-isopropylcholestane ratios compared with younger and older samples (Moldowan *et al.*, 2001, unpublished data).

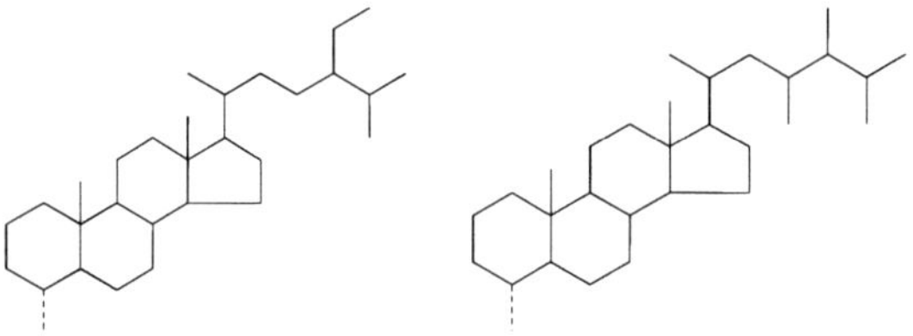

4α-Methyl-24-ethylcholestane   **4α,23,24-Trimethylcholestane**

Figure 13.46. Examples of the two major classes of 4-methylsteranes.

C-4 and C-24 (e.g. the C_{30} compound is 4α-methyl-24-ethylcholestane) and (2) the C_{30} dinosteranes (e.g. 4α,23,24-trimethylcholestanes) (Figure 13.46).

The origin of the 4,24-substituted methylsteranes is unclear. The 4-methylsteranes in petroleum probably originate from 4α-methylsterols in living dinoflagellates (Wolff et al., 1986). However, 4α-methylsterols occur in prymnesiophyte microalgae of the genus *Pavlova* (Volkman et al., 1990). Bird et al. (1971) proposed an additional source from certain bacteria, notably *Methylococcus capsulatus*, but these 4-methylsterols are not alkylated at C-24. Furthermore, the ^{13}C-rich isotopic compositions of C_{30} 4-methyl-24-ethylcholesterane in extracts from evaporitic lacustrine Paleocene-Eocene source rocks in the Jianghan Basin, China (~8‰ enriched in ^{13}C compared with 5α-cholestane) argue against an origin from such methylotrophic bacteria (Grice et al., 1998a). The 4α,23,24-trimethylcholestanes, also known as dinosteranes, originate from dinosterol or dinostanol, compounds that are synthesized by dinoflagellates (Withers, 1983).

The 4-methylsteranes occur in most Mesozoic samples but are absent in the Permian and most Carboniferous samples. Older rocks contain dinosteroids in nearly equal abundance to that in the Mesozoic. This could be related to the diversity of acritarchs, which may be the ancestors of the dinoflagellates. A combined dinoflagellate and acritarch species count appears to mirror the occurrence of the dinosteroids. Nevertheless, these compounds provide age information because when dinosteranes are absent the samples are Triassic or older.

Both 4α-methyl-24-ethyl-cholestanes and dinosteranes occur in marine rocks and related oils with marine dinoflagellate input (Moldowan et al., 1985; Summons et al., 1987; Goodwin et al., 1988). Several non-marine dinoflagellate species are known (Curiale, 1987), although lacustrine sediments appear to contain only the 4α-methyl-24-ethylcholestanes. Extracts from non-marine rocks deposited under hypersaline conditions contain less C_{28} versus C_{29} and C_{30} 4-methylsteranes than those from freshwater and brackish water lacustrine environments in China (Fu Jiamo et al., 1990). Further, samples from freshwater settings contain more total 4-methylsteranes than those from saline settings. They conclude that species differences among dinoflagellates in freshwater, brackish, and hypersaline lacustrine settings account for the different 4-methylsterane compositions and that dinoflagellate blooms occur mainly in the freshwater settings.

Abundant C_{30} 4-methylsteranes occur in both freshwater lacustrine and marine sediments of Mesozoic and Tertiary age. 4α-Methyl-24-ethylcholestanes and 4α,23,24-trimethylcholestanes (dinosteranes) occur in marine dinoflagellate-rich sediments, but the major C_{30} 4α-methylsterane found to date in lacustrine sediments is 4α-methyl-24-ethylcholestane (Goodwin et al., 1988; Summons et al., 1987; Summons et al., 1992). The 4α-methyl-24-ethylcholestanes are less specific for dinoflagellates than dinosteranes because they may originate from 4α-methyl-24-ethylcholesterols in prymnesiophyte algae (Volkman et al., 1990).

Dinosterane has been found only in Triassic and younger crude oils and rock extracts (Summons et al., 1987). This age relationship corresponds with the earliest widespread fossil evidence for dinoflagellates during the Triassic Period, although sporadic occurrences of possible dinoflagellate fossils are indicated into the Paleozoic Era, with the oldest suspected species from rocks of Silurian age (Tappan, 1980). Dinosterane was detected in Paleozoic rocks, although its isomer 24-ethyl-4-methylcholestane occurs in some Early Triassic, Permian, and older samples (Summons et al., 1992). Black shales in the Norian/Rhaetian (Upper Triassic) sequence from Watchet in North Somerset, UK, contain abundant cysts of the earliest positively identified dinoflagellate, *Rhaetogonyaulax rhoetica*, and abundant 4-methylsteranes, including 4,23,24-trimethylcholestanes (Thomas et al., 1989).

The 4α-methylsteranes are an analytical challenge (Figure 13.48) because of their complexity and because their major fragments, except m/z 231 and m/z 232, are the same as the steranes (Figure 8.20). The major homologs are 4α-methyl analogs of cholestane (C_{28}), ergostane (C_{29}), and stigmastane (C_{30}) plus dinosterane (4,23,24-trimethylcholestane). The 4β-methyl group that occurs in thermally immature rock extracts doubles the potential number of stereoisomers compared with the steranes; however, the $4\alpha(CH_3)$-isomers strongly dominate in thermally mature extracts (Wolff et al., 1986). Rubinstein and Albrecht (1975) identified 4β-methylsteranes in the more immature parts of the Toarcian Shale in the Paris Basin. Epimerization at C-23 and C-24 in dinosterane results in four stereoisomers. The large number of 4-methylsterane isomers suggests that these compounds could eventually be as useful for oil–oil and oil–source rock correlation. Like the steranes, the complexity of 4-methylsterane distributions typically requires GCMS/MS for reliable results. The 4-methylsteranes can be analyzed by SIM/GCMS using bench-top quadrupole systems, but only when concentrations of these compounds are high, as in certain immature lacustrine oils (Hwang, 1990). However, analysis of samples containing low 4-methylsteranes can be difficult even using GCMS/MS, particularly with respect to the C_{30} compounds. For example, in some oils m/z 414 → 231 shows 2-, 3-, and 4-methyl-24-ethylcholestanes and their many stereoisomers plus all the epimeric dinosteranes (Figure 13.48). Dinosterane has an additional m/z 98 fragment (see Figure 8.20) from the side chain that is useful for its identification (i.e. m/z 414 → 98) (Summons et al., 1987).

2- and 3-Alkylsteranes

> Potentially highly specific for correlation. Measured using GCMS/MS (M+ → m/z 231).

Two series of compounds with pseudohomologs that range from C_{28} to C_{30}, identified as 3β- and

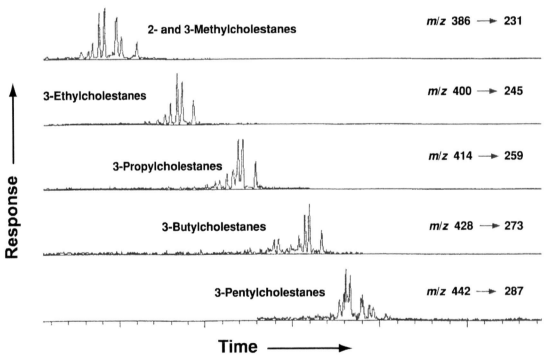

Figure 13.47. Homologous series of cholestane isomers with alkylation at the 3-position are detected in mass chromatograms constructed by gas chromatography/mass spectrometry/mass spectrometry (GCMS/MS) parent–daughter transitions of the saturate fraction of oil from Hamilton Dome, Wyoming. 3-Ethylcholestanes were confirmed by co-injection of authentic standards. Parent-to-daughter m/z 262 transitions (see Figure 8.20) confirmed alkylation in the A-ring for all homologs. Similar homologous series occur for 3-alkyl ergosteranes and 3-alkyl stigmastanes. Intensities are normalized to the largest peak within each chromatogram. Reprinted with permission by ChevronTexaco Exploration and Production Technology Company, a division of Chevron USA Inc.

2α-methylsteranes (Figure 13.47), probably originate from sterols via $\Delta 2$-sterenes through a bacterial alkylation process (Dastillung and Albrecht, 1977), although other origins are possible (Summons and Capon, 1988; Dahl et al., 1995). For C_{28}–C_{30} compounds, the 3β-methylsteranes predominate over 2α-methylsteranes, and their ratio remains constant for oils from source rocks deposited under different conditions. This evidence was used to support a common $\Delta 2$-sterene precursor for the 2- and 3-methylsteranes (Summons and Capon, 1988). There are no known natural product analogs for 2- or 3-methylsterenes in modern organisms.

The 2- and 3-methylsteranes are potentially age-related biomarkers for pre-Mesozoic petroleum. The 2- and 3-methylsteranes are the only nuclear methylated steranes in a suite of Paleozoic and Precambrian oils examined by Summons and Capon (1988). Surveys for 2- and 3-methylsteranes in Mesozoic and Tertiary oils are incomplete, but 4α-methylsteranes are usually predominant in the samples studied. Summons and Capon (1991) also identified 3β-ethyl steranes in sediments and petroleum.

Analysis of 2- and 3-methylsteranes can be difficult because of the large numbers of diastereomers and isomers (i.e. rearranged isomers and 4-methylsteranes). For C_{30} methylsteranes, four series are known (Figure 13.48).

Dahl et al. (1995) identified several novel 3-alkyl sterane and triaromatic steroid series, including 3β-n-pentyl steranes, 3β-isopentyl steranes, 3β-n-hexyl steranes, 3β-n-heptyl steranes, 3,4-dimethyl steranes, 3β-butyl,4-methyl steranes, triaromatic 3-n-pentyl steroids, and 3-isopentylsteroids. They found that the ratio of 3β-n-pentyl steranes to 3β-isopentyl steranes (or 3-n-pentyl triaromatic steroids to 3-isopentyl triaromatic steroids) varies substantially among crude oils and rock extracts and may be useful for geochemical correlation or as indicators of depositional environment. Although no 3-alkyl steroid natural products are known, 3β-alkyl steroids appear to result from bacterial side-chain additions to diagenetic $\Delta 2$-sterenes. High abundance of the C_5 moiety in some samples led Dahl et al. (1992) to postulate that 3-alkyl steranes originate from bacterial addition of a C_5 sugar to sterenes. They believe that this process may be analogous to biosynthesis of bacteriohopanetetrol and related compounds, where microbes couple a hopanoid and a D-pentose (Rohmer, 1993). The resulting 3-alkyl steroids might be suitable as molecular substitutes for bacteriohopanoid membrane components.

Diasteranes/steranes

> Moderately specific for source-rock mineralogy and oxicity, interference due to thermal maturation. Analyzed by GCMS/MS or MRM/GCMS of M+ $\rightarrow$ m/z 217, where M+ = m/z 372, 386, 400, or 414 for C_{27}, C_{28}, C_{29}, or C_{30}, respectively. Alternative ratios, such as C_{27} diasterane/C_{27} sterane, involve fewer peak measurements but are difficult to measure accurately because of interference from C_{28} and C_{29} epimers.

Acidic sites on clays, such as montmorillonite or illite, catalyze the conversion of sterols to diasterenes during diagenesis (Figure 13.49) (Rubinstein et al., 1975; Sieskind et al., 1979). Alternatively, acidic (low pH) and oxic (high Eh) conditions facilitate diasterene formation during diagenesis (Moldowan et al., 1986; Brincat and Abbott, 2001). Diasterenes are ultimately reduced to diasteranes (rearranged steranes) showing $13\beta,17\alpha$(H) 20S and 20R (major isomers) and $13\alpha,17\beta$(H) 20S and 20R (minor isomers) stereochemistries. The diasteranes/steranes ratio is based on $[13\beta,17\alpha$(H) 20S + 20R$]/\{[5\alpha,14\alpha,17\alpha$(H) 20S + 20R$] + [5\alpha,14\beta,17\beta$(H) 20S + 20R$]\}$ for the C_{27}, C_{28}, and C_{29} steranes obtained from GCMS/MS or MRM/GCMS. Occasionally, only one carbon number is used, for example C_{29}, as specified. In most cases, accuracy of the measurements by GCMS/MS or MRM/GCMS is far superior to that from GCMS.

Diasteranes/steranes ratios are commonly used to distinguish petroleum from carbonate versus clastic source rocks (e.g. Mello et al., 1988b). Additional parameters useful for distinguishing crude oils from carbonate and shale source rocks are listed in Table 13.7. Low diasteranes/steranes ratios (m/z 217) in oils indicate anoxic clay-poor or carbonate source rock. During diagenesis of these carbonate sediments, bacterial activity provides bicarbonate and ammonium ions (Berner et al., 1970), resulting in increased water alkalinity. Under these conditions of high pH and low Eh, calcite tends to precipitate and organic matter preservation is improved.

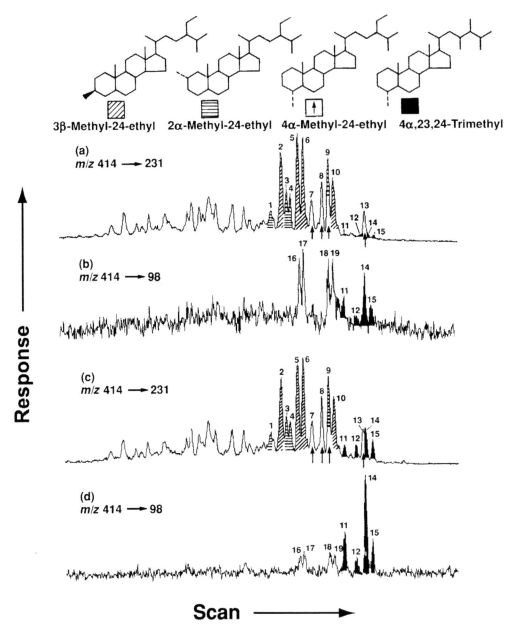

Figure 13.48. Four series of methylsteranes are detected in an Asian oil by gas chromatography/mass spectrometry mass/spectrometry (GCMS/MS) of the saturate fraction using a 60-m DB-1 (J&W Scientific) fused-silica capillary column. (a) Relative retention times of 2α-methyl-24-ethylcholestanes (horizontally striped peaks), 3β-methyl-24-ethylcholestanes (hatched peaks), 4α-methyl-24-ethylcholestanes (arrows), and 4α,23,24-trimethylcholestanes (dinosteranes, black peaks) are indicated on the m/z 414 → 231 chromatogram. (b) Relatively small concentrations of dinosteranes are shown by m/z 414 → 98 peaks that are insignificant for the other methylsteranes. (c) and (d) A co-elution experiment using four dinosteranes confirms their presence in the oil. The four 2α-methyl-24-ethylcholestanes and four 3β-methyl-24-ethylcholestanes were identified by co-elution of 2α-methyl-24-ethylcholestane and 3β-methyl-24-ethylcholestane (courtesy of R. Summons) that had been isomerized over Pd/C catalyst at 260°C (Seifert et al., 1983). The four dinosterane stereoisomers were prepared by reduction of dinosterol isolated from a

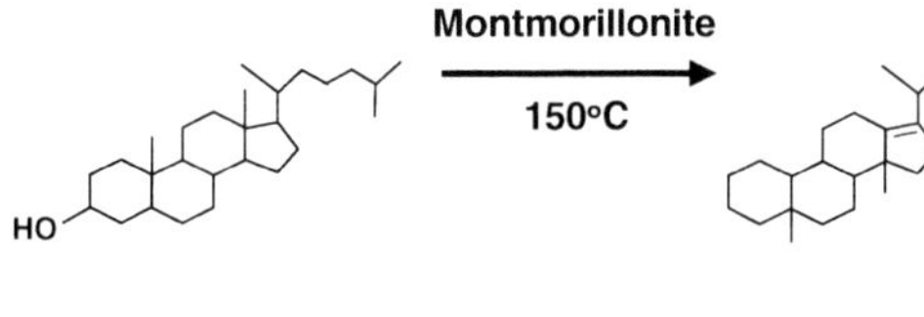

5α-Cholestanol → (Montmorillonite, 150°C) **Diasterene**

Figure 13.49. Laboratory experiments of Rubinstein et al. (1975) support the conclusion that clay-rich source rocks catalyze the formation of diasterane precursors. Diasterenes (rearranged sterenes) are believed to result from clay-catalyzed rearrangement of sterols or sterenes during diagenesis and early catagenesis (Rubinstein et al., 1975). Saturation of the double bond ($\Delta 13$–17) in diasterenes during catagenesis results in diasteranes.

High diasteranes/steranes ratios are typical of petroleum derived from clay-rich source rocks. However, high diasteranes/steranes ratios have also been observed in extracts from organic-lean carbonate rocks from the Adriatic Basin (Moldowan et al., 1991a). These rocks probably originated in low-pH, high-Eh depositional environments. A correlation between low pH, high Eh, and high diasteranes/steranes ratios has been reported for the Toarcian Shale of southwestern Germany (Moldowan et al., 1986). Likewise, Palacas et al. (1984) noted high diasteranes in clay-poor limestones from Florida. Clark and Philp (1989) list several publications where diasteranes were found in carbonates. Van Kaam-Peters et al. (1998) observed that diasteranes/steranes ratios do not correlate directly with clay content but depend on the amount of clay relative to total organic carbon. This correlation may explain the high diasteranes/steranes ratios for some crude oils and extracts from carbonate source rocks, as discussed above.

High diasteranes/steranes ratios in some crude oils can result from high thermal maturity (Seifert and Moldowan, 1978) and/or heavy biodegradation (Seifert and Moldowan, 1979). For example, burial maturation of a series of similar shaly carbonate rocks increases both vitrinite reflectance and the ratio of C_{27} diasteranes/(C_{27} diasteranes + steranes) (Goodarzi et al., 1989). Such correlations between the diasterane ratio and reflectance can be applied only to limited regions, where lithology and organic matter types are similar. At high levels of thermal maturity, rearrangement of steroids to diasterane precursors may become possible, even without clays, due to hydrogen-exchange reactions, which are enhanced by the presence of water (Van Kaam-Peters et al., 1998). Alternatively, diasteranes simply may be more stable and survive thermal degradation better than steranes. The diasteranes/steranes ratio is useful for distinguishing source-rock depositional conditions only when the samples show comparable levels of thermal maturity.

Heavy biodegradation can result in selective destruction of steranes relative to diasteranes. Other evidence supporting heavy biodegradation commonly includes depletion of n-alkanes and isoprenoids or the presence of 25-norhopanes. However, it is possible that non-biodegraded oil might mix with heavily biodegraded oil showing a much higher diasteranes/steranes ratio. In such cases, only careful quantitative assessment

---

Figure 13.48. (cont.) gorgonian (courtesy of R. M. K. Carlson and D. S. Watt), with stereochemical assignments by nuclear magnetic resonance (NMR) (courtesy of R. M. K. Carlson). Assignments of 4α-methyl-24-ethylcholestanes are by comparison of relative retention times with the literature (Summons et al., 1987). Analyses were completed by P. A. Lipton; peak identifications courtesy of B. J. Huizinga. Reprinted with permission by ChevronTexaco Exploration and Production Technology Company, a division of Chevron USA Inc. Identity of peaks:
1. 2α-Methyl-24-ethylcholestane 20S. 2. 3β-Methyl-24-ethylcholestane 20S. 3. 2α-Methyl-24-ethylcholestane 14β,17β(H) 20R*.
4. 2α-Methyl-24-ethylcholestane 14β,17β(H) 20S*. 5. 3β-Methyl-24-ethylcholestane 14β,17β(H) 20R*.
6. 3β-Methyl-24-ethylcholestane 14β,17β(H) 20S*. 7. 4α-Methyl-24-ethylcholestane 20S. 8. 4α-Methyl-24-ethylcholestane 14β,17β(H) 20R*. 9. 2α-Methyl-24-ethylcholestane 20R + 4α-methyl-24-ethylcholestane 14β,17β(H) 20S*.
10. 3β-Methyl-24-ethylcholestane 20R. 11. 4α,23S,24S-Trimethylcholestane 20R. 12. 4α,23S,24R-Trimethylcholestane 20R.
13. 4α-Methyl-24-ethylcholestane 20R. 14. 4α,23R,24R-Trimethylcholestane 20R. 15. 4α,23R,24S-Trimethylcholestane 20R.
16. 4α,23,24-Trimethylcholestane diastereomer**. 17. 4α,23,24-Trimethylcholestane diastereomer**.
18. 4α,23,24-Trimethylcholestane diastereomer**. 19. 4α,23,24-trimethylcholestane diastereomer**.

*20S and 20R designations for 14β,17β(H) compounds may be reversed.
**Based on m/z 414 → 98 response and relative retention time.

536   Source- and age-related biomarker parameters

Table 13.7. *Geochemical properties* differ between non-biodegraded crude oils from marine carbonate, marine shale, and deltaic marine shale source rocks*

Property	Marine carbonate	Marine shale	Deltaic shale
Bulk			
API gravity	10–30	25–40	35–45
Sulfur (wt.%)	>0.6	0.2–0.5	<0.2
Saturates/aromatics	0.3–1.5	1–2	>2
Carbon preference index	<1	1–1.5	>1.5
Biomarkers			
Pristane/Phytane	<1	1.1–1.8	2–4
Phytane/nC_{18}	>0.3	<0.3	<0.1
Steranes	$C_{27} > C_{29}$	$C_{27} < C_{29}$	$C_{27} < C_{29}$
Steranes/hopanes	Low	High	High
Diasteranes/steranes	Low	High	High
C_{24} tetracyclic terpane/C_{26} tricyclic terpane	Medium–high	Low–medium	Low
C_{29}/C_{30} hopanes	High (>1)	Low	Low
C_{35}/C_{30} hopanes	High	Low	Low
Gammacerane/hopane	Low–high	Low	Absent

* Quoted properties encompass most samples, but exceptions occur.

of each biodegradation-sensitive parameter can lead to the correct interpretation.

C_{27}–C_{28}–C_{29} diasteranes

> Highly source-specific. Measured using GCMS/MS of M+ → m/z 217 in saturate fraction. (Although the principal fragment of diasteranes shown in Figure 8.20 is m/z 259, under routine GCMS conditions m/z 217 is more reliable because of its stronger response under GCMS/MS conditions.)

Results for the ternary diasterane plots are normally determined using GCMS/MS or MRM/GCMS. These plots can be used to support oil–oil and oil–source rock correlations based on the analogous sterane ternary diagrams (Figure 13.50). The C_{27}, C_{28}, and C_{29} diasterane distributions can be used when sterane distributions are unreliable, and vice versa, as discussed below. The data include [C_{27} 13β,17α(H) (20S + 20R) diasteranes]/[C_{27} + C_{28} + C_{29} 13β,17α(H)(20S + 20R) diasteranes] and the analogous ratios for the C_{28} and

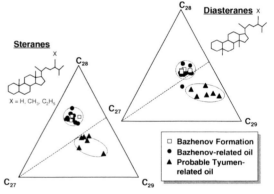

Figure 13.50. Ternary diagrams of sterane and diasterane homologs support oil–oil and oil–source rock correlations indicating two genetically distinct petroleum systems (dotted oval areas) in the West Siberian Basin, Russia. One group of oils correlates closely to extracts from the Upper Jurassic Bazhenov Formation on both diagrams. Data are from metastable reaction monitoring/gas chromatography/mass spectrometry (MRM/GCMS) (Peters *et al.*, 1994). Reprinted by permission of the AAPG, whose permission is required for further use.

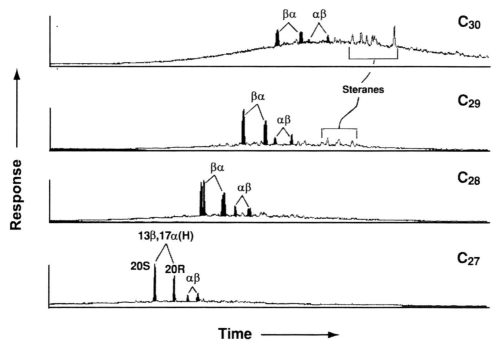

Figure 13.51. Seep oil from Papua New Guinea lost most regular steranes due to heavy biodegradation. However, some C_{29} steranes remain, showing their relative resistance to biodegradation compared with C_{27} and C_{28} steranes. The full suite of C_{30} steranes (24-n-propylcholestanes) remains intact, demonstrating their still greater resistance to biodegradation compared with the C_{27}–C_{29} steranes. The biodegradation rank of this oil (level 8 in Figure 16.11) is consistent with partial removal of 17α-hopanes and all n-alkanes and isoprenoids. Despite biodegradation of the steranes, the unaltered C_{27}–C_{29} diasterane distribution in this oil was used in a ternary diagram to support its correlation with other samples. Reprinted with permission by ChevronTexaco Exploration and Production Technology Company, a division of Chevron USA Inc.

C_{29} compounds. Structures for the diacholestanes are shown in Figure 2.15. Note the two rearranged methyl groups for this compound compared with the steranes in Figure 2.15.

The most important applications of C_{27}, C_{28}, and C_{29} diasterane plots are for (1) heavily biodegraded oils where steranes are altered, but diasteranes remain intact, and (2) some highly mature oils and condensates that show low steranes but more abundant diasteranes. However, some oils from clay-poor source rocks show high steranes, but the diasteranes are not useful for correlation because of low concentrations. Figure 13.51 shows triple quadrupole GCMS/MS results for diasteranes in biodegraded seep oil from Papua New Guinea.

24-Norcholestanes and 24-nordiacholestanes

> Potentially highly temporal specific for eukaryotic (diatom?) input. Measured using GCMS/MS m/z 358 → 217.

Information on C_{26} steranes in petroleum is seldom accessible using conventional SIM/GCMS because concentrations of C_{26} steranes are typically an order of magnitude lower than the C_{27}–C_{29} steranes. Furthermore, their gas chromatographic retention times coincide with the early-eluting C_{27}–C_{29} steranes and diasteranes, resulting in interference. The only practical analysis for these C_{26} steranes is by GCMS/MS.

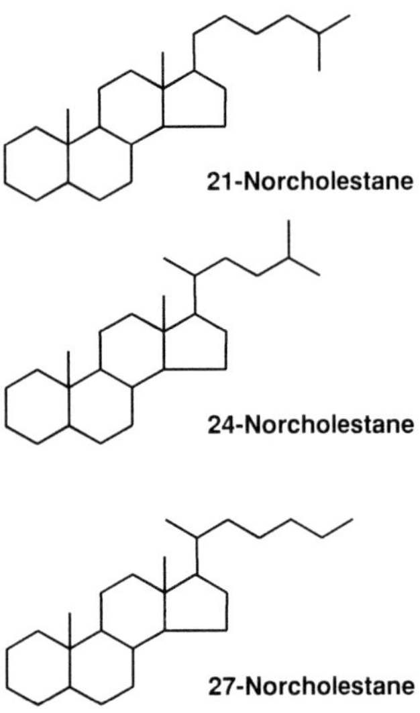

Figure 13.52. Moldowan et al. (1991a) identified three C_{26} sterane structures.

Three series of C_{26} steranes are known, including 21-, 24-, and 27-norcholestanes (Figure 13.52) (Moldowan et al., 1991a). The $5\alpha,14\alpha,17\alpha$(H) 20S + 20R and $5\alpha,14\beta,17\beta$(H) 20S + 20R compounds were identified in the 24- and 27-norcholestanes, but the 20S and 20R isomers do not occur in the 21-norcholestanes due to lack of the methyl group at C-20. Under normal gas chromatography conditions, $5\alpha,14\alpha,17\alpha$(H)- and $5\alpha,14\beta,17\beta$(H)-21-norcholestanes co-elute (Figure 13.53).

The 21- and 27-norcholestanes appear to have no direct sterol precursors but may originate through bacterial oxidation or thermally induced cleavage and loss of a methyl group from larger steroids ($>C_{26}$). On the other hand, traces of 24-norcholesterols occur in living marine algae and invertebrates, suggesting an origin in eukaryotes (Goad and Withers, 1982). All three series of C_{26} steranes occur in both marine and non-marine crude oils.

The ratio of C24/(C24 + C27)-norcholestanes is an effective source-correlation parameter. (Numbers in the ratio represent the nor-position.) For example, it was used to distinguish marine and non-marine crude oils from Upper and Lower Cretaceous source rocks, respectively, in Angola (Figure 13.54), Lower Cretaceous and Permian oils in Wyoming, two non-marine source rocks in the Bohai Basin, China (Moldowan et al., 1991a), and Tertiary and Jurassic non-marine petroleum systems in Qaidam Basin, China (Ritts et al., 1999). Relative concentrations of the 21-norcholestanes can be used to gauge thermal maturity from the middle to late oil-generative window. For example, the ratio of C21/(C21 + C24 + C27)-norcholestanes increases with API gravity and other maturity indicators in a series of oils from Phosphoria (Permian) source rock from Wyoming (Moldowan et al., 1991a). Mature condensates from Wyoming and Angola have a strong predominance of 21-norcholestane over 24- and 27-norcholestanes.

Nordiacholestanes help to distinguish Tertiary from Cretaceous and Cretaceous from older oils. 24-Nordiacholestanes probably originate from diatoms, which evolved during Jurassic time. Figure 13.55 is based on analyses of ~150 crude oils worldwide. 24/(24 + 27)-Nordiacholestane ratios greater than 0.25 and 0.55 typify oils from Cretaceous or younger and Tertiary (generally Neogene) source rocks, respectively.

Greater confidence in age assessment is possible if two age-related biomarker ratios are used together. In Figure 13.56, the Lubna-18 and Dolni Lomna-1 oils from Moravia, Czech Republic, show high oleanane and C_{26} 24-nordiacholestane ratios (Moldowan et al., 1994a; Holba et al., 1998), consistent with a possible Paleogene age source rock. The Zdanice-7 and Damborice-16 oils and Jurassic source rock extracts from the Sedlec-1 and Nemcicky-1 wells lack oleanane and show low 24-nordiacholestane ratios, consistent with a Jurassic age for these oils. The diagonal oval area suggests that Tynec-34 oil from the nearby Vienna Basin is a mixture, which is supported by isotopic and other evidence.

TERPANES AND SIMILAR COMPOUNDS

Cembranoid diterpenes

> Macrocyclic isoprenoids found in resinous plants, soft corals, and insects. First and only reported occurrence in the geosphere limited to ester-bound moieties in an immature Miocene lacustrine kerogen.

Cembranoid diterpanes are macrocyclic compounds formed by cyclization of geranylgeranyl diphosphate,

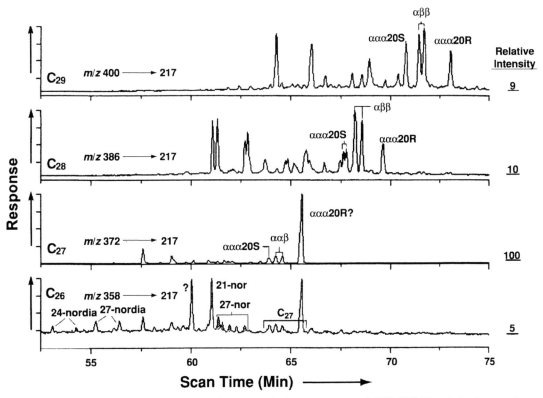

Figure 13.53. Metastable reaction monitoring/gas chromatography/mass spectrometry (MRM/GCMS) analysis of steranes in a condensate from Angola, showing a strong predominance of 21-norcholestanes among the C_{26} steranes. MRM/GCMS gives lower resolution than gas chromatography/mass spectrometry/mass spectrometry (GCMS/MS) run using a multisector mass spectrometer, allowing C_{27} steranes to interfere on m/z 358 → 217 traces. True GCMS/MS data (see Figure 13.54) are not affected by this interference. The mature sterane epimer patterns for the C_{28} and C_{29} steranes agree with this sample being a condensate. However, the C_{27} steranes contain a strong $\alpha\alpha\alpha$20R peak, which could be interpreted to indicate immaturity. Possibly it is $5\alpha,14\alpha,17\alpha$(H) 20R-cholestane picked up from immature organic matter during migration of the mature condensate. Reprinted with permission by ChevronTexaco Exploration and Production Technology Company, a division of Chevron USA Inc.

the common precursor of steroids and terpenoids (Figure 13.57). Resinous conifers, tobacco plants, and soft corals produce various hydrocarbon and functionalized forms of these diterpanes (Wahlberg and Eklund, 1992). Cembranoids also occur in insects (e.g. termites), although they may be not synthesized but accumulated from consumed plant matter.

Barakat and Rullkötter (1993) first reported cembranoids in the geosphere (Figure 13.58) in low-maturity, Miocene lacustrine rocks from the Nördlinger Ries crater in southern Germany. Mild alkaline hydrolysis liberated several cembranoid diterpenes and cembranoid alcohols from the kerogen. They proposed that resinous plants were a likely source of the cembranoids and suggested that they could be markers for arid environments, consistent with the inferred semi-arid, depositional setting of the Nördlinger Ries sediments. However, because of their three double bonds and functional groups, cembranoid diterpenes are highly reactive and are unlikely to be preserved under most conditions. Lack of free cembrenoids in the Nördlinger Ries rock extracts suggests that they were too immature to have

540 Source- and age-related biomarker parameters

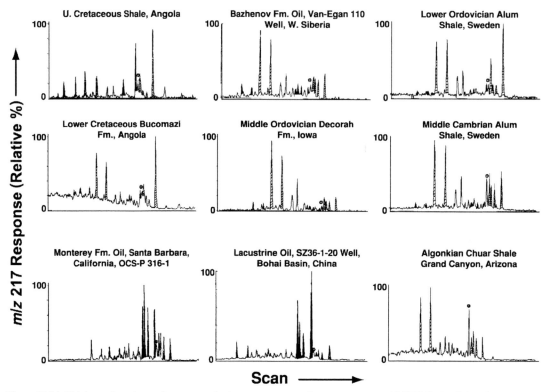

Figure 13.54. Triple quadrupole gas chromatography/mass spectrometry/mass spectrometry (GCMS/MS) of nine oils and rock extracts using the Finnigan MAT TSQ-70, showing various distributions of C_{26} steranes (m/z 358 → 217). The Paleozoic and older rocks show little or no 24-norcholestanes. One lacustrine rock (Bucomazi Formation, Angola) contains little 24-norcholestane, while another (Bohai Basin, China) is dominated by these compounds. The Monterey Formation sample also contains abundant 24-norcholestanes. Black peaks, 24-norcholestane (due to low concentrations, identifications are uncertain in some samples). Hatched peaks, 27-norcholestanes; circle over peak = 21-norcholestane. Refer to Figure 13.52 for structures.

been released from the kerogen or were thermally unstable after release.

Bicyclic sesquiterpanes (eudesmane and drimane)

Specificity unknown; sometimes reported as terrigenous markers. Measured using GCMS m/z 123 (see Figure 8.20).

Several reports show the occurrence of bicyclic sesquiterpanes in crude oils (Bendoraitis, 1974; Seifert and Moldowan, 1979; Philp *et al.*, 1981). The significance of most of these compounds remains unknown due to lack of precise structure assignments. However, Alexander *et al.* (1983a) used synthetic standards to identify 4β-eudesmane and 8β-drimane in oil from the Cormorant Field in the Gippsland Basin, Australia. The source rock for the Cormorant oil contained significant input from higher plants. Furthermore, the carbon skeleton of 4β-eudesmane (Figure 2.15) is clearly related to higher-plant terpenes. Despite its terrigenous origin, 4β-eudesmane is present in very low amounts relative to the other sesquiterpanes in the Cormorant oil.

In contrast to eudesmane, drimane occurs in higher abundance in the Cormorant oil and has structural

Terpanes and similar compounds 541

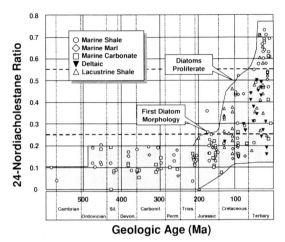

Figure 13.55. C_{26} 24/(24 + 27) nordiacholestanes versus geologic age for 150 crude oils worldwide. Ratios greater than 0.25 and 0.55 (dashed horizontal lines) typify oils from Cretaceous or younger and Oligocene or younger (generally Neogene) source rocks, respectively. Reprinted from Holba et al. (1998). © Copyright 1998, with permission from Elsevier.

Bicyclic and tricyclic diterpanes

High specificity for various microbial, gymnosperm, and angiosperm inputs. Measured using GCMS m/z 123 (see Figure 8.20). Best measured using GCMS/MS: 278–123, 276–247, 262–233, 276–233, 262–219, and others.

The tricyclic diterpanes can be used to evaluate terrigenous input to some crude oils. However, some marine organisms, especially algae, also produce diterpenoids (Simoneit, 1986). More detailed work is needed to differentiate which structures could originate from marine algae versus higher plants. Philp et al. (1981; 1983) analyzed diterpane distributions in various Australian oils from the Gippsland Basin and found a correlation between these distributions and higher-plant input. Richardson and Miller (1982; 1983) recognized diterpanes of terrigenous origin in crude oil from Southeast Asia, and Snowdon (1980b) tentatively identified pimarane-type compounds in Canadian crude oils.

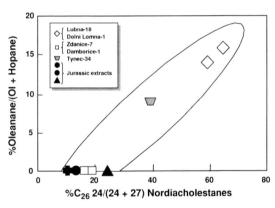

Figure 13.56. Age-related biomarkers support two petroleum systems in the Western Carpathians and their foreland, Moravia, Czech Republic (Picha and Peters, 1998). See also Figures 13.85 and 13.110.

features and a widespread distribution similar to many biomarkers of prokaryotic origin (Alexander et al., 1983a; Volkman, 1988). Thus, we do not recommend the general use of bicyclic sesquiterpanes as markers of terrigenous input to petroleum without rigorous structure identification (Figure 13.59).

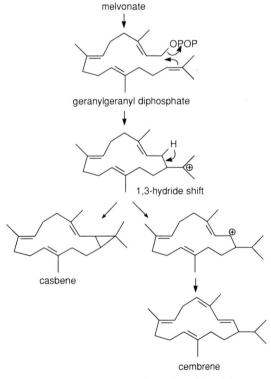

Figure 13.57. Synthetic pathway for cembranoids and the related casbenoids.

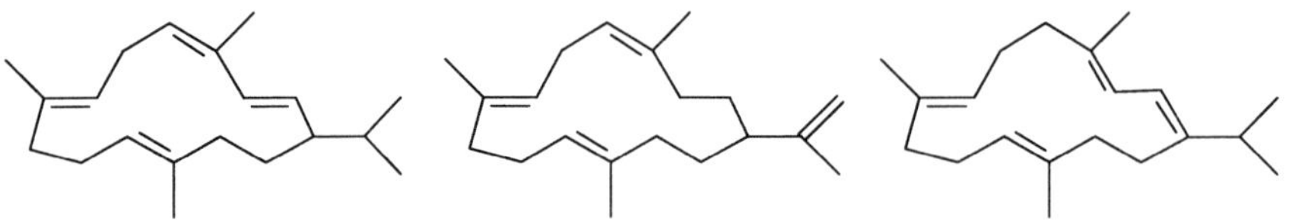

542 Source- and age-related biomarker parameters

cembrene   cembrene-A   cembrene-C

Figure 13.58. Cembrenoid diterpenes liberated from a lacustrine source rock by mild alkaline hydrolysis (from Barakat and Rullkötter, 1993).

Uncertainties in the use of these data arise from lack of structural identification for the compounds. For example, some degraded tricyclohexaprenanes (C_{19}, C_{20}), which could originate from either bacteria or algae, have GCMS fragment ions and retention times similar to the higher-plant diterpanes.

Previous studies suggest a resinite source for some bi- and tricyclic sesqui- and diterpanes (Snowdon, 1980b). Noble (1986) and Noble et al. (1986) identify several bicyclic sesquiterpanes (Figure 13.59) and tricyclic diterpanes (Figure 13.60) and discuss their significance as terrigenous markers. Compounds identified

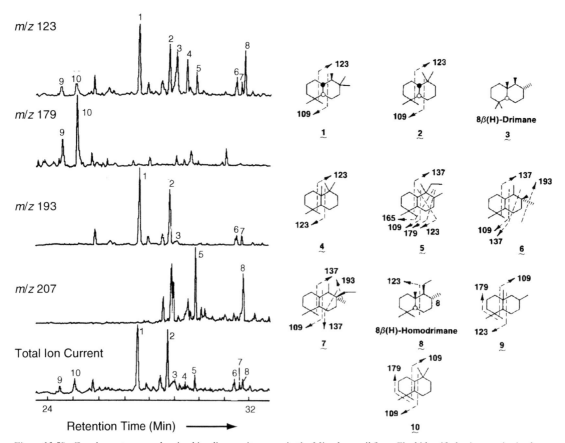

Figure 13.59. Gas chromatograms showing bicyclic sesquiterpanes in the Miandoum oil from Chad identified using synthesized standards and mass spectra (Noble, 1986). The figure also shows structures of the important mass spectral fragments used to identify these compounds. Reprinted with permission by ChevronTexaco Exploration and Production Technology Company, a division of Chevron USA Inc.

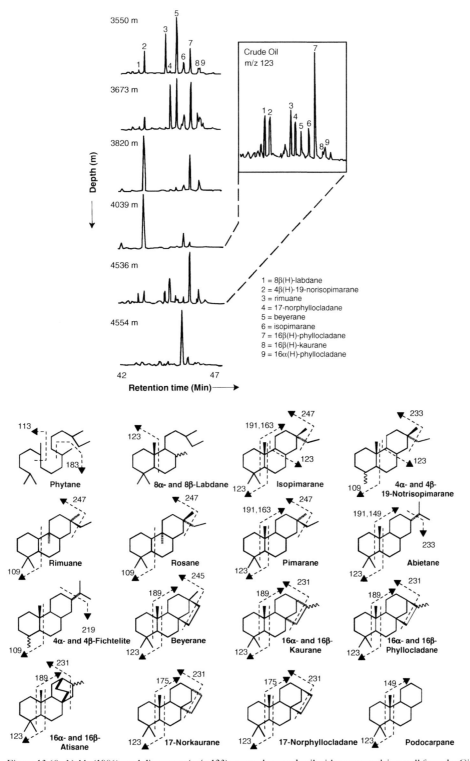

Figure 13.60. Noble (1986) used diterpanes (m/z 123) to correlate crude oil with source rock in a well from the Gippsland Basin, Australia. Structures for tricyclic diterpanes identified in the mass chromatogram show their principal fragment ions. Reprinted with permission by ChevronTexaco Exploration and Production Technology Company, a division of Chevron USA Inc.

in Australian oils include 8β(H)-labdane (bicyclic), 4β(H)-19-norisopimarane, rimuane, and isopimarane (Figure 13.60). Isopimarane [iso(sandaraco)pimarane] was isolated and structurally characterized by Blunt et al. (1988) in seep oils from New Zealand. Restle (1983) and Weston et al. (1989) suggested isopimarane in Maui oil from New Zealand based on mass spectral data.

Tricyclic diterpanes generally show a major fragment at m/z 123 (and m/z 109 for 19-nor compounds), which is useful for identification along with the tetracyclic diterpanes (see below). They are generally found in highest concentrations in conifer (gymnosperm) resins and, when abundant in crude oils, support gymnosperm input to the source rock. However, they are not limited to gymnosperms because suitable precursors occur throughout the plant kingdom. Gymnosperms occur in Permian and younger sediments. Labdane was reported as a series of C_{15}–C_{24} bicyclic alkanes of probable microbial origin in Athabasca tar sand (Dimmler et al., 1984). Weston et al. (1989) used distributions of diterpanes to distinguish among major oil fields in the Taranaki Basin, New Zealand. Fichtelite (below), retene, iosene (phyllocladane), and pimarane (isopimarane?) are common in lignite. Peters et al. (1999a) used the MRM/GCMS transition m/z 276–247 (e.g. see Figure 13.60) to measure rimuane, an isomer called isorimuane (now identified as rosane), and isopimarane to differentiate crude oil samples from eastern Indonesia (Figure 13.61). This family of tricyclic diterpanes occurs widely and is represented in Figure 13.62. They generally can be analyzed as shown by using the m/z 276–247 transition in GCMS/MS or MRM/GCMS. Authentic standards have been used to identify their structures (Blunt et al., 1988; Yongsong et al., 1997; Zinniker, 2004). The MRM/GCMS fingerprint for rimuane, pimarane, rosane, and isopimarane is specific for each oil sample (peaks 1–4 in Figure 13.62, right) and varies over a narrow range for each oil family (see ternary diagram in Figure 13.62). The cause for the variations is uncertain. However, even some marine oil samples without significant terrigenous input (e.g. oil from pre-Silurian source rock) may have these four diterpanes, suggesting that they may originate partly from marine algae.

Hanson et al. (2001) used 4β(H)-19-norisopimarane to indicate relative higher-plant input to the source rocks for crude oils from the northern Qaidam Basin, China. They observed low 4β(H)-19-norisopimarane/[5α(H)androstane + 4β(H)-19-norisopimarane] and 4β(H)-19-norisopimarane/[C_{27} dia + regular steranes + 4β(H)-19-norisopimarane] for oils and related Upper Oligocene hypersaline lacustrine source rock. Only trace amounts of higher-plant organic matter occur in the source rock based on optical study. The oils and rock extracts show low pristane/phytane, low diasteranes, an even-to-odd n-alkane preference, and high β-carotane, γ-carotane, and gammacerane. Another group of lacustrine oils in the basin originated from Jurassic freshwater lacustrine source rock dominated by terrigenous organic matter (Ritts et al., 1999). These freshwater lacustrine rocks and oils have high 4β(H)-19-norisopimarane ratios, high pristane/phytane and diasteranes, and odd-to-even n-alkane predominance, and lack β-carotane, γ-carotane, and gammacerane.

Resinites derived from gymnosperms, i.e. conifers (class I), are enriched in ^{13}C compared with those

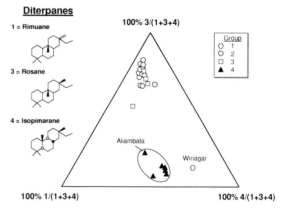

Figure 13.61. Ternary diagram of relative percentages of three tricyclic diterpanes, helping to differentiate oil families in eastern Indonesia (rimuane, rosane, and isopimarane structures at left) based on metastable reaction monitoring/gas chromatograpy/mass spectrometry (MRM/GCMS) (Peters et al., 1999a). The structure of rosane was determined by D. Zinniker (personal communication, 2003). The corners of the triangle represent 100% of the corresponding tricyclic diterpane. See Figure 6.11 for supporting data based on stable carbon isotope ratios of the saturated and aromatic hydrocarbons for these oils. Reprinted by permission of the AAPG, whose permission is required for further use.

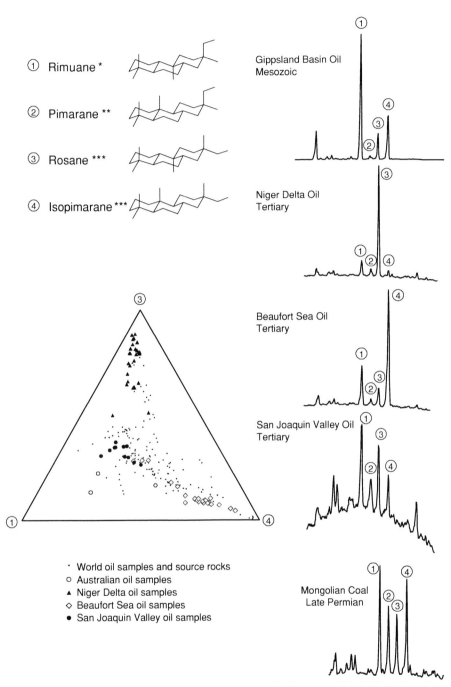

Figure 13.62. Examples of worldwide occurrence of several tricyclic diterpanes and their distributions (courtesy of David Zinniker).
* Peak 1 co-elutes with the hydrogenation product of rimuene. Stereochemistry at C-5 is uncertain.
** Peak 2 co-elutes with a primarane standard formed by hydrogenation of primaradiene.
*** Peaks 3 and 4 co-elute with rosane and isopimarane standards, respectively (Huang et al., 1997; Noble et al., 1986).

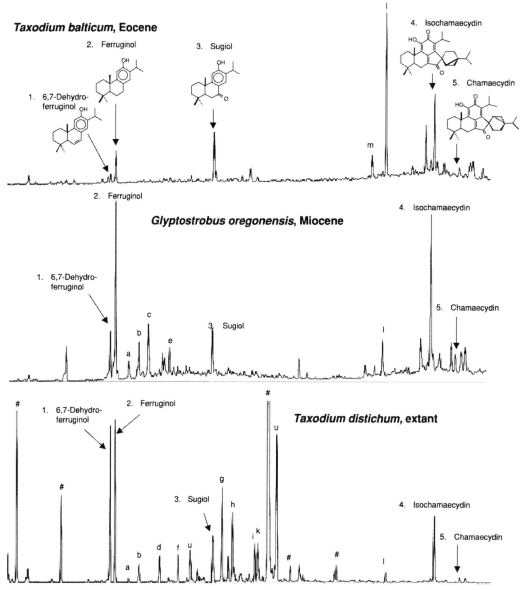

Figure 13.63. Gas chromatography/mass spectrometry (GCMS) traces of total ion current (TIC) for (a) the aromatic compound fraction from the extract of the seed cone of Eocene *Taxodium balticum*, (b) the total extract of the seed cone of Miocene *Glyptostrobus oregonensis*, and (c) the total extract of the seed cone of extant *Taxodium distichum*. Compounds were analyzed after trimethylsilyl (TMS) derivatization: 1 = 6,7-dehydroferruginol; 2 = ferruginol; 3 = sugiol; 4 = isochamaecydin; 5 = chamaecydin; a = taxodione acetate; b = pimaric acid; c = 18- or 19-hydroxyferruginol; d = 7-acetoxy-6,7-dehydroroyleanone; e = communic acid; f = royleanone; g = taxoquinone; h = 6-hydroxytaxo quinone; i = isomer of g; k = taxodone; l = inuroyleanone, 11,14-dioxolambertic acid, 11,14-dioxopisiferic acid, or similar compound; m = isomer of l; u = unknown; # = sugars. Reprinted with permission from Otto *et al.* (2002). © Copyright 2002, American Association for the Advancement of Science.

from angiosperms (class II), and biomarkers derived from these sources are isotopically distinct (Murray et al., 1998). For example, the most abundant diterpane in Tuna-2 oil from the Gippsland Basin is 16β(H)-phyllocladane, which has δ^{13}C (-23.0‰) close to the average for fossil conifer resins (-22.8‰). Isopimarane is the most abundant diterpane in Maui oil from New Zealand, and is slightly enriched in ^{13}C (δ^{13}C $= -25.9$‰) compared with the average for fossil angiosperm resins (-26.4‰).

Mackenzie (1984) suggested that the tricyclic diterpane called fichtelite (Figures 2.15 and 13.60) is a land-plant indicator. Fichtelite, retene, and iosene were known for a long time as crystalline deposits in lignites (Noble, 1986; Simoneit, 1986; Wang and Simoneit, 1990). When fichtelite is heated under vacuum with Pd/C catalyst, it converts rapidly and quantitatively to other compounds (Zinniker, 2004). One saturated compound produced by this process, presumed to be the fichtelite epimer 4β(H)-fichtelite, was identified as the major diterpane peak in the m/z 191 chromatogram of a Colombian oil sample. Fichtelite [4α(H)-fichtelite] was absent or at trace levels in the same oil (Dzou et al., 1999). Thus, the lack of reports of fichtelite in oil is probably due to its low thermal stability compared with other stereoisomers and the lack of structural identification of those stereoisomers until now. Other products produced in the thermal catalytic treatment of fichtelite include retene, which may be produced during diagenesis from fichtelite precursors such as abietenes and intermediates to fichtelite that contain double bonds.

Note: Extracts of fossil conifers from the Eocene Zeitz Formation in Germany and the Miocene Clarkia Formation in the USA contain unaltered terpenoids preserved in seed cones of *Taxodium balticum* and *Glyptostrobus oregonensis*, respectively (Figure 13.63). Terpenoid patterns in the extracts, including the presence of ferruginol, 6,7-dehydroferruginol, and sugiol, are similar to those in extant conifers, especially in the families *Cupressaceae*, *Taxodiaceae*, and *Podocarpaceae*. Resin and clay-rich sediment surrounding these functionalized terpenoids may have retarded diagenesis.

Simoniet et al. (2003) tentatively identified a series of mono- and triaromatic hydroxytriterpenoids ($-$OH at position 1, 2, or 3) of the oleanane, ursane, and lupane classes in extracts of Eocene Geiseltal lignites, Germany. These lignites contain material traditionally called "Affenhaar" (monkey hair) that is composed of the fossilized remains of laticifers (plants having much of their stem volume occupied by cells that secrete and store latex). These novel phenolic triterpenoids were accompanied by known mono-, tri-, and tetra-aromatic forms of oleanane, ursane, and lupane hydrocarbons and their natural product precursors, α-amyrone and β-amyrin.

Cadinanes

Highly specific for resinous input from higher plants, detected using mass spectrometry or various GCMS/MS transitions, such as m/z 412 $\to$ 369.

Cadinane was thought to originate from gymnosperm resins (Simoneit et al., 1986). However, when it occurs with bicadinanes and tricadinanes, it appears to be derived mostly from polycadinene, a resinous polymer produced by angiosperms such as the *Dipterocarpaceae* (van Aarssen et al., 1992).

Grantham et al. (1983) first reported three C_{30}-pentacyclic hydrocarbon compounds in Far Eastern oils, labeled W, T, and R, which now appear to be members of the bicadinane triterpane group. Cox et al. (1986) used X-ray diffraction and NMR to establish the structure for compound T, and van Aarssen et al. (1990a) used NMR to establish that of compound W as *trans-trans-trans*-bicadinane and *cis-cis-trans*-bicadinane, respectively (Figure 13.64). Direct biosynthesis of the bicadinanes from squalene is unlikely, and it was thus postulated (Cox et al., 1986) that bicadinane is a dimer of cadinene.

Van Aarssen and de Leeuw (1989) found cadinanes, bicadinanes, and tricadinanes in Southeast Asian oils and rock extracts (Figure 13.65). Numerous isomers of each oligomer are typically present. Bicadinanes are common in land-plant-derived Tertiary oils from the Surma Basin in northeast Bangladesh (Alam and Pearson, 1990). Oleanane is also abundant but does

548   Source- and age-related biomarker parameters

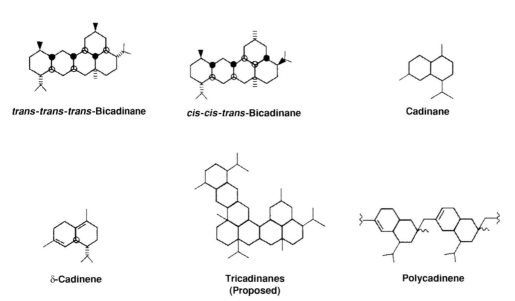

Figure 13.64. Various oligomers of cadinane in crude oils may originate by catagenesis of polycadinene biopolymers in angiosperm dammar resins (Cox et al., 1986; van Aarssen et al., 1990b). The figure shows the two identified stereoisomeric bicadinanes (van Aarssen et al., 1990b). Cadinane has five asymmetric carbons resulting in 32 (or 2^5) possible stereoisomers. A possible structure for tricadinane is suggested (van Aarssen and de Leeuw, 1989).

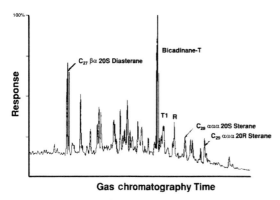

Figure 13.65. The m/z 217 chromatogram for steranes from the saturate fraction of a Southeast Asian crude oil has a dominant peak identified as bicadinane by its mass spectrum and comparison with that in Cox et al. (1986). Other less prominent bicadinane isomers were detected using a collision-activated decomposition/gas chromatography/mass spectrometry/mass spectrometry (CAD/GCMS/MS) m/z 412 → 369 transition, including T1 and R. These and other bicadinane isomers contribute to the complexity of this m/z 217 fingerprint. Reprinted with permission by ChevronTexaco Exploration and Production Technology Company, a division of Chevron USA Inc.

not co-vary with bicadinanes in these oils, suggesting separate land-plant sources for these two triterpanes. Summons et al. (1995) report significant bicadinanes in all of the Jurassic oils that they studied from the Perth Basin. We observed bicadinanes in Permian coals of terrigenous origin (Moldowan, 2002, unpublished data), suggesting that they are not restricted to an origin from angiosperms. Based on pyrolysis of fossil and fresh resins, van Aarssen et al. (1990b) proposed a generalized biopolymer structure called polycadinene that fragments to form the various cadinene oligomers (Figure 13.64). Their work indicates that polycadinene comprises the angiosperm dammar resins of some Southeast Asian higher plants. However, one- and two-dimensional NMR studies of modern and fossil dammar resin suggest that the generalized structural model for polycadinenes is inadequate and requires revision (Anderson and Muntean, 2000).

The mass spectra of bicadinanes contain prominent m/z 191 and 217 fragments and, therefore, peaks can appear in the corresponding chromatograms used in the analysis of hopanes and steranes (e.g. Figure 13.65). Because the bicadinane mass spectrum has an intense m/z 369 for loss of iso-propyl and a strong molecular ion (m/z 412), it can be monitored conveniently with little interference using the m/z 412 → 369 transition (Figure 13.66). Hopanes show the same transition but typically give additional but weaker peaks on this chromatogram. However, the hopanes are distinguished easily because they show much higher retention times than the bicadinanes under normal analytical gas chromatography conditions.

Murray *et al.* (1994) tentatively identified bicadinanes and methylbicadinanes (Figure 13.67) based on comparisons of mass spectra and relative retention times with known, identified bicadinanes W and T (Cox *et al.*, 1986; van Aarrsen *et al.*, 1990b) as follows: W, V, W1, W2, and MeW as *cis-cis-trans* configurations and T, T1,

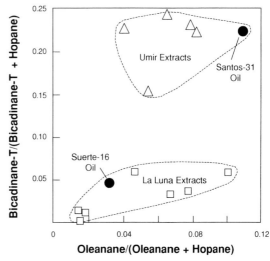

Figure 13.67. The ratio bicadinane-T/(bicadinane-T + hopane) distinguishes Umir from La Luna Formation source rocks and related oils in the Middle Magdalena Basin, Colombia (Rangel *et al.*, 2002). The figure confirms that oleanane and bicadinane relative abundances are independent of each other. Reprinted by permission of the AAPG, whose permission is required for further use.

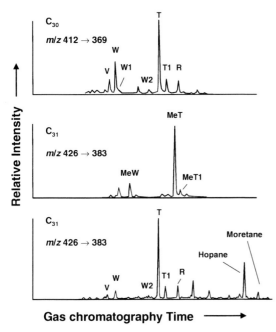

Figure 13.66. Metastable reaction monitoring/gas chromatography/mass spectrometry (MRM/GCMS) analyses of bicadinanes and methylbicadinanes in crude oils. Bicadinane distribution from the MRM/GCMS m/z 412 → 369 transition (top) and putative methylbicadinane distribution from the m/z 426 → 383 transition (middle) for oil from the Tarakan Basin, Kalimantan, Indonesia, which was generated from a fluvial deltaic source rock (Murray *et al.*, 1994). Previously unpublished MRM/GCMS analysis for oil from North Sumatra, Indonesia (bottom), includes bicadinanes, C_{30} 17α-hopane, and moretane. The bicadinane-T/(bicadinane-T + hopane) ratio obtained from this analysis is a useful source parameter for terrigenous-influenced crude oils (see Figure 13.67).

R, MeT, MeT1, and MeR as *trans-trans-trans* configurations (Figure 13.64). The ratio bicadinanes/hopane distinguishes many fluvial deltaic oils (high values) from those with source rocks deposited in either lacustrine or marine depositional settings. Rangel *et al.* (2002) used the modified ratio bicadinane-T/(bicadinane-T + hopane) to characterize and distinguish the La Luna Formation from the Umir Formation and related oils in the Middle Magdalena Basin, Colombia (Figure 13.67). Hopane can be measured on the same GCMS/MS m/z 412 → 369 transition as the bicadinane (Figure 13.66).

Tetracyclic diterpanes (beyerane, phyllocladane, kaurane)

> Many are highly specific for terrigenous organic matter input. Measured using M+ (m/z 274) in combination with various fragment ions (see Figure 13.60).

Many tetracyclic diterpanes are believed to be age-related markers for conifers. For example, Noble *et al.*

(1985a; 1985b; 1986) report a family of bridged tetracyclic diterpanes in Australian coals, sediments, and oils. They identified compounds including *ent*-beyerane, 16α- and 16β-phyllocladane, *ent*-16α- and *ent*-16β-kaurane (Figure 13.60), and a 17-nortetracyclic diterpane thought to be related to diterpanes abundant in leaf resins of conifers belonging to the *Podocarpaceae*, *Araucariaceae*, and *Cupressaceae* families. Phyllocladane was identified in resins from *Podocarpus* and *Dacrydium* (species of *Podocarpaceae*), while kaurane was detected in *Agathis* (a species of *Araucariaceae*). Both of these families are currently restricted to the southern hemisphere. Weston *et al.* (1989) also reported many of these markers in oils from the Taranaki Basin, New Zealand, and Disnar and Harouna (1994) found many of these compounds in lower Cretaceous coals from Niger. Killops *et al.* (1995) used the differential abundance of these compounds and higher-plant triterpanes to distinguish Mesozoic and Tertiary rocks in New Zealand.

While tetracyclic diterpenes are especially abundant in southern hemisphere conifers, all land plants produce trace amounts of tetracyclic diterpenes. This is because of the ubiquitous occurrence of the tetracyclic diterpenoid hormones known as gibberellins in land plants (Crozier, 1983). Gibberellins play an important role in (1) cell division and cell elongation, (2) spore and seed germination, (3) flowering and/or the formation of gametes, and (4) production of chlorophyll (Crozier, 1983). The members of the biochemical pathway used to synthesize gibberellin are enantiomeric, with stereochemistry like that in the hopanes and steranes (signified by the prefix *ent*-). The pathway includes the *ent*-labdane and *ent*-kaurene skeletons as intermediate products and the *ent*-pimarane and *ent*-beyerane skeletons as carbocation intermediates. Therefore, the seemingly more complex tetracyclic diterpanes that lie near the end of the gibberellin synthesis pathway (Figure 13.68) tend to dominate over tricyclic diterpanes in Devonian and Carboniferous coals or plant fossils (Sheng *et al.*, 1992; Thomas, 1990). More derived structures or diversions from this pathway appear to be synthesized and accumulated by other plant species, as evidenced by the predominance of a host of tricyclic diterpane structures that tend to dominate the tetracyclic diterpanes in Mesozoic and Tertiary rocks and oils. In Figure 13.69, the *m/z* 274–123 chromatogram

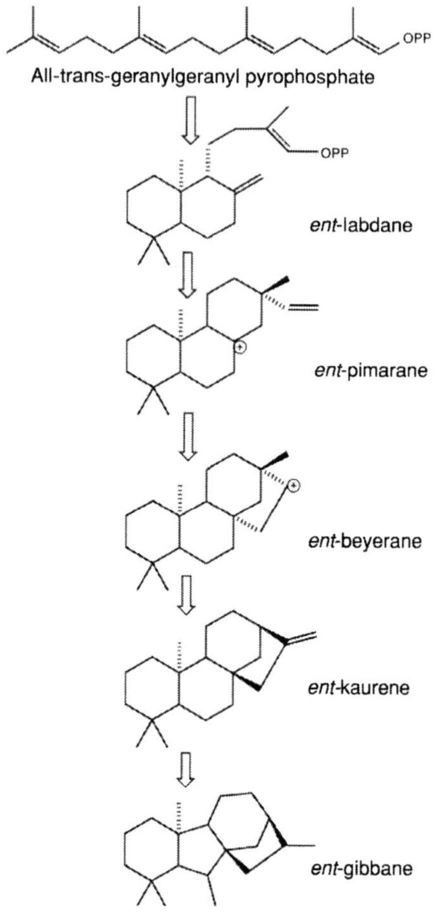

Figure 13.68. Diterpane skeletons in the gibberelin biosynthetic pathway (courtesy of David Zinniker).

shows a typical tetracyclic diterpane distribution for a Carboniferous coal extract, where abundant tetracyclic diterpanes are present. Kaurene rearranges by acid catalysis during diageneis to the products atisene and beyerene. This is proposed based on laboratory simulation experiments treating kaurene with clay, and the observation that beyerane and atisane often dominate this array of tetracyclic diterpanes in oils or coals with higher diasteranes/steranes ratios (D. Zinniker, J.-M. Trendel, and others, unpublished). Atisane occurs as its 16α(H)- and 16β(H)-stereoisomers, one of which co-elutes with 16β(H)-kaurane under typical GCMS conditions (Figure 13.69).

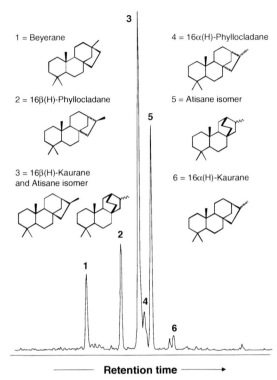

Figure 13.69. Parent–daughter m/z 274–123 transition for extract of Carboniferous coal in the Upper Pennsylvanian Mattoon Formation, New Calhoun, Illinois (courtesy of David Zinniker).

Papanicolaou *et al.* (2000) compared the petrographic and geochemical characteristics of low-maturity coals with distinct plant inputs from the Florina, Lava, Moschopotamos, and Kalavryta basins in Greece. Saturate fractions from Lava and Moschopotamos lignites are rich in C_{23}–C_{33} *n*-alkanes, consistent with petrographic evidence for mainly non-woody, angiosperm vegetation. In contrast, saturate fractions from Florina and Kalavryta lignites are rich in diterpanes and sesquiterpenoids. Gymnosperm remains dominate the Florina and Kalavryta lignites. The Miocene lignite from Florina contains phyllocladane, a diterpane that implies input from *Cupressaceae* and/or *Taxodiaceae*. Diterpenoids in the Pliocene lignite from Kalavryta originate from abietic and pimaric acid, which are found in *Pinaceae*. These different biomarker compositions appear to be related to differences in gymnosperm input resulting from climatic change.

Although difficult to distinguish petrographically, limnic coals from the Saar District are enriched in *ent*-beyerane, while paralic marine coals of the Ruhr district in Germany contain more kauranes (Schulze and Michaelis, 1990). Because these coals are of Late Carboniferous age and *Cupressaceae*, *Araucariaceae*, and *Podocarpaceae* did not evolve until Late Triassic time, the authors conclude that the phyllocladane-type compounds in their coals were derived from early conifers, the *Voltziales*, which may already have been able to biosynthesize these compounds. The *Voltziales* appear to have evolved during Late Carboniferous time.

Thermodynamic equilibrium is reached before the beginning of oil generation where the ratio of isomers $16\alpha/16\beta$ for *ent*-kaurane and *ent*-phyllocladane reach endpoints of <0.1 and 0.3, respectively. Marine organisms, especially algae, also produce diterpenoids (Simoneit, 1986). More detailed work is needed to differentiate which structures could originate from marine algae versus higher plants. Noble (1986) used distributions of tetracyclic and other diterpanes to correlate a crude oil to one horizon within a series of Gippsland Basin rocks extending over a 1 km depth range (Figure 13.60). Villar *et al.* (1988) identified resin-derived diterpenoids with the phyllocladane and kaurane skeleton in Tertiary coals and shales from Argentina, indicating input from conifers.

The various tetracyclic diterpane isomers can be analyzed and distinguished by GCMS using their mass spectra and by selective ion monitoring of the major diagnostic fragments at m/z 123, 231, 245, 259, and 274 (Figure 8.20).

Terpane m/z 191 fingerprint

> Specific for correlation/depositional environment when examined in detail, but affected by interfering peaks. Measured using SIM/GCMS.

Many terpanes in petroleum originate from bacterial (prokaryotic) membrane lipids (Ourisson *et al.*, 1982). These bacterial terpanes include several homologous series, including acyclic, bicyclic (drimanes), tricyclic, tetracyclic, and pentacyclic compounds. The following is a brief overview of more detailed discussions of these compounds found elsewhere in this book.

Bicyclic terpanes of the drimane series are ubiquitous in sediments and crude oils, and for this reason they are thought to be of microbial origin (Alexander et al., 1983a). Noble and Alexander (1989) proposed a mechanism for their formation by oxidation of Δ11(12)-bacteriohopenetetrol during diagenesis. The $\delta^{13}C$ of drimanes and 1,2,5-trimethylnaphthalene are comparable to homohopanes in torbanites, consistent with a common cyanobacterial hopanoid precursor (Grice et al., 2001).

The tricyclic terpanes (Connan et al., 1980; Aquino Neto et al., 1983) extend from C_{19} to at least C_{54} because of their isoprenoid side chains (Moldowan et al., 1983; De Grande et al., 1993). The tricyclic terpanes ($<C_{30}$) appear to originate from a regular C_{30} isoprenoid, such as tricyclohexaprenol (Aquino Neto et al., 1983), and could be constituents in prokaryote membranes (Ourisson et al., 1982). However, high concentrations of tricyclic terpanes correlate with Tasmanites-rich rocks, suggesting they may be related to these primitive algae (Volkman et al., 1989; Azevedo et al., 1992). The C_{28} and C_{29} tricyclics were used extensively in correlations of oils and rock extracts (Seifert et al., 1980; Seifert and Moldowan, 1981). Some tricyclic terpanes (Figure 13.60) are terrigenous indicators (Noble, 1986). Tricyclic diterpanes (C_{19}–C_{20}) originate mainly from diterpenoids, such as abietic acid (Figure 11.14), which are produced by vascular plants (Barnes and Barnes, 1983). Walters and Cassa (1985) show that the ratio of tricyclic diterpanes to the sum of sesterterpane and triterpane tricyclics is a sensitive indicator of source input for offshore Gulf Coast oils. Figure 13.73 shows an example of the tricyclic terpanes (C_{29} extended ent-isocopalanes) (Aquino Neto et al., 1983).

Based on structural studies (Trendel et al., 1982), the C_{24}–C_{27} tetracyclic terpanes appear to be degraded hopanes (17,21-secohopanes, see Figure 2.15). Tetracyclic terpanes appear to be more resistant to biodegradation and maturation than the hopanes.

Pentacyclic triterpenoids (see Figure 2.15), including precursors of the hopanes, occur in prokaryotes and higher plants but appear to be absent in eukaryotic algae. Bacteria are the major source for sedimentary hopanoids. The extended hopanes (C_{31} or more) are related to specific bacteriohopanepolyols in bacteria, such as bacteriohopanetetrol (Figure 2.30), while the lower pseudohomologs (C_{30} or less) may also be re-

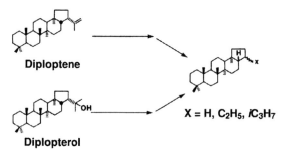

Figure 13.70. Diploptene and diplopterol are common in nearly all hopanoid-producing bacteria and represent likely sources for hopanes containing 30 or fewer carbon atoms. The extended hopanes (C_{31} or more) are most likely derived from bacteriohopanepolyols (see Figure 2.15). Reprinted with permission by ChevronTexaco Exploration and Production Technology Company, a division of Chevron USA Inc.

lated to C_{30} precursors, such as diploptene or diplopterol (Figure 13.70) found in nearly all hopanoid-producing bacteria (Rohmer, 1987).

Hopanes are pentacyclic triterpanes commonly containing 27–35 carbon atoms in a naphthenic structure composed of four six-member rings and one five-member ring (Van Dorsselaer et al., 1977). Hopanes originate from precursors in bacterial membranes (Ourisson et al., 1979). Hopanoids in these membranes, such as bacteriohopanetetrol (Figure 2.30), originate by cyclization of squalene precursors (Figure 3.15) (Rohmer, 1987).

Nytoft and Bojesen-Koefoed (2001) assigned, or tentatively assigned, many of the small peaks remaining to be identified on typical m/z 191 chromatograms as various hopanoids. They showed the existence of the $17\alpha,21\alpha(H)$-hopanes by co-elution experiments using authentic C_{30} $\alpha\alpha$-hopane (Figure 13.71). This C_{30} $\alpha\alpha$-hopane was resolved from C_{30} $\beta\alpha$-hopane, while earlier work (Bauer et al., 1983) was unable to resolve C_{29} $\alpha\alpha$-hopane from co-elution with C_{29} $\beta\alpha$-hopane. Subsequently, the separation of C_{29} $\alpha\alpha$-hopane from C_{29} $\beta\alpha$-hopane was accomplished using a polar column (P. Nytoft, personal communication, 2002). A tentative assignment of C_{30} 18α-neohopane (C_{30}Ts) is also given based on relative retention time and statistical covariance with C_{29} 18α-30-norneohopane (C_{29}Ts).

Tricyclics, tetracyclics, hopanes, and other compounds contribute to the terpane fingerprint (m/z 191) and are commonly used to relate oils and source rocks (Seifert et al., 1980). Examples of terpane fingerprints

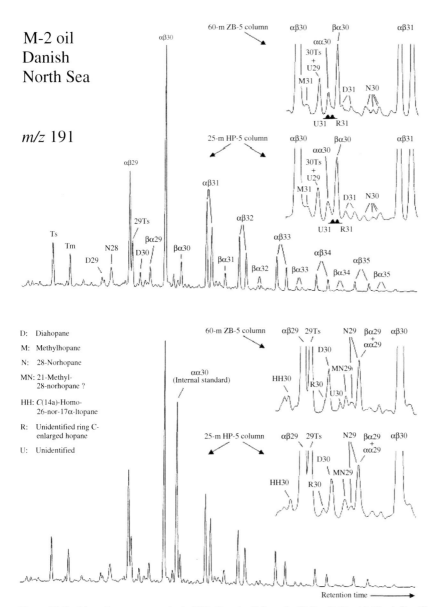

Figure 13.71. Mass chromatograms (m/z 191) of crude oil from the M-2 well, Danish North Sea. Top: no internal standard. Bottom: 0.02 mg $\alpha\alpha$ hopane/72.4 mg oil. The two large chromatograms were obtained using a 25-m HP-5 column. Insets show details of the chromatograms before and after elution of C_{30} $\alpha\beta$ hopane. Partial chromatograms obtained on a 60-m ZB-5 column are shown for comparison. On the 60-m ZB-5 column, two minor C_{31} compounds co-elute with $\alpha\alpha$ hopane. Their retention times are indicated by small black peaks below the chromatograms. Reprinted from Nytoft and Bojesen-Koefoed (2001). © Copyright 2001, with permission from Elsevier.

for oils from different types of source rocks are shown in Figure 13.72 (see also other figures showing m/z 191 mass chromatograms). Terpane fingerprints reflect source rock depositional environment and organic matter input. Because bacteria are ubiquitous in sediments, terpanes occur in nearly all oils, and oils from different source rocks deposited under similar conditions may show similar terpane fingerprints.

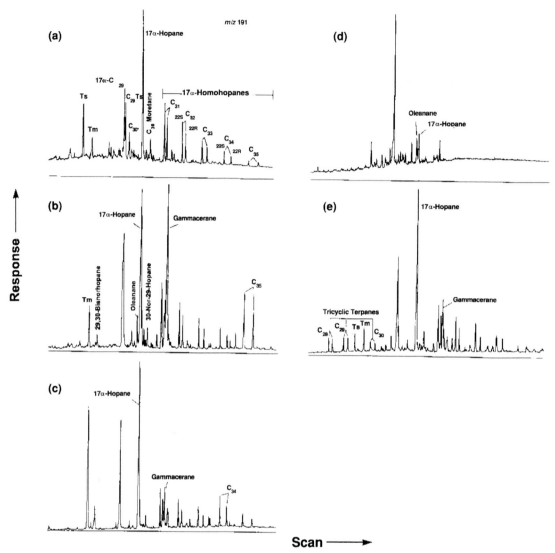

Figure 13.72. Examples of terpane fingerprints (m/z 191) for petroleum from various types of source rock depositional environments. Reprinted with permission by ChevronTexaco Exploration and Production Technology Company, a division of Chevron USA Inc. (a) Oil from the northwestern shelf of Australia has a mass chromatogram typical of source rock deposited under marine shelf conditions. The regular stair-step progression of C_{31}–C_{35} homohopanes is consistent with suboxic bottom waters during deposition. The $C_{30}{}^*$ peak represents a rearranged hopane (17α-diahopane) probably related to hopanes, which resulted from clay-mediated acid catalysis reactions on hopenes. Absence of oleanane in oil from a shelf or deltaic source rock suggests, but does not prove, that the source is older than Late Cretaceous in age. (b) Prinos oil from Greece (Seifert et al., 1984) probably originated from carbonate source rock deposited in an anoxic basin, as indicated by strong preservation of the C_{35} homohopanes. The 30-norhopane series, typified by 17α-29,30-dinorhopane and 17α-30-nor-29-homohopane, is present as in many carbonate-derived oils. The strong gammacerane peak may signify hypersaline water during deposition of the source rock. The presence of oleanane indicates Tertiary-Late Cretaceous age. (c) Ravni Kotari-3 oil from Yugoslavia (Moldowan et al., 1992) is another oil from an anoxic to slightly suboxic hypersaline source rock. Its terpane fingerprint features gammacerane and a predominance of the C_{34}-homohopanes. (d) Miocene oil from the Kenai reservoir, Swanson River Unit at Cook Inlet, Alaska (Peters et al., 1986), yields relatively little response in the C_{27}–C_{35} terpane range. The only identified terpanes are labeled. This unusual liquid is thought to originate from resinitic coaly source rock. (e) Oil from offshore Angola originated from Lower Cretaceous lacustrine source rock. Gammacerane is typical, but not diagnostic, of saline to hypersaline lake deposition. The C_{28}–C_{30} tricyclic terpanes (cheilanthanes) are also present in this m/z 191 fingerprint.

For example, although the C_{29} (norhopane) and C_{30} (hopane) $17\alpha,21\beta(H)$ hopanes are the dominant triterpanes in many oils, their relative abundance is not very useful for separating them into genetically related families. Most crude oils show C_{30} hopane/C_{29} hopane m/z 191 peak ratios greater than 1. The ratio measured using m/z 191 is more sensitive to C_{30} hopanes than C_{29} hopanes because the C_{30} hopanes undergo two fragmentations yielding m/z 191 fragments, while C_{29} hopanes undergo only one fragmentation. Crude oils from organic-rich carbonate-evaporite rocks (e.g. calcite, halite, gypsum, anhydrite) generally show larger peaks on the m/z 191 chromatogram for the C_{29} compared with the C_{30} hopane (Zumberge, 1984; Connan et al., 1986; Clark and Philp, 1989). Brooks (1986) observed that high-C_{29} hopanes characterize many crude oils from source rocks rich in terrigenous organic matter. However, this was probably due to preferential biodegradation of C_{30} over C_{29} hopanes in heavily biodegraded oils and not due to source organic matter type.

Tricyclic terpanes (cheilanthanes) and tricyclics/17α-hopanes

> Widespread in oils and bitumens derived from lacustrine and marine source rocks. Measured using m/z 191 fragmentogram (see Figure 8.21).

Tricyclic terpanes or cheilanthanes were first observed in extracts from the Green River Formation (Anders and Robinson, 1971; Gallegos, 1971). The most prominent tricyclic terpanes are 14-alkyl-13-methylpodocarpanes (Figure 13.73). Moldowan et al. (1983) showed that the tricyclic terpane series extends to C_{45} in oil from California, while De Grande et al. (1993) analyzed various oils and rocks to extend the range to at least C_{54}. Tricyclic terpanes are used to correlate crude oils and source-rock extracts, to predict source-rock characteristics, and to evaluate the extent of thermal maturity and biodegradation (Seifert et al., 1980; Seifert and Moldowan, 1981; Zumberge, 1987a; Peters and Moldowan, 1993). Palacas et al. (1984) found that tricyclic terpanes were the most useful series of biomarkers for differentiating potential from effective source rocks in the South Florida Basin.

The structures of several homologs of the tricyclic terpanes were proven by synthesis (Aquino Neto et al.,

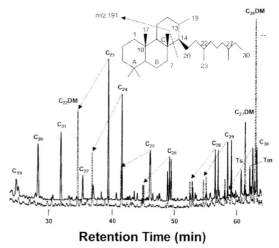

Figure 13.73. Mass chromatogram (m/z 191, solid trace) showing C_{19}–C_{30} tricyclic terpanes in the saturated hydrocarbon fraction of heavily biodegraded crude oil from the Rubiales Field, Well 12 (2603–2608 feet (793.4–794.9 m), 6.9° API), Llanos Basin, Colombia. C_{30} tricyclohexaprenane (inset) is a widespread higher homolog in geological samples. Doublets eluting after C_{24} are stereoisomers resulting from the asymmetric center at C-22. In-reservoir biodegradation resulted in C-10 demethylation of the tricyclic terpanes and formation of the corresponding 17-nor-tricyclic terpanes (DM, m/z 177 dotted trace), where examples are indicated by diagonal arrows. Demethylation slightly favors the second-eluting tricyclic terpane peak, especially for the C_{26}, C_{28}, and C_{29} homologs (e.g. Alberdi et al., 2001). The sample was treated with molecular sieves to remove n-alkanes. Ts = 18α-22,29,30-trisnorneohopane, Tm = 17α-22,29,30-trisnorhopane. Reprinted from Peters (2000). © Copyright 2000, with permission from Elsevier.

1982; Ekweozor and Strausz, 1982; Heissler et al., 1984; Sierra et al., 1984). All four isomers at C-13 and C-14 ($\beta\alpha, \alpha\alpha, \alpha\beta,$ and $\beta\beta$) occur in immature rocks with $\beta\alpha$ and $\alpha\alpha$ predominating, but with increasing maturity the $\beta\alpha$ isomer becomes dominant (Chicarelli et al., 1988).

Extended tricyclic terpanes ($>C_{24}$) contain a regular isoprenoid side chain at C-14 (Aquino Neto et al., 1982), as evidenced by lower abundance of the C_{22}, C_{27}, C_{32}, C_{37}, and C_{42} homologs, which require cleavage of two carbon–carbon bonds to form from higher homologs (e.g. Figures 13.73 and 16.42). Based on their results for the C_{20} and C_{21} tricyclic terpanes, Chicarelli et al. (1988) infer that higher homologs in thermally mature source-rock extracts and crude oils also show

13β,14α stereochemistry. The C_{25}–C_{29} tricyclic terpanes occur as diastereomeric doublet peaks on m/z 191 mass chromatograms, which arise by epimerization at C-22 in the side chain. The C_{25} tricyclic terpanes are commonly poorly resolved (Figure 13.73). Splitting of these two peaks into four peaks should occur for the C_{30}–C_{34} homologs because of the additional asymmetric center at C-27. However, under routine chromatographic conditions, this is not observed (Peters, 2000). For example, Moldowan et al. (1983) did not observe splitting of the first- and second-eluting peaks in each doublet until C_{35} and C_{38}, respectively.

Chromatographic elution times of the tricyclic terpanes are non-uniform due to the isoprenoid character of the side chain. Figure 13.74 shows three distinct linear trends in a plot of tricyclic terpane mass versus log retention time. The C_{24} and C_{29} tricyclic terpanes contain terminal isopropyl groups where addition of a methyl group to form the next homolog activates asymmetric carbon atoms at C-22 and C-27, respectively. These different trends were validated by molecular mechanics calculations, which show distinct low-energy conformations for the tricyclic terpanes within each linear segment of the figure (Peters, 2000).

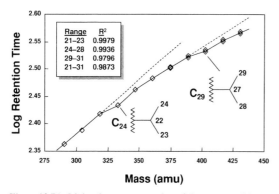

Figure 13.74. Molecular mass versus log of chromatographic retention time for tricyclic terpanes in a non-biodegraded west African oil (30.5° API) has at least two distinct breaks in the linear trend at C_{24} and C_{29}. Data points for the 22S and 22R diastereomers of the C_{25}–C_{28} tricyclic terpanes plot in nearly identical locations due to the log retention time scale. The HP-5 Ultra chromatographic column was slow-programmed to 325°C at 0.1°C/min. An HP Ultra 2 column programmed to 325°C at 2°C/min generates comparable breaks in the linear trend. Reprinted from Peters (2000). © Copyright 2000, with permission from Elsevier.

Regular polyisoprenols, such as C_{30} tricyclohexaprenol in bacterial membranes and malabaricatrienes from algae or bacteria, could be intermediates in the biosynthetic pathway that accounts for many tricyclic terpanes in petroleum (Ourisson et al., 1982; Aquino Neto et al., 1983; Heissler et al., 1984; Behrens et al., 1999). Higher homologs may originate from C_{40} tricyclooctaprenol (Azevedo et al., 1998) or larger precursors. High concentrations of tricyclic terpanes and their aromatic analogs commonly correlate with high paleolatitude *Tasmanites*-rich rocks, suggesting an origin from these algae (Volkman et al., 1989; Azevedo et al., 1992; Simoneit et al., 1993). However, stable carbon isotopic data suggest that other algal or bacterial sources are possible (Revill et al., 1994). Kruge et al. (1990b) and De Grande et al. (1993) noted prominent tricyclic terpanes in highly mature oils and extracts from saline lacustrine and marine carbonate source rocks, suggesting that the precursor organisms lived in moderate salinity conditions. However, these interpretations are suspect because tricyclic terpanes are more stable than many other terpanes and thus are generally more abundant in highly mature petroleum, regardless of source-rock organic matter input (Peters and Moldowan, 1993).

Tricyclics/17α-hopanes is primarily a source parameter that compares a group of bacterial or algal lipids (tricyclics) with markers that arise from different prokaryotic species (hopanes). One such ratio includes the sum of four tricyclic terpane peaks, 22R and 22S doublets, representing the C_{28} and C_{29} pseudohomologs of tricyclohexaprenane (29,30-bisnorcyclohexaprenane and 30-norcyclohexaprenane, respectively), for the numerator of the ratio. The sum of the C_{29}–C_{33} 17α-hopanes comprises the denominator of the ratio. Other common tricyclic terpane ratios include the sum of the four C_{28} and C_{29} tricyclic terpanes divided by hopane and the C_{23} tricyclic terpane/hopane. Figure 13.75 is an example of an early correlation of extract from Triassic Shublik Shale source rock with crude oil produced from the Sag River sandstone reservoir. Later work supports dominantly Shublik input to Sag River oil (e.g. Masterson, 2001).

Note: For tricyclic terpanes containing 25 or more carbon atoms, C-22 is an asymmetric center resulting in two stereoisomers or a doublet on m/z 191 mass chromatograms. The second-eluting peak in each C_{26}–C_{29} doublet is biodegraded more

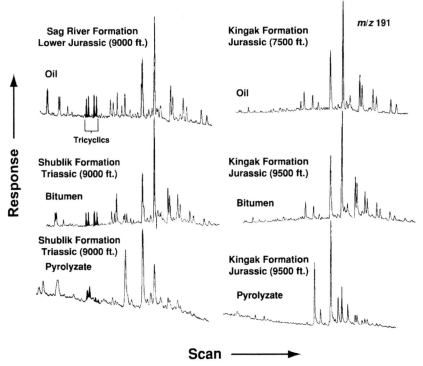

Figure 13.75. Relative amounts of tricyclic terpanes on these m/z 191 fragmentograms (black peaks) support major input from the Triassic Shublik source rock (9000 feet (2743.2 m)) to the Sag River oil (9000 feet (2743.2 m)) (Seifert et al., 1980). Low tricyclic terpanes in the Kingak oil (7500 feet (2268 m)) (Kavearak 32-25 well, Milne Field; see Figure 18.117) and Kingak Shale (9500 feet (2895.6 m)) support an oil–source rock correlation for these samples. The indigenous nature of the bitumen in the source rocks is confirmed by comparing m/z 191 fragmentograms for pyrolyzates of these rocks, which show the presence and absence of tricyclic terpanes in the Shublik and Kingak samples, respectively. Reprinted with permission by ChevronTexaco Exploration and Production Technology Company, a division of Chevron USA Inc.

readily (Alberdi et al., 2001). Molecular mechanics suggests that each 22R epimer elutes after the corresponding 22S epimer (Peters, 2000). For tricyclic terpanes containing 30 or more carbon atoms, C-27 is also an asymmetric center, and at least four stereoisomers are expected for each compound. The tricyclic terpane series commonly lacks, or has low concentrations of, C_{22}, C_{27}, C_{32}, C_{37}, and C_{42} homologs because of the isoprenoid structure of the side chain (methyl substitution every fourth carbon atom). For example, to generate the C_{22} tricyclic terpane requires the cleavage of two carbon–carbon bonds rather than one bond at C-22. As described earlier, the C_{17} acyclic isoprenoid is low or absent in petroleum for a similar reason. Nonetheless, the C_{22} tricyclic terpane can be abundant compared with the C_{21} homolog in certain oils generated from carbonate source rocks (Figure 13.76).

The widespread occurrence of tricyclic terpanes and molecular properties of tricyclohexaprenol suggest that they originate from prokaryotic membranes (Ourisson et al., 1982). These authors propose tricyclohexaprenol as the parent compound for tricyclic terpanes containing 30 or fewer carbon atoms. However, high concentrations of tricyclic terpanes in Tasmanites rock extracts indicate a possible origin from these algae (Volkman et al., 1989; Azevedo et al., 1992). Permian Tasmanite oil shale contains ~29 wt.% TOC, where the only recognizable microscopic structures are compressed disks of Tasmanites (Simoneit et al., 1993). Abundant extended C_{18}–C_{25} tricyclic terpanes in the extract and kerogen from Tasmanite oil shale are

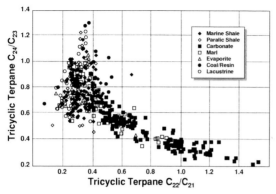

Figure 13.76. Oils from carbonate source rocks can be distinguished by high C_{22}/C_{21} and low C_{24}/C_{23} tricyclic terpanes. The figure is based on more than 500 worldwide crude oil samples used as a database to predict source-rock depositional environments. Sources for the samples are as follows: Lacustrine – Campos, Recôncavo, Beibu, Pearl River, Natuna, Sunda, Sumatra, Penyu, Mekong, Thailand, Inner Moray Firth, west Africa, San Jorge, Uinta. Marine shale – Ghadames/Illizi, Sirte, Llanos, Zagros, North Sea, Maranon, west Siberia, Anadarko, Tarragona, Tunisia. Paralic marine shale – Nigeria, Western Desert, Maturin, Mahakam Delta. Marine carbonate – Oriente, Maracaibo, Williston, and Powder River basins, and the Middle East, Gulf of Suez, and Northwest Palawan. Marine marl – Oriente and Maturin basins, Middle and Upper Magdalena Valleys, East Siberia, Oman, Zagros, Monterey Formation. Evaporite – Ceará, Prinos, Sicily, Great Salt Lake. Figure courtesy of GeoMark Research, Inc. (J. E. Zumberge, 2000, personal communication).

unusually enriched in ^{13}C ($\delta^{13}C = -9.9$ to $-12.2‰$) compared with pristane, phytane, and the n-alkanes (-18 to $-22‰$) and appear to originate from *Tasmanites* (Simoneit et al., 1993). Oils and bitumens from carbonate rocks have low concentrations of tricyclic terpanes above C_{26} compared with those from other depositional environments where the C_{26}–C_{30} and C_{19}–C_{25} homologs have similar concentrations (Aquino Neto et al., 1983).

Evidence suggests a higher-plant source for some C_{19} and C_{20} tricyclic terpanes (Figure 13.60). Zumberge (1983) indicates that C_{19}–C_{20} tricyclics may also be produced by the thermal cleavage of the alkyl side chain in sester- and triterpanes. The C_{23} homolog is typically the most prominent tricyclic terpane (Connan et al., 1980; Aquino Neto et al., 1983; Sierra et al., 1984).

Because of their extreme resistance to biodegradation, tricyclic terpanes permit correlation of intensely biodegraded oils (Seifert and Moldowan, 1979; Palacas et al., 1986). They are also more resistant to thermal maturation than hopanes, although the lower-carbon-number homologs are favored at high thermal maturity (Peters et al., 1990). Tricyclic terpanes crack from the kerogen at relatively higher thermal maturity than hopanes (Aquino Neto et al., 1983; Peters et al., 1990). Sofer (1988) used tricyclic terpanes, stable carbon isotopes of saturated and aromatic hydrocarbons, and other data to classify genetic groups of Gulf Coast oils showing wide variations in thermal maturity.

C_{22}/C_{21}, C_{24}/C_{23}, and C_{26}/C_{25} tricyclic terpane ratios

> Ratios of various tricyclic terpanes by carbon number can be useful in order to distinguish marine, carbonate, lacustrine, paralic, coal/resin, and evaporitic oils. Measured using m/z 191 fragmentogram (see Figure 8.21).

The C_{22}/C_{21} and C_{24}/C_{23} tricyclic terpane ratios help to identify extracts and crude oils derived from carbonate source rocks (Figure 13.76). The C_{26}/C_{25} tricyclic terpane ratio is useful as a supporting method to distinguish lacustrine from marine oils (Figure 13.77).

Extended tricyclic terpane ratio

> Age-related parameter to distinguish Triassic from Jurassic oil samples. Measured using m/z 191 chromatograms.

Interfering effects of thermal maturity on source interpretation can be reduced by using the extended tricyclic terpane ratio (ETR) (= $(C_{28} + C_{29})/(C_{28} + C_{29} + Ts)$) (Holba et al., 2001) rather than tricyclics/17α-hopanes. Ts is thermally more stable than the hopanes, thus making ETR less dependent on maturity than tricyclics/17α-hopanes, although it also rises at very high maturity.

Early results suggest that ETR can be used to differentiate crude oils generated from Triassic, Lower Jurassic, and Middle-Upper Jurassic source rocks (Table 13.8, Figure 13.1). Holba et al. (2001) studied a suite of worldwide crude oil samples with Triassic or Jurassic source rocks based on geochemical and

Table 13.8. *The extended tricyclic terpane ratio (ETR) can be used to distinguish crude oils that originated from Triassic, Lower Jurassic, and Middle-Upper Jurassic source rock (from Holba et al., 2001)*

	Jurassic			Triassic		
	Upper	Middle	Lower	Upper	Middle	Lower
No. Oils	47	26	19	28	8	3
ETR $\geq$ 2.0	0	0	2	25	8	3
ETR $>$ 1.2	6	4	17	0	0	0

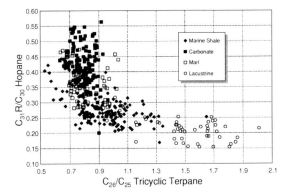

Figure 13.77. Unlike most marine crude oils, oils from lacustrine source rocks generally show high C_{26}/C_{25} tricyclic terpane and low C_{31} 22R/C_{30} hopane. This figure is based on more than 500 worldwide crude oil samples used as a database to predict source-rock depositional environments. Sources for the samples are as follows: Lacustrine – Campos, Recôncavo, Beibu, Pearl River, Natuna, Sunda, Sumatra, Penyu, Mekong, Thailand, Inner Moray Firth, west Africa, San Jorge, Uinta. Marine shale – Ghadames/Illizi, Sirte, Llanos, Zagros, North Sea, Maranon, west Siberia, Anadarko, Tarragona, Tunisia. Marine carbonate – Oriente, Maracaibo, Williston, and Powder River basins, and the Middle East, Gulf of Suez, and Northwest Palawan. Marine marl – Oriente and Maturin basins, Middle and Upper Magdalena Valleys, East Siberia, Oman, Zagros, Monterey Formation. Figure courtesy of GeoMark Research, Inc. (J. E. Zumberge, 2000, personal communication).

geological evidence. All Triassic oil samples (39) had ETR $\geq$ 2.0, except for three Late Triassic samples from the Adriatic Basin, which probably originated from a hypersaline source-rock setting. Early Jurassic oil samples (17) had ETR $\leq$ 2.0 except for two samples with slightly higher values. Middle or Late Jurassic oil samples (73) had ETR $\leq$ 2.0, with most <1.2 (63). The sharp drop in ETR at the end of the Triassic Period corresponds to a major mass extinction (Table 17.1, Figure 17.25), which may have had an impact on the principal biological sources of tricyclic terpanes (e.g. *Tasmanites*?). A subsequent lesser extinction in the Toarcian Age may have further decimated these biological sources of tricyclic terpanes, resulting in low ETR for crude oils generated from Middle-Upper Jurassic source rocks.

C_{24} tetracyclic terpane ratio

> Specificity unknown. Abundant C_{24} tetracyclic may indicate carbonate or evaporite depositional environment. Measured using *m/z* 191. Commonly expressed as C_{24} Tet/hopane, C_{24} Tet/C_{23} tricyclic, and C_{24} Tet/C_{26} tricyclic.

Tetracyclic terpanes of the 17,21-secohopane series (Figure 13.78) are structurally similar to hopanes (Figure 8.20) (Trendel *et al.*, 1982) and occur in most oils and rock extracts. Aquino Neto *et al.* (1983) show that tetracyclic terpanes range from C_{24} to C_{27} with tentative evidence for homologs up to C_{35}. Some of these compounds are identified as peaks on the *m/z* 191 chromatogram shown in Figure 8.21 (peaks 4, 5, and 10). They are thought to originate by thermal or microbial

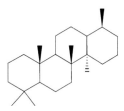

Figure 13.78. Des-E-hopane, a C_{24} 17,21-secohopane.

rupture of the E-ring (Figure 2.20) in hopanes or precursor hopanoids, although an independent biosynthetic route to the tetracyclic terpanes may exist in bacteria. The stable carbon isotopic composition of the C_{24} tetracyclic terpane is fairly constant among torbanites from boghead coals ($\delta^{13}C = -25.5$‰), suggesting an origin from hopanoids or terrigenous precursors with a constant isotopic composition (Grice et al., 2001).

Ratios of tetracyclic terpanes to hopanes increase in more mature source rocks and oils, indicating greater stability of the tetracyclic terpanes. Tetracyclic terpanes also are more resistant to biodegradation than the hopanes (Aquino Neto et al., 1983). For these reasons, they are used occasionally in correlations of altered crude oils (Seifert and Moldowan, 1979).

The C_{24} tetracyclic terpane/hopane, C_{24} tetracyclic/C_{23} tricyclic terpane, and C_{24} tetracyclic/C_{26} tricyclic terpane ratios are common source parameters. The C_{24} tetracyclic terpane (Figure 8.20, X = H) has the most widespread occurrence, followed by C_{25}–C_{27} homologs. Abundant C_{24} tetracyclic terpane in petroleum (e.g. see Figures 2.15 and 14.6) appears to indicate carbonate and evaporite source-rock settings (Palacas et al., 1984; Connan et al., 1986; Connan and Dessort, 1987; Mann et al., 1987; Clark and Philp, 1989). However, this compound also is present in Australian oils believed to originate from terrigenous organic matter (Philp and Gilbert, 1986) and is common in most marine oils generated from mudstone to carbonate source rocks. The C_{25}–C_{27} tetracyclic terpanes also have been reported in carbonate and evaporite samples (Connan et al., 1986).

Tetracyclic polyprenoid ratio

Highly specific for lacustrine organic matter input since at least Devonian time, the tetracyclic polyprenoid ratio (TPP) is measured using the GCMS/MS transition of M+ (414) → 259 for the tetracyclic polyprenoid and M+ (358) → 217 for the 27-norcholestanes. TPP probably originates from precursors in freshwater algae.

The tetracyclic polyprenoid (TPP) ratio (Holba et al., 2000) compares the GCMS/MS peak area for one of the two C_{30} tetracyclic polyprenoids with the peak areas of the four C_{26} 27-norcholestanes (Holba et al., 1998):

$$\text{TPP} = (2 \times \text{peak a})/[(2 \times \text{peak a}) + (\Sigma 27 - \text{norcholestanes})]$$

The two TPP peaks represent 21S and 21R stereoisomers of the C_{30} homolog in a series that ranges from C_{26} to $>C_{40}$ (Figure 13.79). An analogous series of

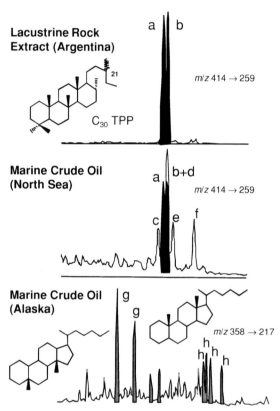

Figure 13.79. The C_{30} tetracyclic polyterpenoid (TPP, top inset) elutes as two peaks on gas chromatography/mass spectrometry/mass spectrometry (GCMS/MS) mass chromatograms of m/z 414 → 259, where peaks a and b are the 21R and 21S isomers, respectively. Top: source-rock extract from Triassic lacustrine Cachueta Shale, Cuyo Basin, Argentina. Middle: Jurassic marine oil from Ekofisk Field, offshore Norway. Bottom: m/z 358 → 217 norcholestanes from Jurassic marine oil, Cook Inlet, Alaska. Insets at bottom show 27-nordiacholestane (left) and 27-norcholestane (right). Peak c = 20S $\alpha\alpha\alpha$-3β-propylcholestane; d = 20R $\alpha\beta\beta$-3β-propylcholestane; e = 20S $\alpha\beta\beta$-3β-propylcholestane; f = 20R $\alpha\alpha\alpha$-3β-propylcholestane; g = 27-nordiacholestane isomers (20S = left, 20R = right); h = 27-norcholestane isomers, $\alpha\alpha\alpha$20S, $\alpha\beta\beta$20R, $\alpha\beta\beta$20S, $\alpha\alpha\alpha$20R, left to right, respectively (modified from Holba et al., 2000).

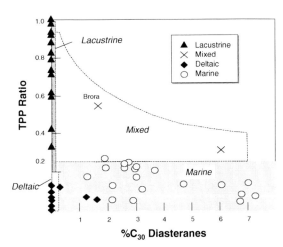

Figure 13.80. Lacustrine crude oils have high tetracyclic polyprenoid (TPP) ratios and lack C_{30} 24-propyldiacholestanes $[C_{30}/(C_{27} + C_{28} + C_{29} + C_{30})]$ (modified from Holba et al., 2000). The 19 lacustrine oil samples are from the Orcadian, Turpan, Songliao, Bohai, Sichuan, Uinta, Recôncavo, Tacatu, Campos, Coastal Gabon, Muglad, Sunda, and central Sumatra basins. The plot location for the oil from Brora Beach, Scotland, supports earlier conclusions that it is a mixture containing lacustrine Devonian and marine Middle Jurassic input (Peters et al., 1999b).

polyprenoid sulfides, which includes two isomers of the C_{30} analog to TPP, is widespread in marine and lacustrine systems (Poinsot et al., 1998). The abundance of TPP in petroleum from lacustrine versus marine source rocks may reflect the predominance of freshwater algal precursors or preferential preservation as saturated hydrocarbons under non-marine conditions (Holba et al., 2000). The area of the first-eluting TPP peak (a) is multiplied by a factor of two to compensate for co-elution of $20S,14\beta,17\beta$-3β-propylcholestane (Summons and Capon, 1991; Dahl et al., 1995) with the later-eluting TPP peak on a moderately polar DB-5 (5% methylphenylsiloxane) chromatographic column. The C_{26} 27-norcholestanes in the denominator of the TPP ratio are low in non-marine oils but common in marine oils.

Plots of the TPP ratio (lacustrine input) versus the 24-propyldiacholestane ratio (marine Chrysophyte algal input) or 4-methylsterane ratio (non-marine dinoflagellate input) readily identify lacustrine, marine, and mixed lacustrine-marine oils (Figure 13.80). Marine oils show low TPP ratios with high C_{30} 24-propyldiacholestanes, while non-marine oils do not contain 24-propyldiacholestanes. Mixed marine and non-marine oils show elevated TPP ratios with non-zero values for C_{30} propyldiacholestanes. For example, Beatrice oil has a high TPP ratio but also contains C_{30} propyldiacholestanes (Figure 13.80), consistent with the conclusion that it is a mixture of Devonian lacustrine and Middle Jurassic oil charges. TPP compounds extend well into the Paleozoic Era, as exemplified by the Beatrice oil.

Holba et al. (2003) observed moderately elevated TPP ratios and high C_{30} steranes for crude oils from certain basins in western and northern South America (Middle Magdalena, Colombia, Maracaibo, Venezuela, and Trinidad Basin). They interpreted these results to indicate (1) influx of freshwater in nearshore marine source-rock settings or (2) input from nearshore shallow marine algae with chemistry similar to that found in lacustrine settings. They also observed high TPP in a nearshore facies of the Chonta Formation, Peru, which also contained abundant Chlorococcalean green algal non-marine palynomorphs but no marine palynomorphs, suggesting that the green algae (Chlorophyta) are a possible source for TPP compounds.

28,30-Bisnorhopane and 25,28,30-trisnorhopanes

> Highly specific as a correlation tool; probable bacterial markers associated with some anoxic depositional environments. Measured using m/z 191, 177, and 163 fragmentograms. Also expressed as C_{28}/H, BNH/H, TNH/H, and BNH/TNH.

28,30-Bisnorhopane (BNH) and 25,28,30-trisnorhopane (TNH) are desmethylhopanes that occur as $17\alpha,18\alpha,21\beta(H)$-, $17\beta,18\alpha,21\alpha(H)$-, and $17\beta,18\alpha,21\beta(H)$-epimers (e.g. Figure 13.81). High concentrations of BNH and TNH are typical of petroleum source rocks deposited under anoxic conditions. For example,

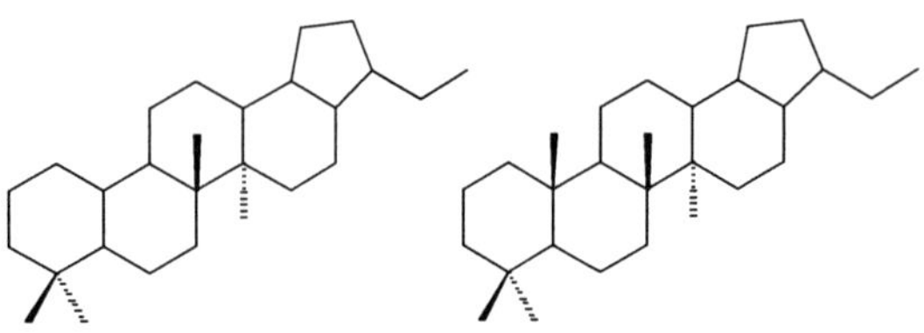

Figure 13.81. 25,28,30-Trisnorhopane (TNH) and 28,30-bisnorhopane (BNH). 28,30-Bisnorhopane is also known as 28,30-dinorhopane (DNH).

extracts of anoxic Upper Cretaceous source rocks from the Brazilian continental margin show abundant BNH and TNH (up to 130 ppm), porphyrins (up to 5700 ppm), and low pristane/phytane (Mello et al., 1990). High BNH and TNH correlate with high benzothiophene and low pristane/phytane for crude oils and extracts from the Monterey Formation and equivalents (Curiale et al., 1985).

BNH/C_{30} hopane is commonly used as a source parameter, although BNH is depleted more rapidly from the source rock as generation progresses. Thus, the ratio decreases dramatically with increasing API gravity of Monterey oils (Curiale et al., 1985). Therefore, BNH/hopane is useful for correlation only when samples have similar thermal maturity.

BNH/TNH is based on all epimers and was reported by Moldowan et al. (1984) for six different California oil–source rock groups. BNH/TNH was used to separate some genetic groups and was unaffected by the range of thermal maturity among the samples. BNH and TNH are not generated from kerogen but are passed from the original free bitumen in the source rock to the oil (Moldowan et al., 1984; Tannenbaum et al., 1986; Noble et al., 1985c). Therefore, the concentrations of BNH and TNH drop as source rocks generate oil during maturation (e.g. note the relative decrease in BNH in Figure 14.2 between Kimmeridge bitumen and related, but more mature, Piper oil). Because BNH and TNH crack at about the same rate, BNH/TNH remains approximately constant during maturation until one or both are depleted. Heavy biodegradation may invalidate BNH/TNH because, like the degradation of 17α-hopanes to 17α-25-norhopanes, BNH may be converted to TNH by microbes.

When abundant in petroleum, BNH and TNH indicate deposition of the source rock under clay-poor, anoxic conditions (Katz and Elrod, 1983; Mello et al., 1988b; Curiale and Odermatt, 1989). However, absence of these compounds does not exclude sedimentation under anoxic conditions. Figure 13.82 shows the relative abundance of BNH compared with clay content in a stratigraphic column spanning the Rincon and Monterey formations at the Naples Beach outcrop, Santa Barbara-Ventura Basin, California (Brincat and Abbott, 2001). The large changes in %BNH with depth are thought to indicate variations in microbial populations in response to changes in sedimentation rate and clastic input.

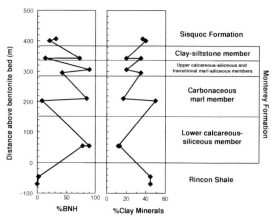

Figure 13.82. Percentage of bisnorhopane relative to total C_{27}–C_{32} hopanoids (hopenes + hopanes) and percentage of clay minerals in outcrop samples from Naples Beach, Santa Barbara-Ventura Basin, California (from Brincat and Abbott, 2001).

BNH is believed to originate from chemoautotrophic bacteria that grow at the oxic–anoxic interface. BNH is not bound to kerogen, suggesting that it is produced as a free, unfunctionalized lipid that should be readily characterized in modern biomass (Noble et al., 1985c). However, the organism responsible for BNH has not been identified. Fossilized bacterial mats in the Monterey Formation led Katz and Elrod (1983) to suggest the sulfide-oxidizing γ-proteobacterium *Thioploca* as a possible diagenetic precursor for BNH in Monterey oils and bitumen extracts (Philp, 1985; Curiale and Odermatt, 1989). However the lack of hopanoids in lipids extracted from *Thioploca* (McCaffrey et al., 1989) indicates that this organism is probably not the source. BNH dominates the terpane distributions of laminated rocks and is less abundant in massive, clay-rich rocks from the Naples Beach section of the Monterey Formation, California (Brincat and Abbott, 2001).

Williams (1984) speculated that BNH originates from *Beggiatoa*, a large (∼200-μm diameter) sulfur-oxidizing bacterium. These and other related microorganisms (e.g. *Thioploca*, *Thiobacillus*, *Thiomicrospira*, *Thiothrix*, and *Thiomargarita*) form dense filamentous mats at the oxic–anoxic (dysaerobic) transition zone in environments supplied with H_2S, often in marine sediments associated with hydrocarbon vents (Jørgensen, 1989; Schoell et al., 1992; Brune et al., 2000). The biochemistry of the sulfide-oxidizing bacteria is very complex and only beginning to be deciphered. These organisms store sulfur and thiosulfate, can simultaneously reduce nitrate to ammonia while oxidizing sulfide, and can switch between chemoautotropic, mixotrophic, and heterotrophic growth. BNH may be produced only under specific nutrient-restricted conditions that force the bacteria to use a specific biochemical pathway. Alternatively, *Beggiatoa* mats are home to many prokaryotic and eukaryotic organisms, most of which have yet to be identified (Bernhard *et al.*, 2000). It is possible that one of these associated organisms is the source for BNH.

Note: The most common lithofacies of the Monterey Formation was deposited mainly as distal organic-rich diatomaceous and phosphatic shales in oxygen-starved deep-marine silled basins (Demaison and Moore, 1980; Pisciotto and Garrison, 1981) or in topographic lows on a transgressed slope (Isaacs, 2001). Anoxic conditions and excess biological oxygen demand associated with upwelling of nutrient-rich waters were reinforced by the basin topography. Sulfate-reducing bacteria in the anoxic water column and surficial sediments generated sulfide. In clay-rich sediments, most sulfide sulfur is sequestered by chemically reactive iron in clay minerals. However, because of low clays, much of this sulfur was incorporated into the Monterey organic matter during diagenesis, resulting in type IIS kerogen (atomic S/C >0.04) and sulfur-rich oils (Orr, 1986; Baskin and Peters, 1992). Monterey crude oils from this lithofacies show high sulfur (>1 wt.% at 30°API, low pristane/phytane (<1) (Orr, 1986), and high BNH, typical of source-rock anoxia. For example, BNH/hopane in Monterey crude oil from the Beta Field in the Los Angeles Basin is 0.89 (Jeffrey *et al.*, 1991). These oils are also characterized by a strong even-predominance in *n*-alkanes, consistent with diatomaceous input under low Eh conditions (Hunt, 1996) and ^{13}C more positive than $-23.5‰$ (Chung *et al.*, 1992). Another lithofacies of Monterey-equivalent source rocks (Puente Formation) occurs along the landward (northern) flank of the Los Angeles Basin. Unlike the more common distal lithofacies, the landward lithofacies is more clay-rich and contains type II and II/III kerogens that yield low-sulfur crude oils with evidence of higher-plant input (Jeffrey *et al.*, 1991; McCulloh *et al.*, 1994).

17α-Diahopane/18α-30-norneohopane

Probable relationship between $C_{30}*$ and oxic–suboxic/clay-rich depositional environments. Measured using *m/z* 191 chromatograms and by parents of *m/z* 191 using GCMS/MS; also expressed as $C_{30}*/C_{29}Ts$. 17α-Diahopane/17α-hopane has similar application.

Two rearranged hopanes were identified in a Prudhoe Bay oil: 17α-diahopane ($C_{30}*$) was characterized by X-ray crystallography and 18α-30-norneohopane ($C_{29}Ts$) by advanced NMR methods (Moldowan *et al.*, 1991c). The peak for $C_{30}*$ is commonly detected in the *m/z* 191 chromatogram of the saturate fraction of crude oils (Figure 13.83). For example, Volkman *et al.* (1983a) found what appears to be $C_{30}*$ in oils and rock extracts from the Barrow Sub-basin of Western Australia, and Philp and Gilbert (1986) used an unknown terpane they called X-C_{30}, which appears to be the same compound, for correlations of oils from many basins. Both papers regard compound $C_{30}*$ as a possible terrigenous marker due to its presence in coals and terrigenous oils.

However, the structure of 17α-diahopane suggests that it rearranged from a hopanoid having functionality on the D-ring (Corbett and Smith, 1969). This rearranged hopane has nearly the same stable carbon isotope ratio as the C_{27}–C_{30} 17α-hopanes, Ts, and $C_{29}Ts$ in the Prudhoe Bay oil, suggesting they all originated from precursors in the same or similar organisms (Moldowan *et al.*, 1991b). Furthermore, the 17α-diahopane homologous series appears to extend to C_{35} (e.g. Figure 13.84), showing stereoisomer doublets in GCMS/MS analyses for C_{30} and higher homologs. This is similar to the distribution of 17α-hopane series, suggesting derivation of 17α-diahopanes from the bacteriohopane biosynthetic route. Thus, the $C_{30}*$ in oils may be related to bacterial hopanoid precursors that have undergone oxidation in the D-ring and rearrangement by clay-mediated acidic catalysis. Many terrigenous source rocks were deposited under oxic to suboxic conditions and are clay-rich. Thus, our results suggest that $C_{30}*$ originates from bacterial input to sediments containing clays deposited

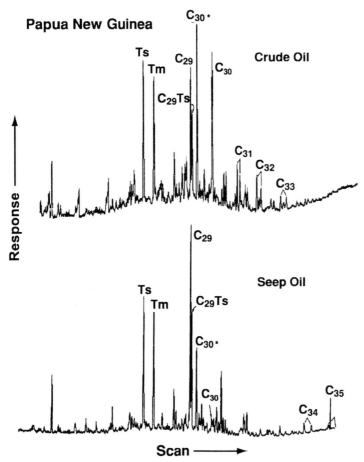

Figure 13.83. Two groups of produced oils and seep oils from Papua New Guinea can be distinguished using the ratio of two rearranged hopanes, 18α-30-norneohopane and 17α-diahopane (C_{29}Ts and C_{30}^*, respectively). The seep oil (bottom) has lost most 17α-hopanes in the range C_{30}–C_{33} due to biodegradation, but C_{29}Ts and C_{30}^* are resistant, and their relative amounts distinguish the seep oil from the crude oil (top). Reprinted with permission by ChevronTexaco Exploration and Production Technology Company, a division of Chevron USA Inc.

under oxic or suboxic conditions. This interpretation is also consistent with the type of highly terrigenous oils where it is found in greatest abundance (Philp and Gilbert, 1986; Volkman et al., 1983a).

Note: Unlike most other biomarkers, the cyclohexyl ring (ring D) in diahopanes is in the boat rather than the usual chair conformation (Dasgupta et al., 1995).

On the other hand, C_{29}Ts has virtually the same geochemical behavior as Ts and is probably as widespread in crude oils, except that poor resolution from C_{29} 17α-hopane may obscure its presence (e.g. see Figure 13.84). It can be used as a mturity marker in the ratio C_{29}Ts/[C_{29}-hopane + C_{29}Ts], which is similar to Ts/(Ts + Tm). (See C_{29}Ts/(C_{29} 17α-Hopane + C_{29}Ts), p. 000.

Certain Proterozoic crude oils contain a pseudohomologous series of peaks in the C_{27}–C_{34} range that appears to include C_{30}^* (Summons et al., 1988a; 1988b). A comparable GCMS/MS analysis of oil from Jordan shows what is probably the same series of peaks (starred series, Figure 13.84). Because these peaks fall in chromatographic succession with Ts, they were thought to be pseudohomologs of Ts. However, the finding that C_{29}Ts is the peak eluting immediately after C_{29}

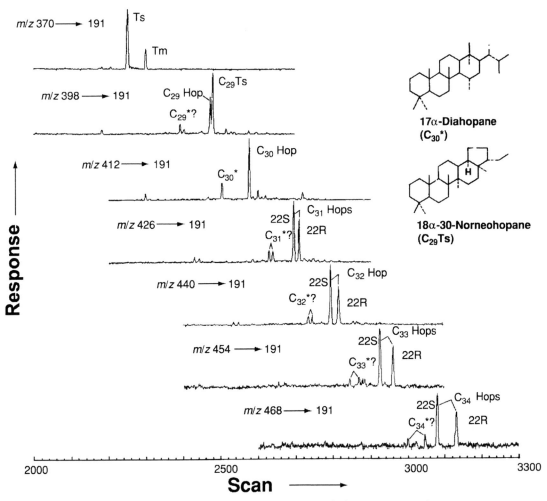

Figure 13.84. Parent-ion analysis of m/z 191 daughter ions by gas chromatography/mass spectrometry/mass spectrometry (GCMS/MS) in the saturate fraction of a Jordanian oil, indicating a pseudohomologous series of C_{29}–C_{34} 17α-diahopanes ($C_{29}*$–$C_{34}*$). The C_{27} and C_{29} 18α-neohopanes (Ts and C_{29}Ts) and C_{27} (Tm) and C_{29}–C_{34} 17α-hopanes are also evident. Each chromatogram is normalized internally to the highest peak. GCMS/MS data were recorded on a Finnigan MAT TSQ-70 triple quadrupole system interfaced to a Varian 3400 gas chromatograph, using a 60-m DB-1 (J&W Scientific) fused-silica capillary column, 0.25-mm internal diameter (ID), 0.25-mm film thickness, H_2 carrier gas, 150°C programmed at 2°C/min to 300°C. Scans are cycled at 1.5 seconds. Reprinted with permission by ChevronTexaco Exploration and Production Technology Company, a division of Chevron USA Inc.

17α-hopane, plus the X-ray structure of $C_{30}*$, shows that the starred peaks in Figure 13.84 are pseudohomologous 17α-diahopanes.

Application of the $C_{30}*/C_{29}$Ts parameter is not well established. However, it appears that relative amounts of $C_{30}*$ and C_{29}Ts depend most strongly on environment of deposition, and that oils derived from shales deposited under oxic-suboxic conditions show higher ratios than those from source rocks deposited under anoxic conditions. Thus, $C_{30}*/C_{29}$Ts effectively distinguished genetic oil groups in Papua New Guinea (Figure 13.83). Both oils in the figure show relatively high $C_{30}*/C_{29}$Ts, suggesting oxic-suboxic source-rock depositional conditions.

Molecular mechanics calculations indicate that compounds of the 17α-diahopane series should be more stable than those of the 18α-neohopane series, which, in turn, should be more stable than those of the 17α-hopane series. Thus, increasing maturity should result in increased ratios of 17α-diahopane to either 18α-30-norneohopane or 17α-hopane, particularly in the late oil window (Moldowan et al., 1991c). The calculated heat of formation for C_{30}^* (76.1 kcal/mol) (D. S. Watt, 1991, personal communication) is 3.7 kcal less than the hypothetical compound 18α-neohopane (C_{30}Ts) and 6.1 kcal less than 17α-hopane (Kolaczkowska et al., 1990). Thus, C_{30}^*/C_{29}Ts should increase with thermal maturity. Horstad et al. (1990) used the ratio $C_{30}^*/[C_{30}^* + 17\alpha$-hopane] along with sterane isomerization [C_{29} ααα 20S/(20S + 20R)] in oils to map maturity gradients in North Sea oil fields (e.g. Cornford et al., 1986).

Homohopane distributions

> Useful in order to assess source-rock redox conditions and for correlation. Measured using m/z 191 chromatograms.

Distributions of the C_{31}–C_{35} extended hopanes (homohopanes) can be used to infer redox conditions during deposition of the source rock. Similar homohopane distributions imply but do not prove genetic relationships among oil samples. For example, elevated C_{35} homohopanes for the Zdanice-7 and Damborice-16 oils (group 1) are characteristic of anoxic conditions during source-rock deposition (Figure 13.85). The Lubna-18

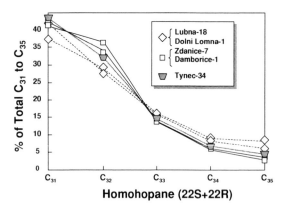

Figure 13.85. Homohopane distributions support two petroleum systems in Moravia, Czech Republic (Picha and Peters, 1998). The Tynec-34 oil is a mixture of these two families, as supported by other geochemical data. See also Figures 13.56 and 13.110.

and Dolni Lomna-1 oils (group 2) show elevated C_{32} homohopanes. Except for the C_{31} homohopanes, the homohopane distribution for Tynec-34 oil is intermediate between groups 1 and 2, consistent with a mixture, as indicated by other data (e.g. Figure 13.56).

Source rocks deposited under highly reducing conditions may preserve extended hopanes beyond C_{35}. For example, Rullkötter and Philp (1981) found extended hopanes up to C_{40} in some samples, but their precursors in organisms are unknown. Grice et al. (1996b) used nickel boride treatment to release C_{36} and C_{37} hopanes from the sulfur-bound fraction of extract from the Permian Kupferschiefer source rock from the Lower Rhine Basin.

C_{35} homohopane index

> Indicator of redox potential in marine sediments during diagenesis. High values indicate anoxia, but also affected by thermal maturity. Measured from m/z 191 chromatograms. Also expressed as C_{35}/C_{34} and $C_{35}S/C_{34}S$ hopanes.

The homohopanes (C_{31}–C_{35}) originate from bacteriohopanetetrol and other polyfunctional C_{35} hopanoids common in prokaryotic microorganisms (Figure 2.30) (Ourisson et al., 1979; Ourisson et al., 1984; Rohmer, 1987). The relative distribution of C_{31}–C_{35} 17α 22S and 22R homohopanes in marine petroleum is used as an indicator of the redox potential (Eh) during and immediately after deposition of the source sediments (e.g. Figure 13.86). High C_{35} homohopanes are commonly associated with marine carbonates or evaporites (Boon et al., 1983; Connan et al., 1986; Fu Jiamo et al., 1986; ten Haven et al., 1988; Mello et al., 1988a; Mello et al., 1988b; Clark and Philp, 1989). However, we interpret high C_{35} homohopanes as a general indicator of highly reducing (low Eh) marine conditions during deposition (Peters and Moldowan, 1991).

The C_{35} homohopane index (homohopane index) is the ratio $C_{35}/(C_{31}$–$C_{35})$ homohopanes, usually expressed as a percentage. For example, the m/z 191 chromatogram for the Baghewala-1 oil from northwestern India has elevated C_{35} homohopanes (22S + 22R doublet) compared with the C_{34} homologs (Figure 13.86). The C_{35} 22S and 22R peaks are labeled for oil from Hamilton Dome, Wyoming in Figure 8.21 (peak

Terpanes and similar compounds 567

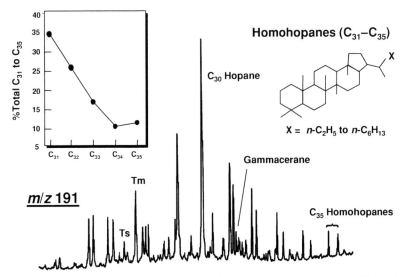

Figure 13.86. Terpane mass chromatogram for Baghewala-1 oil from northwestern India has elevated C_{35} homohopanes, indicating anoxic source-rock depositional conditions with restricted water circulation (Peters et al., 1995). The low Ts/(Ts + Tm) (25%) is consistent with other data (e.g. low diasteranes/(diasteranes + steranes) = 10%) suggesting a clay-poor carbonate source rock. See also Figure 4.25. Note that C_{29}/C_{30} hopane is lower than that for oils from many carbonate source rocks but is within the range observed for this organic facies (e.g. see Figure 13.98). Reprinted by permission of the AAPG, whose permission is required for further use.

numbers 34 and 35). Figure 13.87 (left) shows several examples of homohopane distributions for oils from the Adriatic Basin, Italy, and Yugoslavia. Common variants of the homohopane index include C_{35}/C_{31} and C_{35}/C_{34} hopane ratios.

The C_{29}/C_{30} and C_{35}/C_{34} hopane ratios can be used in tandem to define the source facies of oils (Figure 13.90). The C_{35}/C_{34} hopane ratio in this plot uses the 22S epimer rather than both 22S and 22R to avoid interference. Many crude oils from coal/resin source rocks show lower C_{35}/C_{34} hopanes (<0.6) than marine shale, carbonate, or marine source rocks, consistent with more oxic depositional conditions. Most oils from marine carbonate source rocks show high C_{35}/C_{34} hopane (>0.8) combined with high C_{29}/C_{30} hopane (>0.6).

Oils and bitumens of similar maturity that show high concentrations of C_{33}, C_{34}, or C_{35} homohopanes compared with lower homologs (Figure 13.87) are believed to indicate highly reducing (low Eh) marine source-rock depositional environments with no available free oxygen. When free oxygen is available, the precursor bacteriohopanetetrol is oxidized to a C_{32} acid, followed either by loss of the carboxyl group to C_{31} or, if all the oxygen is used, preservation of the C_{32} homolog. The latter environment is called suboxic or dysoxic, while the former may be oxic or suboxic, depending on the amount of oxygen and its accessibility to the organic matter (Demaison et al., 1983). This situation may be complicated by the bacteriohopane precursors having different (as yet unreported) side-chain hydroxyl substitution or side chain lengths. Thus, C_{33} and C_{34} homohopane predominances might reflect different types of bacterial input (e.g. Obermajer et al., 2000a). Alternatively, preservation of intermediate homologs could indicate mildly suboxic exposure at the time of deposition followed by partial oxidation of the bacteriohopanetetrol side chain. Coaly type III source rocks deposited under oxic conditions yield oils with low C_{32}–C_{35} hopanes and low homohopane indices.

High C_{35} hopane ratios for extracts commonly correlate with high hydrogen indices (e.g. Rangel et al., 2000) in the source rocks due to better preservation of oil-prone organic matter. Paleoenvironmental conclusions based on homohopane distributions should always be supported by other parameters. For example, Moldowan et al. (1986) found a relationship between pristane/phytane and diasteranes/regular steranes in a sequence of Lower Toarcian shales. Both parameters

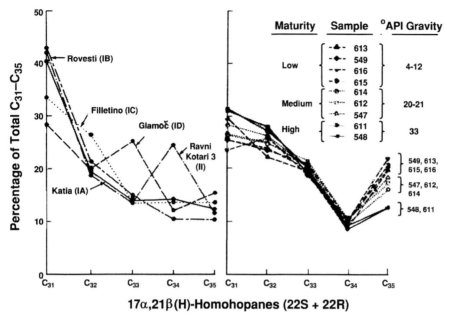

Figure 13.87. (Left) Homohopane distributions for several oils and seeps from the central Adriatic Basin (Italy and Yugoslavia), showing variations used to divide the samples into genetically different groups (Moldowan et al., 1991a). (Right) Related Monterey oils from offshore California, showing similar enrichment in C_{35} homohopanes, typical of organic matter from anoxic depositional settings (Peters and Moldowan, 1991). The example shows that a wide range of thermal maturity can affect homohopane distributions. The C_{35} homohopane index [$\%C_{35}/(C_{31}-C_{35})$] decreases, while $\%C_{31}$ homohopane increases with increasing maturity. The text gives an example of unusually heavily biodegraded oil, where the homohopanes show a reversed distribution in the range $C_{31}-C_{35}$. Reprinted with permission by ChevronTexaco Exploration and Production Technology Company, a division of Chevron USA Inc.

increase in sections of the core where the organic matter was exposed to higher levels of oxidation during deposition, as indicated by lower C_{35} homohopane indices and porphyrin V/(V + Ni).

Homohopane distributions are affected by thermal maturity (Peters and Moldowan, 1991). For example, the homohopane index decreases with maturity in a suite of related oils derived from the Monterey Formation, California, (Figure 13.87, right).

Homohopanes liberated from sulfur complexes in low-mature sulfur-rich oil using Raney nickel show distributions favoring the C_{35} homolog (Schmid, 1986; Trifilieff, 1987; Sinninghe Damsté et al., 1988; de Leeuw and Sinninghe Damsté, 1990; Sinninghe Damsté and de Leeuw, 1990). Köster et al. (1997) confirmed that the degree of sulfurization of hopanoids increases with increasing carbon number from C_{31} to C_{35}. Cleavage of carbon–sulfur bonds in the early oil-generative window can result in the release of sulfur-bound homohopanes. Additional work on the ruthenium tetroxide (RuO_4) oxidation of asphaltenes from sulfur-rich oils shows that some homohopanes are attached to aromatic systems by C–C bonds (Trifilieff et al., 1992). Oxidation by RuO_4 disrupts aromatic systems by producing a carboxyl group at the point of attachment of the side chains or groups. Homohopanoic acids liberated from asphaltenes by this method carry the carboxyl group at the point of original functionality. When cracked from the kerogen during catagenesis, these homohopanes show a mature distribution favoring the C_{31} homolog. Thus, catagenesis of source rock having elevated C_{35} homohopanes (typical of sulfur-bound homohopane species liberated during early maturation) may crack kerogen-bound homohopanes and lead to an increase in the C_{31} homologs later in the oil window.

Diastereomeric pairs (22S and 22R) of side-chain-extended $\alpha\beta$-homohopanes up to C_{40} occur in extracts (Rullkötter and Philp, 1981). The $C_{36}-C_{40}$ homologs are much less abundant than their $C_{31}-C_{35}$ counterparts and are thought to originate during thermal cracking of bacteriohopane precursors that are bound to the kerogen by covalent carbon-carbon bonds.

C_{31}/C_{30} hopane

> Useful in order to distinguish between marine versus lacustrine source-rock depositional environments. Measured using m/z 191 chromatograms (see Figure 8.20). Also expressed as C_{31} 22R/C_{30} hopane (C_{31}R/H).

Unlike crude oils from lacustrine source rocks, oils from marine shale, carbonate, and marl source rocks generally show high C_{31} 22R homohopane/C_{30} hopane (C_{31}R/C_{30} >0.25) (Figure 13.77). Marine and lacustrine crude oils are best distinguished using C_{31}R/C_{30} hopane in combination with other parameters, such as the C_{30} n-propylcholestane and C_{26}/C_{25} tricyclic terpanes, and the canonical variable from stable carbon isotope measurements.

Hexahydrobenzohopanes

> Specificity is unknown but may be diagnostic of carbonate-anhydrite depositional environments. Measured using m/z 191 or molecular ions (see Figure 8.20).

Hexahydrobenzohopanes (Figure 8.20) range from C_{32} to C_{35} and are probably formed in a reducing (low Eh) depositional environment by cyclization of the side chains on extended hopanoids. They appear diagnostic of oil and bitumen from sulfur-rich carbonate-anhydrite source rocks (Connan and Dessort, 1987; Rinaldi *et al.*, 1988). Although hexahydrobenzohopanes (also called hexacyclic hopanoids) can be analyzed using m/z 191 fragmentograms of the saturate fraction, their concentrations are usually low relative to the extended hopanes. These compounds can be monitored using the molecular ions at m/z 438, 452, 466, and 480 (Figure 8.20). Alternatively, the high selectivity of the specific parent-to-daughter transitions (M+ → 191) allows use of linked-scan GCMS/MS approaches (Figure 13.88).

Note: Unlike most other biomarkers, a cyclohexyl ring (ring D) in hexahydrobenzohopanes is in the boat rather than the usually preferred chair conformation.

30-Norhopane/hopane

> High 30-norhopane/hopane is typical of anoxic carbonate or marl source rocks and oils. Measured using m/z 191 chromatograms. Also expressed as C_{29}/C_{30} hopane (C_{29}/H.)

The C_{29} 17α-norhopane (Figure 13.89) rivals hopane as the major peak on m/z 191 mass chromatograms of saturate fractions of many oils and bitumens. C_{29}/C_{30} 17α-hopane (m/z 191 uncorrected peak heights) is greater than 1.0 for many anoxic carbonate or marl source rocks and related oils but generally is less than 1.0 for other samples (Figure 13.90). Norhopane is more stable than hopane at high levels of thermal maturity. Thus, within a group of related oils, 30-norhopane/hopane can increase with thermal maturity.

Methylhopanes

> Potential high specificity as a correlation tool. Prokaryotic source input. Measured using m/z 205.

Ring A and B methylhopanes were first detected in a series of Jurassic oils from the Middle East (Seifert and Moldowan, 1978). Addition of a methyl group to the ring A/B fragment of the hopane series increases the fragment from m/z 191 to m/z 205 (see Figure 8.20). Rohmer and Ourisson (1976) reported 3-methylhopanes in bacteria, while some Precambrian oils show evidence for an additional series of ring A/B 2α-methylhopanes (Figure 13.91) (Summons and Walter, 1990; Summons and Jahnke, 1992).

The 2α-methylhopanes appear to be specific for oxygen-producing cyanobacteria. The 2-methylbacteriohopanepolyol precursors for these compounds are abundant in cultured cyanobacteria and cyanobacterial mats like those that compose stromatolites as old as 2.7 Ga (Summons *et al.* (1999). However, caution is needed because the number of species analyzed is small compared with microbil diversity (C. Blank, 2002, personal communication). Ratios of methylhopanes to hopanes are useful as source input

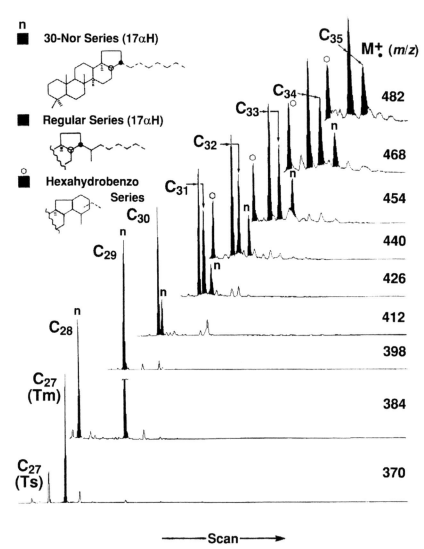

Figure 13.88. Metastable reaction monitoring/gas chromatography/mass spectrometry (MRM/GCMS) analysis of pentacyclic triterpanes in Santa Maria-3 oil, Italy (Moldowan et al., 1991a). Pentacyclic hopanes with 27–35 carbon atoms are analyzed on separate M+ → 191 chromatograms, with M+ ranging every 14 mass units from m/z 370 to m/z 482, respectively. Both the regular 17α-hopane and the 30-nor-17α-hopane series, as well as peaks corresponding to the C_{32}–C_{35} hexahydrobenzohopanes, occur in this sample. The m/z 191 fragments from their molecular ions, m/z 438, 452, 466, and 488, respectively, are recorded here due to the low resolution of the MRM method compared with gas chromatography/mass spectrometry/mass spectrometry (GCMS/MS using a multisector mass spectrometer. The same transitions recorded at higher resolution e.g. with a triple quadrupole system) do not show the hexahydrobenzohopane series. The precise stereochemical structures of these compounds have not been determined. Reprinted with permission by ChevronTexaco Exploration and Production Technology Company, a division of Chevron USA Inc.

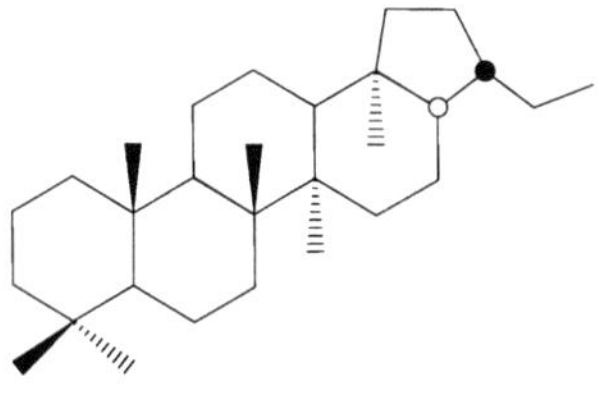

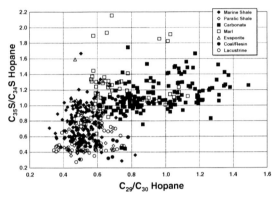

Figure 13.89. 17α-30-Norhopane (C$_{29}$).

Figure 13.90. Crude oils generated from many marine carbonate and marl source rock rocks have high norhopane/hopane and C$_{35}$/C$_{34}$ 22S hopane, consistent with anoxic source-rock depositional settings. The figure is based on more than 500 worldwide crude oil samples used to predict source-rock depositional environments. Carbonate samples are from the Oriente, Maracaibo, Williston, and Powder River basins and the Middle East, Gulf of Suez, and Northwest Palawan, while marl samples are from the Oriente and Maturin basins, the Middle and Upper Magdalena Valleys, East Siberia, Oman, the Zagros area, and the Monterey Formation in California. Figure courtesy of GeoMark Research, Inc. (J. E. Zumberge, 2000, personal communication).

parameters reflecting bacterial populations at the time of sedimentation.

The 2α-methylhopane index is the ratio 2α-methylhopane/(2α-methylhopane + hopane), where the hopane and its 2α-methyl analog are determined using GCMS/MS transitions m/z 412 → 191 and m/z 426 → 191, respectively. Burial temperature strongly affects the 2α-methylhopane index. Immature bitumens, for example, show uniformly low indices, apparently because 2α-methylhopanes require cracking from kerogen. However, thermally mature Proterozoic to Phanerozoic oil and rock samples confirm previous observations (Summons and Jahnke, 1992) that oils from carbonate source rocks have the highest 2α-methylhopane indices (Summons et al., 1999).

Various microorganisms, including several methanotrophic bacteria, synthesize 3-methylhopanoids, which are precursors to 3β-methylhopanes (Neunlist and Rohmer, 1985; Zundel and Rohmer, 1985). Some of these bacteria may synthesize either regular or methylated hopanoids depending on their metabolic state. Summons et al. (1994) found that *Methylococcus capsulatus*, an aerobic methanotroph, synthesized mainly regular hopanoids during exponential growth and 3β-methylated hopanoids during the stationary phase of cell growth. In biogenic mats associated with the Be'eri sulfur mine of Israel, the distribution of 3β-methylhopanes matched that of the regular hopanes (Figure 13.92). These compounds are extremely depleted in ^{13}C, yielding δ^{13}C values as low as −80 to −93‰. Organic matter in this deposit originated primarily from methanotrophic bacteria, which, along with associated archaea and sulfide-oxidizing bacteria, were thriving exclusively on ^{13}C-depleted methane-derived carbon.

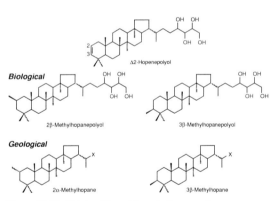

Figure 13.91. Origin of 2- and 3-methylhopanes from a proposed Δ^2-hopenepolyol precursor. The X group in the geological methylhopanes is a saturated *n*-alkyl side chain. The side chain in the biological methylhopanepolyols is probably similar to the polyol side chain in Δ^2-hopenepolyol, as shown. The 2α-methylhopanes, common in some mature sediments and oils, probably originate from the biogenic, less stable 2β-methylhopanes (Summons and Jahnke, 1992). Reprinted with permission by ChevronTexaco Exploration and Production Technology Company, a division of Chevron USA Inc.

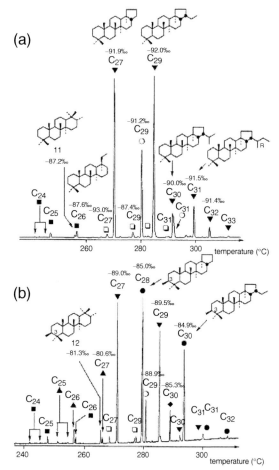

Figure 13.92. Gas chromatograms of the saturated hydrocarbon fractions from (a) the black sandstone and (b) the bacterial mats from the Be'eri sulfur deposit, Israel. Reprinted from Burhan *et al.* (2002). © Copyright 2002, with permission from Elsevier. ▼ = 17β,21β(H)-hopanes; □ = 17α,21β(H)-hopanes; ○ = 17β,21α(H)-hopanes; ● = 17β,21β(H) 3β-Me-hopanes; ■ = 17,21-secohopanes; ▲ = 3β-Me-17,21-secohopanes; ♦ = 17β,21α (H)-3β-Me-hopanes. $\delta^{13}C$ values are indicated above each peak.

17α,18-Dimethyl-des-E-hopane

Possibly specific for methane-oxidizing bacteria. Measured using m/z 191 chromatograms.

Burhan *et al.* (2002) identified 17α,18-dimethyl-des-E-hopane and its 3-methylated equivalent (Figure 13.93) in extracts of biogenic mats from the Be'eri sulfur deposit, Israel. These tetracyclic terpanoid hydrocarbons are believed to result from the cleavage of the C-20 to C-21 bond in the E-ring of hopanoids, a tranformation not reported in other sediments. Organic matter in the Be'eri deposit is extremely depleted in ^{13}C ($\delta^{13}C_{kerogen} = -85‰$) and probably originates from a bacterial ecosystem that oxidizes migrated biogenic methane and sulfides.

Oleanane/C_{30} hopane (oleanane index)

Highly specific for higher-plant input of Cretaceous or younger age. Measured using m/z 191 chromatograms (see Figure 8.20). Also expressed as Ol/H or Ol/(Ol + H).

Oleanane in crude oils and rock extracts is a marker for both source input and geologic age (Figure 13.94). This compound originates from betulins (Figure 11.16) (Grantham *et al.*, 1983), taraxerene (ten Haven and Rullkötter, 1988), and other pentacyclic triterpenoids (Whitehead, 1973; 1974) that are produced by angiosperms (flowering land plants). Based on tentative fossil evidence, such as pollen, leaf, and wood vessel structures, angiosperms probably originated in the Triassic Period or earlier. According to molecular phylogenic projections, the angiosperm lineage should have separated from other seed plants in the late Paleozoic Era (Martin *et al.*, 1989). Taylor *et al.* (2004) show evidence of this based on oleanane occurrence in Permian rocks from China that contain Gigantopterid compression fossils. These Gigantopterids had similar structural morphologies to many angiosperms, except that no flowers or seeds have been found. Undisputed fossil evidence for angiosperms occurs in Upper Jurassic-Lower Cretaceous strata, where *Archaefructus* has fruiting bodies that enclose seeds (Ge Sun *et al.*, 1998). However, angiosperms were rare, with low diversity as late as the

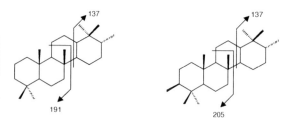

Figure 13.93. 17α,18-Dimethyl-des-E-hopane and its 3β-methylated equivalent.

Terpanes and similar compounds 573

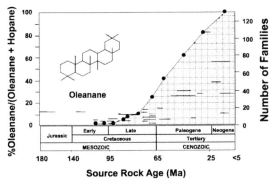

Figure 13.94. Oleanane ratios (horizontal lines) in bitumens extracted from 103 source rocks of various ages. Lengths of horizontal lines indicate uncertainty of biostratigraphic age determination. Oleanane ratios >20% are diagnostic of Tertiary or younger source rocks and related oils. Solid dots and curve indicate the number of fossil pollen reports assigned to extant angiosperm families. Horizontal line at lower left shows data for a Middle Jurassic rock from the Tyumen Formation in West Siberia that contains oleanane. Reprinted with permission from Moldowan et al. (1994a). © Copyright 1994, American Association for the Advancement of Science.

Early Cretaceous Epoch, and did not become prominent until Late Cretaceous time (<100 Ma) (Figures 13.94 and 17.12).

Crude oils from the Tertiary Niger Delta contain abundant oleananes (Ekweozor et al., 1979a), and there is a correlation between the abundance of higher-plant macerals (e.g. vitrinite and resinite) and the oleanane index (Udo and Ekweozor, 1990). Marine shales of the Tertiary Akata Formation, which received substantial input from terrigenous higher plants, are considered to be the main source of the Niger Delta oils. The widespread occurrence of oleananes is now recognized (Grantham et al., 1983; Hoffmann et al., 1984; Riva et al., 1988; Zumberge, 1987a; Czochanska et al., 1988; Ekweozor and Udo, 1988; Zeng et al., 1988; Fu Jiamo and Sheng Guoying, 1989).

Absence of oleanane does not prove that crude oil was generated from Cretaceous or older rocks. Small amounts of oleanane occur in Jurassic crude oil (Peters et al., 1999b) and rock extracts (Moldowan et al., 1994a) and extracts of megafossils from older rocks (Taylor et al., 2004).

18β-Oleanane in immature bitumens can complicate use of the oleanane index (oleanane/C_{30} hopane) to compare samples showing major differences in thermal maturity. Although both 18α- and 18β-oleanane are found in petroleum, the latter is thermally less stable (Riva et al., 1988). Equilibrium between these stereoisomers probably occurs before peak oil generation. Thus, the sum of 18α and 18β isomers should be used in oleanane/C_{30} hopane for purposes of correlation. Based on the relative retention time and mass spectrum, we believe that compound J from the earlier literature (Hills and Whitehead, 1966; Grantham et al., 1983) is 18β-oleanane. Figure 8.28 shows the provisional identification of 18α- and 18β-oleanane in bitumen by co-injection of authentic standards.

Oleanane-rich marine equivalents of more terrigenous, oleanane-poor coals and shales may represent the most effective source rocks in deltaic petroleum systems. Oleanane and related compounds are best preserved in deltaic rocks influenced by marine waters during early diagenesis (Figure 13.95), (Murray et al., 1997a). This explains why oleanane ratios for Beaufort Sea (McCaffrey et al., 1994b) and Mahakam Delta oils (Peters et al., 2000) generally increase in an offshore rather than an onshore direction. Murray et al. (1997a) concluded that diagenetic contact of seawater with higher-plant organic matter in coals and shales dramatically enhances hydrogenation and preservation of oleanane. They found a good correlation of oleanane/hopane in South Sumatra Basin deltaic rocks with indicators of marine influence, such as organic sulfur, dibenzothiophene/phenanthrene, and homohopane

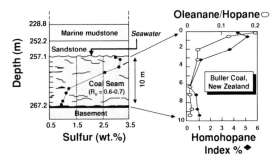

Figure 13.95. Oleanane and related compounds such as C_{35} homohopanes are best preserved in deltaic rocks influenced by marine waters during early diagenesis. The increase in sulfur content toward the top of the Buller Coal, New Zealand, is consistent with more reducing conditions and increased bacterial sulfate reduction due to the influx of seawater in the overlying sands during early diagenesis. Reprinted from Murray et al. (1997b). © Copyright 1997, with permission from Elsevier.

574 Source- and age-related biomarker parameters

ratios. Lowstand systems are expected to more frequently juxtapose permeable sandstones within or on top of the source rock. This increases the potential for both (1) influx of marine waters during source-rock diagenesis and (2) subsequent expulsion of hydrocarbons during catagenesis.

Peters et al. (1999b) confirmed oleanane in certain pre-Cretaceous rocks and crude oils, as noted originally for the Middle Jurassic Tyumen Formation in West Siberia (Moldowan et al., 1994a). Thus, oleanane in crude oil is not conclusive evidence of a post-Jurassic source rock for the oil.

Note: An unusual rounded cobble of brecciated sandstone clasts from a beach near Brora may represent fault gouge from the nearby Helmsdale Fault. Oil cement between the angular clasts in the cobble originated by mixing of marine and lacustrine crude oils. The first oil charge consisted of lacustrine oil generated from organic-rich Devonian carbonate mudstones. This oil charge was heavily biodegraded at shallow depth. The second oil charge originated from marine source rock that contained oleanane. However, the nearest documented Cretaceous or younger rocks occur in subcrop more than 10 km offshore from Brora and are thermally immature. The second charge to the oil cement appears to have originated from an oleanane-bearing marine facies of the Middle Jurassic Brora coal. This hypothesis is supported by positive identification of oleanane in an extract from the Middle Jurassic Brora coal (Figure 13.96). The second charge was biodegraded only mildly. Weathering and transportation of the cobble occurred after oil emplacement. Other papers suggest that the Middle Jurassic is an effective source rock in the general area. For example, oils from the Foinaven Complex to the west of Britain are believed to originate by mixing of oils generated from Upper and Middle Jurassic source rocks (Scotchman et al., 1998).

Ekweozor and Telnaes (1990) suggest that the oleanane ratio increases from low values in immature rocks to a maximum at the top of the oil-generative window, remaining relatively stable at greater depths. Consequently, use of the oleanane index to compare

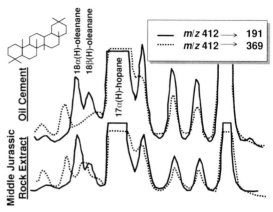

Figure 13.96. Metastable reaction monitoring/gas chromatography/mass spectrometry (MRM/GCMS) confirms oleanane (structure at top left) in oil cementing a sandstone breccia from Brora Beach in Scotland (top) and an extract from Middle Jurassic Brora coal from the Beatrice Field (bottom). The 18α- and 18β-oleanane peaks in the oil cement (m/z 412 → 191, solid) show no m/z 369 response (dotted). Lupane co-elutes with 18α-oleanane on some chromatographic columns but, unlike that compound, shows the m/z 412 → 369 transition (Rullkötter et al., 1994). Some peaks, such as 17α-hopane, are truncated to better view the oleanane peaks. Oleanane was also confirmed by two additional MRM/GCMS analyses on aliquots of the oil cement from different parts of the cobble that were extracted using different batches of ultra-pure solvent. Reprinted from Peters et al. (1999b). © Copyright 1999, with permission from Elsevier.

relative higher-plant input between immature and mature samples is not recommended. Supporting parameters should complement interpretations of terrigenous versus marine organic matter input based on oleanane in mature samples. For example, plots of oleanane versus C_{35} homohopane index or C_{30} sterane index are often useful (Figure 13.42).

Oleanane is difficult to identify at low concentrations using conventional GCMS because of possible interference by other triterpanes with similar retention times and mass spectra. For example, the mass spectra of 17α-hopane and oleanane are similar. Depending on chromatographic column type, lupane (Rüllkotter et al., 1994) and the C_{30} 25-norhomohopane 20S (Dzou et al., 1999) may co-elute with oleanane. In most published work, oleanane is identified by GCMS in SIM mode using m/z 191 and m/z 412 mass chromatograms and occasionally by co-elution experiments with 18α- and 18β-oleanane standards. Unfortunately, both

17α-hopane and oleanane undergo cleavage of the C-ring during ionization, yielding both A/B-ring and D/E-ring fragments that carry the main ion current of the spectrum at m/z 191. The molecular ion, m/z 412, and the M − CH$_3$ (molecular ion minus methyl group) fragment, m/z 397, are only slightly larger for the oleananes than for 17α-hopane. These characteristics are also typical of other known stereoisomeric C$_{30}$-hopanes. Hopanes have a small m/z 369 fragment, which oleananes lack.

Oleanane can be identified by MRM/GCMS using a combination of traits:

- Oleanane occurs as a recognizable doublet at the proper retention times, where the earlier-eluting 18α-peak is larger than the 18β-peak in mature oils.
- Metastable transitions m/z 412 → 397, 412 → 412, and 414 → 191 decrease slightly, in that order, relative to those of 17α-hopane.
- Metastable transition m/z 412 → 369 has no significant peak. A common but unidentified interfering compound yields a singlet peak at m/z 412 → 369 (Figure 13.96). Loss of an isopropyl group from lupane (m/z 43), for example, yields the m/z 369 fragment.

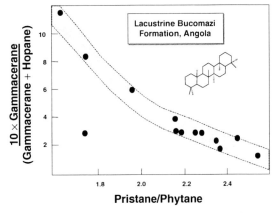

Figure 13.97. Variations in pristane/phytane (Pr/Ph) (redox) and gammacerane index (salinity stratification) for oils derived from lacustrine source rocks in Angola. Inset shows gammacerane. Increased water salinity in the source-rock depositional environment results in higher gammacerane indices. Higher salinity is typically accompanied by density stratification and reduced oxygen content in bottom waters (i.e. lower Eh), which results in lower Pr/Ph. Biomarker evidence suggests a marine co-source for the point that lies off the trend. Figure courtesy of B. J. Huizinga. Reprinted with permission by ChevronTexaco Exploration and Production Technology Company, a division of Chevron USA Inc.

Gammacerane index

> Highly specific for water-column stratification (commonly due to hypersalinity) during source-rock deposition. Measured using m/z 412 or 191 fragmentograms (see Figure 8.20). Also expressed as 10X gammacerane/(gammacerane + C$_{30}$ hopane), gammacerane/C$_{31}$ 22R hopane, and Ga/C$_{31}$R.

Gammacerane indicates a stratified water column in marine and non-marine source-rock depositional environments (Sinninghe Damsté et al., 1995), commonly resulting from hypersalinity at depth. However, stratified water columns can also result from temperature gradients. Oils from lacustrine source rocks in Angola indicate that increased water salinity during deposition of the source rock results in higher gammacerane indices and lower pristane/phytane (Figure 13.97). Similar results were obtained for petroleum from Tertiary rocks, offshore China (Mann et al., 1987). Tertiary lacustrine clayey dolomites from the Biyang Basin, China, show low pristane/phytane and high gammacerane, indicating high salinity during their deposition (Junhong Chen and Summons, 2001). In addition to β-carotane and related carotenoids (see below), gammacerane is a major biomarker in many lacustrine oils and bitumens, including the Green River marl and oils from China (Moldowan et al., 1985; Jiang Zhusheng and Fowler, 1986; Fu Jiamo et al., 1986; Fu Jiamo et al., 1988; Brassell et al., 1988; Grice et al., 1998a). The Green River marl was deposited in a widespread arid or semi-arid lacustrine setting, where the organic matter originated mainly from algae and bacteria. Gammacerane is also abundant in certain marine crude oils from carbonate-evaporite source rocks (Rohrback, 1983; Moldowan et al., 1985; Mello et al., 1988a; Moldowan et al., 1991a).

The origin of gammacerane is uncertain, but it may form by reduction of tetrahymanol (gammaceran-3β-ol) (Venkatesan, 1989; ten Haven et al., 1989). Diagenetic conversion of tetrahymanol to gammacerane most likely proceeds by dehydration to form gammacer-2-ene, followed by hydrogenation. Tetrahymanol is a lipid that replaces steroids in the membranes of certain protozoa (Caspi et al., 1968; Nes and McKean, 1977; Ourisson et al., 1987) and possibly other organisms. Gammacerane may also arise by sulfurization and subsequent cleavage of tetrahymanol (Sinninghe Damsté et al., 1995). Tetrahymanol was first identified as a natural product of the freshwater ciliated protozoan *Tetrahymena* (Mallory et al., 1963), where it replaces certain sterols in the membrane. *Tetrahymena* incorporates sterols from the environment but synthesizes tetrahymanol when sterols are not available (Conner et al., 1971; 1982). Because tetrahymanol is synthesized from squalene in far fewer steps than sterols and may proceed via anaerobic pathways (Kemp et al., 1984), it may originate by a primitive biosynthetic pathway. The oldest known occurrence of abundant gammacerane in Upper Protozeroic rocks of the Chuar Group in the Grand Canyon (~850 Ma) supports this conclusion (Summons et al., 1988b).

Tetrahymanol occurs in sediments from various marine depositional environments (ten Haven et al., 1989), including the Santa Monica Basin and the Palos Verdes shelf, the Santa Barbara Basin and offshore Baja California, the Atlantic shelf, slope, and rise, the Antarctic region, the Peru upwelling region (ODP Leg 112), and Baffin Bay (ODP Leg 105) (Venkatesan, 1989). The common occurrence of tetrahymanol in marine environments implies that organisms other than *Tetrahymena* contain this compound. Tetrahymanol has been identified in a fern (Zander et al., 1969), an anaerobic rumen fungus (Kemp et al., 1984), photosynthetic bacteria (Kleemann et al., 1990), freshwater ciliates (Mallory et al., 1963), and marine ciliates when they consume prokaryotes and their diet is deprived of sterols (Harvey et al., 1997).

The principal source of tetrahymanol appears to be bacterivorous ciliates, which occur at the interface between oxic and anoxic zones in stratified water columns (Sinninghe Damsté et al., 1995). Thus, abundant gammacerane suggests the presence of a stratified water column and possibly salinity stratification during sediment deposition. Tetrahymanol is a principal neutral lipid in eight marine ciliate species, most of which are scuticociliates, a widespread group of protozoa that feed mainly on bacteria (Harvey and McManus, 1991). Tetrahymanol abundance in pure cultures and field samples (sediment traps, water column particulates, and enrichments from coastal and estuarine environments) shows good agreement with ciliate biovolume ($R^2 = 0.89$), suggesting that tetrahymanol is a specific marker for marine ciliates that feed on bacteria. Hopan-3β-ol was also identified positively in several ciliates but did not occur in all species examined. Because of their widespread distribution in modern planktonic and interstitial marine systems, these organisms are a likely source for tetrahymanol in marine sediments. The ciliates are important components of modern aquatic communities. In many marine systems, planktonic ciliates consume ~30–50% of the primary production and may be the dominant group (up to ~100%) of microzooplankton in temperate coastal waters (Pierce and Turner, 1992). Some of the ciliates are, in turn, consumed by larger zooplankton but remain as a significant portion of the sinking biomass.

Although present in at least trace amounts in most crude oils, large amounts of gammacerane indicate highly reducing; hypersaline conditions during deposition of the contributing organic matter (Moldowan et al., 1985; Fu Jiamo et al., 1986). These conditions may favor the organisms that produce tetrahymanol. Although oils and bitumens with high gammacerane ratios (e.g. gammacerane/$\alpha\beta$-hopane) often can be traced to hypersaline depositional environments, these environments do not always result in high gammacerane ratios (Moldowan et al., 1985). Gammacerane is more resistant to biodegradation than the hopanes (Zhang Dajiang et al., 1988).

Gammacerane is useful to distinguish petroleum families. For example, Poole and Claypool (1984) used gammacerane to distinguish oils and bitumens from different source rocks in the Great Basin. Bitumens from the oil shale unit of the lacustrine Elko Formation were related to Elko-derived oils and distinguished from other samples based on high gammacerane and supporting geochemical data. Palmer (1984a) distinguished organic facies within the Elko Formation using gammacerane and other data, such as the distribution of C_{27}–C_{29} steranes (e.g. see Figure 13.37). Gammacerane was present in the oil shale but absent in the lignitic siltstone facies.

Care must be taken to accurately quantify gammacerane using the m/z 191 mass chromatogram. Because of its high degree of symmetry (see Figure 2.27), two identical m/z 191 fragments are generated in the mass spectrometer from gammacerane (see Figure 8.20). Thus, a sizable peak on the m/z 191 mass chromatogram represents a low concentration of gammacerane compared with other terpanes, requiring a correction (Seifert and

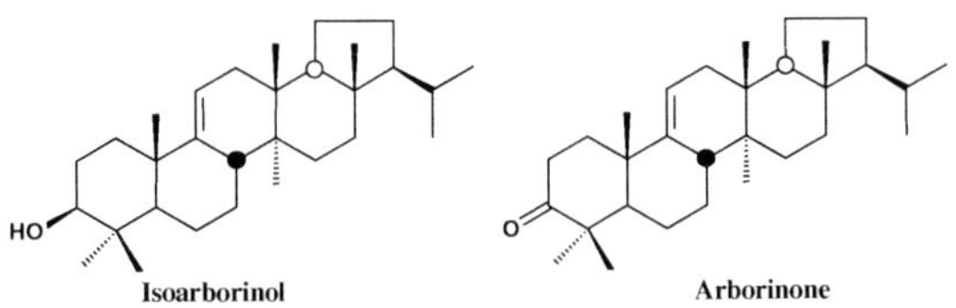

Figure 13.98. Isoarborinol and arborinone are the dominant alcohol and ketone, respectively, in sediments from Lake Valencia, a tropical freshwater lake in Venezuela (Jaffé and Hausmann, 1995).

Moldowan, 1979). Schoell et al. (1992) found high gammacerane in gilsonite from Utah, but it co-eluted with a C_{31} methylhopane. Further, under conditions of poor column performance, gammacerane can nearly co-elute with the 22R epimer of C_{31}-homohopane (M+ = m/z 426). Seifert and Moldowan (1986) suggested that gammacerane is best measured using the m/z 412 (molecular ion) mass chromatogram because it reduces interference from other terpanes with the gammacerane peak that occur on the m/z 191 chromatogram. However, this may not be practical for certain quadrupole instruments with reduced sensitivity above m/z 300. Gammacerane can also be analyzed using the GCMS/MS m/z 412 → 191 transition, which conveniently also quantifies the 17α-hopane required for the gammacerane index.

Isoarborinol and arborinone

> Likely derived from unknown algal/bacterial sources. Possible indicator of high-productivity lacustrine settings. Low thermal stability. Isoarborinol is measured usually as an ester derivative and is characterized by its base mass. Arborinone is sufficiently volatile to be separated by conventional gas chromatography (typically in a ketone fraction).

Many immature sediments contain abundant isoarborinol [arbor-9(11)-en-3β-ol] and arborinone, commonly as the dominant alcohol and ketone, respectively (Figure 13.98). Studies of the lacustrine Messel Shale first identified isoarborinol (Albrecht and Ourisson, 1969) and arborinone (Mattern et al., 1970) in the geosphere. Additional papers from the Strasbourg group identified these compounds in various modern lacustrine and lagoonal sediments (Arpino et al., 1972; Dastillung et al., 1980a; Dastillung et al., 1980b). Certain families of angiosperms (e.g. Rutaceae, Gramineae, and Euphorbiaceae) produce trace amounts of isoarborinol and are its only known biological source. However, high concentrations and the depositional settings of the sediments containing isoarborinol suggest a microbial source. Ourisson et al. (1982) postulated unknown aerobic bacteria as the source for isoarborinol because its biosynthesis involves an enzyme that is intermediate between anaerobic-prokaryotic and aerobic-eukaryotic squalene cyclases.

Jaffé and Hausmann (1995) found abundant arborinone (37 mg/g of dry sediment) and isoarborinol in sediments from Lake Valencia, northern Venezuela. This tropical, freshwater lake has high phytoplankton productivity and year-round anoxic bottom water, suggesting that bacteria or phytoplanktonic organisms are the likely source. Aromatic derivatives tied to isoarborinol occur in the Permian Kupferschiefer of Germany and Triassic black shales from Italy (Hauke et al., 1992a; 1992b; 1995). An angiosperm source of isoarborinol can be dismissed because these rocks predate the emergence of flowering plants, but these authors suggested that unknown algae might be the source.

Fernanes and fernenes

> Likely derived from gymnosperm and microbial precursors. Possible indicator of floral input and age and low thermal stability. Fernenes are measured using m/z 243 and MRM molecular ion M+ → m/z 243. Fernanes are measured using m/z 123 and 191 and M+ → m/z 191 or 259.

As the name suggests, fernenes and the 3β-fernenols (Figure 13.99) were first identified in contemporary ferns and were thought to be unique biomarkers for these land plants (Brassell et al., 1980). The fernenes are unusual because they are biosynthesized directly as hydrocarbons. If fernenes and the saturated fernanes were biomarkers for ferns, then they would be valuable

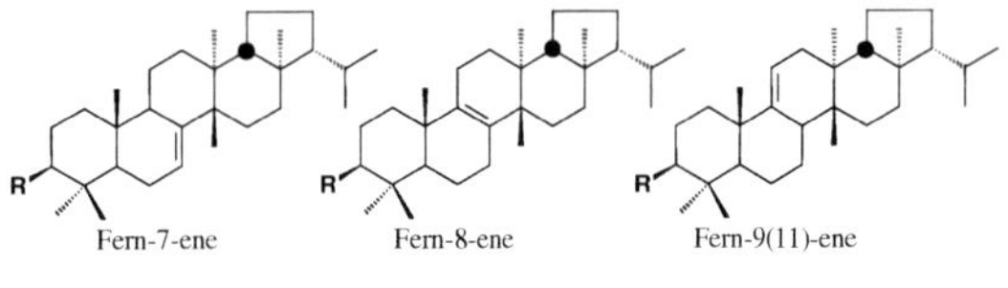

Figure 13.99. Fernenes and 3β-fernenols differ in the location of the double bond.

indicators of floral input and geologic age. Unfortunately, fernenes are found in modern sediments where input from ferns can be discounted, such as numerous deep-sea sites (Brassell and Eglinton, 1983), an Antarctic saline lake (Volkman et al., 1986), and the Ross Sea (Venkatesan, 1988). Volkman et al. (1986) correlated fern-7-ene with high concentrations of methanogenic biomarkers. They suggested that it might originate from purple sulfur bacteria and is a marker for anoxic depositional environments. One report of fernenes in bacteria (Howard, 1980) remains unconfirmed (Rohmer et al., 1992).

A few studies implicate fernenes or their aromatic derivatives as biomarkers for floral input specific to pteridosperms (seed ferns) and coniferophytes, early Gymnospermopsida that were the dominant land plants during the Carboniferous to Jurassic periods (Paull et al., 1998; Vliex et al., 1994; Vliex et al., 1995). Paull et al. (1998) detected C_{29}–C_{31} fernenes and fernanes in Upper Triassic sub-bituminous coals, mudstones, and fossils of the pteridosperm *Dicrodium* from the Leigh Creek coalfield in South Australia. Although the occurrence of these biomarkers correlates with the presence of gymnosperm seeds and pollen, derivation from bacteria cannot be ruled out. Paull et al. (1998) concluded that these compounds are not clear indicators of gymnosperm input, with the possible exception of C_{29} fernane. Incipient aromatization may explain the low quantities of fernanes in these low-rank samples. Saturated hydrocarbon fractions of mature German coals lack fernenes, apparently due to enhanced aromatization above R_o ∼0.5% (Vliex et al., 1994).

Identification of fernenes and fernanes is not difficult if they are abundant, as is common for some immature sediments (Figure 13.100). However, MRM/GCMS is needed to monitor these compounds in rock and coal extracts, where interference by hopenes

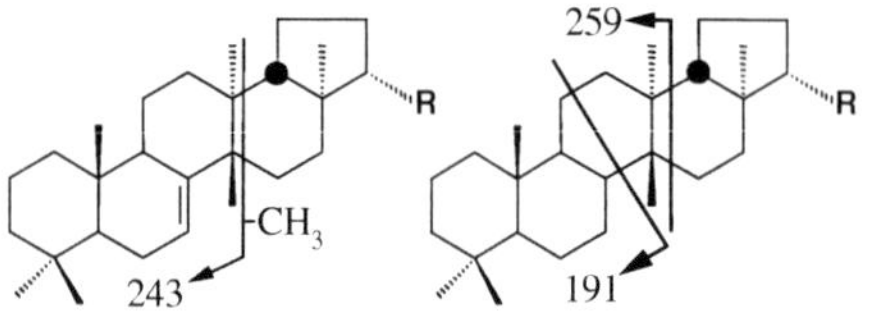

R = H for fernenes, R = OH for fernenols

Figure 13.100. C_{29}–C_{31} fernenes (left) and fernanes (right) yield diagnostic mass spectral fragments. R = nC_2H_5, iC_3H_7, C_4H_9, and C_5H_{11}.

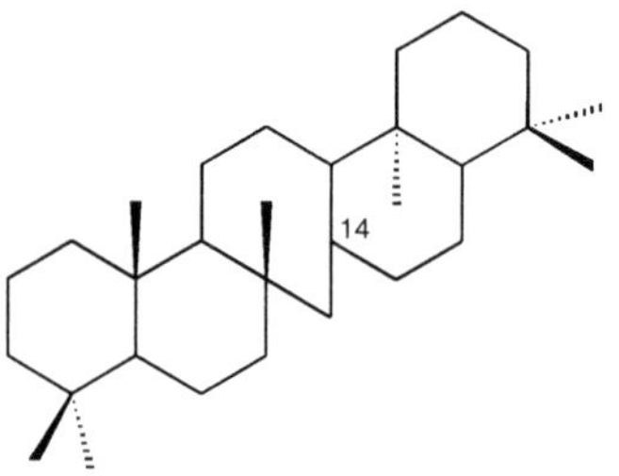

Figure 13.101. Serratanes consist of stereoisomers of a $C_{30}H_{52}$ compound. Kimble et al. (1974b) show mass spectra for serratanes I and II.

and hopanes is likely. The fernenes can be detected by monitoring the molecular ion M+ → 243 transition. Fernanes can be detected by monitoring the transition of their molecular ions to either m/z 191 or 259 (Paull et al., 1996).

Serratane

Origins unknown, but land-plant precursors likely. Possible indicator of paleosalinity in lacustrine environments. Measured using m/z 123 and 191 and MRM m/z 412 → 123

Serratanes are unusual compounds that appear to be structurally related to gammacerane, where the C-ring has opened and a bond formed between the C-8 and C-14 methyl groups (Figure 13.101). Serratanes and gammaceranes, however, do not appear to have common biological origins. The biological precursors for serratane are probably terrigenous land plants. Kimble et al. (1974b) synthesized two isomers of serratane (I and II) that differ at the C-14 asymmetric center. These $C_{30}H_{52}$ compounds yield mass spectra with a strong molecular ion (m/z 412) and predominant peaks at m/z 123 and 191. The isomers can be distinguished from each other by comparing the ratio of the m/z 231 and m/z 259 mass fragments, where m/z 259 > 231 for the 14α isomer (serratane I) and m/z 231 → 259 for the 14β isomer (serratane II).

The only reported geologic occurrence of serratanes was made by Wang et al. (1988) in a suite of Chinese lacustrine oils and rocks that range from Tertiary to Carboniferous age. The identification of serratanes in these samples is questionable because the published mass spectrum is clearly contaminated by co-eluting hydrocarbons. Wang *et al.* (1988) proposed that 14α serratane I and 14β serratane II originate from freshwater and fresh to brackish lacustrine waters, respectively. These findings have not been verified but, if correct, could be very useful in differentiating non-marine sources.

Onocerane

> Associated with rapid deposition in brackish, restricted basins with abundant terrigenous input. Possibly originates mainly from ferns and may be specific for Tertiary or Cretaceous age. Measured using m/z 414 (M+) → 123 or m/z 191 fragmentograms (see Figure 8.20).

The origin of onocerane is uncertain, but it is associated with restricted, terrigenous-dominated deposits in humid subtropical to tropical climates. Onocerane is rare, probably because of the unusual conditions necessary for its formation and preservation. Tropical fern-related plants are the most likely source of onocerane in geological samples. α-Onocerin isolated from *Lycopodium* (club moss), primitive ferns related to early vascular plants and Carboniferous tree ferns such as *Lepidodendron* (Tsuda *et al.*, 1980), is a possible precursor for onocerane. Matsuda *et al.* (1989) extracted α-onoceradiene and related compounds from *Lemmaphyllum*, a Japanese fern.

Onocerane has not been found in living plants but was originally extracted from Miocene angiosperm leaf fossils from the freshwater Clarkia beds in Idaho (Giannasi and Niklas, 1981). They examined extracts from fossil leaves of *Pseudofagus*, a broad-leaved tree typical of warm, humic climates. Onocerane I (Figure 13.102) was reported in Tertiary fluvial and lacustrine environments, including Oligocene shales in the Kishenehn Formation in Montana (Curiale, 1988). It also occurs in Oligocene shales of the Shahejie Formation (Wang *et al.*, 1988) and Paleogene mudstones in several Chinese basins (Fu *et al.*, 1988).

Pearson and Obaje (1999) reported abundant onocerane I in the low-maturity, marine Upper Cretaceous Pindiga Formation from the Upper Benue Trough, Nigeria. These organic-lean rocks (∼2 wt.% TOC) contain abundant vitrinite and inertinite with substantial terrigenous liptinite, apparently deposited

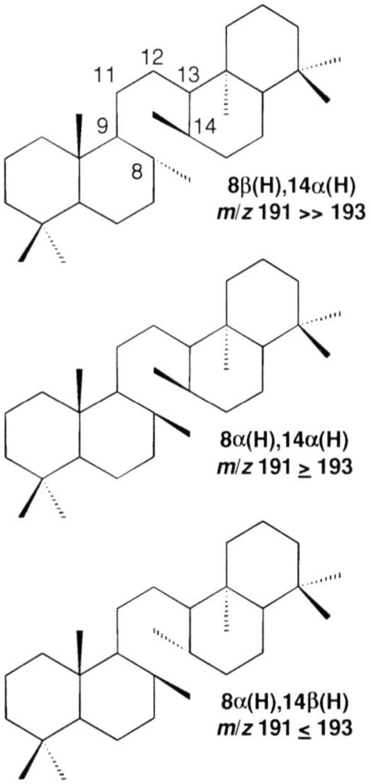

Figure 13.102. Onoceranes are C-ring-opened pentacyclic triterpanes of uncertain origin. Kimble *et al.* (1974b) and Philp (1985) give mass spectra for three onocerane isomers (I–III).

rapidly in a restricted marsh or brackish lagoon near a well-vegetated landmass. The Pindiga Formation lacks the resin (e.g. bicadinanes) and higher-plant markers (e.g. oleanane and lupane) that are characteristic of other terrigenous-dominated source rocks from this region. Pearson and Obaje (1999) proposed that preservation of onocerane required an unusual combination of tropical fern-like biota and rapid burial beneath a restricted water body.

The above reported occurrences of onocerane I suggest that it may be diagnostic of Cretaceous–Tertiary age. However, Wang *et al.* (1988) found onocerane II and III in Chinese oils and rocks as old as Carboniferous. Since α-onocerin (onocera-8(26),14(27)-diene-3β,21α-diol), a possible biological precursor for onocerane, is biosynthesized by both flowering plants and ferns, onocerane may not be age-specific. The mass spectrum published by Wang *et al.* (1988) is clearly contaminated by co-eluting hydrocarbons, and the occurrence of onocerane isomers in samples older than Cretaceous has not been confirmed by other studies.

The three isomers of onocerane (I–III) differ only in the configuration of the methyl groups at C-8 and C-14. These $C_{30}H_{54}$ isomers yield mass spectra with a strong molecular ion (m/z 414) with a predominant base peak of m/z 123 (Kimble et al., 1974b). The isomers differ in the relative intensities of fragment ions at m/z 191 and 193, presumably because the stereochemistry at C-8 and C-14 affects the relative susceptibility to cleavage of bonds 9–11 and 12–13 (Figure 13.102).

Lupane

> Specific for terrigenous input. Lupane has a mass spectrum similar to that of oleanane, and these compounds co-elute with non-polar stationary phases. They can be partially resolved using polar phases, and lupane can be distinguished from oleanane by its m/z 369 mass fragment (e.g. Nytoft et al., 2002).

Lupane is believed to indicate terrigenous organic matter. Various higher-plant lupanoids, such as lupane-$3\beta,20,28$-triol, lup-20(29)-en-$3\beta,28$-diol (betulin), lup-20(29)-en-3β-ol (lupeol), and 3β-hydroxylup-20(29)-en-28-oic acid (betulinic acid), may be possible biological precursors. Lupane is common in extracts of coals and lignites (e.g. Armanios et al., 1995b; Stefanova et al., 1995). Wang and Simoneit (1990) identified lup-20(29)-ene as a possible precursor to lupane in Tertiary brown coals. Poinsot et al. (1995) identified lupane along with several unusual aromatized 12,29-cyclolupa-12,18,20(29)-triene and sulfurized derivatives in Tertiary carbonate-evaporite sediments (Sainte-Cecile, Camargue, France). These terpanoids, along with oleananes and ursanes, provide evidence of terrigenous input into evaporitic deposits. The absence of these compounds in non-evaporitic sediments was attributed to microbially mediated aromatization starting in the A-ring. The diagenetic pathways that give rise to lupane are not understood fully, and likely precursors may rearrange to yield other triterpanes. For example, acid catalysis of lup-20(29)-ene may also produce oleananes and minor amounts of taraxastenes (Rullkötter et al., 1994; Perkins et al., 1995).

Lupane is reported rarely in crude oils. It was identified tentatively in oil in fluid inclusions from the Gippsland Basin (George et al., 1998a) and identified definitively in oils from West Greenland and Beaufort-Mackenzie Delta (Nytoft et al., 2002). Demethylated equivalents are resolved more easily. 24-Norlupane also is present in oil from the Beaufort-Mackenzie Delta (Curiale, 1991; Peakman et al., 1991). The 23,28-bisnorlupanes also originate from lupanes or other higher-plant precursors during diagenesis (Rullkötter et al., 1982). A relationship between the concentrations of bisnorlupanes and oleanane (a higher-plant marker) in crude oils from the Mackenzie Delta, Canada, supports their terrigenous origin (Brooks, 1986). Bisnorlupane/hopane is used commonly in oil–oil and oil–source rock correlation.

The scarcity of reported occurrences of lupane in crude oils may be related to difficulties in its positive identification (Nytoft et al., 2002). For example, the tentative identification of lupane by Hills and Whitehead (1966) was later proved incorrect (Whitehead, 1974). Lupane co-elutes with oleanane on typical apolar chromatographic columns (Figure 13.103) (Rüllkotter et al., 1994)). Lupane also has a mass spectrum similar to oleanane, producing a strong m/z 191 ion fragment. The principal distinguishing characteristic is that lupane produces a prominent m/z 369 fragment resulting from loss of an isopropyl group (m/z 43), while oleanane lacks an isopropyl group and thus yields no m/z 369 fragment. Lupane is best quantified using GCMS/MS methods scanning the m/z 412.4 to 369.3 transition (Figure 13.104). Lupane may be resolved partially from oleanane using gas chromatography capillary columns with polar stationary phases or via HPLC (Nytoft et al., 2002).

AROMATIC BIOMARKERS

Aromatic biomarkers can provide valuable information on organic matter input. For example, aromatic hopanoids originate from bacterial precursors, while tetra- and pentacyclic aromatics with oleanane, lupane, or ursane skeletons indicate higher plants (Garrigues et al., 1986; Loureiro and Cardoso, 1990).

Because living organisms do not biosynthesize aromatic hydrocarbons in appreciable quantities, their ubiquitous occurrence in petroleum is thought to be due to complex transformations of naphthenic and olefinic natural product precursors (Hase and Hites, 1976). These transformations occur during diagenesis and catagenesis (Albrecht and Ourisson, 1971; Johns, 1986; Radke, 1987). It can be difficult to establish

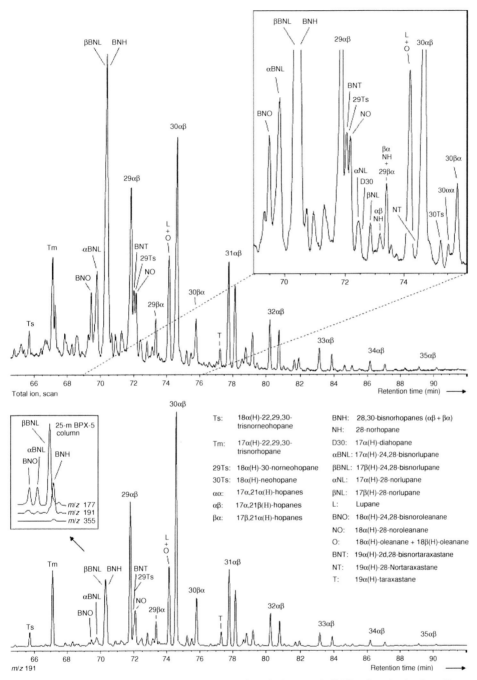

Figure 13.103. Chromatograms of triterpane concentrate from the Marraat-1 oil, West Greenland, using a 60-m non-polar ZB-5 GC column. Top figure is a total ion chromatogram (full scan), with the right insert showing the details of the C_{28}–C_{30} region. Bottom is an m/z 191 ion chromatogram, with the left insert (m/z 177, 191, and 355) showing the separation of 17β(H)-24,28-bisnorlupane and 28,30-bisnorhopanes (αβ,βα) using a 25-m BPX-5 column. Reprinted from Nytoft et al. (2002). © Copyright 2002, with permission from Elsevier.

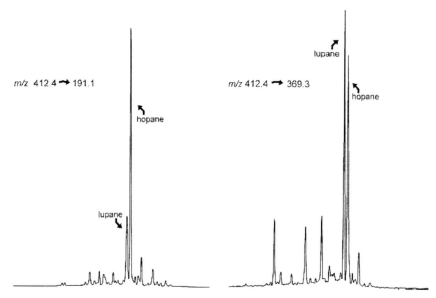

Figure 13.104. Gas chromatography/mass spectrometry/mass spectrometry (GCMS/MS) transitions m/z 412.4 to 191.1 and 369.3, showing the presence of lupane in the Marraat-1 oil, West Greenland. Reprinted from Nytoft et al. (2002). © Copyright 2002, with permission from Elsevier.

genetic relationships between aromatic components in petroleum and natural product precursors. Aromatization can significantly alter precursors by alkylation, dealkylation, isomerization, and ring-opening (Garrigues et al., 1986; Radke, 1987; Püttman and Villar, 1987; Regina et al., 1990; Heppenheimer et al., 1992).

The diagenetic transformations of natural products generally begin with alteration of functional groups. For example, oxygen-containing functional groups are eliminated, usually through decarboxylation or dehydration processes, and unsaturated bonds either are hydrogenated or serve as starting points for progressive aromatization, cleavage, and/or cyclization (Eglinton and Murphy, 1969; Albrecht and Ourisson, 1971; Johns, 1986; Radke, 1987; Rullkötter et al., 1994). Precursors, such as fatty acids and alcohols, may cyclize and aromatize to form equivalent alkylbenzenes, ortho-alkyltoluenes, long-chain alkylnaphthalenes, and other more complex species. With thermal stress, cleavage and alkylation reactions may result in smaller aromatic subunits and/or larger alkylated aromatic species. More complex precursors, such as steroids and terpenoids, may undergo many or all of the aforementioned diagenetic processes. The terpenoids, for example, are susceptible to aromatization during diagenesis (Simoneit, 1986). Aromatization of mono- and sesquiterpenoids

may proceed without loss of structural integrity by disproportionation and dehydrogenation reactions. Examples include p-cymene (Frenkel and Heller-Kallai, 1977; Radke, 1987), dihydro-ar-curcumene (Ellis et al., 1995), and cadalene (Bordoloi et al., 1989), which originate from limonene, bisabolene, and cadinene or cadinol type natural product precursors, respectively.

Many of the A–B ring systems of diterpenoid and triterpenoid natural products show common structural elements, including gem-dimethyl groups or carboxylic acid functions at C-4 and an angular methyl group at C-10. In order for these systems to aromatize, oxidative processes are required, which usually involve loss of one of the gem-dimethyls or the carboxylic acid group and also the angular methyl group (Wakeham et al., 1980). Many aromatic biomarkers in crude oils and rock extracts, such as 1,2,5-trimethylnaphthalene (TMN), 1,2,5,6-tetramethylnaphthalene (TeMN), 9-methylphenanthrene (MP), 1,7-dimethylphenanthrene (DMP), and retene, originate from such terpenoid precursors (Alexander et al., 1992b). Rearrangement and A-ring opening of the carbon skeleton of terpenoids as part of the aromatization process has also been recognized recently (Trendel et al., 1989; Lohmann et al., 1990; Killops 1991; ten Haven et al., 1992; Rullkötter et al., 1994; Ellis et al, 1994; Vliex et al., 1994; Poinsot et al., 1995; Schaeffer et al., 1995a; Ellis, 1995; Ellis et al,

Aromatic biomarkers

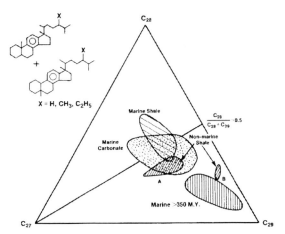

Figure 13.105. Monoaromatic steroids in crude oils give information on source-rock characteristics. The ternary diagram shows the relative abundance of C_{27}, C_{28}, and C_{29} monoaromatic steroids in aromatic fractions of oils determined by gas chromatography/mass spectrometry (GCMS). The labeled areas are a composite of data for oils with known source rocks (Moldowan et al., 1985). Because monoaromatic steroid distributions are more variable than those for steranes (see Figure 13.37), they are more useful to help describe the depositional environments of source rocks for petroleum. Reprinted with permission by ChevronTexaco Exploration and Production Technology Company, a division of Chevron USA Inc.

1996a). The rearrangement reactions of all terpenoids apparently occur via carbocation intermediates, similar to the conversion of sterenes to diasterenes (Rullkötter et al., 1994).

C_{27}–C_{28}–C_{29} C-ring monoaromatic steroids

> High specificity as a correlation tool. Indicates eukaryotic species input. Measured using m/z 253 fragmentograms (see Figure 8.20).

Plot locations of C-ring monoaromatic steroids on C_{27}–C_{28}–C_{29} ternary diagrams (Figure 13.105) were related to various types of source input in a manner similar to the early sterol work of Huang and Meinschein (1979). C-ring monoaromatic steroids may be derived exclusively from sterols with a side-chain double bond during early diagenesis (Riolo et al., 1986; Moldowan and Fago, 1986). In this respect, C-ring monoaromatic steroids may be more precursor-specific than steranes.

Monoaromatic steroids in petroleum plot on the C_{27}–C_{28}–C_{29} diagram in fields associated with terrigenous, marine, or lacustrine input, although overlaps occur in the distributions (Moldowan et al., 1985). Monoaromatic steroid triangular diagrams commonly distinguish oil samples derived from non-marine versus marine shale source rocks. Oils generated from marine shale generally contain less C_{29} monoaromatic steroids than non-marine oils. Typically, more terrigenous organic matter is deposited in non-marine than in marine source rocks, and the non-marine rocks thus contain more C_{29} sterols. Alternatively, non-marine algae may contain relatively higher amounts of C_{29} sterols (e.g. Moldowan et al., 1985; Volkman, 1986). The field for carbonate-derived marine oils extends to higher C_{29} monoaromatic steroids than shale-derived marine oils, although there is considerable overlap. Terrigenous input is poor in C_{27} and C_{28} monoaromatic steroids. Thus, non-marine shales have monoaromatic steroid $C_{29}/(C_{28} + C_{29})$ ratios >0.5. There are non-marine oils that have higher amounts of the C_{27} homologs. These oils generally are related to source rocks deposited under algal-dominated lacustrine settings, with little terrigenous plant input.

Limited data for Lower Paleozoic oils show an increase in the $C_{29}/(C_{27}$–$C_{29})$ monoaromatic steroid ratio with age. The same tendency was observed for steranes (Moldowan et al., 1985; Grantham and Wakefield, 1988). However, several Lower Paleozoic and Precambrian oils have higher C_{27} and/or C_{28} steranes than expected based on this model (R. Summons and D. McKirdy, 1988, personal communication).

The main application of monoaromatic steroid ternary diagrams is correlation. For examples, Figure 13.106 provides supporting evidence for correlation between Piper oil and the Kimmeridge Clay source rock and for a commingled Devonian and Middle Jurassic source for Beatrice oil. Ternary diagrams of both monoaromatic steroids and steranes provide more powerful evidence for correlations than either one alone because they represent compounds of differing origins and provide independent evidence for correlation. Furthermore, plot locations on these diagrams do not change significantly throughout the oil window (e.g. Peters et al., 1989).

Accurate measurement of C-ring monoaromatic steroids requires high-resolution capillary gas chromatography columns (narrow ID, >50-m length) and authentic standards to identify peaks. In mature petroleum, most monoaromatic steroids contain an aromatic C-ring with only a few different combinations

584 Source- and age-related biomarker parameters

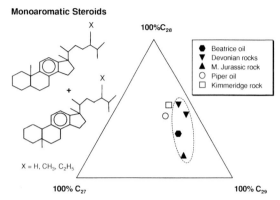

Figure 13.106. Ternary diagram of C_{27}–C_{29} C-ring monoaromatic steroid distributions in aromatic fractions of crude oils and rock extracts, supporting other evidence that Beatrice oil (within dashed oval area) is a mixture of Devonian and Middle Jurassic source input (Peters et al., 1989). See also Figure 13.38. Reprinted by permission of the AAPG, whose permission is required for further use.

of rearranged methyl groups (Figure 13.107). A typical distribution of monoaromatic steroids in marine petroleum is shown in Figure 13.108 (top). However, monoaromatic steroid distributions dominated by the $5\beta(CH_3),10\beta$-isomers are also common (Figure 13.108, bottom).

Ternary diagrams are based on ratios of $C_{27}/(C_{27}$–$C_{29})$, $C_{28}/(C_{27}$–$C_{29})$, and $C_{29}/(C_{27}$–$C_{29})$ monoaromatic steroids, as described above. For each carbon number, six isomeric compounds are used in these ratios, including $5\alpha(20S + 20R)$, $5\beta(20S + 20R)$ and $10\beta \to 5\beta$ methyl-rearranged 20R and 20S isomers. For C_{28} and C_{29}, the number of compounds doubles (12) because of C-24 R and S isomers. However, these R and S isomers are unresolved and, for practical purposes, are ignored.

Dia/(dia + regular) C-ring monoaromatic steroids

Low specificity. Related to source-rock depositional environment. Measured using m/z 253 fragmentogram.

The rearranged monoaromatic (dia-monoaromatic) steroids (Figure 13.107) consist of 10-desmethyl 5α- and 5β-methyl (20S and 20R) diastereomers (Riolo et al., 1985; Riolo and Albrecht, 1985; Moldowan and Fago, 1986). Although the mechanism for their formation is not clear, evidence from a study of Guatemalan samples suggests an influence of clay catalysis on diasterane formation in the source rock (Riolo et al., 1986). Anhydrites formed in evaporitic sabkhas show a strong predominance of dia-monoaromatic steroids, although some clastic source rocks also show this predominance. A study of organic facies in the Toarcian Shale (Moldowan et al., 1986) shows a correlation between C_{27} dia/(dia + regular) monoaromatic steroids and C_{27} dia/(dia + regular) steranes. The dia/(dia + regular) monoaromatic-steroid ratio can also be increased by thermal maturity (Moldowan and Fago, 1986).

We use the C_{27} $5\beta(CH_3)/[5\beta(CH_3) + 5\beta]$ ratio for the 20S isomers to represent the dia/(dia + regular) monoaromatic steroid ratio. The 5α- and $5\beta(CH_3),20S$ (dia) and $5\beta,20S$ (regular) C_{27} monoaromatic steroids show the best resolution by chromatography on the m/z 253 fragmentograms (Figure 8.22). The same isomers for the C_{28} and C_{29} monoaromatic steroids are not generally as well resolved as those for C_{27} (Moldowan and Fago, 1986).

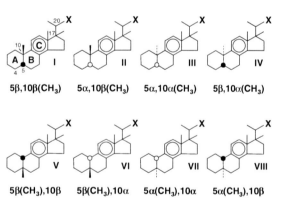

Figure 13.107. Regular (top) and rearranged (bottom) C-ring monoaromatic steroid structures that occur in petroleum. Structures I, II, and V are the most common and are routinely quantified for monoaromatic steroid ternary diagrams (Figure 13.105). Structures V and VII are particularly important in anhydrites from sabkha environments, while structure VII is much less significant in the absence of anhydrite (Riolo et al., 1986; Connan et al., 1986). The other isomers generally occur in lower relative amounts than structures I, II, V, and VII, and more polar gas chromatography columns (e.g. OV-73 and OV-1701) have been used for their analysis.

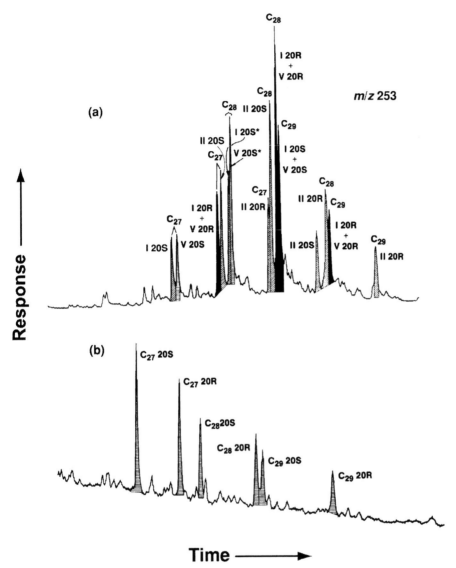

Figure 13.108. Selected ion monitoring/gas chromatography/mass spectrometry (SIM/GCMS) m/z 253 traces of oils from (a) Carneros, California, and (b) Neiber Dome, Wyoming, exemplifying two types of monoaromatic steroid distributions commonly seen in oils. The pattern in (a) shows about equal proportions of regular (see Figure 13.107, structures I and II) and rearranged (structure V) monoaromatic steroids (see Figure 8.22). The pattern in (b) is dominated by rearranged (structure V) monoaromatic steroids with no detectable regular compounds. Gas chromatography conditions: (a) OV-101-coated, 50-m, fused-silica, Hewlett-Packard capillary column programmed from 150 to 320°C at 2°C/minute; (b) 60-m DB-1 J&W Scientific, thick-phase column, programmed as in (a) (Moldowan and Fago, 1986). Reprinted with permission by ChevronTexaco Exploration and Production Technology Company, a division of Chevron USA Inc. Diagonal hatching = structure I or II (see Figure 13.107); horizontal hatching = structure V; solid shading = mixture of structures I and V; * = assignment may be reversed with adjacent peak.

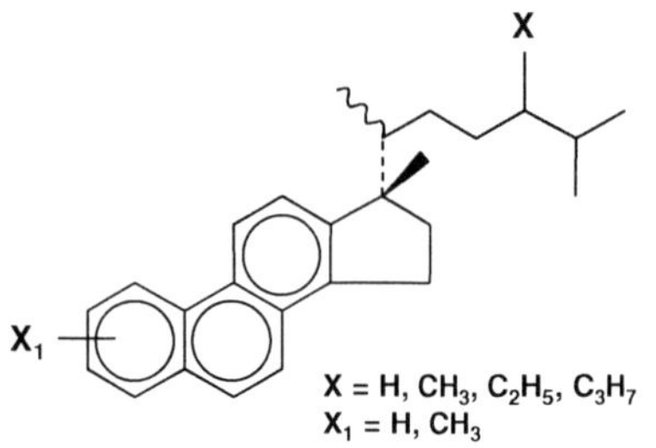

C_{26}–C_{27}–C_{28} triaromatic steroids

> Specificity unknown. Measured using GCMS/MS of M+ → m/z 231 (see Figure 8.20).

Triaromatic steroids can originate by aromatization and loss of a methyl group (-CH_3) from monoaromatic steroids. For example, the C_{29} monoaromatic steroid can be converted to the C_{28} triaromatic steroid (see Figure 14.26). Ratios of $C_{26}/(C_{26}$–$C_{28})$, $C_{27}/(C_{26}$–$C_{28})$, and $C_{28}/(C_{26}$–$C_{28})$ triaromatic steroids are potentially effective source parameters similar to those described above for the C_{27}, C_{28}, and C_{29} monoaromatic steroids. The structures of the triaromatic steroids used in the above parameters are given in Figure 13.109 (e.g. Ludwig *et al.*, 1981). The triaromatic steroid ratios should be more sensitive to thermal maturation than those for monoaromatic steroids or steranes because the triaromatic steroids appear to be maturation products from aromatization of monoaromatic steroids (Mackenzie *et al.*, 1982b). As aromatization proceeds in the early part of the oil window, there may be changes in the triaromatic steroid ratios reflecting the relative ease of aromatization of various monoaromatic precursors and possible additional precursors other than monoaromatic steroids. For example, the ratio of C_{27}/C_{29} monoaromatic steroids does not correlate with the ratio of C_{26}/C_{28} 20S triaromatic steroids in

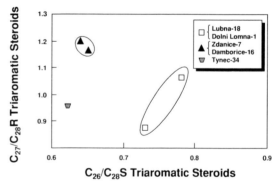

Figure 13.110. Plot of previously unpublished C_{26}/C_{28} 20S versus C_{27}/C_{28} 20R triaromatic steroid ratios helps to distinguish two petroleum systems (circled samples) in Moravia, Czech Republic described by Picha and Peters (1998). The plot location of Tynec-34 is suspect because of low triaromatic steroids, although it does not preclude a mixture of the two oil families as indicated by other data (e.g. see Figures 13.56 and 13.85).

Figure 13.109. Generalized structure for triaromatic steroids common in petroleum. Other side chains and substitution patterns are possible but have not yet been characterized. If X_1 = H, then the major mass spectral fragment is m/z 231 (see Figure 2.15). If X_1 = CH_3, then the major mass spectral fragment is m/z 245. Unlike sterols and steranes, triaromatic steroids lack methyl groups at C-10 and C-13 but show a methyl group at C-17.

m/z 239 Fingerprint

> Specificity unknown. Contains degraded and rearranged monoaromatic steroid hydrocarbons.

The m/z 239 fragmentogram for the monoaromatic hydrocarbon fraction of petroleum has been used successfully to support oil–source rock and oil–oil correlations (Seifert and Moldowan, 1978; Seifert *et al.*, 1980). This fragmentogram is generally complex and contains many peaks representing compounds with unknown structures. Many of these compounds are probably C-ring monoaromatic steroid hydrocarbons that have lost a nuclear methyl group, and some may have undergone rearrangements to aromatic anthrasteroids. Because of the complexity of the m/z 239 fingerprint, its use has been limited compared with other biomarker parameters based on compounds with known structures. Additional research on m/z 239 may improve its usefulness.

m/z 267 Fingerprint

> Specificity unknown.

The m/z 267 fragmentogram is dominated by a series of C-ring monoaromatic steroid hydrocarbons (Figure 8.20), probably derived in part from 2-, 3-, and 4-methylsterols. This fingerprint is little used but has similar potential to the m/z 239 fingerprint (see above).

a study of early mature to mature oils and seeps from Greece (Seifert et al., 1984). The same ratios helped to distinguish oil families in the Western Carpathians and their foreland, Moravia, Czech Republic (Figure 13.110).

Because the C_{26} 20R isomer co-elutes with the C_{27} 20S isomer under all reported gas chromatography conditions using GCMS m/z 231 chromatograms (Figure 8.23), triaromatic $C_{27}/(C_{26}–C_{28})$ and triaromatic $C_{26}/(C_{26}–C_{28})$ cannot be measured readily. Alternatively, a GCMS/MS approach using M+ → 231 (Figure 13.111) can be applied. For example, we used GCMS/MS for triaromatic steroids to successfully show relationships between oils from the Eel River Basin. The triaromatic steroid results for these samples are consistent with other correlation parameters, including C_{27}, C_{28}, and C_{29} sterane distributions. This approach suffers from reduced sensitivity for triaromatic steroids because electron-impact spectra yield low triaromatic steroid molecular ions, leading to weak metastable or collision spectra. For this reason, low-voltage electron-impact, field-ionization, or chemical-ionization methods may be useful in triaromatic steroid analysis. There are additional families of triaromatic steroids having the mass spectral base peak m/z 245, related to the various ring-A methylsterols. However, little work on their structures has been published (Riolo et al., 1986), and they are difficult

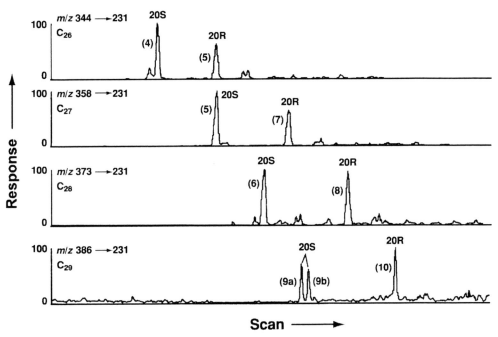

Figure 13.111. Gas chromatography/mass spectrometry/mass spectrometry (GCMS/MS) analysis of triaromatic steroid hydrocarbons (see Figure 13.10, $X_1 = H$) in the aromatic fraction of oil from the Carneros Formation, California using the triple quadrupole (Finnigan MAT TSQ-70) set to record parents of the m/z 231 fragment. Compared with selected on monitoring (SIM) m/z 231 mass fragmentography of the same sample (see Figure 8.23), this method facilitates accurate measurement of all triaromatic steroid epimers. For example, the C_{26} 20R and C_{27} 20S compounds are analyzed separately on the m/z 344 → 231 and m/z 358 → 231 chromatograms, respectively, thus, eliminating most interference problems. Application of ternary diagram relationships and 20S/(20S + 20R) ratios among the C_{26}–C_{28} triaromatic steroid homologs is possible using this method. Small amounts of the C_{29} homologs (peaks 9 and 10) are also recognized and quantified more easily using the GCMS/MS technique. The C_{29} 20S compound is split into a doublet (peaks 9a and 9b) consisting of 24S and 24R epimers (Figure 2.15 shows the steroid-numbering system) of the n-propyl group. Other 24S and 24R epimers co-elute in this analysis. Peak numbers 4–10 refer to identifications in Figure 8.23. Figure modified from Gallegos and Moldowan (1992). Reprinted with permission by ChevronTexaco Exploration and Production Technology Company, a division of Chevron USA Inc.

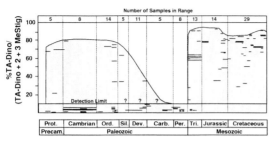

Figure 13.112. Based on analysis of extracts from 112 organic-rich marine rocks from Proterozoic to Cretaceous age, absence of triaromatic dinosteroids (TA-dino) from petroleum indicates a pre-Jurassic source rock (Moldowan et al., 1996; 2001a). The detection limit for the TA-dino/(TA-dino + 2- + 3-methyl-24-ethylcholesteroids) ratio is 10%. The ratio TA-dino/(TA-dino + 4-methyl-24-ethylcholesteroid) has a similar time distribution.

to identify without synthetic standards for co-elution experiments.

Triaromatic dinosteroids

> Age-related biomarkers. Measured using GCMS m/z 245 or GCMS/MS of M+ → m/z 245 of aromatic fraction.

Lack of triaromatic dinosteroids in petroleum generated from marine source rocks indicates a pre-Jurassic source rock (Figure 13.112). Triaromatic dinosteroids are assumed to originate from dinosterol and structurally related 4,23,24-trimethylcholesterols, compounds that characterize modern marine dinoflagellates. Triaromatic dinosteroids are generally abundant in Mesozoic samples but are below detection limits for the vast majority of Paleozoic marine source rocks and petroleum. This age distribution parallels the fossil record of dinoflagellate cysts, with the first undisputed occurrence in Middle Triassic rocks. Pre-Triassic triaromatic dinosteroids correlate temporally with fossil acritarch diversity (e.g. see Figure 13.48) (Moldowan et al., 1996; 2001). Measurement and application of these triaromatic dinosteroids is generally more convenient than dinosteranes because of deconvolution issues in dinosterane measurements (e.g. see Figure 13.48) (Moldowan et al., 2001a).

Benzohopanes

> Moderate specificity for evaporite or carbonate depositional environments. Measured using m/z 191 (see Figure 8.20) or gas chromatography of the aromatic fraction.

Benzohopanes probably form by cyclization of extended hopanoid side chains followed by aromatization (see Figure 2.15) (Hussler et al., 1984a; 1984b). Benzohopanes range in carbon number from C_{32} to C_{35}, consistent with their proposed origin by cyclization of the homohopanoid side chain during early diagenesis. Oils and bitumens from evaporitic and carbonate source rocks show the highest concentrations of benzohopanes, although they occur in trace amounts in most source rocks and crude oils (e.g. He Wei and Lu Songnian, 1990).

Aromatized C_{31} 8(14)-secohopanoid

> A cyclization product of bacteriohopanepolyols that forms under as yet poorly understood depositional conditions. Characterized by its molecular ion at m/z 414 and by fragment ions at m/z 207, 208, and 123. Positive identification can be made only through isolation and ^1H and ^{13}C NMR.

Sinninghe Damsté et al. (1998b) identified a novel C_{31} 8(14)-secohopanoid with a fluorene moiety. Its mass spectrum is dominated by a molecular ion at m/z 414 and by fragment ions at m/z 207, 208 and 123. The compound is believed to originate by cyclization of bacteriohopanepolyols (Figure 13.113). Unlike the benzohopanes that usually exist as a homologous series of four compounds, this C_{31} diaromatized secohopanoid is accompanied only by trace amounts of the C_{32} homolog, which is presumed to result from a methyl addition at C-32. Furthermore, the C_{31} diaromatized secohopanoid occurred in appreciable quantities (150–5000 μg/g TOC) in only 3 of 13 extracts from mudstones in the Kimmeridge Clay Formation, suggesting that its formation requires specific depositional conditions. This compound has not been reported in crude oil.

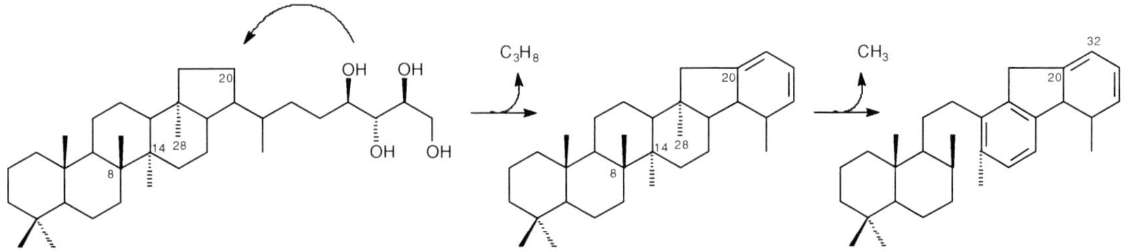

Figure 13.113. Proposed pathway for the formation of C_{31} diaromatic 8(14)-secohopane with a fluorene-like moiety. Cyclization of the polyol side chain is followed by F-ring aromatization and cleavage of a propyl group (Sinninghe Damsté et al., 1998b). The D-ring is then aromatized with a loss of the C-28 methyl and C-ring opening.

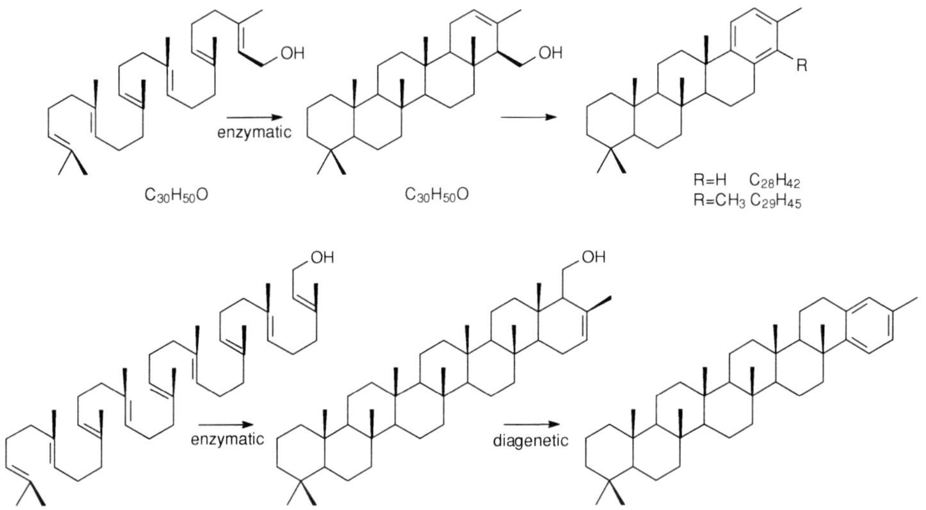

Figure 13.114. Proposed biological and diagenetic formation of monoaromatic polyprenoids (Schaeffer et al., 1994).

Monoaromatic polyprenoids

Unknown specificity. Measured using their molecular ions and/or major terpanoid fragment ions (e.g. m/z 171).

Schaeffer et al. (1994) isolated and identified a series of monoaromatic polyprenoids in extracts of Messel Shale. Tetra-, penta-, hexa-, hepta-, and octacyclic species were detected, all of which appear to originate by biological cyclization of polyprenols followed by diagenetic aromatization of the terminal phenolic ring (Figure 13.114). Mass spectra of these compounds are shown in Figure 13.115.

While isoprenols are distributed widely among living organisms, the cyclized forms have not been identified. Nevertheless, a biological origin is likely because the geologic polyprenoids are optically active and the rings are all in the *trans* configuration (all ring-junction axial methyl groups are on the same side). Optical activity provides strong support for production by enzyme-mediated reactions. Conceivably, optically active compounds could arise by abiogenic (acid-catalyzed) cyclization followed by selective biodegradation to remove all but one configuration, or by additional abiogenic cyclization of biogenic tricycloprenoids. The former is unlikely, and there is no evidence for saturated or aromatic tricycloprenoids in the Messel Shale.

Long-chain isoprenols are common lipids (e.g. C_{45} solanesol, C_{50} castaprenols, C_{65} ficaprenols) that are used in various biochemical reactions in addition to the synthesis of terpenoids (e.g. acylation of proteins, synthesis of vitamins E and K). Solanesol is abundant in tobacco leaves and may be an important precursor for

590 Source- and age-related biomarker parameters

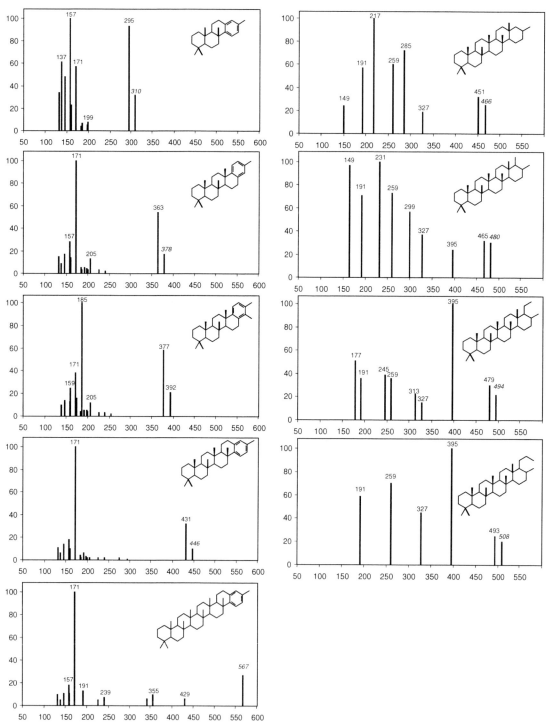

Figure 13.115. Mass spectra of monoaromatic polyprenoids in extracts of Messel Shale and saturated polyprenoids in extracts of Upper Cretaceous sediments, Lameignére quarry, Orthez, France (from Schaeffer et al., 1994).

the tumorigenic polynuclear aromatic hydrocarbons in smoke. Longer polyisoprenols (C_{80}–C_{110}) with one saturated isoprenoid unit (dolichol) are important as carriers of glycosyl in the synthesis of polysaccharides in both eukaryotes and bacteria. These long-chain isoprenols could be cyclized by precursor-specific enzymes, forming bioactive molecules of unknown use. Ourisson and Nakatani (1994) speculated that the cyclic polyprenols could serve as membrane stabilizers, much like hopanols (eubacteria) or sterols (eukaryotes). Alternatively, cyclic polyprenols could arise if enzymes designed for other precursors, such as squalene cyclase, were non-specific in their selection of reactant compounds.

Hexacyclic polyprenoids

> Possibly specific for ostracode input to Cretaceous source rocks. Measured using m/z 191.

Saturated forms of the monoaromatic polyprenoids in Messel Shale occur in Upper Cretaceous sediments from the Lameignére quarry near Orthez, France (Schaeffer et al., 1994). Terpanoids from four to seven rings were identified. Of these, the hexacyclic polyprenoids are unusually abundant. Their mass spectra are shown in Figure 13.115. Because these compounds can yield m/z 191 and 217 mass fragments, they can be detected in standard GCMS ion scans for common biomarkers. The hexacyclic polyprenoids, however, elute after the C_{35} hopanes, and their occurrence may be missed because of analytical limitations.

Riediger et al. (1997) also reported that several hexacyclic C_{34} compounds were abundant in extracts from the Lower Cretaceous Ostracode Zone in southern Alberta. This sedimentary unit contains mixed algal and marine input and, as the name implies, is particularly rich in ostracodes. Because they were not found in other source rocks from the Western Canada Basin, Riediger et al. (1997) suggested that these biomarkers, designated as Q compounds, could be used in oils as an indicator for source contribution from the Lower Cretaceous Ostracode Zone and possibly as a marker specific for ostracodes.

1,1,7,8-Tetramethyl-1,2,3,4-tetrahydrophenanthrene

> A diaromatic biomarker that may originate by the aromatization of tricyclic terpanes or gammacerane. Characterized by a base peak at m/z 223 and a molecular ion at m/z 238.

This alkylated tetrahydrophenanthrene may originate from several different biomarkers. Azevedo et al. (1994) identified this C_{18} compound along with C_{17} and C_{19} homologs in extracts of Permian Tasmanite Oil Shale. They proposed that these tetrahydrophenenanthrenes originated by further aromatizaiton of monoaromatic tricyclic terpanes that were particularly abundant in the rock extracts. In a study of bitumen extracted from the Lower Jurassic Allgäu Formation, Sinninghe Damsté et al. (1998a) offered convincing circumstantial evidence that this diaromatic hydrocarbon originates from gammacerane, which was the dominant saturated hydrocarbon (Figure 13.116). 1,1,7,8-Tetramethyl-1,2,3,4-tetrahydrophenanthrene and gammacerane co-varied in abundance and had similar $\delta^{13}C$ values ($\sim$–29‰) that were distinct from biomarkers associated with algal input (phytane and C_{29} 4α-methyltriaromatic steroid $\sim$–35‰) or anaerobic green sulfur bacteria (isorenieratane $\sim$–18‰).

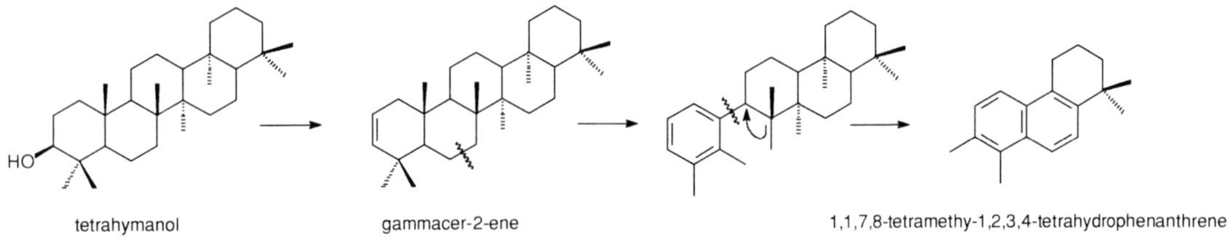

tetrahymanol → gammacer-2-ene → → 1,1,7,8-tetramethy-1,2,3,4-tetrahydrophenanthrene

Figure 13.116. Reaction pathway to the alkylated tetrahydrophenanthrene from tetrahymanol may involve the formation of gammacer-2-ene by dehydration, aromatization of the A ring with concomitant opening of the B ring, aromatization of the C and D rings with elimination of the A and B rings, and methyl transfer (Sinninghe Damsté et al., 1998a).

Aromatic derivatives of the fernane/arborane series precursors

> Aromatic proxies for precursors with the fernane/arborane structure. Determination of specific precursor requires measurement of optical rotation. Hauke et al. (1992a; 1992b) and Jaffé and Hausmann (1995) show mass spectra.

Ancient rocks and coals rarely contain the saturated forms of isoarborinol or fernenes, probably due to rapid B-ring aromatization (Hauke et al., 1995) or degradation of the A ring (Jaffé and Hausmann, 1995) during diagenesis (Figure 13.117). In a study of Upper Carboniferous to Lower Permian coals, concentrations of these aromatic derivatives correspond to increases in the palynomorphs of early gymnosperms (Vliex et al., 1994; 1995). Polynuclear aromatic hydrocarbons also may result from the degradation of the E ring or cleavage of the D ring during degradation of arborane/fernane precursors (Borrego et al., 1997). These polycyclic aromatic hydrocarbons can arise from various unrelated biomarker precursors and can be linked to fernane/arborane precursors only in the context of the other aromatized derivatives described above.

Since the carbon skeletons of the arborane and fernane series are identical, except for orientation around optical centers, the exact precursors of the aromatized derivatives cannot be distinguished by their mass fragmentation alone. Hauke et al. (1992a) found that the $\delta^{13}C$ values of individual monoaromatic hydrocarbons were nearly identical to those of benzohopanes and argued for a prokaryotic source with a fernene skeleton.

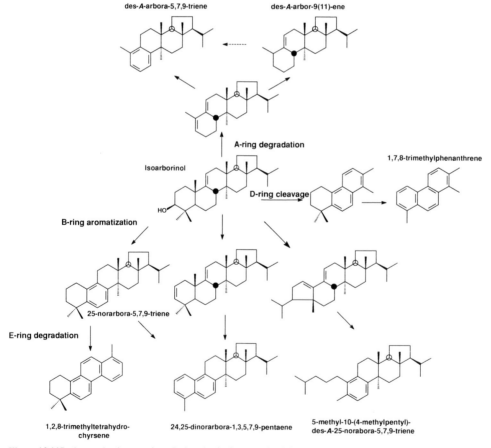

Figure 13.117. Aromatization reactions for isoarborinol summarized from Borrego et al. (1997), Jaffé and Hausmann (1995), and Vliex et al. (1994). The reaction pathways are identical for fernenol and fernenes, except for orientation of optical centers. Product names for the fernane series substitute "fern" for "arbor."

Optical rotation measurements proved later that des-A-triene and the 25-nor-triene in extracts from the Eocene Messel Shale, Permian Kupferschiefer, and Triassic black shales from Italy originated from isoarborinol, while the 24,25-dinorpentene from Carboniferous coal originated from fernenes or fernenols (Hauke et al., 1995). Considering the uncertainty of the biological origins of the precursor compounds, we question the use of aromatic hydrocarbons derived from either the arborane or the fernane series as biomarkers for floral input or depositional environments, even when their optical orientations are known.

Isorenieratane and related compounds

> Specific for *Chlorobiaceae* when enriched in ^{13}C (Koopmans et al., 1996a; 1996b). May co-occur with ^{13}C-rich chlorobactane, methyl iso-butyl maleimide, and/or farnesane (Grice, 2001). Isolated from aromatic hydrocarbon fraction and measured using GCMS.

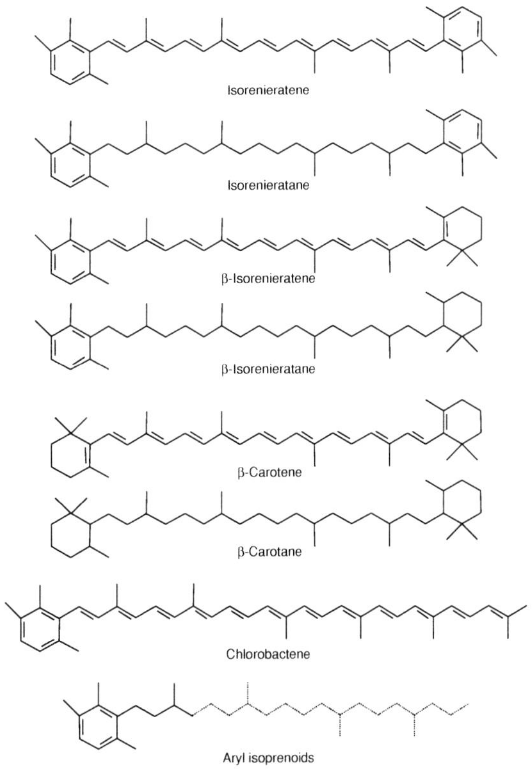

Isorenieratane and many mono-, di-, tri-, and tetra-aromatic compounds originate from the C_{40} diaromatic carotenoid hydrocarbon isorenieratene (Figure 13.118) in green sulfur bacteria (*Chlorobiaceae*) (Grice et al., 1996b). Green sulfur bacteria are anoxygenic phototrophs that fix carbon dioxide using the reverse tricarboxylic acid cycle, leading to biomass that is enriched in ^{13}C (Koopmans et al., 1996a; 1996b). Thus, when ^{13}C-rich isorenieratane and related compounds occur in rock extracts or crude oils, they indicate photic zone anoxia during source-rock deposition (Summons and Powell, 1987; Clifford et al., 1997). These compounds are enriched in ^{13}C compared with biomarkers from phytoplankton by $\sim$15‰ (e.g. Summons and Powell, 1986; Summons and Powell, 1987; Sinninghe Damsté et al., 1993b; Hartgers et al., 1994a; Hartgers et al., 1994b; Grice et al., 1996b; Grice et al., 1997). For example, isorenieratane (-16.4‰), the C_{14} aryl isoprenoid (-17.7‰), the C_{21} and C_{22} diaryl isoprenoids (-12.2‰), and the C_{18} and C_{19} biphenyls (-14.0 to -14.2‰) in Kupferschiefer extract appear to have a common origin from *Chlorobiaceae* based on their structures and ^{13}C-rich isotope composition compared with pristane (-28.4‰), phytane (-28.7‰), and the C_{15}–C_{17} n-alkanes (-28.7‰), which originate mainly from chlorophyll a in cyanobacteria (Grice et al., 1996b). A similar distinct isotopic difference occurs in compounds such as chlorobactane, derived from chlorobactene (Figure 3.28) (Grice et al., 1998a). Stable carbon isotope values for compounds derived from purple sulfur bacteria (*Chromatiaceae*), in particular those related to okenone (Schaeffer et al., 1997), are depleted in ^{13}C (δ^{13}C ~ -45‰) compared with phytoplankton biomarkers (~ -35‰), consistent with *Chromatiaceae* living in deeper parts of the water column, where they use more ^{13}C-depleted inorganic carbon than phytoplankton from the photic zone. Okenone is a C_{40} 1-alkyl-2,3,4-trimethyl-aryl carotenoid.

Figure 13.118. Structures of isorenieratene and other carotenoid derivatives with 1-alkyl-2,3,6-trimethyl substitution, including aryl isoprenoids (2,3,6-trimethylbenzenes, bottom).

Aryl isoprenoids (trimethylbenzenes) in crude oils and rock extracts show a 1-alkyl-2,3,6-trimethyl substitution pattern suggesting an origin from isorenieratane in *Chlorobiaceae* and photic zone anoxia in the source rock (e.g. Summons and Powell, 1986; 1987). However, Koopmans et al. (1996b) showed that aryl isoprenoids originate from at least two sources: isorenieratene or β-isorenieratene in *Chlorobiaceae* and

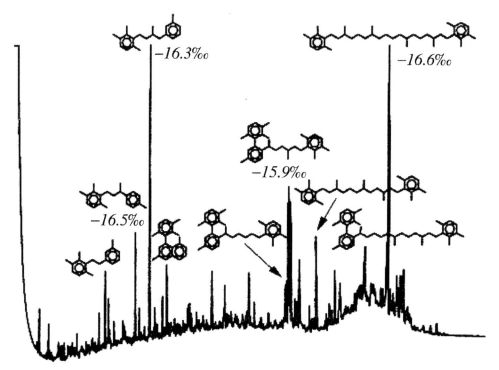

Figure 13.119. Gas chromatogram of the aromatic hydrocarbon fraction from an extract of the Kimmeridge Clay Formation (Van Kaam-Peters et al., 1995b). A series of di- and triaromatic compounds exhibit the same ^{13}C enrichment as isorenieratane, indicating a common origin.

β-isorenieratene from the ubiquitous carotenoid pigment β-carotene. They found that δ^{13}C for β-carotane and β-isorenieratane in a North Sea crude oil are similar ($\sim -26‰$), consistent with a common origin, while isorenieratane is $\sim 15‰$ enriched in ^{13}C, consistent with an origin from *Chlorobiaceae*. The C_{19}–C_{21} aryl isoprenoids in this oil show intermediate isotopic compositions ($\sim -22‰$), suggesting a mixed origin.

Octadecahydro-isorenieratene and β-carotane occur in the desulfurized (Raney nickel) polar fractions of Holocene sediment extracts, ranging in age from present-day to 6200 years old, in the Black Sea (Sinninghe Damsté et al., 1993b). The octadecahydro-isorenieratene is enriched in ^{13}C and originated from photosynthetic green sulfur bacteria growing at the chemocline, i.e. the boundary between oxic and anoxic conditions. Conversely, the β-carotane is consistently depleted in ^{13}C compared with the octadecahydro-isorenieratene and originated from algae living in the upper photic zone. A depth or age profile of the concentrations of these two compounds in the Black Sea sediments shows that photosynthetic green sulfur bacteria have been active in the Black Sea for substantial periods of time and that penetration of the photic zone by anaerobic waters is not a recent phenomenon.

Various related di-, tri-, and higher-ring-number aromatic hydrocarbons may accompany isorenieratane (Figure 13.119). These compounds have ^{13}C-enrichments similar to isorenieratane, indicating a common origin. Koopmans et al. (1996a) identified the structure of many of these compounds and proposed a generalized reaction scheme for their formation (Figure 13.120).

Perylene

Suggested specificity for higher-plant input. Measured using m/z 252 (see Figure 8.20).

The precursor(s) for perylene must have widespread distribution, and its formation requires deposition in highly reducing sediment (Gschwend et al., 1983).

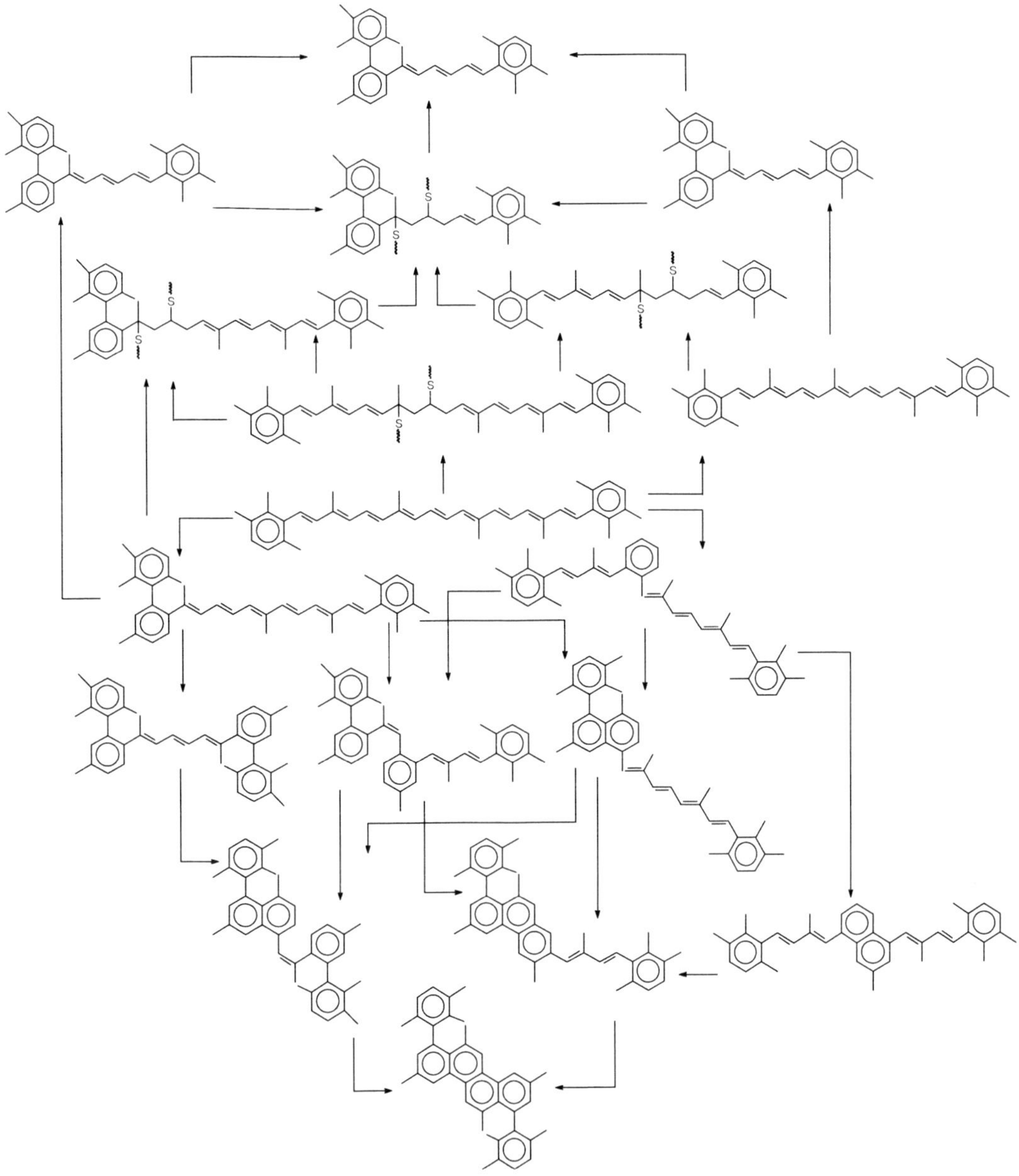

Figure 13.120. Proposed diagenetic pathways for the alteration of isorenieratene (middle, right) involve cyclization, aromatization, sulfurization, and elimination (modified from Koopmans et al., 1996a).

Perylene (see Figure 8.20) may be a land plant source indicator (Aizenshtat, 1973), although its occurrence in Walvis Bay sediment, a site thought to be largely free of terrigenous organic matter input, led Wakeham et al. (1979) to question whether the precursor must be terrigenous.

Fungi may be the major precursor for perylene in sediments (Chunqing Jiang et al., 2000b). These authors found that perylene is a major polyaromatic hydrocarbon (PAH) in low-maturity Lower-Middle Jurassic rocks from the Northern Carnarvon Basin, Australia. The depth/age profiles of perylene are not related to the combustion-derived PAHs produced by paleo-fires, suggesting a diagenetic origin. The concentration of perylene in the rocks is proportional to the amount of terrigenous input, decreasing with distance from the source of land sediments. The carbon isotope composition of perylene is slightly enriched in ^{13}C, but still within the range of the terrigenous PAH, including higher-plant and combustion-derived PAHs.

Silliman et al. (2000) studied the relation of perylene to the amount and type of organic matter in the sediments of Saanich Inlet, a coastal marine anoxic basin. They concluded that perylene originates from more than one precursor, both aquatic and continental organic matter, different microbial processes, or some combination of these possibilities. Organic matter in Saanich Inlet is predominantly marine, but the proportions of marine- and land-derived components varied during deposition. Perylene generally increases with sediment depth relative to TOC, which suggests diagenetic formation of this compound by microbes. However, TOC has a narrow range of δ^{13}C (-21.7 to -21.2‰), while perylene is more variable (-27.7 to -23.6‰) over the same depth interval, suggesting multiple origins for perylene.

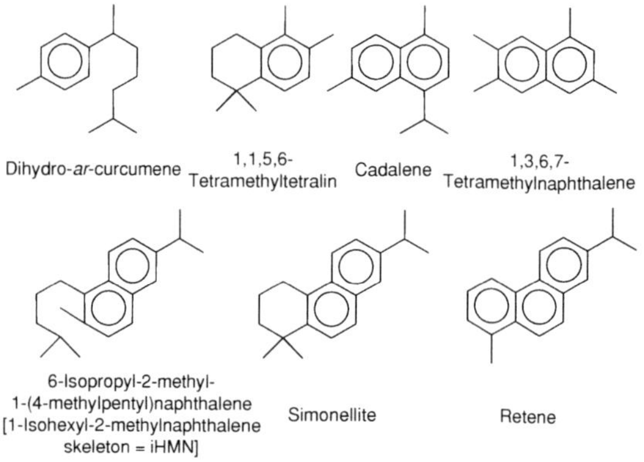

Degraded aromatic diterpanes

> Suggested specificity for higher-plant input. Measured using SIM/GCMS or gas chromatography of aromatic fractions.

Many higher-plant cyclic compounds are unsaturated and undergo aromatization during diagenesis, resulting in reduced source specificity. Alexander et al. (1992a) found that saturated diterpane biomarkers are extremely low in crude oils generated from coal measures in the Cooper and Eromanga basins, Australia. In contrast, the degraded resin precursors are particularly abundant, such as 1,6-dimethylnaphthalene, 1,2,5-trimethylnaphthalene, 1,7-dimethylphenanthrene, 1-methylphenanthrene, and retene. These are probably related mostly to *Araucariaceae* conifer remains in Jurassic-Lower Cretaceous source rocks, which generated the oils in the Eromanga Basin. However, because they can be formed through other processes, they are not particularly specific higher-plant markers. Different contributions of the same compounds occur in Cooper Basin oils derived mostly from remains of pteridosperms (seed ferns) in Permian source rocks. The more degraded aromatics, 1, 2,5-trimethylnaphthalene, 1,7-dimethylphenanthrene, and 1-methylphenanthrene, could originate from

Figure 13.121. Aromatic hydrocarbons have differing specificity for higher-plant input. Dihydro-*ar*-curcumene, 1,1,5,6-tetramethyltetralin, and cadalene are biomarkers for higher plants (Simoneit, 1986; Püttman and Villar, 1987), while retene and simonellite are more specific because they originate from conifers (Simoneit, 1986; Alexander et al., 1988b). A new class of compounds, including 1-isohexyl-2-methylnaphthalene (iHMN), originate from higher plants, but their specific source is unknown (Ellis et al., 1996a; van Aarssen et al., 2000). 1,3,6,7-Tetramethylnaphthalene is a bacterial rather than higher-plant marker (van Aarssen et al., 1996).

pentacyclic triterpanes. However, their collective presence and abundance relative to other aromatics, such as 1,3,6-trimethylnaphthalene and 9-methylphenanthrene (lacking the proper alkyl substitutions to be related to common diterpenes), is strong evidence of their diterpenic origin (Alexander et al., 1992a). Retene (Figure 13.121) is commonly believed to originate during diagenesis by dehydrogenation of abietic acid, a major constituent of coniferous resins (Wakeham et al., 1980). However, Zhou Wen et al. (2000) show that retene can also originate from algal and bacterial precursors.

Higher-plant input and higher-plant parameters

> Specific for land plant input, the higher-plant index (HPI) and higher-plant parameter (HPP) track the global sea-level curve and paleoclimate during the Jurassic Period in Western Australia. Measured using SIM/GCMS or gas chromatography of aromatic fractions.

Several aromatic hydrocarbons, such as dihydro-*ar*-curcumene, 1,1,5,6-tetramethyltetralin, and cadalene, are markers for higher plants, while retene and simonellite are more specific and are thought to originate mainly from conifers (Ellis *et al.*, 1995; Simoneit, 1986; Püttman and Villar, 1987; Alexander *et al.*, 1988b). Figure 8.27 shows the mass spectrum of dihydro-*ar*-curcumene. Three higher-plant biomarkers, including retene, cadalene, and 6-isopropyl-1-isohexyl-2-methylnaphthalene (ip-iHMN) are used in the numerator of the HPI, while the compound 1,3,6,7-tetramethylnaphthalene (1,3,6,7-TeMN) appears to originate from bacteria. Van Aarssen *et al.* (1996) used the HPI to characterize the relative input of higher plants to source rocks and crude oils from the Barrow Sub-basin in Western Australia:

HPI = (retene + cadalene + iHMN)/1,3,6,7-TeMN

Retene, cadalene, and iHMN also occur in marine Oxfordian sedimentary rocks from the Taranaki Basin, New Zealand (Figure 13.121) (van Aarssen *et al.*, 2000), where their relative abundances (higher-plant fingerprint, HPF) were calculated using peak areas from m/z 219, 183, and 197, respectively. The HPF distributions of these compounds were nearly identical at three locations, where retene became more abundant relative to the other two compounds with decreasing age of the Oxfordian rocks. The increase in the abundance of retene relative to that of cadalene during the Oxfordian Age was interpreted to reflect an increase in the contribution of conifers, brought about by climatic change. This was exemplified by measuring the distributions of retene and cadalene, expressed in the HPP for a suite of rocks from the Carnarvon Basin, Western Australia, covering the complete Jurassic Period. HPP is the ratio of retene to the sum of retene plus cadalene. The HPP profile displays three major cycles, each covering a period of at least ten million years. This profile compares well with published paleoclimate data and is remarkably similar to second-order cycles in the global sea-level curve, supporting the proposal that these variations indicate changes in paleoclimate.

Trimethylnaphthalenes

1,2,7-Trimethylnaphthalene indicates angiosperm input. Measured using capillary gas chromatography of aromatic fractions (Roland *et al.*, 1984).

1,2,7-Trimethylhaphthalene (TMN) has been used as a marker of angiosperm input. Two related ratios, TDE-1 and TDE-2, differentiate coal swamp from marine, lacustrine, and deltaic environments (Strachan *et al.*, 1986):

TDE-1 = 1,2,5-TMN/1,2,4-TMN

TDE-2 = 1,2,7-TMN/1,2,6-TMN

1,2,5-Trimethylnaphthalene has $\delta^{13}C$ similar to drimanes in torbanites from boghead coals, suggesting to Grice *et al.* (2001) that both compounds originated from a common cyanobacterial hopanoid precursor. The analyzed torbanites are Permian and Carboniferous in age and predate the evolution of angiosperms.

1,2,5- and 1,2,7-Trimethylnapthalenes can form as diagenetic products of oleanane-type triterpanoids (Figure 13.122) (Chaffee and Johns, 1983; Chaffee *et al.*, 1984). The use and limitations of these compounds as molecular markers for angiosperm input was demonstrated by Strachen *et al.* (1988), who compared the relative abundance of trimethylnapthalene isomers in Southeast Asian samples ranging in age from Permian to Tertiary. They found that the relative concentrations of the 1,2,7-trimethylnapthalenes were appreciably

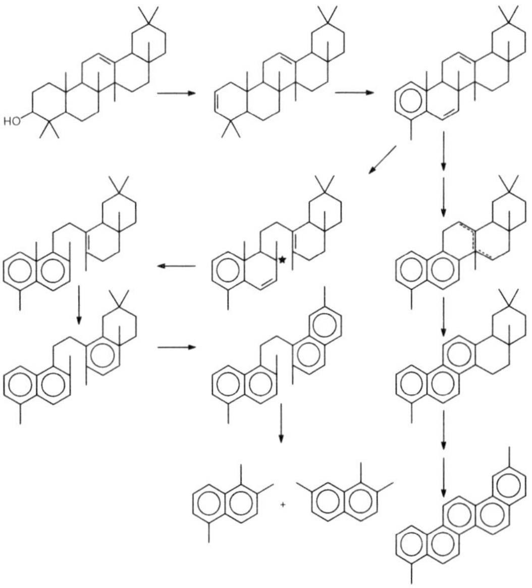

Figure 13.122. Proposed reaction pathways that can give rise to 1,2,7- and 1,2,5-trimethylnapthalenes from oleanane-type triterpanoids (from Strachen *et al.*, 1988; modified from Chaffee and Johns (1983) and Chaffee *et al.* (1984)). The star represents a transition state.

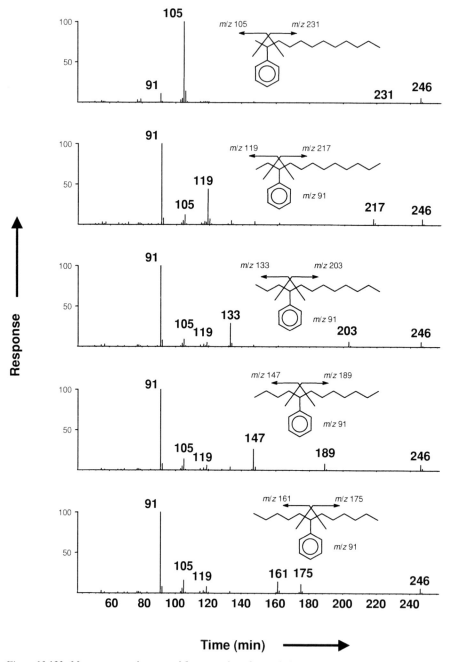

Figure 13.123. Mass spectra and suggested fragmentation of several phenylalkanes (figure courtesy of L. Ellis).

higher in most oils derived from post-Cretaceous source rocks dominated by higher-plant input and in extracts from Cretaceous and younger wellbore samples than in samples of older age or with little or no terrigenous plant input. Oils of this type yielded ratios of 1,2,7-/1,3,7-trimethylnaphthalenes from 0.46 to 1.36, while oils from older source rocks containing higher-plant material yielded ratios from 0.34 to 0.16 and oils from marine source rocks yielded ratios from 0.15 to 0.32.

1,2,5- and 1,2,7-Trimethylnaphthalenes occur in all samples, including oils as old as Ordovician and oils and rock extracts with no input of oleanane-type triterpanoids. These compounds can arise by methyl rearrangement of other trimethylnaphthalene isomers that originate from other precursors (Strachen et al., 1988). Furthermore, not all oils derived from post-Cretaceous source rocks and dominated by higher-plant input yield high 1,2,7-/1,3,7-trimethylnaphthalene ratios. Two deltaic oil samples with high 18α-oleanane from Central Sumatra and the Niger Delta yield 1,2,7-/1,3,7-trimethylnaphthalene ratios of only 0.14 and 0.25, respectively. In these cases, the depositional environment does not appear to facilitate the aromatization and C-ring cleavage reaction needed to produce the trimethylnaphthalenes from oleanane precursors. Hence, while 1,2,7-/1,3,7-trimethylnaphthalene ratios >0.4 appear to indicate oleanoid precursors, low ratios are not diagnostic. These concerns limit the use of the trimethylnaphthalene isomers as age-related parameters. George et al. (1997) demonstrated their utility in the analysis of oils extracted from fluid inclusions.

Phenylalkanes (linear alkylbenzenes)

> Possibly the first aromatic biomarkers for archaea, although they are common surfactants that can contaminate samples. Measured using m/z 91 and m/z 105 fragmentograms.

2-Phenylalkanes (linear alkylbenzenes) (Figure 13.123) in petroleum have been attributed to both bacteria and algae but may be especially useful to identify geologic samples that were exposed to hot subsurface waters, where thermophilic archaebacteria thrive (Ellis et al., 1996b; L. Ellis, 2003, personal communication). Ellis et al. (1996b) isolated 2-phenylalkanes from the obligate acidophilic (pH 2) and thermophilic (60°C) archaebacterium *Thermoplasma acidophilum*, which occurs worldwide in continental and marine hot, acidic vent environments.

Although 2-phenylalkanes have been identified in geologic samples (Figure 13.124), their origin can be uncertain because (1) they are important components in surfactants and (2) they currently lack credible natural product precursors (Eganhouse, 1986; Takada and Ishiwatari, 1990). Some contend that all C_{10}–C_{14} phenylalkanes (C_{16}–C_{20} total carbon atoms) in geologic samples represent anthropogenic contamination (Harvey et al., 1985).

Long-chain alkylnaphthalenes

> High abundance of this long-chain series is specific for algae, such as *Botryococcus braunii* and *Gloeocapsomorpha prisca*. Measured using m/z 141 + 142 fragmentograms.

Laboratory experiments suggest that most long-chain alkylnaphthalenes in the geosphere form by cyclization of unsaturated kerogen moieties, such as alk-1-enylbenzenes, although some may form by clay-catalyzed cyclization of alkylbenzenes (Ellis et al., 1998). Similar heating experiments on clay show ring isomerization of 1-nonylnaphthalene to 2-nonylnaphthalene, with increasing thermal maturation. Long-chain alkylnaphthalenes are generally most abundant in crude oils and rock extracts where the organic matter is dominated by *Botryococcus braunii* or *Gloeocapsomorpha prisca* (Ellis et al., 1999).

Isohexyl alkylaromatics

> Specific for higher-plant terpenoid precursors. Identified by GCMS of the aromatic fraction using the base peak at M+ −71.

Most isohexyl alkylaromatics (Figure 13.125) originate from higher land-plant terpenoids by a novel aromatization-rearrangement mechanism in sediments that leads to opening of the terpenoid A-ring (Figure 13.126). The products of this reaction are characterized by a 4-methylpentyl (isohexyl) substituent on

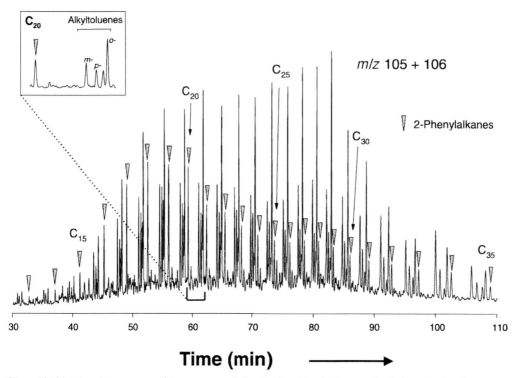

Figure 13.124. Mass chromatogram of the monoaromatic fraction from Rough Range crude oil, Australia, showing the distribution of 2-phenylalkanes extending from C_{13} to at least C_{35} (Figure courtesy of L. Ellis.)

the aromatic ring adjacent to a methyl substituent. Several isohexyl alkylaromatic compounds were identified through laboratory synthesis and dehydrogenation reactions (Ellis *et al.*, 1992; Vliex *et al.*, 1994; Ellis, 1995; Ellis *et al.*, 1996a).

Isohexyl alkylbenzenes in petroleum have both bacterial and higher-plant origins. The presence of isohexyl alkylbenzenes in samples with sources that predate the evolution of land plants can be attributed only to microbial input. Microbial isohexyl alkylbenzenes probably originate from both labdanoid and intermediate 8,14-secohopane type precursors (Wang *et al.*, 1990). 8,14-Secohopanes may originate from microbial and/or thermal geochemical pathways involving ring C-opening of hopanoid precursors in sediments (Bendoraitis 1974; Rullkötter and Wendisch, 1982; Wang *et al.*, 1990). Higher-plant sources of isohexyl alkylbenzenes can be attributed to both triterpenoids and diterpenoids, such as oleanane and labdane-type precursors (Ellis *et al.*, 1996a).

Isohexyl alkylnaphthalenes in crude oil originate mostly from precursors in higher plants (Ellis *et al.*, 1996a). Diterpenoids with tricyclic or tetracyclic ring systems and containing the A–B ring features required for aromatization-rearrangement are common in higher plants. Terpenoid precursors based on podocarpane, pimarane, abietane, and phyllocladane-type carbon skeletons are especially abundant in plant resins and fossil resin deposits and are potential sources of isohexyl alkylnaphthalenes in crude oils. Isohexyl alkylnaphthalenes provide an age-diagnostic capability in that the greatest source of the diterpane biomarkers in most samples has been attributed to conifers. Modern conifer species such as the *Araucariaceae*, *Podocarpaceae*, and *Cupressaceae* began to emerge in the Late Triassic Period and dominated land flora during the Jurassic and Cretaceous period.

Polynuclear aromatic hydrocarbons

Common in carbonaceous chondrites; can be specific for paleo-fires. Measured using SIM/GCMS or gas chromatography on aromatic fractions.

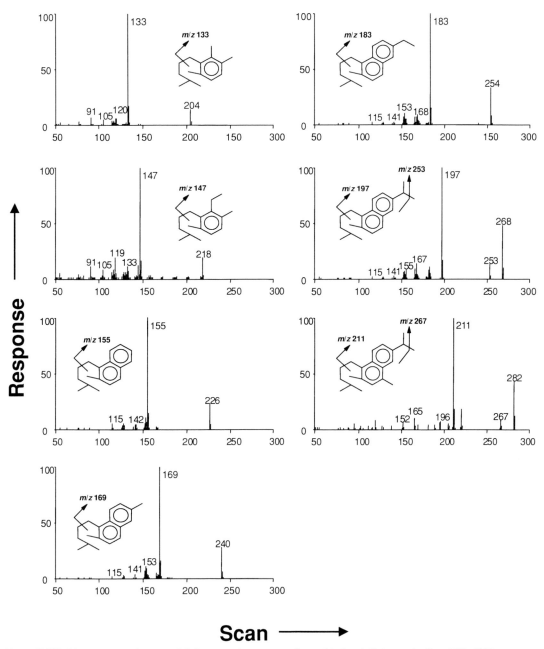

Figure 13.125. Mass spectra and suggested β-fragmentation patterns of several isohexyl alkylaromatics (from Ellis, 1995).

PAHs are markers for biological activity, although they can also form by abiogenic processes. Naraoka et al. (2000) measured carbon isotopic compositions for more than 70 PAHs from the Asuka-881458 carbonaceous chondrite from naphthalene to benzo(ghi)perylene, where fluoranthene and pyrene were the most abundant. Carbon isotopic compositions of individual PAHs range from −26 to 8‰. More condensed PAHs are more

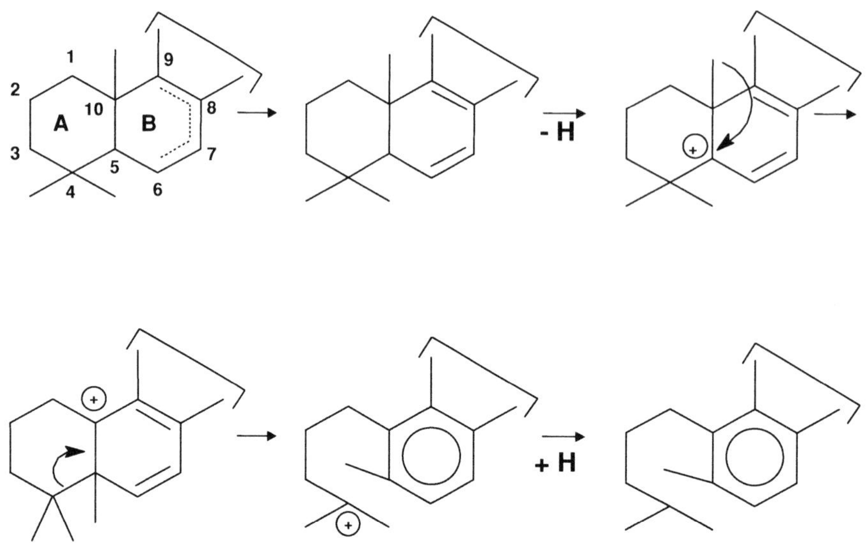

Figure 13.126. Proposed rearrangement-aromatization of terpenoid precursors leading to formation of A-ring opened isohexyl alkylaromatics (from Ellis et al., 1996a).

depleted in ^{13}C as the atomic H/C ratio decreases. The carbon isotope distribution of PAHs containing more than three rings in the Asuka sample is similar to that from the Murchison meteorite but clearly different from that of the terrestrial PAHs. McKay et al. (1996) addressed four aspects of PAHs in the Allen Hills meteorite ALH84001 that suggested their derivation from Martian bacteria. However, as discussed in the text, contamination also could explain these compounds.

Soot from the Cretaceous-Tertiary boundary layer contains pyrosynthetic PAHs, such as retene, coronene, benzoperylene, and benzopyrene (Gilmour and Guenther, 1988; Venkatesan and Dahl, 1989). Arinobu et al. (1999) found an abrupt 1.4–1.8‰ decrease in the δ^{13}C of the sedimentary nC_{29} alkane that parallels sharp increases in various pyrosynthetic PAHs (see Figure 17.26).

Many PAHs generate prominent molecular ions at m/z 252, including 1,2- and 3,4-benzopyrene (see Figure 8.20). Accurate detection and quantitation of PAHs is important in studies of anthropogenic sources for these compounds. Many PAHs in cigarette smoke, such as benzo(a)pyrene, are known carcinogens.

PORPHYRINS

The following is a short discussion of porphyrins to supplement other discussion in the text. More detailed discussions of porphyrins can be found in various publications (Sundararaman et al., 1988a; Sundararaman et al., 1988b; Baker and Louda, 1983; Baker and Louda, 1986; Louda and Baker, 1986; Chicarelli et al., 1987; Gibbison et al., 1995; and Rosell-Melé et al., 1999).

Porphyrins are tetrapyrrolic compounds that occur naturally as metal complexes (e.g. nickel, vanadium, iron) or free-base species. Organometallic porphyrins account for much of the vanadium and nickel in petroleum (Boduszynski, 1987). Deoxophylloerythroetioporphyrin (DPEP) and etioporphyrin structures are the most common tetrapyrroles (Figure 3.24). These compounds are particularly resistant to biodegradation. More details on porphyrins and their use as maturity parameters are discussed below and in Sundararaman et al. (1988a; 1988b).

Porphyrins are complex and show large numbers of isomers (Barwise and Whitehead, 1980). Metalloporphyrins can be isolated from crude oils or rock extracts and demetallated before analysis (Eckardt et al., 1989). An easily reproducible, practical method for fingerprinting these compounds is based on direct HPLC separation of vanadyl porphyrins without demetallation (Sundararaman, 1985).

Routine application of petroporphyrins to geochemical problems has been limited for several reasons. The high molecular weights and low volatilities of these compounds generally preclude their separation using the GCMS approach. Consequently, many analyses involve direct insertion of porphyrins into the mass spectrometer using a solids probe followed by electron-impact ionization and mass analysis (e.g. Quirke et al., 1989; Beato et al., 1991). This requires large amounts of sample and provides no isomer-specific information. Furthermore, it is insufficient to reveal the complexity of natural porphyrin compositions because of interference by fragments from higher-molecular-weight porphyrins. Recent advances in liquid chromatography and tandem mass spectrometry allow rapid, detailed analyses of free-base porphyrins (Rosell-Melé et al., 1999).

New developments in analytical techniques may allow more routine analyses of porphyrins. For example, Gallegos et al. (1991) used commercially available fused-silica columns to chromatographically separate porphyrins and record complete mass spectra without demetallation. For this new approach, the authors employed high-temperature gas chromatography/electron-impact mass spectrometry (HTGC/EIMS) and high-temperature gas chromatography/field ionization mass spectrometry (HTGC/FIMS) to analyze C_{28}–C_{33} etio- and deoxophylloerythroetioporphyrins isolated from Boscan crude oil and a Monterey source rock. Compound-specific isotope analysis of porphyrins provides additional information on their origins (Yu Zhiqiang et al., 2000a).

Porphyrins can be analyzed using liquid chromatography/mass spectrometry (LCMS). Advances in LCMS have been impeded by the problem of delivering the liquid effluent from the column to the vacuum system of the mass spectrometer. A technology called thermospray is the most popular LCMS interface, but it permits only mild chemical ionization of analytes (compounds to be analyzed). Such chemical-ionization spectra do not show the rich fragmentation patterns of electron-impact spectra that make it possible to identify unknown compounds. Newly developed particle-beam LCMS interfaces are more compatible with different ionization sources than other interfaces. Electron-impact, chemical-ionization, and other ionization techniques can be used.

Supercritical fluid chromatography/mass spectrometry (SFCMS) is another technique with potential for porphyrin analysis. This technique has been applied using capillary gas chromatographic columns (Campbell et al., 1988) and may allow many of the benefits obtained using GCMS or GCMS/MS methods for hydrocarbons.

Many alkyl substitution patterns around the porphyrin nucleus were elucidated, and the structures were related to chlorophyll d in green photosynthetic bacteria (*Chlorobiaceae*), chlorophyll c in certain species of eukaryotic algae, or chlorophyll a, which is widespread in eukaryotic algae and higher plants (Ocampo et al., 1984; 1985a; 1985b). Some of the isolated porphyrins from Messel Shale have also been related to specific groups of organisms by stable carbon isotope ratios (Hayes et al., 1987). However, progress in the application of petroporphyrins to source rocks and oils is slow due to the laborious methods necessary for their isolation. Identification of peaks on HPLC chromatograms using authentic standards will be necessary for further progress. New technologies, such as LCMS and SFCMS may ultimately allow routine application of porphyrins to geochemical studies. Rosell-Melé et al. (1996) described the advantages and disadvantages of analysis of metallated versus demetallated porphyrins. They used HPLC/atmospheric pressure chemical ionization (HPLC/APCI) to identify more than 50 significant components in the demetallated vanadyl porphyrin fraction from the Triassic Serpiano oil shale. Trace amounts of cycloalkano porphyrins ($>C_{33}$) indicated the occurrence of photic zone anoxia in the water column during deposition of the oil shale (discussed below).

V/(V + Ni) porphyrins

> Related to redox conditions in source-rock depositional environment. Measured using HPLC methods.

The proportions of vanadyl and nickel porphyrins are used as a source parameter in oil–oil and oil–source rock correlations (Lewan, 1984). Vanadium and nickel are the major metals in petroleum (Boduzynski, 1987) but are not part of the original tetrapyrrole pigments in living organisms. These metals enter into the porphyrin structure by chelation during early diagenesis (Figure 3.24), and the depositional environment strongly influences their relative proportions (Lewan, 1984).

Lewan (1984) proposed that in marine sediments, Ni^{2+} and VO^{2+} in solution in the pore waters compete for chelation with free-base porphyrins. Under normal oxic conditions, nickel reacts more readily than vanadium with free-base porphyrins. However, under low Eh conditions, sulfate-reducing bacteria generate hydrogen sulfide. High sulfide in the pore water of anoxic sediment causes nickel ion (Ni^{2+}) to precipitate as nickel sulfide, leaving vanadyl ion (VO^{2+}) to complex with available free porphyrins. Low $V/(V + Ni)$ porphyrin ratios in marine Toarcian rocks reflect oxic to suboxic conditions, while high ratios reflect anoxic sedimentation (Moldowan et al., 1986).

Nickel porphyrins generally predominate in lacustrine rocks and related oils, while vanadium is low. However, one exceptional lacustrine source rock from the Cretaceous Bucomazi Formation in west Africa is dominated by vanadyl porphyrins in the most organic-rich sections. Sundararaman and Boreham (1991) explain the high $V/(V + Ni)$ porphyrin ratios (up to 0.9) in extracts from parts of the Bucomazi Formation as due to the combined influence low Eh and pH. They observed that two porphyrin ratios decrease uphole in a Bucomazi core: $V/(V + Ni)$ and 3-nor C_{30} DPEP/(3-nor C_{30} DPEP + C_{32} DPEP). Low Eh and pH favor vanadyl porphyrins. The decreasing $V/(V + Ni)$ porphyrin ratios are thought to reflect increasing oxicity during evolution of the lake. However, because the $V/(V + Ni)$ and 3-nor C_{30} DPEP/(3-nor C_{30} DPEP + C_{32} DPEP) ratios show different sensitivities to second-order cycles of deposition, they appear to be controlled by two different factors.

The ratio 3-nor C_{30} DPEP/(3-nor C_{30} DPEP + C_{32} DPEP) is controlled partly by maturity but is sensitive to pH of the source-rock depositional environment. Increased pH apparently suppresses the devinylation reaction that leads from C_{32} DPEP to 3-nor C_{30} DPEP. The results imply that the pH of the lake increased gradually with time, which is supported by more marl and less organic-rich shale and mudstone in the upper parts of the Bucomazi core.

Porphyrin distributions

> Measured using HPLC. Individual porphyrins can be identified by retention time and comparison with standards, or by LCMS.

The complexity of porphyrin fingerprint patterns (Sundararaman, 1985) has largely defied specific application to correlation problems. Some of the porphyrins show specific links to eukaryotes or prokaryotes (Hayes et al., 1987). For example, porphyrins ranging from C_{34} to C_{36} with extended side chains can be identified in HPLC fingerprints and originate from photosynthetic bacteria (Ocampo et al., 1985a; 1985b). Other porphyrins with a rearranged exocyclic five-member ring can be related to algae, such as dinoflagellates (Ocampo et al., 1984).

Michael et al. (1990) correlated oils, source rocks, and heavily biodegraded tar-sand bitumens from the Ardmore and Anadarko basins using various parameters, including tricyclic terpane, C_{24} tetracyclic terpane, hopane, mono- and triaromatic steroid hydrocarbon, and porphyrin distributions. Chicarelli et al. (1987) and Callot et al. (1990) describe sedimentary porphyrins with structures providing clear evidence of specific precursor chlorophylls or bacteriochlorophylls, while others are not obviously related to known pigments. One C_{32} porphyrin isolated from gilsonite contains a methyl-substituted, five-member exocyclic ring. Although of unknown origin, this compound may be a marker for lacustrine settings.

Cycloalkanoporphyrins with extended alkyl substitution

> Analysis of metallo- or free-base porphyrins by HPLC/MS/MS using atmospheric-pressure chemical ionization (e.g. Rosell-Melé et al., 1999). Base peaks in the chemical-ionization spectra of the C_{32}–C_{34} cycloalkanoporphyrins are at m/z 477, 491, 505, and 519, respectively.

High-molecular-weight cycloalkanoporphyrins (CAP) ($>C_{32}$) with extended ($>C_2$) alkyl substitution in crude oils or rock extracts indicate an origin from bacteriochlorophyll d in green sulfur bacteria, thus indicating photic zone anoxia in the source rock (Keely and Maxwell, 1993; Gibbison et al., 1995). Examples include carboxylic acids (series a–c) derived from one of the bacteriochlorophyll d series with ethyl substitution at R_2 (Figure 13.127). Likewise, a C_{34} CAP (Figure 13.127) derived from a second series of bacteriochlorophyll d occurs in marls from the evaporitic Oligocene Mulhouse Basin in Alsace, France. Analysis of these compounds in complex natural porphyrin mixtures is accomplished most reliably using liquid chromatography

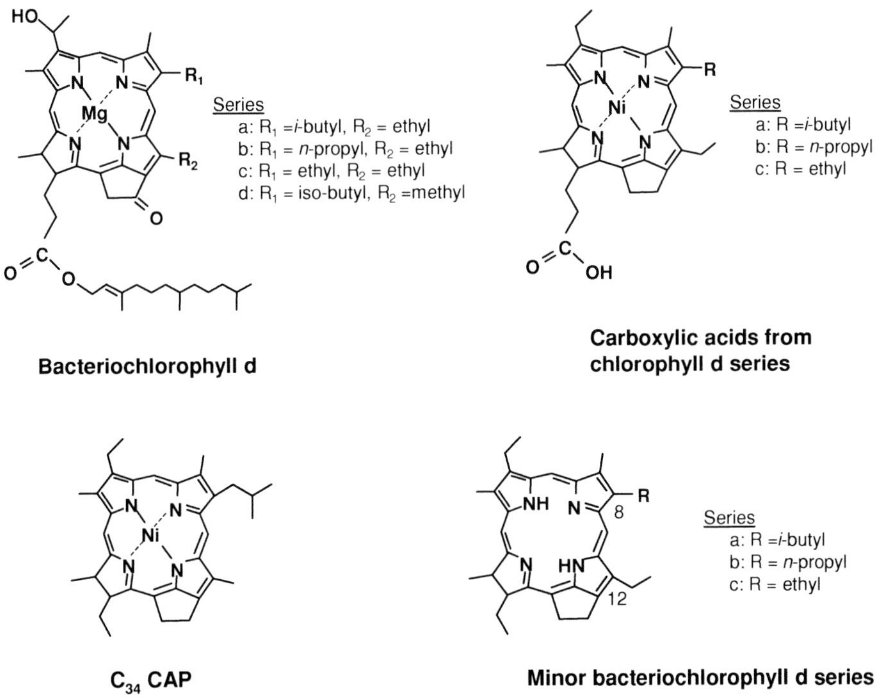

Figure 13.127. The iron porphyrin fraction of the extract from Permian Kupferschiefer source rock contains three porphyrins indicating photic zone anoxic conditions during deposition (lower right, series a–c). The ethyl group at C-12 and i-butyl, n-propyl, or ethyl group at C-8 in the structures indicate a specific origin of these cycloalkanoporphyrins from bacteriochlorophyll d (upper left), specific for green sulfur bacteria (*Chlorobiaceae*), which are obligate anaerobes (modified from Gibbison et al., 1995). Grice (2001) suggests a possible isopropyl group at R_2; see Figure 13.128. Other examples of products from bacteriochlorophyll d include carboxylic acids from the lacustrine Messel oil shale (Eocene, Germany) and a C_{34} cycloalkanoporphyrin. From marls in the evaporitic Mulhouse Basin (Oligocene, France).

(HPLC), atmospheric-pressure chemical ionization (APCI), and tandem mass spectrometry (MS/MS), i.e. HPLC/APCI/MS/MS (Rosell-Melé et al., 1999).

Maleimides

> Isolated from chromic acid oxidation products of porphyrins or from the polar fraction of crude oils or extracts. Typically measured by GCMS as sily-late derivatives. Biomarkers for bacteriochlorophylls c/d/e in green sulfur bacteria and chlorophyll a in phytoplankton (Grice, 2001).

Quirke et al. (1980) used chromic acid oxidation to investigate the alkyl substituents in sedimentary porphyrins. They found that demetallated vanadyl porphyrins from the Cretaceous Boscan crude oil gave a mixture of maleimides (1H-pyrrole-2,5-diones). More recently, maleimides isolated from the polar fraction of the marine Permian Kupferschiefer Shale show a simple distribution dominated by methyl-methyl and methyl-ethyl compounds, apparently due to their origin from the macrocycle of chlorophyll a in phytoplankton (Grice et al., 1996a; 1997). Methyl i-butyl maleimide occurs in relatively lower abundance and is thought to originate from the macrocycle of bacteriochlorophylls from *Chlorobiaceae* (anoxygenic green sulfur bacteria) (Figure 13.128). C_{34} and C_{35} porphyrins consistent with derivation from bacteriochlorophyll d were detected in these samples (Crawford, 1998; Pancost et al., 2002). The methyl ethyl maleimide was 3.5–4.5‰ enriched in ^{13}C compared with phytane from the phytyl side chain of chlorophyll. A similar difference of ~4.5‰ was observed between sedimentary

Table 13.9. *Source-related geochemical data for Monterey crude oils from the Santa Maria Basin*

Oil	Steranes			Monoaromatic steroids			
	%C_{27}	%C_{28}	%C_{29}	%C_{27}	%C_{28}	%C_{29}	$\delta^{13}C_{PDB}$ (‰)
547	41	44	16	28	53	19	−16.94
548	43	42	15	32	44	24	−16.49
549	39	44	18	28	55	17	−17.59
611	41	43	16	31	45	24	−16.51
612	39	45	16	28	53	20	−16.81
613	40	44	16	25	58	17	−17.99
614	38	46	17	29	53	18	−16.89
615	40	43	16	29	55	16	−17.18
616	44	40	16	29	53	18	−16.94

%C_{27} = %$C_{27}/(C_{27} + C_{28} + C_{29})$ regular steranes.

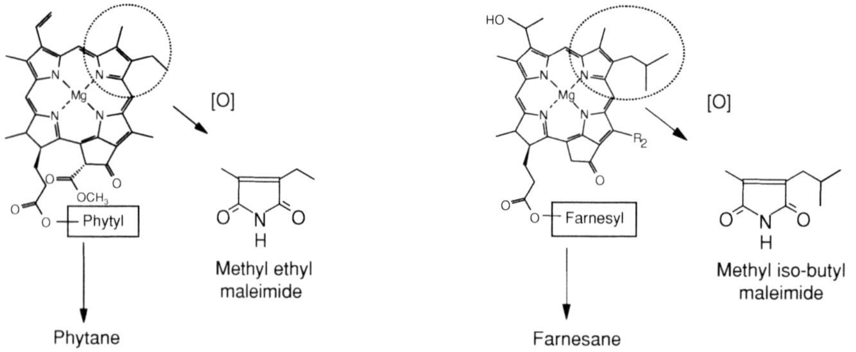

Figure 13.128. Origin of methyl ethyl maleimide (∼−27‰) and phytane (∼−31‰) from chlorophyll a in phytoplankton (algae or cyanobacteria using the C3 fixation pathway, left) and origin of methyl iso-butyl maleimide (∼−15‰) and farnesane (∼−18‰) from bacteriochlorophyll d in *Chlorobiaceae* (green sulfur bacteria) using reversed Krebs tricarboxylic acid cycle (right) (from Grice, 2001).

porphyrins derived from photosynthetic pigments and phytane (Hayes *et al.*, 1990). Methyl *i*-butyl maleimide is enriched in ^{13}C (∼18‰) compared with methyl ethyl maleimide (∼−27‰), which results from carbon fixation by means of the reversed tricarboxylic acid cycle in green sulfur bacteria.

Maleimides may form during various stages of diagenesis. They may be generated within the water column by photo-oxidation or enzymatic oxidation of chlorophylls, which may occur even within a euxinic water column (Magness, 2001). Maleimides also could form during weathering and erosion of source rocks or may even arise by oxidation during sample storage (Grice *et al.*, 1996a). Because of these potential artifacts, use of maleimides as indicators of an oxic or euxinic photic zone should be coupled with supportive biomarkers (e.g. isorenieratenes, porphyrins) and δ^{13}C analysis.

EXERCISE

Generate C_{27}–C_{28}–C_{29} ternary diagrams of sterane and monoaromatic steroid composition for crude oils from the offshore Santa Maria Basin, California in Table 13.9 (Peters *et al.*, 1990; Peters and Moldowan, 1991). Could the oils be related genetically based on their ternary plot locations and the stable carbon isotope data?

Because the relative concentration of specific isomers may depend on more than the initial input of steroids, some ternary diagrams of steroid composition use the sum of several isomers. We use the sum of the $5\alpha,14\alpha,17\alpha(H)$ 20S and 20R and $5\alpha,14\beta,17\beta(H)$ 20S and 20R peaks obtained from GCMS/MS analysis of the four major epimers of the C_{27}, C_{28}, and C_{29} steranes in petroleum.

The accuracy of the $\%C_{27}/(C_{27}-C_{29})$, $\%C_{28}/(C_{27}-C_{29})$, and $\%C_{29}/(C_{27}-C_{29})$ steranes used in the ternary diagrams depends on separating the individual carbon numbers from interfering peaks. It is possible to obtain information on C_{27}, C_{28}, and C_{29} steranes in petroleum using conventional SIM-mode GCMS, but the specificity of sterane analysis improves using GCMS/MS. It is unwise to construct ternary diagrams using data from different instruments (e.g. GCMS versus GCMS/MS) or from the same instrument analyzed at different times, unless the data have been calibrated using standards.

High concentrations of C_{29} steranes (24-ethylcholestanes) compared with the C_{27} and C_{28} steranes may indicate a land-plant source (e.g. Czochanska et al., 1988). This interpretation is based on the work of Huang and Meinschein (1979), who observed high C_{29}-sterol predominance in higher plants and sediments. However, as recommended by Volkman (1986; 1988), caution should be applied in interpreting C_{29}-sterol predominance in recent sediments. For example, Volkman et al. (1981) showed that 24-ethylcholest-5-en-3β-ol (a C_{29} sterol) is a significant component in a mixed diatom culture. The utility of the C_{29} sterol as a terrigenous marker is, thus, questionable. Furthermore, many Paleozoic and older oils and some oils from carbonate source rocks contain high C_{29} steranes but little or no higher-plant input (e.g. Moldowan et al., 1985; Grantham, 1986a; Rullkötter et al., 1986; Vlierbloom et al., 1986; Fowler and Douglas, 1987; Buchardt et al., 1989). Land plants were absent and cannot account for C_{29} steranes in rocks or oils older than Devonian in age. In Precambrian crude oils from southern Oman, Grantham (1986a) attributed strong C_{29}-sterane predominance to algae.

14 · Maturity-related biomarker parameters

> This chapter explains how biomarker analyses are used to assess thermal maturity. The parameters are arranged by groups of related compounds in the order (1) terpanes, (2) polycadinenes and related products, (3) steranes, (4) aromatic steroids, (5) aromatic hopanoids, and (6) porphyrins. Critical information on specificity and the means for measurement are highlighted before the discussion of each parameter.

Thermal maturity describes the extent of heat-driven reactions that convert sedimentary organic matter into petroleum. For example, kerogen in fine-grained source rocks can be converted thermally to oil and gas, which migrate to coarser-grained reservoir rocks (Figure 1.2). Early diagenetic processes convert bacterial and plant debris in sediments to kerogen (insoluble, particulate organic matter) and bitumen (extractable organic matter). Thermal processes generally associated with burial then convert part of this organic matter to petroleum and, ultimately, to gas and graphite. Petroleum is a complex mixture of metastable products that evolve toward greater thermodynamic stability during maturation.

> Note: It is thought that both kerogen and oil are unstable during catagenesis and progressively decompose to pyrobitumen and gases (e.g. Hunt, 1996; Tissot and Welte, 1984). Mango (1991) proposed that hydrocarbons in oil are much more thermally stable than their kerogenous precursors. He believes that oil and gas originate by direct thermal decomposition of kerogen but that hydrocarbons in oils do not thermally decompose to gas in the Earth. This scenario does not exclude some oxidative decomposition of hydrocarbons during thermochemical sulfate reduction (e.g. Orr, 1974; Krouse *et al.*, 1989).

Potential petroleum source rocks are described in terms of the quantity, quality, and level of thermal maturity of the organic matter. A potential source rock contains adequate amounts of the proper type of dispersed kerogen to generate significant amounts of petroleum but is not yet thermally mature. A potential source rock becomes an effective source rock only at the appropriate levels of thermal maturity (i.e. within the oil-generative window).

In general terms, organic matter can be described as immature, mature, or postmature, depending on its relation to the oil-generative window (Tissot and Welte, 1984). Immature organic matter has been affected by diagenesis, including biological, physical, and chemical alteration, but without a pronounced effect of temperature. Mature organic matter has been affected by catagenesis, the thermal processes covering the temperature range between diagenesis and metagenesis. As used in this book, catagenesis is equivalent to the oil-generative window. Postmature organic matter has been heated to such high temperatures that it has been reduced to a hydrogen-poor residue capable of generating only small amounts of hydrocarbon gases.

Recognizing the need to describe accurately the thermal maturity of sedimentary organic matter, organic geochemists developed various thermal maturity parameters. Conventional geochemical methods used to assess source-rock maturity include Rock-Eval pyrolysis, compound class distributions, vitrinite reflectance (R_o), thermal alteration index (TAI) (spore coloration), and carbon preference index (CPI). However, few of these parameters can be applied to crude oils. Molecular parameters based on ratios and distributions of specific biomarkers have found increased use in studies of thermal maturity since 1970, as discussed in the remainder of this chapter.

Two types of thermal maturity parameters exist: (1) generation or conversion parameters used as indices of the stage of petroleum generation (independent of the magnitude of thermal stress) and (2) thermal stress parameters used to describe relative effects of temperature/time. For example, two rocks containing different types of kerogen might generate equivalent amounts of oil at a given atomic H/C, but the vitrinite reflectance of the samples may differ. In this hypothetical case, the atomic H/C is linked to hydrocarbon generation while vitrinite reflectance is linked only to thermal stress (however, see later discussion). For this reason, the reflectance associated with the threshold of oil generation can vary between different rocks. Vitrinite reflectance of about 0.6% is accepted widely as indicating the start of oil generation in most source rocks (Dow, 1977; Peters, 1986).

Figure 14.1 shows the importance of thermal maturity assessment as a predictive tool. All significant oil accumulations occur within 50 km updip of the thermally mature pod of active Devonian-Lower Carboniferous New Albany Shale source rock, as defined by vitrinite reflectance of 0.6% (Demaison, 1984). R_o measures thermal stress rather than generation or conversion. For example, source rock in the Monterey Formation generates significant amounts of oil at R_o as low as 0.3% (Isaacs and Petersen, 1988), while comparable levels of petroleum generation in the Green River marl are not reached until R_o approaches 0.7% (Tissot et al., 1978).

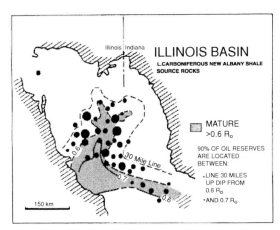

Figure 14.1. Map shows the importance of thermal maturity as measured by vitrinite reflectance (R_o) for oil generation from Devonian-Lower Carboniferous New Albany Shale source rocks (Demaison, 1984). The single point in the upper left portion of the figure may represent part of a different petroleum system. Alternatively, if this oil originated from New Albany Shale, then it may be a key sample for better understanding the distribution of petroleum within this system. Reprinted by permission of the AAPG, whose permission is required for further use.

Reliable assessment of the thermal maturity of organic matter typically requires integrating both biomarker and non-biomarker maturity data. Figure 14.3 shows the general correlation between the biomarker and non-biomarker maturity parameters described below. Because many of the maturity parameters are related to thermal stress rather than generation as defined above, relationships between these parameters and the oil generation window can be only approximate.

CRITERIA FOR BIOMARKER MATURITY PARAMETERS

The ideal molecular maturity parameter is based on measuring the relative concentrations of reactant (A) and product (B) in the following reaction:

$$A \underset{r2}{\overset{r1}{\rightleftharpoons}} B$$

A convenient maturity parameter for expressing the extent of the above reaction is the ratio of the concentration of B to the sum of A and B. Unlike B/A, the ratio B/(A + B) can range only from 0 to 1 (or 0 to 100%) with increasing thermal maturity. The terms $r1$ and $r2$ represent the rates of the forward and reverse reactions, respectively. In the ideal case, the following conditions are satisfied:

- A and B are single compounds, the reaction from A to B is irreversible, or the forward rate of reaction, $r1$, is much greater than the reverse rate, $r2$.
- The initial concentration of B is zero, while easily measured amounts of A are present in all samples before heating.
- A is transformed by heat only to B.
- B is thermally stable and is formed only from A.
- Conversion of A to B occurs in the range of maturity of interest for petroleum generation.

Most biomarker maturity parameters do not satisfy the ideal conditions listed above. In many known reactions, A or B can be derived from or degraded to other reactants and products. Some B may be present before heating begins, and the conversion of A to B may never reach 100%, even at very high maturity. Furthermore, few biomarker maturity parameters can be used throughout the oil-generative window, and all biomarkers are eventually degraded thermally to simpler hydrocarbons.

Proposed equilibrium reactions are commonly employed as biomarker maturity parameters. In contrast to the ideal irreversible reaction of A to B discussed above, the rates of forward and reverse reactions ($r1$ and $r2$) become equal, or reach equilibrium, at a given level of maturity. Thus, after reaching equilibrium, no further information on maturity is available because the ratio $B/(A + B)$ remains constant with further heating.

Compounds not related directly as precursor and product are also used in relative maturity assessments. Their ideal relationship is shown by the following reaction:

$$B \leftarrow C \rightarrow A$$

where A and B are biomarkers related to a common precursor C. During maturation, A and B degrade or change into other compounds at different rates, and B is more stable than A toward these conversions. Although relative maturity can still be expressed using the ratio $B/(A + B)$, there is greater potential for interference with the ratio of products A and B by variables unrelated to maturity than in the simpler case where A and B are the precursor and product, respectively. For example, precursor C might be a family of compounds, and the initial formation of A and B might vary according to source organic matter input, conditions in the depositional environment, and early diagenesis.

MATURITY ASSESSMENT

Caution must be applied when comparing biomarker parameters from different laboratories. For example, biomarker maturity ratios with the same name (e.g. %20S) may be quantified using different peaks by different laboratories. Sample preparation, instrumentation, and column performance can also vary substantially between laboratories.

Assessment of the level of thermal maturity of bitumens and oils assists in correlation studies. Migrated oil in sandstone that is more thermally mature than indigenous bitumen from surrounding shales clearly could not have originated from the shales. Sandstones and other reservoir rocks commonly show anomalously high bitumen/TOC, production index, and S1, indicating migrated oil. Tests to determine whether bitumen is indigenous are described below. For example, biomarker maturity parameters for shales in the Shengli oilfield in China show good correlation with depth, while oils do not (Shi Ji-Yang et al., 1982). This suggests that maturation of the source rock exerts greater control on the oil maturity than the extent of reservoir maturation.

Reduced biomarker concentrations commonly indicate increased maturity (e.g. Mackenzie et al., 1985a) and signal to the interpreter that special care should be taken when interpreting results (e.g. see Figure 8.30). Where biomarkers are low, mass chromatograms should be examined to verify quantified biomarker peak ratios. A general idea of the relative thermal maturity of oils or bitumens can be gained by plotting total steranes versus terpanes or 17α-hopanes (measured in parts per million), especially when the samples are related.

Note: Few oils are so mature that cyclic biomarkers are absent. One possible example is an oil seep from the billion-year-old Nonesuch Shale in the White Pine copper mine, Michigan, which contains no steranes or terpanes (Hoering, 1978; Imbus et al., 1988). However, other oils from the Nonesuch Shale have biomarkers (Pratt et al., 1991). These observations may be explained by oxidative degradation of the biomarkers during ore deposition. Local precipitation of copper sulfides, native copper, and silver in the White Pine mine appears to have occurred by reduction of hot, metalliferous brines and the associated oxidation of organic matter (Ho et al., 1990). Blueberry Debolt oils from the Western Canada Basin are highly unusual because they lack significant amounts of steranes and terpanes but do not appear to be overly mature (Snowdon et al., 1998a).

Some of the more reliable biomarker parameters, such as C_{29}-sterane isomerization ratios, are not useful for samples showing maturities beyond peak oil generation ($R_o \sim 0.9\%$) because the reactions they represent have reached an equilibrium or endpoint. Biomarker ratios that can be used to assess high levels of maturity include the side-chain cleavage ratios for the mono- and triaromatic steroids. Van Graas (1990) described some biomarker parameters that are useful at high thermal maturity (e.g. above the equivalent of vitrinite reflectance 1.0%). These include concentrations of steranes and hopanes, and tricyclics/17α-hopanes, diasteranes/steranes, and $Ts/(Ts + Tm)$ ratios.

Dahl et al. (1999) described a method to estimate the thermal maturity of any liquid hydrocarbon sample

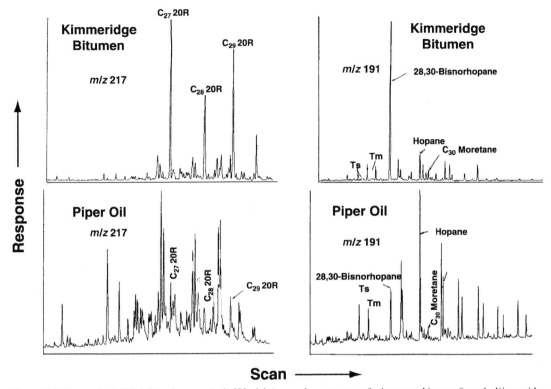

Figure 14.2. Sterane (m/z 217, left) and terpane (m/z 191, right) mass chromatograms for immature bitumen from the Kimmeridge Shale and those for a mature, related oil from the Piper Field in the North Sea. Biomarker and isotopic analyses allowed correlation of these samples despite their differing thermal maturities (Peters et al., 1989). Reprinted with permission by ChevronTexaco Exploration and Production Technology Company, a division of Chevron USA Inc.

based on stigmastane and diamondoid concentrations. This method is particularly useful for highly mature condensates, where other methods can be unreliable. The method relates the extent of cracking (conversion of liquid hydrocarbons to gas and pyrobitumen) to diamondoid concentrations. The extent of cracking can be related further to vitrinite reflectance by a calculation called EASY %R$_o$ (Sweeney and Burnham, 1990; Dahl et al., 1999).

Biomarkers are typically measured using selected ion monitoring (SIM)/GCMS, which is also called multiple ion detection (MID)/GCMS. The SIM approach allows a rapid, general assessment of the thermal maturity of samples. Figure 14.2 shows examples of sterane and terpane mass chromatograms for immature bitumen and mature, related oil. The immature Kimmeridge bitumen in Figure 14.2 has simple mass chromatograms compared with the mature Piper oil. This is due mainly to thermally induced stereoisomerization of the comparatively simple distribution of biological epimers of various biomarkers inherited from organisms to a more complex mixture including geological epimers. For example, the C_{27}–C_{29} steranes in the bitumen (Figure 14.2, top left) are dominated by 20R stereochemistry. Thermal maturation yields a complex mixture of epimers (bottom left).

Note: 28,30-Bisnorhopane (BNH), the dominant peak on the m/z 191 trace of immature Kimmeridge bitumen, behaves somewhat differently during thermal maturation than other compounds in Figure 14.2. BNH is not generated from chemically bound precursors in kerogen but originates as a component of the free bitumen. Thermal maturation generates other terpanes during formation of Kimmeridge oil, which dilute the concentration of BNH measured in the resulting oil (bottom of Figure 14.2). Figure 14.10

gives another example of reduced BNH with increasing maturity.

Biomarker compound classes can also be analyzed using some form of GCMS/MS (see Chapter 8), which reduces interference and allows detection and quantitation at lower concentrations than the SIM approach. Linked scanning is a form of GCMS/MS that can be completed using:

- a double-focusing magnetic sector instrument (e.g. MRM/GCMS);
- a multisector instrument, such as the triple-sector quadrupole (TSQ), by monitoring collision-activated decompositions (CAD);
- an ion trap;
- a hybrid instrument combining magnetic, trap and quad-rupole, or time-of-flight sectors.

Ratios of certain saturated and aromatic biomarkers are some of the most commonly applied thermal maturity indicators. These indicators result from two types of reactions: (1) cracking reactions (including aromatizations) and (2) configurational isomerizations at certain asymmetric carbon atoms. While both types of indicators are used, isomerizations are applied more commonly. For example, one of the more reliable cracking reactions is the conversion of monoaromatic to triaromatic steroids. This reaction is applied less commonly for maturity assessment than isomerizations, partly because of the more tedious quantitation procedures necessary to measure aromatic steroids. The two most commonly used isomerizations are those involving hydrogen atoms at C-22 in the hopanes and C-20 in the steranes, as discussed below.

Several other biomarker thermal maturity parameters supplement the isomerizations. These include the moretane/17α-hopane [17β,21α/17α,21β] ratio, the 18α-trisnorneohopane/17α-trisnorhopane ratio (Ts/Tm, usually presented as Ts/(Ts + Tm)), and ratios of aromatic steroids with short side chains to those with long side chains (side-chain cleavage reactions). In general, these ratios are used to support the more commonly used biomarker ratios or when other maturity indicators are unavailable or unreliable.

Figure 14.3 shows the approximate ranges of various biomarker thermal maturity parameters relative to the oil-generative window. All plots that relate reaction extent to the stage of oil generation are approximate

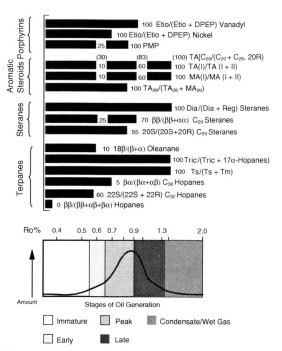

Figure 14.3. Biomarker maturation parameters respond in different ranges of maturity and can be used to estimate the maturity of crude oils or source-rock extracts relative to the oil window. Estimates are approximate because most parameters depend on temperature and time (thermal stress) rather than the amount of petroleum generated (from Peters and Moldowan, 1993). Approximate ranges of biomarker maturity parameters are shown versus vitrinite reflectance and a generalized oil-generation curve (after Mackenzie, 1984). Variations of ± 0.1% reflectance for biomarker ratios are common, and even greater variations can occur. The solid bars indicate the range for each ratio with respect to the stages of oil generation. The ratio reaches a constant percentage value indicated by the number at the end of the bar. This number is a maximum value for most of the parameters, except $18\beta/(18\alpha + 18\beta)$-oleanane, $\beta\alpha/(\alpha\beta + \beta\alpha)$-hopanes ($C_{30}$) and $\beta\beta/(\beta\beta + \alpha\beta + \beta\alpha)$-hopanes, where the numbers 10, 5, and 0, respectively, are minimum values. For a few ratios, intermediate values are shown within the bars to indicate earlier stages of maturity. However, the solid bars are not calibrated scales and do not indicate a linear change in each parameter up to its maximum value (e.g. the reader is cautioned against inferring vitrinite reflectance from a given biomarker ratio). All values and ranges with respect to oil generation and vitrinite reflectance are approximate and may vary with heating rate, lithofacies, and organic facies of the source rock. Reprinted with permission by ChevronTexaco Exploration and Production Technology Company, a division of Chevron USA Inc.

because biomarker thermal maturity parameters depend primarily on temperature and time (thermal stress) rather than the amount of petroleum produced (generation).

Note: To simplify the above discussion, variations in many biomarker maturity ratios were attributed to cracking or isomerization. However, in many cases, the use of these terms is not strictly correct. For example, changes in the ratio TA(I)/TA(I + II) are commonly described as due to side-chain cleavage or cracking of long- [TA(II)] to short- [TA(I)] chain triaromatic steroids. However, the relative abundance of TA(I) and TA(II) is controlled at least partly by differential thermal stability (Beach et al., 1989). The factors controlling each maturity ratio are discussed below.

TERPANES

22S/(22S + 22R) homohopane isomerization

> High specificity for immature to early oil generation (see Figure 14.3). Measured using m/z 191 chromatogram or GCMS/MS typically using the C_{31} or C_{32} homologs.

Isomerization at C-22 in the C_{31}–C_{35} 17α-hopanes (Ensminger et al., 1977) occurs earlier than many biomarker reactions used to assess the thermal maturity of oil and bitumen, such as isomerization at C-20 in the regular steranes. Schoell et al. (1983) showed that equilibrium for the C_{32} hopanes occurs at vitrinite reflectance of ~0.5% in Mahakam Delta rocks. The biologically produced hopane precursors carry a 22R configuration that is converted gradually to a mixture of 22R and 22S diastereomers (Figure 14.4). The proportions of 22R and 22S can be calculated for any or all of the C_{31}–C_{35} compounds. These 22R and 22S doublets in the range C_{31}–C_{35} on the m/z 191 mass chromatogram are called homohopanes (Figure 8.21, peaks 22, 23, 25, 26, 29, 30, 32, 33, 34, 35). For comparison, the reader may wish to identify these peaks on the m/z 191 chromatogram of the mature Piper oil in Figure 14.2.

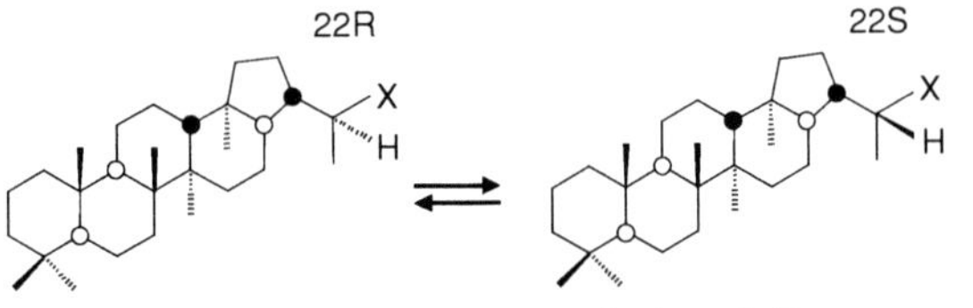

The 22S/(22S + 22R) ratios for the C_{31}–C_{35} 17α-homohopanes may differ slightly. Typically, the C-22 epimer ratios increase slightly for the higher homologs from C_{31} to C_{35}. For example, Zumberge (1987b) calculated the average equilibrium 22S/(22S + 22R) ratios for 27 low-maturity oils at C_{31}, C_{32}, C_{33}, C_{34}, and C_{35} to be 0.55, 0.58, 0.60, 0.62, and 0.59, respectively. In some cases, interference by co-eluting peaks can invalidate certain ratios. For example, the C_{31} homohopane 22S/(22S + 22R) ratio is affected by co-elution of a C_{30} neohopane generated during biodegradation (Subroto et al., 1991). For these reasons, it is useful to:

- check each reported 22S/(22S + 22R) ratio for a given homohopane versus the other homohopanes;
- measure the ratio using other important ions, such as m/z 205 for the C_{31} homohopanes;
- use GCMS/MS of the transition for the appropriate molecular ion to m/z 191.

Typically, C_{31}- or C_{32}-homohopane results are used to calculate the 22S/(22S + 22R) ratio. The 22S/(22S + 22R) ratio rises from 0 to ~0.6 (0.57–0.62 = equilibrium) (Seifert and Moldowan, 1980) during maturation. Samples showing 22S/(22S + 22R) ratios in the range 0.50–0.54 have barely entered oil generation, while ratios in the range 0.57–0.62 indicate that the main phase of oil generation has been reached or surpassed. Some oils exposed to very mild thermal stress apparently can have 22S/(22S + 22R) ratios below 0.50. Philp (1982) described a crude oil from the Gippsland Basin, Australia, with 22S/(22S + 22R) for the C_{31} homohopane <0.5. One possible explanation is leaching or solubilization of homohopanes from immature lignite in contact with the reservoir. However, Hanson et al. (2001) described Upper Oligocene lacustrine oils from the northern Qaidam Basin, northwest China, which have C_{32}, C_{33}, and C_{34} 22S/(22S + 22R) <0.5. Low C_{29} sterane 20S/(20S + 20R) and low Ts/Tm support very low maturity for these oils.

Certain factors, such as lithology, may affect the rate of 17α-homohopane isomerization. For example, Moldowan et al. (1992) found fully isomerized homohopanes in very immature carbonate rocks from the Adriatic Basin. Laboratory simulations of burial

X = n-C_2H_5, n-C_3H_7, n-C_4H_9, n-C_5H_{11}, n-C_6H_{13}

Figure 14.4. Equilibration between 22R (biological epimer) and 22S (geological epimer) for the C_{31}–C_{35} homohopanes.

maturation indicate that free homohopanes in the bitumen isomerize more rapidly than those attached to the kerogen (Peters and Moldowan, 1991). During early maturation, the 22S/(22S + 22R) ratio can be controlled largely by the release of sulfurized hopanoids from the kerogen rather than by isomerization (Köster et al., 1997). For this reason, the 22S/(22S + 22R) ratio should be used with particular caution when applied to extracts and crude oils generated from low-mature, organic sulfur-rich carbonate-marlstone source rocks. Ten Haven et al. (1986) noted that bitumens from immature rocks deposited under hypersaline conditions commonly show mature hopane patterns. These bitumens show hop-17(21)-enes and extended $17\alpha,21\beta(H)$-homohopanes fully isomerized at C-22 (50–60% 22S) typical of immature and mature samples, respectively. Unusual diagenetic pathways for the hopanes in hypersaline environments may account for this discrepancy.

After reaching equilibrium at the early oil-generative stage, no further maturity information is available because the 22S/(22S + 22R) ratio remains constant. However, the inflection point in a plot of 22S/(22S + 22R) versus vitrinite reflectance or other maturity/generation parameters can be used to calibrate these parameters to the onset of oil generation for a given source rock in a basin.

Although not a routine maturity parameter, the 22S/(22S + 22R) ratio for the C_{31}–C_{35} moretanes shows variations similar to those for the C_{31}–C_{35} hopanes (Larcher et al., 1987). As with the 17α-hopanes, the C_{31}–C_{35} moretanes with the presumed 22R configuration of the biological precursor are in much greater concentration in immature rocks. The 22S/(22S + 22R) ratios for 17α-homohopanes and moretanes increase to $\sim$0.6 and 0.4 with thermal maturity, respectively.

Moretanes/hopanes

> High specificity for immature to early oil generation (Figure 14.3). Measured using m/z 191 chromatograms or GCMS/MS using the C_{29} or C_{30} homologs.

The $17\beta,21\alpha(H)$-moretanes are thermally less stable than the $17\alpha,21\beta(H)$-hopanes, and abundances of the C_{29} and C_{30} moretanes decrease relative to the corresponding hopanes with thermal maturity. The biological $17\beta,21\beta(H)$-configuration ($\beta\beta$) of hopanoids in

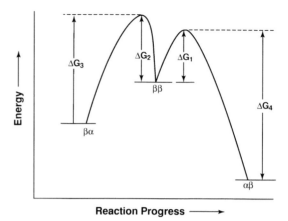

Figure 14.5. Proposed stability relationships between the three major classes of hopanes in petroleum (Seifert and Moldowan, 1980). Reprinted with permission by ChevronTexaco Exploration and Production Technology Company, a division of Chevron USA Inc.

organisms is unstable and is absent in crude oils unless contaminated by immature organic matter. The $\beta\beta$-hopanes readily convert to $\beta\alpha$-(moretane) and $\alpha\beta$-hopane configurations (see Figure 2.29) along a reaction scheme proposed by Seifert and Moldowan (1980). During burial, temperatures are reached where sufficient energy is available to overcome the energy barriers ΔG_1 and ΔG_2 (Figure 14.5), allowing conversion of $\beta\beta$-hopanes to either $\beta\alpha$-moretanes or $\alpha\beta$-hopanes. At this low temperature, the conversion of these compounds back to $\beta\beta$-hopanes is not possible because of high-energy barriers ΔG_3 and ΔG_4. At higher temperatures, the conversion of moretanes back to $\alpha\beta$-hopanes becomes possible through a $\beta\beta$-hopane intermediate. However, the high ΔG_4 energy barrier allows little conversion of $\alpha\beta$-hopanes to $\beta\beta$-hopanes, resulting in an equilibrium mixture favoring $\alpha\beta$-hopanes over $\beta\alpha$-moretanes by $\sim$20 : 1.

The ratio of $17\beta,21\alpha(H)$-moretanes to their corresponding $17\alpha,21\beta(H)$-hopanes decreases with thermal maturity (e.g. see Figure 14.2) from $\sim$0.8 in immature bitumens to <0.15 in mature source rocks and oils to a minimum of 0.05 (Mackenzie et al., 1980a; Seifert and Moldowan, 1980). Based on 234 crude oils, Grantham (1986b) concluded that oils from Tertiary source rocks show higher moretane/hopane (0.1–0.3, with many values between 0.15 and 0.20) than those from older rocks (generally $\leq$0.1).

The C_{30} compounds are used most commonly for moretane/hopane, although this ratio is also quantified using C_{29} compounds (e.g. Seifert and Moldowan, 1980). Others have used both C_{29} and C_{30} compounds for their moretane/hopane ratio (Mackenzie *et al.*, 1980a).

Evidence suggests that moretane/hopane depends partly on source input or depositional environment. For example, Rullkötter and Marzi (1988) noted higher moretane/hopane in bitumens from hypersaline rocks compared with adjacent shales. Isaksen and Bohacs (1995) observed increasing C_{30} moretane ratios from transgressive to highstand systems tracts in Lower-Middle Triassic mudrocks from the Barents Sea, corresponding to increased terrigenous higher-plant input.

Except for the C_{30} pseudohomologs, the three most commonly encountered hopane stereoisomers can be distinguished by their fragmentation patterns in full-scan GCMS. These stereoisomers vary according to the relative intensities of their A + B and D + E ring fragments (Figure 8.20) (Ensminger *et al.*, 1974; Van Dorsselaer, 1974; Seifert and Moldowan, 1980). The $\alpha\beta$-hopane isomers show stronger m/z 191 response for the A/B-ring portion of the molecule than that for the m/z 148 + X (side-chain) fragment for the D/E ring portion (Figure 8.20). For the $\beta\alpha$ and the $\beta\beta$-hopane isomers, the situation is reversed; all show stronger m/z 148 + X than m/z 191 fragments. The $\beta\alpha$ and $\beta\beta$ stereoisomers are distinguished from each other because the ratio of the fragments m/z (148 + X)/191 = ~2 for the $\beta\beta$ isomers but is <1.5 for the $\beta\alpha$ isomers. C_{27} 17β-hopane, which belongs strictly to neither the $\beta\beta$ nor the $\beta\alpha$ series, has a ratio of ~1.5. The A/B- and D/E-ring fragments for the C_{30} hopanes cannot be differentiated using m/z 191. While this method is useful for distinguishing the common hopane series, the same reasoning does not extend to all related hopanoids. For example, the $\alpha\beta$ and $\beta\alpha$-28,30-bisnorhopanes have nearly identical mass spectra.

Tricyclics/17α-hopanes

> Low specificity for mature to postmature range due to interference from source input. Measured on m/z 191 chromatograms or GCMS/MS.

The tricyclics/17α-hopanes ratio increases for related oils of increasing thermal maturity (Seifert and Moldowan, 1978). The ratio increases because proportionally more tricyclic terpanes (i.e. cheilanthanes) (see Figures 8.20 and 13.73) than hopanes are released from the kerogen at higher levels of maturity (Aquino Neto *et al.*, 1983). This is also demonstrated by an increase in the tricyclics/17α-hopanes ratio during hydrous pyrolysis of Monterey Shale (Peters *et al.*, 1990). Figure 14.6 compares terpane mass chromatograms for

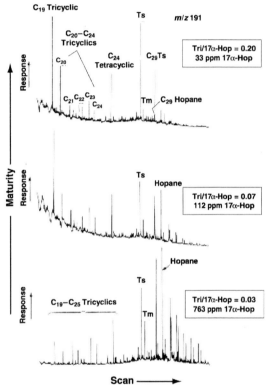

Figure 14.6. Comparison of terpane mass chromatograms (m/z 191) for three related Oman-area oils. Ranking of the oils by thermal maturity was difficult because all are highly mature and contain few biomarkers. Nonetheless, several biomarker parameters were used successfully for this purpose. For example, the tricyclics/17α-hopanes ratio increases and the amounts of 17α-hopanes (C_{29}–C_{33}) decrease with maturity. The figure also demonstrates the use of C_{19}–C_{29} tricyclic terpanes to solve a difficult correlation problem. Based on numerous biomarker and supporting parameters, the two less mature oils (middle and bottom) are closely related. However, the highly mature oil (top) has few if any steranes or hopanes. The similar pattern of the thermally resistant C_{19}–C_{29} tricyclic terpanes for the top and middle oils in the figure and other supporting data were used to show that the most mature oil is related to the other two oils. Reprinted with permission by ChevronTexaco Exploration and Production Technology Company, a division of Chevron USA Inc.

three related Oman-area oils, where the tricyclics/17α-hopanes ratio increases with thermal maturity. The tricyclics/17α-hopanes ratio of expelled oil increases during hydrous pyrolysis (Peters *et al.*, 1990), possibly because tricyclics migrate faster and/or 17α-hopanes show greater affinity for the rock matrix. Laboratory simulations of petroleum migration using an alumina column show that tricyclic terpanes and 5α,14β,17β(H)-steranes elute faster than hopanes and 5α,14α,17α(H) steranes (Jiang Zhusheng *et al.*, 1988). Future research is needed to determine whether this applies to natural expulsion, although Kruge *et al.* (1990b) suggest that more rapid migration of tricyclic terpanes than hopanes partly explains biomarker distributions in black shales of the East Berlin Formation, Hartford Basin, Connecticut.

Because tricyclic terpanes and hopanes originate by diagenesis of different biological precursors (Ourisson *et al.*, 1982), the tricyclics/17α-hopanes ratio can differ considerably between crude oils from different source rocks or different facies of the same source rock. For example, the source effect on the tricyclics/17α-hopanes ratio allows differentiation of Carneros (0.17), Phacoides (0.14), and Oceanic (0.09) oil groups in McKittrick Field, California (Seifert and Moldowan, 1978). Other biomarker ratios show that these oils increase in maturity with average reservoir depth: Carneros (6300 ft (1950.2 m)) < Phacoides (8500 ft (2590.8 m)) < Oceanic (8870 ft (2703.6 m)). The tricyclics/17α-hopanes ratio is highest for the least mature oil (Carneros, 0.17) and lowest for the most mature oil (Oceanic, 0.09), the opposite of that expected for related oils of increasing maturity. The tricyclics/17α-hopanes ratio also helped in the identification of the principal source rocks for various oils from Prudhoe Bay, Alaska (Figure 13.75).

Ts/(Ts + Tm)

Thermal parameter based on relative stability of C_{27} hopanes applicable over the range immature to mature to postmature, but with strong dependence on source. Measured using m/z 191 or GCMS/MS (m/z 370→191). Also expressed as Ts/Tm.

During catagenesis, C_{27} 17α-trisnorhopane (Tm or 17α-22,29,30-trisnorhopane) is less stable than C_{27} 18α-trisnorhopane II (Ts or 18α-22,29,30-trisnor-

Tm (less stable)   Ts (more stable)

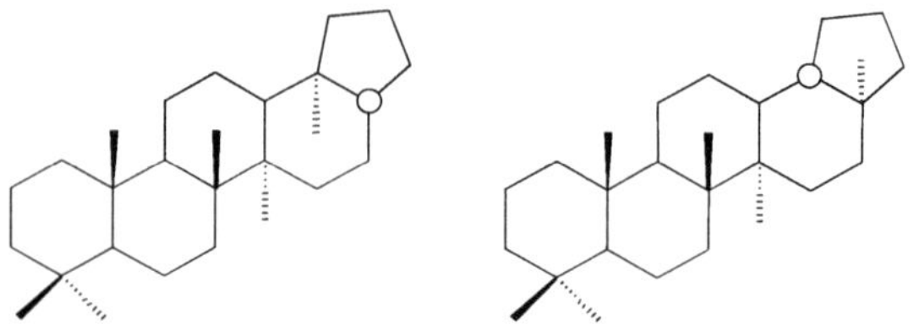

Figure 14.7. Structures of 17α-22,29,30-trisnorhopane (Tm) and 18α-22,29,30-trisnorneohopane (Ts).

neohopane) (Figure 14.7) (Seifert and Moldowan, 1978). This observation was substantiated using molecular mechanics calculations for the formation of various hopanes, including Ts and Tm (Kolaczkowska *et al.*, 1990). It is unknown whether conversion of Tm to Ts may also occur. Figure 14.6 shows an example of maturity effects on the relative amounts of Ts and Tm on m/z 191 mass chromatograms.

The Ts/(Ts + Tm) ratio, sometimes reported as Ts/Tm, depends on both source and maturity (Moldowan *et al.*, 1986). Due to low maturity of shallow rocks in the Linyi Basin, China, Hong *et al.* (1986) included a third compound related to Tm and Ts but showing even lower thermal stability, C_{27} 22,29,30-trisnor-17β-hopane. They found a systematic increase in the relative abundance of Ts and a decrease in C_{27} 17β-trisnorhopane compared with Tm with depth. Concentrations of 17β-trisnorhopanes approach zero at maturity equivalent to the earliest oil window. 17β-Trisnorhopane is more stable than 17β,21β(H)-hopane but less stable than moretane, as shown by a series of Cretaceous shales from the Wyoming Overthrust Belt (Seifert and Moldowan, 1980).

The Ts/(Ts + Tm) ratio is most reliable as a maturity indicator when evaluating oils from a common source of consistent organic facies. The relative importance of lithology and oxicity of the depositional environment in controlling this ratio remains unclear, although some results suggest substantial effects (Figure 14.8). Ts/(Ts + Tm) appears to be sensitive to clay-catalyzed reactions. For example, oils from carbonate source rocks appear to have unusually low Ts/(Ts + Tm) ratios compared with those from shales (McKirdy *et al.*, 1983; McKirdy *et al.*, 1984; Rullkötter *et al.*, 1985; Price *et al.*, 1987). Bitumens from many hypersaline source rocks show high Ts/(Ts + Tm) ratios (Fan Pu

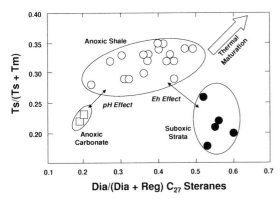

Figure 14.8. Diasteranes/(Dia + Regular) C_{27} steranes versus Ts/(Ts + Tm) is controlled partly by thermal maturity (increasing toward upper right) but can still be used to differentiate extract samples by source-rock depositional environment (Moldowan et al., 1994b). Symbols are for Toarcian rock extracts of the same marginal thermal maturity (R_o = 0.51%). The points cluster according to oxicity and acidity of the depositional environment.

et al., 1987; Rullkötter and Marzi, 1988). The Ts/(Ts + Tm) ratio increases at lower Eh and decreases at higher pH for the depositional environment of a series of Lower Toarcian marine shales from southwestern Germany (Moldowan et al., 1986), but it also decreases in an anoxic (low Eh) carbonate section.

The Ts/(Ts + Tm) ratio should be used with caution. Tm and Ts commonly co-elute with tricyclic or tetracyclic terpanes on m/z 191 mass chromatograms, resulting in spurious Ts/(Ts + Tm) ratios. For example, Rullkötter and Wendisch (1982) show co-elution of a C_{30} tetracyclic triterpane with Tm on their capillary columns. If interference is suspected, then quantifying Tm and Ts from the m/z 370 (molecular ion or M+) mass chromatogram may be helpful (Volkman et al., 1983a). Measurement of the m/z 370 → 191 transition by GCMS/MS gives the most reliable results.

Ts/hopane

Mature to postmature range. Measured on m/z 191 chromatograms.

Volkman et al. (1983a) proposed Ts/C_{30} 17α-hopane as a maturity parameter for very mature oils and condensates (e.g. see Figure 14.6). Cleavage of the side chain of C_{29} and higher-carbon-number hopanoids is believed to produce precursors of Ts by a rearrangement mechanism (Seifert and Moldowan, 1978).

C_{29}Ts/(C_{29} hopane + C_{29}Ts)

Unknown range of specificity. Measured on m/z 191 chromatograms.

18α-30-Norneohopane (C_{29}Ts) (e.g. see Figures 13.83 and 13.84), described for many years as the unknown C_{29} terpane or C_{29}X, elutes immediately after C_{29} 17α-hopane on m/z 191 chromatograms and was identified by advanced NMR methods (Moldowan et al., 1991c). Most reports suggest that the abundance of this compound relative to the 17α-hopane is related to thermal maturity (Hughes et al., 1985; Sofer et al., 1986; Sofer, 1988; Cornford et al., 1988; Riediger et al., 1990c).

Molecular mechanics calculations indicate that C_{29}Ts should be more stable than 17α-30-norhopane by 3.5 kcal/mol, compared with the C_{27} analog Ts, which is more stable than Tm by 4.4 kcal/mol (Kolaczkowska et al., 1990). Thus, the thermal maturity effect on the C_{29}Ts/(C_{29} 17α-hopane + C_{29}Ts) ratio should be comparable with but slightly less than that on Ts/(Ts + Tm). This inference is supported by a study from the Jeanne d'Arc Basin, eastern Canada (Fowler and Brooks, 1990). In this basin, trends in the C_{29}Ts/C_{29}-hopane ratio paralleled those for Ts/Tm, and both parameters increased with thermal maturity, as indicated by other biomarker parameters (Fowler and Brooks, 1990). The highly mature condensate in Figure 14.6 (top); for example, has a very high C_{29}Ts/(C_{29} 17α-hopane + C_{29}Ts) ratio.

18α/(18α + 18β)-Oleananes and oleanane index

Immature to early mature range (see Figure 14.3). Measured on m/z 191 chromatograms (see Figure 8.20).

Apart from being reliable markers for angiosperm input in Cretaceous or younger source rocks, oleanane epimer ratios can be used to indicate thermal maturity (Riva et al., 1988; Ekweozor and Udo, 1988). Oleananes originate from oleanenes and other pentacyclic triterpenes (i.e. taraxerenes) by rearrangements and double-bond isomerizations. Eneogwe et al. (2002) observed three oleanene isomers (olean-13(18)-ene, olean-12-ene, and 18α-olean-12-ene) in onshore and offshore Niger Delta oils. The secondary isomers olean-13(18)-ene and 18α-olean-12-ene formed during diagenesis from the olean-12-ene precursor. They found these olefins only in low-maturity oils, suggesting that the oleanenes originated from the source rocks and migrated with low-maturity, expelled hydrocarbons rather than being extracted from immature strata during migration. The reduction of the $\Delta^{13(18)}$ (or Δ^{18}) double bond intermediates yields a mixture of 18α- and 18β-oleananes. Increased maturation favors the 18α configuration, suggesting it is more stable than 18β. Riva et al. (1988) show correlations between the oleanane epimer ratio and other maturity parameters, including Ts/Tm, vitrinite reflectance, and T_{max} with depth based on data from several wells.

Figure 14.9 compares the relative peak areas for 18α- and 18β-oleanane for two groups of petroleum samples from the Eel River area, California. These two compounds were identified provisionally by coinjection of authentic standards (Figure 13.96). The relative maturity of the samples in Figure 14.9 was established using other biomarker parameters (see Figure 14.29).

Ekweozor and Telnaes (1990) suggest that the oleanane index (oleanane/C_{30} hopane) can be used to delimit the top of the oil-generative window (the boundary between diagenesis and catagenesis). In wells from the Niger Delta, they observed maxima in the oleanane index at the top of the oil-generative window, corresponding to vitrinite reflectance of ~0.55%. Because this approach has not been studied extensively, we recommend that maturity estimates based on the oleanane index be checked using other maturity parameters, such as the hopane isomerization ratio [22S/(22S + 22R)]. At very high maturity, all biomarkers decrease in concentration and hopane decreases more rapidly than oleanane, which can lead to high oleanane/C_{30} hopane.

(BNH + TNH)/hopanes

Highly specific for immature to mature range when bisnorhopane (BNH) and trisnorhopane (TNH) are present. Measured using m/z 191 and 163 fragmentograms (see Figure 8.20). Also expressed as C_{28}/C_{30} hopane.

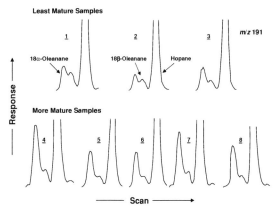

Figure 14.9. Terpane mass chromatograms (m/z 191) of oils and source-rock extracts from the Eel River Basin, California, showing that the relative proportions of 18α- and 18β-oleanane change with thermal maturity. The less mature (top) show proportionally more 18β-oleanane than more mature samples (bottom). Relative maturity among samples (underlined numbers) was established using other biomarker parameters (e.g. see Figure 14.29). Provisional identification of the oleanane epimers in this figure was accomplished using co-injection of authentic standards (see Figure 13.96). Reprinted with permission by ChevronTexaco Exploration and Production Technology Company, a division of Chevron USA Inc.

The (BNH + TNH)/hopanes ratio is defined as (28,30-bisnorhopanes + 25,28,30-trisnorhopanes)/ (C_{29} + C_{30} 17α-hopanes) (Moldowan et al., 1984), but it can be expressed in other ways, such as BNH/[17α-hopane + BNH]. Non-zero values for this ratio indicate contributions from source rocks containing demethylated hopanes and can be useful to distinguish samples from different sources.

The structure of 17α,18α,21β(H)-28,30-bisnorhopane was first proven after its isolation from Monterey Shale by Seifert et al. (1978). This compound is probably the same as the C_{28}-triterpane reported by Petrov et al. (1976) in an East Siberian oil (Seifert, 1980). Later, it was shown that the 17α,21β(H)-isomer co-elutes with the 17β,21α(H)-isomer on most capillary columns and that a special Apiezon-coated column is necessary to separate them (Moldowan et al., 1984). Thus, these two compounds are usually quantified together. Unlike the

17β,21β(H)-isomers of the hopanes, the 17β,21β(H)-isomers of BNH and TNH are stable within the oil-generative window and occur in small amounts relative to the 17α,21β(H)-and 17β,21α(H)-isomers.

As discussed earlier, BNH and TNH do not appear to be generated from the kerogen but are associated with the original, free bitumen in the rock. For example, BNH represents a prominent peak in immature bitumen from the Kimmeridge Shale sample and displays a lower relative concentration in the related, but more mature, Piper oil shown in Figure 14.2. As the source rock generates oil during thermal maturation, products from kerogen cracking dilute pre-existing bitumen and there is a systematic decrease in (BNH + TNH)/hopanes for oils or source rocks that contain these compounds (Moldowan et al., 1984). For example, low-maturity Bazhenov Shale source rock and related oils from Western Siberia contain high concentrations of TNH, but TNH/hopanes decreases to zero in the late oil window (Figure 14.10, bottom) (Peters et al., 1993). Figure 14.10 (top) shows a similar trend for BNH/hopanes for oils from the Monterey Formation in California.

Grantham et al. (1980) showed that bisnorhopane/hopane in North Sea oils decreases with thermal maturity, but other factors, such as organic matter input and depositional environment, may be important. Thus, mature oil showing BNH/hopane of zero is not necessarily unrelated to a less mature potential source rock sample with a non-zero value for the ratio. The presence of BNH and TNH in Kimmeridge source rock and related oils was used to differentiate these samples from other bitumens and oils in the North Sea (Peters et al., 1989).

Unlike Grantham et al. (1980), Hughes et al. (1985) observed an increase in BNH/hopane with maturity. They explain these conflicting observations by recognizing that in diverse samples; source input is the major control on BNH/hopane. However, among samples from a similar source (e.g. Grantham et al., 1980), maturity is the controlling factor.

BNH/hopane can be used to compare the maturity of genetically related crude oils. It is our experience that this ratio decreases with thermal maturity. However, unrelated samples at the same level of maturity can show different BNH/hopane because of different organic matter input and/or depositional environment.

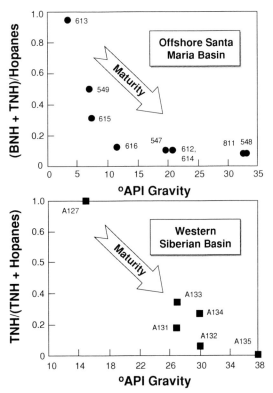

Figure 14.10. Decrease in (bisnorhopane (BNH) + (trisnorhopane (TNH))/hopane with thermal maturity (API gravity) for Monterey oils from the offshore Santa Maria Basin, California (top). Decrease in TNH/(TNH + hopanes) with thermal maturity (API gravity) for oils from Western Siberia (bottom). Reprinted with permission by ChevronTexaco Exploration and Production Technology Company, a division of Chevron USA Inc.

POLYCADINENES AND RELATED PRODUCTS

Several thermal maturity parameters are based on isomeric distributions of the diagenetic products of polycadinene, including bicadinanes, diaromatic secobicadinanes, diaromatic tricadinanes (Figure 14.11), and cadalenes (Alexander et al., 1994; Murray et al., 1994; Murray et al., 1997b; Sosrowidjojo et al., 1993; Sosrowidjojo et al., 1994b; Sosrowidjojo et al., 1996). One advantage of using these compounds is that polycadinene does not thermally decompose until ~80°C (Tegelaar and Noble, 1994). Thus, maturity parameters based on polycadinene diagenesis products are less likely to be affected by microbial and chemical alteration that

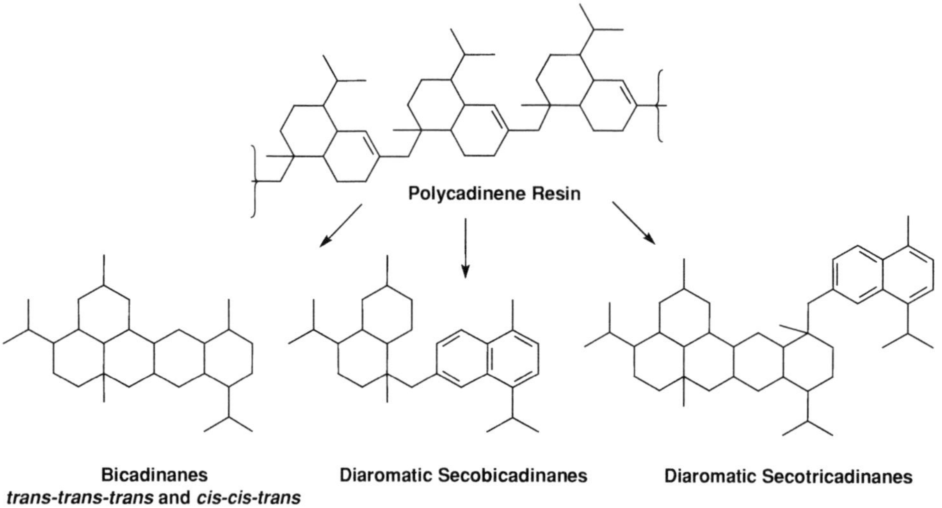

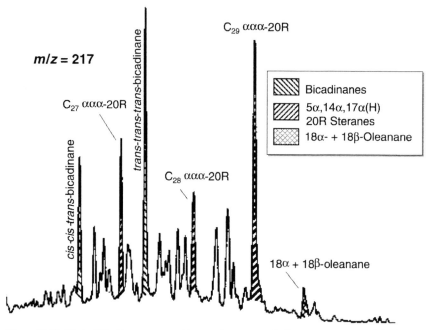

Figure 14.11. Diagenesis of polycadinene resin to form bicadinanes, diaromatic secobicadinanes, and secotricadinanes (Murray et al., 1997b).

Figure 14.12. Typical bicadinanes separated by gas chromatography/mass spectrometry (GCMS) (m/z 217.2), showing co-elution problems. While the *cis-cis-trans* and *trans-trans-trans* bicadinanes are reasonably well resolved, compounds T1 and R are buried in a cluster of C_{27} steranes and C_{29} diasteranes. The ion chromatogram is of the Juragan-1 oil, offshore Brunei Darussalam. Reprinted from Curiale et al. (2000). © Copyright 2000, with permission from Elsevier.

occurs during early diagenesis. A disadvantage is that significant amounts of polycadinene-derived hydrocarbons occur mostly in oils and rock extracts from Southeast Asia. Because these hydrocarbons originate from dammar resins produced by *Dipterocarpaceae* trees (van Aarssen et al., 1990a), their use is limited to Tertiary samples from this region. Low levels of bicadinanes occur elsewhere and in rocks as old as Jurassic (Clifton et al., 1990; Murray et al., 1994; Summons et al., 1995; Armanios et al., 1995a), suggesting that the temporal and geographic range of polycadinene-based maturity parameters could be extended if improved methods of separation and detection are used (Armanios, 1995; Armanios et al., 1995a).

Bicadinane maturity index 1

> Immature to early oil window maturity parameter for Tertiary sequences with *Dipterocarpaceae* higher-plant input. Measured by GCMS using mass fragment m/z 412 or by GCMS/MS using the 412 → 369 transition (preferred).

Murray et al. (1994) proposed various maturity-dependent ratios based on bicadinane isomers. These maturity ratios are most useful in terrigenous dominated crude oils and source rocks that contain abundant bicadinanes and low steranes. Bicadinanes are also more resistant to biodegradation than steranes and hopanes and thus may be useful when other maturity ratios have been altered. Of the bicadinane maturity ratios, the bicadinane maturity index 1 (BMI-1), the ratio of compound T to the sum of compounds T1 and R, is the most useful and reliably measured. Compound T is *trans-trans-trans* bicadinane (Cox et al., 1986). Compounds T1 and R have not been identified fully, but they yield mass spectra similar to that of compound T. They are believed to have the ring configuration of *trans-trans-trans* bicadinane, but they differ in either position or axial/equatorial location of the alkyl groups.

Accurate measurement of the bicadinanes by GCMS can be difficult. The compounds have both m/z 217 and 191 ion fragments, and quantifying the bicadinane peaks, particularly T1 and R, is frequently not possible due to co-elution (Figure 14.12). The bicadinanes can be measured by GCMS using the weaker m/z 412 parent ion. Signal-to-noise ratio can be improved further for the *trans-trans-trans* isomers by using GCMS/MS and monitoring the 412 → 369 transition (Figure 14.13).

Bicadinane maturity index 2

> Immature to early oil window maturity parameter for Tertiary sequences with *Dipterocarpaceae* higher-plant input. Measured by GCMS/MS using the m/z 426 → 383 transition.

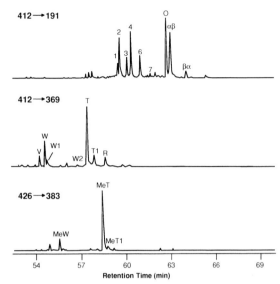

Figure 14.13. Gas chromatography/mass spectrometry/mass spectrometry (GCMS/MS) separation of C_{30} hopane (α,β), moretane (β,α) oleanane (O) and unidentified oleanoid hydrocarbons (1–7) (m/z 412 → 191); *cis-cis-trans* (V, W, W1, and W2) and *trans-trans-trans* (T, T1, and R) bicadinanes (m/z 412 → 369) and *cis-cis-trans* (MeW) and *trans-trans-trans* (MeT, MeT1) methylbicadinanes (m/z 426 → 369). Reprinted from Murray et al. (1994). © Copyright 1994, with permission from Elsevier.

BMI-2 is the equivalent of BMI-1 using methylated compounds (Murray et al., 1994). These compounds have been detected reliably only by GCMS/MS using the m/z 426 → 383 transition, and they are assumed to have the *trans-trans-trans* ring configuration (Figure 14.14). Methylated *trans-trans-trans* bicadinane (MeT) is typically much more abundant than the methylated equivalents of the T1 (MeT1) and R (MeR) compounds, except in thermally immature rocks. Because BMI-2 is defined as MeT/(MeT1 + MeR), the ratio is subject to large errors as the denominator decreases relative to MeT.

Bicadinane maturity index 3

> Immature to early oil window maturity parameter for Tertiary sequences with *Dipterocarpaceae* higher-plant input. Measured by GCMS using mass fragment m/z 412 or by GCMS/MS using the 412 → 369 transition.

BMI-3 is the ratio of bicadinanes with the *cis-cis-trans* (Peaks V, W, W1, and W2) (van Aarssen et al., 1990a)

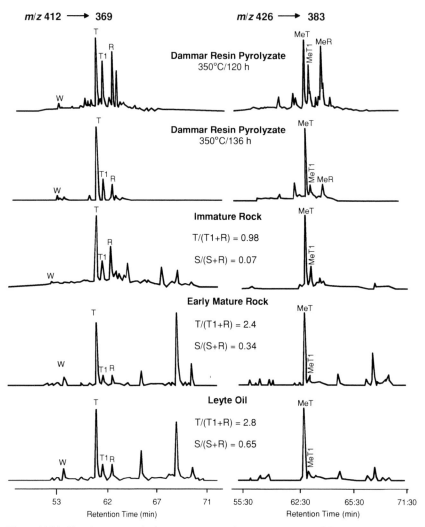

Figure 14.14. Gas chromatography/mass spectrometry/mass spectrometry (GCMS/MS) chromatograms showing the presence of bicadinanes and methylbicadinanes in resin pyrolyzates, rock extracts (Miocene, Visayan Basin), and crude oil (Miocene, Philippines). Bicadinene maturation index 1 (BMI-1) and the C_{29} $\alpha\alpha\alpha$ 20S/(20S + 20R) ratios are shown. Reprinted from Murray et al. (1994). © Copyright 1994, with permission from Elsevier.

to those with the *trans-trans-trans* ring configurations (Murray et al., 1994). In a simpler version, Curiale et al. (1994) found that the ratio of W/T correlates with the Tm/Ts ratio in oils from Myanmar. The bicadinane peaks can be measured by GCMS using the m/z 191 or 217 ion fragments where problems with co-elution can occur, or by using the weaker m/z 412 parent ion. GCMS/MS analysis can improve the signal-to-noise ratio for the *trans-trans-trans* isomers, but the *cis-cis-trans* bicadinanes do not yield a strong m/z 426 → 383 transition. It is possible to determine BMI-3 using different methods optimized for these two types of bicadinanes, provided that calibration samples are run.

Diaromatic secobicadinane ratio

Immature to early oil window maturity parameter for Tertiary sequences with *Dipterocarpaceae* higher-plant input. Measured by GCMS/MS using the m/z 404 → 198 and m/z 362 → 156 transitions.

Van Aarssen et al. (1992) identified C_{27} diaromatic desisopropyl secobicadanes in bicadinane-rich oils from

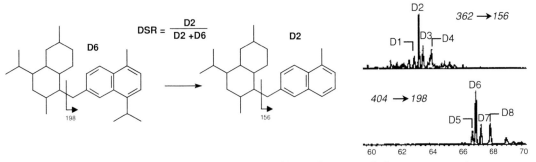

Figure 14.15. Diaromatic secobicadinane ratio (DSR) as determined by gas chromatography/mass spectrometry/mass spectrometry (GCMS/MS). Reprinted from Sosrowidjojo et al. (1996). © Copyright 1996, with permission from Elsevier.

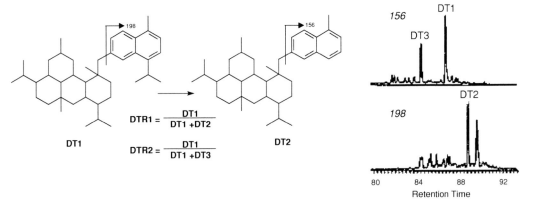

Figure 14.16. Diaromatic tricadinane ratios (DTR1 and DRT2) as determined by gas chromatography/mass spectrometry (GCMS) of the diaromatic fragment. Reprinted from Sosrowidjojo et al. (1996). © Copyright 1996, with permission from Elsevier.

Southeast Asia and postulated a pathway for their formation from polycadalene. Sosrowidjojo et al. (1993; 1994a) identified the C_{30} parent compounds and showed that the diaromatic secobicadinane ratio (DSR) is maturity-dependent (Sosrowidjojo et al., 1994b; 1996). DSR is defined as the ratio of the most abundant C_{27} diaromatic des-isopropyl secobicadinane (D2) over the sum of this compound plus its C_{30} diaromatic secobicadinane parent. The C_{27} and C_{30} diaromatic secobicadinanes are characterized using GCMS/MS of the aromatic hydrocarbon fraction using the m/z 362 → 156 and m/z 404 → 198 transitions, respectively (Figure 14.15).

Diaromatic tricadinane ratios

> Immature to early oil window maturity parameter for Tertiary sequences with *Dipterocarpaceae* higher-plant input. Measured using ion fragments m/z 156 and 198.

Van Aarssen et al. (1992) first identified the C_{42} diaromatic tricadiananes and Sosrowidjojo et al. (1993; 1994a) later identified their C_{45} parent compounds. These compounds are characterized by GCMS using the mass fragment of the diaromatic group. Sosrowidjojo et al. (1994b; 1996) proposed two ratios based on these compounds as maturity parameters (Figure 14.16).

Isocadalene/(isocadalene + cadalene)

> Oil-window maturity parameter for sequences with higher-plant input. Measured using mass fragment m/z 198 in the aromatic hydrocarbon fraction.

Alexander et al. (1994) showed that cadalene (1,6-dimethy-4-isopropylnaphthalene) converts to isocadalene (1,6-3-isopropylnaphthalene) in the presence of

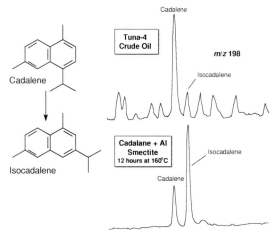

Figure 14.17. Cadalene converts to isocadalene via dealkylation and isomerization reactions. The compounds can be detected in crude oils using the m/z 198 parent ions. Reprinted from Alexander et al. (1994). © Copyright 1994, with permission from Elsevier.

aluminum smectite (Figure 14.17). The only other significant reaction product was 1,6-dimethylnaphthalene. They proposed the ratio of isocadalene to cadalene, both measured by GCMS using the m/z 198 parent ion, as a maturity indicator applicable over the range of oil generation. Although cadalene is most abundant in sediments and oils with input of polycadinenes, the hydrocarbon may also arise from various other higher land-plant biological precursors. Hence, the ratio has broader application than those maturity parameters based on *Dipterocarpaceae* resin input. Alexander et al. (1994) showed that both isocadalene and 1,6-dimethylnaphalene increase with depth in post-Triassic sediments (Figure 14.18).

Comparison of bicadinane with other maturity parameters

Bicadinane maturity indices and sterane and hopane isomerization ratios show systematic trends for extracts from Miocene freshwater lacustrine source rocks in the Visayan Basin, Philippines (Murray et al., 1994). BMI-1 reached maximum values between hopane and sterane isomeric equilibrium, suggesting that the usable range for these parameters ends near the top of the oil window (Figure 14.19a). Sosrowidjojo et al. (1996) found a similar correlation in downhole samples from a well in the South Sumatra Basin (Figure 14.19b). However, BMI-1 continued to increase as the sterane isomerization reached equilibrium. Sosrowidjojo et al. (1996) suggested that the sterane isomerization values reported in Murray et al. (1994) might have been altered by selective biodegradation of the 20R isomer, giving the spurious indication that BM1-1 maximized before sterane equilibrium. A facies effect could also account for the observations. Geochemical logs of thermal maturity indicators from the GK well (South Sumatra Basin) show that sterane isomerization is enhanced in coals compared with marine and shale facies (Figure 14.19). Although the overall trends for the maturity parameters increase with depth, perturbations are attributed to diagenetic effects (Sosrowidjojo et al., 1996).

BMI-2 and BMI-3 appear to increase until sterane equilibrium is obtained (Figure 14.19). In the GK well, neither BMI-2 nor BMI-3 could be determined reliably due to analytical uncertainty associated with small peaks. The diaromatic bicadinane ratios (DSR, DTR-1, and DTR-2) and the isocadalene/(isocadalane + cadalene) ratio generally increase with depth. The useful range for DSR and DRT-1 appears to be above the peak oil window, while DRT-2 and I/(I + C) may extend beyond (Figure 14.20).

Documented use of polycadinene-derived thermal parameters for oils is limited. Murray et al. (1994) showed low BMI-1 for thermally mature oils. Sosrowidjojo et al. (1996) explained that this anomaly was due to analytical error (Figure 14.21). The error was caused by mass spectrometer detector saturation for the *trans-trans-trans* bicadinane when sufficient sample was injected to yield a measurable sterane response. The problem was corrected by conducting two separate analyses that compensated for the large difference in relative abundance of bicadinane compared with the C_{29} steranes. Most of the Southeast Asian oils follow the trend observed in sediment extracts. BMI-1 typically ranges from 3 to 4, but shows little correlation with the sterane isomerization ratio. Two oils from offshore Brunei have anomalous high BMI-1. Sosrowidjojo et al. (1996) suggested that, in this case, the BMI-1 is more consistent with the observed maturity of the oil based on other parameters and that the low sterane isomerization values may be caused by contamination of the oil by low-maturity bitumen.

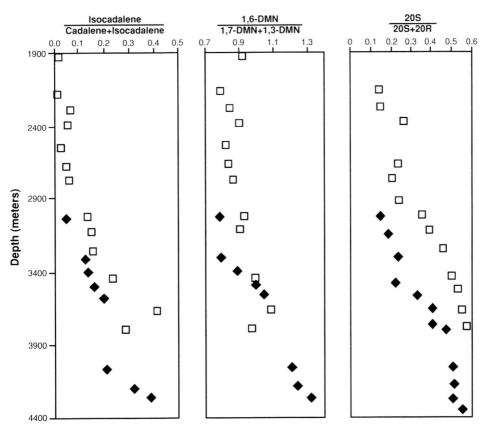

Figure 14.18. Changes in the ratios of isocadalene to cadalene, dimethylnapthalenes, and C_{29} $\alpha\alpha\alpha$ 20S to 20R isomers in two wells from Australia. The top of the oil window in the Grunter-1 (□) and Volador-1 (♦) wells is ~3500 and ~4000 m, respectively. Isocadalene and 1,6-dimethylnapthalene originate from the isomerization and dealkylation of cadalene, respectively. Reprinted from Alexander *et al.* (1994). © Copyright 1994, with permission from Elsevier.

STERANES

20S/(20S + 20R) isomerization

> Highly specific for immature to mature range (see Figure 14.3). Measured using m/z 217 (see Figure 8.20) or preferably by GCMS/MS analysis of C_{29} steranes. Also expressed as %20S and 20S/20R.

Isomerization at C-20 in the C_{29} $5\alpha,14\alpha,17\alpha(H)$-steranes (Figure 14.22) causes 20S/(20S + 20R) to rise from 0 to ~0.5 (0.52–0.55 = equilibrium) with increasing thermal maturity (Figure 14.23) (Seifert and Moldowan, 1986).

Note: The chromatographic peak corresponding to the C_{29} $\alpha\alpha\alpha$ 20S isomer is normally contaminated by other, probably $\alpha\beta\alpha$ C_{29} sterane isomers (Gallegos and Moldowan, 1992), which are effectively ignored because their separation is impractical using routine gas chromatography methods. Also, the 20S and 20R peaks each contain mixtures of 24S and 24R epimers that are usually ignored and measured together due to difficulties in their separation.

Other isomerization reactions for the steranes are shown in Figure 2.31. Only the R configuration at C-20 occurs in steroid precursors in living organisms, and this is gradually converted during burial maturation to a mixture of the R and S sterane configurations. This ratio has been used in kinetic calculations for input to thermal models (Mackenzie and McKenzie, 1983; Beaumont *et al.*, 1985).

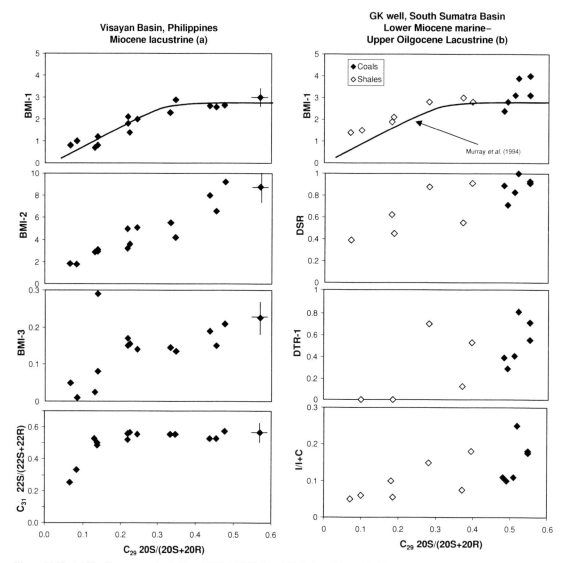

Figure 14.19. (a) Bicadinane maturity indices (BMI-1, BMI-2, and BMI-3) and C_{31} 22S/22R versus C_{29} $\alpha\alpha\alpha$ 20S/20R for extracts of Miocene lacustrine rocks from the Visayan Basin, Philippines (from Murray et al., 1994). Crossbars on the last data point of each curve indicate the standard deviation observed for triplicate measurements. (b) BMI-1, diaromatic bicadinane diaromatic secobicadinane ratio (DSR) and diaromatic tricadinane ratio 1 (DTR1), and isocadalene/(Isocadalene + Cadalene) versus C_{29} $\alpha\alpha\alpha$ 20S/20R for extracts from the GK well in the South Sumatra Basin. The calibration curve in Murray et al. (1994) is plotted for comparison. Reprinted from Sosrowidjojo et al. (1996). © Copyright 1996, with permission from Elsevier.

The sterane isomerization ratios are reported most often for the C_{29} compounds (24-ethylcholestanes or stigmastanes) (Figure 2.15) due to the ease of analysis using m/z 217 mass chromatograms. Isomerization ratios based on the C_{27} and C_{28} steranes commonly show interference by coeluting peaks. However, GCMS/MS measurements allow reasonably good accuracy for C_{27}, C_{28}, and C_{29} 20S/(20S + 20R), all of which have equivalent potential as maturity parameters when measured by this method. The $5\beta,14\alpha,17\alpha$(H)20R stereoisomer

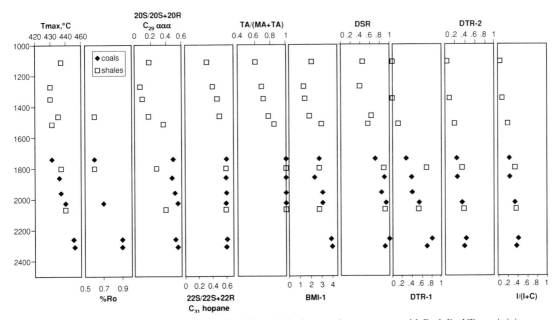

Figure 14.20. Geochemical well profiles compare polycadinane-derived maturation parameters with Rock-Eval T$_{max}$, vitrinite reflectance, and sterane and hopane isomerization ratios. Samples are from the GK well in the South Sumatra Basin (South Palembang Sub-basin) and span the marine Lower Miocene Gumai Formation (1114–1351 m) and Upper Oligocene Talang Akar Formation (1463–2309 m). The top of the oil window is estimated at 2100 m. Reprinted from Sosrowidjojo et al. (1996). © Copyright 1996, with permission from Elsevier.

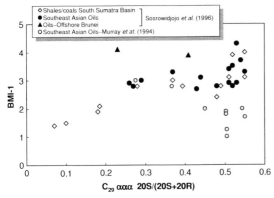

Figure 14.21. Comparison of bicadinane maturity index 1 (BMI-1) and the sterane isomerization ratio for crude oils and rock extracts from Southeast Asia (data from Murray et al., 1994; Sosrowidjojo et al., 1996).

Figure 14.22. Equilibration between 20R (biological epimer) and 20S (geological epimer) for the C$_{29}$ 5α,14α,17α(H)-steranes.

co-elutes with various other epimers used in this ratio. On a polymethylsiloxane capillary gas chromatographic column, C$_{27}$ $\beta\alpha\alpha$20R is poorly resolved from C$_{27}$ $\alpha\alpha\alpha$20R, leading to difficulty in measuring C$_{27}$ 20S/(20S + 20R) sterane ratios, even when using GCMS/MS. C$_{28}$ $\beta\alpha\alpha$20R nearly co-elutes with C$_{28}\alpha\beta\beta$20R, causing problems with its measurement. The C$_{29}$ $\beta\alpha\alpha$20R peak elutes between the $\alpha\beta\beta$20R and $\alpha\beta\beta$20S stereoisomers.

We do not recommend use of ethylcholestane (C$_{29}$) 20S/(20S + 20R) to indicate the onset of petroleum generation unless it is calibrated for each basin and source rock by comparison with other maturity and generation parameters. For example, Huang Difan et al. (1990) show correlations between vitrinite reflectance and two C$_{29}$ sterane isomerization ratios, 20S/(20S + 20R) and $\beta\beta/(\beta\beta + \alpha\alpha)$, specific for the southern part of the

Dagong oil field, China. Mackenzie et al. (1980a) indicate that petroleum generation begins at %20S of ~40%. However, Seifert and Moldowan (1981) observed numerous low-maturity oils with 20S/(20S + 20R) in the range 0.23–0.29. The onset of petroleum generation is currently best estimated using hopane epimer ratios or the porphyrin maturity parameter (see below).

Other factors, such as organofacies differences, weathering, and biodegradation, can affect sterane isomerization ratios. For example, facies effects on C_{29}-sterane 20S/ (20S + 20R) occur in a sequence of Lower Toarcian rocks from southwestern Germany (Moldowan et al., 1986) and in organic-rich lacustrine rocks from offshore west Africa (Hwang et al., 1989). Weathering in Phosphoria Shale outcrops in Utah led to preferential loss of the $\alpha\alpha\alpha$ 20S C_{29}-sterane diastereomer (Clayton and King, 1987). On the other hand, partial sterane biodegradation of oil can result in an increase in C_{27}, C_{28}, and C_{29} sterane $\alpha\alpha\alpha$ 20S/(20S + 20R) to above 0.55, presumably by selective removal of the $\alpha\alpha\alpha$ 20R epimer by bacteria (Rullkötter and Wendisch, 1982; McKirdy et al., 1983; Seifert et al., 1984). Peters et al. (1990) show that this ratio depends partly on the source rock and can decrease at high maturity. Differential stability of epimers, generation of additional material from the kerogen, and other factors may affect this ratio (Dzou et al., 1995).

Immature extracts from hypersaline rocks can appear to be mature due to sterane (and hopane, see p. 614) diagenetic pathways proposed by ten Haven et al. (1986). These immature hypersaline samples contain abundant $5\alpha,14\beta,17\beta$(H) 20R and 20S and $5\alpha,14\alpha,17\alpha$(H) 20R steranes but virtually no $5\alpha,14\alpha,17\alpha$(H) 20S steranes. This contradicts the relationship between these compounds proposed by Mackenzie and Maxwell (1981). Rullkötter and Marzi (1988) and Peakman et al. (1989) indicate that unusually high concentrations of $\alpha\beta\beta$ steranes in immature, hypersaline sediments may result from diagenetic reduction of $\Delta 7$ and $\Delta 8(14),5\alpha$-sterols via the $\Delta 14,5\alpha,17\beta$(H)-sterenes. Care should be used when applying sterane 20S/(20S + 20R) and especially $\beta\beta/(\alpha\alpha + \beta\beta)$ for maturity determinations of samples from hypersaline sources. Bitumens from hypersaline or evaporitic environments are characterized by many of the following characteristics: even-to-odd n-alkane predominance, low pristane/phytane, high gammacerane, low diasteranes/steranes, and preferential preservation of C_{34} and/or C_{35} homohopanes.

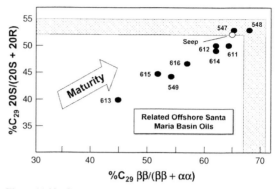

Figure 14.23. Correlation of thermal maturity parameters based on apparent isomerization of asymmetric centers in the C_{29} steranes for oils generated from the Miocene Monterey Formation in the offshore Santa Maria Basin, California (Peters, 1999b). Same samples as in top of Figure 14.11 (right). Endpoints for reactions are indicated by stipled area. Reprinted with permission by ChevronTexaco Exploration and Production Technology Company, a division of Chevron USA Inc.

$\beta\beta/(\beta\beta + \alpha\alpha)$ isomerization

> Highly specific for immature to mature range (see Figure 14.3). Measured using m/z 217 (see Figure 8.20) or preferably by GCMS/MS of C_{29} steranes. Also expressed as $\%\beta\beta$.

Isomerization at C-14 and C-17 in the C_{29} 20S and 20R regular steranes causes an increase in $\beta\beta/(\alpha\alpha + \beta\beta)$ from near-zero values to ~0.7 (0.67–0.71 = equilibrium) with increasing maturity (Figure 14.23) (Seifert and Moldowan, 1986). This ratio appears to be independent of source organic matter input (however, see below) and somewhat slower to reach equilibrium than 20S/(20S + 20R), thus making it effective at higher levels of maturity.

Plots of $\beta\beta/(\alpha\alpha + \beta\beta)$ versus 20S/(20S + 20R) for the C_{29} steranes are particularly effective in order to describe the thermal maturity of source rocks or oils (Seifert and Moldowan, 1986). The plots can be used to check one maturity parameter versus another (Figure 14.23). For example, data for any oils that plot far off the maturity trend line in Figure 14.23 would be re-examined immediately in light of the disagreement between the two maturity parameters. When such disagreements occur, they can sometimes be explained as resulting from some type of analytical error or poor sample quality. Alternatively, such variations can indicate samples that have experienced different heating rates in the subsurface (Mackenzie and McKenzie, 1983) or different levels of clay catalysis (Huang Difan et al., 1990).

Unlike older oils, the sterane isomerization ratios for Tertiary oils are generally not at equilibrium (Grantham, 1986b). This appears to result from insufficient time for complete sterane isomerization in Tertiary rocks, even though the generated oils are thermally mature based on bulk geochemical characteristics, including API gravity, sulfur content, and gross composition.

As for all sterane parameters, GCMS/MS improves the accuracy of the C_{29}-sterane isomerization ratio measurements. In addition, GCMS/MS facilitates more accurate measurement of C_{27} and C_{28} sterane $\beta\beta/(\beta\beta + \alpha\alpha)$, although C_{28} $\beta\beta\alpha$20R co-elutes with $C_{28}\alpha\beta\beta$20R. Comparable data are not available from routine GCMS because of co-eluting compounds on the m/z 217 chromatograms.

Ten Haven et al. (1986) suggested early diagenetic formation of steranes in hypersaline environments. Abundant $\beta\beta$ steranes in these sediments could be due to reaction with sulfur. Schmid (1986), for example, noted that heating $\alpha\alpha\alpha$ cholestane with sulfur gave little isomerization at C-20, but substantial amounts of $\alpha\beta\beta$ 20R and 20S isomers formed. Thus, $\beta\beta$ steranes in these environments may not originate during catagenesis. McKirdy et al. (1983) suggested that the source-rock mineral matrix might affect both C_{29}-sterane maturation parameters. For example, they observed that oils from a probable carbonate source plotted to the left of the Seifert and Moldowan (1981) empirical trend on the C_{29}-sterane diagram. Oils derived from gypsum salt and carbonate-enriched rocks behaved like the carbonate-derived oils (Huang Difan et al., 1990). Rullkötter and Marzi (1988) noted higher $\beta\beta/(\beta\beta + \alpha\alpha)$ for extracts from hypersaline rocks compared with adjacent shales. Laboratory heating experiments indicate that $\beta\beta/(\beta\beta + \alpha\alpha)$ responds differently to different source-rock lithologies (Peters et al., 1990). Despite this finding, $\beta\beta/(\alpha\alpha + \beta\beta)$ is very useful in order to determine maturity for most oils and bitumens.

Biomarker maturation index

Highly specific for immature to mature range. Measured using m/z 217 chromatogram or preferably by GCMS/MS of C_{29} steranes. Use with caution as a migration indicator.

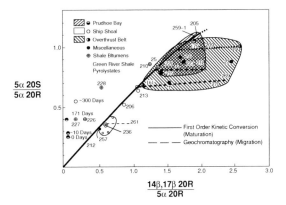

Figure 14.24. Biomarker migration index (BMAI) plot for various oils (Seifert and Moldowan, 1981). This curve was also used to calculate a BMAI. Reprinted with permission by ChevronTexaco Exploration and Production Technology Company, a division of Chevron USA Inc.

24-Ethyl-5α,14α,17α(H)-cholestane 20R is the dominant C_{29} sterane in thermally immature source rocks. Source-rock extracts and oils show greater proportions of 24-ethyl-5α,14α,17α(H)-cholestane 20S and 24-ethyl-5α,14β,17β(H)-cholestane 20R + 20S plus other epimers with increasing maturity. As discussed above, $\alpha\alpha\alpha$ 20S has approximately the same thermal stability as $\alpha\alpha\alpha$ 20R, while $\alpha\beta\beta$ 20R and $\alpha\beta\beta$ 20S are significantly more stable. These thermal stability relationships among epimers are supported by molecular mechanics calculations (Van Graas et al., 1982).

Seifert and Moldowan (1981) derived a theoretical maturity curve (Figure 14.24) that plots the $[5\alpha,14\beta,17\beta(H)\ 20R]/[5\alpha,14\alpha,17\alpha(H)\ 20R]$ versus $[5\alpha,14\alpha,17\alpha(H)\ 20S]/[5\alpha,14\alpha,17\alpha(H)\ 20R]$ for the C_{29} steranes. The distance from the origin along this curve yields a third maturation parameter called the biomarker maturation index (BMAI). This plot is similar to that of $\beta\beta/(\beta\beta + \alpha\alpha)$ versus 20S/(20S + 20R) described above (Figure 14.23). The latter plot gives a straight-line rather than a curved-line relationship.

Based on geological inferences, oils with sterane epimer ratios plotting to the right of the BMAI curve shown in Figure 14.24 (e.g. those on the dashed lines) were explained by geochromatography during migration of the oils through rocks containing clays (Seifert and Moldowan, 1981; 1986). Differences in response of the two ratios to different heating rates of the source rocks was proposed as an alternative explanation (Mackenzie, 1984). No direct evidence for either hypothesis is available. Thus, direct application of the BMAI method to assess relative migration of oils is not advised.

Controversy exists regarding the geochromatographic explanation for data plotting to the right of the BMAI curve shown in Figure 14.24. Although some laboratory simulations suggest that expulsion and migration account for significant changes in biomarker epimer ratios (Fan Zhao-an and Philp, 1987; Jiang Zhusheng et al., 1988), other results for structurally similar biomarkers do not (Hoffmann et al., 1984; Peters et al., 1990; Brothers et al., 1991). For example, Brothers et al. (1991) showed that no consistent redistribution of 5α- versus 5β-cholestane epimers occurred during laboratory migration experiments, although the column may have been too short to simulate nature adequately. Peters et al. (1990) show no significant change in $14\beta,17\beta$(H) 20R/$14\alpha,17\alpha$(H) 20R C_{29}-sterane ratios related to expulsion in heating experiments.

Diasteranes/steranes

> Specificity for early mature to early postmature, but depends partly on depositional environment. Steranes measured on m/z 217 mass chromatograms, or preferably by GCMS/MS (M+ → m/z 217, where M+ = m/z 400, 386, and 372); diasteranes measured using m/z 259 (see Figure 8.20), or preferably by GCMS/MS.

Thermal maturity, lithology, and the redox potential of the source-rock depositional environment affect diasteranes/steranes. As a result, this ratio is useful for maturity determination only when the oils or bitumens being compared are from the same source-rock organic facies. Catalysis by acidic clays has been proposed as the mechanism that accounts for diasterenes in sediments (Rubinstein et al., 1975). Acidic catalysis is necessary to convert sterenes to diasterenes, which are the precursors of diasteranes (see Figure 13.49) (Kirk and Shaw, 1975). Thus, diasteranes/steranes are typically low in carbonate source rocks and oils. However, because diasteranes occur in certain highly calcareous rocks from the Adriatic Sea area that are clay-poor (Moldowan et al., 1991a), other acid mechanisms may be effective. The Adriatic carbonate rocks with high diasteranes are not petroleum source rocks. They contain only small amounts of oxidized organic matter. High Eh during deposition of these sediments may account for the diasteranes. Most reports of low diasteranes in carbonates (McKirdy et al., 1983; Rullkötter et al., 1985) involve organic-rich carbonate source rocks, where the original Eh during deposition was very low (anoxic).

Once formed, diasteranes are more stable than regular steranes. Diasteranes/steranes increase dramatically past peak oil generation, as shown by hydrous pyrolysis experiments (Peters et al., 1990). At these high levels of maturity, rearrangement of steranes to diasteranes may be possible even without clays, probably due to hydrogen-exchange reactions that are enhanced by water (Rullkötter et al., 1984). Alternatively, heating in the postmature range induces destruction of biomarkers, and increased diasteranes/steranes might indicate better survival of diasteranes under high-temperature conditions.

Caution should be used when applying diasteranes/steranes because different versions are applied, depending on the laboratory and the GCMS method. We often use [total C_{27} to C_{29} $13\beta,17\alpha$(H) 20S + 20R diasteranes]/[total C_{27} to C_{29} $5\alpha,14\beta,17\beta$(H) and $5\alpha,14\alpha,17\alpha$(H) 20S + 20R] steranes from GCMS/MS data. However, the ratios at a single carbon number, for example C_{27}, are also used because measurement of C_{27} $\beta\alpha$-diasteranes (20S + 20R) has minimal interference on the m/z 217 chromatogram using GCMS in the SIM mode. However, measurement of the C_{27} steranes is difficult because of interference from the C_{29} diasteranes and some C_{28} epimers. Some laboratories resort to C_{27} diasteranes/C_{29} steranes, but this is a poor compromise because this ratio introduces complications from source input variability.

Diasteranes/steranes determined by SIM or GCMS/MS analysis can result in somewhat different values because the SIM method involves more interfering peaks. The $13\beta,17\alpha$(H) 20S + 20R isomers are the major diasteranes in rocks and petroleum (Ensminger et al., 1978). They can be measured using m/z 259 mass fragmentograms (see Figure 8.20), which are more specific for diasteranes than m/z 217. Minor amounts of $13\alpha,17\beta$(H) 20S + 20R diasterane isomers also occur when the $13\beta,17\alpha$(H) isomers are present, but generally they are not quantified.

The use of GCMS/MS simplifies diasteranes/steranes and improves the accuracy of their measurement by reducing interference. Interference between steranes and diasteranes with different carbon numbers can be eliminated, making it possible to measure accurately the ratio for each carbon number (e.g. C_{27}, C_{28}, C_{29}, C_{30}). Diasteranes/steranes correlate closely at each carbon number for oil and extract samples from a given source rock. Diasteranes/steranes for different carbon numbers can be cross-plotted to differentiate related samples having different maturities. Genetically unrelated samples commonly plot along different vectors on these diagrams, thus allowing source differentiation.

20S/(20S + 20R) 13β,17α(H)-diasteranes

> Mature to highly mature range. Measured by MRM/GCMS m/z 358 → 259 chromatograms (see Figure 8.20).

The 20S/(20S + 20R) ratio for C_{27}–C_{29} diasteranes increases with thermal maturity (Mackenzie et al., 1980a) in a similar fashion to that for the 17α-hopane 22S/(22S + 22R). The steric environment of the C-20 asymmetric carbon in these diasteranes is more similar to the C-22 carbon in the 17α-hopanes than the C-20 carbon in the steranes. C-20 in the 13β,17α(H)-diasteranes equilibrates [20S/(20S + 20R) ~0.6] at about the onset of oil generation (Mackenzie et al., 1980a). However, the 20S epimer forms during the rearrangement of sterenes (Akporiaye et al., 1981) leading to 20S/(20S + 20R) near 0.1, even in thermally immature rocks.

C_{26} 21/(21 + 27)-norcholestane ratio

> Specificity unknown, early mature range. Measured using GCMS/MS m/z 358 → 217 transition.

The 21/(21 + 27)-norcholestane ratio reflects the maturity of crude oils (Moldowan et al., 1991a). Figure 14.25 shows a linear correlation between this parameter and API gravity for oils from central Saudi Arabia (Moldowan et al., 1994b). An extract of Silurian Qusaiba source rock in the Hawtah-1 well correlates with these central Saudi Arabian oils, including 50° API oil from the same well (6153 feet (1875.4 m)). However, Figure 14.25 shows that the Hawtah-1 extract is much less mature than the oil, indicating substantial vertical migration of the oil from deeper, more mature equivalents of the source rock. Qusaiba source-rock extracts from the Stratigraphic-39 (11 650 feet (3550.9 m)) and

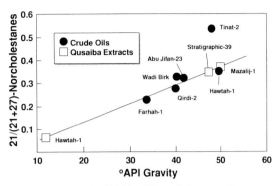

Figure 14.25. The C_{26} 21/(21 + 27)-norcholestane ratio correlates with API gravity and thermal maturity of central Saudi Arabian oils and related Lower Silurian Qusaiba source-rock extracts (Moldowan et al., 1994b). The Tinat-2 oil falls off the trend, possibly due to inaccurate measurements caused by very low C_{26} steranes.

Mazalij-1 (13 505 feet (4116.3 m)) wells also correlate with the Hawtah-1 oil but show sufficient maturity to have generated the oil. Analysis of additional Qusaiba rock samples might allow calibration of burial depth versus oil maturity that could be used to develop models of petroleum migration and generation in the area.

AROMATIC STEROIDS

TA/(MA + TA), monoaromatic steroid aromatization

> Highly specific for immature to mature range (see Figure 14.3). Measured using m/z 253 (monoaromatic steroid, see Figure 8.22) and m/z 231 (triaromatic steroid, see Figure 8.23). (Figure 8.20 shows the principal fragments.)

The aromatization of C-ring monoaromatic (MA) steroids to ABC-ring triaromatic (TA) steroids involves the loss of a methyl group at the A/B ring junction (Figure 14.26). The asymmetric center at C-5 is lost during conversion of the monoaromatic to the triaromatic compounds. Thus, maturation of monoaromatic steroids yields triaromatic steroids with one less carbon. TA/(MA + TA) increases from 0 to 100% during thermal maturation. The ratio has been applied to calibrations of basin models (Mackenzie, 1984). Evidence suggests that this ratio can be affected by expulsion

C_{28}-TA/(C_{29}-MA + C_{28}-TA)

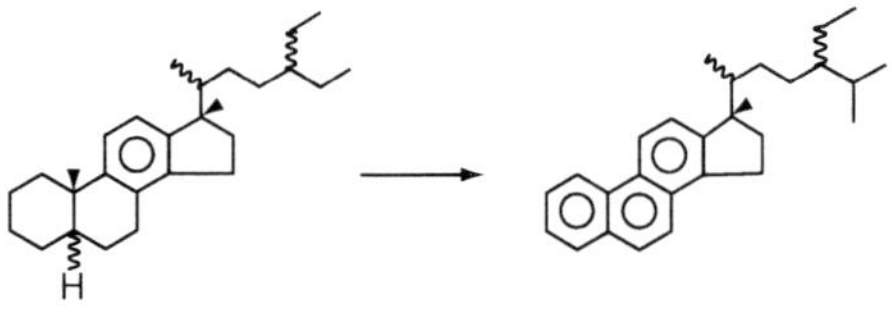

Figure 14.26. Conversion of C_{29}-monoaromatic (MA) with C_{28}-triaromatic (TA) steroids during thermal maturation. Note loss of the methyl group attached to C-10 (A/B-ring juncture) and loss of the asymmetric center at C-5 in this reaction. See Figure 2.20 for nomenclature related to numbering of the carbon atoms.

(Hoffmann et al., 1984; Peters et al., 1990). The more polar triaromatic steroids are retained preferentially in the bitumen compared with the expelled oil.

Several different TA/(MA + TA) maturity parameters are reported in the literature. One of these ratios requires measurement of two C_{29} monoaromatic epimers ($5\alpha,20R$ and $5\beta,20R$) and one C_{28} triaromatic epimer (20R) (Mackenzie et al., 1981a). Palacas et al. (1986) modified the TA/(MA + TA) used by Mackenzie et al. (1981a) to include C_{20} and C_{28} triaromatic steroids. Mackenzie and McKenzie (1983) studied the kinetics of the proposed aromatization reaction. However, additional structure studies on C-ring monoaromatic steroids show that isomer distributions are much more complex than believed previously (Moldowan and Fago, 1986; Riolo and Albrecht, 1985; Riolo et al., 1986). For example, it is likely that monoaromatic steroids with a methyl group rearranged from C-10 to C-5 also generate triaromatic steroids during thermal maturation. These rearranged or dia-monoaromatic steroids are significant, sometimes dominant monoaromatic steroids in oils and source rocks (Figure 8.22). Most dia-monoaromatic steroids and regular monoaromatic steroids cannot be analyzed separately because of co-elution. Thus, C_{29}-monoaromatic steroid in the ratio TA/(MA + TA) appears to also contain some C_{29} dia-monoaromatic steroids.

Use of the TA/(MA + TA) as a maturity parameter requires that the same version of the ratio be used for all samples. The monoaromatic component includes the major known C_{29} monoaromatic steroid peaks (four peaks labeled 12, 13, 15, and 16 from m/z 253 chromatogram) (Figure 8.22) representing 12 compounds: 8 regular monoaromatic steroids ($5\alpha,10\beta(CH_3),20R,24R$, $5\alpha,20R,24S$, $5\alpha,20S,24R$, $5\alpha,20S,24S$, $5\beta,20R24R$, $5\beta,20R,24S$, $5\beta,20S,24R$, and $5\beta,20S,24S$) and 4 dia-$5\beta(CH_3)$ monoaromatic steroids ($5\beta(CH_3)$, $10\beta,20R,24R$, 20R24S, 20S24R, and 20S24S). The triaromatic component includes two C_{28} triaromatic steroid peaks (peaks 6 and 8 from m/z 231 chromatogram) (Figure 8.23), representing four epimeric compounds: 20R24R, 20R24S, 20S24R, and 20S24S. The DB-1 chromatographic column that we use under normal conditions does not resolve the 24R and 24S epimers for the monoaromatic and triaromatic steroids.

One version of TA/(MA + TA) uses the sum of all known C_{27}–C_{29} C-ring monoaromatic steroid peaks (m/z 253) for MA and the sum of all C_{26}–C_{28} triaromatic steroid peaks (m/z 231) for TA in the expression TA/(MA + TA). However, this complex approach is not common. Work from contract laboratories may include any of the above TA/(MA + TA) parameters, or others.

MA(I)/MA(I + II)

Some interference from source input. Early mature to late mature range (see Figure 14.3). Measured using m/z 253 chromatogram (see Figure 8.20).

Apparent side-chain scission (carbon–carbon cracking) with increasing thermal maturity has been documented for aromatic steroids in oils (Seifert and Moldowan, 1978) and rocks (Mackenzie et al., 1981a). MA(I)/MA(I + II) increases from 0 to 100% during thermal maturation (Figure 14.27). It is not known whether this increase is the result of (1) conversion of long-chain to short-chain monoaromatic steroids by carbon–carbon cracking, (2) preferential thermal degradation of the long- versus short-chain series, or (3) both. Moldowan et al. (1986) showed that this parameter is influenced by diagenetic conditions, particularly Eh, in the source sediment. The origin of the short side-chain aromatic steroids is unclear. They could be formed from the C_{27}–C_{29} monoaromatic steroids and/or other precursors. Short-side-chain aromatic steroids could originate from larger aromatic steroids by homolytic (free

MA(I)/MA(I + II)

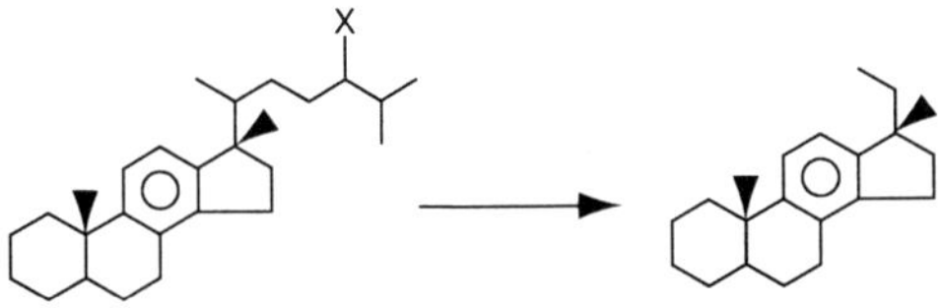

X = H, CH₃, C₂H₅

Figure 14.27. Conversion of group II monoaromatic to group I monoaromatic steroids by side-chain cleavage during thermal maturation. Alternatively, or in addition, short- and long-side-chain steroids may have different natural-product precursors and may undergo differential destruction during maturation. The exact position of the C-19 nuclear methyl group in the short-chain compounds (group I) is not well documented and may reside at C-10 (shown), or at C-5, or as a mixture of the two.

radical) scission of the C-20 to C-22 bond. However, substantial quantities of C_{21} monoaromatic steroids in many samples suggest that they originate from sterols with functionalized side chains (i.e. oxidative cleavage of a C-22 double bond) during diagenesis. There are also C_{21}- and C_{22}- steroid natural products, which could lead to short side-chain monoaromatic steroids.

Mackenzie et al. (1981a) used the C_{28} monoaromatic steroid (mainly the 5β20R isomer) (Moldowan and Fago, 1986) as MA(II) and the C_{21} monoaromatic steroid as MA(I) (Figure 14.27). Our objection to this ratio is that the concentration of C_{28} relative to C_{27} through C_{29} monoaromatic steroids depends partly on source input and the fact that the C_{21} monoaromatic steroid may be derived from any or all of the C_{27}–C_{29} monoaromatic steroids. We use the sum of all major C_{27}–C_{29} monoaromatic steroids as MA(II) and C_{21} plus C_{22} as MA(I) (peaks 4–16 and 1–2 in Figure 8.22 respectively) in order to reduce the source input effect.

The aromatic steroid side-chain scission reactions are based on carbon–carbon cracking and generally require more thermal energy to proceed than isomerizations. They appear to be most useful in the late oil window, when most other biomarkers are no longer effective as maturity indicators.

TA(I)/TA(I + II)

> Some interference from source input. Peak mature to late mature range (see Figure 14.3). Measured using m/z 231 mass chromatogram (see Figure 8.23).

Most of the above discussion on short- and long-chain monoaromatic steroids applies to short- and long-chain triaromatic steroids (Figure 14.28). In addition, the triaromatic steroids are probably derived from the monoaromatic steroids (see discussion on TA/(MA + TA) above). Thus, TA(I)/TA(I + II) has the added advantage of being more sensitive at higher maturity than the same ratio for the monoaromatic steroids. Heating experiments indicate that TA(I)/TA(I + II) increases due to preferential degradation of the long-chain triaromatic homologs rather than conversion of long- to short-chain homologs (Beach et al., 1989).

The TA(I)/TA(I + II) parameter has been measured on two combinations of triaromatic-steroids.

TA(I)/TA(I + II)

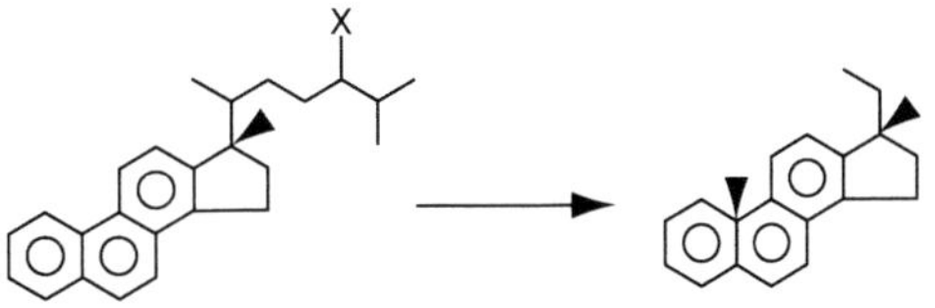

X = H, CH₃, C₂H₅

Figure 14.28. Proposed conversion of group II triaromatic to group I triaromatic steroids by side-chain cleavage during thermal maturation. Beach et al. (1989) show that TA(I)/TA(I + II) increases due to preferential degradation of the long-chain triaromatics rather than conversion of long- to short-chain homologs.

Mackenzie et al. (1981a) suggest the C_{28} triaromatic steroid (20R) as TA(II) and the C_{20} triaromatic steroid as TA(I) (see Figure 14.3). We prefer to use the sum of C_{26}–C_{28} (20S + 20R) triaromatic steroids as TA(II) and the C_{20} and C_{21} triaromatic steroids as TA(I) (peaks 4–8 and peaks 1–2, respectively, in Figure 8.23). As with the monoaromatic steroid scission parameter, we use a summation of carbon numbers for the triaromatic scission parameter to reduce complications caused by different source organic matter. For example, preferential input of C_{27}, C_{28}, or C_{29} sterols in the source sediments could lead to a predominance of C_{26}, C_{27}, or C_{28} triaromatic steroids, respectively, unrelated to thermal maturity.

C_{26} triaromatic 20S/(20S + 20R)

> Highly specific (?) for mature to highly mature range. GCMS/MS analyses of M + → 231.

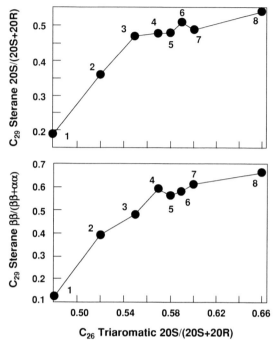

Figure 14.29. The C_{26} triaromatic steroid 20S/(20S + 20R) for oils and bitumens from the Eel River Basin, California, is more sensitive at higher levels of maturity than the two C_{29} sterane biomarker maturity ratios shown. These samples are the same as those numbered in Figure 14.9. Reprinted with permission by ChevronTexaco Exploration and Production Technology Company, a division of Chevron USA Inc.

The first application of this maturity ratio shows good agreement with the two more commonly used maturity parameters based on isomerization in the C_{29} steranes (Figure 14.29). Based on Figure 14.29, C_{26} triaromatic 20S/(20S + 20R) is more sensitive at higher levels of maturity than either C_{29} sterane ratio.

The enhanced sensitivity of C_{26}-triaromatic 20S/(20S + 20R) compared with the more common steroid maturity parameters will be useful for studies of highly mature oils and condensates. Isomerization of C-20 in the triaromatic steroids may be hindered because of an additional methyl group attached to C-17 compared with the steranes. This interpretation is supported by Schmid (1986, p. 158–159), who observed anomalously low 20S/(20S + 20R) for this type of compound in laboratory heating experiments. No other observations of this apparent maturity parameter have been published.

The 20S/(20S + 20R) ratios for the C_{27} and C_{28} triaromatic steroids show trends similar to those for C_{26} triaromatic steroids with maturity, although they are more erratic (Peters and Moldowan, 1993), possibly due to interference. Parameters based on all three carbon numbers (C_{26}–C_{28}) have the same potential for application when measured using GCMS/MS.

AROMATIC HOPANOIDS

Monoaromatic hopanoid parameter

> Unknown specificity. Early mature to late mature range (?). Measured using m/z 191 for benzohopanes and m/z 365 for 8,14-secohopanoids (see Figure 8.20).

He Wei and Lu Songnian (1990) propose a monoaromatic hopanoid (MAH) maturity parameter based on ring-D monoaromatized 8,14-secohopanoids and benzohopanes [8,14-secohopanoids/(8,14-secohopanoids + benzohopanes)] (see Figure 8.20). They observed that aromatic fractions of crude oils and mudstone extracts from Liaohoe Basin, northeastern China, show systematic increases in the MAH ratio with burial depth. Variations in the ratio are attributed to differential generation and destruction of monoaromatic secohopanoids and benzohopanes rather than conversion between the two groups of compounds. Based on comparison of MAH with odd-to-even preference for n-alkanes (OEP) and C_{31} hopane $22S/(22S + 22R)$ for a limited number of samples, they concluded that the threshold of oil generation corresponds to MAH of ~ 0.3. Because both monoaromatic 8,14-secohopanoids and benzohopanes resist biodegradation compared with most saturated biomarkers, they suggest that the MAH maturity parameter can be used for maturity assessment of heavily biodegraded oils. They observed pairs of C_{28}–C_{31} demethylated counterparts of the regular D-ring monoaromatic 8,14-secohopanoids, most probably lacking the C-28 methyl group, in rock and oil samples from the offshore Korea Bay Basin in the Yellow Sea. Trends in the relative distributions of these demethylated compounds parallel those of the regular secohopanoids, suggesting a common origin.

PORPHYRINS

DPEP/etio

> Immature to mature range (as etio/(etio + DPEP) for vanadyl porphyrins) (see Figure 14.3). Measured from total averaged mass spectra or HPLC of demethylated porphyrin concentrates.

The decrease of DPEP/etio porphyrins with thermal maturity (Baker and Louda, 1986) is generally attributed to loss of the isocyclic ring from DPEP. However, the ratio may change as a result of differential stability of the two classes of molecules rather than conversion of DPEP directly to etio compounds. Furthermore, under oxidizing diagenetic conditions, the chlorin precursors to DPEP porphyrins can lose the isocyclic ring (Baker and Louda, 1986). Evidence for loss of the isocyclic ring during early diagenesis has been shown in the Messel Shale (Ocampo et al., 1985a). Sundararaman et al. (1988a) show that etio porphyrins are preferentially released during catagenesis of kerogen, while DPEP porphyrins are dominant in immature bitumen. The DPEP/etio maturity parameter is crude because it involves many compounds. Progress toward improved porphyrin maturity parameters has been slow because of the complex procedures for extraction and isolation of metalloporphyrins.

Porphyrin maturity parameter

> Highly specific for immature to early mature range (see Figure 14.3). Separation and measurement by high-performance liquid chromatography (HPLC).

The porphyrin maturity parameter (PMP) increases with thermal maturity and is tied to generation from the kerogen (Sundararaman et al., 1988a; 1988b). Most other maturity parameters measure the extent of thermally induced chemical reactions not related directly to kerogen breakdown. PMP of 0.2 in bitumen signals the beginning of oil generation. The ratio is determined using a rapid HPLC separation of two vanadyl porphyrins, a C_{28} etioporphyrin and a C_{32} DPEP porphyrin (Figure 14.30). These two porphyrins may or may not be related as precursor and product. For example, the change in $C_{28}E/(C_{28}E + C_{32}D)$ with maturity could be caused by higher relative thermal stability of the C_{28} etio compared with the C_{32} DPEP compound. Detailed structures of these two porphyrins are unknown.

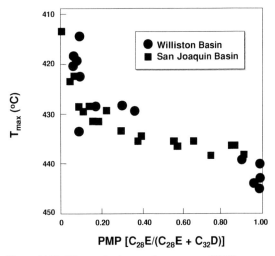

Figure 14.30. The porphyrin maturity parameter (PMP) increases sharply at the onset of oil generation. The plot of Rock-Eval T_{max} versus PMP for source rock samples from the Williston Basin, North Dakota (circles), and San Joaquin Basin, California (squares), indicates that a strong shift in PMP from 0.3 to 1.0 occurs at T_{max} $\sim$435–440°C, consistent with the onset of oil generation (Sundararaman et al., 1988a). Reprinted with permission by ChevronTexaco Exploration and Production Technology Company, a division of Chevron USA Inc.

Table 14.1. *Source-dependent parameters for crude oils from the Czech Republic (Picha and Peters, 1998)*

	Oil 1 (Lubna-18)	Oil 2 (Dolni Lomna-1)	Oil 3 (Zdanice-7)	Oil 4 (Damborice-16)
Depth (m)	1550	2010–2015	961–976	2443–2452
°API	24	Insufficient	28	Insufficient
%Sulfur	0.73	0.27	0.08	0
Saturates/aromatics	1.87	1.98	2.85	3.15
Carbon preference index	–	1.1	1.0	1.1
$\delta^{13}C_{PDB}$	−26.8	−25.6	−26.5	−26.6
Pristane/phytane	~1.09	2.22	1.51	1.41
Pristane/nC_{17}	–	0.92	0.21	0.37
Phytane/nC_{18}	–	0.42	0.97	0.26
Diasteranes/(Dia + Ster)	0.67	0.63	0.67	0.69
Gammacerane index	0.27	0.24	0.41	0.49
Steranes				
%C_{26} 24/(24+27)*	65	59	17	19
%C_{27}/(C_{27}–C_{29})	32	34	42	47
%C_{28}/(C_{27}–C_{29})	42	42	33	31
%C_{30}/(C_{27}–C_{30})	4.2	3.6	4.8	5.0
Diasteranes				
%C_{27}/(C_{27}–C_{29})	22	23	31	33
%C_{28}/(C_{27}–C_{29})	48	45	34	33
Monoaromatic steroids				
%C_{27}/(C_{27}–C_{29})	16	19	20	23
%C_{28}/(C_{27}–C_{29})	47	43	46	37
Triaromatic steroids				
$C_{26}S/C_{28}S$	0.64	0.65	0.78	0.73
$C_{27}R/C_{28}R$	1.2	1.17	1.06	0.88
C_{35} homohopane index	8.2	6.2	3.6	2.8
Oleanane index	0.16	0.14	0.05*	0.04**

* Nordiacholestane ratio (Holba *et al.*, 1998).

** The peak eluting with a retention time equivalent to that of oleanane on m/z 191 mass chromatogram (GCMS) could not be confirmed as oleanane using GCMS/MS.

The mechanism of formation of porphyrins from kerogen during maturation is not clear. Beato *et al.* (1991) pyrolyzed kerogen from the New Albany Shale at 300°C and analyzed the generated nickel and vanadyl porphyrins using tandem mass spectrometry. They conclude that porphyrins are not released by carbon–carbon bond scission at 300°C but are generated by a solubilization or desorption mechanism.

EXERCISE

Tables 14.1 and 14.2 and Appendices 14.1–14.3 show geochemical data for four crude oils from the sub-Carpathian foreland plate in Moravia, Czech Republic. (Appendix A shows gas chromatograms, Appendix B shows GCMS terpane mass chromatograms, and Appendix C shows MRM/GCMS sterane mass

Table 14.2. *Thermal-maturation dependent parameters for crude oils from the Czech Republic (Picha and Peters, 1998)*

	Oil 1 (Lubna-18)	Oil 2 (Dolni Lomna-1)	Oil 3 (Zdanice-7)	Oil 4 (Damborice-16)
Hopanes				
%22S/(22S + 22R)	57	57	57	58
C_{30} moretane/hopane	0.08	0.08	0.09	0.08
Steranes				
%20S/(20S + 20R)	44	49	40	51
%$\beta\beta/(\beta\beta + \alpha\alpha)$	50	56	53	66
Monoaromatic steroids				
MA(I)/MA(I + II)	0.24	0.24	0.28	0.34
Triaromatic steroids				
TA(I)/TA(I + II)	0.11	0.20	0.28	0.25

chromatograms.) Geochemical logs indicate possible Paleogene and Jurassic source rocks in the basin.

1. What are the genetic relationships among the crude oils?
2. Are the oils biodegraded, and what are their relative thermal maturities?
3. What does the geochemistry of the oils indicate about the depositional environment of their source rocks? Were the source rocks marine or non-marine; oxic, suboxic, or anoxic; freshwater, normal salinity, or hypersaline; clay-rich or clay-poor? Do age-related biomarkers suggest a Paleogene or Jurassic source?

APPENDIX A: GAS CHROMATOGRAMS OF CRUDE OILS FROM MORAVIA, CZECH REPUBLIC

APPENDIX B: TERPANE MASS CHROMATOGRAMS OF CRUDE OILS FROM MORAVIA, CZECH REPUBLIC

APPENDIX C: STERANE MASS CHROMATOGRAMS OF CRUDE OILS FROM MORAVIA, CZECH REPUBLIC

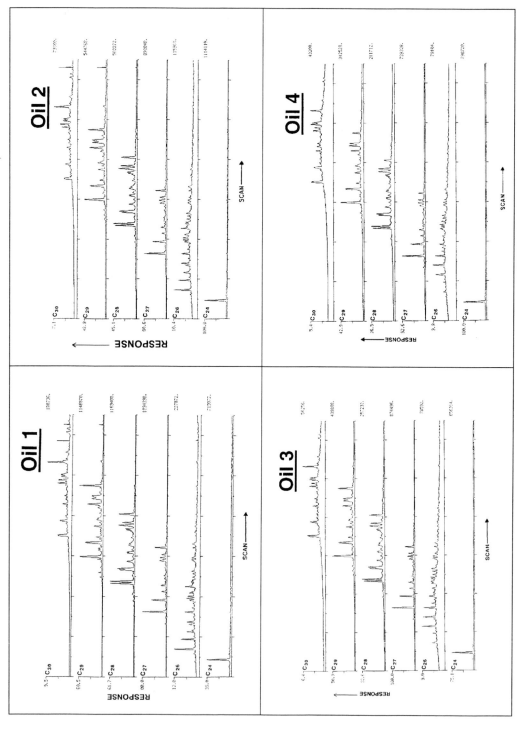

15 · Non-biomarker maturity parameters

This chapter explains how certain non-biomarker parameters, such as ratios involving n-alkanes and aromatic hydrocarbons, are used to assess thermal maturity. Critical information on specificity and the means for measurement are highlighted before the discussion of each parameter.

Various characteristics of petroleum samples can be used to assess their relative level of thermal maturity. For related crude oils of increasing thermal maturity, the n-alkane envelope becomes displaced toward lower-molecular-weight homologs (see Figure 4.23). API gravity and nC_{19}/nC_{31} and saturate/aromatic ratios increase, while sulfur, nitrogen, and isoprenoid/n-alkane ratios decrease. For example, thermal maturity is the principal factor controlling the strong inverse correlation between API gravity and sulfur content for Monterey oils from the Santa Barbara Channel and offshore Santa Maria Basin (Baskin and Peters, 1992). Some of the more commonly applied non-biomarker maturity parameters are discussed below.

ALKANES AND ISOPRENOIDS

Isoprenoid/n-alkane ratios

Specific for maturity, but also affected by other processes, such as source and biodegradation. Measured using peak heights or areas from gas chromatography (e.g. see Figure 2.16).

As discussed above, pristane/nC_{17} and phytane/nC_{18} decrease with thermal maturity as more n-alkanes are generated from kerogen by cracking (Tissot et al., 1971). These isoprenoid/n-alkane ratios can be used to assist in ranking the thermal maturity of related, non-biodegraded oils and bitumens. However, organic matter input (Alexander et al., 1981) and secondary processes such as biodegradation can affect these ratios (Figure 4.26).

Carbon preference index and odd-to-even predominance

Specific for maturity, but also affected by other processes, such as source and biodegradation. Meassured using peak heights or areas from gas chromatography.

The relative abundance of odd versus even carbon-numbered n-alkanes can be used to obtain a crude estimate of thermal maturity of petroleum. These measurements include the carbon preference index (CPI) (Bray and Evans, 1961) and the improved odd-to-even predominance (OEP) (Scalan and Smith, 1970). In practice, the OEP can be adjusted to include any specified range of carbon numbers. Some examples of CPI and OEP variations are shown below.

$$CPI = \left[\frac{C_{25} + C_{27} + C_{29} + C_{31} + C_{33}}{C_{24} + C_{26} + C_{28} + C_{30} + C_{32}} + \frac{C_{25} + C_{27} + C_{29} + C_{31} + C_{33}}{C_{26} + C_{28} + C_{30} + C_{32} + C_{34}} \right] \Big/ 2$$

$$CPI(1) = 2(C_{23} + C_{25} + C_{27} + C_{29})/[C_{22} + 2(C_{24} + C_{26} + C_{28}) + C_{30}]$$

$$OEP(1) = (C_{21} + 6C_{23} + C_{25})/(4C_{22} + 4C_{24})$$

$$OEP(2) = (C_{25} + 6C_{27} + C_{29})/(4C_{26} + 4C_{28})$$

CPI or OEP values significantly above (odd preference) or below (even preference) 1.0 indicate low thermal maturity. Values of 1.0 suggest, but do not prove, that an oil or rock extract is thermally mature. CPI or OEP values below 1.0 are unusual and typify low-maturity oils or bitumens from carbonate (see Table 13.3) or hypersaline environments (e.g. see Figure 4.21, middle).

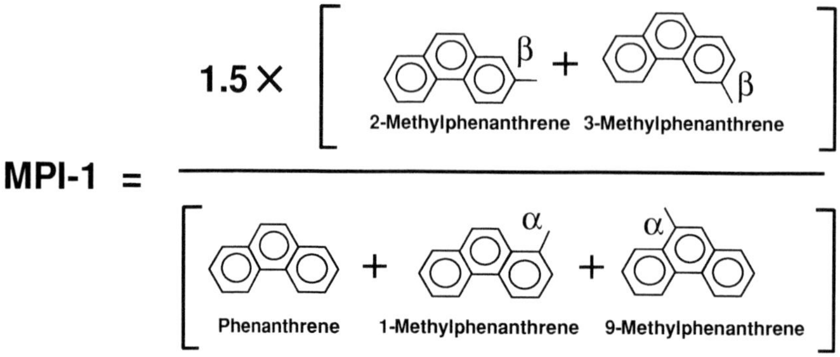

642 Non-biomarker maturity parameters

$$\text{MPI-1} = \frac{1.5 \times [\text{2-Methylphenanthrene} + \text{3-Methylphenanthrene}]}{[\text{Phenanthrene} + \text{1-Methylphenanthrene} + \text{9-Methylphenanthrene}]}$$

Figure 15.1. MPI-1 is a common version of the methylphenanthrene index.

Organic matter input affects CPI and OEP. For example, many Ordovician crude oils show abundant n-alkanes below nC_{20} accompanied by a strong predominance of the odd-numbered homologs in the range nC_{10}–nC_{20} (see Figure 4.21, bottom) (Hatch et al., 1987). This odd-number predominance of n-alkanes is common, even in Ordovician oils showing high API gravities. Thus, traits normally considered to indicate low (odd-number predominance) and high (high API) thermal maturity occur in the same Ordovician oils. This is an example of the effect of organic matter input on these parameters. Estimates of the thermal maturity of these samples can be improved using thermally dependent biomarker parameters.

AROMATIC HYDROCARBONS

The distributions of methylhomologs of naphthalene and phenanthrene are controlled by thermal maturity in the approximate range 0.6–1.7% mean vitrinite reflectance (Rm) (Radke et al., 1982b). The following text describes several of the more common aromatic hydrocarbon maturity parameters.

Methylphenanthrene index

Specific for maturity, but must be calibrated for each petroleum system. Measured using peak heights or areas for phenanthrene and methylphenanthrenes from gas chromatograms or m/z 178 and 192 mass chromatograms.

Various isomer ratios of the methylated aromatic hydrocarbons and sulfur heterocycles were developed as thermal maturity indicators (Radke et al., 1982a; Radke et al., 1986; Radke et al., 1990; Radke and Welte, 1983; Radke, 1987). For example, the methylphenanthrene index (MPI-1) appears to be as useful as vitrinite reflectance for maturity assessment in some field studies (Figure 15.1) (Radke et al., 1982a; Radke, 1988; Farrington et al., 1988b).

Calibrations between methylphenanthrene indices and vitrinite reflectance are available (Radke and Welte, 1983; Boreham et al., 1988). Figure 15.2 shows good positive linear correlation of MPI-1 and Rm in the oil window (0.65–1.35% Rm) and good negative correlation at higher maturity (1.35–2.00% Rm). The two equations in Figure 15.2 can be used to calculate vitrinite reflectance (Rc), depending on which of the two vitrinite reflectance ranges is considered. For example,

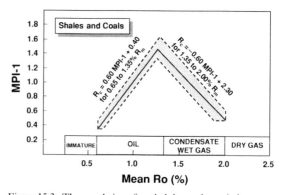

Figure 15.2. The correlation of methylphenanthrene index (MPI-1) with vitrinite reflectance is based on shales and coals containing type III organic matter (Radke and Welte, 1983). %R_c, calculated vitrinite reflectance; %R_m, mean vitrinite reflectance; %R_o, mean vitrinite reflectance under oil.

Rm for a rock sample that lacks vitrinite could be estimated from MPI-1 measured on an extract from the rock. Likewise, MPI-1 for crude oil could be used to estimate Rm of the source rock at the time of expulsion of the oil. Radke *et al.* (1990) used various methylphenanthrene indices for bitumens from coals in a well from Indonesia to estimate the present depth of the source rock for crude oils in the area. If the approximate range of Rm for a sample is unknown, then relationships between Rm and other maturity parameters must be used to decide which of the two equations applies (Radke *et al.*, 1984).

MPI-1 can be calculated using peak areas of phenanthrene and methylphenanthrenes from gas chromatograms or m/z 178 and 192 mass chromatograms, respectively. Calculation of MPI-1 from mass chromatograms requires response factors that are best determined by the analyst. Alternatively, Cassani *et al.* (1988) modified the MPI-1 calculation for peaks measured from m/z 178 and 192 mass chromatograms as follows:

$$\text{MPI-1} = 1.89(2\text{-MP} + 3\text{-MP})/[P + 1.26(1\text{-MP} + 9\text{-MP})]$$

where MP is methylphenanthrene and P is phenanthrene. A similar maturity index was used by Alexander *et al.* (1986b) and modified by Cassani *et al.* (1988):

$$\text{PP-1} = 1\text{-MP}/(2\text{-MP} + 3\text{-MP})$$
$$\text{PP-1}_{\text{modified}} = (1\text{-MP} + 9\text{-MP})/(2\text{-MP} + 3\text{-MP})$$

Several difficulties affect the use of methylphenanthrene indices:

- Samples of different maturity can show identical methylphenanthrene ratios (Figure 15.2).
- Variations in the type of organic matter or lithology in the source rock can affect methylphenanthrene ratios. For example, Cassani *et al.* (1988) noted that high MPI-1 and PP-1 correspond to high carbonate content in La Luna source rocks. Migration may also affect various aromatic hydrocarbon ratios (Radke *et al.*, 1982b).
- Methylphenanthrenes require different separation methods than those used to isolate biomarkers.

Part of the difficulty with MPI may result from the calculation itself. Phenanthrene, the parent and most stable compound, is in the denominator of the ratio along with less stable α-isomers (1-MP and 3-MP), while the β-isomers of intermediate stability (2-MP and 3-MP) are in the numerator. Different rates of generation and destruction of phenanthrene and the α- and β-isomers of the methylphenanthrenes may explain why MPI-1 in Figure 15.2 increases and then decreases with increasing maturity.

Methyl- and ethylnaphthalenes

> Specific for high thermal maturity. Measured using peak heights or areas from gas chromatograms of the aromatic hydrocarbon fraction.

Methylnaphthalene and ethylnaphthalene isomer ratios (MNR and ENR, respectively) (Figure 15.3) are useful maturation indicators (Radke *et al.*, 1982b). For example, thermal rearrangement of of 1-methylnaphthalene results in a pronounced increase in the relative abundance of 2-methylnaphthalene above 0.9% Rm.

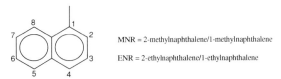

MNR = 2-methylnaphthalene/1-methylnaphthalene
ENR = 2-ethylnaphthalene/1-ethylnaphthalene

Figure 15.3. 1-Methylnaphthalene and the methylnaphthalene (MNR) and ethylnaphthalene (ENR) maturation ratios.

Dimethylnaphthalenes

> Specific for high thermal maturity range. Measured using peak heights or areas from gas chromatograms of the aromatic hydrocarbon fraction.

The ten isomeric dimethylnaphthalenes (DMNs) in crude oils can be separated quantitatively (Alexander, 1983c), and various ratios of these compounds are useful maturity indicators. A relationship exists between the dimethylnaphthalene ratio (DNR) and the mean vitrinite reflectance (Radke *et al.*, 1982b; 1984). Like the methylnaphthalene ratio (MNR), DNR shows a pronounced increase above ~0.9% Rm:

$$\text{DNR-1} = (2{,}6\text{-DMN} + 2{,}7\text{-DMN})/1{,}5\text{-DMN}$$

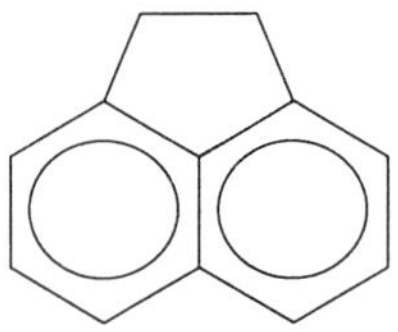

Acenaphthene

Figure 15.4. Acenaphthene may co-elute with 1,8-dimethylnaphthalene (DMN) using certain gas chromatographic columns and conditions. Acenaphthene elutes ~30 seconds after 1,8-DMN using the column and conditions described by Alexander et al. (1983c).

Alexander et al. (1983c) observed a systematic decrease in the abundance of 1,8-dimethylnaphthalene relative to other dimethylnaphthalene isomers in crude oils of increasing maturity. Acenaphthene can co-elute with 1,8-DMN under many chromatographic conditions (Figure 15.4). Several dimethylnaphthalene ratios are used as indicators of thermal maturity as listed below (Alexander et al., 1985), and the ratio −log[1,8-DMN/total DMN] has been used to assess samples associated with the high-maturity condensate zone:

DNR-2 = 2,7-DMN/1,8-DMN
DNR-3 = 2,6-DMN/1,8-DMN
DNR-4 = 1,7-DMN/1,8-DMN
DNR-5 = 1,6-DMN/1,8-DMN

Trimethylnaphthalenes

> Specific for high thermal maturity range. Measured using peak heights or areas from gas chromatograms of the aromatic hydrocarbon fraction.

Trimethylnaphthalene isomers can be separated by capillary gas chromatography. The trimethylnaphthalene ratio (TNR-1) is a useful maturity indicator (Alexander et al., 1985):

TNR-1 = 2,3,6-TMN/(1,4,6-TMN + 1,3,5-TMN)

16 · Biodegradation parameters

> This chapter explains how biomarker and non-biomarker analyses are used to monitor the extent of biodegradation. Compound classes and parameters are discussed in the approximate order of increasing resistance to biodegradation. The discussion covers recent advances in our understanding of the controls and mechanisms of petroleum biodegradation and the relative significance of aerobic versus anaerobic degradation in both surface and subsurface environments. Examples show how to predict the original physical properties of crude oils before biodegradation.

Petroleum biodegradation is the alteration of crude oil by living organisms (e.g. Milner *et al.*, 1977; Connan, 1984; Palmer, 1993; Blanc and Connan, 1994). Numerous eubacteria, fungi, and possibly archaea evolved metabolic pathways to consume saturated and aromatic hydrocarbons. The existence of these pathways in ancient and diverse prokaryotic lineages provides indirect evidence that petroleum systems occurred during the early Precambrian Age. Petroleum biodegradation is primarily a hydrocarbon oxidation process, producing CO_2 and partially oxidized species, such as organic acids. Heteroatomic compounds are less susceptible to biological attack and increase in relative abundance as biodegradation proceeds. This increase is caused primarily by enrichment of pre-existing petroleum NSO compounds due to selective biodegradation of other compounds, but a secondary cause is the direct production of heteroatomic compounds (mainly oxidized species) by microbes. Petroleum quality and net volume thus decrease with increasing biodegradation. API gravity decreases while non-hydrocarbon gases, viscosity, NSO compounds, and trace metals increase. Because biodegradation can impact greatly the economic value and producibility of petroleum, the ability to measure and predict the extent of the alteration is very important in order to evaluate plays and prospects and in field development.

The amount of biodegraded oil worldwide may exceed that of conventional oil (e.g. Tissot and Welte, 1984, p. 480–481). The largest accumulations of petroleum are in sandstones that are saturated with biodegraded oil. The Orinoco Heavy Oil Belt in eastern Venezuela is estimated to contain more than 1.2×10^{12} barrels of oil, while those in western Canada (e.g. Athabasca, Cold Lake, Wabasca, and Peace River) collectively contain a similar amount. In contrast, the supergiant oil field of Ghawar in Saudi Arabia contains $\sim 1.9 \times 10^{11}$ barrels of non-degraded oil (Roadifer, 1987). Subsurface accumulations of biodegraded oil are common throughout the world, particularly in Tertiary deltaic systems, where reservoirs can remain relatively cool to appreciable depths, and in deepwater plays, where only shallow reservoirs are within drilling reach.

> Note: The focus of this chapter is on microbial degradation of petroleum hydrocarbons. We use the term "biodegradation" to indicate this process. Nearly all biological molecules, including biopolymers, can be biodegraded. This is an essential process in order to recycle nutrients in the biosphere. Fungi, for example, are essential for the breakdown of lignin produced by land plants. Consequently, the biochemical pathways for lignin generation and destruction probably have co-evolved. Many synthetic compounds and polymers, such as trinitrotoluene (TNT) (Crawford, 1995; Heiss and Knackmuss, 2002), organophosphorus insecticides, nerve agents (Jokanovi, 2001), and polyester (Abou-Zeid *et al.*, 2001), also are susceptible to biological attack. Enzymes have not evolved spontaneously to degrade these molecules. Rather, enzymes that evolved to degrade naturally occurring compounds exhibit a certain degree of non-specificity that

allows similar compounds to be consumed. Contaminated environments alter the indigenous microbial population by enriching the environment with those organisms that can survive and utilize the contaminant. Most bioremediation projects provide organic and inorganic nutrients to encourage the growth of indigenous microorganisms, rather than introducing microbes with known degrading characteristics.

CONTROLS ON PETROLEUM BIODEGRADATION

The biodegradation of petroleum requires conditions that support microbial life. When these conditions are ideal, large volumes of oil can be degraded in a relatively short time compared with geologic and geochemical processes (Connan, 1984; Palmer, 1993; Blanc and Connan, 1994). In order for petroleum biodegradation to occur, the following apply:

- There must be sufficient access to petroleum, electron acceptors (e.g. molecular oxygen, nitrates, sulfates, ferric iron), and inorganic nutrients (e.g. phosphorus, trace metals). Water must be present.
- The rock fabric must have sufficient porosity and permeability to allow the diffusion of nutrients and bacterial motility.
- Reservoir temperatures must remain within limits that support life. Empirical observations indicate that temperatures less than ~80°C are ideal. Under typical geothermal gradients, such temperatures correspond to depths <2000 m. Transient exposure to higher temperatures may "sterilize" reservoirs.
- Microorganisms capable of degrading hydrocarbon must be present.
- The salinity of the formation water must generally be less than ~100–150 parts per thousand.
- The reservoir must lack H_2S for aerobic microbes or contain no more than ~5% H_2S for anaerobic sulfate reducers to be active.

Nutrients and the aqueous environment

Because hydrocarbons are reduced carbon compounds, they can be viewed as ideal electron donors (food), provided that they are accessible. Most hydrocarbons, however, are nearly insoluble in water. Microorganisms, which must exist in water, have evolved several mechanisms to access hydrocarbons. Some microbes excrete biosurfactants that emulsify hydrocarbons, which can then be transported across cellular membranes. Enzymes also may be excreted that react with hydrocarbons and biopolymers, converting them into water-soluble compounds, which then diffuse to the organism. Some bacteria appear to be able to utilize hydrocarbons directly by existing at the oil–water interface. The uptake mechanism for this assimilation remains unclear.

Chemotrophy of organic compounds requires a terminal electron acceptor. Aerobic organisms use molecular oxygen, converting hydrocarbons to CO_2, H_2O, and biomass. Anaerobic organisms use various inorganic compounds as terminal electron acceptors, such as nitrate, sulfate, and ferric ions. Some microorganisms require a specific electron acceptor, while others can switch acceptors depending on availability. Fermentative organisms use organic compounds, such as acetate, as both the terminal electron donor and acceptor, converting the metabolites to CH_4 and CO_2.

Various inorganic nutrients, such as nitrogen and phosphorus, are needed for the biosynthesis of essential biochemicals, including proteins, phospholipids, and nucleic acids. Trace metals, such as molybdenum, cobalt, and copper, are important components in key enzymes. While these inorganic species are critical for life and their absence may limit biodegradation, only trace amounts are needed to support a stable microbial ecosystem because the elements can be recycled.

Grain size and lithology

Grain size, lithology, porosity, and permeability can greatly influence microbial activity. In general, open rock fabrics offer more favorable conditions than tight rocks by allowing diffusion of nutrients and bacterial motility (Jenneman et al., 1985; Fredrickson et al., 1997; Krumholz, 2000). For example, Brooks et al. (1988) found greater biodegradation in coarse- versus fine-grained reservoir lithologies in heavy oil accumulations in western Canada. On the other hand, McCaffrey et al. (1996) showed that lithology is not the only variable controlling the extent of biodegradation. They found that contours of the extent of biodegradation of Miocene Monterey crude oils in the Cymric Field in California crossed lithologic boundaries.

Most source rocks are fine-grained with pore spaces <4 μm and low Darcy permeability. Consequently, most source-rock extracts are not biodegraded because the organic matter is not readily accessible to microbes. Extracts that exhibit evidence of biodegradation, such as the presence of 25-norhopanes, are commonly from fractured, high-permeability rocks that are stained with migrated oil. Adsorption of petroleum on to clay surfaces may sequester it from microbial attack. For example, Wenger and Isaksen (2002) found that biodegradation does not occur in oil-stained rocks if their concentration is below the saturation threshold.

The preference for bacteria for more open rock fabrics offers the potential for microbially mediated enhanced oil recovery. In many mature fields, water flooding has stripped nearly all the oil from the high-permeability zones. By adding nutrients, microbial growth can be promoted, which might plug these zones, thereby promoting the flushing of low-permeability production zones with higher amounts of residual oil (e.g. Jack, 1993; Zekri *et al.*, 1999).

Temperature

Reservoir temperature is a major factor that limits biodegradation. Empirical observations suggest that microbial degradation of petroleum is optimal at surface or near-surface temperatures and has an upper limit of ~60–80°C. For example, crude oils in reservoirs at depths <1500 m (<~70°C) in the Shengli oilfield, China, show biodegradative removal of *n*-alkanes, while those from greater depths are unaltered (Shi Ji-Yang *et al.*, 1982). Temperature limits of ~60–80°C are consistent with the growth and survival conditions associated with many mesophilic bacteria. However, these temperatures are appreciably lower than the optimal temperatures for thermophilic bacteria and archaea. One extreme thermophile is known as *Pyrolobus fumarii*, a coccoid-shaped archaea isolated from a deep-sea hydrothermal vent at 3650-m water depth on the Mid-Atlantic Ridge (Blöchl *et al.*, 1997). *P. fumarii* grows at temperatures between 90 and 113°C (optimal growth at 106°C), and exponentially growing cultures survived a one-hour autoclaving at 121°C. Lipids from this facultatively aerobic chemolithoautotroph consisted of glycerol-dialkyl glyerol tetraethers (e.g. see Figure 13.12) and traces of 2,3-di-*O*-phytanyl-*sn*-glycerol (e.g. see Figure 13.14).

Note: The discovery of bacteria near "black smoker" hydrothermal vents in the deep ocean (Baross and Deming, 1983) was interpreted to suggest that some organisms might actively degrade petroleum at much higher temperatures (~250°C) than suspected previously. However, these thermophilic (heat-tolerant) archaea are almost certainly not growing at 250°C. They probably were introduced into the samples by strong thermal convection of the water near the vents. The instability of vital macromolecules (e.g. proteins) and monomers (e.g. adenosine triphosphate) in living organisms above ~110°C suggests that this is a good estimate of the upper temperature limit for life. *Pyrodictium brockii* is an anaerobic thermophilic archaebacterium isolated from undersea thermal vents with optimal growth at 105°C (Brock and Madigan, 1991), only 1°C lower than that for *Pyrolobus fumarii*. Recently discovered iron-reducing archaea can survive temperatures as high as 130°C (Kashefi and Lovley, 2003).

Some reservoirs may have been charged with oil at temperatures <80°C but currently are at greater temperatures due to subsidence. These reservoirs are particularly prevalent in rapidly subsiding Tertiary basins. In such situations, shallow oil may be biodegraded before reservoir sterilization occurs at greater depths. For example, biodegraded oil occurs at ~90°C in Pliocene reservoirs in the nearshore South Timbalier Field in the Gulf of Mexico. Basin reconstruction suggests that the reservoir sands were charged at <70°C but later were buried and exposed to higher temperatures within the past one to two million years.

Presence of microorganisms

Until recently, the biosphere was thought to extend only a few meters below the sediment–water interface. Consequently, few microbiologists considered examining deep formation brines for living organisms. In an early study, Nazina and Rozanova (1978) reported thermophilic sulfate-reducing bacteria in oil-bearing strata from Western Siberia. The organism, identified as *Desulfotomaculum nigrificans* by its morphology and biochemical characteristics, grows on various substrates and oxidized forms of sulfur at 40–70°C, with an optimum temperature of ~60°C. The bacterium proved to be a moderate halophile, requiring NaCl for growth with a tolerance of up to 4%. Rozanova and Nazina (1979) also described the thermophilic bacterium *Desulfovibrio thermophilus* in oil-bearing strata from the Apsheron Peninsula, South Caspian. Reservoir temperature was the major factor that determined the distribution of this bacterium in the South Caspian fields.

Belief in deep-subsurface bacteria remained buried in the Russian literature until the 1990s, when a series of papers revealed the occurrence of thermophilic microbes in many subsurface reservoir waters, mostly through the use of 16S rRNA probes. Rosnes et al. (1991a; 1991b) described the isolation of *Thermodesulfobacterium* from North Sea oil field waters and their subsequent growth under simulated reservoir conditions. These bacteria proved to be adapted to high temperatures and pressures, suggesting that they were indigenous to the deep subsurface. Magot et al. (1992) isolated *Desulfovibrio* from wellhead water samples in the Paris Basin, and L'Haridon et al. (1995) described the presence of hyperthermophilic archaea (*Archaeoglobus*) and bacteria (*Thermotoga, Thermoanaerobacter*, and *Thermodesulfobacterium*). This consortium of thermophilic bacteria was found to be common in oil-field waters from around the world (Magot, 1996).

The deep-subsurface biosphere can be viewed as part of a microbial continuum that begins at the sediment surface. We now know that microbial activity can occur in marine sediments at depths below 1000 m (Parkes and Maxwell, 1993; Chapelle and Bradley, 1996). Organisms in the deep subsurface could be descendents of the microbes that were present in the surface sediments or, possibly, could have been transported into the rock pores by meteoric water. For example, Warthmann et al. (2000) found that strains of *Desulfovibrio* from 600-m-deep granitic cores had nearly identical 16S rRNA sequences to strains that grow in sub-seafloor sediments. The ecology and activity of the deep subsurface microbial community, however, are radically different from those in shallow sediments. Increased temperature and salinity and limitations in nutrients and terminal electron acceptors promote the growth of a few species while suppressing others.

There are several reports of viable bacteria and archaea from formation waters >80°C. Rozanova et al. (2001) reported thermophilic sulfate-reducing bacteria from 90°C waters collected from the White Tiger (Bach Ho) Field, Vietnam. These thermophiles oxidize lactate, butyrate, and C_{12}–C_{16} n-alkanes and display optimal growth at ~70°C. Spark et al. (2000) identified apparently indigenous bacteria in cores from North Sea and Irish Basin wells with subsurface temperatures up to 150°C. Indirect evidence for anaerobic bacterial activity (H_2S, pyrite precipitation, and exopolymers) occurs in reservoirs up to 95°C. Discoveries of hyperthermophiles in reservoir rocks (Rothschild and Mancinelli, 2001) and of methanogenic archaeal populations living solely on inorganic H_2 and CO_2 (Stevens and McKinley, 1995; Chapelle et al., 2002) led to speculation that viable microbial ecosystems may exist to the top of the oil window (Parkes et al., 2000) or even beyond (Gold, 1999).

A general correlation exists between reservoir temperature and the extent of biodegradation among oils from Tertiary reservoirs in the Viking Graben and Jurassic reservoirs in the Tampen Spur area, North Sea (Wilhelms et al., 2001). In contrast, oils in reservoirs from the Barents Sea and Wessex basins were not biodegraded, even though several reservoirs are currently appreciably cooler. These basins experienced different geothermal histories. The reservoirs in the Viking Graben and Tampen Spur are currently at their maximum temperatures, whereas the reservoirs in the Barents Sea and Wessex basins have been buried and uplifted. Wilhelms et al. (2001) proposed that the latter reservoirs were sterilized by exposure to >80°C before being charged with oil. Furthermore, once sterilized, the reservoirs were not recolonized by viable, hydrocarbon-degrading organisms. Although the lack of biodegradation in these cooler reservoir rocks could be due to other factors, the sterilization concept supports the theory that active microbes in the deep subsurface are survivors of microbial populations that were present during the deposition of the reservoir sediments. This may account for the low ~80°C limit of petroleum biodegradation and the general lack of extreme thermophilic organisms at higher reservoir temperatures.

Whether microbes can be introduced into reservoirs by subsurface aqueous flow has yet to be resolved. Microbes may be too large to move through many fine-grained rocks. One study documents subsurface transport of microbes. One strain of mesophilic benzoate-degrading sulfate-reducing bacteria occurred in the injection water systems of three Norwegian oil platforms over a period of three years, but not in the *in situ* reservoir water (Beeder et al., 1996). Later, this strain was found in water samples collected from the oil-field production system, showing that the bacteria had penetrated the reservoir with the injection water and eventually reached the production well. Stetter *et al.* (1993) found hyperthermophilic archaea, including some known to occur at deep-sea hydrothermal vents, in four oil reservoirs ~3000 m below the seabottom in the North Sea. Enrichment cultures of these microbes grew at 85°C and 102°C, similar to *in situ* reservoir temperatures. Some of the cultures grew anaerobically in sterilized seawater, with crude oil as the single carbon and energy source. It is likely that these thermophiles entered the reservoirs with injected seawater, although natural inoculation through faults is possible. The investigated reservoirs were flooded with 40 000–127 000 m^3 of seawater each day to displace oil for production. Although treated with biocides, such treatments are not 100% effective, and viable microorganisms likely remain in the injected water.

Note: A major question in oil-field microbiological studies is whether microbes were indigenous to the reservoir or whether they were introduced during drilling and water injection. Several studies argue that recovery of similar microbial communities in different wells, none of which has undergone water injection, indicates that they sampled indigenous populations. However, selection of wells that have not undergone water injection does not guarantee uncontaminated samples. Spark *et al.* (2000) found that many drilling mud samples are already contaminated with bacteria before their use in drilling.

Salinity

Geologists have long known that biodegraded oils tend to occur in reservoirs with low-salinity formation water (<100–150 ppt). Oils in reservoirs with higher salinities are typically non-biodegraded. Wenger and Isaksen (2002) listed 150 g/l of total dissolved solids as an upper limit but noted examples of biodegradation in oil reservoirs above water legs with higher salinities. Such occurrences are particularly important in deepwater areas where the reservoirs are cool and salinity is influenced by the dissolution of nearby salt diapirs.

The precise influence of salinity on biodegradation is uncertain. Representatives of all three domains of life have the ability to regulate internal osmotic pressure by accumulating compatible solutes (Roessler and Müller, 2001). Many halophilic microorganisms (e.g. various species of green algae, photosynthetic sulfur bacteria, and archaea) thrive in waters with salinities over 100–150 ppt. Source-rock deposition occurs in both marine hypersaline and lacustrine high-alkalinity environments (Kirkland and Evans, 1981), suggesting that halotolerant organisms could be present in carbonate and evaporate source and reservoir rocks.

Some microorganisms are capable of surviving, but not necessarily growing, in waters with extreme salinities. For example, a spore-forming bacterium was isolated from an undisturbed brine inclusion in a 250-million-years-old salt crystal from the Upper Permian Salado Formation, New Mexico (Vreeland *et al.*, 2000). *Bacillus* strain 2-9-3 apparently lived in a hypersaline water body during the Late Permian Epoch, was trapped inside a salt crystal, and survived as a spore until released and cultured under strictly sterile conditions. The Salado Formation is approximately the same age as the Phosphoria Formation, a prolific source rock in the western interior USA.

Acid gases and pH

Hydrogen sulfide is a waste product of bacterial sulfate reduction. Some anaerobic microbes have adaptions to cope with relatively high concentrations of this toxic gas. The tolerance of aerobic organisms is substantially lower. Few if any organisms can tolerate concentrations of more than ~5% H_2S. Other toxic chemicals, either produced by bacteria or leached from the reservoir minerals, may inhibit microbial activity.

High acid or base conditions rarely occur in the subsurface, due to buffering effects of minerals. Although formation waters associated with acid gases (H_2S and CO_2) can corrode steel, the pH conditions are relatively mild and can be tolerated by many oil-degrading microbial species. Extreme acidophiles have evolved specialized membranes and ion-transport mechanisms to thrive in low pH waters. An iron-oxidizing archaeon is known to thrive in acid mine drainage waters with pH 0 (Edwards *et al.*, 2000b).

Geopressure

Pressure does not appear to be a significant factor limiting microbial growth. Sharma *et al.* (2002) observed metabolic activity in several strains of bacteria when placed in diamond anvil cells and pressurized up to

1680 MPa. The bacteria resided within fluid inclusions in ice-VI crystals and remained viable when depressurized to atmospheric conditions. Although the findings are controversial (see comments by Yayanos et al. (2002) and reply by Sharma et al. (2002)), they are consistent with geochemical observations. Moldowan et al. (1992) observed a case of biodegraded petroleum in reservoirs at 2500 m in the Adriatic Basin. Walters (1999) showed an example of highly biodegraded oil in reservoirs at ~4000 m in the South Caspian Basin (offshore Azerbaijan). In both cases, biodegradation seemed to have occurred at substantial geopressure, although the reservoir temperatures are <80°C as a result of very low thermal gradients.

RATES OF AEROBIC AND ANAEROBIC BIODEGRADATION

Unfortunately, the rates of petroleum biodegradation are not well known. Empirical evidence from surface or near-surface oil spills suggests that biodegradation occurs relatively quickly in environments that are at least partially oxic with plentiful nutrients (Jobson et al., 1972), while degradation of oil in deep reservoirs is very slow (Larter et al., 2000). Anaerobic bacteria, such as sulfate reducers, can oxidize hydrocarbons but probably do so much more slowly than aerobes. For example, in laboratory studies using estuarine sediments, aerobic degradation rates of individual petroleum hydrocarbons tend to be about ten times faster than anaerobic degradation rates (Yamane et al., 1997). However, this difference in rate may not be significant in terms of geologic time.

PREDICTION OF THE OCCURRENCE AND EXTENT OF BIODEGRADATION

Biodegradation can have a major impact on oil quality and value. Therefore, accurate prediction of the occurrence and extent of biodegradation before drilling is important for exploration risk assessment. A fully explicit model would consider the rate of biodegradation relative to the rate and timing of reservoir charge by petroleum of known composition. Such models are beyond our current ability and knowledge base. At present, some models express the extent of biodegradation as a probability based on statistical data by depth and/or location in a basin or as simple time–temperature correlations (e.g. Scotchmer and Patience, 2002; Larter et al., 2002). For example, low risk for oil biodegradation might be assigned to reservoirs with conditions that are likely to limit microbial growth (e.g. temperatures >80°C, high salinity water), whereas complete absence of these limiting conditions would indicate high risk.

AEROBIC HYDROCARBON DEGRADATION PATHWAYS

Hydrocarbons resist microbial degradation and must be activated if they are to be consumed. In aerobic degradation, oxygen is both the terminal electron acceptor and a necessary reactant for activating hydrocarbons by converting them into oxygenated intermediates. Saturated hydrocarbons are commonly activated by monooxygenases that incorporate oxygen, forming a primary alcohol (Figure 16.1). Oxidation continues at the terminal carbon, forming aldehydes and fatty acids. The latter are cleaved by β-oxidation to form acetyl-CoA. Long-chain n-alkanes in the range $\sim C_{10}$–C_{24} degrade most rapidly. Normal alkanes of greater chain length are more difficult to transport across cell membranes, and n-alkanes $<C_{10}$ are toxic to many microorganisms. The smaller C_3–C_6 n-alkanes may be oxidized subterminally, forming secondary alcohols and then ketones.

Complex branching hinders terminal carbon activation by oxygenating enzymes. For this reason, n-alkanes and isoalkanes are preferentially removed first during microbial degradation of crude oil. These compounds may be partially converted and used as biolipids. However, most appear to be consumed as food or used to build new biomass from two or three carbon units.

Cycloalkanes are fairly resistant to aerobic biodegradation. A specialized pathway evolved for the activation of cycloalkanes that leads to ring-opening and the formation of fatty acids that then can be cleaved by α, β, or ω oxidation (Donoghue and Trudgill, 1975; Atlas, 1981; Prince 2002). The pathway involves a two-step addition of oxygen by a series of monooxygenase enzymes (Figure 16.2). Multi-ring cycloalkanes are highly resistant to biodegradation.

Benzenes and phenolic hydrocarbons are activated by mono- or dioxygenases that result in the formation of catechol. This intermediate may undergo either *ortho-* or *meta-*cleavage to produce various degradation

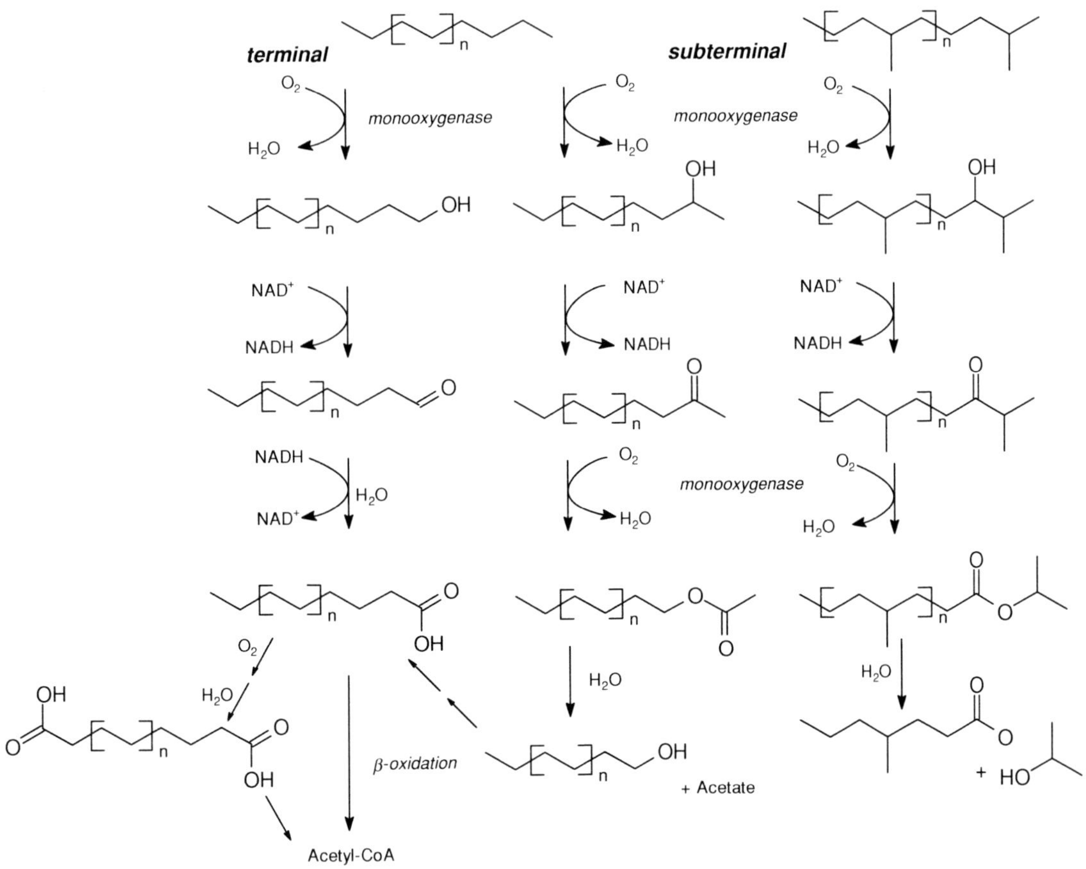

Figure 16.1. Aerobic degradation pathways of normal and branched alkanes (modified from Fritsche and Hofrichter (2000) and Prince (2002)).
Acetyl-CoA, acetyl co-enzyme A; NAD, nicotinamide adenine dinucleotide.

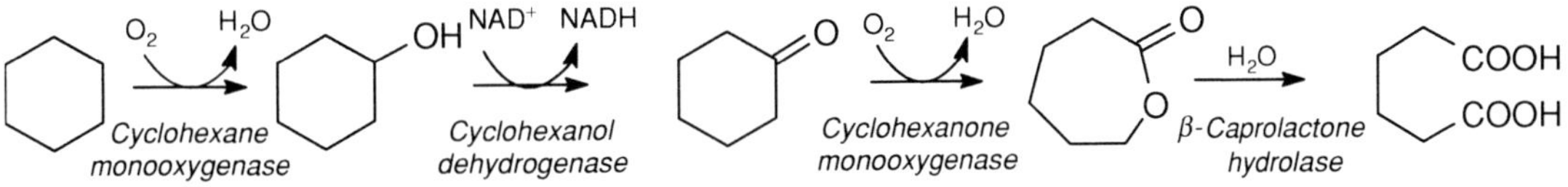

Figure 16.2. Aerobic degradation pathway of cyclohexane (modified from Fritsche and Hofrichter, 2000). NAD, nicotinamide adenine dinucleotide.

products that can be used for energy or biosynthesis (Figure 16.3). The addition of alkyl or other substituent groups on to the aromatic ring may facilitate alternative pathways for opening the aromatic ring. These reactions may also degrade napthalenes, phenanthrenes, dibenzothiophenes, and other condensed hydrocarbons and sulfur-aromatic compounds. Degradation of compounds such as dibenzothiophene involves enzymatic cleavage of C–S bonds and desulfurization of the metabolite (Kirimura et al., 2001).

Organisms capable of aerobic degradation of petroleum

Organisms capable of degrading hydrocarbons are diverse, spanning all domains of life and inhabiting

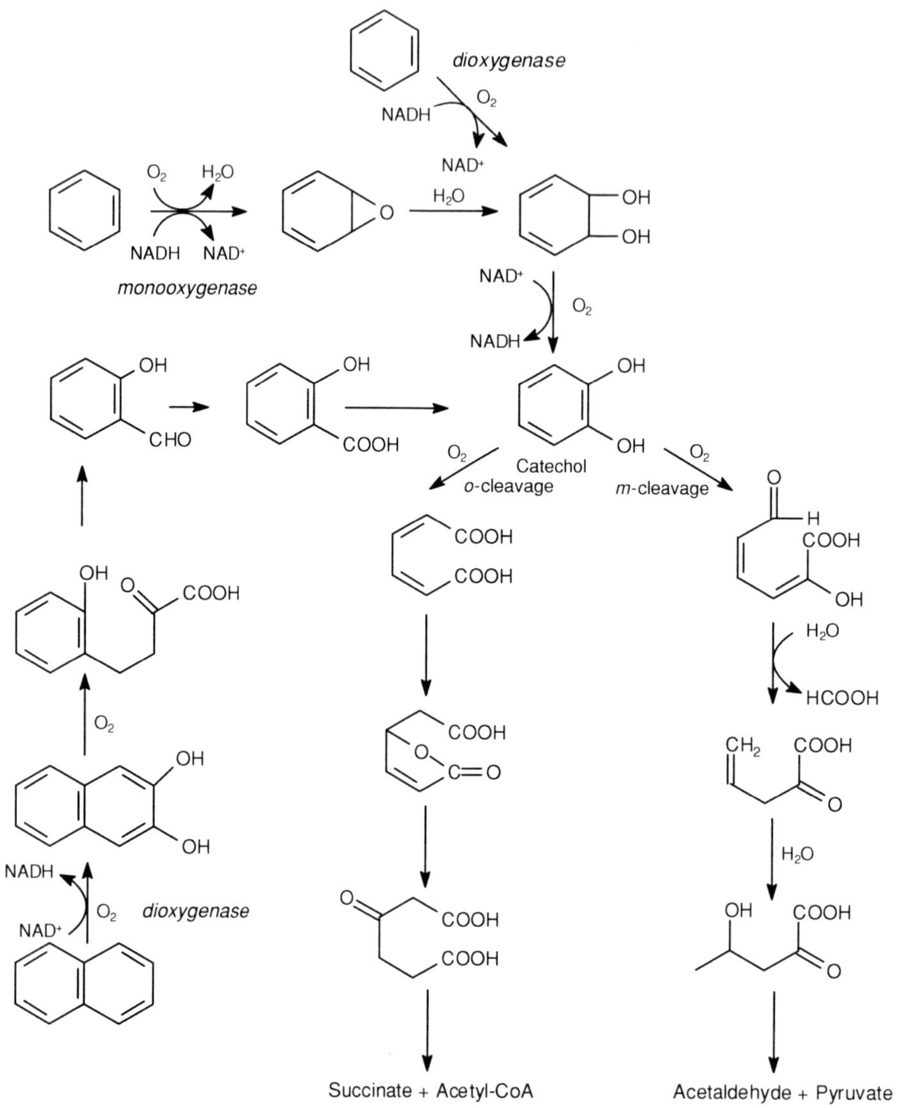

Figure 16.3. Aerobic degradation of benzene and naphthalene (modified from Fritsche and Hofrichter, 2000). Acetyl-CoA, acetyl co-enzyme A; NAD, nicotinamide adenine dinucleotide.

almost all environmental niches (Fritsche and Hofrichter, 2000; Prince, 2002). Under oxic conditions, the dominant degraders of petroleum are heterotrophic bacteria and fungi. Species of *Pseudomonas*, aerobic Gram-negative rod-shaped bacteria that lack fermentative pathways, are the most prolific degraders. Other common Gram-negative bacteria that degrade oil include species of *Acinetobacter*, *Alcaligenes*, *Flavobacterium* of the *Cytophaga* group, and *Xanthomonas*. Common Gram-positive bacteria that degrade hydrocarbons include *Nocardia*, *Mycobacterium*, *Corynebacterium*, *Arthrobacter*, *Rhodococcus*, and *Bacillus*. Two species of aerobic, halophilic archaea also can grow with hydrocarbons as their sole source of carbon. Several algae can degrade hydrocarbons in conjunction with photosynthesis.

Certain species degrade individual hydrocarbons more effectively than others (Prince, 2002). The most common of the alkane monooxygenases can act on a broad range of substrates but is most active on nC_5–C_{12} alkanes. Other microorganisms are very specific for the hydrocarbon substrates that they can metabolize. For example, some aerobic eubacteria can degrade only m-xylene or p-xylene, while others degrade only o-xylene. In natural settings, a mixed population of aerobic bacteria is the most efficient to degrade petroleum. For example, species of *Pseudomonas* readily utilize xylenes as their sole source of carbon, while other species, such as *Nocardia*, can degrade xylenes only as co-metabolites. Microbial diversity changes as hydrocarbons are depleted selectively.

> Note: Bacterial degradation of crude oil may require complex cell-surface adaption to allow adherence to the oil. Norman *et al.* (2002) compared the cell surface characteristics of two *Pseudomonas aeruginosa* strains, U1 and U3, which were isolated from the same crude-oil-degrading community enriched on Bonny Light crude oil. U1 cells demonstrated more lag time for growth on crude oil than U3 cells. During the lag interval, U1 cells showed a smooth morphology that changed gradually to rough morphology when grown on oil. U3 cells exhibited rough morphology thoughout their growth on oil. The rough morphology in both strains was traced to loss of a specific antigen, which resulted in shorter membrane lipopolysaccharides, increased cell-surface hydrophobicity, and increased n-alkane degradation.

Abiotic oxidation of petroleum

Microbial degradation and thermochemical sulfate reduction are the two major processes that oxidize hydrocarbons in reservoirs. Many compounds in petroleum, such as sulfides, are oxidized readily when isolated from crude oil and exposed to air. This process might occur naturally in degraded crude oils, but the compounds susceptible to air oxidation are only minor constituents in petroleum. The concentrations of oxygen-containing polar compounds, such as steroid ketones, benzothiophenic acids, and sulfones, increase with increasing alteration in a suite of genetically related crude oils (Charrié-Duhaut *et al.*, 2000). Oxygen incorporation occurred without any diastereomeric discrimination, indicating a chemical process rather than being mediated by biological enzymatic reactions. They concluded that abiogenic oxidation might be a significant factor in transforming petroliferous compounds into more water-soluble compounds. While these compounds are biodegraded more readily, they also may be more toxic because they are easily taken up by biota.

ANAEROBIC HYDROCARBON DEGRADATION PATHWAYS

Because no compounds can substitute for molecular oxygen in hydrocarbon-activating oxygenase enzymes, it was long thought that hydrocarbons could not be biodegraded under anoxic conditions. Some argued that early experiments that claimed anaerobic degradation of hydrocarbons were not truly anaerobic because oxygen leaked into the system or aerobic microenvironments were present (e.g. Gibson, 1984).

Nevertheless, evidence for anaerobic petroleum degradation of n-alkanes and selected alkylbenzenes by sulfate-reducing bacteria is now overwhelming. Kuhn *et al.* (1985) showed that toluene and xylene are consumed by nitrate-reducing anaerobes in contaminated groundwater and in laboratory experiments. Aeckersberg *et al.* (1991) demonstrated that sulfate-reducing bacteria could degrade n-alkanes. Rueter *et al.* (1994) finally established widespread recognition that anaerobic bacteria could degrade hydrocarbons in crude oils. They showed that a pure culture of moderately thermophilic sulfate reducers preferentially consumed n-alkanes ranging from nC_8 to nC_{34}. Rapid and complete biodegradation of n-alkanes in laboratory incubations produced a residuum that resembled a naturally occurring, severely biodegraded oil, including a large unresolved hump with isoprenoids still resolved. Complete mineralization of n-alkanes was confirmed in experiments using ^{14}C-14,15-octacosane ($C_{28}H_{58}$), with 97% of the radioactivity recovered as $^{14}CO_2$. They also showed that a separate mesophilic sulfate-reducing enrichment culture selectively oxidized alkylbenzenes. These and subsequent studies (e.g. Coates *et al.*, 1996; Coates *et al.*, 1997; Aeckersberg *et al.*, 1998; Caldwell *et al.*, 1998; So and Young, 1999a; So and Young, 1999b; Zengler *et al.*, 1999; Ehrenreich *et al.*, 2000) firmly established anaerobic hydrocarbon degradation. The rapid and extensive biodegradation of *n*-alkanes in oils can no longer be considered a strictly aerobic process (Widdel and Rabus, 2001).

The biochemical pathways used by anaerobes to activate hydrocarbons are clearly different from those used by aerobes (Lovley and Chapelle, 1995; Heider *et al.*, 1998; Widdel and Rabus, 2001). The actual mechanisms of anaerobic alkane oxidation are still not known or understood fully. Two pathways for hydrocarbon activation have been discovered: activation by fumarate and activation of the terminal carbon by a C_1 compound.

As noted above, alkylbenzenes (C_1–C_5) were the first compounds proved to be degraded by anaerobic bacteria (Kuhn *et al.*, 1985; Rabus *et al.*, 1993; Lovley *et al.*, 1995; Harms *et al.*, 1999). Beller *et al.* (1996) identified benzylsuccinate as a metabolite in an enriched culture of toluene-degrading sulfate reducers. This was shown to be the intermediate product formed by the reaction of toluene with fumarate via a double-bond addition that did not require other cosubstrates (Figure 16.4). The benzylsuccinate is metabolized further to form acetyl-CoA and benzoyl-CoA. The benzoyl-CoA undergoes reductive dearomatization and ring cleavage similar to the β-oxidation of fatty acids. Fumarate addition is a common degradation pathway for alkylbenzenes that occurs not only in sulfate-reducing bacteria also but in nitrate- and iron-reducing bacteria, as well as in phototrophic and methanogenic cultures (Widell and Rabus, 2001).

The mechanisms for the degradation of benzene, naphthalene, and phenanthrene are unknown. To date, no pure culture of anaerobic bacteria has been demonstrated to degrade benzene, but benzene can be degraded by enriched cultures of anaerobic bacteria (Phelps *et al.*, 1998). Unlike the alkylbenzenes, benzene biodegradation requires the extraction of hydrogen from the aromatic ring to form a phenyl radical, an energy requirement higher than methane. The degradation pathway is, therefore, not likely to involve fumarate addition, at least in the initial step. Coates *et al.* (1996) conducted experiments where PAHs, such as naphthalene and phenanthrene, were biodegraded by unidentified sulfate-reducing organisms. The experiment used contaminated harbor sediments that liberated $^{14}CO_2$ when placed under sulfate-reducing conditions in the presence of ^{14}C-labeled PAHs. Addition of molybdate, which inhibits sulfate reduction, caused $^{14}CO_2$ generation to cease. The anaerobic degradation of alkyl-naphthalenes is believed to follow the same pathways as for toluene, where activation occurs by fumarate addition to the methyl group (Annweiler *et al.*, 2000). Degradation of naphthalene proceeds by carboxylation followed by a stepwise reduction of the aromatic rings before ring cleavage (Meckenstock *et al.*, 1999). Degradation of alkylated naphthalenes may proceed by the fumarate activation pathway.

Normal alkanes may be activated by the same fumarate mechanism demonstrated to occur for alkylbenzenes (Figure 16.5). Kropp *et al.* (2000) identified dodecylsuccinic acids as trace metabolites produced during the anaerobic biodegradation of *n*-dodecane by an enrichment culture of sulfate reducers. When grown on $C_{12}D_{26}$, the dodecylsuccinic acids retain all of the deuterium atoms, suggesting that the metabolites form by addition across the double bond of fumarate. GCMS

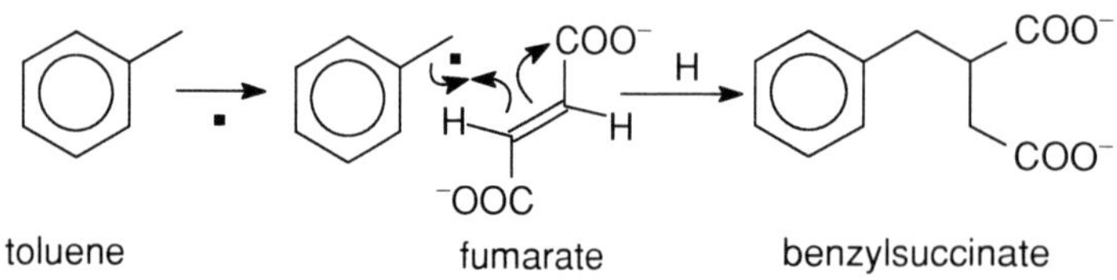

Figure 16.4. Anaerobic activation of toluene by fumarate.

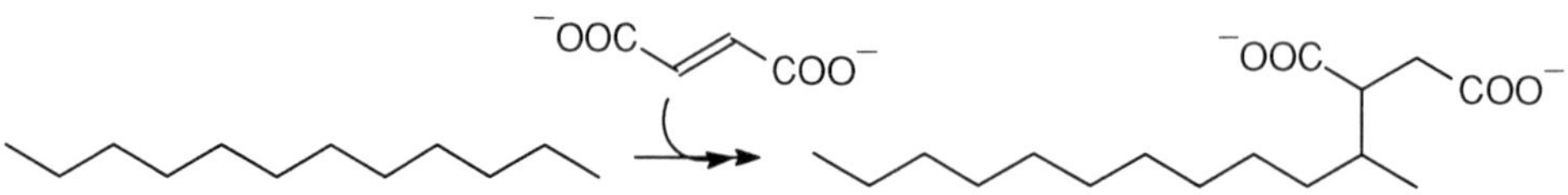

Figure 16.5. Anaerobic activation of *n*-alkanes by fumarate.

analysis indicated that the succinyl moiety in the dodecylsuccinic acid metabolite is attached not at the terminal methyl group of the alkane but at a subterminal position.

An alternative mechanism for alkane activation involves addition of a C_1 moiety to the terminal carbon. Certain sulfate-reducing bacteria generate fatty-acid distributions that suggest that the carbon chains of the precursor n-alkanes are altered by one carbon atom during activation, possibly by terminal addition (Aeckersberg et al., 1998). When one mesophilic strain (Hxd3) was grown on hexadecane, fatty acids with an odd number of carbon atoms were produced. Conversely, the sulfate-reducing bacteria produced fatty acids with an even number of carbon atoms when grown on heptadecane. Similarly, when sulfate-reducing bacteria grew on [1,2 $^{13}C_2$]hexadecane, 2-, 4-, and 6-ethylated fatty acids were detected that contained ^{13}C-labeled carbons at the methyl and adjacent carbon (So and Young, 1999a). This indicates that the methyl carbon was once a terminal carbon. Thus, exogenous carbon was initially added to an alkane at the C-2 position, such that the original terminal carbon of the alkane became a methyl group on the subsequently formed fatty acid (Figure 16.6).

Organisms capable of anaerobic degradation of petroleum

The ability to utilize hydrocarbons appears to be common in several branches of eubacteria (Widdel and Rabus, 2001). These microorganisms rely on electron acceptors other than oxygen, such as nitrate, sulfate, and ferric iron. Other inorganic species, such as manganese (Langenhoff et al., 1997), or organic molecules, such as quinones (Cervantes et al., 2000), may serve as electron acceptors. Nitrate reducers out-compete other anaerobic bacteria for organic nutrients, but nitrate is usually limited in the subsurface unless added intentionally to control well souring (Jackson and McInerney, 1996). Sulfate reducers are the most likely microorganisms that can degrade oil under anoxic conditions. The significance of iron-reducing bacteria is just beginning to be discovered (e.g. Coleman et al., 1993; Schmitt et al., 1996; Prommer et al., 1999).

There is growing evidence that archaea degrade hydrocarbons anaerobically. Anderson and Lovley (2000) proved that ^{14}C-labeled hexadecane was converted to $^{14}CH_4$ and $^{14}CO_2$ in oil-bearing rocks incubated under anaerobic conditions in the absence of nitrate and sulfate. The ratio of $^{14}CH_4$ and $^{14}CO_2$ was consistent

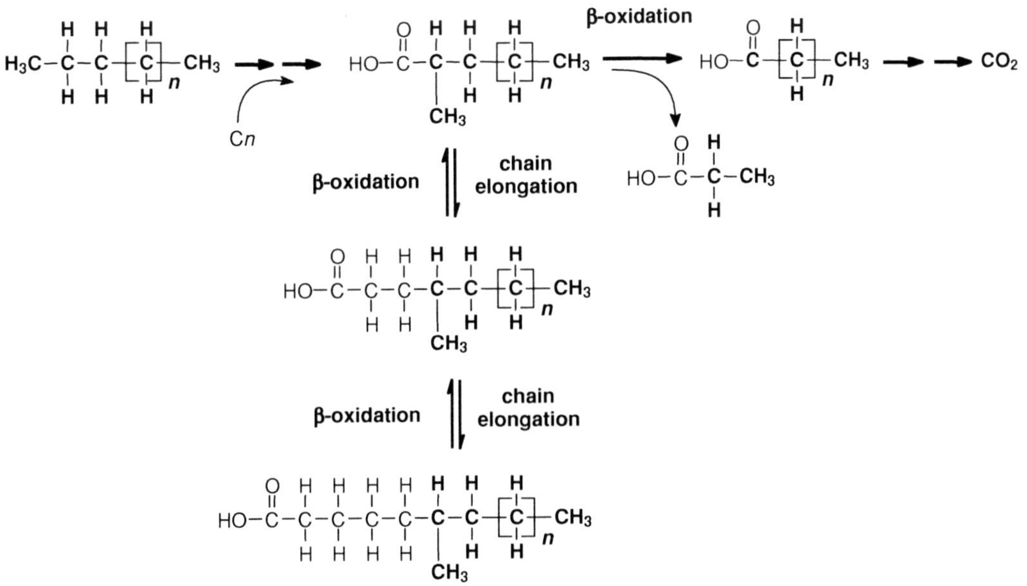

Figure 16.6. Proposed pathway for anaerobic degradation of n-alkanes by a sulfate-reducing bacterium (So et al., 1999a). Degradation pathways may be different in other anaerobic organisms.

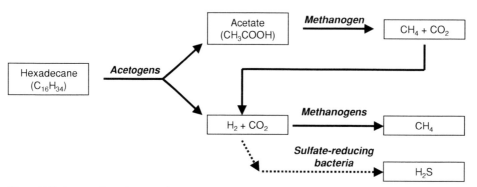

Figure 16.7. A consortium of syntrophic eubacteria and archaea can degrade hexadecane to CH_4, CO_2, and H_2S (adapted from Parkes (1999) and based on Zengler et al. (1999)).

with that expected for hexadecane conversion under methanogenic conditions. The rocks used in the ^{14}C-incubation experiments lacked significant free oxygen, nitrate, or sulfate, and more than 95% of the HCl (0.5 M)-extractable iron was in the Fe(II) rather than the Fe(III) state. It is possible that some oxidation occurred using small amounts of Fe(III) that may have been present in these samples. Nevertheless, Anderson and Lovley (2000) concluded that conversion of alkanes to methane is significant in many anaerobic petroleum reservoirs.

Archaea participate in the anaerobic syntrophic conversion of n-alkanes to CH_4 and CO_2. Jobson et al. (1979) proposed that aerobic microbes degraded petroleum into small functionalized species that methanogens could then consume. However, anaerobic archaeal degradation of hydrocarbons has been demonstrated only recently when Zengler et al. (1999) found that hexadecane could be degraded by a mixed culture under anoxic conditions. The hydrocarbon was first degraded by acetogenic bacteria to acetate and H_2. Acetogens can function only when these waste products are kept at low concentrations by methanogens and sulfate-reducing bacteria (Figure 16.7). No activity was observed when the mixed culture was grown under oxic conditions, indicating that aerobic bacteria did not contribute to the net reaction.

Energy restrictions that apply to microbes grown as pure cultures may not apply for natural assemblages of microbes. For example, in syntrophic metabolism, the metabolites from one species can be used as catabolites by another. This allows anaerobic microbes to exist in systems close to thermodynamic equilibrium (Jackson and McInerney, 2002). There is field evidence that syntrophic reactions occur in the subsurface (Rozanova et al., 1997).

BIODEGRADATION OF METHANE

Methane is more difficult to activate than larger alkanes, and aerobic methanotrophy was long considered the only biological pathway possible for methane consumption. Methanotrophs convert methane to methanol using free oxygen and a specialized enzyme, methane monooxygenase (Whiticar, 1999; Chen et al., 2001). However, while methane generation is confined to sulfate-depleted zones in anoxic marine sediments, methane oxidation occurs within a transition zone where both methane and sulfate decrease (Hoehler et al., 1994). They suggested that a consortium of methanogens and sulfate-reducing bacteria was responsible for anaerobic methane oxidation. ^{13}C-depleted CO_2 in the transition zone provided indirect evidence for anaerobic methane oxidation. The "reverse methanogenesis" reaction $CH_4 + 2H_2O \rightarrow CO_2 + 4H_2$ is energetically favored if the H_2 product can be removed rapidly. Sulfate reducers can utilize the hydrogen, resulting in a net reaction $CH_4 + SO_4^{-2} \rightarrow HCO_3^- + HS^- + H_2O$.

Indirect evidence for reverse methanogenesis was found in archaeal lipids from the transition zone, which were so depleted in ^{13}C ($\delta^{13}C = -90 \pm 10‰$) that methane must be the carbon source, rather than the metabolic product, of the organisms that produced it (Figure 16.8) (Hinrichs et al., 1999; Bian et al., 2001). Isotopic analysis of biomarkers common in the sulfate-reducing bacteria provide evidence that

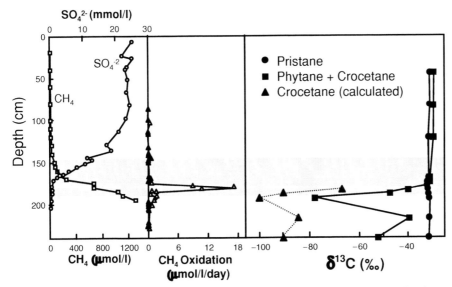

Figure 16.8. Concentrations of dissolved methane and sulfate in pore waters, rates of oxidation of methane per liter of sediment versus depth in a marine sediment core taken from the Kattegat Strait (modified from Bian et al., 2001). Methane oxidation is restricted to a narrow transition zone. Also shown are the $\delta^{13}C$ values of pristane and the gas chromatographic peak containing both phytane and crocetane. Assuming that the $\delta^{13}C$ phytane equals that of pristane, then the $\delta^{13}C$ of co-eluting crocetane, an isoprenoid diagnostic for archaea, can be calculated. © Copyright 2001, American Geophysical Union. Reproduced/modified by permission of American Geophysical Union.

methane-derived carbon is transferred efficiently from the methane oxidizers to the sulfate reducing bacteria (Hinrichs et al., 2000). Genomic 16S rRNA probes indicate that the methanotrophic archaea belong to a new group related most closely to the methanogenic order *Methanosarcinales* (Hinrichs et al., 1999). Boetius et al. (2000) examined fluorescent-labeled phylogenetic strains under the microscope and were able to identify a structured consortium of archaea and sulfate-reducing bacteria. The archaea grow in dense aggregates of ~100 cells surrounded by sulfate-reducing bacteria. The aggregates are abundant in sediments associated with gas hydrates and high rates of methane-based sulfate reduction. These syntrophic consortia mediate anaerobic oxidation of methane. Michaelis et al. (2002) discovered massive 4-m-tall carbonate build-ups and methane seeps in the anoxic waters of the Black Sea. The carbonates were covered with appreciable amounts of biomass composed of the structured consortium.

Orphan et al. (2001b) compared results of rRNA gene surveys and lipid analyses of archaea and bacteria associated with methane-seep sediments from different sites on the Californian continental margin. Two distinct archaea related to the order *Methanosarcinales* were consistently associated with methane-seep marine sediments. Concurrent surveys of bacterial rDNA revealed a predominance of δ-proteobacteria, in particular, some close relatives of *Desulfosarcina variabilis*. Biomarker analyses of the same sediments showed bacterial fatty acids with strong ^{13}C depletion, which are likely products of these sulfate-reducing bacteria (Orphan et al., 2001). The presence of abundant ^{13}C-depleted ether lipids, presumed to be of bacterial origin, but unrelated to ether lipids of members of the order *Desulfosarcinales*, suggests the participation of additional bacterial groups in the methane-oxidizing process. Such studies indicate that other bacteria and archaea are also involved in methane oxidation in marine sediments (Orphan et al., 2002).

How syntrophic bacteria and archaea exchange compounds in reverse methanogenesis is not understood fully, but a possible mechanism is the exchange of acetate (Boetius et al., 2000). Other possible mechanisms for methane oxidation could involve interspecies transfer of H_2, formate, or undefined redox shuttles (Widdel and Rabus, 2001).

BIODEGRADATION OF SUBSURFACE CRUDE OIL: AEROBIC OR ANAEROBIC?

Until recently, the biodegradation of petroleum in surface spills, seeps, and deep reservoirs was assumed to be due to aerobic activity (e.g. Connan, 1984; Palmer, 1993). It was argued that if anaerobic biodegradation of crude oil could occur, then all degraded oil in reservoirs $\leq 80°C$ would be biodegraded. Subsurface anaerobic bacteria were thought to depend on the metabolic residues of aerobes (Bailey et al., 1973a; Bailey et al., 1973b; Jobsen et al., 1979; Nazina et al., 1985). Oxic conditions could be maintained in the subsurface only by transport of oxygen dissolved in meteoritic water, implying that biodegradation could be predicted based on basin hydrology. Microbiological studies appeared to support these conclusions. While there were many reports of aerobic organisms capable of metabolizing hydrocarbons, no anaerobic organisms were known to be capable of this metabolism.

We now know that some prokaryotes can metabolize hydrocarbons under anaerobic conditions and that the biochemical pathways differ from those that require free oxygen. While the diversity of known anaerobic hydrocarbon degraders is limited, it is likely that they are more widespread, including many prokaryotes that have not yet been described. Anaerobic microbial degradation does not require the hydrodynamics needed for aerobic degradation. Furthermore, biodegraded oil need not be ubiquitous in cool, shallow reservoirs if it is anaerobic and limited by factors other than oxygen, such as the availability of alternative terminal electron acceptors.

Aerobic biodegradation clearly dominates in surface petroleum spills and seeps and in shallow, onshore reservoirs with hydrodynamic influx of meteoritic waters. Anaerobic biodegradation may dominate in the anoxic subsurface, but this has yet to be proved. The microbial processes that account for most of the world's biodegraded oil in tar sands are open to debate. Connan et al. (1997) concluded that aerobic biodegradation is a dominant process in shallow petroleum reservoirs and accounts for many tar sands. They found that although deep reservoirs ($>60°C$) contain only anaerobic bacteria, the oils within them are, at most, biodegraded only mildly. Nonetheless, it is difficult to explain how biodegraded accumulations the size of the Orinoco or Alberta tar sands could be degraded solely by aerobic microbes, when small plumes of organic contaminants are sufficient to remove free oxygen from near-surface groundwater (e.g. Baedecker et al., 1993; Chapelle et al., 2002a).

A more reasonable explanation is that biodegraded oils and tar sands may arise from both aerobic and anaerobic microbial activities. The ability of faculative aerobes to use hydrocarbons under suboxic conditions has not been studied sufficiently to rule out aerobic degradation in a largely anoxic environment (Berthe-Corti and Fetzner, 2002). It is possible that trace amounts of oxygen produced by inorganic and organic diagenetic reactions may occur in deep, nutrient-depleted reservoirs, supporting very low metabolic rates (Larter et al., 2000). Since aerobic and anaerobic consumption of hydrocarbons follow different pathways, it is likely that the relative responses of classes of compounds or individual isomers may differ as well. If so, then molecular signatures may exist in degraded oils that indicate the responsible microbial activity.

The following discussion on the effects of biodegradation on the chemical and physical properties of petroleum and molecular distributions of biomarkers is based mostly on observations of surface or near-surface accumulations where aerobic conditions are prevalent.

EFFECTS OF BIODEGRADATION ON PETROLEUM COMPOSITION

The effects of biodegradation on the physical and molecular properties of petroleum are well known (e.g. Connan, 1984; Volkman et al., 1983b; Peters and Moldowan, 1993; Palmer, 1993; Peters et al., 1996b). Because aerobic bacteria require molecular oxygen, active surface recharge waters are necessary for aerobic biodegradation. Water washing typically accompanies biodegradation of petroleum and results in selective loss of light hydrocarbons, especially benzene, toluene, and other aromatics (Bailey et al., 1973a; Palmer, 1984b; Palmer, 1993). Water washing alone induces only minor changes in the chemical and physical properties of oil. However, because biodegradation preferentially consumes hydrocarbons, the residual oil becomes enriched in nitrogen, sulfur, and oxygen (NSO) compounds (polar and asphaltene fractions). Compared with unaltered equivalents, biodegraded oils have lower API gravity, are more viscous, and are richer in sulfur, resins, asphaltenes, and metals (e.g. Ni and V), making them less desirable as refinery feedstocks.

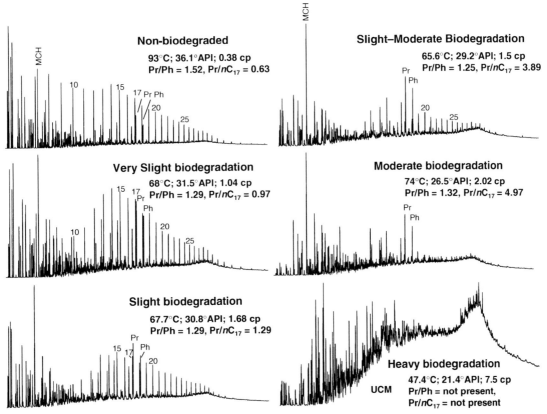

Figure 16.9. Gas chromatograms undergo systematic changes with increasing biodegradation of selected crude oils from Africa. The oils were generated and expelled from the same source rock under comparable thermal conditions. Shown are reservoir temperature, API gravity, viscosity, and pristane/phytane (Pr/Ph) and pristane/nC_{17} (Pr/nC_{17}) ratios (modified from Wenger et al., 2002). MCH, methylcyclohexane; UCM, unresolved complex mixture.

The first indications of oil biodegradation typically occur with the selective removal of C_6–C_{12} normal alkanes (Figure 16.9; however, see also Figure 16.16). As biodegradation proceeds, saturated hydrocarbons outside of the initial range are selectively removed. Normal alkanes degrade at faster rates than mono- and multimethylated alkanes. Light aromatic hydrocarbons (e.g. benzene, toluene, and xylenes) are the first of the aromatic species to be removed, either by micobial degradation or by water washing (e.g. Palmer, 1983). As the major resolved compounds diminish, the chromatographic baseline hump becomes more prominent. This hump is also called the unresolved complex mixture (UCM) and consists of bioresistant compounds, including highly branched and cyclic saturated, aromatic, naphthoaromatic, and polar compounds that are not amenable to routine gas chromatography. Depending on the amounts of high-molecular-weight polar compounds and asphaltenes that are present initially in nondegraded petroleum, the UCM may account for nearly all to less than half of the total mass of a highly degraded oil. Most of the polar and asphaltene compounds do not volatilize and are not detected by whole-oil gas chromatographic analysis.

Since normal and branched paraffins typically constitute ∼35–50% of a non-biodegraded oil, their removal can greatly alter the physical properties and economic value of crude oil. Wenger et al. (2002) illustrated the impact of biodegradation on oil quality by analyzing a suite of petroleum samples that were generated from the same marine shale source facies and were expelled at about the same level of maturity (Figure 16.9).

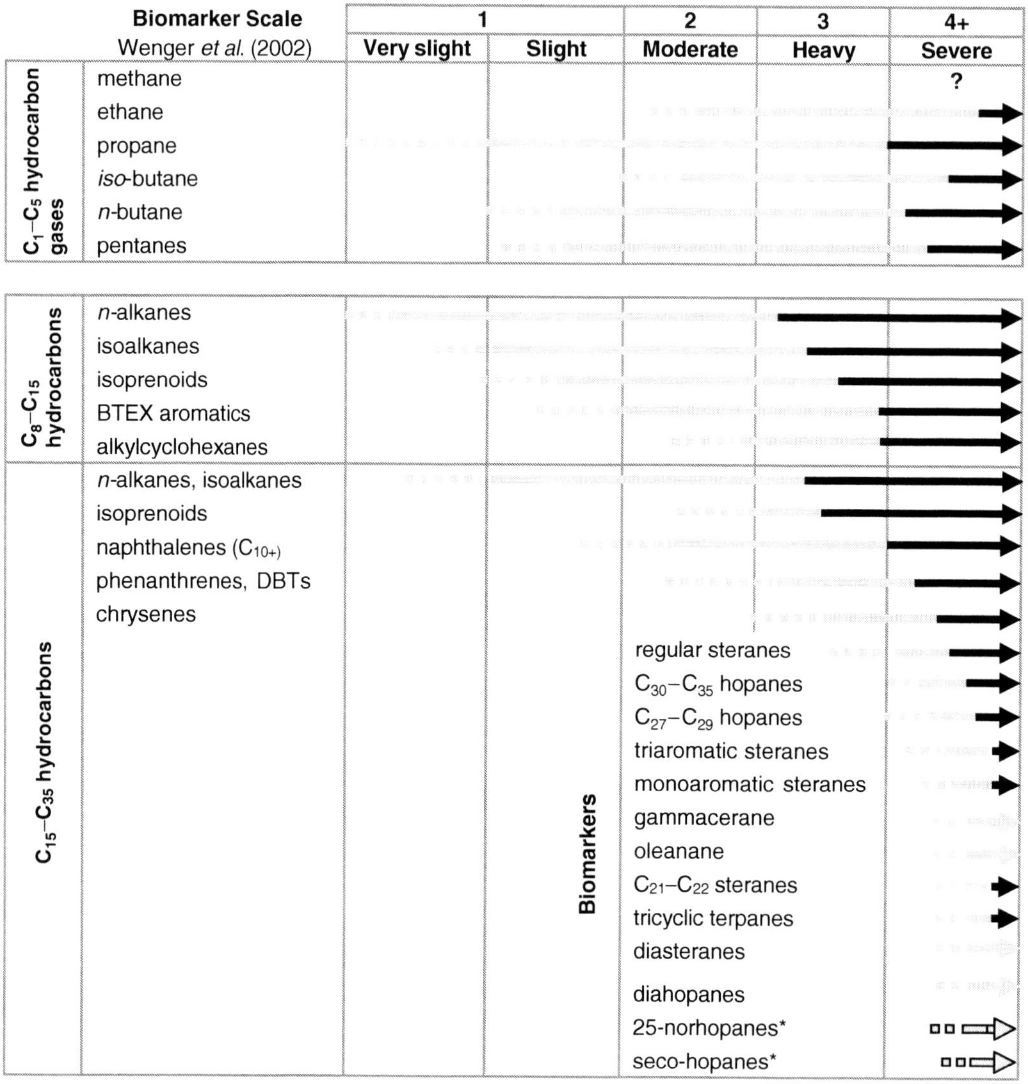

Biodegradation parameters

Biomarker Scale Wenger *et al.* (2002)		1 Very slight	Slight	2 Moderate	3 Heavy	4+ Severe
C_1–C_5 hydrocarbon gases	methane					?
	ethane					→
	propane					→
	iso-butane					→
	n-butane					→
	pentanes					→
C_8–C_{15} hydrocarbons	*n*-alkanes				→	
	isoalkanes				→	
	isoprenoids				→	
	BTEX aromatics				→	
	alkylcyclohexanes				→	
C_{15}–C_{35} hydrocarbons	*n*-alkanes, isoalkanes				→	
	isoprenoids				→	
	naphthalenes (C_{10+})				→	
	phenanthrenes, DBTs				→	
	chrysenes				→	
	Biomarkers regular steranes					→
	C_{30}–C_{35} hopanes					→
	C_{27}–C_{29} hopanes					→
	triaromatic steranes					→
	monoaromatic steranes					→
	gammacerane					
	oleanane					
	C_{21}–C_{22} steranes					→
	tricyclic terpanes					→
	diasteranes					
	diahopanes					
	25-norhopanes*					▫▫⇨
	seco-hopanes*					▫▫⇨

Figure 16.10. Generalized sequence of the removal of selected molecular groups at increasing levels of biodegradation (modified from Wenger *et al.*, 2002). Arrows indicate where compound classes are first altered (dashed lines), substantially depleted (solid gray), and completely eliminated (black).
BTEX, benzene, toluene, ethylbenzene, xylenes; DBTs, dibenzothiophenes.
*25-Norhopanes and seco-hopanes are created by biodegradation.

Correlation of the oils was made using bioresistant cyclic saturated and aromatic hydrocarbons, including many biomarkers, which are largely unaltered at this level of biodegradation. The differences in physical properties between the non-degraded and heavily degraded oils are considerable. API gravity changes from 36.1 to 21.4° API and viscosity changes from 0.38 to 7.5 centipoise, respectively. Even mild to moderate levels of microbial alteration can have a strong influence on whether the reservoir accumulations are deemed suitable for economic development.

Biodegradation is a quasi-stepwise process (Figure 16.10). Saturated and aromatic biomarkers are biodegraded only after consumption of *n*-alkanes, most simple branched alkanes, and some of the alkylated benzenes (Seifert and Moldowan, 1979; Seifert *et al.*,

Effects of biodegradation on petroleum composition 661

1984; Connan, 1984; Moldowan et al., 1992). Laboratory and empirical observations indicate that biomarkers also are consumed in a preferential order. Regular steranes and alkylated aromatics are the most susceptible to biodegradation, followed by hopanes, aromatic steroidal hydrocarbons, diasteranes, and tricyclic terpanes. At advanced stages of alteration, certain biomarkers, such as 25-norhopanes and secohopanes, appear to be created.

The biodegradation scale

Because of their differential resistance to biodegradation, comparisons of the relative amounts of compound classes can be used to rank the extent of biodegradation of petroleum. We developed a scale to assess the extent of biodegradation based on the relative abundance of various hydrocarbon classes (Figure 16.11). The scaling concept was proposed initially by Alexander et al. (1983b) (see also Volkman et al., 1983a; 1983b;

1984) and later was modified by Moldowan et al. (1992). Figure 16.11 shows the effects of various levels of biodegradation on the composition of typical mature oil, ranked from 1 to 10. The biomarker biodegradation scale is based mostly on empirical field observations. These include suites of moderately mature single-source oils that were biodegraded to varying degrees. The relative level of degradation is based partially on changes in bulk chemical and physical properties and partially on the principles of sequential catabolism.

Activation enzymes exhibit varying degrees of substrate specificity, and certain biomarkers may be removed faster than others (either as hydrocarbon classes or as individual isomers). In the subsurface, the microbial ecology is changing continually as individual organisms and syntrophic consortia compete for the pool of available hydrocarbons and inorganic electron acceptors. The quasi-sequential degradation of hydrocarbons

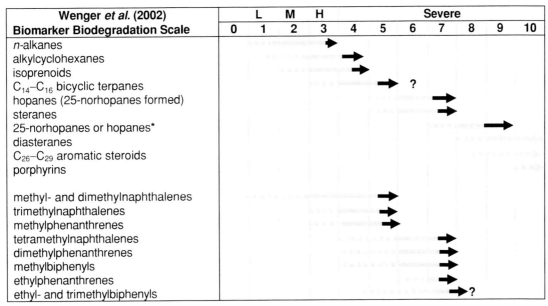

Figure 16.11. The extent of biodegradation of mature crude oil can be ranked on a scale of 1–10 based on differing resistance of compound classes to microbial attack. Biodegradation is quasi-sequential because some of the more labile compounds in the more resistant compound classes can be attacked before complete destruction of less resistant classes (from Peters and Moldowan, 1993). Arrows indicate where compound classes are first altered (dashed lines), substantially depleted (solid gray), and completely eliminated (black). Sequence of alteration of alkylated polyaromatic hydrocarbons is based on work by Fisher et al. (1996b; 1998) and Triolio et al. (1999). Degree of biodegradation from Wenger et al. (2002) reflects changes in oil quality (L, lightly biodegraded; M, moderately biodegraded; H, heavily biodegraded).
*Hopanes degraded without the formation of 25-norhopanes.

does not occur in a true stepwise fashion but reflects differences in the rates of catabolism under varying conditions.

The biomarker biodegradation scale reflects the degradation sequence observed most frequently in subsurface accumulations. It is remarkable that the order of biomarker removal shown in Figure 16.11 is so common. One might anticipate more variable biomarker degradation sequences that reflect differences in reservoir conditions and specific microbial communities. Nonetheless, this ranking scheme must be used cautiously, because there are many known deviations. Some patterns appear to be nearly universal, such as the sharp division that occurs between ranks 5 and 6, after all isoprenoids are removed but before degradation of the steranes. Other sequences of biodegradation are clearly variable. For example, some oils exposed to heavy biodegradation show significant alteration of the hopanes before all steranes are destroyed, although hopanes are generally considered more resistant than steranes to biodegradation. Oil spills that are biodegraded under energetic, aerobic conditions may follow completely different sequences than oil in the subsurface. For example, Wang et al. (2001b) reported a relative degradation order of diasteranes > C_{27} steranes > tricyclic terpanes > pentacyclic terpanes > norhopanes ~C_{29} $\alpha\beta\beta$ steranes in a 24-year-old oil spill that occurred in a marine salt marsh. The early loss of diasteranes is puzzling because these biomarkers are typically highly conserved. Nevertheless, evidence suggests that a biodegradation ranking scale similar to that described above could be used to monitor the progress of bioremediation of refinery wastes (Moldowan et al., 1995).

Qualitative assessment of the degree of biodegradation

Wenger et al. (2002) proposed that the level of biodegradation in crude oils be described using terms that focus on changes that occur before the alteration of biomarkers. Oils are described using the terms "very slight," "slight," "slight to moderate," and "moderately biodegraded" based on the degree of alteration of the n-alkanes. Oils where n-alkanes are mostly absent are heavily biodegraded, while those with altered biomarkers are severely biodegraded (Figure 16.10). Major changes in the bulk physical and chemical properties of crude oils occur well before biomarkers are altered noticeably (Figure 16.12).

In contrast, the main emphasis of the biomarker biodegradation scale is to differentiate the degree of biodegradation that occurs after the removal of n-alkanes. This scale is expressed as a ranking or sequence, and not as numerical values. That is, the numbers are convenient labels that have no quantitative significance. In Peters and Moldowan (1993), descriptive terms were assigned to ranges on the scale: "light" = 1–3, "moderate" = 4–5, "heavy" = 6–7, "very heavy" = 8–9, and "severe" = 10. Similar nomenclature was used in a biomarker degradation scale by Alexander et al. (1983b) and Volkman et al. (1983b; 1984). The extent of biodegradation for oils with no isoprenoids was moderate, with greater degrees of alteration termed "extensive," "very extensive," "severe," and "extreme."

The differences in the qualitative descriptions of the degree of biodegradation depend on whether the scale is based on oil quality issues (producibility, economic value) or on the relative bioresistance of petroleum compounds. Nevertheless, the use of similar terminology leads to confusion. Oils that Wenger et al. (2002) would call severely biodegraded may have unaltered sterane and hopane distributions and would be classified as only moderately biodegraded with respect to biomarkers. For this reason, we suggest that the biomarker biodegradation scale be used without descriptions and that all oils with altered cyclic saturated biomarkers are termed "severely biodegraded," with the degree of severity specified by the biomarker biodegradation scale ranking.

Figure 16.12 illustrates further that bulk physical and chemical measurements can be used effectively to determine the degree of biodegradation, at least up to the point of sterane alteration (rank ~5). Increases in sulfur, nitrogen, and trace metals correspond to decreasing API gravity and logarithmic increases in viscosity. Such systematic changes can be used to predict oil quality, provided the measurements are calibrated using equivalent oils (same source and maturity). Other physical parameters (e.g. pour point and flash point) and bulk chemical measurements (e.g. Conradson carbon residue, aniline point, total acid number (TAN)) that are common refinery assays also may correlate with the degree of biodegradation and can be used to determine economic value.

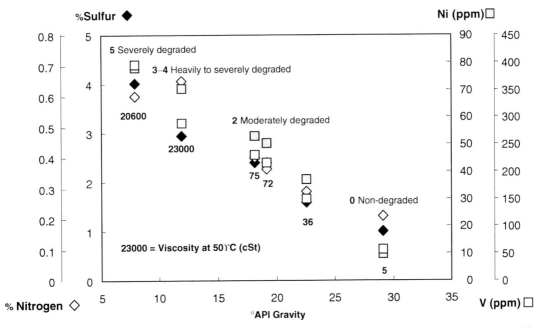

Figure 16.12. Comparison of bulk oil properties for a suite of oils (La Luna source) from Eastern Venezuela. Sulfur, nitrogen, nickel, and vanadium increase proportionally with decreasing API gravity. The degree of biodegradation is indicated by the numerical ranking of the biomarker biodegradation scale and by descriptive terms used by Wenger et al. (2002).

Quantitative assessment of the degree of biodegradation

Biodegradation is primarily a process of hydrocarbon destruction. Petroleum polar molecules and asphaltenes are considered to be highly resistant, and compounds contributed from microbial biomass are considered to be minor or insignificant. Quantitative assessments of biodegradation, therefore, measure the amount of hydrocarbons destroyed. Most of these measurements rely on knowledge of the composition of the unaltered oil. Once the concentration of a conserved compound is determined in non-biodegraded oil, the amount of hydrocarbons removed by microbial activity in a degraded equivalent can then be measured directly or calculated from the enrichment of the conserved component.

Direct measurement of oil composition may involve quantitative gas chromatography using internal or surrogate standards or liquid chromatographic separation into compound classes. Gas chromatographic analysis assumes that all bioreactive compounds are amenable to chromatographic separation, while the non-volatile petroleum components are conserved. The degree of hydrocarbon depletion is measured by comparing the summed detector response of all eluted compounds (normalizing to the added standards) in the biodegraded and unaltered oils. The accuracy of this method is compromised by necessary assumptions about the uniformity of detector response for different hydrocarbons. Liquid chromatographic separation of oils into hydrocarbon and non-hydrocarbon group types and subsequent weighing of the isolated fractions eliminates problems with detector response but is applicable only to the non-volatile portion.

Instead of measuring the amount of hydrocarbons removed, the degree of biodegradation can be assessed by quantifying the enrichment of components that are conserved. Pristane, phytane, or other isoprenoids are conserved only during early stages of biodegradation and are poor reference compounds. Hopane is assumed to be conserved in many environmental and oil spill studies (e.g. Prince et al., 1994; Bragg et al., 1994; Le Dréau et al., 1997; Venosa et al., 1997). However, hopane has been destroyed in many severely biodegraded oils. Minor elements, such as sulfur and nitrogen, appear to be mostly conserved, although biodegradation of the smaller heteroatomic compounds is known. Trace metals, such as nickel and vanadium, appear to be conserved

(Sasaki et al., 1998). These elements occur mostly in highly bioresistant porphyrins and can be measured with relatively high accuracy and precision.

In environmental studies of oil spills and in laboratory culture experiments, changes resulting from biodegradation can be followed over time and compared with the initial oil composition. When dealing with geologic samples, it is impossible to know the exact composition of crude oil before biodegradation. The original composition can be inferred by correlating the biodegraded sample to a non-biodegraded equivalent. However, this inference depends on compensating for all of the usual factors that determine petroleum composition (e.g. source facies, thermal maturity, migration fractionation, and reservoir alteration processes). Volkman et al. (1983b) and Peters et al. (1991) used biodegradation-resistant C_{27} diasterane 20S in genetically related biodegraded and non-biodegraded oils to determine the relative effects of biodegradation on concentrations of various compounds. Many factors may limit the reliability of the correlation (e.g. limited sample availability or incomplete knowledge of the petroleum systems), but the degree of biodegradation is also a major determinant. Oils with unaltered biomarker distributions (rank <4 on the biomarker biodegradation scale) can be correlated easily to their unaltered equivalents. Correlations become more difficult when biomarker and isotopic values are altered by more advanced biodegradation. In the most severely degraded samples, biomarker and isotopic analyses of pyrolyzates of the polar and asphaltene fractions may provide the best means for correlation (Cassani and Eglinton, 1986; Philp et al., 1988b; Dembicki and Mathiesen, 1994; Rooney et al., 1998; Odden et al., 2002).

BIODEGRADATION PARAMETERS

The biomarker biodegradation scale is based on the relative susceptibility of various compound classes to biodegradation. The susceptibility of individual compounds within each class can differ substantially, such that some compounds may be removed before others are affected. These variances may depend on carbon number in homologous sequences or structural or optical isomeric configurations, or they may result from complex and largely unknown microbial relationships. As with the compound classes, biodegradation of individual biomarker compounds follows a quasi-sequential order. The term "quasi-sequential" indicates that some molecules within a given class are more labile than others but may remain while the more resistive molecules begin to be biodegraded.

Molecular differences within various biomarker and non-biomarker compound classes are discussed below, roughly in the order in which they are degraded (Table 16.1). Some of the conclusions described here are based on only one to a few reports and must be considered as tentative pending further studies.

n-Alkanes

As noted above, n-alkanes in the range $\sim C_8$–C_{12} are preferentially removed in the earliest stages of biodegradation (Figure 16.9). This preference is consistent with the substrate specificity of aerobic monooxygenase and, possibly, anaerobic pathways as well. As biodegradation proceeds, n-alkanes $<C_8$ and $>C_9$ begin to be depleted before all C_8–C_{12} n-alkanes are consumed. Biodegradation does not appear to strongly favor either odd- or even-numbered C_{15+} n-alkanes. Some biodegraded oils, particularly marine oil spills and beach strandings, are enriched in C_{35+} n-alkanes. These n-alkanes may be a secondary product produced by bacterial biomass (Connan, 1984), but they are more likely to be enriched components from the parent oil (Heath et al., 1997).

Light C_2–C_6 hydrocarbons exhibit a preferential degradation sequence (Figure 16.10). Propane is altered first, followed by butane, pentane, and C_{6+} hydrocarbons (James and Burns, 1984). Ethane is comparatively resistant, and methane appears to be conserved. Although methane is utilized by aerobic methanotrophs and by anaerobic synthropic communities involving archaea and sulfate-reducing bacteria, such microbial activity has not been demonstrated to occur in reservoir rocks. The net effect of biodegradation is an overall decrease in wet gas components, enrichment in unaltered methane, and an increase in biogenic CO_2 that is produced from the microbial oxidation of reservoir hydrocarbons. The abundance of CO_2 is not a reliable measurement of the degree of biodegradation, as its concentration is highly dependent on numerous inorganic processes. Stable carbon isotopes, however, are diagnostic. Gaseous hydrocarbons exhibit large isotopic fractionations during biodegradation, with the residual compounds becoming enriched in ^{13}C as biodegradation advances. Isotopic shifts in propane relative to

Table 16.1. *Summary of selective biodegradation observed within compound classes*

	Class	biodegradation susceptibility
Least	n-Alkanes	$C_3 \sim C_8–C_{12} > C_6–C_8 \sim C_{12}–C_{15} > C_{6-} \sim C_{15+}$
	Branched alkanes	Monomethyl > polymethyl > highly branched
	Acyclic isoprenoids	Lower molecular weight (e.g. C_{10}) > higher molecular weight (e.g. C_{20})
		Acyclic isoprenoids degraded before major alteration of polycyclic biomarkers
	Alkylated benzenes and polyaromatic hydrocarbons	1-ring > 2-ring > 3-ring > 4-ring
		Methyl and dimethyl > trimethyl or extended alkylated species
	Alkylbiphenyls and alkyldiphenyl-methanes	Alkylation at C-4 > C-2 or C-3
	Hopanes (25-norhopanes present)	When 25-norhopanes are present, microbial attack degrade 22R > 22S for homohopanes and favors $C_{27}–C_{32} > C_{33} > C_{34} > C_{35}$ 17α-hopanes (Peters and Moldowan, 1991). However, Rullkötter and Wendisch (1982) observed that the higher homologs in the series $C_{27}–C_{32}$ were degraded faster than lower homologs
	Steranes (25-norhopanes present)	$\alpha\alpha\alpha$ 20R and $\alpha\beta\beta$ 20R > $\alpha\alpha\alpha$ 20S and $\alpha\beta\beta$ 20S $C_{27} > C_{28} > C_{29} > C_{30}$
	Steranes (25-norhopanes absent)	$\alpha\alpha\alpha 20R(C_{27}–C_{29}) > \alpha\alpha\alpha 20S(C_{27}) > \alpha\alpha\alpha 20S(C_{28}) > \alpha\alpha\alpha 20S(C_{29}) \geq \alpha\beta\beta(20S + 20R)(C_{27}–C_{29})$
	Hopanes (25-norhopanes absent)	When 25-norhopanes are absent, microbial attack favors $C_{35} > C_{34} > C_{33} > C_{32} > C_{31} > C_{30} > C_{29} > C_{27}$ and 22R > 22S
	Diasteranes	$C_{27} > C_{28} > C_{29}$
	Non-hopanoid triterpanes	Gammacerane and oleanane are more resistant to biodegradation than the hopanes
	Aromatic steroids	$C_{20}–C_{21}$ triaromatic (water washing?) > $C_{27}–C_{29}$ 20R monoaromatic $\sim C_{26}–C_{28}$ 20R triaromatic > C_{21}, C_{22} monoaromatic
Most	Porphyrins	No evidence for significant biodegradation of porphyrins (e.g. Sundararaman and Hwang, 1993)

Common order of susceptibility to biodegradation

the more resistant hydrocarbon gases occur in reservoirs with associated oils that are biodegraded only slightly. The difference in $\delta^{13}C$ values between altered and unaltered gases is directly related to the extent of biodegradation.

Carbon isotopic ratios can be used to distinguish slightly biodegraded crude oils (depleted only in $\sim C_8–C_{12}$ n-alkanes) from those that have been moderately biodegraded and then recharged with a non-biodegraded light condensate (Guthrie et al., 2000). In both cases examined by Guthrie et al. (2000), the n-alkanes exhibit a bimodal distribution. In slightly degraded oil, $\delta^{13}C$ values of pentane and hexane were shifted up to 3–4‰. In moderately degraded oil (most of the n-alkanes $<C_{15}$ removed) that was recharged with a non-biodegraded condensate, the $\delta^{13}C$ values of pentane and hexane reflect their unaltered values. These conclusions required prior knowledge of the petroleum system to account for isotopic variations due to source or thermal maturity.

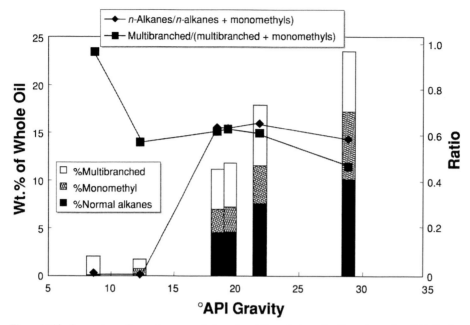

Figure 16.13. Comparison of normal, monomethyl-, and multibranched acyclic alkanes in a suite of oils (La Luna source) from eastern Venezuela (same samples as in Figure 16.12). Their relative susceptibilities to biodegradation are similar until the oils are severely biodegraded.

Branched alkanes

Methyl groups generally inhibit biodegradation. However, a single methyl group appears to lend little or no resistance to microbial removal compared with corresponding normal alkanes (Figure 16.13). The effectiveness of enhanced bioresistance is seen only in the light $\sim C_4$–C_7 hydrocarbons, where the isoalkanes and anteisoalkanes are preserved selectively compared with their normal equivalents (Connan, 1984).

Acyclic isoprenoids with multiple methyl groups are more bioresistant than n-alkanes having similar gas chromatographic elution times. Pirnik et al. (1974) first demonstrated this preferential degradation in laboratory experiments. They found that a pure bacterial culture initially consumed only n-alkanes, but that both n-alkanes and isoprenoids were consumed after the n-alkanes fell below a threshold concentration. Various ratios of acyclic isoprenoids to n-alkanes (e.g. pristane/nC_{17} and phytane/nC_{18}) can be used to measure the relative degree of biodegradation. Slightly to moderately biodegraded oils have higher pristane/nC_{17} and phytane/nC_{18} than related, non-biodegraded oils.

These ratios were first used to indicate the degree of biodegradation for subsurface oils by Winters and Williams (1969) and for surface oil spills by Miget et al. (1969).

Selective biodegradation of the isoprenoid over more bioresistant steranes or hopanes can be used to determine the level of biodegradation in oils ranked 3–4 on the biomarker biodegradation scale. In these heavily to severely degraded oils where n-alkanes are low or absent, isoprenoid hydrocarbons are removed selectively while steranes and hopanes remain unaltered. The ratio of a major isoprenoid, such as pristane or phytane, to a conserved saturated biomarker, such as hopane, indicates relative states of biodegradation within a suite of related oils. The compounds may be sufficiently abundant that the ratio can be determined using gas chromatography of the whole oil (Figure 16.14). Hopane is frequently considered to be conserved in many environmental and oil-spill studies (e.g. Prince et al., 1994; Bragg et al., 1994; Le Dréau et al., 1997; Venosa et al., 1997). As discussed below, hopane is not conserved at higher levels of biodegradation (rank ≥ 6), but it can be considered to be conserved in oils that retain isoprenoid hydrocarbons.

Biodegradation parameters 667

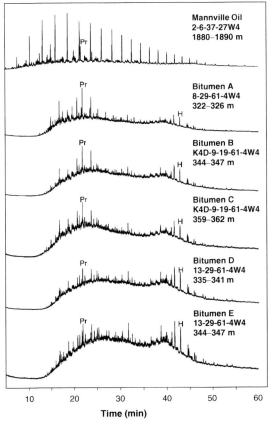

Figure 16.14. Saturate gas chromatogram of Mannville Formation conventional oil (top) and biodegraded heavy oil (bitumen) samples from the Mannville Formation in the Fort Kent thermal project of the Cold Lake Deposit, Alberta. The bitumens are all severely degraded, possessing no normal alkanes, decreasing relative quantities of pristane (Pr), and increasing relative quantities of 17α-hopane (H) and other triterpanes. Reprinted from Peters and Fowler (2002). © Copyright 2002, with permission from Elsevier.

However, the mere presence of isoprenoids is not sufficient to assume that hopane has been conserved because samples may be mixtures of severely biodegraded and less altered oils (Figure 16.15).

Susceptibility of the C_{10}–C_{20} isoprenoids decreases with increasing carbon number. Illich et al. (1977) used the ratio of C_{10}/C_{20} isoprenoids as an indicator of the level of biodegradation in oils from the Marañon Basin, but the difference in volatility between these compounds imposes uncertainty. There are no published studies on the relative bioresistance of >C_{20} linear isoprenoids.

Highly branched isoprenoids and other highly branched alkanes are substantially more bioresistant than alkanes with only methyl groups. These compounds are enriched during biodegradation and are typically part of the unresolved complex mixture (UCM) (Gough and Rowland, 1990; Killops and Al-Juboori, 1990).

Alkylated monocyclic alkanes

Long-chained alkylated cyclopentanes and cyclohexanes are about as susceptible to biodegradation as branched alkanes. Consequently, monocyclic alkanes increase in concentration during early microbial alteration (rank 1–2) with the preferential removal of n-alkanes. Evidence suggests that cyclopentanes are more susceptible than cyclohexanes and that cycloalkanes with branched alkylated side chains are more bioresistant than those with linear side chains. All monocyclic alkanes are absent or in trace quantities by rank ~4–5.

Relative resistance to biodegradation influences many of the parameters based on C_7 hydrocarbon isomers that are commonly used for source correlation and thermal maturation studies (see Chapter 7). Ratios that compare the relative abundance of C_7 alkylated cyclopentanes and methylcyclohexane with n-heptane (e.g. Thompson's heptane ratio) are more susceptible to microbial alteration than those that compare C_7 monocyclic alkanes with branched isomers (e.g. isoheptane ratio).

Little information is available on the relative bioresistance of monocyclic alkanes. Aerobic bacteria appear to degrade low-molecular-weight monocyclic alkanes with little or no isomeric or carbon preference (Solano-Saerena et al., 1999). The same may not be true for anaerobic biodegradation. Monocyclic alkanes were consumed more readily by sulfate reducers than methanogens in sediment-enrichment cultures grown in the presence of a natural gas condensate to promote growth (Townsend et al., 2003). The lower-molecular-weight species (C_0–C_2) were preferentially removed relative to higher homologs (C_3–C_5). Dimethyl-substituted monocyclic alkanes were biodegraded only in the presence of sulfate, which resulted in limited enrichment of specific isomers.

There are a few reports of selective bioresistance for the less volatile monocyclanes. Hostettler and Kvenvolden (2002) reported that anaerobic degradation of n-alkylated cyclohexanes and n-alkanes

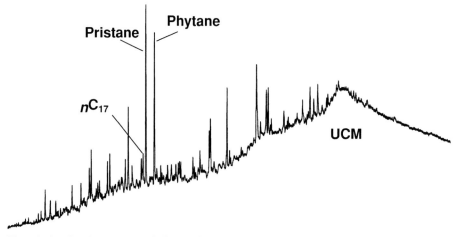

Figure 16.15. Gas chromatogram of oil cementing a sandstone breccia collected from a beach near Brora on the Inner Moray Firth coast of northeastern Scotland has a sharply rising baseline typical of heavily biodegraded oils, where the unresolved complex mixture (UCM) consists of compounds that resist biodegradation. Low n-alkanes (e.g. nC_{17}) combined with apparently unaltered pristane and phytane indicate that a second oil charge mixed with the heavily biodegraded oil and was later mildly biodegraded. Reprinted from Peters et al. (1999b). © Copyright 1999, with permission from Elsevier.

in oil-contaminated sediments resulted in preferential loss of high-molecular-weight homologs. Figure 16.16 shows that n-alkanes degraded from the high-molecular-weight end first, resulting in enhanced low-molecular-weight n-alkanes (nC_{17}) before all n-alkanes were removed. Similar observations were made for spilled diesel fuel (Stout and Lundegard, 1998). The n-alkylcyclohexanes in Figure 16.16 begin to degrade when the n-alkanes are nearly depleted. Beyond this level of biodegradation, n-alkylcyclohexanes increase in abundance compared with the original oil, and the distribution of these compounds progressively favors the low-molecular-weight end. A similar progression of hydrocarbon degradation occurs in chronic diesel fuel spillage at Mandan, North Dakota (Hostettler and Kvenvolden, 2002).

We observed similar enrichments in moderately to severely biodegraded oils. For example, the abundance of cyclohexanes in biodegraded oils (rank ∼3–4) from Cerra Negro, Venezuela, is about a fifth of that of non-degraded equivalent oil. While the overall concentrations of the cyclohexanes decrease with increasing alteration, the distribution of the remaining cyclohexanes shifts slightly. The relative distributions of the surviving cyclohexanes were altered by weathering, which removed all of the volatile compounds ($<C_{15}$), enriched mid-range species ($\sim C_{15}$–C_{25}), and resulted in depletion of the higher-molecular-weight homologs relative to the cyclohexane distribution of the non-degraded equivalent.

The unique progression of anaerobic hydrocarbon degradation described above might contribute to spurious interpretations of processes affecting crude oil or refined products. For example, elevated low-molecular-weight n-alkanes or n-alkylcyclohexanes in crude oils are commonly interpreted to result from admixture of highly mature oil in the reservoir (multiple charge scenario) or diesel or other refined additive (contamination scenario). Such distributions could be the result of anaerobic biodegradation of crude oil. Likewise, spilled diesel and other volatile refinery fuels are defined by the range and distribution of n-alkanes and n-alkylcyclohexanes. If biodegradation progresses beyond n-alkane loss to the stage of enhanced low-molecular-weight n-alkylcyclohexanes, then the altered hydrocarbon pattern could be interpreted erroneously to indicate other lower-range fuels or admixtures of fuels.

Bicyclic terpanes

Alexander et al. (1983b) found that the C_{14}–C_{16} bicyclic terpanes are less susceptible to biodegradation than isoprenoid hydrocarbons and are completely absent before the onset of sterane and hopane biodegradation.

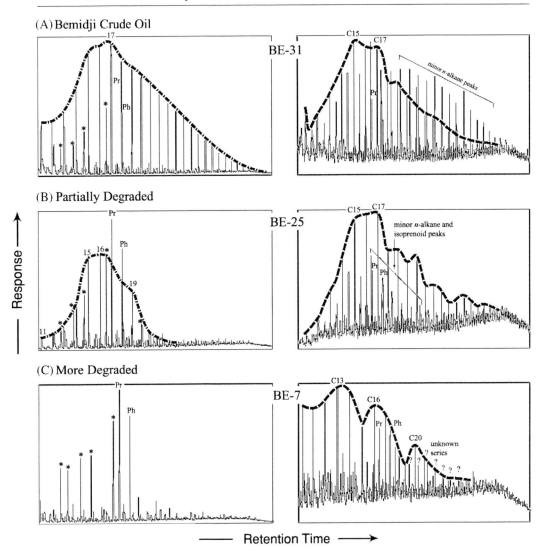

Figure 16.16. Anaerobic biodegradation of crude oil in sediments contaminated by the 1979 rupture of a pipeline in Bemidji, Minnesota, resulted in preferential loss of higher-molecular-weight n-alkanes (m/z 57, left) and n-alkylcyclohexanes (m/z 83, right). Overall n-alkane distributions are emphasized by dashed lines. Samples BE-31, BE-25, and BE-7 are unaltered, partially degraded, and more degraded (e.g. nC_{17}/pristane $= 1.3, 0.64,$ and 0, respectively). Lines on the left and right encompass n-alkanes and cyclohexanes, respectively. Labeled peaks include pristane (Pr) and phytane (Ph) and other isoprenoids (*) and n-alkanes. The unaltered Bemidji oil originated from near Edmonton, Alberta, Canada. From Hostettler and Kvenvolden, 2002. Used with the permission of AEHS, 150 Fearing St, Amherst, MA 01002, USA.

Williams et al. (1986) observed selective removal of 8β(H)-homodrimane and other bicyclic sesquiterpanes compared with 8β(H)-drimane in oils with no evidence of sterane or hopane degradation. Bicyclic terpanes are absent in heavy oils from Alberta, Canada, where the steranes and hopanes have been partially biodegraded (Hoffmann and Strausz, 1986). These observations suggest that the bicyclic terpanes are altered in rank 4 oils and totally removed by rank 5.

However, Kuo (1994b) found that the distribution of bicyclic terpanes could be altered in a laboratory simulation of water washing. The ratio of rearranged drimane to 8β(H)-drimane increased significantly, while the ratio of 8β(H)-drimane to 8β(H)-homodrimane increased slightly. These observations contrasted with the diterpanes and tetracyclic terpanes that exhibited little change with water washing. We urge caution when using the absence of bicyclic terpanes as an indicator of biodegradation rank because there is a high potential for their alteration by water washing, many severely biodegraded oils have lost substantial portions of their semi-volatile hydrocarbons, and preservation of the bicyclic terpanes during sample fractionation is difficult.

Alkylphenols

Alkylphenols are relatively water-soluble and are frequently observed in formation waters associated with oil accumulations. Consequently, their depletion in biodegraded oil may be the result of either microbial removal or water washing. Taylor et al. (2001) studied the distribution of C_0–C_3 alkylphenols in North Sea oil biodegraded with aerobic enrichment culture. They found that the alkylphenols were readily removed with preferential destruction of the C_2 and C_3 compounds. Similar distributions were observed in suites of biodegraded oils from Nigeria (rank 0–5), with the greatest effects occurring in conjunction with the biodegradation of the alkylnaphthalenes. A suite of California (rank 0–4) oils, however, exhibited no clear relationship between alkylphenol abundance and degree of biodegradation. The alkylphenol distributions in this oil suite were enhanced in C_3 compounds and could be explained by water washing. Rolfes and Andersson (2001) describe a highly sensitive gas chromatography/atom-emission detection method to determine alkylphenols after derivatization using ferrocenecarboxylic acid.

Alkylated benzenes and polynuclear aromatic hydrocarbons

The susceptibility of the alkylated benzenes and polynuclear aromatic hydrocarbons (PAHs) to biodegradation has been documented widely in studies of environmental oil spills (see Chapter 10), in microbial culture experiments (see above), and in petroleum. In general, bioresistance increases with increasing aromatic ring number and with increasing number of alkyl substituents. Because PAHs are included in standardized methods of the US Environmental Protection Agency (EPA), oil-spill studies routinely use their shifting distributions as indicators of the level of biodegradation.

The alkylbenzenes are the most labile, in terms of both water solubility and microbial catabolism. These compounds appear to be altered before the near-total depletion of n-alkanes, but they persist beyond the total depletion of isoprenoids. Alkylated naphthalenes are more bioresistant. Minor alteration of methyl- and dimethylnapthalenes occurs during removal of n-alkanes, trimethylnaphthalenes are altered during the removal of the isoprenoids, and tetramethylnaphthalenes persist until steranes are largely depleted (Fisher et al., 1998). Phenanthrenes behave in a similar manner with increasing alkylation, but are generally more resistant to biodegradation than alkylnaphthalenes.

Bastow et al. (1998) found that the relative proportions of dihydro-ar-curcumene and isodihydro-ar-curcumene change upon biodegradation. Dihydro-ar-curcumene is believed to be primarily a biomarker for higher land plants (Ellis et al., 1995). Isodihydro-ar-curcumene is a related compound where the methyl on the aromatic ring is meta to the alkyl group. It is believed to be a rearrangement product of dihydro-ar-curcumene or its precursor (Bastow et al., 1997). Using a chiral chromatographic column that separated these two compound into their R and S isomers, Bastow et al. (1998) studied these compounds in two biodegraded Jurassic oils from Australia (Figure 16.17). Both oils are biodegraded because they lack n-alkanes and acyclic isoprenoids. The less severely degraded oil, estimated at rank ~4, has a dihydro-ar-curcumene and isodihydro-ar-curcumene distribution identical to that of the non-degraded oil (Cretaceous, Gippsland Basin). In the more severely biodegraded oil, estimated at rank ~5, the dihydro-ar-curcumene isomers are absent. The

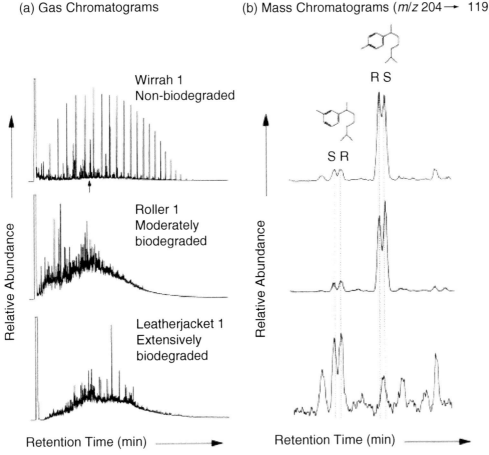

Figure 16.17. (a) Gas chromatograms and (b) partial mass chromatograms (m/z 204→119) show comparative bioresistance of dihydro-*ar*-curcumene (top structure) and isodihydro-*ar*-curcumene (bottom structure) and their enantiomers in non-biodegraded and biodegraded crude oils from Australia. ↑ indicates approximate elution time of dihydro-*ar*-curcumene. Reprinted from Bastow *et al.* (1998). © Copyright 1998, with permission from Elsevier.

persistence of isodihydro-*ar*-curcumene suggests that it is a potentially useful biomarker for land-plant input in severely biodegraded oils. This would be particularly true for condensates that no longer possess C_{15+} biomarkers due to phase separation.

Alkylated PAH isomers show a range of susceptibility to biodegradation, which generally decreases with increasing number of alkyl substituents (e.g. Volkman *et al.*, 1984, their Figure 4; Williams *et al.*, 1986, their Figure 5). Fisher *et al.* (1998) observed similar variations among three alkylnaphthalene ratios during biodegradation for three different sets of samples. The samples included (1) intertidal coastal sediment contaminated with condensate from a North West Shelf field and collected over a three-year period, (2) surficial seafloor sediments collected at varying distances from an offshore platform contaminated by oil-based drilling mud, and (3) biodegraded crude oil. The three alkylnaphthalene biodegradation ratios are as follows:

Dimethylnaphthalene ratio (DBR)
  = 1,6-DMN/1,5-DMN.
Trimethylnaphthalene ratio (TBR)
  = 1,3,6-TMN/1,2,4-DMN.
Tetramethylnaphthalene (TeBR)
  = 1,3,6,7-TeMN/1,3,5,7-TeMN.

Alkylnaphthalene ratios DRB, TBR, and TeBR show comparable variations in petroleum contaminating

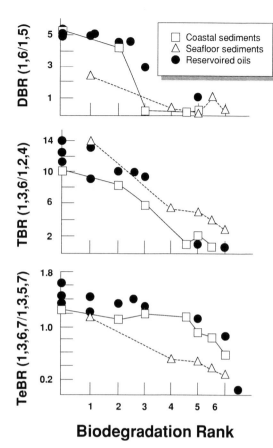

Figure 16.18. Dimethyl-, trimethyl-, and tetramethylnaphthalene biodegradation ratios (DBR, TBR, and TeBR, respectively) for samples from contaminated coastal and seafloor sediments and crude oils from north Western Australia (from Fisher et al., 1998). Values in parentheses indicate the methyl-substitutions for the isomers in each ratio. Coastal sediments are condensate-contaminated intertidal coastal sediments collected over three years. Seafloor sediments are seafloor sediments contaminated with low-toxicity oil-based mud near an offshore production platform. Reservoired oil is crude oil from producing reservoirs.

coastal sediment, oil-based mud contaminating seafloor sediment, and crude oils from reservoirs (Figure 16.18). The value of DBR is altered significantly at biodegradation level 2–3, the TBR between levels 3 and 5, and the TeBR at levels 5–6, indicating depletion of the isomer more susceptible to microbial attack for each case. This biodegradation sequence has not been verified and should be used with caution.

The relative bioresistance of sulfur-containing aromatic species, such as benzothiopehenes and dibenzothiophenes, is greater than or equal to that of PAHs with corresponding ring numbers. Williams et al. (1986) noted nearly complete removal of both methylphenanthrenes and methyldibenzothiophenes in biodegraded oils (rank <5). Connan et al. (1992) reported the removal sequence for aromatic hydrocarbons in biodegraded Dead Sea asphalts as alkylbenzenes, naphthalenes, benzothiophenes, phenanthrenes, and, finally, dibenzothiophenes. We found in other sample suites that the benzothiophenes and dibenzothiophenes are unaltered and become increasingly enriched up to rank 5, and only become depleted at biodegradation levels associated with sterane destruction. Part of this variability may be due to differences in aqueous solubility and the degree of water washing that accompanied microbial alteration.

Alkylbiphenyls and alkyldiphenylmethanes

The relative abundances and distributions of alkylbiphenyls and alkyldiphenylmethanes (Figure 16.19) change systematically with biodegradation in reservoirs and surface sediments, where isomers with alkyl substituents at C-4 increase compared with others (Trolio et al., 1999). For example, 4-ethylbiphenyl increases relative to 2- and 3-ethylbiphenyls at biodegradation level 3–4, and it is the only isomer remaining in Gippsland Basin oils biodegraded to level 4–5. The most biodegraded petroleum in the study (level 7) lacks methylbiphenyls, dimethylbiphenyls, and methyldiphenylmethanes, suggesting that the maximum extent of biodegradation where these compounds are of use is ~level 4–5. However, some ethylbiphenyls and trimethylbiphenyls remain in heavily biodegraded oils. Unlike the C_1 and C_2 biphenyls and the diphenylmethanes, the trimethylbiphenyls remain in oil at biodegradation level 7. Different susceptibilities to microbial attack as well as microbially mediated synthesis of these aromatic compounds may account for changes in the distributions of these compounds.

Xanthones

Xanthone, methylxanthones, and dimethylxanthones are oxygenated aromatic hydrocarbons of uncertain origin (Figure 16.20). They occur in oils and source rocks from offshore Norway, the Western Desert of Egypt,

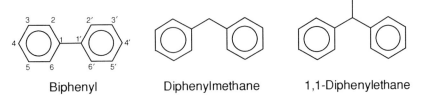

Figure 16.19. Structures for biphenyl and alkyldiphenylalkanes (diphenylmethane and 1,1-diphenylethane). Gas chromatographic elution orders for all methylbiphenyl, ethylbiphenyl, dimethylbiphenyl, and trimethylbiphenyl isomers are in Trolio et al. (1999).

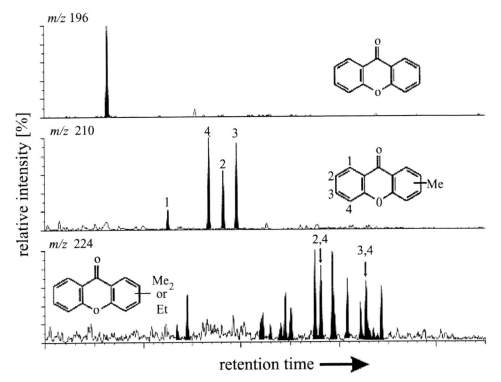

Figure 16.20. Gas chromatographic separation of xanthone, methylxanthones, and dimethylxanthones (from Oldenburg et al., 2002). This elution order reflects the shielding effect of methyl substituents to the polar groups, the keto, and the ether function. Reprinted from Oldenburg et al. (2002). © Copyright 2002, with permission from Elsevier.

and in Carboniferous coals of northern England, but they are absent in carbonate source rocks and oils from Campeche, the Gulf of Suez, and the algal-dominated Posidonia Shale (Oldenburg et al., 2002). These authors concluded that because xanthones correlate with the occurrence of terrigenous organic matter (but only within the thermal maturity range of $R_o = 0.5–1.0\%$), they probably result from diagenetic transformation of biogenic xanthones or xanthone-precursors. Such compounds are common in higher plants, fungi, and lichens. Alternatively, xanthones may form from non-specific precursors and may not be true biomarkers.

Thermal maturity is the principal factor that determines the relative abundance of xanthone and its alkyalted homologs. Xanthone/methylxanthones and methyl-/dimethylxanthones increase within increasing thermal stress. Partitioning during biodegradation or secondary migration may alter isomeric distributions of the alkylated xanthones. The ratios 1-methyl-/Σ

methylxanthones, 4-methyl-/3-methylanthone, and (1 + 4)-/(2 + 3)-methylxanthones correlate well with phytane/nC_{18} for a suite of Gullfaks oils having varying degrees of biodegradation (0 to ~3–4 on the biomarker biodegradation scale) (Oldenburg *et al.*, 2002). These authors argued that both partitioning during migration (water washing) and biodegradation would yield similar effects because the non-shielded isomers are more soluble and, hence, more bioavailable (Tomasek and Crawford, 1986).

Steranes and diasteranes

Microbial alteration and removal of the regular steranes and 4α-methylsteranes from petroleum occurs after the complete removal of C_{15}–C_{20} isoprenoids and before or after the hopanes, depending on circumstances described below. In general, sterane susceptibility to microbial degradation is as follows: $\alpha\alpha\alpha$ 20R $\gg$ $\alpha\beta\beta$ 20R $\geq$ $\alpha\beta\beta$ $\geq$ 20S $\alpha\alpha\alpha$ 20S $\gg$ diasteranes and C_{27} > C_{28} > C_{29} > C_{30} (e.g. Seifert and Moldowan, 1979; McKirdy *et al.*, 1983; Mackenzie *et al.*, 1983c; Sandstrom and Philp, 1984; Seifert *et al.*, 1984; Chosson *et al.*, 1991).

Where partial biodegradation of the steranes has occurred, the C_{27}–C_{30} $\alpha\alpha\alpha$ 20R steranes, which are in the biological configuration, are more susceptible to destruction than other isomeric forms. This selectivity has been documented by field observations (e.g. Rullkötter and Wendisch, 1982; Seifert *et al.*, 1984; Landeis and Connan, 1986), environmental studies of oil spills (e.g. Mille *et al.*, 1998, Wang *et al.*, 2001b), and bacterial culture experiments (Goodwin *et al.*, 1983; Chosson *et al.*, 1991). The relative susceptibility of the $\alpha\beta\beta$ 20R, $\alpha\beta\beta$ 20S, and $\alpha\alpha\alpha$ 20R steranes varies considerably. These isomers seem to be removed at nearly equal rates in some severely biodegraded oils. In other cases, selective removal may first deplete the $\alpha\beta\beta$ 20R, or the $\alpha\beta\beta$ 20R and $\alpha\beta\beta$ 20S isomers together, or the $\alpha\alpha\alpha$ 20S isomer. Some of these differences in sequential bacterial alteration appear to depend on bacterial populations and reservoir conditions. Sterane isomers undergo selective degradation of $\alpha\alpha\alpha$ 20R > $\alpha\beta\beta$ 20R + 20S > $\alpha\alpha\alpha$ 20S in oils that were also enriched in 25-norhopanes (Volkman *et al.*, 1983b). Conversely, Seifert *et al.* (1984) found complete removal of C_{27}–C_{29} $\alpha\alpha\alpha$ 20R isomers, partial removal of C_{27} > C_{28} > C_{29} $\alpha\alpha\alpha$ 20S sterane isomers, and no loss of $\alpha\beta\beta$ 20S + 20R isomers in seep oils from Greece that had intact hopanes and no 25-norhopanes. Chosson *et al.* (1992) found that the preferential removal of steranes proceeded by $\alpha\alpha\alpha$ 20R > $\alpha\beta\beta$ 20R $\geq$ $\alpha\beta\beta$ 20S > $\alpha\alpha\alpha$ 20S in aerobic bacterial culture experiments that degraded hopanes without forming 25-norhopanes.

The susceptibility of steranes to biodegradation typically decreases with increasing carbon number for each isomeric configuration. Selective depletion of C_{27} > C_{28} > C_{29} steranes occurs in subsurface crude oils (e.g. Rullkötter and Wendisch, 1982; Seifert *et al.*, 1984; Zhang Dajiang *et al.*, 1988), oil spills (Wang *et al.*, 2001b), and laboratory culture experiments (Goodwin *et al.*, 1983; Chosson *et al.*, 1991). The C_{30} steranes appear to be even more bioresistant than lower homologs (Lin *et al.*, 1989). Another example (see Figure 13.51) shows complete loss of C_{27} and C_{28} steranes, partial loss of C_{29} steranes with no apparent stereoisomer preference, and complete preservation of the C_{30} steranes.

Little work has been published on the relative susceptiblity to biodegradation of the C_{26} norcholestanes or methylsteranes. Chosson *et al.* (1991) found that the 4α-methylsteranes respond like their desmethylsterane equivalents when degraded by pure cultures of aerobic bacteria. Jiang Zhusheng *et al.* (1990) reported the preservation of C_{25} and C_{26} regular steranes in severely biodegraded oil from the Zhungeer Basin, China, that had lost all other regular steranes. Such preservation is unexpected, as in most cases the analogous C_{27} steranes are degraded before the C_{28} and C_{29} steranes.

Biodegradation can affect the sterane isomers used in thermal maturation parameters. For example, oils in the studies mentioned above show preferential removal of $\alpha\alpha\alpha$ 20R steranes during biodegradation. In the absence of other biodegradation indicators, this might be interpreted as increased maturation (i.e. higher 20S/(20S + 20R)), although in the examples cited 20S/(20S + 20R) were near 1.0. Values of 20S/(20S + 20R) above 0.56 (endpoint) are not observed without selective sterane biodegradation.

C_{20}–C_{21} steranes

Pregnane and homopregnane have high resistance to biodegradation, comparable to the diasteranes (Figure 16.21). The difference in susceptibility between steranes with or without extended alkyl side chains at

C-17 suggests that attack of the alkly group is an important step in the microbial degradation of steranes. Biodegradation of the C_{20}–C_{21} steranes is limited to enzymatic attack of the core structure.

Diasteranes

Diasteranes are particularly resistant to biodegradation. Evidence suggests that the C_{27}–C_{29} steranes are destroyed completely before diasterane alteration (Seifert and Moldowan, 1979; McKirdy *et al.*, 1983; Seifert *et al.*, 1984; Connan, 1984; Requejo *et al.*, 1989). Even in heavily biodegraded oils where steranes and hopanes are totally removed and no 25-norhopanes are present (rank 9), some diasteranes remain (Seifert and Moldowan, 1979).

Diasteranes may exhibit preferential bioresistance with increasing carbon number (Figure 16.21). Seifert and Moldowan (1979) observed that biodegradation of diasteranes results in stereoselective loss of C_{27} 13β,17α(H) 20S over C_{27} 13β,17α(H) 20R epimers. Because diasteranes are more resistant to biodegradation than most other common saturated biomarkers, they have been used as internal standards to measure the comparative loss of less resistant biomarkers.

Hopanes and 25-norhopanes

The relative susceptibility of hopanes, steranes, and individual hopane isomers is problematic, as various bacterial culture experiments and empirical observations of naturally degraded petroleum yield conflicting results. The relative extent of sterane and hopane biodegradation in oils appears to be highly dependent on the occurrence of specific microbial processes (e.g. environmental conditions and the microbial population). In some severely biodegraded oils, hopanes are removed before steranes, while in other, severely biodegraded oils, hopanes are removed only after steranes are highly altered.

25-Norhopanes (also called 10-desmethylhopanes, demethylated hopanes, and degraded hopanes) are a series of C_{26}–C_{34} compounds that are structurally equivalent to the regular hopanes, except for the absence of the methyl group at the A/B ring junction. Reed (1977) first observed degraded hopanes in asphalts from the Uinta Basin. He postulated that they formed by loss of the methyl group at the C-4 position in hopanes. This proved incorrect, as NMR spectrometry of isolated 25-norhopane showed that the methyl group is lost exclusively at the C-10 position (Rullkötter and Wendisch, 1982). Trendel *et al.* (1990) established the complete structure and stereochemistry via single-crystal X-ray analysis of two C_{28} demethylated hopanes isolated from sandstone that was heavily impregnated with biodegraded oil (Loufika outcrop, Congo). The dominant C_{28} compound was identified as 5α,17α,21α-25,30-bisnorhopane; the minor C_{28} compound was similar, except that the methyl group attached to C-18 migrated to the C-17 position (analogous to C_{27} Ts or 18α-22,29,30-trisnorneohopane) (Figure 16.22).

Separation and identification of the 25-norhopanes follow similar methodology as that for the hopanes, except that the m/z 191 A/B ring fragment of the hopanes shifts to m/z 177 in the 25-norhopanes (Figure 16.23). The distribution of the 25-norhopanes usually reflects that of the hopane series, shifted downward by one carbon number. Thus, the single epimer of C_{30} 17α,21β(H)-hopane corresponds to C_{29} 25-nor-17α,21β(H)-hopane, while each pair of the C_{31}–C_{35} 17α–hopane (22S + 22R) epimers corresponds to two C_{30}–C_{34} 25-norhopane epimers. The separation by ion fragments is not clean, and there are potential problems in quantifying the 25-norhopanes. Although m/z 191 is their base peak, all hopanes yield m/z 177 ions as well. C_{29} norhopane produces a significant m/z 177 ion response from its C/D/E-ring fragment and typically dominates m/z 177 ion chromatograms if demethylated hopanes are not present. The extended hopanes produce comparatively minor responses on m/z 171, but they can interfere with co-eluting 25-norhopanes. Similarly, the C_{29} 25-norhopane yields an m/z 191 ion response from its C/D/E-ring fragment. C_{30} 20S-25-norhomohopane may co-elute with oleanane, potentially causing problems in determining the true abundance of this age-related biomarker (Dzou *et al.*, 1999).

In general, 25-norhopanes occur in oils where the hopanes were preferentially removed, but are absent where the hopanes show greater bioresistance than the steranes. For example, severely biodegraded oil from west Siberia shows substantial depletion of 17α-hopanes and corresponding enrichment of 25-norhopanes without sterane degradation (Peters and Moldowan, 1991). Similarly, Malagasy asphalt contains partially biodegraded steranes along with 17α-hopanes and 25-norhopanes (Rullkötter and Wendisch,

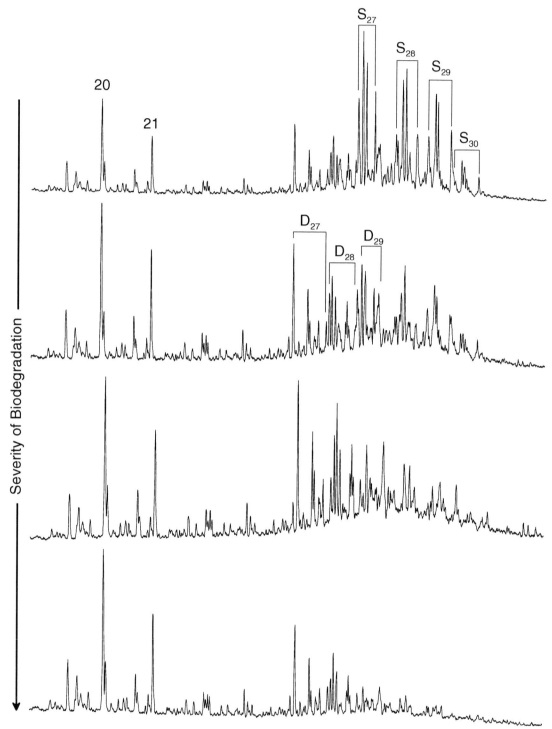

Figure 16.21. Oils from Colombia, showing selective preservation of diasteranes (D), pregnane (20), and homopregnane (21) over the regular steranes (S).

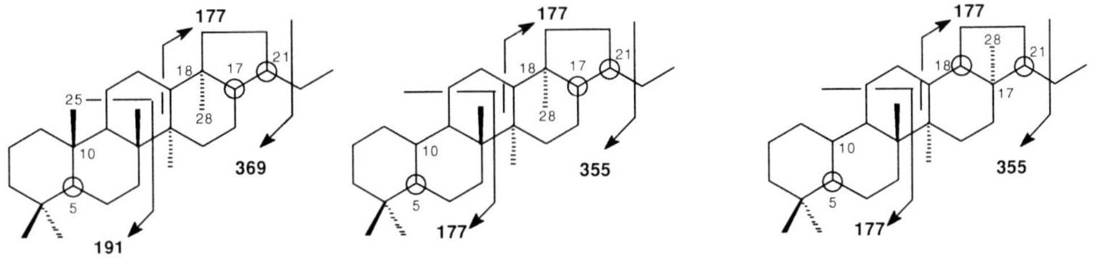

C_{29} 5α,17α,21α(H)-30-norhopane C_{28} 5α,17α,21α(H)-25,30-bisnorhopane C_{28} 5α,18α,21α(H)-25,30-bisnorhopane

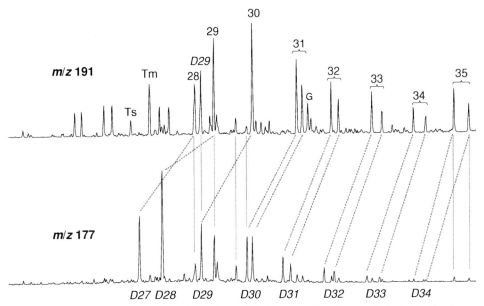

Figure 16.22. 25-Norhopanes have the same optical configuration as hopanes and differ only by the removal of the methyl group attached to the C-10 carbon. An isomeric form of the 28,30-bisnorhopane is known where the methyl group has migrated from the C-18 to the C-17 position (Trendel *et al.*, 1990).

Figure 16.23. Mass chromatograms at m/z 191 and m/z 177 indicate that 25-norhopanes (indicated by D-carbon number) occur in severely biodegraded oil from eastern Venezuela. Ion traces are scaled proportionally to their relative response. The 25-norhopanes are believed to originate by loss of a methyl group from C-10 in hopanes. Thus, the single epimer of C_{30} 17α,21β(H)-hopane (top) has been altered partially to C_{29} 25-nor-17α-hopane (bottom), while each of the C_{31}–C_{35} 17α-hopane (22S + 22R) epimers corresponds to two C_{30}–C_{34} 25-norhopane epimers. Vertical lines indicate some peaks that yield both m/z 191 and m/z 177 ions.

1982). Many biodegraded oils contain abundant 25-norhopanes (e.g. Reed, 1977; Seifert and Moldowan, 1979; Rullkötter and Wendisch, 1982; Goodwin *et al.*, 1983; Seifert *et al.*, 1984; Volkman *et al.*, 1984; Trendel *et al.*, 1990), and high abundance is evidence for severe biodegradation (rank ≥ 6).

On the other hand, hopanes can be biodegraded after steranes. Typically, such oils lack 25-norhopanes. For example, Seifert *et al.* (1984) observed that the least altered seep oils from western Greece show partial loss of steranes without biodegradation of the hopanes, while the most altered samples show complete loss of steranes and only partial loss of hopanes. No 25-norhopanes were detected in any of the seep-oil samples from Greece (see the discussion below on oil seeps). Documented occurrences of hopane destruction with no 25-norhopanes are reported worldwide, including in Texas Gulf Coast oils (Seifert and Moldowan, 1979), asphalts from Switzerland (Goodwin et al., 1983; Connan, 1984), and Yugoslavian oils (Moldowan et al., 1991a). In Athabasca tar sands, hopanes are usually degraded without the formation of 25-norhopanes. However, 25-norhopanes occur in one suite of samples in close proximity to significant subsurface water flow, thus allowing different groups of possibly aerobic microbes to carry out the degradation (Brooks et al., 1988).

From such studies, Brooks et al. (1988) conclude that two pathways for hopane biodegradation occur in nature: one where 25-norhopanes begin to form before sterane alteration and another where steranes are altered before hopanes and 25-norhopanes do not form. Bost et al. (2001) found that laboratory cultures can biodegrade hopanes under aerobic conditions without production of 25-norhopanes or degradation of steranes. These results suggest that the relative biodegradation of steranes and hopanes and production (or enrichment) of 25-norhopanes do not progress along specific biochemical pathways but are results of a complex interplay of multiple microbial reactions.

Origins of 25-norhopanes
The origin of the 25-norhopanes and their enrichment in some severely biodegraded oils have been topics of controversy for several decades. There are three possible origins for these compounds (Figure 16.24):

1. Microorganisms produce 25-norhopanes, hopanes, or their biological precursors. 25-Norhopanes and hopanes can be generated and expelled as petroleum, although 25-norhopanes are less abundant than hopanes. Severe biodegradation removes hopanes but not 25-norhopanes. The 25-norhopanes are, therefore, "unmasked" as they become enriched.
2. 25-Norhopanes originate by microbial demethylation of hopanes.
3. Microorganisms that are significant contributors of sedimentary organic carbon do not produce 25-norhopanes or their biological precursors. Rather, microbes that are responsible for the severe biodegradation of petroleum produce 25-norhopanes. Degradation of hopanes is unrelated to the formation of 25-norhopanes.

The relative abundance of these compounds in biodegraded crude oils provides little insight as to their origins because all three hypotheses could yield similar ratios of 25-norhopanes to hopanes as a function of increasing biodegradation.

Microbiology provides little direct support for any of these hypotheses. No microorganisms are known to biosynthesize 25-norhopanes or their likely biological precursors. No pure or enrichment cultures of bacteria that degrade hopanes produce 25-norhopanes (e.g. Goodwin et al., 1983; Chosson et al., 1992; Bost et al., 2001; Frontera-Suau et al., 2002). De Lemos Scofield (1990) and Nascimento et al. (1999) observed C-10 hopanoid acids, possible intermediates between hopanes and 25-norhopanes, in biodegraded crude oils. However, the origin of these compounds is uncertain, and it is unclear whether these hopanoic acids would degrade thermally to 25-norhopanes under the thermal regime experienced by oils in shallow reservoirs.

Chosson et al. (1992) and Blanc and Connan (1992) argued that 25-norhopanes are present in some oils before biodegradation and are unmasked only when the hopanes are removed. 25-Norhopanes occur in some source-rock extracts (Noble et al., 1985c; Chosson et al., 1992; Blanc and Connan, 1992), although none has been observed in kerogen or asphaltene pyrolyzates. Consequently, it is unclear whether these compounds are syngenetic with indigenous bitumen or are contributions from migrated oil. For example, Noble et al. (1985c) observed 28,30-bisnorhopane, 25,28,30-trisnorhopane, 25,30-bisnorhopane, and other 25-norhopanes in some Western Australian shales. These compounds were apparently not produced by thermal breakdown of kerogen in the shales, indicating that they occur as free hydrocarbons. This could indicate an unusual type of bacterial reworking of the organic matter or the inclusion of paleo-seepage oil into the source rock at the time of deposition.

The major problem with the unmasking hypothesis is that it generally fails mass–balance considerations when the triterpanoid hydrocarbons are quantified rather than expressed as ratios (Volkman et al., 1983b; Peters et al., 1996b). In order for 25-norhopanes to reach their observed concentrations, the enrichment would require the loss of more hydrocarbons than is indicated by other biomarkers, such as diasteranes, aromatic steroidal hydrocarbons, or porphyrins that are believed to be conserved as well as or better than the 25-norhopanes.

Circumstantial evidence strongly supports the origin of most 25-norhopanes by direct microbial

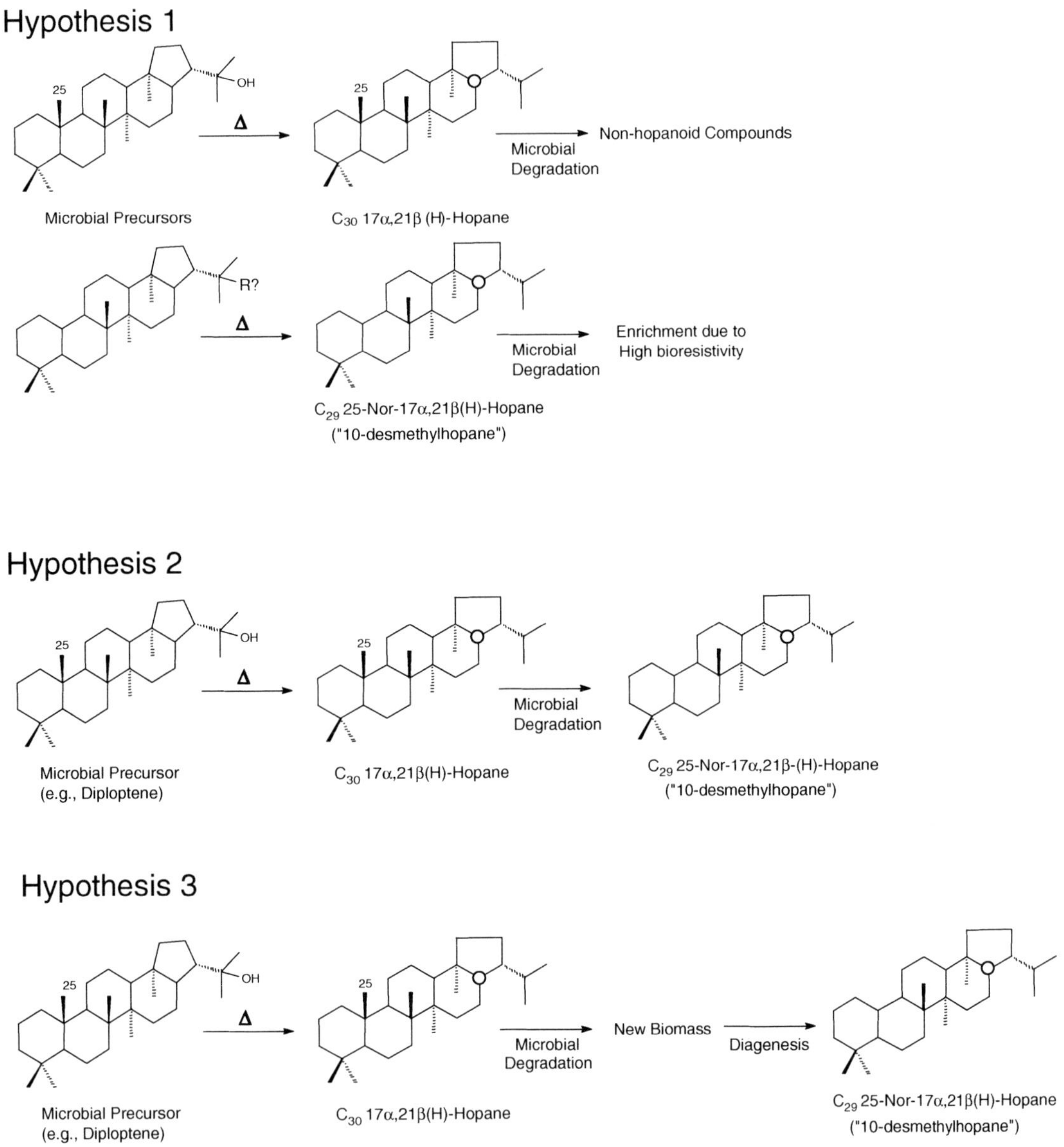

Hypothesis 1

Figure 16.24. Three hypothetical pathways for the origins of 25-norhopanes.

degradation of hopanes (Peters et al., 1996b). Later work by McCaffrey et al. (1996) determined biomarker concentrations in sidewall cores from wells located in the Cymric Field, Kern County, California. Oil in the Cymric Field originates from deep, synclinal siliceous shales of the Antelope Formation. The oil accumulates in shallow diatomite-porcellanite reservoirs of the antelolpe shale, where it has been severely biodegraded. The degree of biodegradation corresponds roughly with depth and ranges from rank 6 (complete loss of n-alkanes and isoprenoids with substantial alteration of steranes) to rank 9 (complete loss of steranes and hopanes). When the quantities of individual hopanes and their 25-nor equivalents are compared, the two are inversely proportional, indicating strongly that the hopanes were

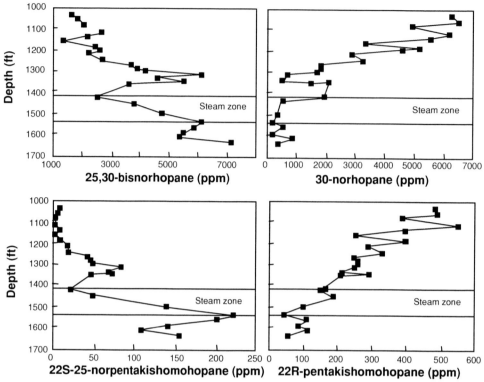

Figure 16.25. Concentrations of selected hopanes and demethylated (25-nor) hopanes in sidewall cores from a well (1815S, Sec 1Y) from the Cymric Field, Kern County, California (McCaffrey et al., 1996). The decrease in concentration of hopanes is roughly proportional to the increase in 25-norhopane equivalents, suggesting strongly that microbes are demethylating hopanes in the reservoir. Reprinted by permission of the AAPG, whose permission is required for further use.

demethylated by microbial activity in the reservoir (Figure 16.25).

The third hypothesis adapts aspects of the first two but currently lacks proof. Like the first hypothesis, it postulates that 25-norhopanes originate from as yet unidentified microorganisms. While these microbes may be present in sediments during diagenesis and contribute small amounts of 25-norhopanes to source rocks, they would be most active under certain, undefined reservoir conditions that promote the biodegradation of hopanes over steranes. Hence, 25-norhopanes would be produced as hopanes are consumed, not as a result of direct conversion but as a consequence of the enhancement of certain microbial populations.

Selective biodegradation of homohopanes

The relative susceptibly of the homohopanes and biodegradation remains controversial. In most biodegraded oils that have both 17α-hopanes and 25-norhopanes, no preferential biodegradation of individual homohopanes by carbon number is evident (e.g. Chosson et al., 1992). In some oils, however, individual hopanes appear to be biodegraded selectively. The C_{31} and C_{32} homohopanes were more susceptible to biodegradation than C_{30} hopane in asphalts from Madagascar (Rullkötter and Wendisch, 1982). In soils exposed to an oil spill for eight years, the biodegradation of homohopanes proceeded systematically by carbon number, with ~25% removal of the C_{35} pentakishomohopanes and only ~3% removal of the C_{31} homohopanes (Munoz et al., 1997). These field observations are consistent with the results of the laboratory culture experiments conducted by Goodwin et al. (1983).

The reverse susceptibility, where the higher-hopane homologs are preferentially bioresistant (particularly the C_{35} pentakishomohopanes), has been reported in oils from the Monterey Formation (Seifert et al., 1984; Requejo and Halpern, 1989; Moldowan and

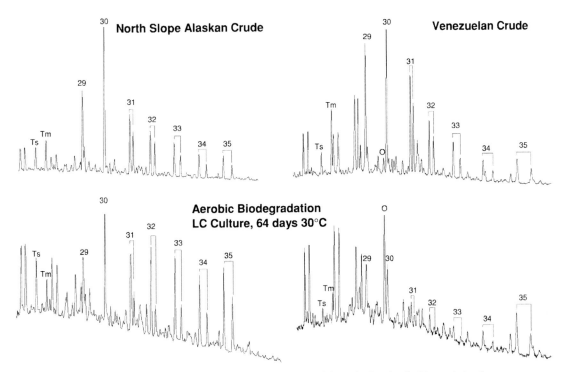

Figure 16.26. Biodegradation of hopanes by an enrichment culture of aerobic bacteria. Reprinted with permission from Frontera-Suau et al. (2002). © Copyright 2002, American Chemical Society. LC culture, light crude culture of creosote-contaminated soil (Fairhope, AL); T_m and T_s, C_{27} trisnorhopanes.

McCaffrey, 1995) and from the Bazhenov Formation of western Siberia (Peters and Moldowan, 1991; Peters et al., 1996b). Bost et al. (2001) and Frontera-Suau et al. (2002) observed such enrichments in laboratory studies using bacterial enrichment cultures grown on whole crude oils under aerobic conditions (Figure 16.26).

These variations suggest that the microbial biodegradation of hopanes proceeds along two different pathways: (1) oxidation of the alkyl side chain and (2) alteration of the cyclic core (Peters et al., 1996b). The first pathway would favor selective biodegradation of the higher-molecular-weight homologs, while the second may be hindered by longer alkyl side chains. Several studies hypothesize that the types of bacterial populations that alter oil control the reaction specificity (Brooks et al., 1988; Requejo and Halpern, 1989; Peters et al., 1996b). Tritz et al. (1999) found evidence for selective microbial degradation of the ring structure. Tritium-labeled hopane and bacteriohopane were incubated in the presence of the *Arthrobacter simplex* (also known as *Nocardioides simplex*), a common aerobic soil bacterium that degrades cholesterol and hopanes. The corresponding 17(21)-hopenes and 17,21-epoxides formed from the saturated hydrocarbon, although no degradation of the *n*-alkyl chain of bacteriohopane was observed.

The selectivity in the biodegradation of homohopanes provides additional evidence for the C-10 demethylation of hopanes during reservoir biodegradation. Peters and Moldowan (1991) observed that C_{26}–C_{34} 25-norhopanes in severely biodegraded western Siberian oil were accompanied by nearly complete removal of the C_{27}–C_{32} 17α-hopanes and partial removal of the C_{33}–C_{35} homohopanes. In less degraded oil, the 17α-hopane distribution was in an intermediate state, with nearly equal abundances of C_{31}–C_{35} homohopanes and a skewed distribution of C_{26}–C_{34} 25-norhopanes (Figure 16.27). Non- or slightly biodegraded equivalent oils (Bazhenov source) had no detectable 25-norhopanes. When the normalized distributions of these oils are compared, the m/z 191 and 177 ions for the biodegraded oils are nearly equivalent to those in the m/z 191 distributions of the

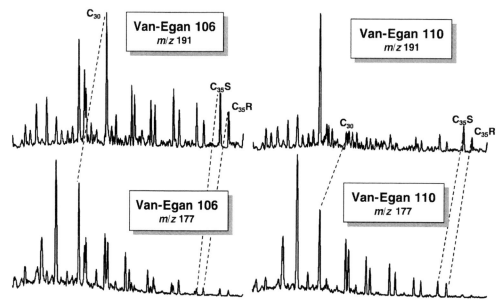

Figure 16.27. Hopane (m/z 191, top) and 25-norhopane (m/z 177) mass chromatograms for two severely biodegraded oils from west Siberia. The Van-Egan 106 oil has no clear decrease in hopanes with increasing carbon number, unlike less biodegraded oils from the same field. More extensive biodegradation of the Van-Egan 110 oil yields a reversed hopane distribution, where abundance increases from C_{31} to C_{35} hopanes. The relative distributions of extended hopanes and 25-norhopanes support preferenial demethylation of low-rather than high-molecular-weight hopanes to 25-norhopanes. Reprinted from Peters et al. (1996b). © Copyright 1996, with permission from Elsevier.

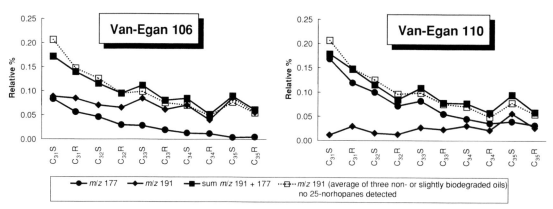

Figure 16.28. Relative distributions of C_{31}–C_{35} homohopanes (m/z 191) and corresponding C_{30}–C_{34} 25-norhopanes (m/z 177) for severely degraded oils from West Siberia. When the normalized abundances are summed, they are equivalent to the homohopane distribution in genetically related Bazenhov oils that are either non-degraded or slightly degraded. Reprinted from Peters et al. (1996b). © Copyright 1996, with permission from Elsevier.

unaltered oils (Figure 16.28).The shifting relative distribution between hopanes and 25-norhopanes with increasing biodegradation and their overall conservation of the original homohopane distribution indicate that hopanes are converted by microbial activity within the reservoir to 25-norhopanes via C-10 demethylation.

Moldowan and McCaffrey (1995) found a similar situation among severely biodegraded oils from Cyrmic Field, California (Figure 16.29). Severely biodegraded oils lost nearly all of their 17α-hopanes and extended 17α-homohopanes, except for the C_{35} pentakishomohopanes, which appear to be considerably

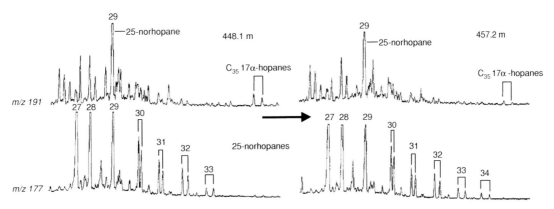

Figure 16.29. Distribution of 17α-hopanes and 17α, 25-norhopanes in extracts of sidewall cores from a well in the Cymric Field. Each ion trace is enlarged ×2 relative to the highest peak to better show the homohopanes. Hopanes have been removed in the severely degraded oil from 448.1 m, except for the C_{35} 17α-pentakishomohopanes, which appear to be the most bioresistant under these conditions of microbial degradation. These compounds have been removed in the 457.2-m sample and appear to have been converted to C_{34} 25-norhopanes. Reprinted from Moldowan and McCaffrey (1995). © Copyright 1995, with permission from Elsevier.

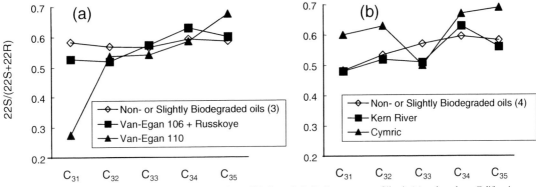

Figure 16.30. 22S/(22S + 22R) for homohopanes in suites of biodegraded oils from western Siberia (a) and onshore California. Averages are shown for non- to slightly biodegraded oils (number of samples). Reprinted from Peters et al. (1996b). © Copyright 1996, with permission from Elsevier.

more bioresistant under these reservoir conditions. The C_{34} 25-norhopanes, presumed C-10 demethylation products of C_{35} pentakishomohopanes, were absent in these samples. In samples that were more biodegraded, even the C_{35} pentakishomohopanes were removed, and, concurrent with their depletion, their product C_{34} 25-norhopanes were detected.

Selective biodegradation of 22R and 22S homohopane epimers

Similar to the steranes, the 22R epimers of the 17α-homohopanes are more susceptible to biodegradation than their 22S counterparts (e.g. Hoffmann and Strausz, 1986; Requejo et al., 1989; Lin et al., 1989). These field observations are consistent with laboratory culture experiments that showed the 22R isomer to be selectively degraded by aerobic microorganisms for all C_{31}–C_{35} homologs (Goodwin et al., 1983).

In mature oils and source rock extracts, an equilibrium ratio for each homohopane pair is achieved with 22S/(22S + 22R) ~0.59. Small variations are observed, but these are usually attributed to coelution of unresolved compounds with either the 22S or 22R isomers. Peters et al. (1996b) noted selective biodegradation of the homohopane isomers that varied by carbon number (Figure 16.30). In the most altered oil from West Siberia (Van-Egan 110), the 22S isomer is comparatively abundant in the

C_{35} pentakishomohopanes but depleted in the C_{31} homohopanes. In the most severely biodegraded oil from the Cymric Field in California (Cymric), the 22S isomer is comparatively abundant in all but the C_{33} trishomohopanes. These effects are subtle and vary between sample suites. Care is needed to prove that these observations are not experimental artifacts. For example, the C_{31} 22R homohopane may co-elute with the C_{31} 22S 25-norhopane and C_{31} 30-norhopane (Subroto et al., 1991) under standard chromatographic separations. Note that the non- to slightly biodegraded oils from California have non-equilibrium ratios and vary consistently by carbon number. Zumberge (1987) observed similar variations in the 22S/(22S + 22R) homohopane ratios in immature oils.

Peters et al. (1996b) hypothesized that the size and shape of the extended hopanes influence either their rates of transport across cellular membranes or access to the enzyme responsible for A/B-ring demethylation. Molecular mechanics calculations and quantum structure–activity relationships indicate that the minimum energy configuration of the 22R isomers results in extension of the alkyl side chains away from the E ring ("rail" orientation), while the 22S isomers favor an orientation with the alkyl side chains folded back over the cyclic core ("scorpion" orientation).

Relative bioresistance of hopanes and 25-norhopanes

25-Norhopanes are clearly more bioresistant than 17α-hopanes under certain reservoir conditions. This conclusion is reached regardless of which hypothesis is responsible for the presence of high concentrations of 25-norhopanes in biodegraded oils. Under conditions where hopanes are degraded with no formation of 25-norhopanes (rank >7), the relative bioresistances of these two triterpanoids are probably equivalent. Bost et al. (2001) studied the aerobic biodegradation of hopanes and 25-norhopanes in two Venezuelan crude oils after a five-week incubation period using a microbial enrichment culture. The microbes simultaneously degraded hopanes and 25-norhopanes, suggesting similar fates for these compounds in oxic surface environments. Degradation caused preferential removal of the 22R epimer for both extended hopanes and 25-norhopanes. The C_{35} 17α-hopane and 18α-oleanane were conserved. None of the nine microorganisms isolated from this culture had the ability to degrade 17α-hopane (C_{30}) in pure culture using crude oil as the sole carbon source. Frontera-Suau et al. (2002) reported similar results using different enrichment cultures.

One interesting aspect of these studies is that the degradation of the triterpanoid hydrocarbons occurred only when the enrichment cultures were grown at 30°C and not at either 15°C or 37°C. This temperature effect may explain why earlier laboratory culture studies failed to observe hopane degradation under laboratory conditions (e.g. Rubinstein et al., 1977; Connan et al., 1980; Teschner and Wehner, 1985; Prince et al., 1994).

The 25-norhopane ratio

Peters et al. (1996b) proposed that the ratio of the total C_{30}–C_{34} 25-norhopanes (m/z 177) to the sum of these compounds plus the C_{31}–C_{35} homohopanes (m/z 191) could be used to assess the extent of microbial alteration among severely biodegraded oils (rank ~6–9). In two suites of oils and sidewall core extracts, this ratio co-varied with the C_{35} hopane index because of the enhanced bioresistance of the C_{35} pentakishomohopanes compared with lower homologs (Figure 16.31).

The 25-norhopane ratio cannot be used in isolation to rank the extent of biodegradation. As noted above, severely biodegraded oils may have altered hopanes without the formation of 25-norhopanes. Furthermore, oils with similar 25-norhopane ratios are not necessarily

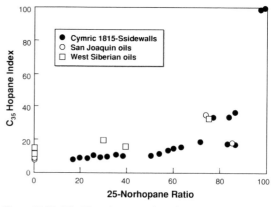

Figure 16.31. The 25-norhopane ratio and the C_{35} hopane index ($100 \times C_{35}/C_{31-35}$) can be used to rank the extent of biodegradation in related, biodegraded oils from onshore Californian and western Sibera (from Peters et al., 1996b).
25-Norhopane ratio = $100 \times C_{30-34}$ 25-norhopanes (m/z 177) / ΣC_{31-35} hopanes (m/z 191) + C_{30-34} 25-norhopanes (m/z 177).

degraded to the same extent with respect to other compound classes (e.g. diasteranes and tricyclic terpanes). Peters *et al.* (1996b) recommended that this ratio be used as a second-order parameter for use within appropriate homogeneous suites of related oils.

Sterane versus hopane biodegradation
The relative extent of sterane and hopane biodegradation in oils depends on various factors, including the type of biodegradation, the environmental conditions, and the microbial population. For example, Malagasy asphalt contains only partially biodegraded steranes, but the 17α-hopanes were partially converted to 25-norhopanes (Rullkötter and Wendisch, 1982). Biodegraded oil from west Siberia shows evidence for substantial conversion of 17α-hopanes to 25-norhopanes without sterane degradation (Peters and Moldowan, 1991). On the other hand, biodegraded seep oil from Greece shows partial loss of steranes without biodegradation of the hopanes (Seifert *et al.*, 1984). Other seeps in the region show complete sterane degradation, with partial loss of hopanes but no formation of 25-norhopanes. This evidence suggests early bacterial attack of the 17α-hopanes compared with steranes in cases where 25-norhopanes are formed, but selective biodegradation in cases where 25-norhopanes do not occur (Brooks *et al.*, 1988; Peters and Moldowan, 1991).

Occurrence of 25-norhopanes in "non-degraded" oils
25-Norhopanes are generally interpreted as indicators of severe biodegradation, especially when they occur as a complete homologous series. However, many oils contain 25-norhopanes but appear to be non-biodegraded or only slightly biodegraded, as indicated by the predominance of *n*-alkanes and acyclic isoprenoids. Volkman *et al.* (1983b) proposed that such oils were mixtures of biodegraded oil residues that were dissolved by non-degraded oil during accumulation in the reservoir. Early studies described such oils as anomalies (e.g. Alexander *et al.*, 1983b; Philp, 1983; Talukdar *et al.*, 1986; Talukdar *et al.*, 1988; Sofer *et al.*, 1986). We now know that such mixed oils are common in basins with shallow (<80°C) reservoirs. These oils may develop either as a continuous process, where the rates of biodegradation are comparable to the rates of reservoir charging, or as episodic events, where a reservoir is charged and biodegraded and then recharged with non-degraded oil. In the latter scenario, the recharge fluid may originate from the same source rock at the same or higher level of thermal maturity, or it may originate from a completely different source. Mixed oils also may result from convection or diffusion within a thick oil column that was biodegraded to varying degrees (Mason *et al.*, 1995). More complex scenarios can be modeled involving the mixing of biodegraded and non-degraded oils coupled with other processes (e.g. differential migration, thermal cracking, phase separation, and solids deposition).

Geochemistry can be used with other methods (e.g. fluid inclusions, thermal modeling, seismic) to describe the filling history of reservoirs. For example, most fields in the Jeanne d'Arc Basin, offshore eastern Canada, have stacked reservoirs that were connected by faults at various times in the past. API gravity generally increases with depth, and nearly all reservoirs shallower than 2000 m (<80°C) contain biodegraded oil. Many shallow oils are undergoing biodegradation today. Some reservoirs toward the depocenters at depths >2000 m contain mixtures of biodegraded and non-biodegraded oils, where present-day temperatures are too high for biodegradation. For example, saturate gas chromatograms and gross compositions of liquids collected from three drill stem tests (DSTs) in separate Cretaceous reservoirs show mixing of biodegraded and non-biodegraded oil in the Mara M-54 well (Figure 16.32) (Von der Dick *et al.*, 1989; Fowler *et al.*, 1998b; Shimeld and Moir, 2001). Biomarker and other data indicate that the two deeper reservoirs in the well originally received a pulse of lower-maturity crude oil than that in the shallowest reservoir. Gross compositional data indicate that the two deeper oils are also more biodegraded, although *n*-alkanes remain. This low-maturity oil (~25°API) was biodegraded when the two deeper reservoirs were <1000 m deep (Shimeld and Moir, 2001). Upon further burial, these reservoirs received a second pulse of more mature oil. More of this high-maturity oil occurs in the DST-2 than the DST-1 reservoir based on the lower API gravity and the gross composition of DST-1. Because of the greater depth of the reservoirs when the second pulse of oil arrived, these oils were not significantly biodegraded. The shallowest reservoir (DST-3) did not receive the original oil pulse and hence contains only oil from the second pulse. The higher maturity of this oil is evident from biomarker parameters (Figure 16.32). Conditions in the shallow reservoir (<2000 m) are conducive to microbial

Biodegradation parameters

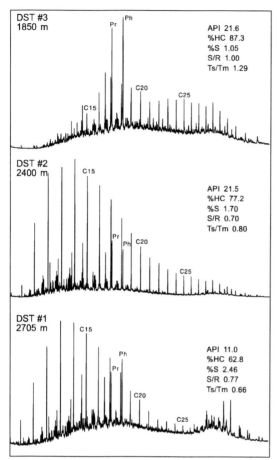

Figure 16.32. Saturate gas chromatogram of three oils from Mara M-54 well, Jeanne d'Arc Basin, offshore eastern Canada. Reprinted from Peters and Fowler (2002). © Copyright 2002, with permission from Elsevier.
%HC, weight percentage saturated and aromatic hydrocarbons; %S, weight percentage sulfur; S/R, C_{29} $5\alpha,14\alpha,17\alpha(H)$ 20S/20R steranes; Ts/Tm, 18α-trisnorneohopane/17α-trisnorhopane.

growth, and the accumulated oil is biodegraded. Differences in pristane/phytane among these oils (0.9, 1.2, and 0.7 from top to bottom in Figure 16.32) are probably due to secondary processes rather than differences in source input. All three oils have similar biomarker compositions indicating an origin from the Kimmeridgian Egret Member of the Rankin Formation.

A scenario similar to that at the Mara M-54 well is thought to occur nearby at the larger Hebron accumulation. Optical fluorescence indicates three populations of oil inclusions in the Ben Nevis Formation in the Hebron area. The first oil to be entrapped had intermediate gravity (~25–30°API). Later entrapped oil had higher gravity (~35–45°API). Both oils were subsequently biodegraded to heavy oil (15–20°API). Detailed Late Cretaceous-Cenozoic biostratigraphy, apatite fission track data, and thermal modeling support this scenario of two phases of oil generation and migration separated by uplift, erosion, and biodegradation. Three-dimensional seismic analysis revealed an Upper Cretaceous surface with features suggesting subaerial exposure. This facilitated a risk assessment model for biodegradation in the Jeanne d'Arc Basin (Shimeld and Moir, 2001).

Selective biodegradation of C_{27}–C_{30} hopanes

The C_{28}–C_{30} 17α-hopanes are typically biodegraded in the same manner and at approximately the same rate as the C_{31}–C_{35} extended hopanes. Preferential biodegradation of individual C_{28}–C_{30} hopanes was reported in several studies. For example, the C_{29} 17α-norhopane was preferentially removed relative to the C_{30} 17α-hopane in Eocene oils from South Texas (Williams et al., 1986). Walters (1993) discussed oils from a steam flood zone in South Belridge Field, onshore California, where the C_{30} and C_{28} hopanes were selectively removed (see Hopanoids in paraffin dirt, pp. 687–9).

28,30-Bisnorhopane (BNH) is demethylated during biodegradation to 25,28,30-trisnorhopane (TNH) (Seifert and Moldowan, 1978; Moldowan et al., 1984). However, TNH also occurs in non-biodegraded petroleum and source-rock extracts, usually with some BNH (Figure 16.33) (Rullkötter et al., 1982). Neither BNH nor TNH is incorporated into kerogen, but they occur as free hydrocarbons (Noble et al., 1985c). Thus, both appear to be products of early diagenetic bacterial reworking in sediments (Curiale et al., 1985). It is possible that BNH is produced and then degraded. A method to rank oil maturities using percentage $17\alpha,21\beta$-TNH/($\alpha\alpha + \beta\alpha + \beta\beta$)-TNH is useful for heavily biodegraded oils where other biomarkers are degraded (Moldowan et al., 1984).

C_{27} Ts and Tm are generally unaffected by even severe biodegradation. If biodegraded, these two isomers are removed at approximately the same rate, preserving their initial ratio. For example, severely biodegraded oils from the Permian Phosphoria Formation in Wyoming have altered distributions of steranes, hopanes, and

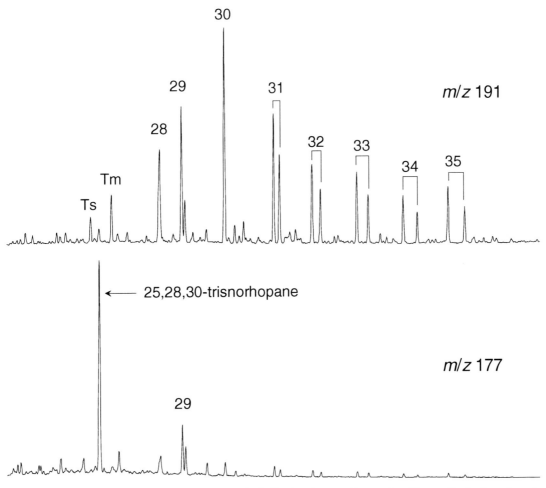

Figure 16.33. Ion chromatograms of crude oil enriched in 25,28,30-trisnorhopane with an unaltered hopane distribution and no other 25-norhopanes.

mono- and triaromatic steroids, but Ts and Tm remain unaltered (Sundararaman and Hwang, 1993). In severely biodegraded oils where the 25-norhopane series occur, C_{26} equivalents of Ts and Tm are not always present (Figure 16.33). Because the C_{27} Ts and Tm hopanes and tricyclic terpanes are highly resistant to biodegradation, they may dominate the m/z 191 ion chromatogram of severely biodegraded oils upon removal of the higher hopane homologs (see Figure 16.42).

In laboratory experiments using enrichment cultures of aerobic bacteria, Bost et al. (2001) found that both Ts and Tm are degraded, with Tm degrading slightly faster than Ts. Using the same enrichment culture, Frontera-Suau et al. (2002) verified that the Ts / (Ts + Tm) ratio increased. However, cultures from other locations that grew under identical conditions and biodegraded higher-hopane homologs had little or even the opposite effect on the Ts/(Ts-Tm) ratio.

Hopanoids in paraffin dirt

In 1901, Patillo Higgins drilled at Spindletop based on active gas seeps and associated "paraffin dirt" (Milner, 1925) – soil with a distinct gummy texture (Halbouty, 2000). Two years later, the Baston oil field was discovered based solely on the presence of paraffin dirt (Smith, 2002). These and other discoveries at Sour Lake

(1901) and Humble (1905) started the East Texas oil boom. Paraffin dirt has been used ever since as a surface indicator of hydrocarbon accumulations (Schumacher, 1996; 1999).

Paraffin dirt is a yellow to brown, gummy to waxy organic substance found in moist soils and outcrops associated with gas seeps (Davis, 1952; 1967). The substance is poorly named because lipids typically account for <3% of the total mass. Paraffin dirt is not migrated hydrocarbons but consists mainly of the by-products of bacterial oxidation of gaseous hydrocarbons. It is primarily the remains of methanotrophic bacteria and associated bacteria and fungi supported by the methanotrophs.

There are numerous reports of paraffin dirt associated with gas seeps along the onshore US Gulf Coast, Colombia, Romania, and Burma (Davis, 1967). Modern geochemical studies were performed on paraffin dirt from Chile (Simoneit and Didyk, 1978; 1986; 1992) and Tanzania (Mpanju and Philp, 1991; 1994). Organic matter in paraffin dirt from Siglia (Antofagasta Province, Chile) proved to be primarily insoluble kerogenous material, humic and fulvic acids, with lesser amounts of carbohydrates, proteins, and lipids. 17β-Trinorhopane

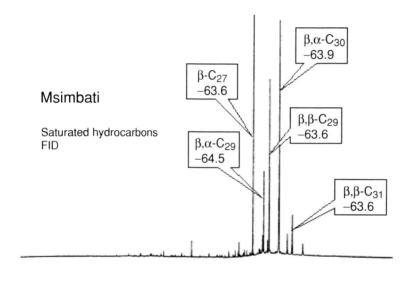

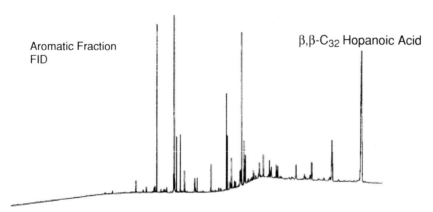

Figure 16.34. Gas chromatograms (flame ionization detector, FID) of the saturated and aromatic fractions of extracted lipids from paraffin dirt at Msimbati, Tanzania. The saturated fraction is dominated by $\beta\beta$-(‰)hopanes with very negative δ^{13}C and $\beta\alpha$-. The aromatic fraction contains hopanoid acids. Reprinted from Mpanju and Philp (1994). © Copyright 1994, with permission from Elsevier.

and β,α-norhopane dominate the saturated hydrocarbon fraction (Simoneit and Didyk, 1978). Isoprenoid and steroidal hydrocarbons are absent.

Saturated hydrocarbons from the Tanzanian paraffin dirt are dominated by ^{13}C-poor hopanoids and are nearly devoid of n-alkanes and steranes (Mpanju and Philp, 1994). In several samples, the saturated hopanoids consist exclusively of the immature β,β- and β,α-isomers and are accompanied by the β,β-isomer of C_{32} hopanoid acid (Figure 16.34). The molecular and isotopic signature of the lipids in the paraffin dirt is unlike that observed in either non-degraded or biodegraded crude oil. Rather, these hopanoids likely originated from methanotrophic bacteria that are consuming thermogenic methane seepage.

The saturated hydrocarbon fraction of the Wingayongo seep sample from Tanzania analyzed by Mpanju and Philp (1994) is composed almost entirely of C_{28}, C_{30}, and C_{32} α,β(H)-hopanes. Unlike the paraffin dirt samples described above, the hopanes in this seep consist of mature isomers with δ^{13}C ($\sim$−32‰) only slightly more negative than that of saturated hydrocarbons in nearby produced oils ($\sim$−29‰). Mpanju and Philp (1994) concluded that the Wingayongo seep is highly biodegraded and that the unusual hopane distribution is the fingerprint of a microbial community that utilized the exposed bitumen as a carbon source. We found that the polar fraction of the Wingayongo seep liberates n-alkanes, isoprenoids, and C_{35} homohopanes upon reaction with Raney nickel. The liberated n-alkanes exhibit an even carbon number preference and are depleted in ^{13}C (δ^{13}C $\sim$−35‰). These observations suggest that the polar fraction is a mixture of thermogenic petroleum and recent bacterial biomass (Figure 16.35).

The hopanes in the paraffin dirts of Tanzania offer an explanation for unusual hopane distributions observed in severely biodegraded oils from the South Belridge Field (Figure 16.36) (Walters, 1993). The South Belridge oils are highly biodegraded and have no n-alkanes or isoprenoids. Steranes also were altered significantly, and the samples contain no demethylated hopanes. Hopanes and tricyclic terpanes are nearly identical in all samples, except for C_{28} bisnorhopane and C_{30} hopane. While the concentration of C_{29} norhopane remains relatively constant ($\sim$500 ppm), the concentration of C_{30} hopane decreases from 1100 ppm in the Lower D sample to 50 ppm in the Upper B sample. C_{28} bisnorhopane decreases from 370 ppm to nearly zero.

The selective biodegradation of the South Belridge oils is the opposite of the hopane enrichment in the Wingayongo tar sand. We hypothesize that while there are methanotrophic bacteria that selectively produce even-carbon-numbered hopanoids, there are associated organisms that selectively consume these hydrocarbons. The Wingayongo tar sands and the South Belridge oils represent end-member cases where only one of these organisms is present or active. In the case of the tar sand, the surface environment is largely oxic, with anoxic microenvironments imparted by methane seepage. In the case of South Belridge, the subsurface environment is largely anoxic, with oxic conditions imparted by steam flooding. This suggests that while the methanotrophs are facultative aerobes, the organisms that selectively consume hopanes are facultative anaerobes.

Irregular hopanoid triterpanes

In addition to the ubiquitous hopanes and the common 25-norhopanes, other series of irregular hopanes and C-ring-modified triterpanoids have been found in biodegraded oils. Most of these compounds have limited documented occurrences, and their origins, geologic distributions, and relative bioresistance are largely unknown.

28-Norhopanes and 25,28-bisnorhopanes

The 28,30-bisnorhopanes and related 25,28,30-trisnorhopanes are usually the only members of the 28-norhopane series readily identified in certain oils and rock extracts. These triterpanoids can be the predominant saturated biomarkers and may have δ^{13}C that differs from the associated C_{29} and C_{30} hopanes (Schoell et al., 1992). Nytoft et al. (2000) identified a complete series of 28-norhopanes (C_{26} and C_{28}–C_{34}) in oils and rock extracts from west Greenland and the North Sea (Figure 16.37). They believe that 28,30-bisnorhopane originates from different biological precursors than the other 28-norhopane homologs. δ^{13}C values for the entire 28-norhopane series were comparable to these of the regular hopanes.

Nytoft et al. (2000) also found that the C_{26} and C_{28}–C_{34} 28-norhopanes are less bioresistant than the regular hopanes. In biodegraded oils, the 28-norhopanes appear to be converted to C_{25} and C_{27}–C_{31}

Biodegradation parameters

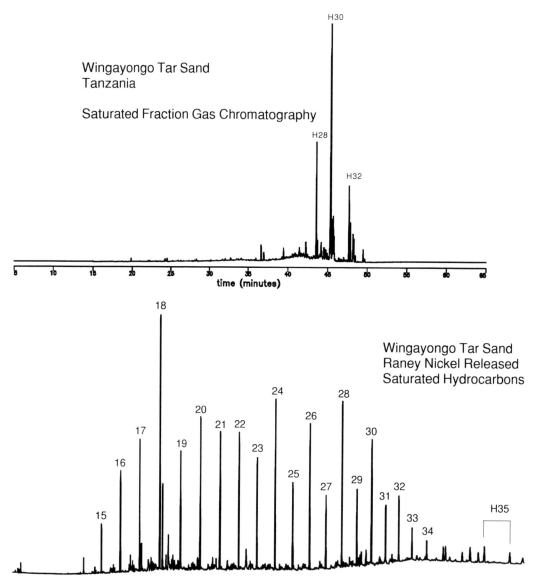

Figure 16.35. Gas chromatogram of the saturated fraction from the Wingayongo tar sand from Tanzania and the saturated hydrocarbons released by the reaction of Raney nickel with the polar fraction.

25,28-bisnorhopanes (Figure 16.38). Lower-molecular-weight homologs are biodegraded more readily than the extended 28-norhopanes.

C_{28}–C_{34} 30-nor-17α-hopanes

In heavily biodegraded oils where the hopanes have been altered, 30-nor-17α-hopanes can provide supplementary information useful in correlation studies. The 30-nor-17α-hopanes are common in petroleum from carbonate source rocks. Moldowan et al. (1992) found these compounds in related, non-biodegraded to severely biodegraded oils from the Adriatic Basin. They used distributions of the C_{30}–C_{34} 30-nor-17α-hopanes analyzed by GCMS/MS to correlate the heavily biodegraded Adriatic oils. These compounds also occur in oils from Greece (Seifert et al., 1984).

The 30-nor-17α-hopanes probably constitute a series of compounds ranging from C_{28} to C_{34} (see

Sand	Well #	Depth (ft)	So %	Sw %	Porosity %
Upper B	9143-3	583	43.4	47.0	36.9
Upper C	9143-3	732	49.2	48.4	35.3
Lower D	4183-3	707	58.4	40.2	35.0
Lower D	4183-3	820	35.4	61.9	36.1

(Note: The reported saturations are from Dean Stark analyses of sands immediately adjacent to those listed.)
Upper B sand sample within a meter of a steam zone.

Sand	Viscosity (cp)	API Gravity	%S
Upper B	2248	14.8	1.000
Upper C	8190	12.7	1.010
Lower C	1090	15.8	0.842
Lower D	670	16.5	0.734

			C_{15+} Chemical Group-Type Composition			
Sand	%Volatile	%Recovered	%Sat.	%Aro.	%Polars	%Asph.
Upper B	6.9	87.0	28.7	34.9	32.8	3.6
Upper C	0.4	97.3	28.9	36.6	32.1	2.4
Lower C	2.3	88.2	29.6	35.7	32.0	2.7
Lower D	6.4	85.9	31.0	31.7	33.3	4.0

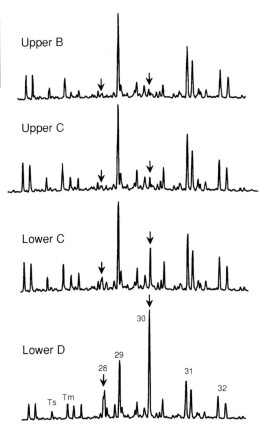

Figure 16.36. Bulk properties of four oils from the South Belridge Field, California, and ion chromatogram (*m/z* 191) showing hopane distributions (from Walters, 1993). All oils were highly biodegraded and contain similar amounts of C_{29} norhopane (~500 ppm). Concentrations of C_{28} bisnorhopane and C_{30} hopane decrease from 370 ppm to nearly zero and from 1100 to 50 ppm from the Lower D to Upper B samples, respectively. Dean Stark analyses measure fluid saturations in core samples by distillation extraction. S_o, oil saturation; S_w, water saturation.

Figure 13.88). Because the highest homolog in the series is C_{34}, precursors for the 30-nor-17α-hopanes are probably about the same size as the hopanoids and serve similar functions in the cell membranes of prokaryotes. Like the 30-nor-17α-hopanes, with which it appears to be related, the C_{28}-hopane pseudohomolog 29,30-bisnor-17α-hopane occurs mainly in oils from carbonate–source rocks. Thermal cracking of this compound from 17α-hopanes having a C-22 methyl branch is not favored because cleavage of two carbon–carbon bonds at C-22 is required (Seifert *et al.*, 1978).

C(14α)-homo-26-nor-17α-hopanes

Trendell *et al.* (1993) discovered an unusual series of C_{27}–C_{35} pentacyclic triterpanoids in severely biodegraded oil extracted from sandstone (Loufika outcrop, Congo). These compounds, identified as C(14α)-homo-26-nor-17α-hopanes by NMR and mass spectrometry, resemble serratanes (Kimble *et al.*, 1974b), possessing a seven-carbon C-ring (Figure 16.39). In addition to the C(14α)-homo-26-nor-17α-hopanes, Trendell *et al.* (1993) identified a minor, equivalent series that appears to involve the C-27 methyl group instead of the C-26 group, as well as a C(14α)-homogammacerane (serratane skeleton).

The origin and utility of C(14α)-homo-26-nor-17α-hopanes are unknown. Trendell *et al.* (1993) considered it unlikely that they are biosynthesized by bacteria. Rather, they may have formed by the oxidation of the C-26 methyl to an alcohol, either in the reservoir or during early diagenesis, followed by a concerted rearrangement and migration of the C8(14) bond. The enlarged C-ring in the serratane-like triterpanoid hydrocarbons may be substantially more thermally stable,

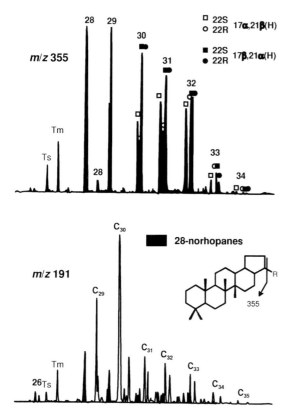

Figure 16.37. Ion chromatograms showing the presence of 28-norhopanes and regular hopanes in the Equaluik oil of west Greenland. The 28-norhopanes are characterized by the m/z 355 mass fragment. Reprinted from Nytoft et al. (2000). © Copyright 2000, with permission from Elsevier.

as well as more bioresistant, than the corresponding 17α-hopanes.

8,14-Secohopanes

The 8,14-secohopanes appear to be highly resistant to biodegradation. C_{27}–C_{31} isomers were first identified in biodegraded Nigerian crude oil (Schmitter et al., 1982) and biodegraded asphalts from Madagascar (Rullkötter and Wendisch, 1982). Extended 8,14-secohopanes up to C_{35} were reported later in biodegraded bitumens from the Aquitaine Basin, France (Dessort and Connan, 1993) and in biodegraded oil seeps from Pakistan (Fazeelat et al., 1994; 1999). 8,14-Secohopanes elute as a complex mixture of optical isomers (Fazeelat et al., 1995). In addition to the optical carbon centers at C-17, C-21, and C-22, the opening of the C-ring creates two more optical carbon centers at C-8 and C-14 (Figure 16.40).

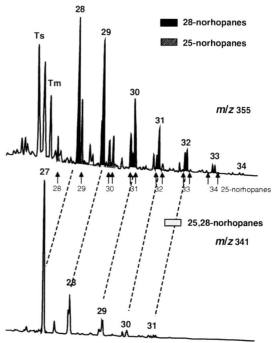

Figure 16.38. 28-Norhopanes in severely biodegraded oil from the Oseberg area, North Sea, biodegrade to 25,28-norhopanes, which are characterized by the m/z 341 ion fragment. Elution position of the 25-norhopanes indicated by arrows. Reprinted from Nytoft et al. (2000). © Copyright 2000, with permission from Elsevier.

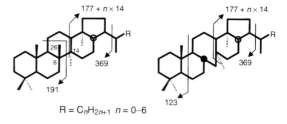

Figure 16.39. Comparison of structure and mass fragmenation pattern of the 17α-hopanes and the C(14α)-homo-26-nor-17α-hopanes.

The occurrence and enrichment of 8,14-secohopanes in biodegraded oils (e.g. Jiang Zhusheng et al., 1990) suggest that these compounds are highly bioresistant and/or may be formed by microbial alteration of reservoir petroleum (Wenger and Isaksen, 2002). 8,14-Secohopanes were reported in nonbiodegraded extracts from Chinese Tertiary argillaceous

8β(H),14α(H),17α,21β(H)-secohopanes   8β(H),14α(H),17β,21α(H)-secohopanes   8α(H),14β(H),17α,21β(H)-secohopanes

8α(H),14α(H),17α,21β(H)-secohopanes   8α(H),14α(H),17β,21α(H)-secohopanes   8α(H),14β(H),17β,21β(H)-secohopanes

Figure 16.40. Examples of 8,14-secohopanes (Fazeelat et al., 1995).

rock (Lu et al., 1985), Chinese Jurassic coals (Wang et al., 1990), and Spanish Puertollano oil shale (del Rio et al., 1994), indicating that these biomarkers either have a direct biological precursor or may form during early diagenesis.

Demethylated D-ring aromatized 8,14-secohopanes

8,14-Secohopanoids with an aromatized D-ring (see Figure 8.20) occur in most crude oils and source rocks, but they are particularly abundant in carbonate-derived oils and bitumens (Hussler et al., 1984b; Connan et al., 1986; Songnian Lu et al., 1988; Rinaldi et al., 1988; del Rio et al., 1994; Chunqing Jiang et al., 2001). These compounds are presumed to be highly resistant to biodegradation, but their relative susceptibility compared with other biomarker has not been determined.

Killops (1991) first identified the demethylated forms of the D-ring aromatized 8,14-secohopaoids in Jurassic lacustrine rocks. Nytoft et al. (2000) also reported that these compounds were present in both non-biodegraded and biodegraded Jurassic oils from the Danish North Sea. Clearly, these compounds can be formed during diagenesis, probably by loss of the C-28 methyl group during aromatization of the 8,14-secohopanes. Their enhanced abundance in biodegraded oils is due mostly to selective preservation. Nytoft et al. (2000) found that the demethylated aromatic 8,14-secohopanoids can lose an additional methyl group (probably C-25) during severe biodegradation (Figure 16.41). The full utility of these compounds in

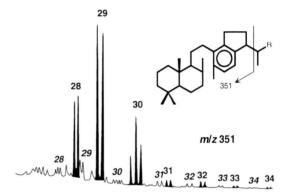

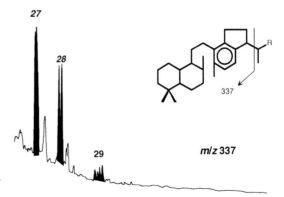

Figure 16.41. The complete C_{28}–C_{34} series of C-28 demethylated aromatic 8,14-secohopanoids (m/z 351) and their C-25 demethylated equivalents (m/z 337) occur in a severely biodegraded oil from the Oseberg area, Norway. Reprinted from Nytoft et al. (2000). © Copyright 2000, with permission from Elsevier.

694 Biodegradation parameters

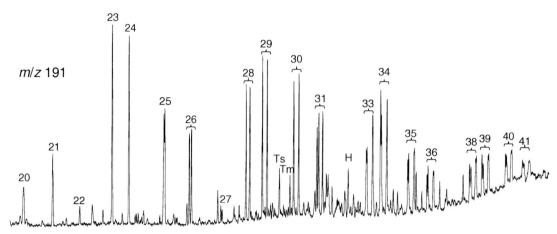

Figure 16.42. Mass chromatogram (m/z 191) showing preservation of tricyclic terpanes in severely biodegraded oil from Venezuela. No demethylated tricyclic terpanes are present. Note the low quantities of C_{27}, C_{32}, and C_{37} tricyclic terpanes. H, hopane.

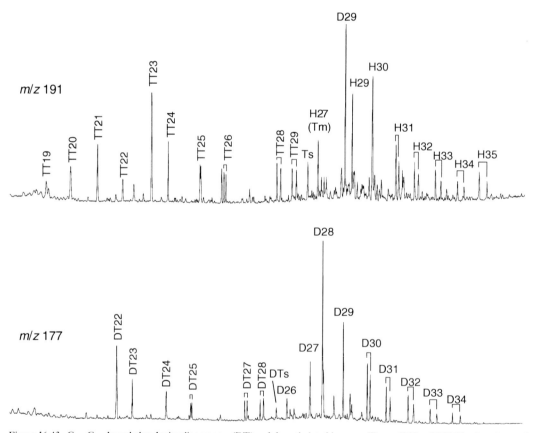

Figure 16.43. C_{22}–C_{28} demethylated tricyclic terpanes (DT) and demethylated hopanes (D; e.g. D27 = 25,28,30-trisnorhopane) occur in the m/z 177 ion chromatogram of a severely biodegraded oil from Colombia that still contains a full suite of tricyclic terpanes (TT) and hopanes (H), as seen in the m/z 191 ion chromatogram (see Howell et al., 1984). Ts = 18α-22,29,30-trisnorneohopane.

the evaluation of the degree of biodegradation has yet to be determined.

Tricyclic terpanes

The C_{19}–C_{45} tricyclic terpanes (cheilanthanes) are highly resistant to biodegradation, surviving even when hopanes are removed (Reed, 1977; Seifert and Moldowan, 1979; Connan et al., 1980) (Figure 16.42). Despite severe biodegradation, seep oils from western Greece still correlate with subsurface oils using a combination of tricyclic terpane and aromatic steroid distributions and stable carbon isotope ratios (Palacas et al., 1986). Similarly, Stojanovic et al. (2001) effectively used the distribution of tricyclic terpanes to evaluate source and maturity relationships in severely biodegraded oil from Sakhalin Island, where other saturated biomarkers were altered.

In some severely biodegraded crude oils, the m/z 191 mass chromatogram is dominated by tricyclic terpanes with no or only trace amounts of hopanes. Such oils may arise by selective biodegradation or high thermal maturity. Tricyclic triterpanes can be altered at about the same rank of biodegradation as the diasteranes (>8) and eventually are removed completely (Connan, 1984; Lin et al., 1989).

Exceptions to these observations have been noted in studies of oil spills and laboratory culturing experiments. Wang et al. (2001b) reported the tricyclic terpanes were removed before hopanes in the *Metula* tanker oil spill. The original oil contained low tricyclic terpanes relative to hopanes, with only the lower-molecular-weight ≤C_{24} tricyclics readily distinguishable. The selective removal of these compounds, therefore, may be due to volatilization and degradation over 24 years of surface exposure. Bost et al. (2001) reported an unusual preference for C_{28} tricyclic terpanes compared with C_{29} tricyclic terpanes in aerobic enrichment culture experiments.

Demethylated tricyclic terpanes

Howell et al. (1984) first reported demethylated tricyclic terpanes in severely biodegraded oils from Colombia (Figure 16.43). The demethylated C_{19} and C_{20} tricyclic terpanes were absent, suggesting that they did not form by microbial alteration of reservoir oils but rather are products of biogenic precursors or early diagenetic reactions. Indeed, the series of demethylated tricyclics does not extend below C_{22} in most documented occurrences, such as Colombia (Howell et al., 1984; Dzou et al., 1999), Venezuela (Cassani and Gallango, 1988; Alberdi et al., 2001), and west Africa (Blanc and Connan, 1992). However, C_{17}–C_{18} tricyclic terpanes occur in heavily biodegraded oil from the Kelamayi oilfield, Zhungeer Basin, northwest China (Jiang Zhusheng et al., 1990). The distribution of these demethylated tricyclic terpanes mirrors that of the regular tricyclanes, suggesting an origin from these compounds.

Alberdi et al. (2001) presented convincing evidence that the demethylated tricyclic terpanes form by loss of the methyl group at C-10 in a manner similar to the microbial alterations that give rise to the 25-norhopanes from hopanes. C_{25}–C_{29} tricyclic terpanes occur as diastereomeric pairs due to an asymmetric carbon at C-22 (Figure 16.44). The elution order of the isomers is unknown, although molecular mechanics calculations predict that the 22R epimers elute after 22S (Peters, 2000). Higher homologs have two asymmetric carbons, C-22 and C-28, resulting in four diastereomers that are resolved only partially by typical chromatographic methods. Analysis of severely biodegraded oil from the Bolivar Coastal Fields of Venezuela revealed that the early-eluting C-22 and C-28 diastereomers are

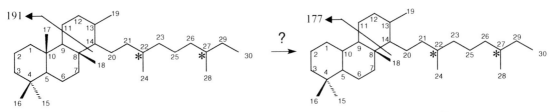

Figure 16.44. Molecular structure of tricyclic terpanes, showing position of optical carbons at C-22 and C-28 and likely demethylated forms.

696 Biodegradation parameters

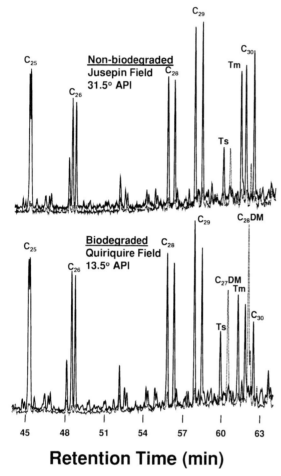

Figure 16.45. Mass chromatogram (m/z 191, solid trace) showing C_{25}–C_{30} tricyclic terpanes in the saturated hydrocarbon fraction of non-biodegraded (top) and heavily biodegraded (bottom) crude oils from Jusepin Field, Well 110, and Quiriquire Field, Well 449 (unknown depths), Maturin Basin, Venezuela. Relative peak heights show that biodegradation favors the second-eluting peak for the C_{28}, C_{29}, and C_{30} tricyclic terpanes (top versus bottom) without formation of 17-nor-tricyclic terpanes (DM; m/z 177, dotted trace). Although four C_{30} tricyclic terpane peaks are possible due to the asymmetric carbon at C-27, only two peaks occur due to co-elution. Reprinted from Peters (2000). © Copyright 2000, with permission from Elsevier.

more bioresistant than the late-eluting diastereomers. Conversely, the late-eluting peaks were dominant over the early-eluting peaks in the 17-nor-tricyclic series. From such data, Alberdi *et al.* (2001) concluded that tricyclic terpane demethylation occurs during sterane

destruction and hopane demethylation, although at a slower rate.

As with the hopanes, tricyclic terpanes can be degraded without the formation of demethylated species. Peters (2000) found evidence for the selective removal of the late-eluting diastereomer in the C_{28}–C_{29} tricyclic terpanes in severely biodegraded oil from the Maturin Basin, Venezuela (Figure 16.45). Although demethylated hopanes are abundant in this oil, demethylated tricyclic terpanes are absent.

De-A-steroidal tricyclic terpanes

Jiang Zhusheng *et al.* (1990) show a series of tricyclic compounds with a base peak at m/z 219 in heavily biodegraded oil from the Kelamayi oilfield, Zhungeer Basin, northwest China, where steranes were removed (Figure 16.46). The mass spectra of these compounds are characterized by molecular ions at m/z 388 and 402, with a base peak at m/z 219, indicating that they originate from C_{28} and C_{29} tricyclanes. The latter fragment suggested that the peaks were de-A-steroidal hydrocarbons. The ratio of the C_{28}/C_{29} tricyclanes is similar to that of C_{28}/C_{29} steranes in non-degraded oil from the same field, supporting a genetic relationship between these compound classes. Full utility of these compounds has not been documented, but the de-A-steroidal tricyclics may be useful for correlation of oils when steranes have been removed by biodegradation.

Non-hopanoid triterpanes

Non-hopanoid triterpanes, including gammacerane, diahopanes, and various biomarkers of angiosperm input,

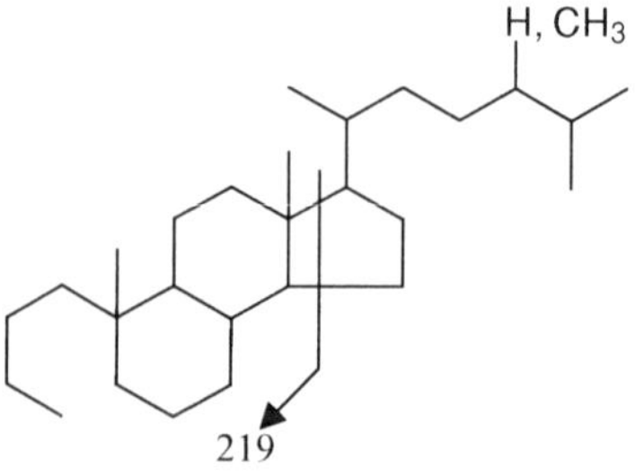

Figure 16.46. Proposed structure of de-A-steroidal tricyclics in a severely biodegraded oil from Zhungeer Basin, northwest China (Jiang Zhusheng *et al.*, 1990).

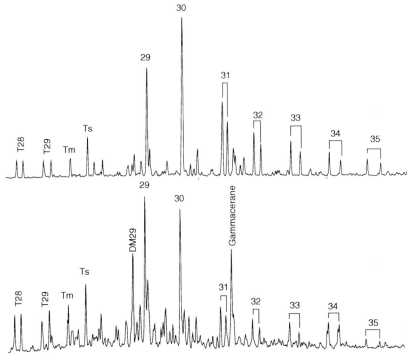

Figure 16.47. Comparison of ion chromatograms (m/z 191) of a non-degraded and equivalent but severely biodegraded oil from Venezuela. Gammacerane, the tricyclic terpanes, Ts, and Tm are enriched relative to the hopanes, which have been partially degraded to 25-norhopanes.

such as oleananes, ursanes, and lupanes, are highly bioresistant. These hydrocarbons persist after many of the saturated biomarkers have been removed (e.g. Seifert et al., 1984; Zhang Dajiang et al., 1988) (Figure 16.47). For example, Fowler et al. (1988) showed ion chromatograms for biodegraded bitumens that lacked steranes and triterpanes yet still retained gammacerane and some of the more resistant compounds, such as tricyclic terpanes and a few unidentified C_{30} pentacyclic compounds. The diahopanes behave in a similar manner. Wenger and Isaksen (2002) showed that these compounds persist and become progressively enriched in oil seeps, even beyond the point where the tricyclic terpanes have been removed.

Oleanane is quite resistant to biodegradation. Figure 16.48 shows another example of hopane biodegradation without 25-norhopane formation for oil from Colombia. In this example, oleanane is clearly more resistant to biodegradation than the C_{27}–C_{35} 17α-hopanes. The bioresistance of oleanane compared with the regular hopanes has been demonstrated in microbial culture experiments (Figure 16.49) (Frontera-Suau et al., 2002).

As noted above, accurate detection of oleanane in severely biodegraded oils can be problematic. In addition to co-elution with lupane (Nytoft et al., 1997), the 22S C_{30} 25-norhomohopane co-elutes with 18α-oleanane under typical chromatographic conditions. Careful chromatographic separation can resolve 18α-oleanane from the 20S C_{30} 25-norhomohopane (Dzou et al., 1999).

Curiale (1991; 1995) identified unusual biomarkers associated with angiosperm plant input in severely biodegraded oil from Tertiary sandstones in the Beaufort Sea, Canada. The oil was biodegradation rank 8, having no steranes or regular hopanes and diminished 25-norhopanes and diasteranes. In addition to the saturated triterpanes oleanane, dinorlupanes, and norlupanes, a series of olefins that included olean-12-ene, 18α-olean-12-ene, and urs-12-ene also are present, along with possible lupanoid ketones (Figure 16.50). Unsaturated, non-aromatic hydrocarbons are rare in

698   Biodegradation parameters

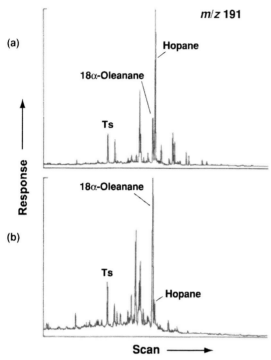

Figure 16.48. Saturated fraction of a very heavily biodegraded (rank 8; see Figure 16.2) seep oil (a) yields a different m/z 191 chromatogram compared with that in a related biodegraded (rank 5) seep oil (b) from Colombia. Biodegradation has nearly completely removed the $C_{27}-C_{35}$ 17α-hopanes from seep No. 851 (a), but 18α-oleanane remains high due to its resistance to biodegradation compared with the hopanes. These seeps do not contain 25-norhopanes (m/z 177), but in some oils the C_{30} 25-norhopane 22S epimer may co-elute with oleanane. Reprinted with permission by ChevronTexaco Exploration and Production Technology Company, a division of Chevron USA Inc.

crude oils and are usually associated with either low maturity or generation via radiolysis (Frolov et al., 1996) or short-term pyrolysis (Li et al., 1998b). Curiale (1995) suggested that the olefinic biomarkers were indications of low-temperature generation, but an alterative explanation is that they form during biodegradation. Tritz et al. (1999) found that an aerobic bacterial culture degraded hopane and bacteriohopane to the corresponding 17(21)-olelfins and 17,21-epoxides.

Aromatic steroids

Aromatized steroidal hydrocarbons remain unaltered in all but the most severely biodegraded crude oils

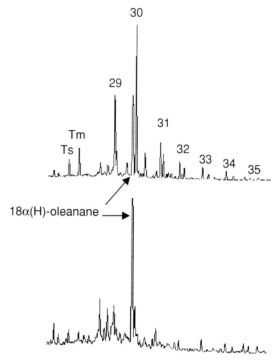

Figure 16.49. Laboratory culture experiments show the relative bioresistance of oleanane compared with the hopanes. A Nigerian oil, initially non-degraded and naturally enriched in oleanane, was degraded using an aerobic enrichment culture grown at 30°C. After 64 days, nearly all hopanes were removed, leaving the m/z 191 ion chromatogram dominated by oleanane. Reprinted with permission from Frontera-Suau et al. (2002). © Copyright 2002, American Chemical Society.

(rank 10). As such, they are particularly useful biomarkers for determining correlations and thermal maturation where the saturated hydrocarbons have been altered or even completely removed. We have found only a few examples where the aromatized steroidal hydrocarbons were altered (e.g. seeps from Madagascar and Nigeria).

Before Wardroper et al. (1984), no biodegradation of triaromatic or ring-C monoaromatic steroids had been reported (Connan, 1984) attesting to their bacterial resistance. Wardroper et al. (1984) indicate that $C_{20}-C_{21}$ triaromatic steroids are among the first aromatic steroids to be depleted during degradation of petroleum, but it is not certain whether this is due to water washing and/or evaporation or to biodegradation. Using biodegraded oils from the North Sea and the Molasse Basin with severely degraded steranes, diasteranes, and hopanes, Wardroper et al. (1984) confirmed degradation of the

Biodegradation parameters 699

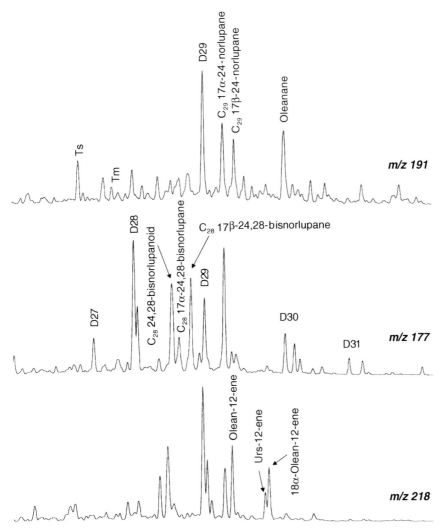

Figure 16.50. Saturated and olefinic angiosperm biomarkers in a severely biodegraded oil from Beaufort Sea. Reprinted from Curiale (1995). © Copyright 1995, with permission from Elsevier.

C_{26}–C_{28} triaromatic steroids. The Molasse Basin oil also shows preferential degradation of the 20R versus 20S epimers of the C_{27}–C_{29} monoaromatic steroids. The low-molecular-weight (C_{21}, C_{22}) monoaromatic steroids in these oils are more resistant to biodegradation than heavier homologs. Laboratory biodegradation experiments support these results (Wardroper et al., 1984)

Adamantanes (diamondoids)

In addition to their high thermal stability, adamantanes appear to be highly bioresistant. Williams et al.

(1986) noted that alkylated adamantanes in biodegraded crude oils (rank 1–4) remained unaltered and suggested that these compounds could be useful for correlation of severely biodegraded oils. To test this hypothesis, they examined severely biodegraded oil where hopanes were nearly absent and 25-norhopanes were abundant (rank 7) and found abundant and apparently unaltered C_2-adamantanes. Wingert (1992) suggested that the ratio of adamantanes to n-alkanes could be used as a measure of biodegradation, but this would be applicable only for oils of low rank (<4). Grice et al. (2000) observed that the ratio of the sum of the 1- and 2-methyladamantanes

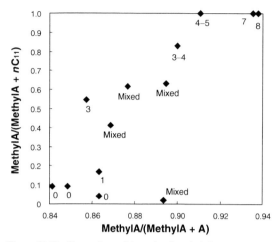

Figure 16.51. Comparison of the ratio of methyladamantanes (methylA) to nC_{11} versus the ratio of methyladamantanes to adamantane (A) in crude oils labeled according to their biomarker biodegradation rank (data from Grice et al., 2000). Mixed indicates a mixture of biodegraded and non-biodegraded oil.

to adamantane remains unchanged from initial values in unaltered oils of equivalent maturity until all n-alkanes were removed (Figure 16.51). At advanced levels of biodegradation, adamantane is preferentially lost relative to the methyladamantanes. Adamantane was still present in the most severely degraded oil studied (rank 8), suggesting that the ratio may have utility at even higher levels of biodegradation. Grice et al. (2000) also noted that the ratio of methyldiamantanes to diadamantane increased in only the most severely degraded oil, suggesting that it could be used to refine interpretations of the level of biodegradation in oils of rank >8.

Porphyrins

Porphyrins are perhaps the most resistant biomarkers to biodegradation. For example, Barwise and Park (1983) showed that related, non-biodegraded and extensively biodegraded oils of nearly equivalent maturity show similar porphyrin distributions (e.g. DPEP/etio). Lin et al. (1989) observed no significant biodegradation of porphyrins in tar-sand bitumens from the Ardmore and Anadarko basins, Oklahoma.

Evidence for the biodegradation of porphyrins is scarce. A study of Colombian oils suggests that nickel porphyrins might be selectively destroyed by severe biodegradation, but that vanadyl porphyrin distributions remain unaltered and could serve as good correlation tools (Palmer, 1983). Vanadyl porphyrin distributions were shown to remain unaltered in heavily biodegraded oils from the Alberta oil sands (Strong and Filby, 1987) and the Permian Phosphoria Formation in Wyoming (Sundararaman and Hwang, 1993).

Organic acids

Organic acids are minor constituents in most crude oils, but they are particularly abundant in low-maturity and biodegraded crude oils. It has been proposed that these compounds react with minerals in carrier beds and reservoir rocks, increasing secondary porosity (McMahon and Chapelle, 1991), although others disagree (Giles et al., 1994). Organic acids are corrosive to metals and can be a major problem for refineries (Robbins, 1998). Because organic acids are marginally water-soluble and toxic, their release in waste-waters is monitored closely (e.g. Holowenko et al., 2002). These detrimental factors initiated some of the earliest studies of petroleum biomarkers, including the discovery of steroidal acids in severely biodegraded Californian crude oil (Seifert 1973; Seifert et al., 1979). Considerable advances have been made in analytical procedures for the extraction, separation, and detection of high-molecular-weight organic acids (e.g. Koike et al., 1992; Warton et al., 1999; Meredith et al., 2000; Galimberti et al., 2000; Tomczyk et al., 2001).

Primary organic acids are produced during kerogen degradation (Mackenzie et al., 1983c). Most of these acids have low molecular weights (C_1–C_4) and partition readily into the aqueous phase (Barth and Bjørlykke, 1993; Barth et al., 1996; Dias et al., 2002b). Low concentrations of more hydrophobic acids that remain in the petroleum phase are released during petroleum generation. Because low-maturity oils tend to have higher total acid numbers (TANs) than equivalent oils of higher maturity, it is reasonable to assume that these carboxylic acids existed either as carboxylated moieties within the kerogen or as readily oxidized precursors. The maximum amounts of organic acids in formation waters associated with oil fields occur consistently in the temperature range 80–140°C. Below 80°C, microbes consume acids, while decarboxylation occurs above 140°C (Carothers and Kharaka, 1978). Because few microbes survive above >80°C in subsurface reservoirs, this temperature distribution suggests that biodegradation is unimportant in the production of organic acids (Lundegard and Kharaka, 1994). Short-chain C_2–C_5 organic acids in oilfield waters generally become isotopically depleted in ^{13}C with increasing carbon number but are consistently more ^{13}C-rich than co-produced crude oils (Franks et al., 2001). These authors show that waters associated with Miocene Monterey and Eocene Kreyenhagen crude oils can be distinguished readily based on the carbon isotope compositions of aliphatic carbon atoms in the organic acids.

Most oils with elevated TAN (>1) are biodegraded, and all high-TAN (>2) crude oils are biodegraded, usually to a severe degree. High acid concentrations may result from the depletion of hydrocarbons and corresponding enrichment of primary organic acids or from newly formed acids. Most carboxylic acids in petroleum are naphthenic or naphthenoaromatic acids. They are believed to originate from the microbial alteration of C_{18} or greater cycloalkanes, alkylaromatic, naphthenoaromatic, sulfur-containing compounds, and possibly polar resins and asphaltenes (Thorn and Aiken, 1998). These compounds may be partially oxidized petroleum compounds, extracellular biosurfactants, or products of dead microbial biomass (Jaffé and Gallardo, 1993; Behar and Albrecht, 1984; Nascimento et al., 1999; Meredith et al., 2000). The generation of microbial acids may accelerate the biodegradation process. Schmitt et al. (1996) proposed that organic acids could serve as ligands for complexing insoluble Fe(III) oxides, making them bioavailable to hydrocarbon-degrading, iron-reducing bacteria. This process, however, may not be needed because some bacteria have direct enzymatic interactions with iron mineral surfaces (Lower et al., 2001) or release quinones to act as an electron shuttle (Childers et al., 2002).

Meredith et al. (2000) showed good agreement between the extent of biodegradation based on the biomarker criteria described above and the concentration of carboxylic acids, as determined by a novel non-aqueous ion-exchange extraction method and gas chromatography of the corresponding methyl esters (Figure 16.52). TAN, as measured by KOH titration, responds to other molecules in addition to carboxylic acids; consequently, it should not be used as a surrogate for total carboxylic acids.

Normal fatty acids are typically minor constituents of the total carboxylic acids in petroleum. Little correlation exists between the concentration of n-fatty acids

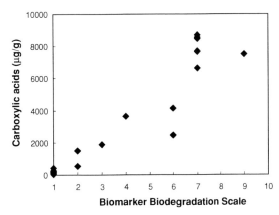

Figure 16.52. Correlation between degree of biodegradation and total carboxylic acid concentration. Reprinted from Meredith et al. (2000). © Copyright 2000, with permission from Elsevier.

with TAN or the degree of biodegradation (Behar and Albrecht, 1984; Jaffé and Gallardo, 1993; Meredith et al., 2000). Some degree of similarity exists between n-alkane and n-fatty acid distributions, indicating a likely precursor–product relationship (Koike et al., 1992; Barakat and Rulkötter, 1995; El-Sabagh and Al-Dhafeer, 2000). C_{16} and C_{18} are frequently much more abundant than the other n-fatty acids, suggesting that they may be indicators of recent microbial activity, including contamination introduced by microbes growing on improperly stored samples.

Note: "Naphthenic acid" is a term confined largely to the petroleum industry that generally refers to cyclic hydrocarbons with one or more carboxylic acid groups. A more formal definition of naphthenic acid indicates compounds with the general formula $C_nH_{2n-1}COOH$. Confusion can arise because the term is sometimes used to refer to all petroleum acids or to all petroleum acids that are in the naphtha distillate fraction. Although high-acid crude oils require special treatment at refineries, naphthenic acids are valuable by-products that are used to formulate many products, including paints, detergents, solvents, lubricants, de-icers, and wood preservatives.

Several organic acids have been described that are clearly partially oxidized biomarker compounds. Phytanic acid [3(R),7(R),11(R)-tetramethylhexadecanoic acid] may be a marker for the early stages of oil biodegradation (Mackenzie et al., 1983c). Hopanoid acids are known biological precursors to hopanes and are common in immature bitumens (Bennett and Abbott, 1999; Stefanova and Disnar, 2000). Hopanoic acids, particularly the biological $17\beta,21\beta$-isomers, also are relatively abundant in biodegraded oils (e.g. Behar and Albrecht, 1984; Mpanju and Philp, 1991). Jaffé et al. (1988a; 1988b) and Jaffé and Gardinali (1990) suggested that these compounds were extracted from immature and moderately mature strata as oils migrated through them. Mass–balance considerations, however, make it unlikely that the concentrations of hopanoic acids in degraded oils could arise from such a process. More likely, $\beta\beta$-hopanoic acids originate from bacterial biomass as a result of reservoir biodegradation (Figure 16.53). The α-, β-, and β,α-hopanoic acids may result from the partial oxidation of petroleum hopanes by microbes (Watson et al., 1999).

Aquifers contaminated by petroleum contain aromatic acids, indicating anaerobic respiration (Jones et al., 2002). Trace amounts of benzylsuccinic acids are metabolic intermediary compounds. Benzoic acids also occur and are interpreted as residual metabolites. These compounds are quite soluble in water and have not yet been discovered in the acid fractions of biodegraded crude oils.

The complexity imposed by the presence of both primary and de novo synthesized carboxylic acids in reservoir oils interferes with their use as biomarkers. Several studies attribute their relative concentrations and distributions to thermal maturation (Mackenzie et al., 1983c; Jaffé and Gardinali 1990) or migration (Jaffé et al., 1988a; Jaffé et al., 1988b; Jaffé et al., 1992; Galimberti et al., 2000; Rodrigues et al., 2000). However, secondary reservoir processes can completely obscure these signatures. Water washing, which is generally associated with biodegradation, strips lower-molecular-weight carboxylic acids from petroleum, leaving a residuum enriched in higher-molecular-weight acids. Biodegradation results in a mixture of thermally stable primary and secondary acids and thermally unstable secondary acids. The latter compounds can remain in steady-state concentrations only during active biodegradation, and they diminish once microbial activity ends. The thermal destruction of carboxylic acids relative to more stable hydrocarbons may explain the occurrence of severely biodegraded oils with low to moderate TANs. Differences in the rates of the thermal destruction of carboxylic acid isomers may provide a measure of the time when active biodegradation occurred (Meredith et al., 2000).

BIODEGRADATION OF COALS AND KEROGEN

Coals and kerogens are generally highly bioresistant, although they are subject to oxidation and degradation by aerobic microbes (Petsch et al., 2001). Biodegradation of bitumen associated with fine-grained source rocks and coals is rare, presumably due to limited nutrients or, in the case of oil-generating strata, thermal sterilization. However, several studies report biodegradation of bitumens associated with coals. Curry et al. (1994) suggested that high concentrations of fluorenes and methylfluorenes indicate extensive oxidation of Permian coals from the Cooper Basin of Australia. Ahmed et al. (1999) examined Permian Moura coals from the Bowen Basin of Australia and found evidence for an unusual sequence of removal of hydrocarbons attributed to biodegradation. Normal alkanes in the saturated fraction of Moura coal extracts were predominant from $\sim nC_{20}$ to nC_{30} but were in trace amounts below nC_{20}. Monocycloalkanes and bicyclic C_{14}–C_{16} sequiterpanes predominate over the lower-molecular-weight saturated hydrocarbons. When compared with nearby non-biodegraded Permian coals, the aromatic hydrocarbons in the Moura coals showed preferential enrichments with increasing ring number (biodegradation susceptibility $\sim$ napthalenes > fluorenes > phenanthrenes > dibenzothiophenes > pyrenes) and with increasing alkyl substitution (biodegradation susceptibility $\sim$parent > methyl- > dimethyl- > trimethyl- > tetramethyl-). Isomers with 1,6-dimethyl-substitution appear to be more susceptible to biodegradation than others within the C_2–C_4 alkylnaphthalenes. These results are similar to those found in crude oils and marine sediments (Volkman et al., 1984; Fisher et al., 1998; Triolio et al., 1999).

The Moura coal extracts violate the established sequence of biodegradation because the aromatic hydrocarbons were biodegraded more readily than n-alkanes $>nC_{20}$. This situation is probably the result of the environment of the coal matrix. Because Permian coals of the Bowen Basin have vitrinite reflectance $\sim 1.0\%$, biodegradation is believed to have occurred after uplift and hydrocarbon generation. Preferential degradation of the lower-molecular-weight n-alkanes and PAHs may

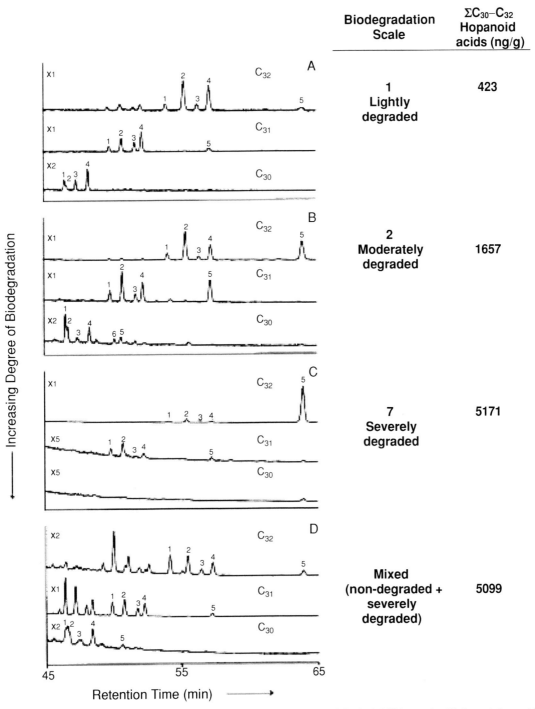

Figure 16.53. Ion chromatograms showing the C_{30} (m/z 235), C_{31} (m/z 249), and C_{32} (m/z 263) hopanoic acids (as methyl esters) in oils of increasing biodegradation. Isomers are designated as 1 = 17α,21α-(22S), 2 = 17α,21β-(22R), 3 = 17β,21α-(22S), 4 = 17β,21α-(22R), 5 = 17β,21β-(22R), and 6 = 17β,21β-(22S). The increase in the C_{32} ββ-22R hopanoic acid with increasing biodegradation is believed to be due to input from bacterial biomass. Reprinted from Meredith et al. (2000). © Copyright 2000, with permission from Elsevier.

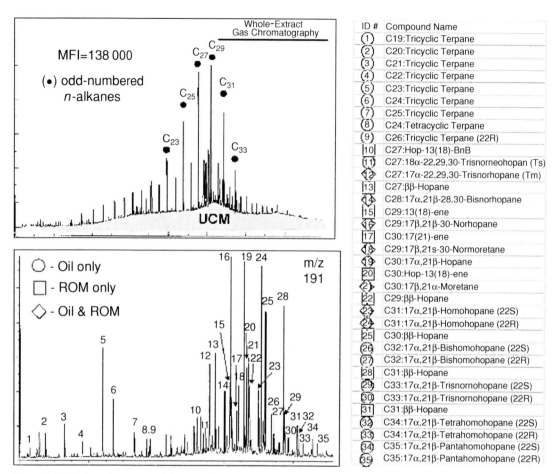

Figure 16.54. Hydrocarbons extracted from a seabottom core containing a mixture of slightly biodegraded oil with unaltered triterpane biomarkers and recent organic matter (ROM). Although the thermogenic biomarkers have been overprinted with contributions from ROM, they can still be used with care to characterize the migrated oil. Reprinted from Wenger and Isaksen (2002). © Copyright 2002, with permission from Elsevier.

be related to aqueous solubility and/or water transport through the coal matrix.

BIODEGRADATION OF OIL SEEPS

Most petroleum systems leak. The majority of the world's known reserves are associated with surface expressions of hydrocarbons. Various surface geochemical techniques were developed to detect trace amounts of thermogenic hydrocarbons in marine and terrigenous sediments, water bodies, groundwater, and the atmosphere.

In the case of seabottom cores, migrated hydrocarbons mix with recent marine organic matter (Figure 16.54). Depending on their relative concentrations, rate of charge, and surface conditions, thermogenic hydrocarbons can vary from non- or slightly biodegraded though severely biodegraded. Wenger and Isasken (2002) found that hydrocarbons in macroseeps (those with a high concentration and/or flux of migrated oil) are biodegraded more heavily than those in microseeps. Several possible reasons were advanced to explain this phenomenon:

- Oil may have to reach a threshold concentration before it can support an oil-degrading microbial population or trigger the necessary enzymatic activity.

- Low concentrations of oil may be sequestered on clay and mineral surfaces where it is not available to biological activity. Only when the absorptive capacity of these mineral surfaces is exceeded can biodegradation occur.
- Bacterial mobility and activity are limited in fine-grained rocks. Small amounts of migrated hydrocarbons that infiltrate marine sediments are not available to biological activity, while those in large pore spaces are degraded more readily.

When biodegradation occurs in seabottom cores, it appears to correlate with the activity of sulfate-reducing bacteria. To eliminate the possibility of modern contaminants, the upper portions of most sediment cores are not analyzed. Oil stains in deeper parts of these cores are usually well below the oxic zone, and the occurrence of H_2S is frequently noted on the sampling logs. Wenger and Isaksen (2002) showed that a core from deep water offshore West Africa contained severely biodegraded oil (rank ~4) at a depth of 1 m, only slightly biodegraded oil (rank ~1) at ~1.5 m, and non-degraded oil at ~2 m. The degree of biodegradation correlated with the amount of available sulfate, which decreased from 500 ppm to 10 ppm over the depth interval. The microbial population also changed. *Moraxella*, an opportunistic oil degrader, is dominant in shallow sediments associated with high sulfate and biodegraded oil, but it is only a minor microbial constituent in deeper sediments that contain non-degraded oil. These findings suggest that the abundance of oxygen, sulfate, and other nutrients near the sediment–water interface support different bacterial populations and promote different degradation pathways compared with oils in deeper reservoirs.

Wenger and Isasken (2002) reported that 25-norhopanes do not form during biodegradation of petroleum in shallow marine sediments. These compounds have not been observed over years of intensive study of the oil spilled in Prince William Sound. They concluded that the microbial processes responsible for biodegradation in marine seabottom sediments must differ from those in the subsurface. This lack of 25-norhopanes in seep samples provides further evidence that hopane demethylation occurs in the reservoir and that 25-norhopanes do not form during diagenesis and are not simply concentrated by biodegradation. Presumably, if oil was biodegraded in a reservoir and leaked to the surface, then 25-norhopanes could occur in the seep, indicating severely biodegraded oil in the deep subsurface.

PREDICTION OF PHYSICAL PROPERTIES

Accurate prediction of the original physical properties of crude oils that have been altered by secondary processes can be an economic necessity. For example, pre-drill estimates of crude-oil API gravity, viscosity, or sulfur content may be necessary to justify drilling and production costs. In one example, seismic data showed a flat spot suggesting an oil–water contact at ~3200 m in an Asian basin. Biodegradation of petroleum trapped in the structure would severely reduce the viability of the prospect as a drilling target. A calibration of pristane/nC_{17} alkane ratios for produced oils from the area indicates that biodegradation is unlikely at depths greater than 2000 m (Figure 16.55).

For oils generated from the same source rock, thermal maturity is a major factor that determines many of the physical (e.g. specific gravity and viscosity) and chemical (%sulfur and trace metals) properties that influence oil productivity and value. In poorly explored basins, the quality of subsurface oil accumulations can be inferred from biodegraded oil seeps, provided that a suitable calibration oil suite is available. For example, Moldowan et al. (1992) determined the relationship between specific gravity and thermal maturity in non-biodegraded oils from the Adriatic Basin using the aromatization of the steroidal hydrocarbons (Figure 16.56). Because these compounds were not altered by biodegradation, the original gravity of severely biodegraded oils could be inferred. This study also illustrates some of the limitations of such extrapolations. The implied relationship is not very accurate for oil with TA/(TA + MA) values >0.7, and the number of oils in the training set is insufficient to rigorously define error.

EXERCISE

Table 16.2 shows nine offshore Santa Maria Basin crude oils and a surface seep sample from a nearby beach (Peters et al., 1990; Peters and Moldowan, 1991). Source-related geochemical parameters in an earlier

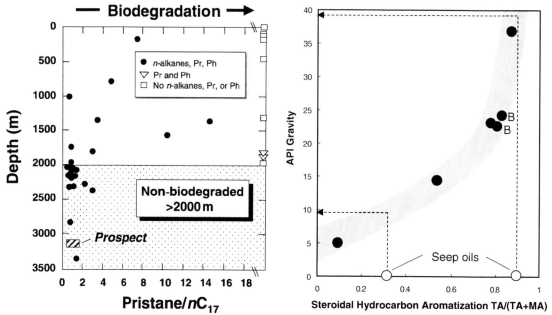

Figure 16.55. Pristane/nC_{17} alkane versus depth for produced oils and seep samples from a basin in the Far East, indicating that biodegradation has not affected oils below 2000 m. The prospect at 3200 m in the study area has low risk for biodegradation.

Figure 16.56. Steroidal aromatization ratio, showing a relationship with API gravity for oils (•) from the Adriatic Basin. Two of the oils (B) are slightly biodegraded (rank <2). Based on this relationship, the unaltered API gravity of oil seeps can be determined with varying degree of confidence. Data from Moldowan et al. (1992).

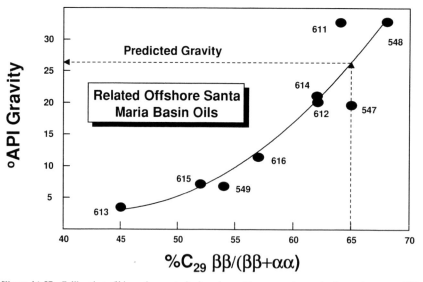

Figure 16.57. Calibration of biomarker maturity based on a C_{29}-sterane isomerization ratio versus API gravity for non-biodegraded crude oils generated from the Miocene Monterey Formation in the offshore Santa Maria Basin, California. The dashed line shows how measured biomarker maturity for biodegraded but genetically related seep oil might be used to predict the original API gravity before alteration. Data from Table 16.2.

Table 16.2. *Geochemical data for nine Monterey crude oils and a hypothetical seep sample from the Santa Maria Basin, California*

Oil	Depth (feet)	°API	S (wt.%)	Steranes[1] %20S	%ββ	%Dia	%22S	Terpanes[2] %βα	%Tric	%Ts	Aromatics[3] %MA(I)	%TA(I)	%TA	Porphyrins %PMP[4]
547	7060–7190	19.6	3.5	53	65	12.3	56	5.0	32.2	35	6	14	75	27
548	–	32.9	1.1	53	68	9.0	56	5.5	65.1	30	24	31	97	74
549	–	6.9	7.5	44	54	6.5	56	5.0	16.1	27	3	8	64	12
611	7870–8060	32.8	0.9	50	64	11.5	56	4.0	56.5	28	15	15	95	75
612	5300–5390	20.3	3.8	50	62	9.9	56	5.0	33.3	21	5	10	79	24
613	6355–6370	3.5	12.7	40	45	9.1	55	6.5	10.8	24	2	7	13	4
614	6700–7000	20.9	3.2	49	62	9.9	56	5.0	34.7	25	5	12	75	25
615	5810–6040	7.3	6.6	45	52	8.3	57	5.0	20.0	28	3	8	64	7
616	–	11.5	6.7[5]	47	57	7.4	56	4.0	16.1	23	3	10	31	13
Seep	Surface	5.2	9.0	52	65	12.0	57	5.0	56.0	37	18	25	96	74

[1] %20S = 100*20S/(20S + 20R), where 20S and 20R are epimers at C-20 in the C_{29} $5\alpha,14\alpha,17\alpha$ steranes. %ββ = 100*ββ/(ββ+αα), where ββ = C_{29} $5\alpha,14\beta,17\beta$ 20S + 20R steranes and αα = C_{29} $5\alpha,14\alpha,17\alpha$ 20S + 20R steranes. %Dia = 100*Diasteranes/(diasteranes + regular steranes) for C_{27}–C_{29} $13\beta,17\alpha$ 20S + 20R diasteranes and C_{27}–C_{29} $5\alpha,14\alpha,17\alpha$ and $5\alpha,14\beta,17\beta$ 20S + 20R steranes.

[2] %22S = 100*22S/(22S + 22R), where 22S is C_{32} $17\alpha21\beta$ 22S hopane and 22R is C_{32} $17\alpha21\beta$ 22R hopane. %βα = 100*βα/(βα+αβ), where βα and αβ are C_{30} hopanes and moretanes, respectively. %Tric = 100*Tric/(Tric + Hop), where Tric is C_{28}–C_{29} tricyclic terpanes and Hop is C_{29}–C_{33} $17\alpha21\beta$ hopanes. %Ts = 100*Ts/(Ts + Tm), where Ts is 18α-22,29,30-trisnorneohopane and Tm is 17α-22,29,30-trisnorhopane.

[3] %MA(I) = 100*MA(I)/(I + II), where MA(I) = C_{21}–C_{22} C-ring monoaromatic steroids and MA(II) = C_{27}–C_{29} C-ring monoaromatic steroids, including $5\alpha,5\beta$ 20S + 20R and CH_3 (10-5β) rearranged isomers. %TA(I) = 100*TA(I)/(I + II), where TA(I) is C_{20}–C_{21} triaromatic steroids and TA(II) = C_{26}–C_{28} triaromatic 20S + 20R isomers. %TA = TA/(TA + MA), where TA = TA_{28} triaromatic 20S + 20R isomers and MA = MA_{29} C-ring monoaromatic steroids, including $5\alpha,5\beta$ 20S + 20R and CH_3 (10-5β) rearranged isomers.

[4] %PMP = 100*vanadyl $C_{28}E/(C_{28}E + C_{32}D)$ by high-performance liquid chromatography (HPLC) analysis, where E = etio- and D = DPEP.

[5] Sample may be contaminated by diesel, which could reduce the measured sulfur content from true value.

DPEP, deoxylphyloerythro-etioporphyrins; E, etioporphyrins.

exercise indicate that all ten oils originated from the Miocene Monterey Formation (see Table 13.9).

1. What is the principal control on the inverse relationship between API gravity and sulfur content among the nine non-biodegraded oils? Does biodegradation move the seep oil off the trend of API gravity versus sulfur? Why?
2. Rank the numbered offshore Santa Maria Basin Monterey oils in order of increasing thermal maturity.
3. Biodegradation readily alters many physical properties of crude oils, adversely affecting their producibility and economic value. Assume that seismic evidence indicates a likely reservoir below the location of a 5°API seep oil and that, to be profitable, any oil discovered by drilling must show at least 25°API gravity and <3.0 wt.% sulfur. Is the prospect below the seep worth drilling?

Figure 16.57 shows a calibration of API gravity for genetically related, non-biodegraded crude oils generated from the Miocene Monterey Formation in California (Table 16.2). The $14\alpha,17\beta$-sterane isomerization ratio is not readily affected by biodegradation, and m/z 217 mass chromatograms for the oils show no evidence for degradation of the steranes. Gas chromatograms of the oils show no evidence for biodegradation of the n-alkanes or isoprenoids. Consequently, the main control on API variations among these oils is thermal maturity. The dashed line in Figure 16.57 was constructed by measuring the $14\alpha,17\beta$-sterane isomerization ratio for a heavily biodegraded seep oil. The predicted API gravity of this oil before biodegradation (>25°API) is much higher than that at the seep (~5°API), making unaltered equivalents of the seep at depth highly desirable. Similar calibrations can be constructed to predict sulfur content, viscosity, or other relevant geochemical parameters.

17 · Tectonic and biotic history of the Earth

> Petroleum systems and their associated biomarkers occur in the context of our evolving planet. The purpose of this chapter is to provide a brief tectonic history of the Earth in relation to the evolution of major life forms and the occurrence of petroleum systems. This chapter is linked closely to Chapter 18, which provides key examples of petroleum systems, also arranged approximately by age of the source rock.

BIRTH OF THE SOLAR SYSTEM (~4.7–4.6 GA)

Many theories have been proposed for the origin of the solar system, but the one advanced by German philosopher Immanuel Kant in 1755 in *Universal Natural History and Theories of the Heavens* is acknowledged as essentially correct. Kant proposed that the planets originated from a rapidly rotating disk of matter orbiting the sun, which coalesced to form larger and larger objects, eventually creating planets. A French mathematician, Pierre-Simon de Laplace, refined Kant's model. In the Kant–Laplace proto-planetary or accretion disk hypothesis, the solar system began as a dense cloud of interstellar gas and dust. About 4.7 billion years ago (4.7 Ga), a shock wave from an exploding supernova triggered the gravitational collapse of the interstellar cloud into a rapidly rotating disk (Figure 17.1). Dust grains attracted by electrostatic forces clumped together, settling in the mid-plane of the disk, where they eventually grew to the size of pebbles. By processes that are poorly understood, the pebbles aggregated to form kilometer-sized planetesimals with sufficient mass to rapidly accrete moon-size bodies. Accretion of these bodies to form the inner planets produced the most intense phase of impacts in the history of the solar system. At least one collision involved the proto-Earth and a Mars-size body, blasting out the material that condensed to form the Moon and dissipating any early oceans and atmosphere that may have formed (Taylor, 1994). Further accretion by impacts of smaller planetesimals and comets continued until ~4.0 Ga.

One significant modification to Figure 17.1 was suggested recently. Reassessment of radioactive decay constants led Scherer *et al.* (2001) to conclude that the first continental crust formed on Earth's surface ~4.3 Ga, at least 200 million years earlier than thought previously. This remnant crust apparently survived the intense meteorite bombardment that ended ~3.9–4.0 Ga.

ORIGINS OF THE OCEANS, ATMOSPHERE, AND LIFE: THE HADEAN EON (~4.6–3.8 GA)

The Hadean Eon is defined as the time before the oldest known rocks on Earth (Figure 17.1). As such, there is no fossil or chemical record of this time to examine. The early Earth was molten and lacked surface volatiles because the accretion disk at Earth's orbit was near 1000 Kelvin and planetesimal impacts and core formation contributed additional heat. Solid crust probably formed within a few million years after the main phase of accretion. The inner planets and moons received a continual rain of comets that may have provided the water and volatiles for the oceans and atmosphere (Delsemme, 1996). Based on the cratering history of the Moon, comets could contribute enough water to account for Earth's oceans if they represented only 10% of the impacts (Oró and Lazcano, 1996). Comets also could contribute the light elements and perhaps more complex molecules necessary for life. Life probably originated after heavy impact bombardment ended ~4.0 Ga.

In the 1920s, Russian biochemist Alexander Oparin and English geneticist J. B. S. Haldane proposed independently that life arose spontaneously from abiotic (abiogenic) processes. Miller (1953) was first to test this

710   Tectonic and biotic history of the Earth

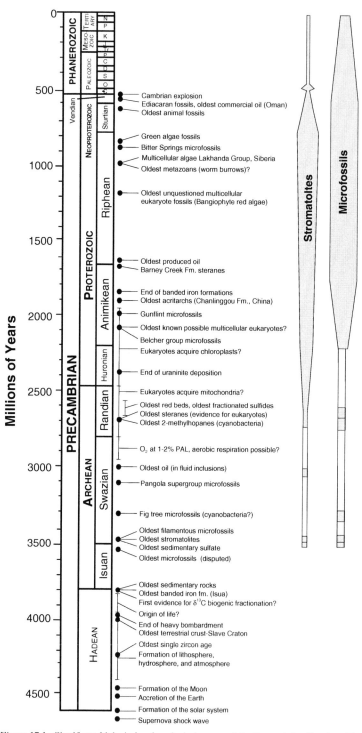

Figure 17.1. Significant biological and geological events of the Precambrian Eon (modified from Schopf et al., 1983).

hypothesis. Miller showed that amino acids could form by electric discharge in a reducing atmosphere like that proposed for the early Earth by Urey (1951). Scientists now know that the early atmosphere was probably oxidizing rather than reducing and that abiotic amino acids, hydrocarbons, and other simple molecules essential for life are common throughout the cosmos. Nonetheless, the Miller–Urey experiment marked the first modern attempt to simulate prebiotic chemistry. Experiments soon demonstrated that many biomolecules could be formed using various processes and energy sources. Fox (1969) found that heating proteins could form microspheres, demonstrating a potential pathway to cell formation. Orgel (1973) assembled short nucleotide strands from simpler organic compounds and salts. Later work showed that certain minerals catalyze the synthesis of nucleotide strands that are large enough (up to 55 monomers) to carry meaningful genetic information (Ferris et al., 1996).

Life requires reproduction and the transfer of genetic information. Reproduction involves complex interactions between proteins and encoded genetic information (e.g. deoxyribonucleic acid, DNA). Models proposing that life started with these processes quickly degrade into circular arguments of which came first, proteins or DNA. Cairns-Smith (1982; 1985) postulated that life began with the replication of layers of clay. The processes that started out as completely inorganic became modified with the absorption of organic species. Over time, the information preserved as crystalline defects was replaced by organic genetic material. Wächtershäuser (1988) hypothesized that life began as autocatalytic, metabolic processes within organic monolayers bound to mineral surfaces. These two-dimensional systems could evolve into cell-like structures as organic systems replaced mineral-catalyzed reactions. Eventually, cells emerged that were able to reproduce without mineral support. Ribonucleic acid (RNA) (Cech, 1987) theories compete with the mineral origin theories. Artificial RNA molecules can act as catalysts, excising portions of themselves or of other RNA molecules, although no RNA has yet been synthesized that is capable of catalyzing RNA replication. The catalytic properties of RNA range from amino acid transfer reactions (Loshe and Szostak, 1996), to peptide bond formation (Zhang and Cech, 1997), to nucleotide synthesis (Unrau and Bartel, 1998). Early life, therefore, could have been based on RNA enzymes (ribozymes) that not only carried genetic information but also provided the machinery needed to reproduce. The RNA world (Gesteland et al., 1998) would be replaced eventually by the DNA–RNA–protein systems common to all extant life (Gilbert, 1986).

Life also requires energy. The first living cells may have been heterotrophs that consumed abiotic molecules from organic-rich "primordial soup." Eventually, autotrophic organisms might have developed as this food source diminished. Others argue that autotrophic systems evolved first as autocatalyzed reactions, which evolved into more complex metabolic pathways (Woese and Wächtershäuser, 1990). The first life may have consisted of chemosynthetic organisms associated with deep-sea hydrothermal vents (Corliss et al., 1981). Deep-sea vents are attractive sites for life's origin because they receive a more or less continuous flux of energy for chemical reactions, they are enriched in metals and sulfides, and they have sharp temperature gradients that may quench metastable species. Complex ecosystems centered on these deep-sea vents are dominated by thermophilic prokaryotes that may be closest to the common ancestor of all life (Corliss et al., 1979). Iron-nickel sulfides can act as catalysts to change methylthiol and carbon monoxide in vent discharges into methylthioacetate (Huber and Wächtershäuser, 1997), which resembles acetyl-CoA, the key enzyme for carbon fixation in organisms. Miller and Bada (1988) suggested that life did not originate near hydrothermal vents because the high temperatures tend to decompose proto-biochemicals. Cold environments may be more conducive to preservation of precursor compounds synthesized on Earth or introduced by extraterrestrial objects (Bada et al., 1994).

AGE OF THE PROKARYOTES: THE ARCHEAN EON (~3.8–2.5 GA)

Geological history began when solid rock formed on the Earth. However, no crustal material remains intact from the age of intense meteoritic bombardment (~4.5–4.0 Ga). Recycled zircon grains imbedded in metasediments contain evidence for early crust formation. The oldest of these terrestrial zircons from the Narryer Gneiss Complex of Australia yield ages up to 4.4 Ga (Wilde et al., 2001) and have hafnium isotopic compositions similar to those of carbonaceous chondrites. Over half of these zircons formed by remelting of older crust (Amelin et al., 1999). The oldest examples of terrestrial crust are 4.1 Ga gneisses from the Slave Craton, Canada

(Figure 17.1) (Bleeker et al., 1999). The Slave Craton may have been a nucleus for the earliest proto-continent, pieces of which are now scattered from Zimbabwe to Wyoming. Whether plate tectonics was active at this early date is unclear (Zimmer, 1999). Isotopic evidence suggests that the mantle differentiated earlier than 3.8 Ga and that melting in deep mantle plumes provided a mechanism to create thick buoyant oceanic plateaus that acted as nuclei for proto-continents (Blichert-Toft et al., 1999). Parts of the Kaapvaal and Pilbara craton contain remains of pristine crust dating from 3.6 to 2.7 Ga. Similarities among these rocks and their overlying sequences suggest that they were part of a larger supercontinent (Zegers et al., 1998). This supercontinent, called Vaalbara, is older than 3.1 Ga and fragmented as early as 2.7 Ga.

The oldest microfossils may occur in the oldest known metasedimentary rocks, the 3.8 Ga amphibolite facies carbonates and banded iron formations of Isua, Greenland (Figure 17.1). However, early reports of yeast-like microfossils by Pflug and Jaeschke-Boyer (1979) are generally thought to be metamorphic inclusions. Graphite that is depleted in the heavy isotope of carbon (^{13}C) occurs in association with magnetite-siderite banded iron formations. While these results suggest life, they are considered ambiguous because the graphite is in isotopic equilibrium with associated carbonate (Walters, 1981; Hayes et al., 1983; Schildowski et al., 1983). Ion-microprobe studies found ^{13}C-depletion ($\delta^{13}C = -21$ to $-49‰$) for graphite globules of bacterial size (Rosing, 1999) and for graphite in apatite grains from Isua and from the nearby Akilia Island that may be as old as 3.9 Ga (Mojzsis et al., 1996). The latter results are intriguing because encapsulation of carbon within apatite eliminates kinetic or thermodynamic fractionation that might abiotically reduce inorganic carbon. However, the radiometric age of the Akilia Island apatites has been questioned (Sano et al., 1999).

Some of the earliest microfossils of putative biological origin are from the Apex Chert of the Warrawoona Group, northwestern Australia (Figure 17.2). These coccoidal and filamentous structures, presumed to be cyanobacteria, are ~3.46 Ga and are morphologically similar to extant prokaryotes (Schopf, 1993; Schopf and Packer, 1987). Conical stromatolites in the 3.5 Ga Strelly Pool Chert from the Warrawoona Group show morphologies attributed to microbially mediated

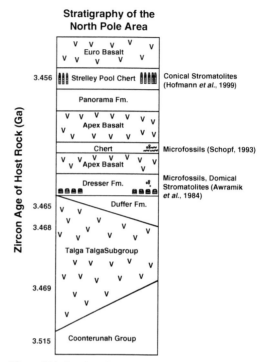

Figure 17.2. Stratigraphy of the Warrawoona Group, showing the age relationships of the earliest known stromatolites and microfossils. Based on Awramik et al. (1984) and Schopf (1999).

accretionary growth (Hofmann et al., 1999). Brasier et al. (2002; 2004) questioned whether the cherts contain the fossilized remains of microorganisms. They believe that the cherts are hydrothermal in origin and that the so-called microfossils are artifacts formed from amorphous graphite. The rock fabric and the association of the cherts with igneous intrusives is consistent with hydrothermal activity. Basier et al. found numerous "microfossils" with morphologies that are difficult to attribute to biogenic origins. Laser-Raman spectroscopy was used to support both biotic (Schopf et al., 2002) and abiotic (Brasier et al., 2002) origins for these structures. Slightly older microfossils from the North Pole location (Awramik et al., 1983; Awramik et al., 1984; Vasconcelos and McKenzie, 1997) are considered suspect (Buick, 1984a; 1984b), as are microfossils reported by Muir and Grant (1972) and Walsh and Lowe (1985) from the Onverwacht Group of South Africa (Schopf and Walter, 1983).

Note: Cyanobacteria trap and bind sedimentary particles of limestone and chert and play a critical role in the accretion, lamination, and early lithification of stromatolites. Because bound particles absorb light, the photosynthetic bacteria move upward during growth, resulting in a laminated texture that generally lacks preserved organic matter. Stromatolites have morphologies that vary from flat-wavy laminations (stratiform) to straight columns, branched columns, and domes ranging in size from millimeters to tens of meters. Stromatolites are the dominant fossils in the Archean Eon and, especially, the Proterozoic Eon. Metazoan grazing animals contributed to the decline of stromatolites. Consequently, most modern stromatolites, such as those in Hamelin Pool, Shark Bay, Australia, occur in restricted to hypersaline or hot spring environments that are unfavorable for metazoan grazers. Some living stromatolites occur in a normal salinity, open marine setting on the margins of Exuma Sound, Bahamas (Reid et al., 2000). Growth of these stromatolites represents a dynamic balance between sedimentation and intermittent lithification of the cyanobacterial mats. During rapid sediment growth, the stromatolites are dominated by gliding filamentous cyanobacteria. During hiatal intervals, surface films of exopolymer and thin crusts of microcrystalline carbonate form during heterotrophic bacterial decomposition. Climax communities, including endolithic coccoid cyanobacteria, form thicker, lithified laminae during prolonged hiatal intervals.

The major anaerobic, heterotrophic, and autotrophic metabolic pathways probably were established early in the Archean Eon, but there is no definitive geochemical or fossil evidence. Archean microfossils are morphologically similar to Paleozoic prokaryotic organisms and even extant cyanobacteria, but they could be unrelated species with totally different biochemistry. Circumstantial evidence suggests that these microfossils were photosynthetic organisms. Kerogens (insoluble organic matter) showing stable carbon isotopic ratios less than $-30‰$ occur in the Onverwacht and Warrawoona sequences, consistent with an Archean carbon cycle driven by photosynthesis under elevated partial pressures of carbon dioxide (Hayes et al., 1983). Such isotopic values could also arise from the consumption of microbial methane by methanotrophs. Some of these early microfossils are in stratigraphic, but rarely concurrent, association with biogenic stromatolites, suggesting photosynthesis. Metabolic processing among the earliest life forms was probably anaerobic and included fermentation, methanogenesis, chemosynthesis, and photosynthesis using photosystem I (H_2S, H_2, or organic molecules as electron donors). Aerobic photosynthesis using photosystems I and II (H_2O as an electron donor producing O_2) probably evolved later but still early in the Archean Eon (Schopf et al., 1983).

The emergence of cyanobacteria and the evolution of photosystem II, which fixes CO_2 and produces O_2, was the most significant biochemical event after the origin of life. Microfossils and stromatolites suggest that cyanobacteria may be as old as $\sim$3.5 Ga and almost certainly as old as $\sim$3.0 Ga (Figure 17.1). Oxygen is toxic to anaerobic life, producing superoxide anions, hydrogen peroxide, and hydroxyl radicals. In response, cyanobacteria evolved free-radical scavengers, such as vitamins C and E, and enzymes, such as superoxide dimutase and catalase, to protect themselves from their own waste products. During the Archean Eon, most of the oxygen produced by cyanobacteria was consumed by oxidation of organic matter, reducing volcanic gases, sulfides, or dissolved ferrous iron that was abundant in the early ocean. However, not all of the oxygen was immediately sequestered in the Archean atmosphere. As early as 2.8 Ga, free oxygen at $\sim$1–2% present atmospheric level (PAL) may have been present in many microenvironments, allowing aerobic respiration (Towe, 1990). Organisms that evolved to exploit oxygen as an electron acceptor in their electron-transport chains gained a tremendous advantage in energy production. This energy-efficient metabolic pathway evolved independently in several prokaryotic lineages that had already split from the universal ancestor.

At one time, the eukaryotes were thought to have evolved after establishment of an oxygen-rich atmosphere. The eukaryotic lineage is known to be as old as the prokaryotes. The earliest proto-eukaryotes were anaerobic and probably chemosynthetic. Microenvironments with free oxygen allowed the first eukaryotes to evolve in the early Archean Eon. Biosynthesis of the sterols in eukaryotic membranes requires oxygen for epoxidation of squalene and removal of the 14α-methyl group from steroid precursors (Figure 3.15) (Chapman and Schopf, 1983). This reaction is needed to make cholesterol planar, which is an important feature of this critical eukaryotic lipid. In addition to oxygen, sterol synthesis requires the cytochrome P450 CYP51 (lanosterol 14α-demethylase). In this and many other aerobic pathways, the oxygen-requiring steps were appended to a series of anaerobic reactions.

The oldest indigenous biomarkers yet discovered occur in 2.5–2.7 Ga laminated kerogenous shales of the Fortescue and lowermost Hamersley Groups from the Pilbara Craton in northwestern Australia (Brocks *et al.*, 1999). Photosynthetic oxygen production had already evolved by the time these rocks were deposited because they contain abundant 2α-methylhopanes (see Figure 13.91), characteristic of cyanobacteria that generate oxygen during photosynthesis (Figure 17.1) (Summons *et al.*, 1999). Excretion of metabolic oxygen by the cyanobacteria is the most likely explanation for the associated oxide-facies banded iron formation. These iron ores resulted from oxidation of water-soluble iron and precipitation of insoluble iron during Earth's early history, when the atmosphere and water column contained little free oxygen. The gradual accumulation of oxygen in the oceans and atmosphere ultimately made Earth hospitable for aerobic organisms. By the Late Archean Eon, the first red beds formed as oxygen levels increased dramatically in the oceans and previously soluble iron was oxidized to insoluble iron oxides. Red beds consist of clastic detritus, usually quartz grains, coated by oxidized iron minerals.

The Pilbara Craton rocks also contain C_{26}–C_{30} steranes, their aromatic counterparts, and A-ring-methylated steranes that originate from sterols characteristic of eukaryotes (Brocks *et al.*, 1999). These results are somewhat controversial because proof that the extracted biomarkers are syngenetic with the shales is based on the lack of such extractable compounds in the surrounding non-source rocks. Extractable organic matter in rock samples may not be indigenous and requires careful evaluation (Table 4.1). If these biomarkers are indigenous, they support the existence of eukaryotes 500 million to 1 billion years before the first documented microfossil evidence for eukaryotes from 1.8 to 1.9 Ga (Knoll, 1992).

THE OXYGEN HOLOCAUST: THE PROTEROZOIC EON (~2.5–1.0 GA)

The oldest continental rocks that have not been reheated to reset their age originated in the earliest Proterozoic Eon, when the Earth cooled sufficiently to allow stable continents to form and plate tectonics to operate much as it does today. The continents presumably collided and broke up several times, but paleo-plate reconstruction extends only to the latest Proterozoic Eon.

During the early Proterozoic Eon, the sinks that kept the global concentrations of free oxygen low were becoming saturated. By ~2.4 Ga, the deposition of uraninite largely ceased, and banded iron formations with alternating beds of silica and hematite became common. Oxygen continued to increase, locally reaching 15% PAL by 1.9 Ga, as indicated by paleo-weathering of siderite surfaces (Towe, 1990). About this time, banded iron formations ceased to be deposited. The rise in atmospheric oxygen in the early Proterozoic Eon probably initiated a mass extinction of many obligate anaerobes and may have also triggered global glaciation as greenhouse gases, such as methane, were oxidized. For example, extensive tillites from ~2.3 Ga occur throughout North America.

While the rise in oxygen was a holocaust for many anaerobes, it allowed diversification of oxygen respiring microorganisms. Microfossil assemblages, such as those in the Gunflint Banded Iron Formation in Canada, where the first Precambrian fossils were described by Tyler and Barghoorn (1954), showed increased diversity but were still dominated by cyanobacteria. Stromatolites became abundant, geographically widespread, and morphologically diverse. Oceanic bottom waters, however, may have remained anoxic throughout much of the Proterozoic Eon (Canfield, 1998).

The first definitive eukaryotic fossils are spheroidal acritarchs from the Chaunlinggou Formation of China dating around 1.8–1.9 Ga (Knoll, 1992). Acritarchs are single-cell, closed-vesicle, acid-resistant, organic-walled microfossils that cannot be assigned to other taxa. They have been interpreted as dinoflagellate cysts (hystrichospheres), but they lack some dinoflagellate features, such as the archaeopyle (the pore through which the dinoflagellate exits the cyst). Acritarchs by definition are of unknown affinity. However, regardless of their affinity, acritarchs are eukaryotes. Increased oxygen allowed multicellular eukaryotic organisms to develop. Disputed multicellular eukaryotic fossils date from ~ 2.1 Ga (Han and Runnegar, 1992). *Grypania* is a large (2 mm wide and up to 80 cm long) organic tube fossil found in ~1.1–1.4 Ga rocks from China, India, and North America (Walter *et al.*, 1990). *Grypania* is unquestionably eukaryotic, but its taxonomic affinity is uncertain.

Note: The eukarya arose in the Late Archean or Early-Middle Proterozoic eons when the atmosphere contained little oxygen and the risk of DNA damage for exposed organisms was high due to excessive ultraviolet radiation. Because deep water eliminates radiation exposure, early eukaryotes may have evolved there as facultative anaerobes. The euxinic Santa Barbara Basin may represent a modern analog for the environment where primeval eukaryotes diversified (Bernhard *et al.*, 2000). *Beggiatoa* bacterial mats in these euxinic deepwater sediments support a prolific community of protists and metazoa, many of which contain previously unknown prokaryotic symbionts. These organisms offer a new opportunity to test the endosymbiotic hypothesis for the origin of eukaryotes.

RISE OF THE EUKARYOTES: THE NEOPROTEROZOIC ERA (1000–544 MA)

The Neoproterozoic Era is the earliest time that allows reasonable estimates of paleogeography. Paleogeographic reconstruction combines evidence from paleomagnetics, paleobiogeography, paleoclimatology, and stratigraphy with basic principles of plate tectonics and geologic history. Uncertainties increase and paleo-reconstruction becomes more speculative with greater age.

During the Late Proterozoic Eon, the continental plates coalesced into a supercontinent called Rodinia. The size and orientation of Rodinia is unknown, but Laurentia (North America) probably formed its core. Rodinia began to break apart ~750 Ma. About half of Rodinia, including Laurentia, migrated toward the South Pole. The other half migrated northward, where it collided with the Congo plate and formed a new supercontinent, Pannotia (Figure 17.3). Near the end of the Neoproterozoic Era, Pannotia fragmented into four new continents, Gondwana, Laurentia, Baltica, and Siberia.

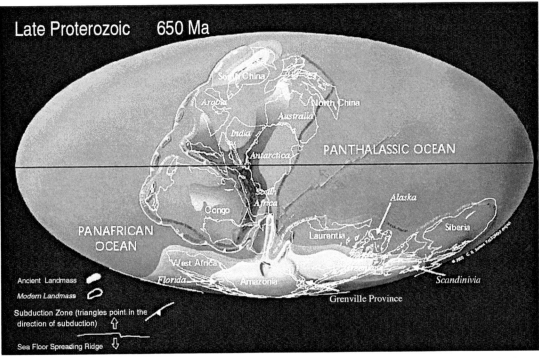

Figure 17.3. Paleo-plate reconstruction of the Late Proterozoic Earth (C. R. Scotese, PALEOMAP project). Pannotia consisted mainly of the Congo plate, with additional South African, Anarctica, and Indian land mass.

Widespread glacial deposits or tillites characterize the Neoproterozoic Era, when a substantial portion of the landmass resided at the southern pole. Detailed stratigraphy and isotopic dating show that several glaciations took place at 850–800, ~750, 720–670, and 610–590 Ma. Geologic evidence and large isotopic excursions indicate that glaciers advanced to within 11 degrees of the Equator (Evans, 1988).

Note: Lake Vostok in eastern Antarctica is comparable in size to Lake Ontario, yet it has never been seen directly. It resides beneath ~4 km of glacial ice, as imaged by ground-penetrating radar and satellite altimeters (Kapitsa et al., 1996), and is the largest of more than 70 similar Antarctic lakes (Siegert et al., 1996). Lake Vostok may represent the closest analog to the subglacial environments that existed during the Neoproterozoic Era on Earth or exobiotic environments on Mars or Europa, the ice-covered moon of Jupiter. The cold ($-3°C$), high-pressured (350 atmospheres), and permanently dark waters of Lake Vostok are believed to have remained liquid and isolated for more than ~400 000 years and may be among the least favorable places for life to exist on Earth (Siegert et al., 2001). Nutrients are severely limited, and there are no hydrothermal systems to drive a chemosynthetic ecosystem (Jean-Baptiste et al., 2001). Nevertheless, Lake Vostok is believed to contain a viable microbial community. Various microorganisms, including bacteria, yeast, fungi, algae, and unidentified species, some of which are viable, occur within the overlying glacial ice (Abyzov et al., 1998). Between the glacial ice and the liquid lake water (>3539 m deep), there is an ice transition formed from refrozen lake water (Jouzel et al., 1999). Microorganisms closely related to extant members of the α- and β-*Proteobacteria* and the *Actinomycetes* occur in this lake ice, and many are viable (Karl et al., 1999; Priscu et al., 1999). Ice coring ceased in 1998 within 120 m of the water surface until protocols for sampling can be established to prevent contamination by organisms foreign to the lake.

Atmospheric oxygen levels were sufficient by 1000 Ma to support multicellular organisms. The first undisputed multicellular eukaryotic fossils are bangiophyte red algae preserved in silicified carbonates on Somerset Island, Canada, dated from 1260–950 Ma (Butterfield et al., 1990). Megafossil remains of multicellular *Chlorophycean* plants occur in 1000-Ga Suket Shales in Central India (Kumar, 2001). Fossilized multicellular algae and slime molds are well preserved in 1000-Ma mudstones of the Lakhanda Group, Siberia, and in younger 750-Ma shales of the Svanbergfjellet Formation, Spitsbergen (Knoll, 1992). Fossils in the Bitter Springs Chert, Australia (~900 Ma), may contain fungal hyphae and algae in various stages of subdivision (Schopf and Blacic, 1987). Siliceous scales of chrysophyte algae from ~610 Ma and calcareous algal fossils in slightly younger rocks indicate that biomineralization was established by the Precambrian-Cambrian boundary. Acritarchs and protists evolved from simple spheroidal morphologies to highly diverse forms with spines and flanges. Acritarch diversity maximized after the Vargner Ice Age (590–610 Ma) but declined with the rise of Ediacara fauna. The latest Proterozoic acritarchs returned to simple, spheroidal forms.

Single-celled eukaryotes diversified rapidly during the Neoproterozoic Era. However, the fossil and biochemical records indicate that these organisms evolved in the Archean Eon and acquired chloroplasts by 1.9 Ga. The reasons for this delay are unclear, although the rapid diversification corresponds approximately to increased atmospheric oxygen and the development of sexual reproduction (Knoll, 1992). Rapid carbon burial near the end of the Neoproterozoic Era increased the oxygen levels to near PAL.

Multicellular animals were among the last organisms to differentiate during the Neoproterozoic Era, probably before 1.0 Ga. The oldest, unequivocal preserved fossils are impressions with radial symmetry from the Mackenzie Mountains, Canada, dating at ~610 Ma (Hofmann et al., 1990). The Ediacara fauna consisted of radically symmetrical, soft-body organisms that dominated the Late Neoproterozoic Era (575–543 Ma). Their fossils are named after the Ediacara Hills of South Australia, where they are superbly preserved. These organisms may represent ancestors that evolved later into invertebrate animals or early experiments in multicellular animal design that failed to survive (Narbonne, 1998). Ediacara fauna probably consisted mainly of bottom-dwelling organisms, many attached permanently to the seafloor. Increased diversification of these organisms occurred during the last 20 million years of the Neoproterozoic Era but ended abruptly. While there are isolated reports of Ediacara in the Cambrian Period (Conway, 1993; Jensen et al., 1998), these unique early animals disappeared from the fossil record at the Precambrian–Cambrian boundary. Their apparent extinction may have resulted from competition and predatory grazing by skeletal animals (McMenamin, 1986), global climatic and geochemical perturbations (Bartley et al., 1998), or changes in the depositional environment (elimination of bacterial mats) that prevented subsequent preservation (Jensen et al., 1998).

Ediacara fauna were not the only animals to evolve in the Late Neoproterozoic Era. Minor trace fossils of burrows that accompany the Ediacara fossils indicate bilateral, burrowing animals (Narbonne et al., 1994). Cnidaria (anemones) (Gehling, 1988), porifera (sponges) (Gehling and Rigby, 1996; Li et al., 1998c), and echinoderms (e.g. sea lilies, sea urchins, and starfish) (Gehling, 1987) all occur in the Neoproterozoic Era (Figure 17.4).

The fossil record indicates that animals began to evolve in the Late Neoproterozoic Era and underwent major divergence during the Cambrian Period. Some studies question this long-standing theory and suggest that animals evolved much earlier in the Proterozoic Eon. One line of evidence for an early divergence is from molecular clocks (Figure 17.5). By examining genome databases for specific proteins, the genetic sequence of individual proteins can be compared across phyla. Some genes appear to mutate at a nearly constant rate. This rate is calibrated using an evolutionary divergence that is well established in the fossil record, such as the time when the reptilian ancestors split into bird and mammal lineages (310 Ma). Using molecular clocks based on the analysis of seven genes, Wray et al. (1996) estimated that metazoan divergence of protostomes and deuterostomes occurred ~1.2 Ga and that chordates diverged from the echinoderms ~1.0 Ga. Animals are classified as protostomes or deuterostomes based on embryonic development of the initial opening of the primitive digestive tract (the archenteron). In protostomes, this initial opening develops into the mouth and a later opening becomes the anus. In deuterostomes, this is reversed. These conclusions were challenged by Ayala et al. (1998), who examined the sequences of 18 different genes, six of which overlap with the Wray et al. study. They concluded that the divergence of protostomes and deuterostomes occurred in the Late Neoproterozoic Era, around 544–700 Ma, and that the divergence of echinoderms and chordates slightly preceded the Cambrian Period. Wang et al. (1999) examined 50 genome sequences and concluded that the plants, fungi, and animals diverged at 1576 ± 88 Ma. Further up the phylogenic tree, the nematodes diverged from the arthropods and chordates by 1177 ± 79 Ma. The basal animal phyla (e.g. porifera and cnidaria) must have diverged between these times. Differences in assumptions

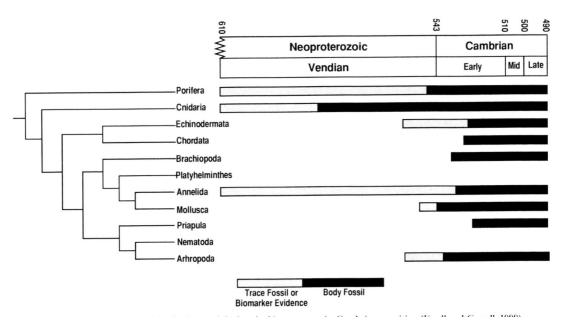

Figure 17.4. Animal diversity and the fossil record during the Neoproterozoic–Cambrian transition (Knoll and Carroll, 1999).

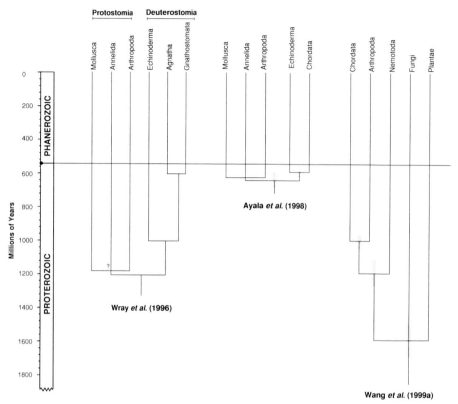

Figure 17.5. Estimated divergence of selected metazoan phyla based on genetic sequences.

and statistical methods used to determine the rate of mutation are responsible for these widely varying times of divergence. The major point of disagreement is whether such evolution rates are heterogeneous between phyla.

These molecular-clock studies suggest that the Cambrian explosion is an artifact of the fossil record. Mineralized animals certainly developed in the Early Cambrian Period and exploded in diversity. The Ediacara fauna were soft-bodied, but they may have had tough, warty surfaces that allowed preservation under the proper conditions (Narbonne, 1998). However, pre-Vendian animals (see Figure 17.4) were likely to have been small and soft-bodied and thus less likely to be preserved. The metazoan diversity recorded in the latest Neoproterozoic Era and Cambrian Period may result from a combination of ideal fossilization conditions, the advent of hard parts, and larger bodies that allowed better preservation (Bromham et al., 1998). The molecular-clock studies predict that the metazoan phyla

diverged before the Vendian Period, a fact now supported by the fossil record. The divergence of the major metazoans certainly occurred before the Cambrian explosion, because sponges and even triploblast bilateralians, such as *Kimberlella*, are known from the Late Neoproterozoic Era (Li et al., 1998c; Xiao et al., 1998; Fedonkin and Waggoner, 1997).

The metazoan phyla probably diverged gradually within the Proterozoic Eon, and gaps in the fossil record greatly reduce the chances of finding remains of particular types of organisms in certain periods (Bromham et al., 1998). Seilacher et al. (1998) reported trace fossils in Chorhat Sandstone of India dating ~1.0 Ga. These bedding plane features are interpreted to be the burrows of worm-like animals grazing below microbial mats. Such trace fossils are controversial but suggest a Proterozoic divergence for the animal lineages. Rasmussen et al. (2002) found discoidal impressions and trace fossils in tidal sandstone from the Stirling Range Formation of southwestern Australia that are dated at >1.2 Ga. The discoidal structures resemble fossils linked to motile Ediacara biota.

Global Anoxia at the Precambrian–Cambrian Boundary

A negative $\delta^{13}C$ anomaly occurs worldwide at the Precambrian–Cambrian boundary. This isotopic excursion was once thought to correspond with the Cambrian explosion, marking the extinction of soft-bodied Ediacaria biota and the rise of shelled fauna. These faunal changes are now recognized to have been much more gradual, ranging from the Late Neoproterozoic Era to the Early Cambrian Period (Bromham et al., 1998; Babcock et al., 2001). Kimura and Watanabe (2001) found that the Precambrian–Cambrian $\delta^{13}C$ anomaly correlates in many locations to low Th/U ratios that indicate oxygen-deficient conditions during deposition. They suggested that these data support a global oceanic anoxic event (OAE). Hence, the mass extinction of Ediacara fauna could be attributed to environmental conditions, rather than to evolutionary pressure.

Proterozoic–Neoproterozoic source rocks

Because of extreme age and the likelihood of deep burial or leakage, preservation of volumetrically significant petroleum generated from Proterozoic or Neoproterozoic source rocks is unlikely. Nonetheless, some crude oils originate from very old source rocks (see Figures 18.6 and 18.10), such as Yurubchen oil from eastern Siberia (see Figure 18.8) and certain oils from Oman (e.g. see Figure 18.12) and the McArthur Basin in Australia (see Figure 18.7).

AGE OF THE METAZOANS: THE PHANEROZOIC EON (544 MA–PRESENT)

Cambrian Period (544–505 Ma)

With the break up of Pannotia, continental landmasses moved from the southern pole to more temperate latitudes (Figure 17.6). Laurentia and Siberia remained around the Equator for most of the Cambrian Period, and the fauna and lithologies reflect warm-water, carbonate-platform environments. Baltica, located in middle to high southern latitudes, had diverse

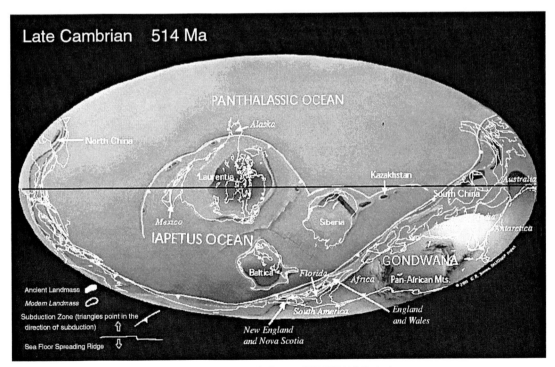

Figure 17.6. Paleo-plate reconstruction of the Cambrian (C. R. Scotese, PALEOMAP Project).

cold-water fauna and sediments ranging from those of nearshore to deep water. Gondwana, the surviving core of Pannotia, extended from the low northern latitudes to the high southern latitudes. Sedimentary rocks and fauna associated with Gondwana reflect a wide range of climates and environments. Sandstones and shales were deposited in many of the terranes. Terranes are fault-bounded fragments of crust that have histories very different from those of neighboring crustal segments. Tectonic movements can merge terranes of varying ages into a single landmass. Rapid changes in sea level, marked throughout the Cambrian Period by extensive unconformities, may have contributed to the extinction of many shelf organisms.

Following the massive ice ages of the Neoproterozoic Era, widespread warm-water carbonate deposits suggest a warm, equitable climate during the Cambrian Period. The absence of Cambrian glacial deposits contrasts with their widespread occurrence in the upper Precambrian age. Glaciers became common again during the Ordovician Period as Gondwana moved over the South Pole. The lack of either land or landlocked seas at the Cambrian poles prevented the accumulation of polar ice caps.

Metazoans diversified rapidly during the Cambrian Period, particularly fauna with hard body parts (Knoll and Carroll, 1999). All present-day marine animal phyla appeared in the Cambrian fossil record, except Bryozoa, which first appeared in the Early Ordovician Period. In the early Cambrian Period, calcified archaeocyathan sponges and mollusks are common. Archaeocyathids formed the first skeletal framework reefs and became extinct in the Late Cambrian Period. Algal stromatolites staged a brief resurgence in the Late Cambrian Period, possibly due to a decline in grazers, such as gastropods. Beginning in the Lower Cambrian Period (Atdabanian Stage), trilobites underwent adaptive radiation and diversified quickly.

Note: Diaminopimelic acid (DAP) is a biomarker for eubacteria, including cyanobacteria. It is part of peptidoglycan, a component of the bacterial cell wall. DAP has two asymmetric carbons, which allows it to assume three different stereochemistries: L-, D-, and meso-configurations. Most bacteria use the meso-form within the muropeptide. DAP has been reported to occur in recent marine sediments and soils, but until recently it was not thought likely to survive early diagenesis.

$$H_2N-\underset{\underset{OH}{|}}{\overset{\overset{H}{|}}{C}}-CH_2-CH_2-\underset{\underset{OH}{|}}{\overset{\overset{H}{|}}{C}}-NH_2$$
$$O=CC=O$$

Borruat et al. (2001) reported on the presence of DAP in modern stromatolites from Lake Thetis, Australia, and ancient Precambrian stromatolites (890 Ma) from the Taoudenni Basin, Mauritania. The concentration of DAP in the Lake Thetis sample was $\sim$50 µmol/gram of rock, equivalent to $\sim 9 \times 10^9$ *Escherichia coli* cells/gram of rock. About 86% of the DAP was in the meso-form, essentially the same as recent marine sediments. In contrast, the ancient stromatolite contained only $\sim$42 pmol/g, equivalent to $\sim 7 \times 10^6$ *E. coli* cells/gram of rock. Only $\sim$70% of the DAP was in the meso-form, with 30% in either the L- or D-configuration. Borruat et al. believe that this 70:30 ratio represents the isomeric equilibrium associated with the diagenetic conditions of this specific calcareous rock. Although appropriate procedures assured that the DAP in the Mauritanian stromatolite was not surface contamination, there was no way to determine whether the DAP was syngenetic or migrated into the rock at a later date.

Cambrian source rocks

Most Cambrian source rocks occur in the Baltic region (Alum Shale) (see Figure 18.17) and in the UK (Clara Group, Stockingford Shales), North Africa, Australia (see Figure 18.16), and Russia. These source rocks were deposited in a semi-restricted tropical to semi-tropical ocean mainly between the continental blocks of Baltica, Gondwana, and Laurentia (Figure 17.6). In particular, Late Cambrian deposition occurred near the peak of a first-order eustatic sea level rise that began in Tommotian (Early Cambrian) and peaked in Llanverian (Middle Ordovician) time (see Figure 17.25).

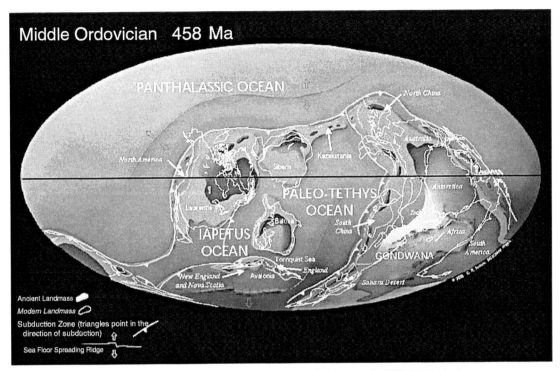

Figure 17.7. Paleo-plate reconstruction of the Middle Ordovician Period (C. R. Scotese, PALEOMAP Project).

Ordovician Period (505–438 Ma)

Throughout the Cambrian and Ordovician Periods, Laurentia and Siberia remained in the tropics and were largely submerged. Baltica moved into southern low latitudes. Gondwana, the largest landmass, extended from the southern pole to the tropics. Toward the end of the Ordovician Period, the portion of Gondwana that is the modern northern and central parts of Africa drifted close to the south pole, resulting in glaciation and lowering of sea level by as much as 100 m.

Restricted ocean circulation resulted in vast areas of tropical shelf seas with anoxic bottom waters during the Ordovician, which helped to preserve organic matter (Figure 17.7). Graptolites appeared to prefer to feed in dysoxic or suboxic waters and consequently are common fossils in petroleum source rocks. Upwelling zones occurred on the westward edges of Laurentia, Siberia, and the Middle East–South China region during much of the Ordovician Period. Restricted circulation within the Paleo-Tethys Sea also favored source-rock deposition on the bounding landmasses.

Carbonate deposition occurred in the equatorial regions, commonly associated with organic-rich shale facies dominated by *Gloeocapsomorpha prisca* (Douglas *et al.*, 1991; Hoffmann *et al.*, 1987), an extinct alga with unusual petrographic and geochemical characteristics (see Table 18.6) that is believed to be related to *Botryococcus* (Metzger and Largeau, 1994). Most siliciclastic rocks formed in the Southern Hemisphere in nontropical marine shelf environments with cooler surface waters. Source rocks were less likely to be deposited here because the waters were better mixed and more oxygen-rich.

A mass extinction of trilobites at the end of the Cambrian Period opened marine niches to new life forms. Brachiopods increased in diversity. Bryozoans, crinoids, tabulates, and tetracorals appeared for the first time. Jawless armored fish, which first appeared in the Late Cambrian Period, evolved in nearshore environments. Trilobites re-established themselves, particularly in clastic, non-tropical shelf seas. Spores suggestive of land plants occurred in the Middle Ordovician Period. A mass extinction affecting all marine animals

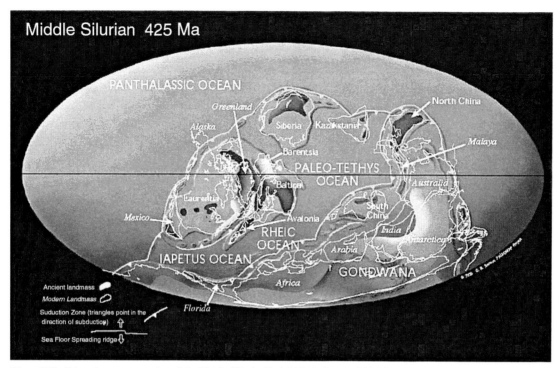

Figure 17.8. Paleo-plate reconstruction of the Middle Silurian Period (C. R. Scotese, PALEOMAP Project).

occurred in the Late Ordovician Period when global sea levels dropped as much as 170 m.

Ordovician source rocks

Petroleum source rocks occur in most Ordovician sedimentary strata in North America, Australia, and the Baltic area (see Table 18.5). Paleo-plate reconstruction places these depositional settings in the tropics. Many of the source rocks are condensed shales within carbonate-evaporite sequences. Compared with other geologic periods, little oil originates from Ordovician source rocks (see Figure 1.8) (Klemme and Ulmishek, 1991), but they are important sources for some petroleum systems. For example, the Albion-Scipio Field, the only giant field in the Michigan Basin, contains petroleum generated from Ordovician Trenton-Black River source rocks (see Figure 18.23). Upper Ordovician (mostly Caradocian) black shales (Vinini, Utica, Sylvan) are widespread in North America, as are the *Dicellograptus* and *Clinograptus* shales in Northern Europe.

Silurian Period (438–408 Ma)

During the Silurian Period, the supercontinent of Gondwana was centered over the South Pole and partially buried under an ice cap (Figure 17.8). A ring of equatorial to mid-latitude continents and terranes surrounded the Tethys Sea. Climate in the Southern Hemisphere was dominated by seasonal monsoons. A large upwelling zone extended along Gondwana's northern boundary. These conditions favored deposition of graptolitic shale source rocks across northern Africa and Arabia. Laurentia, much of which was flooded, straddled the Equator. Clastic source rocks were deposited in several basins (e.g. Permian and Anadarko) on the flooded shelf. Barentsia, a microcontinent that includes present-day Svalbard, collided with the part of Laurentia that is now eastern Greenland. A narrow passage of the Iapetus Ocean separated Baltica and Laurentia. Avalonia, a microcontinent consisting of present-day portions of northern Europe and coastal New England and Canada, joined with Baltica. Source rocks were deposited on the Avalonia and Baltica carbonate shelf.

Age of the metazoans: the Phanerozoic Eon (544 Ma–present)   723

Siberia moved into northern mid-latitudes, but was rotated 180 degrees from its present orientation.

The relatively mild climate of the Silurian Period marks a time of faunal recovery following the Late Ordovician mass extinction. Small fluctuations (30–50 m) in global sea level continued throughout the Silurian Period but had little influence on the diversity of shelf fauna, and only minor extinction and radiation events occurred. Land plants became established by the Middle Silurian Period, and the first air-breathing animals (millipedes and other insects) evolved.

Silurian source rocks

Lower Silurian source rocks and related crude oils are widespread in North Africa (Tanezzuft and equivalents) (see Figure 18.30), the Middle East (Qusaiba Member) (see Figure 18.31) and Europe (*Rastrites* Shale, *Crytograptus* Shale; see Figure 9.13), South America (Trombetas Formation), and North America (Ordovician-Silurian Utica Shale). Source rock deposition was associated with a major second-order eustatic cycle that began in the Early Silurian Period and maximized by the end of the Silurian Period (see Figure 17.25).

Devonian Period (408–355 Ma)

Near the beginning of the Devonian Period, Baltica and Laurentia collided, forming the Caledonian Mountains and causing extensive volcanic activity (Figure 17.9). This Euramerica supercontinent is sometimes called Laurussia or the Old Red Continent because of the prevailing reddish-colored clastic sediments deposited in England, Scotland, the Ardennes, and the Rhenish Mountains. Several of the Asian microplates joined together when Siberia and Kazakhstania collided to form the Ural Mountains. Greenhouse climatic conditions dominated in the Devonian period. Polar ice was greatly reduced, and numerous evaporite basins in the Northern Hemisphere along with widespread deserts and carbonate reefs indicate warm, equable climates. Sea level was high worldwide, and much of the land lay under shallow seas, where tropical reefs flourished. A deep ocean covered the rest of the planet.

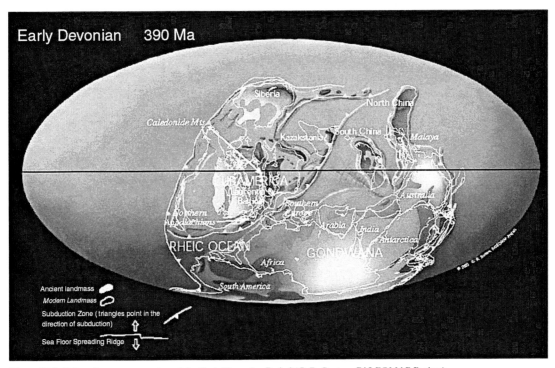

Figure 17.9. Paleo-plate reconstruction of the Early Devonian Period (C. R. Scotese, PALEOMAP Project).

Rapid evolutionary diversification of new fish forms occurred in the Devonian Period, including the Placodermi (primitive rays and sharks), Sarcopterygii (lobe-finned fish and lungfish), and Actinpterygii (conventional bony fish or ray-finned fish). The increase in fish and cephalopod predators may account for the decline in trilobites as well as their tendency toward evolving into giant forms with increased defensive armor. Corals, echinoderms, crinoids, brachiopods, and conodonts were common marine organisms. Sponges evolved siliceous forms, many of which are similar to extant species. Brachiopods reached their zenith in numbers and diversity. Ammonoids first appeared, and most of the other mollusks, such as gastropods, bivalves, and nautiloids, continued with little change from the Silurian Period. A major mass extinction occurred during the Late Devonian Period, which severely affected tropical marine organisms.

While the Devonian Period is frequently termed the age of fish, a more accurate description would be the age of land plants. All major groups of land plants, except angiosperms, existed by the end of the Devonian Period (Figure 17.10). Forests with considerable biomass developed, leading to a significant change in the carbon cycle and new types of organic matter being contributed to the sedimentary environment. Various insects and fish-like amphibians rapidly exploited the new land habitats. Although the megafossil record indicates that the Devonian was a time for major diversification of land plants, the bacteria, fungi, and burrowing metazoans had long since colonized the terrigenous environment, developing the soils needed to support plant life.

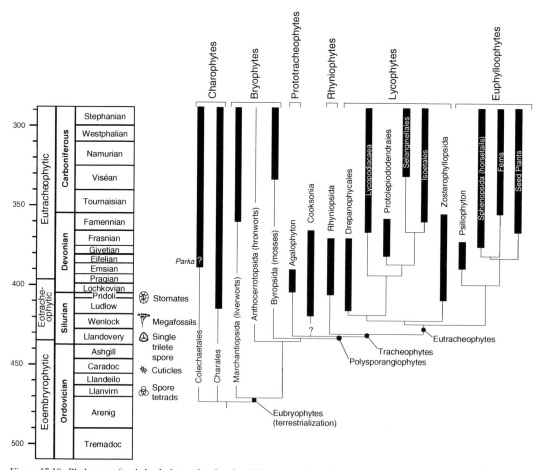

Figure 17.10. Phylogeny of early land plants, showing the minimum stratigraphic ranges based on megafossils (thick bars) and the minimum implied range extensions based on genetic sequences and microfossils (modified from Kenrick and Crane, 1997).

Devonian source rocks

Several marine transgressions and regressions occurred during the Devonian Period, many associated with extinction events and widespread development of anoxic source rocks. The expansion of land plants with deep root systems during the Middle–Late Devonian Period may have resulted in increased chemical weathering and run-off of land-derived nutrients into paralic marine settings (Algeo and Scheckler, 1998; Algeo *et al.*, 2001). This enhanced productivity may have caused excess biological oxygen demand and increased anoxia in bottom waters, thus improving organic matter preservation. Source-rock deposition was particularly pronounced during the Late Devonian Period (mostly Frasnian) (see Figure 17.23), which is associated with a first-order eustatic sea-level rise and the most widespread epeiric seas of the Phanerozoic Eon.

Major Upper Devonian source rocks were deposited throughout the tropics in semi-restricted epeiric seas, especially along the western edges and within the shallow inland sea of Laurentia. These source rocks and related crude oils occur in the present-day Timan-Pechora, Caspian, and Volga-Ural basins (Domanik Formation) (see Figure 18.36), the Dnieper-Donets and Pripyat basins in East Siberia, Alberta Basin (Duvernay and Exshaw shales) (see Figure 18.37), the Northwest Territories (Imperial and Canol shales), and the carbonate settings of North America (Woodford, Chattanooga, New Albany shales) (e.g. see Figure 18.44). In high southern latitudes, strong upwelling deposited algal-rich source rocks on the western edge of Gondwana (Barreirinha and Pimenteriras formations and equivalents in the Amazon and Maranhao basins and the Tomachi Formation in Bolivia) (Figure 4.19).

Evolution of land plants

The transition by plants from marine to terrigenous habitats required new mechanisms for moisture preservation, nutrient uptake, reproduction, structural support, and protection from solar radiation. Cyanobacteria in stromatolitic colonies developed some mechanisms to survive exposure to the atmosphere, but they were unable to grow or reproduce in non-aquatic environments. Nearly all land-plant adaptations involved the management of water. In response to desiccation, plants evolved cuticle, a waxy covering of cutin that reduces water loss to the air. Stomata regulate gas exchange (transpiration) while minimizing the loss of water vapor. The need to transport water and nutrients from the soil to aboveground tissue demands short distances, limiting height (as in the Byrophytes), or cellular differentiation to form a vascular system (tracheids). Symbiotic relationships between plants and mycorrhizal fungi assisted uptake of nutrients. Without the buoyancy of an aqueous environment, plants required new substances, such as cellulose and lignin, for structural support. Reproduction on land required the evolution of spores covered with another new substance, sporopollenin, to allow dispersal and preservation.

Note: These new evolutionary adaptations are the principal reasons for the fundamentally different chemical compositions of marine planktonic algae and terrigenous higher plants. Marine plankton consist mainly of proteins (up to $\sim$50%) and lipids (5–25%), with less than $\sim$40% carbohydrates, while land plants are mainly cellulose (30–50%) and lignin (15–25%) (Tissot and Welte, 1984). Because land-plant organic matter is more aromatic, poor in hydrogen, and rich in oxygen, it has generally lower oil-generative potential (type III) (Figure 4.4) than marine planktonic organic matter (type I or II).

Before the Devonian Period, most biomass consisted of algae and bacteria without significant input from higher plants (e.g. lignin and its derivatives), cuticles, spores, pollen, diatoms, or forams. The *n*-alkanes with more than $\sim$25 carbon atoms originate mainly from waxy coatings on leaves. These *n*-alkanes are rare in pre-Devonian rock extracts and lack a carbon-number predominance in this range. Higher-plant biomarkers, such as oleanane, are absent.

In some classifications, all green plants fall into two major super-phyla: Stretophyta (land plants and their closest green algal relatives, the green algal charophytes) and Chlorophyta (the remaining green algae). Genetic sequencing identifies a flagellated green alga, *Mesostigma viride*, as representing a lineage that emerged before the Stretophyta–Chlorophyta split (Lemieux *et al.*, 2000). Comparisons with extant organisms suggest that land plants evolved from charophytes (the green algal stoneworts), whose fossil record dates only to the Late Silurian Period, although they evolved earlier. Some present-day charophytes (e.g. *Fritschiella*) are remarkably plant-like, having an erect stem on a rhizoidal base. In addition to a high degree of similarity between nuclear DNA and ribosomal RNA, there are many developmental and biochemical affinities that relate the charophytes to land plants (Kranz et al., 1995; Manhart and Palmer, 1990; Mishler et al., 1994). Both produce β-carotene and use chlorophyll b inserted into thylakoid membranes. Unlike other green algae, charophytes have cell walls composed of 20–25% cellulose, critical for structural support. Mechanisms of cell division in the charophytes resemble land plants in that both use microtubules to move vesicles into position to form the plate that divides the cells. The reproductive behavior of charophytes differs from most algae by allowing a zygote to develop into an embryo while remaining attached to the parent plant. The zygote divides, producing a multicellular sporophyte with many spores from a single zygote.

Genome sequencing shows that all extant land plants evolved from a common charophyte ancestor (Qiu and Palmer, 1999). Some extinct land plants may have evolved from different ancestors. Several enigmatic non-vascular plant fossils are of uncertain origin. *Protosalviania* and *Nematothallus* may have evolved from Phaeophyta (brown algae) (Stewart and Rothwell, 1993). These early, non-charophyte land plants evolved and thrived until the end of the Devonian Period, when they could no longer compete with charophytic plants. *Prototaxites*, for example, is a large trunk-like plant up to 1 m in diameter and 2 m in length found in Upper Silurian to Upper Devonian strata. *Prototaxites* is anatomically distinct from tracheophytes, Bryophytes, algae, and fungi. Their closest affinity is to *Nematothallus*.

Bacteria, lichens, mycorrhizal fungi, and Bryophytes (non-vascular land plants, e.g. mosses and liverworts), were essential for developing the soil required for mineral uptake among the Tracheophytes (vascular plants, e.g. club mosses, ferns, horsetails, psilophytes, and seed plants). Recent molecular data and fossils suggest that mutualistic associations between phototrophs and fungi (e.g. lichens) arose repeatedly in branches of parasitic, mycorrhizal, or free-living saprobic fungi (Gargas et al., 1995). Some of these associations are ancient, perhaps predating charophyte-derived land plants (Selosse and Le Tacon, 1998). Various morphological and genetic markers indicate that the Hepatophyta (liverworts) were the first group of land plants to evolve from charophytes (Mishler et al., 1994; Qui et al., 1998). Anthocerophyta (hornworts) and Bryophyta (mosses) evolved shortly after the liverworts.

Note: Symbionts are different organisms that live together to their mutual benefit. For example, lichens consist of algae and fungi, which grow together as symbionts. Root nodules of leguminous plants, such as soybeans, peas, and alfalfa, commonly contain symbiotic bacteria of the genus Rhizobium. These nodules are sites of nitrogen fixation, where atmospheric N_2 is converted into combined organic nitrogen, as found in the chlorophyll and amino acids that are critical to sustain plant life. Without the ability to fix nitrogen, legumes would be less likely to survive in unfertilized soils. Neither legumes nor Rhizobium alone are able to fix nitrogen.

All land plants reproduce sexually by alternating between haploid (gametophyte) and diploid (sporophyte) forms. The haploid stage is generated by meiosis, the diploid stage by fertilization, and both are capable of cellular reproduction via mitosis. Originally, the diploid and haploid generations of the ancestral land plant had similar morphologies. As land plants evolved, two branches developed as the morphologies of the diploid and haploid generations differentiated. In Byrophytes, the gametophyte became dominant, with the sporophyte stage parasitically dependent on the gametophyte. The other lineage, the Polysporangia, developed as the sporophyte generation increased in size relative to the gametophyte stage. Polysporangia differentiated into several extinct and extant lineages. In Spermatophytes (seed-bearing vascular plants), the haploid generation is completely dependent on the sporophyte. The female gametophyte is reduced to just seven cells, one of which is the egg, and the male gametophyte is the pollen grain of just three cells when mature.

The fossil record for the charophytes (land plant precursors) dates from the Late Silurian Period for the Charales (Figure 17.10). Fossilized remains of Colechaetales date from the Middle Devonian Period if one accepts *Parka*, an enigmatic plant fossil, as an ancestral Colechaetales. The liverworts were the earliest land plants, yet their megafossil record dates only to the Late Devonian Period. Traditional studies of early plant evolution focus on the Tracheophytes and their radiation at the Silurian–Devonian boundary (e.g. Knoll and Niklas, 1987). This disparity in the fossil record between Tracheophytes and Bryophytes is due mainly to enhanced preservation of megafossils with lignin compared with cutin. Gray (1993) found land-plant spores in marine strata as old as Middle Ordovician (~476 Ma). As for all land plants, the Bryophytes produce spores encapsulated in sporopollenin that are more likely to be preserved than floral tissues.

By considering fossil records of both spores and megafossils, Gray (1993) proposed that land-plant evolution can be divided into three bio-events: Eoembryophytic, Eotracheophytic, and Eutracheophytic (e.g. see Figure 17.25). These bio-events are decoupled from conventional stratigraphy that is based mostly on marine organisms. The Eoembryophytic Epoch (Middle Ordovician–Early Silurian) is characterized exclusively from fossilized tetrad spores and a few plant fragments. During this period, the first terrigenous plants evolved from charophycean algae and the land-plant lineage was established. Liverworts were the first land plants to split from that common lineage. The Eotracheophytic Epoch (Early Silurian to Early Devonian) is marked by a transition from the tetrad spore to single trilete spores and the earliest megafossils that are characteristic of basal land plants (e.g. hornworts, mosses, and early vascular plants). The Eutracheophytic Epoch (Early Devonian to Middle Permian) is characterized by increased spore diversity and megafossils of vascular plants.

Note: Vitrinites are a group of organic particles in sedimentary rocks that originate from terrigenous higher plants. Vitrinite reflectance is a petrographic method for measuring thermal maturity that quantifies the percentage of incident light reflected from polished vitrinite particles (Figure 4.12). Because land-plant communities were well developed by Devonian time, vitrinites are an important constituent in Devonian and younger sedimentary rocks.

Because the megafossil record of the early diversification of land plants is incomplete, the oldest known fossils are not necessarily the most primitive. The oldest land-plant megafossil, *Cooksonia*, dates from the Middle Silurian Period (Edwards *et al.*, 1992). *Cooksonia* was a small (~2 cm) quasi-vascular land plant with dichotomous branching and terminal sporangia. The phylogeny of *Cooksonia* is uncertain, but it may relate most closely to the protracheophytes, lycophytes, or rhyniophytes (Kenrick and Crane, 1997). The later group of basal vascular plants is named after the Rhynie Chert, a lowermost Devonian (408–380 Ma) hydrothermal silicate located ~40 km northeast of Aberdeen, Scotland, where early plants are preserved so perfectly that cellular structures remain. The rhyniophytes (Figure 17.10) had the dichotomous branching and terminal sporangia of *Cooksonia* but grew to more than 50 cm in height. They exhibit advanced features, such as stomata, and, although lacking true roots, they had single-cell rhizoids for nutrient uptake. Also within the Rhynie Chert are protracheophytes (e.g. *Aglaophyton* and *Horneiophyton*), which are transitional between the bryophytes and the rhyniophytes. The protracheophytes have some of the characteristics that are unique to vascular plants, such as polysporangia and trilete spores, but lack other distinguishing features, such as tracheids. Discoveries of the gametophyte stages of some of the earliest land plants within the Rhynie Chert enabled the recognition of primitive land-plant lifecycles (Remy *et al.*, 1993). Cody *et al.* (2003) examined fossilized *Asteroxylon*, an extinct early plant from the Rynie Chert, using scanning transmission X-ray microscopy (STXM), a technique that provides chemical information within a spatial context. They found chemically differentiated cell walls consistent with lignin and speculated that the colonization of land may have resulted from a single genetic mutation that allowed primitive plants to make this structural biopolymer.

Drepanophytes are other early megafossils that date from the Upper Silurian Period. These short (~20–30 cm) plants had spirally arranged spines (microphylla) along a central stem that probably served to increase the surface area for carbon dioxide uptake. The spines were not needed for defense because animal herbivores had yet to evolve. For many years, the drepanophyte *Baragwanathia* (named after the fossil beds of Victoria, Australia) was accepted as the oldest land plant. Compared with the contemporaneous megafossils of *Cooksonia*, *Baragwanathia* is more complex and clearly associated with the lycophytes (club mosses) that dominated the Early Devonian Period.

The Early Devonian fossil record is characterized by a strong radiation of land plants into four major groups: rhyniophytes, zosterophyllophytes, lycophytes, and trimerophytes (Figure 17.10). Zosterophyllophytes where short (~15 cm) plants characterized by dichotomously branching axes bearing sporangia on short lateral stalks rather than in a terminal position,

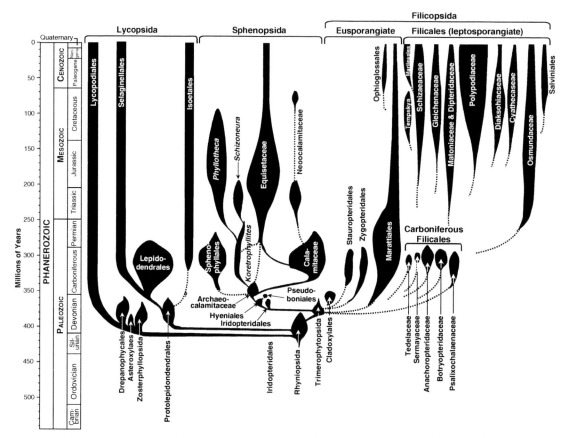

Figure 17.11. Summary of suggested origins of the Lycopsida, Sphenopsida, and Filicopsida phyla of higher plants (Stewart and Rothwell, 1993).

as in the rhyniophytes. Some zosterophyllophytes had microphylla and a curious H-shaped branching habit. The zosterophyllophytes are considered intermediate between the rhyniophytes and the lycophytes. The lycophytes have dichotomous branching microphylla that are usually arranged, spirally and they have single sporangia born by the upper surface of the microphylla. Tree-form lycophytes evolved by the Late Devonian Period and went extinct in the Permian Period. The trimerophytes were constructed on a central axis with dichotomous lateral branches dedicated to photosynthesis and terminally positioned clusters of sporangia. Psilophyton were diverse trimerophytes common in the lower Devonian Period. One genus, *Pertica*, reached heights over 3 m.

The trimerophytes represent the lineage that gave rise to the sphenophytes, cladoxylophytes, ferns, and seed-bearing plants, all of which appeared by the Late Devonian Period (Figure 17.11). The sphenophytes gave rise to a variety of genera common throughout the late Paleozoic and Mesozoic eras, but they were restricted to the genus *Equisetum* (horsetails) by the Tertiary Period. Cladoxylates appeared in the Early Devonian Period and extend into the Lower Carboniferous Epoch. These plants are proto-ferns and were some of the first plants with true leaves (megaphylla). True ferns appeared in the upper Devonian Period and were common tree forms in the Carboniferous Age.

Progymnosperms, such as *Archaeopteris*, are ancestors of modern gymnosperms. The progymosperms had a frond-like axis system with fan-shaped leaflets and wood remarkably similar to that of primitive conifers. They were heterosporous, having small micro- and much larger macrospores, which led to the evolution of seeds. Seeds are embryonic plants contained within protective coats that represent a major advance in plant reproduction. Seeds can be dispersed over long distances and can remain viable for long periods. The first seed plants, the pteridosperms or seed ferns, appeared toward the end of the Devonian Period. By the Late Devonian Period, the earliest forest stands consisted of tree-formed lycophytes, ferns, and gymnosperms.

After the rapid radiation of land plants in the Devonian Period, the Carboniferous Age is characterized by diversification within existing groups, rather than the appearance of new phyla, and by increased biomass. Although Devonian coals are known, substantial peat accumulations formed in tropical swamps during the Carboniferous Age in Europe and North America. The tree-form lycophytes, such as *Lepidodendron* and *Sigillaria*, were the most common and widespread plants and account for up to 70% of biomass in Carboniferous coals. Despite containing little wood, these plants reached heights of 40 m, with trunks up to 2 m in diameter. Their root systems consisted of shallow, hollow tube-like structures, which restricted the lycophytes to wetland environments. Other important groups of the Carboniferous coal swamps were calamites (sphenopsids), pteridosperms, ferns, and cordaites (early conifer-like gymnosperms). The pteridosperms and cordaites could grow in drier environments.

By the end of the Carboniferous Age, the tree-like lycophytes largely became extinct, with few species persisting locally into the Early Permian Period (Figure 17.12). Tree ferns, seed ferns, cordaites, and the earliest true conifers largely replaced them. Many of

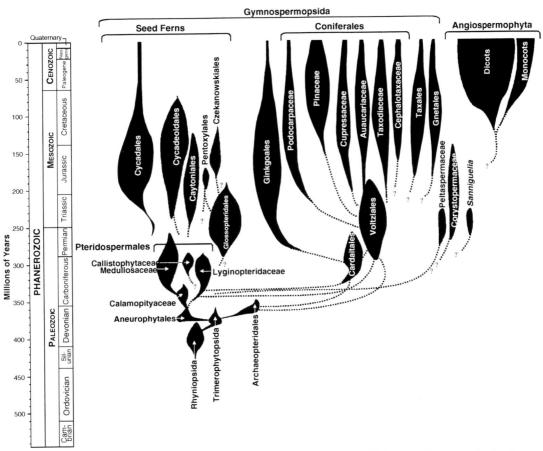

Figure 17.12. Summary of suggested origins of the gymnosperm and angiosperm phyla of higher plants (Stewart and Rothwell, 1993).

the pteridosperms and sphenopsids became extinct and were replaced by gymnosperms, such as Voltziales (transition conifers), cycads, cycadeoids, and Glossopterids. Ginkgophytes appeared in the Permian Period and became increasingly diversified in the Triassic Period. Conifers diversified with the evolution of the compound seed cone.

The angiosperms were the last major land plants to evolve (Figure 17.12). Angiosperm-like pollen is known from the Triassic Period (Cornet, 1989). Unequivocal evidence for both monocotyledon (monocot) and dicotyledon (dicot) angiosperms (leaves, pollen, and flowers) date from the Lower Cretaceous Period. However, the angiosperms were a small component of terrigenous biomass until the Upper Cretaceous and Tertiary periods.

The evolution of land plants is a complex and important issue to biomarker research. Most of the bacterial, algal, and protozoan biochemical pathways evolved by the early Paleozoic Era, whereas land-plant evolution spans the Phanerozoic Eon. The changing biochemistry of the land plants could result in useful biomarkers for age assessment, as exemplified by oleanane as a marker for angiosperms. Since the Silurian Period, type III (terrigenous, gas-prone) organic matter has become an increasingly significant component in sedimentary sequences deposited near land. Further research in land-plant specific biomarkers is needed to extend their utility. The reader is encouraged to examine books by Niklas (1996), Stewart and Rothwell (1993), and Taylor and Taylor (1993) for further details on land-plant evolution.

Carboniferous Period (Mississippian-Pennsylvanian, 355–290 Ma)

During the Early Carboniferous Period, terranes continued to accrete on to Laurentia and Baltica, forming Euramerica (Laurussia) (Figure 17.13). Regional subsidence of the East-European Platform occurred due to compression stress at the margins. Gondwana occupied the Southern Hemisphere, but little of it was over the South Pole. Gondwana continued to drift northward, eventually colliding with Laurasia by the Late

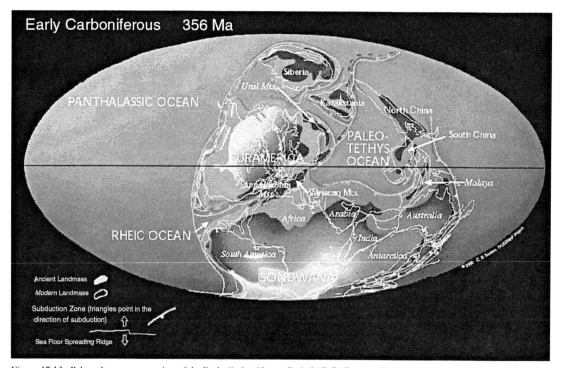

Figure 17.13. Paleo-plate reconstruction of the Early Carboniferous Period (C. R. Scotese, PALEOMAP Project).

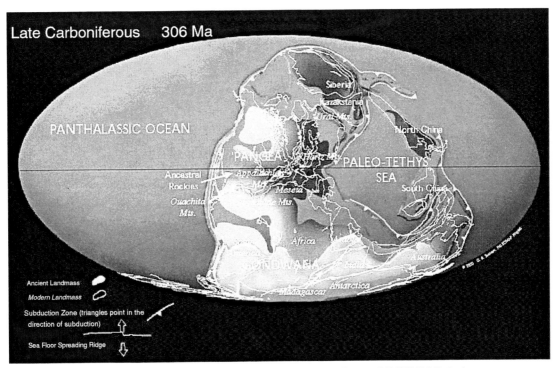

Figure 17.14. Paleo-plate reconstruction of the Late Carboniferous Period (C. R. Scotese, PALEOMAP Project).

Carboniferous Period (Figure 17.14), forming fold belts and the Hercynides Mountains in Central Europe, the Appalachians in eastern North America, the Mauritanides in North America, and the Ouachita (Marathon) Mountains in southern North America. With the merging of these two continents, Pangea emerged as a supercontinent. Siberia, Kazakhstania, China, and Southeast Asia remained as separate micro-continents.

The Mississippian Period was mostly warm and equitable. The equatorial continents were low-lying and frequently swamped. A pronounced cooling and southern glaciation occurred during the Pennsylvanian Period. The global climate had a steep latitude-controlled gradient, with very cold high latitudes. Glaciers covered much of Gondwana, extending as far as 30 degrees south latitude and persisting for millions of years. Gymnosperms began to colonize the cooler hinterlands and exhibit seasonal growth rings. Warm, moist wetland forests dominated the tropical low latitudes. Stands of tall tree-form lycophytes, ferns and early gymnosperms, with a lower canopy of seed ferns and true ferns, eventually became major coal deposits. Sphenopsids occupied drier upland environments. Eustatic sea level changes resulting from glacial advance and retreat probably caused the cyclothems of the Upper Carboniferous coal swamps (DiMichele and Phillips, 1995).

In the oceans, coral reefs, brachiopods, ammonoids, and echinoderms diversified. Various conventional fish and sharks replaced the armored placoderm and ostracoderm fish. Among the invertebrates, the trilobites and nautiloids decreased in variety. Calcareous and agglutinate foraminifera and calcareous algae were well represented. Fusulinids, protozoans that secreted a tightly coiled, chambered calcareous test, appeared in the Upper Carboniferous Period and dominated fossil assemblages through the end of the Permian Period. The fusulinids radiated rapidly, and the diversified forms are commonly used to correlate late Paleozoic strata.

On land, many types of insects, spiders, and other arthropods evolved, some reaching huge sizes. The dragonfly-like *Meganeura* had a wingspan of 75 cm, while the stocky-bodied millipede-like *Arthropleura* was 1.8 m long. Labyrinthodont amphibians emerged as a dominant life form, and many different types inhabited the rivers, ponds, and swamps of the Carboniferous tropics. The first reptiles appeared, adapted to a life on land, but they remained minor until the end of the Carboniferous Period.

Carboniferous source rocks
Late Carboniferous (mostly Westphalian) marine source rocks occur in North America (Atoka Limestone, Chainman, Cherokee, and Excello shales), South America (Amazon Basin), and China (Junggar Basin). These source rocks were deposited mainly during a second-order eustatic sea-level rise that began in Westphalian and peaked in Stefanian time (see Figure 17.25).

Permian Period (290–250 Ma)

During the Permian period (Figure 17.15), Gondwana collided with Euramerica, to which the Angaran sector of Siberia was subsequently fused. This rejoined most of the land masses into one supercontinent, Greater Pangea, for the first time since the Late Proterozoic Eon (Rodinia). By the Middle Permian Period, a single mountainous land mass extended across all climatic zones from one pole to the other. An immense world ocean, Panthalassa, surrounded Greater Pangea. The Tethys Sea was bounded to the east by various microcontinents, island arcs, oceanic plateaux, and trenches. It covered much of what is now southern and central Europe.

The climatic conditions of the Late Carboniferous Period continued into the Early Permian Period. Extensive glaciation persisted in the Early Permian Period, largely in what is now India, Australia, and Antarctica but also in Siberia near the North Pole. The tropics were warm and covered by swampy forests. Continued warming brought on hot, dry conditions. By the Late Permian Period, glaciers retreated and deserts became widespread in the topics and subtropics. Due to the formation of Greater Pangea, sea levels dropped dramatically, drying up much of the previously inundated shallow seas.

The warming Earth at first encouraged evolutionary expansion of shallow-water marine fauna. Many new families or suborders of foraminifera, ammonoids,

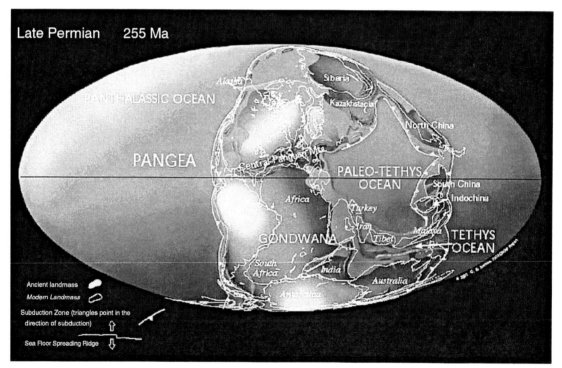

Figure 17.15. Paleo-plate reconstruction of the Late Permian Period (C. R. Scotese, PALEOMAP Project).

brachiopods, bryozoans, and bivalves evolved, and the marine fauna was highly diverse. The shelf fauna on the east and west coasts of Greater Pangea were largely isolated and evolved independently. By the end of the Permian Period, tropical marine fauna suffered a major extinction. Trilobites died out, and only a few genera of bryozoans and brachiopods survived.

Land plants in the Early Permian Period continued the ecosystems of the Late Carboniferous Period, with polar tundra regions and warm, wet tropical swamp forests. The warmer, drier climate eventually eliminated the tropical swamp forests. Tree-form lycopsids and sphenopsids became extinct, and only the small herbaceous (shrub-form) species survived. Ferns, seed ferns, conifers, and ginkgos were better adapted to the drier climate. The gymnosperm *Glossopteris* flourished in Gondwana but was gradually replaced by the seed fern *Dicroidium* as the climate dried. Proto-angiosperms may have emerged in response to the drying conditions.

Increasingly arid conditions greatly influenced the evolution of the vertebrates. Amphibians declined as most swampy environments dried. Reptiles, which emerged in the Carboniferous Period, became the dominant land animals. The Synapsida were the largest and most diverse of the reptile lineages and consisted of two orders: the pelycosaurs and the therapsids. The Early Permian pelycosaurs were both carnivorous and herbivorous. Many early pelycosaurs developed long spines on their vertebrae, which seem to have supported a membrane or sail used to regulate body temperature. The therapsids were diverse and wide-ranging ancestors of the mammals; they had many mammalian traits in dentition, bone structure, and metabolism. The therapsids extended into the Triassic Period, and it is difficult to determine when these mammalian-like reptiles evolved into true mammals. Insects also exploded in diversity in response to the increasing diversification of land plants and animals.

Permian source rocks
Upper Permian (mostly Kungurian) marine source rocks (see Figure 18.15) and related oils occur mainly in North America (Phosphoria, Meade Peak Member, Cache Creek, Leonard Shale) (see Figure 18.48), the North Sea and the Barents Sea (Kupferschiefer) (see Figure 18.49), and basins in South America and Australia (see Figure 18.54). Permian lacustrine source rocks and oils occur in China (Tarim, Junggar basins) (see Figure 18.52) and South America (Paraná Basin).

Triassic Period (250–205 Ma)
The Triassic Period opened with Greater Pangea stretching from pole to pole. Panthalassa covered the rest of the world (Figure 17.16). Scattered across Panthalassa, within 30 degrees of the Triassic Equator, were islands, seamounts, and volcanic archipelagoes, some associated with reef carbonates, that would later be driven into the lithospheric plates on either side of the Panthalassa spreading center as displaced terranes. In the Middle–Late Triassic Period, Greater Pangea began to tear apart, with continental rifts developing between North America and the African portion of Gondwana. Rift basins from the Carolinas to Nova Scotia and along Northern Africa initially received lacustrine sediments but later became restricted marine grabens with thick clastic and evaporite sequences. As rifting continued, a seaway linked the Tethys Sea with Panthalassa by Middle–Late Jurassic time. Triassic mountain building was restricted to the Pacific coastal margin of North America and to China and Japan. Tectonic and igneous activity was prevalent in much of the circum-Pacific and the northeastern portion of the Tethys Sea.

The Triassic climate continued the warm, dry pattern established in the Late Permian Period. The formation of Greater Pangea greatly reduced the shallow sea shelf and led to widespread deserts. Minor coal deposits formed in temperate and tropical latitudes in areas with relatively high rainfall and poor drainage.

The lycopsids, sphenopsids, and tree ferns responsible for the great Paleozoic coal swamps gave way to complete dominance by conifers and other gymnosperms. In Gondwana, *Glossopteris*, which was so prominent in the Permian Period, became extinct and was replaced by the seed fern *Dicroidium* and other gymnosperms. In Laurasia, the flora was a mixture of conifers, cycads, and ginkgos.

Up to 95% of all marine invertebrate species disappeared during the mass extinction at the end of the Permian Period. The Triassic Period opened with many ecological niches ready to be exploited by new animals. During this time, most of the modern invertebrate orders were established. Diverse marine reptiles ruled the warm seas. On land, the ectothermic (cold-blooded) archosaurs attained prominence over

734   Tectonic and biotic history of the Earth

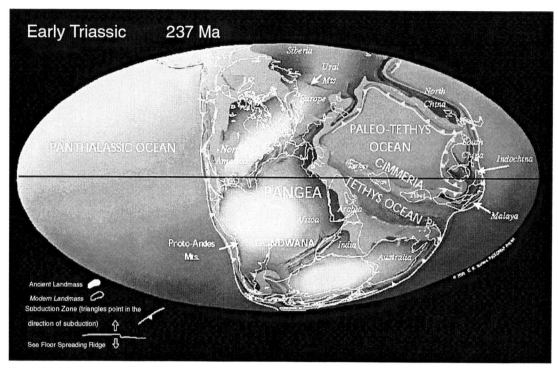

Figure 17.16. Paleo-plate reconstruction of the Early Triassic Period (C. R. Scotese, PALEOMAP Project).

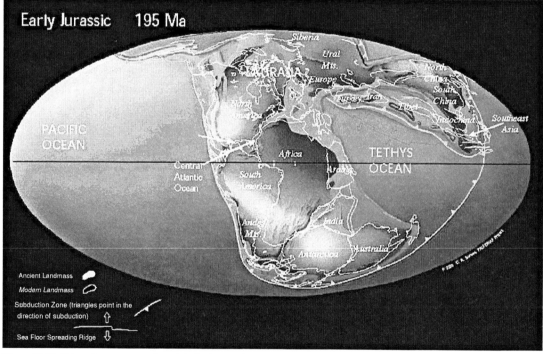

Figure 17.17. Paleo-plate reconstruction of the Early Jurassic Period (C. R. Scotese, PALEOMAP Project).

Age of the metazoans: the Phanerozoic Eon (544 Ma–present) 735

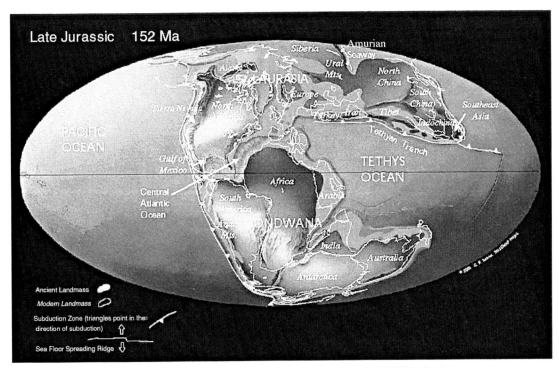

Figure 17.18. Paleo-plate reconstruction of the Late Jurassic Period (C. R. Scotese, PALEOMAP Project).

the endothermic (warm-blooded) therapsids. The first proto-dinosaurs and the first true mammals evolved at this time.

Triassic source rocks

Marine source rocks were deposited during two second-order eustatic cycles during the Middle–Late Triassic (Ladinian–Carnian) Period, primarily in the Arctic. Triassic source rocks and related crude oils occur on the Arctic Slope and Barents Sea (Shublik Formation) (see Figure 18.57), the Eastern Arctic Islands and Western Canada Basin (Doig) in Canada, and to a lesser extent in Eastern Europe (Jura-Bresse).

Jurassic Period (205–140 Ma)

At the start of the Jurassic Period, Greater Pangea was still relatively intact, although divergence along the Atlantic axis had already begun (Figure 17.17). Throughout the Jurassic Period, various plates of the supercontinent began to rotate and move in different directions. As Laurasia drifted westward, the southern part of North America split from Gondwana, forming the Central Atlantic Ocean, a seaway connecting the Tethys and the Pacific (Panthalassic) Oceans (Figure 17.18). The restricted marine environments created with the opening of the paleo-Gulf of Mexico produced the Louanne salt and the Smackover carbonate source rocks. Plate collisions took place around the Pacific rim. The westward movement formed the North American Cordillera, including the Sierra Nevada and the Rocky Mountains. The western margin of South America accreted several microplates and exotic terranes and continued building the Andes, which had started to form in the Triassic Period. Mountain building took place when the eastern margin of Asia collided with the Pacific Plate and when Cimmeria collided with Laurasia.

In the south, the separation of South America from Africa, which began in the Triassic Period with a release of massive flood basalts, continued with the formation of numerous rift basins accompanied by marine transgressions. Eventually, a long narrow seaway, the proto-south Atlantic, separated the two plates. Conditions were ideal for the deposition of lacustrine and restricted marine source rocks. Additional flood basalts marked the separation of Antarctica–India–Australia from southern Africa.

Both poles were land-free during the Jurassic Period, resulting in a generally warm, equitable climate. Rising sea levels established shallow seas in Europe, Western Siberia, Central America, and Arabia. The latter was situated near the Equator in a paleo-upwelling zone within the circulating Tethys Ocean, providing ideal conditions for deposition of marine carbonate and clastic source rocks. The breakup of Europe from North America created various restricted marine circulation patterns, resulting in deposition of the North Sea Kimmeridge Clay Formation and age equivalents (Ziegler, 1988).

Dinoflagellates, radiolarians, coccolithophorids, acritarchs, and foraminifera flourished in the warm tropical seas, accompanied by marine invertebrates, including sponges, coral, bryozoa, gastropods, bivalves, ammonoids, and belemnite cephalopods. Some brachiopods and crinoids survived from the Paleozoic Era. Modern shark groups evolved, and marine reptiles proliferated.

The gymnosperms dominated the land, and cycads were particularly abundant. Conifers continued to be highly diverse, and the extant families were established. Ginkgos were prominent in middle to high latitudes. Seed ferns were successful, but the lycopsids and sphenopsids were relatively insignificant. By the end of the Jurassic Period, only *Equisetum* survived. Coal deposits were extensive in northwestern North America, southwestern Europe, much of Siberia, Asia (including China), and Australia. Many of the continental deposits resulted from Middle-Late Jurassic sea-level fluctuations along the margins of the epicontinental seas, which, in some areas, formed cyclothems. Numerous groups of herbivorous insects evolved, and plants responded with new morphological and biochemical defenses.

The lineage that became birds split and evolved with a dramatic decrease in body size, while the sauropod lineage evolved with an equally dramatic increase in body size. Although the dinosaurs hold a near-universal fascination for both scientists and the general public, their radiation, evolution, and demise have little relevance to biomarker studies. Grazing pressure on the vegetation must have been tremendous, but there is insufficient evidence to clearly show co-evolution of dinosaurs and angiosperms (Sereno, 1999). While the dinosaurs ruled, a few labyrinthodont amphibians struggled on, with most amphibians representing modern orders. Crocodiles were particularly successful, while mammals remained small and limited in numbers.

Jurassic source rocks

Jurassic source rocks account for more than 25% of the total estimated ultimate recovery of petroleum worldwide (see Figure 8.1) (Klemme and Ulmishek, 1991). Figure 18.56 gives some examples of Jurassic source rocks through time. Lower-Middle Jurassic marine source rocks and related crude oils occur in Oman (Marrat Shale), the Arabian Platform and Zagros Fold Belt (Sargelu Shale) (see Figures 18.89 and 18.92). Europe (Lias, Schistes Carton, and Posidonia Shale) (Figure), and Australia (Bridhead Formation). Deposition of these source rocks was tied closely to tectonic breakup along the Tethys–Central Atlantic–Gulf of Mexico rift system. For example, Toarcian source rocks in Europe were deposited during a marine transgression and opening of a seaway linking the Arctic and Tethys oceans (Ziegler, 1988).

Middle–Upper Jurassic source rocks in Saudi Arabia generated substantial amounts of crude oil (see Figures 18.94 and 18.95). Upper Jurassic (Kimmeridgian) marine source rocks are widespread. Rich source rocks and related crude oils occur in the offshore Nova Scotia (Egret Member) (see Figure 18.120), the Gulf of Mexico (see Figure 18.110), the Barents Sea, Timan-Pechora, West Siberian (Bazhenov Formation) (see Figure 18.121), South America (see Figure 18.127), the North Sea and Norwegian Sea (Kimmeridge Clay or Draupne Shale) (see Figure 18.119), Western Europe (Paris Basin) (see Figure 18.63), and Northwest Australia (Dingo Formation) (Figure 18.108).

Cretaceous Period (140–64.5 Ma)

The breakup of Pangea that started in the Jurassic Period continued throughout the Cretaceous Period (Figure 17.19). The Atlantic Ocean widened as Africa and South America split, until marine sequences occurred on both continental margins. Africa moved

Age of the metazoans: the Phanerozoic Eon (544 Ma–present) 737

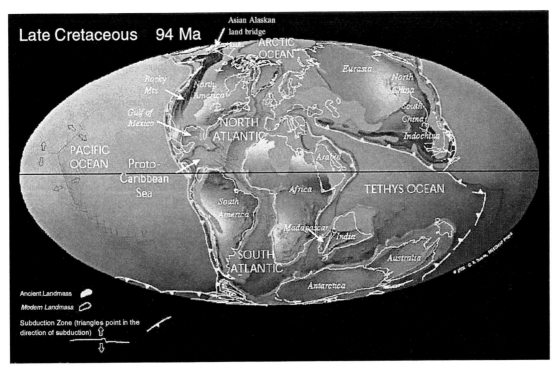

Figure 17.19. Paleo-plate reconstruction of the Late Cretaceous Period (C. R. Scotese, PALEOMAP Project).

northward, closing the eastern Tethys Sea. India broke from Gondwana and also began to drift northward. The Alps formed in southern Europe. High sea levels influenced much of the topography of the Cretaceous continents. Shallow seas covered most of the continental shelves and interior lows. Inland seas inundated much of Europe, Asia, North Africa, and western North America.

The worldwide climate of the Cretaceous Period was much warmer than today, perhaps the warmest in Earth's post-Hadean history. Temperatures were lowest at the start of the period, maximized during the latest Albian Age, and then declined slightly. The poles were free of ice, resulting in only a mild temperature gradient from the poles to the Equator. Extensive evaporites in the tropics indicate that arid conditions prevailed, whereas coals in the mid-latitudes indicate more humid environments with seasonal rainfall. Warm ocean waters and low temperature gradients resulted in sluggish circulation and global bottom-water anoxia, facilitating source-rock deposition.

In the oceans, ammonites, belemnites, and other mollusks were abundant. Diatoms evolved and radiated throughout the Cretaceous seas. Many modern crustaceans, corals, and gastropods appeared, and modern teleost fish became widespread. These fish competed with the ichthyosaurs, which were reduced to a single family by the end of the Cretaceous Period.

On land, the Jurassic ferns, cycads, and conifers continued but were being replaced rapidly by angiosperms. Insects co-evolved with the flowering plants, and many of the modern groups, such as butterflies and ants, appeared and diversified. Dinosaurs were the dominant megafauna on land during the Cretaceous Period, including impressive therapods such as *Tyrannosaurus rex*. Many new dinosaurs and birds evolved and also diversified. New mammalian groups appeared, including the three living groups of mammals (placentals, marsupials, and monotremes). The end of the Cretaceous Period (Figure 17.20) was marked by a major mass extinction, as discussed below.

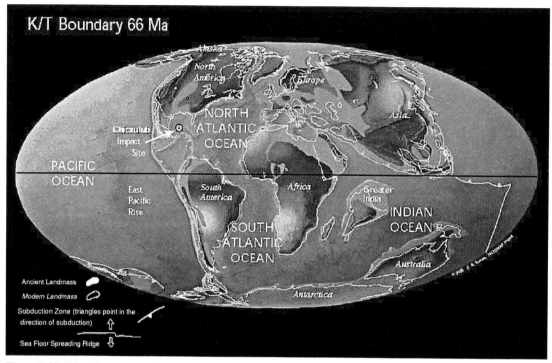

Figure 17.20. Paleo-plate reconstruction at the Cretaceous–Tertiary boundary (C. R. Scotese, PALEOMAP Project).

Cretaceous source rocks

Widespread deposition of marine source rocks occurred during the Cretaceous Period (especially the Aptian–Turonian), and this time period accounts for about 29% of the total estimated ultimate recovery of petroleum worldwide (see Figure 8.1) (Klemme and Ulmishek, 1991). Organic-rich marine shales and limestones were deposited in the North and South America seaways, the Gulf of Mexico, along the South Atlantic basins, North Africa, Southern Europe, the Crimea, Western Siberia, and the North Slope of Alaska. Crude oils generated from Cretaceous source rocks include examples from the western interior of North America (Mowry Shale, Greenhorn Limestone, Frontier Formation, Skull Creek Shale), Gulf of Mexico (see Figure 18.114), North American Arctic ("pebble shale unit," Torok Shale), South America (La Luna Formation (see Figure 18.143), Napo Formation (see Figure 18.144), Villeta Formation (see Figure 18.145), Caballos Formation, Chonta Formation, Ray Formation, Inoceramus Shale)), North Africa (Sirte Formation, Libya (see Figure 18.146), Brown Limestone, Egypt (see Figure 18.147)), western and southern Africa (Nkalagu Formation and offshore Gabon Basin), Arabia (Nahrum Shale, Nahr Umr, Kazhdumi), the UK (Black Band, Mount Holland Shale), Europe (Plenus Marl, Bonarelli Horizon), and Australia (Toolebuc Limestone, Latrobe Formation (see Figure 18.149)). Major lacustrine source rocks developed in China (e.g. Qingshankou Formation (see Figure 18.138)).

Origin of the angiosperms

Angiosperms or flowering plants exploited nearly all terrigenous and non-marine aquatic niches and are highly diversified (>250 000 cataloged species and probably more than one million total species). Angiosperms have flowers and bear seeds surrounded by fleshy fruit. The two angiosperm subphyla, monocots and dicots, bear one and two seed leaves, respectively. A seed leaf (cotyledon) is a leaf-like structure of the seed embryo that provides nutrients during germination. Monocots are generally herbaceous, lacking secondary wood growth, while dicots are both herbaceous and tree-form.

Early theories considered the Magnoliidae as the subclass containing the most primitive living angiosperms. However, subsequent fossil discoveries suggested that the suborder Laurineae and the family Chloranthaceae are more primitive because they have small simple flowers compared with the large cone-like flowers of the family Magnoliaceae (Doyle and Donohue, 1987). Based on this cladistic analysis of morphologic features, the Gnetales were once considered to be a potential sister group of the angiosperms. More recent genome analyses indicate that Nymphaeales (water lilies) are the closest sister clade (Zanis *et al.*, 2002). The genomic molecular data also place the Gnetales within the gymnosperms, indicating that they are not a sister group of the angiosperms (Qiu *et al.*, 1999). These studies show that genomic data for living plants can assist interpretation of the fossil record of extinct plants having similar traits (Soltis *et al.*, 1999).

Fossilized pollen, leaves, and flowers suggest evolutionary radiation of the angiosperms from a few species in the Early Cretaceous Period. These early angiosperms appear to have originated in tropical northern Gondwana and then migrated. All modern angiosperm taxa can be traced back to this evolutionary event (Doyle and Hickey, 1977). Vast gymnosperm forests that dominated the Mesozoic Era were largely replaced by the angiosperms during the Late Cretaceous Period (Figure 17.12). This evolutionary takeover may be linked with the development of symbiotic relationships between angiosperms and animals that facilitated pollination and seed dispersal, and to the evolving seed coat that provided better protection and regulated the timing of germination.

However, recent studies question these observations because older angiosperm fossils are now being discovered. Several angiosperm and angiosperm-like fossils were found in the earliest Cretaceous–Upper Jurassic Period within China (Ge Sun *et al.*, 1998; Shuying, 1998) and the Oxfordian Age in France (Cornet and Habib, 1992). The most spectacular discovery is *Archaefructus sinensis*, discovered in 125-Ma lake deposits of Liaoning Province in northeastern China (Figure 17.21) (Ge Sun *et al.*, 2002). The fossils preserve clear flower-like traits, and phylogenetic analyses indicate that *Archaefructus* is the sister group for all extent angiosperms, including *Amborella*. Ren (1998) reported short-horned flies in Upper Jurassic rocks that are specialized pollinators of angiosperms. These new findings suggest that angiosperms arose from paleoherbs or water plants rather than trees or shrubs (Taylor and Hickey, 1996). The implication of the paleoherb hypothesis is that the key innovations of flowers and a rapid lifecycle were present in the earliest angiosperms. More controversial reports of angiosperm-like fossils in older strata include pollen from the Late Triassic Richmond Rift Basin and Late Triassic fossilized remains of *Sanmiguelia lewisii*, with leaves and flowers that resemble monocot palms (Cornet, 1989; Cornet and Olsen, 1990).

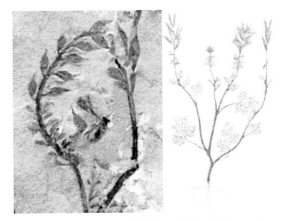

Figure 17.21. Paratype fossil of 25-cm-high *Archaefructus* and reconstruction. Reprinted with permission from Ge Sun *et al.* (2002). © Copyright 2002, American Association for the Advancement of Science.

Bicadinanes in Jurassic (Armanios *et al.*, 1995a) and oleanane in pre-Mesozoic strata (Moldowan *et al.*, 1994a) and crude oils (Peters *et al.*, 1999b) provide indirect evidence for early divergence of the angiosperms (e.g. see Figure 13.94). While the Early Cretaceous Period is still recognized as the age of angiosperm radiation, these plants appear to have split as a separate group in the early Mesozoic Era or even earlier (Taylor and Hickey, 1996). Moldowan *et al.* (2001b) conclude that angiosperms are related more closely to Gnetales than conifers, but that they diverged from Gnetales and other seed plants as early as the Paleozoic Era. They combined morphologic phylogeny, genome cladistics, and pyrolytic generation of oleanoids to suggest that angiosperms may be related closely to gigantopterids.

740 Tectonic and biotic history of the Earth

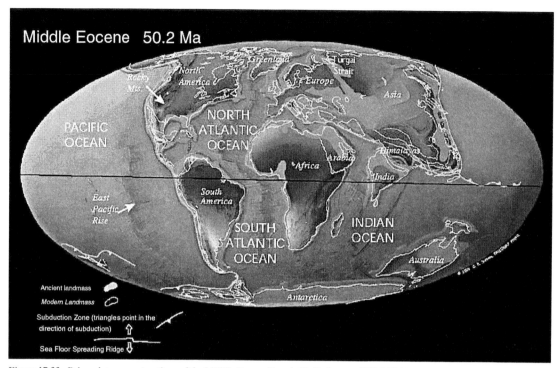

Figure 17.22. Paleo-plate reconstructions of the Middle Eocene Epoch (C. R. Scotese, PALEOMAP Project).

Paleogene Period (Paleocene, Eocene, Oligocene, 64.5–23.7 Ma)

The breakup of Pangea that began in the Triassic Period continued through the Cenozoic Era, causing new oceanic circulation patterns, climates, and migration patterns for terrigenous fauna. India continued to move northward until colliding with Asia in the Middle Eocene Epoch, reducing the Tethys to a seaway. Australia split from Anarctica in the late Paleocene Epoch. Mixing of Arctic and Atlantic waters began in the Early Paleocene Epoch and was completed by the inundation of the Greenland–Scotland Ridge in the Eocene Epoch (Figure 17.22). Before this, animals migrated freely between Europe and North America. Inundation of the Ural Trough ended with a regional uplift at the end of the Eocene Epoch, allowing Eurasian fauna to migrate into Western Europe, an event known as the *Grande Coupure* (Big Break) among vertebrate paleontologists.

The Earth remained warm into the Early Eocene Epoch, with subtropical conditions and heavy rainfall extending into high latitudes. Global cooling began in the Middle–Late Eocene Epoch, accelerating rapidly across the Eocene–Oligocene boundary. The widening seaway between Australia and Antarctica altered oceanic circulation, establishing the circum-Antarctic current that thermally isolated Antarctica from warmer waters and triggered the initial glaciation of Antarctica in the Early Oligocene Epoch.

The extinction event at the Cretaceous–Tertiary boundary terminated many land and marine families, with dinosaurs, large marine reptiles (e.g. mosasaurs ichthyosaurs, and plesiosaurs), and several phyla of marine invertebrate fauna (rudists, belemnites, ammonites) going extinct. Only ~13% of the coccolithophorids and foraminifera families endured. The deep-sea community, however, survived relatively intact until ~54 Ma, when about half the benthic protozoans died out. The cause of these extinctions at the Paleocene–Eocene boundary is unknown but may be related to changing oceanic circulation and warming.

In contrast to the marine environment, the Cretaceous–Tertiary boundary was not marked by significant change in terrigenous flora. The Paleocene

Age of the metazoans: the Phanerozoic Eon (544 Ma–present)   741

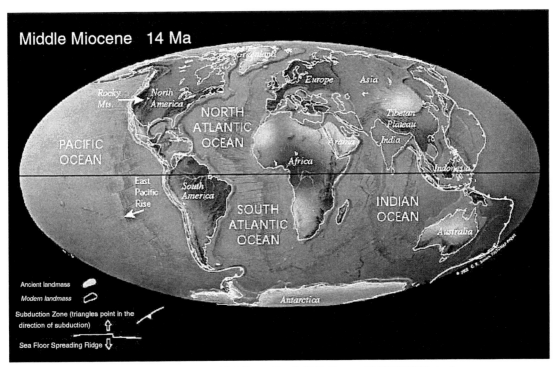

Figure 17.23. Paleo-plate reconstruction of the Middle Miocene Epoch (C. R. Scotese, PALEOMAP Project).

terrigenous flora resembled that of the Cretaceous Period more than of modern times. Angiosperms continued the radiation that began in Cretaceous time. Grasses were present by late Paleocene time but did not expand to form the upland grasslands and prairies until Late Oligocene and Miocene time (Freeman and Colarusso, 2001). With the extinction of the dinosaurs, mammals began to fill the vacated niches. The first lemurs, horses, and rodents appeared in the Paleocene Epoch. Whales may have first appeared in the Late Cretaceous Period, but they were certainly present in the Paleocene Epoch. The Oligocene Epoch was a time of rapid evolution and diversification for mammals. Many of the older megaforms became extinct and were replaced by the modern mammals.

Paleogene source rocks
Eocene (commonly Lutetian) source rocks occur mainly in the circum-Pacific area, including western North America (Green River Formation) (see Figure 18.158), southeast China (Pearl River Mouth Basin), and India (Bombay, Krishna-Goodavari basins). The Eocene Pabdeh Formation generated crude oil in the Dezful Embayment of Iran (see Figure 18.161). Lacustrine oils were generated from the Eocene-Oligocene Brown Shale in central Sumatra (see Figure 18.173) and Oligocene lacustrine shales, offshore Vietnam (see Figure 18.175).

Neogene Period (Miocene, Pliocene, 23.7–2.5 Ma)

Continents in the Neogene Period were similar to their modern configuration (Figure 17.23). Widespread continental uplift and climatic cooling marked the Neogene Period. The Tethys Sea remained open until ∼18 Ma, when the Arabian plate collided with Asia, forming the Zagros Mountains. By ∼ 5.5 Ma, the western portion of the Tethys (proto-Mediterranean) was completely isolated from all oceanic waters. The basin dried up over a period of 0.5 Ma, depositing massive evaporites, and subsequently filled with water from the Atlantic. Organic-rich sapropels were deposited during these events. The tectonic forces that produced the Pyrennes Mountains in the Paleogene Period continued to produce the Alps until ∼9 Ma, when mountain

742 Tectonic and biotic history of the Earth

formation stopped. The Himalayas began to form during the Miocene Epoch as the Indian plate continue to plunge under the Tethys trench, forcing nappes of metamorphic rocks to be thrust over the Indian land mass.

Modern forests and open-vegetation settings, including deserts, tundra, and grasslands, were established. These new habitats allowed further expansion of grazing animals. Most of the animals from the Neogene Period were related closely to extant species. The earliest hominids appeared in east Africa ~5–6 Ma.

Neogene source rocks

Most Neogene source rocks originated in deltaic settings that accumulated mixtures of marine and terrigenous organic matter, including the Kutei (see Figure 18.189) and other Far East basins, the Niger Delta (Figure 18.186), the Mackenzie Delta, and parts of the Gulf of Mexico. However, the San Joaquin and offshore California (e.g. Monterey Formation) (see Figure 18.181)) and Caspian basins contain thick, dominantly marine diatomaceous source-rocks of Miocene age, which was the latest time of major source rock deposition. With a few exceptions, most Miocene source rocks (commonly Langian-Serravalian) occur in the circum-Pacific area, including western North America (see above), western and northern South America (e.g. Oligo-Miocene Heath Formation (see Figure 18.166)), Japan, Korea, southeast China, eastern Russia (Sakhalin), and Southeast Asia, but also in the Caribbean, Spain (see Figure 18.184), southeastern Europe, Pakistan, and Iran.

Quaternary Period (Pleistocene–Holocene, 2.5 Ma–recent)

Sea-level ice sheets developed on west Antarctica during the Early Oligocene Epoch and covered much of the continent by the Middle Miocene Epoch. About 3 Ma, the Isthmus of Panama emerged, separating the Atlantic and Pacific oceans. This event allowed animal migration between North and South America and had profound effects on global climate. The westward North Equatorial Current was now deflected northward, enhancing the Gulf Stream and warming the high northern

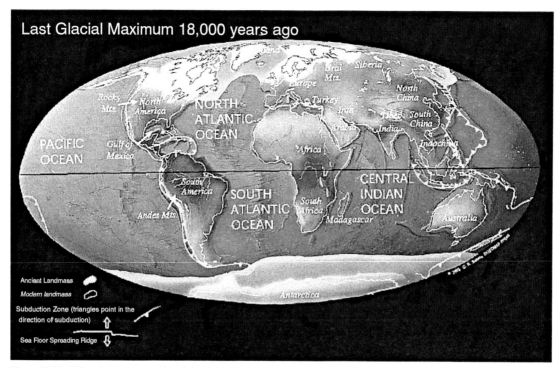

Figure 17.24. Paleo-plate reconstruction of the Pleistocene Epoch (C. R. Scotese, PALEOMAP Project).

Table 17.1. *Intensity of the five major Phanerozoic mass extinctions. Species level estimates based on a rarefaction technique (Hallam and Wignall, 1997)*

Mass extinction event	Families		Genera	
	Observed extinction (%)	Calculated species loss (%)	Observed extinction (%)	Calculated species loss (%)
Late Ordovician Period	25	84	60	85
Late Devonian Period	22	79	57	83
End of Permian Period	51	95	82	95
End of Triassic Period	22	79	53	80
End of Cretaceous Period	16	70	47	76

latitudes. It also contributed to greater precipitation in eastern Canada and Greenland, leading to the development of the Arctic ice cap. The Bering Land Bridge joining Siberia and Alaska was breached only in the past 2.5 million years, allowing the transit of cold-water currents from the Arctic into the Pacific. Only within the past 600 000 years did the Himalayas become the highest mountains on Earth. Repeated glacial advances and retreats in the Northern Hemisphere characterize this time. The earliest members of the genus *Homo* appeared ~2 Ma, and the distribution of land masses was similar to that of today (Figure 17.24).

MASS EXTINCTIONS

Up to 90% of the species that have populated the Earth are now extinct. Rates of extinction, however, have been uneven. Large numbers of species (>70%) perished suddenly (within a few million years) at various times in Earth history. Catastrophes must have occurred that directly impacted over half of the Earth's surface, resulting in the collapse of entire ecosystems (Raup, 1992). Mass extinctions exert control on the evolution of life because new species proliferate and occupy vacated ecological niches in their aftermath.

The Phanerozoic fossil record indicates five major mass extinctions (Table 17.1, Figure 17.25) and as many as 22 lesser mass extinctions (Sepkoski, 1986). However, many of the proposed minor mass extinctions, where 25–70% of the species are terminated, are based on extrapolations and statistical manipulations that are suspect. Recognition of mass extinctions depends on the completeness and chronometric accuracy of the fossil record. Many proposed minor mass extinctions in the Paleozoic Era might be statistical aberrations. For example, Sepkoski (1986) proposed up to five minor mass extinctions during the Cambrian Period. Trilobite, brachiopod, and conodont diversity was severely impacted, but whether these events were caused by global crises is unknown.

The greatest ecosystem collapse recorded in the fossil record occurred at the end of the Permian Period, with the extinction of ~80–95% of all species and >50% of all families. It is the closest that all multicellular life came to extinction. However, the extinction was selective because it targeted species with narrow geographic distributions. Only 35% of the species with global distributions became extinct. The Permian extinction may have occurred in two stages separated by about five million years (Hoffman, 1989).

The second largest mass extinction occurred toward the end of the Ordovician Period (late Ashgill), resulting in the demise of most of the reef-building fauna, one-third of all bryozoan and brachiopod species, and groups of conodonts, trilobites, and graptolites. Collectively, more than 100 families of marine organisms were eliminated. The extinction occurred in two phases separated by about 0.5–1 million years (Brenchley et al., 2001). The major extinction at the Frasnian–Famennian boundary devastated marine life, particularly reef-building organisms. About 75% of the marine invertebrate species were exterminated, but terrigenous plant species were largely unaffected. Several studies have noted that while the Late Ashgill and the Frasnian-Famennian events appear comparable in terms of marine species extinctions, the permanent effects on the global ecology were appreciably different. While the Late Ashgill event produced minimal change, the

744 Tectonic and biotic history of the Earth

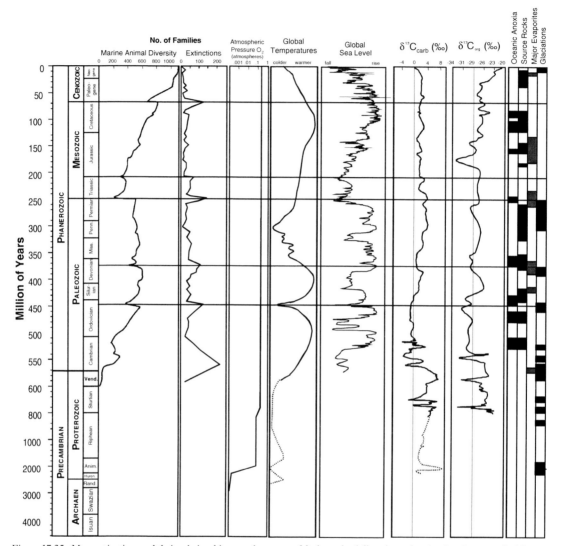

Figure 17.25. Mass extinctions and their relationships to paleo-events. Marine animal diversity and extinction data compiled from Benton (1993b). Atmospheric oxygen curve from Kerr (1999). Paleozoic sea level curve from Ross and Ross (1987). Mesozoic–Cenozoic sea level curve from Haq et al. (1987; 1988). Paleo-temperature curve from Crowley and North (1991). δ^{13}C data from Hayes et al. (1999). Source-rock events from Klemme and Ulmishek (1991).

Frasnian–Famennian event resulted in a complete restructuring of the marine ecosystem (Droser et al., 2000; Brenchley et al., 2001).

The end of the Triassic Period marked the termination of labyrinthodont amphibians, conodonts, and most marine reptiles. Brachiopods, gastropods, mollusks, thecodonts, and mammal-like reptiles also were reduced drastically in diversity (Benton, 1993a). The Triassic–Jurassic boundary marks a major faunal mass extinction, with more than 95% of the megafauna being affected. This extinction allowed the dinosaurs to radiate into vacated terrigenous niches. Extinctions on land appear to have preceded those in the sea by several hundred thousand years (Palfy et al., 2000). The stratigraphic record for the end of the Triassic Period is, however, very poor, and some have questioned whether a true global catastrophe actually occurred or whether the fossil record is more consistent with a gradual decline extending over time (Hallam, 2002; Lucas *et al.*, 2002).

The most notorious Phanerozoic mass extinction at the Cretaceous–Tertiary (K–T) boundary was actually the least severe of the five major extinctions. A meteorite impact is postulated to have killed the dinosaurs, but the ammonoids, rudist bivalves, marine reptiles, belemnites, and pterosaurs also were terminated, along with many species of land plants (Alvarez *et al.*, 1980). Many planktonic species, including foraminifera, calcareous nanoplankton, diatoms, dinoflagellates, and shallow-water brachiopods, mollusks, echinoids, and fish were severely affected. Remarkably, most mammals, birds, and the common extant reptiles and amphibians, ferns, and angiosperms emerged relatively unaffected.

Causes of mass extinctions

Many theories have been proposed to explain mass extinctions and some of their common features (Stanley, 1988). For example, small morphologically simple taxa tend to survive compared with larger and specialized taxa. Certain groups of animals, such as trilobites and ammonoids, experienced repeated episodes of mass extinction. Temperate and arctic species generally resist extinction more effectively than tropical species. On land, the extinction rates for animals are much higher than for plants. These commonalties have led some to suggest that simple triggering events, such as a meteor impact or glaciation, might explain all mass extinctions. However, the proposed single agents either fail to occur during all extinctions or also occur at other times without mass extinctions. Most mass extinctions probably occurred due to multiple cosmological or terrestrial agents that converged in time, resulting in loss of habitat and collapse of biological productivity (Erwin, 1994a; 1994b).

Cosmic rays

Bombardment of the Earth by cosmic rays from a nearby supernova or radiation from a strong solar flare are the only proposed triggering events that might lead directly to mass extinction. The theory was first proposed by Schindewolf (1954), but it fails to explain observed isotopic variations and the preferential extinction of certain species. Nevertheless, a nearby supernova could affect life adversely, and the theory is periodically revived (Ellis and Schramm, 1995; Detre *et al.*, 1998). Short-term solar fluctuations might cause global cooling or warming and ecosystem collapse, but there is little evidence that changes in solar output precipitated past mass extinctions.

Bolide impacts

Large extraterrestrial bodies that impact the Earth can leave craters, trace metal anomalies, shocked quartz, and tektites. Every year, new asteroids in Earth-crossing orbits are discovered, and many cometary bodies are believed to exist in the Oort Cloud. The 1994 impact of the Shoemaker-Levy 9 comet and the long-term disruption of the Jovian atmosphere provided a stunning modern example of a bolide impact. Given sufficient mass and velocity, a bolide impact on Earth could flash-vaporize the oceans, effectively sterilizing the surface so that only subterranean life might survive (Zahnle and Sleep, 1996). Uniformitarianism, where present-day processes are used to explain past geologic events, cannot be applied strictly to meteor impacts because the time represented by human experience is so brief compared with that between major impacts.

Anomalous levels of iridium in sediments at the K–T boundary provided the first direct evidence for a meteor impact that may have caused a mass extinction (Alvarez *et al.*, 1980). The discovery of the buried Chicxulub crater in Yucatan (see Figure 17.20) (Hildebrand *et al.*, 1991) and a piece of the meteorite itself (Kyte, 1998) prove that a catastrophic impact occurred at the K–T boundary. Fullerenes (Heymann *et al.*, 1996a,b), ^{3}He trapped in fullerenes (Becker *et al.*, 2000), and ^{3}He trapped in quartz within boundary clays (Mukhopadhyay *et al.*, 2001), along with a host of shocked minerals and diagnostic osmium and chromium isotopic ratios, provide convincing indirect evidence for a major bolide (see Alvarez (2003) for a summary).

While the shock wave and resulting tsunami were devastating to local flora and fauna, the major impact on the biosphere resulted from disruption of the atmosphere. The bolide collided with a marine carbonate platform, resulting in massive release of CO_2, SO_2, and dust. The impact also triggered a global wildfire that injected massive quantities of soot into the atmosphere (Anders, 1988; Gilmour *et al.*, 1989; Wolbach *et al.*, 1990, Wolbach *et al.*, 2003) (see also

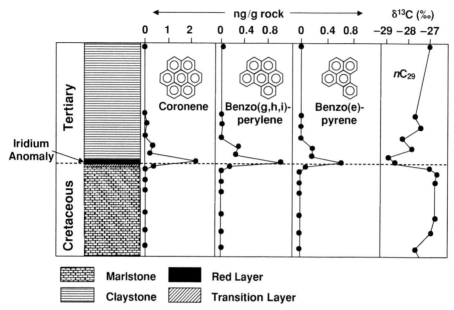

Figure 17.26. Rocks immediately above a chaotic transition layer at the Cretaceous–Tertiary boundary from Caravaca, Spain, are characterized by high iridium and polycyclic aromatic hydrocarbon concentrations and an abrupt decrease in the $\delta^{13}C$ of the C_{29} n-alkane (modified from Arinobu et al., 1999).

Figure 17.26). Decreased light penetration and acid rain caused rapid collapse of marine and land-plant productivity, which led to the mass extinction.

In contrast with the K–T, the evidence that bolide impacts caused other mass extinctions is much less convincing (see Alvarez, 2003). Large craters in Russia (Bottomley et al. 1997) and the Chesapeake Bay (Poag, 1996) (Table 17.2) were formed at ~36.5 Ma, suggesting multiple (synchronous?) strikes from a comet storm (Farley et al., 1998). This Late Eocene event is associated with iridium (Alvarez et al., 1982) and ^{3}He (Farley et al., 1998) anomalies, and shocked minerals (Glass and Wu, 1993; Glass and Liu, 2001). However, unlike the K–T Chicxulub strike, mass extinctions failed to occur immediately following the cometary impacts. Extinction rates of land mammals are unremarkable (Alroy, 2003), and mass extinction of marine organisms occurred later and is related to global warming (Poag, 2002).

The Woodleigh Crater in the Carnarvon Basin of Western Australia is associated with iridium anomalies and microtektites and is dated at the end of the Devonian Period (Table 17.2). This has led some researchers to argue that a bolid impact triggered the Devonian extinction (McLaren, 1983; Wang, 1993; Usyal et al., 2001), but others strongly disagree (McGhee et al., 1984; Fowell et al., 1994).

Evidence that bolide impacts occurred during other mass extinctions is equivocal. Iridium anomalies occur at the Ordovician–Silurian boundary, but these are attributed to terrestrial diagenetic processes (Wang et al., 1992). Olsen et al. (1987) propose that the impact that caused the large Manicouagan crater in Quebec caused the mass extinction at the Triassic–Jurassic boundary, where there is evidence for shocked quartz (Bice et al., 1992) However, the Triassic–Jurassic boundary lacks trace metal anomalies, and the Manicauagan impact has been dated at 12 million years after the end of the Triassic Period (Hodych and Dunning, 1992).

Evidence for a bolide impact at the Permian–Triassic boundary and its role in the greatest mass extinction is equivocal (Erwin, 2003). Becker et al. (2001) and Poreda and Becker (2003) reported the occurrence of C_{100}–C_{400} fullerenes in sediments from the Permian–Triassic boundary that contain trapped helium and argon with isotopic distributions similar to carbonaceous chondrites. However, until recently large-impact craters were unknown from the end of the Permian Period, and others have concluded that a bolide impact

Table 17.2. *Locations of the largest known terrestrial impact craters, their approximate ages, and possible relationships with extinction events*

Impact structure	Location	Diameter (km)	Age (Ma)	Extinction event
Vredefort	Kaapvaal, South Africa	300	~1970	?
Sudbury	Ontario, Canada	250	~1850	?
Chicxulub	Yucatan, Mexico	180	65	Cretaceous-Tertiary
Woodleigh*	Carnarvon Basin, Australia	120	~360*	Devonian-Carboniferous?
Manicouagan	Quebec, Canada	100	212	Triassic-Jurassic?
Popigai	Russia	100	35	Eocene-Oligocene?

*The Woodleigh Crater was originally tied to the Permian-Triassic extinction event (Mory et al., 2000), but more recent findings suggest that it occurred at the end of the Devonian Period (Hadfield, 2002).

is inconsistent with the complex nature of the observed shifts in $\delta^{13}C$, which are explained more readily by the coincidence of the Siberian flood basalts (Wignall, 2001; Erwin, 2003). Becker et al. (2004) identified a candidate for an end-Permian impact crater offshore Australia.

Cyclicity of extinctions and astronomical periodicities

Raup and Sepkoski (1984; 1986) proposed that the marine fossil record indicates mass extinctions throughout the Phanerozoic Eon at a regular spacing of ~26 million years. Regular perturbation of comets might occur as the solar system passes through the galactic plane (Rampino and Strothers, 1984; Schwartz and James, 1984) or as an unknown planet (Nemesis) in an eccentric orbit passes through the Oort Cloud (Davis et al., 1984; Whitmire and Jackson, 1984). Statistical analysis of terrestrial impact craters suggests ~28-million-year periodicity (Alvarez and Muller, 1984) and that meteor impacts may trigger all mass extinctions (Raup, 1992; Shoemaker, 1994). However, many paleontologists discount the statistical analysis of Raup and Sepkoski as evidence for periodic extinction (Benton, 1995; Hallam and Wignall, 1997; Hoffman, 1985; Hoffman, (1989).

Terrestrial agents

Oxygen

The earliest and possibly most devastating mass extinction may have occurred between ~2.2 and 1.8 Ga. As iron and other oxygen sinks became exhausted, atmospheric oxygen levels increased. Many anaerobic microorganisms must have perished as oceanic surface waters became oxic.

Glaciation

Glaciation caused by global cooling or movement of continental landmasses to high latitudes is frequently evoked as a trigger for mass extinctions (Stanley, 1988). A cooler Earth would diminish tropical habitats and explain the preferential loss of warm-water species during mass extinctions. Glaciation favors a global lowering of sea level, decreased shelf environments, and disrupted nutrient supplies. The general correlation between major marine regressions and mass extinctions supports glaciation as a trigger (Newell, 1967).

Several minor mass extinctions occurred throughout the Cambrian Period that correspond to glaciations. The Ediacara fauna may have suffered a mass extinction at the end of the Precambrian Period, although it is unclear whether some taxa of these soft-bodied invertebrates survived into the Cambrian Period. Olnellids, the oldest group of trilobites, and archaeocyathids, the primary reef-building organisms, became extinct during the Cambrian Period. The archaeocyathids left biomarkers that are useful age indicators (see Figure 17.25). Three minor mass extinctions occurred at the end of the Cambrian Period, severely affecting trilobites, brachiopods, and conodonts. Anomalies in the stable carbon isotopic record of carbonates indicate that global glaciation occurred during these extinction events (Saltzman et al., 1995; Strauss et al., 1997).

Glaciation may have triggered the major mass extinction in the Late Ordovician Period when Gondwana migrated into polar latitudes (Wang, 1993). Freezing of large volumes of water resulted in a worldwide lowering of sea levels, and the shallow seas and continental shelves dried up (Brenchley et al., 1993; Marshall et al., 1997).

Table 17.3. *Major Phanerozoic flood basalts*

Flood basalt event	~Volume (km^3 × 10^6)	Eruption duration (Ma)	Eruption age (Ma)	References
Siberian Traps	1–5(?)*	0.9 ± 0.8	250 ± 1.6	Renne and Basu (1991), Reichow *et al.* (2002)
Central Atlantic Magmatic Province	~7	~2.0	200	Mazoli *et al.* (1999)
Paraná basalts	1.3	~1.0	133 ± 1	Renne *et al.* (1992)
Deccan Traps	2.6	>1.0	65.6 ± 0.3	Allegre *et al.* (1999)

*Flood basalts of the Siberian Traps currently cover only ~3.4 × 10^5 km^2 of northwest Siberia. Its original size has been estimated to be as much as ~5 × 10^6 km^2, but many consider sizes ranging from ~1 to 2 × 10^6 km^2 to be more likely (Wignall, 2001).

Cold bottom waters then ventilated the oceans, largely eliminating the dysoxic zone favored by graptolites. A second wave of extinction occurred ~500 000 years later, when global warming resulted in a transgression, re-establishing stratified, euxinic oceans, and inundating the continental shelves. The mass extinction of warm-water marine species suggests that the Frasnian-Famennian mass extinction was caused by global cooling like that in the Late Ordovician Period (McGhee, 1988).

Hypercarnia or CO_2 poisoning is another mechanism for mass extinction that may be precipitated by global cooling. When cold, oxygenated waters ventilate anoxic bottom waters that are saturated with CO_2 and H_2S, the oceans can overturn rapidly, releasing massive amounts of dissolved gases into the atmosphere. In such situations, many organisms would die instantly, while more CO_2-tolerant species, such as fish, would be relatively unaffected (Knoll *et al.*, 1996). A small-scale version of this process occurred in August 1986, when the bottom waters of Lake Nyos, Cameroon, overturned, causing regional hypercarnia (Kling *et al.*, 1994).

The effects of glaciation on the biosphere appear to have been much greater during the Proterozoic Eon and Paleozoic Era than in later times. None of the major post-Carboniferous mass extinctions and only a few of the minor mass extinctions correspond to glaciations, except for the protracted Late Eocene-Early Oligocene ice age.

Volcanism

The magnitudes of many paleo-igneous events far exceed any modern volcanic eruption. In the past, massive quantities of basalt flooded on to the continents within a short time (about one million years), ejecting large quantities of dust, CO_2, and SO_2 into the atmosphere. The global affects of large-scale volcanism were nearly identical to those of large meteor impacts: both caused short-term darkness, acid rain, and long-term global warming. These two causative agents may be coupled if a large impact could trigger massive igneous flows (Boslough *et al.*, 1996).

Strothers (1993) proposed that all mass extinctions are linked to eruptions of flood basalt. The three largest known volcanic eruptions occurred at about the same times as the three major post-Paleozoic mass extinctions (Table 17.3). Wignall (2001) expanded this concept to minor mass extinctions associated with the Emeishan-Panjal Volcanics (mid-Permian) and the Karoo-Ferrar Traps (early Toarcian). However, large-scale volcanism does not always trigger mass extinction. The South American Paraná basalts are comparable in magnitude to the Siberian Traps, yet evidence is lacking for mass extinction during their eruption. Also, the onset of the eruptions appears to slightly post-date the main phase of extinctions, although this could be offset easily by the uncertainties in absolute age dating (Wignall, 2001).

Transgression and oceanic anoxia

Global warming melts polar ice, causing a rise in sea level. Warming of the oceans decreases oxygen solubility, alters circulation patterns, and favors worldwide anoxic events. As with other triggering agents, the correlation between anoxic oceans and mass extinctions is imperfect. Early Paleozoic events, such as the Ordovician Ashgill mass extinction, appear to correspond to rapid fluctuations between global warming and cooling and changes in ocean oxicity. However, no mass extinction occurred during the mid-Cretaceous anoxic event.

Multiple triggers

Multiple factors probably play a role in most major mass extinctions. Until recently, the massive Permian extinction was thought to be caused by multiple terrestrial rather than extraterrestrial agents. Three phases can be identified that contributed to the Permian extinction (Erwin, 1994a). During the Early–Middle Permian Period, all major cratons merged into Pangea, a single supercontinent. This tectonic configuration resulted in severe climatic fluctuations, arid interiors, and glaciation at the poles. The enormous volume of the single oceanic basin lowered sea level to an unrivaled global lowstand, drying out many marine sub-basins and severely reducing shallow-water areas on continental shelves. The worldwide regression accelerated the second phase by triggering the release of gas hydrates and increasing shelf erosion. Marine regressions exposed gas hydrates to weathering and lower pressure, causing release of greenhouse gases into the atmosphere (Kvenvolden, 1993; Erwin, 1994a). The sudden volume increase of sediments deposited into shallow marine settings would further promote regression (Bratton, 1999). Eruption of flood basalt in Siberia released massive quantities of sulfates, ash, and CO_2 into the atmosphere and also contributed to an elevated greenhouse effect. Global warming produced a stratified anoxic ocean, resulting in the extinction of many marine species (Isozaki, 1997; Wignall and Twitchett, 1996). The final phase of the extinction involved the destruction of nearshore terrigenous habitats during the rapid, Early Triassic marine transgression.

Throughout the Middle Permian Period, sulfur isotope ratios of sedimentary sulfates are generally low ($\delta^{34}S$ = −39 to −25‰, Canyon Diablo Troilite (CDT)), but they systematically increase beginning in the Upper Permian Period and persisting into the Lower Triassic Period (−20 to −2‰, CDT). A remarkable excursion occurs at the end of the Permian Period during the extinction event, where the $\delta^{34}S$ reverts to low values (−41 to −23‰). The sulfur isotopic ratios suggest the development of a largely stagnant, anoxic, stratified ocean, which presumably began to form in the lower Upper Permian Period and persisted into the Lower Triassic Period, followed by massive mixing and overturn at the Permian–Triassic boundary (Kajiwara et al., 1994). The carbon isotopic record of organic matter also exhibits a sudden shift at the end of the Permian Period. Kerogen $\delta^{13}C$ shifts from ∼−29‰ in the Upper Permian Period to −33‰ at the boundary, and then back to −29‰ in the Lower Triassic Period. Reduced surface-water primary productivity following the mass extinction at the boundary is largely responsible for the observed $\delta^{13}C$ (Scholle, 1995; Wang et al., 1994).

A bolide impact (Becker et al., 2001) or supernova radiation (Ellis and Schramm, 1995) may explain the Permian extinction event. Becker et al. (2001) report the noble gases helium and argon trapped in fullerenes extracted from rock deposited at the Permian–Triassic boundary. The isotopic compositions of these gases are similar to those of meteorites, suggesting a meteorite impact. Nickel-rich Permian–Triassic-boundary rocks from southern China exhibit remarkable sulfur and strontium isotope excursions and contain impact-metamorphosed grains and clay minerals with unusual trace-element compositions, consistent with an asteroid or comet impact (Kaiho et al., 2001). The large sulfur and strontium isotope excursions at the Permian–Triassic boundary could arise from an ocean impact that caused a massive release of mantle sulfur.

BIOMARKERS AND MASS EXTINCTIONS

With the exception of the K-T event, biomarker studies related specifically to mass extinctions are rare. The marine and terrigenous animals that characterize mass extinctions were generally minor contributors to the preserved biomass. Nonetheless, the Cambrian extinction of the archaeocyathids appears to be responsible for the decrease in 24-isopropylcholestanes in source rocks and crude oils from that time (Figure 17.25). Likewise, the mass extinction at the end of the Triassic Period corresponds to an abrupt decrease in extended tricyclic terpane ratios for crude oils generated from source rocks of that age.

The development of extreme environmental conditions in continental shelf habitats may be tied to the Permian–Triassic extinctions. In one study based on biomarker distributions, Dahl et al. (1993) showed that four different organic facies could be distinguished within the Permian Phosphoria Formation. This was attributed to development of a chemocline that resulted in anoxia and higher salinity in a stratified water column. However, as discussed above, recent results suggest that, like the K–T event, a bolide impact may also be responsible for the Permian-Triassic mass extinction (Becker et al., 2001). Although not strictly biomarkers, fullerenes (C_{60}–C_{200}) from rocks at the Permian-Triassic boundary contain trapped helium and argon with isotope ratios similar to those of the planetary component of carbonaceous chondrites, but unlike those of helium and argon in the Earth's atmosphere. The gas-filled fullerenes are thought to have formed in stars or collapsing gas clouds.

Unlike the older extinction events, the K–T bolide impact received considerable attention from organic geochemists. Several studies of the K–T boundary sediments at Kawaruppu, Hokkaido, Japan, have reported changes in the distribution of n-alkanes and isoprenoids (Mita and Shimoyama, 1999a), n-alkylcyclohexanes and methyl-n-alkylcyclohexanes (Shimoyama and Yabuta, 2002), and polynuclear thiophenes (Katsumata and Shimoyama, 2001a) that are believed to be related to the extinction event. Polynuclear aromatic hydrocarbons (PAHs) (Mita and Shimoyama, 1999b) and various mono- and bicyclic saturated hydrocarbons (e.g. trimethylcyclohexanes, methylethylcyclohexanes, decalins, methyladamantanes, and methyldiamantanes) do not change across the boundary (Shimoyama and Yabuta, 2002).

All sampled K–T boundary sediments are enriched in soot and charcoal that parallels the iridium anomaly (Wolbach et al., 1990). The elemental carbon in soot is evidence for a global wildfire ignited by the impact. The soot contains pyrosynthetic PAHs such as retene, coronene, benzoperylene, and benzopyrene, (Gilmour and Guenther, 1988; Venkatesan and Dahl, 1989). Arinobu et al. (1999) found an abrupt 1.4–1.8‰ decrease in the $\delta^{13}C$ of the sedimentary nC_{29} alkane immediately above the K–T boundary that parallels increased pyrosynthetic PAHs (Figure 17.26). Combustion of 18–24‰ of the total terrigenous biomass would be needed to deplete the atmospheric CO_2 to yield such isotopically depleted n-alkanes. C_{60} and C_{70} fullerenes also have been detected within the soot layers and are believed to originate from the global wildfire (Becker et al., 1995; Heymann et al., 1996a; Heymann et al., 1996b). Abundant soot at the K–T boundary, however, is not proof of global wildfires. Fires were an integral part of mire ecosystems through the latest Cretaceous Period and into the early Tertiary Period (Scott et al., 2000). High concentrations of elemental carbon (3.6 mg/cm) and soot (1.8 mg/cm) in K–T marine sediments from the central Pacific (DSDP Site 465) that were far removed from potential continental sources prove that there was global atmospheric transport of soot at 65 Ma (Wolbach et al., 2003).

18 · Petroleum systems through time

> This chapter defines petroleum systems and provides examples of the geology, stratigraphy, and geochemistry of source rocks and crude oils through geologic time. Gas chromatograms, sterane and terpane mass chromatograms, stable isotope compositions, and other geochemical data are provided for representative crude oils generated from many worldwide source rocks.

PETROLEUM SYSTEM NOMENCLATURE

Traditional exploration focuses on subsurface traps and the play concept in sedimentary basins that are described according to tectonic style. A play consists of prospects and fields with similar geology (e.g. reservoir, cap rock, trap type). Plays use the characteristics of discovered accumulations to predict similar undiscovered accumulations. Focus by interpreters on a particular play type, as in anticline or pinnacle reef trends, may limit creative ideas on other potential play types. Furthermore, although generalizations can be made about many variables affecting basins of a given tectonic style, such as field size, heat flow, and the effectiveness of traps for retaining petroleum, source-rock richness and volumes are related only weakly to tectonic style. Therefore, tectonic classifications are of little value in order to forecast petroleum volumes (Demaison and Huizinga, 1994). The key elements needed to forecast petroleum volumes, such as source, reservoir, and seal rock, and adequate generation, migration, and accumulation factors, were incorporated into the petroleum system concept, as discussed below.

The petroleum system concept was first expressed in terms of oil–source rock correlation (Dow, 1974), as petroleum systems (Perrodon, 1980), generative basins (Demaison, 1984), hydrocarbon machines (Meissner et al., 1984), and independent petroliferous systems (Ulmishek, 1986). Magoon and Dow (1994a) describe these early publications in detail. Magoon (1988; 1989) formalized the concept of the specific elements needed to define a petroleum system by establishing criteria for their identification, mapping, and nomenclature. This work culminated with publication of *The Petroleum System – From Source to Trap* (Magoon and Dow, 1994b). Originally, the petroleum system concept included only mature provinces with produced oil or gas. The concept was expanded to include plays and prospects where potential petroleum systems may occur (Magoon, 1995). By combining known petroleum systems with those that may remain undiscovered, the total petroleum system for a basin can be defined (Magoon, 1995; Magoon and Beaumont, 1999).

Magoon and Beaumont (1999) documented the process required to define a petroleum system. Briefly, to fully characterize a petroleum system, one must identify the source, reservoir, seal, and overburden rocks, and the processes of trap formation, generation, migration, and accumulation of hydrocarbons. It is critical not only that mature source rocks occur within the basin but also that an effective reservoir system be present at the proper time to accumulate and retain expelled petroleum.

A petroleum system encompasses a pod of active or once-active source rock, all related oil and gas, and all geologic elements and processes that are essential for petroleum accumulations to exist (Perrodon, 1992; Magoon and Dow, 1994a). The well-defined petroleum system includes:

- a formal name consisting of the source rock name followed by a hyphen, the principal reservoir formation name, and an indication of the certainty of the correlation, with symbols (?), (.), and (!) indicating speculative, hypothetical, and known genetic relationships, respectively (Table 18.1);
- a summary of geochemical correlations used to derive the level of certainty;
- burial history chart;

Table 18.1. *The degree of confidence in petroleum to source-rock correlation depends on many factors, especially source-rock depositional environment. Gas–source rock correlations in deltaic and coaly sequences are commonly speculative or hypothetical, whereas oil–source rock correlations in marine and lacustrine sequences are defined better*

	Speculative (?)	Hypothetical (.)	Known (!)
Marine			Oil
Deltaic		Gas	
Coal			
Lacustrine			Oil

- petroleum system map;
- petroleum system cross-section;
- events chart;
- table of hydrocarbon accumulations;
- determination of generation-accumulation efficiency.

Geochemistry is the key to petroleum systems because it is required to:

- clearly establish the genetic link between petroleum and the pod of active source rock (oil-source rock correlation);
- map the extent (volumetrics) of the pod of source rock and assess the timing of generation–migration–accumulation relative to trap formation.

Proper definition of a petroleum system requires identification of petroleum source rocks, correlation of petroleum (oil and/or gas) to the source, and correlation of petroleum to petroleum (Figure 18.1). A petroleum system identified only by geological and/or geophysical inference can be classified only as speculative (?) or hypothetical (.). Geochemical correlation is needed for a system to be termed known (!). The degree of confidence in gas–source rock correlations is commonly lower than that for oil–source rock correlations (Table 18.1). The tools available for correlation of hydrocarbon gases are limited to isotopic ratios of gas homologs (e.g. stable carbon or hydrogen isotopes of methane, ethane, etc.) and homolog composition (e.g. gas wetness), whereas crude oils commonly contain many biomarkers and other compounds that can be analyzed for genetic comparisons with prospective source rocks.

Biomarker and other maturation indicators also are essential in order to formulate or verify the petroleum system on the burial history chart. Geothermal histories can be calculated using geophysical information, such as heat-flow measurements and/or models, or geological information, such as thermal conductivities of lithologies and estimates of overburden eroded at unconformities. The uncertainty of the thermal reconstruction can be reduced greatly by applying geochemical temperature and maturity measurements (e.g. vitrinite reflectance, apatite fission tracks, fluid inclusion homogenization temperatures, estimates of fractional conversion from Rock-Eval S2 and T_{max}, and biomarker and isotopic maturity parameters).

Marine and lacustrine rocks that are oil-prone generate crude oils with abundant biomarkers that facilitate correlation to source-rock samples (Table 18.1). Deltaic and coaly rocks that are gas-prone generate hydrocarbon gases and/or condensates that are poor in biomarkers and are more difficult to correlate. In many deltas, the proposed source rocks are deep and not commonly penetrated for sampling. There are many exceptions to the general observations in Table 18.1. For example, oil-prone coals are correlated readily to related crude oils using biomarkers.

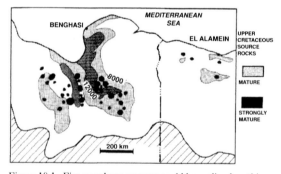

Figure 18.1. Five petroleum systems could be outlined on this map of north Africa (Demaison, 1984). Although the Upper Cretaceous source rocks may be stratigraphically equivalent, they may be named differently in different areas or countries, resulting in different petroleum system names. This also applies to reservoir rocks. Petroleum generated from the source rock can migrate into different reservoir rocks, also yielding different petroleum system names. Reprinted by permission of the AAPG, whose permission is required for further use.

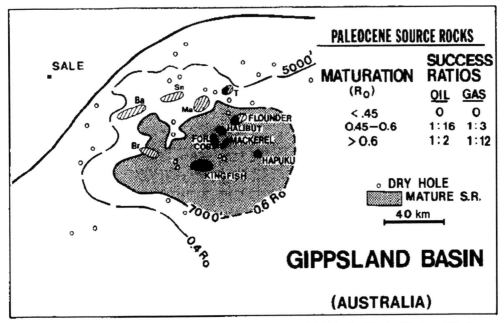

Figure 18.2. The unusual distribution of shallow gas fields over deeper oil fields in the Gippsland Basin could be interpreted to indicate distinct source rocks. However, the oil and gas in both groups of reservoir intervals originated from Paleocene source rock (Demaison, 1984). Their vertical differentiation is the result of pressure–volume–temperature (PVT) relationships. Reprinted by permission of the AAPG, whose permission is required for further use.

The principal objective of the petroleum system approach is to show the geographic boundaries of oil and gas occurrence. A valuable by-product of this approach is identification of limits in our knowledge of the generation, migration, and accumulation of petroleum in each study area. It facilitates the identification of new plays and allows us to identify more readily the additional data, training, and skills needed to allocate resources properly. For example, geochemical confirmation of a petroleum system by oil–oil correlation allows us to focus on defining migration pathways with the expectation that we can find traps that have not yet been identified (e.g. Terken and Frewin, 2000). Figure 18.2 is another example that demonstrates the importance of geochemical identification of the origin of different petroleum accumulations. Both the oil and gas in the Gippsland Basin originate from the same source rock, thus better constraining exploration strategies.

The petroleum system folio sheet consists of five charts that define a systematic method to assess the regional, stratigraphic, and temporal distributions of petroleum. Figure 18.3 describes the hypothetical Deer-Boar(.) petroleum system, where the name includes the source rock (Deer Shale), the major reservoir rock (Boar Sandstone), and a symbol (.) expressing the level of certainty in the genetic relationship between the source and the trapped petroleum as hypothetical (Magoon and Dow, 1994a). The symbols (?), (.), and (!) indicate speculative, hypothetical, and known genetic relationships, respectively. The first chart on the folio sheet is a cross-section showing the extent of the petroleum system at the critical moment, i.e. a snapshot in time that best depicts the generation–migration–accumulation of hydrocarbons. This chart is useful because the present-day distribution of discovered hydrocarbons can obscure important information needed to understand a petroleum system and to predict the location of undiscovered reserves. A map shows the extent of the Deer-Boar(.) petroleum system, including the pod of active source rock and the discovered petroleum accumulations at the critical moment. A table of accumulations for the Deer-Boar(.) petroleum system relates oil and gas fields to their key geochemical and reserves characteristics. The burial history chart shows the critical

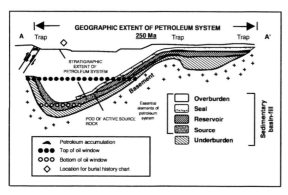

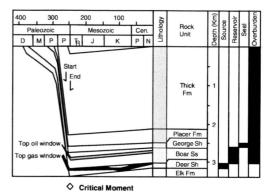

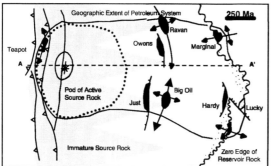

Table of accumulations for Deer-Boar(.) petroleum system.					
Field Name	Date discovered	Reservoir rock	API Gravity (°API)	Cumulative oil production (×10⁶ bbl)	Remain reserves (×10⁶ bbl)
Big oil	1954	Boar Ss	32	310	90
Raven	1956	Boar Ss	31	120	12
Owens	1959	Boar Ss	33	110	19
Just	1966	Boar Ss	34	160	36
Hardy	1989	Boar Ss	29	85	89
Lucky	1990	Boar Ss	15	5	70
Marginal	1990	Boar Ss	18	12	65
Teapot	1992	Boar Ss	21	9	34

Figure 18.3. Folio sheet for the hypothetical Deer-Boar(.) petroleum system, containing five charts that define a systematic method to assess the regional, stratigraphic, and temporal distributions of petroleum (modified from Magoon and Dow, 1994a). Reprinted by permission of the AAPG, whose permission is required for further use.

moment and the timing of oil generation. The petroleum system events, or timing-risk, chart shows timing of the elements and processes in the petroleum system.

Geochemistry is the key to petroleum systems because it is required to:

- clearly establish the genetic link between petroleum and the pod of active source rock (oil–source rock correlation);
- map the extent (volumetrics) of the pod of source rock;
- assess the timing of generation–migration–accumulation relative to trap formation.

Maps of the extent of the active source-rock pod and estimates of generated volumes of petroleum require input from multiple disciplines, including geochemistry, seismic sequence stratigraphy, and well log analysis (e.g. Demaison and Huizinga, 1994; Creaney *et al.*, 1994a). Some geochemical innovations that contribute to constructing these maps include Rock-Eval pyrolysis and geochemical logs (Espitalié et al., 1984; Espitalié et al., 1987; Peters and Cassa, 1994), Δlog R (Passey et al., 1990), and calibrated basin modeling (Welte et al., 1997), including custom kerogen kinetic measurements (Braun et al., 1991). For example, the Δlog R method allows prediction of total organic carbon (TOC) profiles in wells that lack measured TOC by using the separation between scaled transit-time and resistivity curves from conventional well logs (Creaney and Passey, 1993). Predictions of TOC from Δlog R must be calibrated using wells where measured TOC values are available in representative lithologies.

PETROLEUM SYSTEMS

Table 18.2 and Figure 18.4 describe the major petroleum systems outside of the USA. The grand total of petroleum volumes in Table 18.2 ($\sim$2 390 000 MMBOE) accounts for >95% of the cumulative production and discovered reserves outside of the USA. Figure 18.5 compares the volumes of the 50 largest petroleum systems outside of the USA. The three largest petroleum systems dominated by oil consist of the Cretaceous-Tertiary in Zagros-Mesopotamia region (1 in Table 18.2), the Tuwaiq/Hanifa-Arab on the Arabian Peninsula (2), and the Bazhenov-Neocomian in west Siberia (4). Geochemical data for representative crude oils from these three petroleum systems are given in Figures 18.96, 18.94, and 18.121, respectively. The composite Mesozoic system in northern west Siberia (3) is dominantly hydrocarbon gas. Table 18.3 includes some common petroleum volume and energy conversion factors.

In the following discussion, examples of petroleum systems are arranged from oldest to youngest. For our purpose, the discussion is brief and focuses on the source rock, kerogen type, and estimated ultimate recoverable reserves when available. Complete petroleum system names are used only where documented clearly in the literature.

ARCHEAN PETROLEUM SYSTEMS

Since all Archean rocks were metamorphosed at least once, no petroleum systems survived from this time to produce commercial quantities of petroleum. Nevertheless, there is convincing evidence that there were Archean source rocks that generated and expelled oil (Figure 18.6). Archean petroleum systems, however, are largely incomplete, as there is no evidence that this oil was ever trapped in reservoirs that now represent significant accumulations.

Pilbara Craton, Australia

The Pilbara Craton in northwestern Australia contains the most complete (3.5–2.5 Ga) and best-preserved sequence of Archean sedimentary rocks (Buick et al., 1998). It has never been metamorphosed above prehnite-pumpellyite facies ($\sim$250°C maximum). This extraordinary state of preservation may be due to stable continental crust that avoided major collisions with other tectonic plates. Shales within the Pilbara Craton contain more than 1 wt.% TOC and have fully expended any of their original source potential (Strauss and Moore, 1992).

Relict Archean petroleum systems are identified in the Pilbara Craton by pyrobitumen nodules of migrated oil. These formed by radiogenic immobilization (cross-link polymerization) of fluid hydrocarbons around detrital grains of uraninite, thorite, and monazite, or by flash maturation of kerogen by hydrothermal silicification (Buick et al., 1998). These nodules occur at several stratigraphic levels within the Pilbara Craton, including the hydrothermal Warrawoona Group (>3.46 Ga), deltaic sediments of the Mosquito Creek Formation ($\sim$3.25 Ga), and non-marine intevals in the Lalla Rookh Formation ($\sim$3.0 Ga) and Fortescue Group ($\sim$2.75 Ga).

Dutkiewicz et al. (1998) found oil preserved in fluid inclusions in sandstones from the Pilbara Craton ($\sim$3.0 Ga). The inclusions contained up to four fluid phases, including water, liquid CO_2, gaseous CO_2, and oil. The oil was characterized by UV-fluorescence and Fourier transform infrared (FTIR) spectroscopy. Most of the inclusions were in microfractures confined to individual quartz grains that healed during an Archean metamorphic event. Therefore, the oil in the inclusions was generated from an older Archean source rock, migrated, and was encapsulated before the metamorphism. The inclusions acted as inert pressure vessels, protecting the oil from subsequent degradation. These ancient fluid inclusions may still contain biomarkers providing information on the status of biotic evolution.

Table 18.2. *Summary of petroleum systems outside of the USA ranked by known petroleum volume (modified from Magoon and Schmoker, 2000). Numbers at far left refer to locations in Figure 18.4. Gas petroleum systems are shaded*

	Petroleum system name*	Code	Region name	Known oil (MMBO)	Known gas (BCFG)	Known NGL (MMBNGL)	Total (MMBOE)
>100 billion BOE							
1	Zagros–Mesopotamian Cretaceous–Tertiary	203001	Middle East, North Africa	372 226	493 238	1562	455 995
2	Arabian Sub-Basin Tuwaiq/Hanifa-Arab	202102	Middle East, North Africa	198 995	275 295	9205	254 083
3	Northern west Siberian Mesozoic Composite	117403	Former Soviet Union	10 185	1 167 288	2375	207 107
4	Bazhenov–Neocomian	117401	Former Soviet Union	117 972	97 582	764	135 000
						Subtotal =	1 052 185
20–100 billion BOE							
5	Cretaceous Thamama/Wasia	201901	Middle East, North Africa	71 231	102 100	1901	90 148
6	Silurian Qusaiba	201903	Middle East, North Africa	555	452 030	13 815	89 708
7	Volga–Ural Domanik-Paleozoic	101501	Former Soviet Union	63 872	96 458	1096	81 044
8	Kimmeridgian Shales	402501	Europe	43 895	158 914	5971	76 351
9	La Luna/Maracaibo	609901	Central, South America	49 072	26 701	43	53 565
10	Tertiary Niger Delta (Agbada/Akata)	719201	Sub-Saharan Africa, Antarctica	34 522	93 811	2842	53 000
11	Pimienta–Tamabra	530501	North America	44 412	50 822	95	52 977
12	Querecual	609801	Central, South America	26 756	112 296	528	46 001
13	Paleozoic North Caspian	101601	Former Soviet Union	10 808	156 976	8890	45 861
14	Sirte–Zelten	204301	Middle East, North Africa	37 072	37 767	129	43 496
15	Amu-Darya Jurassic-Cretaceous	115401	Former Soviet Union	766	230 614	1175	40 377
16	Carboniferous-Rotliegend	403601	Europe	2872	222 159	172	40 071
17	Jurassic Hanifa/Diyab-Arab	201 902	Middle East, North Africa	19 013	77 512	592	32 523
18	Shahejie-Shahejie/Guantao/Wumishan	312701	Asia Pacific	24 553	15 672	88	27 253
19	Oligocene-Miocene Maykop/Diatom	111201	Former Soviet Union	17 438	35 994	503	23 941
20	Paleozoic–Permian/Triassic	203002	Middle East, North Africa	0	131 220	925	22 795
21	Tanezzuft-Benoud	205805	Middle East, North Africa	88	105 050	4935	22 531
						Subtotal =	841 642

(cont.)

Table 18.2. (cont.)

	Petroleum system name*	Code	Region name	Known oil (MMBO)	Known gas (BCFG)	Known NGL (MMBNGL)	Total (MMBOE)
5–20 billion BOE							
22	Central Arabia Qusaiba-Paleozoic	202101	Middle East, North Africa	6376	79 115	343	19 905
23	Domanik-Paleozoic	100801	Former Soviet Union	13 069	36 632	716	19 890
24	Qingshankou-Putaohua/Shaertu	314401	Asia Pacific	15 570	1598	0	15 836
25	Brown Shale-Sihapas	380801	Asia Pacific	13 217	3866	11	13 872
26	South, North Barents Triassic–Jurassic	105001	Former Soviet Union	51	78 143	100	13 175
27	Brunei-Sabah	370101	Asia Pacific	6898	36 200	180	13 111
28	Duvernay-Leduc	524302	North America	9459	14 357	1025	12 877
29	Eocene-Miocene Composite	804301	South Asia	8440	24 193	267	12 739
30	Togur-Tyumen	117402	Former Soviet Union	11 756	5180	3	12 623
31	Tanezzuft-Oued Mya	205401	Middle East, North Africa	10 843	8973	0	12 338
32	Tanezzuft-Illizi	205601	Middle East, North Africa	3670	45 061	898	12 078
33	Kutei Basin	381701	Asia Pacific	2879	45 473	1273	11 731
34	Dnieper-Donets Paleozoic	100901	Former Soviet Union	1611	59 098	200	11 660
35	Congo Delta Composite	720303	Sub-Saharan Africa, Antarctica	9745	9443	39	11 357
36	Lagoa Feia-Carapebus	603501	Central, South America	10 056	6244	10	11 107
37	Sudr-Nubia	207101	Middle East, North Africa	9810	5995	41	10 850
38	Dingo-Mungaroo/Barrow	394801	Asia Pacific	1149	48 245	991	10 181
39	Gacheta-Mirador	609601	Central, South America	5402	15 314	451	8405
40	Tanezzuft-Ghadames	205403	Middle East, North Africa	4538	16 484	1011	8296
41	East Natuna	370202	Asia Pacific	20	45 045	0	7528
42	Second White Speckled Shale-Cardium	524306	North America	2688	26 449	365	7462
43	Baikal-Patom Fold Belt Riphean–Craton Margin Vendian	121001	Former Soviet Union	2006	30 210	360	7401
44	Sarawak Basin	370201	Asia Pacific	797	37 119	379	7363
45	Oligocene-Miocene Lacustrine	370301	Asia Pacific	3017	24 248	136	7194
46	Lucaogou-Karamay/Ulho/Pindequan	311501	Asia Pacific	6624	2248	0	6998
47	Mesozoic-Cenozoic	604101	Central, South America	6601	1616	0	6871
48	Upper Cretaceous/Tertiary	609802	Central, South America	3447	15 822	172	6256

(*cont.*)

Table 18.2. (cont.)

	Petroleum system name*	Code	Region name	Known oil (MMBO)	Known gas (BCFG)	Known NGL (MMBNGL)	Total (MMBOE)
49	South Mangyshlak	110902	Former Soviet Union	5241	5707	46	6239
50	Latrobe	393001	Asia Pacific	3860	9775	701	6190
51	Neuquén Hybrid	605501	Central, South America	2370	20 704	338	6159
52	Exshaw–Rundle	524303	North America	1728	21 568	836	6158
53	Sembar-Goru/Ghazij	804201	South Asia	180	35 373	63	6139
54	North Sakhalin Neogene	132201	Former Soviet Union	2182	22 383	165	6077
55	Upper Jurassic Spekk	401701	Europe	2660	15 662	702	5973
56	North Oman Huqf/'Q'-Haushi	201401	Middle East, North Africa	2028	20 339	507	5925
57	Bampo-Cenozoic	382201	Asia Pacific	674	25 559	926	5860
58	Mannville–Upper Mannville	524305	North America	0	30 731	447	5569
59	Lower Inoceramus	605901	Central, South America	1216	24 807	211	5562
60	Transylvanian Composite	405701	Europe	0	30 731	0	5122
61	Madbi Amran/Qishn	200401	Middle East, North Africa	1933	17 120	312	5098

0.2–5 billion BOE   Subtotal = 375 175

	Petroleum system name*	Code	Region name	Known oil (MMBO)	Known gas (BCFG)	Known NGL (MMBNGL)	Total (MMBOE)
62	Dysodile Schist–Tertiary	406102	Europe	4115	5048	11	4968
63	Terek-Caspian	110901	Former Soviet Union	3490	7840	29	4826
64	Miocene Coaly Strata	370302	Asia Pacific	592	23 901	186	4761
65	Bou Dabbous–Tertiary	204801	Middle East, North Africa	2114	15 509	44	4743
66	Lahat/Talang Akar-Cenozoic	382801	Asia Pacific	2429	10 204	56	4186
67	North Oman Huqf-Shu'aiba	201601	Middle East, North Africa	2685	7633	58	4015
68	D-129	605801	Central, South America	3346	3565	2	3942
69	Lower Cruse	610301	Central, South America	20	22 600	80	3867
70	Los Monos-Machareti	604501	Central, South America	320	16 985	548	3699
71	Azov-Kuban Mesozoic–Cenozoic	110801	Former Soviet Union	524	18 704	56	3696
72	Sylhet-Kopili/Barail-Tipam Composite	803401	South Asia	2536	6534	5	3630
73	Combined Triassic/Jurassic	524304	North America	1296	10 580	350	3409
74	La Luna-La Paz	609001	Central, South America	2426	5002	68	3328
75	Jatibarang/Talang Akar-Oligocene/Miocene	382402	Asia Pacific	1908	7322	175	3303

(cont.)

Table 18.2. (cont.)

	Petroleum system name*	Code	Region name	Known oil (MMBO)	Known gas (BCFG)	Known NGL (MMBNGL)	Total (MMBOE)
76	Late Jurassic/Early Cretaceous–Mesozoic	391301	Asia Pacific	46	17 960	200	3239
77	Porto Garibaldi	406001	Europe	14	18 475	5	3098
78	Stavropol-Prikumsk	110903	Former Soviet Union	820	13 499	3	3073
79	Lodgepole	524404	North America	2778	893	37	2964
80	Greater Hungarian Plain Neogene	404801	Europe	1089	9657	24	2722
81	Cretaceous Natih	201602	Middle East, North Africa	1852	3100	100	2469
82	Eocene-Miocene Composite	804801	South Asia	716	10 020	63	2449
83	Buzuchi Arch, Surrounding Areas Composite	115001	Former Soviet Union	2286	943	0	2443
84	Azile-Senonian	720302	Sub-Saharan Africa, Antarctica	2114	1835	10	2430
85	Egret-Hibernia	521501	North America	1582	2406	81	2393
86	Keg River-Keg River	524301	North America	1011	7073	100	2290
87	Moesian Platform Composite	406101	Europe	1793	2217	66	2228
88	Cretaceous–Tertiary	608101	Central, South America	1672	2892	0	2154
89	Jurassic/Early Cretaceous–Mesozoic	391003	Asia Pacific	502	7526	385	2141
90	Jenam/Bhuban-Bokabil	804703	South Asia	4	12 289	51	2104
91	Melania-Gamba	720301	Sub-Saharan Africa, Antarctica	1752	843	15	1908
92	Ordovician/Jurassic-Phanerozoic	315401	Asia Pacific	704	5780	189	1856
93	Maokou/Longtang-Jialingjiang/Maokou/Huanglong	314201	Asia Pacific	0	11 072	0	1846
94	Mesozoic/Paleogene Composite	404702	Europe	923	5393	6	1828
95	Zala-Drava-Sava Mesozoic/Neogene	404802	Europe	1101	3929	63	1818
96	Isotopically Light Gas	404701	Europe	1	10 660	3	1782
97	Paleozoic Qusaiba/Akkas/Abba/Mudawwara	202301	Middle East, North Africa	1213	1661	25	1515
98	Banuwati-Oligocene/Miocene	382401	Asia Pacific	1259	614	0	1362
99	Locker-Mungaroo/Barrow	394802	Asia Pacific	6	8102	0	1356
100	Keyling/Hyland Bay-Permian	391002	Asia Pacific	0	5730	50	1005
101	Cretaceous Composite	730301	Sub-Saharan Africa, Antarctica	0	6015	0	1003
102	Maqna	207102	Middle East, North Africa	155	3527	256	999

(cont.)

Table 18.2. (cont.)

	Petroleum system name*	Code	Region name	Known oil (MMBO)	Known gas (BCFG)	Known NGL (MMBNGL)	Total (MMBOE)
103	Yenisey Fold Belt Riphean–Craton Margin Riphean	120701	Former Soviet Union	34	5052	60	936
104	Taiyuan/Shanxi-Majiagou/Shihezi	312802	Asia Pacific	0	5590	0	932
105	Neocomian–Turonian Composite	602901	Central, South America	692	1407	3	929
106	Mesozoic Composite	521502	North America	0	4224	107	804
107	Tanezzuft–Timimoun	205801	Middle East, North Africa	0	4200	0	700
108	Patala–Namal	802601	South Asia	324	1981	41	696
109	Yanchang–Yanan	312801	Asia Pacific	683	20	0	686
110	Tanezzuft–Ahnet	205802	Middle East, North Africa	1	3117	90	610
111	Cretaceous Composite	718301	Sub-Saharan Africa, Antarctica	236	1949	36	596
112	Tanezzuft–Sbaa	205803	Middle East, North Africa	284	1490	10	542
113	Guaratiba–Guaruja (Cretaceous) Composite	603601	Central, South America	285	1125	44	517
114	Belsk Basin	101502	Former Soviet Union	65	2638	5	509
115	Jurassic Gotnia/Barsarin/Sargelu/Najmah	202302	Middle East, North Africa	474	100	0	490
116	Meride/Riva di Solto	406002	Europe	332	412	27	428
117	Jurassic–Cretaceous Composite	204802	Middle East, North Africa	103	1301	23	343
118	North Ustyurt Jurassic	115002	Former Soviet Union	80	1450	8	330
119	Upper Jurassic–Neocomian	611701	Central, South America	297	96	0	313
120	Jurassic Coal-Jurassic/Tertiary	311502	Asia Pacific	210	342	0	267
121	Cambrian/Silurian Shale-Dengying/Lower Paleozoic	314204	Asia Pacific	1	1350	0	226
						Subtotal =	*123 702*
<200 MMBOE							
122	Cretaceous–Paleogene	608302	Central, South America	171	158	0	197
123	Cretaceous Composite	603401	Central, South America	130	222	1	169
124	Daanzhai-Daanzhai/Lianggaoshan	314202	Asia Pacific	109	140	0	132

(*cont.*)

Table 18.2. (cont.)

	Petroleum system name*	Code	Region name	Known oil (MMBO)	Known gas (BCFG)	Known NGL (MMBNGL)	Total (MMBOE)
125	Cuanza Composite	720304	Sub-Saharan Africa, Antarctica	107	59	0	116
126	Cenomanian–Turonian	602101	Central, South America	63	0	0	63
127	Neogene	608301	Central, South America	4	340	0	61
128	Transcarpathian Neogene	404804	Europe	0	261	0	43
129	Danube Neogene	404803	Europe	0	247	0	41
130	Paleozoic Composite	404703	Europe	1	233	0	39
131	Tanezzuft-Melrhir	205402	Middle East, North Africa	6	125	9	36
132	Jurassic Coal-Denglouku/Nongan	314402	Asia Pacific	5	145	0	29
133	Lower Cretaceous Marine	606301	Central, South America	10	75	2	25
134	Milligans-Carboniferous/Permian	391001	Asia Pacific	15	48	0	23
135	Tertiary-Parigi	382403	Asia Pacific	0	119	0	20
136	Cretaceous-Tertiary Composite	701301	Sub-Saharan Africa, Antarctica	10	49	0	19
137	Brightholme	524402	North America	14	11	0	16
138	Hungarian Paleogene	404806	Europe	11	16	0	14
139	Tobago Trough Paleogene	610701	Central, South America	11	22	0	14
140	Bakken	524403	North America	7	0	0	7
141	Yeoman	524401	North America	1	0	0	1
						Subtotal =	1214
						Grand total =	2 393 918

* Some petroleum systems in this table have the same or similar names but reside in different areas (e.g. 2 and 17; 6 and 22; 7, 13, and 23). BCFG, billion cubic feet of gas; MMBNGL, millions of barrels of natural gas liquids; MMBO, millions of barrels of oil; MMBOE, millions of barrels of oil equivalent at one barrel oil/6000 cubic feet of gas.

Code = Total petroleum system code on CD (Magoon and Schmoker, 2000). Also available on the web at http://greenwood.cr.usgs.gov/energy/WorldEnergy/DDS-60/. Known petroleum volumes are cumulative production plus remaining reserves (IHS/Petroconsultants S.A., 1996). For gas petroleum systems (shaded rows), the gas-to-oil ratio (GOR) >20 000 cubic feet of gas/barrel of oil. GOR = 1000 × BCFG/(MMBO + MMBNGL).

Table 18.3. *Some petroleum volume and energy conversion factors*

B To convert A to B, multiply by factor: A	1 billion m^3 NG	1 billion ft^3 NG	1 million tons oil	1 million tons LNG	1 million tons coal	1 trillion BTU	1 petajoule	1 MMBOE
1 billion m^3 NG	1	35.3	0.90	0.73	1.35	36	38	6.29
1 billion ft^3 NG	0.028	1	0.026	0.021	0.036	1.03	1.08	0.18
1 million tons oil	1.11	39.2	1	0.81	1.52	40.4	42.7	7.33
1 million tons LNG	1.38	48.7	1.23	1	1.86	52.0	55.0	8.68
1 million tons coal	0.74	26.1	0.86	0.54	1	26.7	28.1	4.66
1 trillion BTU	0.028	0.98	0.025	0.02	0.038	1	1.06	0.17
1 petajoule	0.026	0.93	0.023	0.019	0.036	0.95	1	0.17
1 MMBOE	0.16	5.61	0.14	0.12	0.21	5.8	6.04	1

BTU, British thermal unit; LNG, liquid natural gas; MMBOE, million barrels of oil equivalent; NG, natural gas.

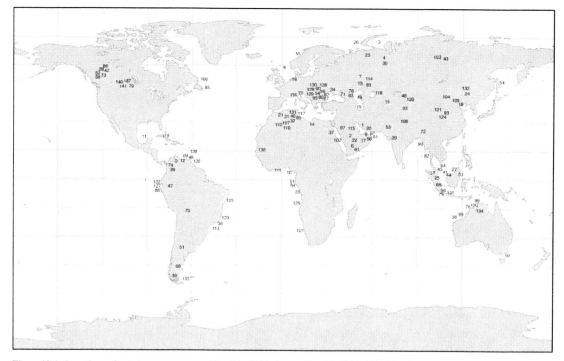

Figure 18.4. Locations of petroleum systems outside of the USA that are listed in Table 18.2 (data from Magoon and Schmoker, 2000).

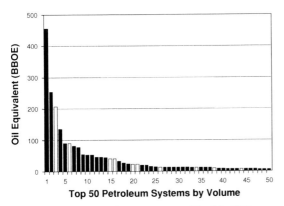

Figure 18.5. Top 50 petroleum systems outside of the USA ranked by volume (data from Magoon and Schmoker, 2000). Numbers on x-axis refer to petroleum systems in Table 18.2. The largest petroleum system (1) consists of the Zagros-Mesopotamian Cretaceous-Tertiary, with ~456 trillion barrels of oil equivalent (BBOE). White and dark bars indicate gas and oil systems based on gas-to-oil ratio (GOR) greater than or less than 20 000, respectively.

Archean biomarkers were reported in ~2.7-Ga shales from the Pilbara Craton (Brocks et al., 1999). The shales contained 2α-methylsteranes, which are characteristic of cyanobacteria, and C_{27}–C_{29} steranes, suggesting that eukaryotes had evolved by this early date. Sequential extraction of adjacent samples suggests that these biomarkers are indigenous to the shales, but it is possible that these hydrocarbons migrated from younger strata. The manner in which biomarkers survive low-grade metamorphism is unknown because their occurrence has not been coupled with fluid-inclusion observations.

Kaapvaal Craton, South Africa

The Kaapvaal Craton of South Africa is similar to the Pilbara Craton in tectonic style, sedimentary sequences, and paleo-source potential. Fluid inclusions in sandstone contain preserved oil (Dutkiewicz et al., 1998) and pyrobitumen nodules in association with radiogenic minerals occurring in the metasediments from the Kaapvaal (Buick et al., 1998). These observations prove that oil was generated from organic-rich shales during the Archean Eon and migrated before metamorphism destroyed fluid pathways during the Late Archean to Early Proterozoic eons. Shales in the Witwatersrand Supergroup (~3.2–2.6 Ga) were metamorphosed to lower greenshist facies (~350°C). They currently contain only small amounts of graphitic carbon, but they may have contained at least 2 wt.% TOC before maturation (Cornford, 2001).

Buick et al. (1998) proposed that methane in the Witwatersrand placer deposits is a vestige of the Archean petroleum system. The gas is mostly methane (~80%), nitrogen (~12%), helium (~6%), and minor amounts of CO_2 and argon. Traces of natural gas are widespread throughout the basin, with the greatest accumulations in structural highs along the unconformity between the Witwatersrand and the overlying Permian Karoo. Although it is possible that the methane originated from the Karoo, Buick et al. (1998) argued for

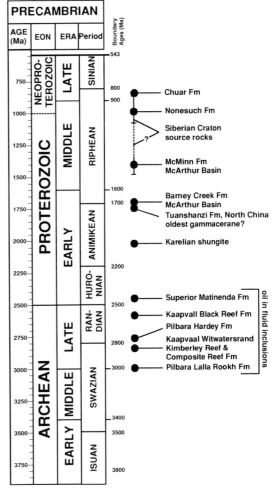

Figure 18.6. Archean and Proterozoic source rocks.

an Archean source. This conclusion rests on observations of methane in mines with no overlying Karoo, unfavorable stratigraphic position of the Karoo (updip from the accumulations), compositional differences compared with gas from the Karoo (~98% methane, ~1% nitrogen, traces of ethane, CO_2, helium, and argon), and relative abundance. One mine produced ~1.8×10^8 m^3 methane from the Witwatersrand over a 16-year period that was used to power a uranium treatment plant.

EARLY PROTEROZOIC PETROLEUM SYSTEMS

The transition between the Archean Eon and the Proterozoic Eon is marked by the stabilization of continental crust, modern-style plate tectonics, and the preservation of sedimentary rocks that have not experienced metamorphism. These rocks contain abundant microfossils and hydrocarbons (e.g. Burlingame et al., 1965; Oró et al., 1965), although it is difficult to prove whether soluble compounds are indigenous or contamination. Evidence suggests that the largely prokaryotic biomass was sufficient to produce organic-rich source rocks during the Early Proterozoic Eon. Veins of pyrobitumen provide clear evidence for oil migration (Mancuso et al., 1989), and it is likely that accumulations of petroleum did occur in reservoirs with modern styles of stratigraphic and structural traps (McKirdy and Imbus, 1992). Nevertheless, there are no Early Proterozoic petroleum systems that produce commercial quantities of oil.

Karalian shungite, Russia

The Upper Zaonezhskaya Formation near Lake Onega, northwest (Karalia) Russia, is the richest known accumulation of organic matter from the Early Proterozoic Eon (~2.0 Ga). It is a 600-m-thick zone that contains shungite, a nearly pure carbonaceous material, averaging ~25 wt.% TOC. The formation is estimated to contain more than 23×10^{10} metric tons of carbon within an area of 9000 km^2. The deposit was discovered in 1879 and has been studied extensively by Soviet and Russian geochemists since the 1980s, but it received little attention in English-language publications until an extensive review by Melezhik et al. (1999).

Shungite occurs as disseminated organic matter (0.1–50 wt.% TOC), as coal-like seams and layered oil shales (50–75 wt.% TOC), and as lustrous veins (80–98 wt.% TOC) (Melezhik et al., 1999). Shungite consists of autochthonous kerogen mixed with migrated oil that is now pyrobitumen. Organic matter was initially deposited in brackish, non-euxinic lagoonal waters, where productivity was enhanced by nutrients from nearby volcanoes. The kerogen and pyrobitumen are highly mature because they experienced greenshist metamorphism ~1.8 Ga. This advanced state of thermal stress is reflected in very low H/C ratios (<0.2) and in highly variable δ^{13}C ranging from −45 to −17‰, with a bimodal distribution and maxima at −28 and −39‰.

Note: Fullerenes occur in meteorites (Harris et al., 2000; Pizzarello et al., 2001), shocked strata associated with bolide impacts (Becker et al., 1994; Mossman et al., 2003), and sediments deposited at major extinction events, such as the Cretaceous–Tertiary (Heymann et al., 1998) and Permian–Triassic (Becker et al., 2001; Chijiwa et al., 1999; Poreda and Becker, 2003) boundaries. Precambrian shungites are the first rocks where fullerenes were reported that are not associated with an impact event (Buseck et al., 1992). The origin of these C_{60} and C_{70} fullerenes remains a mystery, and even their presence in the samples has been questioned (Mossman et al., 2003). Fullerenes have since been reported in the solid bitumen associated with pillow lava in the Bohemian Massif (Jehlycka et al., 2000) and in Tertiary coals from Yunnan Province, China (Fang and Wong, 1996). Fisher et al. (1996a) showed that oil shales produce fullerenes when pyrolyzed at 400°C under helium. It is possible that fullerenes formed with the shungite as the ancient oil shales were exposed to metamorphic temperatures.

MIDDLE PROTEROZOIC PETROLEUM SYSTEMS

Several studies have found soluble hydrocarbons in Mesoproterozoic sedimentary rocks (see reviews by Hayes et al. (1983; 1992) and Summons and Walter (1990). Methyl-branched alkanes and isoprenoids are common biomarkers, while tricylcanes, steranes, and hopanes are absent or found only in small quantities

Table 18.4. *Middle Proterozoic source rocks in the McArthur Basin, Northern Territory, Australia (from Summons et al., 1988a)*

Formation	Environment	Maturity	TOC (wt.%)	Hydrogen index	Atomic H/C	Extractable hydrocarbons (ppm)
McMinn	Marine shelf	Marginal	0.7–2.9	64–484	–	81–881
Velkerri	Marine outer shelf	Mature	0.9–7.2	40–641	0.76–1.01	39–7982
Yalco	Lacustrine or lagoonal	Marginal	0.8–5.4	156–588	0.92–1.20	57–1799
		Marginal	0.6–10.4	200–740	1.20–1.66	722–1119
Barney Creek	Lacustrine or lagoonal	Mature	0.8–7.6	190–740	0.70–1.54	97–2337
		Postmature	0.2–3.2	<100	<0.7	0–192

in most Precambrian rocks. Other biomarkers, such as gammacerane, were reported in only a few samples (Peng *et al.*, 1998). The origin of these soluble hydrocarbons is difficult to prove. Although many of the analyzed rocks contain microfossils, most have negligible organic carbon, with little remaining petroleum-generative potential. Only rarely can a correlation be made between the biomarkers in extracts and those liberated from kerogen by artificial chemical or thermal maturation (Hoering and Navale, 1987).

These concerns are not an issue in at least two cases, the McArthur Basin of Australia and the Siberian Craton of Russia. Lacustrine and marine shales from these basins have source potential comparable to that of later Phanerozoic strata. While petroleum systems were present in the Early Proterozoic Eon, and possibly in the Archean Eon, it was not until the Middle Proterozoic Eon that complete petroleum systems were preserved as potentially commercial sources of produced oil. Summons and Walter (1990) tabulated biomarkers in Proterozoic rocks and oils from worldwide localities and described how these compounds can be used to infer the depositional environment, mineralogy, and type of organic matter input in their source rocks.

McArthur and Roper Group shales, Australia

Largely unmetamorphosed lacustrine and shallow marine Mesoproterozoic (∼1.8–1.4 Ga) sedimentary successions are distributed widely in several major basins across northern Australia (McArthur, Mount Isa, South Nicholson, and Victoria-Birrindudu basins, and the Ashburton and Davenport Provinces of the Tennant Inlier). Except for the Davenport Province, these successions are deformed only gently, and their stratigraphy is relatively continuous (Jackson *et al.*, 1988). The McArthur Basin in the Northern Territory of Australia contains a sedimentary sequence consisting of the Tawallah, McArthur, Nathan, and Roper Groups. The Tawallah is postmature and lacks sedimentary source rocks. The McArthur and Nathan Groups contain stromatolitic and carbonate evaporites with interbedded shales dating from ∼1.7 Ga. Separated by an unconformity, the Roper Group contains alternating quartz arenites and shales dating from ∼1.4 Ga (Jackson *et al.*, 1986). Unconformably overlying the Roper Group is a sequence of probable Neoproterozoic age containing the Jamison Sandstone (Lanigan *et al.*, 1994).

The McArthur and Roper Groups contain several shales with relatively high concentrations of organic carbon. Of these, the Barney Creek, Velkerri, and McMinn (Kyalla Member) formations have the most significant potential (Table 18.4) (Crick *et al.*, 1988; Summons *et al.*, 1988a; Warren *et al.*, 1998). These combined source sequences cover more than 150 000 km^2 and are >1500 m thick. Drilling has not yet defined the full extent of the source formations beneath the Cambrian overburden. Depending on location, the Mesoproterozoic shales range from marginally mature with significant remaining source potential to postmature with largely inert carbon. Thermal maturity cannot be determined using vitrinite reflectance because these rocks predate the evolution of land plants. Therefore, maturity assessments are based on kerogen H/C or Rock-Eval hydrogen index (HI), the methylphenanthrene index, and ratios of alkylphenanthrenes and alkylnaphthalenes cross-calibrated for use in Proterozoic source rocks (e.g. George and Ahmed, 2000).

Barney Creek Formation (~1.7 Ga)

The Barney Creek Formation contains organic-rich shales deposited within numerous sub-basin grabens. These shales and the overlying Lynott Caranbirini Member and the Yalco shales were deposited under shallow-water, evaporitic conditions consistent with either lacustrine or lagoonal marine environments. The remaining source potential for marginally mature samples is impressively high, with TOC exceeding 10 wt.% and Rock-Eval HI >700 mg HC/g TOC, typical of Phanerozoic type I kerogen.

Velkerri Formation (~1.4 Ga)

The Velkerri Formation is composed of couplets of laminated black organic-rich mudstones (~4–7 wt.% TOC) and laminated gray-green glauconitic, organic-lean mud shales (TOC <2 wt.%) (Crick et al., 1988; Crick, 1992,). These rocks originated as distal deltaic marine muds exposed to periodic anoxia in a restricted marine basin. Three black organic-rich mud shales in the Middle Velkerri Member are tens of meters thick, but they have low porosity and permeability. Extract and Rock-Eval pyrolysis data indicate more efficient hydrocarbon expulsion from the upper two units (Warren et al., 1998). Shallow, immature Middle Velkerri Formation rocks contain type II kerogen with HI >500 mg HC/g TOC. Deeper equivalents of the formation have HI ranging from 150 mg HC/g TOC near peak oil generation to 50 mg HC/g TOC or less in the late mature to postmature stage (George and Ahmed, 2000).

McMinn Formation (~1.4 Ga)

The McMinn Formation consists of coarse sandstone and oolitic ironstones deposited in a shallow-water inner shelf environment, grading upward into a quiescent outer shelf that gave rise to sandstones, siltstones, and shales (Jackson et al., 1988). The shales (Kyalla Member) generally have lower petroleum source potential then the underlying Middle Velkerri Member, but they are positioned stratigraphically to charge overlying Jamison sandstone reservoirs.

The McArthur Basin contains the oldest known freely flowing oil. Several stratigraphic test wells drilled by the Australian Bureau of Mineral Resources from 1979 to 1985 encountered oil shows in the McArthur Group and freely flowing oil from black mudstones of the Velkerri Formation (e.g. BMR Urapunga No. 4, Muir et al., 1980). Beginning in 1988, several exploration wells in the Beetaloo Sub-basin encountered oil in the Jamison sandstones (Lanigan et al., 1994). These oils correlate to the Middle Velkerri and Kyalla Member shales (Jackson et al., 1986; Kontorovich et al., 1996b; Warren et al., 1998; Zaunbrecher, 1988). The oils are composed mostly of $<C_{22}$ saturated hydrocarbons (~70‰) that are mainly n-alkanes, with minor amounts of n-alkylcycloalkanes, monomethylalkanes, and isoprenoids, and traces of tricyclic terpanes and steranes (Figure 18.7) (Summons et al., 1988a). This distribution indicates that prokaryotes contributed most of the organic matter to the source rocks. The presence of steranes and specific isoprenoids proves that some eukaryotic and archaeal microorganisms were also present. Subtle differences in the biomarkers between the marine and lacustrine rocks in this sequence indicate that depositional environments can be distinguished using these compounds, even in the Proterozoic Eon (Jackson et al., 1986).

Nonesuch Formation, central North America

The Middle Proterozoic Nonesuch Formation (~1.1 Ga) consists mainly of unmetamorphosed lacustrine siltstones and shales deposited in a euxinic, aborted rift in central North America. This mid-continental rift system began as a proto-oceanic rift but failed after reaching up to 90 km of extension in the Lake Superior area. It extends 1300 km from Lake Superior to northeastern Kansas. The formation hosts a large copper deposit at White Pine, Michigan, where black lacustrine siltstone generated oil seeps (Mauk and Hieshima, 1992). Thermal modeling results are consistent with in situ oil generation at White Pine (Price et al., 1996). Based on 183 outcrop and core samples from northern Wisconsin and Michigan, TOC varies from 0 to 2.5 wt.% and correlates strongly with depositional environment and the petrographic character of associated kerogen (Imbus et al., 1988). Biomarker stereoisomer ratios in extracts of Nonesuch rocks are consistent with moderate thermal maturity, except for very high maturity in the mineralized zone near White Pine (Pratt et al., 1991).

Riphean shales, eastern Siberia

The Siberian Craton contains thick sedimentary sequences (up to 12–14 km) of Riphean to Vendian (~1600–540 Ma) terrigenous carbonate successions

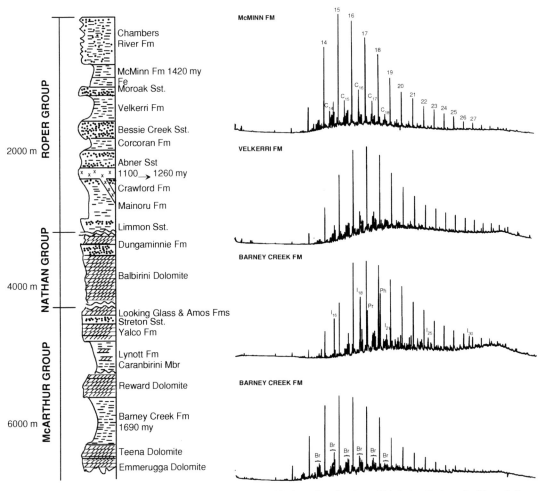

Figure 18.7. Representative gas chromatograms of C_{15+} saturated hydrocarbon fractions of McArthur Basin rocks. Normal alkanes indicated by carbon number, Reprinted from Summons *et al.* (1988a). © Copyright 1988, with permission from Elsevier. Br, cluster of monomethylalkanes; C_x, *n*-alkylcyclohexanes; I_x, isoprenoids.

(Khudoley *et al.*, 2001). More than 30 large accumulations of oil and gas were discovered in Vendian and Riphean reservoir rocks as old as ~1.4 Ga (Kontorovich *et al.*, 1998b). Reserve estimates at Verkhne-Chonskaya (1.5 billion barrels of oil) and Koviktinskoye (35.5 trillion cubic feet) in the Irkutsk region (Thompson and Voropanov, 1997), and the Yurubchen-Tokhomo oil and gas zone of the Kamovsk Arch in the Baykit High (>3.0 BBO), suggest the potential for the development of world-class supergiant fields. The reserves in the Yurubchen-Tokhomo structure may exceed the sum of all other discoveries in the Siberian Craton, occupying an area of ~16 000 km² with a 40-m oil column and a 90-m gas cap (Kontorovich *et al.*, 1997).

Precambrian oils from the Siberian Craton originated from source rocks with similar organic facies (Kashirtsev *et al.*, 1999; Kashirtsev and Philp, 1997; Kontorovich *et al.*, 1998b). They are light, low-sulfur crude oils dominated by saturated hydrocarbons. Pristane/phytane and CPI are typically less than one. The oils have unusually negative $\delta^{13}C$ (~−32 to −34‰) (Kashirtsev and Philp, 1997) and high relative proportions of mid-chain monomethylalkanes and ethylcholestane (Fowler and Douglas, 1987; Kontorovich *et al.*, 1997).

Although not correlated definitively, the source(s) of the Precambrian oils in the eastern Siberian basins are believed to be Riphean. There are several units within the Riphean sequence with high TOC, and there is little evidence for source rocks in the overlying Vendian or Cambrian strata (Ulmishek, 2001b). Black shales and marls occur throughout the Riphean argillaceous, argillaceous-siliceous, and argillaceous-carbonate formations of the Siberian Platform rims. These potential source rocks were deposited in deep, anoxic waters of a back-arc basin. Present-day TOC ranges from ~0.5 to 8 wt.%, and most samples have low HI (<160 mg HC/g TOC) because of high thermal maturity. The regional distribution of these potential source rocks is poorly known, and they were identified in only a few wells.

The Iremeken Formation (Upper Riphean) contains the highest-quality source rocks and generated crude oil (Figure 18.8). This formation is believed to occur only in structural depressions, but it has been encountered in only one well. The penetrated formation contains a ~10-m-thick dark brown shale with TOC >8 wt.% and HI >724 mg HC/g TOC (Filiptsov *et al.*, 1999). These shales entered the oil window after deposition of the Cambrian salt seal.

Any petroleum generated from potential source units before deposition of the Cambrian salt was likely destroyed by the Baikalian tectonic event (~850–820 Ma). From outcrops, the Shuntar Formation contains >1000 m of organic-rich black shales and argillaceous carbonates, with up to 8.7 wt.% TOC and >4% for more than a third of the formation (Kontorovich *et al.*, 1996a). These shales are presently metamorphosed and have low HI. Shales in the Vedreshev and Madrin formations contain up to 2.4 wt.% TOC, with HI as high as ~160 mg HC/g TOC, and are presently in the upper gas window. These shales expended most of their generative potential before deposition of the Cambrian salt seal. In some locations, oil generation could have occurred during the early Paleozoic Era.

Other potential Riphean source rocks

Romeiro Silva *et al.* (1998) proposed that the São Francisco Basin of central Brazil might contain a petroleum system with commercial accumulations. This intracratonic sag basin contains Riphean sedimentary rocks ranging from continental to restricted marine. The uppermost Traíras Group contains black shales with original TOC up to 4–6 wt.%. Seeps and subcommercial flow of thermogenic gas from the Vendian Bambuí fractured carbonates suggest that the Traíras was once an active source in the São Francisco Basin, where gas migrated into Vendian reservoir rocks capped by organic-lean shales and carbonates in the Upper Bambuí Formation.

The seven Purana basins of India (Cuddapah, Vindhyan, Chattishgarh, Bastar, Pranhita-Godavari, Bhima, and Kaladgi) are interior cratonic basins that may be related to the McArthur Basin of Australia before the breakup of a Pre-Rodinian supercontinent (Punati, 1999). These basins contain Precambrian sedimentary rocks ranging from 0.5 to 9 km thick and include mature clastic-carbonate-shale suites. Specific source rocks have not been identified by geochemical measurements. The occurrence of complete petroleum systems is highly speculative, although non-commercial flows of wet gas occur in the Vindhyan Basin (Padhy, 1997).

Note: Deposition of black shales appears to have peaked several times during the Precambrian Period (Figure 18.9). The most prominent peak occurred at 2.0–1.7 Ga, with lesser peaks in the Late Archean Eon (2.7–2.5 Ga) and the Late Neoproterozoic Era (800–600 Ma). Condie *et al.* (2001) proposed that enhanced deposition of these potential source rocks corresponds to periods of mantle superplume events and supercontinent formation. The superplumes could have introduced high concentrations of CO_2 into the atmosphere, resulting in global warming and an increase in the rate of carbon deposition. Productivity and preservation would have been enhanced by the increased supply of inorganic nutrients and the development of anoxic waters resulting from hydrothermal vents, disruption of oceanic currents, and an increase in the number of restricted marine basins.

LATE PROTEROZOIC PETROLEUM SYSTEMS

The Neoproterozoic Era was punctuated by several global glaciations that were so extreme that the surfaces of the oceans may have frozen completely. This so-called snowball Earth would have stressed most life

O.I.L.S.
Oil Information Library System

GEOMARK RESEARCH, INC.
9748 Whithorn Drive Tel: (281) 856-9333
Houston, Texas 77095 Fax: (281) 856-2987
info@geomarkresearch.com

GEOCHEMICAL SUMMARY SHEET
Country: **Russia**   Depth: **2463m**   02-Jul-99
Basin: **Baykit**   Age: **Riphean**   Sample ID: **ES001**
Field: **Yurubchen**   Formation: **Kamov Group**   LAT: 60.5767
Well: **24**   R   LONG: 96.8497

BULK PROPERTIES   API Gravity: **40.2**   % S: **0.23**   ppm V: **8.0**
                       % < C15: **30.7**   ppm Ni: **4.0**

C15+ Composition
% Sat: **64.9**
% Aro: **22.2**
% NSO: **13.0**
% Asph: **0.0**
Sat/Aro= **2.92**
n-Paraffin/Naphthene= **0.12**

Stable Carbon Isotope Composition
δ per mil PDB

C15+ Saturate: **-33.48**
C15+ Aromatic: **-33.20**
Canonical Variable: **-0.65**

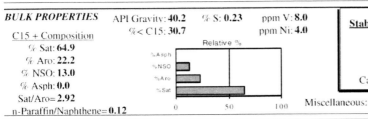

Miscellaneous:

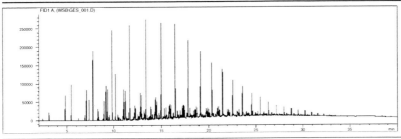

WHOLE CRUDE GAS CHROMATOGRAPHY

Pr/Ph= **0.87**
Pr/n-C17= **0.15**
Ph/n-C18= **0.23**
n-C27/n-C17= **0.08**
CPI= **1.084**

BIOMARKERS   ppm C30 Hopane: **57**

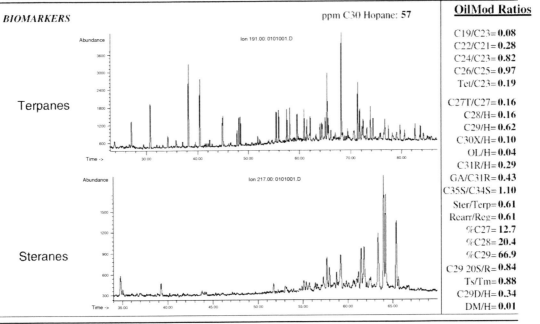

Terpanes

Steranes

OilMod Ratios

C19/C23= **0.08**
C22/C21= **0.28**
C24/C23= **0.82**
C26/C25= **0.97**
Tet/C23= **0.19**

C27T/C27= **0.16**
C28/H= **0.16**
C29/H= **0.62**
C30X/H= **0.10**
OL/H= **0.04**
C31R/H= **0.29**
GA/C31R= **0.43**
C35S/C34S= **1.10**
Ster/Terp= **0.61**
Rearr/Reg= **0.61**
%C27= **12.7**
%C28= **20.4**
%C29= **66.9**
C29 20S/R= **0.84**
Ts/Tm= **0.88**
C29D/H= **0.34**
DM/H= **0.01**

COMMENTS: Precambrian (Upper Riphean) Iremeken Shale

Figure 18.8. Geochemical data for crude oil from Yurubchen Field generated from the Upper Riphean Iremeken Formation, Baykit, Russia (courtesy of GeoMark Research, Inc.).

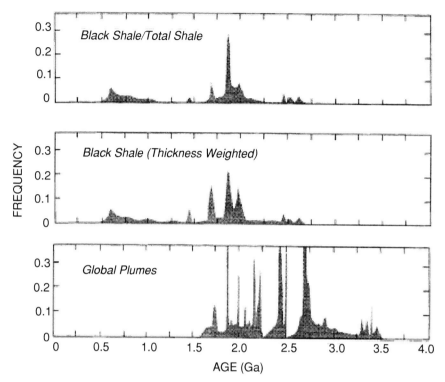

Figure 18.9. Time series of the frequency of black shale and global plumes. Reprinted from Condie *et al.* (2001). © Copyright 2001, with permission from Elsevier.

forms and greatly lowered total biomass. Consequently, source-rock deposition occurred mainly between Proterozoic glaciations (Figure 18.10).

Centralian Superbasin, Australia

During the Neoproterozoic Era, the interior of Australia consisted of one large depositional system, the Centralian Superbasin. This basin was disrupted ~580–600 Ma and reassembled during the late Paleozoic Era into numerous major (Amadeus, Georgina, and Officer) and minor (Aralka, Birrindudu, Ngalia, and Savory) sub-basins. Several individual petroleum systems occur within these sub-basins (Jackson *et al.*, 1984; Summons and Powell, 1991). The oldest source rocks for these petroleum systems are Neoproterozoic (~800–750 Ma) carbonates, marls, and shales of the Bitter Springs, Albinia, and Browne formations. Oil and gas shows were reported in the Magee No. 1 well. Postglacial marine shales from ~650 Ma of the Rinkabeena Formation are responsible for gas in the Ooraminna No. 1 and Davis No. 1 wells. Exoil's 1963 Ooraminna No. 1 well is historically significant because it was the first exploration well to discover hydrocarbons in Precambrian rocks. The well penetrated the Proterozoic strata at 1525 feet (465 m) and continued to a depth of 6105 feet (1861 m), where it encountered salt. Shales of moderate source quality and wet gas from a DST in dolomitic limestone proved the existence of a complete petroleum system. A younger Neoproterozoic petroleum system is derived from ~600-Ma postglacial marine shales of the Pertatatak and Rodda formations, which are believed to be source rocks for the Dingo gas field.

At present, the Neoproterozoic source rocks of the Centralian Superbasin typically contain <1 wt.% TOC with HI below 100 mg HC/g TOC. The low generative potential reflects the fact that these rocks are postmature. Less mature samples from the upper Neoproterozoic Era have HI of 200–400 mg HC/g

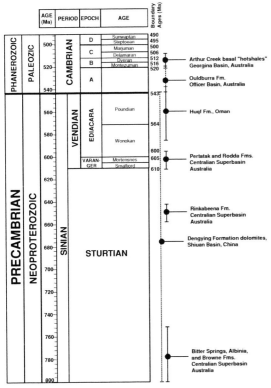

Figure 18.10. Neoproterozoic and Lower Paleozoic source rocks.

TOC. Although the kerogen has been characterized as type III (Jackson et al., 1984), the original kerogen was type I/II. Biomarker and isotopic distributions in organic matter from the Centralian source rocks indicate algal marine input (Logan et al., 1997)

Chuar Group, Utah

The Walcott Member of the Kwagunt Formation in the Upper Proterozoic Chuar Group from southern Utah contains organic-rich black shales (Dehler et al., 2001). Outcrops of these shales in the Grand Canyon have 3–9 wt.% TOC, hydrogen indices up to 255 mg HC/g TOC, and maximum maturity within the oil window (Uphoff, 1997). Modeling indicates a potential 150-square-mile (390 km^3) area with a minimum generative potential of 2700 MBO. The proposed petroleum system consists of Chuar source rock, unconformably overlain by Tapeats Sandstone reservoir, conformably overlain by Bright Angel Shale seal. A potential complication is the age of the source (∼850 Ma) and the long duration of uplift and erosion represented by the unconformity with the overlying Tapeats Sandstone reservoir.

Two wells tested the Tapeats sandstones in southern Utah and encountered oil shows that were chemically distinct from oil produced from upper Paleozoic sandstones and the tar sands common to the region. The compositions of the Tapeats oil shows, however, were similar to extracts from the Chuar Group (Lillis et al., 1995). Summons et al. (1988b) studied the biomarker distribution of extracts from the Walcott.

Infracambrian Huqf Supergroup, Oman

The Infracambrian (Late Vendian–Early Cambrian) Huqf Supergroup of Oman contains several clastic and carbonate source rocks of exceptional quality (Grantham et al., 1988; Pollastro, 1999; Terken et al., 2001). The oldest source rocks are shales and marls in the pre-rift Buah and Shuram formations (Nafun Group, Late Vendian) (Richard et al., 1998). The overlying Lower Cambrian Ara Group is a carbonate-evaporite rift sequence that was deposited in geographically restricted basins where stratified, anoxic conditions prevailed and organic-rich sediments and salts accumulated (Figure 18.11). Source rocks within the Ara Group are the U-Shale Formation and the Athel silicilyte (Al Shomou Formation). The Athel silicilyte is both source and reservoir because this laminated chert unit is completely encapsulated by thick halite (Amthor et al., 1998; 1999).

Oil generated from the Huqf source rock is typically light (25–45°API), sour (1–2 wt.% sulfur), and unusually depleted in ^{13}C (δ^{13}C = −40 to −35‰). Marmoul oil from South Oman is produced from the Permian Haushi Sandstone (Figure 18.12). This oil has lower API gravity (12° API) than many other Huqf-sourced oils because it was partially biodegraded. Huqf oils from the northern Ghaba Salt Basin are enriched in ^{13}C compared with those from the southern Oman Salt Basin, possibly due to higher thermal maturity. Biomarker distributions of these Infracambrian oils are characterized by a strong predominance of C_{29} steranes, a high ratio of 24-isopropyl/n-propyl C_{30} steranes, and the presence of the so-called X-compounds (Grantham et al., 1988; 1990). The latter are a homologous series of mid-chain methylalkanes (Klomp, 1986) that also occur in other Precambrian-Cambrian oils (Fowler and Douglas, 1987; Summons et al., 1988a; Summons et al.,

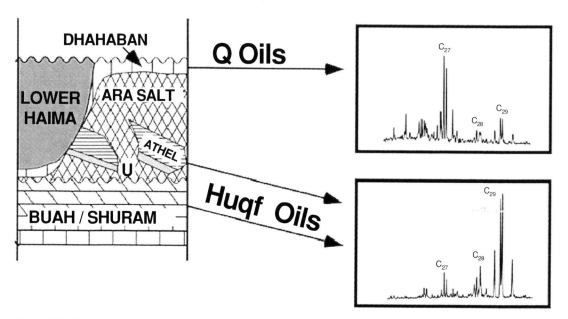

Figure 18.11. Stratigraphic relationship of Infracambrian source rocks from Oman and typical sterane mass chromatograms of related oils (from Terken et al., 2001). Reprinted by permission of the AAPG, whose permission is required for further use.

1988b) and are thought to originate from symbiotic bacteria living within desmosponges (Thiel et al., 1999a).

Crude oils in the Permian Gharif Formation in central and northern Oman and in the Cretaceous Shu'aiba Formation in the Ghaba Salt Basin probably originated from Infracambrian source rock (Grantham, 1986a; Grantham et al., 1988; Grantham et al., 1990). These so-called Q-oils differ from other Infracambrian oils in having low sulfur content (<0.5%), a predominance of C_{27} steranes and tricyclic terpanes, and a peak ('A') of unknown structure in the m/z 217 ion trace (Figures 18.13 and 18.14). Q-oils are isotopically enriched in ^{13}C ($\delta^{13}C$ ~ -32 to $-28\permil$) compared with the Huqf oils. They have the mid-chain monomethylalkanes (X compounds) that are common among Late Precambrian-Cambrian oils. Note that the absence of mid-chain monomethylalkanes does not preclude an Infracambrian source. For example, the Infracambrian Baghewala-1 oil from India lacks these compounds (Peters et al., 1995). Terken and Frewin (2000) and Terken et al. (2001) correlated the Q-oils to post-salt Dhahaban source rocks. The Dhahaban is the cap rock for the Huqf Supergroup. It contains a sequence of carbonate rocks with high source potential, interbedded with dolomitized limestones and anhydrites. The amorphous type I/II organic matter is mostly dispersed but also occurs as discrete lenses and layers.

Note: China, along with Siberia and Oman, is frequently cited as containing large reserves of Precambrian oil and gas. The large Weiyuan gas field in the Sichuan Basin in southwestern China produces from the Sinian Dengying Formation dolomites and was probably generated from algal organic matter within the reservoir dolomites (Korch et al., 1991). However, Precambrian oil in China is more elusive. Riphean Era rocks of the Yanshan fold belt are reported to have good source potential, with conditions favorable for a prospective petroleum system in the fold belt and the nearby Bohai Bay Basin. The giant Renqiu Field in the Bohai Bay Basin in northern China produces oil from the Riphean Wumishan Formation dolomites but was generated by overlying Oligocene Shahejie Formation shales (Hao Shisheng and Guangdi Liu, 1989). Although there are known occurrences of migrated oils and bitumens in the Precambrian Eon of China (e.g. Wang and Simoneit, 1995; Peng et al., 1998), correlation of produced oil to Precambrian source rocks has not been documented.

O.I.L.S.
Oil Information Library System

GEOMARK RESEARCH, INC.
9748 Whithorn Drive Tel: (281) 856-9333
Houston, Texas 77095 Fax: (281) 856-2987
info@geomarkresearch.com

GEOCHEMICAL SUMMARY SHEET

Country: **Oman** Depth: 02-Jul-99
Basin: **South Oman** Age: **Permian** Sample ID: **OM012**
Field: **Marmul** Formation: **Haushi** LAT: 18.1
Well: **Sandstone** LONG: 55.2

BULK PROPERTIES API Gravity: **12.4** % S: **1.59** ppm V: **71.0**
%< C15: **4.0** ppm Ni: **18.0**

C15 + Composition
% Sat: **31.8**
% Aro: **48.2**
% NSO: **17.2**
% Asph: **2.8**
Sat/Aro= **0.66**
n-Paraffin/Naphthene= **0.24**

Stable Carbon Isotope Composition
δ per mil PDB

C15+ Saturate: **-35.59**
C15+ Aromatic: **-35.89**
Canonical Variable: **-1.28**

Miscellaneous:

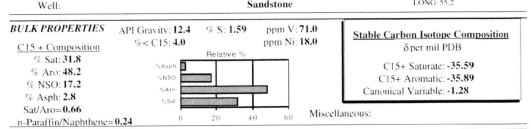

WHOLE CRUDE GAS CHROMATOGRAPHY

Pr/Ph= **0.52**
Pr/n-C17= **1.16**
Ph/n-C18= **2.36**
n-C27/n-C17= **0.22**
CPI= **1.081**

BIOMARKERS ppm C30 Hopane: **1396**

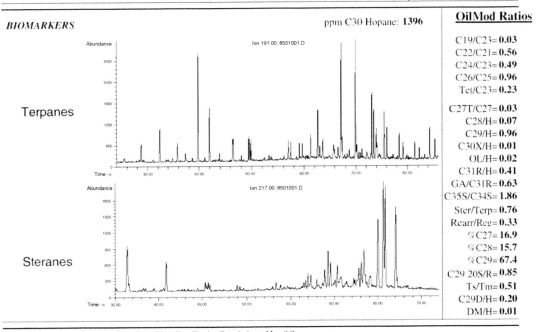

OilMod Ratios

C19/C23= **0.03**
C22/C21= **0.56**
C24/C23= **0.49**
C26/C25= **0.96**
Tet/C23= **0.23**

C27T/C27= **0.03**
C28/H= **0.07**
C29/H= **0.96**
C30X/H= **0.01**
OL/H= **0.02**
C31R/H= **0.41**
GA/C31R= **0.63**
C35S/C34S= **1.86**
Ster/Terp= **0.76**
Rearr/Reg= **0.33**
%C27= **16.9**
%C28= **15.7**
%C29= **67.4**
C29 20S/R= **0.85**
Ts/Tm= **0.51**
C29D/H= **0.20**
DM/H= **0.01**

COMMENTS: Infracambrian (Late Vendian/Early Cambrian) Huqf Supergroup

Figure 18.12. Geochemical data for crude oil from Marmoul Field generated from the Infracambrian Huqf Formation (courtesy of GeoMark Research, Inc.).

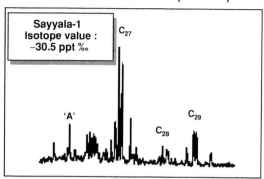
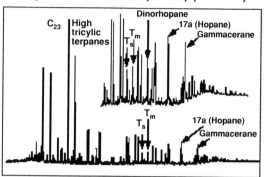

Figure 18.13. Ion chromatograms showing the distribution of steranes and terpanes in a typical Q oil (from Terken and Frewin, 2000). Reprinted by permission of the AAPG, whose permission is required for further use.

PHANEROZOIC PETROLEUM SYSTEMS

The occurrence of petroleum source rocks during Phanerozoic time appears to be controlled largely by tectonically driven interactions of eustasy, subsidence, and climate. Several episodes of widespread organic-rich source rock deposition can be documented, as listed below. Most of these episodes correspond to times of eustatic sea-level rise and widespread basin formation (Figure 17.25). Most of the global petroleum resource originated from effective source rocks that were deposited during major first-order rises and highstands of sea level in the Jurassic-Cretaceous (see Figure 18.56) and, to a lesser extent, Silurian-Devonian (Figure 18.15) periods. Global marine transgressions result in larger areas of continental shelves and interiors under stable water columns, which may become anoxic. At these times, the supercontinents were also breaking apart into smaller land masses, and global climates were warm, leading to poorly connected and ventilated oceans that were prone to anoxia.

The largest volumes of oil-prone source rocks were deposited during the Middle Jurassic to Early Cretaceous periods in the Middle East and Arctic regions. Significant amounts of lacustrine source rocks were deposited in China (Permian-Eocene), Southeast Asia (Tertiary), and the South Atlantic (Mesozoic). Finally, large volumes of source rocks with significant land-plant input were deposited during the Tertiary Period in paralic, low to middle latitude regions.

PALEOZOIC ERA: CAMBRIAN SOURCE ROCKS

Most plate tectonic reconstructions place the Iapetus Ocean at a southern latitude within 30° of the equator during the Cambrian Period. Deposition in the earliest Late Cambrian Period occurred near the peak of a first-order eustatic sea-level rise (Figure 17.25). Organic-rich rocks were deposited under tropical to subtropical conditions on shallow marine platforms in the Baltic region (Alum Shale or Alunskiffer) (Figure 18.15) and parts of the present-day UK (Stockingford Shales, Clara Group) within the Iapetus Ocean.

Middle Cambrian Arthur Creek Formation, Australia

Following the breakup of the Centralian Superbasin, the prolific basal hot shale of the Middle Cambrian Arthur Creek Formation was deposited in the southern Georgina Basin, the largest of the Neoproterozoic-Paleozoic basins on the North Australian Craton (Ambrose et al., 2001). These pyrite-rich shale source rocks were deposited on a gently undulating unconformity under anoxic conditions that graded upward into a more oxic carbonate facies (Figure 18.16). The principal source facies is as thick as 60 m, with 0.5–16 wt.% TOC and HI in the range 600–800 mg HC/g TOC. Thinner source beds extent into the Arthur Creek Formation up to 100 m above the basal hot shale. The source shales cover ~80 000 km^2.

O.I.L.S.
Oil Information Library System

GEOMARK RESEARCH, INC.
9748 Whithorn Drive Tel: (281) 856-9333
Houston, Texas 77095 Fax: (281) 856-2987
info@geomarkresearch.com

GEOCHEMICAL SUMMARY SHEET

Country: **Oman** Depth: **4213-4219'** 02-Jul-99
Basin: **South Oman** Age: **Permian** Sample ID: **OM015**
Field: **Sayyala** Formation: **Gharif** LAT: 19.55
Well: **1** **Sandstone** LONG: 55.77

BULK PROPERTIES API Gravity: **43.4** % S: **0.10** ppm V: **14.0**
%< C15: **54.8** ppm Ni: **12.0**

C15+ Composition
% Sat: **75.4**
% Aro: **15.6**
% NSO: **9.0**
% Asph: **0.0**
Sat/Aro= **4.83**
n-Paraffin/Naphthene= **0.12**

Stable Carbon Isotope Composition
δ per mil PDB

C15+ Saturate: **-30.64**
C15+ Aromatic: **-30.32**
Canonical Variable: **-1.44**

Miscellaneous:

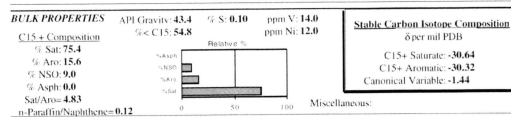

WHOLE CRUDE GAS CHROMATOGRAPHY

Pr/Ph= **1.09**
Pr/n-C17= **0.63**
Ph/n-C18= **0.63**
n-C27/n-C17= **0.17**
CPI= **1.103**

BIOMARKERS ppm C30 Hopane: **20**

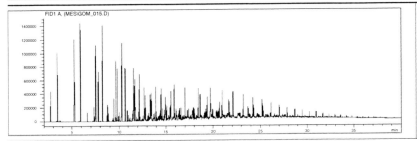

Terpanes

Steranes

OilMod Ratios

C19/C23= **0.07**
C22/C21= **0.38**
C24/C23= **0.88**
C26/C25= **1.02**
Tet/C23= **0.10**

C27T/C27= **0.07**
C28/H= **0.42**
C29/H= **0.65**
C30X/H= **0.22**
OL/H= **0.13**
C31R/H= **0.28**
GA/C31R= **3.86**
C35S/C34S= **1.50**
Ster/Terp= **2.01**
Rearr/Reg= **0.53**
%C27= **74.0**
%C28= **8.6**
%C29= **17.3**
C29 20S/R= **0.86**
Ts/Tm= **2.26**
C29D/H= **0.50**
DM/H= **0.24**

COMMENTS: Infracambrian "Q" Source

Figure 18.14. Geochemical data for Q-type crude oil from Sayyala Field generated from the Infracambrian Dhahaban Formation (courtesy of GeoMark Research, Inc.).

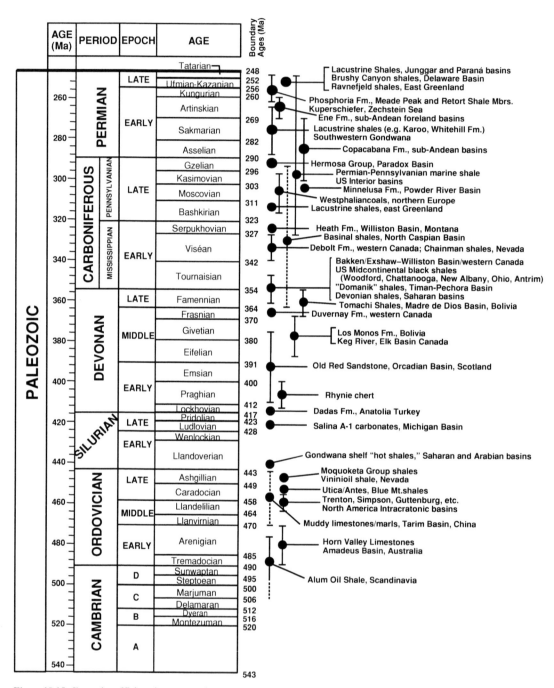

Figure 18.15. Examples of Paleozoic source rocks.

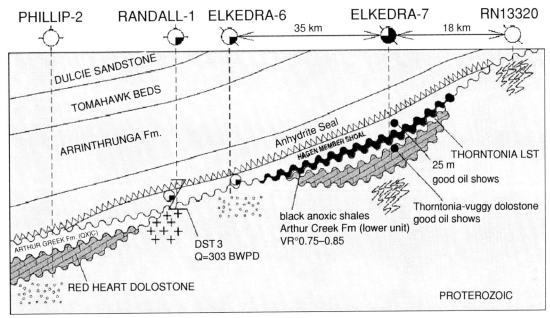

Figure 18.16. Stratigraphic plays within Middle-Upper Cambrian strata of the Georgina Basin, Australia (from Ambrose et al., 2001). The basal Arthur Creek organic-rich shales are the source and bottom seal. Reprinted by permission of the AAPG, whose permission is required for further use. VR, vitrinite reflectance.

Middle Cambrian–Lower Ordovician Alum Shale, Baltic Scandinavia

Thermal maturity of the Alum Shale in Sweden and the surrounding Baltic countries ranges from immature to postmature. Where immature, the Alum Shale and its time equivalents, such as the *Dictyonema* Shale in Estonia, contain ~10–20 wt.% TOC, with atomic H/C ratios in the range 1.0–1.2 (HI ~600 mg HC/g TOC) (Lewan and Buchardt, 1989). Nevertheless, minor production of oil generated from the Alum Shale is limited to the island of Gotland (Figure 18.17), due mainly to unfavorable timing of maturation relative to trap formation and poor reservoir preservation (Dahl et al., 1989).

The Alum Shale is chemically unusual because the kerogen originated from algae but is rich in aromatic moieties. This is attributed to post-depositional irradiation damage induced by high concentrations of uranium in the shale. With increasing uranium content, the atomic H/C of Alum Shale kerogen decreases. Higher uranium content corresponds to condensate-like oils with increasing proportions of aromatic and polar compounds and decreasing proportions of saturated hydrocarbon, including biomarkers such as hopanes (Dahl et al., 1988; Lewan and Buchardt, 1989; Dahl, 1990; Bharati et al., 1995). The Alum Shale also contains chemically distinct, vitrinite-like macerals that did not originate from land-plant material (Lewan and Buchardt, 1989; Buchardt and Lewan, 1990). It is possible that the aromatic character of the Alum Shale kerogen results from macromolecular input from marine green microalgae. *Chlorella marina* contains algaenan that yields aromatic-rich pyrolyzates (Derenne et al., 1996).

ORDOVICIAN SOURCE ROCKS

Widespread organic-rich source rocks were deposited during the Late Ordovician Period (Caradocian Stage) (see Figure 17.10) as part of an extensive marine transgression before the major glaciation (Figure 17.25). Upper Ordovician black shales are common in North America (Sylvan, Utica, Vinini, and equivalents) and northern Europe (*Dicellograptus* and *Clinograptus* shales). Ordovician source rocks also occur in Russia, South Africa, and central Australia. Examples of worldwide Ordovician source rocks are given in Table 18.5.

O.I.L.S.
Oil Information Library System

GEOMARK RESEARCH, INC.
9748 Whithorn Drive   Tel: (281) 856-9333
Houston, Texas 77095   Fax: (281) 856-2987
info@geomarkresearch.com

GEOCHEMICAL SUMMARY SHEET

Country: **Sweden**   Depth: 02-Jul-99
Basin: **Gotland Island**   Age: **Ordovician**   Sample ID: **SW003**
Field: **Gotland Oil**   Formation: LAT: 57.73
Well: LONG: 18.7

BULK PROPERTIES   API Gravity: **36.1**   % S: **0.21**   ppm V: **38.0**
% < C15: **45.4**   ppm Ni: **8.0**

C15+ Composition
% Sat: **52.9**
% Aro: **35.9**
% NSO: **8.1**
% Asph: **3.1**
Sat/Aro= **1.47**
n-Paraffin/Naphthene= **0.38**

Stable Carbon Isotope Composition
δ per mil PDB
C15+ Saturate: **-30.73**
C15+ Aromatic: **-29.76**
Canonical Variable: **0.03**

Miscellaneous:

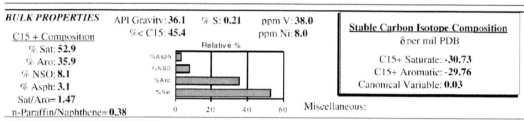

WHOLE CRUDE GAS CHROMATOGRAPHY
Pr/Ph= **2.04**
Pr/n-C17= **0.80**
Ph/n-C18= **0.43**
n-C27/n-C17= **0.20**
CPI= **1.004**

BIOMARKERS   ppm C30 Hopane: **15**

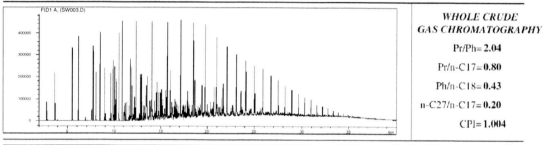

Terpanes

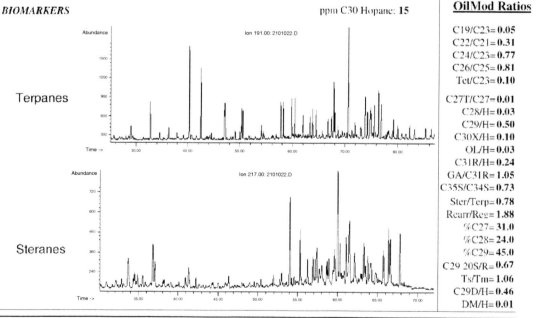

Steranes

OilMod Ratios
C19/C23= **0.05**
C22/C21= **0.31**
C24/C23= **0.77**
C26/C25= **0.81**
Tet/C23= **0.10**
C27T/C27= **0.01**
C28/H= **0.03**
C29/H= **0.50**
C30X/H= **0.10**
OL/H= **0.03**
C31R/H= **0.24**
GA/C31R= **1.05**
C35S/C34S= **0.73**
Ster/Terp= **0.78**
Rearr/Reg= **1.88**
%C27= **31.0**
%C28= **24.0**
%C29= **45.0**
C29 20S/R= **0.67**
Ts/Tm= **1.06**
C29D/H= **0.46**
DM/H= **0.01**

COMMENTS: Upper Cambrian/Lower Ordovician Alum Shale

Figure 18.17. Geochemical data for Gotland crude oil generated from the Cambrian/Ordovician Alum Shale, Sweden (courtesy of GeoMark Research, Inc.).

Table 18.5. *Worldwide examples of Ordovician source rocks*

	Formation	Basins	References
Assemblage A	Decorah (Guttenburg Member), Glenwood Shales	Iowa, Illinois	See text
	Kukersite	Baltic	See text
	Winnipeg, Icebox Shale, Bighorn Group	Williston	Williams (1974), Jarvie (2001)
	Yeoman	Williston	Stasiuk (1991), Stasiuk et al. (1993), Osadetz and Snowdon (1995)
	Red River	Williston	Jarvie (2001)
	Simpson	Forest City, Permian, Anadarko, Salina-Sedgwick	Hatch et al. (1987), Wang and Philp, (1997a), Hatch and Newell (1999), Newell and Hatch (2000)
	Trenton Limestone	Michigan	Reed et al. (1986)
	Goldwyrn	Canning	Foster et al. (1986), Hoffmann et al. (1987)
Assemblage B	Collingwood Shale, Point Pleasant	Ontario, Ohio	See text
	Utica/Antes Shale	Michigan, Appalachian	See text
	Maquoketa Group	Illinois	See text
	Horn Valley Siltstones	Amadeus	See text
	Carbonate muds	Tarim	See text
	Womble Shale	Ouachita Mountains	Fowler and Douglas (1984)
	Vinini	Roberts Mountain, Nevada	Fowler and Douglas (1984)

Ordovician source rocks contain two distinct kerogen assemblages: Assemblage A is telalginite, derived primarily from *Gloeocapsomorpha prisca*. Assemblage B is amorphous kerogen, derived from mixtures of marine algae (Table 18.6) (Hatch et al., 1987 Jacobson et al., 1988). Each assemblage may dominate a source unit or occur as mixed interbedded laminations within the same source unit.

The biological affinity of *G. prisca* has a long history of debate (Fowler, 1992). Zalessky (1917) first characterized the organism in Estonian kukersite, which consists nearly entirely of *G. prisca* remains, and named the colonial organism based on a morphological similarity to the extant cyanobacteria *Gloeocapso* (*Gloeocapsomorpha prisca* Zalessky 1917 is the complete name for *G. prisca*). Because the oil generated from *G. prisca* source rocks contains low concentrations of pristane, phytane, and steranes, Reed et al. (1986) concluded that the microorganism was a benthic non-photosynthetic mat-forming bacterium. However, Hoffmann et al. (1987) showed that pristane, phytane, and steroidal hydrocarbons occur in pyrolyzates of *G. prisca*-dominated source rocks from Australia and concluded that the microorganism was a phototrophic, planktonic alga. While the samples used by Hoffmann et al. (1987) were predominately assemblage A, they contain varying amounts of assemblage B. Thus, the biomarkers diagnostic of photosynthetic eukaryotes cannot be correlated exclusively with *G. prisca*. Foster et al. (1989; 1990) suggested that morphological and biogeochemical characteristics of *G. prisca* are very similar to those of *Entophysalis major*, a mat- and sometimes stromatolite-forming cyanobacterium.

Microscopy and spectroscopy revealed that *G. prisca* exhibits three distinctly different morphologies that differ chemically. *G. prisca* macerals varied from

Table 18.6. *Petrographic and geochemical characteristics of* Gleocapsamorpha prisca *Assemblages A and B*

	Assemblage A	Assemblage B
Kerogen in transmitted light	Yellow to reddish brown "platy" telaglinite; *G. prisca* exhibit varying morphologies from thin-walled small conglomerates, to thick-walled larger conglomerates, to a stromatolitic form; possibly due to different life stages (Stasiuk and Osadetz, 1990)	Gray to brownish yellow, amorphous "fluffy" masses or sheet-like particles with a spongy ragged edge; fewer microfossils than Assemblage A; small amounts of *G. prisca* may be present
Kerogen in fluorescent light	Bright yellow to yellow-orange	Dull yellow brown; fluorescent liptodetrinite, commonly associated with degraded microplankton
TOC (wt.%)	Typically 5–10%; up to 77% in some beds	Typically <3%; beds with higher TOC are usually mixed A + B assemblages
Hydrogen index (HI)	~750–1000 mg HC/g TOC	~300–500 mg HC/g TOC; higher HI are usually mixed A + B assemblages; HI may be lower due to oxidation
Oxygen index	<50 mg CO_2/g TOC	<50 mg CO_2/g TOC; may be lower due to oxidation

TOC, total organic carbon.

thin-walled small conglomerates, to thick-walled larger conglomerates, to stromatolitic forms (Foster *et al.*, 1989). Each morphology differs in chemical composition (Derenne *et al.*, 1990; Stasiuk *et al.*, 1993). These morphologies may represent lifecycle stages induced by changing water depth, oxygen supply, and/or salinity (Stasiuk and Osadetz, 1990). Stasiuk and Osadetz (1990) considered these morphological changes in *G. prisca* to be consistent with a cyanobacterium. Conversely, Derenne *et al.* (1992) showed that the biochemical adaptations to increasing salinity made by the normally freshwater green algae *Botyroccoccus braunii* are similar to those in *G. prisca*. Derenne *et al.* (1992) concluded that *G. prisca* might be a related organism. Metzger and Largeau (1994) and Pancost *et al.* (1998) provide additional biomarker and geochemical evidence suggesting that *G. prisca* microfossils are the selectively preserved algaenan cell walls from photosynthetic, eukaryotic microalgae. However, while *Botyroccoccus* is a freshwater green alga that can adapt to saline conditions, *G. prisca* was clearly an obligate marine organism. The resorcinolic lipids in *G. prisca* that are suggestive of *Botyroccoccus* may be polymerized cyanobacterial sheath material excreted as an antioxidant and/or UV filter (Blokker *et al.*, 2001).

G. prisca flourished during the Ordovician Period. There is scattered fossil evidence for *G. prisca* in the Lower Cambrian Ouldburra Formation, Officer Basin, Australia (Kamali, 1995; Michaelsen *et al.*, 1995) and the Middle Cambrian Mt Cap Formation, Northern Interior Plains, Canada (Wielens *et al.*, 1990; Dixon and Stasiuk, 1998). It may have extended into the Silurian Period (Abrams *et al.*, 1998). However, during the Ordovician Period, *G. prisca* was widespread throughout the epicontential seas, and in some environments it was the only organism contributing primary biomass.

Pancost *et al.* (1998) provide an explanation for *G. prisca*'s ability to dominate an ecological niche. Detailed molecular and isotopic analysis of core samples from above, within, and below the Guttenberg Member of the Decorah Formation (Figure 18.18) revealed changes in depositional redox conditions that correspond with the occurrence of *G. prisca* facies. The Guttenberg Member, and to a lesser degree the underlying Spechts Ferry Member, is dominated by organic matter (up to 40.8 wt.% TOC) derived from *G. prisca*. Biomarkers indicate that oxic conditions prevailed during their deposition. A sharp increase in ^{13}C-enriched aryl isoprenoids and gammacerane at the base of the Spechts Ferry Member indicates

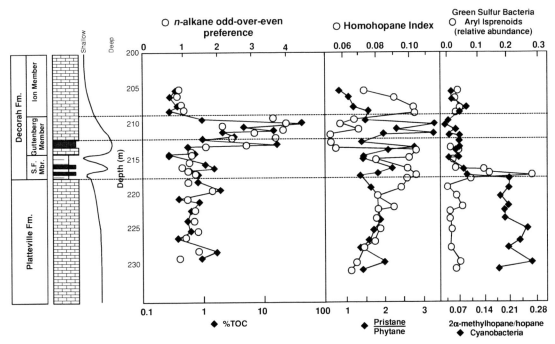

Figure 18.18. Geochemical profiles for core samples through the Platteville and Decorah formations (Cominco SS-9 core, Millbrook Farms, Jackson Co., Iowa). Wt.% total organic carbon (TOC), plotted on a log scale, is roughly proportional to the contribution of *Gloeocapsomorpha prisca*, as indicated by the odd-to-even carbon preference of mid-range n-alkanes [$(C_{17} + C_{19})/(2 \times C_{18})$]. Changes from anoxia in the Platteville and Specht Ferry to oxic conditions in the Guttenberg Member are indicated by shifts in pristane/phytane and homohopane index ($C_{35}/\Sigma C_{31}$–C_{35}). The relative abundance of aryl isoprenoids and 2α-methylhopanes are proxies for green-sulfur bacteria and cyanobacteria, respectively. Reprinted from Pancost *et al.* (1998). © Copyright 1998, with permission from Elsevier.
S.F. Mbr., Spechts Ferry Member.

development of a low-Eh photic zone. Under such dysoxic to anoxic conditions, the occurrence of *G. prisca* is limited, while 2α- and 3β-methylhopane become relatively abundant. These A-ring methylated hopanes can originate from various eubacteria. Based on δ^{13}C of -32 to -28‰, which are consistent with the values of pristane and phytane, Pancost *et al.* (1998) concluded that cyanobacteria rather than methylotrophic bacteria were the primary source of organic matter in the older Platteville Formation strata.

Preservation of abundant organic matter under oxic conditions appears to contradict conventional theories of source-rock deposition. Pancost *et al.* (1998) suggest that *G. prisca* played an important role in organic matter preservation by depositing a thick layer of refractory cellular material that limited diffusion of electron acceptors (oxygen and sulfate) from the bottom waters into the sediments. This could have rendered the underlying sediments anoxic, severely limiting bioturbation.

G. prisca may have grown either as seasonal planktonic blooms or as benthic mats, resulting in discrete organic-rich layers 1–2 cm thick that are separated by thin layers of shale or carbonate mud.

Crude oils generated from Mid-Continent Ordovician source rocks (Figure 18.19) exhibit a wide range of stable carbon isotopic values ($\delta^{13}C_{sat} \sim -34$ to -25‰, $\delta^{13}C_{aro} \sim -34$ to -24‰). Hatch *et al.* (1987) found that extracts from Middle Ordovician shales exhibited similar ranges in δ^{13}C, with more negative and positive values in the lower (Platteville Formation) and upper (Guttenberg Member) source units, respectively. Hatch *et al.* (1987) suggested that the widespread isotopic excursion reflected a global change in the carbon cycle, resulting in high carbon sequestration and restricted water circulation. Compound-specific isotopic analysis of hydrocarbons and TOC from the Cominco SS-9 core suggest that the positive δ^{13}C Middle Ordovician excursion is not the result of changes in atmospheric CO_2 or

O.I.L.S.
Oil Information Library System

GEOMARK RESEARCH, INC.
9748 Whithorn Drive   Tel: (281) 856-9333
Houston, Texas 77095   Fax: (281) 856-2987
info@geomarkresearch.com

GEOCHEMICAL SUMMARY SHEET

Country: **USA**   Depth: 02-Jul-99
Basin: **Williston**   Age: **Ordovician**   Sample ID: **MT004**
Field: **Raymond**   Formation: **Red River**   LAT: 48.875
Well: **State 16**   LONG: -104.657

BULK PROPERTIES

API Gravity: **40.8**   % S: ppm V:
% < C15: **39.2**   ppm Ni:

C15 + Composition
% Sat: **53.0**
% Aro: **31.4**
% NSO: **14.3**
% Asph: **1.3**
Sat/Aro= **1.69**
n-Paraffin/Naphthene=

Stable Carbon Isotope Composition
δ per mil PDB

C15+ Saturate: **-28.31**
C15+ Aromatic: **-28.11**
Canonical Variable: **-2.43**

Miscellaneous:

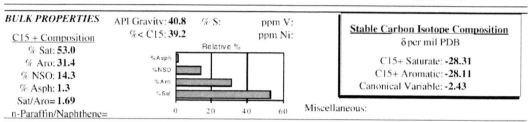

WHOLE CRUDE GAS CHROMATOGRAPHY

Pr/Ph= **0.58**
Pr/n-C17= **0.06**
Ph/n-C18= **0.33**
n-C27/n-C17= **0.06**
CPI= **1.148**

BIOMARKERS   ppm C30 Hopane: **169**

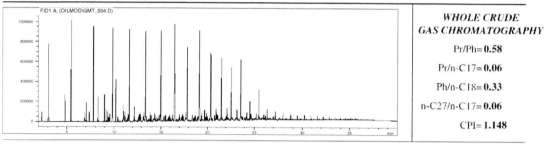

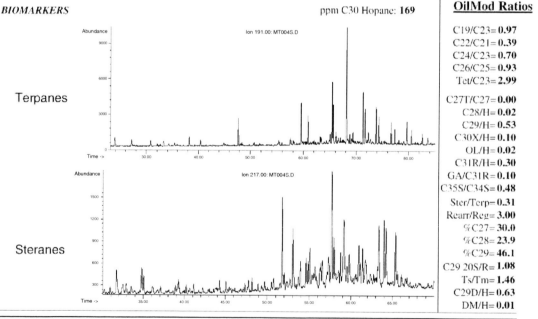

OilMod Ratios

C19/C23= **0.97**
C22/C21= **0.39**
C24/C23= **0.70**
C26/C25= **0.93**
Tet/C23= **2.99**
C27T/C27= **0.00**
C28/H= **0.02**
C29/H= **0.53**
C30X/H= **0.10**
OL/H= **0.02**
C31R/H= **0.30**
GA/C31R= **0.10**
C35S/C34S= **0.48**
Ster/Terp= **0.31**
Rearr/Reg= **3.00**
%C27= **30.0**
%C28= **23.9**
%C29= **46.1**
C29 20S/R= **1.08**
Ts/Tm= **1.46**
C29D/H= **0.63**
DM/H= **0.01**

COMMENTS: Ordovician Red River Formation

Figure 18.19. Geochemical data for crude oil from Raymond Field generated from the Ordovician Red River Formation, Williston Basin, USA (courtesy of GeoMark Research, Inc.).

global conditions but reflects a shift from normal marine to *G. prisca* facies (Pancost *et al.*, 1999). Hydrocarbons attributed to *G. prisca* are ~7‰ enriched in ^{13}C compared with those from other algal sources. Thus, the isotopic excursion observed by Hatch *et al.* (1987) may be caused by local differences in sequestered biomass rather than a global change in isotope mass balance.

Lower Ordovician Horn Valley Siltstone, Australia

Petroleum in the two largest fields under production in the Amadeus Basin of central Australia – the Mereenie oil and Palm Valley gas fields – was generated primarily from the Lower Ordovician Horn Valley Siltstone (Do Rozario, 1990; Havord, 1993). The formation consists of euxinic, pyritic siltstones and shales with thinly interbedded limestone and dolomite that were deposited under shallow marine conditions. Present-day TOC averages ~0.96%, with maxima of ~2.3% (Gorter, 1984; Jackson *et al.*, 1984). Measured HI are low, but the original generative potential is thought to be considerable as the source unit is nearly through the oil window in the location of the Mereenie Field and well within the zone of gas catagenesis around the Palm Valley gas field. Mereenie Field oil and Horn Valley Siltstone extracts have the odd-to-even preference in C_{15}–C_{20} *n*-alkanes that is diagnostic of the *G. prisca* organic facies. They typically have low pristane/phytane ratios (~0.8–1.2) and quite negative δ^{13}C (−30 to −31‰).

Middle Ordovician Galena Group, Iowa Shelf, and Illinois Basin

Many Middle Ordovician sequences in North American intracratonic basins have carbonate source sequences with Assemblage A kerogen. These approximately age-equivalent strata include the Winnipeg Formation (Williston Basin), the Trenton-Black River Formation (Michigan Basin), the Simpson Group in central Kansas and Oklahoma, and the Glenwood and Decorah formations. Deposition of the latter two formations was controlled by the transcontinental arch and corresponds to a semi-arid belt running through Illinois, Iowa, and Wisconsin.

Source rocks with exceptional potential occur in the Glenwood Shales (up to 25 wt.% TOC) and the Guttenberg Member of the Decorah Formation (up to 43 wt.% TOC) (Figure 18.20). HI for immature source

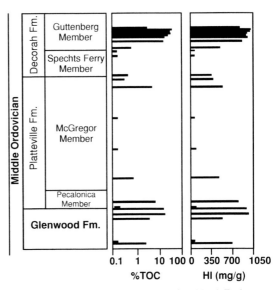

Figure 18.20. Down-hole profiles of TOC and Rock-Eval hydrogen index (HI) from core samples from the E.M. Greene Well, Washington County, Iowa (from Hatch *et al.*, 1987). Reprinted by permission of the AAPG, whose permission is required for further use.

rocks typically is in the range 950–1000 mg HC/g TOC (Hatch *et al.*, 1987; Jacobson *et al.*, 1988). These source rocks have been used in many biomarker and isotopic studies because the Guttenberg Member is the best representative of immature, nearly pure *G. prisca* kerogen with platy colonial morphology, and core material is readily available (Fowler and Douglas, 1984; Jacobson *et al.*, 1986; Jacobson *et al.*, 1988; Derenne *et al.*, 1990; Eglinton, 1994; Pancost *et al.*, 1998; Blokker *et al.*, 2001).

Crude oils derived from Assemblage A kerogen are recognized easily by their *n*-alkane distributions (see Figure 4.21), including a strong odd preference from nC_{11} to nC_{19}, low relative amounts of nC_{20}–nC_{35}, and a C_{40+} waxy component. The waxes have been documented rarely, but it may be common. The oils have very low concentrations of isoprenoid hydrocarbons, and pristane/phytane ratios, when measurable, may be relatively high (>2–3). Sterane and hopane concentrations are usually low, and may approach limits of detection.

Middle Ordovician kukersite, Baltic Basin

Estonian kukerite oil shale has been mined and retorted for fuel oil or burned directly for electricity since the

1920s. From 1950 to the late 1980s, it was the major energy source for the northwestern Soviet Union, reaching maximum production of >31 × 10^9 tons in 1980. Mining operations were reduced drastically since 1990 due to changing politics and economics. Kukersite oil shale occurs in a 20–30-m-thick sequence that contains up to 50 individual organic-rich beds alternating with argillaceous, biomicritic limestone (Bauert, 1994; Arro et al., 1998). The kukersite beds are typically 10–40 cm thick, but they may be as thick as 2.4 m. Individual beds are laterally continuous over 250 km and cover ~50 000 km^2 of the Baltic Basin (Hoffmann et al., 1987). The organic-rich beds typically contain 40–45 wt.% TOC to a maximum of ~70 wt.%.

Telalginite kerogen in kukersite has the thick-walled, large colonial morphology of *G. prisca* (Assemblage A), and its chemical structure has been studied intensively (Figure 18.21) (Derenne et al., 1990; Derenne et al., 1992; Stasiuk et al., 1993; Metzger and Largeau, 1994; Largeau et al., 2001; Bajc et al., 2001; Blokker et al., 2001). The primary chemical difference from the Guttenberg-type *G. prisca* facies is that kukersite has significantly higher oxygen content (O/C ~0.14 versus ~0.10). This oxygen is bound primarily as poly-*n*-alkyl resorcinols (5-*n*-alkyl-benzene-1,3-diol), which are phenolic moieties unlike those in lignin (Derenne et al., 1990; Blokker et al., 2001).

Chemical differences between Guttenberg and kukersite kerogens might be caused by differences in lifecycle stages of *G. prisca*, contributions from other unidentified organisms, or differences in diagenesis (Derenne et al., 1990). Of these possibilities, the most favored is that the chemistry of kukersite results from an adaptative response of *G. prisca* to different environmental conditions. Such lifecycle adaptations occur in *Botryococcus braunii*, a freshwater green alga, when grown under saline conditions (Derenne et al., 1992; Metzger and Largeau, 1994). This leads to the conclusion that *G. prisca* might have genetic affinities to *B. braunii* and that kukersite results from the selective preservation of bioresistant macromolecular material in the thick outer walls. However, Blokker et al. (2001) noted that *n*-alkyl resorcinols in *B. braunii* differ from those in kukersite. They speculated that thick-walled microfossil remains in kukersite are not alginites but sheath material excreted by extinct cyanobacteria, although resorcinols are not known to occur in such organisms.

Middle-Upper Ordovician Trenton Group, North America

The Middle-Upper Ordovician Trenton Group consists of limestones and calcareous shales deposited on a

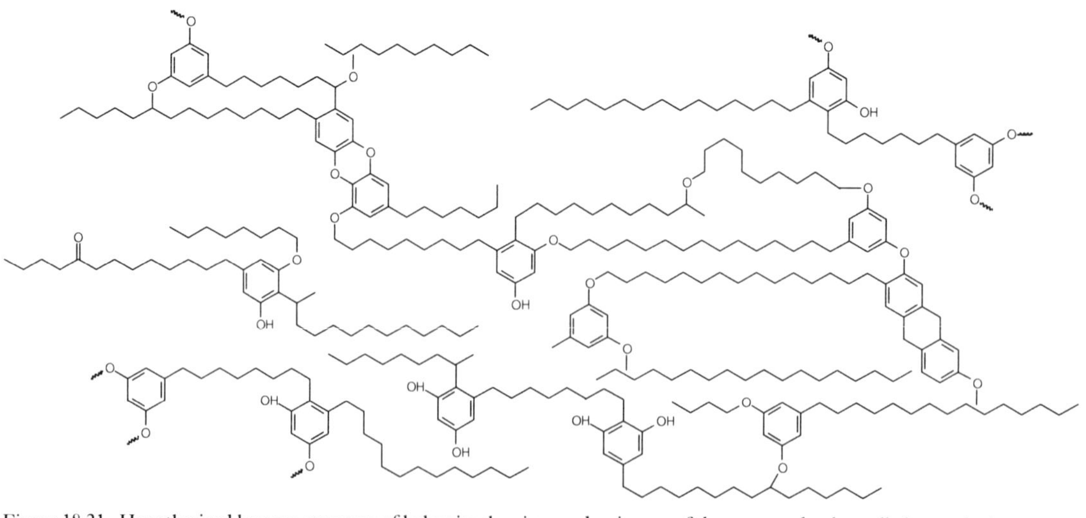

Figure 18.21. Hypothesized kerogen structure of kukersite showing predominance of the suggested poly-*n*-alkyl resorcinol structure, i.e. 5-*n*-alkyl-benzene-1,3-diol subunit. Reprinted from Blokker et al. (2001). © Copyright 2001, with permission from Elsevier.

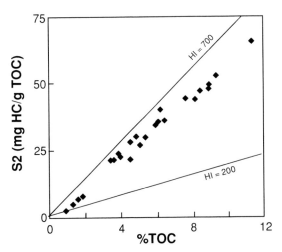

Figure 18.22. Total organic carbon (TOC) (wt.%) versus Rock-Eval S2 for Collingwood Member calcareous shales from the Georgian Bay area, Ontario (data from Obermajer et al., 1999). Reprinted by permission of the AAPG, whose permission is required for further use.

widespread carbonate platform over much of the North American craton. The Collingwood Member (Lindsey Formation) and its stratigraphic equivalent, the Point Pleasant Formation, are the source rocks for crude oils in Ontario (Powell et al., 1984) and Ohio (Drozd and Cole, 1994), respectively. The source potential of the Collingwood calcareous shales varies substantially (Figure 18.22). TOC ranges from ~0.7 to >11 wt.%, while HI ranges from ~170 to 620 mg HC/g TOC (Obermajer et al., 1999). The kerogen is type I/II telalginite composed mostly of planktonic algal biomass. *G. prisca* occurs in minor quantities and is usually associated with acritarchs. The Collingwood Member is typically only 2–4 m thick, with a maximum of ~10 m.

Crude oil from the Trenton/Black River Formation generated from the Middle-Upper Ordovician Trenton Formation, Michigan Basin (Figure 18.23), is typical of that generated from source rock dominated by *G. prisca*. The *n*-alkanes display the characteristic mid-range odd-number carbon preference, and acyclic isoprenoids occur as minor components. Saturated and aromatic hydrocarbon $\delta^{13}C$ values for this oil are within the range of Paleozoic oils. The saturated biomarker distributions indicate a thermally mature source with exclusively algal and bacterial input. Terrigenous plant triterpanoids (e.g. oleanane), C_{30} steranes, bisnorhopane, and gammacerane are absent. The C_{27} and C_{29} rearranged and regular isomers dominate the sterane distribution.

Middle-Upper Ordovician limestone muds and marls, Tarim Basin, China

The identity and distribution of Lower Paleozoic source rocks in the Tarim Basin, the largest and least explored basin in China, are not understood fully. The prospectivity of the basin is controversial. While Lower Paleozoic rocks occur throughout the basin and may exceed 12 000 m thickness in the interior, much of the section is postmature and deposition of source rocks may have been restricted to local depocenters.

Two Paleozoic intervals have been suggested as having source potential. Lower to Middle Cambrian strata in the eastern Tarim Basin include marls and mudstones deposited in a starved marine basin. Intervals that contain >1.0 wt.% TOC occur in ~65% of the source sequence, which has a thickness of 120–415 m. There is a lagoonal facies within Cambrian evaporites on the Bachu Uplift that has a maximum TOC of 2.14 wt.%. Water-column stratification has been implicated in source-rock deposition. The Cambrian strata are postmature where tested. The Middle-Upper Ordovician section contains carbonate muds ~80–100 m thick that appear to once have had excellent source potential. Where measured, present-day TOC averages 0.43 wt.%, with a maximum value of 6 wt.%. Rock-Eval HI is low, ranging from ~50 to 170 mg HC/g TOC, due to high maturity and oxidation in outcrop. Original generative potential is estimated to have been ~4–6 wt.% TOC with HI of 600–800 mg HC/g TOC (Graham et al., 1990). These source rocks may reflect deposition in a high-productivity upwelling zone restricted to the slope of the Tazhong and Tabei uplifts and the Majaer Depression (Wang et al., 2002). Zhang et al. (2000) used biomarkers to identify the Middle-Upper Ordovician marls and muddy limestones as the source of Paleozoic oils in the Tarim Basin (Figure 18.24).

Upper Ordovician-Silurian Utica/Antes shales, North America

A thick blanket of Upper Ordovician shales marks the end of carbonate deposition in the epicontinental seas of the North American craton. These ~50–150 m-thick

O.I.L.S.
Oil Information Library System

GEOMARK RESEARCH, INC.
9748 Whithorn Drive   Tel: (281) 856-9333
Houston, Texas 77095   Fax: (281) 856-2987
info@geomarkresearch.com

GEOCHEMICAL SUMMARY SHEET

Country: **USA** Depth: **3210'** 02-Jul-99
Basin: **Michigan** Age: **Ordovician** Sample ID: **MI001**
Field: **Keel** Formation: **Trenton/Black River** LAT: 42.0155
Well: **#1-21** LONG: -85.007

BULK PROPERTIES

API Gravity: % S: ppm V:
% < C15: **15.9**   ppm Ni:

C15+ Composition
% Sat: **47.3**
% Aro: **41.6**
% NSO: **9.1**
% Asph: **2.0**
Sat/Aro= **1.14**
n-Paraffin/Naphthene=

Stable Carbon Isotope Composition
δ per mil PDB

C15+ Saturate: **-28.92**
C15+ Aromatic: **-28.50**
Canonical Variable: **-1.75**

Miscellaneous:

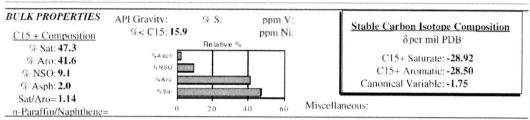

WHOLE CRUDE GAS CHROMATOGRAPHY
Pr/Ph= **1.46**
Pr/n-C17= **0.10**
Ph/n-C18= **0.15**
n-C27/n-C17= **0.12**
CPI= **1.027**

BIOMARKERS   ppm C30 Hopane: 199

OilMod Ratios
C19/C23= **0.46**
C22/C21= **0.27**
C24/C23= **0.58**
C26/C25= **0.95**
Tet/C23= **1.35**
C27T/C27= **0.02**
C28/H= **0.03**
C29/H= **0.54**
C30X/H= **0.14**
OL/H= **0.01**
C31R/H= **0.40**
GA/C31R= **0.12**
C35S/C34S= **0.69**
Ster/Terp= **0.34**
Rearr/Reg= **4.19**
%C27= **28.3**
%C28= **21.4**
%C29= **50.3**
C29 20S/R= **1.21**
Ts/Tm= **0.78**
C29D/H= **0.55**
DM/H= **0.00**

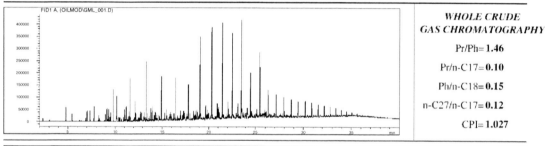
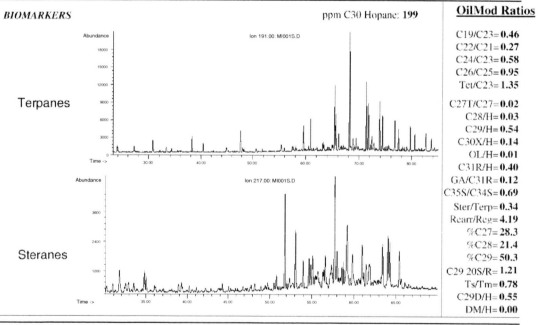

COMMENTS: Middle/Late Ordovician Trenton Formation

Figure 18.23. Geochemical data for crude oil from Keel Field generated from the Middle-Upper Ordovician Trenton Formation, Michigan Basin, USA (courtesy of GeoMark Research, Inc.).

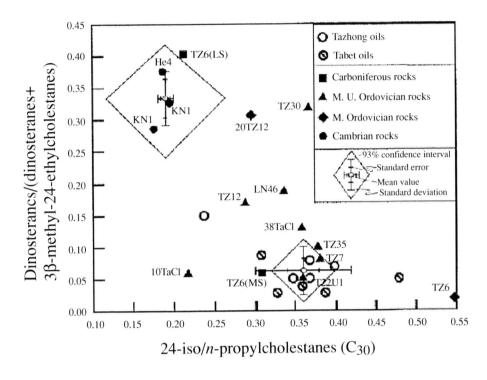
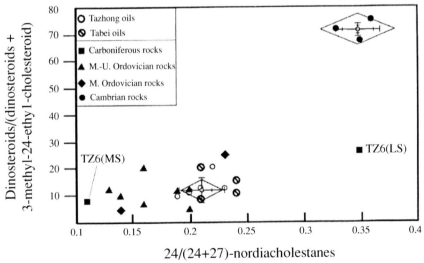

Figure 18.24. Comparison of age-related biomarker ratios (24-isopropycholestanes/24-n-propylcholestanes versus C_{30} methylsterane dinosterane ratio, top, and the nordiacholestane ratio versus triaromatic C_{29} methylsteroid ratio, bottom) for Tarim Basin oils and rock extracts. These age-related parameters clearly distinguish Cambrian and Ordovician extracts from each other and correlate the Tazhong and Tabei oils with the Ordovician extracts. Reprinted from Zhang et al. (2000). © Copyright 2000, with permission from Elsevier.

shales, known as the Utica or Antes shales, cover large portions of the northern Appalachian (Ohio, Pennsylvania, northern West Virginia, and New York) and the Michigan basins. Ryder et al. (1998) correlated oils from the Cambro-Ordovician Knox Dolomite in central and eastern Ohio to these Utica/Antes shales and considered the strata to be the likely source of oils in Ordovician reservoirs throughout the region. TOC typically ranges from ~0.3 to 4.3 wt.%, with median values of ~1.8 wt.%. The richest intervals are at the base of Utica/Antes shales. Rock-Eval HI typically ranges from 200 to 400 mg HC/g TOC, similar to those reported by Obermajer et al. (1999) for stratigraphically equivalent Blue Mountain Shale of Ontario (Figure 18.25).

Upper Ordovician Maquoketa Shale, Illinois Basin

The Upper Ordovician (Cincinnatian) Maquoketa Shale was deposited across a shallow epicontinental sea within the Illinois Basin. Guthrie and Pratt (1994) recognized two depositional cycles resulting in organic-rich, laminated shales. Shales in the Scales and Brainard formations follow the typical pattern for marine transgressional sequences, with the lowermost shales having

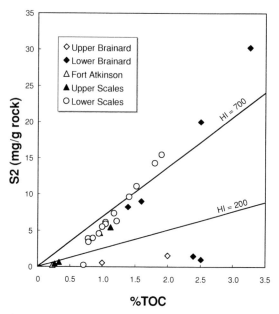

Figure 18.26. Total organic carbon (TOC) (wt.%) versus Rock-Eval S2 for Maquoketa Group shales from New Jersey Zinc J-9 Parrish core, Illinois Basin (data from Guthrie and Pratt, 1994). The Lower Scales and Lower Brainard shales have high source potential (Assemblage B kerogen). HI, hydrogen index.

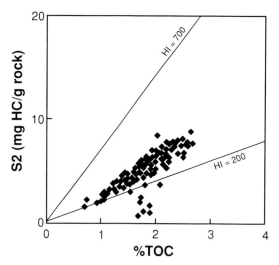

Figure 18.25. Total organic carbon (TOC) (wt.%) versus Rock-Eval S2 for Blue Mountain Shale, Ontario (data from Obermajer et al., 1999). These shales are stratigraphic equivalents of the Utica/Antes shales. Reprinted by permission of the AAPG, whose permission is required for further use. HI, hydrogen index.

the highest source potential (Figure 18.26). These dark brown to black shales have relatively high TOC (~1–3 wt.%), with appreciable generative potential (HI ~500–1000 mg HC/g TOC). Assemblage B kerogen occurs within the Maquoketa shales, but microfossils of G. prisca are absent. The organic facies is considered to be relatively uniform across the basin, except for the occurrence of graptolite-rich zones. The organic-rich shales resulted from high productivity and sluggish water circulation, which favored suboxic to anoxic bottom waters and enhanced preservation. These rich sequences grade into oxidized bioturbated shales with substantially lower potential.

The Maquoketa Shale was the source rock for much of the oil in Ordovician-age reservoirs within the Illinois Basin (Guthrie and Pratt, 1995). This correlation is somewhat unexpected because the Maquoketa Shale overlies the Assemblage A source units of the Middle Ordovician Galena Group (Glenwood shales and Guttenberg Member). Hatch et al. (1991) had concluded earlier that the Glenwood Shale rather than the Guttenberg Member was the source rock for these oils.

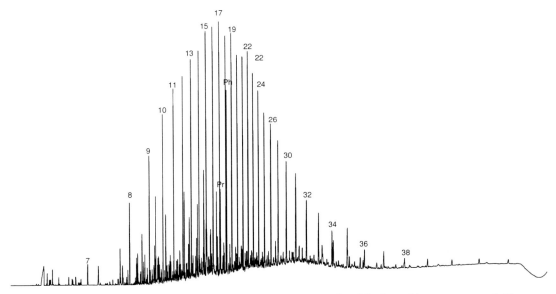

Figure 18.27. Gas chromatogram of 34° API gravity oil from the Cambrian Deadwood Sandstone (chromatogram supplied by D. Jarvie, Humble Instruments & Services, Inc.).

However, compound-specific isotope analysis of individual biomarkers (e.g. steranes, aryl isoprenoids) identifies the Maquoketa Shale as the source (Guthrie and Pratt, 1995; Guthrie, 1996).

Oil from the Newporte Impact Crater, Williston Basin

The Newporte Field, located in the Williston Basin just south of the US–Canada border, produces from fractured Precambrian gneiss and Cambrian Deadwood Sandstone formed by a Late Cambrian–Early Ordovician meteorite impact (Castaño et al., 1994; Forsman et al., 1996). The oil is unlike other early Paleozoic oils from the region because of high wax content, low pristane/phytane (Pr/Ph = 0.47), and no odd-carbon preference in the mid-range n-alkanes (Figure 18.27) (Jarvie, 2001). The source rock for this oil is unknown. It is geochemically unlike Ordovician-sourced Red River or Winnipeg oils, which generally have low concentrations of Pr and Ph and Pr/Ph > 1.

Middle–Upper Ordovician Simpson-Ellenburger(.), North America

The Simpson-Ellenburger(.) petroleum system is a significant petroleum system (>1 × 10^{11} kg) HC with high generation–accumulation efficiency (10.4%) because the source, reservoir, seal, and traps are in close proximity and migration distances are short (Magoon and Valin, 1994). Several factors limited the size of this system, including relatively lean source rock (∼1.7 wt.% TOC) and only moderately oil-prone organic matter (HI = 425 mg HC/g TOC). Furthermore, the proportion of generated hydrocarbons preserved in traps is low because the oils are light (∼44° API) and the critical moment for this system occurred during the Triassic Period.

Upper Ordovician Vinini Formation, Nevada, USA

Oil shales from the Upper Ordovician Vinini Formation are among the oldest source rocks attributed to deposition in an upwelling zone. Exposures of the formation in the Roberts Mountains of central Nevada were displaced tectonically by as much as 100 km (Meissner et al., 1984). The Vinini Formation is ∼9 m thick and consists of alternating 1–25-cm beds of blocky, siliceous mudstones and 2–13-cm beds of graptolitic fetid mud shale. Source potential can be exceptional, with TOC up to 25 wt.% and HI >600 mg HC/g TOC (Berry and Cooper, 2001), although other analyses suggest lower overall potential (Poole and Claypool, 1984). These

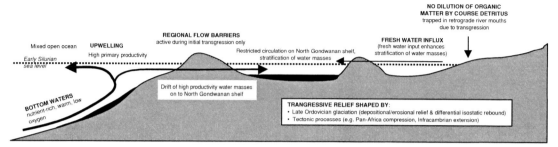

Figure 18.28. Paleoecological scheme showing processes that may have contributed to deposition of the Lower Silurian hot shales in North Africa/Arabia. An initial transgression flooded continental depressions, thus producing silled basins with restricted water circulation and riverine clastic influx. Upwelling along the northern shelf resulted in high productivity and oxygen-minimum zones. These waters drifted into the silled basins, resulting in algal blooms. Influx of cold, fresh water from glacial melt may have enhanced water stratification. Reprinted from Lüning et al. (2000). © Copyright 2000, with permission from Elsevier.

shales contain high concentrations of metals, and phosphatic nodules are common. Vinini source rocks are believed to have generated and expelled oil during the late Paleozoic–early Mesozoic Era, but most of this oil was destroyed or remigrated during the Tertiary Basin-and-Range tectonic event (Meissner et al., 1984)

SILURIAN SOURCE ROCKS

Lower Silurian graptolitic hot shales, Saharan and Arabian basins

Klemme and Ulmishek (1991) identified the lower Silurian Period as one of the six most significant temporal intervals for source-rock deposition, having generated ∼9% of the world's reserves. Most of this petroleum was generated from one source unit, the lowermost Silurian hot shales that occur across the northern African and the Arabian cratons (Tissot et al., 1984).

Global deglaciation at the end of the Ordovician Period (Semtner and Klitzsch, 1994) caused a major marine transgression that flooded a vast portion of the northern Gondwana shelf that was positioned in high southern latitudes (>40°). The transgression established depositional conditions that gave rise to enhanced productivity and preservation, resulting in a regional source-rock of exceptional quality. The detailed geological factors that were the most significant controls on source rock deposition remain unclear. The position of the tectonic plates favored an upwelling zone across the entire shelf of Gondwana, suggesting high productivity and probable offshore oxygen minimum zones (Moore et al., 1993; Parrish, 1982). However, Lüning et al. (2000) point out that upwelling persisted long after deposition of the hot shale (Moore et al., 1993) and that other factors may have been equally important. They suggest that flooding of paleodepressions controlled deposition of the hot shales (Figure 18.28). These depressions would have restricted water circulation and could receive only limited riverine clastic input.

Basinal hot shales (Lowermost Silurian) in North Africa typically have 4–6 wt.% TOC, with maxima as high as 17 wt.%, and type II kerogen (Boote et al., 1998) derived mainly from marine prasinophytes (Tyson, 1995). The overlying Silurian shales have substantially lower source potential, with TOC in the range 1–3 wt.%. However, localized, younger (Telychian? = lower Wenlock) organic-rich hot shales occur in wells from the Algerian part of the Ghadames Basin and Iraq (Figure 18.29) (Lüning et al., 2000). Such source-rock distributions occur in many marine transgressive sequences, where high basinal TOC values correspond to maximum flooding surfaces (Creaney and Passey, 1993; Wignall and Maynard, 1993). In most basins, the Silurian shales are now postmature and consequently have low HI. Original generative potential was considerably higher, with HI estimated to be ∼600 mg HC/g TOC. Although the entire Silurian shale succession may reach a thickness of 700–1000 m, the hot shale horizon is usually <30 m thick, suggesting that the anoxic conditions that gave rise to the exceptional source quality may have lasted no more than a few million years (Lüning et al., 2000).

Major petroleum accumulations derived from Lower Silurian hot shales occur in the northern Sahara,

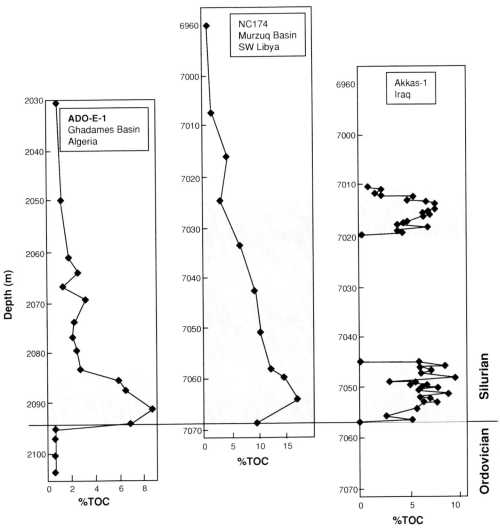

Figure 18.29. Total organic carbon (TOC) well profiles showing the distribution of Silurian hot shales in basins from North Africa and Arabia. Reprinted from Lüning et al. (2000). © Copyright 2000, with permission from Elsevier.

where it is commonly known as the Mokattam Shale, Gotlandien Argilleux, Aquinet Ouernine, and Tanezzuft (Figure 18.30) (Klett, 2000a; 2000b; 2000c), and in Saudi Arabia, where it is known as the Qusaiba Member of the Qalibah Formation (Figure 18.31) (Cole et al., 1994a; Cole et al., 1994b; Jones and Stump, 1999). Less extensive petroleum systems occur in Sudan and Chad (Bedo Formation), Jordan (Bata Formation), Iraq (Akkas Formation), Oman (Sarriah Member of the Safiq Formation), Qatar (Shaly member of the Sharawra Formation), and the United Arab Emirates (Rann Formation). The geographic distribution of the Lower Silurian hot shales is not known completely, although outcrop and/or subsurface samples were retrieved from Mauritania, Iran, and Sardinia. Portions of Nova Scotia may once have been attached to northwestern Morocco and may contain equivalent Silurian hot shales (Lüning et al., 2000).

Upper Silurian Dadas Formation, Turkey

The Upper Silurian Dadas Formation (Tanf Formation in Syria) is one of several source rocks in the southern

O.I.L.S.
Oil Information Library System

GEOMARK RESEARCH, INC.
9748 Whithorn Drive Tel: (281) 856-9333
Houston, Texas 77095 Fax: (281) 856-2987
info@geomarkresearch.com

GEOCHEMICAL SUMMARY SHEET

Country: **Algeria** Depth: **11200'** 02-Jul-99
Basin: **Trias** Age: **Cambrian** Sample ID: **AL013**
Field: **Hassi Messaoud** Formation: **R-1** LAT: 31.711
Well: LONG: 5.984

BULK PROPERTIES API Gravity: **38.8** % S: **0.06** ppm V: **13.0**
C15+ Composition % < C15: **45.3** ppm Ni: **7.0**
% Sat: **51.5**
% Aro: **34.6**
% NSO: **13.2**
% Asph: **0.7**
Sat/Aro= **1.49**
n-Paraffin/Naphthene= **0.27**

Stable Carbon Isotope Composition
δ per mil PDB
C15+ Saturate: **-29.99**
C15+ Aromatic: **-29.35**
Canonical Variable: **-0.93**

Miscellaneous: D/H = **-111**

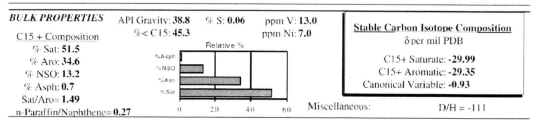

WHOLE CRUDE GAS CHROMATOGRAPHY
Pr/Ph= **1.58**
Pr/n-C17= **0.40**
Ph/n-C18= **0.31**
n-C27/n-C17= **0.10**
CPI= **1.046**

BIOMARKERS ppm C30 Hopane: **8**

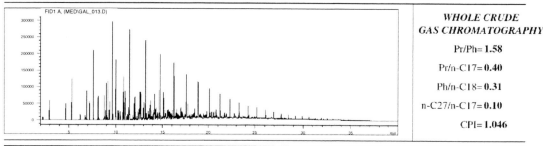
Terpanes

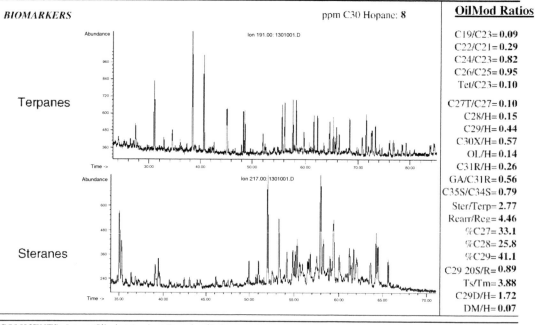
Steranes

OilMod Ratios
C19/C23= **0.09**
C22/C21= **0.29**
C24/C23= **0.82**
C26/C25= **0.95**
Tet/C23= **0.10**
C27T/C27= **0.10**
C28/H= **0.15**
C29/H= **0.44**
C30X/H= **0.57**
OL/H= **0.14**
C31R/H= **0.26**
GA/C31R= **0.56**
C35S/C34S= **0.79**
Ster/Terp= **2.77**
Rearr/Reg= **4.46**
%C27= **33.1**
%C28= **25.8**
%C29= **41.1**
C29 20S/R= **0.89**
Ts/Tm= **3.88**
C29D/H= **1.72**
DM/H= **0.07**

COMMENTS: Lower Silurian Aquinet Ouernine Shale

Figure 18.30. Geochemical data for crude oil from Hassi Messaoud Field generated from the Lower Silurian Tanezzuft Shale, Algeria (courtesy of GeoMark Research, Inc.).

O.I.L.S.
Oil Information Library System

GEOMARK RESEARCH, INC.
9748 Whithorn Drive   Tel: (281) 856-9333
Houston, Texas 77095   Fax: (281) 856-2987
info@geomarkresearch.com

GEOCHEMICAL SUMMARY SHEET

Country: **Saudi Arabia**   Depth: **7953-8134'**   02-Jul-99
Basin: **Western Platform**   Age: **Permian**   Sample ID: **SA116**
Field: **Raghib**   Formation: **Unayzah**   LAT: 23.66
Well: **1**   **Sandstone**   LONG: 47.1

BULK PROPERTIES   API Gravity: **47.3**   % S: **0.04**   ppm V: **11.0**
%< C15: **52.8**   ppm Ni: **13.0**

C15 + Composition
% Sat: **62.2**
% Aro: **25.1**
% NSO: **12.7**
% Asph: **0.0**
Sat/Aro= **2.48**
n-Paraffin/Naphthene= **0.20**

Stable Carbon Isotope Composition
δ per mil PDB
C15+ Saturate: **-30.05**
C15+ Aromatic: **-28.99**
Canonical Variable: **0.02**

Miscellaneous:

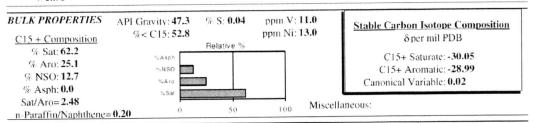

WHOLE CRUDE GAS CHROMATOGRAPHY
Pr/Ph= **2.05**
Pr/n-C17= **0.53**
Ph/n-C18= **0.35**
n-C27/n-C17= **0.06**
CPI=

BIOMARKERS   ppm C30 Hopane: **7**

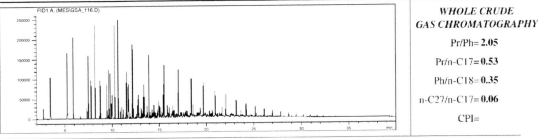
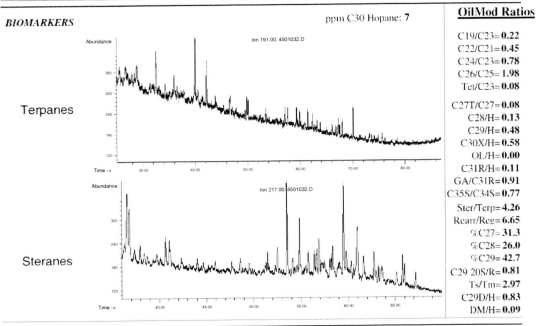

OilMod Ratios
C19/C23= **0.22**
C22/C21= **0.45**
C24/C23= **0.78**
C26/C25= **1.98**
Tet/C23= **0.08**

C27T/C27= **0.08**
C28/H= **0.13**
C29/H= **0.48**
C30X/H= **0.58**
OL/H= **0.00**
C31R/H= **0.11**
GA/C31R= **0.91**
C35S/C34S= **0.77**
Ster/Terp= **4.26**
Rearr/Reg= **6.65**
%C27= **31.3**
%C28= **26.0**
%C29= **42.7**
C29 20S/R= **0.81**
Ts/Tm= **2.97**
C29D/H= **0.83**
DM/H= **0.09**

COMMENTS: Lower Silurian Qusaiba Shale

Figure 18.31. Geochemical data for crude oil from Raghib Field generated from the Lower Silurian Qusaiba Member, Western Platform, Saudi Arabia (courtesy of GeoMark Research, Inc.).

Anatolia region of Turkey (Soylu, 1987). Having a thickness of ~150 m, the Dadas Formation is divided into three members, with organic-rich clastic rocks restricted to the lowermost (Dadas-I) and uppermost (Dadas-III) members. The organic-rich intervals range from 2 to 5 wt.%, with an average of ~3.5 wt.% TOC. Rock-Eval HI averages ~450 mg HC/g TOC where immature. Gürgey (1991) correlated biomarker distributions of oils in the Northern Diyarbakir area to Dadas Formation source rocks.

Upper Silurian Salina A-1 carbonates, Michigan Basin

Oil produced from Middle-Upper Silurian Niagaran pinnacle reefs in southwestern Ontario and the Michigan Basin was generated from Upper Silurian Salina A-1 carbonates. The oils have several characteristics indicating hypersaline source-rock depositional conditions, including abundant high-molecular-weight n-alkanes with an even-to-odd carbon preference, isoprenoids (pristane/phytane ~0.8), gammacerane, and C_{34}-C_{35} homohopanes (Vogler et al., 1981; Illich and Grizzle, 1983; Powell et al., 1984; Rullkötter et al., 1986). Using IR spectroscopy and carbon isotopic analysis of distillate fractions, Gardner and Bray (1984) correlated the Michigan Basin pinnacle-reef oils with the Salina A-1 carbonate. This correlation was verified later by biomarkers (Rullkötter et al., 1986; Rullkötter 1992; Dunham et al., 1988; Obermajer et al., 2000a). A similar facies may occur in the Middle Silurian Niagaran (Eramosa) dolomites, which are a secondary source for these oils (Powell et al., 1984; Gardner and Bray, 1984).

The source facies within the Salina A-1 carbonates is a brown, laminated dolostone deposited during recurrent back-reef lagoonal and inter-biohermal conditions. The source rock intervals are thin and contain from 0.5 to 3.5 wt.% TOC. They contain type II kerogen that is mostly amorphous, with minor amounts of structured alginite. Extracts from Salina source rocks closer to the reefs correlate more closely to the crude oils than more basinal facies (Obermajer et al., 2000a).

> **Note:** Biomarker analyses were conducted on a suite of oil and rock samples from the Middle Silurian Guelph-Salina interval in southern Ontario (Obermajer et al., 2000a). Oils in the Guelph reef reservoirs in southern Ontario, Canada, have distinct biomarker compositions (e.g. high gammacerane and acyclic isoprenoids, low pristane/phytane, and prominent C_{34} and C_{35} homohopanes), indicating one oil family that originated from carbonate source rock deposited under hypersaline, strongly reducing conditions. Extracts of organic-rich, laminated dolostones from the stratigraphically adjacent Salina A-1 Formation are geochemically similar to the reef-hosted oils.

DEVONIAN SOURCE ROCKS

Lower-Middle Devonian Old Red Sandstone, Scotland

The Old Red Sandstone Formation was deposited in the Orcadian Basin, a deep intermontane basin northeast of the Scottish Highlands, during much of the Early and Middle Devonian Period. The formation is >4000 m thick and consists of >100 cycles of continental sedimentary sequences. The sequences consist of four lithologies that reflect changes in water depth in the Orcadian Lake. Thin beds (~0.5 mm) of organic-rich carbonates and organic-poor clastics were deposited under anoxic deep-water conditions. This lithofacies contains three discrete layers reflecting a seasonal cycle of carbonate precipitation, followed by algal blooms and then clastic deposition during the rainy season. The next lithofacies consists of siltstones and shales deposited under shallower water conditions. These <3-mm-thick beds may be rich in organic matter, resulting from high productivity. Above this facies are ~10-mm-thick organic-rich shale and coarse sandstone laminae that were deposited under shallower conditions. The final lithofacies consists of >100-mm-thick shales, siltstones, and sandstones that were deposited under alternating wet to dry conditions. The cycle then repeats in the reverse direction, returning to deep-water conditions (Donovan, 1980).

Source rocks in the Old Red Sandstone contain type I kerogen. Selected outcrop samples contain ~1–5 wt.% TOC, with Rock-Eval HI ~300–930 mg HC/g TOC and OI <60 mg CO_2/g TOC (Peters et al., 1989; Bailey et al., 1990). Figure 18.32 shows an example of geochemical data for Beatrice oil, which is dominantly derived from this source rock. Peters et al. (1989) determined through biomarkers that these rocks contributed to the composition of the Inner Moray Firth Beatrice Field oil. Unlike other North Sea petroleum, the Beatrice

O.I.L.S.
Oil Information Library System

GEOMARK RESEARCH, INC.
9748 Whithorn Drive   Tel: (281) 856-9333
Houston, Texas 77095   Fax: (281) 856-2987
info@geomarkresearch.com

GEOCHEMICAL SUMMARY SHEET

Country: UK   Depth: 02-Jul-99
Basin: **Inner Moray Firth**   Age: Sample ID: **NWE034**
Field: **Beatrice (11-30B)**   Formation: LAT: 58.1204
Well: LONG: -3.087

BULK PROPERTIES   API Gravity: **38.0**   % S: **0.04**   ppm V:
%< C15: **27.1**   ppm Ni:

C15+ Composition
% Sat: **65.8**
% Aro: **24.8**
% NSO: **8.2**
% Asph: **1.2**
Sat/Aro= **2.65**
n-Paraffin/Naphthene= **0.70**

Stable Carbon Isotope Composition
δ per mil PDB
C15+ Saturate: **-32.27**
C15+ Aromatic: **-30.72**
Canonical Variable: **1.79**

Miscellaneous:

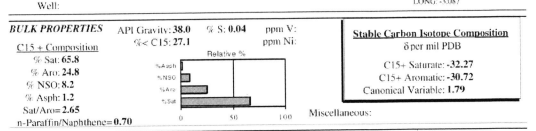

WHOLE CRUDE GAS CHROMATOGRAPHY
Pr/Ph= **2.17**
Pr/n-C17= **0.25**
Ph/n-C18= **0.12**
n-C27/n-C17= **0.54**
CPI= **1.143**

BIOMARKERS   ppm C30 Hopane: **398**

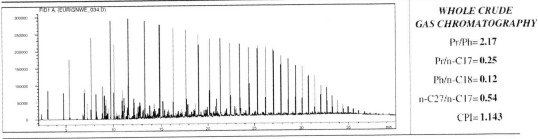

Terpanes

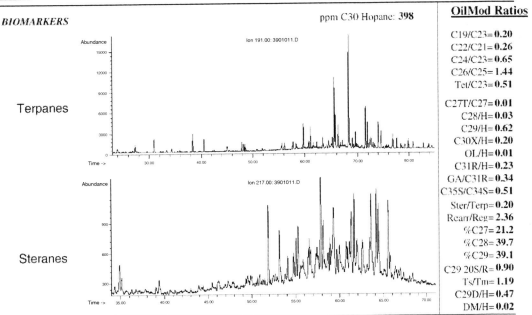

Steranes

OilMod Ratios
C19/C23= **0.20**
C22/C21= **0.26**
C24/C23= **0.65**
C26/C25= **1.44**
Tet/C23= **0.51**

C27T/C27= **0.01**
C28/H= **0.03**
C29/H= **0.62**
C30X/H= **0.20**
OL/H= **0.01**
C31R/H= **0.23**
GA/C31R= **0.34**
C35S/C34S= **0.51**
Ster/Terp= **0.20**
Rearr/Reg= **2.36**
%C27= **21.2**
%C28= **39.7**
%C29= **39.1**
C29 20S/R= **0.90**
Ts/Tm= **1.19**
C29D/H= **0.47**
DM/H= **0.02**

COMMENTS: Early/Middle Devonian Lacustrine Old Red Sandstone

Figure 18.32. Geochemical data for crude oil from Beatrice Field generated mainly from Lower-Middle Devonian lacustrine source rock in the Old Red Sandstone Formation, Inner Moray Firth, offshore UK (courtesy of GeoMark Research, Inc.).

Field oil contains perhydro-β-carotene, a marker for lacustrine sources, as well as C_{30} steranes indicative of marine sources. Based on the mixed biomarker and intermediate isotopic values, Peters *et al.* (1989) suggested that the Beatrice oil had multiple sources. Bailey *et al.* (1990) argued that the Beatrice oil originated exclusively from Devonian shales and that the C_{30} steranes were derived from occasional marine incursions (Marshall, 1996). However, additional evidence for oil being generated and migrated from both Devonian and Middle Jurassic coaly sources was found in a Helmsdale Fault brecciated sandstone that is held together by a bitumen matrix (Peters *et al.*, 1999b).

> Note: The Pragian-age Rhynie cherts occur within the lowermost Old Red Sandstone. These cherts formed from hot springs and geysers related to volcanic activity and contain an exquisitely preserved ecosystem of early land plants, fungi, and insects (see Figure 17.10).

Middle Devonian Los Monos Shale, South America

The Middle Devonian Los Monos Shale is thought to be the major source of oil and gas in the Santa Cruz-Tarija Province that spans Bolivia, Argentina, and Paraguay (Lindquist, 1998a). TOC averages only ~2 wt.%. HI is typically 100–300 mg HC/g TOC, but it can reach as high as 500 mg HC/g TOC. OI is near 10–50 mg CO_2/g TOC, but it can be as high as 300 mg CO_2/g TOC (Dunn *et al.*, 1995). The kerogen is mixed type II and III, although much of the reduced oil-generative potential is due to oxidation of marine organic matter input. The underlying Silurian shales are geochemically similar and may also serve as a petroleum source unit. Up to 4 km of these marine shales were deposited, compensating for the marginal generative potential of the kerogen.

Oils produced from these marine shales tend to be mature to postmature, with abundant diasteranes (Dunn *et al.*, 1995). Carboniferous siliciclastics in the Machareti Formation are the main reservoir rocks, although secondary reservoirs occur in Ordovician, Cretaceous, and Tertiary units. Santa Cruz oils in the younger reservoirs have a wide range of gravities (27–71°API) and were altered by migration–fractionation (Illich *et al.*, 1981). The lightest condensates are depleted in *n*-alkanes and light aromatic hydrocarbons and enriched in iso- and cycloalkanes. Illich *et al.* (1981) attributed these distributions to cross-stratigraphic movement of hydrocarbons in an aqueous phase followed by exsolution of the hydrocarbons in shallower reservoir rocks.

Middle Devonian Keg River/Winnipegosis Formation, Elk Point/Williston basins

Basinal carbonates of the Middle Devonian Keg River Formation are the primary source for oils in the co-eval Keg River pinnacle reef reservoir rocks in northwestern Albert and in the southeastern Saskatchewan, where it is equivalent to the Winnipegosis Formation in the Williston Basin (Powell, 1984; Creaney *et al.*, 1994b). The basinal laminites vary from <1 to 15 m thick. They contain type I/II kerogens consisting largely of well preserved *Tasmanites* and akinete algal cells. TOC ranges from 3 to 16 wt.%, with HI ~650 mg HC/g TOC (Figure 18.33). Isolated samples may contain up to 46 wt.% TOC, with HI up to ~800 mg HC/g TOC (Clegg *et al.*, 1997).

The Keg River Formation contains two distinct organic facies. The lower member was deposited under anoxic water established following a marine transgression. The upper member, also known as the Muskeg, was deposited under high saline conditions following a regression (Chow *et al.*, 1995). Biomarkers clearly differentiate the two facies in rock extracts and in generated oils (Clegg *et al.*, 1997; Behrens *et al.*, 1998). The Lower Keg River has pristane/phytane ratios ranging from 0.4 to 0.9 and abundant tricyclic

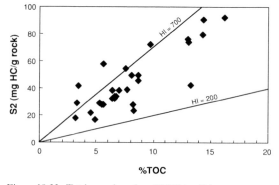

Figure 18.33. Total organic carbon (TOC) (wt.%) versus Rock-Eval S2 for Lower Keg River carbonates. Reprinted from Clegg *et al.* (1997). © Copyright 1997, with permission from Elsevier.

HI, hydrogen index.

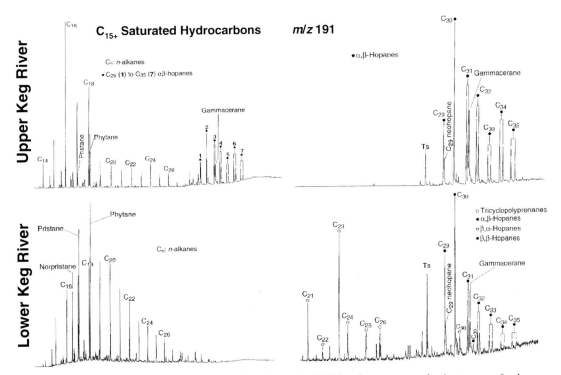

Figure 18.34. Gas chromatograms of free saturated hydrocarbons and *m/z* 191 ion chromatograms, showing terpanes of rock extracts from the Upper and Lower Keg River source rocks. Reprinted from Behrens *et al.* (1998). © Copyright 1998, with permission from Elsevier.

terpanes and aryl isoprenoids, while the Upper Keg River has pristane/phytane ratios ranging from 0.3 to 0.4 and abundant homohopanes and gammacerane (Figures 18.34 and 18.35). Diaromatic carotentoids and derived aryl isoprenoids are diagnostic for *Chlorobiacaea* (green sulfur photosynthetic bacteria). In the Lower Keg River, their abundance is proportional to organic richness, suggesting that a substantial portion of the photic zone was anoxic. Samples from the upper Keg River have reduced indications of *Chlorobiacaea*, suggesting that the photic zone was mostly oxic during deposition (Clegg *et al.*, 1997; Behrens *et al.*, 1998).

Keg River oils in sub-basins of northwestern Alberta can be described as mixtures of two families that can be best differentiated by their triterpane distributions. One end member is characterized by abundant homohopanes, while the other contains abundant 8,14-secohopanes (Li *et al.*, 1999). The diversity of biomarker distributions in the oils indicates a wide range of depositional environments and source facies within each sub-basin.

Global deposition of source rocks at the Devonian–Mississippian boundary

Following the Frasnian-Famennian mass extinction that greatly impacted marine life, an event at the Devonian–Carboniferous boundary resulted in one of the most severe disturbances of the Phanerozoic biosphere (Sepkoski, 1996; Walliser, 1996). This comparatively short period, termed the Hangenberg Event after the Hangenberg black shales of Germany (Walliser, 1996), gave rise to deposition of some of the world's most prolific source rocks. The Hangenberg Event did not result in as many species extinctions as the five greatest extinctions (see Table 17.1), but it severely altered the floral and faunal distributions at all trophic levels in both terrigenous and marine environments.

The factors that gave rise to the Hangenberg bioevent are disputed, but organic-rich, marine shales covered >20% of the depositional area at that time and the resulting source rocks account for ~8% of the world's petroleum reserves (Klemme and Ulmishek, 1991). Asteroid impacts, climate cooling, oceanic

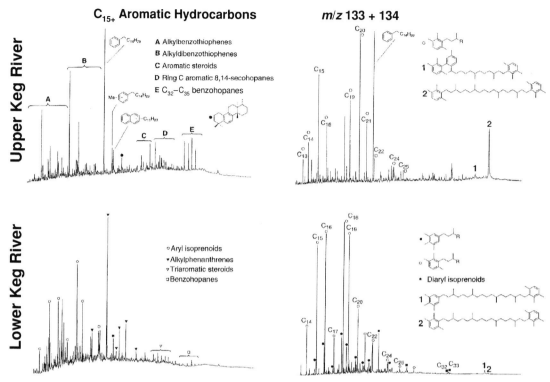

Figure 18.35. Gas chromatograms of free aromatic hydrocarbons and m/z 133 + 134 ion chromatograms, showing aryl isoprenoids in rock extracts from the Upper and Lower Keg River source rocks. Reprinted from Behrens et al. (1998). © Copyright 1998, with permission from Elsevier.

overturn, eutrophication, climatic overheating, marine nutrient deficiency, and eustatic effects have all been proposed as triggers for Devonian–Carboniferous boundary extinctions (Caplan and Bustin, 1999). The widespread, global deposition of Devonian–Carboniferous source rocks has been at the center of the productivity versus preservation controversy. Parrish (1995) argued that upwelling and enhanced productivity can account for about two-thirds of the Upper Devonian source rocks, while Ormiston and Oglesby (1995) argued that upwelling was important in only a few Upper Devonian source rocks and that the vast majority arose from enhanced preservation of organic matter in shallow, anoxic seas. Algeo et al. (2001) suggest that a link exists between evolving vascular plants, increased continental weathering, and enhanced anoxia and preservation in nearshore sediments during the Middle to Late Devonian Period.

Upper Devonian-Mississippian rocks with high generative potential occur worldwide (Table 18.7). The most significant for oil generation are the Domanik Shale of western Russia and the marine shales and novaculites deposited on lower Paleozoic carbonate platforms and in sag basins that covered most of central and western North America. Examples of these source rocks are discussed below. Devonian marine shales are a primary source of light oils and gas in the North African Illizi-Ghadames basins. Determination of their original generative yields is difficult because these rocks are presently postmature. Distinguishing oils and condensates from the Saharan basins is difficult because the organic facies are nearly identical, and many potentially characteristic biomarkers were severely thermally altered (Peters and Creaney, 2003).

Devonian-Mississippian New Albany Shale, Illinois Basin

The New Albany-Chesterian(!) petroleum system in the Illinois Basin has been studied extensively because nearly 90 000 wells penetrate the Chesterian (Upper

Table 18.7. *Major Upper Devonian–Mississippian source rocks are widespread*

Major source formations	Locations	References
North America		
Woodford, New Albany, Chattanooga, Ohio, Millboro, Burket, Geneseo, Antrim, Arkansas Novaculite	US mid-continent – carbonate platform	See below
Bakken, Exshaw	Williston Basin, western Canada	See below
Russian Platform		
Domanik	Volga-Ural, Timan-Pechora, Pricaspian, Pripyat, and Dneiper-Donets basins	See below
North Africa	Illizi-Ghadames Basins	Klett (2000a)
Immature or lesser sources, Europe		
Porsquen Shale	Brittany	Ormiston and Oglesby (1995)
Montagne Noire	Southern France	Paproth *et al.* (1991)
Hangenberg Shale (and equivalents)	Rheinishches Germany	Walliser (1996)
	Carnic Alps	Schönlaub (1986)
	Bohemian region	Chlupac (1988)
	Yugoslavia	Krstic *et al.* (1988)
Siberian Platform	Western Siberia Basin	Ormiston and Oglesby (1995)
	Eastern Siberia Basin	
Middle East		
Pre-Unayzah clastics	Widyan Graben, Saudi Arabia	Al-Laboun (1986), Husseini (1992)
Asia		
Luijiang	Guangxi Province	Ormiston and Oglesby (1995)
Changshun Shale	Chuxiong Basin	Caplan and Bustin (1999)
Australia		
Gogo	Canning Basin	Bishop (1999c)
Ningbing	Bonaparte Basin	
South America		
Curuá	Amazon Basin	Mosmann *et al.* (1984)

Mississippian) reservoir section. Based on material balance calculations, the total petroleum charge from this system is 78 billion bbl, of which 11.4 billion barrels occur as in-place petroleum, 9 billion barrels occur as residual migration losses in carrier beds, and 57.6 billion barrels were lost by erosion (Lewan *et al.*, 2002).

Upper Devonian (Frasnian) Domanik shales, Russia

The Domanik deep marine facies are the primary source rocks in the Timan-Pechora Basin, northwestern Arctic Russia (Lindquist, 1999b). Widespread deposition occurred in the Frasnian Age, with limited deposition in regional depocenters through the earliest Carboniferous (Tournaisian) Period (Ulmishek, 1988). The Domanik source rocks, composed of laminated black shales, siliceous shales, limestones, and marls, were deposited under 100–400 m of water in the deep basinal facies off shelf-edge reefs. Organic matter was preserved when anoxic conditions formed in waters with restricted circulation and low sedimentation (Abrams *et al.*, 1999; Alsgaard, 1992). The Domanik Formation is up to

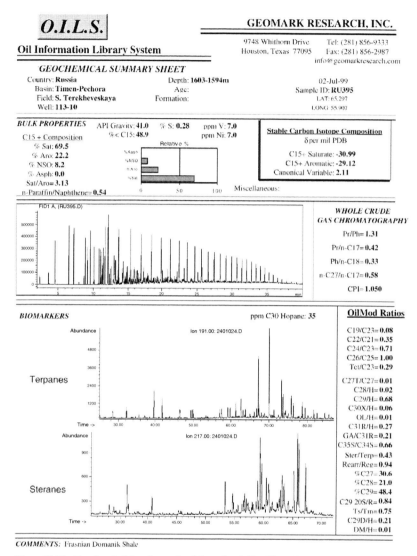

Figure 18.36. Geochemical data for crude oil from South Terekheveskaya Field generated from the Frasnian Domanik Shale, Timan-Pechora Basin, Russia (courtesy of GeoMark Research, Inc.).

500-m thick, and the source potential, particularly in the Middle Frasnian age, is exceptional. The source rocks contain mostly amorphous type I and II organic matter, with average TOC of 5 wt.%, but this can reach 30 wt.% (Abrams *et al.*, 1999; Alsgaard, 1992; Pairazian, 1993). HI for immature source strata range from 500 to 700 mg HC/g TOC.

Timan-Pechora oils typically exhibit low maturity and high sulfur (1–4.5 wt.%), with biomarker distributions characteristic of marly source rocks (Figure 18.36). These oils are rich in aromatic and polar compounds and have pristane/phytane of 1–2, and norhopane is commonly equal to or greater than hopane. Biomarker and isotopic analyses ($\delta^{13}C \sim -29\%_0$) prove that nearly all of the petroleum ($\sim$20 BBOE known recoverable reserves) in the Timan-Pechora Basin was generated from Domanik source rocks (Ulmishek, 1991; Abrams *et al.*, 1999).

Note: Biomarkers are sensitive indicators of depositional environments among different subfacies of a Devonian carbonate complex in Poland and were used to refine classical sedimentologic facies analyses of platform, reef, and off-platform shelf-basinal systems (Marynowski et al., 2000). For example, peritidal dolostones deposited under anoxic to suboxic conditions show little sedimentological evidence for elevated salinity but have the highest potential for organic matter preservation among other platform/reefal facies. Their biomarker compositions include several compounds indicating hypersaline conditions and anoxia, whereas others, in particular gammacerane, indicate that the water column was stratified during sedimentation. The Frasnian/Famennian mass extinction apparently had no impact on the biomarker distribution in the shelf-basinal system across the stage boundary.

Frasnian Duvernay calcareous shales, Western Canada Basin

Shales and marls of the Duvernay Formation are the source rocks for most of the oil and gas within Upper Devonian reservoirs of the Western Canada Basin (Creaney and Allan, 1990; Allan and Creaney, 1991). The Duvernay is the basinal time equivalent of the Frasnian Leduc reef carbonates. The formation consists of dark brown to black bituminous shale and limestones that has been subdivided into three units. The lowermost unit is ~0.3–20 m thick and consists of brown to greenish-gray argillaceous limestones. The middle unit is composed of reef-derived skeletal debris. The upper unit ranges from ~16 to 60 m thick and is the primary source facies, consisting of dark brown to black bituminous shales and brown deep-water limestones (Switzer et al., 1994).

Duvernay shales and marls were deposited under marine, deep-water conditions with slow sedimentation rates. The setting was low-energy and euxinic to anoxic, leading to the absence of bottom fauna, preservation of oil-prone marine organic matter, and the prevalence of bacterial sulfate reduction in the upper sediments. Organic carbon varies from <1 wt.% TOC in bioturbated oxic strata up to 20 wt.% in anoxic laminites. HI typically ranges from ~100 to 560 mg HC/g TOC.

On average, the Duvernay contains ~4 wt.% TOC with HI of ~400 mg HC/g TOC (Creaney et al., 1994b).

Biomarker distributions for Duvernay rock extracts and crude oils are typical for anoxic, marine shales (Allan and Creaney, 1991). Diasteranes are greater than or equal to normal steranes, and the homohopanes express a normal decrease with carbon number (Figure 18.37) (Fowler et al., 2001). Isoprenoids may be abundant, with pristane/phytane ranging from ~1.4 to 2.4. The Duvernay is noted for C_{13}–C_{31} aryl isoprenoids as well as C_{40} perhydrodiaromatic carotenoids that have the 2-alkyl-1,3-4-trimethyl substitution pattern characteristic of the diaromatic carotenoids in *Chlorobiaceae*, a family of photosynthetic sulfur bacteria (Requejo et al., 1992). Upon kerogen pyrolysis, these compounds produce an unusually high abundance of 1,2,3,4-tetramethylbenzene that is believed to arise from β-cleavage of diaromatic carotenoid moieties incorporated in the kerogen structure. Products derived from the diaromatic carotenoids are substantially enriched in ^{13}C relative to *n*-alkanes of algal origin, which is consistent with their biosynthesis by photosynthetic green sulfur bacteria (Hartgers et al., 1994a). These isotopic enrichments are evident in either the whole extract or kerogen, suggesting that, despite the prevalence of pyrolyzates derived specifically from the diaromatic carotenoids, photosynthetic green sulfur bacteria contributed only a small portion of organic matter to the source sediments.

Duvernay source rocks are ideal for studies of thermal alteration because of readily available core samples with maturity that spans from immature through the gas window (Requejo, 1994; Dieckmann et al., 2000). Duvernay samples are used frequently to test maturation and expulsion trends in compound distributions (e.g. Requejo et al., 1997; Li et al., 1997).

Upper Devonian Tomachi Formation, Bolivia

The Upper Devonian Tomachi Formation in the Madre de Dios Basin contains ~200 m of prodeltaic shales with exceptional source potential (Moretti et al., 1995; Peters et al., 1997a). The Tomachi Formation is lithologically equivalent to the Colpacucho Formation (northern Altiplano) and the Iquiri Formation (Cordillera Oriental and Subandino Sur) (Isaacson et al., 1995). TOC ranges from 3 to 16 wt.%, with HI ranging from

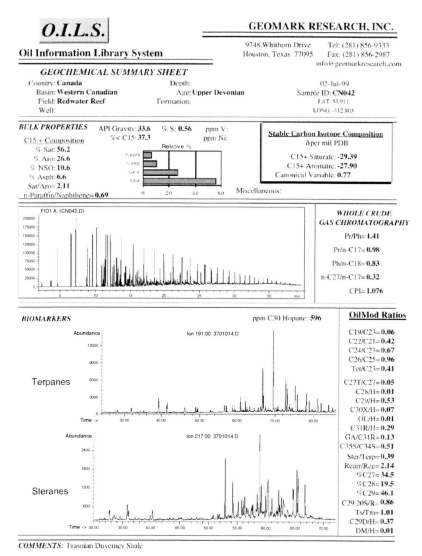

Figure 18.37. Geochemical data for crude oil from Redwater Reef Field generated by the Frasnian Duvernay Formation (courtesy of GeoMark Research, Inc.).

400 to 600 mg HC/g TOC (see Figure 4.19) (Peters et al., 1997a). The kerogen is mostly algal with abundant prasinophyte remains (*Tasmanite*-like). Unlike many of the other Upper Devonian-Mississippian highstand source rocks, the Tomachi shales were deposited in high latitudes. The exceptional source potential probably arose from enhanced productivity caused by upwelling and enhanced preservation under cold, anoxic conditions resulting from the marine transgression (Peters et al., 1996).

Frasnian-Famennian Barreirinha Formation, Amazonas Basin

The basal black shales of the Barreirinha Formation in the Curuá Group are the most organic-rich strata in the Amazonas Basin (Gonzaga et al., 2000). The lowermost section is 30–40 m thick and has high gamma-ray and resistivity response on well logs. These shales contain ~3–8 wt.% TOC, with HI ranging from 100 to 400 mg HC/g TOC. The upper section is 30–150 m thick and

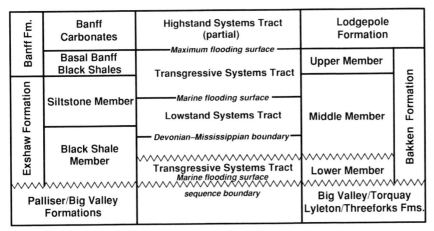

Figure 18.38. Comparison of sequence stratigraphy of the Bakken and Exshaw/Banff formations (from Smith and Bustin, 2000). Reprinted by permission of the AAPG, whose permission is required for further use.

has lower generative potential. TOC ranges from ~1 to 2 wt.%, and HI is typically <200 mg HC/g TOC.

Devonian-Mississippian Bakken/Exshaw-Banff formations, North America

Upper Devonian-Lower Mississippian Bakken Formation organic-rich shales were deposited contemporaneously with the Exshaw/Banff Formation in the Western Canada Basin and other source shales in the Appalachian Basin (e.g. Woodford, Chattanooga) and other sag basins (e.g. Antrim Shale in the Michigan Basin). The Bakken Formation consists of three members – the lower and upper members of finely laminated organic-rich marine mudstones, and a middle member of organic-lean mudstones and sandstones (Smith and Bustin, 2000). In Canada, the Exshaw Formation is equivalent to the Lower and Middle Bakken Formation, while the basal shales of the Banff Formation are equivalent to the Upper Bakken Formation (Figure 18.38). In both systems, these strata are overlain by the carbonates (Lodgepole and Upper Banff), with significant generative potential. Richards (1989b) recognized that black-shale deposition was tied to changes in relative sea level. Smith and Bustin (2000) refined this concept by placing the Bakken/Exshaw-Banff formations into a sequence stratigraphic model involving three distinct systems tracts: a lower highstand sequence, a middle lowstand sequence, and an upper trangressive sequence. Thermal maturation of these source rocks led to petroleum generation from the Exshaw Formation in Canada and the Bakken Formation in the Williston Basin (Figure 18.39).

The geochemistry of the Bakken/Exshaw-Banff source rocks has been characterized extensively (Webster, 1984; Price *et al.*, 1984; Leenheer, 1984;

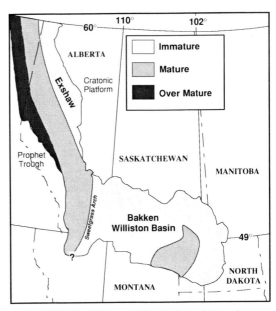

Figure 18.39. Map showing the distribution and thermal maturity of Bakken/Exshaw-Banff source rocks (data from Price *et al.*, 1984; Creaney *et al.*, 1994b; Smith and Bustin, 2000).

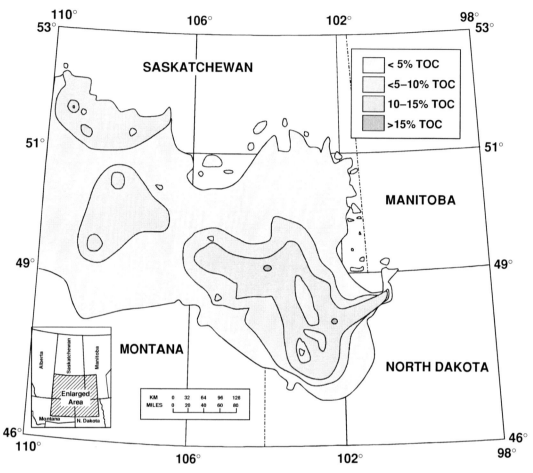

Figure 18.40. Total organic carbon (TOC) distribution for the lower member of the Bakken Formation (from Smith and Bustin, 2000). Reprinted by permission of the AAPG, whose permission is required for further use.

Osadetz et al., 1992; Osadetz and Snowdon, 1995). The lower and Bakken/Exshaw black shales average ~2 m (20 m maximum) and ~3 m thick (7 m maximum), respectively. The lower Bakken/Exshaw contains ~8 wt.% TOC, with maxima ~20 wt.% in the deep basin interior (Figure 18.40), while the upper Bakken/Banff averages ~10 wt.% TOC, with maxima >35 wt.% and individual laminations as high as 63 wt.% (Figure 18.4). All source intervals contain kerogen derived almost entirely of marine algae. Rock-Eval pyrolysis HI averages ~615 mg HC/g TOC for the lower Bakken/Exshaw and ~565 mg HC/g TOC for the upper Bakken/Banff. HI >1000 mg HC/g TOC occurs in both intervals (Smith and Bustin, 2000).

Crude oils produced from the Bakken Formation and Madison Group are geochemically similar. Obermajer et al. (2000b) separated four families of oils from the Williston Basin based on gasoline-range hydrocarbons (iC_5H_{12}–nC_8H_{18}). Their gasoline-range data support different groups established previously using n-alkanes and biomarker distributions (Williams, 1974; Zumberge, 1983; Leenheer and Zumberge, 1987; Osadetz et al., 1992; Osadetz, et al., 1994). Each oil group is confined to one of the following reservoirs: Bakken, Lodgepole, Winnipegosis, or Red River. These oil groups probably originated from Upper Devonian-Mississippian Exshaw/Bakken, Lower Mississippian Lodgepole, Middle Devonian Winnipegosis,

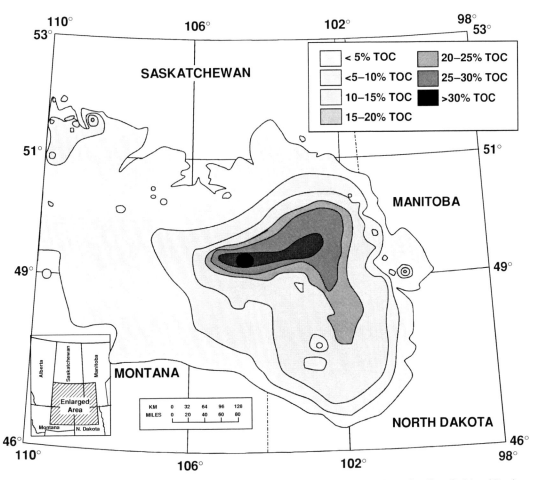

Figure 18.41. Total organic carbon (TOC) distribution for the upper member of the Bakken Formation (from Smith and Bustin, 2000). Reprinted by permission of the AAPG, whose permission is required for further use.

and Ordovician Winnipeg or Bighorn Group source rocks, respectively. Oils from Bakken and Lodgepole (Madison Group) reservoirs could not be distinguished from each other by Obermajer et al. (2000b). Jarvie (2001) also used light hydrocarbons to distinguish Williston Basin oils and found that separating Bakken and Lodgepole oils was problematic. Some Madison oils plot with the Bakken oils, probably because they originate from the Bakken source rock (see Figure 7.37). Large amounts of crude oil generated within the pod of active Bakken source rock in northeastern Montana and northwestern North Dakota and migrated updip over large distances into thermally immature reservoir rocks in Canada (Figure 18.42).

Upper Devonian–Lower Mississippian Woodford Shale, USA

Organic-rich shales with prolific generative potential were deposited in a tectonically stable epicontinental sea that covered much of North America at the end of the Devonian Period. These shales were deposited far from the paleo-shoreline along a broad carbonate shelf extending from southern Texas and Oklahoma (Woodford) and into the Appalachian Basin (Chattanooga). They are time-equivalent to adjacent shales deposited in slightly more restricted sag basins. The Upper Devonian shales are transitional from carbonate deposition that characterized the early Paleozoic

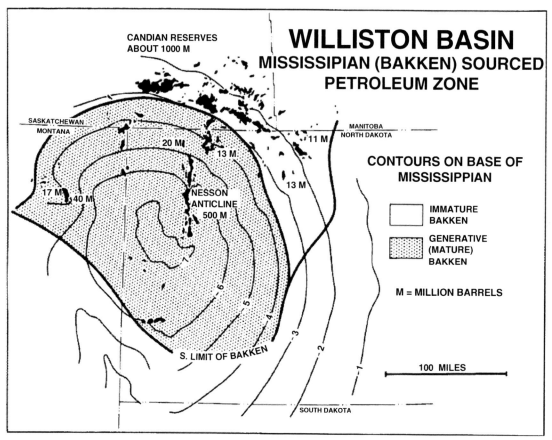

Figure 18.42. About half of the discovered oil in the Williston Basin was trapped within the generative depression, while the other half migrated >100 km toward the northeast (Demaison, 1984). The Nesson Anticline contains ~500 million billion barrels of oil. Reprinted by permission of the AAPG, whose permission is required for further use.

Era to clastic deposition that dominated the late Paleozoic Era.

The middle member of the Woodford Shale is pyritic and rich in radiogenic elements. The shales contain abundant oil-prone kerogen (predominantly type II) deposited under shallow, anoxic waters. TOC typically ranges from 2 to 6 wt.%, but it may be >13 wt.%. HI averages ~350 mg HC/g TOC, with maximum values of ~500 mg HC/g TOC (Figure 18.43) (Lambert, 1993; Lambert et al., 1994; Wang and Philp, 1997b). The original generative potential was higher because these measurements were made on marginally mature to mature shales. The Woodford Shale is typically 40–70 m thick. Novaculite facies are also potential sources, although the organic matter was diluted by siliceous input compared with that in the shales.

The widespread availability of Woodford Shale in core and outcrops facilitated many studies of oil generation (e.g. Lewan, 1983; Michels et al., 1995) and weathering (e.g. Leo and Cardott, 1994). Several studies correlated biomarkers in Woodford source rocks to produced oils in the Anadarko Basin of Oklahoma (Jones and Philp, 1990; Reber, 1989; Wang and Philp, 1997b; Zemmels and Walters, 1987) and in the Permian Basin of southwest Texas (e.g. Jarvie et al., 2001b). Figure 18.44 shows geochemical data for a typical crude oil generated from the Woodford Shale that was produced from the Golden Trend Field in the Anadarko Basin. Gas is produced from fractured Woodford/Chattanooga shale source rocks (Montgomery, 1990), and the formation is commonly used to calibrate well logs (e.g. Guidry et al., 1996).

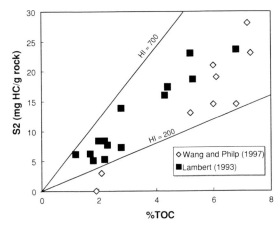

Figure 18.43. Total organic carbon (TOC) (wt.%) versus Rock-Eval S2 for Woodford Shale from Oklahoma and Kansas (Lambert, 1993; Wang and Philp, 1997b).
HI, hydrogen index.

Upper Paleozoic black shale/marls, Kazakhstan

The source rocks for petroleum in the North Caspian (Pricaspian or Pre-Caspian) Basin are believed to be several sequences of Paleozoic (Devonian to Lower Permian) black shales and marls deposited contemporaneously with carbonate reservoir rocks (Ulmishek, 2001a; Pairazian, 1999; Gürgey, 2002). Carbonate reefs, atolls, and platform rocks deposited in shallow, oxic waters along the basin margin became the primary reservoir formations. Muddy limestones, marls, and shales deposited in deep, anoxic waters formed the basinal facies. This depositional setting remained in place for most of the Upper Paleozoic Era, resulting in the formation of multiple source units through time with similar organic facies. Because drilling is confined largely to the carbonate flanks, little is known about the identity and distribution of these source rocks. Where sampled, basinal facies have highly variable TOC up to 10 wt.%. Organic-rich black shales from the Upper Devonian, Middle Carboniferous, and Lower Permianeras are source rocks for petroleum in the Pricaspian Basin. The source rocks contain primarily algal-marine type II kerogen, with lesser amounts of terrigenous type III kerogen. Rock-Eval HI reaches ~400 mg HC/g TOC (see Ulmishek, 2001a). Based on variations in the chemistry of pre-salt (Kungurian) oils, isolated occurrences of high-sulfur type IIS kerogen exist along the eastern flank.

Gürgey (2002) found two major populations of crude oils on the southern margin of the Pricaspian Basin in Kazakhstan. Source-specific biomarker ratios and oil–source rock correlation indicated that the oils originated from Artinskian (Early Permian) marine siliciclastic and carbonate facies immediately below Kungurian salt. The oils that originated from the carbonate facies were non-biodegraded and had generally higher sulfur, C_{24} tetracyclic/C_{26} tricyclic terpanes, and C_{29}/C_{30} hopane, and lower diasteranes/steranes, than the siliciclastic oils. Carboniferous and Devonian shales and carbonates (e.g. Domanik Formation) could also be important source rocks (Pairazian, 1999), but little information on potential correlations with the Pricaspian oils has been published.

CARBONIFEROUS SOURCE ROCKS

Mississippian Chainman Shale, Nevada

Middle Mississippian Chainman Shale is the major petroleum source rock in the Great Basin, Nevada (Poole and Claypool, 1984). These rocks are equivalent to the Delle Phosphatic Member in the Oquirrh Basin, Utah (Sandberg and Gutschick, 1984). The shales were deposited in relatively deep (>200 m), anoxic waters off a carbonate shelf where upwelling may have enhanced primary productivity. The source unit typically contains ~3–8 wt.% TOC, but some cores have up to 13.3 wt.% TOC (Morris and Lovering, 1961). The Chainman Shale contains variable amounts of type II and III kerogen. HI may exceed 400 mg HC/g TOC, but it typically ranges from 100 to 300 mg HC/g TOC.

Mississippian (Viséan) Debolt Formation, Western Canada Basin

The Tournaisian-Viséan Rundle Group in the Western Canada Basin is an overall shallowing-upward succession of carbonate-platform lithofacies that formed during several transgressive/regressive cycles (Richards, 1989a). The Debolt Formation is one of several regressive units within the Rundle Group. The Lower and Upper Debolt are reservoir units dominated by clean platform carbonates deposited under restricted shelf conditions. The Middle Debolt has significant source potential in argillaceous ramp carbonates and shales that were deposited on a restricted shelf. These source rocks contain up to 9.5 wt.% TOC of predominantly

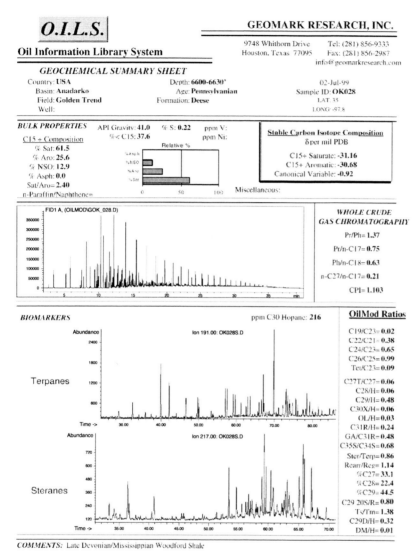

Figure 18.44. Geochemical data for crude oil from Golden Trend Field generated from the Upper Devonian-Lower Mississippian Woodford Shale, Anadarko Basin, USA (courtesy of GeoMark Research, Inc.).

type III to type II/III kerogens (HI up to 400 mg HC/g TOC) (Snowdon et al., 1998b). Rock extracts from the Debolt Formation in the Dunvegan area correlate with some of the oil produced from the Debolt. The oils and extracts exhibit a strong predominance of C_{34} tetrakishomohopanes and low relative abundance of tricyclic terpanes, similar to oils derived from Winnipegosis Formation source rocks in Saskatchewan (Snowdon et al., 1998a; 1998b). The distribution is inferred to indicate source-rock deposition under hypersaline, carbonate-evaporite conditions.

Oils produced from the Debolt Formation in other areas may have originated from other source rocks (Snowdon et al., 1998a; 1998b). Oils from the Royce area have many geochemical characteristics similar to oils derived from the Exshaw Formation (Creaney et al., 1994a), including low pristane/phytane, moderate concentrations of tricyclic terpanes no homohopane predominance, and low C_{24} tetracyclic terpane. Oils from the Blueberry area are highly unusual because they lack steranes and terpanes but do not appear to be highly mature. It is unclear whether this lack of biomarkers is due to an unusual source rock or to secondary processes (Snowdon et al., 1998a; 1998b).

Viséan-Tournaisian Milligans Formation, Bonaparte Gulf Basin

The Bonaparte Gulf Basin, located between Western Australia and East Timor, contains several source rock units, the oldest being the marine shales and carbonates in the Mississippian Milligans Formation (Bishop, 1999c; Edwards et al., 2000a). Organic-rich basinal shales formed during a transgression and rapid basin subsidence (McConachie et al., 1996). Measured source potential is fairly lean in mature samples, with TOC rarely exceeding 2 wt.% and HI typically ranging from 10 to 100 mg HC/g TOC (Jefferies, 1988). Initial source potential may have been higher because the kerogen originated from a mixture of terrigenous and marine organic matter. Rapid sedimentation and anoxic conditions enhanced preservation.

Upper Mississippian Heath Formation, Central Montana and Williston Basin

The Upper Mississippian Heath Formation is comprised mostly of nearshore marine calcareous shales and carbonates with minor anhydrite and coal beds. The black shales and limestones have excellent source potential averaging ~7.6 wt.% TOC, with maxima >17 wt.% TOC. Rock-Eval HI averages ~575 mg HC/g TOC (Cole et al., 1990). The Heath Formation is generally immature, except in the deeper portion of the Williston Basin, a few deeply buried areas in the central Montana trough, and in the Montana thrust belt (Longden et al., 1988). Where mature, the Heath has been correlated to oils produced from the overlying Tyler sandstones (Rinaldi, 1988; Cole et al., 1990; Cole and Drozd, 1994). The Heath-Tyler(!) petroleum system has high generation–accumulation efficiency (36.3%) because the reservoir rock incises into the source rock and migration distances are short.

Pennsylvanian-Permian shales, North America

Beginning in the latest Mississippian Period, tectonic deformations produced a series of basins throughout Oklahoma, Texas, New Mexico, Colorado, and Utah (Figure 18.45). Each basin shares similar sedimentary sequences, tectonic features, geometry, and structural elements, suggesting that they developed from a common cause involving synchronous basin subsidence and basement uplift. It is not understood fully why these deformations occurred, but they are believed to involve plate collisions that sutured North America to South America–Africa, resulting in the Ouachita orogeny, the Marathon orogeny, and finally the Ancestral Rocky Mountains (Kluth, 1986; Ye et al., 1996). Consequently, deep basins developed in the Ouachita foreland during the Late Mississippian Period (e.g. Black Warrior Basin), followed by the formation of deep, sediment-starved basins to the west. The fill in these basins is commonly ~700–1800 m thick, but it is more than 5000 m thick in the Anadarko Basin and more than 6000 m thick in the Arkoma Basin. A series of basins then developed along the Marathon orogeny throughout Texas (e.g. Delaware, Fort Worth, and Palo Duro). By the Middle Pennsylvanian (Desmoinesian) Period, the Ancestral Rocky Mountains were well developed, and numerous restricted basins had formed by tectonic uplift. Deposition of source rocks continued into the Permian Period, although this was largely restricted to the Delaware Basin, forming the Wolfcampian, Leonardian, and Guadalupian series of source rocks.

The Lower Pennsylvanian strata of the Oklahoma Basin are mostly marine shales associated with sandstone, limestone, conglomerate, and coal. They contain varying amounts of organic matter (<0.5 to >25 wt.% TOC) that is predominantly type III and gas-prone. Isolated strata contain mixtures of type II and III organic matter and were the source of considerable volumes of petroleum (Johnson and Cardott, 1992; Wang and Philp, 1997b). Most of these strata are postmature, and it is difficult to assess original source potential (Guthrie et al., 1986). Less altered Pennsylvanian-Permian source rocks that better reflect their original generative potential are preserved in other western interior US basins. Three of these – the Middle Minnelusa shales in the Powder River Basin, the Gothic and Chimney Rock shales from the Paradox Basin, and the Brushy Canyon from the Permian Basin – are described below.

Hermosa Group Pennsylvanian shales, Paradox Basin

Tectonic activity created a silled precursor of the Paradox Basin in the Early Pennsylvanian Period, which restricted the influx of marine waters and resulted in 29 cyclical sequences of alternating black shales,

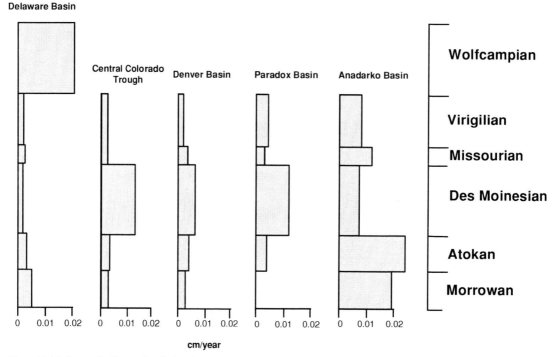

Figure 18.45. Rates of sedimentation during the Pennsylvanian-Early Permian Period for selected basins from the western interior USA (modified from Kluth, 1986). The graphs oversimplify the complexities of basin geometry and changes in sedimentation rates, but they illustrate the general pattern of tectonic events that developed the basins in the Ancestral Rocky Mountains. Reprinted by permission of the AAPG, whose permission is required for further use.

carbonates, and evaporites. Two of the units, the Gothic and Chimney Rock shales, were identified as major source rocks for oil in the Paradox Basin (Hite et al., 1984). These source units formed when marine waters spilled over the sill, either mixing with residual brines or dissolving evaporites to produce hypersaline waters. Oil-prone kerogen was preserved in the deeper, anoxic depocenters within the carbonate shelf.

The Gothic and Chimney Rock units have approximately equal source potential, ranging from 0.43 to 7.71 wt.% TOC, with an average of ~2.53 wt.% TOC. Moderately mature samples have residual HI of 171–460 mg HC/g TOC (Tischler, 1995). There is considerable debate concerning the exact nature of the depositional environment and of the biotic input to the source kerogen (Hite et al., 1984; Tischler, 1995; Tuttle et al., 1996; Van Buchem, 2000). Bacterially altered terrigenous plant matter, halophilic bacteria, cyanobacteria, and normal marine algae and zooplankton have all been suggested as major contributors.

Pennyslvanian lacustrine rocks, east Greenland

Stemmerik et al. (1990) described Westphalian (Upper Bashkirian) organic-rich lacustrine shales in east Greenland. These Permian sedimentary rocks represent the final stage of deposition of a 9-km-thick syn-rift sequence that started during the Early Devonian Period (Ziegler, 1988). The lacustrine basins are believed to be ~10–15 km wide and ~20–100 km in length along the rift axis. Deep-water claystones contain 2–10 wt.% TOC, with HI in the range 300–900 mg HC/g TOC. The shales contain type I kerogen that is mostly amorphous, degraded algal material. Remains of *Botryococcus* and wind-blown spores and pollen also are present. The high generative potential is attributed to high productivity during seasonal algal blooms, low rates of sedimentation, and preservation in deep anoxic waters. The source facies grade upward into rocks deposited in shallow-water oxic environments with negligible source potential. Onshore outcrops range from immature to marginally mature. Stemmerik et al. (1990) speculated that Westphalian lacustrine shales could be an overlooked source rock in the east Greenland basins and offshore Norway.

Pennsylvanian coals, Southern Gas Basin

The Carboniferous Period is characterized by extensive peat accumulations throughout Europe, North America, and Asia. The largest of these is the Paralic Basin, which extends from Ireland to Poland. Subsidence during the Early Pennsylvanian (Namurian and Westphalian) Period nearly equaled sedimentation rate, allowing vast portions of the tropical lowlands to be flooded, thus forming extensive bogs. Although extensive, marine transgressions were short-lived and resulted in minor marine intercalations. The collision of Gondwana with Eurasia at the end of Westphalian time altered the topography and climate, essentially ending the deposition of the Westphalian coals and resulting in widespread extinction of the tree-like lycopsids, the major biomass for the Pennsylvanian coals.

Westphalian coal is believed to be the major source of the massive amount of gas in the Southern Gas Basin (Cornford, 1984). However, commonly associated condensate has led to speculation that there are additional sources of hydrocarbons, such as the Permian Kupferschiefer Formation and/or the Namurian shales (Cameron, 1993). In addition to multiple sources of gas and condensate, the complex tectonic history of the Southern Gas Basin, particularly the inversion of some parts of the basin during the Late Cretaceous Period and the early part of the Tertiary Period, complicated the occurrence of gas and condensate in the basin (Besley, 1998). The significant impact of such inversions was the loss of gas and condensate that had been accumulated in uplifted reservoirs. The gas and condensate generated from the Westphalian coals and non-marine source rocks subsequently migrated into these inverted reservoirs. Finally, some dry gas deposits may originate by long-distance migration of methane generated from the Westphalian coals.

Pennsylvanian Minnelusa shales, Powder River Basin

Middle Pennsylvanian (Desmoinesian) shales blanketed much of the US central plains from Oklahoma to North

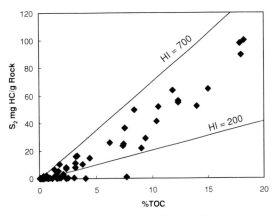

Figure 18.46. Total organic carbon (TOC) (wt.%) versus Rock-Eval pyrolysis S2 yields for Middle Minnelusa and age-equivalent rocks from central US interior basins (data from Clayton and Ryder, 1984).
HI, hydrogen index.

Dakota. These shales are relatively thin (<20 m) and have highly variable in source potential. Some intervals have negligible generative potential, while others have >25 wt.% TOC, with HI >700 mg HC/g TOC (Figure 18.46) (Clayton and Ryder, 1984). In the Powder River Basin, these rocks occur in the Middle Minnelusa Formation, where they were deposited under shallow marine conditions that occasionally became highly saline due to restricted water circulation. Aryl isoprenoids in extracts of these rocks provide evidence for euxinic conditions within the photic zone during deposition of the source rock (Clayton et al., 1988).

Upper Pennsylvanian-Permian Copacabana Formation, Sub-Andean basins

The Copacabana Formation is a shallow-water, marine platform succession that oocurs throughout the Marañon, Ucayali, and Madre de Dios Basins of Peru and Bolivia. Where not eroded, the formation may comprise up 800 m of limestone, shale, and sandstone (Mathalone and Montoya, 1995; Baby et al., 1995). Its source potential is best expressed in the Pando X-1 well in the Madre de Dios Basin (Isaacson et al., 1995; Peters et al., 1997a). TOC ranges from 1 to 9 wt.%, with maximum HI ~440 mg HC/g TOC. Kerogens within the shales are mixtures of type II and type III organic matter.

Carboniferous Kuna-Lisburne interval, Alaska

The Lisburne Group (see Figure 18.28) may be the source rock for limited oil accumulations, primarily in the central and western Brooks Range of Alaska (Huang et al., 1985). North of the Brooks Range, the Lisburne consists mainly of platform carbonates with low TOC (average ~0.7 wt.%) interbedded with a few thin organic-rich shales (Sedivy et al., 1987). Some of these shales contain up to 14 wt.% TOC in the area of the Prudhoe Bay Field (Masterson, 2001). In the Brooks Range south of the Colville Basin, the Lisburne Group contains generally organic-rich deep marine shales (up to ~7 wt.% for highly mature samples) (Brosgé et al., 1981) that are described locally as the Kuna Formation (Mull et al., 1982; Magoon and Bird, 1988).

Only a few crude oils have been attributed to the Kuna-Lisburne source rock. For example, non-biodegraded oils from South Barrow-17 and -19 have low API gravities (19–21°API) and high sulfur (1.7–1.9 wt.%) (Magoon and Claypool, 1984), suggesting a reducing, clay-poor source-rock depositional environment.

Upper Carboniferous-Permian torbanite oil shales

The organic matter in torbanite oil shales (see Figure 4.8) is believed to originate from strains of the extant freshwater alga *Botryococcus braunii* or closely related organisms. Torbanites occur in strata of Carboniferous age in Scotland, of Lower Permian age in South Africa and Australia, and of Upper Permian age in the Sydney Basin in Australia. Surprisingly, botryococcane (see Figure 13.25) and related-biomarkers (see Figures 13.4 and 13.32) associated with the "B" strain of *B. braunii* have not been identified in torbanites (Derenne et al., 1988). Their absence suggests that the fossil algae in torbanites are not related to *B. braunii*, originate from strains of *B. braunii* that do not produce botryococcane, or are so highly degraded by bacteria that *B. braunii* lipids were altered completely (Grice et al., 2001).

PERMIAN SOURCE ROCKS

Lower Permian lacustrine source rocks, southwestern Gondwana

Lower Permian lacustrine shales occur in the Whitehill Formation (Karoo Basin, southern Africa), Huab Formation (Karoo Supergroup, South Africa/Namibia), Lower Irati Formation (Paraná Basin, Brazil), and Black Rock Member of the Port Sussex Formation (Falkland Islands). Where immature, these rocks typically contain ~4–5 wt.% TOC, with HI ranging from 300 to 1000 mg HC/g TOC and OI <30 mg CO_2/g TOC (Faure and Cole, 1999). Algal organic matter was preserved under anoxic, fresh to brackish water conditions. The kerogens from these Lower Permian formations exhibit ^{13}C-rich isotope ratios (δ^{13}C ~−17‰) compared with formations above and below (δ^{13}C ranges from −26 to −22‰, consistent with Paleozoic type III kerogens). Faure and Cole (1999) speculated that these formations were deposited in a single proto-rift basin in southwestern Gondwana. This vast lake basin (~5 × 10^6 km^2) had seasonal algal blooms, possibly the most extensive in Earth's history. Occasional marine transgressions may have allowed sulfate reduction and pyrite deposition.

Middle Permian Ene Formation, Peru

The Ene Formation is one of the major petroleum source rocks in the foreland basins of Peru (Figure 18.47) (Fabre and Alvar, 1993; Mathalone and Montoya, 1995). It is a hypersaline regressive sequence that originally covered much of the Marañon, Ucayali, and Ene basins before being partially eroded in the Triassic Period. Marine shales within the evaporitic sequence are organic-rich and up to 300 m thick. The source potential of the Ene has not been described fully and is inferred by correlation with oils and outcrop samples. Where measured, Ene marine shales have an average of 2–3 wt.% TOC, with a maximum of ~7 wt.% TOC. The rocks contain mixtures of type I and II kerogen, indicating that the Ene Formation had considerable generative potential.

Lower Permian Leonard-Wolfcamp units, Permian Basin

The Permian Basin of west Texas and southeast New Mexico includes the Delaware and Midland basins and the intervening Central Basin Platform. Over 35 BBOE has been produced from the Permian Basin, which contains several source rocks, including the Ordovician Simpson Group, the Upper Devonian-Lower Mississipian Woodford Formation, and the

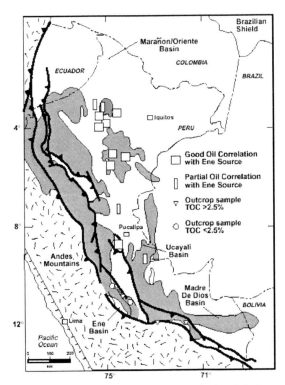

Figure 18.47. Distribution of the Ene Formation in sub-Andean foreland basins (from Mathalone and Montoya, 1995). Reprinted by permission of the AAPG, whose permission is required for further use.

Permian Wolfcampian, Leonardian, and Guadalupian series (Jarvie *et al.*, 2001b). The prolific Lower Permian Leonard-Wolcamp source rocks in the Permian Basin contain organic-rich micritic shales with up to 10 wt.% TOC and type II–III kerogen.

Williams (1977) analyzed 504 crude oils from the Permian Basin representing 391 reservoirs and 297 fields. He found nine major oil groups generated from the following four pre-Permian and five Permian source rocks:

1. Ordovician Simpson Group (Ellenburger reservoirs).
2. Devonian-Mississippian Woodford Shale (Silurian-Devonian reservoirs).
3. Pennyslvanian Strawn or Canyon intervals (Pennsylvanian reservoirs).
4. Wolfcampian (Wolfcamp, Dean, Spraberry reservoirs).

5. Upper Leonardian in Midland Basin only (San Andres reservoirs).
6. Lower Leonardian in Midland Basin only (Clearfork reservoirs).
7. Leonardian in Delaware Basin only (San Andres reservoirs).
8. local Pennsylvanian source in northern Delaware Basin only (Pennsylvanian, Wolfcamp, and Abo reservoirs).
9. basinal Guadalupian in Delaware Basin only (Queen reservoirs).

Group 5 accounts for the largest volumes of oil in the Permian Basin, which occur mainly in San Andres oolitic carbonate reservoirs and originated from the Leonardian Bone Spring Formation. Groups 5–7 probably represent different organofacies of the Bone Spring source rock. Group 9 oil accumulations are volumetrically less significant than the Leonard-Wolcamp oils and originate from the Brushy Canyon Formation, as discussed below. Hill *et al.* (2003) used biomarker, isotope, and sulfur data to confirm the oil groups of Williams (1977) and to recognize the Mississippian Barnett Shale as an additional source rock and oil type.

Permian Brushy Canyon Formation, Delaware Basin

The Brushy Canyon Formation is the lowermost portion of the Guadalupian Delaware Mountain Group in the Delaware Basin, located in the western Permian Basin. The Brushy Canyon Formation consists of submarine canyon fill, slope, and basin-floor deposits primarily of fine-grained arkosic and subarkosic sandstones and siltstones that might be considered unlikely source rocks (Sageman *et al.*, 1998). However, ~20–25% of the lower Brushy Canyon Formation consists of organic-rich siltstones and shales with sufficient generative potential to act as the source for most of the oil produced from the Brushy Canyon sandstones. The siltstones and shales range from < 1 to 12 m thick and have TOC in the range 1–4 wt.%. They are widespread marker beds that are proven source rocks for oils produced to the east (Sageman *et al.*, 1998). Similar source strata occur in the deeper Pipeline Shale and in the overlying Cherry Canyon Formation (Hays, 1992; Carroll *et al.*, 2000).

The organic-rich units within the Brushy Canyon Formation are considered to be suspension fallout deposits representing cycle-bounding condensed sections. They typically contain only ~0.5–1.8 wt.% TOC, with measured HI ranging from only 100 to 200 mg HC/g TOC (Armentrout et al., 1998; Sageman et al., 1998; Justman and Broadhead, 2000). The richest interval occurs within a 3-m-thick third-order flooding surface at the top of the formation. Isolated strata contain as much as 5.3 wt.% TOC, with maximum HI rarely exceeding 300 mg HC/g TOC. The organic-rich units contain mostly amorphous, sapropelic, and herbaceous kerogen.

Middle-Upper Permian Phosphoria Formation, western interior USA

During the Late Carboniferous-Early Permian Period, the western shelf of the North America craton broke into several discrete basins. The Middle-Upper Permian Phosphoria Formation developed in the Sublett Basin, a foreland basin on the edge of the continental shelf within the Cordilleran seaway (Wardlaw et al., 1995). The Phoshoria Formation is noted for its cherts and phosphatic, organic-rich shales that were deposited on the foreslope of persistent upwelling zones (Parrish, 1982; Maughan, 1993; Jewell, 1995). Unlike most modern upwelling zones that occur on the slopes of continental margins, upwelling at the time of Phosphoria Formation deposition occurred near a broad, shallow, evaporative sea. This environment favored cyclic occurrences of salinity stratification and anoxia, which affected biomarker distributions (Dahl et al., 1993). Depositional environments incorporating both upwelling and salinity stratification are rare, but these conditions favor the high productivity and preservation that appear to be responsible for organic-rich shales (up to ~33 wt.% TOC) (Stephens and Carroll, 1999).

Two major source units occur within the Phosphoria Formation: the Meade Peak and the Retort Shale Members. Meade Peak phosphatic shales were deposited through most of the Sublett Basin and over a broad range of environments, including outer-ramp, mid-ramp, and lagoonal inner-ramp settings (Maughan, 1984; Hiatt, 1997; Hiatt and Budd, 2001). Organic content within the Meade Peak Member typically ranges from 1 to 4 wt.% TOC and averages ~2.4 wt.%, with the greatest richness near the Wyoming/Idaho border, where it averages ~9 wt.% TOC (Claypool et al., 1978; Maughan, 1984). In contrast, the Retort Shale member was deposited under hypersaline conditions, reflected by high gammacerane (Dahl et al., 1993). The organic content averages ~4.9 wt.% TOC and is highest near the Idaho/Montana border, where it averages ~10 wt.% TOC (Maughan, 1984). Where immature, Phosphoria shales have H/C ratios of ~1.2–1.3 (Claypool et al., 1978; Lewan et al., 1986). Kerogen within the Phosphoria Formation is considered to be type II, although isolated facies may be rich in sulfur. This is reflected in a lower distribution of activation energies than for typical type II kerogen (Lewan et al., 1986).

Phosphoria crude oils have low pristane/phytane, as exemplified by gas chromatograms of samples from the Dillinger Ranch and Dry Piney fields, Wyoming (see Figure 4.23). Figures 3.20 and 18.48 show quantitative distributions of various compound fractions, biomarker classes, and individual biomarkers in Phosphoria oil from Hamilton Dome, Wyoming. The Hamilton Dome oil has elevated gammacerane and C_{35} hopanes, consistent with elevated salinity and anoxic conditions during source-rock deposition (Figure 8.21).

Permian Kupferschiefer/Zechstein, northern Europe

During the Permian Period, non-marine Rotliegendes Group red beds and sandstones were deposited in a subsiding, tectonic depression across northern Europe. The region was >700 m below sea level when a breach occurred during the Late Permian Period, catastrophically flooding the depression (van Wess et al., 2000). Two large intracratonic basins formed, which blanketed most of northern Europe (Southern Permian Basin) and the middle of the Viking Graben (Northern Permian Basin) (Ziegler, 1988). Under these conditions, a thin, copper-bearing source rock called the Kupferschiefer Formation was deposited across northern Europe. The richest strata occur at the base of the formation, and abundant fish fossils suggest mass mortality caused by anoxic bottom waters and restricted water circulation. These basins were largely isolated from the open ocean, and the arid climate helped to elevate the salinity, giving rise to the Zechstein evaporites.

Concentrations of organic carbon and copper typically are greatest at the base of the Kupferschiefer

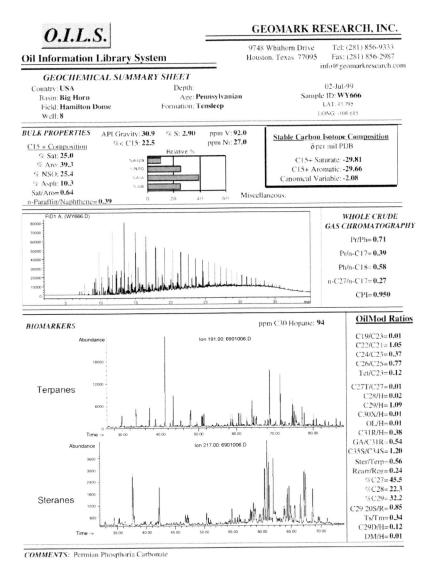

Figure 18.48. Geochemical data for crude oil from the Hamilton Dome Field generated from the Permian Phosphoria Formation, Big Horn Basin, USA (courtesy of GeoMark Research, Inc.).

Formation (Figure 18.49). Copper accumulation in the Kupferschiefer Formation has been attributed to the action of thermochemical sulfate reduction (TSR) on the associated organic matter (Sun, 1996; Sun, 1998; Sun and Püttmann, 1996; Sun and Püttmann, 1997; Sun and Püttmann, 2000; Sun and Püttmann, 2001; Bechtel et al., 2001). A three-step process of metal accumulation was proposed (Sun and Püttmann, 1996). During deposition and diagenesis of the sediment, framboidal pyrite and pyrite precursors precipitated by bacterial sulfate reduction (BSR) and were largely replaced by mixed Cu/Fe minerals and by chalcocite. In the section of the Kupferschiefer Formation with high copper (>8 wt.%), reduced sulfur from iron sulfides was insufficient to precipitate all copper and other trace metals from ascending solutions. In this part of the profile, TSR occurred. By analyzing soluble saturated and aromatic hydrocarbons and kerogen, Sun and Püttmann (2000) estimated that ~60% of the copper in the formation was precipitated by reaction with H_2S generated by TSR.

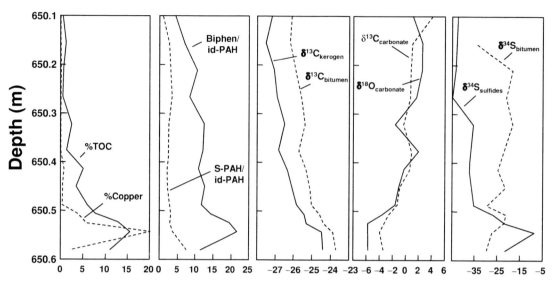

Figure 18.49. Distribution of copper and total carbon, selected polycyclic aromatic hydrocarbon (PAH) ratios, $\delta^{13}C$ of kerogen and bitumen, $\delta^{13}C$ and $\delta^{18}O$ of carbonate, and $\delta^{34}S$ of sulfides and bitumen for a core of Zechstein limestone (~650.1–650.35 m) and Permian Kupferschiefer (650.35–650.6 m) from the Sangerhausen Basin, Germany. Reprinted from Sun and Püttmann (2000) and Bechtel et al. (2001). © Copyright 2000, 2001 with permission from Elsevier.
TOC, total organic carbon.

Maturity assessment in the Kuperferschiefer black shales is difficult because vitrinite is rare and more than 90% of the organic matter consists of bituminite (Koch, 1997). The random reflectance of bituminite is less than vitrinite at low maturity. However, the difference in reflectance between bituminite and vitrinite is small at ~1.25% and becomes nearly identical in the range 2–4% reflectance.

TSR was limited to the highly mineralized zones within the Kupferschiefer Formation. Microscopic evidence for TSR includes elevated vitrinite reflectance (0.95% R_r in the mineralized zone versus 0.83% R_r for overlying strata), and the presence of pyrobitumen, saddle dolomite, and calcite spurs (Sun and Püttmann, 1997). The bituminite filling pores and fractures is highly reflective (0.8–1.2% R_r). Estimates of the maximum temperatures in the mineralized zones based on reflectance, Rock-Eval pyrolysis, and hydrocarbon thermal indicators were downgraded from 150°C (Sun and Püttmann, 1996) to 130°C (Sun and Püttmann, 1997; 2000).

Sun and Püttmann (2000) compared the distributions of saturated and aromatic hydrocarbons in Kupfesrchiefer rock samples and found variations between the zones with copper deposits from BSR and pyrite replacement and the highly mineralized zones attributed to TSR. Samples from the highly mineralized TSR zones are depleted in n-alkanes and isoprenoids, and selectively enriched in biphenyls, phenanthrenes, sulfur-containing PAHs, and dibenzofuran. The comparisons assume homogeneity of organic facies between the BSR and TSR zones and proportional abundance between TOC and concentrations of individual hydrocarbons. The depletion in saturated hydrocarbons and hydrogen content of the kerogen was attributed to TSR. The enrichment in specific aromatic and sulfur-aromatic hydrocarbons, however, may be due to generation or selective uptake of PAHs transported from basin formation waters. The vertical profile of PAHs and copper argue for the latter.

Bechtel et al. (2001) examined the isotopic composition of a 0.58-m-thick succession of Kupferschiefer and Zechstein limestone. The bottom section of the profile has up to 20 wt.% copper. The different processes leading to high-grade copper mineralization are distinguished by contrasting carbon–sulfur–iron elemental relationships, HI, and stable isotopic compositions of organic matter, carbonates, and sulfides

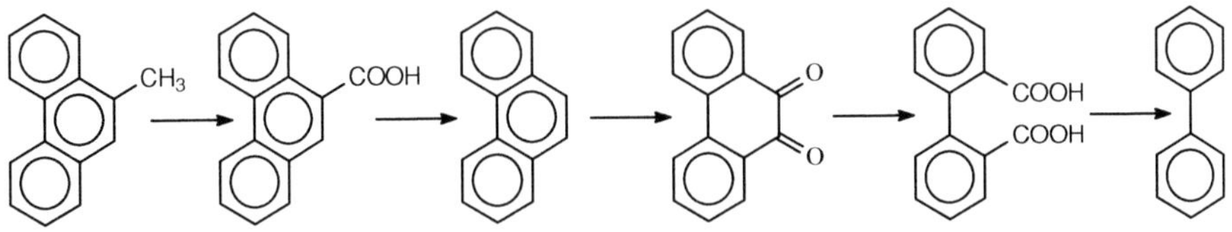

Figure 18.50. Generalized scheme for sequential oxidation reactions in the mineralized Kupferschiefer zone (from Sun and Püttmann, 2000).

(Figure 18.49). The isotopic data correlate with various facies-related parameters (TOC, atomic C/S, and HI) only in those parts of the profile that are not enriched in copper, suggesting precipitation of metal sulfides (mostly pyrite) by BSR with subsequent pyrite replacement. In the high-grade mineralized section (>8 wt.% Cu), $\delta^{13}C$ and $\delta^{18}O$ values of carbonates decrease by up to 4‰ compared with Kupferschiefer sections that are not enriched in copper. The opposite shift to enrichment in ^{13}C occurs in the kerogen and bitumen. $\delta^{34}S$ of sulfide varies in most parts of the profile between −44.7 and −35.2‰. Higher $\delta^{34}S$ values up to −8.4‰ occur in copper sulfides from high-grade mineralized samples near the base of the Kupferschiefer Formation. The results of carbon and sulfur isotopic analyses from this part of the Kupferschiefer Formation provide evidence for oxidative degradation of organic matter and its participation in metal sulfide precipitation through TSR.

Sun and Püttmann (2000) examined this sample suite using FTIR and GCMS for evidence of organic matter oxidation. The mineralized Kupferschiefer zone contained high concentrations of oxidized saturated (mostly long-chain primary alcohols and acids) and aromatic (phthalic acids and esters) species. They attributed the relatively high concentrations of napthalene, phenanthrene, and biphenyl to sequential oxidation reactions (Figure 18.50).

Upper Permian lacustrine rocks, Junggar Basin, northwest China

Upper Permian subsidence in the Junggar Basin of northwest China produced more than 5000 m of lacustrine sediments, including more than 1000 m of organic-rich mudstones. These rocks are the source for oil in the giant Karamay Field in northwestern Junggar and for discoveries elsewhere in the basin (Carroll, 1990). Lacustrine strata may be classified as underfilled, balanced-fill, or overfilled, based on the balance between the supply of water and sediment and its accommodation by basin subsidence (Carroll and Bohacs, 1999; Bohacs et al., 2000). Carroll and Bohacs (2001) used this classification for lacustrine sediments in the Junggar Basin and examined the development of organic-rich source facies in a stratigraphic framework.

The Jingjingzigou Formation, lowermost in the Permian lacustrine sequence, represents the underfilled facies and is characterized by dolomitic mudstones, siltstones, and sandstones. The mudstones typically contain 1–2 wt.% TOC where laminated and <1.0 wt.% where non-laminated. Rock-Eval HI ranges from ~130 to 475 mg HC/g TOC. These rocks have entered the oil window, so the original generative potential was higher. Amorphous kerogen, presumably of algal and bacterial origin, dominates the mudstones of the Jingjingzigou Formation. Extracts contain abundant β-carotane, which has been linked to hypersaline deposition (ten Haven et al., 1988). Karamay Field oils also contain abundant β-carotane (Figures 18.51 and 18.52), suggesting that the Jingjingzigou Formation is the primary source rock (Carroll, 1998).

The Lucaogou Formation represents the balanced-fill facies, resulting in one of the richest and thickest lacustrine source rocks in the world. The formation is dominated by laminated mudstone sequences 1–4 m thick and with highly variable source potential. TOC and HI range from <0.5 to 34 wt.% and ~480 to 760 mg HC/g TOC, respectively (Carroll, 1990). The sampled source rocks are in the oil window, so initially their generative potential was higher. The amorphous kerogen is mostly algal. Although the oil shales within the Lucaogou Formation typically exceed 20 wt.% TOC, their richness is not the result of enhanced productivity (Carroll, 1998). Deposition occurred in middle paleolatitudes (39–43°N) rather than in the tropics, and the lake had limited nutrient supplies. The high source potential appears to result from stable salinity

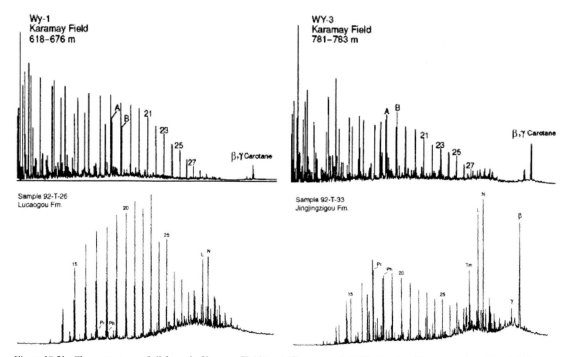

Figure 18.51. Chromatograms of oil from the Karamay Field (top) (Clayton et al., 1997). Reprinted by permission of the AAPG, whose permission is required for further use). Rock extracts from the Lucaogou and Jingjingzigou formations (bottom) (Carroll, 1998). © Copyright 1998, with permission from Elsevier.

stratification in deep water and low inorganic sedimentation rates. Modest levels of β-carotane and gammacerane in these rocks are consistent with deposition under brackish water conditions.

The overlying Hongyanchi Formation represents lacustrine overfilled facies. It was deposited in fresh oxic to suboxic waters and contains a greater amount of kerogen from higher plants than the balanced-fill facies. Consequently, the Hongyanchi Formation contains significant amounts of organic carbon (1–5 wt.% TOC) but has low generative potential. The influx of terrigenous organic matter is reflected in the biomarker distributions, including abundant C_{29} steranes (Figure 18.53) (Carroll, 1998).

Upper Permian Ravnefjeld Formation, east Greenland

The Upper Permian Ravnefjeld Formation of east Greenland contains two major units with significant source potential (Christiansen et al., 1993). Synrift basins formed during the Late Devonian-Carboniferous Period followed by Permian-Mesozoic strata that originated due to thermal subsidence. Upper Permian strata consist of three depositional sequences. The Ravnefjeld Formation is the second of these sequences, which, in turn, can be subdivided into five units. Two of these units are organic-rich laminated shales from the basinal anoxic facies and three are bioturbated shales representing oxic facies. Anoxic conditions are thought to arise from changes in sea level, with the organic-rich layers corresponding to highstand conditions.

The organic-rich layers typically have cumulative thickness of ~15–20 m, TOC ~4–5 wt.%, and HI in the range 300–400 mg HC/g TOC. The laminated basinal shales have geochemical characteristics typical of marine anoxic deposition. Pristane/phytane ratios range from ~1 to 2.5. Extracts of basal shales in the vicinity of carbonates have low pristane/phytane and abundant homohopanes (including $C_{35}>C_{34}$) and C_{23} tricyclic terpanes.

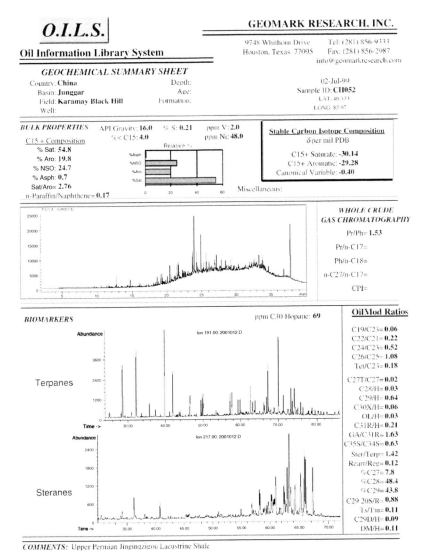

Figure 18.52. Geochemical data for crude oil from Karamay Black Hill Field generated from the Upper Permian Jingjingzigou Formation, Junggar Basin, China (courtesy of GeoMark Research, Inc.).

Upper Permian Irati shales, Paraná Basin, Brazil

The Upper Permian calcareous shales of the Irati Formation (Paraná Basin, Brazil) and its age equivalents in Argentina and Uruguay have exceptional generative potential. They contain ~6 wt.% TOC, although values >15 wt.% are not uncommon. The Irati Formation is thermally immature, except for areas exposed to diabase intrusions where small oil deposits occur (Sousa *et al.*, 1997; Araújo *et al.*, 2000). However, Brazil has a long history of extracting oil from the shale. A large, modern retort operation currently processes ~7000 metric tons of oil shale/day, yielding ~4000 barrels/day of shale oil.

Kerogen in the Irati Formation is mostly amorphous, degraded algal remains that are rich in hydrogen with average HI ~600 mg HC/g TOC (Correa da Silva and Cornford, 1985; Mello *et al.*, 1993). These calcareous black shales, up to 30 m thick, were deposited in the basinal facies of a wide, epicontinental sea. A

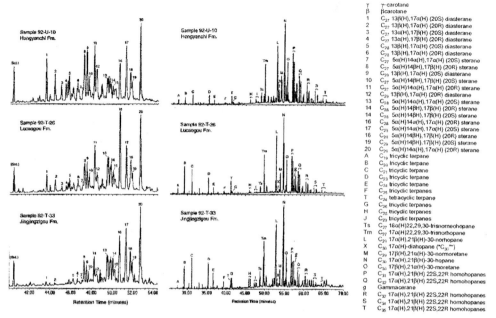

Figure 18.53. Sterane (m/z 217) and terpane (m/z 191) mass chromatograms showing the distribution of biomarkers in extracts from Upper Permian lacustrine source rocks in the Junggar Basin, China. Reprinted from Carroll (1998). © Copyright 1998, with permission from Elsevier.

broad, shallow-water carbonate shelf established hypersaline conditions, resulting in a density-stratified water column. Abundant gammacerane in extracts from these rocks indicates saline to hypersaline conditions during deposition.

Upper Permian Toolachee Formation, Cooper/Eromanga basins, Australia

The Upper Permian coal measures and coaly shales of the Toolachee Formation are a major source of hydrocarbons in the Cooper Basin of central Australia (Figures 18.54 and 18.55). Type III kerogen derived from higher plants dominates the organic matter in these rocks. Although predominantly gas-prone, the source rock still has the potential to generate hydrocarbon liquids (Curry et al., 1994). The oils are characterized by high wax content, high pristane/phytane (>3), and a predominance of terpenoid hydrocarbons characteristic of higher land plants (e.g. phyllocladane, kaurane) (Philp and Gilbert, 1986; Powell et al., 1986; Alexander et al., 1988c; Alexander et al., 1992b).

Shales and coals in the Lower Permian Patchawarra Formation may also contribute to Cooper Basin petroleum, but their source potential is more limited than that of the Upper Permian rocks. Oil and gas from the Upper Permian source rocks may have also migrated into the Eromanga Basin by means of a topographically driven, groundwater flow system that spans nearly 1200 km (Toupin et al., 1997). However, geochemical studies have not clarified whether the light oil and gas in the Eromanga Basin migrated long distances (>50 km) from the Cooper Basin (e.g. Boreham and Summons, 1999), originated from Jurassic source rocks within the Eromanga Basin (e.g. Boult et al., 1998), or represents a mixture of these two possibilities (Arouri et al., 2000).

MESOZOIC ERA: TRIASSIC SOURCE ROCKS

Figure 18.56 shows the temporal distribution of many Mesozoic source rocks.

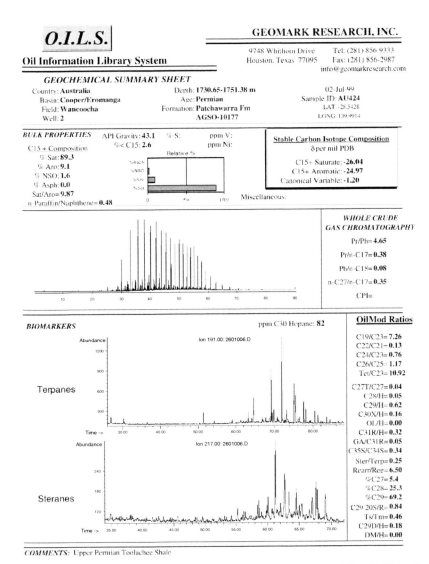

Figure 18.54. Geochemical data for crude oil from Wancoocha Field generated from the Upper Permian Toolachee Formation, Cooper-Eromanga Basin, Australia (courtesy of GeoMark Research, Inc.).

Triassic Eagle Mills Formation

See Source rocks in the Gulf of Mexico (p. 877) and Figure 18.109.

Lower-Middle Triassic Locker Shale, Northwest Shelf, Australia

The Triassic Locker Shale generated gas and condensate in the Northwest Shelf, Australia, and is recognized most commonly in regions outside of sub-basins where the overlying Jurassic Dingo Claystone source rocks are immature (Bishop, 1999a). The Locker Shale was deposited as a quiescent marine transgressive facies across the entire shelf, reaching thicknesses of 200–1000 m. Source potential is highly variable, with TOC ranging from ~1 to 3 wt.% and HI up to ~300 mg HC/g TOC, although it is typically in the range 100–200 mg HC/g TOC (Scott, 1994; Warris, 1993).

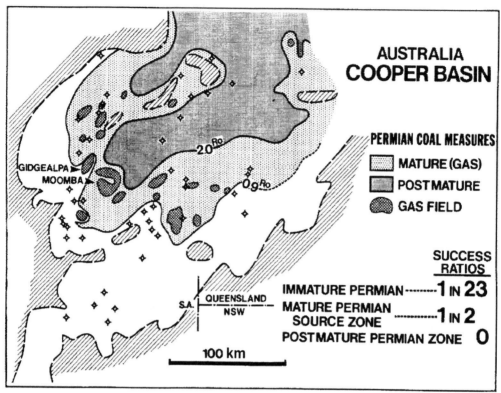

Figure 18.55. The distribution of hydrocarbon gas generated from the Permian coaly source rock in the Cooper Basin, Australia, is controlled largely by thermal maturity, as measured by vitrinite reflectance (Demaison, 1984). Reprinted by permission of the AAPG, whose permission is required for further use.

Kerogen in the Locker Shale is primarily terrigenous type III with minor amounts of algal marine type II. Preferential preservation may have occurred in local sags on the underlying Permian surface (Scott, 1994).

Lower-Middle Triassic shales, Barents Sea

Triasssic rocks are believed to be the source for petroleum in the Barents Sea (Leith et al., 1993; Ferriday et al., 1995; Lindquist, 1999c). These rocks were deposited in a shallow epicontinental sea. Most of the organic matter is coaly and gas-prone, but marine, oil-prone organic matter was deposited during marine transgressions, particularly in areas with local upwelling and/or restricted water circulation. The generative potential of these strata varies considerably (2–8 wt.% TOC, with most HI in the range 200–500 mg HC/g TOC). The coaly shales can have >20 wt.% TOC, but they are characterized by HI <100 mg HC/g TOC. Much of the Triassic strata in the Barents Sea region is buried more than 5 km deep and is in the gas window. The rapid sedimentation (as high as 1 m/1000 years) may have promoted preservation of organic matter, but the rocks are organic-lean due to clastic dilution.

Middle Triassic Meride limestones/Upper Triassic Riva di Solto shales, Italy

Following the breakup of Pangea, a series of shallow-water carbonate platforms formed off the Adriatic microplate system between Africa and Europe. In the Po Valley province, anoxic conditions were most prevalent

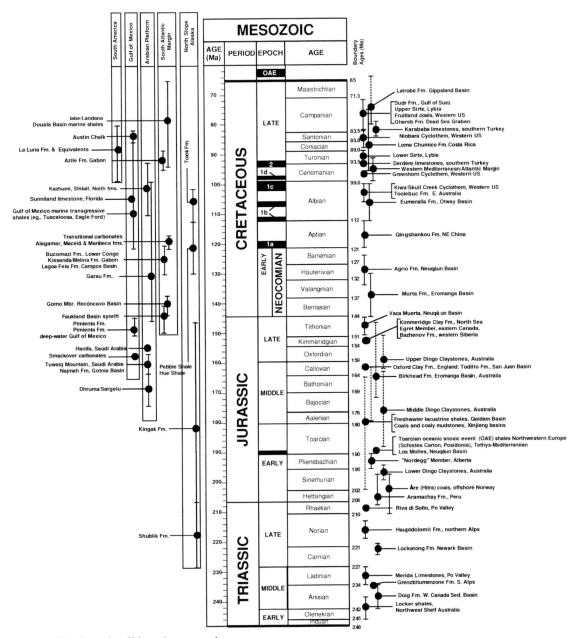

Figure 18.56. Examples of Mesozoic source rocks.

during deposition of the Ladinian Meride limestones and Rhaetian Riva di Solto shales, resulting in the accumulation of organic-rich strata (Lindquist, 1999a; Stefani and Burchell, 1990; Stefani and Burchell, 1993). The extent and thickness of the source units are coupled closely to changes in sea level and are highly variable over short distances. The Meride limestones average ~0.8 wt.% TOC with HI of 513 mg HC/g TOC. The kerogen is mostly amorphous, with relatively high sulfur near 4.5 wt.% (type II/IIS). The Riva di Solto

Shale contains type II/III kerogen and an average of ~1.3 wt.% TOC, with HI of ~251 mg HC/g TOC (Riva et al., 1986). The lower generative potential of the Riva di Solto kerogen is offset by the thickness of the formation, which can reach up to 2 km (Stefani and Burchell, 1990).

Oil generation from the Triassic source rocks of the Po Valley province began in the Mesozoic Era. Most of this oil was lost due to inadequate trap formation, except locally (Lindquist, 1999a), as in the Villafortuna-Gaggiano complex, and the Malossa and Cavone fields. Oils from these fields are chemically distinct, expressing variations in organic facies and thermal maturity (Mattavelli and Novelli, 1990; Riva et al., 1986; Stefani and Burchell, 1990). Oils from the Villafortuna-Gaggiano complex (34–40°API) and condensates from the Malossa Field (47–53°API) are low in sulfur, with pristane/phytane of 1.0–1.2 and $\delta^{13}C$ of -30.9 to $-29.2‰$ for saturated and -30.3 to $-28.5‰$ for aromatic C_{15+} hydrocarbons. These oils contain significant amounts of diasteranes, hopanes, and moretanes. In contrast, the low-maturity crude oils from the Carone Field have low gravity (20–36°API) and high sulfur (up to 4 wt.%), with abundant polar compounds and asphaltenes, as well as steranes and hopanes. Low maturity and their type IIS source facies is reflected in negligible diasteranes and low Ts/Tm ratios (Mattavelli and Novelli, 1990; Riva et al., 1986)

Middle-Upper Triassic carbonates, Arabian Platform

During the Triassic Period, the interior of Pangea was hot and dry, resulting in the deposition of non-marine sands on the Arabian Plate. However, shallow, marginal marine lagoonal shales and marls were deposited along the coast under conditions not that different from those that gave rise to prolific source rocks in the Jurassic and Cretaceous periods. Loutfi and Abdel Satter (1987) reported that some Middle to Upper Triassic restricted shelf and lagoonal carbonate-evaporite sequences have significant organic content in areas around eastern Saudi Arabia, the United Arab Emirates, and Oman. The extent and richness of these strata have not investigated fully, and these Triassic rocks may eventually be recognized as source rocks.

Middle-Upper Triassic carbonates, Swiss Alps

Organic-rich calcareous shales and dolomites occur in the southern and northern Alps. Outcrops are known for their abundance and excellent preservation of fossil vertebrates and are sometimes called "Ichtyol" shales (Tintori, 1991). In the southern Alps, the Anisian-Ladinian Grenzbitumenzone Formation occurs as a 16-m-thick sequence of finely laminated dolomites (up to 10 wt.% TOC) and shales (up to 40 wt.% TOC) (Bernasconi and Riva, 1994). In the northern Alps, similar rocks occur in the Norian Hauptdolomit Formation (up to 50 wt.% TOC) (Köster et al., 1988; Köster, 1989). High organic richness resulted from enhanced preservation in a mesosaline, permanently anoxic environment, which promoted incorporation of sulfur into the kerogen while restricting input of clastics and terrigenous organic matter. Two organic facies occur in these oil shales: (1) an algal, planktonic facies that developed in restricted lagoonal settings and (2) a bacterial, ^{13}C-poor ($\delta^{13}C \sim -30$ to $-32‰$), hopane-rich facies that developed primarily in small slump features on the carbonate platform.

Upper Triassic rift sequence shales, east coast North America

A series of elongate, fault-bounded Triassic rift basins along the eastern coast of North America from Nova Scotia to South Carolina formed during the early fragmentation of Pangea (Olsen, 1986; Olsen, 1990; Olsen et al., 1996). These rift valley lacustrine deposits, collectively known as the Newark Supergroup, consist of transgressive–regressive cycles. Within these sequences are numerous thin, fine-grained, organic-rich lacustrine shales. Deposition of the shales generally corresponded to periods when water depths were deepest (<200 m) and stratified, anoxic bottom waters developed. However, high-resolution stratigraphy suggests that cyclic deposition of the shales reflects periodic changes in water level that are tied to climatic changes forced by variations in the Earth's orbit, i.e. Milankovitch climate forcing (Olsen, 1986; Olsen and Kent, 1996).

Organic-rich lacustrine shales probably occur in all of the eastern coast rift basins. Shales from the Hartford and Newark basins were characterized extensively because they are readily available from outcrops and shallow cores (Pratt *et al.*, 1985; Pratt *et al.*, 1986; Pratt *et al.*, 1988; Pratt and Burruss, 1988; Kotra *et al.*, 1988; Spiker *et al.*, 1988; Walters and Kotra, 1989; Kruge *et al.*, 1990a; Kruge *et al.*, 1990b). In the moderately mature Lower Jurassic units from the Newark Basin, TOC typically exceeds 1 wt.% and can reach 4 wt.% in the dark-colored shales. Currently, HI exceeds 400 mg HC/g TOC and visual kerogen studies indicate an abundance of well-preserved type I kerogen in the dark shales. Considerable variations occur in the abundance and distributions of gammacerane, β-carotane, triterpanes, and other biomarkers, consistent with cyclic deposition in balanced-fill and underfilled settings.

Because of igneous activity and alteration by hydrothermal fluids, these shales range from marginally mature to postmature (R_o >2%) within the same stratigraphic horizon. Consequently, the samples were used to test the effects of heating on various saturated biomarker and aromatic hydrocarbons (e.g. Walters and Kotra, 1990; Kruge *et al.*, 1990a).

Upper Triassic–Lower Jurassic Aramachay Formation, Peru

The Aramachay Formation, part of the Pucará Group, contains organic-rich limestones and mudstones (Ichpachi Member) that are a likely source for oil in the western Marañon and Ucayali basins of Peru (Mathalone and Montoya, 1995). These source rocks were deposited in an oxygen-minimum zone produced by a classical westward-facing, open-shelf setting resulting in upwelling and quasi-estuarine circulation (Loughman and Hallam, 1982). Where exposed, the organic-rich shales and shaly limestones are thicker than 50 m and contain up to 5 wt.% TOC (Loughman, 1984). These rocks are currently postmature, but their original source potential was likely to be high because the kerogen is dominated by degraded, amorphous sapropel with little terrigenous plant matter.

The Pucará Group is well known for its zinc-lead ore deposits. These minerals are associated intimately with organic matter, and their formation was based on sulfide produced by thermochemical sulfate reduction (Spangenberg *et al.*, 1996; Spangenberg *et al.*, 1999; Spangenberg and Macko, 1998).

Triassic source rocks, circum-Arctic

Tectonic, climatic, and biological factors limited source-rock deposition during the Triassic Period. Greater Pangea stretched from pole to pole, and a vast desert spanned the low latitudes. During this time, organic-rich, transgressive marine shales were deposited throughout the Arctic, probably in association with upwelling (Parrish, 1995). Lower–Middle Triassic source rocks were deposited in Svalbard and the Barents Sea (Sassendalen Group) and in the Sverdrup Basin (Schei Point and Blaa Mountain groups) (Leith *et al.*, 1992; 1993). There are several organic-rich Triassic source rocks within the Western Canada Basin. The basal shales of the Middle Triasssic (Anisian) Doig Formation are prolific source rocks containing up to 11 wt.% TOC, with HI up to 480 mg HC/g TOC (Riediger *et al.*, 1990a; Brooks *et al.*, 1991). Source potential of the carbon-rich facies from the Lower Triassic Montney Formation (up to 5 wt.% TOC) and the Upper Triassic Pardonet Formation (up to 6.5 wt.% TOC) is understood less well because these units are now highly mature (Creaney *et al.*, 1994a).

Middle–Upper Triassic Shublik-Otuk interval, Alaska

The Middle-Upper Triassic Shublik Formation (see Figure 18.28) is one of the major source rocks for oil on the North Slope of Alaska (Magoon and Bird, 1985). The Ellesmerian(!) petroleum system of Magoon *et al.* (1987) contains crude oil derived mainly from the Shublik Formation, with variable admixtures from the Hue Shale and Kingak Shale (Bird, 2001). More than 90% of the oil in northern Alaska is assigned to the Ellesmerian(!) system (Bird, 2001).

The Shublik Formation is a heterogeneous source rock consisting of marine carbonate, marl, and phosphorite facies deposited under transgressive, upwelling conditions at a paleolatitude of $\sim$50°N (Kupecz, 1995). The Otuk Formation in the central and western Brooks Range is the age-equivalent of the Shublik Formation (Mull *et al.*, 1982; Bird, 1994) and is thought to be the source rock for oil seeps in the central Brooks Range foothills (Lillis *et al.*, 2002).

Oil in the Kuparuk Field originated mainly from the Shublik Formation, with possible minor input from the Kingak Shale in the southwestern part of the field (Masterson *et al.*, 1997; Masterson, 2001). The Barrow–Prudhoe oil family of Magoon and Claypool (1981) originated mainly from Shublik source rock. However, North Slope reservoirs generally have complex filling histories that involve multiple source rocks. Many fields contain a mixture of Shublik and other oil types (Seifert *et al.*, 1980; Seifert *et al.*, 1981; Claypool and Magoon, 1985; Sedivy *et al.*, 1987; Masterson, 2001). For example, the giant Prudhoe Bay Field on the Barrow Arch, exemplified by oil samples from the Sag River and Sadlerochit Sandstones, contains mixtures of oil generated mainly from the Shublik Formation, but with additional input from the "Post-Neocomian" (Cretaceous "pebble shale," gamma-ray zone (GRZ), Torok, and Hue Shale), and the Jurassic Kingak Formation (Seifert *et al.*, 1980; Seifert *et al.*, 1981; Masterson *et al.*, 1997; Masterson *et al.*, 2000; Masterson *et al.*, 2001). Prudhoe Bay main field oils contain 59:28:13 contributions from Shublik, pebble–GRZ–Torok, and Kingak source rocks, respectively, based on average sulfur contents in representative main field (1%), Kuparuk Field (1.6%), Kingak (0.2%), and pebble shale–GRZ–Torok oils (0.2%) (Masterson, 2001). Figure 13.75 is an example of evidence for Shublik input to the Sag River oil.

Much of the Shublik Formation is currently mature to postmature, complicating assessment of original source potential. In a study of samples from six wells, Bird (1994) reported 2.3 wt.% TOC for the mean, with a range of 0.49–6.73 wt.% TOC and thickness ranging from ∼24 to 149 m. He selected a conservative value for original HI of 200 mg HC/g TOC for calculating reserve estimates. Robison *et al.* (1996) conducted a comprehensive geochemical and lithostratigraphic study of a continuous core through the Shublik Formation. They found that the lower source intervals of the Shublik Formation were deposited during a marine transgression. The laminated marls and shales contain mixtures of fluorescent amorphous kerogen, marine alginite, and other exinites. They observed that the calcareous Shublik facies in the Phoenix-1 well north of the Colville Delta had average TOC near 4 wt.% and HI up to 965 mg HC/gm TOC. Source potential lessens within a highly condensed section near the maximum flooding surface and regressive, more bioturbated shales. These rocks still contain marine, oil-prone kerogen, but with increasing amounts of vitrinite and inertinite higher in the section.

Masterson (2001) compared the biomarker characteristics of extracts of Shublik calcareous and shaly facies with each other and other North Slope source rocks. All Shublik source-rock extracts contain high tricyclic terpanes (e.g. see Figure 13.75). Calcareous Shublik facies extracts have higher average dibenzothiophene/phenanthrene (∼0.6), C_{29}/C_{30} hopane (∼0.7), C_{23}/C_{30} hopane (∼0.6), and C_{35}/C_{34} hopane (∼1.2) than shaly Shublik extracts (∼0.4, 0.5, 0.4, and 0.9, respectively), consistent with sulfur-rich carbonate source rock. The values of these ratios for the calcareous Shublik extracts closely match the average values for Kuparuk Field oil, supporting an oil–source rock correlation. Compared with the Shublik samples, Kingak and GRZ source-rock extracts have comparatively low dibenzothiophene/phenanthrene (∼0.2) and C_{23}/C_{30} hopane (∼0.1).

Most crude oils generated from the carbonate-rich Shublik-Otuk source rock unit have low API gravity and high sulfur compared with other North Slope oils from clastic source rocks, such as oils in the Tarn (pebble shale–GRZ–Torok source) and Alpine (Kingak source) fields. For example, 16°API gravity Shublik oil from the Antares Field has 2.4 wt.% sulfur, elevated C_{22}/C_{21} and low C_{24}/C_{23} tricyclic terpanes, and elevated C_{35} hopanes (Figure 18.57), consistent with an origin from marine carbonate or marl source rock (see Figure 13.75). Shublik-Otuk oils from the Kuparuk Field average 23°API and 1.6 wt.% sulfur and the $\delta^{13}C$ of saturated and aromatic hydrocarbons (average ∼−30.0 and −29.6‰, respectively) corresponds to the range of overlap between extracts from calcareous and shaly Shublik source-rock extracts (Masterson, 2001).

Table 18.11 summarizes geochemical differences between Shublik-Otuk oils and oils generated from Jurassic Kingak-Blankenship and Cretaceous pebble–GRZ–Torok source rock units. For example, Shublik oils have higher sulfur, dibenzothiophene/phenanthrene, C_{29}/C_{30} hopane, and C_{35}/C_{34} hopane, and lower API gravity, than the other oils. The Shublik and Kingak-Blankenship oils have low C_{26}

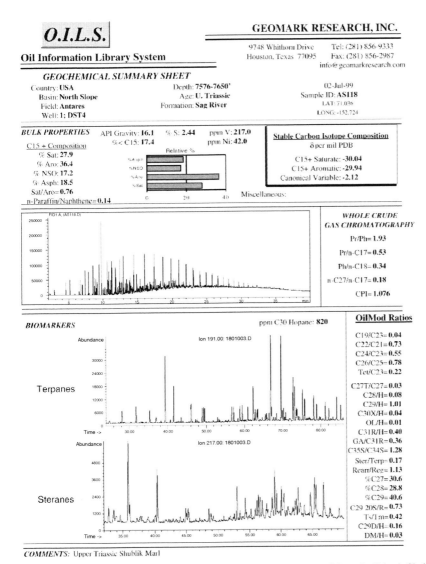

Figure 18.57. Geochemical data for crude oil from Antares-1 well generated from the Triassic Shublik Formation, North Slope, Alaska (courtesy of GeoMark Research, Inc.).

nordiacholestane ratios (0.14), consistent with Jurassic age source rock.

and type II kerogen deposited in a shelf-margin setting (Brosse et al., 1988).

Upper Triassic Streppenosa Formation, Sicily

The Upper Triassic Streppenosa Formation source rock from the Ragusa Basin in Sicily consists of millimeter- to meter-scale intercalations of clay, marls, and carbonates. The highest petroleum potential is associated with clay-rich laminites that contain up to 10 wt.% TOC

Rhaetian to Sinemurian Åre (Hitra) coals, Haltenbanken, Norway

Three source rock units are recognized in the Haltenbanken area, offshore Norway. The Spekk, equivalent to the Draupne or Kimmeridge Clay formations, is the principal source rock, with an average of ~5 wt.%

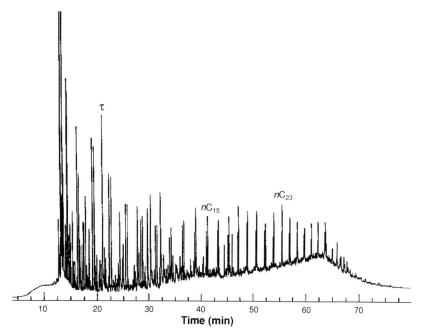

Figure 18.58. Pyrolysis gas chromatography of a sample of Åre coaly shale from the Haltenbanken area (Isaksen, 1995). The sample contained 12.8 wt.% total organic carbon (TOC), with hydrogen index (HI) of 246 mg HC/g TOC. The kerogen consisted of 70% herbaceous, 25% algal(?) amorphous, and 5% woody-inertinitic macerals. Reprinted by permission of the AAPG, whose permission is required for further use.
τ, toluene.

TOC and HI $\sim$390 mg HC/g TOC (Forbes et al., 1991; Karlsen et al., 1995). The Melke, equivalent to the Heather Formation, is a secondary source with lower potential, averaging $\sim$2 wt.% TOC with HI $\sim$200 mg HC/g TOC. The Åre (formerly Hitra), equivalent to the Statfjord Formation in the North Sea and the Kap Stewart Formation in eastern Greenland, also has secondary source potential. The Åre is regionally extensive and contains up to 500 m of shales, coals, and sandstones deposited in shallow-marine, deltaic, and paralic environments. These coals average >35 wt.% TOC, with initial HI $\sim$275 mg HC/g TOC (Hvoslef et al., 1988; Forbes et al., 1991). Pyrolyzates of these coals have significant amounts of paraffins (Figure 18.58).

JURASSIC SOURCE ROCKS

Tithonian source rocks, Gulf of Mexico

See Source rocks in the Gulf of Mexico (p. 878) and Figure 18.109.

Oxfordian Lower Smackover source rocks

See Source rocks in the Gulf of Mexico (p. 877) and Figure 18.109.

Lower Jurassic Toarcian shales, northwestern Europe

The Early Toarcian Age, especially the time represented by the lower part of the *Harpoceras falciferum* (ammonite) zone, was characterized by high paleotemperatures, high sea level, mass extinction of benthic fauna, and deposition of organic-rich black shales throughout the world, particularly in northwestern Europe. Just before the deposition of these shales, there was a major perturbation in the global carbon budget that produced a stable carbon isotopic shift toward more negative values in both organic matter and carbonates (Figure 18.59). Two broad hypotheses were advanced to explain this early isotopic shift:

1. A change in ocean circulation resulted in the release of ^{13}C-depleted CO_2, which built up below

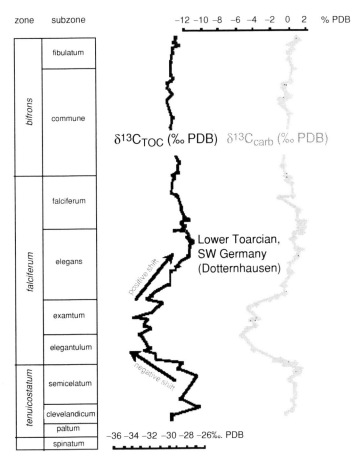

Figure 18.59. δ^{13}C of total organic carbon (TOC) and carbonate carbon in the Posidonia Shale (Dotternhausen). The fractionation between the organic and inorganic carbon is nearly constant, indicating a common carbon source. The positive isotopic shift corresponds to the rapid sequestration of ^{13}C-depleted carbon and is ascribed to widespread deposition of source rocks during an oceanic anoxic event (OAE). The cause of the negative shift just before source-rock deposition is unclear. Reprinted from Schmid-Röhl et al. (2000). © Copyright 2000, with permission from Elsevier.

a deep pycnocline (e.g. Küspert, 1982; Röhl et al., 2001).

2. Increased atmospheric CO_2 from flood basalts in the southern Gondwana Karoo and Ferrar provinces (Palfy and Smith, 2000) triggered a catastrophic release of methane from hydrates (e.g. Hesselbo et al., 2000; Padden et al., 2001).

Carbon isotope ratios for organic matter shifted to more positive values ∼183 Ma, corresponding to widespread deposition of marine, organic-rich shales and the rapid (0.5–1 million years) sequestration of ^{12}C-enriched organic matter (Jenkyns, 1985; Jenkyns, 1988; Jenkyns and Clayton 1997). This synchrony of global source-rock deposition and isotopic change establishes the earliest, well-documented oceanic anoxic event (OAE).

The Early Toarcian anoxic event deposited organic-rich, marine shales across an epicontinental platform that spanned the southern North Sea and most of northwestern Europe. These source rocks include the Jet Rock in Yorkshire (Hallam, 2001), the Posidonia Shale of the southern North Sea, the Posidonienschiefer in Germany and Switzerland, the Schistes

Carton in the Paris Basin (see below), and many lesser deposits (Baudin, 1995). Mixing of waters from the cold Arctic and warm Tethys seas (Ziegler, 1988) and nutrient-rich waters from continental run-off (Loh et al., 1986) resulted in enhanced productivity (Vetö et al., 1997). Anoxic conditions enhanced preservation of this organic matter. The Toarcian source rocks contain 2–20 wt.% TOC, with an average of ~5 wt.% TOC. The kerogen is well-preserved marine algae and plankton, with HI typically between 400 and 700 mg HC/g TOC. There is abundant biomarker evidence to support deposition in the presence of an anoxic photic zone.

Note: About 15 million years ago, a small bolide impact formed the Nördlinger Ries Crater in southern Germany. The impact ejected sedimentary rocks, including immature Posidonia Shale. Hofmann et al. (2001) showed that shale clasts in the ejected Bunte Breccia were slightly more mature than samples of undisturbed Posidonia Shale away from the impact. Biomarker distributions indicated negligible thermal alteration of the shale in the ejected breccia. Unsaturated sterenes predominate over saturated steranes, which included minor amounts of the metastable $5\beta,14\alpha,17\alpha$-isomers. The hopanoids were also immature, with almost equal abundances of $\alpha\beta$-, $\beta\alpha$-, and $\beta\beta$-isomers as well as hopenes. Small variations in alginite fluorescence and vitrinite reflectance were detected between the ejected breccia (R_r ~0.32–0.35%) and the reference sample (R_r ~0.25%) and were attributed to thermal alteration by the impact event. However, the nearly identical biomarker indicators of thermal stress suggest that Posidonia Shale fragments in the Bunte Breccia originated from the outer, cooler portion of the crater rather than from the central part, which reached high temperatures due to the impact.

Organic-rich shales were also deposited in the Mediterranean-Tethyan region during the Lower Toarcian anoxic event (Ziegler, 1988). Intense tectonic activity and subsidence created a complex paleogeography of small carbonate platforms surrounded by basins of varying depth. The source potential of the strata deposited within these basins depends on local conditions, particularly the organic matter input. On average, Tethyan basins have lower TOC than those in northwestern Europe, averaging ~1 wt.%, and many contain mixtures of type II and III kerogens (Baudin, 1995). Deep basins within the Apulian block (Italy, Yugoslavia, and Greece) have strata with exceptional source potential. In the southern Alps, Toarcian black shales occur in the Lombardy and Julian basins, on the Trento Plateau, and in the Belluno Trough. TOC averages ~1 wt.%, with maxima up to 4.3 wt.%. HI typically ranges from 400 to 650 mg HC/g TOC where immature (Figure 18.60) (Farrimond et al., 1988; Farrimond et al., 1989; Farrimond et al., 1994; Baudin, 1995). In the Umbrian Basin in central Italy, shales contain up to 1.6 wt.% TOC, with HI up to 390 mg HC/g TOC (Baudin, 1995). In the Ionian Basin of western Greece, Toarcian shales contain up to 5.2 wt.% TOC (average ~1.2 wt.%), with HI up to 650 mg HC/g TOC (Baudin, 1995). These basins may have been isolated or in only partial communication with the Tethyan Ocean. Biomarker analysis indicates that much of the primary algal organic matter was partly degraded and transported through an oxic water column (Farrimond et al., 1988). In contrast to shales of northwestern Europe, the Tethyan basins were relatively oxic, with anoxic conditions occurring only in the deep basin waters (Farrimond et al., 1989; 1994).

Lower Toarcian transgressive marine source rocks and their relationships with the OAE are characterized less well outside of Europe. Some of these source rocks include the Poker Chip Shale in the Fernie Formation of Alberta and eastern British Columbia (Poulton et al., 1994) and the organic-rich shales of the Karoo Formation in Madagascar (Figure 18.61) (Hiller, 2000). Many other Lower Toarcian formations contain type III kerogen and/or coals (Baudin, 1995).

Schistes Carton, Paris Basin

Toarcian shales in the Paris Basin were used to define the behavior of type II kerogen using Rock-Eval pyrolysis (Espitalié et al., 1977) and are still used as the calibration standard. The Paris Basin shales are ideal for this application because they are fairly homogeneous

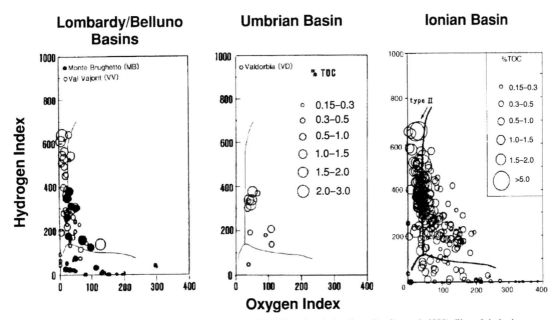

Figure 18.60. Modified van Krevelen diagrams for Apulian block Toarcian shales (from Baudin et al., 1990). Size of circles is proportional to total organic carbon (TOC) (inset at right).

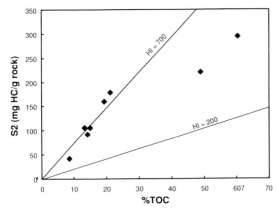

Figure 18.61. Total organic carbon (TOC) (wt.%) versus Rock-Eval S2 for Toarcian oil shales from Madagascar (data from Hiller, 2000).
HI, hydrogen index.

and samples are available that span the range of early oil maturity (vitrinite reflectance ~0.58% at the surface to ~0.73% at 2450 m depth). For the same reasons, much of the early work on biomarker maturity indicators and kinetics was calibrated on Paris Basin Toarcian shale (Ensminger et al., 1977; Mackenzie et al., 1980a; Mackenzie et al., 1980b; Mackenzie et al., 1981a; Mackenzie et al., 1981b; Mackenzie et al., 1982c; Rullkötter and Marzi, 1988; Marzi et al., 1990).

The source potential of the Schistes Carton shales has been studied extensively (e.g. Espitalié et al., 1987; Katz, 1995a; Disnar et al., 1996). TOC in these shales is concentrated in a concentric pattern typical of many anoxic marine silled basins (Figure 18.62). The Lower Toarcian shales are one of several Lower Jurassic organic-rich intervals within the Paris Basin and in Germany (e.g. Posidonia Shale) (Mann and Müller, 1988; Kockel et al., 1994). A series of stacked sequences correlate with four transgressive–regressive cycles and organic-rich intervals corresponding to maximum flooding surfaces (Bessereau et al., 1995).

Crude oils generated from the Schistes Carton Formation have compositions and biomarker distributions typical of Mesozoic transgressive marine source rocks that were deposited in suboxic to anoxic bottom water

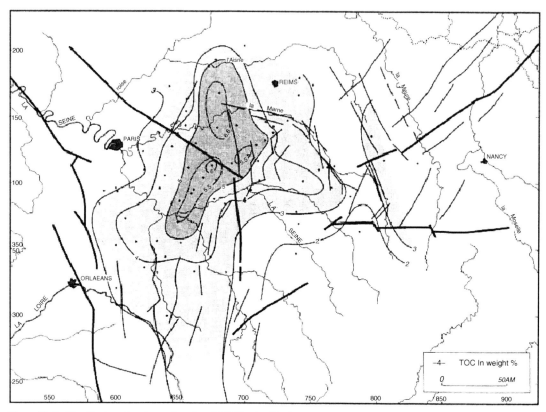

Figure 18.62. Distribution of total organic carbon (TOC) in the Liassic (Lower Jurassic) strata of the Paris Basin (from Bessereau et al., 1995). Reprinted by permission of the AAPG, whose permission is required for further use.

with an overlying oxic photic zone (Figure 18.63). This is evident from the low sulfur concentration, pristane/phytane of ~1.5–2.0, and a homohopane distribution that decreases with carbon number. Significant algal marine input is evident based on the presence of n-propylcholestanes and the high concentration of steranes relative to hopanes. The relatively high proportion of diasteranes to regular steranes indicates a clastic source rock.

Lower Jurassic Nordegg Member, Western Canada Basin

The lowermost Jurassic (Pliensbachian) rocks in the subsurface of the Fernie Basin are believed to be stratigraphically equivalent to the outcropping Nordegg Member. These strata were deposited during the initial Jurassic marine invasion that formed a narrow seaway along the Alberta Trough. The subsurface Nordegg represents two transgressive–regressive events (Creaney and Allan, 1990) and consists of ~20–36 m of dark gray to black, variably phosphatic limestone, marls, and calcareous mudstones. Strata at the top and base of the Nordegg Member have high source potential and high concentrations of radiogenic elements. The organic-rich source intervals are separated by silty limestone that is usually oil-stained. TOC ranges from 5–15 wt.%, with values as high as 28 wt.% for certain laminites. The kerogen is type I/II, with HI up to 800 mg HC/g TOC measured in immature cores. Residual HI typically ranges from 350 to 480 mg HC/g TOC for Nordegg samples that have entered the oil window (Riediger et al., 1990a; Riediger et al., 1990b; Riediger et al., 1991; Riediger, 1992).

Oils generated from the Nordegg have biomarker distributions typical of marine marls deposited under

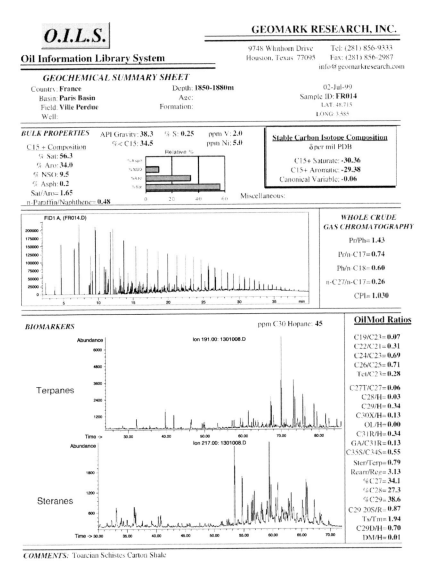

Figure 18.63. Geochemical data for crude oil from Ville Perdue Field generated from the Toarcian Schistes Carton Shale, Paris Basin, France (courtesy of GeoMark Research, Inc.).

anoxic, carbonate-evaporitic conditions. They are enriched in sulfur (up to 4 wt.%), have low relative amounts of diasteranes, and have high C_{29} norhopane and C_{35} homohopanes (Creaney et al., 1994a).

Middle Jurassic Tuxedni Group, Cook Inlet, Alaska

Fossiliferous beds of graywack and siltstone of the Middle Jurassic Tuxedni Group are the major source of oil in Cook Inlet, Alaska (see Table 4.3) (Magoon and Anders, 1992; Magoon, 1994). These strata were deposited in a back-arc setting, blanketing most of the underlying volcanics with more than a kilometer of sediments. Most Tuxedni rock samples are highly mature, and only a few immature samples from well and outcrop locations have been analyzed. These immature samples have only marginal generative potential, with ~1.7 wt.% TOC and an average HI of ~225 mg HC/g TOC. HI of individual samples may exceed 400 mg HC/g TOC. The kerogen originated from a mixture of marine and terrigenous organic matter deposited under oxic to dysaerobic conditions.

Crude oil from the Sunfish #1 well in Cook Inlet has a complex biomarker distribution that resulted from overprinting of migrated Jurassic Tuxedni oil by less mature Tertiary reservoir bitumen (Hughes and Dzou, 1995). These authors estimated the relative proportions of Jurassic and Tertiary sources in the Sunfish #1 oil by measuring concentrations of anthracene and simonellite in (1) a proposed Tuxedni oil from the McArthur River K-1RD well, (2) bitumen extracted from Tertiary coal interbedded with the reservoir sandstones, and (3) Sunfish#1 oil. Anthracene and simonellite occur in (2) and (3), but not in (1). The calculations show that ~95% of the Sunfish #1 oil originated from Jurassic Tuxedni source rock, while ~5% represents low-maturity bitumen extracted from the interbedded Tertiary coals. Because biomarker concentrations in the high-maturity Tuxedni condensate (41°API gravity) are low, mixing with small amounts of low-maturity, biomarker-rich bitumen significantly affected the observed biomarker distributions.

Middle Callovian Todilto Formation, San Juan Basin

The Todilto Formation is an unusual sequence of limestone and gypsum in the San Juan Basin, New Mexico. The formation was deposited during a brief 20 000-year interval in a landlocked, saline lake (salina) that spanned nearly 90 000 km^2 (Lucas and Kietzke, 1985). Within the formation is an undisturbed ~2–4-m-thick unit composed of repetitions of three distinct laminae: limestone, organic matter, and a discontinuous layer of clastic grains. The three laminae constitute a sedimentary cycle that is repeated ~72 times per centimeter. Anderson and Kirkland (1960) interpreted these laminations to represent an annual cycle of varve sedimentation.

Although landlocked, water for the Todilto Lake was drawn from the Curtis Sea in eastern Utah. Consequently, the kerogen, derived from halophilic algae and bacteria, incorporated sulfur. Immature Todilto limestones have HI and atomic S/C ratios near 600 mg HC/g TOC and 0.42, respectively. Where mature, they are the source of oil in the underlying Entrada dune sands. The Todilto gypsum seal blocks vertical migration.

CRETACEOUS SOURCE ROCKS

Albian Sunniland Formation

See Source rocks in the Gulf of Mexico (p. 879) and Figure 18.109.

Cenomanian-Turonian Tuscaloosa and Eagleford formations

See Source rocks in the Gulf of Mexico (p. 881) and Figure 18.109.

Mesozoic source rocks of the South Atlantic margins

The Mesozoic basins along the east coast of South America and the west coast of Africa (Figure 18.64) formed as a result of closely related tectonic events. Consequently, these basins share many similarities, and it is convenient to discuss the source rocks of the South Atlantic margin together.

The South Atlantic Ocean began in the Late Jurassic Period as an incipient rift basin situated between the African, Antarctic, and South American plates (Figure 18.65). Early and late synrift fluvio-lacustrine sediments of Jurassic to Valanginian age in the North Falkland Basin provide evidence for the formation of this small sea (Richards and Hillier, 2000). Deep-Sea cores suggest that marine black shales were also deposited under conditions of restricted water circulation (Zimmerman et al., 1987).

The rifting of the South American and African plates was not a simple, continuous separation but occurred in three discrete phases (Karner et al., 1997; Norvick and Schaller, 1998; Szatmari, 2000). The first phase, known as the Rio da Serra Brazilian stage, began in the Late Tithonian Age or Early Neocomian Epoch and formed a rift valley complex from the Tucano-Recôncavo basins to northern Angola. During this time, the Potiguar, Araripe, and Jatobá basins formed along a second rift parallel to the Trans-Brazilian Shear that extended northward, possibly as far as the Hoggar

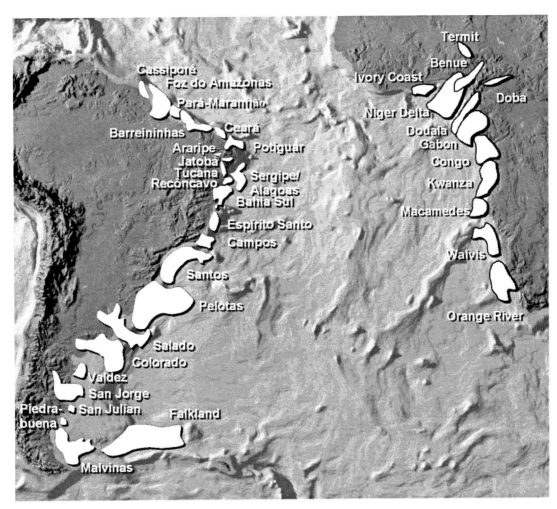

Figure 18.64. South Atlantic margin basins.

Mountains in Algeria. Basalts flooded the Paraná and Etendeka (Namibia) regions, and extensive volcanism occurred in the Espírito Santo, Campos, Santos, and southern Kwanza basins.

The second phase of rifting, the Buracica and Jiquia stages, began during the Late Barremian Age and continued until the Early Aptian Age. Rifting rotated ∼20 degrees clockwise, and a new, second set of rift basins developed from Cameroon to Santos, running roughly parallel with the future Atlantic margin. Rifting along the Trans-Brazilian Shear was abandoned, and the onshore basins in northeastern Brazil were uplifted.

The third phase, or Alagoas stage, began in the Early Aptian Age and continued until the earliest Albian Age. During this time, oceanic crust was emplaced and sag basins formed between the separating plates. This was followed by deposition of a thick layer (∼2 km) of salt from the Kwanza and Santos basins in the south to the Gabon and Sergipe-Alagoas basins in the north. The evaporites may have formed from transitional marine incursions in the final stages of rifting or from the desiccation of a small ocean basin. The emerging South Atlantic was still a largely isolated water body, separated from the Indian Ocean by the Rio Grande Rise/Walvis Ridge. A series of small rift basins formed in the north,

836 Petroleum systems through time

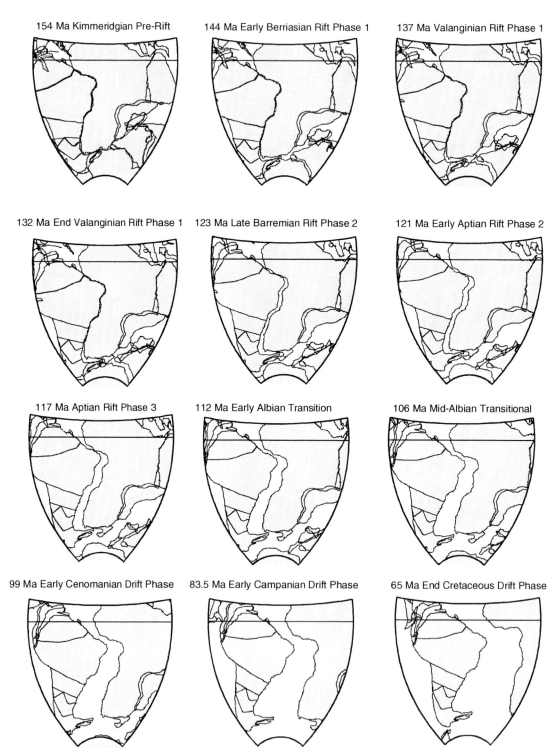

Figure 18.65. Reconstructed plate tectonics views showing the formation of the South Atlantic. Drawings generated from the GEOMAR/ODSN (Ocean Drilling Stratigraphic Network) website, www.odsn.de/odsn/services/paleomap/paleomap.html, based on data in Hay *et al.* (1999).

including the Barreirinhas, Ceará, and Potiguar basins in Brazil and the Benue, Termit, and Doba (Chad) basins in western Africa (Norvick and Schaller, 1998). Rifting was completed in the Late Albian Age, with plate separation around Nigeria.

With completion of the rift phase, the tectonic style of the South Atlantic shifted to a drift-subsidence phase. A gap finally formed during the Albian Age between the eastward spreading Falkland Plateau and the southern tip of Africa, connecting the South Atlantic to the global ocean system. A shallow-water carbonate platform developed along the African and Brazilian margins. The marine transgressive megasequence continued in the Late Cretaceous Period, drowning the carbonate platform and depositing marine clastics. Regressive marine clastics characterize the Tertiary megasequence.

Sedimentation was controlled largely by tectonic evolution and followed a broad stratigraphic megasequence of pre-rift, synrift, transitional (or evaporitic), transgressive marine, and regressive marine facies (Brice *et al.*, 1982). Pre-rift sequences resulted from the stretching of the continental crust, which produced block-faulted troughs filled mostly with sandy fluvial-lacustrine sediments and mafic volcanics. Synrift sequences resulted in graben and half-graben troughs filled with lacustrine sequences. Transitional sequences consist of carbonates and evaporites deposited under saline to hypersaline conditions. Deep-water, clastic source rocks were deposited under anoxic conditions during the transgressive sequence. Few source rocks were deposited during the Tertiary regressive megasequence except in deltaic settings, such as the Niger Delta.

Lacustrine source rocks

The rates of influx and accommodation define sedimentation and the conditions of organic matter preservation within lake systems (Carroll and Bohacs, 1999; Bohacs *et al.*, 2000). The rate of influx is defined as the net amount of sediment and water that a lake receives during a given period of time. It is roughly proportional to the ratio of precipitation to evaporation, and it is controlled strongly by climate. Influx rates are high in humid, high-rainfall environments and low in arid, low-rainfall environments. The rate of accommodation is the net volume change that the lake system experiences over the same time interval. Accommodation is controlled primarily by tectonic subsidence.

The interplay between influx and accommodation produces different conditions of continental sedimentation (Figure 18.66). Fluvial conditions occur where the influx far exceeds the topographic relief needed to retain the water in lakes. Coals and lacustrine deposits form only under special conditions where the influx of organic matter approximately equals accommodation (Bohacs and Suter, 1997). Aeolian and playa deposits form where accommodation far exceeds influx (precipitation ≪ evaporation).

The balance between influx and accommodation determines depositional conditions and the type and amount of organic matter that is preserved. If influx exceeds accommodation, then the lake will have open hydrodynamics. In overfilled lake systems, the water is typically fresh and sedimentation is dominated by clastic progradation (Figure 18.67). Lake fauna and flora are diverse, but deposition of source beds is limited by poor preservation due to oxic water and dilution by rapid

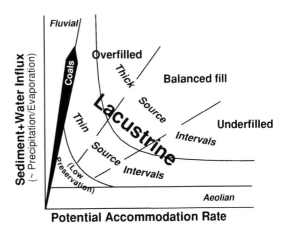

Figure 18.66. Lakes occur where the rate of influx of sediments and water is approximately equal to that of potential accommodation (from Carrol and Bohacs, 1999). Influx is controlled mostly by the climate and is proportional to the rates of precipitation and evaporation. Accommodation is controlled mostly by tectonic subsidence. Lacustrine sedimentation and generative potential depends on the system hydrodynamics. Overfilled systems with open hydrology occur when influx exceeds accommodation and are generally poor for source-rock deposition. Balanced-fill and underfilled conditions result in closed hydrology and may lead to rich source rocks.

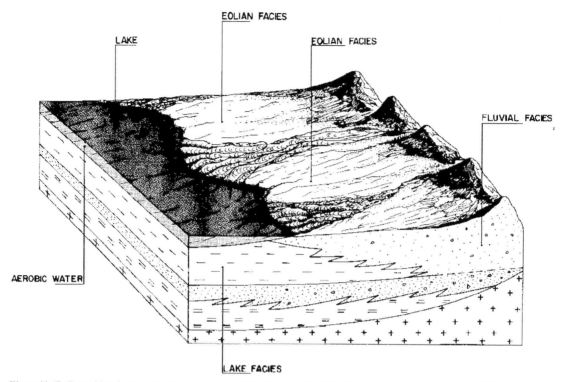

Figure 18.67. Depositional setting for a fluvial/overfilled lake (Mello *et al.*, 1995). Reprinted by permission of the AAPG, whose permission is required for further use.

clastic sedimentation. If accommodation equals or exceeds influx, then the lake will be a closed system. When these factors are equal, the lake is a balanced-fill system (Figure 18.68). The lake waters fluctuate between fresh and brackish conditions and may become stratified. The term "brackish" is used to indicate water chemistry with more dissolved solids than fresh water and is preferred over the term "*saline*," with the inference that the dissolved solids are salts. Balanced-fill lakes fluctuate between clastic and chemical sedimentation, and surviving biota must tolerate changing conditions. Balanced-fill lakes may produce source rocks with exceptional petroleum-generative potential. These rocks are typically deposited in deep lakes where the water is density stratified. The oxic, photic zone is fresh and highly productive. Seasonal algal blooms may occur in response to temperature and nutrient inflow. The organic matter is preserved in deep, suboxic to anoxic bottom water. Underfilled lakes occur where accommodation exceeds influx (Figure 18.69). Under these conditions, the lakes may be ephemeral, i.e. filling during the rainy season and evaporating to dryness during the dry season. Under these conditions, chemical precipitation dominates over clastic sedimentation, the lake biota is limited to halotolerant organisms, and the water may be oxic or anoxic. Rocks with exceptional petroleum-generative potential can be deposited in underfilled conditions.

When Gondwana pulled apart during Late Jurassic–Early Cretaceous time, accommodation increased episodically along the rift axis, and a series of lakes formed that gave rise to source rocks in both west Africa and Brazil. The two continental margins, however, were not mirror images. Most of the discovered oil in Brazil originated from lacustrine source rocks. In contrast, lacustrine source rocks are major contributors

Cretaceous source rocks 839

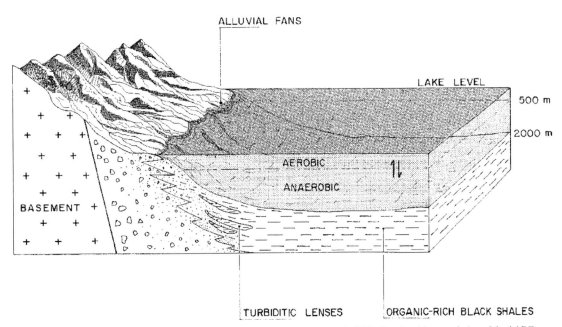

Figure 18.68. Depositional setting for a deep-water, balanced-fill lake (Mello *et al.*, 1995). Reprinted by permission of the AAPG, whose permission is required for further use.

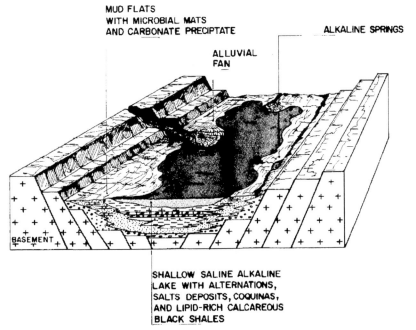

Figure 18.69. Deposition setting for a shallow-water, underfilled lake (Mello *et al.*, 1995). Reprinted by permission of the AAPG, whose permission is required for further use.

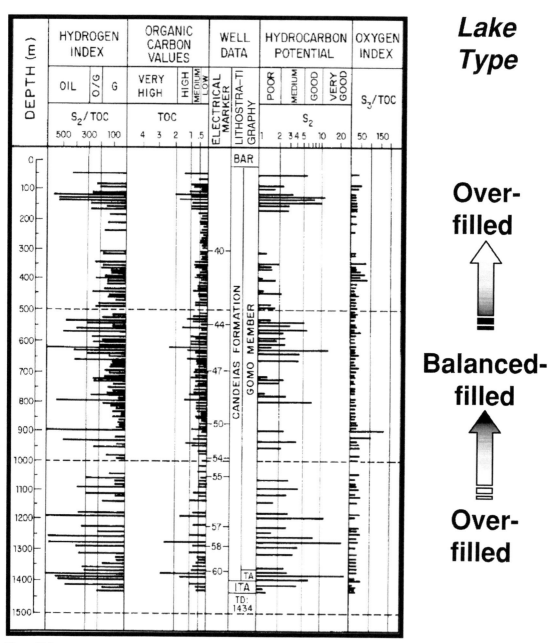

Figure 18.70. Organic geochemical log of lacustrine/fluvial sediments in the Recôncavo Basin of Brazil (Figueiredo et al., 1994). Comparison of biomarkers in the source rocks and oils indicates that the Gomo Member of the Upper Candeias Formation (balanced-fill) facies is most responsible for oil generation in the Recôncavo Basin (Mello and Maxwell, 1990). Reprinted by permission of the AAPG, whose permission is required for further use.

in only the lower Congo/Kwanza basins and are minor contributors in Gabon and several other African basins, where they give rise to presalt accumulations.

Climate exerted major control on the character of the lacustrine basins and the associated organic matter for a given rift phase. Tropical conditions with high rainfall and low evaporation tend to result in overfilled and balanced-fill lakes. Increasingly arid conditions result in a progression from overfilled, to balanced-fill, to underfilled lakes, and finally to restricted marine conditions.

Upper Jurassic lacustrine source rocks, Falkland Basin

The oldest Mesozoic source rocks in the South Atlantic may be in the North Falkland Basin (Richards and Hillier, 2000). Early synrift Middle Jurassic-Tithonian fluvial and overfilled lacustrine rocks have low potential (0.2–1.6 wt.% TOC, HI 12–390 mg HC/g TOC). The low generative potential reflects poor preservation due to oxic conditions during deposition. Late synrift (Tithonian-Berriasian) lacustrine deposits fluctuated between overfilled and balanced-fill conditions. During the latter, source rocks with up to 6 wt.% TOC and HI >700 mg HC/g TOC were deposited. Post-rift sediments reflect the transition to marginal marine and fully open-marine conditions. The source potential of these strata is also considerable, with TOC up to 8.7 wt.% and HI >1000 mg HC/g TOC.

Gomo Member, Recôncavo Basin, Brazil

Lacustine shales of the synrift Gomo Member (Candeias Formation) are source rocks for oils in the Recôncavo Basin (Mello and Maxwell, 1990; Mello et al., 1994a; Penteado et al., 2000). The depositional environments of these shales varied from pre-rift fluvial in the Late Jurassic Period, to synrift lacustrine in the Berriasian Age, and back to fluvial deltaic beginning in the Valanginian Age (Figure 18.70). The Upper Berriasian Gomo Member shales were deposited primarily in balanced-fill, brackish water conditions. TOC ranges from negligible to 10 wt.%. Penteado et al. (2000) showed that the Gomo Member contains primarily type I algal kerogen that has been preserved to varying degrees. Samples from the southern portion of the basin have greater potential (TOC >2 wt.%, HI ~650–850 mg HC/g TOC) than those in the north (TOC <2 wt.%, HI ~400–575 mg HC/g TOC). Figure 18.71 shows geochemical data for a typical crude oil in the Recôncavo Basin that was generated from the Gomo Member source rock.

Bucomazi Formation, Lower Congo Basin of West Africa

The Barremian-Aptian Bucomazi and the overlying Chela formations are the major source rocks for coastal oil in the Lower Congo Basin (Figure 18.72) (Burwood et al., 1990; Burwood et al., 1992; Burwood et al., 1995; Cole et al., 2000; Schoellkopf and Patterson, 2000). The sequence progresses from fluvial, to balanced-fill, to underfilled, and finally to transitional marine (Figure 18.73, Table 18.8). The Lower Bucomazi Formation represents freshwater fluvial to isolated, overfilled lake systems with poorly integrated drainages. Organic matter was preserved poorly in the fluvial deposits but was preserved in the freshwater lakes when sufficiently deep to promote anoxic bottom water. Further rifting merged these isolated lakes into a larger, fresh to alkaline deep-water lake. This phase (Zone D) was balanced-fill and resulted in the deposition of widespread organic-rich Middle Bucomazi source beds. The deep-water clastic facies of the Middle Bucomazi Formation gave way to more calcareous strata deposited under shallow-water conditions (Zone C). The Bucomazi lake was still balanced-fill, and source rocks with excellent generative potential were deposited. This facies transitions into a shallower, brackish water facies in the Upper Bucomazi Formation (Zone B). Water depth was insufficient to prevent bottom water mixing, and conditions fluctuated between oxic and dysoxic. During this phase, the lake remained balanced-fill, and organic matter was preserved when low oxygen conditions occurred. The uppermost Bucomazi Formation continued the development into a broader and shallower, but still balanced-fill, alkaline lake (Zone A). The water column was oxic for most of this facies, resulting in generally lower source potential, although organic-rich strata were deposited during low-oxygen conditions. The Upper Barremian Toca carbonates, an important reservoir unit, were deposited along the lake rim (Harris, 2000). The Aptian Chela Formation marks the time when the lake sills were breached, resulting in the transition from a shallow, balanced-fill

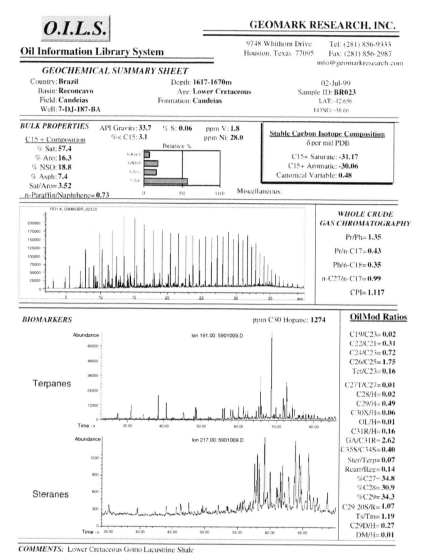

Figure 18.71. Geochemical data for crude oil from Candeias Field generated from the Lower Cretaceous Gomo Formation, Recôncavo Basin, Brazil (courtesy of GeoMark Research, Inc.).

lake, to a marine-connected lagoon, to a shallow gulf or seaway. Conditions were generally oxic, and TOC in the Chela Formation is typically low.

Lagoa Feia Formation, Campos Basin

The Hautervian-Aptian Lagoa Feia Formation is the principal source rock for crude oils in the Campos Basin (see Table 4.3) (Mello, 1988; Mello and Maxwell, 1990; Mello *et al.*, 1994a; Trindade *et al.*, 1994; Mohriak *et al.*, 1990; Guardado *et al.*, 2000). The Lagoa Feia represents mainly lacustrine sediments deposited under arid conditions (Figure 18.74). The lowermost section was deposited under fluvial conditions that evolved to an overfilled lake in the Early Barremian Age. These strata (Talc-Stevensitic sequence) have relatively low generative potential, typically with >1 wt.% TOC and HI ~300 mg HC/g TOC. Deposition occurred under dysaerobic conditions, resulting in partially oxidized organic matter.

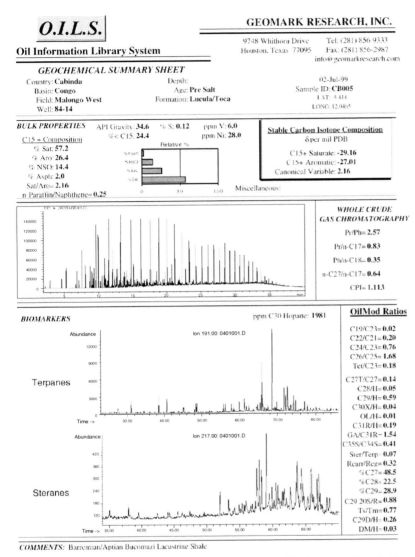

Figure 18.72. Geochemical data for crude oil from Malongo West Field generated from the Barremian-Aptian Bucomazi Formation (courtesy of GeoMark Research, Inc.).

Sedimentation changed to a balanced-fill lake environment in the Barremian Age, followed by an underfilled system in the Aptian Age (Coquinas sequence). The richest source rocks were deposited under balanced-fill conditions in a large, alkaline lake. Organic-rich strata typically exceed 5 wt.% TOC, with individual samples having up to 12 wt.%. The type I–I/II kerogen originated from phytoplankton, particularly dinoflagellates. HI varies greatly from ~250 to >750 mg HC/g TOC. The lake sills were repeatedly breached in the Aptian Age, resulting in marine incursions and a transitional lagoonal facies. This facies contains biomarkers characteristic of both lacustrine (e.g. tetracyclic polyprenoids, β-carotane) and marine (e.g. n-propylcholestanes) biotic input (Mello et al., 1994a). Figure 18.75 shows a typical crude oil from the Garoupa Field in the Campos Basin, Brazil that was generated from the Lagoa Feia source rock.

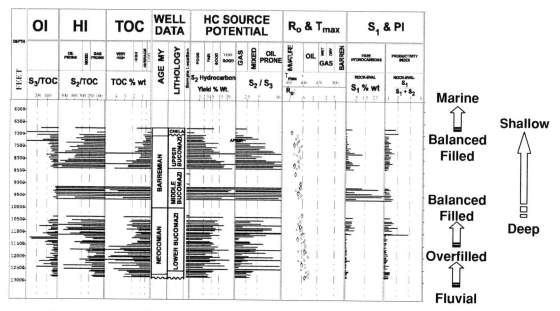

Figure 18.73. Geochemical log of fluvial/lacustrine/marine sediments in the Angola Basin, Cabinda (Schoellkopf and Patterson, 2000). Balanced-fill conditions were established in the Lower Bucomazi Formation and persisted into the Upper Bucomazi Formation. Reprinted by permission of the AAPG, whose permission is required for further use.
HI, hydrogen index; OI, oxygen index; TOC, total organic carbon.

Neocomian Kissenda/Melania Formation, Gabon Basin

The Kissenda/Melania lacustrine shales are the main source of petroleum in pre-salt reservoirs in the Gabon Basin (Figure 18.76) (Teisserenc and Villemin, 1989; Katz et al., 2000c). The lacustrine sequence began with the Kissenda Shale, which was deposited in deep-water, overbalanced conditions. These shales and siltstones contain ~1.7 wt.% TOC composed of mixed type II and III kerogens. HI typically is below 300 mg HC/g TOC, but it may exceed 500 mg HC/g TOC in organic-rich intervals near the top of the formation, where balanced-fill conditions prevailed. The Kissenda Shale has a maximum thickness of 1000 m, although the organic-rich sections occur in the upper 100–200 m.

The lower Melania Formation was deposited under deep-water, balanced-fill conditions, resulting in excellent potential source rocks. These black shales average 6.1 wt.% TOC, with some intervals as high as 20 wt.%. Kerogen is dominated by *Botryococcus* algal remains, resulting in HI in the range ~400–900+ mg HC/g TOC. Variable terrigenous plant material is included, depending on proximity to fluvial input. The upper Melania Formation was deposited under shallow, oxic conditions as thick detrital fans with low source potential. The organic-rich lower Melania is ~500 m thick. However, deposition was contemporaneous with graben and horst block faulting, and sedimentation differed between the blocks.

Other South Atlantic lacustrine source rocks

Lacustrine source units occur in most South Atlantic basins (Table 18.9). These source rocks generated the oil and gas in pre-salt reservoirs as well as some post-salt accumulations where petroleum migrated through the regional Aptian salt. The source beds deposited in overfilled to balanced-fill lake conditions tend to occur in the northern basins, reflecting a humid, high-rainfall climate. Source beds deposited in balanced-fill to underfilled lake conditions tend to occur in the southern basins, reflecting arid, low-rainfall conditions.

Table 18.8. *Summary of source characteristics of the Bucomazi and Chela formations (Well 86–1)*

Source formation	Facies	Mean TOC	Mean HI (mg HC/g TOC)	Kinetic parameters Ea/A (kcal/mol)/s^{-1}	Biomarker signature
Chela	Lagoonal	(1.5)	400	52 2.86×10^{13}	25,30-Bisnorhopane present
Upper Bucomazi (Zone A)	Shallow-water alkaline lake (balanced-fill)	1.3	(570)	52 5.24×10^{13}	Gammacerane present
Upper Bucomazi (Zone B)	Shallow-water alkaline lake (balanced-fill)	2.9	(500+)	52 3.88×10^{13}	Gammacerane and ethylcholestane prominent
Middle Bucomazi (Zone C)	Shallow-water alkaline lake (balanced-fill)	4.2	700+	54 1.035×10^{14}	25,28,30-Trisnorhopane, 28,30-bisnorhopane, 30-norhopane, prominent Low Ts/Tm ratio
Middle Bucomazi (Zone D)	Deep-water fresh to alkaline lake (balanced-fill)	7.9	700+	56 4.852×10^{14}	Abundant cholestane
Lower Bucomazi*	Fluvial to freshwater lake, overfilled	2.5	300	–	Uppermost Bucomazi approaching balanced-fill conditions

*Lower Bucomazi values from the wells in Block 0 from Schoellkopf and Patterson (2000). Kinetic data were not available.
HI, hydrogen index; TOC, total organic carbon.
HI values in parentheses are mean for samples with S2 >5 kg/t.
Well 86–1 data from Burwood *et al.* (1995)

Generative potentials are similar to those of the source rocks described above.

Transitional carbonate source rocks

As noted above, the sills that isolated many of the lacustrine rift lakes were intermittently breached, allowing temporary influx of marine waters. With increased rifting, the breaches became permanent, resulting in narrow seaways. An arid climate produced high rates of evaporation and low clastic influx. When isolated by topographic barriers, the transitional basins formed hypersaline pools that precipitated evaporites (Figure 18.77). The saline to hypersaline conditions produced density-stratified water columns with permanently anoxic bottom waters. High salinity limited the biota to halophilic microorganisms, which flourished in the nutrient-rich brines. High productivity and conditions favoring preservation led to deposition of thin organic-rich source beds.

The organic-rich source beds are composed mostly of calcareous shales and marls that occur typically at the base of the marine incursions before the onset of halite/anhydrite precipitation. Organic richness is variable, with TOC exceeding 15 wt.% and HI up to ~750 mg HC/g TOC. Kerogen in these source rocks consists of mixtures of amorphous algal and bacterial debris (45–60%) and terrigenous macerals. The relatively high input of higher-plant debris (herbaceous, woody,

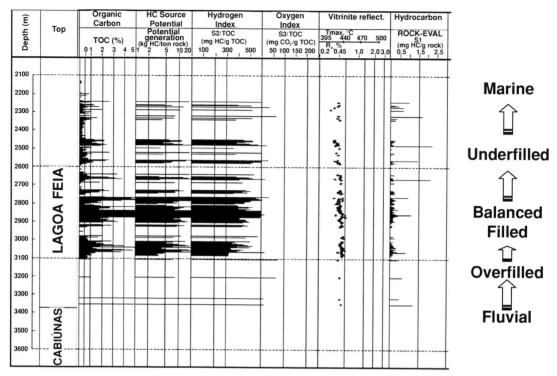

Figure 18.74. Geochemical log of fluvial/lacustrine/marine sediments in the Campos Basin, Brazil (Guardado et al., 2000). Source rocks with excellent source potential were deposited primarily in balanced-fill and underfilled conditions. Reprinted by permission of the AAPG, whose permission is required for further use.

and coaly macerals) is somewhat unexpected in such a depositional setting. This may be due to the increased density of the hypersaline waters suspending organic matter for prolonged periods, allowing bacteria time to rework all but the most resistant macerals (Mello et al., 1994a).

Aptian Alagamar Formation, Potiguar Basin, Brazil

The Alagamar Formation is an example of a hypersaline source sequence. The formation is the major source rock for oils in the Potiguar Basin and spans the transition from rift lacustrine (Upanema Member) to marine-evaporites to restricted marine (Ponta do Tubarão Member). Organic-rich units occur throughout the sequence (Mello et al., 1988b). Lacustrine shales contain up to ~4 wt.% TOC, composed mostly of hydrogen-rich kerogen with HI up to 700 mg HC/g TOC. The restricted marine sequence contains marls and shale with TOC up to ~6 wt.% and HI ~500 mg HC/g TOC. Oils generated from these mature, offshore source facies migrated considerable distances to onshore reservoirs. The chemistry of these oils indicates that mixing occurred during migration and reservoir filling (Trindade et al., 1992; Santos Neto and Hayes, 1999; Souto Filho et al., 2000). Figure 18.78 shows geochemical data for crude oil from Curimã Field in the Ceará Basin, Brazil, that was generated from the Alagamar source rock.

Maceió and Muribeca formations, Sergipe-Alagoas Basin

The Maceió and Muribeca formations lie immediately below and above the Aptian salt, respectively. These transitional calcareous shales and marls are the main source rocks for oils in the Sergipe-Alagoas Basin (Figure 18.77). They average ~6 wt.% TOC with HI ~500 mg HC/g TOC (Mello et al., 1988b). Similar

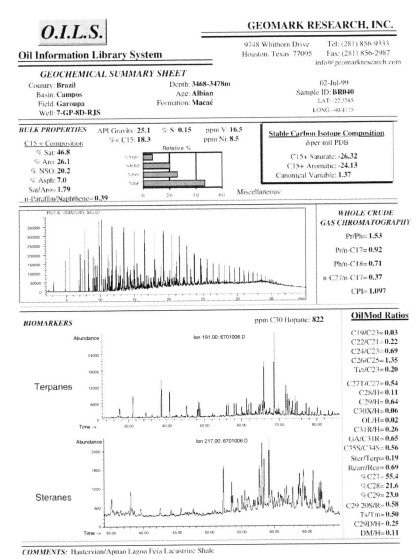

Figure 18.75. Geochemical data for crude oil from the Garoupa Field generated from the Hauterivian-Aptian Lagoa Feia Formation, Campos Basin, Brazil (courtesy of GeoMark Research, Inc.).

source facies exist in the Potiguar and Ceará basins (Mello *et al.*, 1994a) and in the Gabon Basin, although the occurrence of the latter is poorly defined.

Albian marine carbonate sequence

Seafloor spreading formed a permanent seaway between Brazil and West Africa during the Albian Age. The proto-South Atlantic was a narrow epiric sea with restricted circulation. Water influx was high enough to prevent halite precipitation but low enough to promoted carbonate precipitation. The saline to hypersaline, anoxic conditions were ideal for deposition of carbonates and marls with organic-rich laminations dominated by sulfur-rich algal kerogen (Figure 18.79). Source beds, such as those in the Sergipe/Alagoas Basin of Brazil, have up to ~6 wt.% TOC with maximum HI >700 mg HC/g TOC (Koutsoukos *et al.*, 1991; Mello *et al.*, 1996). Oils generated from Albian carbonate source rocks occur in the Cassiporé, Pará-Maranhão, and Bahia Sul basins (Mello *et al.*, 1994a).

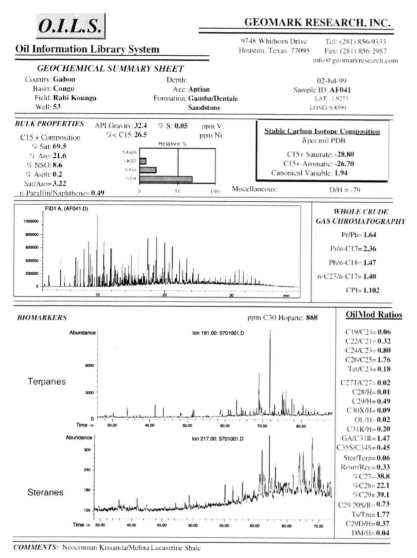

Figure 18.76. Geochemical data for crude oil from Rabi Kounga Field generated from Neocomian Kissenda/Melania lacustrine shales, Gabon (courtesy of GeoMark Research, Inc.).

This facies can be distinguished from the evaporite source facies by higher relative concentrations of dinosteranes and C_{30} steranes and lower concentrations of gammacerane, β-carotane, regular steranes, and hopanes.

Upper Cretaceous open marine shales, transgressive sequence, South Atlantic

Clastic deposition became dominant during Cenomanian time as the basins between the South Atlantic continents deepened with increasing plate separation. A generally high sea level with a major marine transgression gave rise to persistent and widespread deepwater anoxic conditions (Figure 18.80). Anoxia was further promoted and maintained by high productivity, increased salinity, and restricted water circulation. These conditions were interrupted only when continued spreading established a deep connection with the North Atlantic, causing ventilation of the bottom waters. Sedimentation varied from predominantly calcareous

Table 18.9. *Comparison of South Atlantic lacustrine source rocks. Basins are arranged vertically and horizontally in their approximate rift positions relative to one another (modified from Moore et al., 1995)*

	Brazil			West Africa	
Basin	Formation	Environment	Basin	Formation	Environment
Ceará	Mundau	I to II	–	–	–
Potiguar	Pendência	I to II	–	–	–
Jatobá-Tucano	Lower Cretaceous Shales	I to II	–	–	–
Recôncavo	Gomo Member	I to II	Gabon	Kissenda/Melania	I to II
Sergipe/Alagoas	Barro do Itiuba/Ibura	I to II to III			
Bahia Sul	Itaúnas	I to II	–	–	–
Espírito Santo	Juquia	II to III	Congo	Marne de Pointe Noire	III
Campos	Lagoa Feia	II to III	Lower Congo	Bucamozi	II to III
Santos	Guaratiba	II to III	Kwanza	Rio Cuvo	II to III

I, overfilled; II, balanced-fill; III, underfilled.

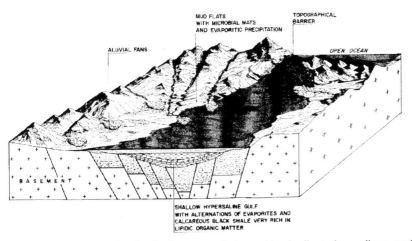

Figure 18.77. Depositional setting for the source rocks in transitional, saline-to-hypersaline evaporitic environments (from Mello *et al.*, 1995). Reprinted by permission of the AAPG, whose permission is required for further use.

mudstones during semiarid climatic conditions to siliciclastic shales during more humic conditions (Mello *et al.*, 1995).

Turonian Azile Formation, Gabon Basin

Marine shales of the Azile Formation are the primary source rocks for oils in post-Aptian salt reservoirs of the offshore Gabon Basin (Figure 18.81) (Teisserenc and Villemin, 1989; Katz *et al.*, 2000c). These calcareous shales were deposited over the Sibang Limestone and represent the earliest source rocks of the transgressive sequence. They contain an average of 3–5 wt.% TOC, with highly variable generative potential. Kerogen consists of mixed marine algal and terrigenous higher-plant matter, resulting in HI in the range 100–500 mg HC/g TOC, and >700 mg HC/g TOC in exceptionally rich intervals. Teisserenc and Villemin (1989) mapped these variations in the offshore south of Port Gentil and

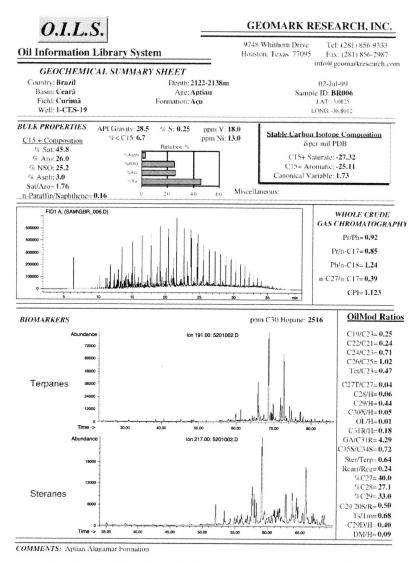

Figure 18.78. Geochemical data for crude oil from Curimã Field generated from the Aptian Alagamar Formation, Ceará Basin, Brazil (courtesy of GeoMark Research, Inc.).

showed that generative potential was expressed in regional facies. The richest facies occur in the southern area. Slightly lower generative potential occurs in the northern area, while the central area has substantially lower potential. Marine shales similar to the Azile Formation occur in the Coniacian-Santonian Anguille and the Campanian Point Clairette formations, but they are less important source rocks because of their limited extent and generally lower maturity.

Iabe/Landana marine shales, Lower Congo and Angola Basin

The Upper Cretaceous Iabe marine shales are the primary source rocks for oils in the offshore Congo Delta (Burwood, 1999; Cole et al., 2000; Schoellkopf and Patterson, 2000). TOC and HI typically exceed 2 wt.% and 300 mg HC/g TOC, respectively (Figure 18.82). Particularly rich intervals can exceed 10 wt.% TOC. A regional unconformity separates the Cenomanian

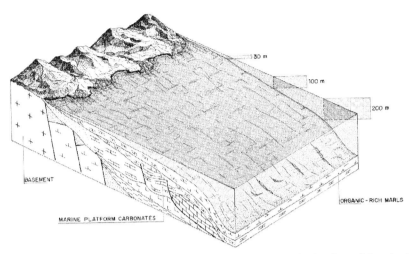

Figure 18.79. Depositional setting for the marine carbonate source rocks. Anoxic conditions developed in the deeper-water slope facies and in topographic lows within the platform (Mello *et al.*, 1995). Reprinted by permission of the AAPG, whose permission is required for further use.

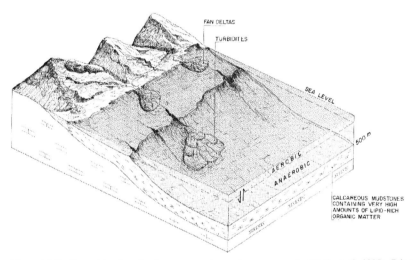

Figure 18.80. Depositional setting for transgressive marine source rocks (Mello *et al.*, 1995). Calcareous marls and siliciclastic shales predominate during semi-arid and humid conditions, respectively. Anoxic conditions developed in the deeper-water facies and persisted through much of the Late Cretaceous Period. Reprinted by permission of the AAPG, whose permission is required for further use.

to Maastrichtian Iabe from the overlying Paleocene to Eocene Landana Formation. The depositional settings and generative potentials for these source rocks are nearly identical. Kerogen in these source rocks originated mostly from a diverse assemblage of algal marine phytoplankton with secondary contributions of terrigenous plant materials, as supported by the geochemistry of related crude oils (Figure 18.83).

Upper Cretaceous/Lower Tertiary marine shales, Douala Basin, Cameroon

The Turonian-Paleocene marine shales of the Douala Basin have historic significance for geochemistry because early calibrations for the maturation of type III kerogen were based on measurements from the wells in the Logbaba gas field (Tissot and Welte, 1984). Similar shales occur in the adjacent Niger Delta (Figure 18.84)

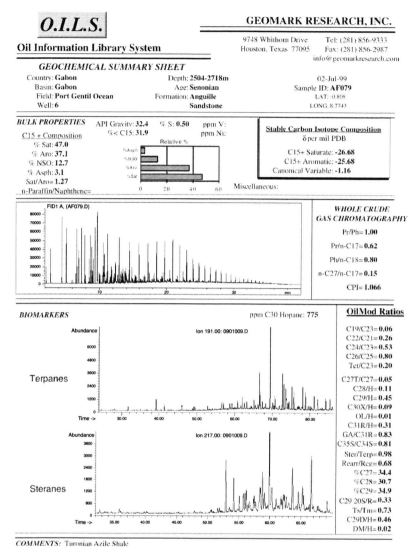

Figure 18.81. Geochemical data for crude oil from Port Gentil Ocean Field generated from the Turonian Azile Formation, Gabon (courtesy of GeoMark Research, Inc.).

Haack *et al.*, 2000). Although these shales contain intervals with oil-prone kerogen (Ackerman *et al.*, 1993), they are generally gas-prone. Offshore Cameroon oils and condensates were likely generated from older Lower Cretaceous evaporites and marls (Pauken, 1992).

Brazilian trangressive marine source rocks
Transgressive shales and marls with high source potential were deposited along most of the Brazilian continental margin. These sediments, however, are immature, except in the Santos, Espírito Santo, and perhaps Campos basins.

Tertiary regressive marine shales
The marine regressive mega-sequence from the latest Cretaceous Period to present produced few source rocks along the South Atlantic margins. Exceptions include several Tertiary deltaic systems (e.g. Niger, Congo, Foz do Amazonas, and Campos), some of which are discussed below. Most of the water column became well

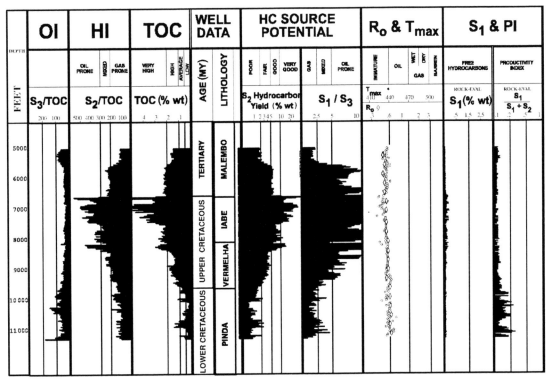

Figure 18.82. Geochemical log of a well from offshore Cabinda, Angola (Schoellkopf and Patterson, 2000). Similar Cabinda logs are seen in Peters and Cassa (1994). Reprinted by permission of the AAPG, whose permission is required for further use. HI, hydrogen index; OI, oxygen index; TOC, total organic carbon.

oxygenated during expansion of the South Atlantic Ocean, and organic productivity was low, except for upwelling areas along the African coast.

Upper Jurassic-Miocene lacustrine shales, East African interior rift basins

Upper Jurassic-Lower Cretaceous synrift lacustrine sediments were deposited in a series of east African rift basins that formed in response to the break-up of the African and South American plates. More lacustrine sediments were deposited during the Cretaceous and Eocene-Oligocene periods, possibly in response to the breakup of the Afro-Arabian plate, and during the Miocene Epoch as a result of basin subsidence.

Upper Jurassic-Lower Cretaceous lacustrine shaly to marly source rocks occur in southern Egypt (Dolson et al., 2001), Sudan (Hwang et al., 1994; Mohamed et al., 1999), and Ethiopia (Cayce and Carey, 1979; Du Toit et al., 1997). Tertiary lacustrine source rocks occur in Kenya (Tiercelin et al., 2001; Jarvie et al., 2001c), and possibly in Uganda and Tanzania (Mpanju and Philp, 1994). The petroleum-generative potentials of many of these units are poorly documented. In one description of basins in the Lake Turkana region of northwestern Kenya, >250 m of Miocene (?) shales have TOC up to ~7 wt.%, with HI >600 mg HC/g TOC (Tiercelin et al., 2001). The organic matter is dominated by phytoplankton, particularly the green algae *Pediastrum* and *Botryococcus*, typical of freshwater, balanced-fill lakes.

South Atlantic Mesozoic oils

Petroleum systems along the continental margins of South America and Africa may have multiple source units. Lacustrine source-rock facies range from freshwater to hypersaline. Large volumes of oil from these sources (e.g. Bucomazi or Lagoa Feia formations)

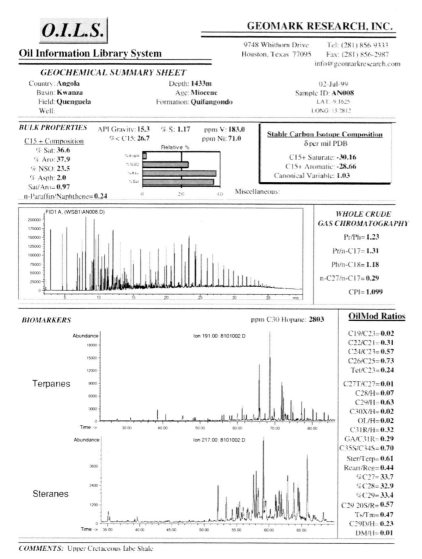

Figure 18.83. Geochemical data for crude oil from Quenguela Field generated from the Upper Cretaceous Iabe Formation, Kwanza Basin, Angola (courtesy of GeoMark Research, Inc.).

(see Table 4.3) are confined largely to reservoirs below the Aptian salt, but they may occur in younger strata where salt diapirs provide vertical migration pathways or where the salt is absent. Marine source-rock facies range from hypersaline and euxinic to normal saline and oxic environments. Post-salt marine rock may be present but it lacks sufficient overburden to be thermally mature.

The potential complexity of petroleum sources in the South Atlantic has led to several attempts to systematically correlate oils to source facies according to specific biomarker, isotopic, and other geochemical criteria. Table 18.10 summarizes the criteria developed by Mello and co-workers in their long-term study of the Brazilian basins. Schiefelbein et al. (1999; 2000) used statistical analysis to correlate oils from both margins (Figure 18.85). These and other modern studies (e.g. Katz and Mello, 2000) tend to rely on specific biomarker parameters that are particularly useful in order to distinguish oils from different source facies

Cretaceous source rocks

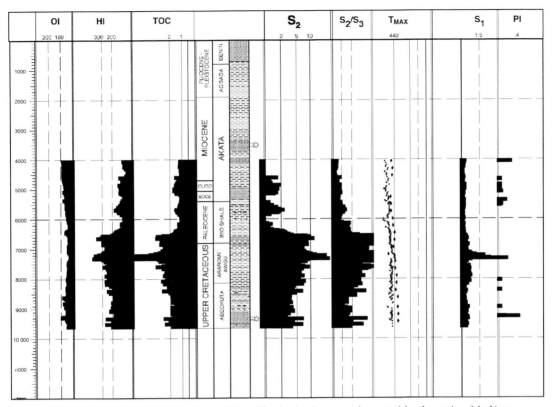

Figure 18.84. Geochemical log of the Epiya-1 well, offshore Nigeria, showing generative potential and maturity of the Upper Cretaceous/Lower Tertiary Period (Haack et al., 2000). Reprinted by permission of the AAPG, whose permission is required for further use.
HI, hydrogen index; OI, oxygen index; TOC, total organic carbon.

(e.g. ratios based on various tricylic terpanes, polyprenoids, oleanane, gammacerane, C_{26} and C_{30} steranes, dinosteranes, and methylsteranes) (Figure 18.86) compared with other parameters (e.g. sulfur, pristane/phytane).

Mesozoic carbonates of the Arabian platform

The Mesozoic carbonates of the Arabian platform are collectively the most productive petroleum source rocks worldwide. The region contains about two-thirds of the world's producible reserves, commonly trapped below tight evaporite seals in high-quality reservoirs that directly overlie organic-rich source rocks. The generative potential of each source rock is not particularly unusual. Rather, the key element that accounts for the enormous Arabian reserves is that the source rocks were deposited in vast, nearly flat intershelf basins ($\gg 10\,000\,km^2$) (Murris, 1980) throughout the Jurassic and Cretaceous periods and into Eocene time, giving rise to widespread, multiple source formations within individual sub-basins (Beydoun, 1991b).

From the Permian–Triassic Period through the early Tertiary Period, the Arabian Plate was appended to the Nubian Shield and evolved as a broad carbonate shelf on the southwestern side of the Southern Tethys Sea (Beydoun, 1991a). A series of sub-basins developed as a result of changes in sea level, stable structural elements, and local tectonic effects (Figure 18.87). These are: Lurestan in northern Iraq/Iran, Khuzestan in the Dezful Embayment of Iran, the Gotnia or Northern Arabian Basin of southern Iraq and Kuwait, the Arabian Basin of Saudi Arabia, the Southern Gulf Basin (also known as the Rub Al Khali Basin) of Qatar and the United Arab Emirates, and the Fahud Salt Basin of Oman. Source rocks of varying quality and thickness

Table 18.10. *Geochemical characteristics of Brazilian oils and source rocks by their inferred environment of deposition (from Mello et al., 1995)*

	Lacustrine, fresh water	Lacustrine, saline	Marine, evaporitic	Marine, carbonate	Marine, calcareous	Marine, siliciclastic	Marine, deltaic
°API (oils)	30–39	24–32	20–30	25–30	34–40	—	42–44
%Saturates (oils)	60–73	45–65	30–59	20–60	50–80	—	60–70
%Sulfur (oils)	<0.1	0.2–0.4	0.3–1.5	0.4–0.7	0.1–0.2	—	0.3–0.4
V/Ni (oils)	<0.05	0.3–0.4	0.2–0.3	0.4–0.5	—	—	0.8–1.0
%R_o (rocks)	0.4–0.7	0.4–0.8	0.5–0.7	0.4–0.6	0.4–0.6	0.5–0.7	0.5–0.6
%Saturates (rocks)	40–60	25–55	25–40	20–45	22–34	25–44	27–30
%Sulfur (rocks)	0.2–0.3	0.1–0.5	0.3–2.5	0.2–0.6	0.4–0.5	0.3–0.7	0.6–0.7
%$CaCO_3$ (rocks)	<7	2–30	5–25	15–65	18–48	6–20	50–70
$\delta^{13}C$ (PDB ‰, oil)	<−28	−23 to −27	−25 to −27	−26 to −28	−25 to −27	−26 to −27	−24 to −26
n-Alkane maximum	nC_{23}	nC_{19}	nC_{18}	nC_{20}–C_{22}	nC_{20}	nC_{17}	nC_{20}–C_{22}
Odd/even	≥1	≥1	≤1	≤1	≤1	>1	≤1*
Pristane/phytane	>1.3	>1.1	<1.0	<1	≤1	<1	<1
i-C_{25} + i-C_{30} (ppm)	<170	70–700	300–1500	100–500	10–100	40–180	10–300
β-Carotane (ppm)	Not detected	10–200	100–400	20–60	10–30	Not detected	Not detected
C_{21} + C_{22} steranes (ppm)	Trace	10–30	10–60	10–60	10–30	25–35	30–50
C_{27} steranes (ppm)	10–50	50–150	500–4000	50–300	50–200	20–400	50–350
C_{27}/C_{29} steranes	1.5–4.0	1.5–2.5	1.0–2.2	1.1–1.5	0.8–1.2	1.5–25	1.3–1.8
Diasterane index	20–40	10–50	6–20	20–30	10–30	30–80	30–60
C_{30} steranes, dinosteranes	Not detected	Not detected	Low	High	Medium	High	Medium
4α-Methylsterane index	30–50	30–150	30–80	30–80	20–60	10–20	<10
Hopane/steranes	5–30	5–30	0.4–2.0	0.9–3.0	0.5–5.0	1.5–8.0	0.5–3.0
Tricyclic index	30–100	100–200	10–60	60–200	50–100	70–100	60–180
C_{34}/C_{35} hopanes	>1	>1	<1	≤1	≤1	>1	<1
Bisnorhopane index	0	3–15	15–40	10–30	20–1000	1–5	0
18α(H)-Oleanane index	0	0	0	0	0	0	20–40
Ts/Tm	>1	<1	≤1	<1	<1	>1	>1
C_{30} αβ hopanes (ppm)	200–500	200–1600	300–2000	80–300	10–70	50–800	100–250

(*cont.*)

Table 18.10 (*cont.*)

Gammacerane index	20–40	20–70	70–120	10–20	0–25	0–5
%Amorphous kerogen	55–65	85–90	46–60	50–60	60–70	85–95
%Herbaceous kerogen	25–35	5–10	15–25	10–15	5–10	2–10
%Woody + coaly kerogen	5–10	5–10	10–25	20–30	20–25	10–15
						15–25
$i\text{-}C_{25} + i\text{-}C_{30}$	Sum of 2,6,10,14,18- and/or 2,6,10,15,19-pentamethylicosane ($i\text{-}C_{25}$) and squalane ($i\text{-}C_{30}$) peak areas in RIC normalized to sterane standard.					
C_{27}/C_{29} steranes	Peak area of 20R 5α,14α,17α(H)-cholestane over 20R 5α,14α,17α(H)-ethylchostane in *m/z* 217 chromatogram.					
Diasterane index	Sum of peak areas of C_{27} 20R 13β,17α-diasteranes in *m/z* 217 chromatogram over sum of peak areas of C_{27} 20R and 20S 5α,14α,17α-cholestane × 100. Low, <30; medium, 30–100; high, >100.					
4-Methylsterane index	Sum of peak areas of all C_{30} 4-methylsteranes in *m/z* 217 chromatogram and *m/z* 414 chromatogram over sum of peak areas of C_{27} 20R and 20S 5α,14α,17α(H)-cholestane × 100. Low, <60; medium, 60–80; high, ∼80.					
Hopane/steranes	Peak areas of C_{30} 17α,21β(H)-hopane in *m/z* 191 chromatogram over sum of peak areas of C_{27} 20R and 20S 5α,14α,17α(H)-cholestane in *m/z* 217 chromatogram. Low, <4; medium, 4–7; high, >7.					
Tricyclic index	Sum of peak areas of C_{19}–C_{29} (excluding C_{22}, C_{27}) tricyclic terpanes *m/z* 191 chromatogram over peak area of C_{30} 17α,21β(H)-hopane × 100. Low, <50; medium, 50–100; high, >100.					
C_{34}/C_{35} hopanes	Peak areas of C_{34} 22R and 22S 17α,21β(H)-hopanes in *m/z* 191 chromatogram over peak areas of C_{35} counterparts. Low, <1; high, >1.					
Bisnorhopane index	Peak areas of 28,30-bisnorhopane over C_{30} 17α,21β(H)-hopane × 100 in *m/z* 191 chromatogram. Low, <10; medium, 10–50; high, >50.					
Oleanane index	Peak areas of 18α-oleanane in *m/z* 191 chromatogram over peak area of C_{30} 17α,21β(H)-hopane × 100 in *m/z* 191 chromatogram.					
Ts/Tm	Peak area of 18α-trisnorneohopane (Ts) over peak area of 17α-trisnorhopane (Tm) in *m/z* 191 chromatogram.					
Gammacerane index	Peak area of gammacerane in *m/z* 191 chromatogram over that of C_{30} 17α,21β(H)-hopane × 100. Low, <50; medium, 50–60; high, >60.					
Tetracyclic index	Peak area of C_{24} tetracyclic over C_{30} 17α,21β(H)-hopane × 100 in *m/z* 191 chromatogram.					

* Although Mello *et al.* (1995) report that Brazilian marine deltaic oils have *n*-alkanes exhibiting an even carbon number preference, it is our experience that this is generally not true. RIC, reconstructed ion chromatogram.

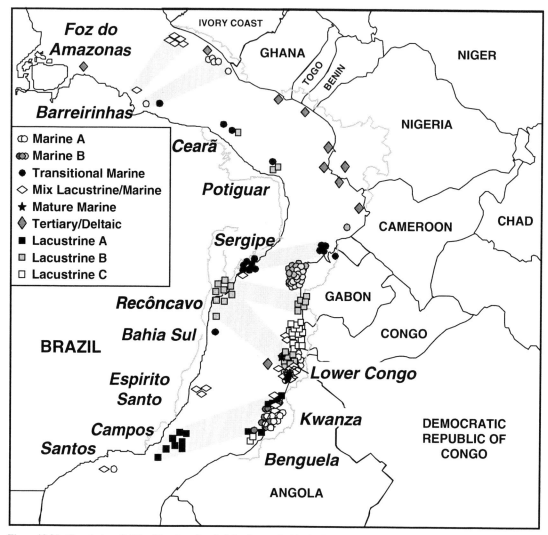

Figure 18.85. Correlation of oil families along South Atlantic marginal basin. Lacustrine A oils originated from source rocks deposited in balanced-fill to underfilled conditions. Lacustrine B oil originated from source rocks deposited in balanced-fill freshwater conditions. Lacustrine C oils originated from the Bucomazi Formation. Marine A oils were derived from late rift transgressional marine shales and marls. Marine B oils, primarily located offshore Gabon and Kwanza Basins, originated from early rift calcareous shales. Modified from Schiefelbein et al. (2000). Reprinted by permission of the AAPG, whose permission is required for further use.

developed at different times within these sub-basins (Figure 18.88).

During the Triassic Period, the interior of Pangea was hot and dry, resulting in deposition of non-marine sands on the Arabian Plate. However, shallow, marginal marine lagoonal shales and marls were deposited along the coast under conditions similar to those that gave rise to prolific source rocks in the Jurassic and Cretaceous periods. Loutfi and Abdel Satter (1987) reported that some Middle-Upper Triassic restricted shelf and lagoonal carbonate-evaporite sequences have significant organic content in areas of eastern Saudi Arabia, the United Arab Emirates, and Oman. Alsharhan and Kendall (1986) suggested that several small oil fields in Iran had

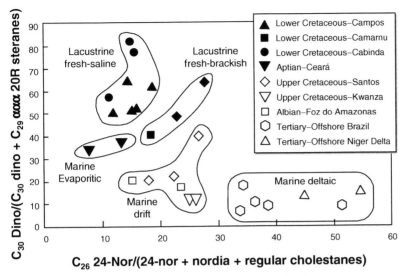

Figure 18.86. Separation of South Atlantic margin oils by %C_{26} and %C_{30} dinosterane ratios (Katz and Mello, 2000). The lacustrine facies are enriched in dinosteranes, reflecting input from freshwater dinoflagellates. Upper Cretaceous/Tertiary deltaic oils are enriched in C_{26} 24-norcholestanes, reflecting input from diatoms. Reprinted by permission of the AAPG, whose permission is required for further use.

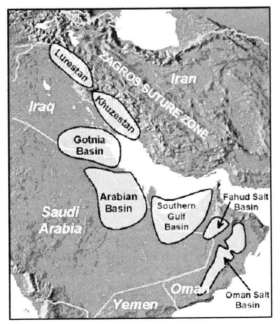

Figure 18.87. Petroliferous basins of the Arabian platform. These are the major intershelf basins responsible for generating most of the world's producible oil. Major source rocks were deposited from Jurassic to Eocene time. The Oman Salt Basin contains mainly Infracambrian source rock.

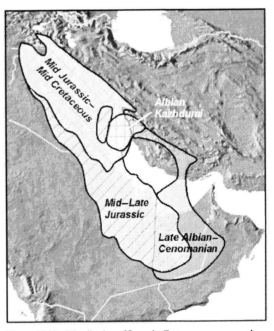

Figure 18.88. Distribution of Jurassic-Cretaceous source rocks on the northeastern margin of the Arabian platform (modified from Beydoun et al., 1992).

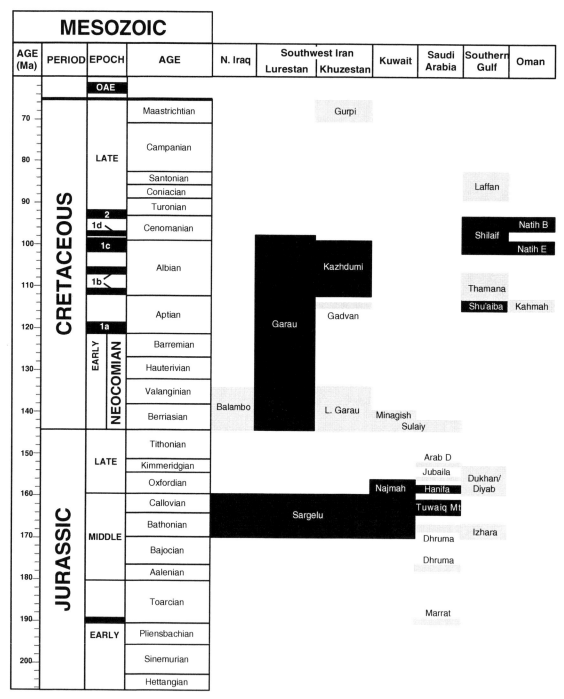

Figure 18.89. Stratigraphic relationships of Jurassic-Cretaceous source rocks on the Arabian platform (modified from Stoneley (1987) and Beydoun (1991b)).

likely Triassic source rocks. These include the paper shales within Kurra Chine limestone, and the Gahkum, Elika, and Kashaf formations. Also within Abu Dhabi, the Jilh Formation contains terrigenous kerogen and may be a minor source of gas. The extent and richness of these strata have not been studied fully, and these Triassic rocks may eventually be recognized as significant source rocks.

At the end of the Triassic Period, climatic conditions became favorable for source-rock deposition. These environments were vast and persistent, resulting in the repeated deposition of organic-rich carbonates and marls under essentially similar conditions (Figure 18.89). The region's climate changed from arid to humid and remained tropical from the Jurassic Period through much of the Tertiary Period. For much of this period, an epiric sea covered the Arabian Platform, and shorelines fluctuated with changes in eustatic sea level and local subsidence. Sediments included mainly high-energy aerobic limestones with low-energy argillaceous limestones in the depressions. Carbonate deposition was interrupted by periods of clastic input from the Nubian Shield, mainly during the Aptian and Albian ages and the Early Miocene Epoch, or by thick regional evaporites at the end of the Jurassic Period and during the Early Miocene Epoch.

Van Buchem *et al.* (2002) conducted a sequence stratigraphic analysis of the Cenomanian Natih Formation in northern Oman. These rocks serve as a model for all Mesozoic intershelf deposits. Carbonate sequences formed in two types of depositional systems: a carbonate-clay ramp and a carbonate-ramp intershelf environment.

Alternating beds of carbonate and clay were deposited under the carbonate-clay ramp conditions. Clastics washed on to the shelf in an early transgressive stage, resulting in a mixed carbonate-clay ramp deposit. The clay-rich layers are interpreted as being deposited in open lagoonal settings, under fairly shallow (5–20 m) water. Clay contents as high as 70% both diluted and inhibited carbonate production. These layers are highly bioturbated and have no source-rock potential. The carbonate beds in this environment consist of wackestones, packstones, and grainstones that contain a diverse faunal assemblage. The setting was relatively open-marine, but the water depth was no greater than that of the clay layers. Deposition of these carbonate beds occurred under oxic conditions and may simply represent decreased clay input.

With increasing transgression, the clays dropped out nearshore and the carbonate ramp stepped back. Sedimentation in the shallow margins kept pace with the rising sea level, but was slower in the intershelf depressions, resulting in a deeper-water column and onset of the carbonate ramp–intershelf system. Four distinct environments include the inner ramp, mid-ramp, outer ramp, and intershelf basin (Figure 18.90). The inner ramp is composed of wackestones to packstones deposited in shallow, high-energy subtidal to intertidal environments (0–5 m water depth). It is dominated by rudists and may be intensively bioturbated. The mid-ramp includes a low-energy bioclastic facies that is relatively clastic-rich and bioturbated and a high-energy, cross-bedded facies. These sediments were deposited in shallow water, at depths that were influenced by fair-weather waves. The outer ramp is the transitional facies between the wave-influenced carbonates and the deeper-water sediments of the intershelf basin. This facies ranges from highly bioturbated to non-bioturbated with few faunal fossils. The intershelf basins are characterized by two facies: couplets of organic-rich/organic-poor limestones and mudstones/wackestones with abundant chert. Within the limestone couplets, the organic-poor beds dominated by oyster fossils are commonly bioturbated. The organic-rich beds are thin ($\sim$2–40 cm), contain few faunal fossils, and are commonly laminated. Paleowater depth was approximately at the storm-wave base ($\sim$30–40 m).

A combination of high productivity, low rates of sedimentation, and anoxic to euxinic conditions led to deposition of organic-rich carbonate source rocks in the intershelf basins. Marine productivity was high because of upwelling along the Arabian platform, and large volumes of nutrient-rich freshwater flowed off the Nubian Plate. During periods of high sea level, the intershelf basins were sufficiently deep (50–100 m) to facilitate an anoxic, saline water layer under the highly productive photic zone. A ridge of topographic highs and barrier islands (proto-Zagros) hindered water circulation.

The type of deposition that occurred on the Arabian platform was determined by the interplay of eustatic, tectonic, and environmental forces. Van Buchem *et al.* (2002) described three discrete settings for

862 Petroleum systems through time

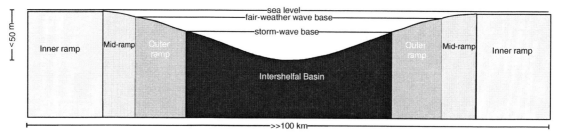

Figure 18.90. The carbonate ramp–intershelf environment of the Arabian platform. The vertical scale is exaggerated.

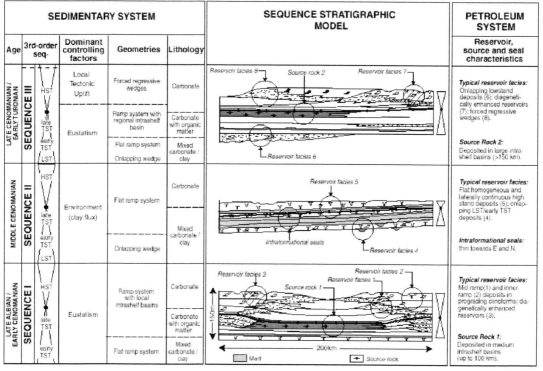

Figure 18.91. Summary of third-order sequences of the Natih Formation (from van Buchem et al., 2002). Sequence I was influenced mostly by eustatic changes. Sequence II was dominated by the influx of clay. Sequence III was controlled mainly by local tectonism. Reprinted by permission of the AAPG, whose permission is required for further use.
HST, highstand systems tract; LST, lowstand systems tract; TSR, transgressive systems tract.

depositional sequences of the Natih Formation in Oman (Figure 18.91). The first sequence, controlled mainly by changes in eustatic sea level, resulted in the Natih E source rocks. The second sequence was influenced most by factors that controlled the influx of clay. In the third sequence, when the Natih B source rocks were deposited, regional tectonic movement was the most influential factor. Because these carbonate sequences contain source rocks that generated most of the world's produced oil, a more detailed description follows.

Jurassic source rocks

Organic-rich carbonate source rocks of Jurassic age occur in several basins in the Middle East (Stonely, 1987).

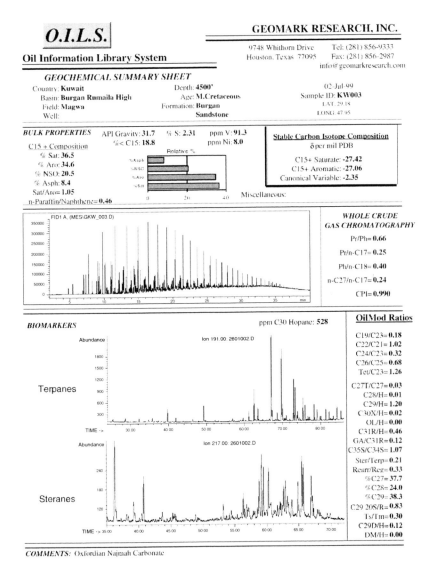

Figure 18.92. Geochemical data for crude oil from Magwa Field generated from the Oxfordian Najmah Formation, Burgan Rumaila High, Kuwait (courtesy of GeoMark Research, Inc.).

These include carbonates and shales of the Madbi in Yemen, the Sargelu and Najmah in Iraq and Kuwait, the Hanifa and Tuwaiq Mountain in Saudi Arabia, and the Izhara and Dukham/Diyab in Qatar and the United Arab Emirates (Beydoun, 1991b). The Jurassic is poorly sampled in the Arabian platform because most of the potential Jurassic source rocks are deeper than reservoir rocks targeted for drilling in the central basins or near the overthrust margins.

Sargelu Formation

The Sargelu source rock was deposited in the Lurestan, Khuzestan, and Gotnia basins (Figure 18.87) during the Middle Jurassic (~Bajocian–Callovian) Period. Source potential is best developed in northeastern Iraq, where 100–200 m of paper shale containing ~1.5–4.5 wt.% TOC is present (Bordenave and Burwood, 1990). Although the Sargelu Formation has excellent source potential, its contribution to commercial accumulations is limited because it is isolated from reservoir rocks by thick evaporites.

Najmah Formation

The Sargelu Formation grades into the Oxfordian Najmah Formation in the Gotnia Basin, where it consists mainly of limestones and calcareous shales. Source beds within the Najmah Formation had significant generative potential and were major contributors of oil to younger reservoirs (Abdullah, 2001). Individual strata contain abundant TOC (up to 31.4 wt.%) composed of type II/III kerogen. Mature samples (T_{max} ~443°C) still yield HI as high as 367 mg HC/g TOC (Abdullah and Connan, 2002). Figure 18.92 shows geochemical data for crude oil from Magwa Field in the Burgan Rumaila High, Kuwait, that was generated from the Najmah source rock.

Tuwaiq Mountain and Hanifa formations

Several Jurassic source rocks occur in the Arabian Basin. Of these, the Tuwaiq Mountain (Callovian) and Hanifa (Oxfordian) formations are the most significant (Table 4.3 Figure 18.93), while contributions of hydrocarbons from the Marrat (Toarcian), Dhruma (Aalenian-Callovian), Jubaila (Oxfordian), and Arab D (Kimmeridgian) are believed to be minor (Ayres et al., 1982; Carrigan et al., 1995). Deposition of the Tuwaiq Mountain source rocks corresponds to a major rise in sea level that began in the Late Callovian Age and drowned much of the interior platform, forming a vast inter-shelf basin. During this time, the Arabian Basin was separated from the Gotnia Basin to the north and the Southern Gulf Basin to the southeast. Restricted water circulation led to euxinic conditions, which, coupled with high productivity, resulted in the deposition of organic-rich Tuwaiq Mountain carbonates. Layers containing organic matter average ~3 wt.% and may be as high as 13 wt.% TOC. The kerogen is mostly hydrogen-rich lamalginite, with HI ~640 mg HC/g TOC. The formation exceeds 150 m in the center of the basin and thins laterally. The source potential of the lower portion of the Hanifa Formation is lower than that of the Tuwaiq Mountain. TOC averages ~3 wt.%, and the kerogen is a variable mixture of marine type II lamalginite and terrigenous type III/IV macerals of vitrinite and inertinite. Greater terrigenous input reflects deposition closer to the paleo-shoreline. Figure 18.94 shows geochemical data for crude oil from Bakr Field in the Central Arabian Basin, Saudi Arabia, that was generated from the Tuwaiq Mountain/Hanifa source rock.

Note: High sulfur content in Arabian Heavy oil is consistent with its origin from sulfur-rich kerogen in Upper Jurassic Tuwaiq Mountain and Hanifa Formation source rocks in the central Gulf area (Murris, 1984). These carbonate-evaporite source rocks originated in sediment-starved intrashelf depressions that became euxinic during a major marine transgression (Alsharhan and Kendall, 1986).

Dukham/Diyab Formation

Oxfordian to Kimmeridgian carbonates in the Southern Gulf Basin are secondary sources of petroleum in onshore/offshore United Arab Emirates and Qatar (Beydoun 1991b; Alsharhan, 2002). The formation grades from onshore carbonate inner ramp deposits to offshore intershelf basins. The organic content varies according to facies. High-energy carbonates have <1 wt.% TOC and little generative potential. However, argillaceous lime mudstones in the shelf interior contain 0.3–5.5 wt.% TOC, range in thickness from ~250 to 450 m, and typically contain oil-prone kerogen. Figure 18.95 shows geochemical data for crude oil from Upper Zakum Field in the United Arab Emirates that was generated from the Jurassic Diyab Formation.

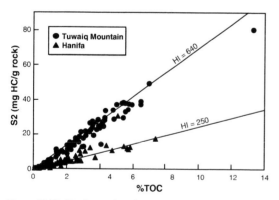

Figure 18.93. Total organic carbon (TOC) (wt.%) versus Rock-Eval S2 for samples from the Tuwaiq Mountain and Hanifa formations, Saudi Arabia (from Carrigan et al., 1995).

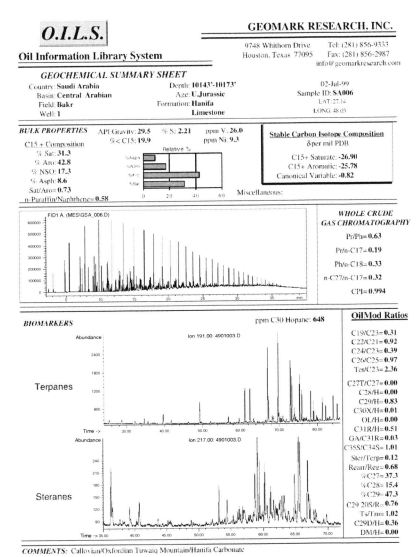

Figure 18.94. Geochemical data for crude oil from Bakr Field generated from the Callovian-Oxfordian Tuwaiq Mountain/Hanifa Formation, Central Arabian Basin, Saudi Arabia (courtesy of GeoMark Research, Inc.).

Lower Cretaceous source rocks

A marine transgression deposited carbonates and clays over much of the eastern Arabian Platform, particularly in the Lurestan and the Gotnia basins (Beydoun et al., 1992). The Garau Formation in Lurestan was deposited in a long-lived depression (Valanginian-Aptian, up to Coniacian in the axial portions of the basin). The Garau Formation reached a total thickness of >2000 m, but only a fraction of this unit has significant generative potential. The formation includes organic-rich laminated, pyritic marls that alternate with fine-grained argillaceous limestones. TOC varies from 2 to 9 wt.% for the marls and from 1 to 2 wt.% for the limestones (Bordenave and Burwood, 1990). The extent of the source facies within the Garau Formation is defined poorly due to limited data.

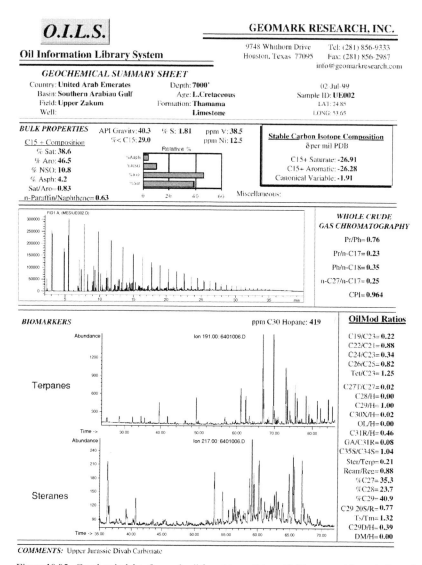

Figure 18.95. Geochemical data for crude oil from Upper Zakum Field generated from the Jurassic Diyab Formation, United Arab Emirates (courtesy of GeoMark Research, Inc.).

Basal source beds equivalent to the Garau Formation occur in the Gotnia Basin (Sulaiy and Minagish formations) and Khuzestan Basin (Lower Garau Formation). Abdullah and Connan (2002) correlated rock extracts from the Sulaiy and Minagish carbonates to oils in Kuwait. These rocks have substantially lower source potential than the underlying Sargelu and Najmah formations, having TOC ~1 wt.% and HI <400 mg HC/g TOC. Gadvan marls (~Barremian) may be source rocks in Khuzestan (Ala et al., 1980).

Regressive facies dominate the end of the Early Cretaceous Period. A drop in sea level and the emergence of the Numbian shield resulted in widespread deposition of continental to marginal deltaic sandstones and marine shales across the eastern Arabian platform. Several of these clastic units, such as the Ratawi Shale Member and Burgan Formation, contain low concentrations of type III kerogen and may have minor source potential.

Middle Cretaceous source rocks

Several source rocks were deposited on the Arabian platform during the Cretaceous Period in a series of marine transgressions. Sea level generally rose over this time, and each transgression was followed by a less developed regression. Carbonate ramp deposition characterized nearly the entire shelf by the end of the mid-Cretaceous Period. During this time, the Garau Formation in the Lurestan, the Kazhdumi Formation in Khuzestan, the Shilaif Formation in the Southern Gulf, and the Natih Formation in the Fahud Salt Basin were deposited in intershelf basins under euxinic conditions. Deposition of these source beds corresponded with several global oceanic anoxic events (OAEs).

Kazhdumi Formation

The major source unit for oils from the Khuzestan Basin (Dezful Embayment of southwest Iran) is the Kazhdumi (Khazdumi) Formation (Bordenave and Burwood, 1990; 1995). A large, subtle depression developed in the area during the Albian Age, and up to 300 m of organic-rich marls and lime mudstones were deposited under euxinic conditions. During deposition of the Kazhdumi Formation, abundant clastics were carried from the exposed Nubian Shelf into the basins due to the tropical climate and high rainfall. However, transgressive conditions trapped most of the clastics in nearshore delatic facies that generally lack source potential. Source-rock richness increases toward the centers of the intershelf basins, typically from ~1 to 5 wt.% to maxima >11 wt.% TOC. Residual HI for marginally mature samples are typically ~230–320 mg HC/g TOC. Original generative potential was estimated to be ~450–600 mg HC/g TOC. Figure 18.96 shows geochemical data for crude oil from Marun Field in the United Arab Emirates that was generated from the Kazhdumi source rock.

Shu'aiba and Shilaif Formations

The Cretaceous Shu'aiba and Shilaif (Khatiyah) formations are major source rocks for oils in the mostly offshore fields of the United Arab Emirates (Azzar and Taher, 1993; Taher, 1997; Milner, 1998) and Abu Dhabi (Lijmback et al., 1992). The Shu'aiba Formation consists of >30 m of organic-rich argillaceous lime mudstones at the base of the Thamama Group, with up to 10 wt.% TOC, mostly as oil-prone type II kerogen. Source quality decreases in shallower facies, and the Shu'aiba is only a lean, marginal source in Oman. Other argillaceous layers within the Thamama Group have lower source potential. The Shilaif Formation is the basin facies of the Wasia Group. Depositional conditions and organic matter content are similar to those of the Shu'aiba Formation, typically with 1–5 wt.% TOC. Figure 18.97 shows geochemical data for crude oil from Zakum Field in the United Arab Emirates that was generated from the Cretaceous Shilaif Formation.

Natih Formation

The Natih E and B strata are the source rocks for oils in northeastern Oman (Terken et al., 2001). These strata were deposited during the Late Albian and Cenomanian ages and are temporal equivalents of the Shilaif Formation in Abu Dhabi. The Natih petroleum system is relatively small and is limited to only ~20 000 km^2 in the Fahud Salt Basin. However, the source rock has very high generative potential and is responsible for more than eight billion barrels of discovered reserves, which may be only 9% of the generated total (Terken, 1999). The Natih Formation has up to 14 wt.% TOC, with HI up to 650 mg HC/g TOC (Figure 18.98; van Buchem et al., 2002).

Upper Cretaceous source rocks

Uplift and erosion dominate much of the Late Cretaceous tectonics the Arabian Plate, where strata of this age are commonly absent. Later, a general transgression resulted in a return to carbonate deposition, although only marginal source rocks occur along the northeastern margin. For example, the Laffan and Gurpi formations typically contain <3 wt.% TOC and primarily gas-prone type III kerogen (Ala et al., 1980; Bordenave and Burwood, 1990). Small intershelf basins and grabens developed along the northwestern (e.g. Turkey, Syria, Jordan) and southern (Yemen) margins. Oil shales in Jordan contain up to 40 wt.% TOC and are best

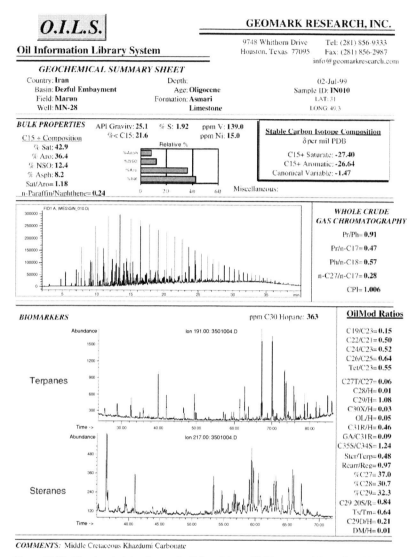

Figure 18.96. Geochemical data for crude oil from Marun Field generated from the mid-Cretaceous Kazhdumi Formation, United Arab Emirates (courtesy of GeoMark Research, Inc.).

described as bituminous marls (Abed and Amireh, 1983). These rocks are immature at present and have not generated or expelled oil.

Lower-Middle Jurassic coals and lacustrine source rocks, Central Asia

Organic-rich, Lower and Middle Jurassic continental strata are source rocks in many Central Asian basins. Examples were reported from Mongolia and Russia (Vakhrameyev and Doludenko, 1977) and northwestern China (e.g. Ulmishek, 1984; Graham et al., 1990). Many of these sedimentary basins have complex Paleozoic histories, but in the Mesozoic Era they were foreland basins in a back-arc position relative to the pre-Himalayan convergent margin of southern Asia (Graham et al., 1990). The basins were deformed by the Cenozoic Himalayan orogenic collision, but they were separated throughout

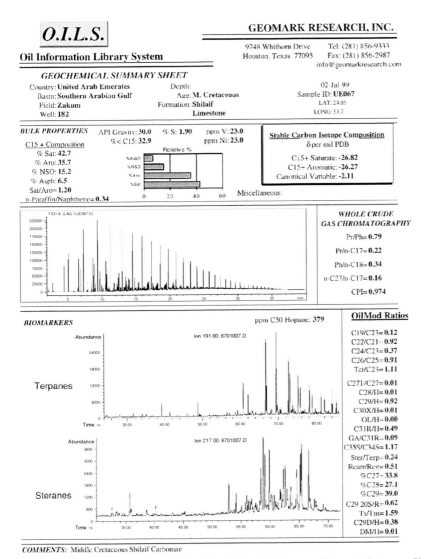

Figure 18.97. Geochemical data for crude oil from Zakum Field generated from the Cretaceous Shilaif Formation, United Arab Emirates (courtesy of GeoMark Research, Inc.).

the Mesozoic Era. Sediment deposition was controlled largely by climatic conditions. During humid times, such as the Early Jurassic Period, large lakes and coal swamps filled the basins, giving rise to vast deposits of coal and organic-rich lacustrine mudstones. These climatic conditions are attributed to monsoon circulation patterns induced by the configuration of the Pangean landmass (Parrish, 1993).

Jurassic coals in the Xinjiang Uygur, northwest China

In the Tarim, Junggar, and Turpan-Hami basins of Xinjiang Uygur Autonomous Region, northwest China, Hendrix (1992) and Hendrix et al. (1995) showed that coals and coaly shales have significant petroleum-generative potential and are responsible for

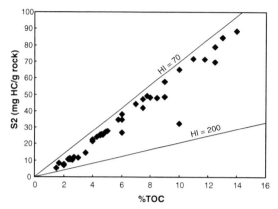

Figure 18.98. Total organic carbon (TOC) (wt.%) versus Rock-Eval S2 for thermally immature Natih B source rocks in the Natih-68 production well (van Buchem et al., 2002). HI, hydrogen index.

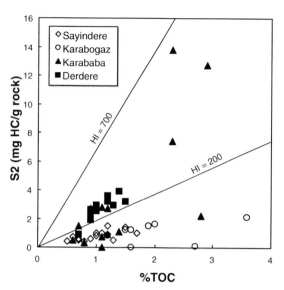

Figure 18.99. Total organic carbon (TOC) (wt.%) versus Rock-Eval S2 for Upper Cretaceous rocks from the Adiyaman region, southeastern Turkey (data from Demirel et al., 2001). HI, hydrogen index.

producible oil. Organic petrography identified vitrinite and inertinite as the dominant macerals in these rocks, with up to 38% hydrogen-rich exinite. HI typically ranges from 50 to 300 mg HC/g TOC, but it may exceed 400 mg HC/g TOC (Figure 18.100). However, elemental analysis does not indicate that these coals and coaly shales are as oil-prone as indicated by Rock-Eval pyrolysis (Figure 18.101). Rock-Eval pyrolysis tends to overestimate the hydrocarbon generative potential of coals (e.g. Black Hawk coal in Figure 4.4) because much of the pyrolyzate consists of volatile polar materials rather than true hydrocarbons (Peters, 1986).

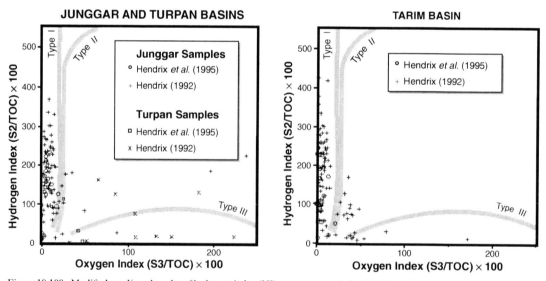

Figure 18.100. Modified van Krevelen plot of hydrogen index (HI) versus oxygen index (OI) for coals and lacustrine mudstones from three Xinjiang basins of northwestern China (from Hendrix et al., 1995). Reprinted by permission of the AAPG, whose permission is required for further use.

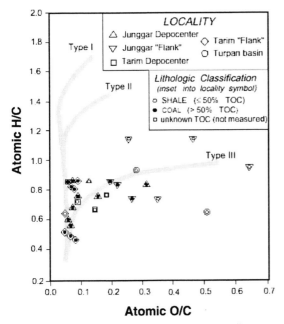

Figure 18.101. Atomic H/C and O/C (van Krevelen) diagram for selected coal samples from three Xinjiang basins in northwestern China (Hendrix et al., 1995). In contrast to the Rock-Eval data, elemental analyses indicate that the coals are predominantly type III. Reprinted by permission of the AAPG, whose permission is required for further use.
TOC, total organic carbon.

light aromatics and relatively abundant naphthalenes (Figure 18.102). Biomarker distributions are typical for coals and sediments dominated by terrigenous higher-plant input.

The distribution of steranes by carbon number was used to correlate the Jurassic coals or older Permian source rocks with produced oils (Figure 18.103) (Hendrix et al., 1995). Although we do not dispute this correlation, the differences in sterane carbon-number distributions between the Permian and Jurassic sources are not sufficient to make such a conclusion without supporting data. Hendrix et al. (1995) compared their samples with published data from Permian oil shales and Karamay oils. Although the illustration appears to support two genetic groups, the variance is no more than ~10%. This variance could also be due to differences in analytical methods and data-processing techniques. Ideally, comparisons of biomarker data should be made on samples that were analyzed using the same instrument and laboratory procedures.

Lower-Middle Jurassic lacustrine shales, Qaidam Basin, northwest China

The Qaidam Basin, to the southeast of the Tarim Basin and north of the Tibet Plateau, contains organic-rich freshwater, fluvial-lacustrine source rocks dating from the Early (Xiaomeigou Formation) and (Chaishiling and Dameigou formations) Middle Jurassic Period. These rocks have a wide range of generative potential, depending on depositional facies and thermal maturity (Figure 18.104) (Ritts et al., 1999). Outcrop samples have vitrinite reflectance in the range 0.77–1.05%, so

Nevertheless, these Jurassic coaly source rocks contain sufficient quantities of hydrogen-rich kerogen to generate highly paraffinic oils. Pyrolysis/gas chromatographic analysis of samples enriched in exinitic macerals (resins, cuticles, and spores) yields high-molecular-weight alkene/alkane doublets along with

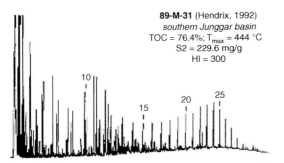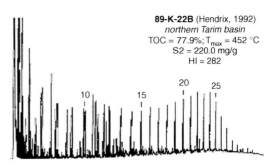

Figure 18.102. Examples of pyrolysis/gas chromatograms of oil-prone coals from the Junggar and Tarim basins (from Hendrix et al., 1995). Reprinted by permission of the AAPG, whose permission is required for further use.
HI, hydrogen index; TOC, total organic carbon.

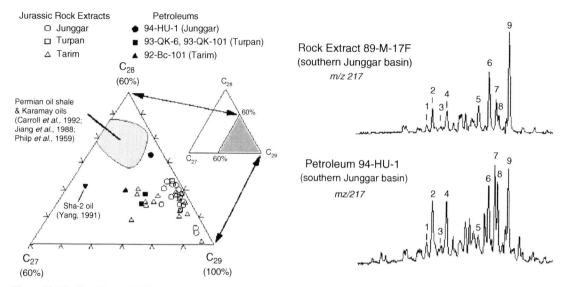

Figure 18.103. Hendrix *et al.* (1995) used the distribution of steranes by carbon number to differentiate oils from Jurassic coals from those generated from older Permian source rocks. Jurassic coals are characterized by abundant C_{29} $\alpha\alpha\alpha$-20R sterane (peak 8) compared with the C_{27} $\alpha\alpha\alpha$-20R (peak 1) and C_{28} $\alpha\alpha\alpha$-20R (peak 5) steranes. Differences between the rock extract and the produced oil are attributed to thermal maturation. The more mature oil is enriched in normal C_{29} steranes with 20S and $\alpha\beta\beta$-isomeric configurations (peaks 6–8) and C_{29} diasteranes (peaks 2, 4). Reprinted by permission of the AAPG, whose permission is required for further use.

they have expended some of their original generative potential. Carbon content is determined mostly by the rock's depositional setting. Lacustrine shales have 1.1–22.1 wt.% TOC, marginal lacustrine/fluvial mudrocks have 5.7–49.5 wt.% TOC, and coals have 50–86.6 wt.% TOC. Most samples have HI of <200 mg HC/g TOC, although a few lacustrine shales of Aalenian age have HI up to 672 mg HC/g TOC.

Biomarkers correlate oils produced in the Qaidam Basin to either Jurassic freshwater or Tertiary hypersaline lacustrine source rocks. The Jurassic oils and source rocks are enriched in C_{29} steranes and terrigenous diterpanes and depleted in gammacerane compared with the Tertiary oils and source rocks. The Qaidam Basin lacks the potentially interfering Paleozoic coaly source rocks that are common in the Xinjiang Uygur to the north. Therefore, correlation of Jurassic oils to freshwater source rocks is straightforward. Differences in diterpane distributions within the Jurassic samples can be related to their depositional setting and reflect variations in biogenic input. Deep-water, lacustrine shales contain abundant norisopimarane, abundant fichtelite isomers, variable concentrations of phylocladane and beyerane, with rimuane, isopimarane, and abietanes present. The nearshore lacustrine shales have more abundant diterpanes and high relative abundances of phylocladane, isopimarane, norisopimarane, and beyerane.

Middle Jurassic synrift coals, Seychelles microplate

Paleotectonic reconstructions place the Seychelles along a portion of Gondwana that was exposed to the Tethyan Gulf during the Mesozoic Era. Eastern Gondwana began to fragment, initially with the separation of Antarctica–Australia ~120 Ma, then with separation of Seychelles–India from Madagascar ~100 Ma, and finally the Seychelles separated from India ~85 Ma. The Seychelles microplate is now isolated completely and appears to be an unlikely location for petroleum systems. However, during the Middle Jurassic Period, synrift coals were deposited along the coastline (Plummer 1992; Plummer *et al.*, 1998). These liptinitic coals have HI ~290 mg HC/g TOC and are enriched in sulfur (~11 wt.%), consistent with deposition in a marginal marine setting, such as a coastal plain or abandoned delta.

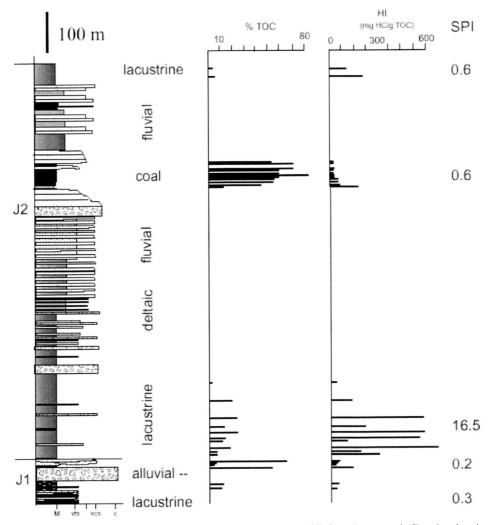

Figure 18.104. Stratigraphic column and geochemical log for Lower-Middle Jurassic strata at the Dameigou location, northeast Qaidam Basin (from Ritts et al., 1999). Reprinted by permission of the AAPG, whose permission is required for further use. HI, hydrogen index; SPI, source potential index (Demaison and Huizinga, 1994); TOC, total organic carbon.

Approximately 6–8 km of Jurassic-Cretaceous sediments were deposited on the Seychelles microplate. Samples from the few available exploration wells have not revealed prospective source rocks other than the Jurassic coals. However, tar balls stranded on beaches suggest that multiple oil-prone source rocks exist. Plummer (1996) described the chemistry of these tar balls, which he believed were indigenous to the Seychelles and not from tanker spillage. Based on biomarker distributions, he identified three genetic families. A Middle Jurassic carbonate source rock was inferred for one group of tar balls, which lacks oleanane and has high sulfur, norhopane, and tetracyclic terpanes. An Upper Cretaceous clastic source rock was inferred for another group based on minor amounts of oleanane and bicadinane and a predominance of diasteranes and tricyclic terpanes. A third group of tar balls from Coetivy Island contains minor amounts of oleanane, bicadinane, and botryoccocane, with high concentrations of pristane and tricyclic terpanes. An Upper

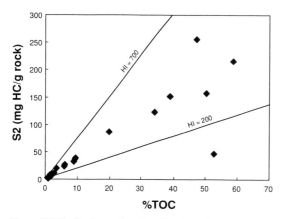

Figure 18.105. Total organic carbon (TOC) (wt.%) versus Rock-Eval S2 for samples from the Birkhead and equivalent formations from the western portion of the Eromanga Basin, southern Australia (data from Michaelsen and McKirdy, 1996). HI, hydrogen index.

Cretaceous–Lower Tertiary clastic deltaic source rock was proposed for this group.

Middle–Upper Jurassic Birkhead Formation, Eromanga Basin

Much of the petroleum in the Eromanga Basin is attributed to long-distance migration from Permian coals in the Cooper Basin (Figure 18.54). However, the Birkhead and Murta formations may be source rocks within the Eromanga Basin. Their contribution to the petroleum system remains controversial (Alexander et al., 1988c; Alexander et al., 1992b; Jenkins, 1989; Powell et al., 1989). The Birkhead Formation consists of siltstones, mudstones, and sandstones with thin (<0.3 m) lenticular coal seams that were deposited in a low-energy lake to swamp (Paton, 1986). The generative potential of Birkhead rocks is variable, with <1 to 58.7 wt.% TOC and HI averaging ~350 mg HC/g TOC (Figure 18.105) (Michaelsen and McKirdy, 1996). The kerogen consists of mixtures of oil-prone, resinitic coals and freshwater algae. HI >700 mg HC/g TOC has been measured in individual samples.

Middle Callovian Oxford Clay, southern England

The Oxford Clay is an organic-rich marine mudstone (Figure 18.106) that underlies much of southern England but is best known from a narrow band of outcrops running from Weymouth to the Yorkshire coast. The formation is noted for its preservation of marine fossils, including ammonites, fish, invertebrates, and marine reptiles (Smith, 1997).

Stratigraphic equivalents of the Oxford Clay occur from the Isle of Skye in Scotland to Boulogne, France. Biomarkers, particularly isorenieratane derivatives in these Callovian strata, suggest that a gradation in depositional conditions existed (Kenig et al., 1994; 2000). Organic-rich, laminated sediments from Skye are enriched in isorenieratane, suggesting deposition in a stratified water column with an anoxic photic zone. Pyrite is common, biomarkers for land plants are abundant, but sulfur-bound biomarkers occur in trace amounts. Oxford Clay of England (Peterborough Member) shows evidence of high-frequency alternations between oxic and anoxic deposition. These sediments contain mostly marine and bacterial biomarkers (Sinninghe Damsté and Schouten, 1997) and abundant polysulfide-bound macromolecules. In northern France, preservation of organic carbon is poor, and stratification of the water column was a rare event. Kenig et al. (2000a) hypothesized that a discharge of fresh water, possibly from the east coast of Greenland, into the Callovian seaway established a permanently stratified water column in Scotland, recurring stratification through central England, and rare stratification in northern France. The model is consistent with the narrow topography of the seaway in the north, broadening to the south. The distribution of pyrite and sulfur-bound macromolecules

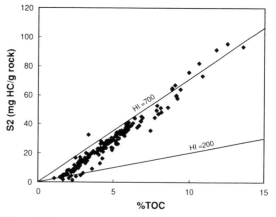

Figure 18.106. Total organic carbon (TOC) (wt.%) versus Rock-Eval S2 for samples of Oxford Clay, UK. HI, hydrogen index.

implies that iron was more abundant in Skye than in central England and that Skye was closer to the source of the freshwater discharge.

The Oxford Clay is not mature enough to have generated substantial oil, although it may have contributed minor input to seeps and subsurface accumulations in southern England (Selley and Stoneley, 1987). The Kimmeridge Bay Field produces from fractures in Middle Jurassic Cornbrash limestones and Oxford Clay. This oil was generated from Lower Toarcian shales but may have extracted hydrocarbons from the Oxford Clay (Evans *et al.*, 1998).

Jurassic Dingo Claystone, Northwest Shelf, Australia

The Dingo Claystone is the primary source rock for oils and gas in the Northwest Shelf, Australia (Figure 18.107) (Bishop, 1999a; Bradshaw *et al.*, 1997). During the Jurassic Period, this petroleum province was separated into a string of *en echelon*, faulted sub-basins. Subsidence of these sub-basins allowed for the deposition of three thick sequences of Dingo Claystone: lower (Sinemurian/Pliensbachian), middle (Toarcian-Bathonian), and upper (Callovian-Kimmeridgian) (Wulff, 1992). Restricted marine circulation favored anoxia and deposition of source rocks. Water depth during deposition (500–1000 m) and total thickness of the formation (maximum in the Barrow Sub-basin at 3–4 km) depended on the subsidence rate within the sub-basins. Secondary source rocks in the Northwest Shelf province include the Triassic Locker Shale (Scott, 1994) and possibly Lower Jurassic limestones at the base of the Lower Dingo Claystone, or the Cunaloo Member, a thin limestone unit that occurs at the base of the Locker Shale (George *et al.*, 1998b).

Source potential within the Dingo Claystone Formation is highly variable. Marine shales deposited during transgressions and periods of maximum subsidence range from ∼1 to 5 wt.% TOC, with HI of 100–400 mg HC/g TOC in the Upper Dingo, and from 0.2 to 3 wt.% TOC, with HI of 100–250 mg HC/g TOC for the Middle Dingo (Bradshaw *et al.*, 1994; Scott, 1994). The kerogen is type II/III, although terrigenous organic matter may dominate some intervals. Associated Jurassic coal measures may have significant oil and gas potential. Figure 18.108 shows geochemical data for crude oil from Pasco Field in the Carnarvon/Barrow Basin, Western Australia, that was generated from the Dingo source rock.

Source rocks in the Gulf of Mexico

Early studies postulated that Upper Tertiary shales were the source rocks for oil in the offshore Gulf of Mexico (Clark and Rouse, 1971). However, these strata are generally immature and have poor source potential for oil. The later discovery of intraslope anoxic depressions, such as the Orca Basin, suggested that Upper Tertiary source rocks were deposited in these basins (Dow and Pearson, 1975; Dow, 1984; Curtis, 1987). It was postulated that the source rocks were deposited where oxygen-minimum zones impinged on the slope or where topography favored a stratified water column. These proposed source rocks could be deep enough to be thermally mature. The temporal and spatial distribution of intraslope basins is not known, and their ability to form source beds is disputed (Bissada *et al.*, 1990). The modern Orca Basin does not conform rigorously to this model because stratification and anoxia resulted from a pycnocline through dissolution of a salt diapir, not from topographic restrictions on water circulation. Furthermore, organic matter in the Orca Basin is mostly transported and oxidized terrigenous debris with little generative potential. The contribution of intraslope source beds to the Gulf of Mexico is now believed to be insignificant, but such source rocks may be important in other Tertiary deltas, such as the Niger Delta.

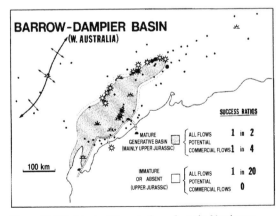

Figure 18.107. Much of the petroleum from the Northwest Shelf of Australia originates from the Upper Jurassic Dingo Formation (Demaison, 1984). Reprinted by permission of the AAPG, whose permission is required for further use.

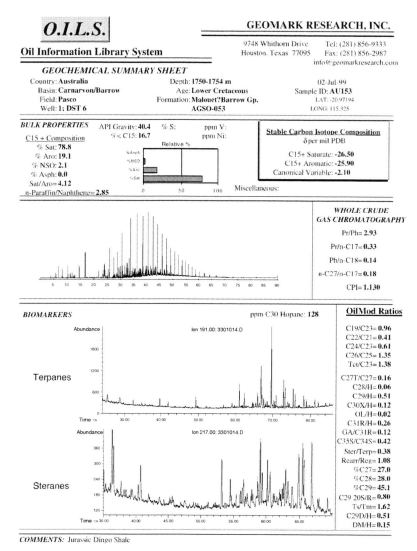

Figure 18.108. Geochemical data for crude oil from Pasco Field generated from the Jurassic Dingo Formation, Carnarvon/Barrow Basin, Western Australia (courtesy of GeoMark Research, Inc.).

In recent years, a consensus has emerged that the source rocks in the Gulf of Mexico are restricted to the Mesozoic Era and early Tertiary Period. This conclusion is based on geochemical analyses of surface seeps, produced oils, and gases (e.g. Walters and Cassa, 1985; Kennicutt et al., 1992; Wenger et al., 1994). Supporting evidence includes seismic facies and migration pathway interpretations (e.g. Hood et al., 2001), thermal history modeling (e.g. Bissada et al., 1990; Stover et al., 2001), and analyses of rock samples (e.g. Sassen, 1990; McDade et al., 1993; Wagner et al., 1994). The Gulf of Mexico is similar to other Mesozoic rift basins because sedimentation progressed through sequence of synrift lacustrine, evaporites, restricted marine, transgressive marine, and regressive marine source rocks. The distribution of oils generated from these source rocks is controlled largely by Tertiary overburden and associated thermal maturation. Oils from Triassic-Jurassic source rocks occur along the northern, outer rim of the Gulf Coast. To the south, the oils originated from increasingly younger

Mesozoic source rocks

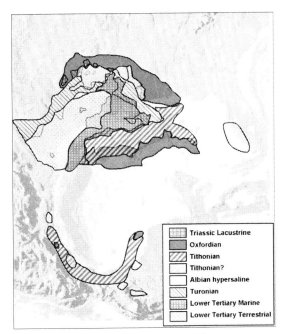

Figure 18.109. Distribution of oils generated from Mesozoic-Lower Tertiary source rocks in the Gulf of Mexico (modified using maps from Hood et al. (2001) and Guzmán-Vega et al. (2001)).

source rocks with increasing overburden. Older sources then come into play as the Tertiary deltaic overburden thins in deep water. Bands of genetically distinct oil groups correlate roughly with source maturity and, in many places, overlap (Figure 18.109).

Triassic lacustrine shales, Eagle Mills Formation, northern Gulf of Mexico

Oils of unusual composition occur in Jurassic reservoirs along the Mexia-Talco fault zone in Cass County of East Texas (Schumacher and Parker, 1988; 1990) and along northern bounding faults of the Gulf rim in Louisiana (James et al., 1993). These oils contain abundant methylalkanes and isoprenoid hydrocarbons up to C_{40}. Although postmature where sampled, the source rocks for these oils are believed to be carbonaceous lacustrine shales of the Triassic Eagle Mills Formation (James et al., 1993). The distribution of this facies is described poorly but is probably of limited areal extent.

Oxfordian Lower Smackover Formation

Oxfordian calcareous mudstones of the Lower Smackover Formation are the major source rock for oil and gas in a Jurassic petroleum system in the southeastern USA (see Table 4.3). In the Gulf of Mexico, the Smackover Formation overlies the Norphlet Formation, which consists of alluvial red beds and aeolian sandstones. The Norphlet Formation overlies the Louanne salt. The Smackover Formation is overlain by the Buckner anhydrite, so that oil generated from the lower Smackover Formation is confined largely to accumulations in either the Norphlet Formation or portions of the upper Smackover Formation where porosity was preserved.

The source facies of the lower Smackover Formation consists of brown to black laminated lime mudstones that were deposited under anoxic, and possibly hypersaline, conditions. These conditions arose in restricted lagoons or coastal subkha environments. The laminations are characteristic of algal and bacterial mats, and the kerogen is largely amorphous. Herbaceous and woody input was insignificant, except in the Manilla Embayment (Mancini, 2000). When mature, the source potential of the lower Smackover lime mudstones is poor. Oehler (1984) reported that the Lower Smackover Formation typically contained only 0.05–1.0 wt.% TOC, with an average of 0.48 wt.% and a maximum of 2.52 wt.% for mature samples from onshore wells. In a larger study involving 537 core samples, Sassen and Moore (1988) reported similar values for the Lower Smackover Formation. Average TOC is ~0.54 wt.%, with a maximum of 5.4 wt.%. HI is typically below 250 mg HC/g TOC. In most cases, the Smackover Formation is mature to postmature for oil generation, and these values represent residual source potential. Where immature (e.g. Arkansas outcrops), the Lower Smackover Formation has TOC >1 wt.% and HI >600 mg HC/g TOC. Klemme (1994) estimated the average pre-generative source potential for the Smackover Formation to be 2–4 wt.% TOC, with an average HI of 500–550 mg HC/g TOC and a thickness of 120 m. The thickness is probably overestimated because the source facies are confined largely to condensed sections (Mancini et al., 1993).

Crude oil generated from the Smackover Formation has geochemical characteristics typical of oil generated from anoxic marine carbonate source rock (Figure 18.110). For example, these high-sulfur oils have low pristane/phytane, low diasteranes, elevated C_{35} homohopanes, and high C_{29}/C_{30} hopanes.

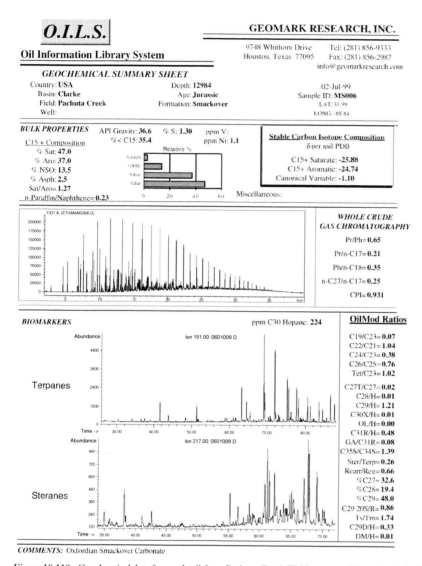

Figure 18.110. Geochemical data for crude oil from Pachuta Creek Field generated from the Oxfordian Smackover Formation, Gulf of Mexico (courtesy of GeoMark Research, Inc.).

Tithonian source rocks

Tithonian source rocks generated oils in the Mexican Coastal Basin and along the northern slope and rim of the US Gulf Coast (Wenger et al., 1994; Guzmán-Vega and Mello, 1999; Hood et al., 2001). These oils differ from younger oils in their elevated sulfur content, which varies considerably depending on source facies, thermal maturity, and extent of reservoir alteration. Tithonian and other Jurassic oils can be separated by their C_{26}-sterane distributions, where the Tithonian oils have higher 21-norcholestanes (Guzmán-Vega et al., 2001).

Pimienta-Tamabra(!) is a giant (~83.6×10^6 BOE), supercharged petroleum system in the southern Gulf of Mexico and adjacent onshore Mexico that accounts for most oil fields in the Tampico-Misantla, Salina, Chiapas-Tabasco, and Bay of Campeche areas (Morelos Garcia, 1996; Magoon et al., 2001). Although the entire Upper Jurassic section has potential source rock, the Tithonian Pimienta Formation was selected for the

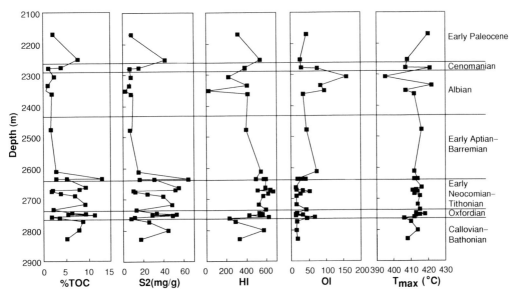

Figure 18.111. Total organic carbon (TOC) and Rock-Eval measurements of sidewall cores and cuttings from the Amerada Hess Garden Banks 754 well, Gulf of Mexico (data from D. Jarvie, Humble Instruments, Inc.). Strata with excellent source potential occur in the interval spanning the Middle Jurassic–Early Cretaceous Period.
HI, hydrogen index; OI, oxygen index.

petroleum system name because it is the richest, particularly in the Tampico and Tuxpan areas. These Upper Jurassic calcareous shales include thin-bedded, dark gray to black limestone, argillaceous limestone, and dark gray shale with at least 2 wt.% TOC and HI >500 mg HC/g TOC that were deposited under anoxic conditions. The source potential index (SPI) of the Upper Jurassic interval is ~15 metric tons HC/m^2.

Guzmán-Vega et al. (2001) found four oil families in the southern Gulf of Mexico area. Oil families 1 and 2 belong to the Pimienta-Tamabra(!) petroleum and originated from Oxfordian and Tithonian source rocks, respectively, while the two other families originated from Lower Cretaceous and Tertiary source rocks. Families 1 and 2 are closely related, but family 2 is widespread and contributed the vast majority of oil, including most oil in the Salina, Villahermosa, and Bay of Campeche areas. For example, oil from the Salina area has a biomarker composition very similar to extracts of Tithonian organic-rich marl from the Tuxpan area, more than 500 km to the northwest. Characteristics of the family 2 oils include low pristane/phytane and Ts/Tm (1), abundant extended hopanes, high C_{35}/C_{34} hopanes, abundant C_{29} versus C_{27} steranes, and presence of 17α-29,30-bisnorhopane. Family 2 has $δ^{13}C$ in the range −26.4 to −28.1‰, while family 1 ranges from −25.4 to −25.7‰.

Jarvie et al. (2002) reported samples of relatively immature Tithonian source rocks recovered from the Garden Banks 754 well (Figure 18.111), but full characterization of these samples has not been published.

Albian Sunniland Formation, south Florida
The Sunniland Formation of south Florida is a self-contained petroleum system within the updip portion of the Bahamas Basin (Mitchell-Tapping, 1987). Sandwiched between layers of anhydrite, the karsted middle member and high-porosity zones of the upper member received oil generated from organic-rich marls and carbonates primarily in the lower member (Palacas et al., 1984). The source facies has HI near 700 mg HC/g TOC (Figure 18.112) and was deposited after a marine transgression under euxinic conditions. Because there is little or no iron-bearing clay within the Sunniland Formation, bacterially generated sulfide was incorporated into the sedimentary organic matter during diagenesis. The resulting type IIS kerogen generated and expelled a high-sulfur (~3.2 wt.%) crude oil

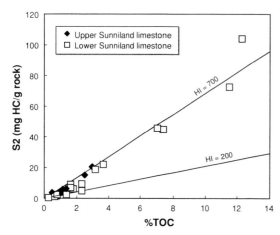

Figure 18.112. Total organic carbon (TOC) (wt.%) versus Rock-Eval S2 for core samples from the upper and lower Sunniland limestones (data from Palacas et al., 1984). HI, hydrogen index.

under relatively mild thermal conditions [20S/(S + R) <0.4].

Biomarker distributions in extracts from the organic-rich lower-member marls and limestones correlate well with produced oils. The C_{31}–C_{35} homohopane distributions are typical of anoxic marine conditions, decreasing in abundance with increasing carbon number (Figure 18.113). In contrast, extracts from some of the organic-lean-upper member limestones are enriched in homohopanes, particularly C_{32} and C_{34}. C_{35} homohopanes are enriched in source rocks deposited under hypersaline conditions. The unusual distribution that favors C_{32} or C_{34} homohopanes may arise from the partial oxidation of bacteriohopanetetrol or other bacterial lipid precursors, which is consistent with deposition under rapidly alternating oxic to dysoxic condition, likely associated with sabkha-type deposits. Figure 18.114 shows geochemical data for crude oil generated from the Sunniland Formation.

Oils from Lower Cretaceous hypersaline source rocks occur in Veracruz, Chiapas-Tabasco, and Sierra de Chiapas Sub-basins along the Mexican coast and in Guatemala (Guzmán-Vega et al., 2001). The sources of these oils are equivalent in facies, if not age, to the Sunniland Formation. The distribution of this hypersaline facies in coastal Mexico is erratic, and the beds are thin and laterally discontinuous. Samples from Sierra de Chiapas wells contained ~1.5 wt.% TOC, with HI ~450 mg HC/g TOC.

Upper Cretaceous marine shales and carbonates

Cretaceous source rocks were deposited primarily during marine transgressions in the Gulf of Mexico. Lower Cretaceous source rocks probably exist but are not well documented. Organic-rich shales within the Pearsall Group are a likely source for early Cretaceous oil in the East Texas Basin (Wescott and Hood, 1994). In contrast, several source rocks are known from the Late Cretaceous Period, especially Turonian time. Lower Cenomanian

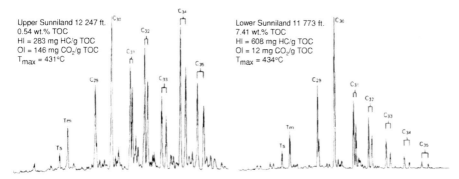

Figure 18.113. Reconstructed m/z 191 ion chromatograms of extracts from upper and lower Sunniland source rocks (from Palacas et al., 1984). The lower member has appreciably greater source potential, and the hopane distribution of the rock extract correlates with oil produced from Middle and Upper Sunniland reservoir rocks. The upper-member carbonate contains partially oxidized organic matter. The unusual enrichment in C_{32} and C_{34} hopanes may be due to partial oxidation or from contribution of unknown bacterial precursors. Reprinted by permission of the AAPG, whose permission is required for further use.
HI, hydrogen index; OI, oxygen index; TOC, total organic carbon.

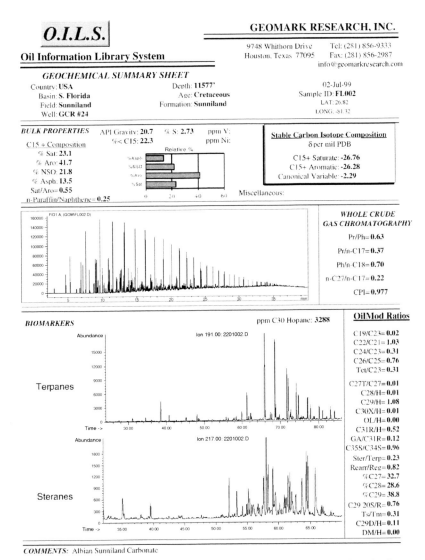

Figure 18.114. Geochemical data for crude oil from Sunniland Field generated from the Cretaceous Sunniland Formation, south Florida (courtesy of GeoMark Research, Inc.).

shales from the offshore Mississippi Canyon 84 well generally contain <1 wt.% TOC of gas-prone structured or recycled kerogen (Wagner *et al.*, 1994). However, a few samples are enriched in oil-prone amorphous organic matter (maximum 5.2 wt.% TOC, HI 423 mg HC/g TOC).

Cenomanian-Turonian marine shales of the Tuscaloosa and Eagleford formations occur across much of Louisiana and Texas. Figure 18.115 shows geochemical data for crude oil produced from the Upper Cretaceous Woodbine Formation in the East Texas Field that was generated from the Upper Cretaceous Eagleford Formation. The source potential and organic facies of these rocks are tightly coupled with sequence stratigraphy (Miranda and Walters, 1992). Most of the Tuscaloosa Shale is postmature in Louisiana, and its occurrence is inferred from the geochemistry of produced oil and gas (Koons *et al.*, 1974; Walters and

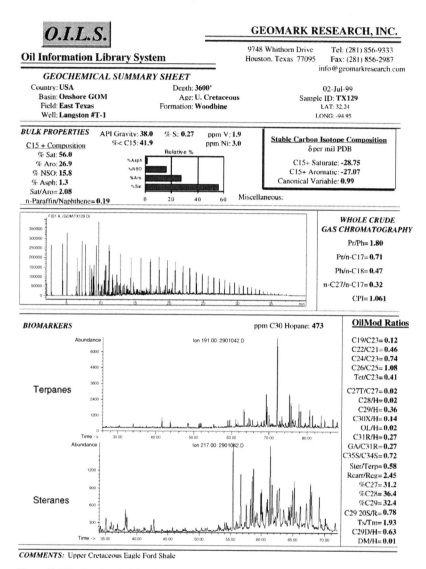

Figure 18.115. Geochemical data for crude oil from the East Texas Field generated from the Upper Cretaceous Eagleford Formation, onshore Gulf of Mexico (courtesy of GeoMark Research, Inc.).

Dusang, 1988; Wenger *et al.*, 1988). In south central Texas, Cenomanian-Turonian Eagleford calcareous shales have generally lower maturity, and original generative potential is determined more easily. Organic richness is highly variable (Grabowski, 1984; 1994). Core samples average ~4 wt.% TOC (maximum ~6 wt.%), with HI ranging from ~90 to 562 mg HC/g TOC. The Santonian Austin Chalk overlies the Eagleford Formation and has generative potential in the lowermost sections with an average of ~3.7 wt.% TOC and HI ~480 mg HC/g TOC. These calcareous source rocks were deposited in restricted basins that preserved both marine algal and terrigenous higher-plant input.

Lower Tertiary marine shales, Gulf of Mexico

Paleocene-Eocene prodelta and transgressional shales of the Midway, Wilcox, and Sparta formations are the source rocks for onshore and near offshore oils of the northern Gulf Coast (Walters and Dusang, 1988; Sassen, 1990; McDade *et al.*, 1993). These

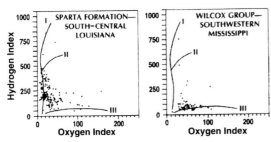

Figure 18.116. Rock-Eval oxygen index (mg CO_2/g total organic carbon (TOC)) versus hydrogen index (mg HC/g TOC) for Lower Tertiary source rocks from the US Gulf Coast (from Sassen, 1990). Reprinted by permission of the AAPG, whose permission is required for further use.

formations are dominantly non-source clastics with low (<1 wt.%) TOC. Some strata, however, contain shales with high generative potential, where TOC ranges from 3 to 6 wt.% and HI averages ~400 mg HC/g TOC (Figure 18.116). The kerogens in these intervals are mixtures of marine algal and terrigenous higher-plant material. In a few cases, the kerogen is amorphous, yielding HI >600 mg HC/g TOC. In general, the Lower Tertiary strata become more gas-prone updip of the depocenters toward Texas. The transition from oil-prone to gas-prone reflects increased contribution of terrigenous plant material from Texas river systems. Lower Tertiary oils and source-rock extracts are characterized by elevated oleanane and bisnorhopane.

Note: The major K-T meteorite impact that resulted in extinction of the dinosaurs occurred at Chicxulub on the Yucatan carbonate platform. This event has been implicated in the formation of some reservoir and seal rocks (Grajales-Nishimura et al., 2000), but it may also have produced carbonate facies with high petroleum-generative potential. The impact created a large, intracratonic depression with restricted water circulation, especially during the Oligocene–Miocene Epoch. These conditions may have been ideal for deposition of organic-rich source beds in a saline-hypersaline setting. The depression finally filled with sediments during the Late Miocene Epoch (Galloway et al., 2000). Future drilling in the Chicxulub Crater may determine whether such source rocks occur.

Jurassic Kingak-Blankenship Interval, Alaska

The widespread organic-rich Kingak Formation (see Figure 18.28) on the North Slope of Alaska contains a mixture of marine and terrigenous organic matter deposited in a siliciclastic setting with little or no carbonate (Magoon and Claypool, 1984). The lowermost 15 m of the Kingak Shale in the Kalubik-1 well averages 5 wt.% TOC, with HI ~400 mg HC/g TOC (Masterson, 2001). The Blankenship Member of the Otuk Formation is the southern, distal equivalent of the Kingak Formation (Mull et al., 1982) and was combined with the Kingak by Bird (1994) for his treatment of the Ellesmerian(!) petroleum system.

The Kingak Formation was a significant contributor of oil to the Prudhoe Bay Field based on oil – source rock correlation (Seifert et al., 1980; Claypool and Magoon, 1985; Premuzic et al., 1986; Sedivy et al., 1987; Masterson, 2001). Seifert et al. (1980) showed that Kingak oil (Kavearak 32–25 well, Milne Point Field) (Figure 18.117) originated from the Kingak Shale partly based on low tricyclic terpanes (see Figure 13.75). This 35°API gravity oil has high diasteranes/steranes (2.34), low sulfur (0.2 wt.%), and C_{22}/C_{21} and C_{24}/C_{23} tricyclic terpane ratios (0.41 and 0.69, respectively) consistent with a clay-rich source rock (see Figure 13.76). Alpine Field, located just to the east of the National Petroleum Reserve in Alaska (NPRA), is the largest accumulation of predominantly Kingak oil, which averages nearly 40°API gravity (Masterson, 2001).

Kingak Shale core extracts are geochemically distinct from Shublik Formation extracts. For example, Kingak extracts have lower dibenzothiophene/ phenanthrene, C_{29}/C_{30} hopane, C_{23} tricyclic terpane/ hopane, and C_{35}/C_{34} hopane than Shublik extracts, consistent with a more siliciclastic, terrigenous source rock (Masterson, 2001).

Kingak-Blankenship oils from the Kavearak Point and Alpine fields differ substantially from Shublik-Otuk oils from Kuparuk Field (Table 18.11). Table 18.11 also compares Kingak-Blankenship oils and those generated from Cretaceous pebble–GRZ–Torok source rock. For example, the Kingak-Blankenship oils are similar to pebble shale–GRZ–Torok oil in many respects because both originated from similar marine siliciclastic source rocks. However, their stable carbon isotope and C_{26} nordiacholestane ratios (NDR) differ. Low NDR

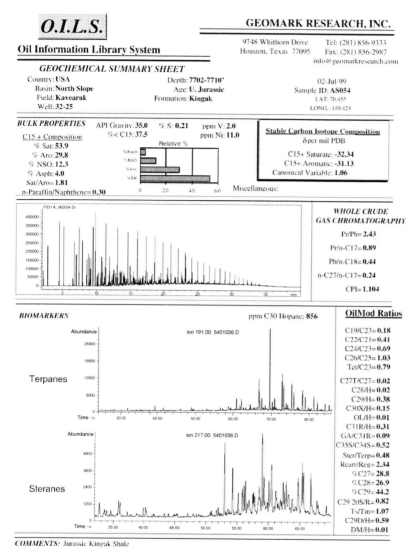

Figure 18.117. Geochemical data for crude oil from Kavearak Field generated from the Jurassic Kingak Formation, North Slope, Alaska (courtesy of GeoMark Research, Inc.).

for Kingak–Blankenship and Shublik–Otuk oils (0.14) and high NDR for pebble shale–GRZ–Torok oils (0.48) is consistent with Jurassic- and Cretaceous-age source rocks, respectively (see Figure 13.55).

Kimmeridigian transgressive shales

A period of maximum trangression occurred during the Kimmeridgian Stage (Haq *et al.*, 1987), resulting in several source rocks of exceptional quality and scale. During this time, the principal source rocks for oil in the North Sea (Kimmeridge Clay Formation), offshore eastern Canada (Egret Member), western Siberia (Bazhenov Formation), and others were deposited.

Kimmeridge Clay Formation, North Sea

The geohistory of the North Sea oil province is dominated by a Late Jurassic episode of crustal extension and accelerated basin subsidence leading to the

Table 18.11. *Average and standard deviation for selected geochemical data from North Slope oils generated from Jurassic Kingak, Triassic Shublik, and Cretaceous gamma-ray zone (GRZ) source rocks. (data from Masterson, 2001)*

Oils by source[1]	S (wt.%)	°API	$\delta^{13}C$ (‰) Saturated	Aromatic	DBT/P	C_{29}/C_{30} Hopane	C_{23} Tri/C_{30} Hopane	C_{35}/C_{34} Hopane	NDR
Kingak	0.3	37	−31.4	−30.4	0.17	0.43	0.12	0.6	0.14
±	0.1	2	0.9	0.4	0.05	0.02	0.04	0.1	0.01
Shublik	1.6	23	−30.0	−29.6	0.67	0.86	0.62	1.2	0.14
±	0.1	1.5	0.1	0.1	0.07	0.3	0.06	0.1	0.01
GRZ	0.2	37	−28.8	−27.9	0.11	0.42	0.11	0.8	0.48
±	0.1	1	0.3	0.5	0.04	0.06	0.04	0.3	0.06

[1] DBT/P, dibenzothiophene/phenanthrene; NDR, C_{26}-nordiacholestane ratio.
Kingak, Kavearak Point (1) and Alpine (3) oils generated from Kingak–Blankenship source rock. Shublik, Kuparuk Field (23) oils generated from Shublik–Otuk source rock. GRZ, Umiat-4, Kuukpik-3, and Tarn-2 oils generated from pebble–GRZ–Torok source rock. Kuukpik-3 oil was excluded from the calculation for sulfur and API gravity because it is slightly biodegraded.

formation of the Central Graben, the Viking Graben, and the Moray Firth rift systems (Ziegler, 1988). Kimmeridge Clay shales were deposited within the rift, which became mature along the rift axis following the deposition of Cretaceous-Tertiary overburden (Johnson and Fisher, 1998). The Kimmeridge Clay Formation is the primary source rock for the major oil accumulations (see Table 4.3). Other sources in the province include the Lower Jurassic Dunlin shales and Middle Jurassic Brent coals, which were deposited in a shallow-water terrigenous setting and tend to be gas-prone. Upper Jurassic Heather shales were deposited immediately before the Kimmeridge Clay and tend to be leaner in potential. The Kimmeridge Clay Formation contains mainly marine shales deposited in saline to transient hypersaline conditions. Source deposition occurred in a stratified sea when cold polar waters flowed over warm saline bottom waters. Source deposition ceased when slight climatic changes overturned the water column (Miller, 1990).

The Kimmeridge Clay Formation is also known as the Mandal or Draupne Formation in the North Sea oil province and is time-equivalent to the Borglum Formation in the Norwegian–Danish Basin. Many regional studies have attempted to map the source potential of the Kimmeridge Clay Formation (Barnard and Cooper, 1981; Barnard *et al.*, 1981; Goff, 1983; Cornford, 1984; Thomas *et al.*, 1985; Harris and Fowler, 1987; Cornford, 1994; Cooper and Barnard, 1995; Cooper *et al.*, 1995; Isaksen and Ledje, 2001). This work is difficult because much of the mature, oil-generative Kimmeridge Clay occurs in the deep grabens that have not been sampled adequately. Furthermore, the formation contains highly variable distributions of organic matter both vertically and regionally. Comparing biomarker, isotopic, and petrographic data, Cooper *et al.* (1995) identified at least eight different kerogen types in the Kimmeridge Clay Formation. Some of these differences include liptinitic versus vitrinitic land-plant macerals, variations in marine algal type I/II kerogens, and varying quantities of sulfur incorporation that, in some cases, result in high-sulfur type IIS kerogens.

Because of this facies variability as well as the sampling limitations, it is difficult to describe the overall source potential of the Kimmeridge Clay Formation in simple terms. For modeling purposes, Ungerer *et al.* (1984; 1985) selected 7 wt.% TOC with an HI of 310 mg HC/g TOC to represent average Kimmeridge Clay Formation source rock. Miller (1990) proposed average values of 4–5 wt.% TOC, with HI of 350–450 mg HC/g TOC, and with a thickness of ~150 m. In reality, TOC ranges from <0.5 to >40 wt.% and HI ranges from <50 to >800 mg HC/g TOC for different facies. Within a single transect, Herbin *et al.* (1993) identified short periods of cyclic deposition with periodicities of ~25 000

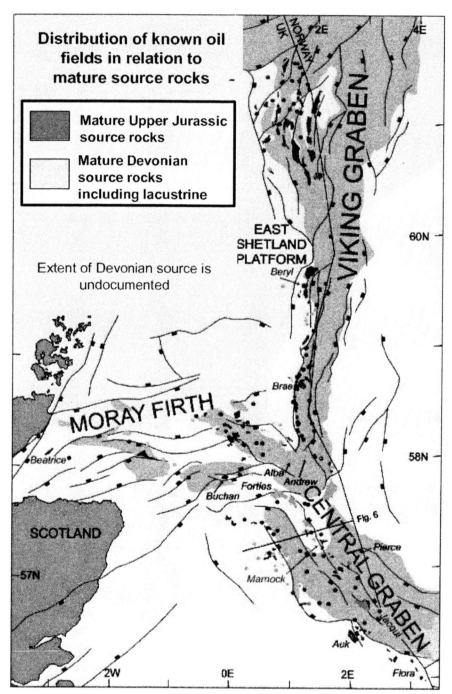

Figure 18.118. Principal reservoirs for oil and gas fields in the North Sea oil province (Brooks *et al.*, 2001). The Kimmeridge Clay Formation was deposited over much of the North Sea. It is mature only along the axis of the rift system. Reprinted by permission of the AAPG, whose permission is required for further use.

and 280 000 years, resulting in alternating shales with high and low source potential. While the general composition of marine zoo- and phytoplanktonic input was fairly consistent, quantities varied significantly, and differences in generative potential arose due to fluctuations in oxygen levels as the cyclic deposition passed from oxic to anoxic conditions.

Most crude oils in the North Sea originate from thermally mature Kimmeridge Clay Formation located in the deep Viking and Central grabens (Figure 18.118).

However, the compositions of these fluids can be highly variable, reflecting differences in source facies, thermal maturation, and reservoir alteration. For example, the Kimmeridge Clay Formation contains varying amounts of type IIS, type II, and mixed type II/III kerogens. In addition to Kimmeridgian oil, many North Sea reservoirs also received migrated oil from the overlying Heather (type II/III) and underlying coals (type III). North Sea oil generated from Kimmeridge Clay marine shales is shown in Figure 18.119. The bulk

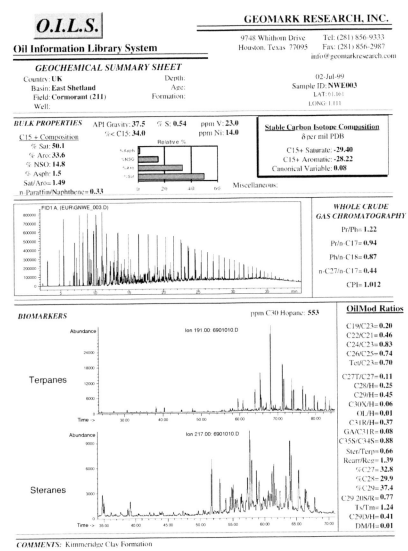

Figure 18.119. Geochemical data for crude oil from Cormorant Field generated from the Kimmeridge Clay Formation, East Shetland Basin, UK (courtesy of GeoMark Research, Inc.).

chemistry and biomarker distributions are typical of oils generated from Mesozoic open-marine shale source rocks deposited under anoxic bottom water. Sulfur contents are moderate (~0.5 wt.%), indicating diagenesis of organic matter within the sulfate reduction zone, but most of the sulfide was tied up with iron. High input of algal marine organic matter is inferred by the high relative abundance of C_{30} n-propylcholestanes and the high sterane/hopane ratio. High diasteranes/steranes is consistent with clastic deposition. Bisnorhopane is relatively abundant, a common feature in many North Sea oils. There is a slight enrichment in the C_{35} homohopanes, which indicates anoxic deposition.

Egret Member, offshore eastern Canada

Calcareous marine marls of Egret Member (Rankin Formation) are the primary source rocks for oil in the Jeanne d'Arc Basin (e.g. Hibernia, Ben Nevis, and Terra Nova fields) located ~100 km off the east coast of

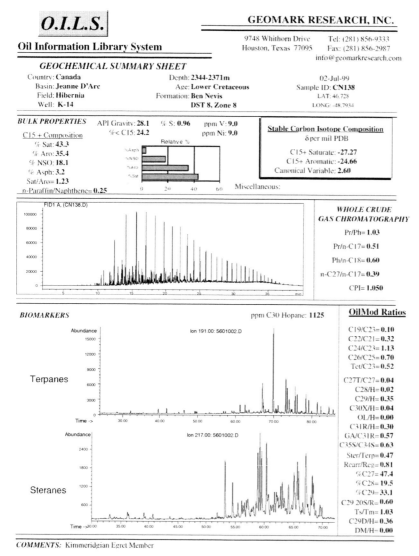

Figure 18.120. Geochemical data for crude oil from the Hibernia K-14 well generated from the Kimmeridgian Egret Member of the Rankin Formation, Jeanne d'Arc Basin, offshore eastern Canada (courtesy of GeoMark Research, Inc.).

Newfoundland, Canada (von der Dick, 1989; von der Dick et al., 1989). The Egret Shale was syndepositional with the Kimmeridge Clay Formation in essentially the same environment, i.e. the deep graben of a failed Mesozoic rift basin. Restricted water circulation allowed development of anoxic bottom waters with cyclic changes in salinity. Deposition of the Egret source facies ceased in the Late Kimmeridgian Age, with complete flooding of the basin, resulting in the disruption of euxinic conditions and increased input of coarse clastics and terrigenous higher-plant debris.

Because the depositional conditions of the Kimmeridge Clay and the Egret Member were nearly identical, it is not surprising that they exhibit similar organic facies variability (Bateman, 1995; Huang et al., 1994). Huang et al. (1996) recognized three different orders of cyclic deposition and noted that strata with high TOC correspond to low sedimentation rates. They concluded that the organic-rich layers were deposited during warmer periods with high sea-level stands.

The Egret Member ranges from 55 to >200 m thick. The source intervals average ~4.6 wt.% TOC, with HI for immature samples in the range ~500–700 mg HC/g TOC (von der Dick, 1989; Huang et al., 1994; Bateman, 1995; Fowler and McAlpine, 1995). The kerogen is mainly amorphous and consists of marine phytoplanktonic organic matter (mainly dinoflagellates) that was heavily reworked by bacteria. Laminated dark brown shales deposited under anoxic, highstand, and warm climatic conditions contain higher TOC and more oil-prone organic matter (type I–II) than other lithofacies within the Egret Member (Bateman, 1995).

A sample from the Hibernia Field in the Jeanne d'Arc Basin, offshore eastern Canada, is typical of crude oils generated from the Egret Member (Figure 18.120).

Bazhenov shales, Western Siberian Basin

Upper Jurassic Bazhenov Formation shales are the major source rock for crude oil (>80%) in the Western Siberian Basin (see Table 4.3 (Peters et al., 1994b; Kontorovich et al., 1998a). They were deposited in a deep-water marine basin with high productivity, slow sedimentation, and anoxic bottom water that led to enhanced preservation and incorporation of bacterial sulfide. The Bazhenov Formation averages ~5 wt.% TOC; however, richness varies considerably, from ~2–3 wt.% TOC along the margins to >10 wt.% TOC in the deep basin (see Figure 1.5 (Kontorovich, 1984). Thermally mature Bazhenov shales have residual HI of ~350–450 mg HC/g TOC, indicating that these source rocks had significantly higher initial HI (>600 mg HC/g TOC). Some organic facies may be exceptionally rich. One anomalous sample from the Ugut 31 well contains 40.2 wt.% TOC, with HI of 840 mg HC/g TOC (Peters et al., 1994b). The HI for immature Bazhenov source rock near the Salym Field is ~550 mg HC/g TOC (Klemme, 1994). Table 4.7 and the related discussion (pp. 97–100) show how original TOC can be calculated for highly mature samples using mass-balance assumptions.

Bazenhov source rocks contain mostly fluorescent amorphous kerogen (type II) with significant amounts of solid bitumen. Anoxic bottom waters and the accompanying bacterial sulfate reduction resulted in near-complete microbial recycling and incorporation of sulfur into the kerogen (2–3 wt.% sulfur on average, >5% in basinal facies). Biomarker distributions reflect oil generated from the sulfur-rich facies with CPI <1, pristane/phytane <1.5 and typically <1.0, and well-developed homohopanes (Figure 18.121).

Cretaceous oceanic anoxic events

The Cretaceous Period was characterized by unusually high production of oceanic crust at spreading centers and by the eruption of flood basalts, resulting in high rates of volcanic CO_2 outgassing and increased global warming. Wide climatic swings occurred, from the extreme greenhouse conditions during the Turonian Age, when global surface temperatures were more than $10°C$ warmer than today, to the cool climates of the Aptian and Masstrichtian ages, when polar icecaps may have existed. Changes in climate, sea level, and ocean circulation influenced the evolution and extinction of both terrigenous and marine biota and the deposition of organic-rich marine sediments.

During certain times, Earth's oceans became stratified, and anoxic bottom-water conditions were widespread. OAEs occurred during the early Toarcian Age (see above) and throughout the Cretaceous Period (Figure 18.122). These events are well established at the Deep Sea Drilling Program (DSDP) and Ocean

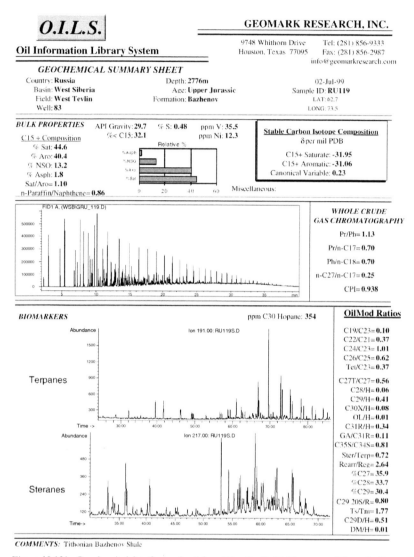

Figure 18.121. Geochemical data for crude oil from West Tevlin Field generated from the Upper Jurassic Bazhenov Formation, West Siberia, Russia (courtesy of GeoMark Research, Inc.).

Drilling Program (ODP) sites that record the syngenetic deposition of organic-rich sediments in all ocean basins at various paleodepths and latitudes (Jenkyns, 1980). When OAEs were first proposed (Schlanger and Jenkyns, 1976; Scholle and Arthur, 1976), considerable debate emerged over questions of productivity versus preservation and whether OAEs were truly global. Cretaceous deep-sea cores indicate that anoxic conditions fluctuated in intensity and might be caused by conditions other than global anoxia (Waples, 1983). Isotopic and geochemical evidence indicate that at least five OAEs occurred during the Cretaceous Period in 121–119 Ma (OAE 1a in the early Aptian Age), 112–109 Ma (OAE 1b in the early Albian Age), 102–98 Ma (OAE 1c–1d in the Late Albian Age), and 93–94 Ma (OAE 2 at the Cenomanian–Turonian boundary). Whether the early Toarcian OAE was global or regional is still debated. OAEs likely occurred before the Toarcian event, but there is little evidence for widespread ocean anoxia in Mesozoic DSDP/ODP cores.

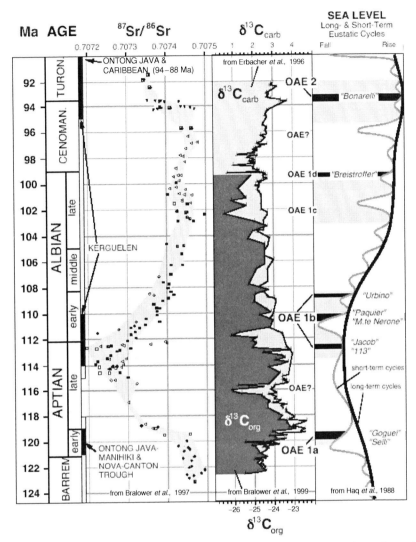

Figure 18.122. The Cretaceous oceanic anoxic events (OAEs) as recorded in the carbon and strontium isotopic record at the Blake Nose, North Atlantic Deep Sea Drilling Program (DSDP) Site 390 in relation to changes in global sea level and the emplacement of large igneous, submarine provinces (from Leckie et al., 2002). © Copyright 2002, American Geophysical Union. Reproduced/modified by permission of the American Geophysical Union.

The causes of OAEs are unclear, but they are related to warm, global temperatures that shut down ocean circulation. Throughout much of the Tertiary Period, and at present, deep ocean waters have been cold and well oxygenated. These conditions arise as sea ice forms at the poles and cold brines sink to the ocean floors. These deep brines circulate toward the Equator from the North and South Atlantic. Upwelling of these deep, nutrient-rich waters enhances marine productivity.

The scenario for creation of the Cretaceous OAEs requires greenhouse conditions, probably resulting from the emission of volcanic CO_2 from flood basalts and the opening of the northern Atlantic Ocean. Warming was probably most pronounced at high latitudes, such that the surface waters at the poles would no longer freeze. Surface waters at low latitudes would become more saline due to enhanced evaporation. Because of the decreased latitudinal salinity gradient, polar waters

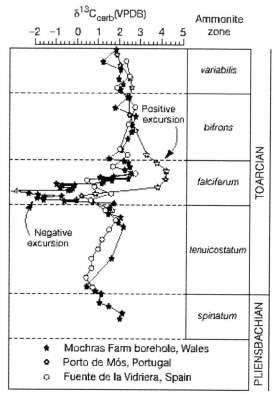

Figure 18.123. Carbonate carbon-isotope data through selected European Toarcian sections (from Hesselbo et al., 2000).

would not sink readily and ocean circulation would decrease dramatically.

Mesozoic OAEs are characterized by a strong isotopic shift in organic-lean marine carbonates toward more positive $\delta^{13}C$ (Figure 18.123). This is attributed to high biological productivity and rapid sequestration of ^{12}C-rich carbon in organic-rich sediments, which shifted the remaining global carbon pool toward ^{13}C enrichment (Arthur et al., 1987; 1988). Diagenesis was once viewed as the cause of negative isotopic excursions in marine carbonates (Jenkyns and Clayton, 1997). Paradoxically, a sharp negative isotopic excursion precedes several prominent OAEs (Figure 18.124) (Hesselbo et al., 2000; Jahren et al., 2001). This negative spike is attributed to massive destabilization of methane hydrates, which increased temperatures rapidly and triggered the OAE. Methane hydrate destabilization is hypothesized more by the elimination of other mechanisms than by direct evidence. For example, oxidation of biomass is insufficient to account for the OAE. Up to three times the total biomass would be needed to account for the magnitude of the negative excursion. Weathering of organic-rich sediments and coals is not fast enough to match the abruptness of the isotopic excursion. Furthermore, although intense volcanism characterized the Cretaceous Period, the emitted CO_2 is relatively enriched in ^{13}C ($\delta^{13}C \sim -6‰$). Warming ocean temperatures could result in destabilization of methane hydrates. The released methane could have contributed further to global warming, destabilizing more methane hydrates. This mechanism can account for the rapidity and magnitude of the global negative excursions that precede OAEs.

Warming oceans, possibly up to 10–15°C hotter than today, led to OAEs by several mechanisms (Jenkyns, 1980; Jacobs and Linberg, 1998; Larson and Erba, 1999). Sluggish water circulation results in lower oxygen flux into the deep ocean. The warmer waters can carry ~15–20% less dissolved oxygen. More importantly, bacterial metabolic oxygen demand doubles for every 5°C. Benthic organisms would also have a higher oxygen demand. Collectively, warm waters during Cretaceous greenhouse conditions resulted in a four- to eightfold increase in oxygen demand at the sediment–water interface. Oxidation of organic matter would consume much of the dissolved oxygen, leading to anoxic bottom waters. Euxinic conditions may have developed upward through the water column into the photic zone (Sinninghe Damsté and Köster, 1998; Schouten et al., 2000b).

Anoxic conditions alone may not account for the deposition of organic-rich shales. Considerable debate exists on the level of biotic productivity during greenhouse conditions. The reduced temperature and density gradients between surface and bottom waters during OAEs would facilitate upwelling, resulting in enhanced vertical mixing, warmer waters, and increased productivity (Erbacher et al., 1996; Erbacher et al., 2001; Wilson et al., 1999; Wilson and Morris, 2001). On the other hand, average wind speeds were probably less than today, resulting in less active surface currents, which are needed to induce upwelling. Sea level was considerably higher during greenhouse conditions, allowing vast areas of the continental shelf to be flooded and providing more surface area for marine productivity. Nutrient levels may have increased by leaching of these flooded lowlands due to an accelerated hydrological cycle. Conversely, continental weathering could be reduced due to decreased

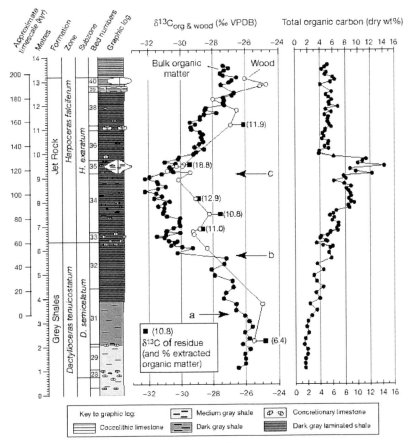

Figure 18.124. Stable carbon isotope composition (δ^{13}C) of bulk organic matter and wood through the lower Whitby Mudstone, Hawsker Bottoms, Yorkshire. The negative δ^{13}C excursion begins at level a, decreases abruptly at level b, and terminates at level c (from Hesselbo et al., 2000).

topographic relief, resulting in decreased nutrient flow into the marine environment.

Oceanic biotas were affected strongly by these events. In addition to extinctions of benthic and planktonic species, OAEs may have dramatically influenced the populations of marine bacteria and archaea. Kuypers et al. (2001) suggested that marine, non-thermophilic archaea contributed up to 80 wt.% of the TOC within an Albian OAE 1b deposit.

Mesozoic petroleum systems of Argentina

Petroleum in the sedimentary basins of Argentina originated from many, sometimes stacked, Mesozoic source rocks (Villar et al., 1998). The earliest source rocks are Upper Triassic–Pliensbachian pre- and synrift lacustrine sequences that generated early-mature, high-wax oils in the Cuyo Basin (Villar et al., 1998) and possibly the Neuquén Basin (Gomez-Perez et al., 2001). Overlying these rocks are four Pliensbachian–Eocene transgressive-regressive sedimentary cycles (Urien and Zambrano, 1994; Urien et al., 1995; Vergani et al., 1995). Two of these cycles contain transgressive marine shales that are the major source rocks for produced hydrocarbons in the Neuquén Basin. The first cycle (Cuyo) began in the Late Pliensbachian–Early Toarcian Age with deposition of the basinal black shales of the Los Molles Formation. These shales may contain >5 wt.% TOC, mostly terrigenous in origin, grading upward into organic-lean facies (Urien and Zambrano, 1994). Most

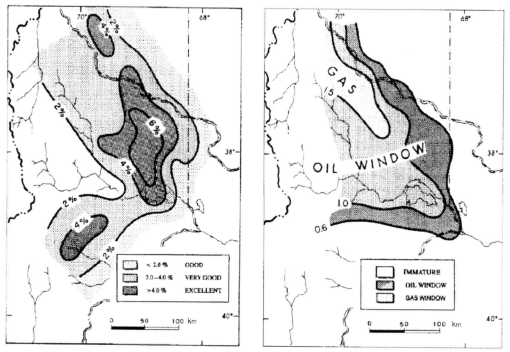

Figure 18.125. Distribution of average total organic carbon and vitrinite reflectance for the Vaca Muerta and Quintuco formations (from Urien and Zambrano, 1994). Reprinted by permission of the AAPG, whose permission is required for further use.

of these shales are now postmature, and the Los Molles is considered to be a minor source of produced liquids. The third transgressive-regressive cycle (Andic) contains the main source rocks in the Neuquén Basin: the Lower Tithonian basal shales of the Vaca Muerta Formation and the Upper Valanginian-Barremian Agrio Formation (Superior and Inferior members). These source rocks are responsible for most of the petroleum produced in Argentina and are discussed in detail below. Source rocks of equivalent age but with less generative potential were deposited in the Austral Basin. Lower Cretaceous fresh to brackish lacustrine shales are source rocks in the San Jorge Basin, and shallow-water algal carbonates are source rocks in the Northwest Late Cretaceous basins.

Lower Cretaceous transgressive shales of the Neuquén Basin, Argentina

The Neuquén Basin contains many shale source rocks that were deposited during the Andic transgressive-regressive cycle on the slope and shelf under euxinic marine conditions. Basal facies within the Vaca Muerta and the Agrio formations corresponding to the maximum flooding surface are the most significant, but many thinner shales (e.g. Quintuco and Loma Montosa formations) were deposited during recurrent transgressions.

The mostly Tithonian Vaca Muerta Formation consists of thermally mature black shales, marls, and lime mudstones and ranges from 30 to 1200 m thick. Organic-rich strata typically have 2–3 wt.% TOC, with maxima >8 wt.%, and with HI from 200 to 675 mg HC/g TOC (Figure 18.125) (Urien and Zambrano, 1994). The richest strata are in a condensed section at the base of the Vaca Muerta that was deposited on the Early Tithonian maximum flooding surface (Lara *et al.*, 1996). TOC generally decreases in the overlying strata as the sequence shallows upward, but some organic-rich intervals correspond to minor changes in sea level. Lateral variations in organic richness reflect changes in depositional settings. The distal facies consists of stacked organic-rich condensed sequences. The shelf-source facies is restricted to a single condensed section, resulting in a relatively thin organic-rich section.

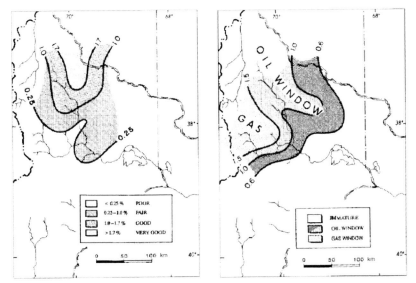

Figure 18.126. Distribution of average total organic carbon and vitrinite reflectance for the Agrio Formation (from Urien and Zambrano, 1994). Compared with the Vaca Muerta, the richest area has shifted toward the northwest, which is related to the regressive stage of the Andic sedimentary cycle. Reprinted by permission of the AAPG, whose permission is required for further use.

After the transgression, the Andic cycle entered into a regressive phase, but two source rocks occur as members within the Upper Valanginian-Barremian Agrio Formation (Figure 18.126). The non-marine Avilé Member, deposited during a Hauterivian lowstand, separates the lower Inferior and the upper Superior member source rocks. Like the Vaca Muerta, the Agrio source rocks were deposited under restricted marine conditions, with organic-rich strata occurring in condensed sections during periods of transgression. The Agrio Inferior formation contains up to 5 wt.% TOC, with HI up to 700 mg HC/g TOC (Cruz et al., 1998). Source potential is greatest at the base and generally decreases upward in the Agrio Inferior formation, while the Agrio Superior formation typically has lower overall source potential.

Figure 18.127 shows geochemical data for crude oil from the Charco Bayo Field in the Neuquén Basin, Argentina, that was generated from the Vaca Muerta source rock.

Neocomian Murta Formation, Eromanga Basin

The Neocomian Murta Formation and the Middle-Upper Jurassic Birkhead Formation are potential sources of petroleum in the Eromanga Basin, southern Australia. The Murta Formation is composed mostly of thinly interbedded siltstones and shales deposited in lacustrine and fluvial settings. Marine transgressions may have influenced the upper portion of the Murta Formation (Gorter, 1994). Michaelsen and McKirdy (1989) and Powell et al. (1989) examined the source potential of the Murta Formation. Most of the core samples have HI <200 mg HC/g TOC and mixtures of gas-prone type III and inert type kerogen. A smaller subset of the core samples contain varying quantities of Botyroccocus, telalginite, and land-plant spores and pollen (up to 60%) plus non-generative macerals. TOC varies from 0.8 to 2.6 wt.%, with HI ~175–540 mg HC/g TOC.

Cretaceous pebble shale–gamma-ray zone–Torok Interval, Alaska

Three organic-rich Cretaceous marine shales from the North Slope of Alaska, including the "pebble shale" (informal name), the gamma-ray or highly radioactive zone (GRZ or HRZ), and the lower part of the Torok Formation (Figure 18.128), are commonly considered to be a single source rock (e.g. Claypool and Magoon, 1985) because of similar kerogen composition (Magoon and Bird, 1988). These marine shales

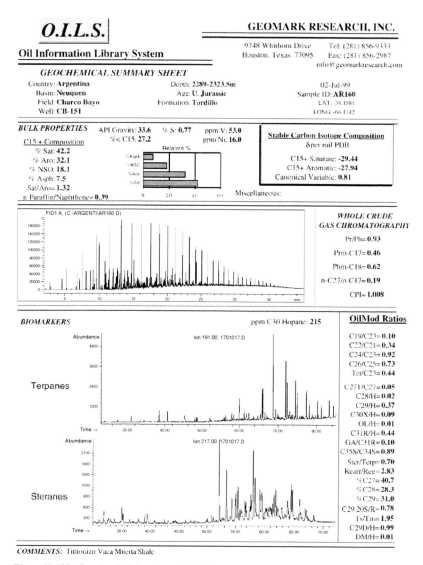

Figure 18.127. Geochemical data for crude oil from the Charco Bayo Field generated from the Tithonian Vaca Muerta Formation, Neuquén Basin, Argentina (courtesy of GeoMark Research, Inc.).

are important possible source rocks in the Arctic National Wildlife Refuge (ANWR) (e.g. Jago and Point Thomson-Flaxman Island areas), although their contribution to the giant accumulation at Prudhoe Bay is less than that of Triassic-Jurassic sources (Seifert et al., 1980; Lillis et al., 1999; Magoon et al., 1999). Masterson (2001) estimated that oil in the Prudhoe Bay area consists of 59% Shublik (Triassic), 28% GRZ, and 13% Kingak contributions. Carmen and Hardwich (1983) deduced that the most likely Cretaceous source rocks contributing to Prudhoe Bay accumulations consisted of the condensed shales of the GRZ within the pebble shale. To the east of the National Petroleum Reserve in Alaska (NPRA), the GRZ is commonly included in the lower part of the overlying Hue Shale rather than in the pebble shale (Molenaar et al., 1987). The Hue-Thomson(!) petroleum system is represented by the Point Thomson gas-condensate field, which originated

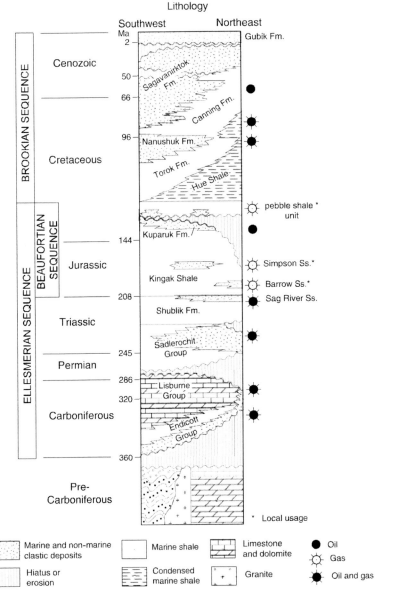

Figure 18.128. Generalized stratigraphic column for the North Slope of Alaska (Bird and Houseknecht, 2002). The principal source rocks identified in the column include the Carboniferous Lisburne Group, the Triassic Shublik Formation, the Jurassic Kingak Shale, and the Cretaceous pebble shale–gamm-ray zone (GRZ)–Torok interval.

from the Hue Shale source rock (Magoon et al., 1999; Bird, 2001). The Torok-Nanushuk(.) petroleum system has also been postulated (Bird, 2001).

The pebble shale is ~6–82 m thick and contains 1.5–3.8 wt.% TOC (Magoon et al., 1987; Keller et al., 1999). The lower portion of the Hue Shale (GRZ) is ~45–105 m thick and contains 1.9–3.9 wt.% TOC (Keller et al., 1999). Individual outcrops may average >12 wt.% TOC (Magoon et al., 1987). The upper portion of the Hue Shale is thicker but has considerably less generative potential. Earlier assessments of these units suggested that they were gas-prone in and around ANWR, but they were inferred to be more oil-prone in the Prudhoe Bay area. Many of these analyses, however, were conducted on weathered outcrops.

Recent high-resolution studies of a continuous core from the Mobil-Phillips Mikkelsen Bay State #1 in Prudhoe Bay found that the pebble shale averages ~4 wt.% TOC, with HI of 366 mg HC/g TOC. The GRZ of the Hue Shale averages ~3.5 wt.% TOC, with HI of 266 mg HC/g TOC (Keller and Macquaker, 2001; Keller et al., 2002). The generative potential of individual samples from the Mikkelsen Bay State #1 core is variable (Figure 18.129) due to differences in clastic dilution, primary productivity, and preservation. Masterson (2001) found a range of TOC (2–7 wt.%) and HI (150–400 mg HC/g TOC) for several GRZ cores near the Prudhoe Bay Field.

Pebble shale–GRZ–Torok and Kingak-Blankenship source-rock extracts have similar biomarker ratios, including high pristane/phytane and low dibenzothiophene/phenanthrene, C_{29}/C_{30} hopane, C_{23} tricyclic terpane/C_{30} hopane, and C_{35}/C_{34} hopane compared with Shublik extracts, because both are from marine shales (Masterson, 2001). However, GRZ extracts have higher $C_{28}/(C_{27}-C_{29})$ steranes (30% versus 24–25%) and are enriched in ^{13}C by at least 0.7‰ in the aromatic hydrocarbons compared with Kingak or Shublik extracts. The age-related C_{26}-nordiacholestane ratio (NDR) averages 0.4 for GRZ extracts and less than 0.2 for Shublik and Kingak extracts.

Table 18.11 compares geochemical data for pebble–GRZ–Torok oils with those generated from Jurassic Kingak-Blankenship and Triassic Shublik-Otuk source rocks. The pebble shale–GRZ–Torok and Shublik-Otuk oils differ markedly because they originated from dominantly siliciclastic and carbonate source rocks, respectively. Pebble shale–GRZ–Torok and Kingak-Blankenship oils are similar because both originated from similar marine siliciclastic source rocks. However, their stable carbon isotope and C_{26}-NDRs differ. Low NDR (0.14) for Kingak-Blankenship and Shublik-Otuk oils and high NDR (0.48) for pebble shale–GRZ–Torok oils is consistent with Jurassic and Cretaceous age source rocks, respectively (see Figure 13.55). North Slope oils with NDR less than 0.2 probably lack input from GRZ source rock.

The Tarn Field in the NPRA contains 37°API, low-sulfur crude oil (0.1 wt.%) believed to originate solely from the pebble–GRZ–Torok interval (Masterson, 2001). Two broad oil types from the North Slope include the Umiat-Simpson and Barrow-Prudhoe families, generated mainly from pebble shale and Shublik source rocks, respectively (Magoon and Claypool, 1981).

Neocomian Kissenda/Melania formations, Gabon

Organic-rich facies within the Berriasian-Valanginian Kissenda and Barremian Melania shales are the principal source rocks for non-marine, pre-salt oils in onshore Gabon (Kuo, 1994a). These strata in N'Komi rift and their equivalents in the nearby Interior Basin were deposited on top of fluvial/alluvial (overfilled) Upper Jurassic synrift sandstones. Greater accommodation during deposition of the Kissenda Formation resulted in oscillating overfilled to balanced-fill conditions and the preservation of organic matter. Overfilled conditions returned during deposition of Hauterivian fluvial sands and siltstones of the Lucina and lower Melania formations. These are overlain by organic-rich Melania shales, indicating a return to a balanced-fill state. There is no evidence that underfilled, evaporitic conditions ever occurred in the Gabon synrift basins or

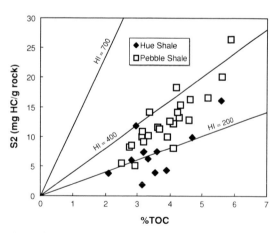

Figure 18.129. Total organic carbon (TOC) (wt.%) versus Rock-Eval S2 for core samples of pebble shale and Hue Shale (gamma-ray zone, GRZ) from the Mobil-Phillips Mikkelsen Bay State #1 (from Keller and Macquaker, 2001; Keller et al., 2002). HI, hydrogen index.

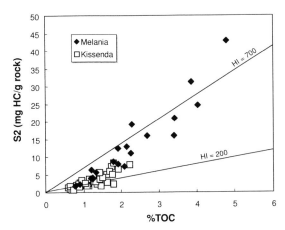

Figure 18.130. Total organic carbon (TOC) (wt.%) versus Rock-Eval S2 for Melania and Kissenda lacustrine shales, onshore Gabon (data from Kuo, 1994a).
HI, hydrogen index.

that a marine transgression occurred before the deposition of the Aptian salt.

Kerogens in the Melania and Kissenda shales are similar because they contain input from freshwater algae and minor contributions from higher plants. The Melania Formation has greater source potential than the Kissenda Formation, due mostly to better preservation and less dilution by sediment rather than different organic matter input (Figure 18.130). The kerogen and expelled oils are low in sulfur.

Cretaceous shales of the Western Interior Seaway

The Western Interior Seaway (WIS) was a broad epicontinental sea that extended ~6000 km, covering most of central North America between the emerging Cordilleran thrust belt to the west and the stable Mid-Continent craton to the east (Figure 18.131). The seaway was present throughout much of the Cretaceous Period during a first-order transgression that resulted in the highest global sea levels during the Mesozoic Era (Kauffman and Caldwell, 1993). The degree of connectivity between the Boreal Ocean and the Gulf of Mexico varied with sequential subcycles of transgression and regression. During highstands, the WIS ran the full length of North America, and the cold, arctic waters from the north mixed with the warm, tropical waters from the Gulf of Mexico. During lowstands, the WIS was not continuous and may have been isolated.

Paleo-oceanographic circulation in the WIS was generally restricted and unlike open oceanic basins. A stratified water column was present throughout much of the seaway's history (Kauffmann and Sageman, 1990), giving rise to benthic communities adapted to dysoxic to anoxic bottom-water conditions. There are no modern analogs for this oceanographic situation, although it is likely that the WIS responded in a manner similar to a large estuarine system (Slingerland et al., 1996; 1998).

Numerous marine shales, marls, and limestones were deposited in the elongated foreland basin. Those with source potential were deposited mostly during three subcycles of transgression and/or basin deepening. These source units typically contain <1–4 wt.% TOC comprised of type II or type II/III kerogen, with HI averaging ~400 mg HC/g TOC (Meissner et al., 1984). The best preservation of organic matter tended to occur in the center of the seaway, where the waters were deepest, and in localized areas with restricted circulation. Coals and sediments with mainly type III kerogen were deposited during the regressive

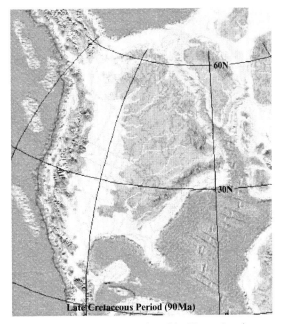

Figure 18.131. Paleo-reconstruction of the Western Interior Seaway during the Late Cretaceous Period (~90 Ma). Figure prepared by Ron Blakey and used with his permission (http://vishnu.glg.nau.edu/rcb/RCB.html).

subcycles. These sequences are mostly gas-prone, although some, such as the Fruitland Formation, have generated oil (Michael et al., 1993).

There is no consistent nomenclature for WIS rock units, and various names are assigned to age- and facies-equivalent strata in the many basins throughout the central and western USA and Canada. Absolute geochronology of the formations is hampered further by the lack of age-related fossils in many of the source beds. In our discussion of the WIS source rocks, below, we group the sources by depositional subcycles of transgression using nomenclature adapted from Kauffman (1977).

Upper Albian Kiowa-Skull Creek cyclothem

WIS shales with moderate source potential were deposited during the Upper Albian Age. These dark-colored, fine-grained rocks underlie the Mowry Formation and include the Kiowa, Skull Creek, and Thermopolis units in the USA and the Fish Scales Zone in Canada. The Kiowa–Skull Creek shales represent the first of the marine transgressions into the Rocky Mountain area, reaching as far south as northernmost New Mexico (Meissner et al., 1984; Burtner and Warner, 1984). They average ~0.8–3.6 wt.% TOC and tend to be richest in the middle of the paleo-seaway, particularly in Canada (Creaney et al., 1994b). The kerogen is mainly terrigenous or oxidized marine in origin, with HI rarely exceeding 200 mg HC/g TOC. The generative potential of the Upper Albian shales tends to be substantially lower than that of the overlying Mowry shales (e.g. Dyman et al., 1996).

Greenhorn cyclothem shales

The oldest major source rocks in the WIS Greenhorn cyclothem are the Lower Cenomanian Graneros/Mowry/Aspen/Colorado shales. These strata were deposited during a major transgression at the end of

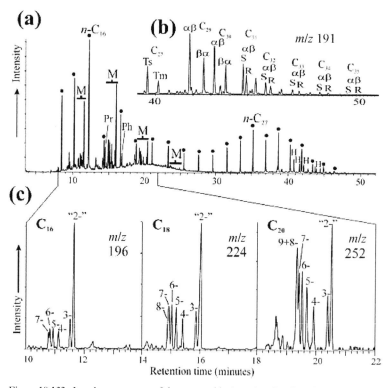

Figure 18.132. Ion chromatograms of the saturated hydrocarbon fraction of basal Graneros Shale, Park District of Pueblo, Colorado. (a) • = n-alkanes, M = clusters of monomethyl- and monoethylalkanes; (b) m/z 191 ion chromatogram showing hopane distributions; (c) partial ion chromatogram (m/z 196 + 224 + 252) showing the C_{16}, C_{18}, and C_{20} monoethylalkanes. Numbers on peaks indicate positions of the ethyl group. Reprinted from Kenig et al. (2001). © Copyright 2001, with permission from Elsevier.

Albian time, when the WIS extended from the circumpolar ocean to southern Colorado. Thickness of these Lower Cenomanian source rocks within the foreland basins varies from ~30 m to more than 200 m, increasing toward the Cordilleran thrust belt to the west. The source potential of the Mowry Shale has been studied extensively within the USA (e.g. Nixon, 1973; Seifert and Moldowan, 1981; Burtner and Warner, 1984; Momper and Williams, 1984; Davis *et al.*, 1989; Dyman *et al.*, 1996). The Mowry Formation is dominantly siliceous shale that is richest in the center of the seaway toward the northern US boundary and into Canada. TOC averages 1–4 wt.%, with maxima near 7 wt.%, and with HI <400 mg HC/g TOC.

Dahl *et al.* (1994) used the distribution of C_{27}–C_{30} steranes to map organic facies within the Mowry Shale. In rock extracts, the relative abundances of C_{28} and C_{30} steranes were indicators of marine algal input and were proportional to the relative abundance of type II kerogen macerals. In contrast, the relative abundance of C_{29} steranes correlated with the abundance of vitrinite macerals in the same rock sequence. Similar relationships were found for the diasteranes. Using this information, the quality of the source rock can be extrapolated from the biomarker distributions in oils, allowing interpreted generative potential and organic facies to be mapped in areas lacking rock samples. This approach is valid only in basins where lateral migration is limited.

Kenig *et al.* (2001) used biomarkers to determine the depositional conditions of selected WIS source units. By studying the molecular and isotopic distributions of a series of monomethyl- and monoethylalkanes in the basal Graneros Shale from Colorado, they determined that cyanobacteria growing in fresh to brackish waters were the most likely source for these compounds (Figure 18.132). They proposed that the basal Lower Cenomanian shales were deposited during the earliest phase of the transgression, when full marine conditions did not yet exist.

Deposition of the Greenhorn/Mancos/Frontier shales occurred during the Late Cenomanian–Early Turonian Age, corresponding to maximum transgression and OAE 2. Other formations associated with the transgressive event include the Belle Fourche/Frontier and the Second White Speckled (Second White Specks) Shale in Canada. In the central US basins, the Greenhorn Formation is divided into three members: the Lincoln Shale, the Hartland Shale, and the Bridge Creek Limestone. All units have generative potential, but the richest source rocks are in the Hartland Shale and Bridge Creek Limestone members (Figures 18.133 and 18.134).

High generative potential resulted from high productivity and excellent conditions for preservation of organic matter during deposition of the Greenhorn Formation. Bottom waters during deposition of most of the Greenhorn cyclothem were anoxic. Euxinic conditions existed in the photic zone based on the widespread presence of isorenieratene derivatives in Greenhorn strata (Figure 18.135). Isorenieratene is produced by anaerobic, photosynthetic green bacteria and occurs in Greenhorn samples as far south as New Mexico (Simons and Kenig, 2001) and as far north as the Manitoba Escarpment (Schröder-Adams *et al.*, 2001). The pervasive stratigraphic occurrence of isorenieratene derivatives indicates that water-column stratification and photic-zone anoxia occurred throughout the WIS during the Late Cretaceous Period. Stratification probably arose from the mixing of warm waters from the Tethys Sea and cold waters from the Boreal Sea, and an influx of freshwater from continental runoff (Hay *et al.*, 1993), although the detailed mechanism is not understood fully. For example, paleooceanographic models fail to predict the establishment of a stable, stratified water column and anoxic bottom waters (Slingerland *et al.*, 1996; Kump and Slingerland, 1999).

The Carlile Shale was deposited during the regressive end of the Greenhorn cyclothem. The basal Fairport Chalk Member was deposited under distal offshore conditions at estimated water depths of 200–300 m (Sageman *et al.*, 1997). Fairport Chalk contains ~3–5 wt.% TOC and type II kerogen (Forster *et al.*, 1999). The Blue Hill Shale Member, the uppermost portion of the Carlile Shale, contains lower TOC (~2.4 wt.%) and is a mixture of terrigenous and marine kerogen. The geochemistry of these shales indicates that source rocks of modest generative potential can be deposited during regressive phases where local conditions promote organic matter preservation.

Niobrara cyclothem

The third transgressive subcycle deposited organic-rich limestones and shales throughout the WIS. Source rocks with excellent generative potential occur in the Coniacian-Santonian Smoky Hill Member of the

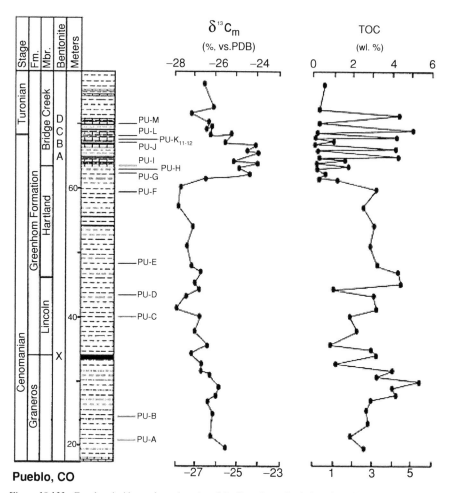

Figure 18.133. Geochemical log and stratigraphy of the Greenhorn Cyclothem based on an outcrop near Pueblo, Colorado (data from Pratt et al., 1985; figure from Simons and Kenig, 2001). The Greenhorn Formation is fully exposed by several road cuts in Pueblo Dam State Park and was sampled extensively for geochemical study.

Niobrara Formation and its Canadian equivalent, the First White Speckled Shale, and in the Campanian Sharon Springs Member of the Pierre Shale (Rice, 1984; Gautier et al., 1984; Creaney et al., 1994b; Dean and Arthur, 1998). These calcareous shales and limestones generally contain lower amounts of predominantly marine algal kerogen than source rocks in the earlier Greenhorn cyclothem (Figure 18.136)

Campanian Fruitland coals

Numerous coals were deposited within the WIS during the regressive stages and are now targets for coalbed methane development (e.g. Takahaski, 2001). For example, the Fruitland coals were deposited at the end of the Niobrara cyclothem. Swamps in the San Juan Basin occurred between perennial streams resulting in coastal plain peats within a 30-mile-wide zone (Nance, 1998). Fruitland coals commonly contain >90% vitrinite, yet they can generate some oil based on HI in the range 200–400 mg HC/g TOC for immature to moderately mature samples (Figure 18.137). Highly mature coals generally have HI below 200 mg HC/g TOC in Figure 18.137. Samples from this formation span a wide range of thermal maturity (0.42–1.49% R_m) for a relatively

Mesozoic source rocks

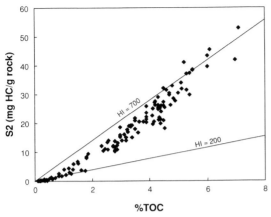

Figure 18.134. Total organic carbon (TOC) (wt.%) versus Rock-Eval S2 for samples of Bridge Creek Limestone, Greenhorn Formation. Samples are from the Amoco Rebecca K. Bounds #1 well (Greeley County, Kansas) and the USGS Escalante #1 well (Garfield County, Utah) (data from Dean and Arthur, 1998).
HI, hydrogen index.

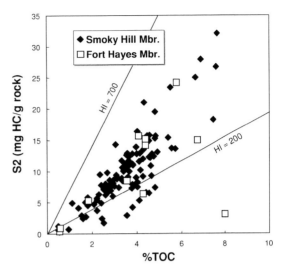

Figure 18.136. Total organic carbon (TOC) (wt.%) versus Rock-Eval S2 for samples from the Niobrara Formation in the Amoco Rebecca K. Bounds #1 well (Greeley County, Kansas) and the Coquina Oil Corporation Berthoud State #3 and #4 wells (Fort Collins, Colorado) (data from Dean and Arthur, 1998).
HI, hydrogen index.

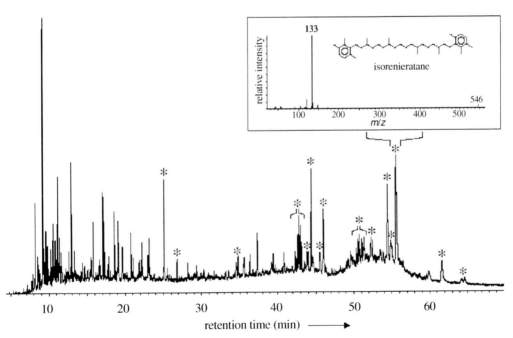

Figure 18.135. Reconstructed m/z 133 ion chromatogram of the aromatic fraction extracted from the Assiniboine Member of the Favel Formation, a Turonian source rock from the Manitoba Escarpment. Isorenieratene derivatives are indicated by *. Reprinted from Schröder-Adams et al. (2001). © Copyright 2001, with permission from Elsevier.

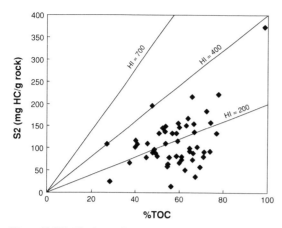

Figure 18.137. Total organic carbon (TOC) (wt.%) versus Rock-Eval S2 for Fruitland coals, San Juan Basin (data from Michael et al., 1993).
HI, hydrogen index.

homogeneous organic facies, thus providing an ideal suite to test molecular and isotopic-derived maturity indicators (Michael et al., 1993).

Although the Fruitland coals are predominantly gas-prone, small amounts of liquids can be recovered from the coal and associated sandstone reservoirs (Clayton et al., 1991). These crude oils are characterized by high pristane/phytane ratios, relatively high proportions of isoprenoid hydrocarbons, high-molecular-weight n-alkanes, and abundant methylcyclohexane and light aromatic hydrocarbons.

Upper Aptian Qingshankou Formation, Songliao Basin, northeast China

The Songliao Basin is one of several basins situated on the pre-Mesozoic fold belt in northeastern China. Intracontinental rifting began in the Late Jurassic Period, leading to development of synrift lacustrine basins in the Early Cretaceous Period. The synrift sequences in 6 of these basins account for at least 22 oil and gas fields (Xue and Galloway, 1993; Dou, 1997). Thick, post-rift lacustrine sequences in the Songliao Basin contain the source and reservoir rocks of the Daqing Field (Figure 18.138), the largest oil field in China and one of the few super giant fields in the world that produces lacustrine oil (Yang Wanli, 1985). During the Early Cretaceous Period, individual rift depressions merged into one large, rapidly subsiding basin. By the Late Aptian Age, a deep-water lake covered ~87 000 km². The lacustrine system continued to expand, depositing deep-water reservoir and source facies through the Late Albian Age, when the lake covered >100 000 km². Deposition continued through the Quaternary Period in the western portions of the basin.

The Upper Aptian lacustrine Qingshankou Formation is the principal source rock for oil in the Songliao Basin (Yang Wanli et al., 1985; Li Desheng et al., 1995). During this time, the climate was subtropical and humid, leading to seasonal algal blooms and lush plant growth along the lake shores. Organic-rich mudstones were deposited in the lake primarily under deep-water, euxinic conditions. Freshwater algae and bacterially altered terrigenous plant matter produced lipinite-rich kerogens near the basin center. TOC within the lower Qingshankou member, which varies from 25 to 112 m thick, ranges from 0.5 to 4% wt.%, with maxima as high as 8.4 wt.%. The upper Qingshankou member, which varies from 53 to 552 m thick, is associated with a deltaic facies and has lower source potential. TOC rarely exceeds 2 wt.% and averages ~0.7%. HI for the deep-water facies ranges from ~200 to 850 mg HC/g TOC, with maxima approaching 1000 mg HC/g TOC. Bog and swamp facies occur around the paleo-lake margins. These typically have HI <200 mg HC/g TOC and contain type III terrigenous organic matter.

Biomarkers provide evidence for marine transgressions during deposition of the Qingshankou Formation (Hou et al., 1999; 2000). Relative high concentrations of 4,22,23,24-tetramethylcholestanes, C_{30} desmethylsteranes, and C_{31} 4α-methylsteranes are present in source-rock extracts and produced oils. These biomarker distributions are characteristic of dinoflagellates from brackish to saline environments.

Albian Eumeralla Formation, Otway Basin

The Otway Basin of southern Australia is one of several basins formed during the Mesozoic rifting of the Australian and Antarctic plates. Several sedimentary sequences were identified as having some source potential, but the basal coals of the Eumeralla Formation are recognized as the most likely source of hydrocarbons in the Otway Basin. The lower Eumeralla Formation consists of fluvial and flood-basin sediments with overbank coals and lacustrine deposits laid down under

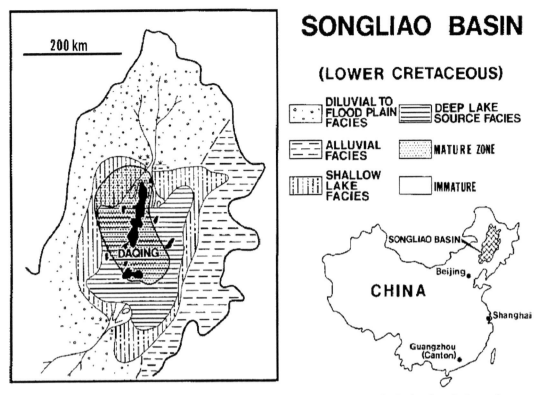

Figure 18.138. The giant Daqing oil field lies updip from the center of the generative Songliao Basin, where the Lower Cretaceous deep lake source rock facies is thermally mature (Demaison, 1984). The maximum distance of horizontal migration is <40 km. Reprinted by permission of the AAPG, whose permission is required for further use.

overfilled to balanced-fill conditions. Cool temperate forests and peat swamps were the primary sources of organic matter (Struckmeyer and Felton, 1990; Tupper et al., 1993). Coaly rocks within the Eumeralla Formation average ~31 wt.% TOC, with HI <100 mg HC/g TOC. However, <1-m-thick coal seams in the basal portion have a lipinitic component and average HI ~244 mg HC/g TOC, with individual samples as high as 380 mg HC/g TOC (Hill, 1995). Collectively, these coals make up ~30% of the total source interval. Edwards et al. (1999) correlated oils in the Otway Basin to these coals (Figure 18.139).

Middle-Upper Albian Toolebuc Formation, eastern Australia

The Toolebuc Formation in the Eromanga and Carpentaria basins contains black, organic-rich calcareous mudstones with abundant fish remains. The carbonate originated from marine planktonic algae, which also contributed most of the organic matter (Sherwood and Cook, 1986; Boreham and Powell, 1987). Biotic input from cyanobacterial mats also has been proposed (Glikson and Taylor, 1986). The Toolebuc Formation is recognized easily in well logs by a strong gamma-ray response caused by high uranium in the organic matter and phosphatic fish remains (Moore et al., 1986; Pitt, 1986). This organic-rich condensed section was deposited in an anoxic epicontinental sea during a maximum transgression (Moore et al., 1986). TOC is highly variable throughout the formation but is typically 5–10 wt.%, with maxima >20 wt.%. Rock-Eval HI is also variable, ranging from <100 to >700 mg HC/g TOC (Boreham and Powell, 1987).

The Toolebuc Formation is thermally immature onshore, where it has been used in oil-shale retorting operations. More mature equivalents of these rocks may be responsible for asphaltic beach strandings,

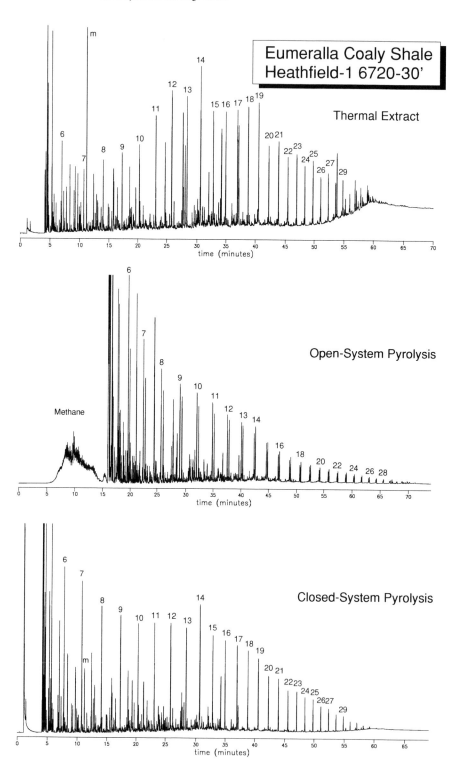

Figure 18.139. Thermal extraction, open–system, and closed–system pyrolysis gas chromatograms of the oil-prone coal from the Heathfield-1 well, Otway Basin, Australia. m = methylcyclohexane; n-alkanes indicated by carbon number.

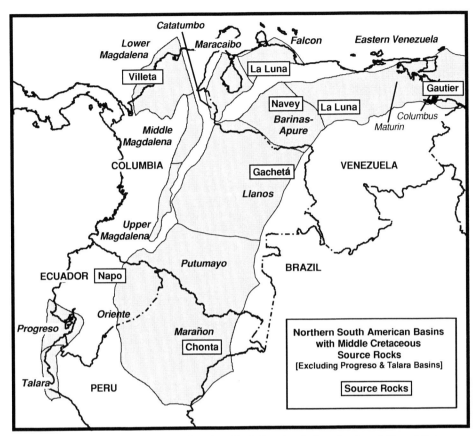

Figure 18.140. Map showing the location of northern South American basins, oil and gas fields, and Cretaceous source rocks (from Schenk *et al.*, 1999).

which are common along the Southern Margin of Australia (Edwards *et al.*, 1998). When the chemistry of the asphaltites was compared with a global database (GeoMark Research, Inc.), the beach standings were most similar to oils generated from the West Siberian Basin and the Arabian platform (Summons *et al.*, 2001). The Toolebuc Formation is the closest analog to these source rocks in the study area, suggesting that a deepwater Cretaceous equivalent facies may be thermally mature offshore. This research illustrates the advantage of applying statistical analysis to a large, internally consistent, geochemical database.

Cretaceous La Luna and equivalents, sub-Andean South America

Cretaceous source rocks in the sub-Andean basins of northern South America are some of the most prolific in the world (Figure 18.140). These source rocks were once considered to be fairly homogeneous marine carbonate source rocks deposited under similar anoxic conditions along a passive continental margin. However, several recent stratigraphic, paleogeographic, and geochemical studies have shown that these source units were deposited in different environments that controlled both the type of organic matter and the preservation conditions (e.g. Escandon *et al.*, 1993; Perez-Teller, 1995; Olivares *et al.*, 1996; Perez-Infante *et al.*, 1996; Alberdi-Genolet and ToCco, 1999; Erlich *et al.*, 1999a; Erlich *et al.*, 1999b; Erlich *et al.*, 2000; Rangel *et al.*, 2000; Davis and Pratt, 2002). Nevertheless, these source rocks share some general traits, which allow them to be discussed together.

The sub-Andean basins have a complex tectonic history that can be traced back to at least the Early Cretaceous Period. The oldest sedimentary rocks are

Table 18.12. *Summary of mid-Cretaceous source rocks along the northern passive margin of South America*

Formation	Basin/sub-basin	County	Key references
Gautier	Columbus	Trinidad	Rodrigues (1988), Persad *et al.* (1993), Requejo *et al.* (1994)
Querecual	Eastern Venezuela/Maturin	Venezuela	Talukdar *et al.* (1988), Alberdi and Lafargue (1993), Parnaud *et al.* (1995)
Navay	Barinas/Apure	Venezuela	Anka *et al.* (1998), Erlich *et al.* (2000)
La Luna	Maracaibo	Venezuela	Talukdar and Marcano (1994), Perez-Infante *et al.* (1996)
La Luna-Capacho	Catatumbo	Colombia	Yurewicz *et al.* (1998)
La Luna	Middle Magdalena Valley	Colombia	Ramón and Dzou (1999), Rangel *et al.* (2000), Zumberge (1984)
Callabos-Villeta (Hondita)	Upper Magdalena Valley	Colombia	Buitrago (1994), Mann and Steine (1997), Higley (2001)
Gachetá	Llanos	Colombia	Ramón *et al.* (2001)
Gachetá	Putumayo	Colombia	Ramón *et al.* (2001), Higley (2001)
Napo	Oriente	Ecuador	Mello *et al.* (1994b), Mathalone and Montoya (1995)
Chonta/Raya	Marañon	Peru	Vargas (1988), Mathalone and Montoya (1995), Higley (2001)

pre-Aptian to Aptian rift deposits that overlie older basement metamorphic rocks. The earliest potential source rocks were deposited in continental and shallow-marine tidal environments during the Aptian–Albian Age. These strata include the Caballos Formation in the central Colombia basins (Ramón *et al.*, 2001), the Hollin Formation in the Oriente Basin (Mello *et al.*, 1994a), and the Raya Formation in the Marañon Basin (Mathalone and Montoya, 1995). Conditions changed to fully marine during the Cenomanian-Turonian transgression and OAE 2. During this period, a broad, continuous marine margin extended from eastern Venezuela to Peru. Thick deposits (>1000 m) of shallow-marine limestones, marls, and shales were deposited under various environmental conditions (dysaerobic, anoxic, and euxinic), resulting in organic-rich source beds throughout as much as one-third of the sedimentary package (Table 18.12). The Santonian-Maastrichtian regressive sequence consists mostly of marine and coastal plain deposits with negligible source potential.

In general, the most prolific source rocks follow the distribution of known oil production (Figure 18.140). Organic-rich facies of the La Luna Formation can exceed 15 wt.% TOC, with HI typical of type II kerogen (~350–650 mg HC/g TOC) (e.g. Perez-Infante *et al.*, 1996). Values of 3–5 wt.% TOC are more representative of the widespread occurrence of La Luna source rocks in eastern Venezuela, Colombia, and Ecuador (e.g. Zumberge, 1984; Mello *et al.*, 1994a; Yurewicz *et al.*, 1998). The Navay Formation of the Barinas/Apure basins (average ~2.6 wt.% TOC) (Erlich *et al.*, 2000) and the Chonta/Raya formations of Peru (average ~0.5–1 wt.% TOC) (Vargas, 1988; Mathalone and Montoya, 1995) represent the limit of the oil-prone source facies. Age-equivalent strata to the south and inland of the production zones are dominated by continental clastics with terrigenous organic matter.

The deposition of Cretaceous source and non-source facies can be explained using a sequence stratigraphic framework (Figure 18.141). Sediments rich in oil-prone kerogen are deposited during transgressional systems tracts, culminating at the maximum flooding surface (Pasley *et al.*, 1993). Mann and Steine (1997) showed how the organic facies of mid-Cretaceous sediments in the Upper Magdalena Valley fit this model (Figure 18.142). Rangel *et al.* (2000) examined the organic facies within the La Luna of the Middle Magdalena Valley and showed distinct system–tract correlations between generative potential, isotopic ratios, and biomarker distributions. Similar trends occur in other basins of the region (e.g. Alberdi and Lafargue, 1993; Mello *et al.*, 1994a).

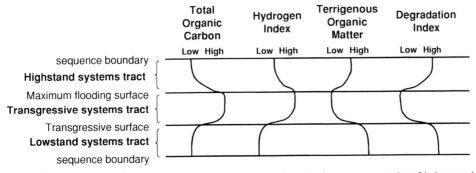

Figure 18.141. Systematic changes in the amount and type of organic matter in a sequence stratigraphic framework (from Mann and Steine, 1997) (modified from Pasley et al., 1993). Petrographic analysis is used to measure relative abundance of land plant macerals and for determining the degradation index, which is the ratio of well-preserved to highly degraded or amorphous phytoclasts. Reprinted by permission of the AAPG, whose permission is required for further use.

The widespread deposition of La Luna and equivalent source rocks (Table 18.12) is a result of many factors that enhanced primary productivity and preservation. Sea-level change is only one of these factors. High rates of productivity and preservation can be attributed partially to marine transgression and OAE 2, but they do not account for local variations in generative potential. Villamil and Arango (1996) explain the temporal differences in paleoproductivity as a result of shifting oceanic currents from the northern migration of the South American Plate. The margin was south of the paleoequator during Jurassic and Early Cretaceous time, where ocean currents favored downwelling. When the passive margin shifted to a position north of the paleoequator, conditions were favorable for upwelling. This model can account for the overall change in facies from non-calcareous shales to calcareous marls, to siliceous shales, and finally to bedded cherts (e.g. Perez-Infante et al., 1996). Conditions that promoted the preservation of organic matter are as important as those that promote high paleoproductivity. The development of paleobathymetric barriers that caused limited water circulation and turnover were key to the local deposition of organic-rich intervals (Erlich et al., 2000). High rates of evaporation and low rates of precipitation further promoted anoxic, hypersaline conditions.

Upper Cretaceous La Luna-Misoa(!) petroleum system

The Upper Cretaceous La Luna Formation is composed of organic-rich carbonate rocks of monotonous appearance, with good to excellent potential for oil generation in outcrops along the eastern flank of the Middle Magdalena Basin, Colombia (Rangel et al., 2000). However, geochemical variations are sufficient to differentiate organic facies in the Salada, Pujamana, and Galembo members. C_{35}/C_{34} hopane correlates with HI for these rocks, suggesting that changes in HI reflect redox conditions more strongly than carbonate content. Certain biomarker ratios within the La Luna Formation characterize the depositional environment. For example, average diasteranes/steranes are <1, Ts/Tm averages <0.33, C_{35}/C_{34} hopane is >0.92, and oleanane/hopane ranges from 0.02 to 0.19. In regressive carbonate shelf settings, HI and TOC tend to increase and decrease, respectively, whereas in regressive siliciclastic shelf settings, both TOC and HI decrease. Some biomarker ratios (oleanane/hopane, C_{20}/C_{23} tricyclic, Ts/Tm) increase during times of decreasing sea level.

Four geographically distinct crude oil families can be distinguished based on sulfur content, pristane/phytane, dibenzothiophene/phenanthrene, oleanane content, and terpane distributions (Ramón and Dzou, 1999). These families originated locally from different organic facies or mixtures of the La Luna Formation rather than having migrated from the region of the eastern Cordillera. The oil compositions indicate dominantly anoxic marine carbonate facies in the northwestern versus more siliciclastic facies in the eastern portions of the Middle Magdalena Basin.

The La Luna-Misoa(!) is a giant petroleum system (8160 bkg recoverable hydrocarbons) (see Table 4.3) because of four factors (Magoon and Valin, 1994): (1) the micritic shale source rock from a shelf-margin setting is organic-rich (~5–9 wt.% TOC) and oil-prone

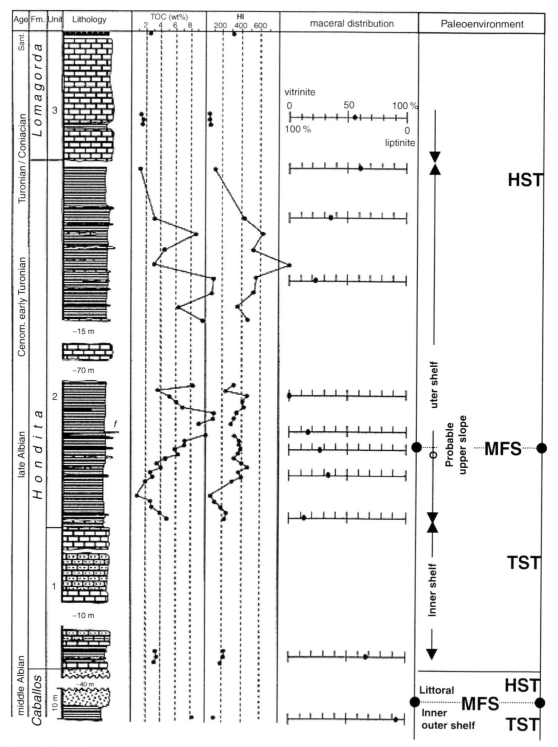

Figure 18.142. Variations of total organic carbon (TOC), hydrogen index (HI), and maceral distribution in a sequence stratigraphic framework for the Ocal section, Upper Magdalena Valley (modified from Mann and Steine, 1997). Reprinted by permission of the AAPG, whose permission is required for further use.
HST, highstand systems tract; MFS, maximum flooding surface; TST, transgressive systems track.

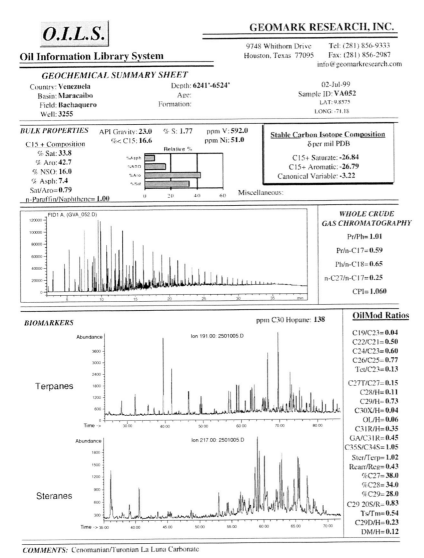

Figure 18.143. Geochemical data for crude oil from Bachaquero Field generated from the Upper Cretaceous La Luna Formation, Maracaibo Basin, Venezuela (courtesy of GeoMark Research, Inc.).

(HI^o = 650 mg HC/g TOC, type II kerogen); (2) the volume of mature source rock is relatively large; (3) comparatively little oil was destroyed because the critical moment is present-day; and (4) much of the generated oil reached viable traps.

Figure 18.143 shows geochemical data for crude oil generated from the La Luna Formation in the Maracaibo Basin, Venezuela. Figures 18.144 and 18.145 show geochemical data for crude oil generated from the Napo Formation in the Oriente Basin, Ecuador, and the Villeta Formation in the Upper Magdalena Basin, Colombia, respectively, which are equivalents of the La Luna Formation.

Cenomanian-Turonian rocks, western Mediterranean and Atlantic Margin

Organic-rich strata occur throughout the western Mediterranean and the adjacent Atlantic Margin that correspond to the major OAE 2. Examples include

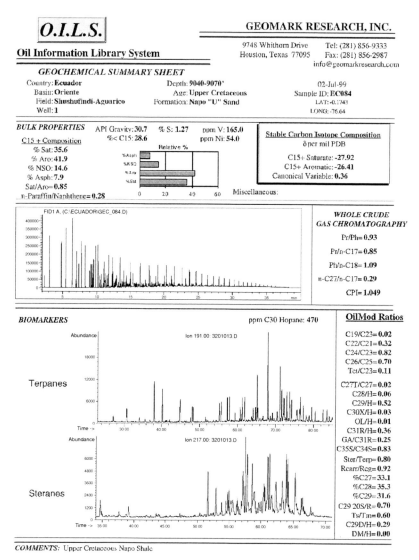

Figure 18.144. Geochemical data for crude oil from the Shushufindi-Aguarico Field generated from the Upper Cretaceous Napo Formation, Oriente Basin, Ecuador (courtesy of GeoMark Research, Inc.).

DSDP cores and well penetrations in Italy (Apennines, southern Alps), Tunisia (Bahloul), Algeria, Morocco (Rif Mountains, Atlas Mountains, Tarfaya Basin), Gibraltar arch, Spain (Betics, Bay of Biscay, Galicia margin), Senegal (Cape Verde Basin, Casamance), and Nigeria (Benue, Calabar flank) (Kuhnt et al., 1990; Kuhnt and Wiedmann, 1995). Most of these rocks are immature with respect to oil generation.

Pre-Cenomanian rocks in the region typically have low TOC (<1 wt.%) and are dominated by terrigenous plant debris. With the onset of OAE 2 in the Cenomanian Age, organic richness increased, reaching a maximum near the Cenomanian–Turonian boundary, with TOC up to 40 wt.%. Biotic input switched from terrigenous to marine and HI typically increased to >500 mg HC/g TOC. These organic-rich rocks commonly represent three different depositional settings: shelf-basin marls, continental margin marls and limestones, and deep-water shales and turbidites. Ocean waters returned to oxic conditions after the Turonian Age, resulting in

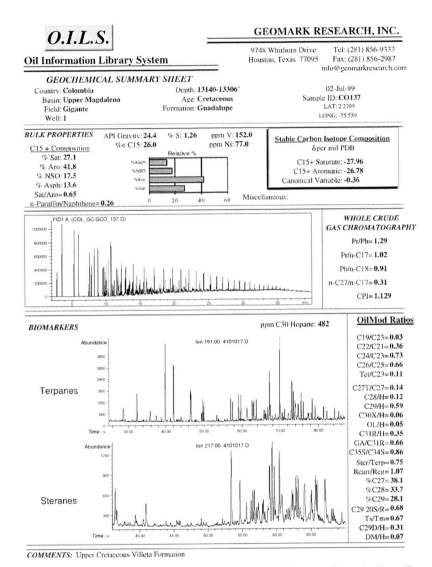

Figure 18.145. Geochemical data for crude oil from Gigante Field generated from the Upper Cretaceous Villeta Formation, Upper Magdalena Basin, Colombia (courtesy of GeoMark Research, Inc.).

decreased deposition of organic-rich sediments and lower generative potential.

The Mesozoic Tarfaya Basin along the coast of southwestern Morocco contains Upper Cretaceous organic-rich marls and limestones (Leine, 1986; Kuhnt et al., 1997; El Albani et al., 1999). The richest samples, including oil shales, are from Late Cenomanian marls and phosphatic chalks (2–16 wt.% TOC, with HI up to 750 mg HC/g TOC) and Turonian chalks (up to 16 wt.% TOC, with HI up to 800 mg HC/g TOC) that consist almost exclusively of biogenic carbonate from planktonic foraminifera. Overlying Coniacian-Santonian and Campanian marls have lower source potential (maximum 4–5 wt.%, with maximum HI typically <520 mg HC/g TOC but sometimes >900 mg HC/g TOC).

Organic-rich Cenomanian-Turonian strata in the Tarfaya Basin and elsewhere correspond in part to OAE 2, but the persistent occurrence throughout the Late Cretaceous Period of organic enrichment is attributed to various factors (Holbourn et al., 1999; El Albani et al., 1999). The Late Cretaceous Period was characterized by high global sea level and major transgressions during the early Turonian and early Campanian ages, which led to widespread flooding of the continents and major changes in water-circulation patterns and biotic productivity. During most of the Late Cretaceous Period, the Tarfaya Basin was located in an area of major upwelling at low latitudes. Hence, conditions were ideal for high productivity of marine planktonic organisms and good preservation of organic matter in the distal portions of the basin where stratified, anoxic waters were common.

Upper Cretaceous Sirte Shale, Sirte Basin, Libya

The Upper Cretaceous Sirte Shale is the source rock for most or all of the oil and gas in the Sirte Basin, one of the world's largest petroleum systems, with known

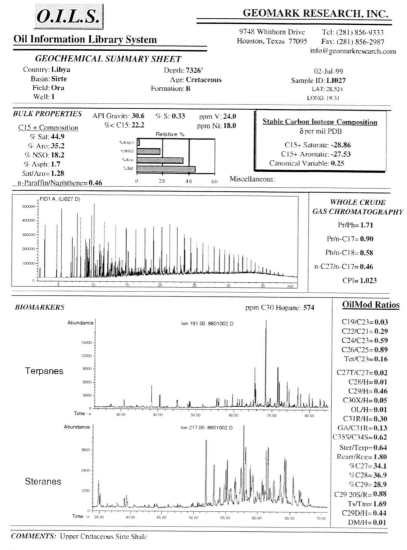

Figure 18.146. Geochemical data for crude oil from Ora Field generated from the Upper Cretaceous Sirte Formation, Libya (courtesy of GeoMark Research, Inc.).

reserves of 43.1 BBOE (Ahlbrandt, 2001). Oils generated from Sirte Shale are typical of those generated from clastic marine source rocks, which have moderate gravity (30–43°API) and low sulfur (~0.3 wt.%) (Figure 18.146). The Sirte Basin formed when Tethyan rifting commenced in the Early Cretaceous Period and terminated in the early Tertiary Period, resulting in a triple junction within the basin (Ambrose, 2000). The source rock is differentiated into Upper Sirte Shale (Campanian) and Lower Sirte Shale (Turonian), separated by the Coniacian/Santonian Tagrifet limestone (Mansour and Magairhy, 1996). All members are source rocks and collectively may range in thickness from a few hundred to nearly 1000 meters in the deep troughs. TOC ranges from 0.5 to 4 wt.%, with an average of ~1.9 wt.% (Parsons et al., 1980) and maxima near 8 wt.% (El-Alami, 1996). The kerogen in the Sirte Shale is predominantly type II, with HI ranging from 300 to 600 mg HC/g TOC (El-Alami, 1996; Baric et al., 1996).

Upper Cretaceous (Campanian) Sudr Formation, Brown/Duwi members

The pre-rift Campanian Brown Limestone/Duwi member of the Upper Cretaceous Sudr Formation is the major source rock in the Gulf of Suez and the northern Red Sea (Robison, 1995; Lindquist, 1998b). It is a 25–70-m-thick, uraniferous, phosphatic limestone with an average of 2.6 wt.% TOC (Lelek et al., 1992) and as much as 21 wt.% TOC (Abdine et al., 1992). The Brown Limestone was deposited during the initial marine transgression of the northeastern margin of Gondwana. Source rocks formed under restricted marine-evaporitic conditions from Libya to Syria (Robison, 1995). The kerogen is type II/IIS, with HI ~450–600 mg HC/g TOC and OI <60 mg HC/g TOC (Alsharhan and Salah, 1997).

The Brown Limestone member generates classic carbonate oils with high sulfur (>1 wt.%), pristane/phytane, and CPI, and nickel/vanadium ratios <1 (Figure 18.147). Diagnositic biomarkers include high gammacerane, C_{30} steranes, and C_{35}/C_{34} hopanes (Alsharhan and Salah, 1997; Barakat et al., 1996; Barakat et al., 1997). Maturity parameters show that expulsion occurred throughout the oil window (Barakat et al., 1998).

Barakat et al. (1997) found two genetic families based on biomarker analyses of oils from the Gulf of Suez, Egypt. Oils from the Ras Fanar and East-Zeit wells have high gammacerane, low diasteranes, and high C_{35}/C_{34} hopanes, consistent with an origin from the Brown Limestone. Oils from the Gama and Amal-9 wells have low gammacerane, high diasteranes, and oleanane indices >20%, indicating an angiosperm-rich Tertiary siliciclastic source rock, probably the Lower Miocene Rudeis Formation.

Seeps from the Gulf of Suez representing both of the above oil groups were used to prepare ancient Egyptian mummies. For example, seep oils from Abu Durba and Gebel Zeit (see Figure 11.2) are geochemically similar to bitumen used to mummify Cleopatra and Pasenhor, respectively (Harrel and Lewan, 2002). Terpane mass chromatograms for the East-Zeit and Amal-9 oils (Figure in Barakat et al., 1997) are remarkably similar to those for Abu Durba and Gebel Zeit, respectively (see Figure 11.3).

Upper Cretaceous limestones, Adiyaman region, southeastern Turkey

Following deposition of the major Cretaceous source rocks in the Persian Gulf, relatively minor source rocks were deposited on the Arabian platform throughout the Late Cretaceous Period. The oil province of southeastern Turkey is one such intrashelf basin lying along the northern passive margin of the Arabian Plate. Demirel et al. (2001) investigated the Cretaceous sequence in the Adiyaman region, identifying shales, mudstones, and carbonates in the Derdere (Late Cenomanian-Early Turonian) and Karababa (Santonian-Early Campanian) formations of the Mardin group as possible source rocks for the crude oil (Figure 18.148). Present-day TOC is typically 1–2 wt.%, with HI between 200 and 300 mg HC/g TOC. A few relatively immature samples from the Karababa Formation exhibit substantially higher TOC and pyrolyzate yields and may be more indicative of the original generative potential of the formation. The source potential of the overlying Karabogaz and Sayindere formations is generally poor, but the Karabogaz Formation may have contributed some petroleum.

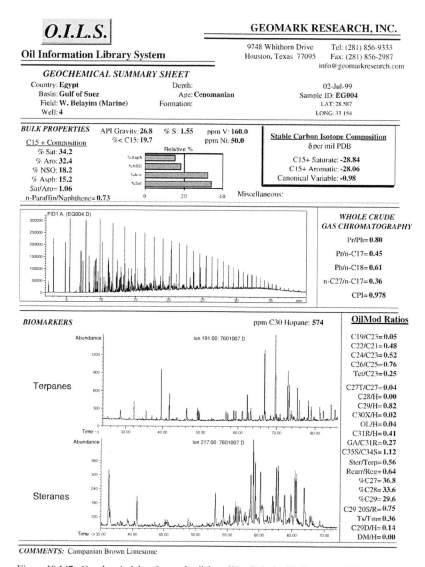

Figure 18.147. Geochemical data for crude oil from West Belayim Field generated from the Campanian Brown Limestone, Gulf of Suez, Egypt (courtesy of GeoMark Research, Inc.).

Upper Cretaceous-Paleocene Latrobe Formation, Gippsland Basin

The Latrobe Formation of the Gippsland Basin is the primary source and reservoir for oils in the Gippsland Basin (Bishop, 1999b). The source rocks are coals and coaly shales deposited in a lower coastal plain setting behind a wave-dominated shoreline (Burns and Emmett, 1992). TOC may exceed 70 wt.%, and HI typically ranges from ~200 to 350 mg HC/g TOC, with maxima near 400 mg HC/g TOC (Fielding, 1992; MacGregor, 1994; Moore et al., 1992; Shanmugam, 1985). These oil-prone coals formed from Araucarian coniferous rainforests that flourished in a raised bog setting. The coniferous vegetation provided large quantities of hydrogen-rich liptinite (e.g. cutinite and resinite) that was enriched further by microbial degradation of woody tissues (e.g. lignite and cellulose). High rainfall, raised groundwater level, low oxygen, high acidity, cold climate, and low-nutrients contributed to

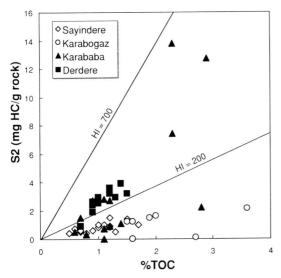

Figure 18.148. Total organic carbon (TOC) (wt.%) versus Rock-Eval S2 for Upper Cretaceous rocks from the Adiyaman region, southeastern Turkey (data from Demirel et al., 2001). HI, hydrogen index.

nated to massively bedded siliceous shale, cherts, ash fall tuff, volcanic glass beds, and volcaniclastic sandstones. Much of the formation is low in carbon (<0.5 wt.% TOC). Organic-rich layers (<1 to ~30 m thick) contain some of the world's richest marine source rocks. These organic-rich layers average >15 wt.% TOC, with maxima of >33 wt.% (Figure 18.150) (Erlich et al., 1996). The kerogen is consistently hydrogen-rich. Samples that contain more than 2 wt.% TOC have average HI >800 mg HC/g TOC. These source rocks were deposited in middle-shelf to outer-slope environments near active subaerial volcanoes. Both upwelling and volcanism provided inorganic nutrients, including silica, which promoted high biotic productivity, particularly episodic blooms of diatoms, radiolarian, and dinoflagellates. The extract of one organic-rich facies is dominated by the C_{25} high-branched isoprenoid produced by diatoms (see Figure 13.16). Anaerobic bacteria extensively reworked the marine phytoplankton (Erlich et al., 1996) and archaea (Walters et al., 1993).

the enrichment and preservation (Shanmugam, 1985). The largely offshore Upper Cretaceous-Paleocene oil-prone facies grades into onshore Middle Eocene-Upper Miocene brown coals with limited oil-generative potential. The Victorian brown coals (Traralgon Formation) are among the largest deposits of sedimentary carbon on Earth (~10^{11} tons) (Holdgate et al., 2000).

The low-sulfur waxy oils generated from the oil-prone coals and coaly shales have distinctive biomarker distributions, including high pristane/phytane (4–10+) and hopane/sterane ratios (Figure 18.149). They contain gymnosperm-derived diterpanes, abundant diasteranes, oleanane and diahopane, and small amounts of tricyclic terpanes (Burns and Emmett, 1992; Noble et al., 1985a; Philp and Gilbert, 1982; Philp and Gilbert, 1986; Philp et al., 1981).

Upper Cenomanian-Campanian Loma Chumico Formation, Costa Rica

The Upper Cretaceous Loma Chumico Formation was deposited in basins along the Pacific flank of Costa Rica in a typical island arc setting. The formation is at least 630 m thick in the Tempisque Basin, consisting of lami-

Campanian-Maastrichtian Ghareb Formation, Dead Sea graben

The Upper Cretaceous ("Senonian") Ghareb Formation contains calcareous oil shale or phosphatic chalk that is the source for Dead Sea asphalts (see Figure 11.2). The formation is exposed in various basins in the region and occurs in downwarped blocks within the Dead Sea graben (Spiro et al., 1983; Minster et al., 1997). The Oil Shale Member of the Ghareb has excellent generative potential (TOC may exceed 25 wt.%) and has been used in retort operations. The organic matter consists of preserved marine planktonic algae (mostly foaminifera) and degraded algal remains. HI typically ranges from 400 to 600 mg HC/g TOC and may exceed 800 mg HC/g TOC for some samples. Deposition occurred under hypersaline conditions resulting from restricted water circulation during closing of the Tethys Ocean (Rullkötter et al., 1984). Dead Sea asphalt has biomarkers indicative of these diagenetic conditions, including high gammacerane and C_{35} hopanes, low diasteranes, low 18α-30-neonorhopane, and little or no oleanane (see Figure 11.3). Kerogen in the Ghareb oil shale is enriched in sulfur incorporated during early diagenesis under euxinic conditions. Immature samples of the Ghareb source rock were used in artificial

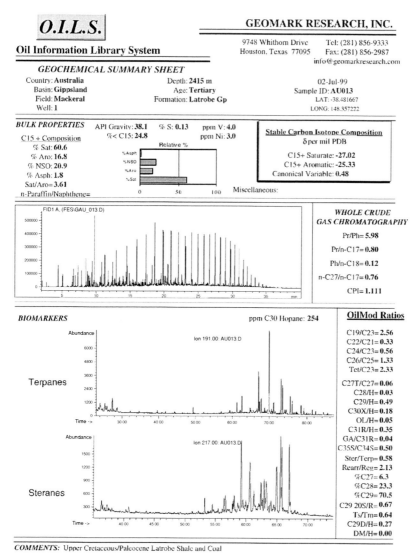

Figure 18.149. Geochemical data for crude oil from Mackeral Field generated from the Upper Cretaceous Latrobe Formation, Gippsland Basin, Australia (courtesy of GeoMark Research, Inc.).

maturation (Clegg et al., 1998; Koopmans et al., 1998) and pyrolysis/chemical degradation experiments (Van Kaam-Peters et al., 1995a; Kohnen et al., 1990b; Koopmans et al., 1999; Dias et al., 2002a).

TERTIARY SOURCE ROCKS

Tertiary source rocks are widespread and generally better preserved than older source rocks (Figure 18.151).

Paleocene–Eocene Midway, Wilcox, and Sparta formations

See Source rocks in the Gulf of Mexico (p. 883) and Figure 18.109.

Paleogene Kapuni coals, Taranaki Basin

Based on biomarker correlations and basin models, Paleocene coals of the Farewell Formation or Eocene

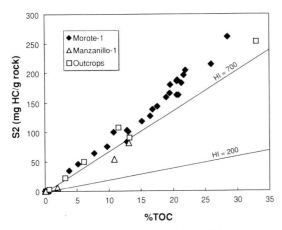

Figure 18.150. Total organic carbon (TOC) (wt.%) versus Rock-Eval S2 for the samples from the Loma Chumico Formation, Costa Rica (data from Erlich et al., 1996). HI, hydrogen index.

coals of the Mangahewa/Kaimiro Formation are the source of petroleum in the Taranaki Basin, New Zealand (Collier and Johnston, 1991; Johnston et al., 1991; Killops et al., 1994; Killops et al., 1996; Armstrong et al., 1996). The exceptions are oils around the Maui Field that correlate with the Upper Cretaceous Rakopi Formation. The Taranaki coals are oil-prone, like those that occur in the Gippsland Basin, which was adjacent to the Taranaki Basin before Late Cretaceous rifting of Gondwana (Curry et al., 1994; Killops et al., 1994). The coals have ~65 wt.% TOC, with HI up to ~300 mg HC/g TOC.

The Late Paleocene thermal maximum: a near-miss ocean anoxic event

About 60 Ma, the Earth entered a period of long-term global warming that weakened ocean circulation for about five million years. At the end of the Paleocene Epoch (~55.5 Ma), ocean temperatures spiked abruptly (Kennett and Stott, 1991; Zachos et al., 1993). This Late Paleocene thermal maximum (LPTM) increased deep-ocean-water temperatures by ~5°C in <10 000 years. Surface-water temperatures increased by 4–8°C, with the greatest impact at high latitudes. The LPTM lasted only ~200 000 years, but it produced low-oxygen conditions that nearly qualify as an OAE. The result was a mass extinction of marine organisms (see Figure 17.25), including up to half of all species of deep-sea foraminifera, many of which had survived the K–T extinction.

Initiation of the LPTM was marked by a sudden decrease in carbon isotopic values in both the oceans and the atmosphere (Figures 18.152 and 18.153). A decrease of ~2.5–4‰ occurs for organic and inorganic carbon in both marine and terrigenous sources (Koch et al., 1992b), probably due to the catastrophic dissociation of gas hydrates and the release of massive quantities of methane into the atmosphere (Dickens et al., 1995; Katz et al., 1999; Katz et al., 2000b). Similar isotopic excursions in Mesozoic OAEs may also be the result of dissociation of methane hydrates.

Carbon dioxide emissions from large flood basalts may have triggered OAEs in the Toarcian Age (Karoo-Ferra basalts) and Cretaceous Period (e.g. the Ontong-Java Plateau basalts in the Pacific Ocean). The LPTM originated due to greenhouse conditions induced by elevated CO_2 (Thomas et al., 1999). Whether global warming progressed gradually until methane hydrate dissolution occurred, or whether it was triggered by massive volcanic eruptions, as suggested by Bralower et al. (1997), remains to be determined (Bains et al., 1999). The rapidity of recovery is perhaps as remarkable as the excursion itself. Carbon isotopic ratios returned quickly to their pre-LPTM values, suggesting an immediate biogenic response. Paleo-productivity spiked after the LPTM, indicating that the enhanced deposition of organic matter in the deep-sea sediments efficiently cooled the greenhouse climate by rapid removal of excess CO_2 (Bains et al., 2000).

Note: Crouch et al. (2001) found that the global abundance of dinoflagellate cysts rose dramatically and then returned to normal in the Late Paleocene Epoch at the time of the $\delta^{13}C$ excursion. Some species of dinoflagellates produce neurotoxins. Blooms of these phytoplankton produce "red tides" that often result in the mass mortality of marine life. Red tides on a global scale during the Late Paleocene Epoch may have resulted in mass mortality that rapidly contributed carbon to the sediments. During this period, mammals were evolving rapidly (e.g. primates, whales, bats, and hoofed species). Whether the thermal event played a role in mammalian evolution is unknown.

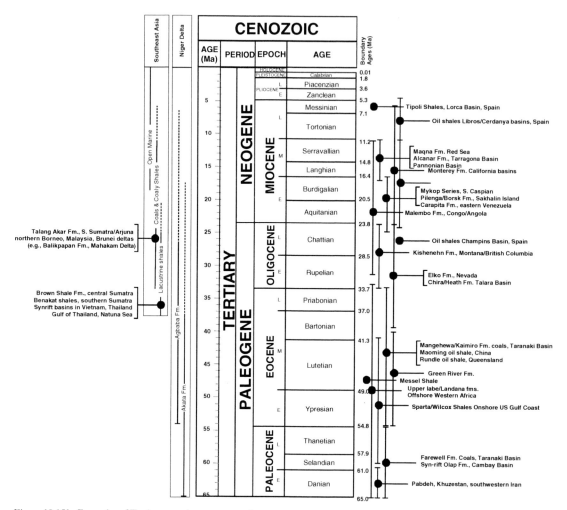

Figure 18.151. Examples of Tertiary petroleum source rocks.

Tertiary deltaic shales, Bangladesh

Source rocks for the Bangladesh Basin are not known with certainty. Over 20 km of largely deltaic Tertiary sediments were deposited, giving rise to multiple potential sources. Curiale *et al.* (2002) list the units identified to date (Table 18.13). All are considered to be gas-prone, and the Jenam Formation is believed to be the most active.

Paleocene–Eocene shales of the Cambay Basin, India

Rifting of the Seychelles and northward movement of the Indian plate resulted in intensive volcanism and deposition of the Deccan Trap flood basalts near the Cretaceous–Tertiary boundary. A sag basin formed within the Cambay rift, and Tertiary sediments were deposited upon the basalt (Biswas, 1982; 1987). Strata with source potential occur in the Paleocene Olpad Formation deposited a continental rift-fill, the Lower Eocene transgressive marine Cambay Shale, and the Upper Eocene regressive coaly shales of the Kalol Formation (Banerjee and Rao, 1993; Biswas *et al.*, 1994; Samanta *et al.*, 1994; Banerjee *et al.*, 2002). Terrigenous organic matter with marginal average generative potential dominates all of these source rocks. TOC averages 2–3 wt.%, while HI averages <150 mg HC/g TOC. However, some immature samples of shales from the lacustrine

Tertiary source rocks   921

Table 18.13. *Potential source rocks in the Bangladesh Basin (Curiale et al., 2002)*

Formation	Age	TOC (wt.%)	Comments
Tipam, Boka Bil, Upper Bhudan	Pliocene/Late Miocene	0.2–1.5	HI = 104–225 mg/g TOC
Lower Bhudan	Middle Miocene	1.76 (average)	S2 = 2–3 mg/g
Middle Miocene (possibly including Upper Jenam)	Oligocene/Middle Miocene	0.4–1.2	HI < 155 mg/g TOC
Upper Jenam	Oligocene	1.4–2.7	HI = 121–166 mg/g TOC
Kopili and Cherra	Eocene	<16	Gas-prone
Gondwana coal measures	Mesozoic	<60	Gas-prone

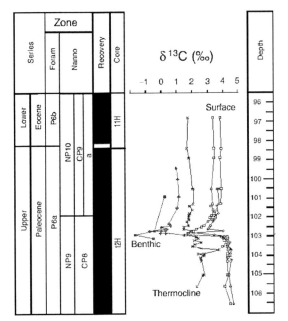

Figure 18.152. Biozonations of a core from ODP Hole 865C (Allison Guyot, Mid-Pacific Mountains) and carbon isotope profiles of individual species of foraminifera from the Late Paleocene–Eocene Epoch. Large isotopic shifts occur for both surface and benthic forams. Reprinted from Kelley et al. (1998). © Copyright 1998, with permission from Elsevier.

Olpad Formation and the marine Cambay Shale have HI >650 mg HC/g TOC and 250 mg HC/g TOC, respectively (Biswas et al., 1994; Banerjee et al., 1998; Banerjee et al., 2000).

Biomarker and isotopic characterization of oils and source-rock extracts define two distinct petroleum systems within the Cambay Basin (Mathur et al., 1987; Pande et al., 1993; Pande et al., 1997; Samanta et al., 1994; Banerjee et al., 2002). In North Cambay, the oils have characteristics of a lacustrine source and correlate with organic facies in the Olpad Formation and lower Cambay Shale. In South Cambay, the oils have characteristics of a deltaic source with high angiosperm input and correlate with upper Cambay regressive coals (see Table 4.3). These inferences are consistent with the geological framework and basin models (Pande et al., 1997; Yalcin et al., 1988).

Lower-Middle Eocene Green River Formation, USA

Approximately 2500 m of Green River Formation lacustrine sediments were deposited in the depressions adjacent to the Laramide Uplift (Roehler, 1992; Fouch et al., 1994) that make up the Greater Green River Basin (Figure 18.154). Two of these early Tertiary lakes included Lake Uinta in the Uinta-Piceance Basin of northwestern Colorado and northeastern Utah and Lake Gosiute in northwestern Colorado and southwestern Wyoming (Fouch et al., 1994). The Green River Formation accounts for ~19 billion barrels of petroleum (Anders and Gerrild, 1984) and is a large potential oil shale resource.

Studies of the origin of the Green River Formation have a long history, partly because of its importance as an oil shale resource (e.g. Henderson, 1924; Bradley, 1925; Bradley, 1931). Many of the earliest biomarker studies were conducted on these organic-rich shales (e.g. Burlingame et al., 1965; Robinson et al., 1965; Murphy et al., 1967; Haug et al., 1967; Burlingame and Simoneit, 1968; Anderson et al., 1969). Readily available outcrop samples have abundant immature to marginally mature extractable hydrocarbons with a relatively simple suite

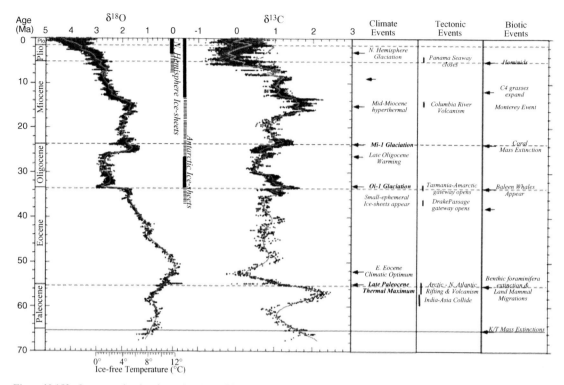

Figure 18.153. Oxygen and carbon isotopic values of foraminifera over Cenozoic time and their relationships to various climatic, tectonic, and biotic events. Reprinted with permission from Zachos et al. (2001). © Copyright 2001, American Association for the Advancement of Science.

of major acyclic and cyclic isoprenoid hydrocarbons. Several key biomarkers were first identified in Green River Formation, including perhydro-β-carotene (e.g. Murphy et al., 1967), various steranes and triterpanes (e.g. Gallegos, 1971) and their aromatic forms (Anders et al., 1973; Gallegos, 1973), and gammacerane (Hills et al., 1966). Geochemists continue to use Green River samples in modern biomarker, maturation, and isotopic investigations (e.g. Ruble et al., 1994; Ruble et al., 2001; Schoell et al., 1994; Kralert et al., 1995; Sinninghe Damsté et al., 1995; Koopmans et al., 1997; Coleman, 2001).

The Green River Formation is not a single organic facies but a collection of members and facies deposited under different conditions. There are several models for the deposition of these sediments, but we have adapted the description of Carroll and Bohacs (2001), who described the generative potential of the formation in Wyoming (Figure 18.154).

The thinnest and earliest lacustrine deposit in the Green River Formation is the Luman Tongue of the Washakie Basin and the laterally equivalent Niland Tongue of the Wasatch Formation, Uinta Basin. These sediments were deposited in fresh to mildly brackish waters under fluvial to overfilled conditions in a humid climate. The Luman and the Niland tongues have highly variable generative potential (Figure 18.155) and were deposited under oxic conditions as profundal lake mudstones and coaly lake margin sediments (Horsfield et al., 1994; Carroll and Bohacs, 2001). Most samples have 2–8 wt.% TOC, but some of the thin coal seams have >50% TOC. Rock-Eval pyrolysis identifies two distinct populations of kerogen. Samples with HI <500 mg HC/g TOC are mixtures of algal and terrigenous plant matter. The other population averages ~4.9 wt.% TOC and contains homogeneous, algal-derived kerogen, with an average HI of ~805 mg HC/g TOC. The differences in generative potential are

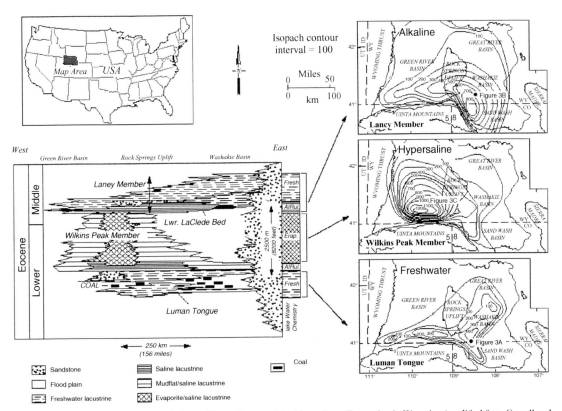

Figure 18.154. Location, sedimentary facies, and isopach maps of the Green River Formation in Wyoming (modified from Carroll and Bohacs, 2001). Reprinted by permission of the AAPG, whose permission is required for further use.

attributed to sediment dilution rather than to variations in biotic input.

The Wilkins Peak Member was deposited under evaporitic, underfilled conditions (Figure 18.155). It is equivalent to the Douglas Creek and Middle Green River Formation in the Uinta Basin. The member has a wide variety of lithofacies (alluvial fans, sandstones, bedded trona and halite, and laminated shales). Sedimentary cycles of organic-rich mudstones deposited during flooding and carbonate and evaporite facies are evident in the basin center. Kerogen in the Wilkens Peak Member mudstones is amorphous and originated from algal and bacterial organic matter. While the kerogen is fairly homogeneous within the member, its concentration depends on lithofacies. TOC is highest (up to 18 wt.%) where the lake was deep and lowest in shallow and alluvial settings. Biomarkers in extracts of the Wilkins Peak Member are consistent with deposition under hypersaline conditions (e.g. high gammacerane).

A shift toward balanced-fill conditions marks deposition of the LaClede Bed (Lower Laney Member) in Wyoming and the Parachute Creek Member in Utah and Colorado. This unit includes the frequently studied Mahogany oil shale bed in the Uinta Basin. Sediments show well-defined parasequences of flooding and dessication with organic-rich deposits occurring in calcareous shales (Figure 18.156). The Parachute Creek Member of the Mahogany Zone in the Uinta Basin has >30 wt.% TOC, with HI typically in the range 800–1000 mg HC/g TOC (Katz, 1995b; Dean and Anders, 1991). The kerogen within these members is similar, consisting almost entirely of alginite with an average HI of ~830 mg HC/g TOC. The parasequences, however, have distinct biomarker signatures (Horsfield et al., 1994).

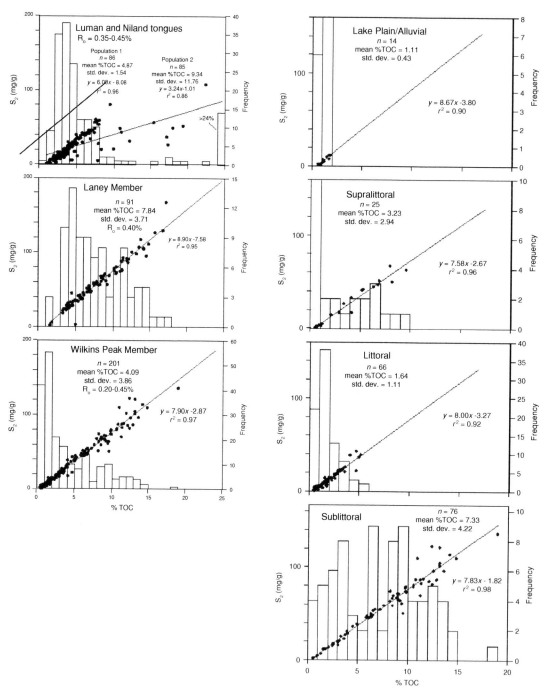

Figure 18.155. Total organic carbon (TOC) (wt.%) versus Rock–Eval S2 (points) and TOC histograms for member units of the Wyoming Green River Formation (left) and interpreted depositional environments for facies within the Wilkins Peak Member (Carroll and Bohacs, 2001). Vitrinite reflectance (R_o) is indicated for each population. Average hydrogen index (HI) is the slope of the linear regression × 100. Reprinted by permission of the AAPG, whose permission is required for further use.

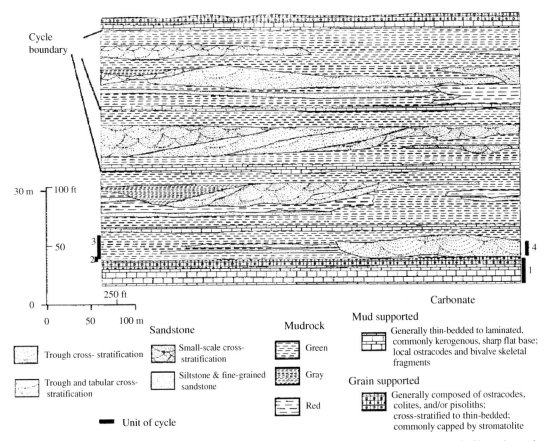

Figure 18.156. Depositional cycles (parasequences) of the Green River Formation, showing strata from a marginal lacustrine setting exposed in the Unita Basin (Fouch et al., 1994). This sequence is typical of Lower-Middle Eocene strata, which formed in a marginal lacustrine setting crossed by numerous streams. Reprinted by permission of the AAPG, whose permission is required for further use.

Much of the Green River Formation is thermally immature (Figure 18.157). However, in the northern part of the Uinta Basin, lacustrine source rocks were buried sufficiently (>3000 m) to generate ~500 million bbl of recoverable, low-sulfur, paraffinic crude oil as produced from the Altamont-Bluebell trend (Figure 18.158) (Fouch et al., 1994; Ruble et al., 2001). Marginal and open-lacustrine source rocks at this location are dominated by carbonate oil shales that contain up to 60 wt.% type I kerogen.

The distinct biomarker signatures of individual source-rock facies, distinguishable by simple multivariate statistical analysis, are preserved in the expelled oils (Mueller, 1998). For example, oils from the Altamont Field correlate (see Figure 9.4) with hydrous pyrolyzates from the type I kerogen in offshore open-lacustirne source rocks like the mahogany shale and upper carbonate marker horizons of the Green River Formation (see Figure 9.5) (Ruble et al., 2001). These low-maturity crude oils have a strong odd-to-even n-alkane predominance and abundant acyclic isoprenoids and β-carotane. Hydrous pyrolyzates from the basal Green River black shale facies (see Figure 9.6) correlate with deeper crude oils from the Altamont Field (see Figure 9.4). These moderate- to high-maturity samples are solid waxes at room temperature and are dominated by n-alkanes with no odd-to-even predominance, lack β-carotane, and have low acyclic isoprenoids.

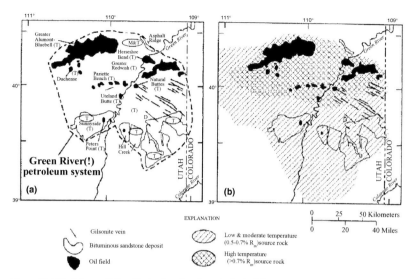

Figure 18.157. Maps of Uinta Basin showing the principal oil fields, bituminous sandstone deposits, and gilsonite veins in the Green River(!) petroleum system (see Table 4.3) and maturity zones for Eocene lacustrine source rocks (from Fouch et al., 1994). Reprinted by permission of the AAPG, whose permission is required for further use.

Eocene Maoming oil shale, China

The Eocene Maoming Shale (Guangdong Province) is the largest deposit of oil shale in China. Other Tertiary deposits, such as Fushun (Liaoning Province), Huadian (Jilin Province), and Haungxian (Shandong Province), are more than an order of magnitude smaller. During the period 1960–90, the Maoming oil shale was exploited in a massive open-pit mining and retorting operation. In addition to its high content of organic matter, the Maoming Shale contains a rich fossiliferous assemblage of megaflora and fauna.

The Maoming Shale contains a typical lacustrine sequence with highly variable organic facies. A basal member was deposited under overfilled conditions. With increased accommodation, balanced-fill conditions promoted the development of anoxic bottom waters and enhanced preservation of organic matter. The lake system later advanced into underfilled conditions, with seasonal periods of desiccation. Variations in organic facies reflect these depositional changes (Brassell et al., 1985; 1988). The basal member is a lignite and vitrinitic coal, whereas the organic-rich shale (up to $\sim$20 wt.% TOC, with HI $\sim$600 mg HC/g TOC) (Kuangzong, 1988) contains mixtures of freshwater planktonic algae. Dinoflagellates dominate some zones, yielding extracts enriched in dinosteranes and 4-methylsteranes. *Botryoccoccus* dominates other zones, yielding extracts enriched in botryoccocane and related isoprenoids (Summons et al., 2002).

Eocene Shahejie Formation, Bohai Bay, China

Eocene shales of the Shahejie Formation (Dongying Depression) have excellent source potential (Figure 18.159). A series of faulted depressions formed during the Paleogene Period, which latter condensed into the Bohai Bay Basin as the entire area subsided. Li et al. (2003) documented the source potential of Eocene shales within one of the depressions. These potential source rocks are immature but have contributed biomarkers to migrating, mature oil.

Middle Eocene Messel shale, Sprendlinger Horst, Germany

The Messel oil shale pit is one of six occurrences of Eocene ($\sim$48 Ma) oil shale in the Sprendlinger Horst, a structure of uncertain origin east of Darmstadt, Germany. These lacustrine shales have high TOC

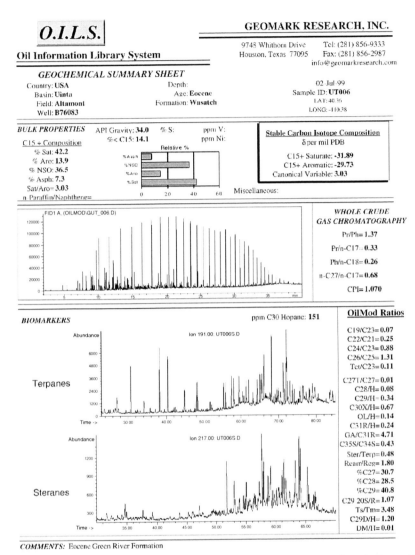

Figure 18.158. Geochemical data for crude oil from Altamont Field generated from the Eocene Green River Formation, Uinta Basin, USA (courtesy of GeoMark Research, Inc.).

(~30 wt.% TOC), most of which occurs in annual laminations of alginite from *Tetraedon*, a unicelluar green alga (Goth et al., 1988). Messel shales are barely lithified, and fossilized *Tetraedon* are morphologically identical to extant organisms. Abundant molecular and petrologic evidence shows that the shales were deposited in a deep freshwater lake under subtropical, balanced-fill conditions with anoxic bottom water and a euxinic photic zone.

Although world-renowned for its animal and plant fossils, the Messel Shale has provided an equally impressive treasure of biomarkers. Many individual biomarkers and biomarker classes were first identified in Messel Shale extracts or kerogen degradation products. These include 4-methylsterols (Mattern et al., 1970), homohopanes (Ensminger et al., 1972; 1974), tricyclic and tetracyclic terpanes (Kimble et al., 1974a), aromatized hopanes (Greiner et al., 1977; Schaeffer et al., 1995b),

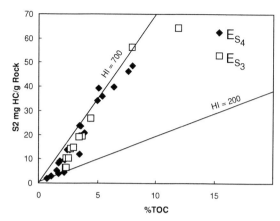

Figure 18.159. Total organic carbon (TOC) versus Rock-Eval S2 for the samples from the Eocene shales (E_{S_4} and E_{S_3} Members) of the Shahejie Formation, Bohai Bay Basin, China (data from Li et al., 2003).
HI, hydrogen index.

archael lipids (Michaelis and Albrecht, 1979; Chappe et al., 1979; Chappe et al., 1980), and numerous porphyrins (Prowse and Maxwell, 1989; Ocampo et al., 1992). Extracts of Messel Shale also provided some of the earliest $\delta^{13}C$ reported for individual biomarkers (Hayes et al., 1987; 1990).

> Note: Since the discovery of a fossilized crocodile in 1875, the exceptionally rich and well-preserved fossils of Messel Shale have provided scientists with a window into the flora and fauna of central Europe during the Eocene Epoch. The Messel pit near Darmstadt, mined for its oil shale between 1859 and 1971, was the source for many famous fossil discoveries, including ancient horses, tapirs, bats, crocodiles, snakes, and turtles, many of which have preserved soft-tissue and stomach contents. Plans drawn up in the 1980s to convert the Darmstadt open-pit mine into a waste dump were abandoned in recognition of scientific value of the strata. The United Nations Educational, Scientific, and Cultural Organization (UNESCO) declared the area a World Heritage site in 1995.

Eocene Rundle oil shale, Queensland, Australia

Several fault-bounded (half-graben), elongated basins along the coast of central Queensland contain organic-rich lower Tertiary lacustrine rocks. The Rundle oil shale in the Narrows Graben is one of the richest and most extensive of these rocks. The bulk of the kerogen in the Rundle oil shale consists of alginite, which is composed of bundles of 10–60-nm-thick ultralaminae based on transmission electron microscopy (Derenne et al., 1991). The ultralaminae are similar to algaenan, the non-hydrolyzable macromolecules that occur in the outer cell walls of microalgae, particularly Chlorophyceae (green algae). Rundle oil shale averages >13 wt.% TOC, with HI >800 mg HC/g TOC, but it is thermally immature, except near igneous intrusions (Gilbert et al., 1985; Saxby and Stephenson, 1987).

Eocene Pabdeh Formation, Iran (Khuzestan/Lurestan basins)

The Pabdeh Formation contains the only Tertiary source rocks in the Persian Gulf region (Bordenave and Burwood, 1990). Deposition of marls under anoxic conditions occurred in a relatively deep trough running from the northern Fars region through the Dezful Embayment (southwest Iran) and into the Lurestan Basin (northeast Iraq) (see Figure 18.87). Sedimentation was continuous from the Late Cretaceous Period to the Oligocene Epoch, but anoxic conditions did not develop until the Middle–Late Eocene Epoch. In Iran, ~150–250 m of organic-rich Pabdeh marls contain ~3–6 wt.% TOC, with HI in the range ~375–600 mg HC/g TOC (Figure 18.160). In southwest Lurestan, the source unit

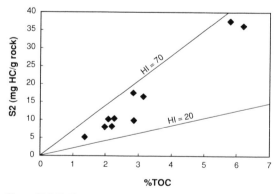

Figure 18.160. Total organic carbon (TOC) (wt.%) versus Rock-Eval S2 for Pabdeh Formation in a well from the Dezful Embayment, Iran (Bordenave and Burwood, 1990).
HI, hydrogen index.

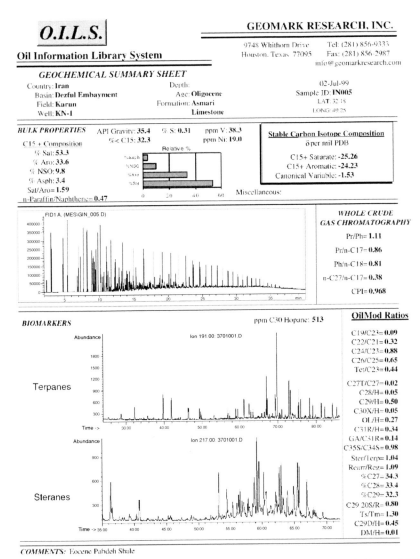

Figure 18.161. Geochemical data for crude oil from Karun Field generated from the Eocene Pabdeh Formation, Dezful Embayment, Iran (courtesy of GeoMark Research, Inc.).

is thinner but richer (80–120 m, up to 11.5 wt.% TOC). Anoxic conditions persisted into the Oligocene Epoch in this area, producing ~100 m of comparatively lean marls (1.5–3.5 wt.% TOC).

Figure 18.161 shows geochemical data for crude oil produced from the Oligocene Asmari Limestone from Karun Field in the Dezful Embayment, Iran, that was generated from the Eocene Pabdeh Formation. The oil contains abundant oleanane, consistent with a Tertiary source rock (e.g. see Figure 13.94).

Eocene–Oligocene Elko Formation, northeast Nevada

Palmer (1984a) described the source potential of the Eocene–Oligocene Elko Formation, one of many lacustrine deposits in the Basin and Range (Figure 18.162). Two organic facies are present – an organic-rich calcareous oil shale and a lignitic siltstone. Biomarkers are consistent with the former being deposited in balanced-fill to underfilled lake conditions,

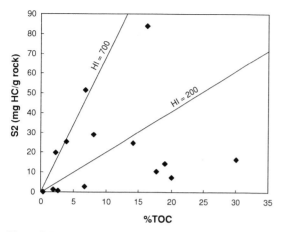

Figure 18.162. Total organic carbon (TOC) (wt.%) versus Rock-Eval S2 for oil shale and lignitic siltstone facies of the Elko Formation, Nevada (data from Palmer, 1984a). HI, hydrogen index.

while the latter was deposited in overfilled to fluvial environments.

> Note: Robert M. Catlin opened the main shaft to begin mining Elko oil shale in 1916. The operation was successful for a brief period (1922–24), but eventually it could not provide a product that was cheaper than conventional oil. The facilities are now in ruins (Figure 18.163), but they are commemorated by Nevada Historical Marker 229 (located on US Highway 40, opposite Elko Airport), which ends with the words: "Easily 50 years ahead of his time, Catlin did, for a few years, give Elkoans and Nevadans a dream and the community an oil boom in the Roaring Twenties."

Eocene Chira and Oligocene–Miocene Heath formations, Talara Basin, Peru

The Talara Basin formed in an Eocene trench-slope setting that was once part of a larger Paleocene forearc basin resulting from the collision of the Andean and Pacific tectonic plates. Progressive accretion to the margin resulted in a complex geologic history that included deposition of source rocks (Jaillard et al., 1995). The Talara Basin and the Progreso Basin to the north have numerous small oil and gas fields that collectively sum to >1.7 billion barrels of produced hydrocarbons (Higley, 2002). Upper Cretaceous Redondo Shale, which is commonly cited as the source rock for these oils, was deposited under upwelling, dysaerobic conditions (Pindell and Tabbutt, 1995). However, the oils are enriched in ^{13}C (δ^{13}C ~−22‰) and contain abundant oleanane, suggesting Tertiary source rocks. Possible source rocks include the Eocene San Cristobal Formation, the Chacra Group, and the lower Talara and Chira formations, Lower Eocene Palegreda neritic marine shales, the Oligocene-Miocene Heath Formation, and the Paleocene Balcones Shale (Higley, 2002).

From a limited data set, we found that samples from the Chira and Heath formations have some generative source potential (Figure 18.164) and correlate well with produced oils (Figure 18.165). Figure 18.166 shows geochemical data for crude oil from Carrizo Field generated from the Heath source rock in the Talara Basin, Peru.

Oligocene Kishenehn Formation, Montana and British Columbia

The Kishenehn Basin in northwestern Montana and southeastern British Columbia is one of many intermontane basins in the Rocky Mountain Overthrust that contain thick (~3–5 km) Tertiary sediments deposited directly on Proterozoic Belt rock (Fields et al., 1985). The Oligocene Kishenehn Formation consists of fluvial and lacustrine calcareous shales and lignites. Deposition occurred under overfilled to balanced-fill conditions, allowing excellent preservation of both land-plant and freshwater algal organic matter (Figure 18.167). Samples with type I kerogen have an average of ~6 wt.% TOC and HI >900 mg HC/g TOC (Curiale et al., 1988).

Curiale (1987; 1988) and Curiale and Lin (1991) studied the biomarker distributions in the Kishenehn Formation. Nine distinct steroidal hydrocarbons families were identified in extracts: regular steranes and sterenes, diasteranes and diasterenes, spirosteranes and methylspirosterenes, 4-methylsteranes, C-ring monoaromatic steroids, and B-ring monoaromatic anthrasteroids. In addition to these steranes, onocerane (indicative of angiosperm input) and botrycoccane (indicative of the green alga *Botryococcus*), along with the usual pentacyclic terpanoids, were also detected. The biomarker distribution is consistent with the Kishenehn lake system, which consisted of lowlands that contained eutrophic alkaline lakes, swamps, marshes, and braided streams surrounded by thickly wooded highlands (Curiale et al., 1988).

Figure 18.163. Photograph of Catlin oil shale plant in 1923. Used with permission of the Northeastern Nevada Museum Photograph Collection (document # 46–77).

Upper Oligocene Talang Akar Formation, South Sumatra and Ardjuna basins

Coals and coaly shales of the Talang Akar Formation are the major source rock in the Ardjuna Basin of the North Sumatra Basin, Java, and are secondary sources in basins in southern Sumatra (Gordon, 1985; Noble *et al.*, 1991; Noble *et al.*, 1997; Nugrahanto and Noble, 1997; Bishop, 2000a). Source facies within the Talang Akar Formation were deposited both in the grabens (>600 m thick) and on the flanks and outside of the grabens (<125 m thick). The oldest basal shales deposited within the grabens are nearly identical to the older lacustrine source units and contain primarily algal-derived organic matter. Younger strata reflect a shift from balanced-fill to overfilled conditions, resulting in the deposition of coals and coaly shales. The rocks contain 40–70 wt.% TOC and tend to be oil-prone, with HI in the range of 200–400 mg HC/g TOC.

Oligocene–Lower Miocene Maykop series, South Caspian

The South Caspian Basin contains more than 22 km of sediment resting on Lower Jurassic oceanic crust. The South Caspian depression accumulated marine

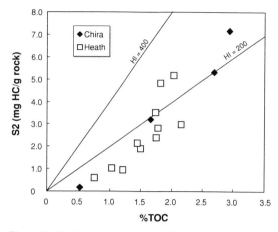

Figure 18.164. Total organic carbon (TOC) (wt.%) versus Rock-Eval S2 for well cuttings and core samples of the Eocene Chira and Oligocene-Miocene Heath formations from the Talara Basin. HI, hydrogen index.

sediments as part of the Tethyan Sea since the Middle Jurassic Period. The Middle Eocene Kuma, Upper Oligocene–Lower Miocene Maykop, and Middle Miocene Diatomaceous Zone have petroleum source potential. During the Miocene Epoch, collision of the Arabian and Eurasian plates began to close off the Caspian Sea. Although it became a lacustrine setting, the resulting water body was vast enough to have open marine characteristics. By the Late Miocene–Early Pliocene Epoch, tectonic activity uplifted the Caucasus Mountains and formed many of the horst and graben structures that host present-day oil accumulations. Rapid basin subsidence and deposition of eroded sediment from the Caucasus region resulted in extremely high sedimentation rates.

Maykop Shale and, to a lesser extent, the Miocene Diatomaceous Zone are the principal source rocks for oil in the Azerbaijan and offshore Turkmenistan areas of

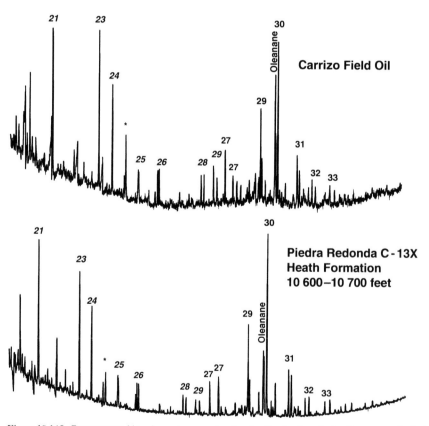

Figure 18.165. Reconstructed ion chromatograms showing the distribution of tricyclic terpanes (italics) and pentacyclic terpanes (normal font) for oil from the Carrizo Field and a rock extract of the Heath Formation. Both contain oleanane and an unusual tetracyclic terpane (*).

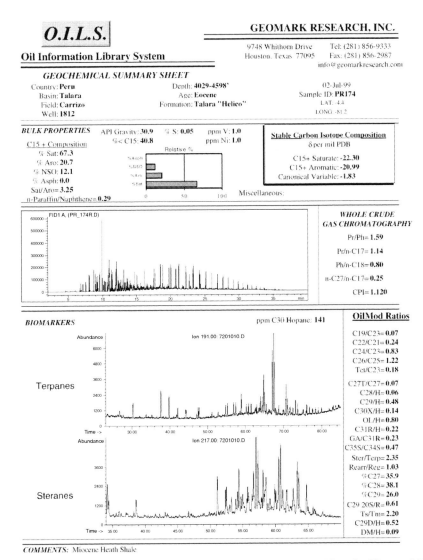

Figure 18.166. Geochemical data for crude oil from Carrizo Field generated from the Oligocene-Miocene Heath Formation, Talara Basin, Peru (courtesy of GeoMark Research, Inc.).

the South Caspian (Abrams and Narimanov, 1997; Inan et al., 1997; Saint-Germes et al., 1997; Wavrek et al., 1998; Katz et al., 2000d). This interpretation is based on thermally immature rock samples from onshore outcrops, offshore wells, and rare, thermally mature ejecta from mud volcanoes (e.g. see Figure 4.32). Maykop Shale tends to have ~1–4 wt.% TOC, although individual samples may be up to ~16 wt.% (Figure 18.168). Rock-Eval HI indicates varying mixtures of type II and III or bacterially degraded organic matter, with the high-TOC samples tending to have the higher-quality kerogen (Figure 18.169). The kerogen is mainly amorphous organic matter, with some terrigenous and marine macerals.

Saint-Germes et al. (2000) showed a correlation between ^{13}C-rich Maykop kerogens and higher HI. They attributed this relation to a reduction in ^{13}C of the marine component, while the amount of ^{13}C in the land-plant component remained unchanged. This suggests that the late Tertiary global isotopic shift in marine organic matter occurred during the Upper Rupelian Age (Early Oligocene Epoch) in this area.

934 Petroleum systems through time

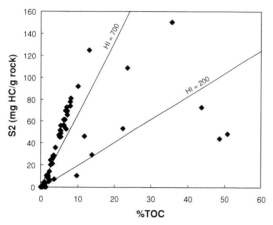

Figure 18.167. Total organic carbon (TOC) (wt.%) versus Rock-Eval S2 for samples from the Kishenehn Fomation, Montana (data from Curiale *et al.*, 1988). HI, hydrogen index.

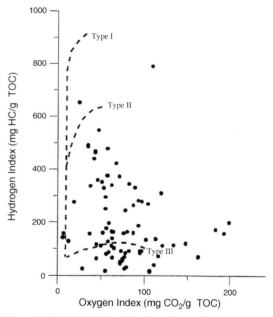

Figure 18.169. Modified van Krevelen diagram for Maykop mudstones. Reprinted from Katz *et al.* (2000d). © Copyright 2000, with permission from Elsevier.

The source potential of the Middle Miocene Diatomaceous Zone is understood less well than that of the Maykop. Although it contains essentially the same organic facies as the older Maykop, most of the available samples have lower source potential, with <1 wt.% TOC. Nonetheless, individual samples suggest that the Diatomaceous Zone may also be a viable source rock. One sample ejected from a mud volcano had 2.62 wt.% TOC, with HI of 721 mg HC/g TOC (Inan *et al.*, 1997).

Note: Exploitation of natural seepage in the Baku Peninsula occurred as early as ∼700 BC. The army of Alexander the Great used Baku oil from hand-dug wells during the siege of Persia in 331 BC. Marco Polo wrote in his Silk Route memoirs of fountains of oil that were barreled and loaded onto camel caravans for trade. Mechanical drilling opened vast untapped reservoirs. The first oil well was drilled in Baku in 1848, 11 years before Colonel Edwin L. Drake's well in Pennsylvania. Baku was the center of the first oil boom, with >200 wells producing by 1870 (Figure 18.170).

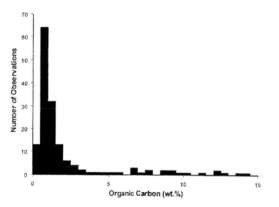
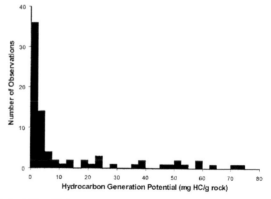

Figure 18.168. Distribution of total organic carbon (TOC) and Rock-Eval S1 + S2 for Maykop outcrop and oil-field samples. Reprinted from Katz *et al.* (2000d). © Copyright 2000, with permission from Elsevier.

Figure 18.170. Hand-digging wells in Baku (left). The Nobel brothers' oil wells in Balakhani, a suburb of Baku (right). The derricks were spaced closely to maximize short-term production rates. Both figures from Mir-Babayev, M. Y. (2002) Azerbaijan's oil history: a chronology leading up to the Soviet Era. www.azer.com/aiweb/categories/magazine/ai102_folder02_articles/102_oil_chronology.html

The potential for vast wealth at Baku attracted established financiers, such as the Rothschilds, as well as upstart entrepreneurs, such as the Nobel brothers. When Immanuel Nobel, a Swedish engineer, went bankrupt in St Petersburg in 1859, he and his wife returned with their sons, Alfred and Emil, to Stockholm, while their other sons Robert and Ludvig remained in Russia. Robert Nobel saw the potential in drilling wells while on a trip to the South Caspian and, in 1876, convinced Ludvig to jointly start BraNobel, a Baku-based oil company. While Robert returned to Sweden, Ludvig remained and amassed a fortune. Alfred Nobel, who made his own fortune in explosives and chemicals, was the largest single shareholder in BraNobel. Five years after Alfred's death, the Nobel Prize was established in 1901, with an endowment of 31 million Swedish crowns, of which ~12% came from shares in the oil fields of Baku.

Middle Oligocene–Upper Miocene Malembo Formation, Congo/Angola basins

The Tertiary Malembo Formation is a minor source unit in the offshore Congo and Angola region (Duval et al., 1995; Reynaud et al., 1998; Schoellkopf and Patterson, 2000) compared with the overlying Upper Cretaceous–Lower Tertiary Iabe/Landana marine shales (Figure 18.171). The Malembo Shale is immature throughout much of the region but typically contains ~1–2 wt.% TOC and type II–III kerogen, with HI of ~250 mg HC/g TOC. The highest generative potential occurs in the basal Oligocene shales, possibly reflecting high sea level or upwelling conditions or association with regional faulting (Schoellkopf and Patterson, 2000). Where mature in deep Tertiary troughs, the Malembo Shale has generated and expelled oil, which commonly migrates into adjacent Pinda structures or Pliocene-Pleistocene sandstones. Malembo oils contain higher oleanane than Iabe oils.

Eocene–Oligocene lacustrine shales, Sumatra

A series of back-arc basins formed in Sumatra, Indonesia, during the Late Cretaceous–Early Tertiary subduction of the Indian Ocean plate under the Southeast Asian plate. Deep freshwater lakes formed in extensional half-grabens that favored deposition of thick source beds with excellent generative potential. A period of basin subsidence and relative quiescence occurred in the Late Oligocene–Miocene Epoch, followed by widespread basement compression from Pliocene to recent time. The occurrence of giant petroleum accumulations within these basins depends mostly on the deposition of lacustrine or reservoir sandstones and on complex structural features and faulting resulting from local tectonics.

Brown Shale Formation

The Pematang Group of the Central Sumatra Basin is divided into lower and upper fluvial red beds that encase coals and organic-rich lacustrine shales deposited

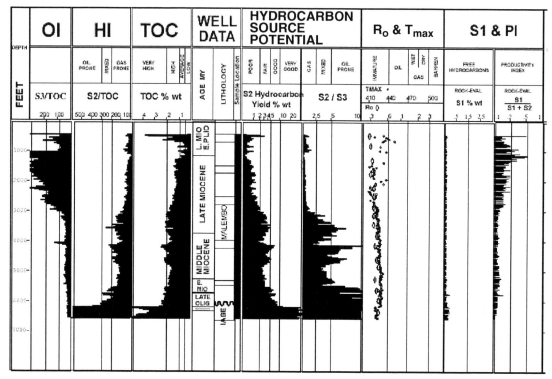

Figure 18.171. Geochemical well log comparing the source potential of the Tertiary Malembo compared with the Upper Cretaceous–Lower Tertiary Iabe/Landana formations (from Schoellkopf and Patterson, 2000). Reprinted by permission of the AAPG, whose permission is required for further use.
HI, hydrogen index; OI, oxygen index; TOC, total organic carbon.

under overfilled to balanced-fill conditions. These coals and lacustrine shales form the Brown Shale, which is the source rock for all oils in the region (Kelley et al., 1994). The source rocks provided the overlying Miocene Sihapas marine sandstones with >60 billion barrels of oil, including the giant Minas and Duri fields (see Table 4.3).

The source facies of the Brown Shale was deposited in deep, anoxic lacustrine systems. Samples with >4 wt.% TOC contain abundant alginite and have HI >700 mg HC/g TOC (Figure 18.172). Samples with <4 wt.% TOC have highly variable generative potential, with HI ranging from <100 to nearly 1000 mg HC/g TOC. The measured variability may be caused by degradation of organic matter during oxic or dysoxic deposition or loss of generative potential due to thermal maturation. The alginite consists mostly of the green microalga *Botryoccocus braunii*. Various isoprenoids characteristic of this organism occur in

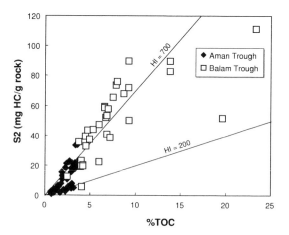

Figure 18.172. Total organic carbon (TOC) (wt.%) versus Rock-Eval S2 for Brown Shale well samples from Central Sumatra (data from Kelley et al., 1994).
HI, hydrogen index.

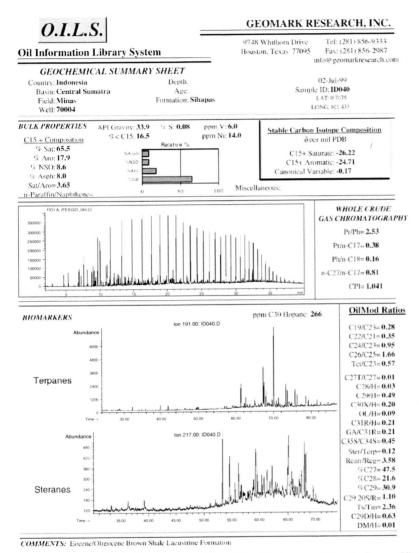

Figure 18.173. Geochemical data for crude oil from Minas Field generated from the Eocene-Oligocene Brown Shale, Central Sumatra, Indonesia (courtesy of GeoMark Research, Inc.).

source-rock extracts and oils generated from this organic facies (Seifert and Moldowan, 1980; Summons et al., 2002).

Crude oil from Minas Field is typical of petroleum generated from the Eocene-Oligocene Brown Shale (Figure 18.173): Minas oil is highly paraffinic because most of its hydrocarbons originate from *Botryococcus*. The oil has prominent high-molecular-weight n-alkanes that exhibit a bimodal distribution. This biological input is also reflected in diagnostic biomarkers, such as botryococcane and polymethylsqualanes. Steroidal hydrocarbons dominated by diasteranes, low relative abundance of C_{28} steranes, and absence of C_{30} n-propylsteranes are consistent with a non-marine clastic source. The distribution of terpanoids is characteristic of many Tertiary lacustrine shales. Low concentrations of tricyclic terpanes and extended homohopanes, a small (but reliably measureable) amount of oleanane, and a predominance of C_{27}–C_{30} hopanes are characteristics common to many Tertiary

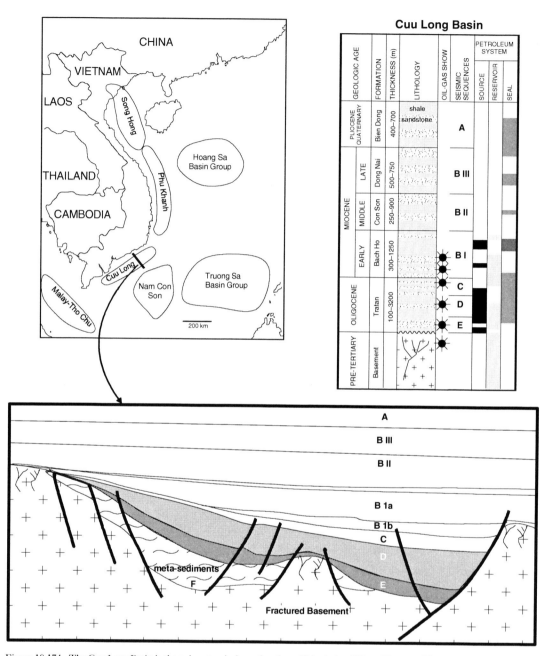

Figure 18.174. The Cuu Long Basin is the only extensively explored synrift basin in offshore Vietnam. The source rocks are primarily Oligocene–Lower Miocene lacustrine shales. Middle Miocene coals may have made a minor contribution to the petroleum system. Oil is produced primarily from Oligocene and Lower Miocene sandstones and fractured granites that immediately underlie the Oligocene source rocks. Modified from figures available from the Vietnam Oil and Gas Corporation, www.petrovietnam.com.vn/main.htm.

lacustrine shales deposited under an oxic water column. The $\delta^{13}C$ values for the Minas oil yield a negative canonical variable, which is usually associated with marine oils.

Benakat Shale

The South Sumatra Basin has two principal source units, the lacustrine Benakat Shale of the Lahat (Lemat) Formation and the overlying Talang Akar coals and coaly shales (Sarjono and Sardjito, 1989; Suseno et al., 1992; Bishop, 2000a). The Benakat Shale typically has 1.7–8.5 wt.% TOC, with individual samples as high as 16 wt.%. HI is comparatively low, ranging from 130 to 290 mg HC/g TOC, reflecting a mixture of degraded algal and terrigenous plant input. The poorer generative potential of the Benakat Shale compared with other equivalent shales in the region may be attributed to deposition under oxic, underfilled conditions.

Banuwati Shale

These shales are the principal source unit for oil in the Sunda and Asri basins (Wickasono et al., 1992; Noble et al., 1997; Bishop, 2000b). The shales are equivalent to the Brown Shale unit of central Sumatra and were deposited in deep, anoxic freshwater lakes. Balanced-fill conditions promoted algal growth and preservation resulting in alginite-rich source rocks with ∼2–8 wt.% TOC. Hydrogen indices are ∼570–640 mg HC/g TOC, but these measurements are from partially mature strata and the original generative potential was higher.

Oligocene lacustrine shales, synrift basins of Vietnam

A series of extensional grabens and half-grabens formed along the coastline of Vietnam during the Paleogene Period, giving rise to the deposition of non-marine, clastic deposits. The Song Hong, Phu Khanh, and Cuu Long basins are near the coastline, while the Nam Con Son, Malay-Tho Chu, Hoang Sa, and Truong Sa basins are in deeper water (Figure 18.174). Most of these basins are explored poorly, but significant reserves have been produced from the Cuu Long Basin. The Song Hong, Nam Con Son, and Malay-Tho Chu basins have significant petroleum potential (Todd et al., 1997; Matthews et al., 1997; Petersen et al., 2001). The synrift basins of Vietnam share many of the characteristics of other petroliferous basins in Southeast Asia. All are defined by high sedimentation rates, abrupt changes in facies, numerous unconformities, and poor lateral continuity in sedimentary sequences. Source rocks were deposited mainly as Oligocene-Early Miocene lacustrine shales, with TOC in the range ∼ 1–7 wt%. Early Miocene carbonate mudrocks and Late Miocene coals also may contribute minor amounts of hydrocarbons to the petroleum system.

The Bach Ho (White Tiger) Field in the Cuu Long Basin was discovered in 1975; it contains waxy terrigenous oils in Miocene sandstone reservoirs (Figures 18.174 and 18.175). Additional reserves were discovered in Oligocene sandstones and fractured basement granites in 1985 and 1986, respectively (Areshev et al., 1992; Canh et al., 1994). The latter occurrence gave rise to the theory that the granites were source rocks that expelled petroleum into overlying sediments (Gavrilov, 2000). However, the lacustrine shales are generally accepted as the source of the oil. For example, Oligocene cuttings from 9740–9930 feet (2970–3027 m) in the Bach Ho-1 well contained an average of 2.75 wt.% TOC, with HI ∼590 mg HC/g TOC. The Bach Ho-1 oil (Figure 18.175) has many geochemical characteristics that are diagnostic of lacustrine oil, such as low $C_{31}R/C_{30}$ hopane (0.24) and high C_{26}/C_{25} tricyclic terpane (1.69) ratios (e.g. see Figure 13.77), low sulfur (0.25 wt.%), and low V/(V+Ni) (0.17).

Petersen et al. (2001) documented the source potential of outcrop Oligocene lacustrine shales at Dong Ho in northern Vietnam. These shales are thought to be analogs for offshore source rocks in the Song Hong Basin. Two organic facies with high generative potential occur: lacustrine mudstones (∼6.5–17 wt.% TOC, HI ∼472–690 mg HC/g TOC) and oil-prone humic coals (HI ∼200–242 mg HC/g TOC). The mudstones were deposited under balanced-fill, oxygen-deficient conditions with reduced input of siliciclastics. Kerogen in these shales is dominated by freshwater algae (e.g. *Botryoccoccus*) and bacterially degraded land-plant matter.

Tertiary lacustrine shales, Gulf of Thailand/Natuna Sea rift basins

A series of north-south trending half-graben rift basins formed in the Gulf of Thailand and Natuna Sea

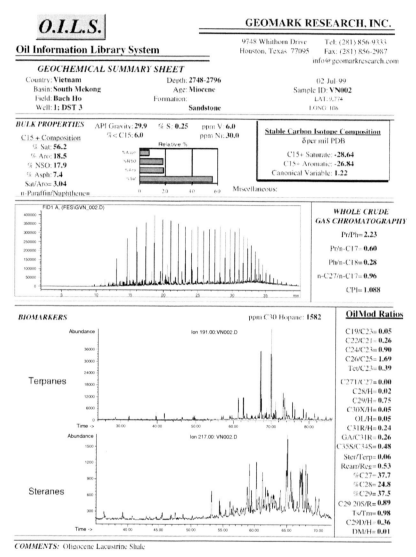

Figure 18.175. Geochemical data for crude oil from the Bach Ho Field generated from Oligocene lacustrine shales, South Mekong Basin, Vietnam (courtesy of GeoMark Research, Inc.).

in response to crustal extension caused by the northward collision of the Indian Plate with Eurasia. These basins typically contain two potential source rocks: Upper Eocene-Upper Oligocene synrift lacustrine shales and Miocene post-rift coals and coaly shales. Both source units are common in most Gulf of Thailand basins. In some basins, high rates of sedimentation and high geothermal gradients have pushed both source units into the oil window. In other basins, only the older lacustrine shales have entered the oil window.

Documentation of the source potential of the Eocene-Oligocene lacustrine shales and the Miocene coaly sources is limited by few wells and high thermal maturity. For example, Oligocene deep-water balanced-filled lacustrine shales in the Pattani Basin are believed to be the main source for oil, but only the marginal facies have been sampled, where the average TOC is <1% (Jardine, 1997). The Miocene coals and coaly shales are mostly gas-prone, with HI ~200 mg/g, although a few sections are more oil-prone, with HI up to ~500 mg/g.

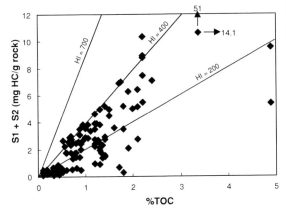

Figure 18.176. Total organic carbon (TOC) versus Rock-Eval (S1 + S2) for Oligocene lacustrine shales, Khmer Trough (data from Okui et al., 1997). Okui et al. (1997) reported the sum of S1 and S2 to indicate original generative potential. This applies only if the rocks have not expelled bitumen and were not stained by migrated petroleum.
HI, hydrogen index.

The Miocene section is thermally immature in the Pattani Basin.

In the Natuna Sea, source rocks occur in three sequences: (1) basal, (2) early synrift deep-water and uppermost late synrift lacustrine shales of the Oligocene Belut Formation, and (3) basal, syn-inversion coals and coaly shales of the Miocene Upper Arang Formation (Phillips et al., 1997; Michael and Bond, 1997). The synrift lacustrine shales are considered to be the major source rocks in the basin. Early synrift source rocks are poorly sampled but consist of typical algal-dominated, deep-freshwater shale facies deposited during balanced-filled conditions. The late synrift source rocks were deposited during underfilled conditions in a broad, shallow lacustrine plain. Petroleum potential of this source unit is greatest in the axial portions of the basin, where TOC typically ranges from 4 to 9 wt.% and HI may exceed 600 mg HC/g TOC. Organic richness decreases, and the source rock becomes more gas-prone toward the basin margins.

The Khmer Trough in the eastern Gulf of Thailand contains Oligiocene lacustrine shales that generated highly paraffinic oils (Okui et al., 1997). Although one sample exceeded 14 wt.% TOC, most have <2 wt.%, with HI ranging from ~150 to 400 mg HC/g TOC (Figure 18.176). These samples are mostly from the basin flanks. Okui et al. (1997) assumed that the generative potential in the center of the trough was higher. Miocene fluvial-deltaic coals have significantly higher potential but are thermally immature.

Oligocene Pilenga/Miocene Borsk formations, Sakhalin Island

Marine siliciclastic facies within the Upper Oligocene Pilenga and Lower Miocene Borsk formations are thought to be the source rocks for crude oils onshore and offshore Sakhalin Island. This conclusion is inferred mostly from the geochemical characteristics (biomarkers and isotopes) rather than from direct observations on rock samples (Popovich and Kravchenko, 1995; Bazhenova and Arefiev, 1997; Chakhmakhchev et al., 1997; Peters et al., 1997b). The oils are similar but can be differentiated into two families, representing proximal and distal deltaic facies of the same general source (Popovich and Kravchenko, 1995; Peters et al., 1997b). Limited analyses of rock samples indicate that the Lower Miocene Borsk (Pilsk) Formation has reasonably good generative potential, with ~1–2 wt.% TOC and HI in the range 300–500 mg HC/g TOC. The Upper Oligocene Pilenga (Daekhurin) Formation has lower potential (Figure 18.177) (Kodina and Vlasova, 1989; Kodina et al., 1989; Peters et al., 1997b).

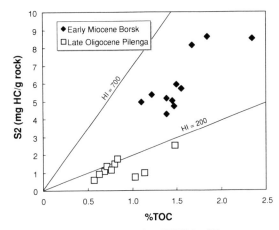

Figure 18.177. Total organic carbon (TOC) (wt.%) versus Rock-Eval S2 of Oligo-Miocene siliciclastic source rocks from Sakhalin Island (data from Peters et al., 1997b).
HI, hydrogen index.

Although of similar age, the Sakhalin Island source rocks are unlike those of the Monterey Formation in California (Peters et al., 1997b; Chakhmakhchev et al., 1997). Extracts and oils from these sources exhibit some similarities (e.g. ^{13}C-rich isotope values, predominance of C_{27} steranes) that reflect an open late Tertiary marine environment dominated by diatomaceous organic matter. The Sakhalin source rocks, however, were deposited under oxic to suboxic conditions, as opposed to the Monterey source facies, which were deposited under suboxic to anoxic conditions. These differences are reflected clearly in a number of biomarker ratios (e.g. pristane/phytane, bisnorhopane/hopane, and C_{35}/C_{34} homohopane ratios).

Oligo-Miocene Carapita Formation, Eastern Venezuela Basin

The Carapita Formation in the Maturin Basin of eastern Venezuela and its equivalents to the south in the Ofinica Basin are marine shales deposited during the late Tertiary Period in a foreland basin. These shales have marginal source potential compared with the Upper Cretaceous Querecual Formation (equivalent to the La Luna outside of eastern Venezuela). Nevertheless, the Carapita Formation generated local petroleum accumulations (e.g. Tocco et al., 1994; Stoufer et al., 1998; Vivas et al., 1998) and contributed hydrocarbons to the Orinoco tar belt, the largest single oil accumulation in the world, with ~ 1.2–2×10^{12} barrels (Roadifer, 1987). The relative proportion of oil from Cretaceous and Tertiary source rocks in the Orinoco tar belt remains unclear (Cassani et al., 1993; Galarraga et al., 1996).

Relatively low TOC and HI characterize the Carapita Formation (Figure 18.178) (Talukdar et al., 1988; Tocco et al., 1994). The kerogen is composed of terrigenous macerals and amorphous material, the latter being a mixture of degraded terrigenous and marine organic matter. The low generative potential is attributed to both sediment dilution and diagenetic oxidation.

Oligo-Miocene coals and coaly shales, northern Borneo, Malaysia, Brunei

The island of Borneo contains several major petroleum provinces associated with Miocene deltaic systems. The Mahakam, Tarakan, and Sandakan deltas lie along the eastern coast, while the Sabah, Baram, and Balingian

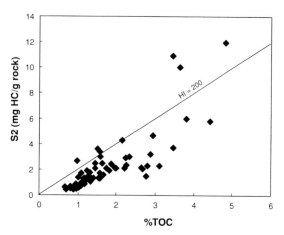

Figure 18.178. Total organic carbon (TOC) (wt.%) versus Rock-Eval S2 for Carapita Formation shales (data from Tocco et al., 1994).
HI, hydrogen index.

deltas lie along the northeast–southwest trending coastline and empty into the South China Sea. All but the Sandakan Delta have major oil and gas reserves and production. The sources of oils along the eastern coast are oil-prone coals and coaly shales that originated as shallow marine tidal and coastal plain deposits. Differences in organic facies are attributed to variations in the relative input of marine versus higher land-plant organic matter (e.g. Dipterocarpaceae and other angiosperms). These variations can be determined qualitatively by comparing the relative abundance of n-propylsteranes, bicadinanes, and oleananes, respectively. The source potential of these source rocks, particularly offshore, is poorly documented. Most coals and coaly shales are predominantly vitrinitic and considered to be generally gas-prone. Some coals and coaly shales contain high proportions of bituminite and exinite macerals and are considered to be oil-prone. These samples have HI ~ 300–450 mg HC/g TOC and produce pyrolyzates with high wax content.

Wan Hasiah (1999) described the source potential of coals and other organic-rich strata in the Upper Oligocene-Lower Miocene strata of the Nyalau Formation, onshore Sarawak. The Nyalau Formation is mostly immature onshore (0.42–0.72% R_o), but it is believed to be a major source rock offshore. Coals and carbonaceous argillites with high generative potential contain 15–35% oil-prone macerals, with HI up to 400 mg HC/g TOC. Microscopic examination of

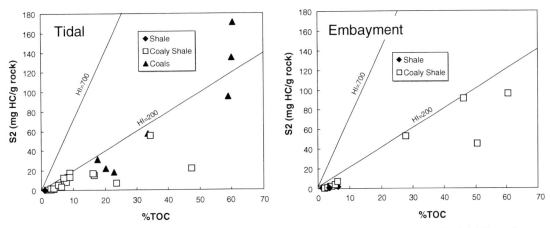

Figure 18.179. Total organic carbon (TOC) (wt.%) versus Rock-Eval S2 for immature outcrop samples of the Belait Formation, onshore Brunei (from Curiale et al., 2000). Coals deposited in nearshore marine settings have the highest petroleum-generative potential. HI, hydrogen index.

these samples provides visual evidence for their generative potential, including exudinite extruding from bituminite.

Curiale et al. (2000) reported on the petroleum potential of the Miocene Belait Formation, onshore Brunei. Here, too, the onshore facies is believed to extend into the more mature offshore area and is considered to be a major source of oil. The analyzed samples typically had low generative potential, but a relationship was found with depositional setting. Shales were uniformly low in organic matter. Coals and, to a lesser extent, coaly shales were richest in nearshore marine environments (tidal, lagoonal, embayment, and shoreface areas), with some tidal coals approaching HI of 300 mg HC/g TOC (Figure 18.179).

Miocene Monterey Formation, California

The Miocene Monterey Formation is the major source rock in the Neogene basins of California. Its depositional setting was once thought to be analogous to modern southern Californian borderland basins (e.g. Isaacs, 1984; Isaacs and Peterson, 1988; Bohacs, 1993). In these basins, steeply dipping horst blocks form a series of silled basins prograding offshore. The inner basins capture most terrigenous clastics, while the outer basins receive hemipelagic mineral and biogenic aggregates that are enriched in marine organic matter enhanced by high productivity caused by upwelling. However, inconsistencies with the "borderland basin" depositional model led to an alternative depositional framework described as a dynamic prograding margin (Isaacs et al., 1996; Isaacs, 2001). With the exception of the San Joaquin Basin, organic-rich strata of the Monterey Formation are thought to have originated during transgressive highstand sequences in sediment-starved depressions on a gentle, prograding slope. Minor topographic irregularities imposed considerable control on lithofacies.

The Monterey Formation varies considerably in lithology and source potential (Table 18.14). The dynamic progradation margin model explains deposition of the Monterey Formation at mid-bathyal depths on a low-gradient open continental margin and in intervening sediment-starved, semi-restricted basins (Isaacs, 2001). The model attributes the calcareous, siliceous, and phosphatic facies of the Monterey Formation to (1) moderate productivity in the deep offshore, (2) high diatom productivity nearer shore, and (3) sediment-starved condensed zones, respectively.

The Monterey Formation contains phosphatic diatomite source rock with TOC typically in the range 2–17 wt.% (some samples >25 wt.%) and type II, IIS, and III kerogens (Isaacs and Rullkötter, 2001). Rock-Eval HI reflects this variability in source potential. Typical Monterey oil-prone strata have HI ~400–500 mg HC/g TOC, with maxima exceeding 830 mg HC/g TOC (Price et al., 1999). Most of the organic matter originated from bacterially reworked diatoms

Table 18.14. *Average total organic carbon (TOC), biogenic silica, and mass accumulation rates for Monterey and associated formations (from Isaacs, 1987; 2001a)*

Age			Rock unit	Lithology		Biogenic silica (%)	~TOC (wt.%)	Total MAR (mg/cm^2/year)
Miocene	Late		Sisquoc Formation	Diatomaceous clay ooze		~30	2	–
		Monterey Formation	Clayey-siliceous member	Laminated diatomaceous ooze and massive diatomaceous clay ooze, deep offshore		25–40	8	14–58
	Mid		Upper calcareous-siliceous transitional marl-siliceous members	Laminated coccolith-foraminiferal ooze	Nearshore	40–60	6	7.5–60
			Organic-rich phosphatic marls		Sediment-starved	5–15	13	1.8–4.9
	Early		Lower calcareous-siliceous member	Massive coccolith-foraminifera and diatom ooze, deep offshore		20–30	6	7.4–24*
			Rincon Shale	Clay ooze		~10	4	–

* Excludes San Joaquin Basin.
MAR, mass accumulation rate.

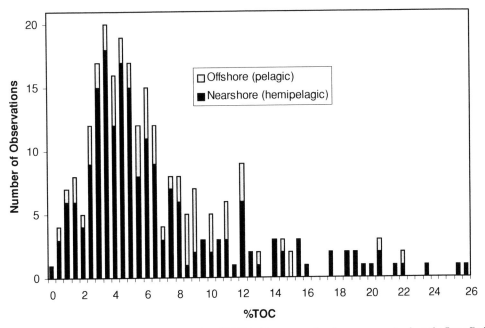

Figure 18.180. Distribution of total organic carbon (TOC) in Monterey well and outcrop samples from the Santa Barbara-Ventura Basin (from Bohacs, 1993). Reprinted by permission of the AAPG, whose permission is required for further use.

associated with slow accumulation and little sediment dilution, rather than high productivity or preservation (Bohacs, 1993; Katz and Royle, 2001). Organic matter in sequences with high biotic productivity is diluted by the increased input of biogenic silica and carbonate. This classic marine petroleum source rock has a complex distribution of organic carbon (e.g. Figure 18.180) that suggests generative yields ranging from 1 to 40×10^6 tons oil/km^3 (Katz and Royle, 2001).

Highly specific biomarkers for Monterey rocks and oils include 28,30-bisnorhopane, possibly related to *Beggiatoa* bacterial mats, and the C_{25} highly branched isoprenoid from diatoms (Curiale *et al.*, 1985; Brincat and Abbott, 2001; Schouten *et al.*, 2001). High 28,30-bisnorhopane/hopane, C_{35}/C_{34} hopanes >1, sulfur content >1%, pristane/phytane <1, and V/Ni >1 in Monterey oils indicate a marine calcareous siliciclastic source rock deposited under highly reducing conditions with abundant reduced sulfur produced by sulfate-reducing bacteria (Michael, 2001). Figure 18.181 shows geochemical data for crude oil from Cat Canyon Field that was generated by the Monterey source rock, onshore Santa Maria Basin, California.

Tertiary Catalan oil shales, western Mediterranean rift system

A series of small extensional lacustrine basins (<50 km^2) developed in the grabens of the western Mediterranean rift system (northeastern Spain). About 100–250 m of immature oil shales occur in the Campins (Late Oligocene), Ribesalbes and Rubielos de Mora (Early–Middle Miocene), Libros (Late Miocene), and Cerdanya basins (Late Miocene) (Anadon *et al.*, 1989). Source quality varies between the basins, with hydrogen-rich algal kerogens in the Rubielos (HI ~850 mg HC/g TOC) and Ribesalbes (HI ~700 mg HC/g TOC), and higher terrigenous plant input in the Campins (HI ~500 mg HC/g TOC), Cerdanya (HI ~440 mg HC/g TOC) and Libros (HI ~340 mg HC/g TOC) basins (Anadon *et al.*, 1988). *Botryococcus braunii* was a major contributor of carbon to these sediments (Metzger and Largeau, 1994). Organic-rich layers were deposited in a cyclic sequence of alternating oxic–anoxic waters in the marginal zones of a meromictic lake. This is attributed to cyclic changes in lake water volume. Oil shales in the Catalan basins are thermally immature.

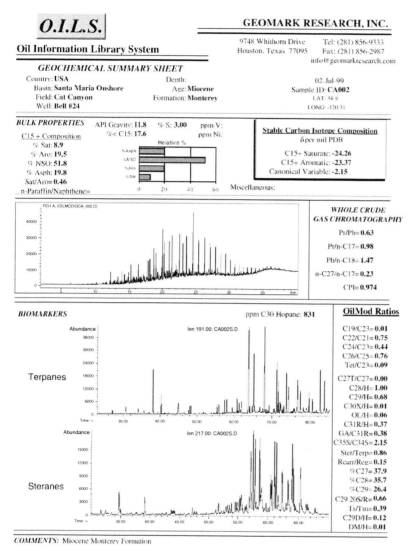

Figure 18.181. Geochemical data for crude oil from Cat Canyon Field generated by the Miocene Monterey Formation, onshore Santa Maria Basin, California, USA (courtesy of GeoMark Research, Inc.).

Kerogen in the Ribesalbes and Cerdanya basins is of special geochemical interest. Erosion of previously deposited anhydrite allowed bacterial sulfate reduction to occur in the anoxic sediments. Sulfides then reacted with the algal organic matter to produce type IS kerogen (Sinninghe Damsté et al., 1993a) and several unusual sulfur biomarkers (de las Heras et al., 1997; Sinninghe Damsté and Rijpstra, 1993). Other occurrences of type I-S kerogen include the Wilkins Peak Member of the Green River Formation (Wyoming), the Jingjingzigou Formation (Junggar Basin, China), the Jianghan and Qaidam basins (China), and the Blanca Lila Formation (Argentina) (Carroll and Bohacs, 2001).

Middle Miocene Maqna Formation, Red Sea Basin

Lower-Middle Miocene synrift and post-rift strata in the Red Sea have varying source potential for oil and gas, with 1–2 wt.% TOC and HI ranging from 100 to

700 mg HC/g TOC (Alsharhan and Salah, 1997). Of these strata, the Middle Miocene Maqna Formation is the most prolific (Lindquist, 1998b). Total organic carbon averages ∼1–2% (maximum 14 wt.%), with HI averaging 200–300 mg HC/g (Cole et al., 1995). Oils from the Miocene shales reflect the high terrigenous input of the source kerogen and contain abundant oleanane.

Middle Miocene Alcanar Formation, Tarragona Basin, Spain

The Alcanar Group is a transgressive unit of carbonate and clastic sediments deposited during Miocene subsidence of the Tarragona Trough. By examining the benthic fossils and lithologic textures, Demaison and Bourgeois (1984) showed that the marly facies of the Alcanar Formation was deposited in a low-energy, suboxic environment, while the chalk facies was a high-energy, oxic environment. The source potential of Alcanar marls and chalks was tied to the energy level of the depositional conditions (Figure 18.182). Extracts from the organic-rich marls correlate with low-sulfur (<1 wt.%) oils produced from the region (Figure 18.183).

Crude oils from Amposta and nearby wells are unlike those from the Tarragona Basin, which were generated from the Alcanar Formation. Amposta oils have high sulfur (∼5.8 wt.%) and biomarker distributions characteristic of marine evaporitic source rocks (e.g. pristane/phytane = 0.61, high concentrations of isoprenoids, $\beta\beta$-steranes, and gamacerane) (Albaigés et al., 1986). The unusual chemistry of the Amposta oils prompted detailed molecular studies of the origin of monoaromatic steroids (Seifert et al., 1983), archaeal isoprenoids (Albaigés et al., 1985), isoprenoid cyclohexanes, and isoprenoid alkylbenzenes (Sinninghe Damsté et al., 1991). The source rocks for these oils are unknown.

Middle Miocene of the Pannonian Basin, Central Europe

The Pannonian Basin of Central Europe is not a single basin but a system of 14 small, deep basins that occur within a low region surrounded completely by the Carpathian Mountains and the Dinaric Alps. The extensional, back-arc basin formed within the past 10–17 Ma during closure of the Tethys Ocean, when the European plate subducted beneath the Carpathian Mountains. More than 7000 m of Miocene-Holocene

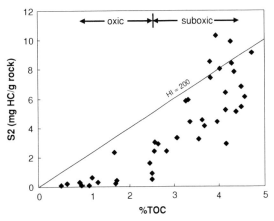

Figure 18.182. Total organic carbon (TOC) (wt.%) versus Rock-Eval S2 for well cores from the Casablanca Field, Tarragona Basin. These samples have entered the oil window (T_{max} = 434–9 °C) and the original generative potential for the marls deposited under suboxic conditions is consistent with marine type II kerogen.
HI, hydrogen index.

sediments were deposited on Proterozoic-Paleozoic metasediments and Mesozoic sediments, some of which are source rocks. The Upper Triassic Kössen Marl Formation in western Hungary is particularly rich in organic matter (>6 wt.% and up to 31 wt.% TOC, with HI up to 900 mg HC/g TOC) (Brukner-Wein and Vetö, 1986; Hetenyi, 1989; Clayton and Koncz, 1994). Lower and Middle Jurassic deltaic and prodeltaic facies in the Vienna Basin contain mostly oxidized and recycled organic matter but generated minor amounts of gas and condensate (Ladwein, 1988). Upper Jurassic marls, however, were deposited in a restricted environment and are the main source rocks for petroleum in the Vienna Basin. These marls average ∼1.7 wt.% TOC, with HI ∼300 mg HC/g TOC (Ladwein, 1988).

During initial extensional tectonics that formed the Pannonian basins, fluvial and lacustrine sediments were deposited under overfilled to balanced-fill conditions. The latter facies has been documented to have significant source potential in many of the Pannonian sub-basins. These occurrences include the Little Plains Basin in Hungary (Mattick et al., 1996), Zala Basin (Clayton and Koncz, 1994), Békés Basin (Clayton et al., 1994a; 1994b), and the Kisalföld Basin, and in Croatia the Sava (Baric et al., 2000), Eastern Drava,

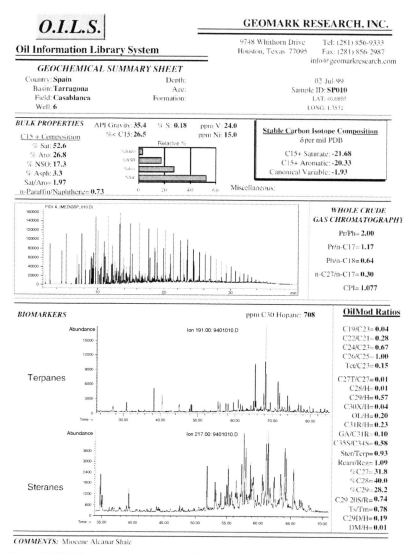

Figure 18.183. Geochemical data for crude oil from Casablanca Field generated from the Miocene Alcanar Formation, Tarragona Basin, Spain (courtesy of GeoMark Research, Inc.).

and Slavonija-Srijem depressions (Alajbeg et al., 1996). These Middle Miocene rocks typically contain 2–5 wt.% TOC, with HI <250 mg HC/g TOC, reflecting a predominance of terrigenous organic matter. Some strata, however, are much richer and reflect input from freshwater algae (e.g. Botryoccoccus).

Source potential also has been documented in Oligocene (Brukner-Wein et al., 1990) and Pliocene lacustrine strata (Derenne et al., 1997; 2000). Brukner-Wein et al. (2000) examined the organic matter in Pliocene oil shale from two Hungarian, maar-type basaltic tuff craters (Egyházaskeszö and Várkeszö). Although these deposits were laid down under identical paleoclimatic conditions and have nearly identical lithologies, the organic matter differed substantially. Kerogens in the Várkeszö crater originated from algal (Botryococcus braunii) input, with minor amounts of land-plant material preserved under anoxic bottom-water conditions. In contrast, kerogens from the Egyházaskeszö crater contain more organic sulfur, pyrite, and land-plant material and have lower source potential. Much more of the primary organic matter was

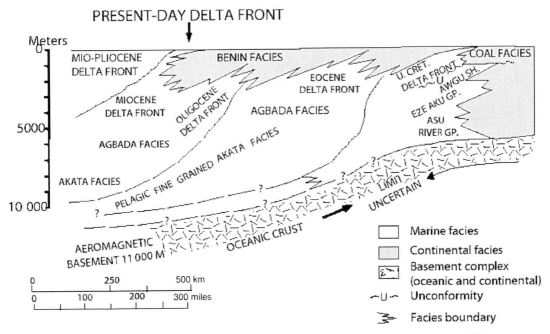

Figure 18.184. Regional cross-section through the Niger Delta (from Whiteman (1982), as used in Tuttle et al. (1999)).

altered bacterially, allowing sulfur incorporation during diagenesis.

Much of the sedimentary fill in the Pannonian Basin occurred within the past 10 Ma. Consequently, the Middle Miocene (12–17 Ma) source units were buried rapidly, and oil generation commenced within the past 7 Ma. The rapid heating rates experienced by Pannonian source rocks attracted attention in several studies of kerogen and biomarker kinetics (Mackenzie et al., 1982b; Mackenzie et al., 1984; Sajgó et al., 1988).

Tertiary Akata and Agbada formations, Niger Delta

The Niger Delta is the twelfth largest petroleum province, with >23 billion barrels of recoverable oil and condensate and 124 trillion cubic feet of gas (Haack et al., 2000). The source rocks for this petroleum occur in the Akata and lowermost Agbada formations (Ekweozor and Daukoru, 1994; Tuttle et al., 1999; Haack et al., 2000). Stratigraphy in the Niger Delta is not temporal but is based on rock facies, primarily sand/shale ratios. Three formations are recognized that represent prograding depositional facies (Kulke, 1995).

The Akata Formation at the base of the sequence is composed of thick, prodeltaic marine shales and turbidites. These marine shales generated significant volumes of petroleum from type II/III organic matter (see Table 4.3). The overlying Agbada Formation is a thick sequence of deltaic siliciclastics and contains the main reservoir rocks for the region. The Benin Formation is a thick deposit of alluvial and upper coastal plain sandstones that overlie the Agbada Formation. Deposition began for the Akata Formation in the Paleocene Epoch, for the Agbada Formation in the Eocene Epoch, and for the Benin formation in the latest Eocene Epoch; deposition continues for each formation today (Figure 18.184).

Most of the Tertiary sedimentary sequence of the Niger Delta contains an average of ~1.5 wt.% TOC, with HI <150 mg HC/g TOC. The kerogen is mostly vitrinitic and gas-prone, leading some to question whether there are Tertiary-age source rocks in the Niger Delta (Bustin, 1988). However, the Akata Formation shales are represented poorly by samples. Nonetheless, several studies show that these shales have sufficient source potential to account for all petroleum reserves in the province (Figure 18.185). Agbada/Akata shales typically contain >2 wt.% TOC, with HI in the range ~200–300 mg HC/g TOC and maxima over

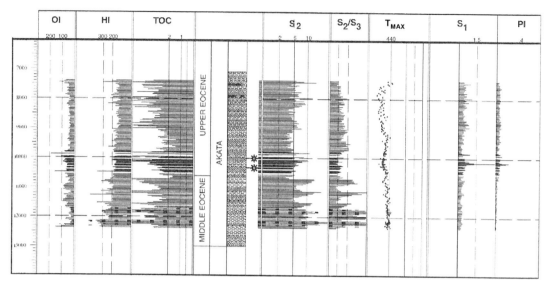

Figure 18.185. Geochemical log of the Akata Formation in the Aroh-2 well, onshore Niger Delta (Haack et al., 2000). Cuttings were extracted with solvent before total organic carbon (TOC) and Rock-Eval analysis to remove drilling additives. The top of the oil window is ~10 500 feet (3200 m). Reprinted by permission of the AAPG, whose permission is required for further use. HI, hydrogen index; OI, oxygen index.

400 mg HC/g TOC (Ekweozor and Daukoru, 1994; Tuttle et al., 1999; Haack et al., 2000).

Three types of depositional environments account for the source rocks in the Niger Delta: (1) shallow-water paralic, (2) upper slope impinged by the oxygen-minimum zone, and (3) deep-water, prodeltaic. Paralic facies were deposited along the coastal plain and are thickest across regional growth faults. Preservation was enhanced on the coastal plain, where large amounts of higher-plant material induced dysaerobic conditions. Source rocks were deposited on the slope near the shelf edge where the slope intersects the oxygen-minimum zone or in deep, oxygen-poor bottom waters. Preservation can be enhanced along the axes of major fan lobes where terrigenous organic matter is rapidly buried. Dysaerobic conditions may be promoted further by bathymetric restrictions resulting in limited water circulation and channeling of clastics.

Because source-rock samples are rare in most deltaic settings, oil chemistry has been used as a proxy to map changes in source facies and generative potential (e.g. Peters et al., 2000; Haack et al., 2000). Oils in the Niger Delta share many common features, such as high concentrations of oleanoids (Ekweozor et al., 1979a; Ekweozor et al., 1979b; Ekweozor and Udo, 1988), which indicate terrigenous plant matter (Figure 18.186). There are sufficient variations among the Niger Delta oils, mainly in the relative amounts of biomarkers characteristic of terrigenous and marine input, to identify the above three source facies (Figure 18.187).

Middle Miocene Balikpapan Formation coals and shales, Mahakam Delta

The prolific Kutei Basin in Kalimantan, which contains the Tertiary Mahakam Delta, has estimated ultimate recovery of ~16 billion barrels of oil equivalent (BBOE) (IHS/Petroconsultants S.A., 1998). The first oil in the delta was discovered in 1898 in the Louise-1 well, located on the westernmost, inboard anticlinal trend (Figure 18.188). The Mahakam Delta was one of the first locations where terrigenous organic matter in coals was shown to have expelled liquid hydrocarbons (Durand and Paratte, 1983; Huc et al., 1986). Most geologists now agree that deltaic coals can generate large quantities of oil and gas (e.g. Law and Rice, 1993). Major oil accumulations were linked to hydrogen-rich coaly source rocks, especially in

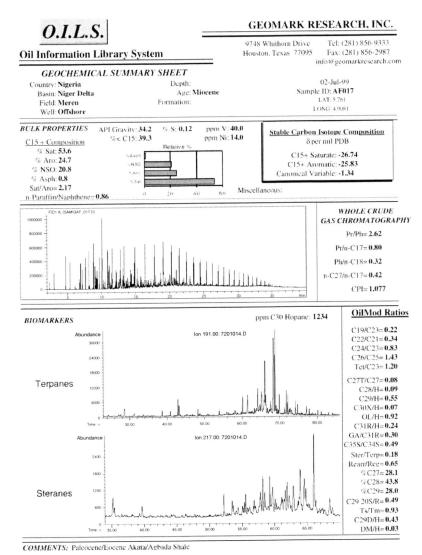

Figure 18.186. Geochemical data for crude oil from Meren Field generated from the Tertiary Akata/Agbada formations, Niger Delta, Nigeria (courtesy of GeoMark Research, Inc.).

Upper Jurassic-Paleogene rocks of Australia and New Zealand and low-latitude (<20°) Tertiary deltaic-coastal deposits throughout Southeast Asia (Scott and Fleet, 1994). Figure 18.189 shows geochemical data for crude oil from onshore Tambora Field that was generated from the Miocene Balikpapan Formation. An equivalent of this oil was interpreted to correlate with a major family of "highstand systems-tract oils," as discussed below (see sample T-19 in Figure 12.2) (Peters et al., 2000).

Previous stratigraphic model

Total and the Institut Français du Pétrole (IFP) produced the first significant chronostratigraphic correlation model for the Mahakam Delta (Burrus et al., 1992; Duval et al., 1992a). Their early stratigraphic and geochemical work on the Mahakam Delta (e.g. Combaz and de Matharel, 1978; Durand and Oudin, 1979; Verdier et al., 1980) provided the basis for a regional hydrocarbon-charging model (Duval et al., 1992b). The Total–IFP model was instrumental in

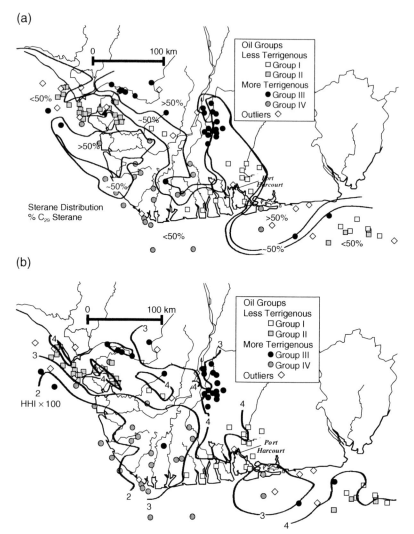

Figure 18.187. Contour maps showing inferred source-facies distributions based on the biomarkers in produced oils from the Niger Delta (modified from Haack et al., 2000). The top map is based on high (>50%) versus low (≤ 40 %) C_{29} steranes, which indicates the relative terrigenous versus marine input. The bottom map is based on the homohopane index (C_{35} homohopane index (HHI) × 100), where low values suggest more oxic, gas-prone source-rock facies than high values. Reprinted by permission of the AAPG, whose permission is required for further use.

successful efforts toward reversing production declines; initially, it included Tunu Field reservoir characterization (Duval et al., 1992b), followed by discoveries of the Northwest Peciko and Sisi fields.

While the Total–IFP model was a major improvement over previous understanding, it downgrades the potential for commercial deep-water petroleum accumulations and fails to explain recent oil discoveries on the outer shelf (Burrus et al., 1992; Duval et al., 1992b). The model shows most of the past shelf margins near or just seaward of Tunu Field (triangles in Figure 18.190, top), which explains why many believe that Middle Miocene coastal-plain coal and shale source rocks occur mainly landward of Tunu Field (Burrus et al., 1992). Those who adhere to the model assume that all oils in the area are genetically related to this updip shelf source

Tertiary source rocks 953

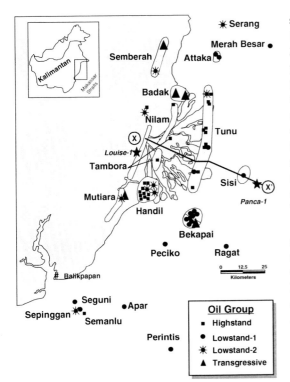

Figure 18.188. Map showing the nearly north–south anticlinal trends (elliptical patterns) and locations of oil samples in the Mahakam-Makassar area, Kutei Basin, Indonesia (Peters *et al.*, 2000). Insert at upper left shows the location of the area in eastern Kalimantan. Heavy and light stippling within the anticlines indicate oil and gas trends, respectively. Genetic groups with symbols defined in the inset at lower right are based on statistical analysis of multivariate geochemical data. Stars indicate locations of the Louise-1 and Panca-1 wells. Reprinted by permission of the AAPG, whose permission is required for further use.

(e.g. Paterson *et al.*, 1997). Because the outer shelf and slope are distant from the presumed source rocks, they are presumed to carry significant exploration risk. By this interpretation, Miocene rocks in deep water have low petroleum potential due to oxidation of coaly source material during transport across the shelf break, deep burial and high maturity (~6 km) Figure 18.190), and diagenetic cementation of reservoirs at great depth.

Highstand-systems-tract coals and shales

The generally accepted source rocks for the Mahakam Delta crude oils are highstand-systems-tract coals and shales from the Miocene Balikpapan Group (Robinson, 1987; Schoell *et al.*, 1983) that were buried deeper than 2600 m (Perrodon, 1983). The Paleogene section is overpressured and believed to be too deep to contribute to petroleum accumulations. The post-Miocene section is not buried sufficiently to have entered the oil window. According to the Total–IFP model, the best source rocks consist of lower delta-plain coals (Duval *et al.*, 1992a) that contain mainly huminite with minor exinite macerals (Burrus *et al.*, 1992). Most workers believe that the Middle Miocene coastal-plain coals and coaly shales (15.5–10.5 Ma) are the primary source interval in the Tunu kitchen (e.g. Burrus *et al.*, 1992).

The Balikpapan Group contains up to 175 m and 1750 m of cumulative coal and shale, respectively (Thompson *et al.*, 1985). Although most Balikpapan Group coals contain abundant vitrinite-like macerals, HI of the coals commonly exceeds 300 mg HC/g TOC and OI is usually <15 mg CO_2/g TOC (Thompson *et al.*, 1985). HI >200 mg HC/g TOC is necessary for significant oil-generative potential from coals (Hunt, 1996).

New geochemical-stratigraphic model

The new model (Figure 18.190, bottom) places the interpreted shelf margins farther seaward than the Total–IFP model, allowing source-rock deposition in less explored outer shelf and slope regions, and the source rocks have more favorable oil-window maturity. This new interpretation allows development of lowstand kitchens, i.e. thick depocenters seaward of the coastal plain that are closer to the slope margin than allowed by the Total–IFP model. Because the lowstand kitchens are downdip of the co-eval shelf margin, the terrigenous organic matter is believed to consist of transported rather than *in situ* coals or coaly shales.

In the new model, the deep-water Middle Miocene interval is not buried as deeply as believed previously (3–4 km) (Figure 18.190), which impacts favorably the expected thermal maturity of the proposed source rocks and cementation of sandstone reservoirs. The new model predicts that the source rocks are now within the oil window based on regional seismic reinterpretation and thermal modeling using source-specific kerogen kinetics (Peters *et al.*, 2000). During lowstand-system-tract time, downdip depocenters received terrigenous organic matter by a process similar to that

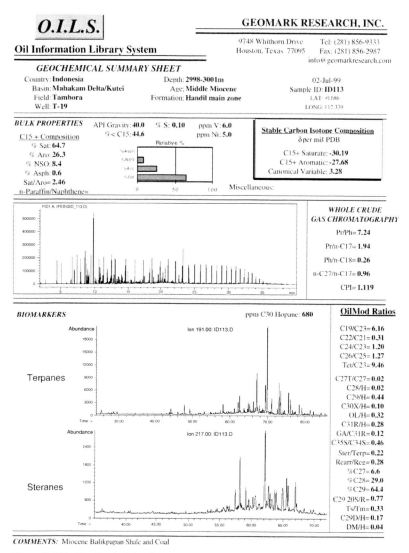

Figure 18.189. Geochemical data for crude oil generated from Tambora Field generated from the Miocene Balikpapan Formation (courtesy of GeoMark Research, Inc.).

responsible for gravity-flow sandstones on the outer shelf and slope. Oxidation of this organic matter was minimized due to proximity of the shelf break to the depocenters.

Geochemical evidence for the new model

Biomarker and isotope analyses for 61 crude oils confirm that genetically distinct petroleum accumulations (see Figure 12.2) originated from local kitchens between anticlinal structural trends in the Mahakam Delta area (Figure 18.188) (Peters et al., 2000). The results confirm that source-rock facies other than the well-documented Middle Miocene coastal-plain coals and coaly shales exist in the Mahakam-Makassar area. Two major (1–2) and two minor (3–4) petroleum systems dominated by terrigenous type III organic matter were recognized:

1. Waxy highstand oils occur mainly onshore in Middle Miocene-Pliocene reservoirs. These oils originated from Middle–Upper Miocene coal and shale source

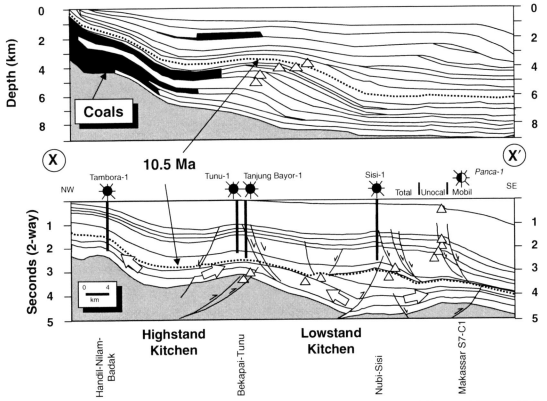

Figure 18.190. Comparison of the old (top) (Burrus et al., 1992; Duval et al., 1992a) and new (bottom) (Peters et al., 2000) models for the Mahakam Delta. Location of X–X' cross-section is shown in Figure 18.188. Triangles indicate shelf break during specified intervals of deposition. White arrows indicate present-day direction of petroleum migration from local kitchens. Burial depth for the 10.5-Ma top Middle Miocene source-rock surface (dotted line) in the lowstand kitchen area for the new model is ~3–4 km, depending on the two-way transit time versus depth conversion that is used, but is ~6 km based on the old model. Reprinted by permission of the AAPG, whose permission is required for further use.

rocks, deposited in coastal-plain highstand kitchens now near the peak of the oil window.
2. Less waxy lowstand-1 oils occur offshore in Middle–Upper Miocene reservoirs. These oils originated from Middle–Upper Miocene coaly source rocks, deposited in deepwater lowstand kitchens now mostly in the early oil window. Their low maturity is inconsistent with the Total–IFP model.
3. Lowstand-2 oils are similar to the lowstand-1 oils but occur mainly onshore in Lower–Middle Miocene reservoirs. These oils are generally more mature than lowstand-1 oils and originated from Lower–Middle Miocene coaly source rocks.
4. Nonwaxy-transgressive oils occur mainly onshore in Middle–Upper Miocene reservoirs. They were generated at low thermal maturity from Middle Miocene suboxic marine shales, deposited near maximum flooding surfaces.

The highstand, lowstand-1, lowstand-2, and transgressive system-tract oils account for ~46%, 31%, 15%, and 8% of the 61 oil samples and ~45%, 32%, 11%, and 12% of the estimated ultimate recoverable reserves from the fields represented by these samples, respectively. These fields account for ~13 of the 16 BBOE estimated ultimate recoverable reserves in the Kutei Basin.

The 61 oil samples were used as a training set to construct a K-Nearest Neighbor (KNN) statistical model of oil families. This KNN model was used to establish

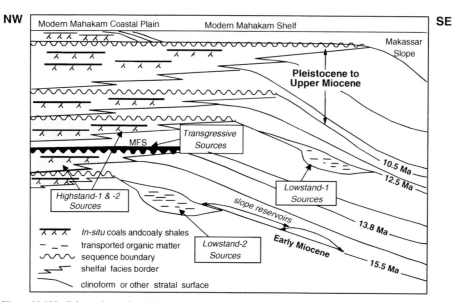

Figure 18.191. Schematic geochemical-stratigraphic model and predicted distribution of source rocks near the Mahakam Delta based on oil–source rock correlation (Peters et al., 2000). Faults are not shown. NW–SE refers to X–X' in Figure 18.188. Reprinted by permission of the AAPG, whose permission is required for further use.
HST, highstand system tract; LST, lowstand system tract; MFS, maximum flooding surface; TST, transgressive system tract.

genetic oil–source rock correlations based on the geochemical compositions of extracts from organic-rich source-rock candidates. The systems tract (e.g. highstand, lowstand, transgressive, discussed below) and geologic age for each source-rock sample, and by inference the related oils, was determined using biostratigraphic and seismic sequence stratigraphic data (Figure 18.191).

Lowstand-system-tract coaly shales
In the new model, terrigenous organic matter accumulates in depocenters seaward of the co-eval coastal plain in outer shelf to slope settings during lowstand systemtracts. Enhanced erosion of coastal-plain sediments during rapid falls in relative sea level could provide large quantities of terrigenous organic matter to the lowstand kitchens (Figure 18.191).

Few downdip wells penetrate marine sections with high terrigenous organic matter in the area. For example, the best shale source rocks in the deep-water Perintis-1 and Sisi-1 wells (Figure 18.118) contain only 1–3 wt.% TOC. However, sampling bias is possible because most exploration wells are drilled on structural highs, where less terrigenous organic matter may have been deposited. The best evidence for terrigenous-rich deep-water source rocks offshore Mahakam Delta comes from observations of updip erosion and transport of terrigenous organic matter. For example, cores of Middle Miocene channel and incised-valley fills show that TOC and HI can be high (Snedden et al., 1996). Seismic evidence for valley fills elsewhere in the delta supports this view.

These terrigenous-rich lowstand depocenters generally lie above the 15.5 and 12.5 Ma sequence boundaries and include periods of extensive updip erosion, downdip subsidence, and deposition (Figure 18.91). Thermal modeling shows that two lowstand depocenters are within the window for hydrocarbon generation and thus are potential kitchens. The presence of two lowstand kitchens is supported by geochemical data, which identify two distinct subgroups of lowstand oils (see Figure 12.2). The lowstand-1 and lowstand-2 oils differ in both source- and maturity-related geochemical parameters, implying origins from source rocks of different age and thermal exposure. The oil–source rock correlation indicates that the lowstand-1 and lowstand-2 oils originated from Middle–Upper and Lower–Middle Miocene source rocks, respectively.

Transgressive-system-tract source rocks
Thick and laterally extensive coals and organic-rich shales can form during both highstand and lowstand system tracts, given the proper balance of peat production and accommodation (Bohacs and Suter, 1997). Little algal organic matter is contributed to these sediments under conditions of high water turbidity, terrigenous organic matter input, and clastic deposition typical of the Mahakam Delta. However, coals are rare during transgressive-system-tract time because swamp sediments commonly are drowned before substantial organic matter can accumulate. Reduced water turbidity and clastic deposition can occur during the peak transgressive events marked by maximum flooding surfaces (MFS) (Figure 18.191). Seismic evidence indicates several MFS events during the Middle Miocene Epoch, especially within the 15.5- and 13.8-Ma sequences.

The geochemistry of a small but distinct group of oils among the 61 samples in the study suggests a transgressive, marine-influenced source rock with significant algal input mixed with the terrigenous organic matter. Furthermore, these oils correlate with extract from source rock deposited in an inner shelf environment as part of a transgressive-system tract (13.8 Ma). While our highstand and lowstand oil samples show genetic affinities for both coals and shales, the transgressive system-tract oils show affinities only for shales.

Messinian Tripoli Unit shales, Lorca Basin, southeastern Spain

The Lorca Basin is one of several intermontane depressions within the Internal Zone of the Betic Cordillera

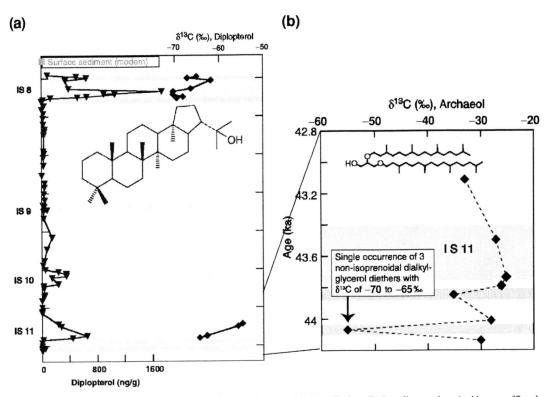

Figure 18.192. (a) Abundance and isotopic composition of diplopterol in Santa Barbara Basin sediments deposited between 37 and 44.2 ka. Laminated sediments deposited during relatively warm interstadials are indicated by light shading. Periods with foraminifera $\delta^{13}C$ excursions are indicated by darker shading. (b) $\delta^{13}C$ of archaeol in sediments from the earliest isotopic excursion (44.2–43 ka). Reprinted with permission from Zachos *et al.* (2001). © Copyright 2001, American Association for the Advancement of Science.

of southeastern Spain. The basin first developed in the Tortonian Age, resulting in deposition of shallow-water, high-energy conglomerates, siliciclastics, and biogenic carbonates. Rapid changes in water depth and salinity in the Messinian Age are marked by intercalation of gypsum layers that preceded major evaporite precipitation when the basin was isolated from the Mediterranean Sea. Thin shale within the Lower Member of the Messinian Tripoli Unit is highly enriched in TOC and sulfur. TOC ranges from 21 to >25 wt.%, with HI ~650–740 mg HC/g TOC and total sulfur of ~6–7.5 wt.% (Benalioulhaj et al., 1994; Benali et al., 1995; Russell et al., 1997). Thick deposits of diatomites and the presence of the biomarker C_{25} HBI throughout the Tripoli Unit indicate high biogenic productivity. These strata, however, are not source rocks, because of low average TOC (<0.2 wt.%) (Rouchy et al., 1998).

The organic-rich shales were deposited under highly saline conditions with extensive bacterial sulfate reduction. Abundant organo-sulfur compounds and elemental sulfur and replacement of sulfates by carbonates provide evidence for hypersaline, euxinic conditions (Rouchy et al., 1998). Unlike the rest of the Tripoli Unit, the organic-rich shales contain no sulfurized C_{25} HBI but yield abundant isoprenoid bithiophenes and C_{40} isoprenoids, which are indicative of archaeal lipids (Sinninghe Damsté et al., 1989b; Russell et al., 1997).

Neogene lacustrine oil shales, intermontane basins, northern Thailand

Uplift during the Late Cretaceous and early Tertiary periods formed a series of intermontane basins in northern Thailand that contain late Tertiary lacustrine shales of exceptional organic richness (Gibling et al., 1985b). The Mae Sot Basin is the largest of these basins, where organic-rich shales cover ~53 km² with in-place oil reserves of ~6.4 billion bbl (Knutson et al., 1988). Lacustrine rocks in the Mae Sot Basin are estimated to be of Late Miocene–Pleistocene age, while those in other northern Thailand intermontane basins may be as old as Eocene (Gibling et al., 1985a; 1985b).

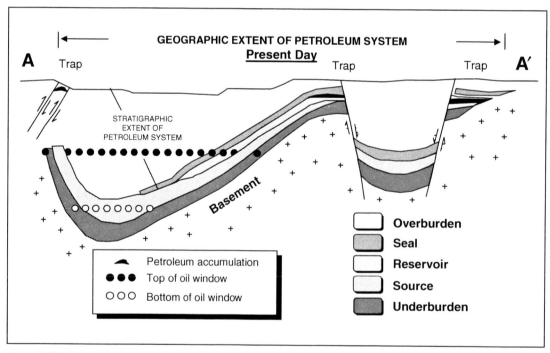

Figure 18.193. Present-day cross-section through the Deer-Boar(.) petroleum system along A–A' (see Figure 18.194), including, from left to right, a thrust belt, foredeep, rift valley, and stable craton. Compare this figure with figure 18.3. How does the present-day cross-section differ from that at the critical moment (250 Ma)?

Lacustrine rocks in the Mae Sot Basin alternate between coals and mudstones, indicating fluctuations between overfilled and balanced-fill conditions. The mudstones contain variable amounts of organic carbon. Individual strata contain up to 22.4 wt.% TOC, most of which is amorphous kerogen or, less frequently, alginite with high hydrogen content (Bjorøy et al., 1988; Curiale and Gibling, 1992; Curiale and Gibling, 1994). These rocks are thermally immature (R_o ~0.36–0.48%), and the steranes in extracts are dominated by C_{27}–C_{29} $\alpha\alpha\alpha$20R isomers, of which ethylcholestanes are dominant. Fluctuating depositional conditions, possibly related to surface-water productivity, imparted large variations in carbon isotopic values and in biomarker distribution (Curiale and Gibling, 1994). Diagnostic biomarkers include β-carotane, dammaranes, norsteranes, and novel distributions of tricyclic terpanes.

Tertiary shales of the Brookian Sequence, Beaufort–Mackenzie Basin

Approximately 12–16 km of paralic, largely terrigenous sediments were deposited in the Canadian Beaufort Sea–Mackenzie Delta area since the Late Cretaceous Period in 11 regionally extensive transgressive-regressive sequences. Because drilling is limited largely to shallow Tertiary reservoirs, the source rock of the produced hydrocarbons has probably not been sampled.

The geochemistry of produced liquids indicates that there are at least four distinct source facies within this sediment package (Dixon et al., 1992). Two types of oils from marine sources are present (McCaffrey et al., 1994b), which are believed to originate from Cenomanian–Middle Maastrichtian Boundary Creek/Smoking Hills sequences that contain thin,

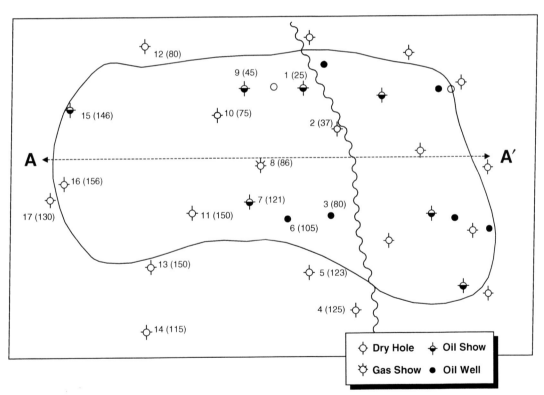

Figure 18.194. Index map showing dry wells, wells with gas or oil shows, oil wells, and the geographical extent of the Deer-Boar(.) petroleum system (enclosed area). The inset applies to Figures 18.194–18.197. The horizontal dashed line A–A′ indicates the location of the cross-section in Figure 18.193. The wavy line is the erosional limit of the Deer Shale source rock, also as indicated in Figure 18.193. Numbers refer to wells in Table 18.15. Numbers in parentheses are thicknesses of the Deer Shale source rock. Use these numbers to generate a contour map of source-rock thickness (isopach).

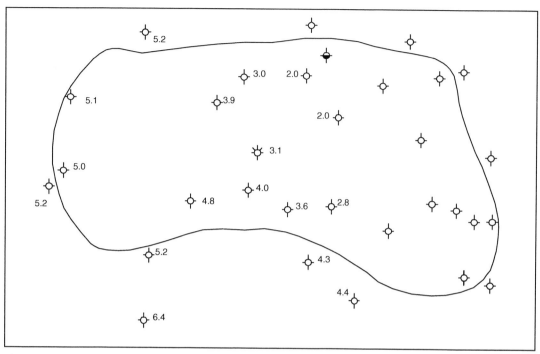

Figure 18.195. Only one hydrocarbon show occurs in the overburden rock based on evaluation of geochemical data from the wells (e.g. Rock-Eval pyrolysis S1 response). Numbers are total organic carbon (TOC) (wt.%) from Table 18.15. Use these data and your interpretation of Figure 18.193 to generate a contour map to TOC.

organic-rich outer-shelf to basinal mudstones (Creaney, 1980; Snowdon, 1980a). Terrigenous-influenced oils can be separated into two genetic families based on the presence of either 28-bisnorhopane or lupane (Snowdon and Powell, 1979; Brooks, 1986, Curiale, 1991; Curiale, 1995). Potential Tertiary sources include the Fish River Shale (Paleocene Tent Island Formation) and shales in the Reindeer Supersequence (Paleocene), Richards Sequence (Eocene), and Kugmallit Sequence (Oligocene). All measured samples of these Tertiary shales have relatively low amounts of dispersed organic matter (1–2 wt.% TOC) derived primarily from gas-prone (HI average <150 mg HC/g TOC) land-plant debris. Of these, the Eocene Richards Sequence may have the greatest generative potential (Dixon et al., 1992). Source-facies mapping based on oil chemistry indicates that the Tertiary deltaic source(s) become substantially more oil-prone in the distal portion than in the region closer to the paleoshoreline (McCaffrey et al., 1994b).

Biomarker evidence for Pleistocene methane hydrate destabilization

Gas hydrates represent some of the most recent hydrocarbon accumulations on Earth. Kennett et al. (2000) conducted a detailed study of the δ^{13}C and δ^{18}O of benthic foraminfera from Santa Barbara Basin over the past 60 000 years. They concluded that millennial-scale shifts in δ^{13}C of ~5‰ record periods of interstadial and Holocene warming and reflect conditions of increased outgassing from gas-hydrate dissociation. Within the record are several very brief but large (up to 6‰) negative excursions that they hypothesized corresponded to periods of massive releases of methane induced by switches in thermohaline circulation. This hypothesis was proved recently by Hinrichs et al. (2003) by examining the δ^{13}C of diagnostic biomarkers within this sedimentary record (Figure 18.192). Diplopterol, a hopanoid formed by various aerobic bacteria, including methanotrophs, was found to have δ^{13}C ~60‰,

Exercise

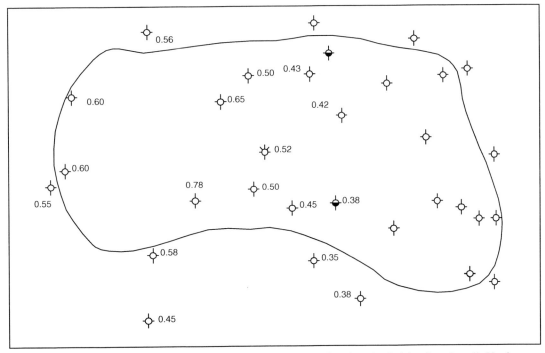

Figure 18.196. Few hydrocarbon shows occur in the seal rock based on evaluation of geochemical data from the wells. Numbers are vitrinite reflectance (%) from Table 18.15. Use these data to generate a contour map of R_o.

indicating the use of methane as a carbon source during its biosynthesis. Within one ~1000-year interval at around 44.1 ka, ^{13}C-depleted archaeol, an ether lipid diagnostic for microbial consortia that oxidize methane anaerobically, was also found. This interval also contained planktonic foraminiferal carbon showing maximum isotopic depletion. These findings support strongly the concept that a very large emission of methane occurred within a brief period of time. This large emission of methane supported methanotrophic oxidation within the entire water column and created anoxic bottom waters.

EXERCISE

Deer-Boar(.) petroleum system evaluation

Using the cross-section and index maps (Figures 18.193–18.197) and Table 18.15, prepare the following:

- thickness map of Deer Shale source rock (Figure 18.194);
- TOC contour map of Deer Shale (Figure 18.195);
- isoreflectance map (at the base of the Boar reservoir rock) (Figure 18.196);
- residual SPI map of Deer Shale (Figure 18.197).

What general conclusions can you make about this petroleum system?

- What happened to the source rock east of the wavy unconformity line in Figure 18.193?
- What does the TOC map indicate about the quantity of organic matter?
- Based on the isoreflectance map, where is the pod of active source rock? What direction(s) of migration do you predict? Are these migration paths consistent with the oil-show map of the Boar reservoir rock?
- What factors account for the regional variations in residual SPI? Are these SPI low or high?
- What can you say about the timing of oil generation based on the presence of oil east of the block faults in the eastern basinal area?

Table 18.15. *Geochemical information for the Deer Shale source rock (Magoon and Dow, 1994a; Peters and Cassa, 1994)*

Well	R_o (%)	h^1 (m)	S1 (mg HC/g rk)	S2 (mg HC/g rk)	SPI^2 (tHC/m^2)	TOC (wt.%)	HI (mg HC/g TOC)
1	0.43	25	1.7	8	0.6	2.0	400
2	0.42	37	1.8	8	0.9	2.0	400
3	0.38	80	2.8	14	3.3	2.8	500
4	0.38	125	4.1	22	7.8	4.4	500
5	0.35	123	4.0	21	7.4	4.3	488
6	0.45	105	3.1	16	4.8	3.6	444
7	0.50	121	1.9	17	5.5	4.0	425
8	0.52	86	3.1	15	3.7	3.1	484
9	0.50	45	2.9	16	2.0	3.0	533
10	0.65	75	3.0	15	3.2	3.9	384
11	0.78	150	2.0	10	4.3	4.8	208
12	0.56	80	5.3	27	6.2	5.2	519
13	0.58	150	5.1	26	11.2	5.2	500
14	0.45	115	6.4	34	11.2	6.4	531
15	0.60	146	5.1	26	10.9	5.1	510
16	0.60	156	5.1	25	11.3	5.0	500
17	0.55	130	5.2	26	9.7	5.2	500

^{1} h = thickness
^{2} SPI = $h\rho$ (average S1 + S2)/1000, assuming density of organic matter $\rho = 2.4$ g/cm^3.
SPI, source potential index.

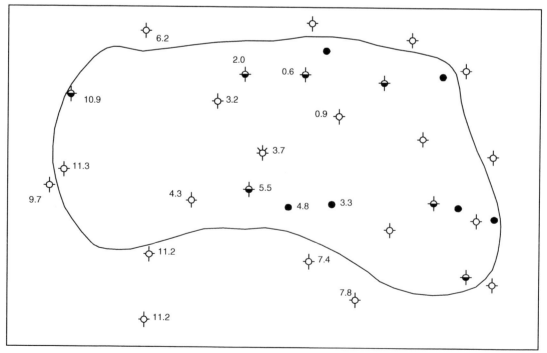

Figure 18.197. Hydrocarbon shows in the reservoir rock interval identify it as the carrier bed for migration. Numbers are residual source potential index (SPI) (tons HC/m^2). Use these data to generate a contour plot of residual SPI.

- Identify an area on the map where biodegradation might adversely affect the preservation of accumulated petroleum.
- Based on the regional geology and the residual SPI map, how would you classify this petroleum system based on the nomenclature of Demaison and Huizinga (1994): (1) supercharged, normally charged, or undercharged, (2) vertically or laterally drained, (3) high impedance or low impedance?

19 · Problem areas and further work

> This chapter describes areas requiring further research, including the application of biomarkers to migration, the kinetics of petroleum generation, geochemical correlation and age assessment, and the search for extraterrestrial life.

Continued advances in analytical instrumentation and molecular chemistry suggest that the use of biomarkers to solve geochemical problems will continue to grow. The following discussion outlines several areas where further biomarker research is likely.

MIGRATION

Two processes appear to affect biomarker distributions during oil migration: solubilization and geochromatography. Solubilization, also called overprinting, is documented poorly, but it involves the incorporation of organic matter from rocks that are unrelated to the migrating petroleum. In most cases, solubilized materials are of lower thermal maturity than those comprising the migrating petroleum (Curiale, 2002). The contamination may become evident through the presence of immature (pre-oil window) biomarkers, such as olefins (Curiale and Frolov, 1998), or as alteration of molecular and isotopic maturation parameters. Solubilization has been observed in various areas, including the Mahakam Delta (Durand, 1983; Hoffmann et al., 1984; Jaffé et al., 1988a; Jaffé et al., 1988b), Australia (Philp and Gilbert, 1982; 1986), Cook Inlet, Alaska (Hughes and Dzou, 1995), Gulf of Mexico (Curiale and Bromley, 1996), offshore Brunei (Curiale et al., 2000), and Angola (Figure 19.1). Evidence of solubilization typically consists of mixed thermal maturity signals in the same crude oil. For example, one parameter might indicate high maturity because the compounds used for the parameter are dominant components in the migrated fraction of the oil. Another parameter might indicate lower thermal maturity because the compounds used for this parameter are dominant in the solubilized contaminants picked up by the migrating oil.

Figure 19.1 shows evidence of solubilization in a crude oil that was generated from Lower Cretaceous lacustrine source rocks in Angola. The oil migrated into an Upper Cretaceous reservoir containing sands with interbedded immature, organic-rich marine shales. The oil appears more mature based on $20S/(20S + 20R)$ for the C_{29} steranes than for the C_{30} steranes. The more mature C_{29} sterane $20S/(20S + 20R)$ for the oil (0.31, bottom) results from combined inputs of mature oil (~ 0.5) and leached, immature bitumen. The less mature C_{30} sterane $20S/(20S + 20R)$ for the oil (0.11, top) reflects the maturity of the immature, leached bitumen only because no C_{30} steranes (24-n-propylcholestanes) are present in lacustrine oils. Also note the prominent C_{30} 5β,20R sterane peak compared with the C_{29} 5β,20R sterane. The C_{30} 5β,20R sterane peak has been contributed largely from the immature bitumen.

Although studied little, few oils are likely to have biomarker distributions that have been altered significantly by solubilization of contaminants in the reservoir or during migration through carrier beds. In general, reservoir and carrier rocks are organic-lean, and the concentrations of biomarkers solubilized from these sources are low compared with those in the migrating oil. However, in some cases, such as migration of a biomarker-poor condensate through organic-rich coals, solubilization can be important.

> Note: Petroleum associated with sulfide-rich sediments at active oceanic spreading centers represents an extreme example of solubilization and admixture of hydrocarbons generated over a wide range of thermal regimes (Kvenvolden and Simoneit, 1990). For example, hopanoid epimer ratios for extracts from the Guaymas Basin and Escanaba Trough indicate low maturity, while ratios of aromatic hydrocarbons (e.g.

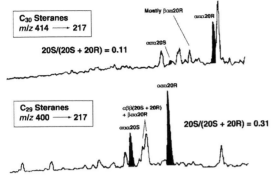

Figure 19.1. Metastable reaction monitoring/gas chromatography/mass spectrometry (MRM/GCMS) sterane chromatograms for an Angolan oil, showing leaching or solubilization of immature bitumen by oil. The oil appears to be more mature based on 20S/(20S + 20R) for the C_{29} steranes (bottom) than for the C_{30} steranes (top). Reprinted with permission by ChevronTexaco Exploration and Production Technology Company, a division of Chevron USA Inc.

benzo(e)pyrene and benzo(a)pyrene)) indicate high maturity. Unlike conventional petroleum, these non-commercial oil stains were generated from organic matter in very young (Quaternary) sediment by hydrothermal activity.

Several studies evoke migration–contamination to explain anomalous biomarker distributions (see Curiale, 2002), although few attempt to quantify the effect or determine its impact on accurate evaluation of petroleum systems. An investigation of oils from offshore Brunei is one such study where incorrect conclusions could have been made, had the phenomenon not been recognized (Curiale et al., 2000). Migration of light oils from deep, unidentified source rocks offshore Brunei resulted in extensive migration–contamination of the tetracyclic and pentacyclic hydrocarbons. The resulting oils contain non-indigenous biomarkers (e.g. bicadinanes and oleananes from carrier and/or reservoir rocks) plus indigenous components (e.g. cholestanes and methylcholestanes from the source rock). Migration contaminants appear to increase in offshore Brunei to the northeast. This contamination resulted in strong correlations between certain molecular maturity indicators and present-day reservoir temperatures.

Curiale and Bromley (1996) studied oils and condensates from the Vermilion 14 Field, offshore Louisiana. The oils yield anomalous maturity parameters (i.e. C_{29} $\alpha\alpha\alpha$ sterane 20S/(20R + 20S) as low as 0.05) and contain unsaturated biomarkers (i.e. diasterenes, oleanenes, and dammarenes). The maturity-dependent biomarker ratios correlated linearly with reservoir depth. Curiale and Bromley (1996) assumed that the maturity-dependent ratios for indigenous biomarker isomers in the oils should have achieved endpoint or equilibrium values before mixing with the reservoir contaminants. They compared measured ratios of these biomarkers in extracts of reservoir rock with the expected endpoint values of the oils to estimate the amount of biomarkers contributed as reservoir contaminants. For example, about 27% of the biomarkers in fluids from 14 000 feet (4267 m), and 53% of the biomarkers in fluids from 10 000 feet (3048 m), were attributed to contaminants. Similarly, Li et al. (2003) showed that "immature" oils from Niuzhuang South Slope of Bohai Bay Basin, China, are mixtures of mature oils and bitumen from immature Eocene shales. They explained the alteration of biomarker maturation parameters using a two-component linear mixing model.

Geochromatography is the hypothesized process by which biomarkers and other compounds migrate at different rates through interstices in the mineral matrix of a rock. Compounds with differing molecular weights, polarities, and stereochemistries should behave differently when exposed to various adsorptive/desorptive processes during movement out of the source rock (primary migration) or through carrier beds (secondary migration). For example, polar molecules are retained more strongly on mineral surfaces. Both field (e.g. Silverman, 1965; Seifert and Moldowan, 1978; Seifert and Moldowan, 1981; Leythaeuser et al., 1984; Mackenzie et al., 1987) and laboratory studies (e.g. Carlson and Chamberlain, 1986; Krooss and Leythaeuser, 1988; Brothers et al., 1991) address geochromatographic effects on organic compounds. Carlson and Chamberlain (1986) address some of the principles controlling migration differences between biomarkers that undergo differential absorption on montmorillonite clay.

Applications of biomarker technology to studies of geochromatography are limited. Although it is not now possible to use biomarkers to quantify migration distances, several attempts were made to use them to distinguish more- versus less-migrated oils (Seifert and Moldowan, 1981; Larter et al., 1996). Detailed studies of alternating sandstone and shale sequences in cores from the North Sea show little if any redistribution among biomarkers, while lower-molecular-weight *n*-alkanes show large effects, probably related to differences in diffusivity and solubility (Leythaeuser *et al.*, 1984; Mackenzie *et al.*, 1988b). However, hydrous pyrolysis experiments indicate that expelled oils and the remaining unexpelled bitumen in pyrolyzed rock chips have different biomarker compositions, paralleling those reported for crude oils and source-rock extracts (Peters *et al.*, 1990). They observed increases in tricyclics/17α-hopanes and diasteranes/steranes in the expelled oils compared with the extracts, but distributions of the hopane and sterane stereoisomers were unchanged. Seifert and Moldowan (1986) also proposed that the ratio of tricyclic terpanes to pentacyclic terpanes is a migration parameter because tricyclics migrate faster than hopanes. More research is needed on geochromatography and its effects on biomarkers.

The non-alkylated benzocarbazoles are measures for the relative distance of secondary oil migration. The ratio benzo[*a*]carbazole/benzo[*a*]carbazole + benzo[*c*]carbazole (the benzocarbazole (BC) ratio) and the concentrations of these compounds decrease with oil migration distance in the North Sea and the Western Canada Basin (Figure 19.2). Larter *et al.* (1996) believed that this was most likely due to different adsorption characteristics of these compounds on mineral surfaces, but more recent molecular dynamics studies suggest that oil/water partitioning may be more important (van Duin and Larter, 2001). Other workers suggest caution because similar effects on benzocarbazole distributions could be caused by varying source input or thermal maturity across the study area (Li *et al.*, 1997; Clegg *et al.*, 1998). Therefore, source and maturity effects must be constrained by other geochemical data before the use of pyrrolic nitrogen data in migration studies. For example, changes in lithology and geometry of migration conduits along migration pathways affect the apparent migration distance of oils of similar maturity and source along the Rimbey–Meadowbrook Trend (Li *et al.*, 1998a).

Note: Relative migration distances based on benzocarbazoles together with migration modeling were used to reconstruct migration paths for Q oils generated from the Infra-Cambrian Dhahaban Formation source rock in Oman (Terken and Frewin, 2000).

BIOMARKER KINETICS

Geochemical basin models attempt to predict the thermal history of sedimentary basins, including the timing and quantities of petroleum generated and migrated (e.g. Welte and Yalcin, 1988; Ungerer, 1990). Accurate assessment of paleotemperatures is critical input for these modeling programs. Many basin models are calibrated using thermally dependent biomarker ratios (Welte and Yalcin, 1987).

Mackenzie and McKenzie (1983) and Mackenzie *et al.* (1984) first advocated the use of thermally dependent biomarker ratios to assess paleotemperatures for basin models. They monitored the progress of several biomarker reactions at depth in basins with known temperature histories. Kinetic expressions, including apparent activation energies (E_a) and frequency factors (A), were derived for proposed steroid aromatization and sterane and hopane isomerizations (Figure 19.3). They observed that the rate expressions for the monoaromatic to triaromatic steroid aromatization are consistent with laboratory heating experiments, while the hopane isomerization is not. Rullkötter and Marzi (1988) recommended adjustments to these apparent activation energies based on the need to account for both laboratory-simulation and natural-evolution kinetics.

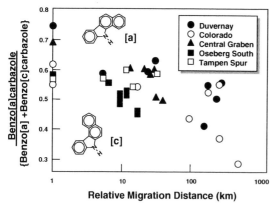

Figure 19.2. The ratio of benzo[*a*]carbazole/benzo[*a*]carbazole + benzo[*c*]carbazole for reservoired oils from five petroleum systems in western Canada and the North Sea versus estimates of secondary migration distance relative to a reference oil nearest the source rock. The reference oil is given an arbitrary migration distance of 1 km. Modified and reprinted with permission from Nature (Larter *et al.*, 1996). © Copyright 1996, Macmillan Magazines Limited.

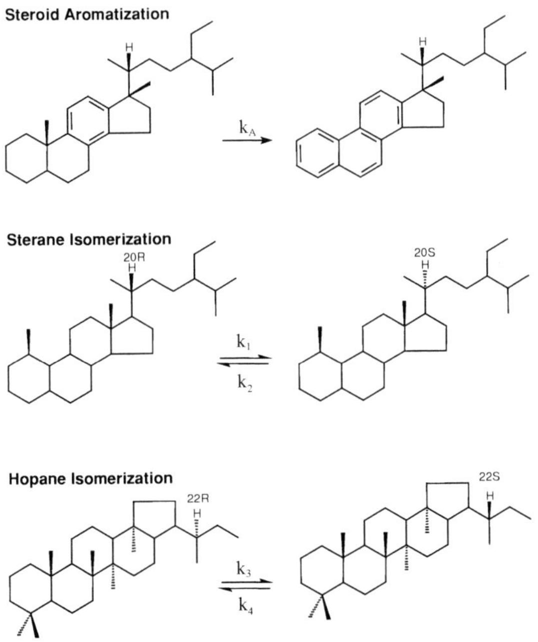

Figure 19.3. Proposed biomarker kinetic reactions used by Mackenzie and McKenzie (1983) to determine the thermal history of extensional sedimentary basins. See Chapter 14 for further discussion.

Biomarker distributions in oils and bitumens generated at varying stages of thermal maturity gave rise to empirical parameters that reflect thermal history (Mackenzie et al., 1980a; Mackenzie et al., 1981a; Mackenzie et al., 1981b; Seifert and Moldowan, 1980). Reaction models were evoked that approximated the conditions required for ideal thermal indicators. For example, during early diagenesis, biological precursors give rise to steranes with only the 20R configuration, which isomerize under geologic conditions to the 20S configuration. Isomerization reactions are presumed to be reversible, so that maturation can result in equilibrium between the various isomers. Hence, sterane 20S/(20S + 20R) ranges from 0 to ~0.5 under thermal conditions associated with oil generation. Homohopane isomerization at C-22 appears to follow a similar reaction model but proceeds at a faster rate. Hopane 22S/(22S + 22R) is initially ~0 in immature sediments, and reaches an apparent equilibrium value of ~0.6 before oil expulsion. The conversion of C-ring monoaromatic (MA) steroid hydrocarbons to triaromatic (TA) equivalents is an irreversible reaction once thought to be ideal for measuring thermal maturity. TA/(MA + TA) ranges from 0 in immature sediment to 1 in postmature oils and rock bitumens.

Mackenzie et al. (1982b) and Mackenzie and McKenzie (1983) observed that the extents of the isomerization and aromatization reactions differ between basins. In young basins with rapid heating rates, aromatization proceeded quickly to completion at about the same stage as hopane isomerization, before the steranes reached equilibrium. Conversely, in old basins with slow heating rates, the hopanes were isomerized completely, while the sterane isomerization and aromatization reactions progressed at about the same rate (Figure 19.4). These observations suggested that the rates of these reactions not only differ but also could be determined from field observations. More importantly, if the rate constants are known, then measurements of the biomarker ratios might be used to constrain the thermal and tectonic histories of basins.

Many assumptions are necessary to derive kinetic rate constants from basin studies. Ideally, the basin thermal and tectonic history is simple and known with a high degree of certainty. Mackenzie and McKenzie (1983) selected extensional sedimentary basins where the heat flow and burial history is constrained largely by the rapid stretching of the lithosphere. The degree of stretching can be defined by a single term, β (McKenzie et al., 1983b; Mackenzie et al., 1984). The reactions are

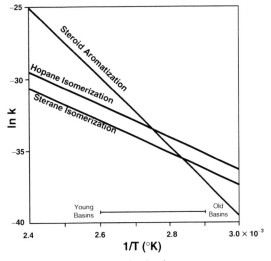

Figure 19.4. Arrhenius plots showing the temperature dependence of biomarker reaction rate constants. Reprinted from Mackenzie et al. (1984). © Copyright 1984, with permission from Elsevier.

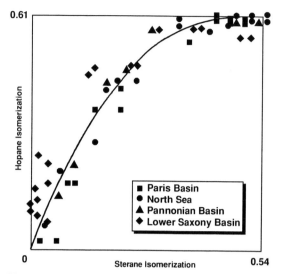

Figure 19.5. The kinetic constants for hopane and sterane isomerization (Mackenzie et al., 1983b) have the same activation energy and differ only in their frequency factor. Consequently, the relationship between hopane and sterane isomerization is independent of heating rate.

modeled using first-order kinetics, a reasonable but unproved assumption, and are assumed to have a temperature dependence defined by the Arrhenius equation, $k = Ae^{-(E/RT)}$. A best fit of the field data provided values for the frequency factor (A) and the activation energy (E).

Mackenzie et al. (1983b) found that the extent of sterane and hopane isomerization was independent of basin thermal history (Figure 19.5). Hence, the activation energies of the isomerization reactions are identical but their frequency factors differ. The activation energy of the aromatization reaction, however, appears to differ substantially from that of the isomerization reactions. Consequently, comparison of the extents of aromatization and isomerization could be used to determine the thermal history of each basin, including the maximum temperature of strata reached before any uplift. Mackenzie et al. (1983b) and Mackenzie et al. (1984) prepared a series of aromatization–isomerization (AI) diagrams based on different basin tectonics for use as guides (Figure 19.6). Thermal histories were modeled successfully using these biomarker kinetic methods in various locations summarized by Lewis (1993), including the Alberta Basin (Beaumont et al., 1985), the Scotian Shelf (Mackenzie et al., 1985b), the Linyi Basin of China (Hong et al., 1986; Zhi-Hua et al., 1986), the Lias Shale of Germany (Mackenzie et al., 1988a), and the Newark Basin (Walters and Kotra, 1990). Combinations of biomarker and vitrinite reflectance kinetic models also were used to predict

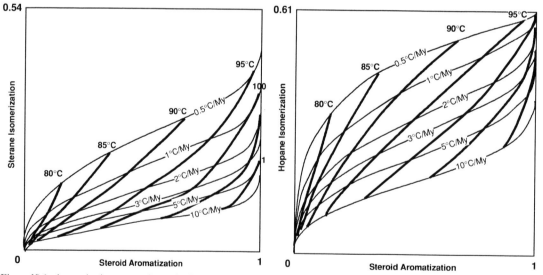

Figure 19.6. Aromatization–sterane isomerization (left) and aromatization–hopane isomerization (right) diagrams based on Arrhenius constants defined by Mackenzie et al. (1983b). Thin lines indicate burial at a constant heating rate; thick lines join points with the same temperature. Reprinted from Mackenzie et al. (1984). © Copyright 1984, with permission from Elsevier. My, million years.

paleoheat flow (Armagnac et al., 1988; Hermansen, 1993). The apparent success of biomarker kinetics led to their use in advanced basin simulators (Welte and Yalcin, 1988).

The ability to model thermal history depends on the accuracy of the biomarker reaction rate constants. After the work by Mackenzie and McKenzie (1983), others attempted to measure rates from laboratory simulations (Suzuki, 1983; 1984), natural samples (Alexander et al., 1986b; Strachan et al., 1989a; Strachan et al., 1989b), or both (Rullkötter and Marzi, 1988; Rullkötter and Marzi, 1989; Marzi and Rullkötter, 1992; Marzi et al., 1990; Ritter et al., 1993a; Ritter et al., 1993b; Ritter et al., 1993c). Differences in the resulting Arrhenius values were attributed to different mechanisms in laboratory and natural systems, uncertainties in the thermal histories, organic facies and mineral matrix effects, and the methods used to derive the rate constants (Figure 19.7). Lewis (1993) reviewed the various experimental methods used to determine Arrhenius constants from field and laboratory data and their limitations.

Unfortunately, inherent error permits a range of A and E for biomarker reactions determined using laboratory or natural data sets. The range of allowable values for Arrhenius constants differs for realistic geologic heating rates and isothermal temperatures used in laboratory simulations (Figure 19.8). If the same reactions are assumed to occur under both conditions, then the allowable Arrhenius constants are constrained to a narrow range. Hence, when Marzi and Rullkötter (1992) and Ritter et al. (1993a; 1993b; 1993c) determined biomarker reaction kinetics by combining natural and laboratory data, the derived Arrhenius constants must fall within this constrained range. Since the Arrhenius constants for the isomerization and aromatization reactions did not differ significantly, Marzi and Rullkötter (1992) suggested that the Mackenzie and McKenzie (1983) and Mackenzie et al. (1984) Arrhenius diagrams were invalid. This conclusion appears to conflict with the observed data.

Subsequent laboratory and natural studies undermine one of the basic assumptions of the Mackenzie and McKenzie model – that sterane and hopane isomerizations occur as single-step, chiral-centered equilibrium reactions. Hydrous pyrolysis simulations by Lewan et al. (1986) and Peters et al. (1990) found unexpected reversals in sterane 20S/(20S + 20R) at high temperatures (>330°C). Pyrolysis experiments by Lu et al. (1989)

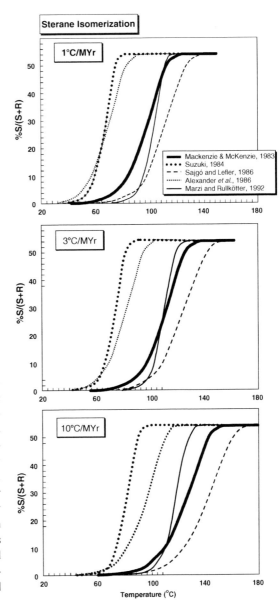

Figure 19.7. Comparison of published sterane-isomerization kinetic models at different heating rates. Models of Suzuki (1984) and Alexander et al. (1986b) are based on laboratory simulations that yield geologically unrealistic results. Models of Mackenzie and McKenzie (1983), Sajgó and Lefler (1986), and Marzi and Rullkötter (1992) are based, at least in part, on geologic field studies and thus yield geologically realistic results. However, these models differ substantially and yield different results at varying heating rates.

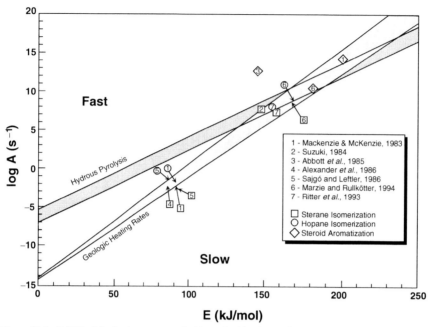

Figure 19.8. Published Arrhenius constants for biomarker kinetic reactions. Values that yield realistic results for geologic heating rates (1–10°C/My) and temperatures typical for hydrous pyrolysis experiments (280–350°C) are indicated. If the same kinetic reactions occur under both conditions, then the Arrhenius constants are constrained to the range where the two shaded areas overlap.

indicated that the thermal release of kerogen-bound biomarkers and their subsequent degradation were far more significant than isomerization. Wang (1990) reached similar conclusions from hydrous pyrolysis experiments. Abbott *et al.* (1990) and Abbott and Stott (1999) showed that deuterated 5α-cholestane does not isomerize in the presence of organic-rich shale under laboratory simulations. These experimental studies suggest strongly that single-step isomerization reactions of saturated steranes and hopanes do not occur.

Several studies using quantitative biomarker data conclude that apparent isomerization reactions involve a complex balance between the relative thermal stabilities and the rates of release of the biomarkers from kerogen. Bishop and Abbott (1993) and Farrimond *et al.* (1996) studied samples of organic-rich Jurassic siltstone from the Dun Caan Shale Member in Scotland. Closely spaced samples in the source rock at varying distances from an intrusive dike allowed them to observe changes in the biomarker distributions and concentrations caused solely by thermal maturity. They found that while isomerization of free steranes and hopanes could not be precluded, release or generation of these compounds from kerogen and/or functionalized moieties and thermal degradation were the dominant factors controlling the biomarker maturity parameters. Requejo (1994) examined a suite of source rocks from the Upper Devonian Duvernay Formation of western Canada that ranged from immature to postmature with respect to hydrocarbon generation. Quantitative assessment indicated that the preferential loss of the 20R isomer, not isomerization, determined 20S/(20S + 20R). In a similar study, Dzou *et al.* (1995) examined coals and vitrinite concentrates of the Lower Kittanning seam (Carboniferous) from western Pennsylvania and eastern Ohio that represented a uniform organic facies at different maturity levels. They found that the free 20R isomer is initially present at much higher concentrations than 20S. With increasing thermal stress, both isomers are released from the kerogen, but the proportion of 20S increases, resulting in a net increase in 20S/(20S + 20R). Farrimond *et al.* (1998) examined homogenous Eocene strata in a Barents Sea well ranging from 1400 to 3000 m depth. They concluded that hopane and sterane isomerization parameters varied according to the interplay of the relative rates of generation or

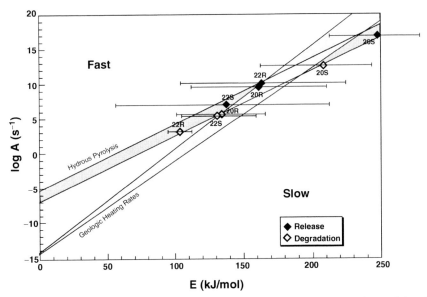

Figure 19.9. Arrhenius constants derived from hydrous pyrolysis experiments for the release and degradation of the 20S + R sterane and 22S + R hopane isomers (Abbott and Stott, 1999). The kinetic rates are consistent with laboratory conditions but not with natural observations. The experimental error, however, is large enough to permit a fit to both conditions.

release and thermal degradation of the different isomers. Again, isomerization of the free biomarkers may occur in the bitumen but was interpreted to be a relatively minor effect.

If maturity parameters based on sterane and hopane isomer ratios result from differences in the rates of their generation and destruction, then kinetic models based on this mechanism might be used in the same manner as Mackenzie and McKenzie proposed for single-step isomerization reactions. However, the kinetics of generation and destruction of isomers are difficult to determine and are likely to vary for different kerogens. Abbott et al. (1990) and Abbott and Stott (1999) tried to determine these kinetic parameters in laboratory simulations (Figure 19.9). The derived Arrhenius constants are consistent with the thermal conditions of hydrous pyrolysis experiments but yield unrealistic results at geologic heating rates. However, the error in these experiments is large, and Arrhenius constants can be found that fit both laboratory and natural conditions.

Although single-step isomerization of sterane and hopane isomers is unlikely, such reactions may occur through multiple steps involving metastable intermediates, such as carbocations and alkenes. Van Duin et al. (1996a; 1997) modeled steroid and hopanoid maturation using a full isomerization scheme involving all likely chiral reactions. Stabilities of the reactants, products, and high-energy intermediates were calculated using molecular mechanics to yield kinetic rate constants for each reaction. The resulting network allowed prediction of biomarker isomerization ratios in both hydrous pyrolysis and natural systems. Van Duin et al. (1997) urge caution in accepting such comparisons as proof of a kinetic model, because good agreement was found even in high-temperature hydrous pyrolysis experiments where the destruction of hopanes is significant.

The early success of the Mackenzie and McKenzie model suggested kinetic control on the sterane and hopane isomerization ratios. However, rather than simple isomerization, these ratios are controlled by the complex interplay of generation and/or release, multistep isomerization reactions through metastable intermediates, and thermal degradation (Figure 19.10). The single-step reaction model can only approximate reality and cannot be used to constrain geothermal history. Furthermore, it is not possible to determine the Arrhenius constants for multistep reactions from observations of natural systems (van Duin et al., 1997).

Our results from hydrous pyrolysis of Monterey Phosphatic and Siliceous rock (Peters et al., 1990) show

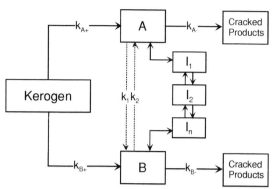

Figure 19.10. Kinetic model for saturated biomarker maturation parameters. The maturation parameter $A/(A + B)$ is no longer believed to involve a single, direct isomeric conversion (k_1/k_2). Rather, it is determined by differences in rates of generation from the kerogen (k_{A+}/k_{B+}) and thermal destruction in the bitumen (k_{A-}/k_{B-}). In addition, interconversion between isomers A and B may occur through a complex series of multistep chiral reactions involving metastable intermediates (e.g. carbocations, alkenes).

that steroid and hopanoid isomerizations are more complex than expected. During heating, kerogen-bound precursors generate steranes and hopanes showing lower levels of thermal maturity based on isomerization than those extracted from the unheated rock. Asymmetric centers in kerogen-bound steroids and hopanoids appear to be protected from isomerization compared with those of free steranes or hopanes in the bitumen. The Phosphatic and Siliceous rocks can show different levels of sterane or hopane isomerization when heated under the same time/temperature conditions. Further, maturity ratios based on these isomerizations unexpectedly decrease at high hydrous pyrolysis temperatures (>330°C) for the Phosphatic but not for the Siliceous samples. Lewan et al. (1986) observed a similar reversal in sterane isomerization for Phosphoria Retort Shale at high hydrous pyrolysis temperature. Further, Strachan et al. (1989b) observed a reversal in the C_{29} sterane $\alpha\alpha\alpha 20S/(20S + 20R)$ at depth in a well in Australia.

Mineralogy clearly affects the kinetics of oil generation and the composition of the products (Tannenbaum et al., 1986). Soldan and Cerqueira (1986) observed variations in biomarker compositions for samples heated under hydrous pyrolysis conditions to the same temperatures using different water salinities.

Unlike sterane and hopane isomerization, steroidal hydrocarbon aromatization appears to occur, although the reaction may not control $TA/(MA + TA)$. In laboratory experiments with pure compounds, Abbott et al. (1985) and Abbott and Maxwell (1988) demonstrated that monoaromatic steroidal hydrocarbons convert to their triaromatic equivalents. The experiments used sulfur to generate free radicals, and the derived kinetics do not apply to the geologic realm. Requejo (1992) suggested that C-ring monoaromatic steroids convert to triaromatic forms based on their concentrations in a suite of oils at varying maturity. Hydrous pyrolysis experiments by Wang (1990) showed increasing concentrations of triaromatic steroids over time at reaction temperatures >330°C, but no such trends were observed at lower temperatures. Bishop and Abbott (1993) studied extracts from rocks that were heated by an igneous intrusive and concluded that changes in $TA/(MA + TA)$ are governed not by aromatization but by the rates of generation/release and thermal destruction. It is likely that $TA/(MA + TA)$ is determined by the interplay of the relative rates of aromatization, generation/release, and thermal degradation.

The kinetics of several other geochemical reactions, or pseudo-reactions, were investigated and applied to constraining geothermal histories. In immature sediments, pristane occurs primarily as the (6R,10S)-diastereoisomer and is converted into racemic mixtures of the (6R,10S) and (6R,10R) + (6S,10S)-diastereoisomers (Patience et al., 1978; Mackenzie et al., 1980a). Abbott et al. (1985) heated (6R,10S)-pristane in the presence of sulfur alone and of sulfur with a shale matrix. Curiously, no isomerization occurred with sulfur only, but isomerization was observed in the presence of sulfur and shale. Thermal degradation of the isoprenoid isomers was significant, and assumptions needed to be made to determine rate constants for the isomerization reaction (Figure 19.11). The Arrhenius values derived from these experiments are not appropriate for geologic conditions.

Based on flash pyrolyzates of Toarcian Shale from the Paris Basin, van Graas et al. (1981) observed that pristane increased while prist-1-ene decreased with thermal maturity. They concluded that pristane evolves under natural conditions from the same kerogen-bound moieties that produce pristane and prist-1-ene by flash pyrolysis. Goossens et al. (1988a; 1988b) developed a maturation indicator, the pristane formation index (PFI) (= pristane/(pristane + prist-1-ene + prist-2-ene)). When calibrated against known geothermal

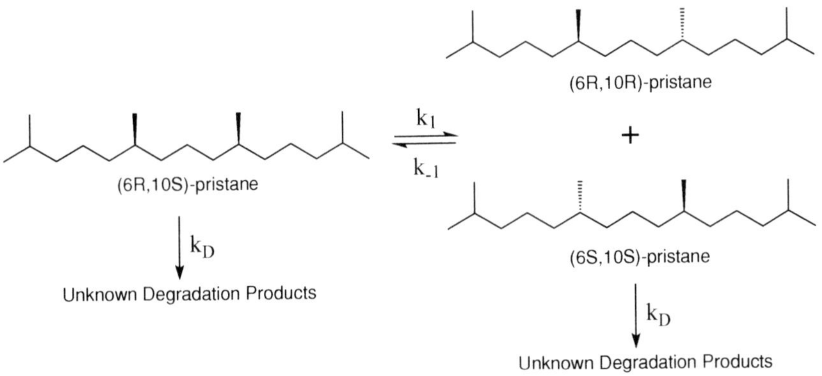

Figure 19.11. Reaction pathways proposed by Abbott *et al.* (1985) for the isomerization of the (6R,10S)-pristane to (6R, 10R) + (6S,10S). Since these diastereoisomers are racemic in mature sediments, the rates k_1 and k_{-1} are assumed to be equal, as are the rates of degradation (K_D).

histories for the Mahakam Delta (Goossens *et al.*, 1988b) and the Paris Basin (Goossens *et al.*, 1988a), the pseudo-kinetics of PFI could be determined. The resulting A and E (2.2×10^7 s^{-1} and 59 kJ/mol, respectively) were appreciably lower than values reported by Mackenzie and McKenzie (1983) for the sterane and hopane isomerization reactions. Burnham (1989) validated the PFI but pointed out that the low Arrhenius constants of Goossens *et al.* (1988b) are unreasonable for reactions involving bond breakage and are an artifact of fitting the data to one activation energy. More realistic values can be derived using a distribution of activation energies. Tang and Stauffer (1995) investigated PFI using both open- and closed-system pyrolysis of Monterey kerogens. They found that two or more first-order reactions are required to explain the origin of pristane. A and E (1.1×10^{14} s^{-1} and 46.1 kJ/mol, respectively) consistent with laboratory and geologic conditions result when the data were forced into a single pseudo-first-order reaction.

Alexander *et al.* (1986a) and (Cumbers *et al.*, 1986) observed that ortho-substituted alkylated biphenyls become depleted relative to the meta- and para-compounds with thermal maturity. One process by which ortho-substituted 2,3-dimethylbiphenyl could be removed is through cyclization to form 1-methylfluorene. Alexander *et al.* (1988a; 1989) and Kagi *et al.* (1989) determined the kinetics of these reactions by heating pure compounds under non-catalytic conditions. Direct application of the derived kinetics, however, is not possible because no systematic changes in the ratio of 1-methylfluorene to 2,3-dimethylbiphenyl occur in rock extracts of increasing maturity. This was attributed to additional, unspecified reactions involving 1-methylfluorene in the geologic environment. To use the ortho-substituted cyclization reactions, the concentration of 3,5-dimethylbiphenyl was compared with 2,3-dimethylbiphenyl. Laboratory simulations showed that 3,5-dimethylbiphenyl is inert under the thermal conditions that converted 2,3-dimethylbiphenyl to 1-methylfluorene (Figure 19.12). The ratio of 3,5- to 2,3-dimethylbiphenyl would thus follow the kinetics determined for the decrease of the 2,3-isomer due to cyclization (assuming no other reaction removes the 2,3-isomer). The derived kinetics are consistent with data from two Australian wells.

Maturity parameters based on isomer distributions of di- and trimethylnaphthalenes (Alexander *et al.*, 1984; 1985) and methylphenanthrenes (Radke and Welte, 1983) mostly reflect their relative thermal stabilities. α-Substituted isomers are more strained and less stable than their β-substituted counterparts. Alexander *et al.* (1986b) proposed that the aromatic maturity

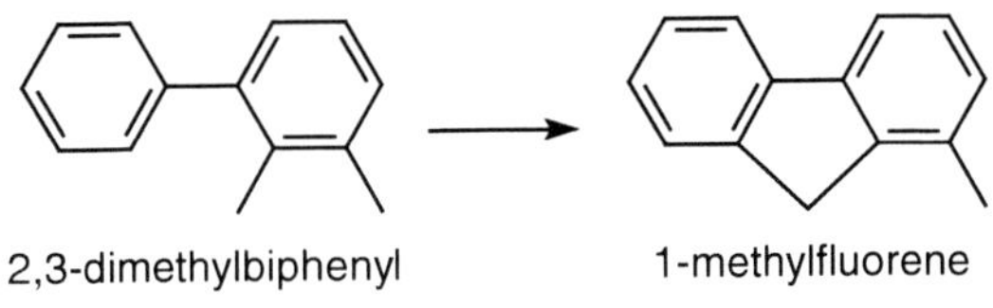

Figure 19.12. Cyclization reaction of 2,3-dimethylbiphenyl to yield 1-methylfluorene. The rate-limiting step is cleavage of the benzylic C–H bond.

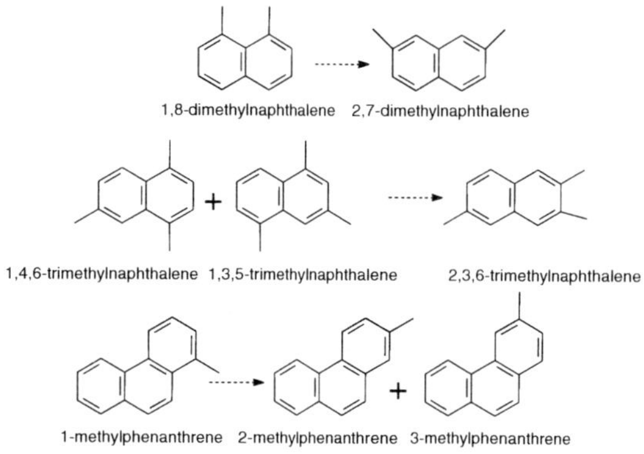

Figure 19.13. Pseudo-reactions based on aromatic maturity indicators. DP1 = 1,8-/2,7-dimethylnapthanene, TP-1 = (1,4,6- + 1,3,5-)/2,3,6-trimethylnaphthalene, PP-1 = 1-/(2- + 3-methylphenanthrene).

indicators could be expressed as pseudo-reactions and their kinetics could be determined from field data. When three aromatic maturity parameters (Figure 19.13) were fitted to data from two Australian wells, they appeared to exhibit different kinetics, which would be ideal for constraining geothermal histories. However, the kinetics differ between the wells, presumably due to different mineral matrices. The shale matrix promoted mainly ionic reactions, while the coal matrix promoted free-radical reactions. Alexander *et al.* (1988b) applied the kinetics of these pseudo-reactions to a study of the geothermal history of the Northwest Shelf of Australia. However, the general utility of these pseudo-reactions for such studies is questionable because the mineral matrix appears to exhibit a strong influence on the kinetics.

Ester decomposition may prove to be the first reliable biomarker reaction for constraining geothermal history. Esters decompose by a consorted reaction into corresponding alkenes and acids (Figure 19.14). The

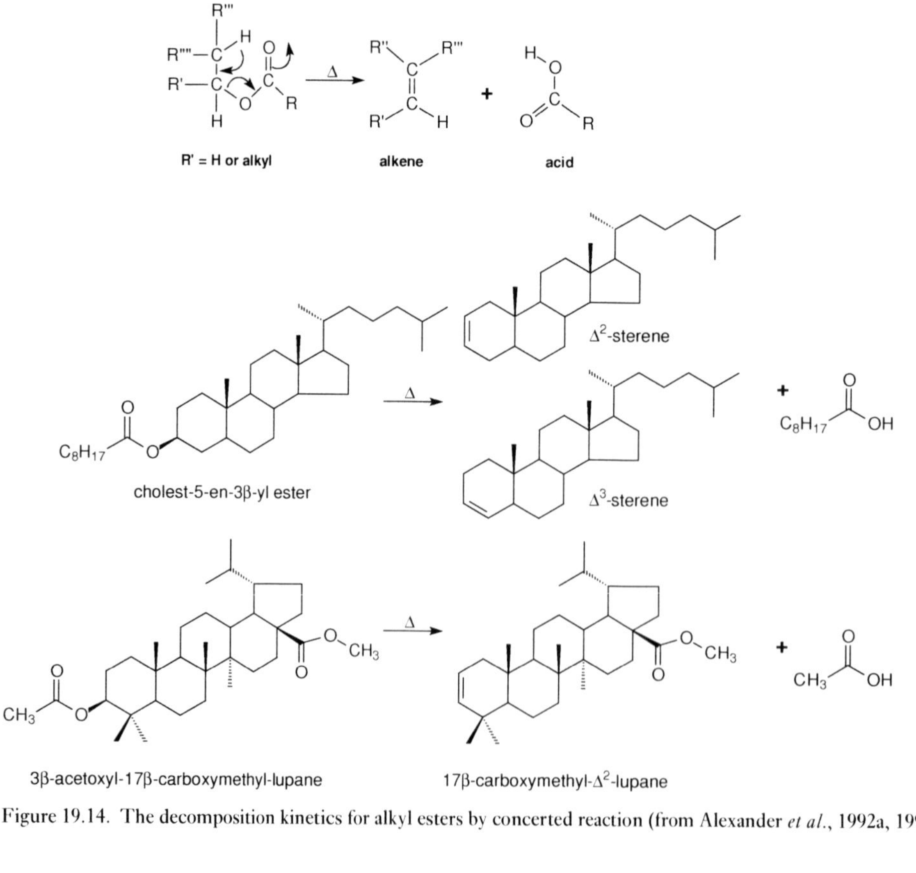

Figure 19.14. The decomposition kinetics for alkyl esters by concerted reaction (from Alexander *et al.*, 1992a, 1997).

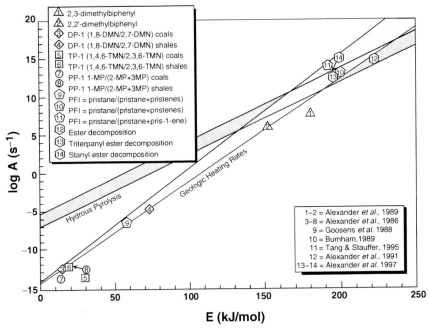

Figure 19.15. Published Arrhenius constants for biphenyl cyclization, aromatic pseudo-reactions, pristane formation index (PFI), and ester-decomposition reactions. Values that yield realistic results for geologic heating rates (1–10°C/My) and temperatures typical for hydrous pyrolysis experiments (280–350°C) are indicated.

reaction follows first-order kinetics and appears to be unaffected by catalysts and free radicals. The decomposition kinetics for long-chain alkyl esters (Alexander et al., 1992a) and for stanyl and triterpanyl esters (Alexander et al., 1997), as measured in laboratory liquid-phase simulations, proved to correspond well with those observed in a sedimentary sequence in the Gippsland Basin. Unfortunately, long-chain alkyl ester decomposition is complete before main-stage oil generation, limiting its utility for petroleum applications (Figure 19.15).

Our present knowledge of biomarker kinetics is inadequate. Uncertainties in the mechanisms and rates of biomarker reactions limit their use to constrain geothermal histories. Currently, vitrinite reflectance, apatite fission tracks, and fluid-inclusion homogenization temperatures are preferred to biomarker reactions for constraining geothermal history. Nevertheless, the concept of constraining geothermal histories by comparing chemical reactions that proceed at different rates is valid. Research in concerted reactions (Alexander et al., 1992a; 1997), biomarker generation/degradation (Abbott and Stott, 1999), and reaction networks via computational chemical mechanics (van Duin et al., 1996; 1997) show promise, and biomarker kinetics could yet emerge as a critical tool in basin modeling. Questions concerning the influence of mineral matrix effects (e.g. Peters et al., 1990), pressure (Costa Neto, 1983; 1991), and salinity (Soldan and Cerqueira (1986) remain unanswered.

Our understanding of the controls on kinetics of proposed biomarker reactions is limited. Rullkötter and Marzi (1988) used kinetic constants derived from the 20S/(20S + 20R) sterane isomerization during hydrous pyrolysis to reconstruct the geothermal history of the Michigan Basin (Rullkötter and Marzi, 1989). Abbott et al. (1990) attempted to derive a kinetic model to rationalize the use of these parameters. They found that direct chiral isomerization at C-20 in the regular steranes appears to be relatively unimportant under hydrous pyrolysis conditions. Marzi et al. (1990) observed wide variations in published kinetic expressions for the isomerization reaction at C-20 in the $5\alpha,14\alpha,17\alpha$-steranes. They suggest that more precise kinetic expressions could be obtained if the results of both laboratory experiments (high temperature, short time) and natural series (low temperature, long time) were combined. This approach assumes that the geologic and paleotemperature histories of the natural samples are known and that the chemical processes are the same in the laboratory and in nature.

Until further research is complete, we recommend that published kinetic expressions for biomarker reactions be used only as rough constraints on paleotemperatures. At least three problems limit the use of biomarker thermal maturity ratios in basin modeling:

- Assumptions of simple precursor–product relationships may be oversimplified (e.g. Mackenzie and McKenzie, 1983).
- Some biomarker maturity ratios show reversals at high maturity (e.g. Lewan et al., 1986; Rullkötter and Marzi, 1988; Strahan et al., 1989b; Peters et al., 1990).
- Small analytical errors in biomarker maturity ratios can have large effects on the kinetic expressions (Marzi et al., 1990).

OIL–OIL AND OIL–SOURCE ROCK CORRELATION

Hydrous pyrolysis of immature source rocks

Biomarker fingerprints of immature source-rock extracts differ considerably from those of mature extracts and oils from more deeply buried lateral equivalents of the same source rock. Mature extracts consist largely of products generated from the kerogen. Before incorporation into the kerogen, these products were largely cell-membrane lipids. Immature extracts, on the other hand, may be derived largely from free lipids in the contributing organisms, which may show a different composition from those incorporated into the kerogen. Thus, petroleum may correlate with extracts from its mature source rock while bearing little resemblance to extracts from an immature equivalent of the same rock.

Hydrous pyrolyzates expelled from potential source rocks are compositionally similar to natural crude oils, except for elevated amounts of aromatic and polar materials. Noble et al. (1991) observed that the gas/oil ratios (GORs) for pyrolyzates from three types of rocks from the Talang Akar Formation, Indonesia, varied according to their petroleum-generative potential and represented reasonable estimates of what might be expected from their subsurface maturation.

Figure 19.16 shows how to evaluate a possible correlation between immature source rock and petroleum using hydrous pyrolysis of the rock. Figure 19.16 shows that the C_{29} homologs dominate the sterane distribution for Monte Prena bitumen. The Monte Prena hydrous pyrolyzate has more C_{27} and C_{28} steranes, similar to the Katia oil. The immature monoaromatic steroids of Monte Prena bitumen also show a change in the hydrous pyrolyzate toward the composition of the Katia oil (Moldowan et al., 1992). Further details on the hydrous pyrolysis procedure are in Lewan (1985), Lewan et al. (1986), and Peters et al. (1990).

Eglinton and Douglas (1988) describe non-systematic differences between source parameters in immature and mature (pyrolyzate) bitumens from Monterey, Kimmeridge, New Albany, and Green River shales. Attempts to correlate immature source rocks with oil (mature) are best approached by pyrolyzing whole or crushed samples of the immature rock rather than extracted rock. Like the oil, the resulting pyrolyzate will thus contain biomarkers from both the original bitumen and those cracked from the kerogen.

Hydrous pyrolysis ternary diagrams for correlation

Sterane and monoaromatic steroid C_{27}–C_{28}–C_{29} ternary diagrams are commonly used to show genetic relationships among crude oils and bitumens extracted from thermally mature source rocks. These plots provide independent evidence for correlation among samples because the compounds in each homologous series have different origins. Proximity of samples to each other on such plots supports a genetic relationship.

Although relationships among crude oils are readily addressed using biomarkers, oil–source rock correlations are more difficult, partly because few thermally mature source-rock extracts are available. Active source rocks are buried deeply and generally occur far from updip, shallower sites of petroleum accumulation. Drilling is costly and normally ends when reservoir objectives are reached at these shallower depths. Although mature source rock may not be available, samples of laterally equivalent, potential (thermally immature) source rock can usually be obtained from outcrops or shallow wells near the periphery of the basin. When thermally mature source rock is unavailable, hydrous pyrolysis (Lewan, 1985; Lewan et al., 1986) is commonly used

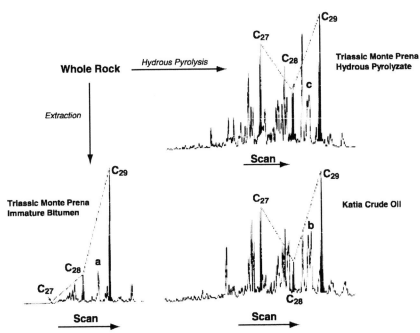

Figure 19.16. Hydrous pyrolysis allows correlation of immature Monte Prena source rock with Katia oil, offshore Adriatic Basin (Moldowan et al., 1992). The immature source rock was pyrolyzed under hydrous conditions (Peters et al., 1990; Lewan et al., 1986) in a pressure vessel at 320°C for 72 hours. The saturate fraction of oil generated by the pyrolysis shows a dramatic shift in sterane composition (m/z 217) from the immature extract. The immature pattern is skewed toward C_{29}-steranes compared with the mature oil pattern, which contains considerable C_{27}- and C_{28}-steranes. The pattern of the C_{27}–C_{29} $5\alpha,14\alpha,17\alpha$(H) 20R-steranes ($\alpha\alpha\alpha$20R) indicated by the dashed line for the Monte Prena pyrolyzate is nearly identical to that for Katia oil. (a) Mostly C_{29} $5\beta,14\alpha,17\alpha$(H) 20R steranes. (b) Mostly C_{29} $5\alpha,14\beta,17\beta$(H) 20R steranes. (c) Mixture C_{29} $5\beta,14\alpha,17\alpha$(H) 20R + C_{29} $5\alpha,14\beta,17\beta$(H) 20S + 20R. Reprinted with permission by ChevronTexaco Exploration and Production Technology Company, a division of Chevron USA Inc.

to artificially mature potential source rocks for geochemical comparison with the oils (e.g. Peters et al., 1990; Moldowan et al., 1992). However, care must be taken when comparing C_{27}–C_{28}–C_{29} ternary diagrams of biomarker compositions for hydrous pyrolyzates with crude oils, as illustrated below.

The distributions of C_{27}–C_{28}–C_{29} steranes for nine non-biodegraded oils generated from the Monterey Formation from the offshore Santa Maria Basin in California do not change (Figure 19.17, top), despite a significant range of thermal maturity as discussed earlier (see Figure 14.23) (Peters et al., 1990; Peters and Moldowan, 1991). API gravities of the oils range from 3.5 to 32.9°, implying a broader range of maturity than for most oil–oil correlations. The C_{27}–C_{28}–C_{29} monoaromatic steroid distributions for the same Monterey oils are similar, except for the least mature (613) and two of the more mature (548 and 611) samples (Figure 19.17, bottom). This represents one of the few cases where otherwise similar oils show different plot locations on a monoaromatic steroid ternary diagram, apparently due to differences in thermal maturity.

We believe that the above trend in monoaromatic steroid composition results from thermal maturity rather than organic facies differences among the Monterey crude oils. For example, the oils show similar stable carbon isotope ratios that become ~1.5‰ more enriched in ^{13}C from the least mature (sample 613, δ^{13}C $= -17.99$‰) to the most mature samples (samples 548 and 611, δ^{13}C $= -16.51$ to -16.49‰, respectively) (Figure 19.18). The similar isotopic values support the sterane distributions and other data indicating a common origin for the samples. Furthermore, the ^{13}C content of the oils increases with thermal maturity, consistent with the expected trend resulting from kinetic isotope fractionation.

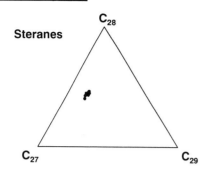

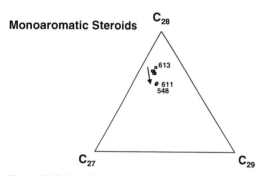

Figure 19.17. C_{27}–C_{28}–C_{29} sterane (top) and monoaromatic steroid (bottom) distributions for nine Miocene Monterey Formation crude oils from the Santa Maria Basin.

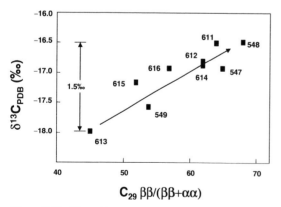

Figure 19.18. The stable carbon isotope ratios for nine Monterey crude oils (see Table 13.9) increase by as much as 1.5‰ due to thermal maturation (data from Peters, 1999b). Oils range in maturity from early oil window to past peak oil window (e.g. samples 613 and 548, respectively). See also Figure 16.57.

We tested the effects of hydrous pyrolysis on the sterane and monoaromatic steroid distributions for thermally immature, organic-rich source rocks from the Siliceous and Phosphatic members of the Monterey Formation and the Green River Formation. Original TOC, hydrogen index, and T_{max} for these rocks were as follows: Siliceous Member = 4.5 wt.%, 501 mg HC/g TOC, 406°C; Phosphatic Member = 11.6 wt.%, 615 mg HC/g TOC, 404°C; Green River = 10.6 wt.% TOC, 798 mg HC/g TOC, 448°C. Figures 19.19–Figure 19.21 show that increasing hydrous pyrolysis temperatures result in systematic trends in sterane and monoaromatic steroid composition similar to those observed in straight-run refinery products (Peters et al., 1992). For example, floating pyrolyzates from heated Siliceous and Phosphatic Shale Members of

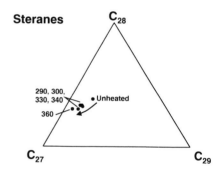

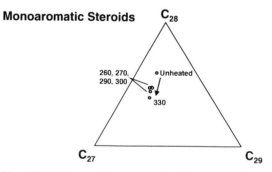

Figure 19.19. C_{27}–C_{28}–C_{29} sterane (top) and monoaromatic steroid (bottom) distributions for bitumen extracted from thermally immature source rock from the Monterey Formation Siliceous Member and floating oil from hydrous pyrolyzates of the same rock. Numbers refer to the temperature (°C) for each 72-hour hydrous pyrolysis experiment. Arrows indicate trends of increasing thermal maturity.

Monterey Phosphatic Member

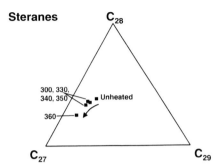

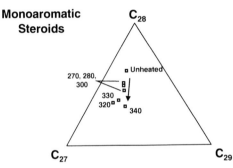

Figure 19.20. C_{27}–C_{28}–C_{29} sterane (top) and monoaromatic steroid (bottom) distributions for bitumen extracted from thermally immature source rock from the Monterey Formation Phosphatic Member and floating oil from hydrous pyrolyzates of the same rock. Numbers refer to the temperature (°C) for each 72-hour hydrous pyrolysis experiment. Arrows indicate trends of increasing thermal maturity.

the Miocene Monterey Formation show similar trends toward increased C_{27} steranes with increasing hydrous pyrolysis temperature (see also Table 13.9). The increase in C_{27} monoaromatic steroids for these members follows a slightly different pathway from that of the steranes. In contrast, the trends for the steranes and monoaromatic steroids for the Green River pyrolyzates differ sharply from those for the Monterey pyrolyzates. Furthermore, the sterane and monoaromatic steroid distributions of the unheated Siliceous and Phosphatic rock extracts resemble the crude oils more closely than any of the pyrolyzates.

SOURCE-ROCK AGE FROM BIOMARKERS

If the age distributions of more biomarkers were known, then determination of the source rocks for oils would be greatly simplified. Because biomarkers represent a molecular record of life, a better understanding of their temporal evolution will necessarily improve our understanding of historical geology. Some major changes in Earth history that might be reflected in biomarker distributions could include:

- evolution of new organisms (e.g. land plants, angiosperms, diatoms, coccolithophorids, and eukaryotes);
- major extinctions or changes in biota (e.g. Cretaceous-Tertiary and Permian-Triassic);
- major geologic events (e.g. oceanic anoxia and glaciations). Table 19.1 shows the limited data for certain biomarkers whose age distribution appears useful. Additional discussion of C_{27}–C_{28}–C_{29} steranes and their age significance is included later.

Green River Formation

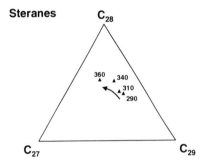

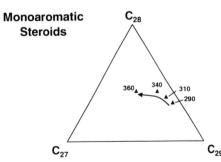

Figure 19.21. C_{27}–C_{28}–C_{29} sterane (top) and monoaromatic steroid (bottom) distributions for floating oil from hydrous pyrolyzates of immature source rock from the Green River Formation. Bitumen from the immature source rock was not analyzed because of insufficient material. Numbers refer to the temperature (°C) for each 72-hour hydrous pyrolysis experiment. Arrows indicate trends of increasing thermal maturity.

Table 19.1. *Age distributions for some biomarkers in petroleum*

Biomarker	Organism	First geologic occurrence[1]
Terpanes		
Oleanane	Angiosperms	Late Cretaceous
Beyerane, kaurane, phyllocladane	Gymnosperms	Ordovician-Devonian
Gammacerane	Protozoa, bacteria	Late Proterozoic
28,30-Bisnorphane	Bacteria	Proterozoic
Steranes		
23,24-Dimethylcholestane	Prymnesiophytes or Coccolithophores?	Triassic
4-Methylsteranes	Dinoflagellates, bacteria	Triassic
Dinosterane	Dinoflagellates	Triassic
24-*n*-Propylcholestane	Marine algae (*Sarcinochrysidales*)	Proterozoic
2- and 3-Methylsteranes	Bacteria?	Pre-Mesozoic?
$C_{29}/(C_{27}$ to $C_{29})$ steranes	Eukaryotes	Varies through time
Isoprenoids		
Botryococcane	*Botryococcus braunii*	Tertiary
Biphytane	Archaea	Proterozoic

[1] These suggested age boundaries are not rigorously documented for the related biomarkers and additional research may necessitate revisions. (See also Table 13.5.)

EXTRATERRESTRIAL BIOMARKERS

McKay *et al.* (1996) presented four lines of evidence that a meteorite called ALH84001 and collected from the Allen Hills in Antarctica contains fossil remnants of Martian life:

- Carbonate globules were formed by liquid water.
- Ovoid and tubular structures within fractures in the carbonate globules are smaller than, but morphologically similar to, bacteria (Figure 19.22).
- Mineral grains of magnetite, pyrrhotite, and greigite that are similar to those produced by some bacteria occur within the carbonate globules.
- Minute quantities of polynuclear aromatic hydrocarbons (PAHs) occur in or near the carbonate globules. If confirmed, then these PAHs are the first known extraterrestrial biomarkers.

Nearly all researchers agree that the meteorite sample ALH84001 began as Martian crust ~4500 Ma and was shocked by an asteroid impact ~4000 Ma. The carbonate globules formed ~3.8 Ma, during a time when Mars was believed to be wetter, but there is no consensus on the temperature of formation. The sample was shocked by meteorite impacts one or more times after the carbonate globules formed. The last impact occurred ~17–12 Ma, resulting in ejection of the sample

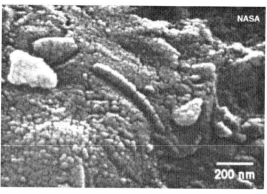

Figure 19.22. Fossils of ancient Martian nanobacteria or shock-induced inorganic mineral shapes? After extensive study, the question remains unanswered. National Aeronautics and Space Administration (NASA) photo. See discussion in McKay *et al.* (1996).

from the Martian surface into space. The sample fell to Earth ~13 000 years ago, where it remained trapped in glacial ice until collection in 1984. The ovoid and tubular objects are not artifacts of sample preparation. Beyond these basic conclusions, there is little agreement.

The organic geochemical evidence that the microscopic objects are fossils is considered by some to be the weakest line of evidence supporting evidence of Martian life. The PAHs in AL84001 could be bacterial relicts, but PAHs could just as easily be abiogenic because these compounds are abundant in carbonaceous chondrites and occur in interstellar space. Perhaps because of this uncertainty, the geochemical studies have focused on whether the organic matter is extraterrestrial or represents terrestrial contamination. Contamination by organic cleaning solvents, disinfectants, and even air-fresheners affects many museum samples of carbonaceous chrondrites (Watson *et al.*, 2003).

McKay *et al.* (1996) addressed four aspects of PAHs in ALH84001 that indicated their derivation from Martian bacteria:

- The PAHs are not contamination from laboratory procedures.
- These compounds did not enter the meteorite while in Antarctica.
- The distribution resembles, but is not identical to, PAHs in carbonaceous chondrites and differs from modern terrestrial PAHs.
- They are consistent with the decomposition of bacteria.

Laboratory contamination was eliminated using ultra-clean analytical procedures and comparison of samples with sample blanks. PAHs in the meteorite did not contain the sulfur-containing PAHs common in modern exhaust and smoke from burning fossil fuels. The concentrations of the PAHs also increased from the exterior to the interior of the meteorite and were higher than those found in preindustrial ice from cores. The authors concluded that these compounds did not enter the meteorite after its arrival on Earth, were indigenous to Mars, and were consistent with degradation of bacteria.

Note: Careful handling and storage of meteorite samples are essential in order to reduce the risk of terrestrial contamination. Pieces of the Orgueil carbonaceous chondrite fell in southern France on May 14, 1864. Organic compounds in the meteorite suggested to some an origin from alien life (Nagy *et al.*, 1961; Meinschein *et al.*, 1963). Notoriety of the meteorite increased after it was discovered that one Orgueil sample was deliberately contaminated with plant and coal fragments (Anders *et al.*, 1964). Compound-specific isotope analysis was used to identify contaminants in certain meteorites, such as Murchison, but the carbon isotope composition of organic matter in Orgueil is not readily distinguishable from terrestrial values (Sephton and Gilmour, 2001). Watson *et al.* (2003) extracted a powdered Orgueil sample using supercritical fluid extraction and pure CO_2. Most of the extracted compounds identified by GCMS are mono- and sesquiterpenoids that fall in a narrow range (C_{10}–C_{15}), suggesting contamination by refined essential plant-oil derivatives. Because of their natural aroma and bacterial action, essential plant-oil derivatives are added to air-fresheners, disinfectants, and cleaning products. Abundant compounds in the Orgueil sample included cadalene, calamenene, 5,6,7,8-tetrahydrocadalene, and curcumene, C_{13}–C_{20} *n*-alkanes, pristane, and phytane. In sediments, cadalene is a diagenetic product derived from cadinene sesquiterpenoids, with calamenene and 5,6,7,8-tetrahydrocadalene as possible intermediates (Simoneit and Mazurek, 1982). Pyrolysis of thermally desorbed and solvent-extracted residues of the Orgueil sample yielded no terpenoid compounds like those in the extract, supporting the contention that the free compounds are terrestrial contaminants. Similar contaminants occur in another Orgueil sample (camphor) and a sample of Murray meteorite (camphene), suggesting that this type of contamination is not an isolated event.

Critics of McKay *et al.* (1996) point out that PAHs are not strictly biomarkers. For example, PAHs readily form by abiogenic reactions that could be catalyzed by the magnetite grains (Anders, 1996), and presumed abiogenic PAHs occur in C2-type carbonaceous chondrites (Bell, 1996). Clemett and Zare (1996) countered that abiogenic processes could have produced the low-mass PAHs, but the high-mass PAHs could have originated only from the thermal breakdown of kerogen or a kerogen-like polymer. Becker et al. (1997) questions whether the PAHs were indigenous to the Martian meteorite and suggested that they leached out of the Antarctic ice and selectively concentrated on carbonate mineral surfaces. They concluded that the organic matter in ALH84001 is a mixture of terrestrial PAH contaminants and fallout from carbonaceous chondrites or interplanetary dust. Clemett et al. (1998) refute much of Becker et al. (1997) by showing that the ice from Allan Hills contains essentially no PAHs, that the carbonate minerals do not preferentially absorb PAHs, and that the spatial distribution of PAHs in ALH84001 is inconsistent with contamination. Furthermore, other Allan Hills meteorites that did not originate from Mars contained little or no PAHs based on the same procedures used to detect PAHs in ALH84001. Bada et al. (1998) demonstrated that all of the amino acids extracted from ALH84001 were terrestrial contaminants, suggesting that other compounds might also represent contamination.

Carbon isotopic studies are not conclusive about the origin of the carbonate and organic carbon in ALH84001. Excitement was generated by early reports of ^{13}C-depleted (δ^{13}C $\sim$32.3‰) and barely detectable amounts of methane (δ^{13}C $\gg$ -60‰) (Wright et al., 1997). Jull et al. (1998) measured the stable and radioactive isotopes of the carbonate and organic carbon associated with the globules. They found that carbonate had little or no ^{14}C and was isotopically unlike terrestrial carbon. In contrast, most of the organic carbon contained ^{14}C and was deemed to be terrestrial contamination. However, the high-temperature pyrolyzate of acid-resistive carbon, which accounts for $\sim$8% of the total, had no ^{14}C and was depleted in ^{13}C. This small amount of non-carbonate carbon was deemed extraterrestrial but not necessarily of biogenic origin. Becker et al. (1999) confirmed at least two types of organic carbon in ALH84001. Terrestrial contamination associated with the carbonate globules yields δ^{13}C of -26‰. After mild acid treatment, the residue yields δ^{13}C of -8‰, which was interpreted to indicate a mixture of the terrestrial contamination at δ^{13}C of -26‰ and carbonate at δ^{13}C of $\sim$36‰. After strong acid treatment, the residue yielded δ^{13}C of -15‰, which was offered as proof for extraterrestrial kerogen. Becker et al. (1999) concluded that this acid-resistant residue was delivered to Mars in carbonaceous meteorites, which returns us to the argument that the PAHs in ALH84001 are not of biological origin (e.g. Bell, 1996).

Other studies of Martian meteorites present additional evidence for and against a biological origin for the bacteria-like structures in ALH84001. Isotopic analysis of the sulfides associated with the carbonate globules found no evidence for sulfur processing by bacteria (Greenwood et al., 1997; Shearer et al., 1996). Stephan et al. (1998) found that PAHs were distributed broadly in the meteorite and not concentrated in the carbonate globules, as indicated by McKay et al. (1996). These results were confirmed by Becker et al. (1999) and further dissociate the PAHs from the bacteria-like structures. On the other hand, McKay et al. (1997) reported biofilms in ALH84001 that resemble sheath-like biological materials in terrestrial sediments.

In summary, although circumstantial evidence suggests that the PAHs in ALH84001 could represent extraterrestrial biomarkers, the question of contamination remains. Stronger evidence for extraterrestrial biomarkers and life on Mars may require samples returned directly from a Mars probe, where contamination is less likely.

SOURCE INPUT AND DEPOSITIONAL ENVIRONMENT

If petroleum explorationists are able to determine the source rock for crude oil, then probable migration pathways or types of petroleum to be expected can be used to develop new plays. Unfortunately, the source rock for trapped petroleum is commonly unknown, and only oil and/or seep samples are available for study. For this reason, the use of biomarkers to predict the character of the source rock is a critical technology.

Current understanding of the effects of depositional and diagenetic processes on geochemical parameters is limited but useful. For example, as discussed earlier in the text, reducing (low Eh) depositional conditions for shale source rocks result in enhanced phytane versus pristane, Tm versus Ts, and vanadyl versus nickel porphyrins, while hindering formation of diasteranes versus steranes and short- versus long-side-chain monoaromatic steroids. Alkaline (high pH) conditions appear to favor formation of Tm over Ts and moretanes over hopanes, while hindering formation of diasteranes and phytane.

Although biomarkers are already powerful tools, further research is needed to improve systematic understanding of source input and sedimentologic controls on biomarker compositions. The results of these studies will build on concepts of controls on organic facies developed by Demaison and Moore (1980) and Jones (1987). For example, diasteranes/steranes in petroleum is commonly used to make qualitative inferences on the relative clay content in the source rock. The application of these and other parameters related to organic matter input and depositional environment is generally based on non-systematic, empirical evidence (experience) and limited support from the literature or internal company memoranda. More systematic, statistical studies could improve the usefulness of these parameters. For example, Peters *et al.* (1986) used multivariate analysis to evaluate the significance of various biomarker and isotopic analyses to distinguish oils derived from organic matter deposited in non-marine shales, marine shales, and marine carbonates. Triangular plots of probabilities were developed that can be used to distinguish between the three source-rock organic-matter end members.

Zumberge (1987a) used factor and stepwise discriminant analysis of tricyclic and pentacyclic terpane data for oils to predict source-rock features and depositional environments. He was able to distinguish five categories of source rocks from the oil compositions: nearshore marine, deep-water marine, lacustrine, phosphatic-rich source beds, and Ordovician source rocks. Telnaes and Dahl (1986) compared 45 oils of probably Kimmeridge source from the North Sea using multivariate analysis of biomarker data. They were able to separate oils from different fields and formations and make inferences about the significance of parameters in regard to depositional environment, maturity, or organic facies. Irwin and Meyer (1990) used multivariate analysis to distinguish organic and mineralogical compositions among four distinct lacustrine depositional settings in the Devonian Orcadian Basin of Scotland. Principal component analysis allowed identification of the biomarker parameters that were most useful in discriminating between the samples by environment or relative thermal maturity.

Sample selection will be a critical factor in further studies of biomarkers for predicting source-rock character. We need well-defined source-rock sequences where depositional and diagenetic processes are understood in order to relate more accurately these processes to measured biomarker parameters. For example, detailed evaluation of the effects of various clay contents on biomarker ratios would require a sample suite where other variables, such as thermal maturity and organic matter input, were relatively uniform. Other sample suites that could be useful include lacustrine rocks showing differences in paleosalinity, marine rocks showing differences in oxicity during deposition, and rocks containing organic matter specific to certain organisms. This work overlaps with the use of biomarkers to predict source-rock age.

Appendix: geologic time charts

EONOTHEM EON	ERATHEM ERA	SYSTEM PERIOD	SERIES EPOCH	STAGE AGE	AGE Ma	+/-
PHANEROZOIC	CENOZOIC	Quaternary	HOLOCENE			
			PLEISTOCENE	Calabrian	0.01	
				Gelasian	1.81	
		NEOGENE	PLIOCENE	Piacenzian	2.58	
				Zanclean	3.60	
					5.32	
			MIOCENE	Messinian	7.12	
				Tortonian	11.2	
				Serravallian	14.8	
				Langhian	16.4	
				Burdigalian	20.5	
				Aquitanian	23.8	
		PALEOGENE	OLIGOCENE	Chattian	28.5	
				Rupelian	33.7	
			EOCENE	Priabonian	37.0	
				Bartonian	41.3	
				Lutetian	49.0	
				Ypresian	55.0	
			PALEOCENE	Thanetian	57.9	
				Selandian	61.0	
				Danian	65.5	0.1
	MESOZOIC	CRETACEOUS	UPPER/LATE	Maastrichtian	71.5	0.5
				Campanian	83.5	0.5
				Santonian	85.8	0.5
				Coniacian	89.0	0.5
				Turonian	93.5	0.2
				Cenomanian	98.9	0.6
			LOWER/EARLY	Albian	112.2	1.1
				Aptian	121.0	1.4
				Barremian	127.0	1.6
				Hauterivian	132.0	1.9
				Valanginian	136.5	2.2
				Berriasian	142.0	2.6
		JURASSIC	UPPER/LATE	Tithonian	150.7	3.0
				Kimmeridgian	154.1	3.3
				Oxfordian	159.4	3.6
			MIDDLE	Callovian	164.4	3.8
				Bathonian	169.2	4.0
				Bajocian	176.5	4.0
				Aalenian	180.1	4.0
			LOWER/EARLY	Toarcian	189.6	4.0
				Pliensbachian	195.3	3.9
				Sinemurian	201.9	3.9
				Hettangian	205.1	4.0
		TRIASSIC	UPPER/LATE	Rhaetian	209.6	4.0
				Norian	220.7	4.4
				Carnian	227.4	4.5
			MIDDLE	Ladinian	234.3	4.6
				Anisian	241.7	4.7
			LOWER/EARLY	Olenekian	244.8	4.8
				Induan	250.0	4.8

EONOTHEM EON	ERATHEM ERA	SYSTEM PERIOD	SERIES EPOCH	STAGE AGE	AGE Ma	+/-
PHANEROZOIC	PALEOZOIC	PERMIAN	LOPINGIAN	Changhsingian	251.4	3.6
				Wuchiapingian	253.4	
			QUADALUPIAN	Capitanian		
				Wordian	265	
				Roadian		
			CISURALIAN	Kungurian		
				Artinskian	283	
				Sakmarian		
				Asselian	292	
		CARBONIFEROUS	PENNSYLVANIAN	Gzhelian		
				Kazimovian		
				Moscovian		
				Bashkirian	320	
			MISSISSIPPIAN	Serpukhovian	327	
				Visean	342	4
				Tournaisian	354	
		DEVONIAN	UPPER/LATE	Famennian	364	
				Frasnian	370	
			MIDDLE	Givetian	380	
				Eifelian	391	
			LOWER/EARLY	Emsian	400	
				Pragian	412	
				Lockhovian	417	
		SILURIAN	PRIDOLI		419	
			LUDLOW	Ludfordian		
				Gorstian	423	
			WENLOCK	Homerian		
				Sheinwoodian	428	
			LLANDOVERY	Telychian		
				Aeronian		
				Rhuddanian	440	
		ORDOVICIAN	UPPER/LATE			
			MIDDLE	Darriwilian	467.5	3
			LOWER/EARLY	Tremadocian	495	
		CAMBRIAN	UPPER/LATE		500	
			MIDDLE		520	
			LOWER/EARLY		545	

EONOTHEM EON	ERATHEM ERA	SYSTEM PERIOD	AGE Ma
PRECAMBRIAN	PROTEROZOIC	NEOPROTEROZOIC	540
		NEOPROTEROZOIC III	
		CRYOGENIAN	650
		TONIAN	850
		STENIAN	1000
	MESOPROTEROZOIC	ECTASIAN	1200
		CALYMMIAN	1400
		STATHERIAN	1600
	PALEOPROTEROZOIC	OROSIRIAN	1800
		RHYACIAN	2050
		SIDERIAN	2300
			2500
	ARCHEAN	NEOARCHEAN	
		MESOARCHEAN	2800
			3200
		PALEOARCHEAN	
		EOARCHEAN	3600

No subdivisions into periods

Geologic time charts

PHANEROZOIC

CENOZOIC

AGE (Ma)	PERIOD	EPOCH	AGE	Boundary Ages (Ma)
		HOLOCENE / PLEISTOCENE	Calabrian	0.01 / 1.8
		PLIOCENE L	Piacenzian	3.6
5		E	Zanclean	5.3
	NEOGENE	MIOCENE L	Messinian	7.1
10			Tortonian	11.2
15		M	Serravallian	14.8
			Langhian	16.4
20		E	Burdigalian	20.5
25	TERTIARY		Aquitanian	23.8
30		OLIGOCENE L	Chattian	28.5
35		E	Rupelian	33.7
	PALEOGENE	EOCENE L	Priabonian	37.0
40			Bartonian	41.3
45		M	Lutetian	49.0
50		E	Ypresian	54.8
55		PALEOCENE L	Thanetian	57.9
60			Selandian	61.0
65		E	Danian	65.0

MESOZOIC

AGE (Ma)	PERIOD	EPOCH	AGE	Boundary Ages (Ma)
70		LATE	Maastrichtian	65 / 71.3
80			Campanian	83.5
			Santonian	85.8
90	CRETACEOUS		Coniacian	89.0
			Turonian	93.5
100			Cenomanian	99.0
110		EARLY	Albian	112
120			Aptian	121
130		NEOCOMIAN	Barremian	127
			Hauterivian	132
140			Valanginian	137
			Berriasian	144
150		LATE	Tithonian	151
			Kimmeridgian	154
160	JURASSIC		Oxfordian	159
		MIDDLE	Callovian	164
170			Bathonian	169
			Bajocian	176
180			Aalenian	180
190		EARLY	Toarcian	190
			Pliensbachian	195
200			Sinemurian	202
210			Hettangian	206
			Rhaetian	210
220	TRIASSIC	LATE	Norian	221
230			Carnian	227
		MIDDLE	Ladinian	234
240			Anisian	242
		EARLY	Olenekian	245
			Induan	248

PALEOZOIC

AGE (Ma)	PERIOD	EPOCH	AGE	Boundary Ages (Ma)
		LATE	Tatarian	248
260			Ufimian-Kazanian	252 / 256
	PERMIAN		Kungurian	260
		EARLY	Artinskian	269
280			Sakmarian	282
			Asselian	290
300			Gzelian	296
	CARBONIFEROUS	LATE	Kasimovian	303
	PENNSYLVANIAN		Moscovian	311
320			Bashkirian	323
			Serpukhovian	327
340	MISSISSIPPIAN	EARLY	Visean	342
			Tournaisian	354
360		LATE	Famennian	364
			Frasnian	370
380	DEVONIAN	MIDDLE	Givetian	380
			Eifelian	391
400		EARLY	Emsian	400
			Praghian	412
420			Lockhovian	417
	SILURIAN	LATE	Pridolian	423
			Ludlovian	428
		EARLY	Wenlockian	
440			Llandoverian	443
		LATE	Ashgillian	449
460	ORDOVICIAN		Caradocian	458
		MIDDLE	Llandeilian	464
			Llanvirnian	470
480		EARLY	Arenigian	485
			Tremadocian	490
		D	Sunwaptan	495
500			Steptoean	500
	CAMBRIAN	C	Marjuman	506
			Delamaran	512
520		B	Dyeran	516
			Montezuman	520
540		A		543

PRECAMBRIAN

AGE (Ma)	EON	ERA	ERA	Boundary Ages (Ma)
750	NEOPROTEROZOIC	LATE	SINIAN	543
1000				800 / 900
1250	PROTEROZOIC	MIDDLE	RIPHEAN	
1500				1600
1750				1700
2000		EARLY	ANIMIKEAN	
2250				2200
2500			HURONIAN	2500
2750			RANDIAN	
3000	ARCHEAN	LATE	SWAZIAN	2800
3250		MIDDLE		3000 / 3400
3500		EARLY	ISUAN	3500
3750				3800

Glossary

abiogenic see Abiotic.

abiotic non-biological, not related to living organisms.

abyssal oceanic depths in the range 3500–6000 m.

accretionary wedge a piece of continental crust that has been accreted or attached to a larger continental mass.

acetogenic bacteria prokaryotic organisms that use carbonate as a terminal electron acceptor and produce acetic acid as a waste product.

acidophile an organism that grows best under acid conditions (as low as pH = 1).

acritarch a unicellular, alga-like eukaryotic microfossil of uncertain biological origins. Acritarchs may be related to dinoflagellates.

activation energy the energy necessary for chemical transformations to take place, usually expressed in kcal/mol. Activation energy distributions are used in maturation modeling to calculate the degree of transformation of kerogen to oil and gas.

active margin a continental margin characterized by uplifted mountains, earthquakes, and igneous activity caused by convergent or transforming plate motion.

active source rock a source rock that is currently generating petroleum due to thermal maturation or microbial activity (e.g. methanogenesis). Source rock that was active in the past is either inactive or spent today.

acyclic straight or branched carbon – carbon linkage in a compound without cyclic structures. See Aliphatic.

adaptation a modification of an organism that improves its ability to exist in its environment or enables it to live in a different environment.

adaptive radiation diversity that develops among species as they adapt to different sets of environmental conditions.

adenosine diphosphate (ADP) a natural product in living cells formed by the hydrolysis of adenosine triphosphate, which is accompanied by release of energy and organic phosphate.

adenosine triphosphate (ATP) adenosine 5′-triphosphoric acid. A natural product in living cells that serves as a source of energy for biochemical reactions.

adsorbent packing used in adsorption chromatography. Silica gel and alumina are the most commonly used adsorbents in liquid chromatography.

adsorption the interaction between the solute and the surface of an adsorbent. The forces in adsorption can be strong (e.g. hydrogen bonds) or weak (van der Waals' forces). For silica gel, the silanol group is the driving force for adsorption, and any solute functional group that can interact with this group can be retained by liquid–solid chromatography on silica.

adsorption chromatography a type of liquid chromatography that relies on adsorption to separate compounds. Silica gel and alumina are the most commonly used supports. Molecules are retained by the interaction of their polar functional groups with the surface functional groups (e.g. silanols of silica).

aeolian wind-deposited, e.g. desert sands.

aerobe an organism that requires molecular oxygen (terminal electron acceptor) to carry out respiratory processes.

aerobic refers to metabolism under oxic conditions using molecular oxygen, as in aerobic respiration (see Table 1.5). Most organic matter is destroyed or severely altered by aerobic microbes when exposed to oxic conditions for extended periods.

alcohol a compound that contains one or more hydroxyl groups (–OH), e.g. methanol, ethanol, bacteriohopanetetrol.

aldehyde a compound that contains the –CHO group.

alga (*pl.* **algae**) a non-vascular uni- or multicellular eukaryotic plant (thallophyta) that contains chlorophyll and is capable of photosynthesis. Blue-green algae are not true algae but belong to a group of bacteria called cyanobacteria.

alginite an oil-prone maceral of the liptinite group composed of morphologically recognizable remains of algae.

aliphatic hydrocarbons in petroleum that contain saturated and/or single unsaturated bonds that elute during liquid chromatography using non-polar solvents. Aliphatic hydrocarbons include normal and branched alkanes and alkenes, but not aromatics.

alkane (paraffin) a saturated hydrocarbon that may be straight (normal), branched (isoalkane or isoparaffin), or cyclic (naphthene).

alkaline having pH > 7.3.

alkene (olefin) a straight- or branched-chain unsaturated hydrocarbon (C_nH_{2n}) containing only hydrogen and carbon and with one or more double bonds, but not aromatic. Alkenes are not abundant in crude oils, but they can occur as a result of rapid heating, as during turbo-drilling, steam treatment, refinery cracking, and laboratory pyrolysis. Ethene (C_2H_4) is the simplest alkene. Refined olefins, such as ethene and propene, are made from oil or natural gas liquids and are commonly used to manufacture plastics and gasoline.

alkylation a refining process that converts light olefins into high-octane gasoline components, i.e. the reverse of cracking.

alkyl group a hydrocarbon substituent with the general formula C_nH_{2n+1} obtained by dropping one hydrogen from the fully saturated compound, e.g. methyl (–CH_3), ethyl (–CH_2CH_3), propyl (–$CH_2CH_2CH_3$), or isopropyl [(CH_3)$_2$CH–]. The letter "R" typically symbolizes alkyl groups attached to biomarkers, such as the alkyl side chain in steranes. In the text, we prefer to use the letter "X" so as not to confuse the reader with the R designation for stereochemistry (e.g. see Figure 2.29).

allochthonous not indigenous, derived from elsewhere. For example, sediments in a marine basin might receive autochthonous macerals from aquatic organisms in the overlying water column and allochthonous macerals from terrigenous plants that were transported into the basin from distal swamps and soils.

alumina a common adsorbent in liquid chromatography. Aluminum oxide (Al_2O_3) is a porous adsorbent that is available with a slightly basic surface.

aluminosilicate a silicate mineral where aluminum substitutes for silicon in the SiO_4 tetrahedra.

amino acid organic compounds that contain amino (–NH_2) and carboxylic acid (–COOH) groups. Certain amino acids, such as alanine, $CH_3CH(NH_2)COOH$, are the building blocks of proteins in living organisms.

amino group an –NH_2 group attached to a carbon skeleton, as in the amines and amino acids.

ammonification liberation of ammonium (ammonia) from organic nitrogenous compounds by the action of microorganisms.

ammonite an extinct group of cephalopods with coiled, chambered shells having septa with crenulated margins.

amorphous non-crystalline, lacking a crystal structure; a solid, such as glass, opal, wood, or coal, that lacks an ordered internal arrangement of atoms or ions.

amorphous kerogen insoluble particulate organic matter that lacks distinct form or shape as observed in microscopy. Amorphous kerogen that fluoresces in ultraviolet light has petroleum-generative potential. However, highly mature oil-prone kerogen and degraded gas-prone organic matter may appear amorphous but do not fluoresce and do not act as a source for petroleum.

amphibians cold-blooded vertebrates with gills for respiration in early life but with air-breathing lungs as adults.

amphipathic refers to organic compounds with polar (hydrophilic) and non-polar (hydrophobic) ends, e.g. cholesterol, bacteriohopanetetrol, phospholipids.

anaerobe an organism that does not require oxygen for respiration but uses other processes, such as fermentation, to obtain energy. Anaerobic bacteria grow in the absence of molecular oxygen.

anaerobic refers to metabolism under anoxic conditions without molecular oxygen, as in anaerobic respiration (see Table 1.5).

anaerobic respiration a metabolic process in which electrons are transferred from organic or, in some cases, inorganic compounds to an inorganic acceptor molecule other than oxygen. The most common acceptors are nitrate, sulfate, and carbonate.

analyte a compound to be analyzed.

angiosperms flowering or higher land plants having floral reproductive structures and seeds in an ovary that became prominent on land during Early Cretaceous time. Oleanane is a biomarker believed to indicate angiosperm input to petroleum from source rocks of Late Cretaceous or younger age.

angular unconformity an unconformity between two groups of rocks where the bedding planes of rocks above and below are not parallel. Usually, the older, underlying rocks dip at a steeper angle than the younger, overlying strata.

anhydrite   see Evaporite.

anion   a negatively charged atom that has gained one or more electrons.

annulus   the space between the drill string and the exposed wall of the well bore, or between the production tubing and the casing.

anoxic   water column or sediments that lack oxygen (see Table 1.5). Demaison and Moore (1980) define anoxic sediments as containing <0.5 ml oxygen/l water, which is the threshold below which the activity of multicellular deposit feeders is significantly depressed. Between ∼0.1 and 0.5 ml oxygen/l, epifauna (non-burrowing) can still survive above the sediment–water interface, but bioturbation by deposit feeders virtually ceases. Organic matter in sediments below anoxic water stands a better chance of preservation and is commonly more hydrogen-rich, more lipid-rich, and more abundant than that under oxic water.

anoxygenic photosynthesis   a type of photosynthesis in green and purple bacteria in which oxygen is not produced.

anteisoalkanes   straight-chain alkanes that have a methyl group attached to the third carbon atom (i.e. 3-methyl alkanes).

anthracene   an aromatic hydrocarbon consisting of three fused benzene rings; an isomer of phenanthrene.

anthracite   coal of the highest thermal maturity. See also Coal.

anthropogenic   derived from human activities.

antibody   protein that is produced by animals in response to the presence of an antigen and that can combine specifically with that antigen.

anticline   a concave-downward fold where strata are bent into an arch.

API gravity   a scale of the American Petroleum Institute that is related inversely to the density of liquid petroleum: API gravity $= [141.5°/(\text{specific gravity at } 16°C) - 131.5°]$. The unusual form of the equation results in a convenient scale; higher API indicates lighter oil. Fresh water has a gravity of 10°API. Heavy oils are <25°API; medium oils are 25–35°API; light oils are 35–45°API; condensates are >45°API.

aquifer   permeable rock that carries moving subsurface water.

aquitard   a relatively impermeable rock unit that retards the flow of groundwater.

archaea   prokaryotic organisms that contain ether-linked lipids built from phytanyl chains and are common in extreme environments. They include extreme halophiles (hypersaline environments), thermoacidophiles (hot, acidic environments), thermophiles (hot environments), and methanogens (methane generated as a product of metabolism). Although generally considered prokaryotes due to their similarities to bacteria, archaea show a number of eukaryotic features indicating that they may represent a third distinct group (Woese *et al.*, 1978).

archaebacteria   an old term for the archaea.

archaeocyathids   extinct marine organisms with double, perforated, calcareous, conical to cylindrical walls.

Archean   part of Precambrian time beginning 3.8 and ending 2.5 billion years ago.

archosaurs   advanced reptiles of a group called diapsids, which includes thecodonts, dinosaurs, pterosaurs, and crocodiles.

argillaceous   composed largely of clay-sized particles or clay minerals, e.g. argillaceous marls or shales contain appreciable clay.

argillite   a low-grade metamorphic rock composed of shaly sedimentary rock and characterized by irregular fractures and lack of foliation.

aromaticity ratio (B)   toluene/*n*-heptane, as in Figure 7.21 (Thompson, 1987).

aromatics   organic compounds with one or more benzene rings in their structure, including pure aromatics, such as benzene and polycyclic aromatic hydrocarbons, plus cycloalkanoaromatics, such as monoaromatic steroids, some cyclic sulfur compounds, such as benzothiophenes, and porphyrins. Because of their high solubility, low-molecular-weight aromatics such as benzene and toluene are readily washed from petroleum by circulating groundwater. Light aromatics are components in unleaded gasoline and are used as feedstocks for petrochemicals.

aromatic steroids   aromatic hydrocarbons that originate from sterols, including monoaromatic and triaromatic steroids.

aromatization parameter (monoaromatic steroid aromatization)   a biomarker maturity parameter based on the hypothesis of irreversible, thermal conversion of C_{29} monoaromatic (MA) to C_{28} triaromatic (TA) steroids [TA/(MA + TA)].

arthropod   Invertebrate with jointed body and limbs (includes insects, arachnids, and crustaceans).

asphalt dark brown to black, solid to semi-solid heavy ends of petroleum that gradually liquefy when heated and are generally soluble in carbon disulfide but insoluble in *n*-heptane. Natural and refined asphalts contain resins, asphaltenes, and heavy waxes, are dominated by carbon and hydrogen, but are also rich in nitrogen, sulfur, oxygen, and complexed metals, such as vanadium and nickel. Refinery asphalts are residues of crude oil distillations and air oxidation of crude oil feedstock.

asphaltenes a complex mixture of heavy organic compounds precipitated from oils and extracts by natural processes or in the laboratory by addition of excess *n*-pentane, *n*-hexane, or *n*-heptane. After precipitation of asphaltenes, the remaining oil or bitumen consists of saturates, aromatics, and NSO compounds. See Asphalt.

associated gas natural gas that occurs with oil, either dissolved in the oil or as a gas cap above the oil.

asteroid small planetary bodies (<800 km in diameter), mostly orbiting the sun between Mars and Jupiter. Asteroids in Earth-crossing orbits can collide with Earth, resulting in major-impact structures and possible extinction events.

asthenosphere the layer of Earth below the lithosphere where isostatic adjustments take place, magmas can be generated, and seismic waves are strongly attenuated, suggesting that the asthenosphere is a zone of convective flow.

ASTM American Society for Testing Materials.

asymmetric carbon (center) a carbon atom surrounded by four different substituents (see Figure 2.21), thus having no plane of symmetry.

atom the smallest unit of an element that retains the physicochemical properties of that element.

atomic mass the mass of the number of neutrons and protons in an atomic nucleus. A Dalton is one atomic mass unit (amu).

atomic number the number of protons in the nucleus of an atom.

atomic weight the average mass number of an element; the mass (in grams) of Avogadro's number of atoms of an element.

autochthonous indigenous, derived from nearby. For example, sediments in a marine basin might receive autochthonous macerals from aquatic organisms in the overlying water column and allochthonous macerals from terrigenous plants that were transported into the basin from distal swamps and soils.

autotroph an organism that uses carbon dioxide as the source for cellular carbon.

azeotrope a mixture of two solvents that boils at a constant boiling point, as if it were a pure compound, e.g. the azeotrope of ethanol : water is a 95 : 5 mixture.

bacillus a bacterium with an elongated, rod shape.

backflushing a column-switching technique whereby a four-way valve placed between the injector and the column allows for mobile phase flow in either direction. Backflushing is used to elute strongly held compounds from the head of the column.

bacteria a general term that includes unicellular, prokaryotic (archaea or eubacteria) organisms and microorganisms.

bacterial sulfate reduction (BSR) bacterially mediated reduction of sulfate to sulfide, accompanied by oxidation of organic matter to carbon dioxide.

bacteriochlorophyll light-absorbing pigment in green sulfur and purple sulfur bacteria.

bacteriohopanetetrol a C_{35}-hopanoid containing four hydroxyl groups found in the lipid membranes of prokaryotes (see Figure 2.29). This compound is presumed to be a major precursor for the hopanes in petroleum.

bacteriophage a type of virus that infects bacteria, often with destruction or lysis of the host cell.

barrel (bbl) the standard unit of oil in the petroleum industry; 42 US standard gallons.

barrel of oil equivalent (BOE) the volume of gas divided by about six and added to volumes of crude oil and natural gas.

basalt an extrusive igneous rock that is rich in ferromagnesian minerals and low in silica.

basement the oldest rocks in an area; commonly igneous or metamorphic rocks of Precambrian or Paleozoic age that underlie other sedimentary formations. Basement generally does not contain significant oil or gas, unless it is fractured and in a position to receive these materials from sedimentary strata.

base peak the largest peak in the mass spectrum of a compound. Typically all other peaks in the spectrum are normalized to the base peak, which is assigned an intensity of 100%. For example, the mass spectrum of cholestane (a C_{27} sterane) has a base peak at m/z 217 (see Figure 8.27), typical of many steranes.

basin a low-lying or depressed area that collects sediments or where sediments collected in the past; also refers to the basin fill.

bathyal oceanic depths in the range 200–3500 m.

BBOE billions of barrels of oil equivalent.

bedding sedimentary rock layers of varying thickness and character that represent the original sediments that were deposited.

bedrock solid rock that underlies unconsolidated surface materials.

belemnites mollusks of the class Cephalopoda that have straight internal shells.

benthic refers to a bottom-dwelling marine or lacustrine organism.

benzene the simplest aromatic hydrocarbon, consisting of a flat, six-member ring with the formula C_6H_6. Some important products manufactured from benzene include phenol, styrene, nylon, and synthetic detergent.

B–F diagram a plot of aromaticity ratio (toluene/n-heptane, B) versus paraffinity ratio (n-heptane/methylcyclohexanes, F) that has trends associated with several reservoir alteration processes (Thompson, 1987).

bioaccumulation build-up of pollutants, such as heavy metals, in organisms at higher trophic levels in a food chain.

biodegradable capable of being decomposed by biological processes.

biodegradation microbial alteration of organic matter. Petroleum can be biodegraded during migration, while in the reservoir, and at surface seep locations. Biodegradation of petroleum occurs at low temperatures (not more than $\sim 80\,^{\circ}C$) and, hence, at shallow depths. Conditions may be anoxic or involve circulating groundwater with dissolved oxygen where water washing accompanies biodegradation. Microorganisms typically degrade petroleum by attacking the less complex, hydrogen-rich compounds first. For example, n-alkanes and acyclic isoprenoids are attacked before steranes and triterpanes.

biofacies distinct spatial or temporal distributions of assemblages of organisms that are controlled by environmental factors in the surrounding environment. For example, biofacies can vary within the same lithologic unit due to differences in water depth, oxicity, or salinity during deposition. Organic facies are analogous to biofacies, except that they refer to distinctive organic matter composition (e.g. including geochemical characteristics) rather than only assemblages of recognizable organisms.

biofilm an adhesive, usually a polysaccharide, that encases some microbial material and assists with attachment to a surface.

biogenic related to, or originating from, organisms. The terms "biogenic," "abiogenic," and "microbial" can be confusing. Although some methane is abiogenic (originating abiotically), most originates as a result of microbial or thermal degradation of (biogenic) organic matter. For this reason, we prefer to use the terms "biogenic" and "abiogenic" when describing the biotic or abiotic origins of methane, and the terms "microbial" and "thermogenic" when describing the origin of most petroleum.

biogeochemistry the study of microbially mediated chemical transformations of geochemical interest, such as nitrogen or sulfur cycling.

biological marker (biomarker, molecular fossil) complex organic compounds composed of carbon, hydrogen, and other elements that are found in petroleum, rocks, and sediments and show little or no change in structure from their parent organic molecules in living organisms. These compounds are typically analyzed using gas chromatography/mass spectrometry. Most, but not all, biomarkers are isopentenoids, composed of isoprene subunits. Biomarkers include pristane, phytane, steranes, triterpanes, and porphyrins.

biological oxygen demand (BOD) the tendency to consume dissolved oxygen during the biological breaking down of organic matter.

biomarker see Biological marker.

biomass the amount of living organic matter in a particular environment or habitat.

bioremediation the process by which organisms remove or detoxify hazardous organic contaminants in soils, water, and the subsurface.

biosphere zones of the atmosphere, water, sediments, and rocks that are capable of supporting life. Most life in these zones depends on energy from the sun, although some organisms living in the deep subsurface or at deep-sea hydrothermal vents obtain their energy from other sources, such as carbon dioxide and hydrogen.

biosynthesis production of cellular constituents from simpler molecules.

biotic caused or induced by living organisms.

bioturbation disturbance of sediment by burrowing, boring, and sediment-ingesting multicellular organisms.

28,30-bisnorhopane a desmethylhopane (two methyl groups removed compared with hopane) that contains 28 carbon atoms and is common in sulfur-rich source rocks, such as the Miocene Monterey Formation. Also called 28,30-BNH and 28,30-dinorhopane (28,30-DNH).

bitumen organic matter extracted from fine-grained rocks using common organic solvents, such as methylene chloride. Unlike oil, bitumen is indigenous to the rock in which it is found (i.e. it has not migrated). If mistaken for bitumen, migrated oil impregnating a proposed source rock could result in an erroneous oil–source rock correlation. Bitumen differs from pyrobitumen, which is less soluble. In archeology, bitumen refers to heavy oil or asphalt.

bitumen ratio (transformation ratio) the ratio of extractable bitumen to total organic carbon (Bit/TOC) in fine-grained, non-reservoir rocks. The ratio varies from near-zero in shallow sediments to $\sim$250 mg/g TOC at peak generation and decreases because of conversion of bitumen to gas at greater depths. Anomalously high values for the bitumen ratio in immature sediments can be used to help show contamination by migrated oil or manmade products.

bituminous coal coal with maturity between lignite and anthracite; see also Coal.

bivalves a class of the phylum Mollusca (also known as the class Pelecypoda).

bleaching whitening of red-colored sandstones along faults by migrating saline brines that convert the immobile Fe^{3+} in hematite to mobile Fe^{2+} (e.g. Chan et al., 2000). These brines were reducing due to interaction with petroleum, organic acids, or hydrogen sulfide.

blue-green algae see Cyanobacteria.

blueschist a high-pressure, low-temperature metamorphic rock characteristic of subduction zones.

BOE barrels of oil equivalent.

boghead coal a liptinite-rich, oil-prone (type I) coal that is dominated by algal remains.

bolide an extraterrestrial object, such as an asteroid or comet, that explodes upon entering the atmosphere or striking the Earth.

bonded-phase chromatography the most popular mode of liquid chromatography, in which a stationary phase chemically bonded to a support is used for the separation. The most popular support and bonded phase are silica gel and an organosilane, such as octadecyl (for reversed-phase chromatography), respectively.

BOPD barrels of oil per day.

botryals C_{52}–C_{64} even-carbon-numbered α-branched, α-unsaturated aldehyldes that represent up to 45% of the lipids in some strains of living *Botryococcus braunii* race A (Metzger et al., 1989).

botryococcane a saturated, irregular isoprenoid biomarker formed from precursors in *Botryococcus braunii*, a fresh/brackish water, colonial Chlorophycean alga found in ancient rocks and still living today. Concentrated deposits of the alga may result in boghead coals or oil shales.

bottoms the heavy fraction of crude oil that does not vaporize during distillation.

bottom-simulating reflector (BSR) a seismic response that marks the interface between higher-sonic-velocity, gas hydrate-bearing sediment (above) and lower-sonic-velocity, free gas-bearing sediment (below). BSRs at depth in the sediment commonly parallel the overlying seafloor topography.

boulder a rock >256 mm in diameter.

BPD barrels per day.

brachiopod marine bivalve (double-shelled) invertebrates that were widespread during the Paleozoic Era but are less common today.

brackish mixed fresh and marine waters with intermediate salinities (<35 parts per thousand).

branched hydrocarbon branched alkanes containing carbon atoms that are linked to more than two other carbon atoms.

breccia clastic sedimentary rock composed mainly of angular fragments of various sizes.

Bryozoa a phylum of attached and encrusting colonial marine invertebrates.

BSR see Bacterial sulfate reduction and Bottom-simulating reflector (BSR is a common abbreviation for both terms; they can be distinguished by the context of the discussion).

BTEX water-soluble aromatic hydrocarbons, including benzene, toluene, ethylbenzene, and xylenes.

Burgess Shale fauna preserved fossil fauna of soft-bodied Cambrian animals discovered in 1910 by Charles Walcott in Kicking Horse Pass, Alberta, Canada.

burial history the depth of a sedimentary layer versus time, usually corrected for compaction.

burial history chart a diagram that shows the time interval for hydrocarbon generation, the essential elements, and the critical moment for a petroleum system.

butadiene a widely used raw material in the manufacture of synthetic rubber.

butane flammable, gaseous hydrocarbons consisting of a mixture of isobutane and normal butane.

^{14}C carbon-14. A radioactive isotope of carbon with atomic mass 14 that can be used to determine the age of materials <50000 years old.

C3 plant most green plants, eukaryotic algae, and autotrophic bacteria use the C3 or Calvin pathway to fix carbon from carbon dioxide during photosynthesis. All resin-producing plants, including gymnosperms and angiosperms, use the C3 pathway (Murray et al., 1998). C3 and C4 plants are isotopically distinct (see Figure 6.2).

C4 plant Most tropical grasses use C4 photosynthesis, which originated during the Paleocene Epoch, to fix carbon from carbon dioxide. C3 and C4 plants are isotopically distinct (see Figure 6.2).

$C_{15}+$ fraction the fraction of oil or bitumen composed of compounds eluting after the C_{15} n-alkane from a gas chromatographic column (boiling point $nC_{15} \sim 271°C$). This fraction is not altered significantly by evaporation or sample preparation and is the most reliable for correlations involving oils and bitumens. For example, after extraction of bitumen from rock, removal of the solvent by rotoevaporation results in loss of the $<C_{15}$ fraction. It is for this reason that oils are commonly topped (distilled) to remove the $<C_{15}$ fraction so that the remaining material ($C_{15}+$) can be compared with bitumens more reliably.

C_{30} sterane index ratio of parts per thousand C_{30} steranes (proposed markers of marine algal input) to total C_{27}–C_{30} steranes.

C_{35} homohopane index percentage of C_{35} 17α-homohopanes among the total C_{31}–C_{35} homohopanes. High values indicate selective preservation of the C_{35} homolog due to low redox potential conditions or complexation of bacteriohopanoids with sulfur.

CAI see Color alteration index.

calcite see Evaporite.

Caledonian Orogeny an early Paleozoic mountain-building episode in Europe that created the Caledonides orogenic belt, extending from Ireland and Scotland northeastward through Scandinavia.

Calvin cycle the biochemical route of carbon dioxide fixation in many autotrophic organisms.

CAM plants Succulents that can use C3 or C4 pathways to fix carbon from carbon dioxide during photosynthesis. CAM plants show $\delta^{13}C$ that covers the range for both C3 and C4 plants.

cannel coal oil-prone (type II) coal that consists mainly of liptinite macerals (i.e. spores and pollen) with little or no alginite.

canonical variable (CV) a statistical parameter based on stable carbon isotopic compositions of the saturated and aromatic hydrocarbons for a collection of terrigenous ($\delta^{13}C = 1.12 \delta^{13}C_{sat} + 5.45$) and marine oils ($\delta^{13}C = 1.10 \delta^{13}C_{sat} + 3.75$). Based on stepwise discriminant analysis of 339 non-biodegraded oils, CV = $-2.53 \delta^{13}C_{sat} + 2.22 \delta^{13}C_{aro} - 11.65$. CV >0.47 indicates mainly waxy terrigenous oils, while CV <0.47 indicates mainly non-waxy marine oils (Sofer, 1984).

capillary tubing metal, plastic, or silica tubing used to connect various parts of a chromatograph. Tubing with inner diameters ranging from 0.2 to 0.5 mm is in common use.

carbohydrate carbohydrates are a class of organic compounds consisting of polyhydroxy aldehydes and ketones or substances that yield these materials when hydrolyzed. These range from monosaccharides (i.e. simple sugars such as glucose, $C_6H_{12}O_6$) to polysaccharides (e.g. cellulose), which are polymers of many monosaccharide units. Complex carbohydrates decompose to water-soluble sugars in shallow sediments.

carbon the main element in all hydrocarbons. Capable of combining with hydrogen to form huge numbers of compounds.

carbonaceous shale organic-rich shales that contain less total organic carbon (TOC) than coals (50 wt.% TOC).

carbonado microcrystalline black diamond thought to originate in the crust rather than the mantle of the Earth.

carbonate carbon carbon present in a rock as carbonate minerals. Commonly measured by converting the carbonate to carbon dioxide.

carbonate rock a sedimentary rock dominated by calcium carbonate ($CaCO_3$). In the text, we define carbonate rocks as containing at least 50 wt.% carbonate minerals, the consolidated equivalent of limy mud, shell fragments, or calcareous sand.

carbonates chemical precipitates formed when carbon dioxide dissolved in water combines with oxides of calcium, magnesium, potassium, sodium, and iron. Carbonate rocks, such as limestone and dolostone, consist mainly of the carbonate minerals calcite ($CaCO_3$) and dolomite $CaMg(CO_3)_2$, respectively.

carbon cycle the geochemical cycle by which carbon dioxide is fixed to form organic matter by photosynthesis or chemosynthesis, cycled through various tropic levels in the biosphere, partially lost to sediments, and finally returned to its original state through respiration or combustion.

carbon fixation conversion of carbon dioxide or other single-carbon compounds to organic matter, such as carbohydrate.

carbonization an antiquated term that refers to the concentration of carbon during fossilization.

carbon isotope ratio see Stable carbon isotope ratio.

carbon preference index (CPI) ratio of peak heights or peak areas for odd- to even-numbered n-alkanes in the range nC_{24}–nC_{34} (Bray and Evans, 1961). If other ranges of n-alkanes are used, then the results are usually referred to as odd-to-even predominance (OEP) (Scalan and Smith, 1970). Petroleum that originates from terrigenous organic matter has high CPI that decreases toward 1.0 with increasing maturation.

carboxyl group a functional group (–COOH) attached to a carbon skeleton, as in fatty acids.

carboxylic acid an acid that contains a carboxyl (–COOH) group, e.g. fatty acids and various naphthenic and aromatic acids.

carcinogen a substance that causes cancer.

carotane (β-carotane, perhydro-β-carotene) a saturated tetraterpenoid biomarker ($C_{40}H_{78}$) typical of petroleum from saline, lacustrine environments.

carotenoid yellow to black pigments that include a class of hydrocarbons (carotenes) and their oxygenated derivatives (xanthophylls). They are characterized by eight isoprenoid units (tetraterpanoid, C_{40}) that are linked so that the two central methyl groups are in 1,6-positional relationship (tail-to-tail) while the remaining non-terminal methyl groups are in a 1,5-positional relationship (head-to-tail).

carrier bed a permeable rock that acts as a conduit for migrating petroleum.

casing steel pipe that is cemented into a well bore to prevent caving and loss of circulating fluids into the formation. In a completed well, the casing is perforated at the proper depths to allow oil or natural gas to flow into the production tube.

catabolism the biochemical breakdown of organic compounds, resulting in the production of energy.

catagenesis thermal alteration of organic matter by burial and heating in the range of ~50–150°C, typically requiring millions of years. Thermal maturity equivalent to the range 0.6–2.0% vitrinite reflectance includes the oil window or principal zone of oil formation and wet gas zone (e.g. see Figure 1.2).

catalyst a substance that promotes a chemical reaction by lowering the activation energy but does not enter into the reaction, e.g. enzymes in living organisms or platinum catalysts in refineries.

catalytic cracking a petroleum-refining process whereby heavy hydrocarbons and other compounds break down (crack) into lighter molecules in the presence of heated catalysts.

cation a positively charged atom that has lost one or more electrons.

cation exchange capacity the total exchangeable cations that a material can adsorb at a given pH.

cation-exchange chromatography a form of ion-exchange chromatography that uses resins or packings with functional groups that can separate cations.

cell the basic unit of living matter.

cell membrane a selectively permeable membrane that surrounds the cytoplasm.

cellulose a complex carbohydrate or glucose polysaccharide with beta-1,4-linkage in the cell walls of woody plants.

cell wall layer outside the cell membrane that supports and protects the membrane and gives the cell shape.

chalk a soft, white, fine-grained variety of limestone that is composed mainly of the calcium carbonate skeletal remains of marine plankton.

channeling occurs when voids in the packing material of a chromatographic column cause mobile phase and solutes to move more rapidly than the average flow velocity, resulting in band broadening. These voids can be caused by poor packing or by erosion of the packed bed.

charge the volume of expelled petroleum available for entrapment.

cheilanthanes see Tricyclic terpanes.

chelate a chemical structure that allows a metal to be held between two or more atoms.

chemoautotroph   an organism that uses inorganic compounds as the source for cellular carbon and energy.

chemocline   the depth at which hydrogen sulfide first appears and oxygen disappears (Sinninghe Damsté *et al.*, 1993b). For example, the present-day chemocline in the Black Sea occurs at 80–100 m.

chemoheterotroph   an organism that uses organic compounds as the source for cellular carbon and energy.

chemosynthesis   the process by which certain microbes create biomass (e.g. carbohydrate) by chemical oxidation of simple inorganic compounds. For example, chemosynthetic bacteria create biomass from carbon dioxide and water by mediating inorganic chemical reactions at hydrothermal vents in the deep sea.

chemotroph   an organism that uses inorganic or organic substances for energy and as a source of carbon. Includes all non-photosynthetic microorganisms, animals, and fungi.

chert   a dense, hard, sedimentary rock composed of chemically or biologically precipitated silica (SiO_2) as microcrystalline quartz. Unless colored by impurities, chert is white while flint is black.

chiral   many molecules containing asymmetric carbon atoms are chiral, i.e. they have the potential to exist as two non-superimposable structures that are mirror images. These mirror-image structures differ in the spatial relationship between atoms in the same manner as right- and left-handed gloves. In solution, chiral molecules rotate plane-polarized light either clockwise or counterclockwise. Certain chiral molecules are not asymmetric.

chiral stationary phase   stationary chromatographic phase designed to separate enantiomeric compounds.

Chlorobiaceae   photosynthetic green sulfur bacteria in which carbon fixation takes place by the reversed tricarboxylic acid cycle, resulting in ^{13}C-rich organic matter (e.g. Grice *et al.*, 1997). *Chlorobiaceae* are obligate anaerobes that require light and hydrogen sulfide (Clifford *et al.*, 1997). Presence of isorenieratane in crude oil or rock extracts indicates that the source rock was deposited under conditions of photic zone anoxia.

chlorophylls   green tetrapyrrole pigments in higher plants, algae, and certain bacteria that are required for photosynthesis. Most porphyrins in petroleum originate from the pyrrole ring in chlorophyll, while much of the pristane and phytane originate from the isoprenoid side chain.

chloroplast   an organelle in eukaryotes that contains chlorophyll, which is necessary for photosynthesis.

cholestane C_{27} saturated sterane derived from cholesterol. A major biomarker in petroleum.

cholesterol (choles-5-en-3β-ol)   a sterol (steroid alcohol) containing 27 carbon atoms that is found in the lipid membranes of eukaryotes (see Figure 2.30). This compound is a precursor for the cholestane in petroleum.

chondrites   stony meteorites that contain rounded silicate grains or chondrules, believed to have formed by crystallization of liquid silicate droplets from the solar nebula.

chordates   animals that develop a notochord during some stage of their development, including vertebrates.

chromatogram   a plot of the detector signal for separated components as peaks versus time or elution volume during chromatography (e.g. see Figure 8.9). Each peak on the chromatogram may represent more than one compound. Comparison of petroleum samples is typically accomplished using peak-height or peak-area ratios. However, internal standards allow direct comparisons of concentrations for specific peaks between samples.

chromatography   separation of mixtures of compounds or ions based on their physicochemical properties. Liquid chromatography and gas chromatography are used routinely to separate petroleum components based on their partitioning between a mobile and a stationary phase in a chromatographic column.

chromosome   the genetic template for a living organism that consists of a threadlike, microscopic body that contains genes. Occurs in the nucleus of eukaryotes but within the cytoplasm in prokaryotes. The number of chromosomes is normally constant for each species.

chronostratigraphic unit   rocks formed during a specific geologic time. Also known as time-stratigraphic and time-rock unit.

cilia (*sing*. cilium)   short, whiplike appendages on some protozoa that are used for propulsion.

ciliate   a protozoan that moves by means of cilia found on the surface of the cell.

circulation   (1) water movement. (2) Movement of drilling mud pumped down through the drill string to the drill bit and back up the annulus during rotary drilling.

cladistic phylogeny   biologic classification in which organisms are grouped according to similar characteristics that were inherited from a common ancestor.

clastic sedimentary rock that consists of fragments or clasts of pre-existing rock, such as sandstone or shale.

clastic wedge an accumulation of mainly clastic sediments deposited adjacent to uplifted areas. Sediments in the wedge become finer and the section thinner in a direction away from the upland source area.

clay a particle <0.004 mm (<4 μm) in diameter, or a group of layer-silicate minerals characterized by poor crystallinity and fine particle size.

claystone a non-fissile sedimentary rock composed mainly of clay-sized particles.

cleavage (1) bond-breaking. (2) Tendency of a mineral or rock to break in preferred directions.

coal a rock containing >50 wt.% organic matter. Most, but not all, coals are of higher-plant origin, plot along the type III (gas-prone) pathway on a van Krevelen diagram, and are dominated by vitrinite-group macerals. Increasing burial maturation of peat results in lignite, bituminous, and anthracite coals.

coalbed methane a methane-rich, sulfur-free natural gas from coal beds.

cobble a rock particle 64–256 mm in diameter.

coccolithophorids planktonic, marine, golden-brown algae that typically secrete discoidal calcareous coverings called coccoliths.

coccus spherical bacterial cell.

coenzyme a chemical that participates in an enzymatic reaction by accepting and donating electrons or functional groups.

co-injection chromatographic technique used to help identify unknown compounds. A synthesized or isolated standard (some are available commercially) is mixed with the sample (a process called spiking) containing the compound to be identified. If the standard and unknown compounds co-elute, then the relative peak intensity of the unknown compound on chromatograms of the mixture is higher than that for the neat (unspiked) sample (e.g. see Figure 13.96). Co-elution supports, but does not prove, the idea that the compounds are identical. Proof of structure might include co-elution of the unknown and standard on various chromatographic columns, identical mass spectra, and nuclear magnetic resonance (NMR) or X-ray structural verification.

coke solid carbonaceous residues that can form during refinery processes.

collinite a structureless vitrinite group maceral that originates from woody organic matter.

collision-activated decomposition (CAD) dissociation of a projectile ion into fragment ions due to collision with a target neutral species. For example, in a typical application of triple quadrupole mass spectrometry, parent ions separated in the first quadrupole collide with an inert collision gas (e.g. argon) and decompose into daughter ions in the second quadrupole (collision cell).

colloid a gas, liquid, or solid that is dispersed so finely as to settle only very slowly.

color alteration index (CAI) measure of thermal stress based on color changes in fossil conodonts (tiny fossil cone-shaped teeth of the extinct eel-like vertebrate). Also called condodont alteration index.

column chromatography use of a column or tube to hold the stationary phase, e.g. open-column chromatography, high-performance liquid chromatography (HPLC), and open-tubular capillary chromatography.

column performance refers to the efficiency of a column as measured by the number of theoretical plates for a test compound.

column switching use of multiple columns connected by switching valves to effect better chromatographic separations or for sample cleanup.

comet an extraterrestrial body consisting of a nucleus of frozen gases (mostly water, but also CO, CO_2, and others) and organic- and silicate-based dust.

commingled production a mixture of petroleum from two or more reservoirs at different depth.

competition rivalry between two or more species or groups for a limiting environmental factor in the environment.

compost a mixture of moist organic materials, with or without added fertilizer and lime, that is allowed to undergo thermophilic decomposition until the original organic materials are altered substantially.

compound-specific isotope analysis (CSIA) a technique that measures isotopic compositions of individual components separated by a gas chromatograph. Compounds that elute from the gas chromatograph are combusted to CO_2 or pyrolyzed to form H_2 or N_2 for measurement of carbon, hydrogen, or nitrogen isotopes, respectively.

compression lateral force or stress (e.g. tectonic) that tends to decrease the volume of, or shorten, a substance.

condensate a light oil with an API gravity greater than 45° API. The term was used originally to indicate petroleum that is gaseous under reservoir conditions but is liquid at surface temperatures and pressures.

condensed rings naphthenic or aromatic molecules containing multiple rings of carbon atoms where adjacent rings are joined by two shared carbon atoms (e.g. cholestane, benzopyrene).

condensed section a thin marine stratigraphic unit, typically deep-water shale, that may be rich in organic matter and/or phosphate that was deposited at very slow sedimentation rates (<1–10 mm/1000 years). High total organic carbon (TOC)-condensed sections are time-synchronous markers that can be correlated across sedimentary basins (Creaney and Passey, 1993).

configuration refers to the arrangement of atomic groups around an asymmetric carbon atom.

conformation the three-dimensional arrangements that an organic molecule can assume by rotating carbon atoms or their substituents around single covalent carbon bonds. Although not fixed, one conformation may be more likely to occur than another. For example, cyclohexane exists in the preferred chair formation in addition to the boat and twisted conformations (see Figure 2.11).

conglomerate a sedimentary rock dominated by rounded pebbles, cobbles, or boulders.

conjugation the process of mating and exchange of DNA between bacterial cells.

conodonts small, toothlike fossils, probably from primitive chordates, that are composed of calcium phosphate and that occur in Cambrian to Triassic rocks. Conodonts are useful for stratigraphic correlation and maturity assessment.

contact metamorphism changes in the mineralogy and texture of a rock caused by heat and pressure from a nearby igneous intrusion.

continental crust the part of the Earth's crust that underlies the continents and continental shelves and ranges in thickness from ~35 km to as much as 60 km under mountain ranges.

continental drift the horizontal movement or rotation of continents relative to one another. The theory proposes that continents move away from each other by seafloor spreading along a median ridge or rift, producing new oceanic areas between the continents. Materials moving away from the ridges consist of thick plates of continental and oceanic crust, which move independently of each other.

continental margin part of the ocean floor that extends from the shoreline to the landward edge of the abyssal plain, including the continental shelf, slope, and rise.

continental rise a broad, gentle slope that rises from the abyssal plain to the continental slope.

continental shelf the gently seaward-sloping submerged edge of a continent that extends to the edge of the continental slope, typically near 200 m depth.

continental slope steeply sloping seafloor between the continental shelf and the continental rise.

convection circulation of a fluid in response to uneven heat distribution.

convergent evolution a process in which genetically different organisms develop similar characteristics during their adaption to similar habitats.

coorongite a low-maturity equivalent of torbanite or boghead coal from Australia that consists mainly of *Botryococcus braunii*.

cordaites a primitive order of treelike plants with long, blade-like leaves and clusters of naked seeds that may be evolutionary intermediates between seed ferns and conifers.

cordillera an extensive mountain range or system of mountain ranges.

Coriolis effect deflection of winds and water currents to the right in the Northern Hemisphere and to the left in the Southern Hemisphere due to the Earth's rotation.

cosmic rays high-energy particles, mostly protons, that move through the space and frequently penetrate the Earth's atmosphere.

covalent (sigma) bond a non-ionic chemical bond in which valence electrons are shared equally between the bonding atoms (for example, C–C or C–H bonds). Because of equally shared electrons, covalent bonds are stronger than ionic bonds (as in NaCl), which are characterized by electrical asymmetry.

CPI see Carbon preference index (CPI).

cracking a thermal process in which large molecules break down into smaller molecules, which occurs during burial maturation and in refineries. Kerogen in source rocks and oil in reservoirs are cracked to generate lighter petroleum products at high temperatures. Catalytic cracking increases the rate and efficiency of generation of light hydrocarbons in refineries.

crater a bowl-shaped depression, commonly surrounding the central vent of a volcano.

craton the old, geologically stable interior of a continent. Commonly composed of Precambrian rocks at the surface or covered only thinly by younger sedimentary rocks.

crinoids stalked marine echinoderms with a calyx composed of regularly arranged plates and radiating arms for gathering food.

critical moment the time that best depicts the generation–migration–accumulation of hydrocarbons in a petroleum system. The geographic and stratigraphic extents of the system are best evaluated using a map and cross-section drawn at the critical moment.

cross-linking during copolymerization of resins, a difunctional monomer is added to form cross-linkages between adjacent polymer chains. The degree of cross-linking is determined by the amount of this monomer added to the reaction. For example, divinylbenzene is a typical cross-linking agent for polystyrene ion-exchange resins.

crude oil natural oil that consists of a mixture of hydrocarbons and other compounds that have not been refined. It commonly contains solution gas and closely resembles the original oil from the reservoir rock. Hydrocarbons are the most abundant compounds in crude oils, but crude oils also contain NSO compounds and widely varying concentrations of trace elements, such as vanadium, nickel, and iron (e.g. National Research Council, 1985).

crust outer part of the Earth, from the surface to the Mohorovicic discontinuity (Moho). Continental crust is thick (30–60 km), light, silica-rich, and older, while oceanic crust is thin (<20 km), dense, silica-poor, and younger.

Crustacea a subphylum of the phylum Arthropoda that includes lobsters and crayfish.

cryogenic trap a cold trap used to remove contaminants, e.g. in compound-specific isotope analysis, a cryogenic trap is commonly used to remove water and other combustion products from the carbon dioxide analyte in the effluent from the gas chromatograph/combustion furnace (see Figure 6.12).

CSIA see Compound-specific isotope analysis.

culture (1) a population of microbes cultivated in an artificial growth medium. Pure cultures grow from one cell, while mixed cultures consist of two or more strains or species growing together. (2) A human population with a distinct set of customs and traditions.

cuticle the waxy layer on the outer walls of epidermal plant cells that represents the precursor to cutinite.

cutinite a maceral derived from the waxy remains of land plants.

CV see Canonical variable (CV).

cyanobacteria prokaryotic, photosynthetic microorganisms with chlorophyll and phycobilins that produce oxygen. Formerly called blue-green algae.

cycadales seed plants that were common during the Mesozoic Era and were characterized by palm-like leaves and coarsely textured trunks marked by leaf scars.

cycloalkane (naphthene, cycloparaffin) a saturated, cyclic compound containing only carbon and hydrogen. One of the simplest cycloalkanes is cyclohexane (C_6H_{12}). Steranes and triterpanes are branched cycloalkanes consisting of multiple condensed five- or six-carbon rings.

cyclohexane the cyclic form of hexane; used as a raw material to manufacture nylon.

cyclothem a vertical package of sedimentary units that reflects environmental events that occurred in a constant order. Multiple cyclothems are particularly characteristic of the Pennsylvanian System.

cyst the resting stage of some bacteria, nematodes, and protozoa, during which the cell becomes surrounded by a protective layer.

cytochrome respiratory pigment that consists of an iron-containing porphyrin ring (e.g. heme) complexed with proteins and serves to transfer electrons from a substrate to a terminal electron acceptor, such as oxygen.

cytoplasm the contents of a cell inside the cell membrane, but excluding the nucleus.

28,30-dinorhopane see 28,30-Bisnorhopane.

Dalton one atomic mass unit (amu).

data processing storage and manipulation of gas chromatography/mass spectrometry (GCMS) data using an electronic computer.

daughter ion an electrically charged product of the reaction of a particular parent ion.

Dean Stark extraction a method used to measure fluid saturations in core samples by distillation extraction. Water in the sample is vaporized by boiling the solvent, condensed, and measured in a calibrated trap. The solvent is condensed repeatedly and flows back through the sample to extract oil. Extraction continues until the condensed solvent is clean or the sample lacks residual fluorescence. The sample is weighed before and after extraction, and the volume of oil can be calculated from the loss in weight of the sample. Water (S_w) and oil (S_o) saturations are calculated from these data.

deasphalting the process by which asphaltenes precipitate from crude oil. Laboratory or refinery deasphalting is used to remove complex components from oil by adding light hydrocarbons, such as pentane or hexane. This process can occur in nature when methane and other gases that escape from deep reservoirs enter a shallower oil reservoir.

decarboxylation loss of one or more carboxyl (–COOH) groups from a compound. For example, decarboxylation of acetic acid (CH_3COOH) results in methane (CH_4) and carbon dioxide (CO_2). Decarboxylation of phytanic acid ($C_{18}H_{36}O_2$) results in pristane ($C_{17}H_{36}$).

Deccan Traps a thick (3200 m) sequence of Upper Cretaceous basaltic lava flows that cover $\sim$500 000 km^2 in India.

deformation folding, faulting, shearing, compression, or extension of rocks due to the Earth's forces.

dehydration loss of one or more molecules of water from a compound. For example, dehydration of ethyl alcohol (CH_3CH_2OH) results in ethylene ($CH_2=CH_2$) and water.

delta a low, nearly flat area near the mouth of a river, commonly forming a fan-shaped plain that can extend beyond the coast into deep water. Deltas form in lakes and oceans when sediment supplied by a stream or river overwhelms that removed by tides, waves, and currents.

delta log R a method used to estimate total organic carbon (TOC) content in rocks based on the difference between the scaled transit-time and resistivity curves from conventional well logs (Creaney and Passey, 1993).

demethylated hopanes demethylated hopanes (25-norhopanes) resulting from microbial removal of the methyl group attached to C-10 during heavy biodegradation of hopanes (Peters et al., 1996b).

denitrification reduction of nitrate or nitrite to molecular nitrogen or nitrogen oxides by microbial activity or by chemical reactions involving nitrite.

deoxyribonucleic acid (DNA) a biomacromolecule that consists of nucleotides connected by a phosphate–deoxyribose sugar backbone. DNA contains the genetic information of the cell and controls protein synthesis.

deposition sedimentation of any material, as in the mechanical settling of sediment from suspension in water, precipitation of mineral matter by evaporation from solution, and accumulation of organic material.

detrital refers to sediment or organic matter deposited by moving water from a source within or outside the depositional basin.

detritus loose rock, mineral, or organic material that is worn off or removed by mechanical means, as by disintegration or abrasion and deposited in a basin, e.g. sand, silt, and clay derived from older rocks and moved from its place of origin.

development a phase in which newly discovered oil or gas fields are put into production by drilling and completing production wells.

development well a well drilled with the intent to produce petroleum from a proven field.

diagenesis chemical, physical, and biological changes that affect sediment during and after deposition and lithification but before significant changes caused by heat. Alteration processes occurring at maturity levels up to an equivalent of 0.6% vitrinite reflectance (e.g. see Figure 1.2). Diagenesis occurs before oil generation but includes the formation of microbial gas.

diahopane hopane in which the methyl group attached to C-14 has been rearranged to C-15, e.g. 17α-15α-methyl-27-norhopane (Dasgupta et al., 1995).

diamondoid fused-ring cycloalkane with a diamond structure that shows high thermal stability and commonly precipitates directly from the gas to the solid phase during production.

diasterane (rearranged sterane) rearrangement product from sterol precursors (e.g. see Figure 2.15) through diasterenes (see Figure 13.49). The rearrangement involves migration of C-10 and C-13 methyl groups to C-5 and C-14 and is favored by acidic conditions, clay catalysis, and/or high temperatures. Diasteranes increase relative to steranes with thermal maturation and they are low in clay-poor carbonate source rocks and related oils.

diasterane index the ratio of rearranged (diasteranes) to regular steranes. Diasterane concentrations in petroleum depend on anoxicity and pH of the depositional environment and clay content and thermal maturity in the source rock.

diastereomer a steroisomer that is not an enantiomer (mirror image). Diastereomers show different physical and chemical properties. An epimer is one type of diastereomer.

diatom microscopic unicellular or colonial alga (chrysophyte) with a siliceous cell wall called a frustule that persists after death.

diatomaceous refers to geologic deposits of fine siliceous material composed mainly of the remains of diatoms.

diazotroph organism that can use diatomic nitrogen as a sole nitrogen source, i.e. capable of N_2 fixation.

diesel a straight-run refinery distillation cut that contains hydrocarbons with ~14–18 carbon atoms. Diesel is a common additive in oil-based drilling muds and can interfere with geochemical analyses. See also Straight-run refinery products.

dike a small, discordant (injected into massive igneous, metamorphic, or across layers of sedimentary rock) body of intrusive igneous rock.

dinoflagellates unicellular marine algae typified by two flagella and a cellulose wall.

diploid describes eukaryotic organism or cell with two chromosome complements, one derived from each haploid gamete.

disproportionation cracking of organic matter resulting in lighter and heavier products, e.g. cracking of petroleum in the reservoir generates gases and a pyrobitumen residue.

distillation heating of crude oil or other materials to separate components according to their relative volatility (see Table 5.1).

diterpanes a class of biomarkers containing four isoprene subunits (e.g. see Figure 2.15) that are mainly bi- and tricyclic compounds. Many diterpanes originate from higher plants.

DNA see Deoxyribonucleic acid (DNA).

DNA fingerprinting molecular genetic techniques for assessing the differences between DNA in samples.

dolomite a rhombohedral carbonate mineral with the formula $CaMg(CO_3)_2$.

dolostone a carbonate sedimentary rock that contains over 50% of the mineral dolomite [$CaMg(CO_3)_2$].

domain a major taxonomic division ranking higher than a kingdom. The three domains of organisms consist of the archaea, eubacteria, and eukarya.

dome a fold in rocks shaped like an inverted bowl. Strata in a dome dip outward and downward in all directions from a central area.

downstream all petroleum activities from refining of crude oil into petroleum products to the distribution, marketing, and shipping of these products.

DPEP deoxophylloerythroetioporphyrins. Common tetrapyrrole pigments in sediments and petroleum. The DPEP/etio porphyrin ratio is used as a thermal maturity parameter.

drill bit a drilling tool that cuts through rock by a combination of crushing and shearing.

drilling mud a mixture of clay, water, and chemicals that is pumped in and out of the well bore during drilling. Circulation of drilling mud flushes rock cuttings produced by the drill bit at the bottom of the well bore to the surface and maintains pressure in the well bore.

dry gas gas that is dominated by methane (>95% by volume) with little or no natural gas liquids and is generally microbial or highly mature. Microbial gas is depleted in ^{13}C compared with highly mature gas.

dry hole a well lacking oil or gas in commercial quantities.

dwell time the time spent by the detector at a given mass during scanning by the mass spectrometer. Longer dwell times result in more accurate measurements of peaks by increasing the signal-to-noise ratio. Dwell times for selected ion monitoring/gas chromatography/mass spectrometry (SIM/GCMS) and full-scan GCMS are ~0.04 and <0.0075 s/ion, respectively.

dysaerobic refers to metabolism under dysoxic conditions with limited molecular oxygen (see Table 1.5).

dysoxic refers to water column or sediments with molecular oxygen contents of 0.2–2 ml/l of interstitial water (see Table 1.5).

early-mature early-mature source rocks have begun to generate some petroleum but have not yet reached the main stage of oil generation. Early mature is commonly assumed to represent ~0.5–0.6% vitrinite reflectance.

earthquake shaking of the ground caused by a sudden release of energy stored in the rocks beneath the Earth's surface, commonly caused by movement along faults.

EASY %R_o a simple model that uses Arrhenius first-order kinetics for a distribution of activation energies to calculate vitrinite reflectance for given time/temperature conditions (Sweeney and Burnham, 1990). The model applies in the range 0.3–4.5% mean random reflectance.

echinoderms marine invertebrates of the phylum Echinodermata, showing fivefold symmetry and with exoskeletons and spines commonly consisting of calcite. Cystoids, blastoids, crinoids, and echinoids are examples of echinoderms.

ecology the science of the interrelationships among organisms and between organisms and their environment.

ecosystem the community of organisms and the environment in which they live.

Ediacaran fauna Late Proterozoic fauna first discovered in Australia but subsequently found in rocks ~600 million years old in many continents.

effective source rock a source rock that is generating and expelling or has generated and expelled oil and gas.

effluent mobile phase used to carry out a chromatographic separation. See also Eluate.

Eh see Redox potential.

electron a subatomic particle with an electric charge of -1 and mass of 9.11×10^{-28} g.

electron acceptor a small organic or inorganic compound that accepts electrons during a redox reaction and is reduced to complete an electron-transport chain.

electron donor a small organic or inorganic compound that donates electrons during a redox reaction and is oxidized to provide an electron for the electron transport chain.

electron impact a gas chromatography/mass spectrometry (GCMS) ionization technique. Compounds eluting from the gas chromatograph enter the ion source of the mass spectrometer through a transfer line, where they are bombarded with energetic electrons, typically at 70 eV (electron-volts). This causes the compounds to ionize into a series of fragment ions characteristic of the structure of the particular compound, e.g. $M + e- \rightarrow M + \cdot 2e-$, where M is an eluting compound.

electron-transport chain the sequence of reactions in biological oxidation that moves electrons through a series of oxidizing agents arranged in order of increasing strength and terminating in oxygen.

eluate combined mobile phase and solute exiting a chromatographic column. Also called effluent.

eluent a liquid used as the mobile phase to carry components to be separated through a chromatographic column. For example, in liquid chromatographic fractionation of petroleum, hexane can be used as the eluent for the saturate fraction while methylene chloride can be used for the polar fraction.

elution the process of passing mobile phase through a chromatographic column to transport solutes.

enantiomers a pair of stereoisomers of a molecule that have the same molecular formula but differ in the arrangement of substituents around every asymmetric carbon atom, thus representing two mirror-image structures (e.g. see Figure 2.21). When only one asymmetric carbon atom is present, the mirror-image structure can be obtained only by transposing (inverting) any two of the four substituents on the asymmetric carbon atom (isomerization). In a molecule containing more than one asymmetric center, inversion of all the centers leads to the enantiomer. Enantiomers are isomers that are non-superimposable mirror images.

endosymbiosis a permanent or long-lasting intracellular incorporation of one or more genetically autonomous biological systems. For example, mitochondria and chloroplasts in eukaryotic cells may have originated as aerobic non-photosynthetic bacteria and photosynthetic cyanobacteria, respectively.

enhanced oil recovery displacement and recovery of oil or gas by injecting gas or water into the reservoir.

enzyme a protein in a living organism that catalyzes specific reactions.

eon the largest division of geologic time consisting of several eras. The Phanerozoic Eon consists of all geologic periods, from the Cambrian to the Quaternary.

EPA United States Environmental Protection Agency.

epifaunal organisms living on the sediment surface.

epimer in a molecule containing more than one asymmetric center, inversion of all the centers leads to the enantiomer (mirror image). Inversion of less than all the asymmetric centers yields an epimer or diastereomer.

epiric sea a shallow water body formed during major transgressions of continental shelf or landmass. Source rocks with high petroleum-generative potential can be deposited in shallow basins within epiric seas. The name originates from the Greek *Epirus*, a region of northwestern Greece bordering the Ionian Sea.

epoch a subdivision of a geologic period. Rocks deposited during an epoch represent the series for that epoch.

era a division of geologic time that is less than an eon and that consists of several periods.

ergostane the C_{28} sterane; also known as methylcholestane.

erosion the movement of unconsolidated materials at the Earth's surface by wind, water, or glacial ice.

essential elements the source, reservoir, seal, and overburden rocks in a petroleum system. Together with the processes of generation–migration–accumulation and trap formation, essential elements control the distribution of petroleum in the lithosphere.

ethane (C_2H_6) a gaseous hydrocarbon that occurs in petroleum.

ethylcholestane see Stigmastane.

etio a class of porphyrins. The deoxyphylloerythroetioporphyrin (DPEP)/etio porphyrin ratio is used as a maturity parameter.

ETR see Extended tricyclic terpane ratio.

eubacteria prokaryotic organisms that contain ester-linked lipids and are not archaea.

eukarya see Eukaryotes.

eukaryotes a phylogenetic domain that contains all higher organisms (eukarya or eukaryotes) with a membrane-bound nucleus and organelles and well-defined chromosomes. Includes virtually all organisms except the prokaryotes (bacteria and cyanobacteria) and the archaea.

euphotic zone see Photic zone.

eustatic refers to worldwide simultaneous changes in sea level, such as might result from change in the volume of continental glaciers.

eutrophic describes a lake containing abundant nutrients that support high primary productivity. Eutrophic lakes commonly have low dissolved oxygen due to high biological oxygen demand. They are generally last in the age sequence oligotrophic, mesotrophic, and eutrophic, before complete infilling of the lake basin and subaerial conditions.

euxinic anoxic, restricted depositional conditions with free hydrogen sulfide (H_2S). Present-day examples include anoxic bottom water and sediments in Lake Tanganyika, the Black Sea, the coast of Peru, and the northeastern Pacific. Anoxic sediments with organic matter and sulfate become euxinic through the activity of sulfate-reducing bacteria. The term originates from the classical Greek name for the Black Sea, *Euxynos*. While the deep Black Sea is euxinic, the name is contradictory because it means "hospitable" or "good to foreigners."

evaporation the process in which a substance passes from the liquid or solid state to the vapor state. Also called vaporization.

evaporite various minerals that precipitate from a water solution during evaporation, e.g. anhydrite ($CaSO_4$), gypsum [($CaSO_4 \cdot 2(H_2O)$)], and halite (NaCl). A typical sequence of minerals deposited with increasing extent of evaporation is (1) calcite ($CaCO_3$), (2) gypsum, (3) halite, and (4) sylvite (KCl).

events chart a petroleum system chart (also called timing-risk chart) that shows the timing of essential elements and processes and the preservation time and critical moment of the system.

evolution genetic adaptation of organisms or species to the environment.

exinite the liptinite group, sometimes called exinite, consists of hydrogen-rich macerals that generate significant quantities of oil during maturation. These macerals include alginite, resinite, sporinite, and cutinite (from algae, terrigenous plant resins, spores, and cuticle, respectively). In our simplified classification, exinite consists of all liptinite macerals, except alginite.

exobiology the branch of biology that deals with the effects of extraterrestrial environments on living organisms.

expulsion the process of primary migration, whereby oil or gas escapes from the source rock due to increased pressure and temperature. Generally involves short distances (meters to tens of meters).

extended tricyclic terpane ratio ETR = $(C_{28} + C_{29})/(C_{28} + C_{29} + Ts)$. Early results by Holba *et al.* (2001) suggest that ETR can be used to differentiate crude oils generated from Triassic, Lower Jurassic, and Middle–Upper Jurassic source rocks (see Table 13.8).

extract bitumen or oil removed from rocks using organic solvents.

extractable organic matter (EOM) see Extract.

extrusive igneous rock that erupted on to the surface of the Earth, such as lava flows, tuff (consolidated pyroclastic material), and volcanic ash.

facies the characteristics of a rock unit that reflect the conditions of its depositional environment.

facultative describes an organism that can carry out both options of mutually exclusive processes (e.g. aerobic and anaerobic metabolism).

facultative anaerobes microbes that can grow under both oxic and anoxic conditions but do not shift from one mode of metabolism to another as conditions change. They obtain energy by fermentation.

fatty acids saturated or unsaturated organic compounds with a terminal carboxyl group (–COOH) that are common membrane components in organisms and are converted to *n*-alkanes and other hydrocarbons during diagenesis and thermal maturation. Fatty acids commonly contain an even number of carbon atoms in the range $\sim C_{14}$–C_{24}, which are synthesized by condensation of malonyl coenzyme A (see Figure 3.5).

fault a fracture or zone of fractures in the Earth's crust along which rocks on one side were displaced relative to those on the other side.

FC43 a standard compound (perfluorotributylamine) used to calibrate the mass scale of a mass spectrometer.

feedstock crude oil, natural gas liquids, natural gas, and refined materials that are used to make gasoline, other refined products, and chemicals.

fermentation an anaerobic oxidation–reduction by which some prokaryotes and eukaryotes (e.g. yeasts) break down organic compounds to yield energy for growth, alcohol, and lactic acid. Some atoms of the energy source become more reduced, while others become more oxidized.

FID see Flame ionization detector.

field an area under which a producing or prospective oil and/or natural gas reservoir lies.

filter feeders animals that filter water to obtain small particles of food.

fingerprint a chromatographic signature used in oil–oil or oil–source rock correlations. Mass chromatograms of steranes or terpanes are examples of fingerprints that can be used for qualitative or quantitative comparison of oils and source-rock extracts.

Fischer–Tropsch an industrial process used to convert carbon monoxide and carbon dioxide from coal into synthetic petroleum. Fischer–Tropsch reactions may occur naturally in the presence of magnetite or other metal oxide catalysts.

fissile the tendency of certain rocks, such as shale, to split into thin plates or lamina.

fissility a property of rocks, mainly shales, that causes them to split into thin pieces parallel to bedding.

fission tracks tiny damage trails in minerals produced by high-energy particles emitted from radioactive elements (e.g. uranium) during spontaneous fission.

fjord a long, U-shaped, glacier-carved inlet from the ocean.

flagellum (*pl.* flagella) a whip-like structure attached to certain microbial cells that allows motility.

flame ionization detector (FID) a detector for gas chromatographs or Rock-Eval pyrolysis that measures ions generated by combustion of eluting compounds. FID responds to all compounds that combust (e.g. hydrocarbons, but not carbon dioxide).

flash point the lowest temperature to which liquid fuel must be heated to produce a vapor/air mixture above the liquid that ignites when exposed to an open flame.

flood basalt a regionally extensive plateau consisting of layers of basalt that originated from repeated fissure eruptions, e.g. the Columbia Plateau in the western USA and the Deccan Traps in India.

floodplain a flat area of unconsolidated sediment near a stream channel that is submerged only during high-flow periods.

fluorescence light emission by organic matter when excited by energy, such as ultraviolet (UV) light. UV fluorescence is diagnostic of low to moderately mature, oil-prone kerogen. UV-excitation fluorescence can be used to identify crude oil in conventional cores.

fluvial sediments or other geologic features formed by streams.

flux the rate of emission, sorption, or deposition of a material from one pool to another.

fold a curve or bend of a formerly planar structure, such as rock strata or bedding planes, that generally results from deformation.

footwall the underlying side of a fault, below the hanging wall.

foraminifera an order of mostly marine, unicellular protozoans that secrete tests (shells) composed of calcium carbonate and that account for most carbonate sediments in the modern oceans.

formation a distinct stratigraphic unit of rock that shares common lithologic features and is large enough to be mapped.

formation water water that occurs naturally in sedimentary rocks.

fossil remains or indications of an organism that lived in the geologic past.

fragment ion an electrically charged product of the fragmentation of a molecule or ion. Fragment ions can dissociate further to generate other, lower-molecular-weight fragments.

fragmentogram see Mass chromatogram.

free energy energy that is, or that can be made, available to do useful work. Defined as the difference between the internal energy of a system and the product of its absolute temperature and entropy. $\Delta G = \Delta H - T\Delta S$. Also called Gibbs free energy, after Josiah Willard Gibbs, who developed the concept.

frustule the siliceous wall and protoplast of a diatom.

fuel oil heavy distillates from oil refining that are used mainly for heating and fueling industrial processes, ships, locomotives, and power stations.

full scan an operational mode in mass spectrometry in which a range of ion mass/charge ratios is monitored repeatedly at a specified rate.

fulvic acid yellow organic material that remains in solution after removal of humic acid by acidification.

functional group a site of chemical reactivity in a molecule that arises from differences in electronegativity or from a pi bond. For example, alcohol (–OH), thiol (–SH), and carboxyl (–COOH) groups, and double bonds (C=C), are common functional groups in organic compounds.

fungus (*pl.* **fungi**) non-phototrophic, eukaryotic microorganisms that contain rigid cell walls.

fusinite an inert maceral of the inertinite group that is rich in carbon and has high reflectance.

fusulinids mainly spindle-shaped foraminifers with coiled, calcareous tests separated into many complex chambers. Fusulinids were abundant during the Pennsylvanian and Permian periods.

gammacerane a C_{30} pentacyclic triterpane in which each ring contains six carbon atoms. Source rocks deposited in stratified anoxic water columns (commonly hypersaline) and related crude oils commonly have high gammacerane indices (gammacerane/hopane).

gamma ray high-frequency electromagnetic wave.

gas cap part of a petroleum reservoir that contains free gas.

gas chromatography an analytical technique designed to separate compounds (e.g. see Figure 8.6), whereby a mobile phase (inert carrier gas) passes through a column containing immobile stationary phase (high-molecular-weight liquid). The stationary phase can be coated on the walls of the column or on a solid support packing material. Compounds separate based on their relative tendencies to partition into the stationary or mobile phases as they move through the column. Various detectors can be used to measure the separated components as they elute from the column. Most gas chromatographs are equipped with flame ionization detectors (FIDs).

gas chromatography/mass spectrometry see GCMS.

gas hydrate a crystalline phase of water that contains gas (mainly methane) in arctic and deep-water settings.

gas injection an enhanced recovery method, whereby natural gas is injected under pressure into a producing reservoir through an injection well to drive oil to the well bore.

gasoline a light fuel used to drive automobiles and other vehicles. The gasoline distillation cut from a refinery includes the approximate range C_5–C_{10}, while commercial gasoline may also contain synthetic additives. See also Straight-run refinery products.

gasoline range the fraction of crude oil boiling between ~15°C and 200°C, including low-molecular-weight compounds, usually containing fewer than 12 carbon atoms; e.g. the C_5–C_{10} hydrocarbons represent the gasoline-range compounds that are low or absent in uncontaminated, thermally immature sediments.

gas-prone describes organic matter that generates predominantly gases with only minor oil during thermal maturation. Type III organic matter is gas-prone.

gas-to-oil ratio (GOR) the amount of hydrocarbon gas relative to oil in a reservoir, commonly measured in cubic feet per barrel.

gas wetness can be expressed in different ways, but generally represents the amount of methane (C_1) relative to the total hydrocarbon gases ($C_1 + C_{2+}$) in a sample. For example, $C_1/(C_1–C_5)$ ratios above and below 98% are dry and wet gases, respectively.

Gaussian curve a symmetrical, bell-shaped error curve; e.g. ideal chromatographic peaks are Gaussian curves.

GC see Gas chromatography. Most gas chromatographs are equipped with a flame ionization detector (FID); see Flame ionization detector. As used in the text, the term "gas chromatography" implies use of FID unless another detector is specified.

GCMS (gas chromatography/mass spectrometry) the gas chromatograph separates organic compounds while the mass spectrometer is used as a detector to provide structural information (see Figure 8.5). Biomarker interpretations depend mainly on GCMS analysis.

GC/MSD (gas chromatograph/mass selective detector) a benchtop instrument that provides separation and detection of organic compounds, including biomarkers.

gel the solid packing used in gel permeation chromatography for separation of macromolecules (e.g. nucleic acids or proteins), which consists of a dispersed medium (solid portion) and dispersing medium (the solvent).

gel permeation chromatography a type of chromatography used to separate and characterize polymers.

gene the unit of heredity in a chromosome consisting of a segment of DNA that codes for a single t-RNA, r-RNA, or protein.

gene pool all of the genes in the living population of a species.

generation–accumulation efficiency (GAE) the percentage of the total volume of trapped (in-place) petroleum for a petroleum system relative to the total volume of petroleum generated from the corresponding pod of active source rock (Magoon and Valin, 1994).

generation–migration–accumulation a petroleum system process that includes the generation and movement of petroleum from the pod of active source rock to the petroleum seep, show, or accumulation.

genetic code information for the synthesis of proteins contained in the nucleotide sequence of a DNA molecule (or, in certain viruses, of an RNA molecule).

genome a complete set of genes for an organism. All of the chromosomal genes in a haploid cell (e.g. prokaryotes and archaea) or the haploid part of chromosomes in eukaryotes.

genotype the complete genetic constitution of an organism.

genus (*pl.* genera) a subdivision of a taxonomic family of plants or animals that usually consists of more than one species; the first name (e.g. *Homo*) of the scientific name (e.g. *Homo sapiens*).

geochemical fossil see Biological marker.

geochromatography chromatographic separation of organic compounds, in which clay minerals and organic matter in the source rock and carrier beds retard heavier, more polar compounds.

geochronology the study of geologic time.

geographic extent the area of occurrence of a petroleum system as defined by a line that encloses the pod of active source rock and all discovered petroleum shows, seeps, and accumulations that originated from that pod. The geographic extent is mapped at the critical moment.

geologic range the span of geologic time between the origin and extinction of an organism.

geology the science of the history of the Earth, including the materials that comprise the planet, the physicochemical changes that occur on and within the Earth, and the evolution of life as recorded in the rocks.

geopolymer a term commonly used to describe fulvic acids, humic acids, and kerogen in rocks and sediments. However, these organic materials are not strictly polymers because they are not composed of repeating subunits like those in organic (e.g. cellulose) or inorganic (e.g. nylon) polymers.

geosphere the solid Earth, including continental and oceanic crust, mantle, and deep interior. The geosphere does not include the oceans, atmosphere, or life.

geothermal gradient the rate of increase in temperature with depth in the Earth, usually measured in °C/km or °F/100 ft. Gradients are sensitive to lithology, circulating groundwater, and the cooling effect of drilling fluids. Worldwide average geothermal gradients range from 24 to 41°C/km (1.3–2.2°F/100 ft), with extremes outside this range.

Gibb's free energy see Free energy.

gilsonite solid bitumen composed mainly of nitrogen, sulfur, and oxygen (NSO) compounds and asphaltenes that occurs as dike complexes in the Uinta Basin, Utah. It is a common drilling mud additive.

glacier a massive body of ice, formed from recrystallized snow, that flows slowly under the influence of gravity.

glauconite a green clay mineral common in marine sandstones and believed to have formed at the site of deposition.

Gloeocapsomorpha prisca a microorganism peculiar to Middle Ordovician rocks that appears responsible for the strong odd *n*-alkane ($<C_{20}$) predominance in oils and source-rock extracts of this age. Ordovician rocks that contain abundant *G. prisca*, such as the Estonian kukersites and the Guttenberg Member of the Decorah Formation (North America), contain a polymeric material consisting mainly of C_{21} and C_{23} *n*-alkenyl resorcinol building blocks that represent selectively preserved cell-wall or sheath components or materials that polymerized during diagenesis (Blokker *et al.*, 2001).

Glossopteris **flora** an assemblage of late Paleozoic to early Mesozoic fossil plants from South Africa, India, Australia, and South America named after the associated seed fern, *Glossopteris*.

glucose a common monosaccharide sugar ($C_6H_{12}O_6$) that occurs in honey and fruit juices. Glucose is also called dextrose because it is optically active and rotates the plane of polarized light in a clockwise direction.

glycerides esters formed between one or more acids and glycerol. For example, fatty-acid esters with glycerol occur in plant oils and animal fats.

Gondwana a large continent in the southern hemisphere during Permo-Carboniferous time composed of parts of modern Australia, South Africa, India, Africa–Arabia, and Antarctica.

GOR see Gas-to-oil ratio.

GPA Gas Processors Association.

graben a down-dropped block of crust bounded by parallel normal faults.

Gram stain a stain that divides bacteria into two groups, Gram-positive and Gram-negative, depending on their ability to retain crystal violet when decolorized with an organic solvent, such as ethanol. The cell wall of Gram-positive bacteria consists chiefly of peptidoglycan and lacks the outer membrane of Gram-negative cells.

granite an intrusive igneous rock with high silica (SiO_2) content typical of continental regions.

graphite a mineral composed mainly of staggered flat layers of carbon atoms with minor amounts of hydrogen. Individual layers are bound weakly to each other and are composed of strongly bonded carbon atoms at the vertices of a network of regular hexagons.

graptolites extinct colonial marine invertebrates, possibly protochordates that range from Late Cambrian to Mississippian in age.

gravity segregation a process whereby heavier and lighter petroleum components accumulate near the bottom and top of the reservoir, respectively, possibly due to movement of gas toward the top of the reservoir.

greenhouse effect a process whereby short-wavelength solar radiation that impinges on the Earth is re-radiated from Earth and cannot escape back into space because the Earth's atmosphere is not transparent to the re-radiated energy, which is in the form of infrared radiation (heat).

greenschist schist containing the minerals chlorite and epidote (which are green) and formed by low-pressure, low-temperature metamorphism of mafic volcanic rocks.

groundwater water that lies below the surface in fractures and pore space in rocks.

growth fault a slump fault sedimentary rock that grows continuously during deposition of sediments, such that strata on the downthrown side of the fault are thicker than equivalent strata on the upthrown side.

GRZ three organic-rich Cretaceous intervals on the North Slope of Alaska, including the pebble shale (informal name), gamma-ray zone (GRZ) (also called the highly radioactive zone, HRZ), and the lower part of the Torok Formation, are considered to be one source rock because of similar kerogen composition (Magoon and Bird, 1988).

guard column a small column placed between the injector and the analytical chromatographic column that protects the latter from contamination by sample particulates and strongly retained species.

gymnosperms flowerless seed plants in which the seeds are not enclosed (naked seeds), e.g. Coniferales, Ginkgoales, Bennettitales, and Cycadales, which were the dominant land plants after their origin in the middle Paleozoic Era until the advent of angiosperms during the Cretaceous Period.

habitat the environment where an organism lives.

half-life the time required for one-half of an original amount of radioactive material to decay to daughter products. The half-life of $_{14}C$ is $\sim$5730 years.

halite see Evaporite.

halocline a layer of water in which the salinity increase with depth is greater than that of the underlying or overlying water.

halophile organism that grows readily in a highly saline environment, e.g. purple halophilic bacteria in modern lakes.

hanging wall the overlying side of a fault, above the footwall.

haploid describes cells with a single set of chromosomes, as in gametes.

HBIs highly branched isoprenoids in petroleum probably originate from diatom precursors (Nichols *et al.*, 1988) and could be markers of Jurassic or younger source rock. The C_{25} HBI is generally more abundant than C_{20} and C_{30} HBI and is also called 2,6,10,14-tetramethyl-7-(3-methylpentyl)-pentadecane.

H/C the atomic hydrogen to carbon ratio, typically used to describe kerogen type on van Krevelen diagrams.

HRZ see GRZ.

headspace gas analysis gas chromatographic analysis of light hydrocarbon gases that collect in the contained space above canned cuttings.

heavy crude thick, viscous crude oil showing <20°API.

heavy oil crude oil that shows low API gravity (<25°API).

hemipelagic describes ocean waters near land masses. Hemipelagic sediments consist of large amounts of lithic materials transported from land by turbidity currents, volcanic activity, and other processes.

heptane ratio a thermal maturity parameter based on the abundance of *n*-heptane versus other gasoline-range hydrocarbons and commonly used with the isoheptane ratio (Thompson, 1983).

herbaceous organic matter cuticular, spore, and pollen components in kerogen (type II).

heterocompounds see NSO compounds.

heterocyst a specialized cyanobacterial cell that carries out fixation of diatomic nitrogen.

heterotroph an organism that uses organic compounds as a source for cellular carbon.

hexane a petroleum liquid found in natural gasoline.

HGT see Horizontal gene transfer.

HI see Hydrogen index.

higher-plant index (HPI) (retene + cadalene + iHMN)/1,3,6,7-TeMN, where iHMN is 1-isohexyl-2-methyl-6-isopropylnaphthalene and TeMN is tetramethylnaphthalene (van Aarssen et al., 1996).

high-performance liquid chromatography (HPLC) a chromatographic method used to separate petroleum into saturate, aromatic, and polar fractions (see Figure 8.3). The method is typically automated. Larger amounts of materials can be separated by column chromatography.

high-sulfur oil see Sour crude.

holocene the time since the last major episode of glaciation. Equivalent to Recent.

homohopane index see C_{35} Homohopane index.

homohopane isomerization a biomarker maturity ratio [22S/(22S + 22R)] describing the conversion of the biological 22R to the geological 22S configuration of homohopane molecules. Typically, this ratio is calculated for the C_{32} 17α-homohopanes, but other carbon numbers in the range C_{31}–C_{35} are used sometimes.

homohopanes C_{31}–C_{35} 17α,21β(H)-hopanes (pentacyclic triterpanes) that elute as isomeric doublets after the C_{30} hopane on m/z 191 mass chromatograms.

homolog one member of a series of organic compounds of the same chemical type that are constructed from discrete chemical subunits. For example, nC_{17} is the seventeenth homolog in the series of n-alkanes (n-paraffins) that begins with methane (nC_1). Each homolog in the n-alkane series differs from the previous and subsequent homolog by one methylene group (–CH$_2$–). Likewise, the homologous series of acyclic isoprenoids includes various compounds, such as norpristane (iC_{18}), pristane (iC_{19}), and phytane (iC_{20}).

homologous series compounds showing similar structures but differing by the number of methylene (–CH$_2$–) groups (homologs). For example, the homologous series of n-alkanes can be described by the formula C_nH_{2n+2}, where $n = 1, 2, 3, 4$, etc. (methane, ethane, propane, butane, etc.). Many biomarkers consist of homologous series.

hopane the C_{30} homolog in the hopane series of pentacyclic hydrocarbons (see Hopanes).

hopanes C_{27}–C_{35} pentacyclic triterpanes that originate from bacteriohopanoids in bacterial membranes and generally dominate the triterpanes in petroleum. Various isomeric series of hopanes include diahopanes, neohopanes, moretanes, demethylated hopanes, and homohopanes. The C_{30} hopane has 17α,21β-stereochemistry and commonly is more abundant than later-eluting homohopanes on m/z 191 mass chromatograms.

horizontal gene transfer (HGT) (also called lateral gene transfer) the acquisition of new genes by organisms. For example, bacteria can incorporate genes during conjugation with other genetically distinct bacteria, or they can absorb genes introduced into the surrounding environment by dead bacteria.

host an organism capable of supporting the growth of a virus or other parasite.

HPI see Higher-plant index.

HPLC see High-performance liquid chromatography.

HTB "high TOC at the base, decreasing upward" units (Creaney and Passey, 1993). Most marine organic-rich shales are composed of these discrete sedimentary units, with TOC values that decrease upward from maxima near their bases.

humic acids complex, high-molecular-weight organic acids that originate from decomposing organic matter and can be extracted from sediments and low-maturity rocks using weak base (e.g. dilute NaOH) and precipitated by acid (pH 1–2). Some kerogen originates from humic acids during diagenesis.

humic coal gas-prone (type III) coal that consists mainly of higher-plant detritus, including vitrinite and inertinite group macerals, with little or no liptinite macerals.

humic substances high-molecular-weight, brown to black substances formed in sediments by secondary synthesis reactions during diagenesis. The generic term describes the colored material or its fractions obtained on the basis of solubility characteristics, such as humic acid or fulvic acid.

hydration a chemical reaction in which water or hydrogen is added to a compound or mineral.

hydrocarbons various organic compounds composed of hydrogen and carbon atoms that can exist as solids, liquids, or gases. Sometimes this term is used loosely to refer to petroleum.

hydrogenation any reaction of hydrogen with an organic compound. Typically, hydrogen reacts with double bonds of unsaturated compounds, resulting in a saturated product. For example, hydrogen reacts with ethylene ($CH_2=CH_2$) to produce ethane.

hydrogen bond a chemical bond between a hydrogen atom of one molecule and two unshared electrons of another molecule.

hydrogen index (HI) a Rock-Eval pyrolysis parameter defined as (S2/TOC) × 100 (TOC is total organic carbon) and measured in mg hydrocarbon/g TOC (see Figure 4.5). The HI is used in van Krevelen-type plots of HI versus oxygen index (OI) to determine organic matter type (see Figure 4.4).

hydrologic cycle the movement of water from the ocean by evaporation, precipitation on land, and transport back to the ocean as surface and ground water.

hydrology study of the distribution and movement of water in the atmosphere and upon and within the Earth.

hydrophilic refers to water-soluble molecules and chromatographic stationary phases that are compatible with water.

hydrophobic refers to molecules with little affinity to water and chromatographic stationary phases that are not compatible with water.

hydrostatic equal pressure in all directions, as under a column of water extending from the surface.

hydrothermal formed by precipitation from hot aqueous solutions.

hydrotreating a refinery process that removes sulfur and nitrogen from crude oil and other feedstocks.

hydrous pyrolysis a laboratory technique in which potential source rocks are heated without air, under pressure, and with water to artificially increase the level of thermal maturity. Oils generated during hydrous pyrolysis may be used for oil–source rock correlation when natural source rock extracts of suitable maturity are not available (see Figure 19.16).

hypercapnia carbon dioxide poisoning, generally caused by rapid overturn of CO_2-rich bottom waters and release of dissolved gases to the atmosphere.

hypersaline refers to brines that show salinities greater than ~35 parts per thousand.

hypha (*pl.* hyphae) a long, commonly branched tubular filament that serves as the vegetative body for many fungi and bacteria of the order Actinomycetes.

icecap a mountain glacier that flows outward in several directions.

ice sheet a large mass of ice covering a significant portion of a continent, as in Greenland and Antarctica.

ichnofossils trace fossils, e.g. tracks, trails, burrows, borings, castings, and other markings made in sediments.

igneous rock a rock formed by cooling or solidification of magma.

illite a general name for three-layer, mica-like clay minerals in argillaceous rocks, especially marine shales and related soils.

immature refers to conditions too cool (or of too short duration) for thermal generation of petroleum, i.e. vitrinite reflectance <0.6%.

I_{max} a thermal maturity parameter based on the maximum wavelength of the fluorescence intensity, e.g. acritarch fluorescence

inactive source rock a source rock with remaining generative potential that has stopped generating petroleum (e.g. due to uplift and reduced thermal stress).

inertinite a maceral group composed of inert, hydrogen-poor organic matter with little or no petroleum-generative potential. Type IV kerogen is dominated by inertinite.

infauna organisms that live within bottom sediments.

infrared (IR) the electromagnetic spectrum with wavelengths from ~0.1 to 10 μm.

inlet the initial part of a chromatographic column, where the solvent and sample enter.

inspissation drying-up. For example, inspissation of seep oils can result in loss of gases and the lighter oil fractions due to reduced pressure (compared with that in the reservoir), evaporation and drying by exposure to sunlight, oxidation, and other related processes.

internal standard a compound that shows similar behavior to the unknown compounds in a mixture to be analyzed but that is absent in the original mixture.

Addition of the internal standard to the mixture allows more reliable measurement of the unknown compounds. For example, 5β-cholane is added to the saturate fractions of oils as an internal standard. This compound is not found in significant amounts in natural oils, but its mass spectral characteristics are similar to those of steranes and other compounds in saturate fractions.

interstitial refers to the pores of a host rock.

intracellular inside the cell.

intrusion the process of emplacement of magma into pre-existing rock.

ion a charged atom or compound that is electrically charged as a result of the loss of electrons (to produce cations) or the gain of electrons (to produce anions).

ionic bond a type of chemical bond in which atoms exchange electrons.

IRM/GCMS see Compound-specific isotope analysis.

island arc a linear or arc-shaped chain of volcanic islands formed at a convergent plate boundary, characterized by volcanic and earthquake activity, and usually near deep oceanic trenches.

isoalkanes straight-chain alkanes that have a methyl group attached to the second carbon atom (i.e. 2-methyl alkanes).

isoheptane ratio a thermal maturity parameter based on the abundance of several C_7 hydrocarbons and commonly used with the heptane ratio (Thompson, 1983).

isomerization (configurational isomerization, stereoisomerization) any rearrangement of the atoms in a molecule to form a different structure. During stereoisomerization, a hydride or hydrogen radical is removed from an asymmetric carbon atom, resulting in formation of a planar carbocation or radical intermediate, followed by reattachment of the hydride or hydrogen radical. Reattachment can occur on the same side of the planar intermediate as removal (thus resulting in no change in configuration) or on the opposite side (resulting in an inverted or mirror-image configuration). Stereoisomerization occurs only when cleavage and reassembly of bonds results in an inverted configuration compared with the starting asymmetric center.

isomers compounds with the same molecular formula but different arrangements of their structural groups (e.g. n-butane and isobutane).

isopach a contour on a map that shows equal thickness of a sedimentary unit.

isopentenoids compounds composed of isoprene subunits (see Figure 2.15). Most biomarkers are isopentenoids. Some biomarkers that are not isopentenoids include certain normal, iso-, and anteisoalkanes.

isoprene (isopentadiene) the basic structural unit composed of five carbon atoms found in most biomarkers, including mono-, sesqui-, di-, sester-, and triterpanes, steranes, and polyterpanes (see Figure 2.15). The saturated analog of isoprene is 2-methylbutane.

isoprenoids hydrocarbons composed of, or derived from, polymerized isoprene units. Typical acyclic isoprenoids include pristane (iC_{19}) and phytane (iC_{20}).

isostacy gravitational balance or equilibrium, comparable to floating, of the units of the lithosphere above the asthenosphere.

isotopes atoms whose nuclei contain the same number of protons but different numbers of neutrons. For example, all carbon atoms have six protons, but there are isotopes containing 6, 7, and 8 neutrons resulting in atomic masses 12, 13, and 14. ^{12}C is the principal naturally occurring isotope (98.89%). ^{13}C is the stable carbon isotope (1.11%) and ^{14}C is the unstable (radioactive) carbon isotope ($\sim 1 \times 10^{-11}$%). Carbon isotopes are commonly used in oil–oil and oil–source rock correlation.

IUPAC the International Union of Pure and Applied Chemistry.

jet rock an organic-rich black gem material of uncertain origin that takes on a bright luster when polished. It is believed to consist of plant remains (possibly related to the *Araucaria* tree) that were deposited under anoxic conditions.

joint a fracture in rock on which no movement has taken place.

kerogen insoluble (in organic solvents) particulate organic matter preserved in sedimentary rocks that consists of various macerals originating from components of plants, animals, and bacteria. Kerogen can be isolated from ground rock by extracting bitumen with solvents and removing most of the rock matrix with hydrochloric and hydrofluoric acids.

kerogen type I, II, IIS, III, and IV see Type I kerogen, Type II kerogen, Type IIS kerogen, Type III kerogen, and Type IV kerogen.

kerosene a distillate from crude oil refining that is used for heating and fuel for aircraft engines. See also Straight-run refinery products.

ketone an organic compound with a carbonyl group (–CO).

kinetics study of the rates of biological, physical, and chemical changes.

Krebs cycle see Tricarboxylic acid cycle.

lacustrine refers to lakes; e.g. lacustrine oils were generated from organic matter originally deposited in lake sediments.

lamella (*pl.* lamellae) a thin plate-like arrangement or membrane.

lamination fine stratification in shale or siltstone, where layers commonly 0.05–1.0 mm thick differ from adjacent layers.

Laurasia a supercontinent that was composed of parts of modern Europe, Asia, Greenland, and North America.

lava magma that reaches the Earth's surface.

leaching the removal of materials in solution; e.g. extraction of soluble materials by water or crude oil percolating through rock.

level of certainty a measure of confidence that petroleum originated from a specific pod of active source rock. Three levels include known (!), hypothetical (.), and speculative (?), depending on the level of geochemical, geophysical and geological evidence.

lichen a symbiotic association of fungi and algae or cyanobacteria.

light hydrocarbons gases that are volatile liquids at standard temperature and pressure (STP) and range from methane to octane, including normal, iso-, and cyclic alkanes, and aromatic compounds.

light oil crude oil that has 35–45° API gravity.

lignans in woody plants, dimers of p-coumaryl, coniferyl, and sinapyl alcohols that serve mainly as supportive fillers or toxins.

lignin a phenolic organic polymer found with cellulose in the cell walls of many plants. Composed primarily of three aromatic precursors, p-coumaryl, coniferyl, and sinapyl alcohols.

lignite a low-rank coal ($\sim$0.2–0.4% R_o) that is commonly used as a drilling mud additive. See also Coal.

limestone see Carbonate rock.

linked-scan mode a gas chromatography/mass spectrometry (GCMS) mode in which two or more quadrupole, electrostatic, and/or magnetic fields are scanned simultaneously, thus, allowing detection of specific parent, daughter, or neutral loss relationships between ions.

lipids oil-soluble, water-insoluble organic compounds, including fatty acids, waxes, pigments, sterols, and hopanoids, and substances related biosynthetically to these compounds. Lipids are major precursors for petroleum (Silverman, 1971).

liptinite a maceral group composed of oil-prone, fluorescent, hydrogen-rich kerogen. Both structured (e.g. resinite, sporinite, and cutinite) and unstructured or amorphous liptinites, sometimes called amorphinite, can occur.

liquid chromatography a preparative column method used to separate oils and bitumens into saturate/aromatic and porphyrin/polar fractions before high-performance liquid chromatography (HPLC).

lithification the solidification of loose sediments to form sedimentary rocks.

lithofacies map a map that shows the lateral variation in lithologic attributes of a stratigraphic unit.

lithosphere the outermost shell of the solid Earth, consisting of $\sim$100 km of crust and upper mantle, that lies above the asthenosphere.

lithotroph an organism that oxidizes an inorganic substrate, such as ammonia or hydrogen, to use it as an electron donor in energy metabolism. Also known as chemoautotroph.

low-sulfur oil see Sweet oil.

Lycopsida leafy plants with simple, closely spaced leaves that carry sporangia on their upper surfaces. Lycopsida include living club mosses and many extinct late Paleozoic Lepidodendrales (scale trees). Also called lycopsids and lycophytes.

lysis rupture of a cell, which results in the loss of cell contents.

maceral microscopically recognizable, particulate organic component of kerogen showing distinctive physicochemical properties that change with thermal maturity. The three main maceral groups are liptinite, vitrinite, and inertinite.

macromolecule a large molecule formed from the connection of a number of small molecules.

mafic igneous rock composed mainly of dark-colored ferromagnesian minerals.

magma molten rock below the Earth's surface.

magnetic sector   a magnetic (versus electric or quadrupole) mass analyzer in mass spectrometers. Ions follow a curved path down the flight tube in the magnetic field (e.g. see Figure 8.15). The degree of curvature is related to the mass and velocity of the ion. Only those ions with a given m/z reach the detector for a given magnetic field strength or accelerating voltage. All other ions collide with the walls of the flight tube.

maleimides   biomarkers for bacteriochlorophylls c, d, and e in green sulfur bacteria and chlorophyll a in phytoplankton (Grice, 2001).

mammoth   extinct elephant of the Pleistocene Epoch.

mantle   the region of the Earth composed mainly of solid silicate rock that extends from the base of the crust (Moho) to the core–mantle boundary at a depth of $\sim$2900 km.

marl   a sedimentary rock containing calcareous clay.

marsupials   mammals of the order Marsupialia. Females have mammary glands and carry their young in a stomach pouch.

mass chromatogram (fragmentogram)   intensity of a specific ion versus gas-chromatographic retention time (e.g. see Figure 8.18). Allows identification of carbon number and isomer distributions for selected compound types.

mass number   the number of protons plus neutrons in the nucleus of an isotope of an element.

mass spectrometer   an instrument that separates ions of different mass but equal charge and measures their relative quantities.

mass spectrometry   a method used to supply information on the molecular structure of compounds, particularly biomarkers. Molecules in the gaseous state (inserted directly into the mass spectrometer or eluting from a gas chromatograph after separation) are ionized, usually by high-energy electrons. The resulting molecular and fragment ions are detected and displayed on the basis of increasing mass/charge ratio (m/z) in a mass spectrum.

mass spectroscopy   see Mass spectrometry.

mass spectrum   depiction of a beam of ions on a plot of the mass/charge (m/z) ratio versus intensity. Mass spectra can be used for provisional compound identification because they represent a fingerprint that is often diagnostic of specific structures.

MA steroid   see Monoaromatic steroid.

mature   describes organic matter that is in the oil-generative window, i.e. at a maturity equivalent to vitrinite reflectance in the approximate range 0.6–1.4% (e.g. see Figure 14.3). Organic matter can also be mature with respect to the gas window ($\sim$0.9–2.0% vitrinite reflectance).

maturity   see Thermal maturity.

meiosis   the process of nuclear division in eukaryotes, in which the change from diploid to haploid occurs, resulting in daughter cells with half the number of chromosomes of the original cell.

mercaptan   sulfur-containing compounds that occur in sour crude and gas and are added to odorless natural gas and natural gas liquids to give them a characteristic smell and thus allow them to be detected. Light mercaptans have a strong, repulsive odor like that of rotten eggs.

meso compound   molecule with two or more asymmetric centers that also has planes of symmetry and thus cannot exist as enantiomers. For example, the 6(R),10(S)- and 6(S),10(R)-configurations of pristane are identical and are called mesopristane.

mesophile   an organism with optimal growth in the approximate range 15–40°C.

mesotrophic   describes a lake at a stage of evolution intermediate between oligotrophic and eutrophic.

metabolism   all biochemical reactions in a cell.

metagenesis   the thermal destruction of organic molecules by cracking to gas, which occurs after catagenesis but before greenschist metamorphism (>200°C) in the range of $\sim$150–200°C (e.g. see Figure 1.2). The level of maturity equivalent to the range 2.0–4.0% vitrinite reflectance.

metamorphic rock   a rock formed from pre-existing rock due to high temperature and pressure in the Earth's crust, but without complete melting.

metamorphism   the solid-state transformation of pre-existing rock into texturally or mineralogically distinct new rock due to high temperatures, pressures, and chemically reactive fluids at depth, as occurs in the roots of mountain chains and adjacent to large intrusive igneous bodies.

metastable ion   an ion that has been accelerated from the ion source and decomposes in one of the field-free regions of the mass spectrometer, producing a broad, diffuse peak in the mass spectrum.

metastable reaction (metastable transition)   decomposition of a metastable ion. In metastable reaction monitoring/gas chromatography/ mass spectrometry (MRM/GCMS, also called metastable reaction monitoring/gas chromatography/

mass spectrometry/mass spectrometry, MRM/GCMS/MS) parent mode, all or selected metastable ions (usually molecular ions) in the mass spectrometer decomposing to a single daughter ion are monitored using a linked-scan method (see Linked-scan mode). A daughter mode is also possible, i.e. monitoring daughter ions from a given parent ion.

metazoa multicellular animals whose cells differentiate to form tissues, i.e. all animals except protozoa.

meteorites stony or metallic extraterrestrial bodies that have impacted the Earth's surface.

meteors solid extraterrestrial materials, mostly small particles, that are heated frictionally to incandescence upon entering the Earth's atmosphere. Most disintegrate, but some reach the surface of the Earth to become meteorites.

methane the main compound in natural gas.

methanogen a methane-producing prokaryote and member of the archaea.

methanogenesis the generation of methane by bacterial fermentation. Occurs only under anoxic conditions, where little sulfate is available (e.g. water column or sediments).

methanotroph a methane-oxidizing microbe.

methylcholestane see Ergostane.

methylotrophic bacteria heterotrophs that use reduced carbon substrates with no carbon–carbon bonds (e.g. methane, methanol, methylated amines, and methylated sulfur species) as their sole source of carbon and energy.

methylphenanthrene index (MPI) a thermal maturity parameter based on the relative abundances of phenanthrene and methylphenanthrenes (three-ring aromatic hydrocarbons). MPI depends on organic facies but is generally most reliable for petroleum generated from type III kerogen.

methylsteranes steranes with a methyl group attached to C-2, C-3, or C-4 on the A-ring. The 4-methylsteranes in petroleum generally originated from dinoflagellates.

microbe a living organism too small to be seen with the naked eye (<0.1 mm), e.g. algae, bacteria, fungi, and protozoans.

microbial biomass total mass of living microorganisms in a volume or mass of sediment or water.

microbial gas microbial gases are dominated by methane (typically >99%) produced by bacteria in shallow sediments. Microbial methane is generally depleted in ^{13}C compared with thermogenic gas.

microbiology the study of microorganisms.

microenvironment the close physical and chemical surroundings of a microorganism.

micrometer 10^{-6} m (one millionth of a meter). A common unit for measuring microorganisms.

microorganism an organism of microscopic size.

microscopic organic analysis (MOA) petrographic analysis of kerogen or rock in transmitted and/or reflected light. General use of the term includes both maceral (relative percentages of phytoclast types) and thermal maturity (vitrinite reflectance, thermal alteration index (TAI), transmittance color index (TCI), color alteration index (CAI)) analysis. MOA provides critical support for interpretations based on elemental analysis (Jones and Edison, 1978) and biomarkers.

MID (multiple ion detection) see Selected ion monitoring.

migration the process whereby petroleum moves from source rocks toward reservoirs or seep sites. Primary migration consists of movement of petroleum to exit the source rock. Secondary migration occurs when oil and gas move along a carrier bed from the source to the reservoir or seep. Tertiary migration is where oil and gas move from one trap to another or to a seep.

Milankovitch effect the proposed long-term effect on world climate caused by three components of the Earth's motion. The combination of these components provides a possible explanation for repeated glacial to interglacial climatic swings.

mineral an element (e.g. gold) or compound (e.g. calcite, $CaCO_3$) that has a definite chemical composition or range of compositions with distinctive properties and form that reflect its atomic structure.

mineralization conversion of an element from an organic to an inorganic state due to microbial decomposition.

mitochondrion (*pl.* mitochondria) the eukaryotic organelle responsible for respiration and oxidative phosphorylation.

mitosis the reproductive process of cell division in eukaryotes in which each of the two daughter nuclei receives exactly the same complement of chromosomes as had existed in the parent nucleus.

MMCFD million cubic feet per day.

mobile phase see Gas chromatography.

Moho the Mohorivicic discontinuity (seismic reflector) at the base of the crust.

Mohorovicic discontinuity a zone that separates the crust of the Earth from the underlying mantle. The so-called Moho occurs at ~70 km below the surface of continents and 6–14 km below the floor of the oceans.

molecular fossil see Biological marker.

molecular ion (M^+ or M^-) an ion formed by the removal (or addition) of one or more electrons from a molecule without fragmentation.

molecular weight the sum of the atomic weights of all the atoms in a molecule. A compound's mass may be expressed by its nominal, accurate, or average molecular weight. Nominal molecular weight is calculated from integer masses (e.g. H = 1, C = 12, O = 16). Accurate molecular weight uses the exact monoisotopic mass values (e.g. $^1H = 1.00783$, $^{12}C = 12$, $^{16}O = 15.99491$). Average molecular weight uses the average of the accurate weight for isotopes as they occur in their natural abundance (e.g. H = 1.01, C = 12.01 O = 16.00). For example, $C_{35}H_{62}$ has a nominal mass of 482, an accurate mass of 482.485152, and an average mass of 482.88 Daltons (atomic mass unit, amu).

molecule two or more atoms combined by chemical bonding.

mollusk a member of the invertebrate phylum Mollusca, which includes cephalopods, pelecypods, gastropods, scaphopods, and chitons.

monoaromatic steroid a class of biomarkers that contain one aromatic ring (usually the C-ring) (e.g. see Figure 13.107), probably derived from sterols.

monoaromatic steroid triangle plot of C_{27}, C_{28}, and C_{29} monoaromatic steroids on a ternary diagram (e.g. see Figure 13.105) used for correlation similar to the sterane triangle (e.g. see Figure 13.38). However, monoaromatic steroids are probably derived from different sterol precursors compared with the steranes.

monomer a simple molecular unit, such as ethylene or styrene, from which a polymer can be made.

monoterpane a class of saturated biomarkers constructed from two isoprene subunits (~C_{10}).

monotremes egg-laying mammals.

montmorillonite a smectite clay that is cation-exchangeable and occurs as a major component in bentonite clay deposits, soils, sedimentary and metamorphic rocks, and some mineral deposits. Catalytic activity of montmorillonite and other clays is believed to account for the origin of diasterenes and disasteranes from sterols (see Figure 13.49).

moraine a deposit of rock and soil (till) left by a retreating glacier.

moretanes C_{27}–C_{35} pentacyclic triterpanes that are stereoisomers of the hopanes and have $17\beta,21\alpha$-stereochemistry (see Figure 2.29). Because of lower stability, moretanes decrease relative to hopanes with increasing thermal maturation (see Figure 14.5).

mosasaurs large marine lizards from the Late Cretaceous Period.

MPI see Methylphenanthrene index.

MRM see Metastable reaction monitoring.

mucilage a gelatinous secretion produced by many microorganisms and plant roots.

mud gas gases in the drilling mud that originate from formations in a well.

mudrock see Mudstone.

mudstone a general term for sedimentary rock made up of clay-sized particles, typically massive and not fissile.

multidimensional chromatography use of two or more chromatographic columns (e.g. two-dimensional gas chromatography) to improve compound separation, either off-line by collecting fractions and re-injecting on to a second column, or on-line by the use of a switching valve.

multiple ion detection (MID) see Selected ion monitoring.

murein (peptidoglycan) an aminosugar polymer that consists of two hexoses, N-acetylglucosamine and N-acetyl-muramic acid, linked by a β-1,4-bond and short oligopeptides, which cross-link the polysaccharide chains in bacterial cell walls.

mutagen a substance that causes genes to mutate.

mutant an organism, gene, or chromosome that differs from the corresponding wild type by one or more base pairs.

mutation an inheritable change in a gene of the DNA of an organism.

mycoplasmas very small bacteria that lack a cell wall and have a single triple-layered membrane that is stabilized by small amounts of sterols. These sterols are apparently not synthesized by mycoplasmas but are obtained from the surrounding environment.

mycorrhiza the symbiotic association between specific fungi with the fine roots of higher plants.

m/z the mass/charge ratio of an ion in mass spectrometry measured in units of Daltons per charge, with positive or negative values denoting cations or anions, respectively, e.g. m/z 217 is a characteristic fragment of steranes (e.g. see Figure 8.15). Older publications may use the term m/e instead of m/z. Incorrect use of mass and Daltons as synonyms for m/z is especially confusing when applied to multiply charged ions.

NADH, NADPH NADH is the reduced form of nicotinamide adenine dinucleotide (NAD). NADH and NADPH (additional phosphate) are key coenzymes for biochemical reactions in living cells (e.g. Figures 3.5 and 3.21).

nafion R a family of perfluorosulfonate ion-exchange membranes (Dupont, Inc.) used in many component-specific isotope analysis (CSIA) instruments to remove H_2O from CO_2 during continuous combustion of the gas chromatography effluent.

naphtha a distillation cut of petroleum showing volatility between gasoline and kerosene ($\sim C_7$–C_{10}). Used as a manufacturing solvent, a dry-cleaning fluid, and a gasoline-blending stock.

naphthalene an aromatic hydrocarbon ($C_{10}H_8$) consisting of two fused benzene rings.

naphthenes see Cycloalkane.

natural gas gaseous petroleum that can consist of a mixture of C_1–C_5 hydrocarbons, CO_2, N_2, H_2, H_2S, and He. When natural gas occurs with oil, it is called associated gas.

NBS-22 a Pennsylvania No. 30 lubricating oil supplied by the National Bureau of Standards as one international standard for stable carbon isotope ratios. The most common standard for carbon is Peedee belemnite (PDB).

NDR C_{26} 24/(24 + 27) nordiacholestanes greater than 0.25 and 0.55 typify oils from Cretaceous or younger and Oligocene or younger (generally Neogene) source rocks, respectively (see Figure 13.55).

Neogene Period a subdivision of the Cenozoic Era that consists of the Miocene and Pliocene epochs.

neohopane a hopane in which the methyl group at C-18 is rearranged to C-17 (i.e. hopane II). Ts is the C_{27}-neohopane.

neritic describes oceanic depths in the range between low tide and 200 m or the approximate edge of the continental shelf.

neutron an electrically neutral subatomic particle of matter with a rest mass of 1 Dalton that occurs with protons in the atomic nucleus of all elements except the mass 1 isotope of hydrogen.

niche the physical and biologic conditions under which an organism can live and reproduce.

nitrogen fixation the conversion of molecular nitrogen (N_2) to ammonia and finally to organic nitrogen.

NMR see Nuclear magnetic resonance.

non-associated gas natural gas in gas accumulations.

non-hydrocarbon gases mainly carbon dioxide (CO_2), nitrogen (N_2), and hydrogen sulfide (H_2S), but also including helium (He), argon (Ar), and hydrogen (H_2).

non-hydrocarbons see NSO compounds.

non-polar not easily dissolved in water due to hydrophobic (water repelling) characteristics.

25-norhopanes see Demethylated hopanes.

normal fault a fault in which the hanging wall appears to have moved downward relative to the footwall, normally occurring in areas of crustal tension.

normal-phase chromatography chromatography using a polar stationary phase and a non-polar mobile phase, e.g. adsorption on silica gel using hexane as a mobile phase.

notochord a rod-shaped cord of cartilage cells forming the primary axial structure of the chordate body. In vertebrates, the notochord is present in the embryo and is later supplanted by the vertebral column.

NSO compounds (resins) a pentane-soluble fraction (or individual compounds) of petroleum that contains various elements in addition to hydrogen and carbon, including nitrogen, sulfur, and/or oxygen. Compounds in this fraction are sometimes called heterocompounds or non-hydrocarbons. Other fractions include saturates, aromatics, and asphaltenes.

nuclear magnetic resonance (NMR) proton NMR (^{1}H NMR) is a spectroscopic method for determining molecular structure. The high resolution of the method commonly distinguishes stereoisomers that cannot be distinguished by mass spectroscopy. ^{13}C NMR requires more sample than proton NMR and can identify the types of carbon atoms (e.g. aromatic versus saturated). New ^{13}C NMR methods allow determination of numbers of hydrogen atoms bound to individual carbon atoms (i.e. methyl, methylene, methine, or quaternary). New two-dimensional NMR techniques allow correlation of ^{13}C and proton NMR data and provide improved resolution and in some cases can rival X-ray diffraction crystallography in structural elucidation.

nucleic acids organic acid polymers of nucleotides that control hereditary processes within cells and make possible the manufacture of proteins from the amino acids ingested by the cells as food.

nucleotide a monomeric unit of nucleic acid that consists of a pentose sugar, a phosphate, and a nitrogenous base.

nucleus a membrane-enclosed structure that contains the genetic material (DNA) organized in chromosomes.

obligate refers to an environmental factor (e.g. oxygen) that is always required for growth.

O/C the atomic oxygen/carbon ratio from elemental analysis of kerogen, typically used to describe kerogen type on van Krevelen diagrams. Because oxygen is not measured readily on kerogens, atomic O/C is not always used directly on van Krevelen plots (Jones and Edison, 1978).

oceanic crust the Earth's crust underlying the ocean basins, which is ~5–10 km thick.

octane number a measure of the resistance of a fuel to engine knock (pre-ignition) when burned in an internal combustion engine. In tests using pure hydrocarbons in gasoline engines, n-heptane and isooctane caused the most and least engine knock and were assigned octane ratings of 0 and 100, respectively. Higher octane numbers for fuels on this scale correspond to less engine knock.

odd-to-even predominance (OEP) the ratio of odd- to even-numbered n-alkanes in a given range (Scalan and Smith, 1970). Immature rock extracts can show high or low OEP, but most mature oils and source rocks show OEP near 1.0.

OEP see Odd-to-even predominance.

OI see Oxygen index.

oil a mixture of liquid hydrocarbons and other compounds of different molecular weights.

oil deadline the depth at which oil no longer exists as a liquid phase in petroleum reservoirs, generally corresponding to gas-to-oil ratio (GOR) >5000 standard cubic feet/barrel or temperatures >150°C (typically in the range 165–185°C).

oil field an area with an underlying oil reservoir.

oil–oil correlation a comparison of chemical compositions to describe the genetic relationships among crude oils based on source-related geochemical data that might include biomarkers, isotopes, and metal distributions, although the source rock may not be defined.

oil-prone organic matter that generates significant quantities of oil at optimal maturity. For example, type I and II kerogens are highly oil-prone and oil-prone, respectively. Oil-prone organic matter is typically also more gas-prone than gas-prone kerogen.

oil shale organic-rich shale that contains significant amounts of oil-prone kerogen and liberates crude oil upon heating, as might occur during laboratory pyrolysis or commercial retorting.

oil–source rock correlation a comparison of chemical compositions to describe the genetic relationships among crude oils and source-rock extracts based on source-related geochemical data that might include biomarkers, isotopes, and metal distributions.

oil–water contact (OWC) the boundary between oil and underlying water in a reservoir.

oil window the maturity range in which oil is generated from oil-prone organic matter (~0.6–1.4% vitrinite reflectance), i.e. within the catagenesis zone (~0.5–2.0% vitrinite reflectance) (e.g. see Figure 14.3).

oleanane index the ratio of oleanane (a C_{30} triterpane marker of angiosperms) to 17α-hopane (a bacterial marker). Oleanane occurs mainly in Late Cretaceous or younger rocks, but its absence cannot be used to prove age.

olefin see Alkene.

oligotrophic describes a lake with few nutrients and abundant dissolved oxygen that supports little primary productivity. Oligotrophic lakes are earliest in the age sequence oligotropic, mesotrophic, and eutrophic, before complete infilling of the lake basin and subaerial conditions.

OPEC the Organization of Petroleum Exporting Countries, which includes Algeria, Indonesia, Iran, Iraq, Kuwait, Libya, Nigeria, Qatar, Saudi Arabia, United Arab Emirates, and Venezuela.

open-tubular column a chromatographic column with small internal diameter, e.g. fused-silica tubing for capillary gas chromatography.

optical activity the tendency of some organic compounds to rotate monochromatic plane-polarized light to the right or left. Biomarkers and other organic materials contain asymmetric carbon atoms resulting in left- or right-handed molecules (enantiomers). Enzymes in living organisms tend to produce one or the other of these optically active compounds. Solutions of these compounds rotate plane-polarized light to varying degrees. Solutions of the same compounds produced without enzymes or living organisms are not optically active because they represent a racemic (50:50) mixture of enantiomers.

organelles various membrane-enclosed bodies specialized for carrying out certain functions in eukaryotic cells.

organic acids organic molecules that contain at least one functional group capable of releasing a proton. Carboxyl groups (–COOH) account for the acidity of most organic acids, e.g. fatty, naphthenic, and aromatic acids. Common monocarboxylic acids include formic (HCOOH), acetic (CH_3COOH), and proprionic (CH_3CH_2COOH) acids (methanoic, ethanoic, and propanoic acids, respectively, in International Union of Pure and Applied Chemistry (IUPAC) nomenclature). Oxalic acid (HOOCCOOH) is the simplest dicarboxylic acid (ethanedioic acid in IUPAC nomenclature).

organic carbon carbon in compounds derived from living organisms rather than inorganic sources.

organic facies a mappable rock unit containing a distinctive assemblage of organic matter without regard to the mineralogy (Jones, 1987). Biofacies are analogous to organic facies, except that biofacies refer specifically to distinctive assemblages of recognizable organisms.

organic matter biogenic, carbonaceous materials. Organic matter preserved in rocks includes kerogen, bitumen, oil, and gas. Different types of organic matter can have different oil-generative potential. See also Biomass.

organic yield a crude estimate (total organic carbon (TOC) is more accurate) of the amount of organic matter in a rock based on the volume of organic material that survives demineralization of the rock with acids during preparation of kerogen.

organotroph an organism that obtains reducing equivalents (stored electrons) from organic substrates. Also known as chemoheterotroph.

orogenic belt extensive tracts of deformed rocks, primarily developed near continental margins by compressional forces accompanying mountain building.

orogeny the process of mountain building; the process whereby structures within fold-belt mountainous areas formed.

ostracodes (also ostracods) small (~0.4–30 mm) aquatic crustaceans with a calcified bivalve carapace. Range from Lower Cambrian to present.

ostracoderms extinct jawless fish from the early Paleozoic Era.

outcrop an exposure of bedrock at the surface.

outer continental shelf the portion of a continental land mass that constitutes the slope down to the ocean floor. The outer continental shelves may contain much of the world's remaining undiscovered oil and gas.

overburden rock sedimentary or other rock that compresses and consolidates the underlying rock. Overburden rock is an essential element of petroleum systems because it contributes to the thermal maturation of the underlying source rock.

overload in chromatography, the mass of sample injected on to the column at which efficiency and resolution begin to be affected adversely if the sample size is increased further.

overmature see Postmature.

oxic refers to water column or sediments with molecular oxygen contents of 2–8 ml/l of interstitial water. Used in reference to a microbial habitat (see Table 1.5). Maximum oxygen saturation in seawater is in the range 6–8.5 ml/l, depending on salinity and water temperature.

oxidation the process of (1) loss of one or more electrons by a compound, or (2) addition of oxygen to a compound, or (3) loss of hydrogen (dehydrogenation). Oxidation and reduction reactions are always coupled (redox reactions). For example, phytol (an alcohol) can be oxidized to phytanic acid, or iron can be oxidized from 2^+ (ferrous) to 3^+ (ferric).

oxidation–reduction (redox) reaction a coupled pair of reactions, in which one compound is oxidized while the other is reduced and takes up the electrons released in the oxidation reaction. See also Redox potential.

oxidation state the number of electrons to be added or removed from an atom in a combined state to convert it to the elemental form.

oxidative phosphorylation the synthesis of adenosine triphosphate, involving a membrane-associated electron-transport chain.

oxygenic photosynthesis the use of light energy to synthesize ATP and NADPH by acyclic photophosphorylation with the production of oxygen from water.

oxygen index a Rock-Eval parameter defined as (S3/TOC) × 100 (where TOC is total organic carbon) and measured in mg carbon dioxide/g TOC (see Figure 4.5). The OI is used on van Krevelen-type plots of hydrogen index (HI) versus OI to describe organic matter type (see Figure 4.4), and there is a general relationship between OI and atomic oxygen/carbon ratio (O/C).

PAH see Polynuclear or polycyclic aromatic hydrocarbons.

PAL present atmospheric level of molecular oxygen (21%).

paleoecology the study of the relationship of ancient organisms to their environment.

Paleogene Period a subdivision of the Cenozoic Era that consists of the Paleocene, Eocene, and Oligocene epochs.

paleogeography the geography as it existed in the geologic past.

paleolatitude the latitude of a location at a particular time in the geologic past.

paleomagnetism remnant magnetization of iron minerals in ancient rocks that allows reconstruction of the orientation of Earth's ancient magnetic field and the former positions of the continents relative to the magnetic poles.

paleontology the study of all ancient forms of life, their fossils, interactions, and evolution.

palynomorph organic-walled, acid-resistant microfossils useful in providing information on age, paleoenvironment, and thermal maturity (e.g. thermal alteration index (TAI)).

Pangea a late Paleozoic supercontinent that included all major present-day continents, which broke up in the Mesozoic Era.

Panthalassa a great ocean that surrounded the supercontinent Pangea before its breakup.

paraffin see Alkane.

paraffinicity ratio (F) n-heptane/methylcyclohexane, as in Figure 7.21 (Thompson, 1987).

parasequence a conformable succession of genetically related progradational beds bounded by marine-flooding surfaces that mark an abrupt increase in water depth.

parasitic a relationship in which an organism lives in intimate association with another, from which it obtains food and other benefits.

paraxylene an aromatic compound used to make polyester fibers and plastic bottles.

partition coefficient the amount of solute in the chromatographic stationary phase relative to that in the mobile phase.

passive margin a continental margin characterized by thick, flat-lying shallow-water sediments and limited tectonic activity.

pathogen an organism that can damage or kill a host that it infects.

pay zone the layer of rock in which significant oil and/or gas are found.

PDB see Peedee belemnite.

peak mature the level of maturity associated with the maximum rate of petroleum generation from the kerogen.

peat unconsolidated, slightly decomposed organic matter accumulated under moist conditions. Related to the early stages of coal formation.

pebble a rock particle 2–64 mm in diameter.

Peedee belemnite (PDB) an international primary stable isotope standard for carbon obtained from the carbonate fossil of a cephalopod, *Belemnitella americana*, in the Cretaceous Peedee Formation. Measurements are given in parts per thousand (‰ or per mil) relative to the standard using the standard delta notation. On the PDB scale, a secondary standard, NBS-22 oil, measures $\sim-29.81‰$. Accurate conversion of PDB to NBS-22 values is not straightforward, especially for gases, and requires a correction.

pelagic describes deep ocean areas far removed from land masses. Pelagic sediments consist of sea-surface materials that settle to the deep ocean floor, including biogenic oozes and aeolian (wind-blown) clays.

pelycosaurs early mammal-like reptiles, e.g. the sail-back animals of the Permian Period.

period the most commonly used unit of geologic time, representing one subdivision of an era.

permeability the capacity of a rock layer to allow water or other fluids, such as oil, to pass through it.

petrochemical a chemical derived from petroleum, hydrocarbon liquids, or natural gas, e.g. ethylene, propylene, benzene, toluene, and xylene.

petroleum a mixture of organic compounds composed predominantly of hydrogen and carbon and found in the gaseous, liquid, or solid state in the Earth, including hydrocarbon gases, bitumen, migrated oil, pyrobitumen, and their refined products, but not kerogen. In European usage, the term is sometimes restricted to refined products only.

petroleum province a geographic area where petroleum occurs. Also referred to as a petroleum basin or basin. For example, the Western Canada Basin and the Zagros Thrust Belt are well-known petroleum provinces.

petroleum system the essential elements and processes and all genetically related petroleum that occurs in shows, seeps, and accumulations and originated from one pod of active source rock. Also called a hydrocarbon system.

petroleum system name a name that includes the pod of active source rock, the reservoir rock containing the largest volume of corresponding petroleum, and the level of certainty in the petroleum system, e.g. the Mandal-Ekofisk(!) petroleum system.

petroleum system processes trap formation and generation–migration– accumulation. Combined with the essential elements, the processes control the distribution of petroleum in the lithosphere.

pH a measure of acidity; the tendency of an environment to supply protons (hydrogen ions) to a base or to take up protons from an acid (negative logarithm of the hydrogen ion concentration). A pH of 7.0 under standard conditions is neutral, while lower and higher values are acidic and basic, respectively. In nature, most environments fall in the pH range 4–9.

Phanerozoic Eon the eon of geologic time when Earth became populated by abundant and diverse life. The Phanerozoic Eon is divided into the Paleozoic, Mesozoic, and Cenozoic eras.

phenanthrene an aromatic hydrocarbon consisting of three fused aromatic rings; an isomer of anthracene.

phenol hydroxybenzenes; oxygen-containing aromatic compounds that are acidic due to the effect of the aromatic ring on the hydroxyl group.

phosphate rock a rock that contains abundant apatite or other phosphatic minerals. Most phosphate minerals are mined from phosphorites.

phospholipid lipids that contain a substituted phosphate group and two fatty acid chains on a glycerol backbone.

phosphorite a sedimentary rock composed mainly of microcrystalline carbonate fluorapatite.

photic zone the uppermost layer of water or soil that receives enough sunlight to permit photosynthesis, e.g. for water, typically <200 m depending on water turbidity.

photoautotroph an organism that uses light as the source of energy and carbon dioxide as the source for cellular carbon.

photoheterotroph an organism that uses light as the source of energy and organic compounds as the source for cellular carbon.

photosynthesis a process in plants that uses light energy captured by chlorophyll to synthesize carbohydrates from carbon dioxide and water. Photosynthesis requires chlorophyll and other light-trapping pigments, such as carotenoids and phycobilins. Photosynthetic eukaryotes include higher green plants, multicellular green, brown, and red algae, and unicellular organisms, such as dinoflagellates and diatoms. Photosynthetic prokaryotes include cyanobacteria, green bacteria, and purple bacteria.

phototroph an organism that uses light as the energy source to drive the electron flow from the electron donors, such as water, hydrogen and sulfide. Chlorophyll and other light-trapping pigments, such as carotenoids and phycobilins, are used by phototrophs in photosynthesis. Phototrophs include green cells of higher plants, cyanobacteria, photosynthetic bacteria, and non-sulfur purple bacteria.

phthalate an ester or salt of phthalic acid common in plasticizers.

phycobilins water-soluble compounds that occur in cyanobacteria and function as the light-harvesting pigments for photosystem II.

phylogeny classification of species into higher taxa and evolutionary trees based on genetic relationships.

phytane (Ph) a branched acyclic (no rings) isoprenoid hydrocarbon containing 20 carbon atoms (see Figure 2.15); a prominent peak eluting immediately after the C_{18} n-alkane in petroleum on most gas chromatographic columns (e.g. see Figure 8.9).

phytoclast an identifiable particle or maceral in kerogen, e.g. phytoclasts of vitrinite are used for measurement of vitrinite reflectance.

phytol a branched acyclic (no rings) isoprenoid alcohol containing 20 carbon atoms (e.g. see Figure 2.26).

phytoplankton unicellular photosynthetic planktonic plants, such as algae, diatoms, and dinoflagellates that inhabit the photic zone of water bodies.

PI see Production index.

placoderms extinct jawed fish from the Paleozoic Era.

plasmids circular DNA molecules that are separate from chromosomal DNA and are exchanged between bacteria during conjugation. They may provide a selective advantage to bacteria, e.g. resistance to antiobiotics. Plasmids are common in bacteria, but they also occur in eukaryotes.

plastid specialized cell organelle that contains pigments.

plates segments of the Earth's lithosphere that ride as distinct units over the asthenosphere.

platform part of a craton that is covered by layered sedimentary rocks and characterized by relatively stable tectonic conditions; or an offshore facility where development wells are drilled.

plankton small, free-floating aquatic organisms.

plateau an elevated area with little internal relief.

plate tectonics a theory that explains the tectonic behavior of the crust of the Earth by means of moving but rigid crustal plates that form by volcanic activity at oceanic ridges and are destroyed along ocean trenches.

playa a desert lake bed that is dry for most of the year.

PMP see Porphyrin maturity parameter.

PNA see Polynuclear or polycyclic aromatic hydrocarbons.

pod of active source rock a contiguous volume of organic-rich rock that generates and expels petroleum at the critical moment and accounts for genetically related petroleum shows, seeps, and accumulations in a petroleum system. A pod of mature source rock may be active, inactive, or spent.

polar compound an organic compound with distinct regions of partial positive and negative charge. Polar compounds include alcohols, such as sterols, and aromatics, such as monoaromatic steroids. Because of their polarity, these compounds are more soluble in polar solvents, including water, compared with non-polar compounds of similar molecular weight.

polymer a macromolecule composed of a large number of repeating subunits (monomers). Biopolymers include proteins (composed of amino acids) and polysaccharides (composed of sugars). Rubber is a polyunsaturated polymer composed of isoprene $[(C_5H_8)_n]$ subunits. Kerogen is not a polymer.

polymerase chain reaction a method used to amplify DNA *in vitro* that involves the use of DNA polymerase to copy target genetic sequences.

polynuclear or polycyclic aromatic hydrocarbons (PAHs or sometimes PNAs) organic compounds containing more than two aromatic rings.

polysaccharide long chains of monosaccharides (simple sugars) that are linked by glycosidic bonds.

polysiloxane see Siloxane.

polyterpanes a class of saturated biomarkers constructed of more than eight isoprene subunits ($\sim C_{40}+$).

pool see Reservoir.

population a group of individuals of the same species that occupy an area at the same time.

porosity the volume percentage of a rock that consists of open or pore space.

porphyrin maturity parameter (PMP) a biomarker maturity parameter for petroleum based on generation. $PMP = C_{28}$ etio/(C_{28} etio + C_{32} DPEP).

porphyrins complex biomarkers characterized by a tetrapyrrole ring, usually containing vanadium or nickel (e.g. see Figure 3.24), which originate from various sources including chlorophyll and heme.

postmature a high level of maturity at which no further oil generation can occur, i.e. vitrinite reflectance $>1.3\%$.

potential source rock a gas- or oil-prone, organic-rich rock that has not yet generated petroleum. A potential source rock becomes an effective source rock when it generates microbial gas at low temperatures or when it reaches the level thermal maturity necessary to generate petroleum.

pour point the temperature below which a crude oil does not flow freely as a liquid. High API gravity oils can show low pour points due to abundant wax content.

ppb parts per billion.

ppm parts per million – the scale for measuring many components in oils, gases, and petrochemicals.

Precambrian Era refers to all of geologic time before the Paleozoic Era.

precipitate to drop out of a saturated solution.

precision the degree of agreement of repeated measurements of a quantity, which may or may not be accurate; i.e. a measurement can be precise but inaccurate.

precolumn a small chromatographic column placed between the pump and the injector to remove particulate matter that may be in the mobile phase and chemically adsorb substances that might interfere with the separation.

prenols alcohols having the general formula $H-[CH_2C(CH_3)=CHCH_2]_nOH$, where the carbon skeleton is composed of one or more isoprene units.

preparative chromatography use of liquid chromatography to isolate sufficient amounts of material for other purposes.

preservation time the time after generation–migration–accumulation of petroleum, including time during which petroleum might be exposed to secondary processes, such as remigration and inspissation, biodegradation, and water washing.

primary migration see Expulsion and Migration.

primary producer an organism that adds biomass to the ecosystem by synthesizing organic molecules from carbon dioxide and simple inorganic nutrients.

pristane (Pr) a branched acyclic (no rings) isoprenoid hydrocarbon containing 19 carbon atoms (see Figure 2.15); a prominent peak eluting immediately after the C_{17} *n*-alkane in petroleum on most gas chromatography columns (e.g. see Figure 8.9).

production index (PI) a Rock-Eval parameter useful for describing thermal maturity of source rocks or indicating contamination. PI = S1/(S1 + S2) (Peters, 1986).

production platform a platform from which development wells are drilled and that carries all of the processing plants and other equipment needed to maintain production of a field.

production string the pipes in a production well through which oil or gas flows from the reservoir to the wellhead.

production well a well used to remove oil or gas from a reservoir.

progradation movement of the shoreline into a sedimentary basin when clastic input exceeds the accommodation space, as might occur due to reduced basinal subsidence or increased erosion and sediment supply.

prokaryotes a phylogenetic domain that consists of organisms (bacteria and cyanobacteria) that lack membrane-bound nuclei and organelles and that maintain their genome dispersed throughout the cytoplasm. Prokarya is one of the three domains of life.

proteins large molecules in all living cells that are composed of chains of amino acids.

protista an old taxonomic term that refers to algae, fungi, and protozoa (eukaryotic protists), and the prokaryotes.

proton a subatomic particle in the nucleus of all atoms with an electric charge of +1 and a mass similar to that of a neutron.

protoplasm all cellular contents, cytoplasmic membrane, cytoplasm, and nucleus. Usually considered the living portion of the cell, thus excluding those layers peripheral to the cell membrane.

protozoan (*pl.* protozoa) unicellular eukaryotic microorganisms.

pseudohomologous series compounds showing similar structures but differing in the length of a branched alkyl group. For example, the tricyclic terpanes (cheilanthanes) represent a pseudohomologous series.

psychrophile an organism able to grow at low temperatures. Optimal growth occurs below 15°C.

pterosaur a flying reptile of the Jurassic and Cretaceous periods.

pure culture a population composed of a single strain of organism.

pycnocline a layer of water in which the density increase with depth is greater than that of the underlying or overlying water. For example, a pycnocline or halocline in an estuary might separate shallow, fresh water from deeper, saline water with higher density.

pyrobitumen thermally altered, solidified bitumen that is insoluble in common organic solvents and has an atomic hydrogen/carbon ratio (H/C) of 0.5 or less.

pyrogram a graph showing detector response to products generated by pyrolysis (e.g. see Figure 4.5). For the Rock-Eval pyroanalyzer, a pyrogram has S1, S2, S3, T_{max}, the programmed temperature trace, and other information (Peters, 1986).

pyrolysis the breakdown of organic matter during heating in the absence of oxygen; as in Rock-Eval pyrolysis (for source-rock evaluation) or hydrous pyrolysis (for simulating oil generation in source rocks). Rock-Eval pyrolysis employs programmed-temperature pyrolysis because the temperature is programmed to increase at a selected rate during analysis. Hydrous pyrolysis typically employs a constant temperature for each experiment.

quadrupole rods an electrical (versus magnetic) mass analyzer in a mass spectrometer (e.g. see Figure 8.17). Using a combination of direct current (DC) and radio frequency (RF) fields on the quadrupole rods serves as a mass filter for ions. Only ions of a given m/z reach the detector for a given magnitude of the DC and RF fields. All other ions collide with the rods before reaching the detector.

quasi-dysaerobic refers to metabolism under suboxic conditions with limited molecular oxygen (see Table 1.5).

R or S configuration the stereochemical arrangement of atomic groups around a asymmetric carbon atom can be assigned by orienting the molecule according to specific conventions and finding whether a circle passes through the remaining groups surrounding the center clockwise (R, rectus) or counterclockwise (S, sinister).

racemic mixture a 50:50 or racemic mixture or racemate of enantiomers in solution results in no optical activity. Racemic mixtures of enantiomers are typical of the non-biologic (abiotic) synthesis of molecules containing asymmetric centers. Many biologically formed compounds are optically active because they consist of only one enantiomer.

radioactive the spontaneous emission of a particle from the atomic nucleus, thus transforming the atom from one element to another.

radiolaria protozoa that secrete a skeleton of opaline silica.

radiometric dating techniques used to determine the age of geologic materials utilizing the known rates of decay of radioactive isotopes.

rearranged hopane see Diahopane and Neohopane.

rearranged sterane see Diasterane.

recharge the replacement of groundwater by infiltrating rain or stream water or the replenishment of water in a lake or ocean.

reconstructed ion chromatogram (RIC) the magnitude of the total ion current during a gas chromatography/mass spectrometry (GCMS) analysis plotted versus scan or retention time. RIC and gas chromatographic traces of petroleum samples are nearly identical. Also known as total ion chromatogram (TIC).

recoverable reserves the amount of oil and/or gas in a reservoir that can be removed using currently available techniques.

recovery the amount of solute (sample) that elutes from a chromatographic column relative to that injected.

red beds red-colored, usually clastic sedimentary deposits.

redox potential (oxidoreduction potential) a measure of the ability of an environment to supply electrons to an oxidizing agent or to take up electrons from a reducing agent. In redox reactions, there is transfer of electrons from an electron donor (the reducing agent or reductant) to an electron acceptor (the oxidizing agent or oxidant). Expressed as the tendency of a reducing agent to lose electrons relative to the standard reduction potential. Standard reduction potential is the electromotive force in volts given by a half-cell in which the reductant and oxidant are present at 1.0-M concentration, 25°C, and pH 7.0, in equilibrium with an electrode, which can reversibly accept electrons from the reductant species. The standard of reference is the reduction potential of the reaction: $H_2 = 2H^+ + 2e^-$, which is set at 0.0 V under conditions in which the pressure of H_2 gas is 1.0 atmosphere, $[H^+]$ is 1.0 M, pH is 7.0, and temperature is 25°C. Systems having a more negative standard reduction potential than the H_2–$2H^+$ couple have a greater tendency to lose electrons than hydrogen, and vice versa. Most redox potentials of seawater, for example, lie in the range between +0.3 V for aerated water to −0.6 V for oxygen-depleted (anoxic) bottom water.

reduction the process of (1) acceptance of one or more electrons by a compound, or (2) removal of oxygen from a compound, or (3) addition of hydrogen (hydrogenation). Reduction and oxidation always occur as coupled reactions (redox reactions). For example, alkenes can be reduced to alkanes by hydrogenation or iron can be reduced from 3^+ (ferric) to 2^+ (ferrous) state by gaining electrons.

reef a wave-resistant organic structure built by calcareous organisms that stands in relief above the surrounding seafloor.

refinery a facility used to separate various components from crude oil and convert them into fuel products or feedstock for other processes.

reflectance see Vitrinite reflectance.

refractory polycyclic aromatic hydrocarbon index (PAH-RI) ratio of $C_{26}(20R) + C_{27}(20S)$ triaromatic steroids (co-eluting compounds that are a major peak on m/z 231 traces) to methylchrysene (Hostettler *et al.*, 1999).

regression withdrawal of the sea from land areas that shifts the boundary between marine and non-marine deposition (or between deposition and erosion) toward the center of a marine basin. Regressions can be caused by tectonic uplift of the land, eustatic lowering of sea level, and progradation of sediments into the basin.

relief the maximum regional difference in elevation.

replication conversion of one double-stranded DNA molecule into two identical double-stranded DNA molecules.

reservoir a porous, permeable sedimentary rock formation that contains oil and/or natural gas enclosed or surrounded by layers of less permeable rock. An oil pool consists of a reservoir or group of reservoirs. However, the term is misleading because petroleum exists not in pools but in pores between rock grains.

reservoir characterization integrating and interpreting geological, geophysical, petrophysical, fluid, and performance data to describe a reservoir.

reservoir rock any porous and permeable rock that contains petroleum, such as porous sandstone, vuggy carbonate, and fractured shale.

resin a solid or semi-solid mixture of complex organic substances with significant nitrogen, sulfur, and oxygen. Resins can be divided into five classes based on chemical composition (Anderson and Muntean, 2000), but classes I and II are the most common. Class I resins originate from gymnosperms (i.e. conifers), especially the family Araucariacea, while class II resins originate mainly from angiosperms, especially tropical hardwoods of the family *Dipterocarpaceae*.

resinite macerals of the liptinite group derived from plant resins, e.g. amber.

resolution the efficiency of a column to separate adjacent chromatographic peaks. Baseline separation indicates that two adjacent peaks are resolved (separated) completely and that the valley between the peaks reaches background (low) levels. In mass spectrometry, resolution refers to the ratio of mass to the difference between two adjacent masses ($M/\Delta M$) that a mass spectrometer can just separate completely. Low and high resolutions are ~ 1000 and >2000 (no units).

respiration oxidative catabolic reactions in living organisms that yield energy (as ATP), where either organic or inorganic substrates are primary electron donors and oxygen or other compounds are the ultimate electron acceptors.

retention time the time required for an injected compound to pass through a chromatographic column. An unknown compound can be identified provisionally if it has the same retention time as a standard compound when injected into the same chromatographic column under the same conditions.

retinoids oxygenated derivatives of 3,7-dimethyl-1-(2,6,6-trimethylcyclohex-1-enyl)nona-1,3,5,7-tetraene.

retrograde condensate a liquid petroleum formed by condensation due to reduced pressure and temperature.

reverse fault a fault in which the hanging wall appears to have moved upward relative to the footwall. Common in compressional regimes.

reverse-phase chromatography the most common high-performance liquid chromatography (HPLC) mode, which uses hydrophobic packing, such as octadecyl- or octylsilane phases bonded to silica, or neutral polymeric beads. The mobile phase is usually water and a water-miscible organic solvent, such as methanol or acetonitrile.

rhizoid a root-like structure that anchors an organism to a substrate.

ribonucleic acid (RNA) a polymer of nucleotides involved in protein synthesis.

ribonucleic acid (RNA) a biochemical consisting of long usually single-strand chains of alternating phosphate and ribose units with the nitrogen bases adenine, guanine, cytosine, and uracil bound to the ribose. Found in all organisms, RNA is used in protein synthesis, transmission of genetic material, and regulation of cellular processes.

ribosomes small organelles consisting of rRNA and enzymes where proteins are manufactured from amino acids in the cytoplasm of living organisms.

ribosomal RNA (rRNA) a type of ribonucleic acid (RNA) found in the ribosome that participates in protein synthesis. 16S rRNA is a large polynucleotide (~ 1500 bases) that is part of the ribosome of prokaryotes. Comparisons of 16S rRNA base sequences are used to show evolutionary relationships among organisms. The eukaryotic counterpart is 18s rRNA.

RIC see Reconstructed ion chromatogram.

rift a narrow depression in a rock caused by cracking or splitting.

rift valley a valley formed by faulting, usually involving a central fault block that moves downward in relation to adjacent blocks.

R_m see Vitrinite reflectance.

RNA see Ribonucleic acid.

R_o see Vitrinite reflectance.

Rock-Eval a commercially available pyrolysis instrument used as a rapid screening tool in evaluating the quantity, quality, and thermal maturity of rock samples.

root nodule a specialized structure in plant roots, such as legumes, where bacteria fix diatomic nitrogen, making it available for the plant.

rRNA see Ribosomal RNA.

rudists specialized Mesozoic bivalves, commonly having one valve in the shape of a horn coral and covered by the other valve in the form of a lid.

ruminant a herbivorous ungulate, e.g. cow.

ruthenium tetroxide degradation oxidation of organic matter by RuO_4 disrupts aromatic systems by producing a carboxyl group at the point of attachment of the side chains or groups. Gas chromatography/mass spectrometry (GCMS) or other analysis of the products of RuO_4 oxidation provides evidence on principal structural components in these complex materials (e.g. Blokker et al., 2001).

S configuration see R or S configuration.

S1, S2, S3 Rock-Eval pyrolysis parameters, where S1 is volatile organic compounds (mg HC/g TOC), S2 is organic compounds generated by cracking of the kerogen (mg HC/g TOC), and S3 is organic carbon dioxide generated from the kerogen up to $390\,°C$ (mg CO_2/g TOC) (TOC is total organic carbon) (see Figure 4.5).

sabkha an evaporitic environment of sedimentation formed under arid to semi-arid conditions, usually on restricted coastal plains. Characterized by evaporite-salt, tidal-flood, and aeolian deposits along many modern coastlines, e.g. the Persian Gulf and the Gulf of California.

salt dome a structural dome in sedimentary strata resulting from the upward flow of a large body of salt.

sand particles measuring 0.05–2.0 mm in diameter.

sandstone sedimentary rock composed of sand-sized particles, typically quartz.

sapropel putrefying organic matter, mostly algae and microbes, deposited in an anoxic setting. The word "sapropel" originates from the Greek *sapros*, meaning rotten.

saprophyte an organism that feeds on dead organic material.

Sarcopterygii lobe-finned bony fish, including air-breathing crossopterygian fish.

saturates (saturate fraction) non-aromatic organic compounds in petroleum. Includes normal and branched alkanes (paraffins) and cycloalkanes (naphthenes).

Saurischia an order of dinosaurs with triradiate pelvic structures, including both the gigantic herbivorous sauropods and the carnivorous theropods.

scan mode a gas chromatography/mass spectrometry (GCMS) operating mode in which the detector records the entire mass range per unit of time (e.g. 50–600 amu/3 s), resulting in a spectrum of masses for each analyzed peak.

schist a foliated metamorphic rock rich in mica.

Schulz–Flory distribution a mixture of hydrocarbons dominated by lighter components, which shows a linear decrease on a semi-log plot of hydrocarbon concentration (e.g. mole%) versus carbon number.

screening rejection of inappropriate samples using rapid, inexpensive analyses to allow upgrading of other samples for more detailed analysis. Large numbers of potential source rocks can be screened using Rock-Eval pyrolysis and total organic carbon (TOC) analysis before further study. Similarly, benchtop gas chromatography/mass spectrometry (GCMS) in selected ion monitoring (SIM) mode can be used to screen petroleum samples for their general biomarker composition before more detailed analysis, such as gas chromatography/mass spectrometry/mass spectrometry (GCMS/MS).

seafloor spreading the process in which oceanic crust rises from deep in the Earth by convective upwelling of magma along mid-oceanic ridges or rift systems and spreads outward at ∼1–10 cm/year, carrying tectonic plates.

seal rock (cap rock) an impervious layer of rock that overlies a reservoir rock, thus preventing leakage of petroleum to the surface.

secondary migration migration of petroleum through permeable rocks (carrier beds) after expulsion from the source rock. Unlike primary migration (expulsion), secondary migration generally involves long distances from tens of meters to hundreds of kilometers.

secondary recovery enhanced recovery of oil or gas from a reservoir beyond the oil or gas that could be recovered by normal pumping operations. Secondary recovery techniques involve maintaining or enhancing reservoir pressure by injecting water, gas, or other substances into the formation.

sediment various materials deposited by water, wind, or glacial ice, or by precipitation from water by chemical or biological action, e.g. clay, sand, carbonate.

sedimentary rock rock formed by lithification of sediment transported or precipitated at the Earth's surface and accumulated in layers. These rocks can contain fragments of older rock transported and deposited by water, air, or ice, chemical rocks formed by precipitation from solution, and remains of plants and animals.

sedimentation the process of deposition and accumulation of sedimentary layers.

seismic exploration an exploration technique that provides information on subsurface structures by monitoring seismic waves from a surface energy source (e.g. dynamite or vibrating heavy equipment), which reflect from the subsurface structures. Used to determine the best places to drill for hydrocarbons.

seismic reflection profiling three-dimensional analysis of structures beneath the Earth's surface using seismic waves. See also Seismic exploration.

selected ion monitoring (SIM) mass spectrometric monitoring of a specific mass/charge (m/z) ratio or limited number of ratios. For example, SIM for m/z 217 results in a mass chromatogram dominated by steranes. Using the SIM method of monitoring one or a few masses results in better sensitivity than can be obtained using the full-scan mode. Also known as multiple ion detection (MID).

series the time-rock term representing rocks deposited or emplaced during a geologic epoch. A series is a subdivision of a system.

serpentinization low-temperature hydration and alteration of mafic minerals, such as olivine, to form serpentine-group minerals.

sesquiterpanes a class of saturated biomarkers constructed from three isoprene subunits ($\sim C_{15}$).

sesterterpanes a class of saturated biomarkers constructed from five isoprene subunits ($\sim C_{25}$).

shale a fine-grained sedimentary rock formed by lithification of mud that is fissile or fractures easily along bedding planes and is dominated by clay-sized particles. In this text, shales, calcareous shales, and limestones contain <25%, 25–50%, and >50 wt.% carbonate, respectively.

shear a frictional force that tends to cause contiguous parts of a body to slide relative to each other in a direction parallel to their plane of contact.

sheath tubular structure formed around a chain of cells or around a bundle of filaments.

shelf the physiographic area between the shoreline and the slope.

silica gel an amorphous, porous packing material common in liquid chromatography composed of siloxane and silanol groups.

silicalite a term that has been used incorrectly to designate high Si/Al ZSM-5 zeolite (Flanigen et al., 1978). ZSM-5 and so-called silicalite are equivalent (Fyfe et al., 1982). Budiansky (1982) describes the patent dispute that arose due to the use of this term. It is now generally recognized that the Si/Al ratio for synthetic ZSM-5 is limited only by Al impurities in the starting materials.

silicate a compound whose crystal structure contains SiO_4 tetrahedra, either isolated or joined through one or more of the oxygen atoms to form groups, chains, sheets, or three-dimensional structures with metallic elements.

silicic rich in silica.

siliciclastic clastic, non-carbonate rocks dominated by quartz or silicate minerals.

silicon tetrahedron an atomic structure in silicate minerals that consists of a central silicon atom linked to four oxygen atoms placed symmetrically around the silicon at the corners of a tetrahedron (SiO_4).

sill (1) a small, concordant (injected between layers of sedimentary rock) body of intrusive igneous rock; (2) a subaqueous elevated area that partially isolates the water in two basins.

siloxane the Si–O–Si bond as in polysiloxane; a principal bond in silica gel or for attachment of a silylated compound or bonded phase.

silt a rock fragment or particle 0.004–0.063 mm in diameter.

SIM see Selected ion monitoring.

slime mold non-phototrophic eukaryotic microbes that lack cell walls.

soil weathered, unconsolidated minerals and organic matter on the surface of the Earth that serves as a growth medium for organisms.

solid bitumen includes pyrobitumen and bitumen formed by non-thermal processes, such as deasphalting and biodegradation (Curiale, 1986).

solute the dissolved component of a mixture that is to be separated in a chromatographic column.

sorting a measure of the size uniformity of particles in a sediment or sedimentary rock.

source rock an essential part of the petroleum system that consists of fine-grained, organic-rich rock that could generate (potential source rock) or has already generated (effective source rock) significant amounts of thermogenic or microbial petroleum. An effective source rock must satisfy requirements as to quantity, quality, and thermal maturity of the organic matter.

sour crude crude oil with high sulfur content (>0.5 wt.%).

sour gas natural or associated gas with abundant hydrogen sulfide.

Soxhlet extraction a common method for removing soluble organic compounds from crushed rock using a hot organic solvent (e.g. dichloromethane) or solvent mixture that is refluxed through the sample.

species a population of individuals sufficiently different from others to be recognized as a distinct group and that in nature breed only with one another.

specific gravity a measure of the density of a material compared with that of water (g/cm^3).

spent source rock source rock that generated and expelled petroleum but that is now postmature, with no further generative potential.

sphenopsids (also called sphenophytes) sponge-bearing plants that were common during the late Paleozoic Era and were characterized by articulated stems with leaves borne in whorls at the nodes.

sphingolipids membrane lipids that function in cell-to-cell communication, signal transduction, immunorecognition, Ca^{2+} mediation, and the physical state of membranes and lipoproteins. Sphingolipids are derivatives of sphingosine, a long-chain unsaturated amino alcohol in nervous tissue and cell membranes. It has a similar structure to a glycerol-based phospholipid, having a polar head group and two hydrophobic hydrocarbon chains (one is the sphingosine and the other is a fatty acid chain).

spiking   see Co-injection.

spore coloration index   see Thermal alteration index.

spores   specialized asexual reproductive cells that germinate without uniting with other cells, as occurs in bacteria, ferns, and mosses.

stable carbon isotope ratio   relative amount of ^{13}C versus ^{12}C (non-radioactive isotopes) in organic matter. Generally used to show relationships between oils or between oils and source rocks.

stage   the time-rock unit equivalent to an age. A subdivision of a series.

standard temperature and pressure (STP)   0 °C and one atmosphere pressure.

stationary phase   see Gas chromatography.

sterane isomerization   stereochemical conversions between the biological and geological configuration at several asymmetric centers, which are used as indicators of thermal maturity. Includes 20S/(20S + 20R) and $\beta\beta/(\beta\beta + \alpha\alpha)$ parameters.

steranes   a class of tetracyclic, saturated biomarkers constructed from six isoprene subunits ($\sim C_{30}$) (see Figure 2.15). Steranes originate from sterols, which are important membrane and hormone components in eukaryotic organisms. Most commonly used steranes are in the range C_{26}–C_{30} and are detected using m/z 217 mass chromatograms.

sterane triangle   plot of %C_{27}, C_{28}, and C_{29} steranes on a ternary diagram used for correlation of oils and bitumens (e.g. see Figure 13.38). Relative location on plot can be used to infer organic matter input. For example, abundant %C_{29} usually, but not always, indicates major input from higher-plant sterols.

stereochemistry   the three-dimensional relationship between atoms in molecules.

stereoisomers   compounds that have the same molecular formula and the same linkage between atoms but different spatial arrangements of the atoms, typically around an asymmetric carbon atom. Stereoisomers include enantiomers (mirror-image structures) and diastereomers (epimers), which differ at certain asymmetric centers but are identical at others.

steroids   similar to steranes, except that other elements in addition to carbon and hydrogen occur, especially oxygen in alcohol groups. Certain structural elements may be rearranged or missing compared with sterols. For example, unlike sterols (see Figure 2.31) triaromatic steroid hydrocarbons lack the α-methyl group at C-10. Further, the methyl group at C-13 in sterols is rearranged to C-17 in triaromatic steroids (see Figure 13.109).

stigmastane   the C_{29} ethylcholestane.

stoichiometry   a branch of science that deals with the laws of definite proportions and conservation of matter during chemical reactions, e.g. the predictable quantities of substances that enter into and are produced by chemical reactions. For example, carbon in saturated organic compounds is always bound to four substituents.

STP   see Standard temperature and pressure.

straight-run refinery products   distillation cuts from crude-oil feedstock that include gasoline (C_5–C_{10}), kerosene (C_{11}–C_{13}), diesel (C_{14}–C_{18}), heavy gas oil (C_{19}–C_{25}), lubricating oil (C_{26}–C_{40}), and residuum ($>C_{40}$) (Hunt, 1996). The carbon number ranges are approximate and differ depending on specific distillation conditions (e.g. see Table 5.1).

strain   (1) change in the shape of a body as a result of stress; a change in relative configuration of the particles of a substance. (2) A population of cells that descended from a single pure isolate.

stratification   layering in sedimentary rocks that results from changes in texture, color, or rock type.

stratigraphy   the study of sedimentary rock strata, including their age, form, distribution, lithology, fossil content, and other characteristics useful in interpreting their environment of origin and geologic history.

stratum   a tabular or sheet-like bed of sedimentary rock that is distinct from other layers above and below.

stromatolites   Precambrian to Recent laminated accumulations of calcium carbonate with rounded, branching, or frondose shape that form as a result of the metabolic activity of marine cyanobacterial mats, commonly in the high intertidal to low supratidal zones.

stromatoporids   extinct reef-building organisms believed to have affinities with the Porifera and noted for the large, often laminated masses constructed by the colonies.

structural isomers   molecules that have the same molecular formula but different linkages between atoms, e.g. *n*-butane and isobutane (see Figure 2.7), and

2-methyl,3-ethyl-heptane and 2,6-dimethyloctane (see Figure 2.14). Stereoisomers are a special form of structural isomer.

subaerial exposed near to or at a sediment surface above sea level.

subduction zone an inclined planar zone defined by a high frequency of earthquakes that is thought to locate the descending leading edge of a moving oceanic plate.

sublimation passing from a gaseous to a solid state without going through the liquid state.

suboxic refers to a water column or sediment with molecular oxygen contents of 0–0.2 ml/l of interstitial water (see Table 1.5).

subsalt refers to rock formations lying beneath long, horizontal layers of salt. These rock formations may contain hydrocarbons.

subsidence sinking or gradual downward settling of the Earth's surface with little or no horizontal motion.

substrate (1) a nutrient or medium required by an organism to live and grow. (2) The surface or underlying rocks to which organisms are attached.

supercritical fluid chromatography a method that uses a supercritical fluid as the mobile phase. This allows separation of compounds that cannot be handled by liquid chromatography because of detection problems or by gas chromatography because of the lack of volatility.

sweet crude crude oil with low sulfur content (<0.5 wt.%).

sweet gas natural gas that contains hydrocarbons with little or no hydrogen sulfide.

sweet oil a low-sulfur (<0.5 wt.%) crude oil.

symbiosis a permanent, mutually beneficial association of different organisms. Symbiosis differs from parasitism, in which one organism benefits to the detriment of its host. See also Endosymbiosis.

syncline a concave-upward fold in rock that contains stratigraphically younger strata toward the center.

synthetic fuels combustible fluids made from coal or other hydrocarbon-containing substances.

synthetic natural gas gases made from coal or other hydrocarbon-containing substances.

synthetic oil liquid fuel made from coal and other hydrocarbon-containing substances.

TAI see Thermal alteration index.

tailing the chromatographic phenomenon where, unlike a normal Gaussian peak, the peak has a trailing edge.

tar a black liquid or solid residue produced by petroleum refining. The terms "tar mat" and "tar sand" refer to natural bitumens.

taxon (*pl.* **taxa**) a unit in the taxonomic classification, e.g. phylum, class, order, family.

taxonomy the science of classifying and naming organisms.

TBR see Trimethylnaphthalene ratio.

tectonics the structural behavior of the Earth's crust.

TCI see Transmittance color index.

teleosts the most advanced of the bony fish, characterized by thin, rounded scales, completely bony internal skeleton, and symmetric tail. Teleosts range from Cretaceous to Recent.

telinite a maceral of the vitrinite group that has a distinct cellular structure of woody tissue.

temperature programming changing the column temperature with time during the gas chromatographic separation.

terminal electron-acceptor an external oxidant, usually oxygen, that accepts electrons as they exit from the electron-transport chain.

terpanes a broad class of complex branched, cyclic alkane biomarkers, including hopanes and tricyclic compounds, commonly monitored using m/z 191 mass chromatograms.

terpenoids (isopentenoids) a broad class of complex branched, cyclic biomarkers composed of isoprene (C_5) subunits, including mono- (two isoprene units), sesqui- (three), di- (four), sester- (five), tri- (six), tetra- (eight), and higher terpenoids. Terpenoids include the saturated terpanes (hydrocarbons) and compounds that may contain double bonds or other elements (in addition to carbon and hydrogen), such as oxygen. Certain structural units may be rearranged or missing compared with terpenes and terpanes.

terrane an area of crust with a distinct assemblage of rocks (as opposed to terrain, which implies topography, such as rolling hills or rugged mountains).

terrestrial pertaining to the Earth. The term "terrestrial" is sometimes used to refer to dry land, rather than marine, fluvial, or lacustrine settings. However, this is confusing because the antonym of terrestrial is extraterrestrial. Organic matter that originated on land is best described using the term "terrigenous."

terrigenous refers to land rather than marine, fluvial, or lacustrine settings. Terrigenous organic matter can be deposited in any of these settings. The term "terrestrial" is commonly used as a synonym for terrigenous. However, this is confusing because the antonym of terrestrial is extraterrestrial.

Tethys Seaway an east–west trending seaway between Laurasia and Gondwana during Paleozoic and Mesozoic time from which arose the Alpine–Himalayan mountain ranges.

tetrapyrrole pigments compounds required for photosynthesis that contain a macrocyclic nucleus composed of four linked pyrrole (nitrogen-containing) rings, i.e. chlorophylls. Porphyrins contain a tetrapyrrole nucleus and are degradation products of these pigments.

tetraterpanes a class of saturated biomarkers constructed from eight isoprene subunits ($\sim C_{40}$).

texture a property of rocks relating to size, size variability, rounding or angularity, and orientation of mineral grains.

thecodonts an order of primarily Triassic reptiles considered to be the ancestral archosaurians.

theoretical plate a concept that relates chromatographic separation to the theory of distillation and is a measure of column efficiency. The length of column relating to this concept is called height equivalent to a theoretical plate (HETP).

therapods the carnivorous saurischian dinosaurs.

therapsids an order of mammal-like reptiles.

thermal alteration index (TAI) various maturity scales based on changes in the color of spores and pollen from yellow to brown to black with thermal maturity. Although this TAI scale does not correspond to that of Staplin (1969), it was designed to correlate linearly with vitrinite reflectance, e.g. 0.4% and 0.7% vitrinite reflectances are 2.4 and 2.7, respectively, on the TAI scale.

thermal maturity the extent of heat-driven reactions that convert sedimentary organic matter into petroleum and finally to gas and graphite. Three levels of maturity are early or low, mid or peak, and late or high. Different geochemical scales including vitrinite reflectance, pyrolysis T_{max}, and biomarker maturity ratios indicate the level of thermal maturity of organic matter. Because common thermal maturity parameters are measured using irreversible reactions, uplifted source rocks may show higher maturity indices than expected based on their current burial depth.

thermocline a layer of water in a lake or ocean where the temperature decrease with depth is greater than that of the underlying and overlying water. Thermoclines may be seasonal or permanent.

thermodynamics the physics and chemistry of reactions and chemical stability.

thermogenic gas hydrocarbon gases generated by the thermal breakdown of organic matter.

thermophile an organism whose optimum temperature for growth is between 45 and 85°C.

thin-layer chromatography (TLC) separation of organic compounds by movement of a solvent through a thin layer of stationary phase coated on a glass plate. The glass plate is positioned vertically in a tray of solvent (mobile phase) so that the solvent rises upward by capillary action. The sample is placed near the bottom edge of the plate, above the original level of solvent. Upward movement of solvent results in partitioning of components between the mobile and stationary phases. Because it is less time-intensive, most laboratories use column chromatography and high-performance liquid chromatography (HPLC) rather than TLC.

thrust fault a low-angle reverse fault, with inclination of fault plane generally <45°.

TIC see Reconstructed ion chromatogram.

till a glacial deposit of unconsolidated, unsorted, unstratified angular fragments of rock.

tillite lithified rock composed mainly of poorly sorted fragments ranging in size and shape (e.g. clay, silt, sand, gravel, and boulders) that were deposited by a glacier.

TLC see Thin-layer chromatography.

Tm see Ts/Tm

T_{max} a Rock-Eval pyrolysis thermal maturity parameter based on the temperature at which the maximum amount of pyrolyzate (S2) is generated from the kerogen in a rock sample. The beginning and end of the oil-generative window approximately correspond to T_{max} of 435°C and 470°C, respectively (see Figure 4.5) (Peters, 1986).

TNR see Trimethylnaphthalene ratio.

TOC see Total organic carbon.

toluene methylbenzene. A key petrochemical and an organic solvent. Toluene and xylene are important components in unleaded gasoline.

topography the configuration of the land surface, including its relief and the position of natural and man-made features.

torbanite an oil-prone coal that consists mainly of algal remains, e.g. Permian torbanites in Australia and Carboniferous torbanites in the Midland Valley,

Scotland. See also Boghead coal. Although many torbanites contain recognizable remains of *Botryococcus braunii*, no biomarkers from *B. braunii* lipids have been reported, possibly due to intense microbial reworking or transformation of reactive precursors (Audino *et al.*, 2001).

total ion chromatogram see Reconstructed ion chromatogram.

total organic carbon (TOC) the quantity of organic carbon (excluding carbonate carbon) expressed as percentage weight of the rock. For rocks at a thermal maturity equivalent to vitrinite reflectance of 0.6% (beginning of oil window), TOC can be described as follows: poor, TOC <0.5 wt.%; fair, TOC = 0.5–1 wt.%; good, TOC = 1–2 wt.%; very good, TOC >2 wt.% (see Table 4.1). TOC decreases with maturity. TOC × 1.22 is commonly equated to total organic matter (TOM). The resulting TOM is approximate because the 1.22 factor assumes certain amounts of oxygen, nitrogen, and sulfur are present.

total organic matter (TOM) see Total organic carbon.

trace fossils see Ichnofossils.

transformation ratio the difference between the original hydrocarbon potential of a sample before maturation and the measured hydrocarbon potential divided by the original hydrocarbon potential. Ranges from 0 to 1.0.

transgression advance of the sea on to land areas that shifts the boundary between marine and non-marine deposition (or between deposition and erosion) toward the edges of a marine basin. Transgressions can be caused by tectonic subsidence of the land or a eustatic rise in sea level.

transmittance color index (TCI) an optical maturity measurement of amorphous kerogen that complements other optical maturity parameters, such as vitrinite reflectance, thermal alteration index (TAI), and CAI. TCI covers the range of petroleum generation and preservation (Robison *et al.*, 2000). TCI is obtained by analysis of white light from a 100-W 6-V tungsten lamp attached to a photometric microscope.

transpiration evaporation of water from plant leaves.

trap a geometric arrangement of reservoir rock and overlying or updip seal rock that allows the accumulation of oil or gas, or both, in the subsurface. Traps may be structural (e.g. domes, anticlines), stratigraphic (pinchouts, permeability changes), or combinations of both (e.g. Levorsen, 1967).

triaromatic steroids a class of biomarkers containing three fused aromatic rings and one five-member naphthene ring (naphthenophenanthrenes). Probably derived from monoaromatic steroids during maturation (e.g. see Figure 13.109).

tricarboxylic acid cycle (Krebs cycle) a series of metabolic reactions in which pyruvate is oxidized to carbon dioxide, also forming NADH, which allows ATP production.

tricyclic terpanes the most prominent tricyclic terpanes (cheilanthanes, 14-alkyl-13-methylpodocarpanes) range from C_{19} to more than C_{54} and consist of three six-member rings with an isoprenoid side chain (Peters, 2000). Abundant tricyclic terpanes commonly correlate with high paleolatitude *Tasmanites*-rich rocks, suggesting an origin from these algae, although other sources are possible. Tricyclic terpane/hopane increases with thermal maturity of petroleum.

trilobites Paleozoic marine arthropods of the class Crustacea that show longitudinal and transverse division of the carapace into three parts, or lobes.

trimethylnaphthalene ratio a maturity parameter defined by different authors as follows: TBR = 1,3,6-TMN/1,2,4-DMN (Fisher *et al.*, 1998); TNR-1 = 2,3,6-TMN/(1,4,6-TMN + 1,3,5-TMN) (Alexander *et al.*, 1985).

triterpanes a class of saturated biomarkers constructed from six isoprene subunits ($\sim C_{30}$) (see Figure 2.15).

trophic level refers to nutrients in various organisms along a food chain, ranging from the primary autotrophs to the predatory carnivorous animals.

Ts/Tm ratio of two C_{27} hopanes: 18α-22,29,30-trisnorneohopane (Ts) (previously called trisnorhopane-II) and 17α-22,29,30-trisnorhopane (Tm). Ts/Tm and Ts/(Ts + Tm) depend on both source and maturity (Moldowan *et al.*, 1986).

tsunami a large sea wave generated by an earthquake or underwater landslide.

tuff volcanic ash that was consolidated into rock.

turbidites sediment deposited from a turbidity current and characterized by graded bedding and generally poor sorting.

turbidity scattering of light due to fine, suspended particulate matter, such as clay, in water.

turbidity current a submarine flow of water and suspended sediment that is denser than surrounding water and thus moves downward.

two-dimensional gas chromatography see Multidimensional chromatography.

type I kerogen highly oil-prone organic matter showing Rock-Eval pyrolysis hydrogen indices over 600 mg hydrocarbon/g total organic carbon (TOC) when thermally immature. Contains algal and bacterial input dominated by amorphous liptinite macerals. Common in, but not restricted to, lacustrine settings.

type II kerogen oil-prone organic matter showing Rock-Eval pyrolysis hydrogen indices in the range 400–600 mg hydrocarbon/g total organic carbon (TOC) when thermally immature. Contains algal and bacterial organic matter dominated by liptinite macerals, such as exinite and sporinite. Common in, but not restricted to, marine settings.

type IIS kerogen composition similar to type II kerogen, but sulfur-rich. Type IIS kerogens contain unusually high organic sulfur (8–14 wt.%, atomic S/C ≥ 0.04) and appear to begin to generate oil at lower thermal exposure than typical type II kerogens with <6 wt.% sulfur (Orr, 1986).

type III kerogen gas-prone organic matter showing Rock-Eval pyrolysis hydrogen indices in the range 50–150 mg hydrocarbon/g total organic carbon (TOC) when thermally immature. Contains higher-plant organic matter dominated by vitrinite macerals. Common in, but not restricted to, paralic marine settings.

type IV kerogen inert organic matter showing Rock-Eval pyrolysis hydrogen indices below 50 mg hydrocarbon/g total organic carbon (TOC) in immature rocks. Contains predominantly higher-plant organic matter that was recycled or oxidized extensively during deposition.

UCM see Unresolved complex mixture.

ultralaminae bundles of very thin lamellae (~10–30 nm) observed under transmission electron microscopy that are widespread in so-called amorphous kerogens (Largeau et al., 1990).

ultramafic igneous rock dominated by mafic minerals, such as augite and olivine.

unassociated gas natural gas in reservoirs that do not contain crude oil.

unconformity a surface that separates an overlying younger rock formation from an underlying formation and represents a break in deposition (non-deposition) or erosion of formerly deposited material, or both (e.g. see Figure 9.10). Unconformities represent gaps in the geologic record.

uniformitarianism a principle that suggests that Earth history can be interpreted using natural laws, i.e. the Earth is old and its geologic features can be explained by the action of present-day geologic processes operated over long periods of time.

unit cell the smallest repeating unit of a crystal that has the properties and symmetry of that crystal.

unresolved complex mixture (UCM) the hump on gas chromatograms of biodegraded (e.g. see Figure 4.22) or low-maturity crude oils (see Figure 4.25). Consists of complex organic compounds not resolved by the column (Gough and Rowland, 1990). Becomes more pronounced in biodegraded oils, probably because it contains abundant T-branched alkanes, such as 2,6,10,14-tetramethyl-7-(3-methylpentyl)-pentadecane (C_{25} highly branched isoprenoid).

unsaturated describes compounds that contain one or more double or triple bonds, such as olefins (alkenes). Saturation of the double bond in ethylene (C_2H_4) results in ethane (C_2H_6). Most unsaturated organic compounds are unstable in natural petroleum, except for aromatics such as benzene.

upstream describes oil and natural gas exploration and production activities.

upwelling upward movement of shallow waters due to wind. Upwelling can carry nutrients required for algal blooms, especially on the western coast of continents (Demaison and Moore, 1980).

urea adduction a procedure used to separate *n*-alkanes from other organic compounds in the saturate fraction of petroleum. Used to concentrate biomarkers for more reliable gas chromatography/mass spectrometry (GCMS) analysis (e.g. see Figure 13.26). The method is typically applied to samples with low biomarkers, such as extremely waxy oils and condensates.

valence the combining capacity of an element for other elements based on the number of electrons available for chemical bonding. For example, carbon has a valence of four, indicating that it will form four covalent bonds with other elements.

van Krevelen diagram a plot of atomic H/C versus O/C originally used to characterize the compositions of coals but now also used to describe different types of kerogen (see Type I kerogen, Type II kerogen, Type III kerogen, and Type IV kerogen) in rocks (see Figure 4.4). Van

Krevelen-type plots can be made using hydrogen index (HI) versus oxygen index (OI) from Rock-Eval pyrolysis (Peters, 1986).

varve a lamina or pair of laminae (thin sedimentary layers) that represents the depositional record of a single year.

vascular plants higher land plants that have a system of vessels and ducts for distributing moisture and nutrients.

virus a submicroscopic infective agent capable of growth only within a living cell. Viruses typically consist of a protein coat surrounding a nucleic acid core.

viscosity the resistance of a fluid to flow. Viscosity decreases with increasing temperature, as measured in centipoise at a given temperature.

vitrinite a group of gas-prone macerals derived from land-plant tissues. Particles of vitrinite are used for vitrinite reflectance (R_o) determinations of thermal maturity.

vitrinite reflectance (R_o) a parameter for determining maturation of organic matter in fine-grained rocks (see Figure 14.3). Average R_o is based on measurements of at least 50–100 randomly oriented vitrinite phytoclasts in a polished slide of kerogen, coal, or whole rock (see Figure 4.12). Each measurement represents the percentage of incident light (546 nm) reflected from a phytoclast under an oil-immersion microscope objective, as measured by a photometer. Some microscopes have rotating stages that allow measurements of anisotropy. Thus, R_m, R_{max}, R_{min}, and R_r indicate mean, maximun, minimum, and random vitrinite reflectance, respectively.

void volume the total volume of mobile phase in the column, where the remainder of the column is taken up by packing material.

V/(V + Ni) porphyrin the ratio of vanadyl to nickel porphyrins, used for correlation of oils and bitumens. The ratio is higher for oils and bitumens derived from marine source rocks deposited under anoxic compared with oxic or suboxic conditions (see Figure 4.11). Low ratios may indicate oxic to suboxic or lacustrine deposition. The ratio is sometimes expressed as Ni/(Ni + V).

waterflooding a method of secondary recovery, whereby water is injected into an oil reservoir to force additional oil into the well bores of producing wells.

water injection a method of enhanced recovery in which water is injected into an oil reservoir to increase pressure and maintain or improve oil production.

water table the upper surface of the zone of water saturation in sediments. Below this surface, the pore space is filled by liquid water.

water washing contact by formation waters in reservoirs or during migration that causes removal of light hydrocarbons, aromatics, and other soluble compounds from petroleum. Biodegradation commonly accompanies water washing of petroleum because bacteria can be introduced from the water (Palmer, 1984; 1993).

wax the solid *n*-alkanes in petroleum. The term "waxy oil" generally indicates oil that contains abundant *n*-alkanes greater than nC_{25} (see Figure 4.21).

weathering the breakdown of rocks and other materials at the Earth's surface caused by mechanical action and reactions with air, water, and organisms. Weathering of seep oils or improperly sealed oil samples by exposure to air results in evaporative loss of light hydrocarbons. Biodegradation and water washing commonly accompany weathering.

well a hole drilled for the purpose of obtaining water, oil, gas, or other natural resources.

wet gas natural gas that contains ethane, propane, and heavier hydrocarbons and <98% methane/total hydrocarbons. The wet-gas zone occurs during catagenesis below the bottom of the oil window and above the top of the gas window (1.4–2.0% vitrinite reflectance).

wildcat an exploration well drilled in unproven territory to discover petroleum, without direct evidence of the contents of the underlying rock structure.

xanthophylls a class of carotenoids consisting of the oxygenated carotenes.

xylene an aromatic compound consisting of a benzene ring with methyl groups at C-1 and C-2 (orthoxylene), C-1 and C-3 (metaxylene), or C-1 and C-4 (paraxylene).

yeast a fungus consisting of single cells that multiply by budding or fission.

zygote the single diploid cell in eukaryotes that results from the fusion of two haploid gametes.

References

Abbott, G. D. and Maxwell, J. R. (1988) Kinetics of the aromatisation of rearranged ring-C monoaromatic steroid hydrocarbons. *Organic Geochemistry*, 13, 881–5.

Abbott, G. D. and Stott, A. W. (1999) Biomarker reaction kinetics during kerogen microscale pyrolysis. In: *Advances in Biochirality* (G. Pályi, C. Zucchi and L. Caglioti, eds.), Elsevier Science, Amsterdam, Netherlands, pp. 231–46.

Abbott, G. D., Lewis, C. A. and Maxwell, J. R. (1985) Laboratory models for aromatization and isomerization of hydrocarbons in sedimentary basins. *Nature*, 318, 651–3.

Abbott, G. D., Wang, G. Y., Eglinton, T. I., Home, A. K. and Petch, G. S. (1990) The kinetics of sterane biological marker release and degradation processes during hydrous pyrolysis of vitrinite kerogen. *Geochimica et Cosmochimica Acta*, 54, 2451–61.

Abdine, A. S., Meshref, W., Shahin, A. N., Garossino, P. and Shazly, S. (1992) Ramadan Field – Egypt, Gulf of Suez Basin. In: *Treatise of Petroleum Geology, Atlas of Oil and Gas Fields, Structural Traps*, Vol. VI (N. H. Foster and E. A. Beaumont, eds.), American Association of Petroleum Geologists, Tulsa, OK, pp. 113–39.

Abdullah, F. H. A. (2001) A preliminary evaluation of Jurassic source rock potential in Kuwait. *Journal of Petroleum Geology*, 24, 361–78.

Abdullah, F. H. and Connan, J. (2002) Geochemical study of some Cretaceous rocks from Kuwait: comparison with oils from Cretaceous and Jurassic reservoirs. *Organic Geochemistry*, 33, 125–48.

Abed, A. M. and Amireh, B. S. (1983) Petrography and geochemistry of some Jordanian oil shales from north Jordan. *Journal of Petroleum Geology*, 5, 261–74.

Abou-Zeid, D.-M., Müller, R.-J. and Deckwer, W.-D. (2001) Degradation of natural and synthetic polyesters under anaerobic conditions. *Journal of Biotechnology*, 86, 113–26.

Abrams, M. and Narimanov, A. A. (1997) Geochemical evaluation of hydrocarbons and their potential sources in the Western South Caspian Depression, Republic of Azerbaijan. *Marine Petroleum Geology*, 14, 451–68.

Abrams, M. A., Stanley, K. O. and Gilbert, D. (1998) Geochemical evaluation of hydrocarbons and their potential sources in the CIS. *American Association of Petroleum Geologists Annual Meeting Abstracts*, 1, A3.

Abrams, M. A., Apanel, A. M., Timoshenko, O. M. and Kosenkova, N. N. (1999) Oil families and their potential sources in the northeastern Timan Pechora Basin, Russia. *American Association of Petroleum Geologists Bulletin*, 83, 553–77.

Abyzov, S. S., Mitskevich, I. N. and Poglazova, M. N. (1998) Microflora of the deep glacier horizons of central Antarctica. *Microbiology*, 67, 451–8.

Ackerman, W. C., Boatwright, D. C., Burwood, R., *et al.* (1993) Geochemical analysis of selected hydrocarbon samples in the Douala Basin, Cameroon: implications for an oil-prone source rock. AAPG International Conference, The Hague, the Netherlands, October 17–20, 1993. *American Association of Petroleum Geologists Bulletin*, 77, 1604.

Adam, P., Schmid, J. C., Mycke, B., *et al.* (1993) Structural investigation of nonpolar sulfur cross-linked macromolecules in petroleum. *Geochimica et Cosmochimica Acta*, 57, 3395–419.

Adam, P., Schneckenburger, P. and Albrecht, P. (2000) Clues to early diagenetic sulfurization processes from mild chemical cleavage of labile sulfur-rich geomacromolecules. *Geochimica et Cosmochimica Acta*, 64, 3485.

Aeckersberg, F., Bak, F. and Widdel, F. (1991) Anaerobic oxidation of saturated hydrocarbons to CO_2 by a new type of sulfate-reducing bacterium. *Archives of Microbiology*, 156, 5–14.

Aeckersberg, F., Rainey, F. A. and Widdel, F. (1998) Growth, natural relationships, cellular fatty acids and metabolic adaptation of sulfate-reducing bacteria that utilize long-chain alkanes under anoxic conditions. *Archives of Microbiology*, 170, 361–9.

Ahlbrandt, T. S. (2001) The Sirte Basin Province of Libya: Sirte-Zelten total petroleum system. *U.S. Geological Survey Bulletin 2202-F*.

Ahmed, M., Smith, J. W. and George, S. C. (1999) Effects of biodegradation on Australian Permian coals. *Organic Geochemistry*, 30, 1311–22.

Ahmed, M., Schouten, S., Baas, M. and de Leeuw, J. W. (2001) Bound lipids in kerogens from the Monterey Formation, Naples Beach, California. In: *The Monterey Formation: From Rocks to Molecules* (C. M. Isaacs and J. Rullkötter, eds.), Columbia University Press, New York, pp. 189–205.

Aizenshtat, Z. (1973) Perylene and its geochemical significance. *Geochimica et Cosmochimica Acta*, 37, 559–67.

Akporiaye, E. E., Farrant, R. D. and Kirk, D. N. (1981) Deuterium incorporation in the backbone rearrangement of cholest-5-ene. *Journal of Chemical Research*, Synopses, 210–11.

Ala, M. A., Kinghorn, R. R. F. and Rahman, M. (1980) Organic geochemistry and source rock characteristics of the Zagros petroleum province. *Journal of Petroleum Geology*, 3, 61–9.

Alajbeg, A., Brodic-Jakupak, Z., Svilkovic, D., *et al.* (1996) Geochemical study of oils and oil source rock from the eastern Drava and Slavonija-Srijem depressions, Pannonian Basin, Croatia. *Geology of Croatia*, 49, 135–43.

Alam, M. and Pearson, M. J. (1990) Bicadinanes in oils from the Surma Basin, Bangladesh. *Organic Geochemistry*, 15, 461–4.

Albaigés, J. (1980) Identification and geochemical significance of long chain acyclic isoprenoid hydrocarbons in crude oils. In: *Advances in Organic Geochemistry 1979* (A. G. Douglas and J. R. Maxwell, eds.), Pergamon, New York, pp. 19–28.

Albaigés, J. and Albrecht, P. (1979) Fingerprinting marine pollutant hydrocarbons by computerized GC-MS. *International Journal of Environmental Analytical Chemistry*, 6, 171–90.

Albaigés, J., Borbon, J. and Walker, W., II (1985) Petroleum isoprenoid hydrocarbons derived from catagenetic degradation of archaebacterial lipids. *Organic Geochemistry*, 8, p. 293–7.

Albaigés, J., Algaba, J., Clavell, E. and Grimalt, J. (1986) Petroleum geochemistry of the Tarragona Basin (Spanish Mediterranean off-shore). *Organic Geochemistry*, 10, 441–50.

Alberdi, M. and Lafargue, E. (1993) Vertical variations of organic matter content in Guayuta Group (Upper Cretaceous), Interior Mountain Belt, Eastern Venezuela. *Organic Geochemistry*, 20, 425–36.

Alberdi, M., Moldowan, J. M., Peters, K. E. and Dahl, J. E. (2001) Stereoselective biodegradation of tricyclic terpanes in heavy oils from the Bolivar Coastal Fields, Venezuela. *Organic Geochemistry*, 32, 181–91.

Alberdi-Genolet, M. and Tocco, R. (1999) Trace metals and organic geochemistry of the Machiques Member (Aptian-Albian) and La Luna Formation (Cenomanian-Campanian), Venezuela. *Chemical Geology*, 160, 19–38.

Albrecht, P. and Ourisson, G. (1969) Triterpene alcohol isolation from oil shale. *Science*, 163, 1192–3.

(1971) Biogenic substances in sediments and fossils. *Angewandte Chemie International Edition*, 10, 209–25.

Albrecht, P., Vandenbroucke, M. and Mandengue, M. (1976) Geochemical studies on the organic matter from the Douala Basin (Cameroon). 1. Evolution of the extractable organic matter and the formation of petroleum. *Geochimica et Cosmochimica Acta*, 40, 791–9.

Alexander, R., Kagi, R. and Woodhouse, G. W. (1981) Geochemical correlation of Windalia oil and extracts of Winning Group (Cretaceous) potential source rocks, Barrow Subbasin, Western Australia. *American Association of Petroleum Geologists Bulletin*, 65, 235–50.

Alexander, R., Kagi, R. and Noble, R. (1983a) Identification of the bicyclic sesquiterpenes, drimane, and eudesmane in petroleum. *Journal of the Chemical Society*, Chemical Communications, 226–8.

Alexander, R., Kagi, R. I., Woodhouse, G. W. and Volkman, J. K. (1983b) The geochemistry of some biodegraded Australian oils. *Australian Petroleum Exploration Association Journal*, 23, 53–63.

Alexander, R., Kagi, R. I. and Sheppard, P. N. (1983c) Relative abundance of dimethylnaphthalene isomers in crude oils. *Journal of Chromatography*, 267, 367–72.

(1984) 1,8-Dimethylnaphthalene as an indicator of petroleum maturity. *Nature*, 308, 442–3.

Alexander, R., Kagi, R. I., Roland, S. J., Sheppard, P. N. and Chirila, T. V. (1985) The effects of thermal maturity on distributions of dimethylnaphthalenes and trimethylnaphthalenes in some ancient sediments and petroleum. *Geochimica et Cosmochimica Acta*, 49, 385–95.

Alexander, R., Cumbers, K. M. and Kagi, R. I. (1986a) Alkylbiphenyls in ancient sediments and petroleums. *Organic Geochemistry*, 10, 841–5.

Alexander, R., Strachan, M. G., Kagi, R. I. and van Bronswijk, W. (1986b) Heating rate effects on aromatic maturity indicators. *Organic Geochemistry*, 10, 997–1003.

Alexander, R., Fisher, S. J. and Kagi, R. I. (1988a) 2,3-Dimethylbiphenyl; kinetics of its cyclisation reaction and effects of maturation upon its relative concentration in sediments. *Organic Geochemistry*, 13, 833–7.

Alexander, R., Kagi, R. I., Toh, E. and van Bronswijk, W. (1988b) The use of aromatic hydrocarbons for assessment of thermal histories in sediments. In: *The North West Shelf, Australia: Proceedings of the Petroleum Exploration Society of Australia, Perth* (P. G. Purcell and R. R. Purcell, eds.), Petroleum Exploration Society of Australia, Perth, Australia, pp. 559–62.

Alexander, R., Larcher, A. V., Kagi, R. I. and Price, P. L. (1988c) The use of plant-derived biomarkers for correlation of oils with source rocks in the Cooper/Eromanga Basin system, Australia. *APEA Journal*, 28, 310–24.

Alexander, R., Kagi, R. I. and Toh, E. (1989) Organic reaction kinetics and the thermal histories of sediments. *Chemistry in Australia*, 56, 74–6.

Alexander, R., Kralert, P. G. and Kagi, R. I. (1992a) Kinetics and mechanism of the thermal decomposition of esters in sediments. *Organic Geochemistry*, 19, 133–40.

Alexander, R., Larcher, A. V., Kagi, R. I. and Price, P. L. (1992b) An oil-source correlation study using age-specific plant-derived aromatic biomarkers. In: *Biological Markers in Sediments and Petroleum* (J. M. Moldowan, P. Albrecht and R. P. Philp, eds.), Prentice-Hall, Englewood Cliffs, NJ, pp. 201–21.

Alexander, R., Kagi, R. I., Singh, R. K. and Sosrowidjojo, I. B. (1994) The effect of maturity on the relative abundances of cadalene and isocadalene in sediments from the Gippsland Basin, Australia. *Organic Geochemistry*, 21, 115–20.

Alexander, R., Kralert, P. G., Sosrowidjojo, I. B. and Kagi, R. I. (1997) Kinetics and mechanism of the thermal elimination of alkenes from secondary stanyl and triterpenyl esters: implications for sedimentary processes. *Organic Geochemistry*, 26, 391–8.

Algeo, T. J. and Scheckler, S. E. (1998) Terrestrial-marine teleconnections in the Devonian: links between the evolution of land plants, weathering processes, and marine anoxic events. *Philosophical Translations of the Royal Society of London*, 353, 113–30.

Algeo, T. J., Scheckler, S. E. and Maynard, J. B. (2001) Effects of the Middle to Late Devonian spread of vascular plants on weathering regimes, marine biotas, and global climate. In: *Plants Invade the Land: Evolutionary and Environmental Perspectives* (P. G. Gensel and D. Edwards, eds.), Columbia University Press, New York, pp. 213–36.

Al-Laboun, A. A. (1986) Stratigraphy and hydrocarbon potential of the Paleozoic succession in both Tabuk and Widyan basins, Arabia. In: *Future Petroleum Provinces of the World* (M. T. Halbouty, ed.), American Association of Petroleum Geologists, Tulsa, OK, pp. 373–97.

Allan, J. and Creaney, S. (1991) Oil families of the Western Canada Basin. *Bulletin of Canadian Petroleum Geology*, 39, 107–22.

Allegre, C. J., Birck, J. L., Capmas, F. and Courtillot, V. (1999) Age of the Deccan Traps using ^{187}Re–^{187}Os systematics. *Earth and Planetary Science Letters*, 170, 197–204.

Aloisi, G., Bouloubassi, I., Heijs, S. K., et al. (2002) CH_4-consuming microorganisms and the formation of carbonate crusts at cold seeps. *Earth and Planetary Science Letters*, 203, 195–203.

Alroy, J. (2003) Cenozoic bolide impacts and biotic change in North American mammals. *Astrobiology*, 3, 119–32.

Alsgaard, P. C. (1992) Eastern Barents Sea Late Palaeozoic setting and potential source rocks. In: *Norwegian Arctic Geology and Petroleum Potential, Petroleum Society (NPF) Special Publication 2*. (T. O. Vorren, E. Bergsager, O. A. Dahl-Stamnes, et al. eds.), Elsevier Science, Amsterdam, pp. 405–18.

Alsharhan, A. S. (2002) Sequence stratigraphy and source rock potential of Upper Jurassic Diyab Formation in the United Arab Emirates. Presented at the Annual Meeting of the American Association of Petroleum Geologists, March 10–13, 2002, Houston, TX.

Alsharhan, A. S. and Kendall, C. G. St. C. (1986) Precambrian to Jurassic rocks of Arabian Gulf and adjacent areas: their facies, depositional setting, and hydrocarbon habitat. *American Association of Petroleum Geologists Bulletin*, 70, 977–1002.

Alsharhan, A. S. and Salah, M. G. (1997) A common source rock for Egyptian and Saudi hydrocarbons in the Red Sea. *American Association of Petroleum Geologists Bulletin*, 81, 1640–59.

Alvarez, W. (2003) Comparing the evidence relevant to impact and flood basalt at times of major mass extinctions. *Astrobiology*, 3, 153–61.

Alvarez, W. and Muller, R. A. (1984) Evidence from crater ages for periodic impacts on the Earth. *Nature*, 308, 718–20.

Alvarez, L. W., Alvarez, W., Asaro, F. and Michel, H. V. (1980) Extraterrestrial cause for the Cretaceous Tertiary extinction. *Science*, 208, 1095–108.

Alvarez, W., Asaro, F., Michel, H. V. and Alvarez, L. W. (1982) Iridium anomaly approximately synchronous with terminal Eocene extinctions. *Science*, 216, 886–8.

Ambrose, G. (2000) The geology and hydrocarbon habitat of the Sarir Sandstone, SE Sirt Basin, Libya. *Journal of Petroleum Geology*, 23, 165–92.

Ambrose, G. J., Kruse, P. D. and Putnam, P. E. (2001) Exploration in a Middle Cambrian carbonate succession, Georgina Basin, Australia. Presented at the Annual Meeting of the American Association of Petroleum Geologists, June 3–6, 2001, Denver, CO.

Amelin, Y., Lee, D.-C., Halliday, A. N. and Pidgeon, R. T. (1999) Nature of the Earth's earliest crust from hafnium isotopes in single detrital zircons. *Nature*, 399, 252–5.

Amthor, J. E., Smits, W. and Nederlof, P. (1998) Prolific oil production from a source rock: the Athel silicilyte source-rock play in South Oman. Presented at the Annual Meeting of the American Association of Petroleum Geologists, May 17–20, 1998, Salt Lake City, UT.

Amthor, J. E., Frewin, N. L. and Ramseyer, K. (1999) Enigmatic origin of a late Pre-Cambrian laminated chert: have bacteria done it? Presented at the Annual Meeting of the American Association of Petroleum Geologists, April 11–14, 1999, San Antonio, TX.

Anadon, P., Cawley, S. J. and Juliá, R. (1988) Oil source rocks in lacustrine sequences from Tertiary Grabens, Western Mediterranean Rift System, Northeast Spain. *American Association of Petroleum Geologists Bulletin*, 72, 983.

Anadon, P., Juliá, R., Cabera, L., Roca, E. and Rosell, L. (1989) Lacustrine oil-shale basins in Tertiary grabens from Northeastern Spain (western European rift system). *Palaeogeography, Palaeoclimatology, Palaeoecology*, 70, A7–28.

Anders, E. (1988) A global fire at the Cretaceous-Tertiary boundary. *Chemical Geology*, 70, 118.

(1996) Evaluating the evidence for past life on Mars. *Science*, 274, 2119–21.

Anders, D. E. and Gerrild, P. M. (1984) Hydrocarbon generation in lacustrine rocks of Tertiary age, Uinta Basin, Utah – organic carbon, pyrolysis yield, and light hydrocarbons. In: *Hydrocarbon Source Rocks of the Greater Rocky Mountain Region* (J. Woodward, F. F. Meissner and J. L. Clayton, eds.), Rocky Mountain Association of Geologists, Denver, CO, pp. 513–529.

Anders, D. E. and Robinson, W. E. (1971) Cycloalkane constituents of the bitumen from Green River Shale. *Geochimica et Cosmochimica Acta*, 35, 661–78.

Anders, E., Dufresne, E. R., Hayatsu, R., et al. (1964) Contaminated meteorite. *Science*, 146, 1157–61.

Anders, D. E., Doolittle, F. G. and Robinson, W. E. (1973) Analysis of some aromatic hydrocarbons in a benzene-soluble bitumen from Green River Shale. *Geochimica et Cosmochimica Acta*, 37, 1213–28.

Anders, D. E., King, J. D. and Lubeck, C. (1985) Correlation of oils and source rocks from the Alaskan North Slope. In: *Alaska North Slope Oil-Rock Correlation Study: Analysis of North Slope Crude. AAPG Studies in Geology 20* (L. B. Magoon and G. E. Claypool, eds.), American Association of Petroleum Geologists, Tulsa, OK, pp. 281–303.

Anders, D. E., Palacas, J. G. and Johnson, R. C. (1992) Thermal maturity of rocks and hydrocarbon deposits, Uinta Basin, Utah. In: *Hydrocarbon and Mineral Resources of the Uinta Basin, Utah and Colorado* (T. D. Fouch, V. F. Nuccio and T. C. Chidsey, eds.), *Utah Geological Association*, Salt Lake City, UT, pp. 55–76.

Anderson, R. Y. and Kirkland, D. W. (1960) Origin, varves, and cycles of Jurassic Todilto Formation, New Mexico. *American Association of Petroleum Geologists Bulletin*, 44, 37–52.

Anderson, R. T. and Lovley, D. R. (2000) Hexadecane decay by methanogenesis. *Nature*, 404, 722–3.

Anderson, K. B. and Muntean, J. V. (2000) The nature and fate of natural resins in the geosphere. Part X. Structural characteristics of the macromolecular constituents of modern dammar resin and Class II ambers. *Geochemical Transactions*, 1, 1–9.

Anderson, P. C., Gardner, P. M., Whitehead, E. V., Anders, D. E. and Robinson, W. E. (1969) The isolation of steranes from Green River oil shales. *Geochimica et Cosmochimica Acta*, 33, 1304–7.

Andrusevich, V. E., Engel, M. H. and Zumberge, J. E. (2000) Effects of paleolatitude on the stable carbon isotope composition of crude oils. *Geology*, 28, 847–50.

Anka, Z., Callejón, A., Hernández, V., *et al.* (1998) The petroleum system in Barinas–Southern Andes, Venezuela: a stratigraphic and geochemical model. *American Association of Petroleum Geologists Bulletin*, 82, 1883–4.

Annweiler, E., Materna, A., Safinowski, M., *et al.* (2000) Anaerobic degradation of 2-methylnaphthalene by a sulfate-reducing enrichment culture. *Applied and Environmental Microbiology*, 66, 5329–33.

Aquino Neto, F. R., Restle, A., Connan, J., Albrecht, P. and Ourisson, G. (1982) Novel tricyclic terpanes (C_{19}, C_{20}) in sediments and petroleums. *Tetrahedron Letters*, 23, 2027–30.

Aquino Neto, F. R., Trendel, J. M., Restle, A., Connan, J. and Albrecht, P. A. (1983) Occurrence and formation of tricyclic and tetracyclic terpanes in sediments and petroleums. In: *Advances in Organic Geochemistry 1981* (M. Bjorøy, C. Albrecht, C. Cornford, *et al.*, eds.), John Wiley & Sons, New York, pp. 659–76.

Araújo, L. M., Trigüis, J. A., Cerqueira, J. R. and da Silva Freitas, L. C. (2000) The atypical Permian petroleum system of the Paraná Basin, Brazil. In: *Petroleum Systems of South Atlantic Margins* (M. R. Mello and B. J. Katz, eds.), American Association of Petroleum Geologists, Tulsa, OK, pp. 377–402.

Areshev, E. G., Dong, T. L., San, N. T. and Shnip, O. A. (1992) Reservoirs in fractured basement on the continental shelf of southern Vietnam. *Journal of Petroleum Geology*, 15, 451–64.

Arinobu, T., Ishiwatari, R., Kaiho, K. and Lamolda, M. A. (1999) Spike of pyrosynthetic polycyclic aromatic hydrocarbons associated with an abrupt decrease in $\delta^{13}C$ of a terrestrial biomarker at the Cretaceous-Tertiary boundary at Caravaca, Spain. *Geology*, 27, 723–6.

Armagnac, C., Kendall, C. G. S. C., Kuo, C., Lerche, I. and Pantano, J. (1988) Determination of paleoheat flux from vitrinite reflectance data, and from sterane and hopane isomer data. *Journal of Geochemical Exploration*, 30, 1–28.

Armanios, C. (1995) *Molecular sieving, analysis and geochemistry of some pentacyclic triterpanes in sedimentary organic matter*. Ph.D. thesis, Curtin University of Technology, School of Applied Chemistry, Perth, Australia.

Armanios, C., Alexander, R., Sosrowidjojo, I. M. and Kagi, R. I. (1995a) Identification of bicadinanes in Jurassic organic matter from the Eromanga Basin, Australia. *Organic Geochemistry*, 23, 837–43.

Armanios, C., Alexander, R., Kagi, R. I., Skelton, B. W. and White, A. H. (1995b) Occurrence of 28-nor-18α-oleanane in the hydrous pyrolysate of a lignite. *Organic Geochemistry*, 23, 21–7.

Armentrout, J. M., Peters, K. E., Sageman, B. B., Murphy, A. E. and Gardner, M. H. (1998) Stratigraphic hierarchy of organic carbon-rich siltstones in deep-water facies, Brushy Canyon Formation (Guadalupian), Delaware Basin, Texas. *American Association of Petroleum Geologists Bulletin*, 10, 1887.

Armstrong, P. A., Chapman, D. S., Funnell, R. H., Allis, R. G. and Kamp, P. J. J. (1996) Thermal modeling and hydrocarbon generation in an active-margin basin: Taranaki Basin, New Zealand. *American Association of Petroleum Geologists Bulletin*, 80, 1216–41.

Arouri, K. R., Yu, X., McKirdy, D. M. and Hill, T. (2000) Petroleum of the Cooper and Eromanga basins, Australia: a model of mixed sources. *American Association of Petroleum Geologists Bulletin*, 84, 1399.

Arpino, P., Albrecht, P. and Ourisson, G. (1972) Studies on the organic constituents of lacustrine Eocene sediments. Possible mechanisms for the formation of some geolipids related to biologically occurring triterpenoids. In: *Advances in Organic Geochemistry 1971* (H. R. von Gaertner and H. Wehner, eds.), Pergamon Press, Oxford, pp. 173–187.

Arro, H., Prikk, A. and Pihu, T. (1998) Calculation of composition of Estonian oil shale and its combustion products on the basis of heating value. *Oil Shale*, 15, 329–40.

Arthur, M. A., Schlanger, S. O. and Jenkyns, H. C. (1987) The Cenomanian/Turonian oceanic anoxic event II. Palaeoceanographic controls on organic matter production and preservation. In: *Marine Petroleum Source Rocks* (J. Brooks and A. J. Fleet, eds.), Geological Society of London, London, pp. 401–20.

Arthur, M. A., Dean, W. E. and Pratt, L. M. (1988) Geochemical and climatic effects of increased marine organic carbon burial at the Cenomanian/Turonian boundary. *Nature*, 335, 714–17.

Atlas, R. M. (1981) Microbial degradation of petroleum hydrocarbon: an environmental perspective. *Microbiology Review*, 45, 180–209.

Audino, M., Grice, K., Alexander, R., Boreham, C. J. and Kagi, R. I. (2001a) Unusual distribution of monomethylalkanes in *Botryococcus braunii*-rich samples. Origin and significance. *Geochimica et Cosmochimica Acta*, 65, 1995–2006.

Audino, M., Grice, K., Alexander, R. and Kagi, R. I. (2001b) Macrocyclic-alkanes: a new class of biomarker. *Organic Geochemistry*, 32, 759–63.

Audino, M., Grice, K., Alexander, R., Boreham, C. J. and Kagi, R. I. (2001c) Origin of macrocyclic alkanes in sediments and crude oils from the algaenan of *Botryococcus braunii*. *Geochimica et Cosmochimica Acta*, 65, 1995–2006.

Avron, M. and Ben-Amotz, A. (1992) *Dunaliella: Physiology, Biochemistry, and Biotechnology*. CRC Press, Boca Raton, FL.

Awramik, S. M., Schopf, J. W. and Walter, M. R. (1983) Filamentous fossil bacteria from the Archean of Western Australia. *Precambrian Research*, 20, 357–74.

(1984) Carbonaceous filaments from North Pole, Western Australia: are they fossil bacteria in Archean stromatolites? A discussion. *Precambrian Research*, 39, 303–9.

Ayala, F. J., Rzhetsky, A. and Ayala, F. J. (1998) Origin of the metazoan phyla: molecular clocks confirm paleontological estimates. *Proceedings of the National Academy of Science, USA*, 95, 606–11.

Ayres, M. G., Bilal, M., Jones, R. W., *et al.* (1982) Hydrocarbon habitat in main producing areas, Saudi Arabia. *American Association of Petroleum Geologists Bulletin*, 66, 1–9.

Azevedo, D. A., Aquino Neto, F. R., Simoneit, B. R. T. and Pinto, A. C. (1992) Novel series of tricyclic aromatic terpanes characterized in Tasmanian tasmanite. *Organic Geochemistry*, 18, 9–16.

(1994) Extended saturated and monoaromatic tricyclic terpenoid carboxylic acids found in Tasmanian tasmanite. *Organic Geochemistry*, 22, 991–1004.

(1998) Extended ketones of the tricyclic terpane series in a Tasmanian tasmanite bitumen. *Organic Geochemistry*, 28, 289–95.

Azevedo, D. A., André Zinu, C. J., Aquino Neto, F. R. and Simoneit, B. R. T. (2001) Possible origin of acyclic (linear and isoprenoid) and tricyclic terpane methyl ketones in a Tasmanian tasmanite bitumen. *Organic Geochemistry*, 32, 443–8.

Azzar, I. N. and Taher, A. K. (1993) Sequence stratigraphy and source rock potential of Middle Cretaceous (Upper Wasia Group) in West Abu Dhabi. In: *Proceedings of the Middle East Oil Technology Conference, Bahrain, April 3–6, 1993*. Society of Petroleum Engineers, Richardson, TX, pp. 475–87.

Babcock, L. E., Zhang, W. and Leslie, S. A. (2001) The Chengjiang biota: record of the Early Cambrian diversification of life and clues to exceptional preservation of fossils. *GSA Today*, 11, 4–9.

Baby, P., Moretti, I., Guillier, B., *et al.* (1995) Petroleum system of the northern and central Bolivian Sub-Andean Zone. In: *Petroleum Basins of South America* (A. J. Tankard, R. Suarez Soruco and H. J. Welsink, eds.), American Association of Petroleum Geologists, Tulsa, OK, pp. 445–58.

Bada, J. L., Bigham, C. and Miller, S. L. (1994) Impact melting of frozen oceans on the early Earth: implications for the origin of life. *Proceedings of the National Academy of Science, USA*, 91, 1248.

Bada, J. L., Glavin, D. P., Mcdonald, G. D. and Becker, L. (1998) A search for endogenous amino acids in Martian meteorite ALH84001. *Science*, 279, 362–5.

Baedecker, M. J., Cozzarelli, I. M., Eganhouse, R. P., Siegel, D. I. and Bennett, P. C. (1993) Crude oil in a shallow sand and gravel aquifer-III. Biogeochemical reactions and mass balance modeling in anoxic groundwater. *Applied Geochemistry*, 8, 569–86.

Bailey, N. J. L., Krouse, H. R., Evans, C. R. and Rogers, M. A. (1973a) Alteration of crude oil by waters and bacteria – evidence from geochemical and isotopic studies. *American Association of Petroleum Geologists Bulletin*, 57, 1276–90.

Bailey, N. J. L., Jobsen, A. M. and Rogers, M. A. (1973b) Bacterial degradation of crude oil: comparison of field and experimental data. *Chemical Geology*, 11, 203–21.

Bailey, N. J. L., Burwood, R. and Harriman, G. E. (1990) Application of pyrolyzate carbon isotope and biomarker technology to organofacies definition and oil correlation problems in North Sea basins. *Organic Geochemistry*, 16, 1157–72.

Bains, S., Norris, R. D. and Cornfield, R. M. (1999) Mechanisms of climate warming at the end of the Paleocene. *Science*, 285, 724–7.

Bains, S., Norris, R. D., Cornfield, R. M. and Faul, K. L. (2000) Termination of global warmth at the Palaeocene/Eocene boundary through productivity feedback. *Nature*, 406, 171–4.

Bajc, S., Amblès, A., Largeau, C., Derenne, S. and Vitorovi, D. (2001) Precursor biostructures in kerogen matrix revealed by oxidative degradation: oxidation of kerogen from Estonian kukersite. *Organic Geochemistry*, 32, 773–84.

Baker, E. W. and Louda, J. W. (1983) Thermal aspects of chlorophyll geochemistry. In: *Advances in Organic Geochemistry 1981* (M. Bjorøy, C. Albrecht, C. Cornford, *et al.*, eds.), John Wiley & Sons, New York, pp. 401–21.

(1986) Porphyrins in the geological record. In: *Biological Markers in the Sedimentary Record* (R. B. Johns, ed.), Elsevier, New York, pp. 125–224.

Banerjee, A. and Rao, K. L. N. (1993) Geochemical evaluation of part of the Cambay Basin, India. *American Association of Petroleum Geologists Bulletin*, 77, 29–48.

Banerjee, A., Sinha, A. K., Jain, A. K., *et al.* (1998) A mathematical representation of Rock-Eval hydrogen index vs. T_{max} profiles. *Organic Geochemistry*, 28, 43–55.

Banerjee, A., Jha, M., Mittal, A. K., Thomas, N. J. and Misra, K. N. (2000) The effective source rocks in the North Cambay Basin, India. *Marine Petroleum Geology*, 17, 1111–29.

Banerjee, A., Pahari, S., Jha, M., *et al.* (2002) The effective source rocks in the Cambay Basin, India. *American Association of Petroleum Geologists Bulletin*, 86, 433–56.

Barakat, A. O. and Rullkötter, J. (1993) Gas chromatography/mass spectrometric analysis of cembrenoid determined in kerogen from a lacustrine sediment. *Organic Mass Spectrometry*, 28, 157–62.

(1995) Extractable and bound fatty acids in core sediments from the Nördlinger Ries, southern Germany. *Fuel*, 74, 416–25.

Barakat, A. O., Omar, M. F., Scholz-Boettcher, B. M. and Rullkötter, J. (1996) A biomarker study of crude oils from the Gulf of Suez, Egypt. In: *212th American Chemical Society National Meeting. Removal of Aromatics, Sulfur and Olefins from Gasoline and Diesel Symposium (Orlando, Florida, 8/25–29/96). ACS Petroleum Chemistry Division Preprints*, 41, 637–41.

Barakat, A. O., Mostafa, A., El-Gayar, M. S. and Rullkötter, J. (1997) Source-dependent biomarker properties of five crude oils from the Gulf of Suez, Egypt. *Organic Geochemistry*, 26, 441–50.

Barakat, A. O., El-Gayar, M. S., Mostafa, A. R. and Omar, M. F. (1998) Occurrence and distribution of bicyclic and tricyclic aromatic hydrocarbons in crude oils from the Gulf of Suez, Egypt. *Petroleum Science and Technology*, 16, 21–39.

Barbanti, S. M., Moldowan, J. M., Mello, M. R., *et al.* (1999) Analysis and occurrence of novel triaromatic 23, 24-dimethylcholestanes in geologic time. Presented at the 19th International Meeting on Organic Geochemistry, September 6–10, 1999, Istanbul.

Barber, C. J., Bastow, T. P., Grice, K., Alexander, R. and Kagi, R. I. (2001a) Analysis of crocetane in crude oils and sediments: novel stationary phases for use in GC-MS. *Organic Geochemistry*, 32, 765–9.

Barber, C. J., Grice, K., Bastow, T. P., Alexander, R. and Kagi, R. I. (2001b) The identification of crocetane in Australian crude oils. *Organic Geochemistry*, 32, 943–7.

Baric, G., Spanic, D. and Maricic, M. (1996) Geochemical characterization of source rocks in NC-157 Block (Zaltan Platform), Sirt Basin. In: *The Geology of Sirt Basin* (M. J. Salem, A. S. El-Hawat and A. M. Sbeta, eds.), Elsevier, Amsterdam, pp. 541–53.

Baric, G., Ivkovic, Z. and Perica, R. (2000) The Miocene petroleum system of the Sava Depression, Croatia. *Petroleum Geoscience*, 6, 165–73.

Barnard, P. C. and Cooper, B. S. (1981) Oils and source rocks of the North Sea area. In: *Petroleum Geology of the Continental Shelf of Northwest Europe* (L. V. Illing and G. D. Hobson, eds.), Heyden, London, pp. 169–75.

Barnard, P. C., Collins, A. G. and Cooper, B. S. (1981) Identification and distribution of kerogen facies in a source rock horizon – examples from the North Sea Basin. In: *Organic Maturation Studies and Fossil Fuel Exploration* (J. Brooks, ed.), Academic Press, London pp. 271–82.

Barnes, M. A. and Barnes, W. C. (1983) Oxic and anoxic diagenesis of diterpenes in lacustrine sediments. In: *Advances in Organic Geochemistry 1981* (M. Bjorøy, C. Albrecht, C. Cornford, *et al.* eds.), John Wiley & Sons, New York, pp. 289–98.

Baross, J. A. and Deming, J. W. (1983) Growth of "black smoker" bacteria at temperatures of at least 250°C. *Nature*, 303, 423–6.

Barth, T. and Bjørlykke, K. (1993) Organic acids from source rock maturation: generation potentials, transport mechanisms and relevance for mineral diagenesis. *Applied Geochemistry*, 8, 325–37.

Barth, T., Andresen, B., Iden, K. and Johansen, H. (1996) Modeling source rock production potentials for short-chain organic acids and CO_2 – a multivariate approach. *Organic Geochemistry*, 25, 427–38.

Bartley, J. K., Pope, M., Knoll, A. H., Semikhatov, M. A. and Petrov, P. Y. (1998) A Vendian-Cambrian boundary succession from the northwestern margin of the Siberian Platform: stratigraphy, palaeontology, chemostratigraphy and correlation. *Geological Magazine*, 135, 473–94.

Barwise, A. J. G. and Park, P. J. D. (1983) Petroporphyrin fingerprinting as a geochemical parameter. In: *Advances in Organic Geochemistry 1981* (M. Bjorøy, C. Albrecht, C. Cornford, *et al*. eds.), John Wiley & Sons, New York, pp. 668–74.

Barwise, A. J. G. and Whitehead, E. V. (1980) Separation and structure of petroporphyrins. In: *Advances in Organic Geochemistry 1979* (A. G. Douglas and J. R. Maxwell, eds.), Pergamon, New York, pp. 181–92.

Baskin, D. K. and Peters, K. E. (1992) Early generation characteristics of a sulfur-rich Monterey kerogen. *American Association of Petroleum Geologists Bulletin*, 76, 1–13.

Bastow, T. P., Alexander, R. and Kagi, R. I. (1997) Identification and analysis of dihydro-*ar*-curcumene enantiomers and related compounds in petroleum. *Organic Geochemistry*, 26, 79–83.

Bastow, T. P., Alexander, R., Kagi, R. I. and Sosrowidjojo, I. B. (1998) The effect of maturity and biodegradation on the enantiomeric composition of sedimentary dihydro-*ar*-curcumene and related compounds. *Organic Geochemistry*, 29, 1297–304.

Bateman, J. A. (1995) Mineralogical and geochemical traits of the Egret Member oil source rock (Kimmeridgian), Jeanne d'Arc Basin, offshore Newfoundland, Canada. M. S. thesis, Dalhousie University, Halifax, Canada.

Baudin, F. (1995) Depositional controls on Mesozoic source rocks in the Tethys. In: *Paleogeography, Paleoclimate, and Source Rocks* (A.-Y. Huc, ed.), American Association of Petroleum Geologists, Tulsa, OK, pp. 191–211.

Baudin, R., Herbin, J.-P., Bassoullet, J.-P., *et al*. (1990) Distribution of organic matter during the Toarcian in the Mediterranean Tethys and Middle East. In: *Deposition of Organic Facies* (A. Y. Huc, ed.), American Association of Petroleum Geologists, Tulsa, OK, pp. 73–91.

Bauer, P. E., Dunlap, N. K., Arseniyadis, S., *et al*. (1983) Synthesis of biological markers in fossil fuels. 1. 17α and 17β isomers of 30-norhopane and 30-normoretane. *Journal of Organic Chemistry*, 48, 4493–7.

Bauert, H. (1994) Kukersite oil shale in Estonia: basin geology, resources and utilization. *American Association of Petroleum Geologists Bulletin*, 78, 101.

Bazhenova, O. K. and Arefiev, O. A. (1996) Geochemical peculiarities of Pre-Cambrian source rocks in the East European Platform. *Organic Geochemistry*, 25, 341–51.

(1997) Source rocks and hydrogen potential, South Sakhalin. Presented at the 59th EAGE Conference, May 26–30, 1997, Geneva, Switzerland.

Beach, F., Peakman, T. M., Abbott, G. D., Sleeman, R. and Maxwell, J. R. (1989) Laboratory thermal alteration of triaromatic steroid hydrocarbons. *Organic Geochemistry*, 14, 109–11.

Beato, B. D., Yost, R. A., van Berkel, G. J., Filby, R. H. and Quirke, M. E. (1991) The Henryville bed of the New Albany Shale. III: tandem mass spectrometric analyses of geoporphyrins from the bitumen and kerogen. *Organic Geochemistry*, 17, 93–105.

Beaumont, C., Boutlier, R., Mackenzie, A. S. and Rullkötter, J. (1985) Isomerization and aromatization of hydrocarbons and the paleothermometry and burial history of Alberta Foreland Basin. *American Association of Petroleum Geologists Bulletin*, 69, 546–66.

Bechtel, A., Sun, Y., Püttmann, W., Hoernes, S. and Hoefs, J. (2001) Isotopic evidence for multi-stage base metal enrichment in the Kupferschiefer from the Sangerhausen Basin, Germany. *Chemical Geology*, 176, 31–49.

Becker, L., Bada, J. L., Winans, R. E., *et al*. (1994) Fullerenes in the 1.85 billion-year-old Sudbury impact structure. *Science*, 265, 642–5.

Becker, L., Bada, J. L. and Bunch, T. E. (1995) Fullerenes in the K/T boundary: are they a result of global wildfires? *Abstracts of Papers Submitted to the 26th Lunar and Planetary Science Conference*, 26, 85–6.

Becker, L., Glavin, D. P. and Bada, J. L. (1997) Polycyclic aromatic hydrocarbons (PAHs) in Antarctic Martian meteorites, carbonaceous chondrites, and polar ice. *Geochimica et Cosmochimica Acta*, 61, 475–81.

Becker, L., Popp, B., Rust, T. and Bada, J. L. (1999) The origin of organic matter in the Martian meteorite ALH84001. *Earth and Planetary Science Letters*, 167, 71–9.

Becker, L., Poreda, R. J. and Bunch, T. E. (2000) Fullerenes (C_{100}–C_{400}): a new extraterrestrial carrier phase of noble gases in nature. *Proceedings of the National Academy of Science USA*, 97, 2979–83.

Becker, L., Poreda, R. J., Hunt, A. G., Bunch, T. E. and Rampino, M. (2001) Impact event at the Permian-Triassic boundary: evidence from extraterrestrial noble gases in fullerenes. *Science*, 291, 1530–3.

Becker, L., Poreda, R. J., Basu, A. R., Pope, K. D., Harrison, T. M., Nicholson, C. and Iasky, R. (2004) Becout: a possible end-Permian impact crater offshore northwestern Australia. *Science*, 304, 1469–76.

Beeder, J., Nilsen, R. K., Thorstenson, T. and Torsvik, T. (1996) Penetration of sulfate reducers through a porous North Sea oil reservoir. *Applied and Environmental Microbiology*, 62, 3551–3.

Behar, F. H. and Albrecht, P. (1984) Correlations between carboxylic acids and hydrocarbons in several crude oils: alteration by biodegradation. *Organic Geochemistry*, 6, 597–604.

Behrens, A., Wilkes, H., Schaeffer, P., Clegg, H. and Albrecht, P. (1998) Molecular characterization of organic matter in sediments from the Keg River Formation (Elk Point Group), western Canada Sedimentary Basin. *Organic Geochemistry*, 29, 1905–20.

Behrens, A., Schaeffer, P., Bernasconi, S. and Albrecht, P. (1999) 17(E)-13α(H)-malabarica-14(27),17,21-triene, an unexpected tricyclic hydrocarbon in sediments. *Organic Geochemistry*, 30, 379–83.

Bell, J. F. (1996) Evaluating the evidence for past life on Mars. *Science*, 274, 2121–2.

Beller, H. R., Spormann, A. M., Sharma, P. K., Cole, J. R. and Reinhard, M. (1996) Isolation and characterization of a novel toluene-degrading, sulfate-reducing bacterium. *Applied and Environmental Microbiology*, 62, 1188–96.

Belt, S. T., Cooke, D. A., Robert, J.-M. and Rowland, S. (1996) Structural characterization of widespread polyunsaturated isoprenoid biomarkers: a C_{25} triene, tetraene and pentaene from the diatom *Haslea ostrearia* Simonsen. *Tetrahedron Letters*, 37, 4655–8.

Belt, S. T., Allard, W. G., Massé, G., Robert, J. M. and Rowland, S. J. (2000) Highly branched isoprenoids (HBIs): identification of the most common and abundant sedimentary isomers. *Geochimica et Cosmochimica Acta*, 64, 3839–51.

Belt, S. T., Massé, G., Allard, W. G., Robert, J. -M. and Rowland, S. J. (2001) Identification of a C_{25} highly branched isoprenoid triene in the freshwater diatom *Navicula sclesvicensis*. *Organic Geochemistry*, 32, 1169–72.

Benali, S., Schreiber, B. C., Helman, M. L. and Philp, R. P. (1995) Characterization of organic matter from a restricted/evaporative sedimentary environment: Late Miocene of Lorca Basin, southeastern Spain. *American Association of Petroleum Geologists Bulletin*, 79, 816–30.

Benalioulhaj, S., Schreiber, B. C. and Philp, R. P. (1994) Relationship of organic geochemistry to sedimentation under highly variable environments, Lorca Basin (Spain): preliminary results. In: *Sedimentology and Paleolimnological Record of Saline Lakes* (R. W. Renaut and W. M. Last, eds.), Society of Economic Paleontologists and Mineralogists, Tulsa, OK, pp. 315–24.

Ben-Amotz, A, Shaish, A. and Avron, M. (1989) Mode of action of the massively accumulated beta-carotene of *Dunaliella bardawil* in protecting the alga against damage by excess irradiation. *Plant Physiology*, 91, 1040–3.

Bendoraitis, J. G. (1974) Hydrocarbons of biogenic origin in petroleum – aromatic triterpenes and bicyclic sesquiterpenes. In: *Advances in Organic Geochemistry 1973* (B. Tissot and F. Bienner, eds.), Editions Technip, Paris, pp. 209–24.

Bennett, B. and Abbott, G. D. (1999) A natural pyrolysis experiment – hopanes from hopanoic acids? *Organic Geochemistry*, 30, 1509–16.

Benton, M. J. (1993a) Late Triassic extinctions and the origin of the dinosaurs. *Science*, 260, 769–70.

(1993b) *The Fossil Record 2*. Chapman & Hall, London.

(1995) Diversification and extinction in the history of life. *Science*, 268, 52–58.

Bernasconi, S. and Riva, A. (1994) Organic geochemistry and depositional environment of a hydrocarbon source rock: the Middle Triassic Grenzbitumenzone Formation, southern Alps, Italy/Switzerland. In: *Generation, Accumulation, and Production of Europe's Hydrocarbons III, Proceedings of the 3rd EAPG Conference, Florence, Italy*, May 26–30, 1991 (A. M. Spencer, ed.), Springer-Verlag, New York, pp. 179–90.

Berner, R. A., Scott, M. R. and Thomlinson, C. (1970) Carbonate alkalinity in the pore waters of anoxic marine sediments. *Limnology and Oceanography*, 14, 544–9.

Bernhard, J. M., Buck, K. R., Farmer, M. A. and Bowser, S. S. (2000) The Santa Barbara Basin is a symbiosis oasis. *Nature*, 403, 77–80.

Berry, W. B. N. and Cooper, J. D. (2001) Late Ordovician oil shale, Vinni Formation, Central Nevada. Presented at the Annual Meeting of the Geological Society of America, November 5–8, 2001, Boston, MA.

Berthe-Corti, L. and Fetzner, S. (2002) Bacterial metabolism of *n*-alkanes and ammonia under oxic, suboxic and anoxic conditions. *Acta Biotechnologica*, 22, 299–336.

Besley, B. M. (1998) Carboniferous. In: *Petroleum Geology of the North Sea: Basin Concepts and Recent Advances*,

4th edn (K. W. Glennie, ed.), Blackwell Science, London, pp. 104–36.

Bessereau, G., Guillocheau, F. and Huc, A.-Y. (1995) Source rock occurrence in a sequence stratigraphic framework: the example of the Lias of the Paris Basin. In: *Paleogeography, Paleoclimate, and Source Rocks* (A.-Y. Huc, ed.), American Association of Petroleum Geologists, Tulsa, OK, pp. 273–301.

Beydoun, Z. R. (1991a) Post-Paleozoic development: formation of the Arabian northeast passive margin. In: *Arabian Plate Hydrocarbon Geology and Potential – A Plate Tectonic Approach*. American Association of Petroleum Geologists, Tulsa, OK, pp. 19–29.

(1991b) The hydrocarbon potential of the Arabian Plate. In: *Arabian Plate Hydrocarbon Geology and Potential – A Plate Tectonic Approach*. American Association of Petroleum Geologists, Tulsa, OK, pp. 49–65.

Beydoun, Z. R., Hughes Clarke, M. W. and Stoneley, R. (1992) Petroleum in the Zagros Basin: a late Tertiary foreland basin overprinted onto the outer edge of a vast hydrocarbon-rich Paleozoic-Mesozoic passive-margin shelf. In: *Foreland Basins and Fold Belts* (R. W. Macqueen and D. A. Leckie, eds), American Association of Petroleum Geologists, Tulsa, OK, pp. 309–39.

Bharati, S., Patience, R. L., Larter, S. R., Standen, G. and Poplett, I. J. F. (1995) Elucidation of the Alum Shale kerogen structure using a multi-disciplinary approach. *Organic Geochemistry*, 23, 1043–58.

Bian, L. (1994) Isotopic biogeochemistry of individual compounds in a modern coastal marine sediment (Kattegat, Denmark and Sweden). Ph. D. thesis, Indiana University, Bloomington, IN.

Bian, L., Hinrichs, K.-U., Xie, T., *et al.* (2001) Algal and archaeal polyisoprenoids in a recent marine sediment: molecular isotopic evidence for anaerobic oxidation of methane. *Geochemistry, Geophysics, Geosystems*, 2, paper 2000GC000112.

Bice, D. M., Newton, C. R., McCauley, S. E., Reiners, P. W. and McRoberts, C. A. (1992) Shocked quartz at the Triassic-Jurassic boundary in Italy. *Science*, 255, 443–6.

Bieger, T., Abrajano, T. A. and Hellou, J. (1997) Generation of biogenic hydrocarbons during a spring bloom in Newfoundland coastal (NW Atlantic) waters. *Organic Geochemistry*, 26, 207–18.

Bird, K. J. (1994) Ellesmerian(!) petroleum system, North Slope, Alaska, USA. In: *The Petroleum System – From Source to Trap* (L. B. Magoon and W. G. Dow, eds.), American Association of Petroleum Geologists, Tulsa, OK, pp. 339–58.

(2001) Alaska: a twenty-first-century petroleum province. In: *Petroleum Provinces of the Twenty-First Century* (M. W. Downey, J. C. Threet and W. A. Morgan, eds.), American Association of Petroleum Geologists, Tulsa, OK, p. 137–65.

Bird, K. J. and Houseknecht, D. W. (2002) *U.S. Geological Survey 2002 Petroleum Resource Assessment of the National Petroleum Reserve in Alaska (NPRA): Play Maps and Technically Recoverable Resource Estimates*. U.S. Geological Survey Open-File Report 02-207.

Bird, C. W., Lynch, J. M., Pirt, F. J., *et al.* (1971) Steroids and squalene in *Methylococcus capsulatus* grown on methane. *Nature*, 230, 473–4.

Bishop, M. G. (1999a) *Total Petroleum Systems of the Northwest Shelf, Australia: the Dingo-Mungaroo/Barrow and the Locker-Mungaroo/Barrow*. U.S. Geological Survey Open-File Report 99-50-E.

(1999b) *Petroleum System of the Gippsland Basin, Australia*. U.S. Geological Survey Open-File Report 99-50-Q.

(1999c) *Total Petroleum Systems of the Bonaparte Gulf Basin Area, Australia: Jurassic, Early Cretaceous-Mesozoic; Keyling, Hyland Bay-Permian; Milligans-Carboniferous, Permian*. U.S. Geological Survey Open-File Report 99-50-P.

(2000a) *South Sumatra Basin Province, Indonesia: The Lahat/Talang Akar-Cenozoic Total Petroleum System*. U.S. Geological Survey Open-File Report 99-50-s.

(2000b) *Petroleum Systems of Northwest Java Province, Java and Offshore Southeast Sumatra, Indonesia*. U.S. Geological Survey Open-File Report 99-50-R.

Bishop, A. N. and Abbott, G. D. (1993) The interrelationship of biological marker maturity parameters and molecular yields during contact metamorphism. *Geochimica et Cosmochimica Acta*, 57, 3661–8.

Bissada, K. K., Katz, B. J., Barnicle, S. C. and Schunk, D. J. (1990) On the origin of hydrocarbons in the Gulf of Mexico Basin – a reappraisal. In: *Gulf Coast Oils and Gases: Their Characteristics, Origin, Distribution and Exploration and Production Significance* (D. Schumacher and B. F. Perkins, eds.), Society of Economic Paleontologists and Mineralogists – Gulf Coast Section Research, Houston, TX, 163–71.

Biswas, S. K. (1982) Rift basins in the western margin of India and their hydrocarbon prospects. *American Association of Petroleum Geologists Bulletin*, 66, 1497–513.

(1987) Regional tectonic framework, structure, and evolution of the western marginal basins of India. *Tectonophysics*, 137, 307–27.

Biswas, S. K., Rangaraju, M. K., Thomas, J. and Bhattacharya, S. K. (1994) Cambay-Hazad(!) petroleum system in South Cambay Basin, India. In: *The Petroleum System – From Source to Trap* (L. B. Magoon and W. G. Dow, eds.), American Association of Petroleum Geologists, Tulsa, OK, pp. 615–24.

Bitner, A. K., Yefimov, A. S., Kasatkin, V. E., *et al.* (1999) Prospects for developing reserves and resource base of Yurubcheno-Tokhom Field. *Petroleum Geology*, 36, 36–8. [Translated from *Neftyanoye Khozyaystvo*, 1998, 6, 2–5.]

Bjorøy, M., Hall, P. B., Loberg, R., McDermott, J. A. and Mills, N. (1988) Hydrocarbons from non-marine source rocks. *Organic Geochemistry*, 13, 221–44.

Blanc, P. and Connan, J. (1992) Origin and occurrence of 25-norhopanes: a statistical study. *Organic Geochemistry*, 18, 813–28.

(1994) Preservation, degradation and destruction of trapped oil. In: *The Petroleum System – From Source to Trap* (L. B. Magoon and W. G. Dow, eds.), American Association of Petroleum Geologists, Tulsa, OK, pp. 237–47.

Bleeker, W., Ketchum, J. W. F. and Davis, W. J. (1999) The Central Slave Basement Complex, Part II: age and tectonic significance of high-strain zones along the basement-cover contact. *Canadian Journal of Earth Sciences*, 36, 1111–30.

Blichert-Toft, J., Albarède, F., Rosing, M. Frei, R. and Bridgwater, D. (1999) The Nd and Hf isotopic evolution of the mantle through the Archean. Results from the Isua supracrustals, West Greenland, and from the Birimian terranes of West Africa. *Geochimica et Cosmochimica Acta*, 63, 3901–14.

Blöchl, E., Rachel, R., Burggraf, S., *et al.* (1997) *Pyrolobus fumarii*, gen. and sp. nov., represents a novel group of archaea, extending the upper temperature limit for life to 113°C. *Extremophiles*, 1, 14–21.

Blokker, P., Van Bergen, P., Pancost, R., *et al.* (2001) The chemical structure of *Gloeocapsomorpha prisca* microfossils: implications for their origin. *Geochimica et Cosmochimica Acta*, 65, 885–900.

Blumer, M. and Snyder, W. D. (1965) Isoprenoid hydrocarbons in recent sediments; presence of pristane and probable absence of phytane. *Science*, 150, 1588–9.

Blumer, M. and Thomas, D. W. (1965) "Zamene," isomeric C_{19} monoolefins from marine zooplankton, fishes, and mammals. *Science*, 148, 370–1.

Blumer, M., Mullin, M. M. and Thomas, D. W. (1963) Pristane in zooplankton. *Science*, 140, 974.

Blunt, J. W., Czochanska, Z., Sheppard, C. M., Weston, R. J., and Woolhouse, A. D. (1988) Isolation and structural characterization of isopimarane in some New Zealand seep oils. *Organic Geochemistry*, 12, 479–86.

Boduszynski, M. M. (1987) Composition of heavy petroleums. 1. Molecular weight, hydrogen deficiency, and heteroatom concentration as a function of atmospheric equivalent boiling point up to 1400°F (760°C). *Energy & Fuels*, 1, 2–11.

Boetius, A., Raveschlag, K., Schubert, C. J., *et al.* (2000) A marine microbial consortium apparently mediating anaerobic oxidation of methane. *Nature*, 407, 577–9.

Bohacs, K. M. (1993) Source quality variations tied to sequence development in the Monterey and associated formations, southwestern California. In: *Source Rocks in a Sequence Stratigraphic Framework* (B. J. Katz and L. M. Pratt, eds.), American Association of Petroleum Geologists, Tulsa, OK, pp. 177–204.

Bohacs, K. M. and Suter, J. (1997) Sequence stratigraphic distribution of coaly rocks: fundamental controls and paralic examples. *American Association of Petroleum Geologists Bulletin*, 81, 1612–39.

Bohacs, K. M., Carroll, A. R., Neal, J. E. and Mankiewicz, P. J. (2000) Lake-basin type, source potential, and hydrocarbon character: an integrated sequence-stratigraphic-geochemical framework. In: *Lake Basins Through Space and Time* (E. H. Gierlowski-Kordesch and K. R. Kelts, eds.), American Association of Petroleum Geologists, Tulsa, OK, pp. 3–34.

Boon, J. J., Hine, S. H., Burlingame, A. L., *et al.* (1983) Organic geochemical studies of Solar Lake laminated cyanobacterial mats. In: *Advances in Organic Geochemistry 1981* (M. Bjorøy, C. Albrecht, C. Cornford, *et al.* eds.), John Wiley & Sons, New York, pp. 207–27.

Boote, D. R. D., Clark-Lowes, D. D. and Traut, M. W. (1998) Palaeozoic petroleum systems of North Africa. In: *Petroleum Geology of North Africa* (D. S. Macgregor, R. T. J. Moody and D. D. Clark-Lowes, eds.), Geological Society, London pp. 7–68.

Bordenave, M. L. and Burwood, R. (1990) Source rock distribution and maturation in the Zagros Orogenic

Belt: provenance of the Asmari and Bangestan reservoir accumulations. *Organic Geochemistry*, 369–87.

(1995) The Albian Kazhdumi Formation of the Dezful Embayment, Iran: one of the most efficient petroleum generating systems. In: *Petroleum Source Rocks, Casebooks in Earth Science* (B. J. Katz, ed.), Springer-Verlag, New York, pp. 183–207.

Bordoloi, M., Shukla, V. S., Nath, S. C. and Sharma, R. P. (1989) Naturally occurring cadinenes. *Phytochemistry*, 28, 2007–37.

Boreham, C. J. and Powell, T. G. (1987) Sources and preservation of organic matter in the Cretaceous Toolebuc Formation, eastern Australia. *Organic Geochemistry*, 11, 433–49.

Boreham, C. J. and Summons, R. E. (1999) New insights into the active petroleum systems in the Cooper and Eromanga basins, Australia. *APEA Journal*, 39, 263–96.

Boreham, C. J., Crick, I. H. and Powell, T. G. (1988) Alternative calibration of the Methylphenanthrene Index against vitrinite reflectance: application to maturity measurements on oils and sediments. *Organic Geochemistry*, 12, 289–94.

Borrego, A. G., Blanco, C. G. and Püttman, W. (1997) Geochemical significance of the aromatic hydrocarbon distribution in the bitumens of the Puertollano oil shales, Spain. *Organic Geochemistry*, 26, 219–28.

Borruat, G., Roten, C.-A. H., Marchant, R., Fay, L.-B. and Karamata, D. (2001) Chromatographic method for diaminopimelic acid detection in calcareous rocks. Presence of a bacterial biomarker in stromatolites. *Journal of Chromatography A*, 922, 219–34.

Boslough, M. B., Chael, E. P., Trucano, T. G. and Crawford, D. A. (1996) Axial focusing of impact energy in the Earth's interior; a possible link to flood basalts and hotspots. In: *The Cretaceous-Tertiary Event and Other Catastrophes in Earth History, Special Paper – Geological Society of America No. 307* (D. L. Campbell, G. Ryder, D. Fastovsky and S. Gartner, eds.) Geological Society of America, Boulder, CO, pp. 541–50.

Bost, F. D., Frontera-Suau, R., McDonald, T. J., Peters, K. E. and Morris, P. J. (2001) Aerobic biodegradation of hopanes and norhopanes in Venezuelan crude oils. *Organic Geochemistry*, 32, 105–14.

Bottomley, R. J., York, D. and Grieve, R. A. F. (1978) ^{40}Ar-^{39}Ar ages of Scandinavian impact structures. I. Mien and Siljan. *Contributions to Mineralogy and Petrology*, 68, 79–84.

Boult, P. J., Lanzilli, E., Michaelsen, B. H., McKirdy, D. M. and Ryan, M J. (1998) A new model for the Hutton/Birkhead reservoir/seal couplet and the associated Birkhead-Hutton(!) petroleum system. *APPEA Journal*, 38, 724–44.

Bourbonniere, R. A. and Meyers, P. A. (1996) Sedimentary geolipid records of historical changes in the watersheds and productivities of Lakes Ontario and Erie. *Limnology and Oceanography*, 41, 352–9.

Bradley, W. H. (1925) A contribution to the origin of the Green River Formation and its oil shale. *American Association of Petroleum Geologists Bulletin*, 9, 247–62.

(1931) *Origin and Microfossils of the Oil Shale of the Green River Formation of Colorado and Utah*. U.S. Geological Survey professional paper 168.

Bradshaw, M. T., Bradshaw, J., Murray, A. P., et al. (1994) Petroleum systems in west Australian basins. In: *The Sedimentary Basins of Western Australia* (P. G. Purcell and R. R. Purcell, eds.), Petroleum Exploration Society of Australia, Perth, Australia, pp. 93–118.

Bradshaw, M. T., Edwards, D., Bradshaw, J., et al. (1997) Australian and eastern Indonesian petroleum systems. In: *Proceedings of the Petroleum Systems of SE Asia and Australasia Conference, May 1997*. Indonesian Petroleum Association, Jakarta, Indonesia, pp. 141–53.

Bragg, J. R., Prince, R. C., Harner, E. J. and Atlas, R. M. (1994) Effectiveness of bioremediation for the *Exxon Valdez* oil spill. *Nature*, 368, 413–18.

Bralower, T. J., Thomas, D. J., Zachos, J. C. et al. (1997) High-resolution records of the late Paleocene thermal maximum and circum-Caribbean volcanism: is there a causal link? *Geology*, 25, 963–6.

Brasier, M. D., Green, O. R., Jephcoat, A. P., et al. (2002) Questioning the evidence for Earth's oldest fossils. *Nature*, 416, 76–81.

Brasier, M., Green, O. Lindsay, J. and Steele, A. (2004) Earth's oldest ($\sim$3.5 Ga) fossils and the "Early Eden Hypothesis": questioning the evidence. *Origins of Life and Evolution of the Biosphere*, 34, 257–69.

Brassell, S. C. and Eglinton, G. (1983) Steroids and triterpenoids in deep sea sediments as environmental and diagenetic indicators. In: *Advances in Organic Geochemistry, 1981* (M. Bjorøy, C. Albrecht, C. Cornford, et al., eds.), John Wiley & Sons, New York, pp. 684–97.

Brassell, S. C., Comet, P. A., Eglinton, G., et al. (1980) The origin and fate of lipids in the Japan Trench. In: *Advances in Organic Geochemistry 1979* (A. G. Douglas and J. R. Maxwell, eds.), Pergamon Press, Oxford, pp. 375–91.

Brassell, S. C., Wardroper, A. M. K., Thompson, I. D., Maxwell, J. R. and Eglinton, G. (1981) Specific acyclic isoprenoids as biological markers of methanogenic bacteria in marine sediments. *Nature*, 290, 693–6.

Brassell, S. C., Eglinton, G. and Fu, J. M. (1985) Biological marker compounds as indicators of the depositional history of the Maoming oil shale. *Organic Geochemistry*, 10, 927–41.

Brassell, S. C., Sheng, G., Fu, J. and Eglinton, G. (1988) Biological markers in lacustrine Chinese oil shales. In: *Lacustrine Petroleum Source Rocks* (A. J. Fleet, K. Kelts and M. R. Talbot, eds.), Geological Society of London, London, pp. 299–308.

Bratton, J. F. (1999) Clathrate eustasy: methane hydrate melting as a mechanism for geologically rapid sea-level fall. *Geology*, 27, 915–18.

Braun, R. L., Burnham, A. K., Reynolds, J. R. and Clarkson, J. E. (1991) Pyrolysis kinetics for lacustrine and marine source rocks by programmed pyrolysis. *Energy & Fuels*, 5, 192–204.

Bray, E. E. and Evans, E. D. (1961) Distribution of *n*-paraffins as a clue to recognition of source beds. *Geochimica et Cosmochimica Acta*, 22, 2–15.

Brenchley, P. J., Carden, G. A. F. and Marshall, J. D. (1993) Environmental changes associated with the "first strike" of the Late Ordovician mass extinction. *Modern Geology*, 20, 69–82.

Brenchley, P. J., Marshall, J. D. and Underwood, C. J. (2001) Do all mass extinctions represent an ecological crisis? Evidence from the Late Ordovician. *Geological Journal*, 36, 329–40.

Brice, S. E., Cochran, M. D., Pardo, G. and Edwards, A. D. (1982) Tectonics and sedimentation of the South Atlantic rift sequence: Cabinda and Angola. In: *Studies in Continental Margin Geology* (J. S. Watkins and C. L. Drake, eds.), American Association of Petroleum Geologists, Tulsa, OK, pp. 5–18.

Brincat, D. and Abbott, G. (2001) Some aspects of the molecular biogeochemistry of laminated and massive rocks from the Naples Beach Section (Santa Barbara-Ventura Basin). In: *The Monterey Formation: From Rocks to Molecules* (C. M. Isaacs and J. Rullkötter, eds.), Columbia University Press, New York, pp. 140–9.

Brock, T. D. and Madigan, M. T. (1991) *Biology of Microorganisms*. Prentice-Hall, Englewood Cliffs, NJ.

Brocks, J. J., Logan, G. A., Buick, R. and Summons, R. E. (1999) Archean molecular fossils and the early rise of eukaryotes. *Science*, 285, 1033–6.

Bromham, L., Rambaut, A., Fortey, R., Cooper, A. and Penny, D. (1998) Testing the Cambrian explosion hypothesis by using a molecular dating technique. *Proceedings of the National Academy of Science, USA*, 95, 12 386–9.

Brooks, P. W. (1986) Unusual biological marker geochemistry of oils and possible source rocks, offshore Beaufort-Mackenzie Delta, Canada. In: *Advances in Organic Geochemistry 1985* (D. Leythaeuser and J. Rullkötter, eds.), Pergamon Press, Oxford, pp. 401–6.

Brooks, J. D., Gould, K. and Smith, J. W. (1969) Isoprenoid hydrocarbons in coal and petroleum. *Nature*, 222, 257–9.

Brooks, P. W., Fowler, M. G. and Macqueen, R. W. (1988) Biological marker and conventional organic geochemistry of oil sands/heavy oils, Western Canada Basin. *Organic Geochemistry*, 12, 519–38.

Brooks, P. W., Macqueen, R. W., Fowler, M. G. and Riediger, C. L. (1991) Source rock organic geochemistry and oil-source correlations, Western Canada Basin. *Bulletin of Canadian Petroleum Geology*, 39, 206.

Brooks, J. R. V., Stoker, S. J. and Cameron, T. D. J. (2001) Hydrocarbon exploration opportunities in the 21st century in the United Kingdom. In: *Future Petroleum Provinces of the 21st Century* (M. W. Downey, J. C. Threet and W. A. Morgan, eds.), American Association of Petroleum Geologists, Tulsa, OK, pp. 167–99.

Brosgé, W. P., Reiser, H. N., Dutro, J. T., Jr, and Detterman, R. L. (1981) *Organic Geochemcial Data for Mesozoic and Paleozoic Shales, Central and Eastern Brooks Range, Alaska*. U.S. Geological Survey Open-File Report 81–551.

Brosse, E., Loreau, J. P., Huc, A. Y., *et al.* (1988) The organic matter of interlayered carbonates and clays sediments; Trias/Lias, Sicily. *Organic Geochemistry*, 13, 433–43.

Brothers, L., Engel, M. H. and Kroos, B. M. (1991) The effects of fluid flow through porous media on the distribution of organic compounds in a synthetic crude oil. *Organic Geochemistry*, 17, 11–24.

Brukner-Wein, A. and Vetö, I. (1986) Preliminary organic geochemical study of an anoxic Upper Triassic sequence from W. Hungary. *Organic Geochemistry*, 10, 113–18.

Brukner-Wein, A., Hetényi, M. and Vetö, I. (1990) Organic geochemistry of an anoxic cycle; a case history from the

Oligocene section, Hungary. *Organic Geochemistry*, 15, 123–30.

Brukner-Wein, A., Sajgó, C. and Hetényi, M. (2000) Comparison of Pliocene organic-rich lacustrine sediments in twin craters. *Organic Geochemistry*, 31, 453–61.

Brune, A., Frenzel, P. and Cypionka, H. (2000) Life at the oxic–anoxic interface: microbial activities and adaptations. *FEMS Microbiology Reviews*, 24, 691–710.

Buchardt, B. and Lewan, M. D. (1990) Reflectance of vitrinite-like macerals as a thermal maturity index for Cambrian-Ordovician Alum Shale, Southern Scandinavia. *American Association of Petroleum Geologists Bulletin*, 74, 394–406.

Budiansky, S. (1982) Research article triggers dispute on zeolite. *Nature*, 300, 309.

Buick, R. (1984a) Carbonaceous filaments from North Pole, Western Australia; are they fossil bacteria in Archean stromatolites? *Precambrian Research*, 24, 157–72.

 (1984b) Carbonaceous filaments from North Pole, Western Australia; are they fossil bacteria in Archean stromatolites? A reply. *Precambrian Research*, 39, 311–17.

Buick, R., Rasmussen, B. and Krapez, B. (1998) Archean oil: evidence for extensive hydrocarbon generation and migration 2.5–3.5 Ga. *American Association of Petroleum Geologists Bulletin*, 82, 50–69.

Buitrago, J. (1994) Petroleum systems of the Neiva Area, Upper Magdalena Valley, Colombia. In: *The Petroleum System – From Source to Trap* (L. B. Magoon and W. G. Dow, eds.), American Association of Petroleum Geologists, Tulsa, OK, pp. 483–97.

Burchardt, B., Christiansen, F. G., Nohr-Hansen, H., Larsen, N. H. and Ostfeldt, P. (1989) Composition of organic matter in source rocks. In: *Petroleum Geology of North Greenland* (F. G. Christiansen, ed.). *Grønlands Geologiske Undersøgelse Bulletin*, pp. 32–39.

Burhan, R. Y. P., Trendel, J. M., Adam, P., *et al.* (2002) Fossil bacterial ecosystem at methane seeps. Origin of organic matter from Be'eri sulfur deposit, Israel. *Geochimica et Cosmochimica Acta*, 66, 4085–101.

Burlingame, A. L. and Simoneit, B. R. (1968) Isoprenoid fatty acids isolated from the kerogen matrix of the Green River Formation (Eocene). *Science*, 160, 531–3.

Burlingame, A. L., Haug, P., Belsky, T. and Calvin, M. (1965) Occurrence of biogenic steranes and pentacyclic triterpanes in an Eocene shale (52 million years) and in an early Precambrian shale (2.7 billion years): a preliminary report. *Proceedings of the National Academy of Sciences, USA*, 54, 1406–12.

Burnham, A. K. (1989) On the validity of the Pristane Formation Index. *Geochimica et Cosmochimica Acta*, 53, 1693–7.

Burnham, A. K., Clarkson, J. E., Singleton, M. F., Wong, C. M. and Crawford, R. W. (1982) Biological markers from Green River kerogen decomposition. *Geochimica et Cosmochimica Acta*, 46, 1243–51.

Burns, B. J. and Emmett, J. K. (1992) Gippsland Basin: regional biomarker oil-source correlation and maturation study. *American Association of Petroleum Geologists Bulletin*, 76, 1093.

Burrus, J., Brosse, E., Choppin de Janvry, G., Grosjean, Y. and Oudin, J. L. (1992) Basin modeling in the Mahakam Delta based upon the integrated 2D model Temispack. In: *Proceedings of the Indonesian Petroleum Association Annual Meeting*. Indonesian Petroleum Association, Jakarta, Indonesia, pp. 23–43.

Burtner, R. L. and Warner, M. A. (1984) Hydrocarbon generation in Lower Cretaceous Mowry and Skull Creek Shales of the Northern Rocky Mountain area. In: *Hydrocarbon Source Rocks of the Greater Rocky Mountain Region* (J. Woodward, F. F. Meissner and J. L. Clayton, eds.), Rocky Mountain Association of Geologists, Denver, CO, pp. 449–67.

Burwood, R. (1999) Angola: source rock control for Lower Congo coastal and Kwanza Basin petroleum systems. In: *The Oil and Gas Habitat of the South Atlantic* (N. Cameron, R. H. Bate and V. Clure, eds.), Geological Society of London, London, pp. 181–94.

Burwood, R., Cornet, P. J., Jacobs, L. and Paulet, J. (1990) Organofacies variation control on hydrocarbon generation: a Lower Congo Coastal Basin (Angola) case history. *Organic Geochemistry*, 16, 325–38.

Burwood, R., Leplat, P., Mycke, B. and Paulet, J. (1992) Rifted margin source rock deposition, a carbon isotope and biomarker study of a West African Lower Cretaceous "lacustrine" section. *Organic Geochemistry*, 19, 41–52.

Burwood, R., de Witte, S. M., Mycke, B. and Paulet, J. (1995) Petroleum geochemical characterisation of the lower Congo coastal basin Bucomazi Formation. In: *Petroleum Source Rocks, Casebooks in Earth Science* (B. J. Katz, ed.), Springer-Verlag, New York, pp. 235–63.

Buseck, P. R., Tsipursky, S. J. and Hettlich, R. (1992) Fullerenes in the geological environment. *Science*, 257, 215–17.

Bustin, R. M. (1988) Sedimentology and characteristics of dispersed organic matter in Tertiary Niger Delta: origin of source rocks in a deltaic environment. *American Association of Petroleum Geologists Bulletin*, 72, 277–98.

Butterfield, N. J., Knoll, A. H. and Swett, K. (1990) A bangiophyte red alga from the Proterozoic of Arctic Canada. *Science*, 250, 104–7.

Cairns-Smith, A. G. (1982) *Genetic Takeover and the Mineral Origins of Life*. Cambridge University Press, Cambridge, UK.

Cairns-Smith, A. G. (1985) *Seven Clues to the Origin of Life*. Cambridge University Press, Cambridge, UK.

Caldwell, M. E., Garrett, R. M., Prince, R. C. and Suflita, J. M. (1998) Anaerobic biodegradation of long-chain n-alkanes under sulfate-reducing conditions. *Environmental Science & Technology*, 32, 2191–5.

Callot, H. J., Ocampo, R. and Albrecht, P. (1990) Sedimentary porphyrins: correlations with biological precursors. *Energy & Fuels*, 4, 635–9.

Cameron, T. D. J. (1993) *Carboniferous and Devonian of the Southern North Sea*. British Geological Survey, Nottingham, UK.

Campbell, R. M., Djordjevic, N. M., Markide, S. K. E. and Lee, M. L. (1988) Supercritical fluid chromatographic determination of hydrocarbon groups in gasolines and middle distillate fuels. *Analytical Chemistry*, 60, 356–62.

Cane, R. F. (1969) Coorongite and the genesis of oil shale. *Geochimica et Cosmochimica Acta*, 33, 257–65.

Canfield, D. E. (1998) A new model for Proterozoic ocean chemistry. *Nature*, 396, 450–3.

Canh, T., Ha, D. V., Carstens, H. and Berstad, S. (1994) Vietnam – attractive plays in a new geological province. *Oil and Gas Journal*, 92, 78–83.

Caplan, M. L. and Bustin, R. M. (1999) Devonian–Carboniferous Hangenberg mass extinction event, widespread organic-rich mudrock and anoxia: causes and consequences. *Palaeogeography, Palaeoclimatology, Palaeoecology*, 148, 187–207.

Carlson, R. M. K. and Chamberlain, D. E. (1986) Steroid biomarker clay mineral adsorption free energies: implications to petroleum migration indices. *Organic Geochemistry*, 10, 163–80.

Carmen, G. J. and Hardwich, P. (1983) Geology and regional setting of Kuparuk oil field, Alaska. *American Association of Petroleum Geologists Bulletin*, 67, 1014–31.

Carothers, W. W. and Kharaka, Y. K. (1978) Aliphatic acid anions in oil-field waters – implications for the origin of natural gas. *American Association of Petroleum Geologists Bulletin*, 62, 2441–53.

Carrigan, W. J., Cole, G. A., Colling, E. L. and Jones, P. J. (1995) Geochemistry of the Upper Jurassic Tuwaiq Mountain and Hanifa Formation petroleum source rocks of eastern Saudi Arabia. In: *Petroleum Source Rocks, Casebooks in Earth Science* (B. J. Katz, ed.), Springer-Verlag, New York, pp. 67–87.

Carroll, A. R. (1990) Biomarker analysis of Upper Permian lacustrine oil shales, Junggar Basin, NW China. *American Association of Petroleum Geologists Bulletin*, 74, 625.

(1998) Upper Permian lacustrine organic facies evolution, Southern Junggar Basin, NW China. *Organic Geochemistry*, 28, 649–67.

Carroll, A. R. and Bohacs, K. M. (1999) Stratigraphic classification of ancient lakes: balancing tectonic and climatic controls. *Geology*, 27, 99–102.

(2001) Lake-type controls on petroleum source rock potential in nonmarine basins. *American Association of Petroleum Geologists Bulletin*, 85, 1033–53.

Carroll, A. R., Brassell, S. C. and Graham, S. A. (1992) Upper Permian lacustrine oil shales, southern Junggar Basin, Northwest China. *American Association of Petroleum Geologists Bulletin*, 76, 1874–902.

Carroll, A. R., Wegner, M., Simo, J. A. T., *et al.* (2000) Deep water organic facies and marine transgression, Cherry Canyon Formation, West Texas. Presented at the Annual Meeting of the American Association of Petroleum Geologists, April 16–19, 2000, New Orleans.

Caspi, E., Zander, J. M., Greig, J. B., *et al.* (1968) Evidence for nonoxidative cyclization of squalene in the biosynthesis of tetrahymanol. *Journal of the American Chemical Society*, 90, 3563–4.

Cassani, F. and Eglinton, G. (1986) Organic geochemistry of Venezuelan extra-heavy oils. 1. Pyrolysis of asphaltenes: a technique for the correlation and maturity evaluation of crude oils. *Chemical Geology*, 56, 167–83.

Cassani, F. and Gallango, O. (1988) Organic geochemistry of Venezuelan heavy and extra heavy crude oils. In: *Fourth UNITAR/UNDP International Conference on Heavy Crude and Tar Sands* (R. F. Meyers and E. J. Wiggins, eds.), Alberta Oil Sands Technology and Research Authority, Edmonton, Canada, pp. 543–53.

Cassani, F., Gallango, O., Talukdar, S., Vallejos, C. and Ehrmann, U. (1988) Methylphenanthrene maturity index of marine source rock extracts and crude oils from the Maracaibo Basin. *Organic Geochemistry*, 13, 73–80.

Cassani, F., Audemard, N. and Eglinton, G. (1993) Geochemistry of the hydrocarbons from the Orinoco Belt, eastern Venezuela Basin. *American Association of Petroleum Geologists Bulletin*, 77, 310.

Castaño, J. R., Clement, J. H., Kuykendall, M. D. and Sharpton, V. L. (1994) Source rock potential of impact craters. *American Association of Petroleum Geologists Annual Meeting, Denver Colorado, June 12–15, 1994, Abstracts*, p. 118.

Cayce, P. W. and Carey, B. D., Jr (1979) Regional hydrocarbon source rock and thermal maturity evaluation of Ogaden Basin, Ethiopia. *American Association of Petroleum Geologist Bulletin*, 63, 431.

Cech, T. R. (1987) RNA as an enzyme. *Scientific American*, 257, 64–75.

Cervantes, F. J., van der Velde, S., Lettinga, G. and Field, J. A. (2000) Quinones as terminal electron acceptors for anaerobic microbial oxidation of phenolic compounds. *Biodegradation*, 11, 313–21.

Chaffee, A. L. and Johns, R. B. (1983) Polycyclic aromatic hydrocarbons in Australian coals. I. Angularly fused pentacyclic tri- and tetraaromatic components of Victorian brown coal. *Geochimica et Cosmochimica Acta*, 47, 2141–55.

Chaffee, A. L., Strachan, M. G. and Johns, R. B. (1984) Polycyclic aromatic hydrocarbons in Australian coals: II. Novel tetracyclic components from Victorian brown coal. *Geochimica et Cosmochimica Acta*, 48, 2037–43.

Chakhmakhchev, A., Suzuki, M., Waseda, A. and Takayama, K. (1997) Geochemical characteristics of Tertiary oils derived from siliceous sources in Japan, Russia and USA. *Organic Geochemistry*, 27, 523–36.

Chan, M. A., Parry, W. T. and Bowman, J. R. (2000) Diagenetic hematite and manganese oxides and fault-related fluid flow in Jurassic Sandstones, Southeastern Utah. *American Association of Petroleum Geologists Bulletin*, 84, 1281–310.

Chapelle, F. H. and Bradley, P. M. (1996) Microbial acetogenesis as a source of organic acids in ancient Atlantic coastal plain sediments. *Geology*, 24, 925–8.

Chapelle, F. H., Bradley, P. M., Lovley, D. R., O'Neill, K. and Landmeyer, J. E. (2002) Rapid evolution of redox processes in a petroleum hydrocarbon-contaminated aquifer. *Ground Water*, 40, 353–60.

Chapman, D. J. and Schopf, J. W. (1983) Biological and biochemical effects of the development of an aerobic environment. In: *Earth's Earliest Biosphere* (J. W. Schopf, ed.), Princeton University Press, Princeton, NJ, pp. 302–20.

Chappe, B., Michaelis, W., Albrecht, P. and Ourisson, G. (1979) Fossil evidence for a novel series of archaebacterial lipids. *Naturwissenschaften*, 66, 522–3.

Chappe, B., Michaelis, W. and Albrecht, P. (1980) Molecular fossils of archaebacteria as selective degradation products of kerogen. In: *Advances in Organic Geochemistry 1979* (A. G. Douglas and J. R. Maxwell, eds.), Pergamon Press, Oxford, pp. 265–74.

Chappe, B., Albrecht, P. and Michaelis, W. (1982) Polar lipids of archaebacteria in sediments and petroleums. *Science*, 217, 65–6.

Charrié-Duhaut, A., Lemoine, S., Adam, P., Connan, J. and Albrecht, P. (2000) Abiotic oxidation of petroleum bitumens under natural conditions. *Organic Geochemistry*, 31, 977–1003.

Chen, J. B., Xia, C. G., Xin, J. Y., Cui, J. R. and Li, S. B. (2001) Review on catalysis mechanism of methane monooxygenase. *Progress in Chemistry*, 13, 376–81.

Chicarelli, M. I., Kaur, S. and Maxwell, J. R. (1987) Sedimentary porphyrins: unexpected structures, occurrence, and possible origins. In: *Metal Complexes in Fossil Fuels* (R. H. Filby and J. F. Branthaven, eds.), American Chemical Society, Washington, DC, pp. 41–67.

Chicarelli, M. I., Aquino Neto, F. R. and Albrecht, P. (1988) Occurrence of four stereoisomeric tricyclic terpane series in immature Brazilian shales. *Geochimica et Cosmochimica Acta*, 52, 1955–9.

Chijiwa, T., Arai, T., Sugai, T., *et al.* (1999) Fullerenes found in the Permo-Triassic mass extinction period. *Geophysical Research Letters*, 26, 767–70.

Childers, S. E., Ciufo, S. and Lovley, D. R. (2002) *Geobacter metallireducens* accesses insoluble Fe(III) oxide by chemotaxis. *Nature*, 416, 767–9.

Chlupac, I. (1988) The Devonian of Czechoslovakia and its stratigraphic significance. In: *Devonian of the World. Proceedings of the Canadian Society of Petroleum Geologists International Symposium* (N. J. McMillan, A. F. Embrey and D. J. Glass, eds.), Canadian Society of Petroleum Geologists, Calgary, Canada, pp. 481–97.

Chosson, P., Lanau, C., Connan, J. and Dessort, D. (1991) Biodegradation of refractory hydrocarbon biomarkers from petroleum under laboratory conditions. *Nature*, 351, 640–2.

Chosson, P., Connan, J., Dessort, D. and Lanau, C. (1992) In vitro biodegradation of steranes and terpanes: a clue to understanding geological situations. In: *Biological Markers in Sediments and Petroleum* (J. M. Moldowan, P. Albrecht and R. P. Philp, eds.), Prentice-Hall, Englewood Cliffs, NJ, pp. 320–49.

Chow, N., Wendte, J. and Stasiuk, L. D. (1995) Productivity vs. preservation controls on two organic-rich carbonate facies in the Devonian of Alberta: sedimentological and organic petrological evidence. *Bulletin of Canadian Petroleum Geology*, 43, 433–60.

Christiansen, F. G., Piasecki, S., Stemmerik, L. and Telnaes, N. (1993) Depositional environment and organic geochemistry of the Upper Permian Ravnefjeld Formation source rock in East Greenland. *American Association of Petroleum Geologists Bulletin*, 77, 1519–37.

Christie, O. H. J. (1992) Multivariate methodology in petroleum exploration. A geochemical software package. *Chemometrics and Intelligent Laboratory Systems*, 14, 319–29.

Christie, O. H. J., Esbensen, K., Meyer, T. and Wold, S. (1984) Aspects of pattern recognition in organic geochemistry. *Organic Geochemistry*, 6, 885–91.

Chung, H. M., Rooney, M. A., Toon, M. B. and Claypool, G. E. (1992) Carbon isotope composition of marine crude oils. *American Association of Petroleum Geologists Bulletin*, Vol. 76, p. 1000–7.

Chunqing, J., Alexander, R., Kagi, R. I. and Murray, A. P. (2000b) Origin of perylene in ancient sediments and its geological significance. *Organic Geochemistry*, 31, 1545–59.

Chunqing, J., Li, M., Osadetz, K. G., *et al.* (2001) Bakken/Madison petroleum systems in the Canadian Williston Basin. Part 2. Molecular markers diagnostic of Bakken and Lodgepole source rocks. *Organic Geochemistry*, 32, 1037–54.

Clark, J. P. and Philp, R. P. (1989) Geochemical characterization of evaporite and carbonate depositional environments and correlation of associated crude oils in the Black Creek Basin, Alberta. *Canadian Petroleum Geologists Bulletin*, 37, 401–16.

Clark, R. H. and Rouse, J. T. (1971) A closed system for generation and entrapment of hydrocarbons in Cenozoic deltas, Louisiana Gulf Coast. *American Association of Petroleum Geologists Bulletin*, 55, 1170–8.

Claypool, G. E. and Magoon, L. B. (1985) Comparison of oil-source rock correlation data for Alaskan North Slope: techniques, results, and conclusions. In: *Alaska North Slope Oil/Source Rock Correlation Study* (L. B. Magoon and G. E. Claypool, eds.), American Association of Petroleum Geologists, Tulsa, OK, pp. 49–81.

Claypool, G. E., Love, A. H. and Maughan, E. K. (1978) Organic geochemistry, incipient metamorphism, and oil generation in black shale members of Phosphoria Formation, western interior United States. *American Association of Petroleum Bulletin*, 62, 98–120.

Clayton, J. L. and King, J. D. (1987) Effects of weathering on biological marker and aromatic hydrocarbon composition of organic matter in Phosphoria Shale outcrop. *Geochimica et Cosmochimica Acta*, 51, 2153–7.

Clayton, J. L. and Koncz, I. (1994) Petroleum geochemistry of the Zala Basin, Hungary. *American Association of Petroleum Geologists Bulletin*, 78, 1–22.

Clayton, J. L. and Ryder, R. T. (1984) Organic geochemistry of black shales and oils in the Minnelusa Formation (Permian and Pennsylvanian), Powder River Basin, Wyoming. In: *Hydrocarbon Source Rocks of the Greater Rocky Mountain Region* (J. Woodward, F. F. Meissner and J. L. Clayton, eds.), Rocky Mountain Association of Geologists, Denver, CO, pp. 231–53.

Clayton, J. L., King, J. D., Threlkeld, C. N. and Vuletich, A. (1987) Geochemical correlation of Paleozoic oils, Northern Denver Basin – implications for exploration. *American Association of Petroleum Geologists Bulletin*, 71, 103–9.

Clayton, J. L., King, J. D., Lubeck, C. M., Leventhal, J. S. and Daws, T. A. (1988) Paleoenvironmental and source rock assessment of black shales of Pennsylvanian age, Powder River and Northern Denver basins. *American Association of Petroleum Geologists Bulletin*, 72, 867.

Clayton, J. L., Rice, D. D. and Michael, G. E. (1991) Oil-generating coals of the San Juan Basin, New Mexico and Colorado, USA. *Organic Geochemistry*, 17, 735–42.

Clayton, J. L., Koncz, I., King, J. D. and Tatar, E. (1994a) Organic geochemistry of crude oils and source rocks, Békés Basin. In: *Basin Analysis in Petroleum Exploration* (P. G. Teleki, R. E. Mattick and J. Kokai, eds.), Kluwer, Dordrecht, pp. 161–85.

Clayton, J. L., Spencer, C. W. and Koncz, I. (1994b) Tólkomlós-Szolnok(.) petroleum system of southeastern Hungary. In: *The Petroleum System: From Source to Trap* (L. B. Magoon and W. G. Dow, eds.), American Association of Petroleum Geologists, Tulsa, OK, pp. 587–98.

Clayton, J. L., King, J. D., Lillis, P. G., Warden, A. and Yang, J. (1997) Geochemistry of oils from the Junggar Basin, northwest China. *American Association of Petroleum Geologists Bulletin*, 81, 1926–44.

Clegg, H., Horsfield, B., Stasiuk, L., Fowler, M. and Vleix, M. (1997) Geochemical characterisation of organic matter in Keg River Formation (Elk Point Group, Middle Devonian), La Crete Basin, Western Canada. *Organic Geochemistry*, 26, 627–43.

Clegg, H., Horsfield, B., Wilkes, H., Sinninghe Damsté, J. and Koopmans, M. P. (1998) Effect of artificial maturation on carbazole distribution, as revealed by the hydrous pyrolysis of an organic-sulfur-rich source rock (Ghareb Formation, Jordan). *Organic Geochemistry*, 29, 1953–60.

Clemett, S. J. and Zare, R. N. (1996) Evaluating the evidence for past life on Mars. *Science*, 274, 2122–3.

Clemett, S. J., Dulay, M. T., Gilette, J. S., *et al.* (1998) Evidence for the extraterrestrial origin of polycyclic aromatic hydrocarbons (PAHs) in the Martian meteorite ALH84001. *Faraday Discussions (Royal Society of Chemistry)*, 109, 417–36.

Clifford, D. J., Clayton, J. L. and Sinninghe Damsté, J. S. (1997) 3,4,5–2,3,6 Substituted diaryl carotenoid derivatives (*Chlorobiaceae*) and their utility as indicators of photic zone anoxia in sedimentary environments. In: *Abstracts from the 18th International Meeting on Organic Geochemistry, September 22–26, 1997, Maastricht, The Netherlands* (B. Horsfield, ed.), Forschungszentrum Jülich, Jülich, Germany, pp. 685–6.

Clifton, C. G., Walters, C. C. and Simoneit, B. R. T. (1990) Hydrothermal petroleums from Yellowstone National Park, Wyoming, USA. *Applied Geochemistry*, 5, 169–91.

Coates, J. D., Anderson, R. T. and Lovley, D. R. (1996) Oxidation of polycyclic aromatic hydrocarbons under sulfate-reducing conditions. *Applied and Environmental Microbiology*, 62, 1099–101.

Coates, J. D., Woodward, J., Allen, J., Philp, P. and Lovley, D. R. (1997) Anaerobic degradation of polycyclic aromatic hydrocarbons and alkanes in petroleum-contaminated marine harbor sediments. *Applied and Environmental Microbiology*, 63, 3589–93.

Cody, G. D., Boyce, C. K., Knoll, A., Wirick, S. and Jacobsen, C. (2003) In situ chemical analysis of ancient microfossils with STXM. Presented at the 225th American Chemical Society National Meeting, March 23–27, 2003, New Orleans, LA.

Cole, G. A. and Drozd, R. J. (1994) Heath-Tyler(!) petroleum system in central Montana, USA. In: *The Petroleum System – From Source to Trap* (L. B. Magoon and W. G. Dow, eds.), American Association of Petroleum Geologists, Tulsa, OK, pp. 371–85.

Cole, G. A., Drozd, R. J. and Daniel, J. A. (1990) Oil-source correlations between the Mississippian Heath shales and the reservoired oils in the Pennsylvanian Tyler sands, Montana. *American Association of Petroleum Geologists Bulletin*, 74, 630–1.

Cole, G. A., Abu-Ali, M. A., Aoudeh, S. M., *et al.* (1994a) Organic geochemistry of the Paleozoic petroleum system of Saudi Arabia. *Fuel*, 8, 1425–42.

Cole, G. A., Halpern, H. I. and Aoudeh, S. M. (1994b) The relationships between iron-sulfur-carbon and gamma-ray response, Silurian basal Qusaiba Shale, northern Saudi Arabia. *Saudi Aramco Journal of Technology*, 95, 9–19.

Cole, G. A., Abu-Ali, M. A., Colling, E. L., *et al.* (1995) Petroleum geochemistry of the Midyan and Jaizan basins of the Red Sea, Saudi Arabia. *Marine and Petroleum Geology*, 12, 597–614.

Cole, G. A., Requejo, A. G., Ormerod, D., Yu, Z. and Clifford, A. (2000) Petroleum geochemical assessment of the Lower Congo Basin. In: *Petroleum Systems of South Atlantic Margins* (M. R. Mello and B. J. Katz, eds.), American Association of Petroleum Geologists, Tulsa, OK, pp. 325–39.

Coleman, D. D. (2001) −70 per mil methane: is it biogenic or thermogenic? *Annual AAPG-SEPM Meeting Abstracts*, 2001, A39.

Coleman, M. L., Hedrick, D. B., Lovley, D. R., White, D. C. and Pye, K. (1993) Reduction of Fe(III) in sediments by sulphate-reducing bacteria. *Nature*, 361, 436–8.

Collier, R. J. and Johnston, J. H. (1991) The identification of possible hydrocarbon source rocks, using biomarker geochemistry, in the Taranaki Basin, New Zealand. *Journal of Southeast Asian Earth Sciences*, 5, 231–9.

Collister, J., Summons, R. E., Lichtfouse, E. and Hayes, J. M. (1992) An isotopic biogeochemical study of the Green River oil shale. *Organic Geochemistry*, 19, 265–76.

Colombo, J. C., Silverberg, N. and Gearing, J. N. (1996) Lipid biogeochemistry in the Laurentian Trough. I – fatty acids, sterols and aliphatic hydrocarbons in rapidly settling particles. *Organic Geochemistry*, 25, 211–25.

(1997) Lipid biogeochemistry in the Laurentian Trough. II – changes in composition of fatty acids, sterols and aliphatic hydrocarbons during early diagenesis. *Organic Geochemistry*, 26, 257–74.

Combaz, A. and de Matharel, M. (1978) Organic sedimentation and genesis of petroleum in the

Mahakam Delta, Kalimantan. *American Association of Petroleum Geologists Bulletin*, 62, 1684–5.

Condie, K. C., des Marais, D. J. and Abbott, D. (2001) Precambrian superplumes and supercontinents: a record in black shales, carbon isotopes, and paleoclimates. *Precambrian Research*, 106, 239–60.

Connan, J. (1974) Diagenese naturelle et diagenese artificielle de la matière organique à element vegetaux predominants. In: *Advances in Organic Geochemistry 1973* (B. P. Tissot and F. Bienner, eds.), Editions Technip, Paris, pp. 73–95.

(1981) Biological markers in crude oils. In: *Petroleum Geology in China* (J. F. Mason, ed.), Penn Well, Tulsa, OK, pp. 48–70.

(1984) Biodegradation of crude oils in reservoirs. In: *Advances in Petroleum Geochemistry*, Vol. 1 (J. Brooks and D. H. Welte, eds.), Academic Press, London, pp. 299–335.

Connan, J. and Dessort, D. (1987) Novel family of hexacyclic hopanoid alkanes (C_{32}–C_{35}) occurring in sediments and oils from anoxic paleoenvironments. *Organic Geochemistry*, 11, 103–13.

Connan, J., Restle, A. and Albrecht, P. (1980) Biodegradation of crude oil in the Aquitaine Basin. In: *Advances in Organic Geochemistry 1979*, Vol. 12 (A. G. Douglas and J. R. Maxwell, eds.), Pergamon Press, Oxford, pp. 1–17.

Connan, J., Bouroullec, J., Dessort, D. and Albrecht, P. (1986) The microbial input in carbonate-anhydrite facies of a sabkha palaeoenvironment from Guatemala: a molecular approach. *Organic Geochemistry*, 10, 29–50.

Connan, J., Nissenbaum, A. and Dessort, D. (1992) Molecular archaeology: export of Dead Sea asphalt to Canaan and Egypt in the Chalcolithic-Early Bronze Age (4th–3rd millennium BC). *Geochimica et Cosmochimica Acta*, 56, 2743–59.

Connan, J., Lacrampe-Couloume, G. and Magot, M. (1997) Anaerobic biodegradation of petroleum in reservoirs: a widespread phenomenon in nature. In: *Proceedings of the 18th International Meeting on Organic Geochemistry*, September 22–26, 1997, Maastricht, The Netherlands, pp. 5–6.

Conner, R. L., Mallory, F. B., Landrey, J. R., et al. (1971) Ergosterol replacement of tetrahymanol in *Tetrahymena* membranes. *Biochemical and Biophysical Research Communications*, 44, 995–1000.

Conner, R. L., Landrey, J. R. and Czarkowski, N. (1982) The effect of specific sterols on cell size and fatty acid composition of *Tetrahymena pyriformis*. *Journal of Protozoology*, 29, 105–9.

Conway, M. S. (1993) Ediacaran-like fossils in Cambrian Burgess Shale – type faunas of North America. *Paleontology*, 36, 593–635.

Cooke, D. A., Barlow, R., Green, J., Belt, S. T. and Rowland, S. J. (1998) Seasonal variations of highly branched isoprenoid hydrocarbons and pigment biomarkers in intertidal sediments of the Tamar estuary, UK. *Marine Environmental Research*, 45, 309–24.

Cooper, B. S. and Barnard, P. C. (1995) Source rocks and oils of the central and northern North Sea. In: *Petroleum Geochemistry and Basin Evaluation* (G. Demaison and R. J. Murris, eds.), American Association of Petroleum Geologists, Tulsa, OK, pp. 303–14.

Cooper, B. S., Telnaes, N. and Barnard, P. C. (1995) The Kimmeridge Clay Formation of the North Sea. In: *Petroleum Source Rocks, Casebooks in Earth Science* (B. J. Katz, ed.), Springer-Verlag, New York, pp. 89–110.

Corbett, R. E. and Smith, R. A. (1969) Lichens and fungi. Part VI. Dehydration rearrangements of 15-hydroxyhopanes. *Journal of the Chemical Society (C)*, 1969, 44–7.

Corliss, J. B., Dymond, J., Gordon, L. I., et al. (1979) Submarine thermal springs on the Galapagos Rift. *Science*, 203, 1073–83.

Corliss, J. B., Baross, J. A. and Hoffman, S. E. (1981) An hypothesis concerning the relationship between submarine hot springs and the origin of life on Earth. *Oceanologica Acta*, 4, 59–69.

Cornet, B. (1989) Late Triassic angiosperm-like pollen from the Richmond Rift Basin of Virginia, USA. *Palaeontographica Abstracts B*, 213, 37–87.

Cornet, B. and Habib, D. (1992) Angiosperm-like pollen from the ammonite-dated Oxfordian (Upper Jurassic) of France. *Review of Palaeobotany and Palynology*, 71, 269–94.

Cornet, B. and Olsen, P. E. (1990) *Early to Middle Carnian (Triassic) Flora and Fauna of the Richmond and Taylorsville Basins, Virginia and Maryland, USA*. Virginia Museum of Natural History, Martinsville, VA.

Cornford, C. (1984) Source rocks and hydrocarbons of the North Sea. In: *Introduction to the Petroleum Geology of the North Sea* (K. W. Glennie, ed.), Blackwell Science, London, pp. 171–204.

(1994) Mandal-Ekofisk(!) petroleum system in the Central Graben of the North Sea. In: *The Petroleum System – From Source to Trap* (L. B. Magoon and W. G. Dow, eds.), American Association of Petroleum Geologists, Tulsa, OK, pp. 537–71.

(2001) The oldest petroleum system in the world? In: *Hydrocarbons in Crystalline Rocks Joint Meeting Abstracts*, Geological Society, London.

Cornford, C., Needham, C. E. J. and de Walque, L. (1986) Geochemical habitat of North Sea oils. In: *Habitat of Hydrocarbons on the Norwegian Continental Shelf* (A. M. Spencer, ed.), Graham and Trotman, London, pp. 39–54.

Cornford, C., Christie, O., Endresen, U., Jensen, P. and Myhr, M.-B. (1988) Source rock and seep oil maturity in Dorset, southern England. *Organic Geochemistry*, 13, 399–409.

Correa da Silva, Z. C. and Cornford, C. (1985) The kerogen type, depositional environment and maturity, of the Irati Shale, Upper Permian of Paraná Basin, southern Brazil. *Organic Geochemistry*, 8, 399–411.

Costa Neto, C. (1983) Theoretical organic geochemistry. I. An alternative model for the epimerization of hydrocarbon chiral centers in sediments. In: *Advances in Organic Geochemistry 1981* (M. Bjorøy, C. Albrecht, C. Cornford, *et al.*, eds.), John Wiley & Sons, New York, pp. 834–8.

(1991) The effect of pressure on geochemical maturation: theoretical considerations. *Organic Geochemistry*, 17, 579–84.

Cox, H. C., de Leeuw, J. W., Schenck, P. A., *et al.* (1986) Bicadinane, a C_{30} pentacyclic isoprenoid hydrocarbon found in crude oil. *Nature*, 319, 316–18.

Crawford, R. L. (1995) The microbiology and treatment of nitroaromatic compounds. *Current Opinion in Biotechnology*, 6, 329–36.

Crawford, N. (1998) *Maleimides-(1H-pyrrole-2,5-diones) from Ancient Sediments as Indicators of Photic Zone Anoxia*. University of Bristol Press, Bristol, UK.

Creaney, S. (1980) The organic petrology of the Upper Cretaceous Boundary Creek Formation, Beaufort-Mackenzie Basin. *Bulletin of Canadian Petroleum Geology*, 28, 112–29.

Creaney, S. and Allan, J. (1990) Hydrocarbon generation and migration in the Western Canada Sedimentary Basin. In: *Classic Petroleum Provinces* (J. Brooks, ed.), Geological Society of London, London, pp. 189–202.

Creaney, S. and Passey, Q. R. (1993) Recurring patterns of total organic carbon and source rock quality within a sequence stratigraphic framework. *American Association of Petroleum Geologists Bulletin*, 77, 386–401.

Creaney, S., Osadetz, K. G., Allan, J., *et al.* (1994a) Petroleum generation and migration in the Western Canada Basin. In: *Geological Atlas of the Western Canada Sedimentary Basin*. (G. D. Mossop and I. Shetsen, eds.), Canadian Society of Petroleum Geologists and Alberta Research Council, Calgary, pp. 455–68.

Creaney, S., Allan, J., Cole, K. S., *et al.* (1994b) Petroleum systems. In: *Geological Atlas of the Western Canada Sedimentary Basin* (G. D. Mossop and I. Shetsen, eds.), Canadian Society of Petroleum Geologists and Alberta Research Council, Calgary, pp. 455–68.

Crick, I. H. (1992) Petrological and maturation characteristics of organic matter from the Middle Proterozoic McArthur Basin, Australia. *Australian Journal of Earth Sciences*, 39, 501–19.

Crick, I. H., Boreham, C. J., Cook, A. C. and Powell, T. G. (1988) Petroleum geology and geochemistry of Middle Proterozoic McArthur Basin, northern Australia. II. Assessment of source rock potential. *American Association of Petroleum Geologists Bulletin*, 72, 1495–514.

Crouch, E. M., Heilmann-Clausen, C., Brinkhuis, H., *et al.* (2001) Global dinoflagellate event associated with the late Paleocene thermal maximum. *Geology*, 29, 315–18.

Crowley, T. J. and North, G. R. (1991) *Paleoclimatology*. Oxford University Press, New York.

Crozier, A. (1983) *The Biochemistry and Physiology of Gibberellins*. Praeger Press, New York.

Cruz, C. E., Kozlowski, E. and Villar, H. J. (1998) Agrio (Neocomian) petroleum systems: main target in the Neuquén Basin Thrust Belt, Argentina. Presented at the ABGP/American Association of Petroleum Geologists International Conference and Exhibition, November 8–11, 1998, Rio de Janeiro, Brazil.

Cumbers, K. M., Alexander, R. and Kagi, R. I. (1986) Methylbiphenyl, ethylbiphenyl and dimethylbiphenyl isomer distribution in some sediments and crude oils. *Geochimica et Cosmochimica Acta*, 51, 3105–12.

Curiale, J. A. (1986) Origin of solid bitumens, with emphasis on biological marker results. *Organic Geochemistry*, 10, 559–80.

(1987) Steroidal hydrocarbons of the Kishenehn Formation, northwest Montana. *Organic Geochemistry*, 11, 233–44.

(1988) Molecular genetic markers and maturity indices in intermontane lacustrine facies: Kishenehn Formation, Montana. *Organic Geochemistry*, 13, 633–8.

(1991) The petroleum geochemistry of Canadian Beaufort Tertiary "non-marine" oils. *Chemical Geology*, 93, 21–45.

(1995) Saturated and olefinic terrigenous triterpenoid hydrocarbons in a biodegraded Tertiary oil of northeast Alaska. *Organic Geochemistry*, 23, 177–82.

(2002) A review of the occurrences and causes of migration-contamination in crude oil. *Organic Geochemistry*, 33, 1389–400.

Curiale, J. A. and Bromley, B. W. (1996) Migration of petroleum into Vermilion 14 Field, Gulf Coast, USA – molecular evidence. *Organic Geochemistry*, 24, 563–79.

Curiale, J. A. and Frolov, E. B. (1998) Occurrence and origin of olefins in crude oils. A critical review. *Organic Geochemistry*, 29, 397–408.

Curiale, J. A. and Gibling, M. R. (1992) Organic geochemistry of Mae Sot Basin oil shales, Thailand: implications for depositional setting and basin reconstruction. *American Association of Petroleum Geologists Bulletin*, 76, 1095.

(1994) Productivity control on oil shale formation: Mae Sot Basin, Thailand. *Organic Geochemistry*, 21, 67–89.

Curiale, J. A. and Lin, R. (1991) Tertiary deltaic and lacustrine organic facies; comparison of biomarker and kerogen distributions. *Organic Geochemistry*, 17, 785–803.

Curiale, J. A. and Odermatt, J. R. (1989) Short-term biomarker variability in the Monterey Formation, Santa Maria Basin. *Organic Geochemistry*, 14, 1–13.

Curiale, J. A., Cameron, D. and Davis, D. V. (1985) Biological marker distribution and significance in oils and rocks of the Monterey Formation, California. *Geochimica et Cosmochimica Acta*, 49, 271–88.

Curiale, J. A., Sperry, S. W. and Senftle, J. T. (1988) Regional source rock potential of lacustrine Oligocene Kishenehn Formation, Northwestern Montana. *American Association of Petroleum Geologists Bulletin*, 72, 1437–49.

Curiale, J. A., Kyi, P., Collins, I. D., *et al.* (1994) The central Myanmar (Burma) oil family – composition and implications for source. *Organic Geochemistry*, 22, 237–55.

Curiale, J., Morelos, J., Lambiase, J. and Mueller, W. (2000) Brunei Darussalam – characteristics of selected petroleums and source rocks. *Organic Geochemistry*, 31, 1475–93.

Curiale, J. A., Covington, G. H., Shamsuddin, A. H. M., Morelos, J. A. and Shamsuddin, A. K. M. (2002) Origin of petroleum in Bangladesh. *American Association of Petroleum Geologists Bulletin*, 84, 625–52.

Curry, D. J., Emmett, J. K. and Hunt, J. W. (1994) Geochemistry of aliphatic-rich coals in the Cooper Basin, Australia and Taranaki Basin, New Zealand: implications for the occurrence of potentially oil-generative coals. In: *Coal and Coal-bearing Strata as Oil-prone Source Rocks?* (A. J. Fleet, ed.), Geological Society of London, London, pp. 149–82.

Curtis, D. M. (1987) The northern Gulf of Mexico Basin. *Episodes*, 10, 267–70.

Czochanska, Z., Gilbert, T. D., Philp, R. P., *et al.* (1988) Geochemical application of sterane and triterpane biomarkers to a description of oils from the Taranaki Basin in New Zealand. *Organic Geochemistry*, 12, 123–35.

Dahl, J. E. (1990) The organic geochemistry of the Alum Shale, Sweden. Ph. D. Thesis, University of California, Los Angeles, CA.

Dahl, J., Hallberg, R. and Kaplan, I. R. (1988) Effects of irradiation from uranium decay on extractable organic matter in the Alum Shales of Sweden. *Organic Geochemistry*, 12, 559–71.

Dahl, J. E., Chen, R. T. and Kaplan, I. R. (1989) Alum Shale bitumen maturation and migration: implication for Gotland's oil. *Journal of Petroleum Geology*, 12, 465–76.

Dahl, J., Moldowan, J. M., McCaffrey, M. A. and Lipton, P. A. (1992) A new class of natural products revealed by 3β-alkyl steranes in petroleum. *Nature*, 355, 154–7.

Dahl, J., Moldowan, J. M. and Sundararaman, P. (1993) Relationship of biomarker distribution to depositional environment; Phosphoria Formation, Montana, USA *Organic Geochemistry*, 20, 1001–17.

Dahl, J. E., Moldowan, J. M., Teerman, S. C., *et al.* (1994) Source rock quality determination from oil biomarkers. I: a new geochemical technique. *American Association of Petroleum Geologists Bulletin*, 78, 1507–26.

Dahl, J., Moldowan, J. M., Summons, R. E., *et al.* (1995) Extended 3β-alkyl steranes and 3-alkyl triaromatic steroids in crude oils and rock extracts. *Geochimica et Cosmochimica Acta*, 59, 3717–29.

Dahl, J. E., Moldowan, J. M., Peters, K. E., *et al.* (1999) Diamondoid hydrocarbons as indicators of natural oil cracking. *Nature*, 399, 54–7.

Dasgupta, S., Tang, Y., Moldowan, J. M., Carlson, R. M. K. and Goddard, W. A., III (1995) Stabilizing the boat conformation of cyclohexane rings. *Journal of the American Chemical Society*, 117, 6532–4.

Dastillung, M. and Albrecht, P. (1977) Δ2-Sterenes as diagenetic intermediates in sediments. *Nature*, 269, 678–9.

Dastillung, M. and Corbert, B. (1975) La géochemie organique des sédiments marine profond. In: *Orgon II, Atlantique-N. W. Brésil* (A. Combaz and R. Pelet, eds.), Centre National de la Recherche Scientifique, Paris, pp. 296–326.

Dastillung, M., Albrecht, P. and Ourisson, B. (1980a) Aliphatic and polycyclic alcohols in sediments: hydroxylated derivatives of hopane and of 3-methylhopane. *Journal of Chemical Research*, Symposium, 168–9.

(1980b) Aliphatic and polycyclic ketones in sediments. C_{27}–C_{35} ketones and aldehydes of the hopane series. *Journal of Chemical Research*, Symposium, 166–7.

Davis, J. B. (1952) Studies on soil samples from "paraffin dirt" bed. *American Association of Petroleum Geologists Bulletin*, 11, 2186–8.

(1967) *Petroleum Microbiology*. Elsevier, New York.

Davis, C. L. and Pratt, L. M. (2002) Organic and inorganic geochemistry of Early and Late Cretaceous black shales from Western Venezuela: implications for the paleoceanographic evolution of northern South America. Presented at the Annual Meeting of the American Association of Petroleum Geologists, March 10–13, 2002, Houston, TX.

Davis, M., Hut, P. and Muller, R. A. (1984) Extinction of species by periodic comet showers. *Nature*, 308, 713–17.

Davis, H. R., Byers, C. W. and Pratt, L. M. (1989) Depositional mechanisms and organic matter in Mowry Shale (Cretaceous), Wyoming. *American Association of Petroleum Geologists Bulletin*, 79, 1103–16.

Dean, W. E. and Anders, D. E. (1991) Effect of occurrence, depositional environment, and diagenesis on characteristics of organic matter in oil shale from the Green River Formation, Wyoming, Utah, and Colorado In: *Geochemical, Biogeochemical, and Sedimentological Studies of the Green River Formation, Wyoming, Utah, and Colorado* (M. L. Tuttle, ed.), U.S. Geologic Survey Washington, DC, pp. F1–16.

Dean, W. E. and Arthur, M. A. (1998) Stratigraphy and paleoenvironments of the Cretaceous Western Interior Seaway, USA. *Concepts in Sedimentology and Paleontology*, 6, 1–10.

De Graaf, W., Sinninghe Damsté, J. S. and de Leeuw, J. W. (1992) Laboratory simulation of natural sulphurization. I. Formation of monomeric and oligomeric isoprenoid polysulphides by low-temperature reactions of inorganic polysulphides with phytol and phytadienes. *Geochimica et Cosmochimica Acta*, 56, 4321–8.

De Grande, S. M. B., Aquino Neto, F. R. and Mello, M. R. (1993) Extended tricyclic terpanes in sediments and petroleum. *Organic Geochemistry*, 20, 1039–47.

Dehler, C. M., Elrick, M., Karlstrom, K. E., *et al.* (2001) Neoproterozoic Chuar Group (ca 800–742 Ma), Grand Canyon: a record of cyclic marine deposition during global cooling and supercontinent rifting. *Sedimentary Geology*, 141–2, 465–99.

De las Heras, F. X. C., Grimalt, J. O., Lopez, J. F., *et al.* (1997) Free and sulphurized hopanoids and highly branched isoprenoids in immature lacustrine oil shales. *Organic Geochemistry*, 27, 41–63.

De Leeuw, J. W. and Sinninghe Damsté, J. S. (1990) Organic sulfur compounds and other biomarkers as indicators of paleosalinity. In: *Geochemistry of Sulfur in Fossil Fuels* (W. L. Orr and C. M. White, eds.), American Chemical Society, Washington, DC, pp. 417–43.

De Leeuw, J. W., Sinninghe Damsté, J. S., Klok, J., Schenck, P. A. and Boon, J. J. (1985) Biogeochemistry of Gavish Sabkha sediments. I. Studies on neutral reducing sugars and lipid moieties by gas chromatography-mass spectrometry. In: *Hypersaline Ecosystems*: Vol. 53 (G. M. Friedman and W. E. Krumbein, eds.), Spinger-Verlag, Berlin, pp. 350–67.

De Lemos Scofield, A. (1990) Nouveaux marqueurs biologiques de sediments et petroles riches en soufre: identification et mode de formation. Ph. D. thesis, L'Universite Louis Pasteur de Strasbourg, Strasbourg, France.

De Long, E. F. (1992) Archaea in coastal marine environments. *Proceedings of the National Academy of Science*, 89, 5685–9.

De Long, E. F., Ying, W. K., Prezelin, B. B. and Jovine, R. V. M. (1994) High abundance of archaea in Antarctic marine picoplankton. *Nature*, 371, 695–7.

Del Río, J. C., Garcia-Molla, J., Gonzalez-Vila, F. J. and Martin, F. (1994) Composition and origin of the aliphatic extractable hydrocarbons in the Puertollano (Spain) oil shale. *Organic Geochemistry*, 21, 897–909.

Delsemme, A. H. (1996) The origin of the atmosphere and oceans. In: *Comets and the Origins of Life* (P. J. Thomas, C. F. Chyba and C. P. McKay, eds.), Springer-Verlag, Berlin, pp. 26–67.

Demaison, G. J. (1984) The generative basin concept. In: *Petroleum Geochemistry and Basin Evaluation* (G. J. Demaison and R. J. Murris, eds.), American Association of Petroleum Geologists, Tulsa, OK, pp. 1–14.

Demaison, G. J. and Bougeois, G. T. (1984) Environment of deposition of Middle Miocene Alcanar carbonate source beds, Casablanca Field, Tarragona Basin, Spain. In: *Petroleum Geochemistry and Source Rock Potential of Carbonate Rocks* (J. G. Palacas, ed.), American Association of Petroleum Geologists, pp. 151–61.

Demaison, G. J. and Huizinga, B. J. (1994) Genetic classification of petroleum systems using three factors: charge, migration, and entrapment. In: *The Petroleum System – From Source to Trap* (L. B. Magoon and W. G. Dow, eds.), American Association of Petroleum Geologists, Tulsa, OK, pp. 73–89.

Demaison, G. J. and Moore, G. T. (1980) Anoxic environments and oil source bed genesis. *American Association of Petroleum Geologists Bulletin*, 64, 1179–209.

Demaison, G., Holck, A. J. J., Jones, R. W. and Moore, G. T. (1983) Predictive source bed stratigraphy; a guide to regional petroleum occurrence. In: *Proceedings of the 11th World Petroleum Congress*, Vol. 2, John Wiley & Sons, London, pp. 1–13.

Dembicki, H., Jr and Mathiesen, M. D. (1994) Biomarkers from asphaltene pyrolysis; an additional tool for oil correlation. Presented at the Annual Meeting of the American Association of Petroleum Geologists, June 3–6, 1994, Denver, CO.

Demirel, I. H., Yurtsever, T. S. and Guneri, S. (2001) Petroleum systems of the Adiyaman region, Southeastern Anatolia, Turkey. *Marine and Petroleum Geology*, 18, 391–410.

Derenne, S., Largeau, C., Casadevall, E. and Connan, J. (1988) Comparison of torbanites of various origins and evolutionary stages. Bacterial contribution to their formation. Cause of lack of botryococcane in bitumens. *Organic Geochemistry*, 12, 43–59.

Derenne, S., Largeau, C., Casadevall, E., *et al.* (1990) Characterization of Estonian kukersite by spectroscopy and pyrolysis; evidence for abundant alkyl phenolic moieties in an Ordovician, marine, type II/I kerogen. *Organic Geochemistry*, 16, 873–88.

Derenne, S., Largeau, C., Casadevall, E., Berkaloff, C. and Rousseau, B. (1991) Chemical evidence of kerogen formation in source rocks and oil shales via selective preservation of thin resistant outer walls of microalgae: origin of ultralaminae. *Geochimica et Cosmochimica Acta*, 55, 1041–50.

Derenne, S., Metzger, P., Largeau, C., *et al.* (1992) Similar morphological and chemical variations of *Gloeocapsomorpha prisca* in Ordovician sediments and cultured *Botryococcus braunii* as a response to changes in salinity. *Organic Geochemistry*, 19, 299–313.

Derenne, S., Largeau, C. and Berkaloff, C. (1996) First example of an algaenan yielding an aromatic-rich pyrolysate. Possible geochemical implications on marine kerogen formation. *Organic Geochemistry*, 24, 617–27.

Derenne, S., Largeau, C., Hetényi, M., *et al.* (1997) Chemical structure of the organic matter in a Pliocene maar-type shale: implicated *Botryococcus* race strains and formation pathways. *Geochimica et Cosmochimica Acta*, 61, 1879–89.

Derenne, S., Largeau, C., Bruker-Wein, A., *et al.* (2000) Origin of variations in organic matter abundance and composition in a lithologically homogeneous maar-type oil shale deposit (Gérce, Pliocene, Hungary). *Organic Geochemistry*, 31, 787–98.

De Rosa, M., Gambacorta, A. and Minale, L. (1974) Cyclic diether lipids from very thermophilic acidophilic bacteria. *Journal of the Chemical Society, Chemical Communications*, 543–4.

De Rosa, M., de Rosa, S. and Gambacorta, A. (1977a) ^{13}C-NMR assignments and biosynthetic data for the ether lipids of *Caldariella*. *Phytochemistry*, 16, 1909–12.

De Rosa, M., de Rosa, S., Gambacorta, A., Minale, L. and Bu'lock, J. D. (1977b) Chemical structure of the ether lipids of thermophilic bacteria of the *Caldariella* group. *Phytochemistry*, 16, 1961–5.

De Rosa, M., de Rosa, S., Gambacorta, A. and Bu'lock, J. D. (1977c) Lipid structures in the *Caldariella* group of extreme thermoacidophile bacteria. *Journal of the Chemical Society, Chemical Communications*, 514–5.

De Rosa, M., Trincone, A., Nicolaus, B. and Gambacorta, A. (1991) Achaebacteria: lipids, membrane structures and adaptation to environmental stresses. In: *Life Under Extreme Conditions* (G. di Prisco, ed.), Spinger-Verlag, Berlin, pp. 61–87.

Dessort, D. and Connan, J. (1993) Occurrence of novel C-ring opened hopanoids in biodegraded oils from carbonate source rocks. In: *Proceedings of the 16th International Meeting on Organic Geochemistry, Stavanger, Norway, September 1993* (K. Oygard, ed.), Falch Hurtigtrykk, Kalbakken, Norway, pp. 485–95.

Detre, C. H., Toth, I., Don, G., *et al.* (1998) A nearby supernova explosion at the Permo-Triassic boundary. In: *23rd Symposium on Antarctic Meteorites* (T. Hirasawa, ed.), National Institute of Polar Research, Tokyo, pp. 23–4.

Dias, R. F., Freeman, K. H. and Franks, S. G. (2002a) Gas chromatography-pyrolysis-isotope ratio mass spectrometry: a new method for investigating intramolecular isotopic variation in low molecular weight organic acids. *Organic Geochemistry*, 33, 161–8.

Dias, R. F., Freeman, K. H., Lewan, M. D. and Franks, S. G. (2002b) $\delta^{13}C$ of low-molecular-weight organic acids generated by the hydrous pyrolysis of oil-prone source rocks. *Geochimica et Cosmochimica Acta*, 66, 2755–69.

Dickens, G. R., O'Neil, J. R., Rea, D. K. and Owen, R. M. (1995) Dissociation of oceanic methane hydrate as a cause of the carbon isotope excursion at the end of the Paleocene. *Paleoceanography*, 10, 965–71.

Didyk, B. M., Simoneit, B. R. T., Brassell, S. C. and Eglinton, G. (1978) Organic geochemical indicators of palaeoenvironmental conditions of sedimentation. *Nature*, 272, 216–22.

Dieckmann, V., Horsfield, B. and Schenk, H. J. (2000) Heating rate dependency of petroleum-forming reactions: implications for compositional kinetic predictions. *Organic Geochemistry*, 31, 1333–48.

Dimichele, W. A. and Phillips, T. L. 1995. The response of hierarchically structured ecosystems to long-term climate change: a case study using tropical peat swamps of Pennsylvanian age. In: *Effects of Past Global Change on Life* (S. M. Stanley, A. H Knoll, and J. P. Kennett, eds.), National Research Council, Studies in Geophysics, Washington, DC, pp. 134–55.

Dimmler, A., Cyr, T. D. and Strausz, O. P. (1984) Identification of bicyclic terpenoid hydrocarbons in the saturate fraction of Athabasca oil sand bitumen. *Organic Geochemistry*, 7, 231–8.

Disnar, J.R. and Harouna, M. (1994) Biological origin of tetracyclic diterpanes, *n*-alkanes and other biomarkers in Lower Carboniferous Gondwana coals (Niger). *Organic Geochemistry*, 21, 143–52.

Disnar, J. R., le Strat, P., Farjanel, G. and Fikri, A. (1996) Organic matter sedimentation in the northeast of the Paris Basin: consequences in the deposition of the lower Toarcian black shales. *Chemical Geology*, 131, 15–35.

Dixon, J. and Stasiuk, L. D. (1998) Stratigraphy and hydrocarbon potential of Cambrian strata, Northern Interior Plains, Northwest Territories. *Bulletin of Canadian Petroleum Geology*, 46, 445–70.

Dixon, J., Dietrich, J. R., Snowdon, L. R., Morrell, G. and McNeil, D. H. (1992) Geology and petroleum potential of Upper Cretaceous and Tertiary strata, Beaufort-Mackenzie Area, Northwest Canada. *American Association of Petroleum Geologists Bulletin*, 76, 927–47.

Dolson, J. C., Shann, M. V., Matbouly, S., *et al*. (2001) The petroleum potential of Egypt. In: *Petroleum Provinces of the Twenty-first Century* (M. W. Downey, J. C. Threet and W. A. Morgan, eds.), American Association of Petroleum Geologists, Tulsa, OK, pp. 453–82.

Donoghue, N. A. and Trudgill, P. W. (1975) The metabolism of cyclohexanol by *Acinetobacter NCIB9871*. *European Journal of Biochemistry*, 6, 1–7.

Donovan, R. N. (1980) Cyclicity in the Orcadian Lake Basin. *Scottish Journal of Geology*, 16, 35–50.

Do Rozario, R. F. (1990) Palm Valley Gas Field–Australia Amadeus Basin, Northern Territory. In: *Structural Traps IV: Tectonic and Nontectonic Fold Traps* (E. A. Beaumont and N. H. Foster, eds.), American Association of Petroleum Geologists, Tulsa, OK, pp. 255–79.

Dou, L. (1997) The lower Cretaceous petroleum system in Northeast China. *Journal of Petroleum Geology*, 20, 475–88.

Douglas, A. G., Sinninghe Damsté, J. S., Fowler, M. G., Eglinton, T. I. and de Leeuw, J. W. (1991) Unique distributions of hydrocarbons and sulphur compounds released by flash pyrolysis from the fossilized alga *Gloeocapsomorpha prisca*, a major constituent in one of four Ordovician kerogens. *Geochimica et Cosmochimica Acta*, 55, 275–91.

Dow, W. G. (1974) Application of oil-correlation and source-rock data to exploration in Williston Basin. *American Association of Petroleum Geologists Bulletin*, 58, 1253–62.

(1977) Kerogen studies and geological interpretations. *Journal of Geochemical Exploration*, 7, 79–99.

(1984) Oil source beds and oil prospect definition in the upper Tertiary of the Gulf Coast. *Transactions of the Gulf Coast Association of Geological Societies*, 34, 329–39.

Dow, W. G. and Pearson, D. B. (1975) Organic matter in Gulf Coast sediments. In: *Proceedings of the Seventh Annual Offshore Technology Conference*, OTC, Houston, TX, pp. 85–94.

Dowling, L. M., Boreham, C. J., Hope, J. M., Murray, A. P. and Summons, R. E. (1995) Carbon isotopic composition of hydrocarbons in ocean-transported bitumens from the coastline of Australia. *Organic Geochemistry*, 23, 729–37.

Doyle, J. A. and Donoghue, M. (1987) The importance of fossils in elucidating seed plant phylogeny and macroevolution. *Review of Palaeobotany and Palynology*, 50, 63–95.

Doyle, J. A. and Hickey, L. (1977) Early Cretaceous fossil evidence for angiosperm evolution. *Botanical Review*, 43, 3–104.

Droser, M. L., Bottjer, D. J., Sheehan, P. M. and McGhee, G. R., Jr. (2000) Decoupling of taxonomic and ecologic severity of Phanerozoic marine mass extinctions. *Geology*, 28, 675–8.

Drozd, R. J. and Cole, G. A. (1994) Point Pleasant-Brassfield(!) petroleum system, Appalachian Basin, USA. In: *The Petroleum System – From Source to Trap* (L. B. Magoon and W. G. Dow, eds.), American Association of Petroleum Geologists, Tulsa, OK, pp. 387–98.

Dujie, H., Maowen, L. and Qinghua H. (2000) Marine transgressional events in the gigantic freshwater Lake Songliao: paleontological and geochemical evidence. *Organic Geochemistry*, 31, 763–8.

Dunham, K. W., Meyers, P. A. and Rullkötter, J. (1988) Biomarker comparisons of Michigan Basin oils. In: *Geochemical Biomarkers* (T. F. Yen and J. M. Moldowan, eds.), Harwood Academic Publishers, Chur, Switzerland, pp. 181–202.

Dunlop, R. W. and Jeffries, P. R. (1985) Hydrocarbons of the hypersaline basins of Shark Bay, Western Australia. *Organic Geochemistry*, 8, 313–20.

Dunn, J. F., Hartshorn, K. G. and Hartshorn P. W. (1995) Structural styles and hydrocarbon potential of the sub-Andean thrust belt of southern Bolivia. In: *Petroleum Basins of South America* (A. J. Tankard, S. R. Suarez, and H. J. Welsink, eds.), American Association of Petroleum Geologists, Tulsa, OK, pp. 523–43.

Durand, B. (1983) Present trends in organic geochemistry in research on migration of hydrocarbons. In: *Advances in Organic Geochemistry 1981* (M. Bjorøy, C. Albrecht, C. Cornford, *et al.*, eds.), John Wiley & Sons, New York, pp. 117–28.

Durand, B. and Oudin, J. L. (1979) Exemple de migration des hydrocarbures dans une série deltaique: le delta de la Mahakam, Kalimantan, Indonesie. *Proceedings of the Tenth World Petroleum Congress*, 2, 3–11.

Durand, B. and Paratte, M. (1983) Oil potential of coals: a geochemical approach. In: *Petroleum Geochemistry and Exploration of Europe* (J. Brooks, ed.), Blackwell Scientific Publishing, Boston, MA, pp. 255–65.

Dutkiewicz, A., Rasmussen, B. and Buick, R. (1998) Oil preserved in fluid inclusions in Archean sandstones. *Nature*, 395, 885–8.

Du Toit, S. R., Kurdy, S., Asfaw, S. H. and Gessesse, A. A. (1997) Dual rifts, dual source systems and strike-slip tectonics: keys to oil and gas in the Ogaden Basin, Ethiopia. Presented at the Annual Meeting of the American Association of Petroleum Geologists, April 6–9, 1997, Dallas, TX.

Duval, B. C., Choppin de Janvry, G. and Loiret, B. (1992a) Detailed geoscience reinterpretation of Indonesia's Mahakam Delta scores. *Oil and Gas Journal*, 90, 67–72.

(1992b) The Mahakam Delta province; an ever-changing picture and a bright future. In: *Proceedings of the 24th Offshore Technology Conference*, OTC, Houston, TX, pp. 393–404.

Duval, B. C., Choppin de Janvry, G., Heidmann, J. C., Pition, J. L. and Ten Haven, H. L. (1995) A hybrid petroleum system on the Western African Margin, an example in offshore Angola. Presented at the Annual Meeting of the American Association of Petroleum Geologists, May 3–5, San Francisco.

Dyman, T. S., Palacas, J. A., Tysdal, R. G., Perry, W. Jr and Pawlewicz, M. J. (1996) Source rock potential of Middle Cretaceous rocks in southwestern Montana. *American Association of Petroleum Geologists Bulletin*, 80, 1177–83.

Dzou, L. I. P., Noble, R. A. and Senftle, J. T. (1995) Maturation effects on absolute biomarker concentration in a suite of coals and associated vitrinite concentrates. *Organic Geochemistry*, 23, 681–97.

Dzou, L. I., Holba, A. G., Ramón, J. C., Moldowan, J. M. and Zinniker, D. (1999) Application of new diterpane biomarkers to source, biodegradation and mixing effects on Central Llanos Basin oils, Colombia. *Organic Geochemistry*, 30, 515–34.

Eckardt, C. B., Wolf, M. and Maxwell, J. R. (1989) Iron porphyrins in the Permian Kupferschiefer of the Lower Rhine Basin, N. W. Germany. *Organic Geochemistry*, 14, 659–66.

Edwards, D., Davies, K. L. and Axe, L. M. (1992) A vascular conducting strand in the early land plant, *Cooksonia*. *Nature*, 357, 683–5.

Edwards, D. S., McKirdy, D. M. and Summons, R. E. (1998) Enigmatic asphaltites from the Southern Australian Margin: molecular and carbon isotopic composition. *Petroleum Exploration Society of Australia Journal*, 26, 106–29.

Edwards, D. S., Struckmeyer, H. I. M., Bradshaw, M. T. and Skinner, J. E. (1999) Geochemical characteristics of Australia's southern margin petroleum systems. *APPEA Journal*, 39, 297–321.

Edwards, D. S., Kennard, I. M., Summons, R. E., *et al.* (2000a) Bonaparte Basin: geochemical characteristics of hydrocarbon families and petroleum systems. *AGSO Research Newsletter*, 14–19.

Edwards, K. J., Bond, P. L., Gihring, T. M. and Banfield, J. F. (2000b) An archaeal iron-oxidizing extreme acidophile important in acid mine drainage. *Science*, 287, 1796–9.

Eganhouse, R. P. (1986) Long-chain alkylbenzenes: their analytical chemistry, environmental occurrence and fate. *International Journal of Environmental and Analytical Chemistry*, 26, 241–63.

(1997) *Molecular Markers in Environmental Geochemistry*. American Chemical Society, Washington, DC.

Eglinton, T. I. (1994) Carbon isotopic evidence for the origin of macromolecular aliphatic structures in kerogen. *Organic Geochemistry*, 21, 721–35.

Eglinton, T. I. and Douglas, A. G. (1988) Quantitative study of biomarker hydrocarbons released from kerogens during hydrous pyrolysis. *Energy & Fuels*, 2, 81–8.

Eglinton, G. and Hamilton, R. J. (1967) Leaf epicuticular waxes. *Science*, 156, 1322–35.

Eglinton, G. and Murphy, M. E. (1969) *Organic Geochemistry: Methods and Results*. Springer-Verlag, Berlin.

Ehrenreich, P., Behrends, A., Harder, J. and Widdel, F. (2000) Anaerobic oxidation of alkanes by newly isolated denitrifying bacteria. *Archives of Microbiology*, 173, 58–64.

Ekweozor, E. M., and Daukoru, E. M. (1994) Northern delta depobelt portion of the Akata-Agbada(!) petroleum system, Niger Delta, Nigeria. In: *The Petroleum System – From Source to Trap* (L. B. Magoon and W. G. Dow, eds.), American Association of Petroleum Geologists, Tulsa, OK, pp. 599–614.

Ekweozor, C. M. and Strausz, O. P. (1982) Tricyclic terpanes in the Athabasca oil sands: their geochemistry. In: *Advances in Organic Geochemistry 1981* (M. Bjorøy, C. Albrecht, C. Cornford, *et al.*, eds.), John Wiley & Sons, New York, pp. 746–66.

Ekweozor, C. M. and Telnaes, N. (1990) Oleanane parameter: verification by quantitative study of the biomarker occurrence in sediments of the Niger Delta. *Organic Geochemistry*, 16, 401–13.

Ekweozor, C. M. and Udo, O. T. (1988) The oleananes: origin, maturation, and limits of occurrence in southern Nigeria sedimentary basins. *Organic Geochemistry*, 13, 131–40.

Ekweozor, C. M., Okogun, J. I., Ekong, D. E. U. and Maxwell, J. M. (1979a) Preliminary organic geochemical studies of samples from the Niger Delta (Nigeria). I. Analyses of crude oils for triterpanes. *Chemical Geology*, 27, 11–28.

(1979b) Preliminary organic geochemical studies of samples from the Niger Delta (Nigeria). II. Analyses of shale for triterpenoid derivatives. *Chemical Geology*, 27, 29–37.

El-Alami, M. (1996) Habitat of oil in Abu Attiffel area, Sirt Basin, Libya. In: *The Geology of Sirt Basin*, Vol. II (M. J. Salem, A. S. El-Hawat and A. M. Sbeta, eds.), Elsevier, Amsterdam, pp. 337–48.

El Albani, A., Kuhnt, W., Luderer, F., Herbin, J. P. and Caron, M. (1999) Paleoenvironmental evolution of the Late Cretaceous sequence in the Tarfaya Basin (southwest of Morocco). In: *The Oil and Gas Habitats of the South Atlantic* (N. R. Cameron, R. H. Bate and V. S. Clure, eds.), Geological Society of London, London, pp. 223–40.

Ellis, L. (1995) Aromatic hydrocarbons in crude oil and sediments: Molecular sieve separations and biomarkers. Ph. D. thesis, Curtin University of Technology, Perth, Australia.

Ellis, J. and Schramm, D. N. (1995) Could a nearby supernova explosion have caused a mass extinction? *Proceedings of the National Academy of Sciences, USA*, 92, 235–8.

Ellis, L., Kagi, R. I. and Alexander, R. (1992) Separation of petroleum hydrocarbons using dealuminated mordenite molecular sieve. I. Monoaromatic hydrocarbons. *Organic Geochemistry*, 18, 587–93.

Ellis, L., Alexander, R. and Kagi, R. I. (1994) Separation of petroleum hydrocarbons using dealuminated mordenite molecular sieve. II. Alkylnaphthalenes and alkylphenanthrenes. *Organic Geochemistry*, 21, 849–55.

Ellis, L., Singh, R. K., Alexander, R. and Kagi, R. I. (1995) Identification and occurrence of dihydro-*ar*-curcumene in crude oils and sediments. *Organic Geochemistry*, 23, 197–203.

(1996a) Formation of isohexyl alkylaromatic hydrocarbons from aromatization-rearrangement of terpenoids in the sedimentary environment – a new class of biomarker. *Geochimica et Cosmochimica Acta*, 60, 4747–63.

Ellis, L., Langworthy, T. A. and Winans, R. (1996b) Occurrence of phenylalkanes in some Australian crude oils and sediments. *Organic Geochemistry*, 24, 57–69.

Ellis, L., Fisher, S. J., Singh, R. K., Alexander, R. and Kagi, R. I. (1998) Identification of alkenylbenzenes in pyrolyzates using GC-MS and GC-FTIR techniques: evidence for kerogen aromatic moieties with various binding sites. *Organic Geochemistry*, 30, 651–65.

Ellis, L., Singh, R. K., Alexander, R. and Kagi, R. I. (1999) Long-chain alkylnaphthalenes in crude oils and sediments. Presented at the 19th International Meeting on Organic Geochemistry, September 6–10, 1999, Istanbul.

El-Sabagh, S. M. and Al-Dhafeer, M. M. (2000) Occurrence and distribution of n-alkanes and n-fatty acids in Saudi Arabian crude oils. *Petroleum Science and Technology*, 18, 743–54.

Elvert, M., Suess, E. and Whiticar, M. J. (1999) Anaerobic methane oxidation associated with marine gas hydrates: superlight C-isotopes from saturated and unsaturated C_{20} and C_{25} irregular isoprenoids. *Naturwissenschaften*, 86, 295–300.

Elvert, M., Suess, E., Greinert, J. and Whiticar, M. J. (2000) Archaea mediating anaerobic methane oxidation in deep-sea sediments at cold seeps of the eastern Aleutian subduction zone. *Organic Geochemistry*, 31, 1175–87.

Eneogwe, C., Ekundayo, O. and Patterson, B. (2002) Source-derived oleanenes identified in Niger Delta oils. *Journal of Petroleum Geology*, 25, 83–96.

Ensminger, A., Albrecht, P. and Oursson, G. (1972) Homohopane in Messel oil shale: first identification of a C_{31} pentacyclic triterpane in nature. Bacterial origin of some triterpanes in ancient sediments? *Tetrahedron Letters*, 36, 3861–4.

Ensminger, A., van Dorsselaer, A., Spyckerelle, C., Albrecht, P. and Ourisson, G. (1974) Pentacyclic triterpenes of the hopane type as ubiquitous geochemical markers: origin and significance. In: *Advances in Organic Geochemistry 1973* (B. Tissot and F. Bienner, eds.), Editions Technip, Paris, pp. 245–60.

Ensminger, A., Albrecht, P., Ourisson, G., and Tissot, B. (1977) Evolution of polycyclic alkanes under the effect of burial (Early Toarcian shales, Paris Basin). In: *Advances in Organic Geochemistry 1975* (R. Campos and J. Goni, eds.), ENADIMSA, Madrid, pp. 45–52.

Ensminger, A., Joly, G. and Albrecht, P. (1978) Rearranged steranes in sediments and crude oils. *Tetrahedron Letters*, 19, 1575–8.

Erbacher, J., Thurow, J. and Littke, R. (1996) Evolution patterns of radiolaria and organic matter variations – a new approach to identify sea-level changes in Mid-Cretaceous pelagic environments. *Geology*, 24, 499–502.

Erbacher, J., Huber, B. T., Norris, R. D. and Markey, M. (2001) Increased thermohaline stratification as a possible cause for an ocean anoxic event in the Cretaceous Period. *Nature*, 409, 325–7.

Erlich, R. N., Astorga, A., Sofer, Z., Pratt, L. M. and Palmer, S. E. (1996) Palaeoceanography of organic-rich rocks of the Loma Chumico Formation of Costa Rica, Late Cretaceous, Eastern Pacific. *Sedimentology*, 43, 691–718.

Erlich, R. N., Macsotay, I. O., Nederbragt, A. J. and Lorente, M. A. (1999a) Palaeoecology, palaeogeography and depositional environments of Upper Cretaceous rocks of western Venezuela. *Palaeogeography, Palaeoclimatology, Palaeoecology*, 153, 203–38.

Erlich, R. N., Palmer-Koleman, S. E. and Lorente, M. A. (1999b) Geochemical characterization of oceanographic and climatic changes recorded in upper Albian to lower Maastrichtian strata, western Venezuela. *Cretaceous Research*, 20, 547–81.

Erlich, R. N., Macsotay, I. O., Nederbragt, A. J. and Lorente, M. A. (2000) Birth and death of the Late Cretaceous "La Luna Sea", and origin of the Tres Esquinas phosphorites. *Journal of South American Earth Sciences*, 13, 21–45.

Erwin, D. H. (1994a) The Permo-Triassic extinction. *Nature*, 367, 231–6.

(1994b) The Permo-Triassic extinction. In: *The Permian of Northern Pangea* (P. A. Scholle, T. M. Peryt and D. S. Ulmer-Scholle, eds.), Springer-Verlag, Berlin, pp. 20–34.

(2003) Impact at the Permo-Triassic boundary: a critical evaluation. *Astrobiology*, 3, 67–74.

Escandon, M., Vallejos, C., Pratt, L. and Gambino, F. (1993) Variations in organic facies and sedimentary environments of the La Luna Formation and Machiques Member in the Perija Mountain Range, Venezuela. *American Association of Petroleum Geologists Bulletin*, 77, 317.

Espitalié, J., Laporte, J. L., Madec, M., *et al.* (1977) Méthode rapide de characterisation des roches mères, de leur potentiel pétrolier et de leur degré d'évolution. *Revue de l'Insitut Francais du Petrole*, 32, 23–42.

Espitalié, J., Marquis, F. and Barsony, I. (1984) Geochemical logging. In: *Analytical Pyrolysis* (K. J. Voorhees, ed.), Butterworths, Boston, MA, pp. 276–304.

Espitalié, J., Marquis, F. and Sage, L. (1987) Organic geochemistry of the Paris Basin. In: *Petroleum Geology of Northwest Europe* (J. Brooks and K. Glennie, eds.), Graham and Trotman, London, pp. 71–86.

Evans, R. (1988) Geological and geochemical determination of generation and migration of hydrocarbons in the South Mississippi Salt Basin; multiple sources, multiple pathways. In: *Gulf Coast Oils and Gases: Their Characteristics, Origin, Distribution, and Exploration and Production Significance. Proceedings of the Ninth Annual Research Conference* (D. Schumacher and B. F. Perkins, eds.), Society of Economic Paleontologists and Mineralogists, Gulf Coast Section Research, Houston, TX, pp. 4–5.

Evans, J., Jenkins, D. and Gluyas, J. (1998) The Kimmeridge Bay Oilfield: an enigma demystified. In: *Development, Evolution and Petroleum Geology of the Wessex Basin* (J. R. Underhill, ed.), Geological Society of London, London, pp. 407–13.

Fabre, A. D. and Alvar, D. U. (1993) The hydrocarbon potential of the Ene Basin (Peru). *American Association of Petroleum Geologists Bulletin*, 77, 317.

Fan, Z.-A. and Philp, R. P. (1987) Laboratory biomarker fractionations and implications for migration studies. *Organic Geochemistry*, 11, 169–75.

Fan, P., King, J. D. and Claypool, G. E. (1987) Characteristics of biomarker compounds in Chinese crude oils. In: *Petroleum Geochemistry and Exploration in the Afro-Asian Region* (R. K. Kumar, P. Dwivedi, V. Banerjie and V. Gupta, eds.), Balkema, Rotterdam, pp. 197–202.

Fang, P. H. and Wong, R. (1996) Evidence for fullerene in a coal of Yunnan, southwestern China. *Materials Research Innovations*, 1, 129–34.

Farley, K. A., Montanari, A., Shoemaker, E. M. and Shoemaker, C. S. (1998) Geochemical evidence for a comet shower in the late Eocene. *Science*, 280, 1250–3.

Farrimond, P., Eglinton, G., Brassell, S. C. and Jenkyns, H. C. (1988) The Toarcian black shale event in northern Italy. *Organic Geochemistry*, 13, 823–32.

(1989) Toarcian anoxic event in Europe: an organic geochemical study. *Marine and Petroleum Geology*, 6, 136–47.

Farrimond, P., Stoddart, D. P. and Jenkyns, H. C. (1994) An organic geochemical profile of the Toarcian anoxic event in northern Italy. *Chemical Geology*, 111, 17–33.

Farrimond, P., Bevan, J. C. and Bishop, A. N. (1996) Hopanoid hydrocarbon maturation by an igneous intrusion. *Organic Geochemistry*, 25, 149–64.

Farrimond, P., Taylor, T. and Telnaes, N. (1998) Biomarker maturity parameters: the role of generation and thermal degradation. *Organic Geochemistry*, 29, 1181–97.

Farrington, J. W., Davis, A. C., Sulanowski, J., *et al.* (1988a) Biogeochemistry of lipids in surface sediments of the Peru upwelling area – 15°S. *Organic Geochemistry*, 10, 607–17.

Farrington, J. W., Davis, A. C., Tarafa, M. E., *et al.* (1988b) Bitumen molecular maturity parameters in the Ikpikpuk well, Alaska North Slope. *Organic Geochemistry*, 13, 303–10.

Faure, K. and Cole, D. (1999) Geochemical evidence for lacustrine microbial blooms in the vast Permian Main Karoo, Paraná, Falkland Islands and Huab basins of southwestern Gondwana. *Palaeogeography, Palaeoclimatology, Palaeoecology*, 152, 189–213.

Fazeelat, T., Alexander, R. and Kagi, R. I. (1994) Extended 8,14-secohopanes in some seep oils from Pakistan. *Organic Geochemistry*, 21, 257–64.

(1995) Molecular structures of sedimentary 8,14-secohopanes inferred from their gas chromatographic retention behaviour. *Organic Geochemistry*, 23, 641–6.

(1999) Effects of maturity on the relative abundances of 8,14-secohopanes in sediments and crude oils. *Journal of the Chemical Society of Pakistan*, 21, 154–63.

Fedonkin, M. A. and Waggoner, B. M. (1997) The late Precambrian fossil *Kimberella* is a mollusc-like bilaterian organism. *Nature*, 388, 868–71.

Ferriday, I. L., Borisov, A., Gudkova, A., *et al.* (1995) Source rock potential and the characterization of finds in the Russian part of the Barents Sea. In: *Petroleum Exploration and Production in Timan-Pechora Basin and Barents Sea* (M. D. Belonin and V. K. Makarevich, eds.), VNIGRI, St Petersburg, Russia, pp. 43–4.

Ferris, J. P., Hill, A. R., Jr, Liu, R. and Orgel, L. E. (1996) Synthesis of long prebiotic oligomers on mineral surfaces. *Nature*, 381, 59–61.

Fielding, C. R. (1992) A review of Cretaceous coal-bearing sequences in Australia. In: *Controls on the Distribution and Quality of Cretaceous Coals* (P. J. McCabe and J. T.

Parrish, eds.), Geological Society of America, Boulder, CO, pp. 303–24.

Fields, R. W., Rasmussen, D. L., Tabrum, A. R. and Nichols, R. (1985) Cenozoic rocks of the intermontane basins of western Montana and eastern Idaho: a summary. In: *Cenozoic Paleogeography of the West-Central United States* (R. M. Flores and S. S. Kaplan, eds.), Rocky Mountain Section Society of Economic Paleontologists and Mineralogists, Denver, CO, pp. 9–36.

Figueiredo, A. M. F., Braga, J. A. E., Zabalga, H. M. C., *et al.* (1994) Recôncavo Basin Brazil: a prolific intracontinental rift basin. In: *Interior Rift Basins* (S. M. Landon, ed.), American Association of Petroleum Geologists, Tulsa, OK, pp. 157–203.

Filiptsov, Y. A., Petrishina, Y. V., Bogorodskaya, L. I., Kontorovich, A. A. and Krinin, V. A. (1999) Evaluation of maturity and oil-and gas-generation properties of organic matter in Riphean and Vendian rocks of the Baykit and Katanga petroleum regions. *Geologiya i Geofizika*, 40, 1362–74.

Fisher, K., Largeau, C. and Derenne, S. (1996a) Can oil shales be used to produce fullerenes? *Organic Geochemistry*, 24, 715–23.

Fisher, S. J., Alexander, R., Ellis, L. and Kagi, R. I. (1996b) The analysis of dimethylphenanthrenes by direct deposition gas chromatography-Fourier transform infrared spectroscopy (GC-FTIR). *Polycyclic Aromatic Compounds*, 9, 257–64.

Fisher, S. J., Alexander, R., Kagi, R. I. and Oliver, G. A. (1998) Aromatic hydrocarbons as indicators of biodegradation in north Western Australian reservoirs. In: *Sedimentary Basins of Western Australia: West Australian Basins Symposium* (P. G. Purcell and R. R. Purcell, eds.), Petroleum Exploration Society of Australia, WA Branch, Perth, Australia, pp. 185–94.

Flanigen, E. M., Bennett, J. M., Grosee, R. W., *et al.* (1978) Silicalite, a new hydrophobic crystalline silica molecular sieve. *Nature*, 271, 512–6.

Forbes, P. L., Ungerer, P. M., Kuhfuss, A. B., Riis, F. and Eggen, S. S. (1991) Compositional modeling of petroleum generation and expulsion: trial application to a local mass balance in the Smorbukk Sor Field, Haltenbanken Area, Norway. *American Association of Petroleum Geologist Bulletin*, 75, 873–93.

Forsman, N. F., Gerlach, T. R. and Anderson, N. L. (1996) Impact origin of the Newporte structure, Williston Basin, North Dakota. *American Association of Petroleum Geologists Bulletin*, 8, 721–30.

Forster, A., Dean, W. E., Schwark, L., *et al.* (1999) Geochemical investigations on cyclicity, paleoenvironment and organic matter deposition of the Middle Turonian Carlile Shale Cretaceous Western Interior Basin, USA. *Geological Society of America Abstracts with Programs*, 31, A–225.

Foster, C. B., O'Brien, G. W. and Watson, S. T. (1986) Hydrocarbon source potential of the Goldwyer Formation, Barbwire Terrace, Canning Basin, western Australia. *Australian Petroleum Exploration Association Journal*, 26, 142–55.

Foster, C. B., Wicander, R. and Reed, J. D. (1989) *Gloeocapsomorpha prisca* Zalessky, 1917: a new study. Part 1: taxonomy, geochemistry, and paleoecology. *Geobios*, 22, 735–59.

(1990) *Gloeocapsomorpha prisca* Zalessky, 1917: a new study. Part 2: origin of kukersite, a new interpretation. *Geobios*, 23, 133–44.

Fouch, T. D., Nuccio, V. F., Anders, D. E., *et al.* (1994) Green River(!) petroleum system, Uinta Basin, Utah, USA. In: *The Petroleum System-From Source to Trap* (L. B. Magoon and W. G. Dow, eds.), American Association of Petroleum Geologists, Tulsa, OK, pp 399–421.

Fowell, S. J., Olsen, P. E. and Cornet, B. (1994) *Geologically Rapid Late Triassic Extinctions: Palynological Evidence from the Newark Supergroup*. Geological Society of America Special Paper 288, Geological Society of America, Boulder, CO, 197–206.

Fowler, M. G. (1992) The influence of *Gloeocapsomorpha prisca* on the organic geochemistry of oils and organic-rich rocks of Late Ordovician age from Canada. In: *Early Organic Evolution: Implications for Mineral and Energy Resources* (M. Schidlowski, S. Golubic, M. M. Kimberley, D. M. McKirdy and P. A. Trudinger, eds.), Springer-Verlag, Berlin, pp. 336–56.

Fowler, M. G. and Brooks, P. W. (1990) Organic geochemistry as an aid in the interpretation of the history of oil migration into different reservoirs at the Hibernia K-18 and Ben Nevis I-45 wells, Jeanne d'Arc Basin, offshore eastern Canada. *Organic Geochemistry*, 16, 461–75.

Fowler, M. G. and Douglas, A. G. (1984) Distribution and structure of hydrocarbons in four organic-rich Ordovician rocks. In: *Advances in Organic Geochemistry 1983* (P. A. Schenck, J. W. de Leeuw and G. W. M. Lijmbach, eds.), Pergamon Press, New York, pp. 105–14.

(1987) Saturated hydrocarbon biomarkers in oils of Late Precambrian age from Eastern Siberia. *Organic Geochemistry*, 11, 201–13.

Fowler, M. G. and McAlpine, K. D. (1995) The Egret Member, a prolific Kimmeridgian source rock from offshore eastern Canada. In: *Petroleum Source Rocks* (B. Katz, ed.), Springer-Verlag, Berlin, pp. 111–30.

Fowler, M. G., Snowdon, L. R., Brooks, P. W. and Hamilton, T. S. (1988) Biomarker characterisation and hydrous pyrolysis of bitumens from Tertiary volcanics, Queen Charlotte Islands, British Columbia, Canada. *Organic Geochemistry*, 13, 715–25.

Fowler, M. G., Stasiuk, L. D., Williamson, M. A., *et al.* (1998) Explanation for the occurrence of heavy oil accumulations in the Grand Banks area, offshore Eastern Canada. Presented at the Geo-triad '98, Joint Meeting of the CSPG, CSEG and CWLS, June 1998, Calgary, Alberta.

Fowler, M. G., Stasiuk, L. D., Hearn, M. and Obermajer, M. (2001) Devonian hydrocarbon source rocks and their derived oils in the Western Canada Sedimentary Basin. *Bulletin of Canadian Petroleum Geology*, 49, 117–48.

Fox, S. W. (1969) Self-ordered polymers and propagative cell-like system. *Naturwissenschaften*, 56, 1–9.

Franks, S. G., Dias, R. F., Freeman, K. H., *et al.* (2001) Carbon isotopic composition of organic acids in oil field waters, San Joaquin Basin, California, USA. *Geochimica et Cosmochimica Acta*, 65, 1301–10.

Fredrickson, J. K., McKinley, J. P., Bjornstad, B. N., *et al.* (1997) Pore-size constraints on the activity and survival of subsurface bacteria in a Late Cretaceous shale-sandstone sequence, northwestern New Mexico. *Geomicrobiology Journal*, 14, 183–202.

Freeman, K. H. (1991) The carbon isotopic compositions of individual compounds from ancient and modern depositional environments. Ph. D. thesis, Indiana University, Bloomington, IN.

Freeman, K. H. and Colarusso, L. A. (2001) Molecular and isotopic records of C_4 grassland expansion in the late Miocene. *Geochimica et Cosmochimica Acta*, 65, 1439–54.

Freeman, K. H., Hayes, J. M., Trendel, J. M. and Albrecht, P. (1990) Evidence from carbon isotope measurements for diverse origins of sedimentary hydrocarbons. *Nature*, 343, 254–6.

Freeman, K. H., Wakeham, S. G. and Hayes, J. M. (1994) Predictive isotopic biogeochemistry: hydrocarbons from anoxic marine basins. *Organic Geochemistry*, 21, 629–44.

Frenkel, M. and Heller-Kallai, L. (1977) Aromatization of limonene – a geochemical model. *Organic Geochemistry*, 1, 3–5.

Fritsche, W. and Hofrichter, M. (2000) Aerobic degradation by microorganisms. In: *Biotechnology*, Vol. 11b (J. Klein, ed.), John Wiley & Sons, New York, pp. 146–64.

Frolov, E. B., Melikhov, V. A. and Smirnov, M. B. (1996) Radiolytic nature of n-alkene/n-alkane distributions in Russian Precambrian and Palaeozoic oils. *Organic Geochemistry*, 24, 1061–4.

Frontera-Suau, R., Bost, F. D., McDonald, T. J. and Morris, P. J. (2002) Aerobic biodegradation of hopanes and other biomarkers by crude oil-degrading enrichment cultures. *Environmental Science & Technology*, 36, 4578–84.

Fu, J. and Sheng, G. (1989) Biological marker composition of typical source rocks and related oils of terrestrial origin in the People's Republic of China; a review. *Applied Geochemistry*, 4, 13–22.

Fu, J., Sheng, G., Peng, P., *et al.* (1986) Peculiarities of salt lake sediments as potential source rocks in China. *Organic Geochemistry*, 10, 119–26.

Fu, J., Sheng, G., and Liu, D. (1988) Organic geochemical characteristics of major types of terrestrial source rocks in China. In: *Lacustrine Petroleum Source Rocks* (A. J. Fleet, K. Kelts and M. R. Talbot, eds.), Blackwell, London, pp. 279–89.

Fu, J., Sheng, G., Xu, J., *et al.* (1990) Application of biological markers in the assessment of paleoenvironments of Chinese non-marine sediments. *Organic Geochemistry*, 16, 769–79.

Fyfe, C. A., Gobbi, G. C., Klinowski, J., Thomas, J. M. and Ramdas, S. (1982) Resolving crystallographically distinct tetrahedral sites in silicalite and ZSM-5 by solid-state NMR. *Nature*, 296, 530–3.

Galarraga, F., Gonzalez, R., and Perez, A. (1996) Geochemical evidences of terrestrial oils from Orinoco Oil Belt, Venezuela, by gas chromatography-mass spectrometry and pyrolysis-GC-MS in whole oil and asphaltenes. Presented at the 30th International Geological Congress, August 4–14, 1996, Beijing, China.

Galimberti, R., Ghiselli, C. and Chiaramonte, M. A. (2000) Acidic polar compounds in petroleum: a new analytical methodology and applications as molecular migration indices. *Organic Geochemistry*, 31, 1375–86.

Gallegos, E. J. (1971) Identification of new steranes, terpanes, and branched paraffins in Green River Shale by combined gas chromatography and mass spectrometry. *Analytical Chemistry*, 43, 1151–60.

(1973) Identification of phenylcycloparaffin alkanes and other monoaromatics in Green River Shale by gas chromatography-mass spectrometry. *Analytical Chemistry*, 45, 1399–403.

Gallegos, E. J. (1976) Analysis of organic mixtures using metastable transition spectra. *Analytical Chemistry*, 48, 1348–51.

Gallegos, E. J. and Moldowan, J. M. (1992) The effect of hold time on GC resolution and the effect of collision gas on mass spectra in geochemical "biomarker" research. In: *Biological Markers in Sediments and Petroleum* (J. M. Moldowan, P. Albrecht and R. P. Philp, eds.), Prentice-Hall, Englewood Cliffs, NJ, pp. 156–81.

Gallegos, E. J., Fetzer, J. C., Carlson, R. M. and Pena, M. M. (1991) High-temperature GC/MS characterization of porphyrins and high molecular weight hydrocarbons. *Energy & Fuels*, 5, 376–81.

Galloway, W. E., Ganey-Curry, P. E., Li, X. and Buffler, R. T. (2000) Cenozoic depositional history of the Gulf of Mexico Basin. *American Association of Petroleum Geologists Bulletin*, 84, 1743–74.

Gardner, W. C. and Bray, E. E. (1984) Oils and source rocks of Niagaran Reefs (Silurian) in the Michigan Basin. In: *Petroleum Geochemistry and Source Rock Potential of Carbonate Rocks* (J. G. Palacas, ed.), American Association of Petroleum Geologists, Tulsa, OK, pp. 33–44.

Gardner, P. M. and Whitehead, E. V. (1972) The isolation of squalane from a Nigerian petroleum. *Geochimica et Cosmochimica Acta*, 36, 259–63.

Gargas, A., Depreist, P. T., Grube, M. and Tehler, A. (1995) Multiple origins of lichen symbiosis in fungi suggested by SSU rDNA phylogeny. *Science*, 268, 1492–5.

Garrigues, P., Saptorahardjo, A., Gonzalez, C., *et al.* (1986) Biogeochemical aromatic markers in the sediments from Mahakam Delta (Indonesia). *Organic Geochemistry*, 10, 959–64.

Gautier, D. L., Clayton, J. L., Leventhal, J. S. and Reddin, N. J. (1984) Origin and source-rock potential of the Sharon Springs Member of the Pierre Shale, Colorado and Kansas. In: *Hydrocarbon Source Rocks of the Greater Rocky Mountain Region* (J. Woodward, F. F. Meissner and J. L. Clayton, eds.), Rocky Mountain Association of Geologists, Denver, CO, pp. 369–85.

Gavrilov, V. P. (2000) Petroleum-bearing granites. *Geologiya Nefti i Gaza*, 6, 44–9.

Ge, S., Dilcher, D. L., Shaoling, Z., and Zhekun, Z. (1998) In search of the first flower: a Jurassic angiosperm, *Archaefructus*, from Northeast China. *Science*, 282, 1692–5.

Ge, S., Qiang, J., Dilcher, D. L., *et al.* (2002) Archaefructaceae, a new basal angiosperm family. *Science*, 296, 899–904.

Gearing, P. J., Gearing, J. N., Lytle, T. F. and Lytle, J. S. (1976) Hydrocarbons in 60 northeast Gulf of Mexico shelf sediments: a preliminary study. *Geochimica et Cosmochimica Acta*, 40, 1005–17.

Gehling, J. G. (1987) Earliest known echinoderm – a new Ediacaran fossil from the Pound Subgroup, South Australia. *Alcheringa*, 11, 337–45.

(1988) A cnidarian of actinian grade from the Ediacaran Pound Subgroup, South Australia. *Alcheringa*, 12, 299–314.

Gehling, J. G. and Rigby, J. K. (1996) Long expected sponges from the Neoproterozoic Ediacara fauna of South Australia. *Journal of Paleontology*, 70, 185–95.

Gelpi, V., Schneider, H., Mann, J. and Oró, J. (1970) Hydrocarbons of geochemical significance in microscopic algae. *Phytochemistry*, 9, 603–12.

George, S. C. and Ahmed, M. (2000) Organic maturity evaluation of the Proterozoic Middle Velkerri Formation, Northern Territory, Australia: use of aromatic hydrocarbon distributions. *American Association of Petroleum Geologists Bulletin*, 84, 1429.

George, S. C., Krieger, F. W., Eadington, P. J., *et al.* (1997) Geochemical comparison of oil-bearing fluid inclusions and produced oil from the Toro sandstone, Papua New Guinea. *Organic Geochemistry*, 26, 155–73.

George, S. C., Eadington, P. J., Lisk, M. and Quezada, R. A. (1998a) Geochemical comparison of oil trapped in fluid inclusions and reservoired oil in Blackback Oilfield, Gippsland Basin, Australia. *PESA (Petroleum Exploration Society of Australia) Journal*, 26, 64–81.

George, S. C., Lisk, M., Summons, R. E. and Quezada, R. A. (1998b) Constraining the oil charge history of the South Pepper oilfield from the analysis of oil-bearing fluid inclusions. *Organic Geochemistry*, 29, 631–48.

Gesteland, R. F., Cech, T. R. and Atkins, J. F. *The RNA World*. Cold Spring Harbor Laboratory Monograph no. 37. CSHL Press, Cold Spring Harbor.

Giannasi, D. E. and Niklas, K. L. (1981) Comparative palaeobiochemistry of some fossil and extant Fagaceae. *American Journal of Botany*, 68, 762–70.

Gibbison, R., Peakman, T. M. and Maxwell, J. R. (1995) Novel porphyrins as molecular fossils for anoxygenic photosynthesis. *Tetrahedron Letters*, 36, 9057–60.

Gibling, M. R., Tantisukrit, C., Uttamo, W., Thanasuthipitak, T. and Hara, M. (1985a) Oil shale sedimentology and geochemistry in Cenozoic Mae Sot Basin, Thailand. *American Association of Petroleum Geologists Bulletin*, 69, 767–80.

Gibling, M. R., Ukakimaphan, Y. and Srisuk, S. (1985b) Oil shale and coal in intermontane basins of Thailand. *American Association of Petroleum Geologists Bulletin*, 69, 760–6.

Gibson, D. T. (1984) *Microbial Degradation of Organic Compounds*. Marcel Dekker, New York.

Gilbert, W. (1986) Origin of life, the RNA world. *Nature*, 319, 618.

Gilbert, T. D., Stephenson, L. C. and Philp, R. P. (1985) Effect of a dolerite intrusion on triterpane stereochemistry and kerogen in Rundle oil shale, Australia. *Organic Geochemistry*, 8, 163–9.

Giles, M. R., De Boer, R. B., and Marshall, J. D. (1994) How important are organic acids in generating secondary porosity in the subsurface? In: *Organic Acids in Geological Processes* (E. D. Pittman and M. D. Lewan, eds.), Springer-Verlag, New York, pp. 449–70.

Gilmour, I. and Guenther, F. (1988) The global Cretaceous-Tertiary fire: biomass or fossil carbon? *Lunar and Planetary Institute (LPI) Contribution*, 673, 60–1.

Gilmour, I., Wolbach, W. S., and Anders, E. (1989) Major wildfires at the Cretaceous-Tertiary boundary. In: *Castastrophes and Evolution: Astronomical Foundations* (S. V. M. Clube, ed.), Cambridge University Press, Cambridge, UK, pp. 195–213.

Glass, B. P. and Liu, S. (2001) Discovery of high-pressure $ZrSiO_4$ polymorph in naturally occurring shock-metamorphosed zircons. *Geology*, 29, 371–3.

Glass, B. P. and Wu, J. (1993) Coseite and shocked quartz discovered in the Australasian and North American microtektite layers. *Geology*, 21, 435–8.

Glikson, M. and Taylor, G. H. (1986) Cyanabacterial mats: major contributors to the organic matter in Toolebuc Formation oil shales. In: *Contributions to the Geology and Hydrocarbon Potential of the Eromanga Basin* (D. I. Gravestock, P. S. Moore and G. M. Pitt, eds.), Geological Society of Australia, Boulder, CO, pp. 273–286.

Goad, L. J. and Withers, N. (1982) Identification of 27-nor-(24R)24-methylcholesta-5,22-dien-3β-ol and brassicasterol as the major sterols of the marine dinoflagellate *Gymnodinium simplex*. *Lipids*, 17, 853–8.

Goff, J. C. (1983) Hydrocarbon generation and migration from Jurassic source rocks in the East Shetland Basin and Viking Graben of the northern North Sea. *Journal of the Geological Society*, 140, 445–74.

Gold, T. (1999) *The Deep Hot Biosphere*. Copernicus, New York.

Gomez-Perez, I., Franzese, J. and Wavrek, D. A. (2001) Deep sedimentary systems and structure in the Neuquén Basin (Argentina). Presented at the Annual Meeting of the American Association of Petroleum Geologists June 3–6, 2001, Denver, CO.

Gonzaga, F. G., Gonçalves, F. T. T., and Coutinho, L. C. E. (2000) Petroleum geology of the Amazonas Basin, Brazil: modeling of hydrocarbon generation and migration. In: *Petroleum Systems of South Atlantic Margins* (M. R. Mello and B. J. Katz, eds.), American Association of Petroleum Geologists, Tulsa, OK, pp. 159–78.

Goodarzi, F., Brooks, P. W. and Embry, A. F. (1989) Regional maturity as determined by organic petrography and geochemistry of the Schei Point Group (Triassic) in the western Sverdrup Basin, Canadian Arctic Archipelago. *Marine and Petroleum Geology*, 6, 290–302.

Goodwin, N. S., Park, P. J. D., and Rawlinson, T. (1983) Crude oil biodegradation. In: *Advances in Organic Geochemistry 1981* (M. Bjorøy, C. Albrecht, C. Cornford, *et al.*, eds.), John Wiley & Sons, New York, pp. 650–8.

Goodwin, N. S., Mann, A. L. and Patience, R. L. (1988) Structure and significance of C_{30} 4-methylsteranes in lacustrine shales and oils. *Organic Geochemistry*, 12, 495–506.

Goossens, H., de Leeuw, J. W., Schenck, P. A. and Brassell, S. C. (1984) Tocopherols as likely precursors of pristane in ancient sediments and crude oils. *Nature*, 312, 440–2.

Goossens, H., de Lange, F., de Leeuw, J. W. and Schenck, P. A. (1988a) The Pristane Formation Index, a molecular maturity parameter. Confirmation in samples from the Paris Basin. *Geochimica et Cosmochimica Acta*, 52, 2439–44.

Goossens, H., Due, A., de Leeuw, J. W., van de Graff, B. and Schenck, P. A. (1988b) The Pristane Formation Index, a new molecular maturity parameter, easily measured by pyrolysis-gas chromatography of unextracted samples. *Geochimica et Cosmochimica Acta*, 52, 1189–93.

Gordon, T. L. (1985) Talang Akar coals – Ardjuna subbasin oil source. In: *Proceedings of the Fourteenth Annual Convention of the Indonesian Petroleum Association*, Vol. 2, Indonesian Petroleum Association, Jakarta, Indonesia, pp. 91–120.

Gorter, J. D. (1984) Source potential of the Horn Valley Siltstone, Amadeus Basin. *Australian Petroleum Exploration Association Journal*, 24, 66–90.

(1994) Sequence stratigraphy and the depositional history of the Murta Member (upper Hooray Sandstone), southeastern Eromanga Basin, Australia: implications for the development of source and reservoir facies. *APEA Journal*, 34, 644–73.

Goth, K., de Leeuw, J. W., Püttmann, W. and Tegelaar, E. W. (1988) Origin of Messel oil shale kerogen. *Nature*, 336, 759–61.

Gough, M. A. and Rowland, S. J. (1990) Characterization of unresolved complex mixtures of hydrocarbons in petroleum. *Nature*, 344, 648–50.

Gough, M. A., Rhead, M. M. and Rowland, S. J. (1992) Biodegradation studies of unresolved complex mixtures of hydrocarbons; model UCM hydrocarbons and the aliphatic UCM. *Organic Geochemistry*, 18, 17–22.

Grabowski, G. J., Jr (1984) Generation and migration of hydrocarbons in Upper Cretaceous Austin Chalk, South-Central Texas. In: *Petroleum Geochemistry and Source Rock Potential of Carbonate Rocks* (J. G. Palacas, ed.), American Association of Petroleum Geologists, Tulsa, OK, pp. 71–96.

(1994) Organic-rich chalks and calcareous mudstones of the upper Cretaceous Austin Chalk and Eagleford Formation, south-central Texas, USA. In: *Petroleum Source Rocks, Casebooks in Earth Science* (B. J. Katz, ed.), Springer-Verlag, Tulsa, OK, pp. 209–34.

Graham, S. A., Brassell, S. C., Carroll, A. R., *et al.* (1990) Characteristics of selected petroleum source rocks, Xianjiang Uygur autonomous region, Northwest China. *American Association of Petroleum Geologists Bulletin*, 74, 493–512.

Grajales-Nishimura, J. M., Cedillo-Pardo, E., Rosales-Domínguez, C., *et al.* (2000) Chicxulub impact: the origin of reservoir and seal facies in the southeastern Mexico oil fields. *Geology*, 28, 307–10.

Grantham, P. J. (1986a) The occurrence of unusual C_{27} and C_{29} sterane predominances in two types of Oman crude oil. *Organic Geochemistry*, 9, 1–10.

(1986b) Sterane isomerization and moretane/ hopane ratios in crude oils derived from Tertiary source rocks. *Organic Geochemistry*, 9, 293–304.

Grantham, P. J. and Wakefield, L. L. (1988) Variations in the sterane carbon number distributions of marine source rock derived crude oils through geological time. *Organic Geochemistry*, 12, 61–73.

Grantham, P. J., Posthuma, J. and DeGroot, K. (1980) Variation and significance of the C_{27} and C_{28} triterpane content of a North Sea core and various North Sea crude oils. In: *Advances in Organic Geochemistry 1979* (A. G. Douglas and J. R. Maxwell, eds.), Pergamon Press, Oxford, UK, pp. 29–38.

Grantham, P. J., Posthuma, J. and Baak, A. (1983) Triterpanes in a number of Far-Eastern crude oils. In: *Advances in Organic Geochemistry 1981* (M. Bjorøy, C. Albrecht, C. Cornford, *et al.*, eds.), John Wiley & Sons, New York, pp. 675–83.

Grantham, P. J., Lijmback, G. W. M., Posthuma, J., (1988) Origin of crude oils in Oman. *Journal of Petroleum Geology*, 11, 61–80.

(1990) Geochemistry of crude oils in Oman. In: *Classic Petroleum Provinces* (J. Brooks, ed.), Geological Society of London, London, pp. 317–28.

Gray, J. (1993) Major Paleozoic land plant evolutionary bio-events. *Palaeogeography, Palaeoclimatology, Palaeoecology*, 104, 153–69.

Greenwood, P. F. and Summons, R. E. (2003) GC-MS detection and significance of crocetane and pentamethylicosane in sediments and crude oils. *Organic Geochemistry*, 34, 1211–22.

Greenwood, J. P., Riciputi, L. R. and McSween, H. Y., Jr. (1997) Sulfide isotopic compositions in shergottites and ALH 84001, and possible implications for life on Mars. *Geochimica et Cosmochimica Acta*, 61, 4449–53.

Greiner, A. C., Spyckerelle, C., Albrecht, P. and Ourisson, G. (1977) Hydrocarbures aromatiques d'origine geologique. V. Derives mono- et di-aromatiques du hopane. [Aromatic hydrocarbons of geologic origin. V. Mono-aromatic and di-aromatic derivatives of hopane.] *Journal of Chemical Research, Miniprint*, 12, 3829–71.

Greinert, J., Bohrmann, G. and Elvert, M. (2002) Stromatolitic fabric of authigenic carbonate crusts: result of anaerobic methane oxidation at cold seeps in 4,850 m water depth. *International Journal of Earth Sciences*, 91, 698–711.

Grice, K. (2001) $\delta^{13}C$ as an indicator of paleoenvironments: a molecular approach. In: *Application of Stable Isotope*

Techniques to Study Biological Processes and Functioning Ecosystems (M. Unkovich, J. Pate, A. McNeill and J. Gibbs, eds.), Kluwer Scientific, Dordrecht, The Netherlands, pp. 247–81.

Grice, K., Gibbison, R., Atkinson, J. E., *et al.* (1996a) Maleimides (^{1}H-pyrrole-2,5-diones) as indicators of anoxygenic photosynthesis in ancient water columns. *Geochimica et Cosmochimica Acta*, 60, 3913–24.

Grice, K., Schaeffer, P., Schwark, L. and Maxwell, J. R. (1996b) Molecular indicators of palaeoenvironmental conditions in an immature Permian shale (Kupferschiefer, Lower Rhine Basin, north-west Germany) from free and sulfide-bound lipids. *Organic Geochemistry*, 25, 131–47.

(1997) Changes in palaeoenvironmental conditions during deposition of the Permian Kupferschiefer (Lower Rhine Basin, northwest Germany) inferred from molecular and isotopic compositions of biomarker components. *Organic Geochemistry*, 26, 677–90.

Grice, K., Schouten, S., Peters, K. E. and Sinninghe Damsté, J. S. (1998a) Molecular isotopic characterisation of hydrocarbon biomarkers in Palaeocene-Eocene evaporitic, lacustrine source rocks from the Jianghan Basin, China. *Organic Geochemistry*, 29, 1745–64.

Grice, K., Schouten, S., Nissenbaum, A., Charrach, J. and Sinninghe Damsté, J. S. (1998b) Isotopically heavy carbon in the C_{21} to C_{25} regular isoprenoids in halite-rich deposits from the Sdom Formation, Dead Sea Basin, Israel. *Organic Geochemistry*, 28, 349–59.

(1998c) A remarkable paradox: sulfurized freshwater algal *(Botryococcus braunii)* lipids in an ancient hypersaline euxinic ecosystem. *Organic Geochemistry*, 28, 195–216.

Grice, K., Alexander, R. and Kagi, R. I. (2000) Diamondoid hydrocarbon ratios as indicators of biodegradation levels in Australian crude oils. *Organic Geochemistry*, 31, 67–73.

Grice, K., Audino, M., Boreham, C. J., Alexander, R. and Kagi, R. I. (2001) Distributions and stable carbon isotopic compositions of biomarkers in torbanites from different palaeogeographical locations. *Organic Geochemistry*, 32, 1195–210.

Gschwend, P. M., Chen, P. H. and Hites, R. A. (1983) On the formation of perylene in recent sediments: kinetic models. *Geochimica et Cosmochimica Acta*, 47, 2115–19.

Guardado, L. R., Spadini, A. R., Brandão, J. S. L. and Mello, M. R. (2000) Petroleum system of the Campos Basin, Brazil. In: *Petroleum Systems of South Atlantic Margins* (M. R. Mello and B. J. Katz, eds.), American Association of Petroleum Geologists, Tulsa, OK, pp. 317–24.

Guidry F. K., Luffel, D. L. and Olszewski, A. J. (1996) Devonian shale formation evaluation model based on logs, new core analysis methods, and production tests. In: *Production from Fractured Shales, SPE Reprint Series 45 Society of Petroleum Engineers*, Richardson, TX, pp. 101–102.

Gürgey, K. (1991) Genetic classification of the SE Turkey crude oils and delineation of source rock types with the use of biological markers. Unpublished Ph. D. thesis, Middle East Technical University, Ankara, Turkey.

(2002) An attempt to recognize oil populations and potential source rock types in Paleozoic sub- and Mesozoic-Cenozoic supra-salt strata in the southern margin of the Pre-Caspian Basin, Kazakhstan Republic. *Organic Geochemistry*, 33, 723–41.

Guthrie, J. M. (1996) Molecular and carbon isotopic analysis of individual biological markers: evidence for sources of organic matter and paleoenvironmental conditions in the Upper Ordovician Maquoketa Group, Illinois Basin, USA. *Organic Geochemistry*, 25, 439–60.

Guthrie, J. M. and Pratt, L. M. (1994) Geochemical indicators of depositional environment and source-rock potential for the Upper Ordovician Maquoketa Group, Illinois Basin. *American Association of Petroleum Geologists Bulletin*, 78, 744–57.

(1995) Geochemical character and origin of oils in Ordovician reservoir rock, Illinois and Indiana, USA. *American Association of Petroleum Geologists Bulletin*, 79, 1631–49.

Guthrie, J. M., Houseknecht, D. W. and Johns, W. D. (1986) Relationships among vitrinite reflectance, illite crystallinity, and organic geochemistry in Carboniferous strata, Ouachita Mountains, Oklahoma and Arkansas. *American Association of Petroleum Geologists Bulletin*, 70, 26–33.

Guthrie, J. M., Rooney, M. A., Chung, H. M. and Unomah, G. (2000) The effects of biodegradation versus recharging on the composition of oils, offshore Nigeria. Presented at the Annual Meeting of the American Association of Petroleum Geologists. April 16–19, 2000, New Orleans, LA.

Guzmán-Vega, M. A. and Mello, M. R. (1999) Origin of oil in the Sureste Basin, Mexico. *American Association of Petroleum Geologists Bulletin*, 83, 1068–95.

Guzmán-Vega, M. A., Ortíz, L. C., Ramán-Ramos, J. R., *et al.* (2001) Classification and origin of petroleum in the

Mexican Gulf Coast Basin: an overview. In: *The Western Gulf of Mexico: Tectonics, Sedimentary Basins, and Petroleum Systems* (C. Bartolini, R. T. Buffler and A. Cantú-Chapa, eds.), American Association of Petroleum Geologists, Tulsa, OK, pp. 127–42.

Haack, R. C., Sundararaman, P., Diedjomahor, J. O., *et al.* (2000) Niger Delta petroleum systems, Nigeria. In: *Petroleum Systems of South Atlantic Margins* (M. R. Mello and B. J. Katz, eds.), American Association of Petroleum Geologists, pp. 213–31.

Hadfield, P. (2002) Destroyer of worlds. *New Scientist*, 173, 11.

Halbouty, M. T. (2000) Spindletop. *Houston Geological Society Bulletin*, 42, 30.

Hall, P. B. and Douglas, A. G. (1983) The distribution of cyclic alkanes in two lacustrine deposits. In: *Advances in Organic Geochemistry 1981* (M. Bjorøy, C. Albrecht, C. Cornford, *et al.*, eds.), John Wiley & Sons, Chichester, UK, pp. 576–87.

Hallam, A. (2001) Environment of the Jet Rock, a classic black shale in the lower Toarcian of Yorkshire, England. Presented at the Annual Meeting of the Geological Society of America, November 5–8, 2001, Boston, MA.
  (2002) How catastrophic was the end-Triassic mass extinction? *Lethaia*, 35, 147–57.

Hallam, A. and Wignall, P. B. (1997) *Mass Extinctions and Their Aftermath*. Oxford University Press, Oxford, UK.

Han, J. and Calvin, M. (1970) Branched alkanes from blue-green algae. *Journal of the Chemical Society, Section D, Chemical Communications*, 22/1970, 1490–1.

Han, T.-M. and Runnegar, B. (1992) Megascopic eukaryotic algae from the 2.1-billion-year-old Negaunee Iron-Formation, Michigan. *Science*, 257, 232–5.

Han, J., McCarthy, E. D. and Calvin, M. (1968) Hydrocarbon constituents of the blue-green algae *Nostoc muscorum*, *Anacystis nidulans*, *Phormidium lubridum*, and *Chlorogloea fritschii*. *Journal of the Chemical Society (C)*, 2785–91.

Hanson, A. D., Ritts, B. D., Zinniker, D., Moldowan, J. M. and Biffi, U. (2001) Upper Oligocene lacustrine source rocks and petroleum systems of the northern Qaidam Basin, northwest China. *American Association of Petroleum Geologists Bulletin*, 85, 601–19.

Hao, S. and Guangdi, L. (1989) Precambrian oil and gas in China. *American Association of Petroleum Geologists Bulletin*, 73, 412.

Haq, B. U., Hardenbol, J. and Vail, P. R. (1987) Chronology of fluctuating sea levels since the Triassic. *Science*, 235, 1156–67.

Haq, B. U., Hardenbol, J. and Vail, P. R. (1988) Mesozoic and Cenozoic chronostratigraphy and cycles of sea-level changes. In: *Sea-Level Changes: An Integrated Approach*, (C. K. Wilgus, B. S. Hastings, C. G. S. C. Kendall, *et al.*, eds.), SEPM special publication no. 42. SEPM, Tulsa, OK, pp. 71–108.

Harms, G., Zengler, K., Rabus, R., *et al.* (1999) Anaerobic oxidation of o-xylene, m-xylene, and homologous alkylbenzenes by new types of sulfate-reducing bacteria. *Applied and Environmental Microbiology*, 65, 999–1004.

Harrell, J. A. and Lewan, M. D. (2002) Sources of mummy bitumen in ancient Egypt and Palestine. *Archaeometry*, 44, 285–93.

Harris, N. B. (2000) Toca carbonate, Congo Basin: response to an evolving rift lake. In: *Petroleum Systems of South Atlantic Margins* (M. R. Mello and B. J. Katz, eds.), American Association of Petroleum Geologists, Tulsa, OK, pp. 341–60.

Harris, J. P. and Fowler, M. R. (1987) Enhanced prospectivity of the Mid-Late Jurassic sediments of the South Viking Graben, northern North Sea. In: *Petroleum Geology of North West Europe* (J. Brooks and K. W. Glennie, eds.), Graham and Trotman, London, pp. 879–98.

Harris, P. J. F., Vis, R. D. and Heymann, D. (2000) Fullerene-like carbon nanostructures in the Allende meteorite. *Earth and Planetary Science Letters*, 183, 355–9.

Hartgers, W. A., Sinninghe Damsté, J. S., Requejo, A. G., *et al.* (1994a) A molecular and carbon isotopic study towards the origin and diagenetic fate of diaromatic carotenoids. *Organic Geochemistry*, 22, 703–25.
  (1994b) Evidence for only minor contributions from bacteria to sedimentary organic carbon. *Nature*, 369, 224–7.

Harvey, H. R. and McManus, G. B. (1991) Marine ciliates as a widespread source of tetrahymanol and hopan-3β-ol in sediments. *Geochimica et Cosmochimica Acta*, 55, 3387–90.

Harvey, G. R., Sinninghe Damsté, J. S. and De Leeuw, J. W. (1985) On the origin of alkylbenzenes in geochemical samples. *Marine Chemistry*, 16, 187–8.

Harvey, H. R., Ederington, M. C. and McManus, G. B. (1997) Lipid composition of the marine ciliates *Pleuronema* sp. and *Fabrea salina*: shifts in response to changes in diet. *Journal of Eukaryotic Microbiology*, 44, 189–93.

Hase, A. and Hites, R. A. (1976) On the origin of polycyclic aromatic hydrocarbons in recent sediments: biosynthesis by anaerobic bacteria. *Geochimica et Cosmochimica Acta*, 40, 1141–3.

Hatch, J. R. and Newell, K. D. (1999) *Geochemistry of Oils and Hydrocarbon Source Rocks from the Forest City Basin, Northeastern Kansas, Northwestern Missouri, Southwestern Iowa and Southeastern Nebraska*. Kansas Geological Survey, Lawrence, KS.

Hatch, J. R., Jacobson, J. R., Witzke, B. J., et al. (1987) Possible Middle Ordovician organic carbon isotope excursion: evidence from Ordovician oils and hydrocarbon source rocks, Mid-Continent, and East-Central United States. *American Association of Petroleum Geologists Bulletin*, 71, 1342–54.

Hatch, J. R., Risatti, J. B. and King, J. D. (1991) Geochemistry of Illinois Basin oils and hydrocarbon source rocks. In: *Interior Cratonic Basins* (M. W. Leighton, D. R. Kolata, D. F. Oltz and J. J. Eidel, eds.), American Association of Petroleum Geologists, Tulsa, OK, pp. 403–23.

Hatcher, H. J., Meuzelaar, H. L. C. and Urban, D. T. (1992) A comparison of biomarkers in gilsonite, oil shale, tar sand and petroleum from Threemile Canyon and adjacent areas in the Uinta Basin, Utah. In: *Hydrocarbon and Mineral Resources of the Uinta Basin, Utah and Colorado* (T. D. Fouch, V. F. Nuccio and T. C. Chidsey, eds.), Utah Geological Association, Salt Lake City, UT, pp. 271–87.

Haug, P. and Curry, D. J. (1974) Isoprenoids in a Costa Rican seep oil. *Geochimica et Cosmochimica Acta*, 38, 601–10.

Haug, P., Schnoes, H. K. and Burlingame, A. L. (1967) Isoprenoid and dicarboxylic acids isolated from Colorado Green River Shale (Eocene). *Science*, 158, 772–3.

Hauke, V., Graff, R., Wehrung, P., et al. (1992a) Novel triterpene-derived hydrocarbons of arborane/fernane series in sediments. Part I. *Tetrahedron Letters*, 48, 3915–24.

(1992b) Novel triterpene-derived hydrocarbons of the arborane/fernane series in sediments. Part II. *Geochimica et Cosmochimica Acta*, 56, 3595–602.

Hauke, V., Adam, P., Trendel, J.-M., et al. (1995) Isoarborinol through geological times: evidence for its presence in the Permian and Triassic. *Organic Geochemistry*, 23, 91–3.

Havord, P. J. (1993) *Mereenie Field – Australia, Amadeus Basin, Northern Territory*. American Association of Petroleum Geologists, Tulsa, OK.

Hay, W. W., Eicher, D. L. and Diner, R. (1993) Physical oceanography and water masses in the Cretaceous Western Interior Seaway. In: *Evolution of the Western Interior Basin* (W. G. E. Caldwell and E. G. Kauffman, eds.), Geological Association of Canada, New Foundland, pp. 297–318.

Hay, W. W., Deconto, R., Wold, C. N., et al. (1999) Alternative global Cretaceous paleogeography. In: *The Evolution of Cretaceous Ocean/Climate Systems* (E. Barrera and C. G. Johnson, eds.), Geological Society of America, Boulder, CO, pp. 1–47.

Hayes J. M., Kaplan, I. R. and Wedeking, K. M. (1983) Precambrian organic geochemistry, preservation of the record. In: *Earth's Earliest Biosphere, Its Origin and Evolution* (J. W. Schopf, ed.), Princeton University Press, Princeton, NJ, pp. 93–134.

Hayes, J. M., Takigiku, R., Ocampo, R., Callot, H. J. and Albrecht, P. (1987) Isotopic compositions and probable origins of organic molecules in the Eocene Messel Shale. *Nature*, 329, 48–51.

Hayes, J. M., Freeman, K. H., Popp, B. N. and Hoham, C. H. (1990) Compound-specific isotopic analyses: a novel tool for reconstruction of ancient biogeochemical processes. *Organic Geochemistry*, 16, 1115–28.

Hayes J. M., Des Marais, D. J., Lambert, I. B., Strauss, H. and Summons, R. E. (1992) Proterozoic biogeochemistry. In: *The Proterozoic Biosphere: A Multidisciplinary Study* (J. W. Schopf and C. Klein, eds.), Cambridge University Press, Cambridge, UK, pp. 81–134.

Hayes, J. M., Strauss, H. and Kaufman, A. J. (1999) The abundance of ^{13}C in marine organic matter and isotopic fractionation in the global biogeochemical cycle of carbon during the past 800 Ma. *Chemical Geology*, 161, 103–25.

Hays, P. D. (1992) Diagenesis in the Delaware Mountain Group: an example of organic-inorganic interaction during burial. In: *American Association of Petroleum Geologists, 1992 Annual Convention Abstracts* (G. Eynon, ed.), American Association of Petroleum Geologists, Tulsa, OK, pp. 53–4.

He, W. and Lu, S. (1990) A new maturity parameter based on monoaromatic hopanoids. *Organic Geochemistry*, 16, 1007–3.

Heath, D. J., Lewis, C. A. and Rowland, S. J. (1997) The use of high temperature gas chromatography to study the biodegradation of high molecular weight hydrocarbons. *Organic Geochemistry*, 26, 769–85.

Heider, J., Spormann, A. M., Beller, H. R. and Widdel, F. (1998) Anaerobic bacterial metabolism of hydrocarbons. *FEMS Microbiology Reviews*, 22, 459–73.

Heiss, G. and Knackmuss, H.-J. (2002) Bioelimination of trinitroaromatic compounds: immobilization versus mineralization. *Current Opinion in Microbiology*, 5, 282–7.

Heissler, D., Ocampo, R., Albrecht, P., Riehl, J. and Ourisson, G. (1984) Identification of long-chain tricyclic terpene hydrocarbons (C_{21}–C_{30}) in geological samples. *Journal of the Chemical Society*, Chemical Communications, 496–8.

Henderson, J. (1924) The origin of the Green River Formation. *American Association of Petroleum Geologists Bulletin*, 9, 662–8.

Hendrix, M. S. (1992) Sedimentary basin analysis and petroleum potential of Mesozoic strata, Northwest China. Ph. D. thesis, Stanford University, Stanford, CA.

Hendrix, M. S., Brassell, S. C., Carroll, A. R. and Graham, S. A. (1995) Sedimentology, organic geochemistry, and petroleum potential of Jurassic coal measures; Tarim, Junggar, and Turpan basins, Northwest China. *American Association of Petroleum Geologists Bulletin*, 79, 929–59.

Heppenheimer, H., Steffens, K., Püttman, W. and Kalkreuth, W. (1992) Comparison of resinite-related aromatic biomarker distributions in Cretaceous-Tertiary coals from Canada and Germany. *Organic Geochemistry*, 18, 273–87.

Herbin, J. P., Muller, C., Geyssant, J. R., *et al.* (1993) Variation of the distribution of organic matter within a transgressive system tract: Kimmeridge Clay (Jurassic), England. In: *Rocks in a Sequence Stratigraphic Framework* (B. J. Katz and L. M. Pratt, eds.), American Association of Petroleum Geologists, Tulsa, OK, pp. 67–100.

Hermansen, D. (1993) Optimization of temperature history – aspects of vitrinite reflectance and sterane isomerization. In: *Basin Modeling: Advances and Applications* (A. G. Dore, J. H. Augustson, C. Hermanrud, D. J. Stewart and O. Sylta, eds.), Elsevier, New York, pp. 119–26.

Hesselbo, S. P., Gröcke, D. R., Jenkyns, H. C., *et al.* (2000) Massive dissociation of gas hydrate during a Jurassic oceanic anoxic event. *Nature*, Vol. 406, pp. 392–95.

Hetényi, M. (1989) Hydrocarbon generative features of the Upper Triassic Kössen marl from W. Hungary. *Acta Mineralogica-Petrographica, Szeged*, 30, 137–47.

Heymann, D., Chibante, L. P. F., Smalley, R. E., *et al.* (1996a) Fullerenes of possible wildfire origin in Cretaceous-Tertiary boundary sediments. In: *The Cretaceous-Tertiary Event and Other Catastrophes in Earth History* (D. L. Campbell, G. Ryder, D. Fastovsky, and S. Gartner, eds.), Geological Society of America, Boulder, CO, pp. 453–64.

Heymann, D., Korochantsev, A., Nazarov, M. A. and Smith, J., (1996b) Search for fullerenes C_{60} and C_{70} in Cretaceous-Tertiary boundary sediments from Turkmenistan, Kazakhstan, Georgia, Austria, and Denmark. *Cretaceous Research*, 17, 367–80.

Heymann, D., Yancey, T. E., Wolbach, W. S., *et al.* (1998) Geochemical markers of the Cretaceous-Tertiary boundary event at Brazos River, Texas, USA. *Geochimica et Cosmochimica Acta*, 62, 173–81.

Hiatt E. E. (1997) A paleoceanographic model for oceanic upwelling in a late Paleozoic epicontinental sea; a chemostratigraphic analysis of the Permian Phosphoria Formation. Ph. D. thesis, University of Colorado, Boulder, CO.

Hiatt, E. E. and Budd, D. A. (2001) Sedimentary phosphate formation in warm shallow waters: new insights into the palaeoceanography of the Permian Phosphoria Sea from analysis of phosphate oxygen isotopes. *Sedimentary Geology*, 145, 119–33.

Higley, D. K. (2001) *The Putumayo-Oriente-Maranon Province of Colombia, Ecuador, and Peru – Mesozoic-Cenozoic and Paleozoic Petroleum Systems*. US Geological Survey Digital Data Series 63, Washington, DC.

(2002) The Talara Basin province of northwestern Peru: Cretaceous-Tertiary total petroleum system. Presented at the Annual Meeting of the American Association of Petroleum Geologists, March 10–13, 2002, Houston, TX.

Hildebrand, A. R., Penfield, G. T., Kring, D. A., *et al.* (1991) Chicxulub Crater: a possible Cretaceous/Tertiary boundary impact crater on the Yucatan Peninsula, Mexico. *Geology*, 19, 867–71.

Hill, A. J. (1995) Source rock distribution and maturity modeling. In: *Petroleum Geology of South Australia*, Vol. 1, (J. G. G. Morton and J. F. Drexel, eds.), South Australia Department of Mines and Energy, Adelaide, Australia, pp. 103–25.

Hill, R. J., Jarvie, D. M., Claxton, B. L., Burgess, J. D. and Williams, J. A. (2003) An investigation of petroleum systems of the Permian Basin, USA. Presented at the *21st International Meeting on Organic Geochemistry*, September 8–12, 2003, Krakow, Poland.

Hiller, K. (2000) *Madagascar: Hydrocarbons – Rohstoffwirtschaftliche Länderstudien*. Bundesanstalt für Geowissenschaften und Rohstoffe, Hanover, Germany.

Hills, I. R. and Whitehead, E. V. (1966) Triterpanes in optically active petroleum distillates. *Nature*, 209, 977–9.

Hills, I. R., Whitehead, E. V., Anders, D. E., Cummins, J. J. and Robinson, W. E. (1966) An optically active triterpane, gammacerane in Green River, Colorado, oil shale bitumen. *Journal of the Chemical Society, Chemical Communications*, 20, 752–4.

Hinrichs, K.-U., Haver, J. M., Sylva, S. P., Brewer, P. G. and Delong, E. F. (1999) Methane-consuming archaebacteria in marine sediments. *Nature*, 398, 802–5.

Hinrichs, K.-U., Summons, R. E., Orphan, V., Sylva, S. P. and Hayas, J. M. (2000) Molecular and isotopic analysis of anaerobic methane-oxidizing communities in marine sediments. *Organic Geochemistry*, 31, 1685–701.

Hinrichs, K.-U., Hmelo, L. R. and Sylva, S. P. (2003) Molecular fossil record of elevated methane levels in Late Pleistocene coastal waters. *Science*, 299, 1214–17.

Hird, S. J. and Rowland, S. J. (1995) An investigation of the source and seasonal variations of highly branched isoprenoid hydrocarbons in intertidal sediments of the Tamar Estuary, UK. *Marine Environmental Research*, 40, 423–38.

Hite, R. J. Anders, D. E. and Ging, T. G. (1984) Organic-rich source rocks of Pennsylvanian age in the Paradox Basin of Utah and Colorado. In: *Hydrocarbon Source Rocks of the Greater Rocky Mountain Region* (J. Woodward, F. F. Meissner and J. L. Clayton, eds.), Rocky Mountain Association of Petroleum Geologists, Denver, CO, pp. 255–74.

Ho, E. S., Meyers, P. A. and Mauk, J. L. (1990) Organic geochemical study of mineralization in the Keweenawan Nonesuch Formation at White Pine, Michigan. *Organic Geochemistry*, 16, 229–34.

Hodych, J. P. and Dunning, G. R. (1992) Did the Manicouagan impact trigger end-of-Triassic mass extinction? *Geology*, 20, 51–4.

Hoefs, M. J. L., Sinninghe Damsté, J. S. and de Leeuw, J. W. (1995) A novel C_{35} highly branched isoprenoid polyene in Recent Indian Ocean sediments. *Organic Geochemistry*, 23, 263–7.

Hoefs, M. J. L., Schouten, S., King, L. L., *et al.* (1997) Ether lipids of planktonic archaea in the marine water column. *Applied and Environmental Microbiology*, 63, 3090–5.

Hoehler, T. M., Alperin, M. J., Albert, D. B. and Martens, C. S. (1994) Field and laboratory studies of methane oxidation in an anoxic marine sediment: evidence for a methanogen-sulfate reducer consortium. *Global Biogeochemical Cycles*, 8, 451–63.

Hoering, T. C. (1971) The conversion of polar organic molecules in rock extracts to saturated hydrocarbons. *Carnegie Institution of Washington Yearbook*, 70, 251–8.

(1972) The benzene-soluble organic matter, humic acids, and insoluble organic matter in a core from the Cariaco Trench. *Carnegie Institute of Washington Yearbook*, 71, 585–92.

(1978) Molecular fossils from the Precambrian Nonesuch Shale. In: *Comparative Planetology* (C. Ponnamperuma, ed.), Academic Press, New York, pp. 243–55.

(1980) Monomethyl, acyclic hydrocarbons in petroleum and source extracts. *Carnegie Institute of Washington Yearbook*, 80, 389–94.

Hoering, T. C. and Freeman, D. H. (1984) Shape-selective sorption of monomethylalkanes by silicalite, a zeolite form of silica. *Journal of Chromatography*, 316, 333–41.

Hoering, T. C. and Navale, V. (1987) A search for molecular fossils in the kerogen of Precambrian sedimentary rocks. *Precambrian Research*, 34, 247–67.

Hoffman, A. (1985) Patterns of family extinction depend on definition and geological time-scale. *Nature*, 315, 659–62.

(1989) Mass extinctions: a view of a skeptic. *Journal of the Geological Society*, 146, 21–35.

Hoffmann, C. F. and Strausz, O. P. (1986) Bitumen accumulation in Grosmont Platform Complex, Upper Devonian, Alberta, Canada. *American Association of Petroleum Geologists Bulletin*, 70, 1113–28.

Hoffmann, C. F., Mackenzie, A. S., Lewis, C. A., *et al.* (1984) A biological marker study of coals, shales, and oils from the Mahakam Delta, Kalimantan, Indonesia. *Chemical Geology*, 42, 1–23.

Hoffmann, C. F., Foster, C. B., Powell, T. G. and Summons, R. E. (1987) Hydrocarbon biomarkers from Ordovician sediments and the fossil alga *Gloeocapsomorpha prisca* Zalessky 1917. *Geochimica et Cosmochimica Acta*, 51, 2681–97.

Hofmann, H. J., Narbonne, G. M. and Aitken, J. D. (1990) Ediacaran remains from intertillite beds in northwestern Canada. *Geology*, 18, 1199–202.

Hofmann, H. J., Grey, K., Hichman, A. H. and Thorpe, R. I. (1999) Origin of 3.45 Ga coniform stromatolites in Warrawoona Group, Western Australia. *Geological Society of America Bulletin*, 111, 1256–62.

Hofmann, P., Leythaeuser, D. and Schwark, L. (2001) Organic matter from the Bunte Breccia of the Ries Crater, southern Germany: investigating possible thermal effects of the impact. *Planetary and Space Science*, 49, 845–51.

Holba, A. G., Dzou, L. I. P., Masterson, W. D., (1998) Application of 24-norcholestanes for constraining source age of petroleum. *Organic Geochemistry*, 29, 1269–83.

Holba, A. G., Tegelaar, E., Ellis, L., Singletary, M. S. and Albrecht, P. (2000) Tetracyclic polyprenoids: indicators of freshwater (lacustrine) algal input. *Geology*, 28, 251–4.

Holba A. G., Ellis, L., Dzou, I. L., *et al.* (2001) Extended tricyclic terpanes as age discriminators between Triassic, Early Jurassic and Middle-Late Jurassic oils. Presented at the 20th International Meeting on Organic Geochemistry, 10–14 September, 2001, Nancy, France.

Holba, A. G., Dzou, L. I., Wood, G. D., *et al.* (2003) Application of tetracyclic polyprenoids as indicators of input from fresh-brackish water environments. *Organic Geochemistry*, 34, 441–69.

Holbourn, A., Kuhnt, W., El Albani, A., *et al.* (1999) Upper Cretaceous paleoenvironments and benthonic foraminiferal assemblages of potential source rocks from the western African margin, Central Atlantic. In: *The Oil and Gas Habitats of the South Atlantic* (N. R. Cameron, R. H. Bate and V. S. Clure, eds.), Geolocial Society, London, pp. 195–222.

Höld, I. M., Brussee, N. J., Schouten, S. and Sinninghe Damsté, J. S. (1998) Changes in the molecular structure of a type II-S kerogen (Monterey Formation, USA) during sequential chemical degradation. *Organic Geochemistry*, 29, 1403–17.

Höld, I. M., Schouten, S., Jellema, J. and Sinninghe Damsté, J. S. (1999) Origin of free and bound mid-chain methyl alkanes in oils, bitumens and kerogens of the marine, Infracambrian Huqf Formation (Oman). *Organic Geochemistry*, 30, 1411–28.

Holdgate, G. R., Wallace, M. W., Gallagher, S. J. and Taylor, D. (2000) A review of the Traralgon Formation in the Gippsland Basin – a world class brown coal resource. *International Journal of Coal Geology*, 45, 55–84.

Holowenko, F. M., Mackinnon, M. D. and Fedorak, P. M. (2002) Characterization of naphthenic acids in oil sands wastewaters by gas chromatography-mass spectrometry. *Water Research*, 36, 2843–55.

Holzer, G., Oro, J. and Tornabene, T. G. (1979) Gas chromatographic-mass spectrometric analysis of neutral lipids from methanogenic and thermoacidophilic bacteria. *Journal of Chromatography*, 186, 795–809.

Hong, Z.-H., Li, H.-X., Rullkötter, J. and Mackenzie, A. S. (1986) Geochemical application of sterane and triterpane biological marker compounds in the Linyi Basin. *Organic Geochemistry*, 10, 433–9.

Hood, K. C., Wenger, L. M., Gross, O. P. and Harrison, S. C. (2001) Hydrocarbon systems analysis of the northern Gulf of Mexico: delineation of hydrocarbon migration pathways using seeps and seismic imaging. In: *Applications of Surface Exploration Methods in Exploration, Field Development and Production* (D. Schumacher, ed.), American Association of Petroleum Geologists, Tulsa, OK, 25–40.

Hopmans, E. C., Schouten, S., Pancost, R. D., Van der Meer, M. J. T. and Sinninghe Damsté, J. S. (2000) Analysis of intact tetraether lipids in archaeal cell material and sediments using high performance liquid chromatography/atmospheric pressure ionization mass spectrometry. *Rapid Communications in Mass Spectrometry*, 14, 585–9.

Horsfield, B., Curry, D. J., Bohacs, K., *et al.* (1994) Organic geochemistry of freshwater and alkaline lacustrine sediments in the Green River Formation of the Washakie Basin, Wyoming, USA *Organic Geochemistry*, 22, 415–40.

Horstad, I., Larter, S. R., Dypvik, H., *et al.* (1990) Degradation and maturity controls on oil field petroleum column heterogeneity in the Gullfaks Field, Norwegian North Sea. *Organic Geochemistry*, 16, 497–510.

Hostettler, F. D. and Kvenvolden, K. A. (2002) Alkylcyclohexanes in environmental geochemistry. *Environmental Forensics*, 3, 293–301.

Hostettler, F. D., Rosenbauer, R. J. and Kvenvolden, K. A. (1999) PAH refractory index as a source discriminant of hydrocarbon input from crude oil and coal in Prince William Sound, Alaska. *Organic Geochemistry*, 30, 873–9.

Hou, D. J., Wang, T. G., Kong, Q. Y., Feng, Z. H. and Moldowan, J. M. (1999) Distribution and characterization of C_{31} sterane from Cretaceous sediments and oils, Songliao Basin, China. *Chinese Science Bulletin*, 44, 560–3.

Hou, D., Li, M. and Huang, Q. (2000) Marine transgressional events in the gigantic freshwater lake Songliao: paleontological and geochemical evidence. *Organic Geochemistry*, 31, 763–8.

Howard, D. L. (1980) Polycyclic triterpenes of anaerobic photosynthetic bacterium, *Rhodomicrobium vannielii*. Ph. D. thesis, University of California at Los Angeles, Los Angeles, CA.

Howell, V. J., Connan, J. and Aldridge, A. K. (1984) Tentative identification of demethylated tricyclic terpanes in nonbiodegraded and slightly biodegraded crude oils from the Los Llanos Basin, Colombia. *Organic Geochemistry*, 6, 83–92.

Huang, W.-Y. and Meinshein, W. G. (1979) Sterols as ecological indicators. *Geochimica et Cosmochimica Acta*, 43, 739–45.

Huang, W. Y., Grizzle, P. L. and Haney, F. R. (1985) Source rock-crude oil correlations: Alaskan North Slope. In: *Alaska North Slope Oil/Source Rock Correlation Study* (L. B. Magoon and G. E. Claypool, eds.), American Association of Petroleum Geologists, Tulsa, OK, pp. 557–70.

Huang, D., Li, J., and Zhang, D. (1990) Maturation sequence of continental crude oils in hydrocarbon basins in China and its significance. *Organic Geochemistry*, 16, 521–9.

Huang, Z., Williamson, M. A., Fowler, M. G. and McAlpine, K. D. (1994) Predicted and measured petrophysical and geochemical characteristics of the Egret Member oil source rock, Jeanne d'Arc Basin, offshore eastern Canada. *Marine and Petroleum Geology*, 11, 294–306.

Huang, Z., Williamson, M. A., McAlpine, K. D., Bateman, J. and Fowler, M. G. (1996) Cyclicity in the Egret Member (Kimmeridgian) oil source rock, Jeanne d'Arc Basin, offshore eastern Canada. *Marine and Petroleum Geology*, 13, 91–105.

Huang Y., Peakman, T. M. and Murray, M. (1997) $8\beta,9\alpha,10\beta$-rimuane, a novel, optically active, tricyclic hydrocarbon of algal origin. *Tetrahedron Letters*, 38, 5363–6.

Huber, C. and Wächterhäuser, G. (1997) Activated acetic acid by carbon fixation on (Fe, Ni)S under primordial conditions. *Science*, 276, 245–7.

Huc, A. Y., Durand, B., Roucachet, J., Vandenbroucke, M. and Pittion, J. L. (1986) Comparison of three series of organic matter of continental origin. *Organic Geochemistry*, 10, 65–72.

Hughes, W. B. (1984) Use of thiophenic organosulfur compounds in characterizing crude oils derived from carbonate versus siliciclastic sources. In: *Petroleum Geochemistry and Source Rock Potential of Carbonate Rocks* (J. G. Palacas, ed.), American Association of Petroleum Geologists, Tulsa, OK, pp. 181–196.

Hughes, W. B. and Dzou, L. I. P. (1995) Reservoir overprinting of crude oils. *Organic Geochemistry*, 23, 905–14.

Hughes, W. B., Holba, A. G., Mueller, D. E. and Richardson, J. S. (1985) Geochemistry of greater Ekofisk crude oils. In: *Geochemistry in Exploration of the Norwegian Shelf* (B. M. Thomas, ed.), Graham and Trotman, London, pp. 75–92.

Hunt, J. M. (1996) *Petroleum Geochemistry and Geology*. W. H. Freeman, New York.

Husseini M. I. (1992) Potential petroleum resources of the Paleozoic rocks of Saudi Arabia. In: *Proceedings of the Thirteenth World Petroleum Congress, Buenos Aires*, Vol. II, John Wiley & Sons, New York, pp. 3–13.

Hussler, G., Albrecht, P. and Ourisson, G. (1984a) Benzohopanes, a novel family of hexacyclic geomarkers in sediments and petroleums. *Tetrahedron Letters*, 25, 1179–82.

Hussler, G., Connan, J. and Albrecht, P. (1984b) Novel families of tetra- and hexacyclic aromatic hopanoids predominant in carbonate rocks and crude oils. *Organic Geochemistry*, 6, 39–49.

Hvoslef, S., Larter, S. R. and Leythaeuser, D. (1988) Aspects of generation and migration of hydrocarbons from coal-bearing strata of the Hitra Formation, Haltenbanken area, offshore Norway. *Organic Geochemistry*, 13, 525–36.

Hwang, R. J. (1990) Biomarker analysis using GC-MSD. *Journal of Chromatographic Science*, 28, 109–13.

Hwang R J., Sundararaman, P., Teerman, S. C. and Schoell, M. (1989) Effect of preservation on geochemical properties of organic matter in immature lacustrine sediments. Presented at the 14th International Meeting on Organic Geochemistry, September 18–22, 1989, Paris, France.

Hwang, R. J., Ahmed, A. S. and Moldowan, J. M. (1994) Oil composition variation and reservoir continuity: Unity Field, Sudan. *Organic Geochemistry*, 21, 171–88.

IHS (Information Handling Services)/Petroconsultants S. A. (1996–99) Petroleum exploration and production database. Available from Petroconsultants, Inc., PO Box 740619, Houston, TX 77274-0619, USA.

Illich, H. A. (1983) Pristane, phytane, and lower molecular weight isoprenoid distributions in oils. *American Association of Petroleum Geologists Bulletin*, 67, 385–93.

Illich, H. A. and Grizzle, P. L. (1983) Comment on "Comparison of Michigan Basin crude oils" by Vogler et al. *Geochimica et Cosmochimica Acta*, 47, 1151–6.

Illich, H. A., Haney, F. R. and Jackson, T. J. (1977) Hydrocarbon geochemistry of oils from Maranon Basin,

Peru. *American Association of Petroleum Geologists Bulletin*, 61, 2103–14.

Illich, H. A., Haney, F. R. and Mendoza, M. (1981) Geochemistry of oil from Santa Cruz Basin, Bolivia: case study of migration-fractionation. *American Association of Petroleum Geologists Bulletin*, 65, 2388–402.

Imbus, S. W., Engel, M. H., Elmore, R. D. and Zumberge, J. E. (1988) The origin, distribution and hydrocarbon generation potential of the organic-rich facies in the Nonesuch Formation, Central North American Rift system: a regional study. *Organic Geochemistry*, 13, 207–19.

Inan, S. I., Yalçin, M. N., Guliev, I. S., Kuliev, K. and Feizullayev, A. A. (1997) Deep petroleum occurrences in the Lower Kura Depression, South Caspian Basin, Azerbaijan: an organic geochemical and basin modeling study. *Marine and Petroleum Geology*, 14, 731–62.

Irwin, H. and Meyer, T. (1990) Lacustrine organic facies. A biomarker study using multivariate statistical analysis. *Organic Geochemistry*, 16, 176–210.

Isaacs, C. M. (1984) Hemipelagic deposits in a Miocene basin, California: toward a model of lithologic variation and sequecnce. In: *Fine-Grained Sediments: Deep-Water Processes and Facies* (D. A. V. Stow and D. J. W. Piper, Jr. eds.), Blackwell, Oxford, UK, pp. 481–96.

(1987) Sources and deposition of organic matter in the Monterey Formation, south-central coastal basins of California: Section II. Characterization, maturation, and degradation. In: *Exploration for Heavy Crude Oil and Natural Bitumen* (R. F. Meyer, ed.), American Association of Petroleum Geologists Tulsa, OK, pp. 193–205.

Isaacs, C. M. (2001) Depositional framework of the Monterey Formation, California. In: *The Monterey Formation: From Rocks to Molecules* (C. M. Isaacs and J. Rullkötter, eds.) Columbia University Press, New York, pp. 461–524.

Isaacs, C. M. and Petersen, N. F. (1988) Petroleum in the Miocene Monterey Formation, California. In: *Siliceous Sedimentary Rock-Hosted Ores and Petroleum* (J. R. Hein, ed.), Van Nostrand Reinhold, New York, pp. 83–116.

Isaacs, C. M. and Rullkötter, R. (2001) *The Monterey Formation: From Rocks to Molecules* (C. M. Isaacs and J. Rullkötter, eds.), Columbia University Press, New York.

Isaacs, C. M., Baumgartner, T. R., Tennyson, M. E., Piper, D. Z. and Ingle, J. C., Jr. (1996) A prograding margin model for the Monterey Formation, California.

Presented at the Annual Meeting of the American Association of Petroleum Geologists and the Society of Economic Paleontologists and Mineralogists, San Diego, CA, May 17–19, 1996.

Isaacson, P. E., Palmer, B. A., Mamet, B. L., Cooke, J. C. and Sanders, D. E. (1995) Devonian-Carboniferous stratigraphy in the Madre de Dios Basin, Bolivia: Pando X-1 and Manuripi X-1 Wells. In: *Petroleum Basins of South America* (A. J. Tankard, R. Suarez Soruco and H. J. Welsink, eds.), American Association of Petroleum Geologists, Tulsa, OK, pp. 501–9.

Isaksen, G. H. (1995) Organic geochemistry of paleodepositional environments with a predominance of terrigenous higher-plant organic matter. In: *Paleogeography, Paleoclimate, and Source Rocks* (A.-Y. Huc, ed.), American Association of Petroleum Geologists, Tulsa, OK, pp. 81–104.

Isaksen, G. H. and Bohacs, K. M. (1995) Geological controls on source rock geochemistry through relative sea level; Triassic, Barents Sea. In: *Petroleum Source Rocks* (B. J. Katz, ed.), Springer-Verlag, New York, pp. 25–50.

Isaksen, G. H. and Ledje, K. H. I. (2001) Source rock quality and hydrocarbon migration pathways within the Greater Utsira High area, Viking Graben, Norwegian North Sea. *American Association of Petroleum Geologists Bulletin*, 85, 861–83.

Isozaki, Y. (1997) Permo-Triassic boundary superanoxia and stratified superocean; records from lost deep sea. *Science*, 276, 235–8.

Jack, T. R. (1993) M.O.R.E. to M.E.O.R.: an overview of microbially enhanced oil recovery. In: *Microbial Enhancement of Oil Recovery – Recent Advances, Proceedings of the 1992 International Conference on Microbial Enhanced Oil Recovery* (E. T. Premuzic and A. Woodhead, eds.), Elsevier Science, Amsterdam, pp. 7–16.

Jackson, B. E. and McInerney, M. J. (1996) Thermophilic denitrifying bacteria isolated from petroleum reservoirs and the environmental factors that influence their metabolism. Presented at the 3rd US DOE Petroleum and Environment Conference, (Albuquerque, NM, September 24–27, 1996.

(2002) Anaerobic microbial metabolism can proceed close to thermodynamic limits *Nature*, 415, 454–6.

Jackson, K. S., Mckirdy, D. M. and Deckelman, J. A. (1984) Hydrocarbon generation in the Amadeus Basin, Central Australia. *Australian Petroleum Exploration Association Journal*, 24, 42–63.

Jackson, M. J., Powell, T. G., Summons, R. E. and Sweet, I. P. (1986) Hydrocarbon shows and petroleum source rocks in sediments as old as 1.7×10^9 years. *Nature*, 322, 727–9.

Jackson, K. S., Sweet, I. P. and Powell, T. G. (1988) Studies on petroleum geology and geochemistry of the Middle Proterozoic McArthur Basin, Northern Australia. I: petroleum potential. *Australian Petroleum Exploration Association Journal*, 28, 283–302.

Jacobs, D. K. and Linberg, D. R. (1998) Oxygen and evolutionary patterns in the sea: onshore/offshore trends and recent recruitment of deep-sea faunas. *Proceedings of the National Academy of Science, USA*, 95, 9296–401.

Jacobson, S. R., Hatch, J. R. and Teerman, S. C. (1986) Microscopical and geochemical observations on *Gloeocapsomorpha* sp. from the St Peter Sandstone to the Guttenberg Limestone Member of the Decorah Formation (Middle Ordovician) of the Midcontinent, USA. Presented at the 19th Annual Meeting of the American Association of Stratigraphic Palynologists New York, October 29–November 1, 1986.

Jacobson, S. R., Hatch, J. R., Teerman, S. C. and Askin, R. A. (1988) Middle Ordovician organic matter assemblages and their effect on Ordovician-derived oils. *American Association of Petroleum Geologists Bulletin*, 72, 1090–100.

Jaffé, R. and Gallardo, M. T. (1993) Application of carboxylic acid biomarkers as indicators of biodegradation and migration of crude oils from the Maracaibo Basin, western Venezuela. *Organic Geochemistry*, 20, 973–84.

Jaffé, R. and Gardinali, P. R. (1990) Generation and maturation of carboxylic acids in ancient sediments from the Maracaibo Basin, Venezuela. *Organic Geochemistry*, 16, 211–18.

Jaffé, R. and Hausmann, K. B. (1995) Origin and early diagenesis of arborinone/isoarborinol in sediments of a highly productive freshwater lake. *Organic Geochemistry*, 22, 231–5.

Jaffé, R., Albrecht, P. and Oudin, J. L. (1988a) Carboxylic acids as indicators of oil migration. I. Occurrence and geochemical significance of C-22 diastereoisomers of the $17\beta(H), 21\beta(H)$ C_{30} hopanoic acid in geological samples. *Organic Geochemistry*, 13, 483–88.

(1988b) Carboxylic acids as indicators of oil migration. II. Case of the Mahakam Delta, Indonesia. *Geochimica et Cosmochimica Acta*, 52, 2599–607.

Jaffé, R., Gardinali, P. and Wolff, G. A. (1992) Evolution of alkanes and carboxylic acids in ancient sediments from the Maracaibo Basin. *Organic Geochemistry*, 18, 195–201.

Jahren, A. H., Arens, N. C., Sarmiento, G., Guerrero, J. and Amundson, R. (2001) Terrestrial record of methane hydrate dissociation in the Early Cretaceous. *Geology*, 29, 159–62.

Jaillard, E., Ordonez, M., Benitez, S., *et al.* (1995) Basin development in an accretionary, oceanic-floored fore-arc setting: southern coastal Ecuador during Late Cretaceous-Late Eocene time. In: *Petroleum Basins of Southern South America*, (A. J. Tankard, R. Suarez Soruco and H. J. Welsink, eds.), American Association of Petroleum Geologists, Tulsa, OK, pp. 615–31.

James, A. T. and Burns, B. J. (1984) Microbial alteration of subsurface natural gas accumulations. *American Association of Petroleum Geologists Bulletin*, 86, 957–60.

James, A. T., Wenger, L. M., Melia, M. B., Ross, A. H. and Kuminecz, C. P. (1993) Recognition of a new hydrocarbon play in a mature exploration area through integration of geochemical, palynologic, geologic, and seismic interpretation (onshore northern rim of Gulf of Mexico). Presented at the Annual Meeting of the American Association of Petroleum Geologists and the Society of Economic Paleontologists and Mineralogists, New Orleans, LA, April 16–19, 2000.

Jardine E. (1997) Dual petroleum systems governing the prolific Pattani Basin, offshore Thailand. In: *Proceedings of the International Conference on Petroleum Systems of SE Asia and Australia* (J. V. C. Howes and R. A. Noble, eds.), Indonesian Petroleum Association, Jakarta, Indonesia, pp. 351–63.

Jarvie, D. M. (2001) Williston Basin petroleum systems: inferences from oil geochemistry and geology. *Mountain Geologist*, 38, 19–42.

Jarvie, D. M., Burgess, J. D., Morelos, A., Mariotti, P. A. and Lindsey, R. (2001a) Permian Basin petroleum systems investigations: inferences from oil geochemistry and source rocks. *American Association of Petroleum Geologists Bulletin*, 85, 1693–4.

Jarvie, D. M., Mbatau, F., Maende, A., Ngenoh, D. and Wavrek, D. A. (2001b) Petroleum systems in northwest Kenya. Presented at the Annual Meeting of the American Association of Petroleum Geologists, Denver, CO, June 3–6, 2001.

Jarvie, D. M., Morelos, A., Sassen, R., Chenet, P. Y. and Brame, J. W. (2002) Hydrocarbon charge assessment,

Gulf of Mexico: rates of oil/gas generation from source rocks and oil asphaltenes. Presented at the Annual Meeting of the American Association of Petroleum Geologists, Houston, TX, March 10–13, 2002.

Jean-Baptiste, P., Petit, J.-R., Lipenkov, V. Y., Raynaud, D. and Barkov, N. I. (2001) Constraints on hydrothermal processes and water exchange in Lake Vostok from helium isotopes. *Nature*, 411, 460–2.

Jeffrey, A. W. A., Alimi, H. M. and Jenden, P. D. (1991) Geochemistry of Los Angeles Basin oil and gas systems. In: *Active Margin Basins* (K. T. Biddle, ed.), American Association of Petroleum Geologists, Tulsa, OK, pp. 197–219.

Jeffries, P. J. (1988) Geochemistry of the Turtle oil accumulation, offshore Southern Bonaparte Basin. In: *The North West Shelf, Australia: Proceedings of the North West Shelf Symposium 1988* (R. R. Purcell, ed.), Petroleum Exploration Society of Australia, Perth, Australia, pp. 563–9.

Jehlýcka, J., Ozawa, M., Slanina, Z. and Osawa, E. (2000) Fullerenes in solid bitumens from pillow lavas of Precambrian age (Mitov, Bohemian Massif). *Fullerene Science and Technology*, 18, 449–52.

Jenkins C. C. (1989) Geochemical correlation of source rocks and crude oils from the Cooper and Eromanga basins. In: *The Cooper and Eromanga Basins, Australia Proceedings of the Cooper and Eromanga Basins Conference, Adelaide, 1989* (B. J. O'Neil, ed.), Petroleum Exploration Society of Australia, Perth, Australia, pp. 525–40.

Jenkyns, H. C. (1980) Cretaceous anoxic events: from continents to oceans. *Journal of the Geological Society London*, 137, 171–88.

(1985) The early Toarcian and Cenomanian-Turonian anoxic event in Europe: comparisons and contrasts. *Geologische Rundschau*, 74, 505–18.

(1988) The early Toarcian (Jurassic) anoxic event: stratigraphic, sedimentary and geochemical evidence. *American Journal of Science*, 288, 101–51.

Jenkyns, H. C. and Clayton, C. J. (1997) Lower Jurassic epicontinental carbonates and mudstones from England and Wales: chemostratigraphic signals and the early Toarcian anoxic event. *Sedimentology*, 44, 687–706.

Jenneman, G. E., McInerney, M. J. and Knapp, R. M. (1985) Microbial penetration through nutrient saturated Berea sandstone. *Applied and Environmental Microbiology*, 50, 383–91.

Jensen, S., Gehling, J. G. and Droser, M. L. (1998) Ediacara-type fossils in Cambrian sediments. *Nature*, 393, 567–9.

Jewell, P. W., (1995) Geologic consequences of globe-encircling equatorial currents. *Geology*, 23, 117–20.

Jiang, Z. and Fowler, M. G. (1986) Carotenoid-derived alkanes in oils from northwestern China. *Organic Geochemistry*, 10, 831–9.

Jiang, Z., Philp, R. P. and Lewis, C. A. (1988) Fractionation of biological markers in crude oils during migration and the effects on correlation and maturation parameters. *Organic Geochemistry*, 13, 561–71.

Jiang, Z., Fowler, M. G., Lewis, C. A. and Philp, R. P. (1990) Polycyclic alkanes in a biodegraded oil from the Kelamayi oilfield, northwestern China. *Organic Geochemistry*, 15, 35–46.

Jianyu, C., Yanpong, B., Jiguo, Z., and Shuafu, L. (1996) Oil-source correlation in the Fulin Basin, Shengli petroleum province, East China. *Organic Geochemistry*, 24, 931–40.

Jobson, A., Cook, F. D. and Westlake, D. W. S. (1972) Microbial utilization of crude oil. *Applied Microbiology*, 23, 1082–89.

(1979) Interaction of aerobic and anaerobic bacteria in petroleum biodegradation. *Chemical Geology*, 24, 355–65.

Johns, R. B. (1986) *Biological Markers in the Sedimentary Record*. Elsevier, New York.

Johnson, K. S. and Cardott, B. J. (1992) Geologic framework and hydrocarbon source rocks of Oklahoma. In: *Source Rocks in the Southern Midcontinent, 1990 Symposium* (K. S. Johnson and B. J. Cardott, eds.), Oklahoma Geological Survey Circular 93, Oklahoma Geological Survey Norman, OK, pp. 21–37.

Johnson, H. D. and Fisher, M. J. (1998) North Sea plays: geological controls on hydrocarbon distribution. In: *Petroleum Geology of the North Sea, Basic Concepts and Recent Advances*, 4th Edn (K. W. Glennie, ed.), Blackwell Science, London, pp. 463–547.

Johnston, J. H., Collier, R. J. and Maidment, A. I. (1991) Coals as source rocks for hydrocarbon generation in the Taranaki Basin, New Zealand, a geochemical biomarker study. *Journal of Southeast Asian Earth Sciences*, 5, 283–9.

Jokanovi, M. (2001) Biotransformation of organophosphorus compounds. *Toxicology*, 166, 139–60.

Jones, R. W. (1984) Comparison of carbonate and shale source rocks. In: *Petroleum Geochemistry and Source Rock Potential of Carbonate Rocks* (J. G. Palacas, ed.), American Association of Petroleum Geologists, Tulsa, OK, pp. 163–80.

(1987) Organic facies. In: *Advances in Petroleum Geochemistry* (J. Brooks and D. Welte, eds.), Academic Press, New York, pp. 1–90.

Jones, R. W. and Edison, T. A. (1978) Microscopic observations of kerogen related to geochemical parameters with emphasis on thermal maturation. In: *Low Temperature Metamorphism of Kerogen and Clay Minerals* (D. F. Oltz, ed.), Society of Economic Paleontologists and Mineralogists, Los Angeles, pp. 1–12.

Jones, P. J. and Philp, R. P. (1990) Oils and source rocks from Pauls Valley, Anadarko Basin, Oklahoma, USA. *Applied Geochemistry*, 5, 429–48.

Jones, P. J. and Stump, T. E. (1999) Depositional and tectonic setting of the Lower Silurian hydrocarbon source rock facies, central Saudi Arabia. *American Association of Petroleum Geologists Bulletin*, 83, 314–32.

Jones, D. M., Meredith, W. and Aiken, C. M. (2002) Carboxylic acids in biodegraded oils and natural bitumens. Presented at the 19th Annual Meeting of the Society of Organic Petrographers: Emerging Concepts in Organic Petrography and Geochemistry, Banff, Canada, August 31–September 4, 2002.

Jörgensen, B. B. (1989) Biogeochemistry of chemoautotrophic bacteria. In: *Autotrophic Bacteria* (H. G. Schlegel and B. Bowien, eds.), Science Tech Publishers, Madison, WI, pp. 117–46.

Jouzel, J., Petit, J. R., Souchez, R., *et al.* (1999) More than 200 meters of lake ice above subglacial Lake Vostok, Antarctica, *Science*, 286, 2138–41.

Jull, A. J. T., Courtney, C. Jeffrey, D. A. and Beck, J. W. (1998) Isotopic evidence for a terrestrial source of organic compounds found in Martian meteorites Allan Hills 84001 and Elephant Moraine 79001. *Science*, 279, 366–9.

Junhong, C. and Summons, R. E. (2001) Complex patterns of steroidal biomarkers in Tertiary lacustrine sediments of the Biyang Basin, China. *Organic Geochemistry*, 32, 115–26.

Justman, H. A. and Broadhead, R. F. (2000) Source rock analysis for the Brushy Canyon Formation, Delaware Basin, southeastern New Mexico. In: *The Permian Basin: Proving Ground for Tomorrow's Technologies*, (D. DeMis, M. K. Nelis and R. C. Trentham, eds. West Texas. Geological Society Midland, TX pp. 211–20.

Kagi, R.I, Alexander, R. and Toh, E. (1989) Kinetics and mechanism of the cyclisation reaction of ortho-methylbiphenyls. *Organic Geochemistry*, 16, 161–6.

Kaiho, K., Kajiwara, Y., Nakano, T., *et al.* (2001) End-Permian catastrophe by a bolide impact: evidence of a gigantic release of sulfur from the mantle. *Geology*, 29, 815–18.

Kajiwara, Y., Yamakita, S., Ishida, K., Ishiga, H. and Imai, A. (1994) Development of a largely anoxic stratified ocean and its temporary massive mixing at the Permian/Triassic boundary supported by the sulfur isotopic record. *Palaeogeography, Palaeoclimatology, Palaeoecology*, 111, 367–79.

Kamali, M. R. (1995) Sedimentology and petroleum geochemistry of the Ouldburra Formation, Eastern Officer Basin, Australia. Ph. D. thesis, University of Adelaide, Adelaide, Australia.

Kapitsa, A. P., Ridley, J. K., Robin, G. D. Q., Siegert, M. J. and Zotikov, I. A. (1996) A large deep freshwater lake beneath the ice of central East Antarctica. *Nature*, 381, 684–6.

Karl, D. M., Bird, D. F., Björkman, K., *et al.* (1999) Microorganisms in the accreted ice of Lake Vostok, Antarctica. *Science*, 286, 2144–7.

Karlsen, D. A., Backer-Owe, K., Nyland, B., *et al.* (1995) *Petroleum Geochemistry of the Haltenbanken, Norwegian Continental Shelf*. Geological Society of London, London, pp. 203–56.

Karner, G. D., Driscoll, N. W., McGinnis, J. P., Brumbaugh, W. D. and Cameron, N. R. (1997) Tectonic significance of syn-rift sediment packages across the Gabon-Cabinda continental margin. *Marine and Petroleum Geology*, 14, 973–1000.

Karner, M. B., Delong, E. F. and Karl, D. M. (2001) Archaeal dominance in the mesopelagic zone of the Pacific Ocean. *Nature*, 409, 507–10.

Kashefi, K. and Lovley, D. R. (2003) Extending the upper temperature limit for life. *Science*, 301, 934.

Kashirtsev, V. A. and Philp, R. P. (1997) Biomarkers and carbon isotopes of oils and natural bitumens in the Precambrian and Paleozoic-Mesozoic formations of the Siberian Platform, Russia. *American Association of Petroleum Geologists Bulletin*, 81, 58.

Kashirtsev, V. A., Kontorovich, A. E., Philp, R. P., et al. (1999) Biomarkers in crude oils of the Eastern Siberian Platform as indicators of paleoenvironment of source-rock deposition. *Geologiya i Geofizika*, 40, 1700–10.

Katsumata, H. and Shimoyama, A. (2001a) Alkyl and polynuclear aromatic thiophenes in Neogene sediments of the Shinjo Basin, Japan. *Geochemical Journal*, 35, 37–48.

(2001b) Thiophenes in the Cretaceous/Tertiary boundary sediments at Kawaruppu, Hokkaido, Japan. *Geochemical Journal*, 35, 67–76.

Katz, B. J. (1995a) The Schistes Carton – the lower Toarcian of the Paris Basin. In: *Petroleum Source Rocks, Casebooks in Earth Science* (B. J. Katz, ed.), Springer-Verlag, Berlin, pp. 51–65.

(1995b) Factors controlling the development of lacustrine petroleum source rocks – an update. In: *Paleogeography, Paleoclimate, and Source Rocks* (A.-Y. Huc, ed.), American Association of Petroleum Geologists, Tulsa, OK, pp. 61–79.

(1995c) Factors controlling the development of lacustrine petroleum source rocks – an update. In: *Paleogeography, Paleochimate, and Source Rocks* (A.-Y. Huc, ed.), American Association of Petroleum Geologists, Tulsa, OK, pp. 61–70.

Katz, B. J. and Elrod, L. W. (1983) Organic geochemistry of DSDP Site 467, offshore California, Middle Miocene to Lower Pliocene strata. *Geochimica et Cosmochimica Acta*, 47, 389–96.

Katz, B. J. and Mello, M. R. (2000) Petroleum systems of South Atlantic marginal basins – an overview. In: *Petroleum Systems of South Atlantic Margins* (M. R. Mello and B. J. Katz, eds.), American Association of Petroleum Geologists, Tulsa, OK, pp. 1–13.

Katz, B. J. and Royle, R. A. (2001) Variability of source rock attributes in the Monterey Formation, California. In: *The Monterey Formation: From Rocks to Molecules* (C. M. Isaacs and J. Rullkötter, eds.), Columbia University Press, New York, pp. 107–30.

Katz, M. E., Pak, D. K., Dickens, G. R. and Miller, K. G. (1999) The source and fate of massive carbon input during the latest Paleocene thermal maximum. *Science*, 286, 1531–3.

Katz, M. E., Miller, K. G., Pak, D. K. and Dickens, G. R. (2000a) The LPTM gas hydrate dissociation hypothesis: new evidence from the western North Atlantic. In: *Early Paleogene Warm Climates and Biosphere Dynamics: Short Papers and Extended Abstracts* (B. Schmitz, B. Sundquist and F. P. Andreasson, eds.), Geological Society of Sweden, Stockholm, pp. 84–85.

Katz, B. J., Dawson, W. C., Liro, L. M., Robinson, V. D. and Stonebraker, J. D. (2000b) Petroleum systems of Ogooué Delta, offshore Gabon. In: *Petroleum Systems of South Atlantic Margins* (M. R. Mello and B. J. Katz, eds.), American Association of Petroleum Geologists, Tulsa, OK, pp. 247–56.

Katz, B., Richards, D., Long, D. and Lawrence, W. (2000c) A new look at the components of the petroleum system of the South Caspian Basin. *Journal of Petroleum Science and Engineering*, 28, 161–82.

Kauffman, E. G. (1977) Geological and biological overview: Western Interior Cretaceous Basin. *Mountain Geologist*, 14, 75–99.

Kauffman, E. G. and Caldwall, W. G. E. (1993) The Western Interior Basin in space and time. In: *Evolution of the Western Interior Basin* (W. G. E. Caldwall and E. G. Kauffman, eds.), Geological Association of Canada Newfoundland, pp. 1–30.

Kauffman, E. G. and Sageman, B. B. (1990) Biological sensing of benthic environments in dark shale and related oxygen restricted facies. In: *Cretaceous Resource, Events and Rhythms* (R. N. Ginsburg and B. Beaudoin, eds.), Kluwer, Amsterdam, pp. 121–38.

Kaufman, R. L., Ahmed, A. S. and Elsinger, R. J. (1990) Gas chromatography as a development and production tool for fingerprinting oils from individual reservoirs: applications in the Gulf of Mexico. In: *Proceedings of the 9th Annual Research Conference of the Society of Economic Paleontologists and Mineralogists* (D. Schumacher and B. F. Perkins, eds.), Society of Paleontologists and Mineralogists, Tulsa, OK, pp. 263–82.

Keely, B. J. and Maxwell, J. R. (1993) The Mulhouse Basin: evidence from porphyrin distributions for water column anoxia during deposition of marls. *Organic Geochemistry*, 20, 1217–25.

Keller, M. A. and MacQuaker, J. H. S. (2001) High resolution analysis of petroleum source potential and lithofacies of Lower Cretaceous mudstone core pebble shale unit and GRZ of Hue Shale), Mikkelsen Bay State #1 well, North Slope Alaska. In: *NPRA Core Workshop: Petroleum Plays and Systems in the National Petroleum Reserve – Alaska* (D. W. Houseknecht, ed.), Society of Economic Paleontologists and Mineralogists, Tulsa, OK, pp. 37–56.

Keller, M. A., Bird, K. J. and Evans, K. R. (1999) Petroleum source rock evaluation based on sonic and resistivity logs. In: *The Oil and Gas Resource Potential of the 1002 Area, Arctic National Wildlife Refuge, Alaska*. U.S. Geological Survey Open-File Report 98–34, pp. SR1–35.

Keller, M. A., MacQuaker, J. H. S. and Lillis, P. G. (2002) *High Resolution Study of Petroleum Source Rock Variation, Lower Cretaceous (Hauterivian and Barremian) of Mikkelsen Bay, North Slope, Alaska*. U.S. Geological Survey Open-File Report 01–480.

Kelley, P. A., Mertani, B. and Williams, H. H. (1994) Brown Shale Formation: Paleogene lacustrine source rocks of Central Sumatra. In: *Petroleum Source Rocks, Casebooks in Earth Science* (B. J. Katz, ed.), Springer-Verlag, Bervry pp. 283–308.

Kelly, D. C., Brabower, T. J. and Zachos, J. C. (1998) Evolutionary consequences of the latest Paleocene thermal maximum for tropical planktonic foraminifera. *Palaeogeography, Palaeoclimatology, Palaeoecology*, 141, 139–61.

Kemp, P., Lander, D. L. and Orpin, C. G. (1984) The lipids of the rumen fungus *Piromonas communis*. *Journal of General Microbiology*, 130, 27–37.

Kenrick, P. and Crane, P. R. (1997) The origin and early evolution of plants on land. *Nature*, 389, pp. 33–39.

Kenig, F., Huc, A.-Y., Purser, B. H. and Oudin, J. L. (1990) Sedimentation, distribution and diagenesis of organic matter in recent carbonate environment, Abu Dhabi, UAE. *Organic Geochemistry*, 16, 735–47.

Kenig, F., Popp, B. N., Hayes, J. M. and Summons, R. E. (1994) An isotopic biogeochemical study of the Oxford Clay Formation, Jurassic, UK. *Journal of the Geological Society, London*, 151, 139–52.

Kenig, F., Sinninghe Damsté, J. S., Kock-van Dalen, A. C., *et al.* (1995a) Occurrence and origin of mono-, di-, and trimethylalkanes in modern and Holocene cyanobacterial mats from Abu Dhabi, United Arab Emirates. *Geochimica et Cosmochimica Acta*, 59, 2999–3015.

Kenig, F., Sinninghe Damsté, J. S., Frewin, N. L., Hayes, J. M. and de Leeuw, J. W. (1995b) Molecular indicators for palaeoenvironmental change in a Messinian evaporitic sequence (Vena del Gesso, Italy). II: high-resolution variations in abundances and ^{13}C contents of free and sulphur-bound carbon skeletons in a single marl bed. *Organic Geochemistry*, 23, 485–526.

Kenig, F., Difrancesco, G. and Simons, D. J. H. (2000) Biomarker constraints on water column structure and oceanographic circulation in an epeiric sea (Callovian, Jurassic, North-Western Europe). Presented at the Research Conference, Goldschmidt September 3–8, 2000, Oxford, UK. *Journal of Conference Abstracts*, 5, 579.

Kenig, F., Simons, D.-J. H. and Anderson, K. B. (2001) Distribution and origin of ethyl-branched alkanes in a Cenomanian transgressive shale of the Western Interior Seaway (USA). *Organic Geochemistry*, 32, 949–54.

Kennett, J. P. and Stott, L. D. (1991) Abrupt deep sea warming palaeoceanographic changes and benthic extinctions at the end of the Palaeocene. *Nature*, 353, 225–9.

Kennett, J. P., Cannariato, K. G., Hendy, I. L. and Behl, R. J. (2000) Carbon isotopic evidence for methane hydrate instability during Quaternary interstadials. *Science*, 288, 128–33.

Kennicutt, M. C., II, Mcdonald, T. J., Comet, P. A., Denoux, G. J. and Brooks, J. M. (1992) The origins of petroleum in the northern Gulf of Mexico. *Geochimica et Cosmochimica Acta*, 56, 1259–80.

Kerr, R. A. (1999) Early life thrived despite Earthly travails. *Science*, 284, 2111–13.

Khudoley, A. K., Rainbird, R. H., Stern, R. A., *et al.* (2001) Sedimentary evolution of the Riphean-Vendian basin of southeastern Siberia. *Precambrian Research*, 111, 129–63.

Killops, S. D. (1991) Novel aromatic hydrocarbons of probable bacterial origin in a Jurassic lacustrine sequence. *Organic Geochemistry*, 17, 25–36.

Killops, S. D. and Al-Juboori, M. A. H. A. (1990) Characterization of the unresolved complex mixture (UCM) in the gas chromatograms of biodegraded petroleums. *Organic Geochemistry*, 15, 147–60.

Killops, S. D., Woolhouse, A. D., Weston, R. J. and Cook, R. A. (1994) A geochemical appraisal of oil generation in the Taranaki Basin, New Zealand. *American Association of Petroleum Geologists Bulletin*, 78, 1560–85.

Killops, S. D., Raine, J. I., Woolhouse, A. D. and Weston, R. J. (1995) Chemostratigraphic evidence of higher-plant evolution in the Taranaki Basin, New Zealand. *Organic Geochemistry*, 23, 429–45.

Killops, S. D., Allis, R. G. and Funnell, R. H. (1996) Carbon dioxide generation from coals in Taranaki Basin, New Zealand: implications for petroleum migration in

Southeast Asian Tertiary Basins. *American Association of Petroleum Geologists Bulletin*, 80, 545–69.

Kimble, B. J., Maxwell, J. R., Philp, R. P., *et al.* (1974a) Tri- and tetraterpenoid hydrocarbons in the Messel oil shale. *Geochimica et Cosmochimica Acta*, 38, 1165–81.

Kimble, B. J., Maxwell, J. R., Philp, R. P. and Eglinton, G. (1974b) Identification of steranes and triterpanes in geolipid extracts by high-resolution gas chromatography and mass spectrometry. *Chemical Geology*, 14, 173–98.

Kimura, H. and Watanabe, Y. (2001) Oceanic anoxia at the Precambrian-Cambrian boundary. *Geology*, 29, 995–8.

Kirimura, K., Furuya, T., Nishii, Y., *et al.* (2001) Biodesulfurization of dibenzothiophene and its derivatives through the selective cleavage of carbon-sulfur bonds by a moderately thermophilic bacterim *Bacillus subtilis* WU-S2B. *Journal of Bioscience and Bioengineering*, 91, 262–6.

Kirk, D. N. and Shaw, P. M. (1975) Backbone rearrangements of steroidal 5-enes. *Journal of Chemical Society Perkin Transactions I*, 22, 2284–94.

Kirkland, D. W. and Evans, R. (1981) Source-rock potential of evaporitic environments. *American Association of Petroleum Geologists Bulletin*, 65, 181–90.

Kissin, Y. V. (1993) Catagenesis of light acyclic isoprenoids in petroleum. *Organic Geochemistry*, 20, 1077–90.

Kleemann, G., Poralla, K., Englert, G., *et al.* (1990) Tetrahymanol from the phototrophic bacterium *Rhodopseudomonas palustris*: first report of a gammacerane triterpane from a prokaryote. *Journal of General Microbiology*, 136, 2551–3.

Klemme, H. D. (1994) Petroleum systems of the world involving Upper Jurassic source rocks. In: *The Petroleum System – From Source to Trap* (L. B. Magoon and W. G. Dow, eds.), American Association of Petroleum Geologists, Tulsa, OK, pp. 51–72.

Klemme, H. D. and Ulmishek, G. F. (1991) Effective petroleum source rocks of the world: stratigraphic distribution and controlling depositional factors. *American Association of Petroleum Geologists Bulletin*, 75, 1809–51.

Klett, T. R. (2000a) Total petroleum systems of the Illizi Province, Algeria and Libya – Tanezzuft-Illizi. *U.S. Geological Survey Bulletin*, 2202-A.

(2000b) Total petroleum systems of the Grand Erg/Ahnet Province, Algeria and Morocco – the Tanezzuft-Timimoun, Tanezzuft-Ahnet, Tanezzuft-Sbaa, Tanezzuft-Mouydir, Tanezzuft-Benoud, and Tanezzuft-Béchar/Abadla. *U.S. Geological Survey Bulletin*, 2202-B

(2000c) Total petroleum systems of the Trias/Ghadames Province, Algeria, Tunisia, and Libya – the Tanezzuft-Oued Mya, Tanezzuft-Melrhir, and Tanezzuft-Ghadames. *U.S. Geological Survey Bulletin*, 2202-C.

Kling, G. W., Evans, W. C., Tuttle, M. L. and Tanyileke, G. (1994) Degassing of Lake Nyos. *Nature*, 368, 405–6.

Klomp, U. C. (1986) The chemical structure of a pronounced series of iso-alkanes in South Oman crudes. *Organic Geochemistry*, 10, 807–14.

Kluth, C. F. (1986) Plate tectonics of the Ancestral Rocky Mountains. In: *Paleotectonics and Sedimentation in the Rocky Mountain Region, United States. Part III. Middle Rocky Mountains* (J. A. Peterson, ed.), American Association of Petroleum Geologists, Tulsa OK, pp. 353–69.

Knights, B. A., Brown, A. C., Conway, E. and Middleditch, B. S. (1970) Hydrocarbons from the green form of the freshwater alga *Botryococcus braunii*. *Phytochemistry*, 9, 1317–24.

Knoll, A. H. (1992) The early evolution of eukaryotes: a geologic perspective. *Science*, 256, 622–7.

Knoll, A. H. and Carroll, S. B. (1999) Early animal evolution: emerging views from comparative biology and geology. *Science*, 284, 2129–37.

Knoll, A. H. and Niklas, K. J. (1987) Adaptation, plant evolution, and the fossil record. *Reveiw of Paleobotony and Palynotology*, 50, 127–49.

Knoll, A. H., Bambach, R. K., Canfield, D. E. and Grotzinger, J. P. (1996) Comparative Earth history and Late Permian mass extinction. *Science*, 273, 452–7.

Knutson, C. F., Dana, G. F., Solti, G., *et al.* (1988) Developments in oil shale in 1988. *American Association of Petroleum Geologists Bulletin*, 73, 375–84.

Koch, J. (1997) Organic petrographic investigations of the Kupferschiefer in Northern Germany. *International Journal of Coal Geology*, 33, 301–16.

Koch, P. L., Zachos, J. C. and Gingerich, P. D. (1992) Coupled isotopic changes in marine and continental carbon reservoirs at the Paleocene-Eocene boundary. *Nature*, 358, 319–22.

Kockel, F., Whener, H. and Gerling, P. (1994) Petroleum systems of the Lower Saxony Basin, Germany. In: *The Petroleum System – From Source to Trap* (L. B. Magoon and W. G. Dow, eds.), American Association of Petroleum Geologists, pp. 573–86.

Kodina, L. A. and Vlasova, L. N. (1989) Geochemistry of organic matter in East Sakhalin siliceous rocks [in Russian]. *Geokhimiya*, 3, 365–75.

Kodina, L. A., Vlasova, L. N., Kuznetsova, L. V., Bazilevskaya, O. L. and Galimov, E. M. (1989) Use of isotope fractionation in diagnosing oil source rocks and oil-oil correlations in the siliceous rocks of East Sakhalin [in Russian]. *Geokhimiya*, 6, 807–15.

Kohnen, M. E. L. (1991) Sulphurised lipids in sediments: the key to reconstruct palaeobiochemicals and their origin. Ph. D. thesis, Universiteit Delft, Delft, the Netherlands.

Kohnen, M. E. L., Sinninghe Damsté, J. S., Kock-Van Dalen, A. C., et al. (1990a) Origin and diagenetic transformations of C_{25} and C_{30} highly branched isoprenoid sulphur compounds: further evidence for the formation of organically bound sulphur during early diagenesis. *Geochimica et Cosmochimica Acta*, 54, 3053–63.

Kohnen, M. E. L., Sinninghe Damsté, J. S., Rijpstra, W. I. C. and De Leeuw, J. W. (1990b) Alkylthiophenes as sensitive indicators of palaeoenvironmental changes: a study of a Cretaceous oil shale from Jordan. In: *Geochemistry of Sulfur in Fossil Fuels* (W. L. Orr and C. M. White, eds.), American Chemical Society Tulsa, OK, pp. 444–85.

Koike, L., Reboucas, L. M. C., Reis, F. D. A. M., et al. (1992) Naphthenic acids from crude oils of Campos Basin. *Organic Geochemistry*, 18, 851–60.

Kolaczkowska, E., Slougui, N.-E., Watt, D. S., Marcura, R. E. and Moldowan, J. M. (1990) Thermodynamic stability of various alkylated, dealkylated, and rearranged 17α- and 17β-hopane isomers using molecular mechanics calculations. *Organic Geochemistry*, 16, 1033–8.

Kontorovich, A. E. (1984) Geochemical methods for the quantitative evaluation of the petroleum potential of sedimentary basins. In: *Petroleum Geochemistry and Basin Evaluation* (G. Demaison and R. J. Murris, eds.), American Association of Petroleum Geologists, Tulsa, OK, pp. 79–109.

Kontorovich, A. E., Izosimova, A. N., Kontorovich, A. A., Khabaraov, E. M. and Timoshina, I. D. (1996a) Geological structure and conditions of the formation of the giant Yurubcheno-Tokhoma zone of oil and gas accumulation in the Upper Proterozoic of the Siberian Platform. *Geologiya i Geofizika*, 37, 166–95.

Kontorovich, A. E., Larichev, A. I., Tukwell, K., et al. (1996b) The oldest oil of Australia. *Geologiya i Geofizika*, 37, 100–15.

Kontorovich, A. E., Izosimova, A. N., Timoshina, I. D., et al. (1997) The giant Yurubchen-Tokhomo zone of oil and gas accumulation in the Riphean of the Siberian Platform. *American Association of Petroleum Geologists Bulletin*, 81, 1391.

Kontorovich, A. E., Danilova, V. P., Kostyreva, E. A., et al. (1998) Main marine oil source formations of the West Siberian petroleum megabasin and their genetic relations to oils. Presented at the Annual Meeting of the American Association of Petroleum Geologists, Salt Lake City, UT, May 17–20, 1998.

Kontorovich, A. E., Izosimova, A. N., Larichev, A. I., et al. (1998c) Hydrocarbon sources in the Precambrian of the Siberian Platform, their genetic relations to oil. Presented at the Annual Meeting of the American Association of Petroleum Geologists, Salt Lake City, UT, May 17–20, 1998.

Koons, C. B., Bond, J. G. and Peirce, F. L. (1974) Effects of depositional environment and postdepositional history on chemical composition of Lower Tuscaloosa oils. *American Association of Petroleum Geologists Bulletin*, 58, 1272–80.

Koopmans, M. P., Köster, J., Van Kaam-Peters, H. M. E., (1996a) Diagenetic and catagenetic products of isorenieratene: molecular indicators for photic zone anoxia. *Geochimica et Cosmochimica Acta*, 60, 4467–96.

Koopmans, M. P., Schouten, S., Kohnen, M. E. L., and Sinninghe Damsté, J. S. (1996b) Restricted utility of aryl isoprenoids as indicators for photic zone anoxia. *Geochimica et Cosmochimica Acta*, 60, 4873–6.

Koopmans, M. P., de Leeuw, J. W., and Sinninghe Damsté, J. S. (1997) Novel cyclised and aromatised diagenetic products of β-carotene in the Green River Shale. *Organic Geochemistry*, 26, 451–66.

Koopmans, M. P., Rijpstra, W. I. C., de Leeuw, J. W., Lewan, M. D. and Sinninghe Damsté, J. S. (1998) Artificial maturation of an immature sulfur- and organic matter-rich limestone from the Ghareb Formation, Jordan. *Organic Geochemistry*, 28, 503–21.

Koopmans, M. P., Rijpstra, W. I. C., Klapwijk, M. M., et al. (1999) A thermal and chemical degradation approach to decipher pristane and phytane precursors in sedimentary organic matter. *Organic Geochemistry*, 30, 1089–104.

Korch, R. J., Mai, H., Sun, Z. and Gorter, J. D. (1991) The Sichuan Basin, southwest China: a Late Proterozoic (Sinian) petroleum province. *Precambrian Research*, 54, 45–64.

Köster, J. (1989) Organische Geochemie und Organo-Petrologie kerogenreicher und bituminoeser Einschaltungen im Hauptdolomit (Trias, Nor) der Noerdlichen Kalkalpen. [Organic geochemistry and organic petrology of kerogenous and bituminous intercalations within the "Hauptdolomite" (Triassic, Norian) of the Northern Calcareous Alps.] Ph. D. thesis, University of Clausthal, Clausthal-Zellerfeld, Germany.

Köster, J., Wehner, H. and Hufnagel, H. (1988) Organic geochemistry and organic petrology of organic rich sediments within the "Hauptdolomit" Formation (Triassic, Norian) of the northern Calcareous Alps. *Organic Geochemistry*, 13, 377–86.

Köster, J., Van Kaam-Peters, H. M. E., Koopmans, M. P., et al. (1997) Sulphurisation of homohopanoids: effects on carbon number distribution, speciation, and 22S/22R epimer ratios. *Geochimica et Cosmochimica Acta*, 61, 2431–52.

Kotra, R. K., Gottfried R. M., Spiker, E. C., Romankiw, L. A. and Hatcher P. A. (1988) Chemical composition and thermal maturity of kerogen and phytoclasts of the Newark Supergroup in the Hartford Basin. In: *Studies of the Early Mesozoic Basins of the Eastern United States* (A. J. Froelich and G. R. Robinson, Jr, eds.), U.S. Geogical Survey Bulletin, 1776, 68–74.

Koutsoukos, E. A. M., Mello, M. R., de Azambuja Filho, N. C., Hart, M. B. and Maxwell, J. R. (1991) The upper Aptian-Albian succession of the Sergipe Basin, Brazil, an integrated paleoenvironmental assessment. *American Association of Petroleum Geologists Bulletin*, 75, 479–98.

Kralert, P. G., Alexander, R. and Kagi, R. I. (1995) An investigation of polar constituents in kerogen and coal using pyrolysis-gas chromatography-mass spectrometry with *in situ* methylation. *Organic Geochemistry*, 23, 627–39.

Kramer, R. (1998) *Chemometric Techniques for Quantitative Analysis*. Marcel Dekker, New York.

Kranz, H. D., Miks, D., Siegler, M. L., et al. (1995) The origin of land plants: phylogenetic relationships among charophytes, bryophytes, and vascular plants inferred from complete small-subunit ribosomal RNA gene sequences. *Journal of Molecular Evolution*, 41, 74–84.

Krooss, B. M. and Leythaeuser, D. (1988) Experimental measurements of the diffusion parameters of light hydrocarbons in water-saturated sedimentary rocks – II. Results and geochemical significance. *Organic Geochemistry*, 12, 91–108.

Kropp, K. G., Davidova, I. A. and Suflita, J. M. (2000) Anaerobic oxidation of *n*-dodecane by an addition reaction in a sulfate-reducing bacterial enrichment culture. *Applied and Environmental Microbiology*, 66, 5393–8.

Krouse, H. R., Viau, C. A., Eliuk, L. S., Ueda, A. and Halas, S. (1989) Chemical and isotopic evidence of thermochemical sulphate reduction by light hydrocarbon gases in deep carbonate reservoirs. *Nature*, 333, 415–19.

Krstic, B., Grubic, A., Ramovs, A. and Filipovic, I. (1988) The Devonian of Yugoslavia. In: *Devonian of the World* (N. J. McMillan, A. F. Embrey and D. J. Glass, eds.), Canadian Society of Petroleum Geologists, Calgary, Alberta, pp. 499–506.

Kruge, M. A. (2000) Determination of thermal maturity and organic matter type by principal components analysis of the distributions of polycyclic aromatic compounds. *International Journal of Coal Geology*, 43, 27–51.

Kruge, M. A., Hubert, J. F., Akes, R. J. and Meriney, P. E. (1990a) Biological markers in Lower Jurassic synrift lacustrine black shales, Hartford Basin, Connecticut, USA. *Organic Geochemistry*, 15, 281–9.

Kruge, M. A., Hubert, J. F., Bensley, D. F., Crelling, J. C. and Akes, R. J. (1990b) Organic geochemistry of a Lower Jurassic synrift lacustrine sequence, Hartford Basin, Connecticut, USA. *Organic Geochemistry*, 16, 689–701.

Krumholz, L. R. (2000) Microbial communities in the deep subsurface. *Hydrogeology Journal*, 8, 4–10.

Kuangzong, Q. (1988) Thermal depolymerization of kerogen and formation of immature oil. *Organic Geochemistry*, 13, 1045–50.

Kuhn, E. P., Colberg, P. J., Schnoor, J. L., et al. (1985) Microbial transformations of substituted benzenes during infiltration of river water to groundwater: laboratory column studies. *Environmental Science & Technology*, 19, 961–8.

Kuhnt, W. and Wiedmann, J. (1995) Cenomanian-Turonian source rocks: paleobiogeographic and paleoenvironmental aspect. In: *Paleogeography, Paleoclimates and Source Rocks*, (A.-Y. Huc, ed.), American Association of Petroleum Geologists, Tulsa, OK, pp. 213–32.

Kuhnt, W., Herbin, J. P., Thurow, J. and Wiedmann, J. (1990) Distribution of Cenomanian-Turonian organic facies in the western Mediterranean and along the adjacent Atlantic margin. In: *Deposition of Organic Facies* (A.-Y. Huc, ed.), American Association of Petroleum Geologists, Tulsa, OK, pp. 133–60.

Kuhnt, W., Nederbragt, A. J. and Leine, L. (1997) Cyclicity of Cenomanian-Turonian organic-carbon-rich sediments in the Tarfaya Atlantic Coastal Basin (Morocco). *Cretaceous Research*, 18, 587–601.

Kulke, H. (1995) Nigeria. In: *Regional Petroleum Geology of the World. Part II: Africa, America, Australia and Antarctica* (H. Kulke, ed.), Gebrüder Borntraeger, Berlin, pp. 143–72.

Kumar, S. (2001) Mesoproterozoic megafossil *Chuaria-Tawula* association may represent parts of a multicellular plant, Vindhyan Supergroup, Central India. *Precambrian Research*, 106, 187–211.

Kump, L. R. and Slingerland, R. L. (1999) Circulation and stratification of the early Turonian Western Interior Seaway: sensitivity to a variety of forcings. In: *Evolution of the Cretaceous Ocean-Climate System* (E. Barrera and C. C. Johnson, eds.), Geological Society of America, Boulder, CO, pp. 18–191.

Kuo, L.-C. (1994a) Lower Cretaceous lacustrine source rocks in northern Gabon: effect of organic facies and thermal maturity on crude oil quality. *Organic Geochemistry*, 22, 257–73.

(1994b) An experimental study of crude oil alteration in reservoir rocks by water washing. *Organic Geochemistry*, 21, 465–79.

Kupecz, J. A. (1995) Depositional setting, sequence stratigraphy, diagenesis, and reservoir potential of a mixed-lithology, upwelling deposit: Upper Triassic Shublik Formation, Prudhoe Bay, Alaska. *American Association of Petroleum Bulletin*, 79, 1301–19.

Küspert, W. (1982) Environmental change during oil-Shale deposition as deduced from stable isotope ratios. In: *Cyclic and Event Stratification* (G. Einsele and A. Seilacher, eds.), Springer Verlag, Berlin, pp. 482–501.

Kuypers, M. M. M., Blokker, P., Erbacher, J., *et al.* (2001) Massive expansion of marine archaea during a Mid-Cretaceous oceanic anoxic event. *Science*, 293, 92–4.

Kvalheim, O. M., Aksnes, D. W., Brekke, T., Eide, M. O. and Sletten, E. (1985) Crude oil characterization and correlation by principal component analysis of carbon-13 nuclear magnetic resonance spectra. *Analytical Chemistry*, 57, 2858–64.

Kvenvolden, K. A. (1993) Gas hydrates – geological persepctive and global change. *Reviews of Geophysics*, 31, 173–87.

Kvenvolden, K. A. and Simoneit, B. R. T. (1990) Hydrothermally derived petroleum: examples from Guaymas Basin, Gulf of California, and Escanaba Trough, Northeast Pacific Ocean. *American Association of Petroleum Geologists Bulletin*, 74, 223–37.

Kyte, F. T. (1998) A meteorite from the Cretaceous/Tertiary boundary. *Nature*, 396, 237–9.

Ladwein, H. W. (1988) Organic geochemistry of Vienna Basin: model for hydrocarbon generation in overthrust belts. *American Association of Petroleum Geologists Bulletin*, 72, 587–99.

Lambert, M. W. (1993) Internal stratigraphy and organic facies of the Devonian-Mississippian Chattanooga (Woodford) Shale in Oklahoma and Kansas. In: *Source Rocks in a Sequence Stratigraphic Framework*, (B. J. Katz and L. M. Pratt, eds.), American Association of Petroleum Geologists, Tulsa, OK, pp. 163–76.

Lambert, M. W., Burkett, P. J., Chiou, W.-A., Bennett, R. H. and Lavoie, D. M. (1994) Kerogen networks and hydrocarbon generation in the Chattanooga (Woodford) Shale of Oklahoma and Kansas. Presented at the Annual Meeting of the American Association of Petroleum Geologists, June 12–15, 1994, Denver, CO.

Landais, P. and Connan, J. (1986) Source rock potential and oil alteration in the uraniferous basin of Lodève (Hérault, France). *Sciences Geologiques Bulletin*, 39, 468–76.

Langenhoff, A. A. M., Nijenhuis, I., Tan, N. C. G., *et al.* (1997) Characterisation of a manganese-reducing, toluene-degrading enrichment culture. *FEMS Microbiology Ecology*, 24, 113–25.

Lanigan, K., Hibbird, S., Menpes, S. and Torkington, J. (1994) Petroleum exploration in the Proterozoic Beetaloo Sub-basin, Northern Territory. *Australian Petroleum Exploration Association Journal*, 34, 674–91.

Lara, M. E., Wavrek, D. A., Vines, R., Laffitte, G. A. and Del Vo, S. (1996) An integrated sequence stratigraphy-organic geochemistry study of the Vaca Muerta petroleum systems, Neuquén Basin, Argentina. *American Association of Petroleum Geologists Bulletin*, 80, 1306.

Larcher, A. V., Alexander, R. and Kagi, R. I. (1987) Changes in configuration of extended moretanes with increasing sediment maturity. *Organic Geochemistry*, 11, 59–63.

Largeau, C., Derenne, S., Casadevall, E., *et al.* (1990) Occurrence and origin of ultralaminar structures in amorphous kerogens of various source rocks and oils shales. *Organic Geochemistry*, 16, 889–95.

Largeau, C., Derenne, S., Metzger, P. Mongenot, T. and Riboulleau, A. (2001) Occurrence and origin of non-lignin aromatic moieties in kerogens. Presented at the 221st National Meeting of the American Chemical Society, April 1–5, 2001, San Diego, CA.

Larson, R. L. and Erba, E. (1999) Onset of the Mid-Cretaceous greenhouse in the Barremian-Aptian: igneous events and the biological, sedimentary, and gechemical consequences. *Paleoceanography*, 14, 663–78.

Larter, S. R., Solli, H., Douglas, A. G., de Lange, F. and de Leeuw, J. W. (1979) Occurrence and significance of prist-1-ene in kerogen pyrolysates. *Nature*, 279, 405–7.

Larter, S. R., Solli, H. and Douglas, A. G. (1983) Phytol-containing melanoidins and their bearing on the fate of isoprenoid structures in sediments. In: *Advances in Organic Geochemistry, 1981, Proceedings of the International Meeting on Organic Geochemistry* (M. Bjorøy, C. Albrecht, C. Cornford, *et al.*, eds.), John Wiley & Sons, New York, pp. 513–23.

Larter, S. R., Bowler, F., Li, M., *et al.* (1996) Benzocarbazoles as molecular indicators of secondary oil migration distance. *Nature*, 383, 593–7.

Larter, S., Koopmans, M. P., Head, I., *et al.* (2000) Biodegradation rates assessed geologically in a heavy oilfield – implications for a deep, slow (Largo) biosphere. Presented at GeoCanada 2000 – The Millennium Geoscience Summit, May 29–June 2, 2000, Calgary, Alberta.

Larter, S., Wilhelms, A., Erdmann, M., Zwach, C. and Aplin, A. (2002) Deterministic modeling of deep subsurface biodegradation–progress and prognosis. Presented at the Annual Meeting of the American Association of Petroleum Geologists, March 10–13, 2002, Houston, TX.

Law, B. E. and Rice, D. D. (1993) *Hydrocarbons from Coal*. American Association of Petroleum Geologists, Tulsa, OK.

Leckie, R. M., Bralower, T. J. and Cashman, R. (2002) Oceanic anoxic events and plankton evolution: biotic response to tectonic forcing during the Mid-Cretaceous. *Paleoceanography*, 17, 1–29.

Le Dréau, Y., Gilbert, F., Doumenq, P., *et al.* (1997) The use of hopanes to track *in situ* variations in petroleum composition in surface sediments. *Chemosphere*, 34, 1663–72.

Leenheer, M. J. (1984) Mississippian Bakken and equivalent formations as source rocks in the Western Canadian Basin. *Organic Geochemistry*, 6, 521–32.

Leenheer, M. J. and Zumberge, J. E. (1987) Correlation and thermal maturity of Williston Basin crude oils and Bakken source rocks using terpane biomarkers. In: *Williston Basin: Anatomy of a Cratonic Oil Province* (M. W. Longman., ed.), Rocky Mountain Association of Geologists, Denver, CO, pp. 287–98.

Leif, R. N. and Simoneit, B. R. T. (1995) Ketones in hydrothermal petroleums and sediment extracts from Guaymas Basin, Gulf of California. *Organic Geochemistry*, 23, 889–904.

Leine, L. (1986) Geology of the Tarfaya oil shale deposit, Morocco. *Geologie en Mijnbouw*, 65, 57–74.

Leith, T. L., Weiss, H. M. and Schou, L. (1992) Biomarker geochemistry of Triassic and Upper Jurassic organic-rich shales and associated oil staining, Barents Sea and Svalbard. Presented at the Annual Meeting of the American Association of Petroleum Geologists, June, 1992, Calgory, Alberta.

Leith, T. L., Weiss, H. M., Mork, A., *et al.* (1993) Mesozoic hydrocarbon source-rocks of the Arctic region. In: *Arctic Geology and Petroleum Potential: Proceedings of the Norwegian Petroleum Society Conference, Norwegian Petroleum Society (NPF) Special Publication 2* (T. O. Vorren, E. Bergsager, O. A. Dahl-Stamnes, *et al.*, eds.), Elsevier, New York, pp. 1–25.

Lelek, J. J., Shepherd, D. B., Stone, D. M. and Abdine, A. S. (1992) October Field, the latest giant under development in Egypt's Gulf of Suez. In: *Giant Oil and Gas Fields of the Decade 1978–1988* (M. T. Halbouty, ed.), American Association of Petroleum Geologists, Tulsa, OK, pp. 231–49.

Lemieux, C., Otis, C. and Turmel, M. (2000) Ancestral chloroplast genome in *Mesostigma viride* reveals an early branch of green plant evolution. *Nature*, 403, 649–52.

Leo, H. B. and Cardott, B. J. (1994) Detection of natural weathering of upper McAlester Coal and Woodford Shale, Oklahoma, USA. *Organic Geochemistry*, 22, 73–83.

Lewan, M. D. (1983) Effects of thermal maturation of stable organic carbon isotopes as determined by hydrous pyrolysis of Woodford Shale. *Geochimica et Cosmochimica Acta*, 47, 1471–9.

(1984) Factors controlling the proportionality of vanadium to nickel in crude oils. *Geochimica et Cosmochimica Acta*, 48, 2231–8.

(1985) Evaluation of petroleum generation by hydrous pyrolysis experimentation. *Philosophical Transactions of the Royal Society of London, A*, 315, 123–34.

Lewan, M. D. and Buchardt, B. (1989) Irradiation of organic matter by uranium decay in the Alum Shale, Sweden. *Geochimica et Cosmochimica Acta*, 53, 1307–22.

Lewan, M. D., Bjorøy, M. and Dolcater, D. L. (1986) Effects of thermal maturation on steroid hydrocarbons as determined by hydrous pyrolysis of Phosphoria Retort Shale. *Geochimica et Cosmochimica Acta*, 50, 1977–87.

Lewan, M. D., Henry, M. E., Higley, D. K. and Pitman, J. K. (2002) Material-balance assessment of the New Albany-Chesterian petroleum system of the Illinois basin. *American Association of Petroleum Geologists Bulletin*, 86, 745–77.

Lewis, C. A. (1993) The kinetics of biomarker reactions. Implications for the assessment of the thermal maturity of organic matter in sedimentary basins. In: *Organic Geochemistry: Principles and Applications* (M. H. Engel and S. A. Macko, eds.), Plenum Press, New York, pp. 491–510.

Leythaeuser, D., Mackenzie, A., Schaefer, R. G. and Bjorøy, M. (1984) A novel approach for recognition and quantification of hydrocarbon migration effects in shale-sandstone sequences. *American Association of Petroleum Geologists Bulletin*, 68, 196–219.

L'Haridon, S., Reysenbach, A.-L., Glénat, P., Prieur, D. and Jeanthon, C. (1995) Hot subterranean biosphere isolated from continental oil reservoir. *Nature*, 377, 223–4.

Li, M. and Larter, S. R. (1995) Biomarkers or not biomarkers; a new hypothesis for the origin of pristane involving derivation from methyltrimethyltridecylchromans (MTTCs) formed during diagenesis from chlorophyll and alkylphenols. Reply to comments by Sinninghe Damsté and de Leeuw. *Organic Geochemistry*, 23, 1085–93.

Li, M., Larter, S. R., Taylor, P., *et al.* (1995a) Biomarkers or not biomarkers? A new hypothesis for the origin of pristane involving derivation from methyltrimethyltridecylchromans (MTTCs) formed during diagenesis from chlorophyll and alkylphenols. *Organic Geochemistry*, 23, 159–67.

Li, D., Jiang, R. and Katz, B. J. (1995b) Petroleum generation in the nonmarine Qingshankou Formation (Lower Cretaceous), Songliao Basin, China. In: *Petroleum Source Rocks* (B. J. Katz, ed.), Springer-Verlag, Berlin, pp. 131–48.

Li, M., Yao, H., Stasiuk, L. D., Fowler, M. G. and Larter, S. R. (1997) Effect of maturity and petroleum expulsion on pyrrolic nitrogen compound yields and distributions in Duvernay Formation petroleum source rocks in central Alberta, Canada. *Organic Geochemistry*, 26, 731–44.

Li, M., Yao, H., Fowler, M. G. and Stasiuk, L. D. (1998a) Geochemical constraints on models for secondary petroleum migration along the Upper Devonian Rimbey-Meadowbrook reef trend in central Alberta, Canada. *Organic Geochemistry*, 29, 163–82.

Li, M., Osadetz, K. G., Yao, H., *et al.* (1998b) Unusual crude oils in the Canadian Williston Basin, southeastern Saskatchewan. *Organic Geochemistry*, 28, 477–88.

Li, C.-W., Chen, J.-Y. and Hua, T.-E. (1998c) Precambrian sponges with cellular structures. *Science*, 279, 879–82.

Li, M., Fowler, M. G., Obermajer, M., Stasiuk, L. D. and Snowdon, L. R. (1999) Geochemical characterisation of Middle Devonian oils in NW Alberta, Canada: possible source and maturity effect on pyrrolic nitrogen compounds. *Organic Geochemistry*, 30, 1039–57.

Li, S., Pang, X., Li, M. and Jin, Z. (2003) Geochemistry of petroleum systems in the Niuzhuang South Slope of Bohai Bay Basin – part 1: source rock characterization. *Organic Geochemistry*, 34, 389–412.

Lijmbach, G. W. M., Buskool Toxopeus, J. M., Rodenburg, T. and Hermans, L. J. P. C. M. (1992) Geochemical study of crude oils and source rocks from offshore Abu Dhabi. In: *Proceedings of the SPE Abu Dhabi Conference*, Society of Petroleum Engineers, Dallas, TX, pp. 395–422.

Lillis, P. G., Palacas, J. G. and Warden, A. (1995) A Precambrian-Cambrian oil play in southern Utah. *American Association of Petroleum Geologists Bulletin*, 79, 921.

Lillis, P. G., Lewan, M. D., Warden, A., Monk, S. M. and King, J. D. (1999) Identification and characterization of oil types and their source rocks. In: *The Oil and Gas Resource Potential of the 1002 Area, Arctic National Wildlife Refuge, Alaska*. U.S. Geological Survey Open-File Report 98-34.

Lillis, P. G., King, J. D., Warden, A. and Pribil, M. J. (2002) Oil-source correlation studies, Central Brooks Range foothills and National Petroleum Reserve, Alaska (NPRA). *American Association of Petroleum Geologists Bulletin*, 86, 1150.

Lin, L. H., Michael, G. H., Kovachev, G., *et al.* (1989) Biodegradation of tar-sands bitumens from the Ardmore and Anadarko basins, Carter County, Oklahoma. *Organic Geochemistry*, 14, 511–23.

Lindquist, S. J. (1998a) *The Santa Cruz-Tarija Province of Central South America: Los Monos – Machareti(!) Petroleum System*. Open-File Report 99-50-C, U.S. Geological Survey, Washington, DC.

(1998b) *The Red Sea Basin Province: Sudr-Nubia(!) and Maqna(!) Petroleum Systems*. Open-File Report 99-50-A, U.S. Geological Survey, Washington, DC.

(1999a) *Petroleum Systems of the Po Basin Province of Northern Italy and the Northern Adriatic Sea: Porto Garibaldi (Biogenic), Meride/Riva di Solto (Thermal), and Marnoso Arenacea (Thermal)*. Denver Open-File Report 99-50-D, U.S. Geological Survey, Washington, DC.

(1999b) *The Timan-Pechora Basin Province of Northwest Arctic Russia: Domanik – Paleozoic Total Petroleum System*. Open-File Report 99-50-G, U.S. Geological Survey, Washington, DC.

(1999c) *South and North Barents Triassic-Jurassic Total Petroleum System of the Russian Offshore Arctic*. Open-File Report 99-50-N, U.S. Geological Survey, Washington, DC.

Logan, G. A., Summons, R. E. and Hayes, J. M. (1997) An isotopic biogeochemical study of Neoproterozoic and Early Cambrian sediments from the Centralian Superbasin, Australia. *Geochimica et Cosmochimica Acta*, 61, 5391–409.

Loh, H., Maul, B., Prauss, M. and Riegel, W. (1986) Primary production, maceral formation and carbonate species in Posidonia Shale of NW Germany. *Mitteilungen Geologisch-Paläontologisches Institut University Hamburg*, 60, 397–421.

Lohmann, F., Trendel, J. M., Hetru, C. and Albrecht, P. (1990) C-29 tritiated β-amyrin: chemical synthesis aiming at the study of aromatization processes in sediments. *Journal of Labeled Compounds in Radiopharmaceuticals*, 28, 377–86.

Longden, M. R., Banowsky, B. R. and Woodward, L. A. (1988) Hydrocarbon generation in Heath Formation (Mississippian) in Montana Thrust Belt in response to tectonic burial. *American Association of Petroleum Geologists Bulletin*, 72, 875.

Longman, M. W. and Palmer, S. E. (1987) Organic geochemistry of mid-continent Middle and Late Ordovician oils. *American Association of Petroleum Geologists Bulletin*, 71, 938–50.

Loshe, P. A. and Szostak, J. W. (1996) Ribozyme-catalysed amino-acid transfer reactions. *Nature*, 381, 442–4.

Louda, J. W. and Baker, E. W. (1986) The biogeochemistry of chlorophyll. In: *Organic Marine Chemistry* (M. L. Sohn, ed.), Vol. 305, American Chemical Society, Washington, DC, pp. 107–41.

Loughman, D. L. (1984) Phosphate authigenesis in the Aramachay Formation (Lower Jurassic) of Peru. *Journal of Sedimentary Petrology*, 54, 1147–56.

Loughman, D. L. and Hallam, A. (1982) A facies analysis of the Pucará Group (Norian-Toarcian carbonates, organic rich shale and phosphate) of Central and North Peru. *Sedimentary Geology*, 32, 161–94.

Loureiro, M. R. and Cardoso, J. N. (1990) Aromatic hydrocarbons in the Paraiba Valley oil shale. *Organic Geochemistry*, 15, 351–9.

Loutfi, G. and Abdel Satter, M. M. (1987) Geology and hydrocarbon potential of the Triassic succession in Abu Dhabi, UAE. In: *Proceedings of the Middle East Oil Technology Conference and Exhibition, Bahrain*, Society of Petroleum Engineers, Richardson, TIX, pp. 717–35.

Lovley, D. and Chapelle, F. H. (1995) Deep subsurface microbial processes. *Reviews of Geophysics*, 33, 365–81.

Lovley, D. R., Coates, J. D., Woodward, J. and Phillips, E. J. P. (1995) Benzene oxidation coupled to sulfate reduction. *Applied and Environmental Microbiology*, 61, 953–8.

Lower, S. K., Hochella, M. F., Jr and Beveridge, M. F. (2001) Bacterial recognition of mineral surfaces: nanoscale interactions between *Shewanella* and -FeOOH. *Science*, 292, 1360–3.

Lu, S. N., Li, W. M., Gu, H. M. and Gao, P. (1985) Effect of biological markers and kerogens in geochemical exploration for oil and gas. *American Association of Petroleum Geologists Bulletin*, 69, 281.

Lu, S. T., Ruth, E. and Kaplan, I. R. (1989) Pyrolysis of kerogens in the absence and presence of montmorillonite. I. The generation, degradation and isomerization of steranes and triterpanes at 200 and 300 degrees C. *Organic Geochemistry*, 14, 491–9.

Lucas, S. G. and Kietzke, K. K. (1985) Todilto Formation: a Jurassic salina and its petroleum potential in east-central New Mexico. *American Association of Petroleum Geologists Bulletin*, 70, 346.

Lucas, S. G., Tanner L. H. and Chapman, M. G. (2002) End-Triassic mass extinction or the compiled correlation effect? Presented at the Annual Meeting of the Geological Society of America, October 27–30, 2002, Denver, CO.

Ludwig, B., Hussler, G., Wehrung, P. and Albrecht, P. (1981) C_{26}–C_{29} triaromatic steroid derivatives in sediments and petroleums. *Tetrahedron Letters*, 22, 3313–16.

Lundegard, P. D. and Kharaka, Y. K. (1994) Distribution and occurrence of organic acids in subsurface waters. In: *Organic Acids in Geological Processes* (E. D. Pittman and M. D. Lewan, eds.), Springer-Verlag, New York, pp. 40–69.

Lüning, S., Craig, J., Loydell, D. K., Štorch, P. and Fitches, B. (2000) Lower Silurian "hot shales" in North Africa and Arabia: regional distribution and depositional model. *Earth-Science Reviews*, 49, 121–200.

MacGregor, D. S. (1994) Coal-bearing strata as source rocks – a global overview. In: *Coal and Coal-bearing Strata as Oil-prone Source Rocks?* (A. C. Scott and A. J. Fleet, eds.), Geological Society, Boulder, CO, pp. 107–16.

Mackenzie, A. S. (1984) Application of biological markers in petroleum geochemistry. In: *Advances in Petroleum Geochemistry* Vol. 1 (J. Brooks and D. H. Welte, eds.), Academic Press, London, pp. 115–214.

Mackenzie, A. S. and Maxwell, J. R. (1981) Assessment of thermal maturation in sedimentary rocks by molecular measurements. In: *Organic Maturation Studies and Fossil Fuel Exploration* (J. Brooks, ed.), Academic Press, London, pp. 239–54.

Mackenzie, A. S. and McKenzie, D. (1983) Isomerization and aromatization of hydrocarbons in sedimentary basins formed by extension. *Geology Magazine*, 120, 417–70.

Mackenzie, A. S., Patience, R. L., Maxwell, J. R., Vandenbroucke, M. and Durand, B. (1980a) Molecular parameters of maturation in the Toarcian shales, Paris Basin, France – I. Changes in the configuration of acyclic isoprenoid alkanes, steranes, and triterpanes. *Geochimica et Cosmochimica Acta*, 44, 1709–21.

Mackenzie, A. S., Quirke, J. M. E. and Maxwell, J. R. (1980b) Molecular parameters of maturation in the Toarcian shales, Paris Basin, France – II. Evolution of metalloporphyrins. In: *Advances in Organic Geochemistry 1979* (A. G. Douglas and J. R. Maxwell, eds.), Pergamon Press, New York, pp. 239–48.

Mackenzie, A. S., Hoffmann, C. F. and Maxwell, J. R. (1981a) Molecular parameters of maturation in the Toarcian shales, Paris Basin, France – III. Changes in aromatic steroid hydrocarbons. *Geochimica et Cosmochimica Acta*, 45, 1345–55.

Mackenzie, A. S., Lewis, C. A. and Maxwell, J. R. (1981b) Molecular parameters of maturation in the Toarcian shales, Paris Basin France – IV. Laboratory thermal alteration studies. *Geochimica et Cosmochimica Acta*, 45, 2369–76.

Mackenzie, A. S., Lamb, N. A. and Maxwell, J. R. (1982a) Steroid hydrocarbons and the thermal history of sediments. *Nature*, 295, 223–6.

Mackenzie, A. S., Patience, R. L., Yon, D. A. and Maxwell, J. R. (1982b) The effect of maturation on the configurations of acyclic isoprenoid acids in sediments. *Geochimica et Cosmochimica Acta*, 46, 783–92.

Mackenzie, A. S., Disko, U. and Rullkötter, J. (1983a) Determination of hydrocarbon distributions in oils and sediment extracts by gas chromatography–high resolution mass spectrometry. *Organic Geochemistry*, 5, 57–63.

Mackenzie, A. S., Wolff, G. A. and Maxwell, J. R. (1983b) Fatty acids in some biodegraded petroleums. Possible origins and significance. In: *Advances in Organic Geochemistry* 1981 (M. Bjorøy, C. Albrecht, C. Cornford, et al. eds.), John Wiley & Sons, New York, pp. 637–49.

Mackenzie, A. S., Beaumont, C. and McKenzie, D. P. (1984) Estimation of the kinetics of geochemical reactions with geophysical models of sedimentary basins and applications. *Organic Geochemistry*, 6, 875–84.

Mackenzie, A. S., Rullkötter, J., Welte, D. H. and Mankiewicz, P. (1985a) Reconstruction of oil formation and accumulation in North Slope, Alaska, using quantitative gas chromatography-mass spectrometry. In: *Alaska North Slope Oil/Source Rock Correlation Study* (L. B. Magoon and G. E. Claypool, eds.), American Association of Petroleum Geologists, Tulsa, OK, pp. 319–77.

Mackenzie, A. S., Beaumont, C., Boutitier, R. and Rullkötter, J. (1985b) The aromatization and isomerization of hydrocarbons and the thermal and subsidence history of the Nova Scotia margin. *Philosophical Transactions of the Royal Society London, Series A*, 315, 203–32.

Mackenzie, A. S., Price, I., Leythaeuser, D., et al. (1987) The expulsion of petroleum from Kimmeridge Clay source-rocks in the area of the Brae Oilfield, UK continental shelf. In: *Petroleum Geology of North West Europe* (J. Brooks and K. Glennie, eds.), Graham and Trotman, London, pp. 865–77.

Mackenzie, A. S., Leythaeuser, D., Altebäumer, F.-J., Disko, U. and Rullkötter, J. (1988a) Molecular measurements of maturity for Lias δ shales in N. W. Germany. *Geochimica et Cosmochimica Acta*, 52, 1145–54.

Mackenzie, A. S., Leythaeuser, D., Muller, P., Quigley, T. M. and Radke, M. (1988b) The movement of hydrocarbons in shales. *Nature*, 331, 63–5.

Magness, S. L. (2001) Distributions and stable carbon isotope characteristics of maleimides (1-*H*-pyrrole-2,5-diones). Unpublished Ph. D. thesis, University of Bristol, Bristol, UK.

Magoon, L. B. (1988) The petroleum system – a classification scheme for research, exploration, and resource assessment. In: *Petroleum Systems of the United States* (L. B. Magoon, ed.), U.S. Geological Survey, Washington, DC, pp. 2–15.

(1989) *The Petroleum System – Status of Research and Methods*. U.S. Geological Survey, Washington, DC, p. 88.

(1994) Tuxedni-Hemlock(!) petroleum system in Cook Inlet, Alaska, U.S.A. In: *The Petroleum System – From Source to Trap*. (L. B. Magoon and W. G. Dow, eds.), American Association of Petroleum Geologists, Tulsa, OK, pp. 359–70.

(1995) The play that complements the petroleum system – a new exploration equation. *Oil and Gas Journal*, 93, 85–7.

Magoon, L. B. and Anders, D. E. (1992) Oil-to-source-rock correlation using carbon-isotopic data and biological marker compounds, Cook Inlet-Alaska Peninsula, Alaska. In: *Biological Markers in Sediments and Petroleum* (J. M. Moldowan, P. Albrecht and R. P. Philp, eds.), Prentice-Hall, Englewood Cliffs, NJ, pp. 241–74.

Magoon, L. B. and Beaumont, E. A. (1999) Petroleum systems. In: *Handbook of Petroleum Geology: Exploring for Oil and Gas Traps* (E. A. Beaumont and N. H. Foster, eds.), American Association of Petroleum Geologists, Washington, DC, pp. 3.1–34.

Magoon, L. B. and Bird, K. J. (1985) Alaskan North Slope petroleum geochemistry for the Shublik Formation, Kingak Shale, pebble shale unit, and Torok Formation. In: *Alaska North Slope Oil/Source Rock Correlation Study* (L. B. Magoon and G. E. Claypool, eds.), American Association of Petroleum Geologists, Tulsa, OK, pp. 31–48.

(1988) Evaluation of petroleum source rocks in the National Petroleum Reserve in Alaska using organic-carbon content, hydrocarbon content, visual kerogen, and vitrinite reflectance. In: *Geology and Exploration of the National Petroleum Reserve in Alaska, 1974 to 1982* (C. Gryc, ed.), U.S. Geological Survey, Washington, DC, 1399, pp. 381–450.

Magoon, L. B. and Claypool, G. E. (1981) Two oil types on the North Slope of Alaska – Implications for future exploration. *American Association of Petroleum Geologists Bulletin*, 65, 644–52.

(1984) The Kingak Shale of North Alaska-Regional variations in organic geochemical properties and petroleum source rock quality. *Organic Geochemistry*, 6, 533–42.

Magoon, L. B. and Dow, W. G. (1994a) The petroleum system. In: *The Petroleum System – From Source to Trap* (L. B. Magoon and W. G. Dow, eds.), American Association of Petroleum Geologists, Tulsa, OK, pp. 3–24.

(1994b) *The Petroleum System – From Source to Trap*. American Association of Petroleum Geologists, Tulsa, OK.

Magoon, L. B. and Schmoker, J. W. (2000) The total petroleum system: natural fluid network that constrains the assessment unit. In: *U.S. Geological Survey World Petroleum Assessment 2000 – Description and Results*, Chapter PS, disk one (U.S. Geological Survey World Energy Asssessment Team), U.S. Geological Survey Digital Data Series DDS-60, Version 1.0, CD-ROM; also available at http://greenwood.cr.usgs.gov/energy/WorldEnergy/DDS-60/.

Magoon, L. B. and Valin, Z. C. (1994) Overview of petroleum system case studies. In: *The Petroleum System – From Source to Trap* (L. B. Magoon and W. G. Dow, eds.), American Association of Petroleum Geologists, Tulsa, OK, pp. 329–38.

Magoon, L. B., Woodward, P. V., Banet, A. C., Griscom, S. B. and Daws, T. A. (1987) Thermal maturity, richness, and type of organic matter of source-rock units. In: *Petroleum Geology of the Northern Part of the Arctic National Wildlife Refuge, Northeastern Alaska* (K. J. Bird and L. B. Magoon, eds.), US Geological Survey, Washington, DC, pp. 127–179.

Magoon, L. B., Bird, K. J., Burruss, R. C., *et al.* (1999) Evaluation of hydrocarbon charge and timing using the petroleum system. In: *The Oil and Gas Resource Potential of the 1002 Area, Arctic National Wildlife Refuge, Alaska*. U.S. Geological Survey Open-File Report 98–34 (Alaska Assessment Team, eds.), US Geological Survey, Washington, DC.

Magoon, L. B., Hudson, T. L. and Cook, H. E. (2001) Pimienta-Tamabra(!) – a giant supercharged petroleum system in the southern Gulf of Mexico, onshore and offshore, Mexico. In: *The Western Gulf of Mexico: Tectonics, Sedimentary Basins, and Petroleum Systems* (C. Bartolini, R. T. Buffler and A. Cantú-Chapa, eds.), American Association of Petroleum Geologists, Tulsa, OK, pp. 83–125.

Magot, M. (1996) Similar bacteria in remote oil fields. *Nature*, 379, 681.

Magot, M., Caumette, P., Desperrier, J. M., *et al.* (1992) *Desulfovibrio longus* sp. nov., a sulfate-reducing bacterium isolated from an oil-producing well. *International Journal of Systematic Bacteriology*, 42, 398–403.

Makushina, V. M., Arefev, O. A., Zabrodina, M. N. and Petrov, A. A. (1978) New relic alkanes of petroleums. *Neftekhimiya*, 18, 847–51.

Mallory, F. B., Gordon, J. T. and Conner, R. L. (1963) The isolation of a pentacyclic triterpenoid alcohol from a protozoan. *Journal of the Americal Chemical Society*, 85, 1362–3.

Mancini, E. A. (2000) Variability in components of the Upper Jurassic Smackover petroleum system, Northeastern Gulf of Mexico. Presented at the Annual Meeting of the American Association of Petroleum Geologists, New Orleans, LA, April 16–19 2000.

Mancini, E. A., Tew, B. H. and Mink, R. M. (1993) Petroleum source rock potential of Mesozoic condensed section deposits of Southwest Alabama. In: *Source Rocks in a Sequence Stratigraphic Framework* (B. J. Katz and L. M. Pratt, eds.), American Association of Petroleum Geologists, Tulsa, OK, pp. 147–62.

Mancuso, J. J., Kneller, W. A. and Quick, J. C. (1989) Precambrian vein pyrobitumen: evidence for petroleum generation and migration 2 Ga ago. *Precambrian Research*, 44, 137–46.

Mango, F. D. (1991) The stability of hydrocarbons under the time-temperature conditions of petroleum genesis. *Nature*, 352, 146–8.

Manhart, F. R. and Palmer, J. D. (1990) The gain of two chloroplast tRNA introns marks the green algal ancestors of land plants. *Nature*, 345, 268–70.

Mann, U. and Müller, P. J. (1988) Source rock evaluation by well log analysis (Lower Toarcian, Hils syncline). *Organic Geochemistry*, 13, 109–19.

Mann, U. and Steine, R. (1997) Organic facies variations, source rock potential, and sea level changes in Cretaceous black shales of the Quebrada Ocal, Upper Magdalena Valley, Colombia. *American Association of Petroleum Geologists Bulletin*, 81, 556–76.

Mann, A. L., Goodwin, N. S. and Lowe, S. (1987) Geochemical characteristics of lacustrine source rocks: a combined palynological/molecular study of a Tertiary sequence from offshore China. In: *Proceedings of the Indonesian Petroleum Association, Sixteenth Annual Convention*, Vol. 1 Indonesian Petroleum Association, Jakarta, Indonesia, pp. 241–58.

Mansour, A. T. and Magairhy, I. A. (1996) Petroleum geology and stratigraphy of the southeastern part of the Sirt Basin, Libya. In: *The Geology of Sirt Basin*, Vol. II (M. J. Salem, A. S. El-Hawat and A. M. Sbeta, eds.), Elsevier, Amsterdam, pp. 485–528.

Marshall, J. E. A., Rogers, D. A. and Whiley, M. J. (1996) Devonian marine incursions into the Orcadian Basin, Scotland. *Journal of the Geological Society of London*, 153, 451–66.

Marshall, J. D., Brenchley, P. J., Mason, P., *et al.* (1997) Global carbon isotopic events associated with mass extinction and glaciation in the Late Ordovician. *Palaeogeography, Palaeoclimatology, and Palaeoecology*, 132, 195–210.

Martin, W., Gierl, A. and Saelter, H. (1989) Molecular evidence for pre-Cretaceous angiosperm origins. *Nature*, 339, 46–8.

Martz, R. F., Sebacher, D. I. and White, D. C. (1983) Biomass measurement of methane forming bacteria in environment samples. *Journal of Microbiological Methods*, 1, 53–61.

Marynowski, L., Narkiewicz, M. and Grelowski, C. (2000) Biomarkers as environmental indicators in a carbonate complex, example from the Middle to Upper Devonian, Holy Cross Mountains, Poland. *Sedimentary Geology*, 137, 187–212.

Marzi, R. and Rullkötter, J. (1992) Qualitative and quantitative evolution and kinetics of biological marker transformations: laboratory experiments and application to the Michigan Basin. In: *Biological Markers in Sediments and Petroleum* (J. M. Moldowan, P. Albrecht and R. P. Philp, eds.), Prentice-Hall, Englewood Cliffs, NJ, pp. 18–41.

Marzi, R., Rullkötter, J. and Perriman, W. S. (1990) Application of the change of sterane isomer ratios to the reconstruction of geothermal histories: implications of the results of hydrous pyrolysis experiments. *Organic Geochemistry*, 16, 91–102.

Mason, G. M., Rudell, L. G. and Branthaver, J. F. (1990) Review of the stratigraphic distribution and diagenetic history of abelsonite. *Organic Geochemistry*, 14, 585–94.

Mason, P. C., Burwood, R. and Mycke, B. (1995) The reservoir geochemistry and petroleum charging histories of Palaeogene-reservoired fields in the Outer Witch Ground Graben. In: *The Geochemistry of Reservoirs* (J. M. Cubitt and W. A. England, eds.), Geological Society of London, London, pp. 281–301.

Masterson, W. D. (2001) Petroleum filling history of central Alaskan North Slope fields. Unpublished Ph. D. thesis, University of Texas at Dallas, Dallas, TX.

Masterson, W. D., Holba, A. and Dzou, L. (1997) Filling history of America's two largest oil fields: Prudhoe Bay and Kuparuk, North Slope, Alaska. Presented at the Annual Meeting of the American Association of Petroleum Geologists, Dallas, TX, April 6–9, 1997.

Masterson, W. D., He, Z., Corrigan, J., Holba, A. G. and Dzou, L. I. P. (2000) Petroleum generation, migration, and filling history models for the Prudhoe Bay, Kuparuk, and West Sak fields, North Slope, Alaska. Presented at the Annual Meeting of the American Association of Petroleum Geologists, New Orleans, LA, April 16–19, 2000.

Masterson, W. D., Dzou, L. I. P., Holba, A. G., Fincannon, A. L. and Ellis, L. (2001) Evidence for biodegradation and evaporative fractionation in West Sak, Kuparuk and Prudhoe Bay field areas, North Slope, Alaska. *Organic Geochemistry*, 32, 411–41.

Mathalone, J. M. P. and Montoya, R. M. (1995) Petroleum geology of the Sub-Andean basins of Peru. In: *Petroleum Basins of South America* (A. J. Tankard, R. Suarez Soruco and H. J. Welsink, eds.), American Association of Petroleum Geologists, Tulsa, OK, pp. 423–44.

Mathur, S., Jain, V. K., Tripathi, G. K., Jassall, J. K. and Chandra, K. (1987) Biological marker geochemistry of crude oils of Cambay Basin, India. In: *Petroleum Geochemistry and Exploration in the Afro-Asian Region*. (R. K. Kumar, P. Dwivedi, V. Banerjie and V. Gupta, eds.), Balkema Publishing, Rotterdam, the Netherlands, pp. 459–73.

Matsuda, K., Shiojima, K. and Ageta, H. (1989) Fern constituents: four new onoceradienes isolated from *Lemmaphyllum microphyllum* Presl. *Chemical and Pharmaceutical Bulletin*, 37, 263–5.

Matsueda, H., Handa, N., Inoue, I. and Takano, H. (1986) Ecological significance of salp fecal pellets collected by sediment trap experiments in the eastern North Pacific. *Marine Biology*, 91, 421–31.

Matsumoto, G. I. and Watanuki, K. (1990) Geochemical features of hydrocarbons and fatty acids in sediments of the inland hydrothermal environments of Japan. *Organic Geochemistry*, 15, 199–208.

Mattavelli, L. and Novelli, L. (1990) Geochemistry and habitat of the oils in Italy. *American Association of Petroleum Geologists Bulletin*, 74, 1623–39.

Mattern, G., Albrecht, P. and Ourisson, G. (1970) 4-Methylsterols and sterols in Messel Shale (Eocene). *Chemical Communications*, 1570–1.

Matthews, S. J., Fraser, A. J., Lowe, S., Todd, S. P. and Peel, F. J. (1997) Structure, stratigraphy and petroleum geology of the SE Nam Con Son Basin, offshore Vietnam. In *Petroleum Geology of Southeast Asia* (A. J. Fraser, S. J. Matthews and R. W. Murphy, eds.), Geological Society of London, London, pp. 89–106.

Mattick, R. E., Teleki, P. G., Phillips, R. L., *et al.* (1996) Structure, stratigraphy, and petroleum geology of the Little Plain Basin, Northwestern Hungary. *American Association of Petroleum Geologists Bulletin*, 80, 1780–99.

Maughan, E. K. (1984) Geological setting and some geochemistry of petroleum source rocks in the Permian Phosphoria Formation. In: *Hydrocarbon Source Rocks of the Greater Rocky Mountain Region* (J. Woodward, F. F. Meissner and J. L. Clayton, eds.), Rocky Mountain Association of Geologists, Denver, CO, pp. 281–94.

  (1993) Phosphoria Formation (Permian) and its resource significance in the Western Interior, USA. Presented at the CSPG Pangeo: Global Environment and Resources Conference, Calgary, August 15–19, 1993.

Mauk, J. L. and Hieshima, G. B. (1992) Organic matter and copper mineralization at White Pine, Michigan. *Chemical Geology*, 99, 189–211.

Maxwell, J. R., Douglas, A. G., Eglinton, G. and McCormick, A. (1968) The botryococcenes – hydrocarbons of novel structure from the alga *Botryococcus braunii* Kutzing. *Phytochemistry*, 7, 2157–71.

Mazoli, A., Renne, P. R., Piccirillo, E. M., *et al.* (1999) Extensive 200-million year old continental flood basalts of the Central Atlantic Magmatic Province. *Science*, 284, 616–18.

McCaffrey, M. A., Farrington, J. W. and Repeta, D. J. (1989) Geochemical implications of the lipid composition of *Thioploca* spp. from the Peru upwelling region – 15°S. *Organic Geochemistry*, 14, 61–8.

McCaffrey, M. A., Moldowan, J. M., Lipton, P. A., *et al.* (1994a) Paleoenvironmental implications of novel C_{30} steranes in Precambrian to Cenozoic age petroleum and bitumen. *Geochimica et Cosmochimica Acta*, 58, 529–32.

McCaffrey, M. A., Dahl, J. E., Sundararaman, P., Moldowan, J. M. and Schoell, M. (1994b) Source rock quality determination from oil biomarkers II – a case study using Tertiary-reservoired Beaufort Sea oils. *American Association of Petroleum Geologists Bulletin*, 78, 1527–40.

McCaffrey, M. A., Legarre, H. A. and Johnson, S. J. (1996) Using biomarkers to improve heavy oil reservoir management; an example from the Cymric Field, Kern County, California. *American Association of Petroleum Geologists Bulletin*, 80, 898–913.

McConachie, B. A., Bradshaw, M. T. and Bradshaw, J. (1996) Petroleum systems of the Petrel Sub-basin – an integrated approach to basin analysis and identification of hydrocarbon exploration opportunities. *APPEA Journal*, 36, 248–68.

McCulloh, T. H., Kirkland, D. W., Koch A. J., Orr, W. L. and Chung, H. M. (1994) How oil composition relates to kerogen facies in the world's most petroliferous basin. Presented at the Annual Meeting of the American Association of Petroleum Geologists, Denver, CO, June 12–15, 1994.

McDade, E. C., Sassen, R., Wenger, L. and Cole, G. A. (1993) Identification of organic-rich lower Tertiary shales as petroleum source rocks, South Louisiana. In: *Transactions of the 43rd Annual Convention Gulf Coast Association of Geological Societies* (N. C. Rosen, A. Sartin and M. Barrett, eds.), GCACS, Austin, TX, pp. 257–67.

McGhee, G. R., Jr (1988) The Late Devonian extinction event: evidence for abrupt ecosystem collapse. *Paleobiology*, 14, 250–7.

McGhee, G. R., Jr, Gilmore, J. S., Orth, C. J. and Olsen, E. (1984) No geochemical evidence for an asteroidal impact at Late Devonian mass extinction horizon. *Nature*, 308, 629–31.

McKay, D. S., Gibson, E. K., Jr, Thomas-Keprta, K. L., *et al.* (1996) Search for past life on Mars: possible relic biogenic activity in Martian meteorite ALH84001. *Science*, 273, 924–30.

McKay, D. S., Gibson, E. K., Thomas-Keprta, K., Romanek, C. S. and Allen, C. C. (1997) Possible biofilms in ALH84001. Presented at the 28th Lunar and Planetary Science Conference, Houston, TX, March 17–21, 1997.

McKenzie, D., Mackenzie, A. S., Maxwell, J. R. and Sajgó, C. (1983) Isomerization and aromatization of hydrocarbons in stretched sedimentary basins. *Nature*, 301, 504–6.

McKirdy, D. M. and Imbus, S. W. (1992) Precambrian petroleum: a decade of changing perceptions. In: *Early Organic Evolution: Implications for Mineral and Energy Resources* (M. Schidlowski, S. Golubic, M. M. Kimberley, D. M. McKirdy and P. A. Trudinger, eds.), Springer-Verlag, Berlin, pp. 176–192.

McKirdy, D. M., Aldridge, A. K. and Ypma, P. J. M. (1983) A geochemical comparison of some crude oils from Pre-Ordovician carbonate rocks. In: *Advances in Organic Geochemistry 1981* (M. Bjorøy, C. Albrecht, C. Cornford, *et al.*, eds.), John Wiley & Sons, New York, pp. 99–107.

McKirdy, D. M., Kantsler, A. J., Emmett, J. K. and Aldridge, A. K. (1984) Hydrocarbon genesis and organic facies in Cambrian carbonates of the Eastern Officer Basin, South Australia. In: *Petroleum Geochemistry and Source Rock Potential of Carbonate Rocks* (J. G. Palacas, ed.), American Association of Petroleum Geologists Geology, Tulsa, OK, pp. 13–31.

McKirdy, D. M., Cox, R. E., Volkman, J. K. and Howell, V. J. (1986) Botryococcane in a new class of Australian nonmarine crude oils. *Nature*, 320, 57–9.

McLaren, D. J. (1983) Bolides and biostratigraphy. *Bulletin of the Geological Society of America*, 94, 312–24.

McMahon, P. B. and Chapelle, F. H. (1991) Microbial production of organic acids in aquitard sediments and its role in aquifer geochemistry. *Nature*, 349, 233–5.

McMenamin, M. A. S. (1986) The garden of Ediacara. *Palaios*, 1, 178–82.

Meckenstock, R. U., Annweiler, E., Michaelis, W., Richnow, H. H. and Schink, B. (1999) Anaerobic naphthalene degradation by a sulfate-reducing enrichment culture. *Applied and Environmental Microbiology*, 66, 2743–7.

Meinschein, W. G., Nagy, B. and Hennessy, D. J. (1963) Evidence in meteorites of former life: the organic compounds in carbonaceous chondrites are similar to those found in marine sediments. *Annals of the New York Academy of Sciences*, 108, 339–66.

Meissner, F. F., Woodward, J. and Clayton, J. L. (1984) Stratigraphic relationships and distribution of source rocks in the Greater Rocky Mountain Region. In: *Hydrocarbon Source Rocks of the Greater Rocky Mountain Region* (J. Woodward, F. F. Meissner and J. L. Clayton, eds.), Rocky Mountain Association of Geologists, Denver, CO, pp. 1–34.

Melezhik, V. A., Fallick, A. E., Filippov, M. M. and Larsen, O. (1999) Karelian shungite – an indication of 2.0-Ga-old metamorphosed oil-shale and generation of petroleum: geology, lithology and geochemistry. *Earth-Science Reviews*, 47, 1–40.

Mello, M. R. (1988) Geochemical and molecular studies of the depositional environments of source rocks and their derived oils from the Brazilian marginal basins. Unpublished Ph. D. thesis, Bristol University, Bristol, UK.

Mello, M. R. and Maxwell, J. R. (1990) Organic geochemical and biological marker characterization of source rocks and oils derived from lacustrine environments in the Brazilian continental margin. In: *Lacustrine Basin Exploration: Case Studies and Modern Analogs* (B. J. Katz, ed.), American Association of Petroleum Geologists, Tulsa, OK, pp. 77–97.

Mello, M. R., Gaglianone, P. C., Brassell, S. C. and Maxwell, J. R. (1988a) Geochemical and biological marker assessment of depositional environments using Brazilian offshore oils. *Marine and Petroleum Geology*, 5, 205–23.

Mello, M. R., Telnaes, N., Gaglianone, P. C., et al. (1988b) Organic geochemical characterization of depositional paleoenvironments in Brazilian marginal basins. *Organic Geochemistry*, 13, 31–46.

Mello, M. R., Koutsoukos, E. A. M., Hart, M. B., Brassell, S. C. and Maxwell, J. R. (1990) Late Cretaceous anoxic events in the Brazilian continental margin. *Organic Geochemistry*, 14, 529–42.

Mello, M. R., Koutsoukos, E. A. M., Santos Neto, E. V. and Silva Telles, A. C., Jr (1993) Geochemical and micropaleontological characterization of lacustrine and marine hypersaline environments from Brazilian sedimentary basins. In: *Source Rocks in a Sequence Stratigraphic Framework* (B. J. Katz and L. M. Pratt, eds.), American Association of Petroleum Geologists, Tulsa, OK, pp. 17–34.

Mello, M. R., Koutsoukos, E. A. M., Mohriak, W. U. and Bacoccoli, G. (1994a) Selected petroleum systems in Brazil. In: *The Petroleum System – From Source to Trap* (L. B. Magoon and W. G. Dow, eds.), American Association of Petroleum Geologists, Tulsa, OK, pp. 499–512.

Mello, M. R., Koutsoukos, E. A. M. and Erazo, W. Z. (1994b) The Napo Formation, Oriente Basin, Ecuador: hydrocarbon source potential and paleoenvironmental assessment. In: *Petroleum Source Rocks* (B. J. Katz, ed.), Springer-Verlag, Berlin pp. 167–81.

Mello, M. R., Telnaes, N. and Maxwell, J. R. (1995) The hydrocarbon source potential in the Brazilian marginal basins: a geochemical and paleoenvironmental assessment. In: *Paleogeography, Paleoclimate, and Source Rocks* (A.-Y. Huc, ed.), American Association of Petroleum Geologists, Tulsa, OK, pp. 233–72.

Mello, M. R., Koutsoukos, E. A. M., Mohriak, W. U., et al. (1996) The petroleum system concept in the Sergipe Basin, Northeastern Brazil. *American Association of Petroleum Geologists Bulletin*, 80, 1314.

Meredith, W., Kelland, S.-J. and Jones, D. M. (2000) Influence of biodegradation on crude oil acidity and carboxylic acid composition. *Organic Geochemistry*, 31, 1059–73.

Metzger, P. and Casadevall, E. (1987) Lycopadiene, a tetraterpenoid hydrocarbon from new strains of the green alga *Botryococcus braunii*. *Tetrahedron Letters*, 28, 3931–4.

Metzger, P. and Largeau, C. (1994) A new type of ether lipid comprising phenolic moieties in *Botryococcus braunii*. Chemical structure and abundance, and geochemical implications. *Organic Geochemistry*, 22, 801–14.

Metzger, P., Casadevall, E., Pouet, M. J. and Pouet, Y. (1985a) Structures of some botryococcenes: branched hydrocarbons from the B-race of the green alga *Botryococcus braunii*. *Phytochemistry*, 14, 2995–3002.

Metzger, P., Berkaloff, C., Casadevall, E. and Coute, A. (1985b) Alkadiene- and botryococcene-producing races of wild strains of *Botroycoccus braunii*. *Phytochemistry*, 24, 2305–12.

Metzger, P., Villarreal-Rosalles, E., Casadevall, E. and Coute, A. (1989) Hydrocarbons, aldehydes and tricylglycerols in some strains of the A race of the green alga *Botryococcus braunii*. *Phytochemistry*, 28, 2349–53.

Metzger, P., Largeau, C. and Casadevall, E. (1991) Lipids and macromolecular lipids of the hydrocarbon-rich microalga *Botryococcus braunii*. Chemical structure and biosynthesis. In: *Fortschritte der Chemie Organischer Naturstoffe 57* (W. Herz, eds.), Springer-Verlag, Berlin, pp. 1–70.

Meunier-Christmann, C. (1988) Geochimie organique de phosphates et schistes bitumineux marocains: étude du processus de phosphatogenese. Ph. D. thesis, L'Universite Louis Pasteur de Strasbourg, Strasbourg, France.

Meyers, P. A. (1997) Organic geochemical proxies of paleooceanographic, paleolimnlogic, and paleoclimatic processes. *Organic Geochemistry*, 27, 213–50.

Michael, G. E. (2001) Geochemical characterization of the Miocene Monterey Formation and oils in the Santa Barbara-Ventura and Santa Maria basins. In: *The Monterey Formation: From Rocks to Molecules* (C. M. Isaacs and J. Rullkötter, eds.), Columbia University Press, New York, pp. 241–267.

Michael, E. and Bond, D. (1997) Integration of 2D modeling, drainage polygon analysis and geochemistry as petroleum systems analysis tools; West Block B PSC, S. Natuna Sea. In: *Proceedings of the International Conference on Petroleum Systems of SE Asia and Australia* (J. V. C. Howes and R. A. Noble, eds.), Indonesian Petroleum Association, Jakarta, Indonesia, pp. 391–401.

Michael, G. E., Lin, L. H., Philp, R. P., Lewis, C. A. and Jones, P. J. (1990) Biodegradation of tar-sand bitumens from the Ardmore/Anadarko basins, Oklahoma – II. Correlation of oils, tar sands, and source rocks. *Organic Geochemistry*, 14, 619–33.

Michael, G. E., Anders, D. E. and Law, B. E. (1993) Geochemical evaluation of Upper Cretaceous Fruitland Formation coals, San Juan Basin, New Mexico and Colorado. *Organic Geochemistry*, 20, 475–98.

Michaelis, W. and Albrecht, P. (1979) Molecular fossils of archaebacteria in kerogen. *Naturwissenschaften*, 66, 420–1.

Michaelis, W., Seifert, R., Nauhaus, K., *et al.* (2002) Microbial reefs in the Black Sea fueled by anaerobic oxidation of methane. *Science*, 297, 1013–15.

Michaelsen, B. H. and McKirdy, D. M. (1989) Organic facies and petroleum geochemistry of the lacustrine Murta Member (Mooga Formation) in the Eromanga Basin, Australia. In: *The Cooper and Eromanga Basins, Australia, Proceedings of the Cooper and Eromanga Basins Conference, Adelaide, 1989.* (B. J. O'Neil, ed.), Petroleum Exploration Society of Australia, Perth, Australia, pp. 541–558.

  (1996) Source rock distribution and hydrocarbon geochemistry. In: *The Petroleum Geology of South Australia*, Vol. 2 (E. M. Alexander and J. E. Hibburt, eds.), South Australia Mines and Energy Petroleum Division, Adelaide, Australia, pp. 101–10.

Michaelsen, B. H., Kamali, M. R. and McKirdy, D. M. (1995) Unexpected molecular fossils from Early Cambrian carbonates of the Officer Basin. In: *Organic Geochemistry: Developments and Applications to Energy, Climate, Environment and Human History* (J. O. Grimalt and C. Dorronsoro, eds.), A.I.G.O.A., San Sebastian, Spain, pp. 218–221.

Michels, R., Landais, P., Torkelson, B. E. and Philp, R. P. (1995) Effects of effluents and water pressure on oil generation during confined pyrolysis and high-pressure hydrous pyrolysis. *Geochimica et Cosmochimica Acta*, 59, 1589–604.

Miget, R. J., Oppenheimer, C. H., Kator, H. I. and Larock, D. A. (1969) Microbial degradation of normal paraffin hydrocarbons in crude oil. In: *Proceedings of the Joint Conference on Prevention and Control of Oil Spills*. American Petroleum Institute, Washington, DC, pp. 327–31.

Mille, G., Munoz, D., Jacquot, F., Rivet, L., and Bertrand, J.-C. (1998) The Amoco Cadiz oil spill: evolution of petroleum hydrocarbons in the Ile Grande salt marshes (Brittany) after a 13-year period. *Estuarine, Coastal and Shelf Science*, 47, 547–59.

Miller, S. L. (1953) A production of amino acids under possible primitive Earth conditions. *Science*, 117, 528–9.

  (1990) A paleoceanographic approach to the Kimmeridge Clay Formation. In: *Deposition of Organic Facies, Studies in Geology 30* (A.-Y. Huc, eds.), American Association of Petroleum Geologists, Tulsa, OK, pp. 13–26.

Miller, S. and Bada, J. (1988) Submarine hot springs and the origin of life. *Nature*, 334, 564.

Milner, H. B. (1925) "Paraffin dirt." Its nature, origin, mode of occurrence, and significance as an indication of petroleum. *Mining Magazine (London)*, 32, 73–85.

  (1998) Source rock distribution and thermal maturity in the Southern Arabian Peninsula. *GeoArabia*, 3, 339–56.

Milner, C. W. D., Rogers, M. A. and Evans, C. R. (1977) Petroleum transformations in reservoirs. *Journal of Geochemical Exploration*, 7, 101–53.

Minster, T., Yoffe, O., Nathan, Y. and Flexer, A. (1997) Geochemistry, mineralogy, and paleoenvironments of deposition of the oil shale member in the Negev. *Israel Journal of Earth Science*, 46, 41–59.

Miranda, R. M. and Walters, C. C. (1992) Geochemical variations in sedimentary organic matter within a "homogeneous" shale core (Tuscaloosa Formation, Upper Cretaceous, Mississippi, USA). *Organic Geochemistry*, 18, 899–911.

Mishler, B. D., Lewis, L. A., Buchheim, M. A., *et al.* (1994) Phylogenetic relationships of the "green algae" and "bryophytes". *Annals of the Missouri Botanical Garden*, 81, 451–83.

Mita, H. and Shimoyama, A. (1999a) Characterization of *n*-alkanes, pristane and phytane in the Cretaceous/Tertiary boundary sediments at Kawaruppu, Hokkaido, Japan. *Geochemistry Journal*, 33, 285–94.

(1999b) Distribution of polycyclic aromatic hydrocarbons in the K/T [Cretaceous/Tertiary] boundary sediments at Kawaruppu, Hokkaido, Japan. *Geochemistry Journal*, 33, 305–15.

Mitchell-Tapping, H. J. (1987) Application of the tidal mudflat model to the Sunniland Formation of South Florida. *Gulf Coast Association of Geological Societies Transactions*, 37, 415–26.

Mohamed, A. Y., Pearson, M. J., Ashcroft, W. A., Iliffe, J. E. and Whiteman, A. J. (1999) Modeling petroleum generation in the southern Muglad rift basin, Sudan. *American Association of Petroleum Geologists Bulletin*, 83, 1943–64.

Mohriak, W. U., Mello, M. R., Dewey, J. F. and Maxwell, J. R. (1990) Petroleum geology of the Campos Basin, offshore Brazil. In: *Classic Petroleum Provinces* (J. Brooks, ed.), Geological Society of London, London, pp. 119–41.

Mojzsis, S.-J., Arrhenius, G., McKeegan, K. D., *et al.* (1996) Evidence for life on Earth before 3,800 million years ago. *Nature*, 384, 55–9.

Moldowan, J. M. (2000) Trails of life. *Chemistry in Britain*, 36, 34–7.

Moldowan, J. M. and Fago, F. J. (1986) Structure and significance of a novel rearranged monoaromatic steroid hydrocarbon in petroleum. *Geochimica et Cosmochimica Acta*, 50, 343–51.

Moldowan, J. M. and Jacobson, S. R. (2000) Chemical signals for early evolution of major taxa: biosignatures of taxon-specific biomarkers. *International Geology Review*, 42, 805–12.

Moldowan, J. M. and McCaffrey, M. A. (1995) A novel microbial hydrocarbon degradation pathway revealed by hopane demethylation in a petroleum reservoir. *Geochimica et Cosmochimica Acta*, 59, 1891–4.

Moldowan, J. M. and Seifert, W. K. (1979) Head-to-head linked isoprenoid hydrocarbons in petroleum. *Science*, 204, 169–71.

(1980) First discovery of botryococcane in petroleum. *Journal of Chemical Society, Chemical Communications*, 912–4.

Moldowan, J. M. and Talyzina, N. M. (1998) Biogeochemical evidence for dinoflagellate ancestors in the Early Cambrian. *Science*, 281, 1168–70.

Moldowan, J. M., Seifert, W. K. and Gallegos, E. J. (1983) Identification of an extended series of tricyclic terpanes in petroleum. *Geochimica et Cosmochimica Acta*, 47, 1531–4.

Moldowan, J. M., Seifert, W. K., Arnold, E. and Clardy, J. (1984) Structure proof and significance of stereoisomeric 28,30-bisnorhopanes in petroleum and petroleum source rocks. *Geochimica et Cosmochimica Acta*, 48, 1651–61.

Moldowan, J. M., Seifert, W. K. and Gallegos, E. J. (1985) Relationship between petroleum composition and depositional environment of petroleum source rocks. *American Association of Petroleum Geologists Bulletin*, 69, 1255–68.

Moldowan, J. M., Sundararaman, P. and Schoell, M. (1986) Sensitivity of biomarker properties to depositional environment and/or source input in the Lower Toarcian of S. W. Germany. *Organic Geochemistry*, 10, 915–26.

Moldowan, J. M., Fago, F. J., Lee, C. Y., *et al.* (1990) Sedimentary 24-*n*-propylcholestanes, molecular fossils diagnostic of marine algae. *Science*, 247, 309–12.

Moldowan, J. M., Lee, C. Y., Watt, D. S., *et al.* (1991a) Analysis and occurrence of C_{26}-steranes in petroleum and source rocks. *Geochimica et Cosmochimica Acta*, 55, 1065–81.

Moldowan, J. M., Fago, F. J., Carlson, R. M. K., *et al.* (1991b) Rearranged hopanes in sediments and petroleum. *Geochimica et Cosmochimica Acta*, 55, 3333–53.

Moldowan, J. M., Fago, F. J., Huizinga, B. J. and Jacobson, S. R. (1991c) Analysis of oleanane and its occurrence in Upper Cretaceous rocks. In: *Organic Geochemistry. Advances and Applications in the Natural Environment* (D. A. C. Manning, ed.), Manchester University Press, Manchester, UK, pp. 195–7.

Moldowan, J. M., Lee, C. Y., Sundararaman, P., *et al.* (1992) Source correlation and maturity assessment of select oils and rocks from the Central Adriatic Basin (Italy and Yugoslavia). In: *Biological Markers in Sediments and Petroleum* (J. M. Moldowan, P. Albrecht and R. P. Philp, eds.), Prentice-Hall, Englewood Cliffs, NJ, pp. 370–401.

Moldowan, J. M., Dahl, J., Huizinga, B. J., *et al.* (1994a) The molecular fossil record of oleanane and its relation to angiosperms. *Science*, 265, 768–71.

Moldowan, J. M., Peters, K. E., Carlson, R. M. K., Schoell, M. and Abu-Ali, M. A. (1994b) Diverse applications of petroleum biomarker maturity parameters. *Arabian Journal for Science and Engineering*, 19, 273–98.

Moldowan, J. M., Dahl, J., McCaffrey, M. A., Smith, W. J. and Fetzer, J. C. (1995) Application of biological marker technology to bioremediation of refinery by-products. *Energy & Fuels*, 9, 155–62.

Moldowan, J. M., Dahl, J., Jacobson, S. R., et al. (1996) Chemostratigraphic reconstruction of biofacies: molecular evidence linking cyst-forming dinoflagellates with pre-Triassic ancestors. *Geology*, 24, 159–62.

Moldowan, J. M., Jacobson, S. R., Dahl, J., et al. (2001a) Molecular fossils demonstrate Precambrian origin of dinoflagellates. In: *Ecology of the Cambrian Radiation* (A. Zhuravlev and R. Riding, eds.), Columbia University Press, New York, pp. 474–93.

Moldowan, J. M., Zinniker, D. A., Dahl, J., Li, H. and Taylor, D. W. (2001b) Molecular paleontology: tracing the evolutionary roots of angiosperms using the molecular fossil oleanane. American Chemical Society, 221st national meeting, April 1–5, Sar Diego, CA.

Molenaar, C. M., Bird, K. J. and Kirk, A. R. (1987) Cretaceous and Tertiary stratigraphy of Northeastern Alaska. In: *Alaskan North Slope Geology* (I. Tailleur and P. Weimer, eds.), Society of Economic Paleontologists and Mineralogists and Alaska Geological Society, Pacific Section, pp. 513–28.

Momper, J. A. and Williams, J. A. (1984) Geochemical exploration in the Powder River Basin. In: *Petroleum Geochemistry and Basin Evaluation*. American Association of Petroleum Geologists, Tulsa, OK, pp. 181–91.

Monson, B. and Parnell, J. (1992) The origin of gilsonite vein deposits in the Unita Basin, Utah. In: *Hydrocarbon and Mineral Resources of the Uinta Basin, Utah and Colorado* (T. D. Fouch, V. F. Nuccio and T. C. Chidsey, eds.), Utah Geological Association, Salt Lake City, UT, pp. 257–70.

Montgomery, S. (1990) Tight gas formation: an ongoing effort by the U.S. Department of Energy. *Petroleum Frontiers*, 7, 26–35.

Moore, P. S., Pitt, G. M. and Dettman, M. E. (1986) The Early Cretaceous Coorikiana Sandstone and Toolebuc Formation: their recognition and stratigraphic relationship in the southwestern Eromanga Basin. In: *Contributions to the Geology and Hydrocarbon Potential of the Eromanga Basin* (D. I. Gravestock, P. S. Moore and G. M. Pitt, eds.), Geological Society of Australia, Sydney, Australia, pp. 97–114.

Moore, P. S., Burns, B. J., Emmett, J. K. and Guthrie, D. A. (1992) Integrated source, maturation and migration analysis, Gippsland Basin, Australia. *Australian Petroleum Exploration Association Journal*, 32, 313–24.

Moore, G. T., Hayashida, D. N. and Ross, C. A. (1993) Late Early (Silurian) Wenlockian general circulation model-generated up-welling, graptolitic black shales, and organic-rich source rocks – an accident of plate tectonics. *Geology*, 21, 17–20.

Moore, G. T., Barron, E. T., Bice, K. L. and Hayashida, D. N. (1995) Paleoclimatic controls on Neocomian-Barremian (Early Cretaceous) lithostratigraphy in northern Gondwana's rift lakes interpreted from a general circulation model simulation. In: *Paleogeography, Paleoclimate, and Source Rocks* (A.-Y. Huc, ed.), American Association of Petroleum Geologists, Tulsa, OK, pp. 173–89.

Morelos Garcia, J. A. (1996) Geochemical evaluation of southern Tampico-Misantla Basin, Mexico: oil–oil and oil–source rock correlations. Ph. D. thesis, University of Texas at Dallas, Richardson, TX.

Moretti, I., Martinez, E. D., Montemurro, G., Aguilera, E. and Perez, M. (1995) The Bolivian source rocks. Sub-Andean Zone, Madre de Dios, Chaco. *Review Institut Français du Petrolé*, 50, 753–77.

Morris, H. T., and Lovering, T. S. (1961) *Stratigraphy of the East Tintic Mountains, Utah*. U.S. Geological Survey Professional Paper 361, U.S. Geological Survey, Washington, DC.

Mory, A. J., Iasky, R. P., Glikson, A. Y. and Pirajno, F. (2000) Woodleigh, Carnarvon Basin, Western Australia: a new 120 km diameter impact structure. *Earth And Planetary Science Letters*, 177, 119–28.

Mosmann, R., Falkenhein, F. U. H., Goncalves, A. and Filho, F. N. (1984) Oil and gas potential of the Amazon Paleozoic basins. In: *Future Petroleum Provinces of the World* (M. T. Halbouty, ed.), American Association of Petroleum Geologists, Tulsa, OK, pp. 207–41.

Mossman, D., Eigendorf, G., Tokaryk, D., et al. (2003) Testing for fullerenes in geologic materials: Oklo carbonaceous substances, Karelian shungites, Sudbury Black Tuff. *Geology*, 31, 255–8.

Mpanju, F. and Philp, R. P. (1991) Geochemical characteristics of bitumens and seeps from Tanzania. *American Association of Petroleum Geologists Bulletin*, 75, 642.

(1994) Organic geochemical characterization of bitumens, seeps, rock extracts and condensates from Tanzania. *Organic Geochemistry*, 21, 359–71.

Mueller, E. (1998) Temporal and spatial source rock variations and the consequence on crude oil composition in the Tertiary petroleum system of the Uinta Basin, Utah, USA. Ph. D. thesis, University of Oklahoma, Norman, OK.

Muir, M. D. and Grant, P. R. (1972) Micropaleontological evidence from the Onverwacht Group, South Africa. In: *The Early History of the Earth* (B. F. Windley, ed.), John Wiley & Sons, New York, pp. 584–604.

Muir, M. D., Armstrong, K. J. and Jackson, M. J. (1980) Precambrian hydrocarbons in the McArthur Basin, NT. *Bureau of Mineral Resources Journal of Australian Geology and Geophysics*, 5, 301–4.

Mukhopadhyay, S., Farley, K. A. and Montanari, A. (2001) A short duration of the Cretaceous-Tertiary boundary event: evidence from extraterrestrial helium-3. *Science*, 291, 1952–5.

Mull, C. G., Tailleur, I. L., Mayfield, C. F., Ellersieck, I. and Curtis, S. (1982) New Upper Paleozoic and Lower Mesozoic stratigraphic units, Central and Western Brooks Range, Alaska. *American Association of Petroleum Geologists Bulletin*, 66, 348–62.

Munoz, D., Guiliano, M., Doumenq, P., *et al.* (1997) Long term evolution of petroleum biomarkers in Mangrove soil (Guadeloppe). *Marine Pollution Bulletin*, 34, 868–74.

Murphy, M. T. J., McCormick, A. and Eglinton, G. (1967) Perhydro-β-carotene in Green River Shale. *Science*, 157, 1040–2.

Murray, A. P., Summons, R. E., Boreham, C. J. and Dowling, L. M. (1994) Biomarker and n-alkane isotope profiles for Tertiary oils: relationship to source rock depositional setting. *Organic Geochemistry*, 22, 521–42.

Murray, A. P., Sosrowidjojo, I. M., Alexander, R., *et al.* (1997a) Oleananes in oils and sediments: evidence of marine influence during early diagenesis? *Geochimica et Cosmochimica Acta*, 61, 1261–76.

Murray, A. P., Sosrowidjojo, I. B., Alexander, R. and Summons, R. E. (1997b) Locating effective source rocks in deltaic petroleum systems: making better use of land-plant biomarkers. In: *Proceedings of an International Conference on Petroleum Systems of SE Asia and Australasia* (J. V. C. Howes and R. A. Noble, eds.), Indonesian Petroleum Association, Jakarta, Indonesia, pp. 939–45.

Murray, A. P., Edwards, D., Hope, J. M., *et al.* (1998) Carbon isotope biogeochemistry of plant resins and derived hydrocarbons. *Organic Geochemistry*, 29, 1199–214.

Murris, R. J. (1980) Middle East: stratigraphic evolution and oil habitat. *American Association of Petroleum Geologists Bulletin*, 64, 597–618.

(1984) Introduction. In: *Petroleum Geochemistry and Basin Evaluation* (G. Demaison and R. J. Murris, eds.), American Association of Petroleum Geologists, Tulsa, OK, pp. i–xii.

Müürisepp, A.-M., Urov, K., Liiv, M. and Sumberg, A. (1994) A comparative study of non-aromatic hydrocarbons from kukersite and *Dictyonema* Shale semicoking oils. *Oil Shale*, 11, 211–16.

Nagy, B., Meinschein, W. G. and Hennessy, D. J. (1961) Mass spectrometric analysis of the Orgueil meteorite: evidence for biogenic hydrocarbons. *Annals of the New York Academy of Sciences*, 93, 25–35.

Nance, H. S. (1998) Late Cretaceous paleogeography and patchwork coalbeds: Fruitland Formation, Northern San Juan Basin, Western Interior Seaway, USA. Presented at the Annual Meeting of the American Association of Petroleum Geologists, Salt Lake City, Utah, May 17–20, 1998.

Naraoka, H., Shimoyama, A. and Harada, K. (2000) Isotopic evidence from an Antarctic carbonaceous chondrite for two reaction pathways of extraterrestrial PAH formation. *Earth & Planetary Science Letters*, 184, 1–7.

Narbonne, G. M. (1998) The Ediacara biota: a terminal Neoproterozoic experiment in the evolution of life. *Geological Society of America Today*, 8, 1–6.

Narbonne, G. M., Kaufman, A. J. and Knoll, A. H. (1994) Integrated chemostratigraphy and biostratigraphy of the Windermere Supergroup, northwestern Canada: implications for Neoproterozoic correlations and the early evolution of animals. *Geological Society of America Bulletin*, 106, 1281–92.

Nascimento, L. R., Rebouças, L. M. C., Koike, L., *et al.* (1999) Acidic biomarkers from Albacora oils, Campos Basin, Brazil. *Organic Geochemistry*, 30, 1175–91.

National Research Council (1985) *Oil in the Sea: Input, Fates, and Effect*. National Academy Press, Washington, DC.

Navale, V. (1994) Comparative study of low and high temperature hydrous pyrolysis products of monoglyceryl diether lipid from archaebacteria. *Journal of Analytical and Applied Pyrolysis*, 29, 33–43.

Nazina, T. N. and Rozanova, E. P. (1978) Thermophillic sulfate-reducing bacteria from oil-bearing strata [in Russian]. *Mikrobiologiia*, 47, 142–8.

Nazina, T. N., Rozanova, E. P. and Kuznetsov, S. I. (1985) Microbial oil transformation processes accompanied by methane and hydrogen sulfide formation. *Geomicrobiology Journal*, 4, 103–30.

Nes, W. R. and McKean, M. L. (1977) *Biochemistry of Steroids and Other Isopentenoids*. University Park Press, Baltimore.

Neunlist, S. and Rohmer, M. (1985) Novel hopanoids from the methylotrophic bacteria *Methylococcus capsulatus* and *Methylomonas methanica*. (22S)-35-aminobacteriohopane-30,31,32,33,34-pentol and (22S)-35-amino-3β-methylbacteriohopane-30,31,32,33,34-pentol. *Biochemistry Journal*, 231, 635–9.

Newell, N. D. (1967) Revolution in the history of life. *Geological Society of America Special Paper*, 89, 63–91.

Newell, K. D. and Hatch, J. R. (2000) A petroleum system for the Salina Basin in Kansas based on organic geochemistry and geologic analog. *Natural Resources Research*, 9, 169.

Nichols, P. D., Mancuso, C. A. and White, D. C. (1987) Measurement of methanotroph and methanogen signature phospholipids for use in assessment of biomass and community structure in model systems. *Organic Geochemistry*, 11, 451–61.

Nichols, P. D., Volkman, J. K., Palmisano, A. C., Smith, G. A. and White, D. C. (1988) Occurrence of an isoprenoid C_{25} diunsaturated alkene and high neutral lipid content in Antarctic sea-ice diatom communities. *Journal of Phycology*, 24, 90–6.

Niklas, K. J. (1996) *The Evolutionary Biology of Plants*. University of Chicago Press, Chicago, IL.

Nixon, R. P. (1973) Oil source beds in Cretaceous Mowry Shale of Northwestern Interior United States. *American Association of Petroleum Geologists Bulletin*, 57, 136–61.

Noble, R. A. (1986) A geochemical study of bicyclic alkanes and diterpenoid hydrocarbons in crude oils, sediments, and coals. Ph. D. thesis, Department of Organic Chemistry, University of Western Australia, Perth, Australia.

Noble, R. A. and Alexander, R. (1989) Origin and significance of drimanes and related bicyclic alkanes in crude oils and ancient sediments. Presented at the 215th National Meeting of the American Chemical Society, Dallas, TX, April 9–14, 1989.

Noble, R. A. and Henk, F. H., Jr (1998) Hydrocarbon charge of a bacterial gas field by prolonged methanogenesis: an example from the East Java Sea, Indonesia. *Organic Geochemistry*, 29, 301–14.

Noble, R. A., Knox, J., Alexander, R. and Kagi, R. I. (1985a) Identification of tetracyclic diterpane hydrocarbons in Australian crude oils and sediments. *Journal of Chemical Society, Chemical Communications*, 32–3.

Noble, R. A., Alexander, R., Kagi, R. I. and Knox, J. (1985b) Tetracyclic diterpenoid hydrocarbons in some Australian coals, sediments and crude oils. *Geochimica et Cosmochimica Acta*, 49, 2141–7.

Noble, R., Alexander, R. and Kagi, R. I. (1985c) The occurrence of bisnorhopane, trisnorhopane, and 25-norhopanes as free hydrocarbons in some Australian shales. *Organic Geochemistry*, 8, 171–6.

Noble, R. A., Alexander, R., Kagi, R. I. and Knox, J. (1986) Identification of some diterpenoid hydrocarbons in petroleum. *Organic Geochemistry*, 10, 825–9.

Noble, R. A., Wu, C. H. and Atkinson, C. D. (1991) Petroleum generation and migration from Talang Akar coals and shales offshore N. W. Java, Indonesia. *Organic Geochemistry*, 17, 363–74.

Noble, R. A., Pratomo, K. H., Nugrahanto, K., et al. (1997) Petroleum systems of Northwest Java, Indonesia. In: *Proceedings of an International Conference on Petroleum Systems of SE Asia and Australasia* (J. V. C. Howes and R. A. Noble, eds.), Indonesian, Petroleum Association, Jakarta, Indonesia, pp. 585–600.

Norman, R. S., Frontera-Suau, R. and Morris, P. J. (2002) Variability in *Pseudomonas aeruginosa* lipopolysaccharide expression during crude oil degradation. *Applied and Environmental Microbiology*, 68, 5096–103.

Norvick, M. S. and Schaller, H. (1998) A three-phase Early Cretaceous rift history of the South Atlantic salt basins and its influence on lacustrine source facies distribution. Presented at the ABGP/AAPG International Conference and Exhibition, Rio de Janeiro, Brazil, November 8–11, 1998.

Nugrahanto, K. and Noble, R. A. (1997) Structural control on source rock development and thermal maturity in the Ardjuna Basin, offshore northwest Java, Indonesia. In: *Proceedings of an International Conference on Petroleum Systems of Southeast Asia and Australasia* (J. V. C. Howes and R. A. Noble, eds.), Indonesian Petroleum Association Jakarta, Indonesia, pp. 631–653.

Nytoft, H. P. and Bojesen-Koefoed, J. A. (2001) 17α,21α(H)-hopanes: natural and synthetic. *Organic Geochemistry*, 32, 841–56.

Nytoft, H. P., Bojesen-Koefoed, J. A., Christiansen, F. G. and Fowler, M. G. (1997) HPLC-separation of coeluting lupane and oleanane – confirmation of the presence of lupane in oils derived from terrigenous source rocks of Cretaceous-Tertiary age. In: *Abstracts from the 18th International Meeting on Organic Geochemistry, September 22–26, 1997, Maastricht, The Netherlands* (B. Horsfield, ed.), Forschungszentrom Jülich, Jülich, Germany, pp. 313–14.

Nytoft, H. P., Bojesen-Koefoed, J. A. and Christiansen, F. G. (2000) C_{26} and C_{28}–C_{34} 28-norhopanes in sediments and petroleum. *Organic Geochemistry*, 31, 25–39.

Nytoft, H. P., Bojesen-Koefoed, J. A., Christiansen, F. G. and Fowler, M. G. (2002) Oleanane or lupane? Reappraisal of the presence of oleanane in Cretaceous-Tertiary oils and sediments. *Organic Geochemistry*, 33, 1225–40.

Obermajer, M., Fowler, M. G. and Snowdon, L. R. (1999) Depositional environment and oil generation in Ordovician source rocks from southwestern Ontario, Canada: organic geochemical and petrological approach. *American Association of Petroleum Geologists Bulletin*, 83, 1426–53.

Obermajer, M., Fowler, M. G., Snowdon, L. R. and Macqueen, R. W. (2000a) Compositional variability of crude oils and source kerogen in the Silurian carbonate-evaporite sequences of the eastern Michigan Basin, Ontario, Canada. *Bulletin of Canadian Petroleum Geology*, 48, 307–22.

Obermajer, M., Osadetz, K. G., Fowler, M. G. and Snowdon, L. R. (2000b) Light hydrocarbon (gasoline range) parameter refinement of biomarker-based oil-oil correlation studies: an example from Williston Basin. *Organic Geochemistry*, 31, 959–76.

Ocampo, R., Callot, H. J., Albrecht, P. and Kintzinger, J. P. (1984) A novel chlorophyll c related petroporphyrin in oil shale. *Tetrahedron Letters*, 25, 2589–92.

Ocampo, R., Callot, H. J. and Albrecht, P. (1985a) Identification of polar porphyrins in oil shale. *Journal of the Chemical Society, Chemical Communications*, 198–200.

(1985b) Occurrence of bacteriopetroporphyrins in oil shale. *Journal of the Chemical Society, Chemical Communications*, 200–1.

Ocampo, R., Bauder, C., Callot, H. J. and Albrecht, P. (1992) Porphyrins from Messel oil shale (Eocene, Germany); structure elucidation, geochemical and biological significance, and distribution as a function of depth. *Geochimica et Cosmochimica Acta*, 56, 745–61.

Odden, W., Barth, T. and Talbot, M. R. (2002) Compound-specific carbon isotope analysis of natural and artificially generated hydrocarbons in source rocks and petroleum fluids from offshore Mid-Norway. *Organic Geochemistry*, 33, 47–65.

Oehler, J. H. (1984) Carbonate source rocks in the Jurassic Smackover trend of Mississippi, Alabama, and Florida. In: *Petroleum Geochemistry and Source Rock Potential of Carbonate Rocks* (J. G. Palacas, ed.), American Association of Petroleuum Geologists, Tulsa, OK, pp. 63–70.

Okui, A., Imayoshi, A. and Tauji, K. (1997) Petroleum system in the Khumer Trough, Cambodia. In: *Proceedings of the International Conference on Petroleum Systems of SE Asia and Australia* (J. V. C. Howes and R. A. Noble, eds.), Indonesian Petroleum Association, Jakarta, Indonesia, pp. 365–71.

Oldenburg, T. B. P., Wilkes, H., Horsfield, B., *et al.* (2002) Xanthones – novel aromatic oxygen-containing compounds in crude oils. *Organic Geochemistry*, 33, 595–609.

Olivares, C., Lorente, M. A. and Cassani, F. (1996) Geochemistry and organic facies of La Luna-Tres Esquinas cycle; maturity, biomarkers and kerogen issues. *American Association of Petroleum Geologists Bulletin*, 80, 1319.

Olsen, P. E. (1986) A 40-million-year lake record of early Mesozoic orbital climatic forcing. *Science*, 234, 842–8.

(1990) Tectonic, climatic, and biotic modulation of lacustrine ecosystems – examples from Newark Supergroup of Eastern North America. In: *Lacustrine Basin Exploration: Case Studies and Modern Analogs* (B. J. Katz, ed.), American Association of Petroleum Geologists, Tulsa, OK, pp. 209–24.

Olsen, P. E. and Kent, D. V. (1996) Milankovitch climate forcing in the tropics of Pangaea during the Late Triassic. *Palaeogeography, Palaeoclimatology, Palaeoecology*, 122, 1–26.

Olsen, P. E., Shubin, N. H. and Anders, M. H. (1987) New Early Jurassic tetrapod assemblages constrain Triassic-Jurassic tetrapod extinction event. *Science*, 237, 1025–9.

Olsen, P. E., Kent, D. V., Comet, B., Witte, W. K. and Schlische, R. W. (1996) High-resolution stratigraphy of the Newark Rift Basin (early Mesozoic, Eastern North

America). *Geological Society of American Bulletin*, 108, 40–77.

Orgel, L. E. (1973) *The Origins of Life: Molecules and Natural Selection*. MacMillan, London.

Ormiston, A. R. and Oglesby, R. J. (1995) Effect of Late Devonian paleoclimate on source rock quality and location. In: *Paleogeography, Paleoclimate, and Source Rocks* (A.-Y. Huc, ed.), American Association of Petroleum Geologists, Tulsa, OK, pp. 105–32.

Oró, J. and Lazcano, A. (1996) The origin of evolution of life. In: *Comets and the Origins of Life* (P. J. Thomas, C. P. McKay and C. F. Chyba, eds.), Springer-Verlag, Berlin, pp. 3–27.

Oró, J., Nooner, D. W., Zlatkis, A., Wikstrom, S. A. and Barghoorn, E. S. (1965) Hydrocarbons of biological origin in sediments about two billion years old. *Science*, 148, 77–9.

Orphan, V. J., House, C. H., Hinrichs, K.-U., McKeegan, K. D. and Delong, E. F. (2001) Methane-consuming archaea revealed by directly coupled isotopic and phylogenetic analysis. *Science*, 293, 479–81.

Orphan, V. J., House, C. H., Hinrichs, K.-U., McKeegan, K. D. and Delong, E. F. (2002) Multiple archaeal groups mediate methane oxidation in anoxic cold seep sediments. *Proceedings of the National Academy of Sciences, USA*, 99, 7663–8.

Orr, W. L. (1974) Changes in sulfur content and isotopic ratios of sulfur during petroleum maturation. Study of Big Horn Basin Paleozoic oils. *American Association of Petroleum Geologists Bulletin*, 58, 2295–318.

 (1986) Kerogen/asphaltene/sulfur relationships in sulfur-rich Monterey oils. *Organic Geochemistry*, 10, 499–516.

Osadetz, K. G. and Snowdon, L. R. (1995) Significant Paleozoic petroleum source rocks in the Canadian Williston Basin: their distribution, richness and thermal maturity (southeastern Saskatchewan and southwestern Manitoba). *Geological Survey of Canada Bulletin*, 487, 60.

Osadetz, K. G., Brooks, P. W. and Snowdon, L. R. (1992) Oil families and their sources in Canadian Williston Basin, (southeastern Saskatchewan and southwestern Manitoba). *Bulletin of Canadian Petroleum Geology*, 40, 254–73.

Osadetz, K. G., Snowdon, L. R. and Brooks, P. W. (1994) Oil families in Canadian Williston Basin, southwestern Saskatchewan. *Bulletin of Canadian Petroleum Geology*, 42, 155–77.

Otto, A., White, J. D. and Simoneit, B. R. T. (2002) Natural product terpenoids in Eocene and Miocene conifer fossils. *Science*, 297, 1543–5.

Ourisson, G. and Nakatani, Y. (1994) The terpenoid theory of the origin of cellular life: the evolution of terpanoids to cholesterols. *Chemistry and Biology*, 1, 11–23.

Ourisson, G., Albrecht, P. and Rohmer, M. (1979) The hopanoids. Palaeochemistry and biochemistry of a group of natural products. *Pure and Applied Chemistry*, 51, 709–29.

 (1982) Predictive microbial biochemistry – from molecular fossils to procaryotic membranes. *Trends in Biochemical Sciences*, 7, 236–9.

 (1984) The microbial origin of fossil fuels. *Scientific American*, 251, 44–51.

Ourisson, G., Rohmer, M. and Poralla, K. (1987) Prokaryotic hopanoids and other polyterpenoid sterol surrogates. *Annual Review of Microbiology*, 41, 301–33.

Øygard, K., Grahl-Hielsen, O. and Ulvøen, S. (1984) Oil/oil correlation by aid of chemometrics. *Organic Geochemistry*, 57, 561–7.

Padden, M., Weissert, M. and De Rafelis, M. (2001) Evidence for Late Jurassic release of methane from gas hydrate. *Geology*, 29, 223–6.

Padhy, P. K. (1997) Proterozoic petroleum systems of Indian sedimentary basins: a perspective. *American Association of Petroleum Geologists Bulletin*, 81, 1403.

Pairazian, V. V. (1993) Petroleum geochemistry of the Timano-Pechora Basin. *First Break*, 11, 279–86.

Pairazian, V. V. (1999) A review of the petroleum geochemistry of the Precaspian Basin. *Petroleum Geoscience*, 5, 361–9.

Palacas, J. G. (1984) Carbonate rocks as sources of petroleum: geological and chemical characteristics and oil-source correlations. In: *Proceedings of the Eleventh World Petroleum Congress 1983*, Vol. 2, John Wiley & Sons, Chichester, UK, pp. 31–43.

Palacas, J. G., Anders, D. E. and King, J. D. (1984) South Florida Basin – a prime example of carbonate source rocks in petroleum. In: *Petroleum Geochemistry and Source Rock Potential of Carbonate Rocks* (J. G. Palacas, ed.), American Association of Petroleum Geologists, Tulsa, OK, pp. 71–96.

Palacas, J. G., Monopolis, D., Nicolaou, C. A. and Anders, D. E. (1986) Geochemical correlation of surface and subsurface oils, western Greece. *Organic Geochemistry*, 10, 417–23.

Palfy, J. and Smith, P. L. (2000) Synchrony between Early Jurassic extinction, oceanic anoxic event, and the Karoo-Ferrar flood basalt volcanism. *Geology*, 28, 747–50.

Palmer, S. E. (1983) Porphyrin distributions in degraded and non-degraded oils from Colombia. Presented at the 186th National Convention of the American Chemical Society, August, 1983, Washington, DC.

(1984a) Hydrocarbon source potential of organic facies of the lacustrine Elko Formation (Eocene/Oligocene), Northeast Nevada. In: *Hydrocarbon Source Rocks of the Greater Rocky Mountain Region* (J. Woodward, F. F. Meissner and J. L. Clayton, eds.), Rocky Mountain Association of Geologists, Denver, CO, pp. 491–511.

(1984b) Effect of water washing on $C_{15}+$ hydrocarbon fraction of crude oils from northwest Palawan, Phillipines. *American Association of Petroleum Geologists Bulletin*, 68, 137–49.

(1993) Effect of biodegradation and water washing on crude oil composition. In: *Organic Geochemistry* (M. H. Engel and S. A. Macko, eds.), Plenum Press, New York, pp. 511–33.

Pancost, R. D., Freeman, K. H., Patzkowsky, M. E., Wavrek, D. A. and Collister, J. W. (1998) Molecular indicators of redox and marine photoautotroph composition in the late Middle Ordovician of Iowa, USA. *Organic Geochemistry*, 29, 1649–62.

Pancost, R. D., Freeman, K. H. and Patzkowsky, M. E. (1999) Organic-matter source variation and the expression of a late Middle Ordovician carbon isotope excursion. *Geology*, 27, 1015–18.

Pancost, R. D., Sinninghe Damsté, J. S., de Lint, S., *et al.* (2000a) Biomarker evidence for widespread anaerobic methane oxidation in Mediterranean sediments by a consortium of methanogenic archaea and bacteria. *Applied and Environmental Microbiology*, 66, 1126–32.

Pancost, R. D., van Geel, B., Baas, M. and Sinninghe Damsté, J. S. (2000b) $\delta^{13}C$ values and radiocarbon dates of microbial biomarker as tracers for carbon recycling in peat deposits. *Geology*, 28, 663–6.

Pancost, R. D., Bouloubassi, I., Aloisi, G., Sinninghe Damsté, J. S. and the Medinaut Shipboard Scientific Party (2001a) Three series of non-isoprenoidal dialkyl glycerol diethers in cold-seep carbonate crusts. *Organic Geochemistry*, 32, 695–707.

Pancost, R. D., Hopmans, E. C., Sinninghe Damsté, J. S. and the Medinaut Shipboard Scientific Party (2001b) Archaeal lipids in Mediterranean cold seeps: molecular proxies for anaerobic methane oxidation. *Geochimica et Cosmochimica Acta*, 65, 1611–27.

Pancost, R. D., Crawford, N. and Maxwell, J. R. (2002) Molecular evidence for basin-scale photic zone euxinia in the Permian Zechstein Sea. *Chemical Geology*, 188, 217–27.

Pande, A., Hazra, P. N., Singh, B. P., *et al.* (1993) Origin and evolutionary histories of crude oils of Cambay Basin through biomarker composition. In: *Proceedings of the Second Seminar on Petroliferous Basins of India* (S. K. Biswas, A. Dave, P. Garg, *et al.*, eds.), Indian Petroleum Publishers, Dehra Dun, India, pp. 137–60.

Pande, A., Singh, B. P., Khan, M. S. R. and Misra, K. N. (1997) Organic geochemical perspective of the origin of oil – a case study of Kadi Formation oils, North Cambay Basin, India. In: *Petrotech 97 Proceedings, January 9–12, New Delhi, India*, Vol.1, India Oil and Natural Gas Corporation Ltd, New Delhi, India, pp. 223–35.

Papanicolaou, C., Dehmer, J. and Fowler, M. (2000) Petrological and organic geochemical characteristics of coal samples from Florina, Lava, Moschopotamos and Kalavryta coal fields, Greece. *International Journal of Coal Geology*, 44, 267–92.

Paproth, E., Feist, R. and Flajs, G. (1991) Decision on the Devonian-Carboniferous boundary stratotype. *Episodes*, 14, 331–5.

Parkes, J. (1999) Cracking anaerobic bacteria. *Nature*, 401, 217–18.

Parkes, J. and Maxwell, J. R. (1993) Some like it hot (and oily). *Nature*, 365, 694–5.

Parkes, R. J., Wellsbury, P., Mather, I. D. and Maxwell, J. R. (2000) Possible deep biosphere energy sources: where does bacterial diagenesis end and thermal alteration begin? *EOS (Transactions AGU)*, 81, 236.

Parnaud, F., Gou, Y., Pacual, J.-C., *et al.* (1995) Petroleum geology of the central part of the Eastern Venezuelan Basin. In: *Petroleum Basins of Southern South America*. (A. J. Tankard, R. Suarez Soruco and H. J. Welsink, eds.), American Association of Petroleum Geologists, Tulsa, OK, pp. 741–56.

Parrish, J. T. (1982) Upwelling and petroleum source beds, with reference to Paleozoic. *American Association of Petroleum Geologists Bulletin*, 66, 750–74.

(1993) Jurassic climate and oceanography of the circum-Pacific region. In: *The Jurassic of the Circum-Pacific, International Geological Correlation Programme Project 171* (G. E. G. Westermann, ed.), Cambridge University Press, New York, pp. 365–79.

(1995) Paleogeography of Corg-rich rocks and the preservation versus production controversy. In: *Paleogeography, Paleoclimate, and Source Rocks* (A.-Y. Huc, ed.), American Association of Petroleum Geologists, Tulsa, OK, pp. 1–20.

Parsons, M. B., Azgaar, A. M. and Curry, J. J. (1980) Hydrocarbon occurrences in the Sirte Basin, Libya. In: *Canadian Society of Petroleum Geologists Memoir 6* (A. D. Miall, ed.), pp. 723–32.

Pasley, M. A., Riley, G. W. and Nummedal, D. (1993) Sequence stratigraphic significance of organic matter variations: example from the Upper Cretaceous Mancos Shale of the San Juan Basin, New Mexico. In: *Source Rocks in a Sequence Stratigraphic Framework* (B. J. Katz and L. M. Pratt, eds.), American Association of Petroleum Geologists, Tulsa, OK, pp. 221–41.

Passey, Q. R., Creaney, S., Kulla, J. B., Moretti, F. J. and Stroud, J.D (1990) A practical model for organic richness from porosity and resistivity logs. *American Association of Petroleum Geologists Bulletin*, 74, 1777–94.

Paterson, D. W., Bachtiar, A., Bates, J. A., Moon, J. A. and Surdam, R. C. (1997) Petroleum system of the Kutei Basin, Kalimantan, Indonesia. In: *Petroleum Systems of SE Asia and Australasia* (J. V. C. Howes and R. A. Noble, eds.), Indonesian Petroleum Association, Jakarta, Indonesia, pp. 709–26.

Patience, R. L., Rowland, S. J. and Maxwell, J. R. (1978) The effect of maturation on the configuration of pristane in sediments and petroleum. *Geochimica et Cosmochimica Acta*, 42, 1871–6.

Paton, I. M. (1986) The Birkhead Formation – a Jurassic petroleum reservoir. In: *Contributions to the Geology and Hydrocarbon Potential of the Eromanga Basin* (D. I. Gravestock, P. S. Moore and G. M. Pitt, eds.), Geological Society of Australia, Sydney, Australia, pp. 195–201.

Pauken, R. J. (1992) Sanaga Sud Field, offshore Cameroon, West Africa. In: *Giant Oil and Gas Fields of the Decade 1978–1988* (M. T. Halbouty, ed.), American Association of Petroleum Geologists, Tulsa, OK, pp. 217–30.

Paull, R., Michaelsen, B. H. and McKirdy, D. M. (1996) GC-MS-MS characterisation of fernenes and fernanes in *Dicroidium*-bearing mudstones and coals. Presented at the Australian Organic Geochemistry Conference, October 2–4, 1996, Fremantle, Australia.

(1998) Fernenes and other triterpenoid hydrocarbons in *Dicroidium*-bearing Triassic mudstones and coals from South Australia. *Organic Geochemistry*, 29, 1331–43.

Pauly, G. G. and van Vleet, E. S. (1986a) Archaebacterial ether lipids: natural tracers of biogeochemical processes. *Organic Geochemistry*, 10, 859–67.

(1986b) Acyclic archaebacterial ether lipids in swamp sediments. *Geochimica et Cosmochimica Acta*, 50, 1117–25.

Peabody, C. E. (1993) The association of cinnabar and bitumen in mercury deposits of the California Coast Ranges. In: *Bitumens in Ore Deposits* (J. Parnell, H. Kucha and P. Landais, eds.), Springer-Verlag, New York, pp. 178–209.

Peakman, T. M., ten Haven, H. L., Rechka, J. R., de Leeuw, J. W. and Maxwell, J. R. (1989) Occurrence of (20R)- and (20S)-$\Delta 8(^{14})$ and Δ^{14} 5α(H)-sterenes and the origin of 5α(H),14β(H),17β(H)-steranes in an immature sediment. *Geochimica et Cosmochimica Acta*, 53, 2001–9.

Peakman, T. M., ten Haven, H. L., Rullkötter, J. and Curiale, J. A. (1991) Characterisation of 24-nor-triterpenoids occurring in sediment and crude oils by comparison with synthesized standards. *Tetrahedron*, 47, 3779–86.

Pearson, M. J. and Obaje, N. G. (1999) Onocerane and other triterpenoids in Late Cretaceous sediments from the Upper Benue Trough, Nigeria: tectonic and palaeoenvironmental implications. *Organic Geochemistry*, 30, 583–92.

Pearson, A., McNichol, A. P., Benitez-Nelson, B. C., Hayes, J. M. and Eglinton, T. I. (2001) Origins of lipid biomarkers in Santa Monica Basin surface sediment: a case study using compound-specific δ^{14}C analysis. *Geochimica et Cosmochimica Acta*, 65, 3123–37.

Peng, P. A., Sheng, D., Fu, J. and Yan, Y. (1998) Biological markers in 1.7 billion year old rock from the Tuanshanzi Formation, Jixian strata section, North China. *Organic Geochemistry*, 29, 1321–9.

Penteado, H. L. D. B. and Behar, F. (2000) Geochemical characterization and compositional evolution of the Gomo Member source rocks in the Recôncavo Basin, Brazil. In: *Petroleum Systems of South Atlantic Margins* (M. R. Mello and B. J. Katz, ed.), American Association of Petroleum Geologists, Tulsa, OK, pp. 179–94.

Perez-Infante, J., Farrimond, P. and Furrer, M. (1996) Global and local controls influencing the deposition of the La Luna Formation (Cenomanian-Campanian), western Venezuela. *Chemical Geology*, 130, 271–88.

Perez-Teller, G. (1995) La Luna Formation of Colombia: lithofacies, organic facies and generation capacity. In: *Proceedings of the VI Colombian Petroleum Congress,*

Colombian Association of Petroleum Engineers, Bogotá, Colombia, pp. 135–44.

Perkins, G. M., Bull, I. D., ten Haven, H. L., et al. (1995) First positive identification of triterpenes of the taraxastane family in petroleums and oil shales: 19α(H)-taraxastane and 24-nor-19α(H)-taraxastane. Evidence for a previously unrecognised diagenetic alteration pathway of lup-20(29)-ene derivatives. In: *Selected Papers from the 17th International Meeting on Organic Geochemistry* (J. O. Grimalt and C. Dorronsoro, eds.), AIGOA, San Sebastián, Spain, pp. 247–9.

Perrodon, A. (1980) *Géodynamique pétrolière. Genèse et répartition des gisements d'hydrocarbures.* Masson–Elf-Aquitaine, Paris.

(1983) Dynamics of oil and gas accumulations. In: *Bulletin des Centres de Resserches Exploration-Production Elf Aquitaine, Memoir 5*. Elf Aquitaine, Paris.

(1992) Petroleum systems: models and applications. *Journal of Petroleum Geology*, 15, 319–326.

Persad, K., Talukdar, S. and Dow, W. G. (1993) Tectonic control in source rock maturation and oil migration in Trinidad and implications for petroleum exploration. In: *Mesozoic and Early Cenozoic Development of the Gulf of Mexico and Caribbean Region: A Context for Hydrocarbon Exploration* (J. L. Pindell and R. F. Perkins, eds.), GCSSEPM foundation, Houston, TX, pp. 237–49.

Peters, K. E. (1986) Guidelines for evaluating petroleum source rock using programmed pyrolysis. *American Association of Petroleum Geologists Bulletin*, 70, 318–29.

(1999) Biomarkers: assessment of thermal maturity. In: *Encyclopedia of Geochemistry* (C. P. Marshall and R. W. Fairbridge, eds.), Kluwer Academic Publishers, Boston, MA, pp. 36–9.

(2000) Petroleum tricyclic terpanes: predicted physicochemical behavior from molecular mechanics calculations. *Organic Geochemistry*, 31, 497–507.

Peters, K. E. and Cassa, M. R. (1994) Applied source rock geochemistry. In: *The Petroleum System – From Source to Trap* (L. B. Magoon and W. G. Dow, eds.), American Association of Petroleum Geologists, Tulsa, OK, pp. 93–117.

Peters, K. E. and Creaney, S. (2004) Geochemical differentiation of Silurian and Devonian oils from Algeria. *Geochemical Investigations in Earth and Space Science: A Tribute to Isaac R. Kaplan* (R. J. Hill, J. Leventhal, Z. Aizenshtat, et al., eds.), Geological Society of America, Boulder, CO, pp. 287–301.

Peters, K. E. and Fowler, M. G. (2002) Applications of petroleum geochemistry to exploration and reservoir management. *Organic Geochemistry*, 33, 5–36.

Peters, K. E. and Moldowan, J. M. (1991) Effects of source, thermal maturity, and biodegradation on the distribution and isomerization of homohopanes in petroleum. *Organic Geochemistry*, 17, 47–61.

(1993) *The Biomarker Guide. Interpreting Molecular Fossils in Petroleum and Ancient Sediments*. Prentice-Hall, Englewood Cliffs, NJ.

Peters, K. E. and Nelson, D. A. (1992) REESA – an expert system for geochemical logging of wells. *American Association of Petroleum Geologists Annual Meeting Abstracts*, 103.

Peters, K. E., Moldowan, J. M., Schoell, M. and Hempkins, W. B. (1986) Petroleum isotopic and biomarker composition related to source rock organic matter and depositional environment. *Organic Geochemistry*, 10, 17–27.

Peters, K. E., Moldowan, J. M., Driscole, A. R. and Demaison, G. J. (1989) Origin of Beatrice oil by cosourcing from Devonian and Middle Jurassic source rocks, Inner Moray Firth, UK. *American Association of Petroleum Geologists Bulletin*, 73, 454–71.

Peters, K. E., Moldowan, J. M. and Sundararaman, P. (1990) Effects of hydrous pyrolysis on biomarker thermal maturity parameters: Monterey Phosphatic and Siliceous Members. *Organic Geochemistry*, 15, 249–65.

Peters, K. E., Kontorovich, A. E., Moldowan, J. M., et al. (1993) Geochemistry of selected oils and rocks from the central portion of the West Siberian Basin, Russia. *American Association of Petroleum Geologists Bulletin*, 77, 863–87.

Peters, K. E., Kontorovich, A. E., Huizinga, B. J., Moldowan, J. M. and Lee, C. Y. (1994) Multiple oil families in the West Siberian Basin. *American Association of Petroleum Geologists Bulletin*, 78, 893–909.

Peters, K. E., Clark, M. E., das Gupta, U., McCaffrey, M. A. and Lee, C. Y. (1995) Recognition of an Infracambrian source rock based on biomarkers in the Bagewhala-1 oil, India. *American Association of Petroleum Geologists Bulletin*, 79, 1481–94.

Peters, K. E., Cunningham, A. E., Walters, C. C., Jiang, J. and Fan, Z. (1996a) Petroleum systems in the Jiangling-Dangyang area, Jianghan Basin, China. *Organic Geochemistry*, 24, 1035–60.

Peters, K. E., Moldowan, J. M., McCaffrey, M. A. and Fago, F. J. (1996b) Selective biodegradation of extended hopanes to 25-norhopanes in petroleum reservoirs. Insights from molecular mechanics. *Organic Geochemistry*, 24, 765–83.

Peters, K. E., Wagner, J. B., Carpenter, D. G. and Conrad, K. T. (1997a) World class Devonian potential seen in eastern Madre de Dios Basin. *Oil and Gas Journal*, 95, 61–5, 84–7.

Peters, K. E., Woods, M., Spencer, G., *et al.* (1997b) Prediction of source-rock organic facies variations using advanced geochemistry: northeastern Sakhalin Island, Russia. *American Association of Petroleum Geologists Bulletin*, 81, 1403.

Peters, K. E., Fraser, T. H., Amris, W., Rustanto, B. and Hermanto, E. (1999a) Geochemistry of crude oils from eastern Indonesia. *American Association of Petroleum Geologists Bulletin*, 83, 1927–42.

Peters, K. E., Clutson, M. J. and Robertson, G. (1999b) Mixed marine and lacustrine input to an oil-cemented sandstone breccia from Brora, Scotland. *Organic Geochemistry*, 30, 237–48.

Peters, K. E., Snedden, J. W., Sulaeman, A., Sarg, J. F. and Enrico, R. J. (2000) A new geochemical-stratigraphic model for the Mahakam Delta and Makassar slope, Kalimantan, Indonesia. *American Association of Petroleum Geologists Bulletin*, 84, 12–44.

Peters, K. E., Scheurman, G. L., Lee, C. Y., *et al.* (1992) Effects of refinery processes on biological markers. *Energy and Fuels*, 6, 560–77.

Petersen, H. I., Andersen, C., Anh, P. H., *et al.* (2001) Petroleum potential of Oligocene lacustrine mudstones and coals at Dong Ho, Vietnam – an outcrop analogue to terrestrial source rocks in the greater Song Hong Basin. *Journal of Asian Earth Sciences*, 19, 135–54.

Petrov, A. A. (1987) *Petroleum Hydrocarbons*. Springer-Verlag, New York.

Petrov, A. A., Tzedilina, A. L., Pustil'Nikova, S. D., *et al.* (1973) Isoprenoid hydrocarbons in oils. *Neftekimiya*, 13, 779–85.

Petrov, A. A., Pustil'Nikova, S. D., Abriutina, N. N. and Kagramonova, G. R. (1976) Petroleum steranes and triterpanes. *Neftekhimiia*, 16, 411–27.

Petrov, A. A., Vorobyova, N. S. and Zemskova, Z. K. (1990) Isoprenoid alkanes with irregular "head-to-head" linkages. *Organic Geochemistry*, 16, 1001–5.

Petsch, S. T., Eglinton, T. I. and Edwards, K. J. (2001) ^{14}C-dead living biomass: evidence for microbial assimilation of ancient organic carbon during shale weathering. *Science*, 292, 1127–31.

Pflug, H. D. and Jaeschke-Boyer, H. (1979) Combined structural and chemical analysis of 3,800 Myr-old microfossils. *Nature*, 280, 483–6.

Phelps, C. D., Kerkhof, L. J. and Young, J. L. (1998) Molecular characterization of a sulfate-reducing consortium which mineralizes benzene. *FEMS Microbiology Ecology*, 27, 269–79.

Phillips, S., Laird, L., Michael, E. and Odell, V. (1997) Sequence stratigraphy of Tertiary petroleum systems in the West Natuna Basin, Indonesia. In: *Proceedings of the International Conference on Petroleum Systems of SE Asia and Australia* (J. V. C. Howes and R. A. Noble, eds.), Indonesian Petroleum Association, Jakarta, Indonesia, pp. 381–90.

Philp, R. P. (1982) Application of computerized gas chromatography/mass spectrometry to fossil fuel research. *Spectra (Finnigan MAT)*, 8, 6–31.

(1983) Correlation of crude oils from the San Jorges Basin, Argentina. *Geochimica et Cosmochimica Acta*, 47, 267–75.

(1985) *Fossil Fuel Biomarkers*. Elsevier, New York.

Philp, R. P. and Brassell, S. (1986) Arguments against abiogenic origin for hydrocarbons. *Chemical and Engineering News*, 64, 2–3, 48, 59.

Philp, R. P. and Gilbert, T. D. (1982) Unusual distribution of biological markers in Australian crude oil. *Nature*, 299, 245–7.

(1986) Biomarker distributions in Australian oils predominantly derived from terrigenous source material. *Organic Geochemistry*, 10, 73–84.

Philp, R. P., Gilbert, T. D. and Friedrich, J. (1981) Bicyclic sesquiterpenoids and diterpenoids in Australian crude oils. *Geochimica et Cosmochimica Acta*, 45, 1173–80.

Philp, R. P., Simoneit, B. R. T. and Gilbert, T. D. (1983) Diterpenoids in crude oils and coals of South Eastern Australia. In: *Advances in Organic Geochemistry 1981* (M. Bjorøy, C. Albrecht, C. Cornford, et al., eds.), John Wiley & Sons, New York, pp. 698–704.

Philp, R. P., Bakel, A., Galvez-Sinibaldi, A. and Lin, L. H. (1988b) A comparison of organosulphur compounds produced by pyrolysis of asphaltenes and those present in related crude oils and tar sands. *Organic Geochemistry*, 13, 915–26.

Philp, R. P., Li, J. and Lewis, C. A. (1989) An organic geochemical investigation of crude oils from Shanganning, Jianghan, Chaidamu and Zhungeer basins, People's Republic of China. *Organic Geochemistry*, 14, 447–60.

Picha, F. J. and Peters, K. E. (1998) Biomarker oil-to-source rock correlation in the Western Carpathians and their foreland, Czech Republic. *Petroleum Geoscience*, 4, 289–302.

Pierce, R. W. and Turner, J. T. (1992) Ecology of planktonic ciliates in marine food webs. *Reviews in Aquatic Science*, 6, 139–81.

Pindell, J. L. and Tabbutt, K. D. (1995) Mesozoic-Cenozoic Andean paleogeography and regional controls on hydrocarbon systems. In: *Petroleum Basins of Southern South America*. (A. J. Tankard, R. Suarez Soruco and H. J. Welsink, eds.), American Association of Petroleum Geologists, pp. 101–28.

Pirnik, M. P., Atlas, R. and Bartha, R. (1974) Hydrocarbon metabolism by *Brevibacterium erythrogenes*: normal and branched alkanes. *Journal of Bacteriology*, 119, 868–78.

Pisciotto, K. A. and Garrison, R. E. (1981) Lithofacies and depositional environments of the Monterey Formation. In: *The Monterey Formation and Related Siliceous Rocks of California* (R. W. Garrison et al., eds.), Society of Economic Paleontology and Mineralogy, Tulsa, OK, pp. 97–122.

Pitt, G. M. (1986) Geothermal gradients, geothermal histories and the timing of thermal maturation in the Eromanga-Cooper basins. In: *Contributions to the Geology and Hydrocarbon Potential of the Eromanga Basin* (D. I. Gravestock, P. S. Moore and G. M. Pitt, eds.), Geological Society of Australia, Sydney, Australia, pp. 323–51.

Pizzarello, S., Huang, Y., Becker, L., *et al.* (2001) The organic content of the Tagish Lake meteorite. *Science*, 293, 2236–9.

Plummer, P. S. (1992) Geochemical analyses may indicate oil kitchen near Seychelles Bank. *Oil and Gas Journal*, 90, 52–4.

(1996) Origin of beach-stranded tars from source rocks indigenous to Seychelles. *American Association of Petroleum Geologists Bulletin*, 80, 323–39.

Plummer, P. S., Joseph, P. R. and Samson, P. J. (1998) Depositional environments and oil potential of Jurassic/Cretaceous source rocks within the Seychelles microcontinent. *Marine and Petroleum Geology*, 15, 385–401.

Poag, C. W. (1996) Structural outer rim of Chesapeake Bay impact crater: seismic and bore hole evidence. *Meteoritics*, 31, 218–26.

(2002) Biospheric consequences of the Chesapeake Bay bolide impact. Presented at the Annual Meeting of the Geological Society of America, October 27–30, 2002, Denver, CO.

Poinsot, J., Adam, P., Trendel, J. M., Connan, J. and Albrecht, P. (1995) Diagenesis of higher plant triterpenes in evaporitic sediments. *Geochimica et Cosmochimica Acta*, 59, 4653–61.

Poinsot, J., Schneckenburger, P., Adam, P., *et al.* (1998) Novel polyprenoid sulfides derived from regular polyprenoids in sediments: characterization, distribution, and geochemical signficance. *Geochimica et Cosmochimica Acta*, 62, 805–14.

Pollastro, R. M. (1999) Ghaba Salt Basin Province and Fahud Salt Basin Province, Oman – geological overview and total petroleum systems. In: *U.S. Geological Survey Bulletin*, 2167, 41.

Poole, F. G. and Claypool, G. E. (1984) Petroleum source-rock potential and crude-oil correlation in the Great Basin. In: *Hydrocarbon Source Rocks of the Greater Rocky Mountain Region* (J. Woodward, F. F. Meissner and J. L. Clayton, eds.), Rocky Mountain Association of Geologists, Denver, CO, pp. 491–511.

Popovich, T. A. and Kravchenko, T. I. (1995) Genetic characteristics of hydrocarbon composition of oils of north Sakhalin oil-gas basin. *Petroleum Geology*, 29, 309–13 (transl. from *Geologiya Nefti i Gaza*, 1995 1, 40–4.

Poreda, R. J. and Becker, L. (2003) Fullerenes and interplanetary dust at the Permian-Triassic boundary. *Astrobiology*, 3, 75–90.

Poulton, T. P., Christopher, J. E., Hayes, B. J. R., *et al.* (1994) Jurassic and lowermost Cretaceous strata of the Western Canada Sedimentary Basin. In: *Geological Atlas of the Western Canada Sedimentary Basin* (G. D. Mossop and I. Shetsen, eds.), Canadian Society of Petroleum Geologists and Alberta Research Council, Calgary, Alberta, pp. 297–316.

Powell, T. G. (1984) Some aspects of the hydrocarbon geochemistry of a Middle Devonian barrier-reef complex, Western Canada. In: *Petroleum Geochemistry and Source Rock Potential of Carbonate Rocks* (J. G. Palacas, ed.), American Association of Petroleum Geologists, Tulsa, OK, pp. 45–61.

Powell, T. G. and McKirdy, D. M. (1973) Relationship between ratio of pristane to phytane, crude oil composition and geological environment in Australia. *Nature*, 243, 37–9.

Powell, T. G., MacQueen, R. W., Barker, F. J. and Bree, D. G. (1984) Geochemical character and origin of

Ontario oils. *Bulletin of Canadian Petroleum Geology*, 32, 299–312.

Powell, T. G., Boreham, C. J. and McKirdy, D. H. (1986) Nature of source material for non-marine oils in Australia. Presented at the 12th International Sedimentological Congress, 1986, Canberra, Australia.

Powell, T. G., Boreham, C. J., McKirdy, D. M., Michaelsen, B. H. and Summons, R. E. (1989) Petroleum geochemistry of the Murta Member, Mooga Formation and associated oils, Eromanga Basin. *APEA Journal*, 29, 114–29.

Pratt, L. M. and Burruss R. C. (1988) Evidence for petroleum generation and migration in the Hartford and Newark basins. In: *Studies of the Early Mesozoic Basins of the Eastern United States* (A. J. Froelich and G. R. Robinson, Jr, eds.), U.S. Geological Survey Bulletin 1776, U.S. Geological Survey, Washington, DC, pp. 74–9.

Pratt, L. M., Vuletich, A. K. and Daws, T. A. (1985) Geochemical and isotopic characterization of organic matter in rocks of the Newark Supergroup. In: *Proceedings of the Second U.S. Geological Survey Workshop on the Early Mesozoic Basins in the Eastern United States* (G. R. Robinson, Jr and A. J. Froelich, eds.), U.S. Geological Survey Circular 946, U.S. Geological Survey, Washington, DC, pp. 74–8.

Pratt, L. M., Vuletich, A. K. and Shaw, C. A. (1986) *Preliminary Results of Organic Geochemical and Stable Isotope Analyses of Newark Supergroup Rocks in the Hartford and Newark Basins, Eastern US.* Geological Survey Open-File Report 86–284.

Pratt, L. M., Shaw, C. A. and Burruss, R. C. (1988) Thermal histories of the Hartford and Newark basins inferred from maturation indices of organic matter. In: *Studies of the Early Mesozoic Basins of the Eastern United States* (A. J. Froelich and G. R. Robinson Jr, eds.), U.S. Geological Survey Bulletin 1776, U.S. Geological Survey, Washington, DC, pp. 58–63.

Pratt, L. M., Summons, R. E. and Hieshima, G. B. (1991) Sterane and triterpane biomarkers in the Precambrian Nonesuch Formation, North American Midcontinent Rift. *Geochimica et Cosmochimica Acta*, 55, 911–16.

Premuzic, E. T., Gaffney, J. S. and Manowitz, B. (1986) The importance of sulfur isotope ratios in the differentiation of Prudhoe Bay crude oils. *Journal of Geochemical Exploration*, 26, 151–9.

Price, L. C., Ging, T., Daws, T., *et al.* (1984) Organic metamorphism in the Mississippian-Devonian Bakken Shale, North Dakota portion of the Williston Basin. In: *Hydrocarbon Source Rocks of the Greater Rocky Mountain Region* (J. Woodward, F. F. Meissner and J. L. Clayton, eds.), Rocky Mountain Association of Geologists, Denver, CO. pp. 83–134.

Price, P. L., O'Sullivan, T. and Alexander, R. (1987) The nature and occurrence of oil in Seram, Indonesia. In: *Proceedings of the Indonesian Petroleum Association. Sixteenth Annual Convention*, Vol. 1, Indonesian Petroleum Association, Jakarta, Indonesia, pp. 141–73.

Price, K. L., Huntoon, J. E. and McDowell, S. D. (1996) Thermal history of the 1.1-Ga Nonesuch Formation, North American Mid-Continent Rift, White Pine, Michigan. *American Association of Petroleum Geologists Bulletin*, 80, 1–15.

Price, L. C., Pawlewicz, M. and Daws, T. (1999) *Organic Metamorphism in the California Petroleum Basins Chapter A – Rock-Eval and Vitrinite Reflectance.* U.S. Department of the Interior, U.S. Geological Survey, Denver, CO.

Prince, R. C. (2002) Petroleum and other hydrocarbons, biodegradation of. In: *Encyclopedia of Environmental Microbiology* (G. Bitton, ed.), John Wiley & Sons, New York, pp. 2402–16.

Prince, R. C., Elmendorf, D. L., Lute, J. R., *et al.* (1994) $17\alpha(H),21\beta(H)$-hopane as a conserved internal standard for estimating the biodegradation of crude oil. *Environmental Science & Technology*, 28, 142–5.

Priscu, J. C., Adams, E. E., Lyons, W. B., *et al.* (1999) Geomicrobiology of subglacial ice above Lake Vostok, Antarctica. *Science*, 286, 2141–4.

Prommer, H., Davis, G. B. and Barry, D. A. (1999) Geochemical changes during biodegradation of petroleum hydrocarbons: field investigations and biogeochemical modeling. *Organic Geochemistry*, 30, 423–35.

Prowse, W. G. and Maxwell, J. R. (1989) Novel polar sedimentary porphyrins. *Geochimica et Cosmochimica Acta*, 53, 3081–3.

Punati, S. R. (1999) Palaeoproterozoic supercontinental fragmentation and amalgamation: evidence from Indian and Australian continents. *Journal of Conference Abstracts*, 4, 118–19.

Püttmann, W. and Villar, H. (1987) Occurrence and geochemical significance of 1,2,5,6-tetramethylnaphthalene. *Geochimica et Cosmochimica Acta*, 51, 3023–9.

Qiu, Y.-L. and Palmer, J. D. (1999) Phylogeny of early land plants: insights from genes and genomes. *Trends in Plant Science*, 4, 26–30.

Qui, Y.-L., Cho, Y., Cox, J. C. and Palmer, J. D. (1998) The gain of three mitochondrial introns identifies liverworts as the earliest land plants. *Nature*, 394, 671–4.

Qiu, Y.-L., Lee, J., Bernasconi-quadroni, F., *et al.* (1999) The earliest angiosperms: evidence from mitochondrial, plastid and nuclear genomes. *Nature*, 402, 404–7.

Quirke, J. M. E., Shaw, G. J., Soper, P. D. and Maxwell, J. R. (1980) Petroporphyrins – II. The presence of porphyrins with extended alkyl substituents. *Tetrahedron*, 36, 3261–7.

Quirke, J. M. E., Cuesta, L. L., Yost, R. A., Johnson, J. and Britton, E. D. (1989) Studies on high carbon number geoporphyrins by tandem mass spectrometry. *Organic Geochemistry*, 14, 43–50.

Rabus, R., Nrdhaus, R., Ludwig, W. and Widdel, F. (1993) Complete oxidation of toluene under strictly anoxic conditions by a new sulfate-reducing bacterium. *Applied and Environmental Microbiology*, 59, 1444–51.

Radke, M. (1987) Organic geochemistry of aromatic hydrocarbons. In: *Advances in Petroleum Geochemistry* (J. Brooks and D. Welte, eds.), Academic Press, New York, pp. 141–207.

  (1988) Application of aromatic compounds as maturity indicators in source rocks and crude oils. *Marine Petroleum Geology*, 5, 224–36.

Radke, M. and Welte, D. H. (1983) The methylphenanthrene index (MPI). A maturity parameter based on aromatic hydrocarbons. In: *Advances in Organic Geochemistry 1981* (M. Bjorøy, C. Albrecht, C. Cornford, *et al.*, eds.), John Wiley & Sons, New York, pp. 504–12.

Radke, M., Schaefer, R. G., Leythaeuser, D. and Teichmüller, M. (1980) Composition of soluble organic matter in coals: relation to rank and liptinite fluorescence. *Geochimica et Cosmochimica Acta*, 44, 1787–800.

Radke, M., Welte, D. H. and Willsch, H. (1982a) Geochemical study on a well in the Western Canada Basin: relation of the aromatic distribution pattern to maturity of organic matter. *Geochimica et Cosmochimica Acta*, 46, 1–10.

Radke, M., Willsch, H. and Leythaeuser, D. (1982b) Aromatic components of coal: relation of distribution pattern to rank. *Geochimica et Cosmochimica Acta*, 46, 1831–48.

Radke, M., Leythaeuser, D. and Teichmüller, M. (1984) Relationship between rank and composition of aromatic hydrocarbons for coal of different origins. *Organic Geochemistry*, 6, 423–30.

Radke, M., Welte, D. H. and Willsch, H. (1986) Maturity parameters based on aromatic hydrocarbons: influence of the organic matter type. *Organic Geochemistry*, 10, 51–63.

Radke, M., Garrigues, P. and Willsch, H. (1990) Methylated dicyclic and tricyclic aromatic hydrocarbons in crude oils from the Handil Field, Indonesia. *Organic Geochemistry*, 15, 17–34.

Raederstorff, D. and Rohmer, M. (1984) Sterols of the unicellular algae *Nematochrysopsis roscoffensis* and *Chrysotila lamellosa*: isolation of (24E)-24-*n*-propylidenecholesterol and 24-*n*-proplycholesterol. *Phytochemistry*, 23, 2835–8.

Ramón, J. C. and Dzou, L. I. (1999) Petroleum geochemistry of Middle Magdalena Valley, Colombia. *Organic Geochemistry*, 30, 249–66.

Ramón, J. C., Hughes, W. B., Dzou, L. I. and Holba, A. G. (2001) Evolution of the Cretaceous organic facies in Colombia: implications for oil composition. *Journal of South American Earth Science*, 14, 31–50.

Rampino, M. R. and Strothers, R. B. (1984) Terrestrial mass extinctions, cometary impacts and the Sun's motion perpendicular to the galactic plane. *Nature*, 308, 709–12.

Rangel, A., Parra, P. and Niño, C. (2000) The La Luna Formation: chemostratigraphy and organic facies in the Middle Magdalena Basin. *Organic Geochemistry*, 31, 1267–84.

Rangel, A., Moldowan, J. M., NiñO, C., Parra, P. and Giraldo, B. N. (2002) Umir Formation: organic geochemical and stratigraphic assessment as cosource for Middle Magdalena Basin oil, Colombia. *American Association of Petroleum Geologists Bulletin*, 86, 2069–87.

Rasmussen, B., Bengtson, S., Fletcher, I. R. and McNaughton, N. J. (2002) Discoidal impressions and trace-like fossils more than 1200 million years old. *Science*, 296, 1112–15.

Raup, D. M. (1992) Large-body impact and extinction in the Phanerozoic. *Paleobiology*, 18, 80–8.

Raup, D. M. and Sepkoski, J. J., Jr (1984) Periodicity of extinctions in the geologic past. *Proceedings of the National Academy of Science, USA*, 81, 801–5.

  (1986) Periodic extinction of families and genera. *Science*, 231, 833–6.

Reber, J. J. (1989) Biomarker characterization of Woodford-type oil and correlation to source rock, Aylesworth Field, Marshall County, Oklahoma. In: *Anadarko Basin Symposium, 1988* (K. S. Johnson, ed.), University of Oklahoma, Norman, OK, p. 271.

Reed, W. E. (1977) Molecular compositions of weathered petroleum and comparison with its possible source. *Geochimica et Cosmochimica Acta*, 41, 237–47.

Reed, J. D., Illich, H. A. and Horsfield, B. (1986) Biochemical evolutionary significance of Ordovician oils and their sources. *Organic Geochemistry*, 10, 347–58.

Regina, M., Loureiro, B. and Cardoso, J. N. (1990) Aromatic hydrocarbons in the Paraiba Valley oil shale. *Organic Geochemistry*, 15, 351–9.

Reichow, M. K., Saunders, A. D., White, R. V., *et al.* (2002) $^{40}Ar/^{39}Ar$ dates from the West Siberian Basin: Siberian Flood Basalt Province doubled. *Science*, 296, 1848–9.

Reid, R. P., Visscher, P. T., Decho, A. W., *et al.* (2000) The role of microbes in accretion, lamination and early lithification of modern marine stromatolites. *Nature*, 406, 989–92.

Remy, W., Gensel, P. G. and Hass, H. (1993) The gametophye generation of some early Devonian land plants. *International Journal of Plant Sciences*, 154, 35–58.

Ren, D. (1998) Flower-associated *Brachycera* flies as fossil evidence for Jurassic angiosperm origins. *Science*, 280, 85–8.

Renne, P. R. and Basu, A. R. (1991) Rapid eruption of the Siberian Traps flood basalts at the Permo-Triassic boundary. *Science*, 253, 176–9.

Renne, P. R., Ernesto, M., Pacca, I. G., *et al.* (1992) The age of Parana flood volcanism, rifting of Gondwanaland, and the Jurassic-Cretaceous boundary. *Science*, 258, 975–9.

Repeta, D. J. (1989) Carotenoid diagenesis in recent marine sediments – II. Degradation of fucoxanthin to loliolide. *Geochimica et Cosmochimica Acta*, 53, 699–707.

Repeta, D. J. and Gagosian, R. B. (1987) Carotenoid diagenesis in recent marine sediments – I. The Peru contenental shelf (15°S, 75°W). *Geochimica et Cosmochimica Acta*, 51, 1001–9.

Requejo, A. G. (1992) Quantitative analysis of triterpane and sterane biomarkers: methodology and applications in molecular maturity studies. In: *Biological Markers in Sediments and Petroleum* (J. M. Moldowan, P. Albrecht and R. P. Philp, eds.), Prentice-Hall, Englewood Cliffs, NJ, pp. 222–40.

(1994) Maturation of petroleum source rocks – II. Quantitative changes in extractable hydrocarbon content and composition associated with hydrocarbon generation. *Organic Geochemistry*, 21, 91–105.

Requejo, A. G. and Halpern, H. I. (1989) An unusual hopane biodegradation sequence in tar sands from the Point Arena (Monterey) Formation. *Nature*, 342, 670–3.

Requejo, A. and Quinn, J. (1983) Geochemistry of C_{25} and C_{30} biogenic alkenes in sediments of the Narragansett Bay estuary. *Geochimicia et Cosmochimica Acta*, 47, 1075–90.

Requejo, A. G., Hollywood, J. and Halpern, H. I. (1989) Recognition and source correlation of migrated hydrocarbons in Upper Jurassic Hareelv Formation, Jameson Land, East Greenland. *American Association of Petroleum Geologists Bulletin*, 73, 1065–88.

Requejo, A. G., Allan, J., Creaney, S., Gray, N. R. and Cole, K. S. (1992) Aryl isoprenoids and diaromatic carotenoids in Paleozoic source rocks and oils from the Western Canada and Williston basins. *Organic Geochemistry*, 19, 245–64.

Requejo, A. G., Wielchowsky, C. C., Klosterman, M. J. and Sassen, R. (1994) Geochemical characterization of lithofacies and organic facies in Cretaceous organic-rich rocks from Trinidad, East Venezuela Basin. *Organic Geochemistry*, 22, 441–59.

Requejo, A. G., McDonald, T. J., Sassen, R., Hieshima, G. B. and Hsu, C. S. (1997) Short-chain (C_{21} and C_{22}) diasteranes in petroleum and source rocks as indicators of maturity and depositional environment. *Geochimica et Cosmochimica Acta*, 61, 2653–67.

Restle, A. (1983) Etude de nouveaux marqueurs biologiques dans des petroles biodegrades: cas naturels et simulations in vitro. Ph.D. thesis, L'Universite Louis Pasteur de Strasbourg, Strasbourg France.

Revill, A. T., Volkman, J. K., O'Leary, T., *et al.* (1994) Hydrocarbon biomarkers, thermal maturity, and depositional setting of tasmanite oil shales from Tasmania, Australia. *Geochimica et Cosmochimica Acta*, 58, 3803–22.

Reynaud, F., Amaral, J., Drapeau, D. and Grauls, D. (1998) Post-salt petroleum systems in the Lower Congo Basin (Congo-Angola). Presented at the ABGP/American Association of Petroleum Geologists International Conference and Exhibition, November 8–11, 1998, Rio de Janeiro, Brazil.

Rice, D. D. (1984) Occurrence of indigenous biogenic gas in organic-rich immature chalks of Late Cretaceous age, eastern Denver Basin. In: *Petroleum Geochemistry and Source Rock Potential of Carbonate Rocks* (J. G. Palacas, ed.), American Association of Petroleum Geolgists Studies, Tulsa, OK, pp. 135–50.

Richard, P. D., Nederlof, P. J. R., Terken, J. M. J. and Al-Ruwehy, N. (1998) Generation and retention of hydrocarbons in the Haushi Play, North Oman. *GeoArabia*, 3, 493–506.

Richards, B. C. (1989a) Uppermost Devonian and Lower Carboniferous stratigraphy, sedimentation, and diagenesis, southwestern District of Mackenzie and southeastern Yukon Territory. *Geological Survey of Canada Bulletin*, 390, 135.

(1989b) Upper Kaskaskia sequence-uppermost Devonian and lower Carboniferous. In: *Western Canadian Sedimentary Basin, A Case History* (B. D. Ricketts, ed.), Canadian Society of Petroleum Geologists, Calgary, Alberta, pp. 165–201.

Richards, P. C. and Hillier, B. V. (2000) Post-drilling analysis of the North Falkland Basin – Part 2: petroleum system and future prospects. *Journal of Petroleum Geology*, 23, 273–92.

Richardson, J. S. and Miller, D. E. (1982) Identification of dicyclic and tricyclic hydrocarbons in the saturate fraction of a crude oil by gas chromatography-mass spectrometry. *Analytical Chemistry*, 54, 765–8.

(1983) Biologically-derived compounds of significance in the saturate fraction of a crude oil having predominant terrestrial input. *Fuel*, 62, 524–8.

Richnow, H. H., Jenisch, A. and Michaelis, W. (1992) Structural investigations of sulphur-rich macromolecular oil fractions and a kerogen by sequential chemical degradation. *Organic Geochemistry*, 19, 351–70.

Riediger, C. L. (1992) Sedimentology and diagenetic features of the Lower Jurassic "Nordegg Member," Western Canada Sedimentary Basin. Presented at the Annual Meeting of the American Association of Petroleum Geologists, Tulsa, OK.

Riediger, C. L., Brooks, P. W., Fowler, M. G. and Snowdon, L. R. (1990a) Lower and Middle Triassic source rocks, thermal maturation, and oil-source rock correlations in the Peace River Embayment area, Alberta and British Columbia. *Bulletin of Canadian Petroleum Geology*, 38a, 218–35.

Riediger, C. L., Fowler, M. G., Snowdon, L. R., Goodarzi, F. and Brooks, P. W. (1990b) Source rock analysis of the Lower Jurassic "Nordegg Member" and oil-source rock correlations, northwestern Alberta and northeastern British Columbia. *Bulletin of Canadian Petroleum Geology*, 38a, 236–49.

Riediger, C. L., Fowler, M. G., Brooks, P. W. and Snowdon, L. R. (1990c) Triassic oils and potential Mesozoic source rocks, Peace River Arch area, Western Canada Basin. *Organic Geochemistry*, 16, 295–305.

Riediger, C. L., Fowler, M. G. and Snowdon, L. R. (1991) The Lower Jurassic "Nordegg Member", Western Canada Sedimentary Basin. *Bulletin of Canadian Petroleum Geology*, 39, 222.

(1997) Organic geochemistry of the Lower Cretaceous ostracode zone, a brackish/non-marine source for some Lower Mannville oils in southeastern Alberta. In: *Petroleum Geology of the Cretaceous Mannville Group, Western Canada*, pp. 93–102.

Rinaldi, G. G. L. (1988) Oil-source studies in central Montana; correlations and migration implications. *Organic Geochemistry*, 13, 373–6.

Rinaldi G. G. L., Leopold, V. M. and Koons, C. B. (1988) Presence of benzohopanes, monoaromatic secohopanes, and saturate hexacyclic hydrocarbons in petroleum from carbonate environments. In: *Geochemical Biomarkers* (T. F. Yen and J. M. Moldowan, eds.), Harwood Academic Publishers, New York, pp. 331–53.

Riolo, J. and Albrecht, P. (1985) Novel rearranged ring C monoaromatic steroid hydrocarbons in sediments and petroleums. *Tetrahedron Letters*, 26, 2701–4.

Riolo, J., Ludwig, B. and Albrecht, P. (1985) Synthesis of ring C monoaromatic steroid hydrocarbons occurring in geological samples. *Tetrahedron Letters*, 26, 2607–700.

Riolo, J., Hussler, G., Albrecht, P. and Connan, J. (1986) Distribution of aromatic steroids in geological samples: their evaluation as geochemical parameters. *Organic Geochemistry*, 10, 981–90.

Risatti, J. B., Rowland, S. J., Yan, D. A. and Maxwell, J. R. (1984) Sterochemical studies of acyclic isoprenoids – XII. Lipids of methanogenic bacteria and possible contributors to sediments. *Organic Geochemistry*, 6, 93–104.

Ritter, U., Myhr, M. B., Aareskjold, K., *et al.* (1993a) Distributed activation energy models from hydrous pyrolysis: towards a second-generation tool of thermal histories. In: *Basin Modeling: Advances and Applications. Proceedings of the Norwegian Petroleum Society* (A. G. Dore, J. H. Augustson, C. Hermanrud, D. J. Stewart and O. Sylta, eds.), Elsevier, New York, pp. 201–8.

Ritter, U., Myhr, M. B., Aareskjold, K. and Schou, L. (1993b) Validation and first application of paleotemperature models of isomerization and NMR aromatization parameters. In: *Basin Modeling: Advances and Applications. Proceedings of the Norwegian Petroleum Society* (A. G. Dore, J. H. Augustson, C. Hermanrud, D. J. Stewart and O. Sylta, eds.), Elsevier, New York, pp. 185–99.

Ritter, U., Aareskjold, K. and Schou, L. (1993c) Distributed activation energy models of isomerization reactions from hydrous pyrolysis. *Organic Geochemistry*, 20, 511–20.

Ritts, B. D., Hanson, A. D., Zinniker, D. and Moldowan, J. M. (1999) Lower-Middle Jurassic nonmarine source rocks and petroleum systems of the northern Qaidam Basin, northwest China. *American Association of Petroleum Geologists Bulletin*, 83, 1980–2005.

Riva, A. Salvatori, T., Cavaliere, R., Ricchiuto, T. and Novelli, L. (1986) Origin of oils in Po Basin, northern Italy. *Organic Geochemistry*, 10, 391–400.

Riva, A., Caccialanza, P. G. and Quagliaroli, F. (1988) Recognition of 18β(H)-oleanane in several crudes and Tertiary-Upper Cretaceous sediments. Definition of a new maturity parameter. *Organic Geochemistry*, 13, 671–5.

Roadifer, R. E. (1987) Size distributions of the world's largest known oil and tar accumulations. Section I. Regional resources. In: *Exploration for Heavy Crude Oil and Natural Bitumen* (R. F. Meyer, ed.), American Association of Petroleum Geologists, Tulsa, OK, pp. 3–23.

Robbins, W. K. (1998) Challenges in the characterization of naphthenic acids in petroleum. Presented at the 215th National Meeting of the American Chemical Society, March 29–April 3, 1998, Dallas, TX.

Robinson, A. (1987) An overview of source rocks in Indonesia. In: *Proceedings of the Indonesian Petroleum Association*, Indonesian Petroleum Association, Jakarta, Indonesia, pp. 97–122.

Robinson, N. and Eglinton, G. (1990) Lipid chemistry of Icelandic hot spring microbial mats. *Organic Geochemistry*, 15, 291–8.

Robinson, W. E., Cummins, J. J. and Dinneen, G. U. (1965) Changes in Green River oil-shale paraffins with depth. *Geochimica et Cosmochimica Acta*, 29, 249–58.

Robison, V. D. (1995) Source rock characterization of the Late Cretaceous Brown Limestone of Egypt. In: *Petroleum Source Rocks* (B. J. Katz, ed.), Springer-Verlag, Berlin, pp. 265–81.

Robison, V. D., Liro, L. M., Robison, C. R., Dawson, W. C. and Russo, J. W. (1996) Integrated geochemistry, organic petrology, and sequence stratigraphy of the Triassic Shublik Formation, Tenneco Phoenix # 1 well, North Slope, Alaska, USA. *Organic Geochemistry*, 24, 257–72.

Robison, C. R., Darnell, L. M. and van, Gijzel, P. (2000) The transmittance color index of amorphous organic matter: a thermal maturity indicator for petroleum source rocks. *International Journal of Coal Geology*, 43, 83–103.

Robson, J. N. and Rowland, S. J. (1986) Identification of novel widely distributed sedimentary acyclic sesterterpenoids. *Nature*, 324, 561–3.

(1988) Synthesis of a highly branched C_{30} sedimentary hydrocarbon. *Tetrahedron Letters*, 29, 3837–40.

(1993) Synthesis, chromatographic and spectral characterisation of 2,6,11,15-tetramethylhexadecane (crocetane) and 2,6,9,13-tetramethyltetradecane: reference acyclic isoprenoids for geochemical studies. *Organic Geochemistry*, 20, 1093–8.

Rodrigues, K. (1988) Oil source bed recognition and crude oil correlation, Trinidad, West Indies. *Organic Geochemistry*, 13, 365–71.

Rodrigues, D. C., Koike, L., Reis, F. de, A. M., et al. (2000) Carboxylic acids of marine evaporitic oils from Sergipe-Alagoas Basin, Brazil. *Organic Geochemistry*, 31, 1209–22.

Roehler, H. W. (1992) *Correlation, Composition, Areal Distribution, and Thickness of Eocene Stratigraphic Units, Greater Green River Basin, Wyoming, Utah, and Colorado*. U.S. Geological Survey Professional Paper 1506-E, U.S. Geological Survey, Washington, DC.

Roessler, M. and Müller, V. (2001) Osmoadaptation in bacteria and archaea: common principles and differences. *Environmental Microbiology*, 3, 743–54.

Röhl, H., Schmid-Röhl, A., Oschmann, W., Frimmel, A. and Schwark, L. (2001) The Posidonia Shale (Lower Toarcian) of SW-Germany: an oxygen-depleted ecosystem controlled by sea level and palaeoclimate. *Palaeogeography, Palaeoclimatology, Palaeoecology*, 165, 27–52.

Rohmer, M. (1987) The hopanoids, prokaryotic triterpenoids and sterol surrogates. In: *Surface Structures of Microorganisms and Their Interactions with the Mammalian Host* (E. Schriner et al., eds.), VCH Publishing, Weinlein, Germany, pp. 227–42.

Rohmer, M. (1993) The biosynthesis of triterpenoids of the hopane series in eubacteria: a mine of new enzyme reactions. *Pure and Applied Chemistry*, 65, 1293–8.

Rohmer, M. and Ourisson, G. (1976) Methyl-hopanes d'*Acetobacter xylinum* et d'*Acetobacter rancens*: une nouvelle famille de composes triterpeniques. *Tetrahedron Letters*, 40, 3641–4.

Rohmer, M., Bisseret, P. and Neunlist, S. (1992) The hopanoids, prokaryotic triterpenoids and precursors of ubiquitous molecular fossils. In: *Biological Markers in Sediments and Petroleum: A Tribute to Wolfgang K. Seifert* (J. M. Moldowan, P. Albrecht and R. P. Philp, eds.), Prentice Hall, Englewood Cliffs, NJ, pp. 1–17.

Rohrback, B. G. (1983) Crude oil geochemistry of the Gulf of Suez. In: *Advances in Organic Geochemistry 1981* (M. Bjorøy, C. Albrecht, C. Cornford, *et al.*, eds.), John Wiley & Sons, New York, pp. 39–48.

Roland, S. J., Alexander, R. and Kagi, R. I. (1984) Analysis of trimethylnaphthalenes in petroleum by capillary chromatography. *Journal of Chromatography*, 294, 407–12.

Rolfes, J. and Andersson, J. T. (2001) Determination of alkylphenols after derivitization to ferrocenecarboxylic acid esters with gas chromatography-atom emission detection. *Analytical Chemistry*, 73, 3073–82.

Romeiro, P. C., Braun, O., Zalán, P. V., de Paula, L. and Martins, M. (1998) São Francisco Basin – exploration frontier for gas in central Brazil. *American Association of Petroleum Geologists Bulletin*, 82, 1959.

Rooney, M. A., Vuletch, A. K. and Griffith, C. E. (1998) Compound-specific isotope analysis as a tool for characterizing mixed oils: an example from the west of Shetlands area. *Organic Geochemistry*, 29, 241–54.

Rosell-Melé, A., Carter, J. F. and Maxwell, J. R. (1996) High-performance liquid chromatography-mass spectrometry of porphyrins by using an atmospheric pressure interface. *Journal of the American Society for Mass Spectrometry*, 7, 965–71.

  (1999) Liquid chromatography/tandem mass spectrometry of free base alkyl porphyrins for the characterization of the macrocyclic substitutents in components of complex mixtures. *Rapid Communications in Mass Spectrometry*, 13, 568–73.

Rosing, M. T. (1999) ^{13}C-depleted carbon microparticles in >3700-Ma sea-floor sedimentary rocks from West Greenland. *Science*, 283, 674–6.

Rosnes, J. T., Graue, A. and Lien, T. (1991a) Activity of sulfate-reducing bacteria under simulated reservoir conditions. *Society of Petroleum Engineers, Production and Engineering*, 5, 217–20.

Rosnes, J. T., Torsvik, T. and Lien, T. (1991b) Spore-forming thermophilic sulfate-reducing bacteria isolated from North Sea oil field waters. *Applied and Environmental Microbiology*, 57, 2302–7.

Rospondek, M. J., Köster, J. and Sinninghe Damsté, J. S. (1997) Novel C_{26} highly branched isoprenoid thiophenes and alkane from the Menilite Formation, Outer Carpathians, SE Poland. *Organic Geochemistry*, 26, 295–304.

Ross, C. A. and Ross, J. R. P. (1987) Late Paleozoic sea levels and depositional sequences. In: *Timing and Depositional History of Eustatic Sequences: Constraints on Seismic Stratigraphy* (C. A. Ross and D. Haman, eds.), Cushman Foundation for Foraminiferal Research, Ithaca, New York, pp. 137–49.

Rothschild, L. J. and Mancinelli, R. L. (2001) Life in extreme environments. *Nature*, 409, 1092–101.

Rouchy, J. M., Taberner, C., Blanc-Valleron, M.-M., *et al.* (1998) Sedimentary and diagenetic markers of the restriction in a marine basin: the Lorca Basin. *Sedimentary Geology*, 121, 25–55.

Rowland, S. J. (1990) Production of acyclic isoprenoid hydrocarbons by laboratory maturation of methanogenic bacteria. *Organic Geochemistry*, 15, 9–16.

Rowland, S. J. and Robson, J. N. (1990) The widespread occurrence of highly branched acyclic C_{20}, C_{25}, and C_{30} hydrocarbons in recent sediments and biota. A review. *Marine Environmental Research*, 30, 191–216.

Rowland, S. J., Yon, D. A., Lewis, C. A. and Maxwell, J. R. (1985) Occurrence of 2,6,10-trimethyl-7-(3-methylbutyl)-dodecane and related hydrocarbons in the green alga *Enteromorpha profilera* and sediments. *Organic Geochemistry*, 8, 207–13.

Rowland, S. J., Hird, S. J., Robson, J. H. and Venkatesan, M. I. (1990) Hydrogenation behaviour of two highly branched C_{25} dienes from Antarctic marine sediments. *Organic Geochemistry*, 15, 215–18.

Rozanova, E. P. and Nazina, T. N. (1979) Distribution of thermophilic sulfate reducing bacteria in the oil-bearing strata of Apsheron and Western Siberia [in Russian]. *Mikrobiologiia*, 48, 1113–17.

Rozanova, E. P., Savvichev, A. S., Miller, Y. M. and Ivanov, M. V. (1997) Microbial processes in a West Siberian oil field flooded with waters containing a complex of organic compounds. *Microbiology (Mikrobiologiya)*, 66, 718–25.

Rozanova, E. P., Borzenkov, I. A., Tarasov, A. L., *et al.* (2001) Microbiological processes in a high-temperature oil field [in Russian]. *Mikrobiologiia*, 70, 118–27.

Rubinstein, I. and Albrecht, P. (1975) The occurrence of nuclear methylated steranes in a shale. *Journal of the Chemical Society, Chemical Communications*, 957–8.

Rubinstein, I., Sieskind, O. and Albrecht, P. (1975) Rearranged sterenes in a shale: occurrence and simulated formation. *Journal of Chemical Society, Perkin Transaction I*, 1833–6.

Rubinstein, I., Strausz, O. P., Spyckerelle, C., Crawford, R. J. and Westlake, D. W. S. (1977) The origin of oil sand bitumens of Alberta. *Geochimica et Cosmochimica Acta*, 41, 1341–53.

Ruble, T. E., Bakel, A. J. and Philp, R. P. (1994) Compound specific isotopic variability in Uinta Basin native bitumens: paleoenvironmental implications. *Organic Geochemistry*, 21, 661–71.

Ruble, T. E., Lewan M. D. and Philp, R. P. (2001) New insights on the Green River petroleum system in the Unita Basin from hydrous pyrolysis experiments. *American Association of Petroleum Geologists Bulletin*, 85, 1333–71.

Rueter, P., Rabus, R., Wilkes, H., *et al.* (1994) Anaerobic oxidation of hydrocarbons in crude oil by new types of sulfate-reducing bacteria. *Nature*, 372, 455–8.

Rullkötter, J. and Marzi, R. (1988) Natural and artificial maturation of biological markers in a Toarcian shale from northern Germany. *Organic Geochemistry*, 13, 639–45.

  (1989) New aspects of the application of sterane isomerization and steroid aromatization to petroleum exploration and the reconstruction of geothermal histories of sedimentary basins. Presented at the 215th National Meeting of the American Chemical Society, April 9–14, 1989, Dallas, TX.

Rullkötter, J. and Philp, R. P. (1981) Extended hopanes up to C_{40} in Thornton bitumen. *Nature*, 292, 616–18.

Rullkötter, J. and Wendisch, D. (1982) Microbial alteration of $17\alpha(H)$-hopane in Madagascar asphalts: removal of C-10 methyl group and ring opening. *Geochimica et Cosmochimica Acta*, 46, 1543–53.

Rullkötter, J., Leythaeuser, D. and Wendisch, D. (1982) Novel 23,28-bisnorlupanes in Tertiary sediments. Widespread occurrence of nuclear demethylated triterpanes. *Geochimica et Cosmochimica Acta*, 46, 2501–9.

Rullkötter, J., Aizenshtat, Z. and Spiro, B. (1984) Biological markers in bitumens and pyrolyzates of Upper Cretaceous bituminous chalks from the Ghareb Formation (Israel). *Geochimica et Cosmochimica Acta*, 48, 151–7.

Rullkötter, J., Spiro, B. and Nissenbaum, A. (1985) Biological marker characteristics of oils and asphalts from carbonate source rocks in a rapidly subsiding graben, Dead Sea, Israel. *Geochimica et Cosmochimica Acta*, 49, 1357–70.

Rullkötter, J., Meyers, P. A., Schaefer, R. G. and Dunham, K. W. (1986) Oil generation in the Michigan Basin: a biological marker and carbon isotope approach. *Organic Geochemistry*, 10, 359–75.

Rullkötter, J., Marzi, R. and Meyers, P. A. (1992) Biological markers in Paleozoic sedimentary rocks and crude oils in the Michigan Basin: reassessment of sources and thermal history of organic matter. In: *Early Organic Evolution: Implications for Mineral and Energy Sources* (M. Schidlowski, S. Golubic, M. M. Kimerley, D. M. McKirdy and P. A. Trudinger, eds.), Springer-Verlag, Heidelberg, pp. 324–35.

Rullkötter, J., Peakman, T. M. and ten Haven, H. L. (1994) Early diagenesis of terrigenous triterpenoids and its implications for petroleum geochemistry. *Organic Geochemistry*, 21, 215–33.

Russell, M., Grimalt, J. O., Hartgers, W. A., Taberner, C. and Rouchy, J. M. (1997) Bacterial and algal markers in sedimentary organic matter deposited under natural sulphurization conditions (Lorca Basin, Murcia, Spain). *Organic Geochemistry*, 26, 605–25.

Ryder, R. T., Burruss, R. C. and Hatch, J. R. (1998) Black shale source rocks and oil generation in the Cambrian and Ordovician of the Central Appalacian Basin, USA. *American Association of Petroleum Geologists Bulletin*, 82, 412–41.

Sageman, B. B., Rich, J., Arthur, M. A., Birchfield, G. E. and Dean, W. E. (1997) Evidence for Milankovitch periodicities in Cenomanian-Turonian lithologic and geochemical cycles, Western Interior USA. *Journal of Sedimentary Research*, 19, 286–302.

Sageman, B. B., Gardner, M. H., Armentrout, J. M. and Murphy, A. E. (1998) Stratigraphic hierarchy of organic carbon-rich siltstone in deep-water facies, Brushy Canyon (Guadalupian), Delaware Basin, West Texas. *Geology*, 26, 451–4.

Saint-Germes, M., Baudin, F., Bazhenova, O. K. and Fadeeva, N. (1997) Organic facies and petroleum potential of the Oligocene–Lower Miocene source rocks from Crimea to Azerbaijan. Presented at the AAPG International Conference and Exhibition, September 7–10, 1997, Vienna, Austria.

Saint-Germes, M., Bocherens, H., Baudin, F. and Bazhenova, O. (2000) Evolution of the δ^{13}C values of organic matter of the Maykop series during Oligocene-Lower Miocene [Evolution des valeurs de δ^{13}C des matieres organiques de la serie de Maykop au cours de L'oligocene-Miocene inferieur.] *Bulletin Société Géologique France*, 171, 13–21.

Sajgó, C. and Lefler, J. (1986) A reaction kinetic approach to the time-temperature history of sedimentary basins. In: *Paleogeothermics*, Vol. 5 (G. Bunterbath and L. Stegena, eds.), Springer-Verlag, Berlin, pp. 119–51.

Sajgó, C., Horvath, Z. A. and Lefler, J. (1988) An organic maturation study of the Hod-I Borehole (Pannonian Basin). In: *The Pannonian Basin: A Study in Basin Evolution* (L. H. Royde and F. Horvath, eds.), American Association of Petroleum Geologists, Tulsa, OK, pp. 297–309.

Saltzman, M. R., Davidson, J. P., Holden, P., Runnegar, B. and Lohmann, K. C. (1995) Sea-level-driven changes in ocean chemistry at an Upper Cambrian extinction horizon. *Geology*, 23, 893–6.

Samanta, U., Mishra, C. S. and Misra, K. N. (1994) Indian high wax crude oils and the depositional environments of their source rocks. *Marine and Petroleum Geology*, 11, 756–9.

Samman, N., Ignasiak, T., Chen, C.-J., Strausz, O. P. and Montgomery, D. S. (1981) Squalene in petroleum asphaltenes. *Science*, 213, 1381–3.

Sandberg, C. A. and Gutschick, R. C. (1984) Distribution, microfauna, and source-rock potential of Mississippian Delle Phosphatic Member of Woodman Formation and equivalents, Utah and adjacent states. In: *Hydrocarbon Source Rocks of the Greater Rocky Mountain Region* (J. Woodward, F. F. Meissner and J. L. Clayton, eds.), Rocky Mountain Association of Geologists, Denver, CO, pp. 135–78.

Sandstrom, M. W. and Philp, R. P. (1984) Biological marker analysis and stable carbon isotopic composition of oil seeps from Tonga. *Chemical Geology*, 43, 167–80.

Sano, Y., Terada, K., Takahasi, Y. and Nutman, A. P. (1999) Origin of life from apatite dating? *Nature*, 400, 127.

Santos Neto, E. V. and Hayes, J. M. (1999) Use of hydrogen and carbon stable isotopes characterizing oils from the Potiguar Basin (onshore), northeastern Brazil. *American Association of Petroleum Geologists Bulletin*, 83, 496–518.

Sarjono, S. and Sardjito, F. (1989) Hydrocarbon source rock identification in the South Palembang Sub-basin. In: *Proceedings of the Indonesian Petroleum Association Eighteenth Annual Convention*, October, 1989, p. 427–467.

Sasaki, T., Maki, H., Ishihara, M. and Harayama, S. (1998) Vanadium as an internal marker to evaluate microbial degradation of crude oil. *Environmental Science & Technology*, 33, 3618–21.

Sassen, R. (1990) Lower Tertiary and Upper Cretaceous source rocks in Louisiana and Mississippi; implications to Gulf of Mexico crude oil. *American Association of Petroleum Geologists Bulletin*, 74, 857–78.

Sassen, R. and Moore, C. H. (1988) Framework of hydrocarbon generation and destruction in Eastern Smackover Trend. *American Association of Petroleum Geologists Bulletin*, 72, 649–63.

Saxby, J. D. and Stephenson, L. C. (1987) Effect of an igneous intrusion on oil shale at Rundle (Australia). *Chemical Geology*, 63, 1–16.

Scalan, R. S. and Smith, J. E. (1970) An improved measure of the odd-to-even predominance in the normal alkanes of sediment extracts and petroleum. *Geochimica et Cosmochimica Acta*, 34, 611–20.

Schaeffer, P., Poinsot, J., Hauke, V., *et al.* (1994) Novel optically active hydrocarbons in sediments: evidence for an extensive biological cyclization of higher regular polyprenols. *Angewandte Chemie International Edition*, 33, 1166–9.

Schaeffer, P., Trendel, J.-M. and Albrecht, P. (1995a) An unusual aromatization process of higher plant triterpenes in sediments. *Organic Geochemistry*, 23, 273–5.

Schaeffer, P., Adam, P., Trendel, J.-M., Albrecht, P. and Connan, J. (1995b) A novel series of benzohopanes widespread in sediments. *Organic Geochemistry*, 23, 87–9.

Schaeffer, P., Adam, P., Werung, P., Bernasconi, S. and Albrecht, P. (1997) Molecular and isotopic investigation of free and S-bound lipids from an actual meromictic lake (Lake Cadagno, Switzerland). In: *Proceedings of the 18th International Meeting on Organic Geochemistry, September 22–26, 1997, Maastricht, The Netherlands*, Forschungszentrum Jülich, Jülich, Germany, pp. 57–8.

Schenk, C. J., Viger, R. J. and Anderson, C. P. (1999) *Maps Showing Geology, Oil and Gas Fields, and Geologic Provinces of South America*. U.S. Geological Survey Open-File Report 97-470D.

Scherer, E., Münker, C. and Mezger, K. (2001) Calibration of the lutetium-hafnium clock. *Science*, 293, 683–7.

Schidlowski, M., Hayes, J. M. and Kaplan, I. R. (1983). Isotopic inferences of ancient biochemistries: carbon, sulfur, hydrogen, and nitrogen. In: *Earth's Earliest Biosphere: Its Origins and Evolution* (J. W. Schopf, ed.), Princeton University Press, Princeton, NJ, pp. 149–86.

Schiefelbein, C. F., Zumberge, J. E., Cameron, N. R. and Brown, S. W. (1999) Petroleum systems in the South Atlantic Margins. In: *The Oil and Gas Habitats of the South Atlantic* (N. R. Cameron, R. H. Bate, R. H. and V.S Clure, eds.), Geological Society of London, London, pp. 169–79.

(2000) Geochemical comparison of crude oil along South Atlantic margins. In: *Petroleum Systems of South Atlantic Margins* (M. R. Mello and B. J. Katz, eds.), American Association of Petroleum Geologists, Tulsa, OK, pp. 15–26.

Schindewolf, O. H. (1954) Über die möglichen Urasachen der grossen erdgeschichtlichen Faunenschnitte. *Neues Jahrbuch für Geologie und Paläontologie Monatshefte*, 1954, 457–65.

Schlanger, S. O. and Jenkyns, H. C. (1976) Cretaceous oceanic anoxic events: causes and consequences. *Geologie en Mijnbouw*, 55, 179–84.

Schmid, J. C. (1986) Marqueurs biologiques soufres dans les petroles. Ph. D. thesis, University of Strausbourg, Strasbourg, France.

Schmid-Röhl A., Röhl, H.-J., Oschmann, W., Frimmel, A. and Schwark, L. (2000) Palaeoenvironmental reconstruction of epicontinental Lower Toarcian black shales (Posidonia Shale, SW-Germany) by means of stable carbon and oxygen isotope analysis: global versus regional controls. *Geobios*, 35, 13–20.

Schmidt, K. (1978) Biosynthesis of carotenoids. In: *Photosynthetic Bacteria* (R. K. Clayton and W. R. Sistrom, eds.), Plenum Press, New York, pp. 729–50.

Schmitt, R., Langguth, H.-R., Püttmann, W., *et al.* (1996) Biodegradation of aromatic hydrocarbons under anoxic conditions in a shallow sand and gravel aquifer of the Lower Rhine Valley, Germany. *Organic Geochemistry*, 25, 41–50.

Schmitter, J. M., Sucrow, W. and Arpino, P. J. (1982) Occurrence of novel tetracyclic geochemical markers: 8,14-seco-hopanes in a Nigerian crude oil. *Geochimica et Cosmochimica Acta*, 46, 2345–50.

Schoell, M., Teschner, M., Wehner, H., Durand, B. and Oudin, J. L. (1983) Maturity related biomarker and stable isotope variations and their application to oil/source rock correlation in the Mahakam Delta, Kalimantan. In: *Advances in Organic Geochemistry 1981* (M. Bjorøy, C. Albrecht, C. Cornford, *et al.*, eds.) John Wiley & Sons, New York, pp. 156–63.

Schoell, M., McCaffrey, M. A., Fago, F. J. and Moldowan, J. M. (1992) Carbon isotopic compositions of 23,30-bisnorhopanes and other biological markers in a Monterey crude oil. *Geochimica et Cosmochimica Acta*, 56, 1391–9.

Schoell, M., Hwang, R. J., Carlson, R. M. K. and Welton, J. E. (1994) Carbon isotopic composition of individual biomarkers in gilsonites (Utah). *Organic Geochemistry*, 21, 673–83.

Schoellkopf, N. B. and Patterson, B. (2000) Petroleum systems of offshore Cabinda, Angola. In: *Petroleum Systems of South Atlantic Margins* (M. R. Mello and B. J. Katz, eds.), American Association of Petroleum Geologists, Tulsa, OK, pp. 361–76.

Scholle, P. A. (1995) Carbon and sulfur isotope stratigraphy of the Permian and adjacent intervals. In: *The Permian of Northern Pangea*, Vol. I (P. A. Scholle, T. M. Peryt and D. S. Ulmer-Scholle, eds.), Springer-Verlag, New York, pp. 133–49.

Scholle, P. A. and Arthur, M. A. (1976) Carbon-isotopic fluctuations in Upper Cretaceous sediments: an indication of paleo-oceanic circulation. Presented at the Annual Meeting of the Geological Society of America, November, 1976, Denver, CO.

Schönlaub, H. P. (1986) Significant geological events in the Paleozoic record of the Southern Alps (Austrian Part). In: *Global Bio-Events: A Critical Approach, Lecture Notes in Earth Science 8* (O. H. Walliser, ed.), Springer-Verlag, Berlin, pp. 163–67.

Schopf, J. W. (1993) Microfossils of the early Archean Apex Chert; new evidence of the antiquity of life. *Science*, 260, 640–6.

Schopf, J. W. and Blacic, J. M. (1987) New microorganisms from the Bitter Springs Formation (late Precambrian) of the north-central Amadeus Basin, Australia. *Journal of Paleontology*, 45, 925–60.

Schopf, J. W. and Packer, B. M. (1987) Early Archean (3.3 billion to 3.5 billion-year-old) microfossils from Warrawoona Group, Australia. *Science*, 237, 70–3.

Schopf, J. W. and Walter, M. R. (1983) Archean microfossils: new evidence of ancient microbes. In: *Earth's Earliest Biosphere* (J. W. Schopf, ed.), Princeton University Press, Princeton, NJ, pp. 214–39.

Schopf, J. W., Hayes, J. M. and Walter, M. R. (1983) Evolution of Earth's earliest ecosystems: recent progress and unsolved problems. In: *Earth's Earliest Biosphere* (J. W. Schopf, ed.), Princeton University Press, Princeton, NJ, pp. 361–84.

Schopf, J. W., Kudryavtsev, A. B., Agresti, D. G., Wdowiak, T. J. and Czaja, A. D. (2002) Laser-Raman imagery of Earth's earliest fossils. *Nature*, 416, 73–6.

Schouten, S., Sinninghe Damsté, J. S., Baas, M., et al. (1995) Quantitative assessment of mono- and polysulphide-linked carbon skeletons in sulphur-rich macromolecular aggregates present in bitumens and oils. *Organic Geochemistry*, 23, 765–75.

Schouten, S., Rijpstra, W. I. C., Sinninghe Damsté, J. S., de Leeuw, J. W. and Schoell, M. (1997a) A molecular stable carbon isotope study of organic matter in immature Miocene Monterey sediments, Pismo Basin. *Geochimica et Cosmochimica Acta*, 61, 2065–82.

Schouten, S., van der Maarel, M. J. E. C., Huber, R. and Sinninghe Damsté, J. S. (1997b) 2,6,10,15,19-Pentamethylicosenes in *Methanolobus bombayensis*, a marine methanogenic archaeon, and in *Methanosarcina mazei*. *Organic Geochemistry*, 26, 409–14.

Schouten, S., Hoefs, M. J. L., Koopmans, M. P., Bosch, H.-J. and Sinninghe Damsté, J. S. (1998a) Structural characterization, occurrence and fate of archaeal ether-bound acyclic and cyclic biphytanes and corresponding diols in sediments. *Organic Geochemistry*, 29, 1305–19.

Schouten, S., Breteler, W. C. M. K., Blokker, P., et al. (1998b) Biosynthetic effects on the stable carbon isotopic compositions of algal lipids: implications for deciphering the carbon isotopic biomarker record. *Geochimica et Cosmochimica Acta*, 62, 1397–406.

Schouten, S., Baas, M., van Kaam-Peters, H. M. E. and Sinninghe Damsté, J. S. (1998c) Long-chain 3-isopropyl alkanes: a new class of sedimentary acyclic hydrocarbons. *Geochimica et Cosmochimica Acta*, 62, 961–4.

Schouten, S., van Kaam-Peters, H. M. E., Rijpstra, W. I. C., Schoell, M. and Sinninghe Damsté, J. S. (2000b) Effects of an oceanic anoxic event on the stable carbon isotope composition of Early Toarcian carbon. *American Journal of Science*, 300, 1–22.

Schouten, S., Hopmans, E. C., Pancost, R. D. and Sinninghe, Damsté, J. S. (2000c) Widespread occurrence of structurally diverse tetraether membrane lipids: evidence for the ubiquitous presence of low-temperature relatives of hyperthomophiles. *Proceedings of the National Academy of Science, USA*, 97, 14 421–6.

Schouten, S., de Loureiro, M. R. B., Sinninghe Damsté, J. S. and de Leeuw, J. W. (2001) Molecular biogeochemistry of Monterey sediments, Naples Beach, California. I: distributions of hydrocarbons and organic sulfur compounds. In: *The Monterey Formation: From Rocks to Molecules* (C. M. Isaacs and J. Rullkötter, eds.), Columbia University Press, New York, pp. 150–74.

Schröder-Adams, C. J., Cumbaa, S. L., Bloch, J., et al. (2001) Late Cretaceous (Cenomanian to Campanian) paleoenvironmental history of the Eastern Canadian margin of the Western Interior Seaway: bonebeds and anoxic events. *Palaeogeography, Palaeoclimatology, Palaeoecology*, 170, 261–89.

Schulze, T. and Michaelis, W. (1990) Structure and origin of terpenoid hydrocarbons in some German coals. *Organic Geochemistry*, 16, 1051–8.

Schumacher, D. (1996) Hydrocarbon-induced alteration of soils and sediments. In: *Hydrocarbon Migration and its Near-Surface Expression* (D. Schumacher and M. A. Abrams, eds.), American Association of Petroleum Geologists, Tulsa, OK, pp. 71–89.

(1999) Surface geochemical exploration for petroleum in exploring for oil and gas traps. In: *Treatise of Petroleum Geology/Handbook of Petroleum Geology: Exploring for Oil and Gas Traps* (E. A. Beaumont and N. H. Foster, eds.), American Association of Petroleum Geologists, Tulsa, OK, pp. 18–27.

Schumacher, D. and Parker, R. M. (1988) Possible Paleozoic or Triassic origin for some Jurassic-reservoired oil and gas, Cass County, Northeast Texas. In: *Geochemistry of Gulf Coast Oils and Gases: Their Characteristics, Origin, Distribution, and Exploration and Production Significance. Proceedings of the Ninth Annual Research Conference, Gulf Coast Section, Society of Economic Paleontologists and Mineralogists Foundation* (D. Schumacher and B. F. Perkins, eds.), Society of Economic Paleontologists and Mineralogists, Tulsa, OK, pp. 10–11.

(1990) Possible pre-Jurassic origin for some Jurassic reservoired oils, Cass Co., Texas. In: *Geochemistry of Gulf Coast Oils and Gases: Their Characteristics, Origin,*

Distribution, and Exploration and Production Significance. Proceedings of the Ninth Annual Research Conference, Gulf Coast Section, Society of Economic Paleontologists and Mineralogists Foundation (D. Schumacher and B. F. Perkins, eds.), Society of Economic Paleontologists and Mineralogists, Tulsa, OK, pp. 59–68.

Schwark, L., Vliex, M. and Schaeffer, P. (1998) Geochemical characterization of Malm Zeta laminated carbonates from the Franconian Alb, SW-Germany (II). *Organic Geochemistry*, 29, 1921–52.

Schwartz, R. D. and James, P. B. (1984) Periodic mass extinctions and the Sun's oscillation about the galactic plane. *Science*, 308, 712–13.

Schwartz, H., Sample, J., Weberling, K. D., Minisini, D. and Moore, J. C. (2003) An ancient linked fluid migration system: cold-seep deposits and sandstone intrusions in the Panoche Hills, California, USA. *Geo-Marine Letters*. 23, 340–50.

Scotchman, I. C., Griffith, C. E., Holmes, A. J. and Jones, D. M. (1998) The Jurassic petroleum system north and west of Britain: a geochemical oil-source correlation study. *Organic Geochemistry*, 29, 671–700.

Scotchmer, J. and Patience, R. (2002) Predicting petroleum biodegradation in undrilled prospects: how far can you go? Presented at the Annual Meeting of the American Association of Petroleum Geologists, March 10–13, 2002, Houston, TX.

Scott, J. (1994) Source rocks of west Australian basins – distribution, character and models. In: *Proceedings of Petroleum Exploration Society of Australia Symposium, Perth, 1994: The Sedimentary Basins of Western Australia* (P. G. Purcell and R. R. Purcell, eds.), Petroleum Exploration Society of Australia, Perth, Australia, pp. 141–58.

Scott, A. C. and Fleet, A. J. (1994) *Coal and Coal-Bearing Strata as Oil-Prone Source Rocks?* Geological Society Special Publication 77, Geological Society of London, London.

Scott, A. C., Lomax, B. H., Collinson, M. E., Upchurch, G. R. and Beerling, D. J. (2000) Fire across the K-T boundary: initial results from the Sugarite Coal, New Mexico, USA. *Palaeogeography, Palaeoclimatology, Palaeoecology*, 164, 381–95.

Sedivy, R. A., Penfield, I. E., Halpern, H. I., et al. (1987) Investigation of source rock-crude oil relationships in the northern Alaska hydrocarbon habitat. In: *Alaskan North Slope Geology* (I. Tailleur and P. Weimer, eds.), Society of Economic Paleontologists and Mineralogists and Alaska Geological Society, Tulsa, OK, pp. 169–179.

Seifert, W. K. (1973) Steroid acids in petroleum; animal contribution to the origin of petroleum. In: *International Symposium on Chemistry in Evolution and Systematics*, (T. Swain, ed.), Crane, Russak and Co., New York, pp. 633–40.

(1980) Impact of Treib's discovery of porphyrins on present day biological marker organic geochemistry. In: *Proceedings of the Treibs International Symposium* (A. Prashnowsky, ed.), Halbigdruck Publishing, Wurzburg, Germany, pp. 13–35.

Seifert, W. K. and Moldowan, J. M. (1978) Applications of steranes, terpanes and monoaromatics to the maturation, migration and source of crude oils. *Geochimica et Cosmochimica Acta*, 42, 77–95.

(1979) The effect of biodegradation on steranes and terpanes in crude oils. *Geochimica et Cosmochimica Acta*, 43, 111–26.

(1980) The effect of thermal stress on source-rock quality as measured by hopane stereochemistry. *Physics and Chemistry of the Earth*, 12, 229–37.

(1981) Paleoreconstruction by biological markers. *Geochimica et Cosmochimica Acta*, 45, 783–94.

(1986) Use of biological markers in petroleum exploration. In: *Methods in Geochemistry and Geophysics* Vol. 24 (R. B. Johns, ed.), Elsevier, Amsterdam, pp. 261–90.

Seifert, W. K., Moldowan, J. M., Smith, G. W. and Whitehead, E. V. (1978) First proof of a C_{28}-pentacyclic triterpane in petroleum. *Nature*, 271, 436–37.

Seifert, W. K., Gallegos, E. J. and Teeter, R. M. (1979) Proof of structure of steroid carboxylic acids in a California petroleum by deuterium labeling, synthesis, and mass spectrometry. In: *Geochemistry of Organic Molecules* (K. A. Kvenvolden, ed.), Dowden, Hutchinson and Ross, Stroudsburg, PA, pp. 212–19.

Seifert, W. K., Moldowan, J. M. and Jones, R. W. (1980) Application of biological marker chemistry to petroleum exploration. In: *Proceedings of the Tenth World Petroleum Congress*, Heyden & Son, Inc., Philadelphia, PA pp. 425–40.

(1981) Application of biologic markers in combination with stable carbon isotopes to source rock /oil correlations, Prudhoe Bay, Alaska. *American Association of Petroleum Geologists Bulletin*, 65, 990–1.

Seifert, W. K., Carlson, R. M. K. and Moldowan, J. M. (1983) Geomimetic synthesis, structure assignment, and geochemical correlation application of monoaromatized petroleum steranes. In: *Advances in Organic Geochemistry 1981* (M. Bjorøy, C. Albrecht, C. Cornford, *et al.*, eds.), John Wiley & Sons, New York, pp. 710–24.

Seifert, W. K., Moldowan, J. M. and Demaison, G. J. (1984) Source correlation of biodegraded oils. *Organic Geochemistry*, 6, 633–43.

Seilacher, A., Bose, P. K. and Pflüger, F. (1998) Triploblastic animals more than 1 billion years ago: trace fossil evidence from India. *Science*, 282, 80–3.

Selley, R. C. and Stoneley, R. (1987) Petroleum habitat in south Dorset. In: *Petroleum Geology of North West Europe* (J. Brooks and K. W. Glennie, eds.), Graham and Trotman, London, pp. 139–48.

Selosse, M.-A. and le Tacon, F. (1998) The land flora: a phototroph-fungus partnership? *Trends in Ecology and Evolution*, 13, 15–20.

Semtner, A. K. and Klitzsch, E. (1994) Early Paleozoic paleogeography of the northern Gondwana margin: new evidence for Ordovician-Silurian glaciation. *Geologische Rundschau*, 83, 743–51.

Sephton, M. A. and Gilmour, I. (2001) Compound-specific isotope analysis of the organic matter constituents in carbonaceous chondrites. *Mass Spectrometry Reviews*, 20, 111–20.

Sepkoski, J. J., Jr (1986) Phanerozoic overview of mass extinctions. In: *Patterns and Processes in the History of Life* (D. M. Raup and D. Jablonski, eds.), Springer-Verlag, Berlin, pp. 277–295.

(1996) Patterns of Phanerozoic extinction: a perspective from global databases. In: *Global Events and Event Stratigraphy in the Phanerozoic* (O. H. Walliser, ed.), Springer-Verlag, Berlin, pp. 35–51.

Sereno, P. C. (1999) The evolution of dinosaurs. *Science*, 284, 2137–46.

Shanmugam, G. (1985) Significance of coniferous rain forests and related organic matter in generating commercial quantities of oil, Gippsland Basin, Australia. *American Association of Petroleum Geologists Bulletin*, 69, 1241–54.

Sharma, A., Scott, J. H., Cody, G. D., *et al.* (2002) Microbial activity at gigapascal pressures. *Science*, 295, 1514–6.

Shearer, C. K., Layne, G. D., Papike, J. J. and Spilde, M. N. (1996) Sulfur isotope systematics in alteration assemblages in Martian meteorite ALH 84001. *Geochimica et Cosmochimica Acta*, 60, 2921–6.

Sheng, G., Simoneit, B. R. T., Lief, R. N., Chen, X. and Fu, J. (1992) Tetracyclic terpanes enriched in Devonian cuticle humic coals. *Fuel*, 71, 523–32.

Sherwood, N. R. and Cook, A. C. (1986) Organic matter in the Toolebuc Formation. In: *Contributions to the Geology and Hydrocarbon Potential of the Eromanga Basin* (D. I. Gravestock, P. S. Moore and G. M. Pitt, eds.), Geological Society of Australia, Sydney, Australia, pp. 255–65.

Shi, J.-Y., Mackenzie, A. S., Alexander, R., *et al.* (1982) A biological marker investigation of petroleums and shales from the Shengli oilfield, the People's Republic of China. *Chemical Geology*, 35, 1–31.

Shiea, J., Brassell, S. C. and Ward, D. M. (1990) Mid-chain branched mono- and dimethyl alkanes in hot spring cyanobacterial mats: a direct biogenic source for branched alkanes in ancient sediments? *Organic Geochemistry*, 15, 223–31.

(1991) Comparative analysis of extractable lipids in hot spring microbial mats and their component photosynthetic bacteria. *Organic Geochemistry*, 17, 309–19.

Shimeld, J.W and Moir, P. N. (2001) *Heavy Oil Accumulations in the Jeanne d'Arc Basin: A Case Study in the Hebron, Ben Nevis, and West Ben Nevis Oil Fields*. Geological Survey of Canada Open-File Report D4012.

Shimoyama, A. and Yabuta, H. (2002) Mono- and bicyclic alkanes and diamondoid hydrocarbons in the Cretaceous/Tertiary boundary sediments at Kawaruppu, Hokkaido, Japan. *Geochemical Journal*, 36, 173–89.

Shoemaker, E. N. (1994) Large-body impacts are a cause of mass extinctions. Presented at the Annual Meeting of the American Association of Petroleum Geologists, June 12–15, 1994, Denver, CO.

Shuying, D. (1998) The oldest angiosperm: a tricarpous female reproductive fossil from western Liaoning Province, NE China. *Science in China. Series D, Earth Sciences*, 41, 14–20.

Siegert, M. J., Dowdeswell, J. A., Gorman, M. R. and McIntyre, N. F. (1996) An inventory of Antarctic sub-glacial lakes. *Antarctic Science*, 8, 281–6.

Siegert, M. J., Ellis-Evans, J. C., Tranter, M., *et al.* (2001) Physical, chemical and biological processes in Lake Vostok and other Antarctic subglacial lakes. *Nature*, 414, 603–9.

Sierra, M. G., Cravero, R. M., Laborde, M. A. and Ruveda, E. A. (1984) Stereoselective synthesis of $(+/-)$-18,19-Dinor-13β(H), 14α(H)-cheilanthane: the most abundant tricyclic compound from petroleums and sediments. *Journal of the Chemical Society, Chemical Communications*, 417–18.

Sieskind, O., Joly, G. and Albrecht, P. (1979) Simulation of the geochemical transformation of sterols: superacid effects of clay minerals. *Geochimica et Cosmochimica Acta*, 43, 1675–9.

Silliman, J. E., Meyers, P. A., Ostrom, P. H., Ostrom, N. W. and Eadie, B. J. (2000) Insights into the origin of perylene from isotopic analyses of sediments from Saanich Inlet, British Columbia. *Organic Geochemistry*, 31, 1133–42.

Silverman, S. R. (1965) Migration and segregation of oil and gas. In: *Fluids in Subsurface Environments*, Vol. 4 (A. Young and G. E. Galley, eds.), American Association of Petroleum Geologists, Tulsa, OK, pp. 53–65.

(1971) Influence of petroleum origin and transformation on its distribution and redistribution in sedimentary rocks. In: *Proceedings of the Eighth World Petroleum Congress*, Applied Science Publishers, London, pp. 47–54.

Simoneit, B. R. T. (1986) Cyclic terpenoids of the geosphere. In: *Biological Markers in the Sedimentary Record* (R. B. Johns, ed.), Elsevier, New York, pp. 43–99.

Simoneit, B. R. T. and Didyk, B. M. (1978) Organic geochemistry of Chilean paraffin dirt. *Chemical Geology*, 23, 21–40.

(1986) Paraffin dirt: bacterial biomass from natural gas, not weathered petroleum. In: *Workshop on Advances in Biomarkers and Kerogens*. Academia Sinica, Institute of Geochemistry, Guiyang, China, pp. 131–2.

(1992) Chilean paraffin dirt: part 3. Systematics of compound specific isotope analysis of bitumen. Presented at the 3rd Latin American Association of Organic Geochemistry (ALAGO) Congress, November, 1992 Manaus, Brazil.

Simoneit, B. R. T. and Mazurek, M. A. (1982) Organic matter of the troposphere – II. Natural background of biogenic lipid matter in aerosols over the rural western United States. *Atmospheric Environment*, 16, 2139–59.

Simoneit, B. R. T., Grimalt, J. G., Wang, T. G., *et al.* (1986) Cyclic terpenoids of contemporary resinous plant detritus and of fossil woods, ambers and coals. *Organic Geochemistry*, 10, 877–89.

Simoneit, B. R. T., Schoell, M., Dias, R. F. and Aquino Neto, F. R. (1993) Unusual carbon isotope compositions of biomarker hydrocarbons in a Permian tasmanite. *Geochimica et Cosmochimica Acta*, 57, 4205–11.

Simoneit, B. R. T., Otto, A. and Wilde, V. (2003) Novel phenolic biomarker triterpenoids of fossil laticifers in Eocene brown coal from Geiseltal, Germany. *Organic Geochemistry*, 34, 121–9.

Simons, D.-J. H. and Kenig, F. (2001) Molecular fossil constraints on the water column structure of the Cenomanian-Turonian Western Interior Seaway, USA. *Palaeogeography, Palaeoclimatology, Palaeoecology*, 169, 129–52.

Sinninghe Damsté, J. S. and de Leeuw, J. W. (1990) Analysis, structure and geochemical significance of organically-bound sulphur in the geosphere: state of the art and future research. *Organic Geochemistry*, 16, 1077–101.

(1995) Biomarkers or not biomarkers: a new hypothesis for the origin of pristane involving derivation from methyltrimethyltridecylchromans (MTTCs) formed during diagenesis from chlorophyll and alkylphenols. Comments on Li *et al.*, 1995 OG 23(2), 159–167. *Organic Geochemistry*, 23, 1085–93.

Sinninghe Damsté, J. S. and Koopmans, M. P. (1997) The fate of carotenoids in sediments: an overview. *Pure and Applied Chemistry*, 69, 2067–74.

Sinninghe Damsté, J. S. and Köster, J. (1998) A euxinic southern North Atlantic Ocean during the Cenomanian/Turonian oceanic anoxic event. *Earth and Planetary Science Letters*, 158, 165–73.

Sinninghe Damsté, J. S. and Rijpstra, W. I. C. (1993) Identification of a novel C_{25} highly branched isoprenoid thiophene in sediments. *Organic Geochemistry*, 20, 327–31.

Sinninghe Damsté, J. S. and Schouten, S. (1997) Is there evidence for a substantial contribution of prokaryotic biomass to organic carbon in Phanerozoic carbonaceous sediments? *Organic Geochemistry*, 26, 517–30.

Sinninghe Damsté, J. S., Kock-Van Dalen, A. C., de Leeuw, J. W., *et al.* (1987b) The identification of mono-, di- and trimethyl 2-methyl-2-(4,8,12-trimethyltridecyl) chromans and their occurrence in the geosphere. *Geochimica et Cosmochimica Acta*, 51, 2393–400.

Sinninghe Damsté, J. S., Rijpstra, I. C., de Leeuw, J. W. and Schenck, P. A. (1988) Origin of organic sulfur compounds and sulfur-containing high molecular weight substances in sediments and immature crude oils. *Organic Geochemistry*, 13, 593–606.

Sinninghe Damsté, J. S., van Koert, E. R., Kock-van Dalen, A. C., de Leeuw, J. W. and Schenck, P. A. (1989a) Characterisation of highly branched isoprenoid thiophenes occurring in sediments and immature crude oils. *Organic Geochemistry*, 14, 555–67.

Sinninghe Damsté, J. S., Rijpstra, W. I. C., de Leeuw, J. W. and Schenck, P. A. (1989b) The occurrence and identification of series of organic sulfur compounds in oils and sediment extracts: II. Their presence in samples from hypersaline and non-hypersaline paleoenvironments and possible application as source, paleoenvironmental and maturity indicators. *Geochimica et Cosmochimica Acta*, 53, 1323–41.

Sinninghe Damsté, J. S., Kock van Dalen, A. C., Albrecht, P. A. and de Leeuw, J. W. (1991) Identification of long-chain 1,2-di-*n*-alkylbenzenes in Amposta crude oil from the Tarragona Basin, Spanish Mediterranean: implications for the origin and fate of alkylbenzenes. *Geochimica et Cosmochimica Acta*, 55, 3677–83.

Sinninghe Damsté, J. S., de las Heras, F. X. C., van Bergen, P. F. and de Leeuw, J. W. (1993a) Characterization of Tertiary Catalan lacustrine oil shales: discovery of extremely organic sulphur-rich type I kerogens. *Geochimica et Cosmochimica Acta*, 57, 389–415.

Sinninghe Damsté, J. S., Wakeham, S. G., Kohnen, M. E. L., Hayes, J. M. and de Leeuw, J. W. (1993b) A 6,000-year sedimentary molecular record of chemocline excursions in the Black Sea. *Nature*, 362, 827–9.

Sinninghe Damsté, J. S., Keely, B. J., Betts, S. E., *et al.* (1993c) Variations in abundances and distributions of isoprenoid chromans and long-chain alkylbenzenes in sediments of the Mulhouse Basin: a molecular sedimentary record of palaeosalinity. *Organic Geochemistry*, 20, 1201–15.

Sinninghe Damsté, J. S., Kenig, F., Koopmans, M. P., *et al.* (1995) Evidence for gammacerane as an indicator of water-column stratification. *Geochimica et Cosmochimica Acta*, 59, 1895–900.

Sinninghe Damsté, J. S., Schouten, S., van Vliet, N. H., Huber, R. and Geenevasen, J. A. J. (1997) A polyunsaturated irregular acyclic C_{25} isoprenoid in a methanogenic archaeon. *Tetrahedron Letters*, 38, 6881–4.

Sinninghe Damsté, J. S., Köster, J., Bass, M., *et al.* (1998a) A sedimentary tetrahydrophenanthrene derivative of tetrahymanol. *Tetrahedron Letters*, 40, 3949–52.

Sinninghe Damsté, J. S., Schouten, S., van Vliet, N. H. and Geenevasen, J. A. J. (1998b) A sedimentary fluorene derivative of bacteriohopanepolyols. *Tetrahedron Letters*, 39, 3021–4.

Sinninghe Damsté, J. S., Rijpstra, W. I. C., Schouten, S., *et al.* (1999) A C_{25} highly branched isoprenoid alkene and C_{25} and C_{27} *n*-polyenes in the marine diatom *Rhizosolenia setigera*. *Organic Geochemistry*, 30, 95–100.

Sinninghe Damsté, J. S., Hopmans, E. C., Pancost, R. D., Schouten, S. and Geenevasen, J. A. J. (2000) Newly discovered non-isoprenoid glycerol dialkyl glycerol tetraether lipids in sediments. *Chemical Communications*, 2000, 1683–4.

Sinninghe Damsté, J. S., Schouten, S., Hopmans, E. C., van Duin, A. C. T. and Geenevasen, J. A. J. (2002a) Crenarchaeol: the characteristic core glycerol dibiphytanyl glycerol tetraether membrane lipid of cosmopolitan pelagic Crenarchaeota. *Journal of Lipid Research*, 43, 1641–51.

Slingerland, R., Kump, L. R., Arthur, M. A., *et al.* (1996) Estuarine circulation in the Turonian Western Interior Seaway of North America. *Geological Society of America Bulletin*, 108, 941–52.

(1998) Estuarine circulation in the Turonian Western Interior Seaway of North America – a reply. *Geological Society of America Bulletin*, 110, 693–4.

Smith, J. B. (1997) Oxford Clay. In: *Encyclopedia of Dinosaurs* (P. J. Currie and K. Padian, eds.), Academic Press, San Diego, CA, pp. 509–10.

(2002) The Batson-Old oilfield. In: *The New Handbook of Texas Online*. The General Libraries at the University of Texas at Austin and the Texas State Historical Association, Austin, TX.

Smith, M. G. and Bustin, R. M. (2000) Late Devonian and Early Mississippian Bakken and Exshaw black shale source rocks, Western Canada Sedimentary Basin: a sequence stratigraphic interpretation. *American Association of Petroleum Geologists Bulletin*, 84, 940–60.

Smith, G. C. and Floodgate, G. D. (1992) A chemical method for estimating methanogen biomass. *Continental Shelf Research*, 12, 1187–96.

Snedden, J. W., Sarg, J. F., Clutson, M. J., *et al.* (1996) Using sequence stratigraphic methods in high-sediment supply deltas: examples from the ancient Mahakam and Rajang-Lupar deltas. *Proceedings of the Indonesian Petroleum Association*, 25, 281–96.

Snowdon, L. R. (1980a) Petroleum source potential of the Boundary Creek Formation, Beaufort-Mackenzie Basin. *Bulletin of Canadian Petroleum Geology*, 28, 46–58.

(1980b) Resinite – a potential petroleum source in the Upper Cretaceous/Tertiary of the Beaufort-Mackenzie Basin. In: *Facts and Principles of World Petroleum Occurrence* (A. D. Miall, ed.), Canadian Society of Petroleum Geologists, Canadian Society of Petroleum Geologists, Calgary, Alberta, pp. 509–21.

Snowdon, L. R. and Powell, T. G. (1979) Families of crude oils and condensates in the Beaufort-Mackenzie Basin. *Bulletin of Canadian Petroleum Geology*, 27, 139–62.

Snowdon, L. R., Beauvilain, J. C. and Davies, G. R. (1998a) Debolt Formation oil-source systems: 1. Crude oil families in the Dunvegan-Blueberry area of Alberta and British Columbia. *Bulletin of Canadian Petroleum Geology*, 46, 266–75.

Snowdon, L. R., Davies, G. R. and Beauvilain, J. C. (1998b) Debolt Formation oil-source systems: 2. Authigenic petroleum source potential. *Bulletin of Canadian Petroleum Geology*, 46, 276–87.

So, C. M. and Young, L. Y. (1999a) Initial reactions in anaerobic alkane degradation by a sulfate reducer, strain AK-01. *Applied and Environmental Microbiology*, 65, 5532–40.

(1999b) Isolation and characterization of a sulfate-reducing bacterium that anaerobically degrades alkanes. *Applied and Environmental Microbiology*, 65, 2969–76.

Sofer, Z. (1984) Stable carbon isotope compositions of crude oils: application to source depositional environments and petroleum alteration. *American Association of Petroleum Geologists Bulletin*, 68, 31–49.

(1988) Biomarkers and carbon isotopes of oils in the Jurassic Smackover Trend of the Gulf Coast States, USA. *Organic Geochemistry*, 12, 421–32.

Sofer, Z., Zumberge, J. E. and Lay, V. (1986) Stable carbon isotopes and biomarkers as tools in understanding genetic relationship, maturation, biodegradation, and migration in crude oils in the Northern Peruvian Oriente (Maranon) Basin. *Organic Geochemistry*, 10, 377–89.

Solano-Saerena, F., Marchal, R., Ropars, M., Lebeault, J.-M. and Vandecasteele, J.-P. (1999) Biodegradation of gasoline: kinetics, mass balance, and fate of individual hydrocarbons. *Journal of Applied Microbiology*, 86, 1008–16.

Soldan, A. L. and Cerqueira, J. R. (1986) Effects of thermal maturation on geochemical parameters obtained by simulated generation of hydrocarbons. *Organic Geochemistry*, 10, 339–45.

Soltis, P. S., Soltis, D. E. and Chase, M. W. (1999) Angiosperm phylogeny inferred from multiple genes as a tool for comparative biology. *Nature*, 402, 402–4.

Songnian, L., Weimin, L. and Wei, H. (1988) Ring D aromatized 8,14-secohopanes in crude oils and source rocks and their geochemical significance. *Acta Sedimentologica Sinica*, 6, 41–9.

Sosrowidjojo, I. B., Alexander, R. and Kagi, R. I. (1993) Identification of novel diaromatic secobicadinanes and tricadinanes in crude oils. In: *Proceedings of the 16th International Meeting on Organic Geochemistry, September 20–24*, Stavanger, Norway (K. Øygard, ed.), Falch Hurtigtrykk, Oslo, pp. 481–4.

(1994a) The biomarker composition of some crude oils from Sumatra. *Organic Geochemistry*, 21, 303–12.

Sosrowidjojo, I. B., Setiardja, B., Zakaria, K. P. G., Alexander, R. and Kagi, R. I. (1994b) A new geochemical method for assessing the maturity of petroleum: application to the South Sumatra Basin. In: *Proceedings of the 23rd Annual Convention of the Indonesian Petroleum Association*, Indonesian Petroleum Association, Jakarta, Indonesia, pp. 433–5.

Sosrowidjojo, I. B., Murray, A. P., Alexander, R., Kagi, R. I. and Summons, R. E. (1996) Bicadinanes and related compounds as maturity indicators for oils and sediments. *Organic Geochemistry*, 24, 43–55.

Sousa, J. J. F., Vugman, N. V. and Neto, C. C. (1997) Free radical transformations in the Irati oil shale due to diabase intrusion. *Organic Geochemistry*, 26, 183–9.

Souto Filho, J. D., Correa, A. C. F., Santos Neto, E. V. and Trindade, L. A. F. (2000) Alagamar-Açu petroleum system, onshore Potiguar Basin, Brazil: a numerical approach for secondary migration. In: *Petroleum Systems of South Atlantic Margins* (M. R. Mello and B. J. Katz, eds.), American Association of Petroleum Geologists, Tulsa, OK, pp. 151–8.

Soylu, C. (1987) Source rock potential of Dadas Formation (Southeast Turkey). Presented at the International Conference on the Petroleum Geochemistry & Exploration in the Afro-Asian Region, November 25–27, 1987, Dehra Dun, India.

Spangenberg, J. E. and Macko, S. A. (1998) Organic geochemistry of the San Vicente zinc–lead district, eastern Pucará Basin, Peru. *Chemical Geology*, 146, 1–24.

Spangenberg, J. E., Fontboté, L., Sharp, Z. D. and Hunziker, J. (1996) Carbon and oxygen isotope study of hydrothermal carbonates in the zinc–lead deposits of the Sand Vicente district, central Peru: a quantitative modeling on mixing processes and CO_2 degassing. *Chemical Geology*, 133, 289–315.

Spangenberg, J. E., Fontboté, L. and Macko, S. A. (1999) An evaluation of the inorganic and organic geochemistry of the San Vicente Mississippi Valley-type Zinc–Lead District, Central Peru: implications for ore fluid composition, mixing processes, and sulfate reduction. *Economic Geology*, 94, 1067–92.

Spark, I., Patey, I., Duncan, B., *et al.* (2000) The effects of indigenous and introduced microbes on deeply buried hydrocarbon reservoirs, North Sea. *Clay Minerals*, 35, 5–12.

Spiker, E. C., Kotra, R. K., Hatcher, P. G., *et al.* (1988) Source of kerogen in black shales from the Hartford and Newark basins, Eastern United States. In: *Studies of the Early Mesozoic Basins of the Eastern United States* (A. J. Froelich and G. R. Robinson, Jr, eds.), U.S. Geological Survey Bulletin 1776, U.S. Geological Survey, Washington, DC, pp. 63–8.

Spiro, B., Welte, D. H., Rullkötter, J. and Schaefer, R. G. (1983) Asphalts, oils, and bituminous rocks from the Dead Sea area: a geochemical correlation study. *American Association of Petroleum Geologists Bulletin*, 67, 1163–75.

Stanley, S. M. (1988) Paleozoic mass extinctions: shared patterns suggest global cooling as a common cause. *American Journal of Science*, 288, 334–52.

Staplin, F. L. (1969) Sedimentary organic matter, organic metamorphism, and oil and gas occurrence. *Canadian Petroleum Geologists Bulletin*, 17, 47–66.

Stasiuk, L. D. (1991) Organic petrology and petroleum formation in Paleozoic rocks of northern Williston Basin, Canada. Ph. D. thesis, University of Regina, Regina, Canada.

Stasiuk, L. D. and Osadetz, K. G. (1990) Progress in the life cycle and phyletic affinity of *Gloeocapsomorpha prisca* Zalessky 1917 from Ordovician rocks in Canadian Williston Basin. In: *Geological Survey of Canada, Current Research, Part D, Paper 90–1D*, National Resources Canada, Ottawa, pp. 127–37.

Stasiuk, L. D., Kybett, B. D. and Bend, S. L. (1993) Reflected light microscopy and micro-FTIR of Upper Ordovician *Gloeocapsomorpha prisca* alginite in relation to paleoenvironment and petroleum generation, Saskatchewan, Canada. *Organic Geochemistry*, 20, 707–19.

Stefani, M. and Burchell, M. (1990) Upper Triassic (Rhaetic) argillaceous sequences in northern Italy: depositional dynamics and source potential. In: *Deposition of Organic Facies* (A.-Y. Huc, ed.), American Association of Petroleum Geologists, Tulsa, OK, pp. 93–106.

(1993) A review of the Upper Triassic source rocks of Italy. In: *Generation, Accumulation and Production of Europe's Hydrocarbons III. Special Publication of the European Association of Petroleum Geoscientists 3*, (A. M. Spencer, ed.), Springer-Verlag, Berlin, pp. 169–78.

Stefanova, M. (2000) Head-to-head linked isoprenoids in Miocene coal lithotypes. *Fuel*, 79, 755–8.

Stefanova, M. and Disnar, J. R. (2000) Composition and early diagenesis of fatty acids in lacustrine sediments, Lake Aydat (France). *Organic Geochemistry*, 31, 41–55.

Stefanova, M., Magnier, C. and Velinova, D. (1995) Biomarker assemblage of some Miocene-aged Bulgarian lignite lithotypes. *Organic Geochemistry*, 23, 1067–84.

Stemmerik, L., Christiansen, F. G. and Piasecki, S. (1990) Carboniferous lacustrine shale in East Greenland – additional source rock in Northern North Atlantic? In: *Lacustrine Basin Exploration: Case Studies and Modern Analogs* (B. J. Katz, ed.), American Association of Petroleum Geologists, Tulsa, OK, pp. 277–86.

Stephan, T., Rost, D., Jessberger, E. K. and Greshake, A. (1998) Polycyclic aromatic hydrocarbons in ALH84001 analyzed with time-of-flight secondary ion mass spectrometry. Presented at the 29th Lunar and Planetary Science Meeting, Lunar and Planetary Institute, Houston, TX.

Stephens, N. P. and Carroll, A. R. (1999) Salinity stratification in the Permian Phosphoria sea; a proposed paleoceanographic model. *Geology*, 27, 899–902.

Stetter, K. O., Huber, R., Blöchl, E., *et al.* (1993) Hyperthermophilic archaea are thriving in deep North Sea and Alaskan oil reservoirs. *Nature*, 365, 743–5.

Stevens, T. O. and Mckinley, J. P. (1995) Lithoautotrophic microbial ecosystems in deep basalt aquifers. *Science*, 270, 450–4.

Stewart, W. N. and Rothwell, G. W. (1993) *Paleobotany and the Evolution of Plants*. Cambridge University Press, Cambridge, UK.

Stojanovic, K., Jovancicevic, B., Pevneva, G. S., *et al.* (2001) Maturity assessment of oils from the Sakhalin oil fields in Russia: phenanthrene content as a tool. *Organic Geochemistry*, 39, 721–31.

Stoneley, R. (1987) A review of petroleum source rocks in parts of the Middle East. In: *Marine Petroleum Source Rocks* (J. Brooks and A. J. Fleet, eds.), Blackwell, Oxford, UK, pp. 263–9.

Stoufer, S., Chigne, N., Santchez, J. H. and Mello, M. R. (1998) Cretaceous and Tertiary petroleum systems of the northwest region of the Eastern Venezuela Basin. *American Association of Petroleum Geologists Bulletin*, 82, 1883–984.

Stout, S. A. and Lundegard, P. D. (1998) Intrinsic biodegradation of diesel fuel in an interval of separate phase hydrocarbons. *Applied Geochemistry*, 13, 851–9.

Stover, S. C., Ge, S., Weimer, P. and McBride, B. C. (2001) The effects of salt evolution, structural development, and fault propagation on Late Mesozoic-Cenozoic oil migration: a two-dimensional fluid-flow study along a megaregional profile in the northern Gulf of Mexico Basin. *American Association of Petroleum Geologists Bulletin*, 11, 1945–66.

Strachan, M. G., Alexander, R. and Kagi, R. I. (1986) Trimethylnaphthalenes as depositional environmental indicators. Presented at the 192nd *Annual American Chemical Society Meeting*, September, 1986, Anaheim, CA.

(1988) Thimethylnaphthalenes in crude oils and sediments: effects of source and maturity. *Geochimica et Cosmochimica Acta*, 52, 1255–64.

Strachan, M. G., Alexander, R., Subroto, E. A. and Kagi, R. I. (1989a) Constraints upon the use of 24-ethylcholestane diastereomer ratios as indicators of the maturity of petroleum. *Organic Geochemistry*, 14, 423–32.

Strachan, M. G., Alexander, R., van, B. W. and Kagi, R. I. (1989b) Source and heating rate effects upon maturity parameters based on ratios of 24-ethylcholestane diastereomers. *Journal of Geochemical Exploration*, 31, 285–94.

Strauss, H. and Moore, T. B. (1992) Abundances and isotopic compositions of carbon and sulfur species in whole rock and kerogen samples. In: *The Proterozoic Biosphere: A Multidisciplinary Study* (J. W. Schopf and C. Klein, eds.), Princeton University Press, Princeton, NJ, pp. 709–98.

Strauss, H., Vidal, G., Moczydlowska, M. and Paczesna, J. (1997) Carbon isotope geochemistry and palaeontology of Neoproterozoic to early Cambrian siliciclastic successions in the East European Platform, Poland. *Geology Magazine*, 134, 1–16.

Strausz, O. P., Chen, H. H., Ekweozor, C. M., Ekwenchi, M. M. and Samman, N. (1982) Organic geochemistry of the Alberta Oil Sands: the nature and distribution of biomarkers. In: *Eleventh International Congress on Sedimentology. Hamilton, Ontario, Canada, Aug. 22–27, 1982, 11* (J. O. Nriagu and R. Troost, eds.), McMaster University, Hamilton, Ontario, p. 89.

Strong, D. and Filby, R. H. (1987) Vanadyl porphyrin distribution in the Alberta oil-sand bitumens. In: *Metal Complexes in Fossil Fuels* (R. H. Filby and J. F. Branhhaven, eds.), American Chemical Society, Washington, DC, pp. 154–72.

Strothers, R. B. (1993) Flood basalts and extinction events. *Geophysical Research Letters*, 20, 1399–402.

Struckmeyer, H. I. M. and Felton, E. A. (1990) The use of organic facies for refining palaeoenvironmental interpretations: a case study from the Otway Basin, Australia. *Australian Journal of Earth Sciences*, 37, 351–64.

Subroto, E. A., Alexander, R. and Kagi, R. I. (1991) 30-Norhopanes: their occurrence in sediments and crude oils. *Chemical Geology*, 93, 179–92.

Sulistyo, G. B. (1994) *Source Rock Evaluation and Oil-Source Rock Correlation of the Upper Pennsylvanian (Virgilian) and Lower Permian (Wolfcampian) in Southwestern Nebraska*. Colorado School of Mines, Golden, CO.

Summons, R. E. (1987) Branched alkanes from ancient and modern sediments: isomer discrimination by GC/MS with multiple reaction monitoring. *Organic Geochemistry*, 11, 281–9.

Summons, R. E. and Capon, R. J. (1988) Fossil steranes with unprecedented methylation in ring A. *Geochimica et Cosmochimica Acta*, 52, 2733–6.

(1991) Identification and significance of 3β-ethyl steranes in sediments and petroleum. *Geochimica et Cosmochimica Acta*, 55, 2391–5.

Summons, R. E. and Jahnke, L. L. (1992) Hopenes and hopanes methylated in ring A: correlation of the hopanoids from extant methylotrophic bacteria with their fossil analogues. In: *Biological Markers in Sediments and Petroleum* (J. M. Moldowan, P. Albrecht and R. P. Philp, eds.), Prentice-Hall, Englewood Cliffs, NJ, pp. 182–200.

Summons, R. E. and Powell, T. G. (1986) *Chlorobiaceae* in Palaeozoic sea revealed by biological markers, isotopes, and geology. *Nature*, 319, 763–5.

(1987) Identification of aryl isoprenoids in a source rock and crude oils: biological markers for the green sulfur bacteria. *Geochimica et Cosmochimica Acta*, 51, 557–66.

(1991) Petroleum source rocks of the Amadeus Basin. In: *Geological and Geophysical Studies in the Amadeus Basin, Central Australia* (R. J. Korsch and J. M. Kennard, eds.), Australian Bureau of Mineral Resources, Canberra, Australia, pp. 511–24.

Summons, R. E. and Walter, M. R. (1990) Molecular fossils and microfossils of prokaryotes and protists from Proterozoic sediments. *American Journal of Science*, 290A, 212–44.

Summons, R. E., Volkman, J. K. and Boreham, C. J. (1987) Dinosterane and other steroidal hydrocarbons of dinoflagellate origin in sediments and petroleum. *Geochimica et Cosmochimica Acta*, 51, 3075–82.

Summons, R. E., Powell, T. G. and Boreham, C. J. (1988a) Petroleum geology and geochemistry of the Middle Proterozoic McArthur Basin, Northern Australia: 3. Composition of extractable hydrocarbons. *Geochimica et Cosmochimica Acta*, 52, 1747–63.

Summons, R. E., Brassell, S. C., Eglinton, G., *et al.* (1988b) Distinctive hydrocarbon biomarkers from fossiliferous sediment of the Late Proterozoic Walcott Member, Chuar Group, Grand Canyon, Arizona. *Geochimica et Cosmochimica Acta*, 52, 2625–37.

Summons, R. E., Thomas, J., Maxwell, J. R. and Boreham, C. J. (1992) Secular and environmental constraints on the occurrence of dinosterane in sediments. *Geochimica et Cosmochimica Acta*, 56, 2437–44.

Summons, R. E., Barrow, R. A., Capon, R. J., Hope, J. M. and Stranger, C. (1993) The structure of a new C_{25} isoprenoid alkene biomarker from diatomaceous microbial communities. *Australian Journal of Chemistry*, 46, 907–15.

Summons, R. E., Jahnke, L. L. and Roksandic, Z. (1994) Carbon isotopic fractionation in lipids from methanotrophic bacteria: relevance for interpretation of the geochemical record of biomarkers. *Geochimica et Cosmochimica Acta*, 58, 2853–63.

Summons, R. E., Boreham, C. J., Foster, C. B., Murray, A. P. and Gorter, J. D. (1995) Chemostratigraphy and the composition of oils in the Perth Basin, Western Australia. *APEA Journal*, 35, 613–31.

Summons, R. E., Nichols, P. D. and Franzmann, P. D. (1996) C-Isotopic fractionation during methylotrophic methanogenesis. Presented at the Annual Meeting of the American Association of Petroleum Geologists, May 19–22, 1996, San Diego, CA.

Summons, R. E., Jahnke, L. L., Hope, J. M. and Logan, G. A. (1999) 2-Methylhopanoids as biomarkers for cyanobacterial oxygenic photosynthesis. *Nature*, 400, 554–7.

Summons, R. E., Logan, G. A., Edwards, D. S., *et al.* (2001) Geochemical analogs for Australian coastal asphaltites – search for the source rock. Presented at the Annual Meeting of the American Association of Petroleum Geologists, June 3–6, 2001, Denver, CO.

Summons, R. E., Metzger, P., Largeau, C., Murray, A. P. and Hope, J. M. (2002) Polymethylsqualanes from *Botryococcus braunii* in lacustrine sediments and oils. *Organic Geochemistry*, 33, 99–109.

Sun, Y. (1996) Geochemical evidence for multi-stage base metal enrichment in Kupferschiefer. Ph. D. thesis, Aachen University of Technology, Aachen, Germany.

(1998) Influences of secondary oxidation and sulfide formation on several maturity parameters in Kupferschiefer. *Organic Geochemistry*, 29, 1419–29.

Sun, Y. and Püttmann, W. (1996) Relationship between metal enrichment and organic composition in Kupferschiefer hosting structure-controlled mineralization from Oberkatz Schwelle, Germany. *Applied Geochemistry*, 11, 567–81.

(1997) Metal accumulation during and after deposition of Kupferschiefer from Niederröblingen, Sangerhausen Basin, Germany. *Applied Geochemistry*, 12, 577–92.

(2000) The role of organic matter during copper enrichment in Kupferschiefer from the Sangerhausen Basin, Germany. *Organic Geochemistry*, 31, 1143–61.

(2001) Oxidation of organic matter in the transition zone of the Zechstein Kupferschiefer from the Sangerhausen Basin, Germany. *Energy & Fuels*, 15, 817–29.

Sundararaman, P. (1985) High-performance liquid chromatography of vanadyl porphyrins. *Analytical Chemistry*, 57, 2204–6.

Sundararaman, P. and Boreham, C. J. (1991) Vanadyl 3-nor C_{30}DPEP: indicator of depositional environment of a lacustrine sediment. *Geochimica et Cosmochimica Acta*, 55, 389–95.

Sundararaman, P. and Hwang, R. J. (1993) Effect of biodegradation on vanadylporphyrin distribution. *Geochimica et Cosmochimica Acta*, 57, 2283–90.

Sundararaman, P., Biggs, W. R., Reynolds, J. G. and Fetzer, J. C. (1988a) Vanadylporphyrins, indicators of kerogen breakdown and generation of petroleum. *Geochimica et Cosmochimica Acta*, 52, 2337–41.

Sundararaman, P., Moldowan, J. M. and Seifert, W. K. (1988b) Incorporation of petroporphyrins into biomarker correlation problems. In: *Geochemical Biomarkers* (T. F. Yen and J. M. Moldowan, eds.), Harwood Academic Publishers, New York, pp. 373–82.

Suseno, P. H. Z., Mujahindin, N. and Subroto, E. A. (1992) Contribution of Lahat Formation as hydrocarbon source rock in South Palembang area, South Sumatra, Indonesia. In: *Proceedings of the Indonesian Petroleum Association Twenty First Annual Convention, October, 1992*. Indonesian Petroleum Association, Jakarta, Indonesia, pp. 325–37.

Suzuki, N. (1983) Estimation of paleomaximum temperature of mudstone by two kinetic parameters (isomerizations of sterane and hopane). Presented at the 4th International Symposium on Water–Rock Interaction, Misasa, Japan.

(1984) Estimation of maximum temperature of mudstone by two kinetic parameters: epimerization of sterane and hopane. *Geochimica et Cosmochimica Acta*, 8, 2273–82.

Suzuki, N., Yoshikazu, S. and Koga, O. (1993) Norcholestane in Miocene Onnagawa siliceous sediments, Japan. *Geochimica et Cosmochimica Acta*, 57, 4539–45.

Sweeney, J. J. and Burnham, A. K. (1990) Evaluation of a simple model of vitrinite reflectance based on chemical kinetics. *American Association of Petroleum Geologists Bulletin*, 74, 1559–70.

Switzer, S. B., Holland, W. G., Christie, D. S., et al. (1994) Devonian Woodbend-Winterburn strata of the Western Canada Sedimentary Basin. In: *Geological Atlas of the Western Canada Sedimentary Basin* (G. D. Mossop and I. Shetsen, eds.), Canadian Society of Petroleum Geologists and Alberta Research Council, Alberta, pp. 165–202.

Szatmari, P. (2000) Habitat of petroleum along the South Atlantic margins. In: *Petroleum Systems of South Atlantic Margins* (M. R. Mello and B. J. Katz, eds.), American Association of Petroleum Geologists, Tulsa, OK, pp. 69–75.

Taher, A. A. (1997) Delineation of organic richness and thermal history of the Lower Cretaceous Thamama Group, East Abu Dhabi. A modeling approach for oil exploration. *Geoarabia*, 2, 56–88.

Takada, H. and Ishiwatari, R. (1990) Biodegradation experiments of linear alkylbenzenes (LABs): isomeric composition of C_{12} LABs as an indicator of the degree of LAB degradation in the aquatic environment. *Environmental Science & Technology*, 24, 86–91.

Takahashi, K. I. (2001) U.S. *Geological Survey Coalbed Methane Field Conference, Casper, WY, United States, May 9–10, 2001*. US Geological Survey Open-File Report, OF 01-0235.

Talukdar, S. C. and Marcano, F. (1994) Petroleum systems of the Maracaibo Basin, Venezuela. In: *The Petroleum System – From Source to Trap* (L. B. Magoon and W. G. Dow, eds.), American Association of Petroleum Geologists Tulsa, OK, pp. 463–81.

Talukdar, S., Gallango, O. and Chin-A-Lien, M. (1986) Generation and migration of hydrocarbons in the Maracaibo Basin, Venezuela: an integrated basin study. *Organic Geochemistry*, 10, 261–79.

Talukdar, S., Gallango, O. and Ruggiero, A. (1988) Generation and migration of oil in the Maturin Subbasin, Eastern Venezuelan Basin. *Organic Geochemistry*, 13, 537–47.

Tang, Y. C. and Stauffer, M. (1995) Formation of pristene, pristane and phytane: kinetic study by laboratory pyrolysis of Monterey source rock. *Organic Geochemistry*, 23, 451–60.

Tannenbaum, E., Ruth, E., Huizinga, B. J. and Kaplan, I. R. (1986) Biological marker distribution in coexisting kerogen, bitumen and asphaltenes in Monterey Formation diatomite, California. *Organic Geochemistry*, 10, 531–6.

Tappan, H. N. (1980) *The Paleobiology of Plant Protists*. W. H. Freeman, San Francisco.

Taylor, P. J. (1994) The scientific legacy of Apollo. *Scientific American*, 271, 40–7.

Taylor, D. W. and Hickey, L. J. (1996) Evidence for and implications of an herbaceous origin for angiosperms, In: *Flowering Plant Origin, Evolution and Phylogeny* (D. W. Taylor and L. J. Hickey, eds.), Chapman and Hall, New York, pp. 232–66.

Taylor, T. N. and Taylor, E. L. (1993) *The Biology and Evolution of Fossil Plants*. Prentice-Hall, Englewood Cliffs, NJ.

Taylor, P., Bennett, B., Jones, M. and Larter, S. (2001) The effect of biodegradation and water washing on the occurrence of alkylphenols in crude oils. *Organic Geochemistry*, 32, 341–58.

Taylor, D. W., Li, H., Dahl, J., et al. (2004) Biogeochemical evidence for late Paleozoic origin of angiosperms. *Paleobiology*, submitted.

Tegelaar, E. W. and Noble, R. A. (1994) Kinetics of hydrocarbon generation as a function of the molecular structure of kerogen as revealed by pyrolysis-gas chromatography. *Organic Geochemistry*, 22, 543–74.

Teisserenc, P. and Villemin, J. (1989) Sedimentary basin of Gabon – geology and oil systems. In: *Divergent/Passive Margin Basins* (J. D. Edwards and P. A. Santogrossi, eds.), American Association of Petroleum Geologists Tulsa, OK, pp. 117–99.

Telnaes, N. and Dahl, B. (1986) Oil-oil correlation using multivariate techniques. *Organic Geochemistry*, 10, 425–32.

Ten Haven, H. L. and Rullkötter, J. (1988) The diagenetic fate of taraxer-14-ene and oleanene isomers. *Geochimica et Cosmochimica Acta*, 52, 2543–8.

Ten Haven, H. L., de Leeuw, J. W., Peakman, T. M. and Maxwell, J. R. (1986) Anomalies in steroid and hopanoid maturity indices. *Geochimica et Cosmochimica Acta*, 50, 853–5.

Ten Haven, H. L., de Leeuw, J. W., Rullkötter, J. and Sinninghe Damsté, J. S. (1987) Restricted utility of the pristane/phytane ratio as a palaeoenvironmental indicator. *Nature*, 330, 641–3.

Ten Haven, H. L., de Leeuw, J. W., Sinninghe Damsté, J. S., *et al.* (1988) Application of biological markers in the recognition of palaeohypersaline environments. In: *Lacustrine Petroleum Source Rocks* (A. J. Fleet, K. Kelts and M. R. Talbot, eds.), Blackwell, London, pp. 123–30.

Ten Haven, H. L., Rohmer, M., Rullkötter, J. and Bisseret, P. (1989) Tetrahymanol, the most likely precursor of gammacerane, occurs ubiquitously in marine sediments. *Geochimica et Cosmochimica Acta*, 53, 3073–9.

Ten Haven, H. L., Peakman, T. M. and Rullkötter, J. (1992) Δ^2-Triterpenes: early intermediates in the diagenesis of terrigenous triterpenoids. *Geochimica et Cosmochimica Acta*, 56, 1993–2000.

Terken, J. M. J. (1999) The Natih petroleum system of north Oman. *Geoarabia*, 4, 157–80.

Terken, J. M. J. and Frewin, N. L. (2000) The Dhahaban petroleum system of Oman. *American Association of Petroleum Geologists Bulletin*, 84, 523–44.

Terken, J. M. J., Frewin, N. L. and Indrelid, S. L. (2001) Petroleum systems of Oman: charge timing and risks. *American Association of Petroleum Geologists Bulletin*, 85, 1817–45.

Teschner, M. and Wehner, H. (1985) Chromatographic investigations of biodegraded crude oils. *Chromatographia*, 20, 407–16.

Thiel, V., Jenisch, A., Wörheide, G., *et al.* (1999a) Mid-chain branched alkanoic acids from "living fossil" desmosponges: a link to ancient sedimentary lipids? *Organic Geochemistry*, 30, 1–14.

Thiel, V., Peckmann, J., Seifert, R., *et al.* (1999b) Highly isotopically depleted isoprenoids: molecular markers for ancient methane venting. *Geochimica et Cosmochimica Acta*, 63, 3959–66.

Thiel, V., Peckmann, J., Richnow, H. W., *et al.* (2001) Molecular signals for anaerobic methane oxidation in Black Sea seep carbonates and a microbial mat. *Marine Chemistry*, 73, 97–112.

Thomas, J. (1990) Biological markers in sediments with respect to geological time. Ph. D. Thesis, University of Bristol. Bristol, UK.

Thomas, B. M., Moller-Pedersen, P., Whitaker, M. F. and Shaw, N. D. (1985) Organic facies and hydrocarbon distributions in the Norwegian North Sea. In: *Petroleum Geochemistry in Exploration of the Norwegian Shelf* (B. M. Thomas, S. S. Eggen, R. M. Larsen, P. C. Home and A. G. Doré, eds.), Graham and Trotman, London, pp. 3–26.

Thomas, J. B., Mann, A. L., Brassell, S. C. and Maxwell, J. R. (1989) 4-Methyl steranes in Triassic sediments: molecular evidence for the earliest dinoflagellates. Presented at the 14th International Meeting on Organic Geochemistry, September 18–22, 1989, Paris.

Thomas, D. J., Bralower, T. J. and Zachos, J. C. (1999) New evidence for subtropical warming during the late Paleocene thermal maximum: stable isotopes from Deep Sea Drilling Project Site 527, Walvis Ridge. *Paleoceanography*, 14, 561–70.

Thompson, K. F. M. (1983) Classification and thermal history of petroleum based on light hydrocarbons. *Geochimica et Cosmochimica Acta*, 47, 303–16.

(1987) Fractionated aromatic petroleums and the generation of gas-condensates. *Organic Geochemistry*, 11, 573–90.

Thompson, S. and Voropanov, V. (1997) The Irkutsk region of Eastern Siberia. Proterozoic and Early Cambrian exploration potential. *American Association of Petroleum Geologists Bulletin*, 81, 1416.

Thompson, S., Cooper, B. S., Morley, R. J. and Barnard, P. C. (1985) Oil-generating coals. In: *Petroleum Geochemistry in Exploration of the Norwegian Shelf.* (B. M. Thomas, S. S. Eggen, R. M. Larsen, et al., eds.). Graham and Trotman, London, pp. 59–73.

Thorn, K. A. and Aiken, G. R. (1998) Biodegradation of crude oil into nonvolatile organic acids in a contaminated aquifer near Bemidji, Minnesota. *Organic Geochemistry*, 29, 909–31.

Tiercelin, J.-J., Potdevin, J.-L. and Talbot, M. R. (2001) Lacustrine source rocks and fluvio-lacustrine reservoirs in northern Kenya. Presented at the Annual Meeting of the American Association of Petroleum Geologists, Denver, CO. June 3–6, 2001.

Tintori, A. (1991) Fish taphonomy and Triassic anoxic basins from the Alps: a case history. *Rivista Italiana di Paleontologia e Stratigrafia*, 97, 393–407.

Tischler, K. L. (1995) Paradox Basin source rock, southeastern Utah: organic geochemical characterization of Gothic and Chimney Rock Units, Ismay and Desert Creek zones within a sequence stratigraphic framework. M. S. thesis, University of Texas, TX.

Tissot, B. P. and Welte, D. H. (1984) *Petroleum Formation and Occurrence*. Springer-Verlag, New York.

Tissot, B., Califet-Debyser, Y., Deroo, G. and Oudin, J. L. (1971) Origin and evolution of hydrocarbons in early Toarcian shales, Paris Basin, France. *American Association of Petroleum Geologists Bulletin*, 55, 2177–93.

Tissot, B. P., Deroo, G. and Hood, A. (1978) Geochemical study of the Uinta Basin: formation of petroleum from the Green River Formation. *Geochimica et Cosmochimica Acta*, 42, 1469–85.

Tissot, B., Espitalié, J. Deroo, G., Tempere, C. and Jonathan, D. (1984) Origin and migration of hydrocarbons in the eastern Sahara (Algeria). In: *Petroleum Geochemistry and Basin Evaluation*, (G. Demaison and R. J. Murris, eds.), American Association of Petroleum Geologists, pp. 315–24.

Tocco, R., Alberdi, M., Ruggiero, A. and Jordan, N. (1994) Organic geochemistry of the Carapita Formation and terrestrial crude oils in the Maturin Subbasin, Eastern Venezuelan Basin. *Organic Geochemistry*, 21, 1107–19.

Todd, S. P., Dunn, M. E. and Barwise, A. J. G. (1997) Characterizing petroleum charge systems in the Tertiary of SE Asia. In *Petroleum Geology of Southeast Asia* (A. J. Fraser, S. J. Matthews, and R. W. Murphy, eds.), Geological Society Special Publication 126, p. 25–47.

Tomasek, P. H. and Crawford, R. L. (1986) Initial reactions of xanthone biodegradation by an *Arthrobacter* sp. *Journal of Bacteriology*, 167, 818–27.

Tomczyk, N. A., Winans, R. E., Shinn, J. H. and Robinson, R. C. (2001) On the nature and origin of acidic species in petroleum. 1. Detailed acid type distribution in a California crude oil. *Energy & Fuels*, 15, 1498–504.

Tornabene, T. G. (1978) Non-aerated cultivation of *Halobacterium cutirubrum* and its effects on cellular squalenes. *Journal of Molecular Evolution*, 11, 253–7.

Tornabene, T. G., Wolfe, R. S., Balch, W. E., *et al.* (1978) Phytanyl-glycerol ethers and squalenes in the archaebacterium *Methanobacterium thermoautotrophicum*. *Journal of Molecular Evolution*, 11, 259–66.

Tornabene, T. G., Langworth, T. A., Holzer, G. and Oró, J. (1979) Squalenes, phytanes, and other isoprenoids as major neutral lipids of methanogenic and thermoacidophilic "archaebacteria". *Journal of Molecular Evolution*, 13, 73–83.

Toupin, D., Eadington, P. J., Person, M., *et al.* (1997) Petroleum hydrogeology of the Cooper and Eromanga basins, Australia. Some insights from mathematical modeling and fluid inclusion data. *American Association of Petroleum Geologists Bulletin*, 81, 577–603.

Towe, K. M. (1990) Aerobic respiration in the Archean? *Nature*, 348, 54–6.

Townsend, G. T., Prince, R. C. and Suflita, J. M. (2003) Anaerobic oxidation of crude oil hydrocarbons by the resident microorganisms of a contaminated anoxic aquifer. *Environmental Science & Technology*. 37, 5213–18.

Trendel, J. M., Restle, A., Connan, J. and Albrecht, P. (1982) Identification of a novel series of tetracyclic terpene hydrocarbons (C_{24}–C_{27}) in sediments and petroleums. *Journal of the Chemical Society, Chemical Communications*, 304–6.

Trendel, J. M., Lohmann, F., Kintzinger, J. P., *et al.* (1989) Identification of des-*A*-triterpenoid hydrocarbons occurring in surface sediments. *Tetrahedron*, 45, 4457–70.

Trendel, J., Guilhem, J., Crisp, P., *et al.* (1990) Identification of two C-10 demethylated C_{28} hopanes in biodegraded petroleum. *Journal of the Chemical Society, Chemical Communications*, 424–5.

Trendel, J. M., Graff, R., Wehrung, P., *et al.* (1993) C(14α)-Homo-26-nor-17α-hopanes, a novel and unexpected series of molecular fossils in biodegraded petroleum. *Journal of the Chemical Society (London), Chemical Communications*, 461–3.

Trifilieff, S. (1987) Etude de la structure des fractions polaires de pétroles (résines et asphalténes) par dégradations chimiques sélectives. Ph. D. thesis, L'Universite Louis Pasteur de Strasbourg, Strasbourg, France.

Trifilieff, S., Sieskind, O. and Albrecht, P. (1992) Biological markers in petroleum asphaltenes: possible mode of incorporation. In: *Biological Markers in Sediments and Petroleum* (J. M. Moldowan, P. A. Albrecht, and R. P. Philp, eds.), Prentice-Hall, Englewood Cliffs, NJ, pp. 350–69.

Trindade, L. A. F., Brassell, S. C. and Santos Neto, E. V. (1992) Petroleum migration and mixing in the Potiguar Basin, Brazil. *American Association of Petroleum Geologists Bulletin*, 76, 1903–24.

Trindade, L. A. F., Dias, J. L. and Mello, M. R. (1994) Sedimentological and geochemical characterization of the Lagoa Feia Formation, rift phase of the Campos Basin, Brazil. In: *Petroleum Source Rocks* (B. J. Katz, ed.), Springer-Verlag, Berlin, pp. 149–65.

Tritz, J.-P., Herrmann, D., Bisseret, P., Connan, J. and Rohmer, M. (1999) Abiotic and biological hopanoid transformation: towards the formation of molecular fossils of the hopane series. *Organic Geochemistry*, 30, 499–514.

Trolio, R., Grice, K., Fisher, S. J., Alexander, R. and Kagi, R. I. (1999) Alkylbiphenyls and alkyldiphenylmethanes as indicators of petroleum biodegradation. *Organic Geochemistry*, 30, 1241–53.

Tsuda, Y., Tabata, Y. and Ichinohe, Y. (1980) *Lycopodium* triterpenoids (10). Triterpenoid constituents of *Lycopodium wightianum* collected in Borneo. *Chemical and Pharmaceutical Bulletin*, 28, 3275–82.

Tupper, N. P., Padley, D., Lovibond, R., Duckett, A. K. and McKirdy, D. M. (1993) A key test of Otway Basin potential: the Eumeralla-sourced play on the Chama Terrace. *Australian Petroleum Exploration Association Journal*, 33, 77–93.

Tuttle, M. L., Klett, T. R., Richardson, M. and Breit, G. N. (1996) Geochemistry of two interbeds in the Pennsylvanian Paradox Formation, Utah and Colorado – a record of deposition and diagenesis of repetitive cycles in a marine basin. In: *Evolution of Sedimentary Basins* (A. C. Huffman, Jr, ed.), U.S. Geological Survey Bulletin 2000-N, U.S. Geological Survey, Washington, DC, pp. N1–86.

Tuttle, M. L. W., Charpentier, R. R. and Brownfield, M. E. (1999) *The Niger Delta Petroleum System: Niger Delta Province, Nigeria, Cameroon, and Equatorial Guinea, Africa*. U.S. Geological Survey Open-File Report 99-50-H. U.S. Geological Survey Denver, CO.

Tyler, S. and Barghoorn, E. S. (1954) Occurrence of structurally preserved plants in pre-Cambrian rocks of the Canadian Shield. *Science*, 119, 606–8.

Tyson, R. V. (1995) *Sedimentary Organic Matter: Organic Facies and Palynofacies*. Chapman and Hall, New York.

Udo, O. T. and Ekweozor, C. M. (1990) Significance of oleanane occurrence in shales of the Opuama Channel Complex, Niger Delta. *Energy & Fuels*, 4, 248–54.

Ulmishek, G. F. (1984) *Geology and Petroleum Resources of Basins in Western China*. Argonne National Laboratory Report ANL/ES-146, Argonne National Laboratory, Argonne, IL.

(1986) Stratigraphic aspects of petroleum resource assessment. In: *Oil and Gas Assessment – Methods and Applications* (D. D. Rice, ed.), Association of Petroleum Geologists, Tulsa, OK, pp. 59–68.

(1988) Upper Devonian – Tournaisian facies and oil resources of the Russian craton's eastern margin. In: *Devonian of the World, Volume I: Regional Syntheses*. (N. J. McMillan, A. F. Embry and D. J. Glass, eds.), Canadian Society of Petroleum Geologists, Calgary, pp. 527–49.

(1991) Volga-Ural Basin, USSR: rich petroleum systems with a single source rock. *American Association of Petroleum Geologists Bulletin*, 75, 685.

(2001a) *Petroleum Geology and Resources of the North Caspian Basin, Kazakhstan and Russia*. U.S. Geological Survey Bulletin 2201-B, U.S. Geological Survey, Washington, DC.

(2001b) *Petroleum Geology and Resources of the Baykit High Province, East Siberia, Russia*. U.S. Geological Survey Bulletin 2201-F, U.S. Geological Survey, Washington, DC.

Ungerer, P. (1990) State of the art of research in kinetic modeling of oil formation and expulsion. *Organic Geochemistry*, 16, 1–25.

Ungerer, P., Bessis, F., Chenet, P. Y., et al. (1984) Geological and geochemical models in oil exploration: principles and practical examples. In: *Petroleum Geochemistry and Basin Evaluation* (G. Demaison and R. J. Murris, eds.), American Association of Petroleum Geologists, Tulsa, OK, pp. 53–77.

Ungerer, P., Chiarelli, A. and Oudin, J. L. (1985) Modeling of petroleum genesis and migration with a bidimensional computer model in the Frigg sector, Viking Graben. In: *Petroleum Geochemistry in Exploration of the Norwegian Shelf* (B. M. Thomas, S. S. Eggen, R. M. Larsen, P. C. Home and A. G. Doré, eds.), Graham and Trotman, London, pp. 121–9.

Unrau, P. J. and Bartel, D. P. (1998) RNA-catalysed nucleotide synthesis. *Nature*, 395, 260–3.

Uphoff, T. L. (1997) Precambrian Chuar source rock play: an exploration case history in Southern Utah. *American Association of Petroleum Geologists Bulletin*, 81, 1–15.

Urey, H. C. (1951) The origin and development of the Earth and other terrestrial planets. *Geochimica et Cosmochimica Acta*, 1, 209–77.

Urien, C. M. and Zambrano, J. J. (1994) Petroleum systems in the Neuquén Basin, Argentina. In: *The Petroleum System – From Source to Trap* (L. B. Magoon and W. G. Dow, eds.), American Association of Petroleum Geologists, Tulsa, OK, pp. 513–34.

Urien, C. M., Zambrano, J. J. and Yrigoyen, M. R. (1995) Petroleum basins of South America: an overview. In: *Petroleum Basins of Southern South America* (A. J. Tankard, R. Suarez Soruco and H. J. Welsink, eds.), American Association of Petroleum Geologists, Tulsa, OK, pp. 63–77.

Usyal, I. T., Golding, S. D., Glikson, A. T., *et al.* (2001) K-Ar evidence from illitic clays of a Late Devonian age for the 120 km diameter Woodleigh impact structure, Southern Carnarvon Basin, Western Australia. *Earth and Planetary Science Letters*, 192, 281–9.

Vakhrameyev, V. A. and Doludenko, M. P. (1977) The Middle-Late Jurassic boundary, an important threshold in the development of climate and vegetation of the Northern Hemisphere. *International Geology Review*, 19, 621–32.

Van Aarssen, B. G. K. and de Leeuw, J. W. (1989) On the identification and occurrence of oligomerized sesquiterpenoid compounds in oils and sediments of Southeast Asia. Presented at 14th International Meeting on Organic Geochemistry, September 18–22, 1989, Paris.

Van Aarssen, B. G. K., Kruk, C., Hessels, J. K. C. and de Leeuw, J. W. (1990a) Cis-*cis*-*trans*-Bicadinane, a novel member of an uncommon triterpane family isolated from crude oils. *Tetrahedron Letters*, 31, 4645–8.

Van Aarssen, B. G. K., Cox, H. C., Hoogendoorn, P. and de Leeuw, J. W. (1990b) A cadinene biopolymer in fossil and extant dammar resins as a source for cadinanes and bicadinanes in crude oils from Southeast Asia. *Geochimica et Cosmochimica Acta*, 54, 3021–31.

Van Aarssen, B. G. K., Hessels, J. K. C., Abbink, O. A. and de Leeuw, J. W. (1992) The occurrence of polycyclic sesqui-, tri-, and oligo-terpenoids derived from a resinous polymeric cadinene in crude oils from South East Asia. *Geochimica et Cosmochimica Acta*, 56, 3021–31.

Van Aarssen, B. G. K., Alexander, R. and Kagi, R. I. (1996) The origin of Barrow Sub-basin crude oils: a geochemical correlation using land-plant biomarkers. *APPEA Journal*, 36, 465–76.

 (2000) Higher plant biomarkers reflect palaeovegetation changes during Jurassic times. *Geochimica et Cosmochimica Acta*, 64, 1417–24.

Van Buchem, F. S. P., Houzay, J.-P., and Peniguel, G. (2000) Variations in distribution and quality of organic matter in Middle Pennsylvanian source rock levels of the Paradox Basin (Utah, USA). In: *Genetic Stratigraphy on the Exploration and Production Scales* (P. W. Homewood and G. P. Eberli, eds.), Elf Aquitaine, Pau, France, pp. 131–7.

Van Buchem, F. S. P., Razin, P., Homewood, P. W., Oterdoom, W. H. and Philip, J. (2002) Stratigraphic organization of carbonate ramps and organic-rich intrashelf basins: Natih Formation (Middle Cretaceous) of northern Oman. *American Association of Petroleum Geologists Bulletin*, 86, 21–53.

Van Dorsselaer, A. (1974) Triterpenes de sediments. Ph. D. thesis, L'Universite Louis Pasteur de Strasbourg, Strasbourg, France.

Van Dorsselaer, A., Albrecht, P. and Ourisson, G. (1977) Identification of novel 17α(H)-hopanes in shales, coals, lignites, sediments and petroleum. *Bulletin de la Sociétè Chimique de France*, 1–2, 165–70.

Van Duin, A. C. T. and Larter, S. R. (2001). Molecular dynamics investigation into the adsorption of organic compounds on kaolinite surfaces. *Organic Geochemistry*, 32, 143–50.

Van Duin, A. C. T., Bass, J. M. A. and van de Graaf, B. (1996) A molecular mechanics force field for tertiary carbocations. *Journal Chemical Society Faraday Transactions*, 92, 353–62.

Van Duin, A. C. T., Sinninghe Damsté, J. S., Koopmans, M. P., van de Graaf, B. and de Leeuw, J. W. (1997) A kinetic calculation method of homohopanoid maturation: applications in the reconstruction of burial histories of sedimentary basins. *Geochimica et Cosmochimica Acta*, 61, 2409–29.

Van Graas, G. W. (1990) Biomarker maturity parameters for high maturities: calibration of the working range up to the oil/condensate threshold. *Organic Geochemistry*, 16, 1025–32.

Van Graas, G., de Leeuw, J. W., Schenck, P. A. and Haverkamp, J. (1981) Kerogen of Toarcian shales of the Paris Basin. A study of its maturation by flash pyrolysis techniques. *Geochimica et Cosmochimica Acta*, 45, 2465–74.

Van Graas, G., Baas, J. M. A., de Graaf, V. and de Leeuw, J. W. (1982) Theoretical organic geochemistry. 1. The thermodynamic stability of several cholestane isomers calculated by molecular mechanics. *Geochimica et Cosmochimica Acta*, 46, 2399–402.

Van Kaam-Peters, H. M. E. and Sinninghe Damsté, J. S. (1997) Characterisation of an extremely organic sulphur-rich, 150 Ma old carbonaceous rock: palaeoenvironmental implications. *Organic Geochemistry*, 27, 371–97.

Van Kaam-Peters, H. M. E., Köster, J., de Leeuw, J. W. and Sinninghe Damsté, J. S. (1995a) Occurrence of two novel benzothiophene hopanoid families in sediments. *Organic Geochemistry*, 23, 607–16.

Van Kaam-Peters, H. M. E., Schouten, S., de Leeuw, J. W. and Sinninghe Damsté, J. S. (1995b) The Kimmeridge Clay Formation biomarker and molecular stable carbon isotope analysis. In: *Organic Geochemistry: Developments and Applications to Energy, Climate, Environment and Human History* (J. O. Grimalt and C. Dorronsoro, eds.), AIGOA, San Sebastian, Spain, pp. 216–18.

Van Kaam-Peters, H. M. E., Heidy, M. E., Köstter, J., *et al.* (1998) The effect of clay minerals on diasterane/sterane ratios. *Geochimica et Cosmochimica Acta*, 62, 2923–9.

Van Wess, J.-D., Stephenson, R. A., Ziegler, P. A., *et al.* (2000) On the origin of the Southern Permian Basin, Central Europe. *Marine and Petroleum Geology*, 17, 43–59.

Vargas, J. M. (1988) The hydrocarbon potential of the Huallaga Basin, Peru. In: *Proceedings of the Third Society of Venezuela Geology and Petroleum Exploration in the Subandean Basins Boilivariano Symposium*. Sociedad Venezolana de Géologs, Caracas, Venezuela, pp. 195–225.

Vasconcelos, C. and McKenzie, J. A. (1997) SEM evidence for diverse Archean microbes in 3.5 Ga dolomite from Pilbara Craton, NW Australia: implications for early life. *Eos, Transactions, American Geophysical Union*, 78, 400.

Vella, A. J. and Holzer, G. (1990) Ether-derived alkanes from sedimentary organic matter. *Geochemistry*, 15, 209–14.

Venkatesan, M. I. (1988) Organic geochemistry of marine sediments in Antarctic region: marine lipids in McMurdo Sound. *Organic Geochemistry*, 12, 13–27.

(1989) Tetrahymanol: its widespread occurrence and geochemical significance. *Geochimica et Cosmochimica Acta*, 53, 3095–101.

Venkatesen, M. I. and Dahl, J. (1989) Organic geochemical evidence for global fires at the Cretaceous/Tertiary boundary. *Nature*, 338, 57–60.

Venosa, A. D., Suidan, M. T., King, D. and Wrenn, B. A. (1997) Use of hopane as a conservative biomarker for monitoring the bio-remediation effectiveness of crude oil contaminating a sandy beach. *Journal of Industrial Microbiology and Biotechnology*, 18, 131–9.

Verbeek, E. R. and Grout, M. A. (1993) *Geometry and Structural Evolution of Gilsonite Dikes in the Eastern Uinta Basin, Utah*. U.S. Geological Survey Bulletin 1787-HH, U.S. Geological Survey, Washington, DC.

Verdier, A. C., Oki, T. and Suardy, A. (1980) Geology of the Handil Field (East Kalimantan – Indonesia) In: *Giant Oil and Gas Fields of the Decade 1968–1978* (M. T. Halbouty, ed.), American Association of Petroleum Geologists, Tulsa, OK, pp. 399–421.

Vergani, G. D., Tankard, A. J., Belotti, H. J. and Welsink, H. J. (1995) Tectonic evolution and paleogeography of the Néuquen Basin, Argentina. In: *Petroleum Basins of Southern South America* (A. J. Tankard, R. Suarez Soruco and H. J. Welsink, eds.), American Association of Petroleum Geologists, Tulsa, OK, pp. 383–402.

Vetö, I., Demény, A., Hertelendi, E. and Hetényi, M. (1997) Estimation of primary productivity in the Toarcian Tethys: a novel approach based on TOC, reduced sulphur and manganese contents. *Palaeogeography, Palaeoclimatology, Palaeoecology*, 132, 355–71.

Villamil, T. and Arango, C. (1996) A plate tectonic-paleoceanographic hypothesis for Cretaceous source rocks and cherts of northern South America. Presented at the Annual Meeting of the *American Association of Petroleum Geologists*, May 19–22, 1996, San Diego, CA.

Villanueva, J., Grimalt, J. O., De Wit, R., Keely, B. J. and Maxwell, J. R. (1994) Sources and transformations of chlorophylls and carotenoids in a monomictic sulphate-rich karstic lake environment. *Organic Geochemistry*, 22, 739–57.

Villar, H. J., Püttmann, W. and Wolf, M. (1988) Organic geochemistry and petrography of Tertiary coals and carbonaceous shales from Argentina. *Organic Geochemistry*, 13, 1011–21.

Villar, H. J., Laffitte, G. A. and Legarreta, L. (1998) The source rocks of the Mesozoic petroleum systems of Argentina: a comparative overview on their geochemistry, paleoenvironments and hydrocarbon generation patterns. Presented at the American Association of Petroleum Geologists International Conference and Exhibition, November 8–11, 1998, Rio de Janeiro, Brazil.

Vink, A., Schouten, S., Sephton, S. and Sinninghe Damsté, J. S. (1998) A newly discovered norisoprenoid, 2,6,15,19-tetramethylicosane, in Cretaceous black shales. *Geochimica et Cosmochimica Acta*, 62, 965–70.

Vivas, M. A. M., Sanchez, H. and Santos Neto, E. (1998) Cretaceous/Tertiary Petroleum System in the Tala-Urica Area, EVB. Presented at the *American Association of Petroleum Geologists International Conference and Exhibition*, November 8–11, 1998, Rio de Janeiro, Brazil.

Vlierboom, F. W., Collini, B. and Zumberge, J. E. (1986) The occurrence of petroleum in sedimentary rocks of the meteor impact crater at Lake Siljan, Sweden. *Organic Geochemistry*, 10, 153–61.

Vliex, M., Hagemann, H. W. and Püttmann, W. (1994) Aromatized arborane/fernane hydrocarbons as molecular indicators of floral changes in Upper Carboniferous/Lower Permian strata of the Saar-Nahe Basin, southwestern Germany. *Geochimica et Cosmochimica Acta*, 58, 4689–702.

(1995) Rekonstruktion des Florensprungs an der Westfal/ Stefan-Grenze im Saar-Nahe-Becken ueber die Analytik von Biomarkern. *Neues Jahrbuch für Geologie und Palaeontologie Abhandlungen*, 197, 225–51.

Vogler, E. A., Meyers, P. A. and Moore, W. A. (1981) Comparison of Michigan Basin crude oils. *Geochimica et Cosmochimica Acta*, 45, 2287–93.

Volkman, J. K. (1986) A review of sterol markers for marine and terrigenous organic matter. *Organic Geochemistry*, 9, 83–99.

(1988) Biological marker compounds as indicators of the depositional environments of petroleum source rocks. In: *Lacustrine Petroleum Source Rocks* (A. J. Fleet, K. Kelts and M. R. Talbot, eds.), Blackwell, London, pp. 103–22.

Volkman, J. K. and Maxwell, J. R. (1986) Acyclic isoprenoids as biological markers. In: *Biological Markers in the Sedimentary Record* (R. B. Johns, ed.), Elsevier, New York, pp. 1–42.

Volkman, J. K., Gillan, F. T., Johns, R. B. and Eglinton, G. (1981) Sources of neutral lipids in a temperate intertidal sediment. *Geochimica et Cosmochimica Acta*, 45, 1817–28.

Volkman, J. K., Alexander, R., Kagi, R. I., Noble, R. A. and Woodhouse, G. W. (1983a) A geochemical reconstruction of oil generation in the Barrow Sub-basin of Western Australia. *Geochimica et Cosmochimica Acta*, 47, 2091–106.

Volkman, J. K., Alexander, R., Kagi, R. I. and Woodhouse, G. W. (1983b) Demethylated hopanes in crude oils and their application in petroleum geochemistry. *Geochimica et Cosmochimica Acta*, 47, 785–94.

Volkman, J. K., Alexander, R., Kagi, R. I., Rowland, S. F. and Sheppard, P. N. (1984) Biodegradation of aromatic hydrocarbons in crude oils from the Barrow Sub-basin of Western Australia. *Organic Geochemistry*, 6, 619–32.

Volkman, J. K., Allen, D. I., Stevenson, P. L. and Burton, H. R. (1986) Bacterial and algal hydrocarbons in sediments from a saline Antarctic lake, Ace Lake. *Organic Geochemistry*, 10, 671–81.

Volkman, J. K., Banks, M. R., Denwer, K. and Aquino Neto, F. R. (1989) Biomarker composition and depositional setting of *Tasmanite* oil shale from northern Tasmania, Australia. Presented at the 14th International Meeting on Organic Geochemistry, September 18–22, 1989, Paris.

Volkman, J. K., Kearney, P. and Jeffrey, S. W. (1990) A new source of 4-methyl and $5\alpha(H)$-stanols in sediments: prymnesiophyte microalgae of the genus *Pavlova*. *Organic Geochemistry*, 15, 489–97.

Volkman, J. K., Barnett, S. M., Dunstan, G. A. and Jeffrey, S. W. (1994) C_{25} and C_{30} highly branched isoprenoid alkenes in laboratory cultures of two marine diatoms. *Organic Geochemistry*, 21, 407–13.

Von der Dick, H. (1989) Environment of petroleum source rock deposition in the Jeanne d'Arc Basin off Newfoundland. In: *Extensional Tectonics and Stratigraphy of the North Atlantic Margins* (A. J. Tankard and H. R. Balkwill, eds.), American Association of Petroleum Geologists, Tulsa, OK, pp. 295–303.

Von der Dick, H., Meloche, J. D., Dwyer, J. and Gunther, P. (1989) Source-rock geochemistry and hydrocarbon generation in the Jeanne d'Arc Basin, Grand Banks, offshore Eastern Canada. *Journal of Petroleum Geology*, 12, 51–68.

Vreeland, R. H., Rosenzweig, W. D. and Powers, W. D. (2000) Isolation of a 250-million-year-old halotolerant bacterium from a primary salt crystal. *Nature*, 407, 897–900.

Wächtershäuser, G. (1988) Before enzymes and templates: theory of surface metabolism. *Microbiological Reviews*, 52, 452–84.

Wagner, B. E., Sofer, Z. and Claxton, B. L. (1994) Source rock in the lower Tertiary and Cretaceous, deep-water Gulf of Mexico. *Gulf Coast Association of Geological Societies Transactions*, 44, 729–36.

Wahlberg, I. and Eklund, A. M. (1992) Cembranoids, pseudopteranoids, and cubitanoids of natural occurrence. In: *Progress in the Chemistry of Organic Natural Products*, 59 (W. Hertz, G. W. Kirby, R. E. Moore, W. Steglich and C. Tamm, eds.), Springer-Verlag, New York, pp. 142–294.

Wake, L. V. and Hillen, L. W. (1981) Nature and hydrocarbon content of blooms of the alga *Botryococcus braunii* occurring in Australian freshwater lakes. *Australian Journal of Marine and Freshwater Research*, 32, 353–67.

Wakeham, S. G. (1990) Algal and bacterial hydrocarbons in particulate matter and interfacial sediment of the

Cariaco Trench. *Geochimica et Cosmochmica Acta*, 54, 1325–36.

Wakeham, S. G. and Beier, J. A. (1991) Fatty acid and sterol biomarkers as indicators of particulate matter source and alteration processes in the Black Sea. *Deep Sea Research. Part A. Oceanographic Research Papers*, 38, 943–68.

Wakeham, S. G. and Ertel, J. R. (1988) Diagenesis of organic matter in suspended particles and sediments in the Cariaco Trench. *Organic Geochemistry*, 13, 815–22.

Wakeham, S. G., Schaffner, C., Giger, W., Boon, J. J. and de Leeuw, J. W. (1979) Perylene in sediments from the Namibian Shelf. *Geochimica et Cosmochimica Acta*, 43, 1141–4.

Wakeham, S. G., Schaffner, C. and Giger, W. (1980) Polycyclic aromatic hydrocarbons in recent lake sediments – II. Compounds derived from biogenic precursors during early diagenesis. *Geochimica et Cosmochimica Acta*, 44, 415–29.

Wakeham, S. G., Freeman, K. H., Pease, T. K. and Hayes, J. M. (1993) A photoautotrophic source for lycopane in marine water columns. *Geochimica et Cosmochimica Acta*, 57, 159–65.

Wakeham, S. G., Lewis, C. M., Hopmans, E. C., Schouten, S. and Sinninghe Damsté, J. S. (2003) Archaea mediate anaerobic oxidation of methane in deep euxinic waters of the Black Sea. *Geochimica et Cosmochimica Acta*, 67, 1359–74.

Walliser, O. H. (1996) Global events in the Devonian and Carboniferous. In: *Global Events and Event Stratigraphy in the Phanerozoic* (O. H. Walliser, ed.), Springer-Verlag, Berlin, pp. 225–50.

Walsh, M. M. and Lowe, D. R. (1985) Filamentous microfossils from the 3,500-Myr-old Onverwacht Group, Barberton Mountain Land, South Africa. *Nature*, 314, 530–2.

Walter, M. R., Du, R. and Horodyski, R. J. (1990) Coiled carbonaceous megafossils from the Middle Proterozoic of Jixian (Tianjin) and Montana. *American Journal of Science*, 290A, 133.

Walters, C. C. (1981) Organic geochemistry of the 3800 million year old metasediments from Isua Greenland. Ph. D. thesis, University of Maryland, College Park, MD.

 (1993) Selective biodegradation of hopanes in oils from the South Belridge Field. In: *Proceedings of the 16th International Meeting on Organic Geochemistry* (K. Øygard, ed.), Falch Hurtigtrykk, Oslo, pp. 16–19.

 (1999) Oil-oil and oil-source rock correlations. In: *Encyclopedia of Geochemistry* (C. P. Marshall and R. W. Fairbridge, eds.), Kluwer Academic Publishers, Dorcrecht, the Netherlands, pp. 442–4.

Walters, C. C. and Cassa, M. R. (1985) Regional organic geochemistry of offshore Louisiana. *Transactions: Gulf Coast Association of Geological Societies*, 35, 277–86.

Walters, C. C. and Dusang, D. D. (1988) Source and thermal history of oils from Lockhart Crossing, Livingston Parish, Louisiana. *Transactions: Gulf Coast Association of Geological Societies*, 38, 37–44.

Walters, C. C. and Kotra, R. K. (1989) Geochemistry of organic matter-rich Mesozoic lacustrine formations of eastern United States. *American Association of Petroleum Geologists Bulletin*, 75, 423–4.

 (1990) Thermal maturity of Jurassic shales from the Newark Basin, USA: influence of hydrothermal fluids and implications to basin modeling. *Applied Geochemistry*, 5, 211–25.

Walters, C. C., Pierce, S. E., Gormly, J. R. and Rooney, M. A. (1993) Loma Chumico Shale: a super-rich source rock with unusual geochemical characteristics. *American Association of Petroleum Geologists Bulletin*, 77, 354.

Wang, G.-Y. (1990) A quantitative and qualitative investigation of hydrocarbon release from kerogens during hydrous pyrolysis. Ph. D Thesis, University of Newcastle upon Tyne, Newcastle upon Tyne, UK.

 (1993) Global events and event stratigraphy in the mid-Paleozoic. Ph. D. thesis, Alberta University, Edmonton, Alberta, Canada.

Wang, H. D. and Philp, R. P. (1997a) A geochemical study of Viola source rocks and associated crude oils in the Anadarko Basin, Oklahoma. In: *Simpson and Viola Groups in the Southern Midcontinent* (K. S. Johnson, ed.), University of Oklahoma, Norman, Oklahoma, pp. 87–101.

 (1997b) Geochemical study of potential source rocks and crude oils in the Anadarko Basin, Oklahoma. *American Association of Petroleum Geologists Bulletin*, 81, 249–75.

Wang, T. G. and Simoneit, B. R. T. (1990) Organic geochemistry and coal petrology of Tertiary brown coal in the Zhoujing mine, Baise Basin, South China. 2. Biomarker assemblage and significance. *Fuel*, 69, 12–20.

 (1995) Tricyclic terpanes in Precambrian bituminous sandstone from the eastern Yanshan region, North China. *Chemical Geology*, 120, 155–70.

Wang, T., Fan, P. and Swain, F. M. (1988) Geochemical characteristics of crude oils and source beds in different continental facies of four oil-bearing basins, China. In: *Lacustrine Petroleum Source Rocks* (A. J. Fleet, K. Kelts and M. R. Talbot, eds.), Geological Society of London, London, pp. 309–25.

Wang, T.-G., Simoneit, B. R. T., Philp, R. P. and Yu, C.-P. (1990) Extended 8β(H)-drimane and 8,14-secohopane series in a Chinese boghead coal. *Energy & Fuels*, 4, 177–83.

Wang, K., Chatterton, B. D. E., Attrep, M., Jr. and Orth, C. J. (1992) Iridium abundance maxima at the latest Ordovician mass extinction horizon, Yangtze Basin, China: terrestrial or extraterrestrial? *Geology*, 20, 39–42.

Wang, K., Geldsetzer, H. H. J. and Krouse, H. R. (1994) Permian-Triassic extinction; organic δ^{13}C evidence from British Columbia, Canada. *Geology*, 22, 580–4.

Wang, D. Y.-C., Kumar, S. and Hedges, S. B. (1999) Divergence time estimates for the early history of animal phyla and the origin of plants, animals, and fungi. *Proceedings of the Royal Society of London, Series B*, 266, 163–71.

Wang, Z., Fingas, M. and Sigouin, L. (2001a) Characterization and identification of a "mystery" oil spill from Quebec (1999). *Journal of Chromatography A*, 909, 155–69.

Wang, Z., Fingas, M. F., Sigouin, L. and Owens, E. H. (2001b) Fate and persistence of long-termed spilled Metula oil in the marine salt marsh environment: degradation of petroleum biomarkers. In: *Proceedings of the 2001 International Oil Spill Conference, Tampa, Florida, March 26–29, 2001*, American Petroleum Institute, Washington, DC, pp. 115–25.

Wang, F., Zhang, B. and Zhang, S. (2002) Anoxia vs. bioproductivity controls on the Cambrian and Ordovician marine source rocks in Tarim Basin, China. Presented at the Annual Meeting of the American Association of Petroleum Geologists, March 10–13, 2002, Houston, TX.

Wan Hasiah, A. (1999) Oil-generating potential of Tertiary coals and other organic-rich sediments of the Nyalau Formation, onshore Sarawak. *Journal of Asian Earth Sciences*, 17, 255–67.

Waples, D. W. (1983) A reappraisal of anoxia and organic richness, with emphasis on Cretaceous of North Atlantic. *American Association of Petroleum Geologists Bulletin*, 67, 963–78.

Waples, D. W. and Curiale, J. A. (1999) Oil-oil and oil-source rock correlations. In: *Exploring for Oil and Gas Traps* (E. A. Beaumont and N. H. Foster, eds.), American Association of Petroleum Geologists, Tulas, OK, pp. 8–1 to 8–71.

Waples, D. W. and Machihara, T. (1991) *Biomarkers for Geologists*. American Association of Petroleum Geologists, Tulsa, OK.

Waples, D. W., Haug, P. and Welte, D. H. (1974) Occurrence of a regular C_{25} isoprenoid hydrocarbon in Tertiary sediments representing a lagoonal-type, saline environment. *Geochimica et Cosmochimica Acta*, 38, 381–7.

Warburton, G. A. and Zumberge, J. E. (1982) Determination of petroleum sterane distributions by mass spectrometry with selective metastable ion monitoring. *Analytical Chemistry*, 55, 123–6.

Wardlaw, B. R., Snyder, W. S., Spinosa, C. and Gallegos, D. M., (1995) Permian of the Western United States. In: *The Permian of Northern Pangea*, Vol. 2 (P. A. Scholle, ed.), Springer-Verlag, New York, pp. 23–40.

Wardroper, A. M. K. (1979) Aspects of the geochemistry of polycyclic isoprenoids. Ph. D. thesis, University of Bristol, Bristol, UK.

Wardroper, A. M. K., Hoffmann, C. F., Maxwell, J. R., *et al.* (1984) Crude oil biodegradation under simulated and natural conditions – 2. Aromatic steriod hydrocarbons. *Organic Geochemistry*, 6, 605–17.

Warren, J. K., George, S. C., Hamilton, P. J. and Tingate, P. (1998) Proterozoic source rocks: sedimentology and organic characteristics of the Velkerri Formation, Northern Territory, Australia. *American Association of Petroleum Geologists Bulletin*, 82, 442–63.

Warris, B. J. (1993) The hydrocarbon potential of the Palaeozoic basin of Western Australia. *Journal of the Australian Petroleum Exploration Association*, 33, 123–37.

Warthmann, R., van Lith, Y., Vasconcelos, C. and McKenzie, J. (2000) Microbial link between the deep subsurface and hypersaline shallow-water environments: a geo-microbiological study of SRB strains. *EOS (Transaction AGU)*, 81, 215.

Warton, B., Alexander, R. and Kagi, R. I. (1997) Identification of some single branched alkanes in crude oils. *Organic Geochemistry*, 27, 465–76.

(1999) Characterisation of the ruthenium tetroxide oxidation products from the aromatic unresolved complex mixture of a biodegraded crude oil. *Organic Geochemistry*, 30, 255–72.

Watson, J. S., Jones, D. M. and Swannell, R. P. J. (1999) Formation of carboxylic acids during biodegradation of crude oil. In: *In Situ Bioremediation of Petroleum*

Hydrocarbon and Other Organic Compounds (B. C. Alleman and A. Leeson, eds.), Battelle, Columbus, OH, pp. 251–6.

Watson, J. S., Pearson, V. K., Gilmour, I. and Sephton, M. A. (2003) Contamination by sesquiterpenoid derivatives in the Orgueil carbonaceous chondrite. *Organic Geochemistry*, 34, 37–47.

Wavrek, D. A., Curtiss, D. K., Guliyev, I. S. and Feizullayev, A. A. (1998) Maikop/Diatom-Productive Series (!) petroleum system, South Caspian Basin, Azerbaijan. Presented at the Annual Meeting of the American Association of Petroleum Geologists, May 17–20, 1998, Salt Lake City, UT.

Webster, R. L. (1984) Petroleum source rocks and stratigraphy of the Bakken Formation in North Dakota. In: *Hydrocarbon Source Rocks of the Greater Rocky Mountain Region* (J. Woodward, F. F. Meissner and J. L. Clayton, eds.), Rocky Mountain Association of Geologists, Denver CO, pp. 57–81.

Weimer, P. (1987). Northern Alaska exploration – the past dozen years. In: *Alaskan North Slope Geology*, Vol. 1 (I. Tailleur and P. Weimer, eds.), Pacific Section Society of Economic Paleontologists and Mineralogists, Los Angeles, CA. pp. 31–7.

Welte, D. H. and Yalcin, M. N. (1987) Formation and occurrence of petroleum in sedimentary basins as deduced from computer aided modeling. In: *Petroleum Geochemistry and Exploration in the Afro-Asian Region* (R. K. Kumar, P. Dwiwedi, V. Banerjie and V. Gupta, eds.), Balkema, Rotterdam, pp. 17–23.

  (1988) Basin modeling – a new comprehensive method in petroleum geology. *Organic Geochemistry*, 13, 141–51.

Welte, D. H., Horsfield, B. and Baker, D. R. (1997) *Petroleum and Basin Evolution*. Springer-Verlag, New York.

Wenger, L. M. and Isaksen, G. H. (2002) Control of hydrocarbon seepage intensity on level of biodegradation in sea bottom sediments. *Organic Geochemistry*, 33, 1277–92.

Wenger, L. M., Sassen, R. and Schumacher, D. (1988) Molecular characterization of Smackover, Wilcox, and Tuscaloosa-reservoired oils in the eastern Gulf Coast. In: *Gulf Coast Oils and Gases: Their Characteristics, Origin, Distribution, and Exploration and Production Significance* (D. Schumacher and B. F. Perkins, eds.), Society of Economic Paleontologists and Mineralogists, Tulsa, OK, pp. 37–58.

Wenger, L. M., Goodoff, L. R., Gross, O. P., Harrison, S. C. and Hood, K. C. (1994) Northern Gulf of Mexico: an integrated approach to source, maturation, and migration. Presented at the *First Joint American Association of Petroleum Geologists/AMGP Research Conference*, October 2–6, 1994, Mexico, Mexico.

Wenger, L. M., Davis, C. L. and Isaksen, G. H. (2002) Multiple controls on petroleum biodegradation and impact on oil quality. *SPE Reservoir Evaluation and Engineering*, 5, 375–83.

Werne, J. P., Hollander, D. J., Behrens, A., *et al.* (2000) Timing of early diagenetic sulfurization of organic matter: a precursor-product relationship in Holocene sediments of the anoxic Cariaco Basin, Venezuela. *Geochimica et Cosmochimica Acta*, 64, 1741–51.

Wescott, W. A. and Hood, W. C. (1994) Hydrocarbon generation and migration routes in the East Texas Basin. *American Association of Petroleum Geologists Bulletin*, 78, 287–306.

West, N., Alexander, R. and Kagi, R. I. (1990) The use of silicalite for rapid isolation of branched and cyclic alkane fractions of petroleum. *Organic Geochemistry*, 15, 499–501.

Weston, R. J., Philp, R. P., Sheppard, C. M. and Woolhouse, A. D. (1989) Sesquiterpanes, diterpanes and other higher terpanes in oils from the Taranaki Basin of New Zealand. *Organic Geochemistry*, 14, 405–21.

Whitehead, E. V. (1973) Molecular evidence for the biogenesis of petroleum and natural gas. In: *Proceedings of Symposium on Hydrogeochemistry and Biogeochemistry*, Vol. II (E. Ingerson, ed.), The Clarke Company, Washington, DC, pp. 158–211.

  (1974) The structure of petroleum pentacyclanes. In: *Advances in Organic Geochemistry 1973* (B. Tissot and F. Bienner, eds.), Editions Technip, Paris, pp. 225–43.

Whiteman, A. (1982) *Nigeria: Its Petroleum Geology, Resources and Potential*. Graham and Trotman, London.

Whiticar, M. J. (1999) Carbon and hydrogen isotope systematics of bacterial formation and oxidation of methane. *Chemical Geology*, 161, 291–314.

Whitmire, D. P. and Jackson, A. A., IV. (1984) Are periodic mass extinctions driven by a distant solar companion? *Nature*, 308, 713–5.

Wickasono, P., Armon, J. W. and Haryono, S. (1992) The implications of basin modeling for exploration – Sunda Basin case study, offshore southeast Sumatra. In: *Proceedings of the Twenty First Annual Convention of the Indonesian Petroleum Association*, Vol. 1. Indonesian Petroleum Association Jakarta, Indonesia, pp. 379–415.

Widdel, F. and Rabus, R. (2001) Anaerobic biodegradation of saturated and aromatic hydrocarbons. *Current Opinion in Biotechnology*, 12, 259–76.

Wielens, J. B. W., von der Dick, H., Fowler, M. G., Brooks, P. W. and Monnier, F. (1990) Geochemical comparison of a Cambrian alginite potential source rock, and hydrocarbons from the Colville/Tweed lake area, Northwest Territories. *Bulletin of Canadian Petroleum Geology*, 38, 236–45.

Wignall, P. B. (2001) Large igneous provinces and mass extinctions. *Earth Science Reviews*, 53, 1–33.

Wignall, P. B. and Maynard, J. R. (1993) The sequence stratigraphy of transgressive black shales. In: *Source Rocks in a Sequence Stratigraphic Framework* (B. J. Katz and L. M. Pratt, eds.), American Association of Petroleum Geologist Tulsa, OK, pp. 35–47.

Wignall, P. B. and Twitchett, R. J. (1996) Oceanic anoxia and the end Permian mass extinction. *Science*, 272, 1155–8.

Wilde, S.A, Valley, J. W., Peck, W. H. and Graham, C. M. (2001) Evidence from detrital zircons for the existence of continental crust and oceans on the Earth 4.4 Gyr ago. *Nature*, 409, 175–8.

Wilhelms, A., Larter, S. R., Head, I., *et al.* (2001) Biodegradation of oil in uplifted basins prevented by deep-burial sterilization. *Nature*, 411, 1034–7.

Williams, J. A. (1974) Characterization of oil types in the Williston Basin. *American Association of Petroleum Geologists Bulletin*, 58, 1243–52.

  (1977) Characterization of oil types in the Permian Basin presented at the *American Association of Petroleum Geologists, Southwest Section Meeting*, March 7, 1977, Abilene, TX.

Williams, L. A. (1984) Subtidal stromatolites in Monterey Formation and other organic-rich rocks as suggested contributors to petroleum formation. *American Association of Petroleum Geologists Bulletin*, 68, 1879–93.

Williams, J. A., Bjorøy, M., Dolcater, D. L. and Winters, J. C. (1986) Biodegradation in South Texas Eocene oils – effects on aromatics and biomarkers. *Organic Geochemistry*, 10, 451–61.

Wilson, P. A. and Morris, R. D. (2001) Warm tropical ocean surface and global anoxia during the Mid-Cretaceous Period. *Nature*, 412, 425–9.

Wilson, P. A., Norris, R. D. and Erbacher, J. (1999) Tropical sea surface temperature records and black shale deposition in the Mid-Cretaceous western Atlantic (Blake Nose and Dermerara Rise). *EOS (Transactions American Geophysical Union)*, 80, F488.

Wingert, W. S. (1992) GC-MS analysis of diamondoid hydrocarbons in Smackover petroleum. *Fuel*, 71, 37–43.

Winters, J. C. and Williams, J. A. (1969) Microbiological alteration of crude oil in the reservoir. *American Chemical Society, Division of Petroleum Chemistry, New York Meeting Preprints*, 14, E22–31.

Wise, S. A. Campbell, R. M., West, W. R., Lee, M. L. and Bartle, K. D. (1986) Characterization of polycyclic aromatic hydrocarbon minerals curtisite, idrialite and pendletonite using high-performance liquid chromatography, gas chromatography, mass spectrometry and nuclear magnetic resonance spectroscopy. *Chemical Geology*, 54, 339–57.

Withers, N. (1983) Dinoflagellate sterols. In: *Marine Natural Products 5* (P. J. Scheuer, ed.), Academic Press, New York, pp. 87–130.

Woese, C. R. and Wächterhäuser, G. (1990) Origins of life. In: *Paleobiology: A Synthesis* (D. E. G. Briggs and P. R. Crowther, eds.), Blackwell Scientific Publications, Oxford, UK, pp. 3–9.

Woese, C. R., Magrum, L. J. and Fox, G. E. (1978) Archaebacteria. *Journal of Molecular Evolution*, 11, 245–52.

Wolbach, W. S., Gilmour, I. and Anders, E. (1990) Major wildfires at the Cretaceous/Tertiary boundary. In: *Global Catastrophes in Earth History: An Interdisciplinary Conference on Impacts, Volcanism, and Mass Mortality*, 247, (V. L. Sharpton and P. D. Ward, eds.), Geological Society of America, Snowbird, UT, pp. 391–400.

Wolbach, W. S., Widicus, S. and Kyte, F. T. (2003) A search for soot from global wildfires in central Pacific Cretaceous-Tertiary boundary and other extinction and impact horizon sediments. *Astrobiology*, 3, 91–7.

Wolff, G. A., Lamb, N. A. and Maxwell, J. R. (1986) The origin and fate of 4-methyl steroid hydrocarbons 1. 4-methyl sterenes. *Geochimica et Cosmochimica Acta*, 50, 335–42.

Wraige, E. J., Belt, S. T., Lewis, C. A., *et al.* (1997) Variations in structures and distributions of C_{25} highly branched isoprenoid (HBI) alkenes in cultures of the diatom, *Haslea ostrearia (Simonsen)*. *Organic Geochemistry*, 27, 497–505.

Wraige, E. J., Belt, S. T., Massé, G., Robert, J.-M. and Rowland, S. J. (1998) Variations in distributions of C_{25} highly branched isoprenoid (HBI) alkenes in the diatom, *Haslea ostrearia*: influence of salinity. *Organic Geochemistry*, 28, 855–9.

Wray, G. A., Levinton, J. S. and Shapiro, L. H. (1996) Molecular evidence for deep Precambrian divergences among metazoan phyla. *Science*, 274, 568–73.

Wright, I. P., Assanov, S., Verchovsky, A. B., *et al.* (1997) Further investigations of isotopically light carbon in Allen Hills 84001. In: *Conference on Early Mars: Geologic and Hydrologic Evolution, Physical and Chemical Environments, and the Implications for Life* (S. M. Clifford, A. H. Treiman, H. E. Newsom and J. D. Farmer, eds.), Lunar and Planetary Institute, Houston, TX, pp. 86–7.

Wulff, K. (1992) Depositional history and facies analysis of the Upper Jurassic sediments in the eastern Barrow Sub-basin. *Journal of the Australian Petroleum Exploration Association*, 32, 104–22.

Xiao, S., Zhang, Y. and Knoll, A. H. (1998) Three-dimensional preservation of algae and animal embryos in a Neoproterozoic phosphorite. *Nature*, 391, 553–8.

Xue, L. and Galloway, W. E. (1993) Genetic sequence stratigraphic framework, depositional style, and hydrocarbon occurrence of the Upper Cretaceous QYN formations in the Songliao lacustrine basin, Northeastern China. *American Association of Petroleum Geologists Bulletin*, 77, 1792–808.

Yalcin, M. N., Welte, D. H., Misra, K. N., *et al.* (1988) 3-D computer aided basin modeling of Cambay Basin, India – a case history of hydrocarbon generation. In: *Petroleum Geochemistry and Exploration in Afro-Asian Region* (R. Kumar, P. Dwivedi, V. Banerjie and V. Gupta, eds.), Balkema, Rotterdam, pp. 417–50.

Yamane, A., Sakakibara, K., Hosomi, M. and Murakami, A. (1997) Microbial degradation of petroleum hydrocarbons in estuarine sediment of Tama River in Tokyo urban area. *Water Science and Technology*, 35, 69–76.

Yang, W. (1985) Daqing oil field, People's Republic of China: a giant field with oil of nonmarine origin. *American Association of Petroleum Geologists Bulletin*, 69, 1101–11.

Yang, W., Li, Y. and Gao, R. (1985) Formation and evolution of nonmarine petroleum in Songliao Basin, China. *American Association of Petroleum Geologists Bulletin*, 69, 1112–22.

Yayanos, A. A., Sharma, A., Scott, J. H., *et al.* (2002) Are cells viable at gigapascal pressures? *Science*, 297, 295a.

Ye, H., Royden, L., Burchfiel, C. and Schuepbach, M. (1996) Late Paleozoic deformation of Interior North America: the Greater Ancestral Rocky Mountains. *American Association of Petroleum Geologists Bulletin*, 80, 1397–432.

Yon, D. A., Maxwell, J. R. and Rybach, G. (1982) 2,6,10-Trimethyl-7-(3-methylbutyl)-dodecane, a novel sedimentary biological marker compound. *Tetrahedron Letters*, 23, 2143–6.

Yongsong, H., Peakman, T. M. and Murray, M. (1997) $8\beta,9\alpha,10\beta$-Rimuane, a novel, optically active, tricyclic hydrocarbon of algal origin. *Tetrahedron Letters*, 38, 5363–6.

Yu, Z., Peng, P., Sheng, G. and Fu, J. (2000a) The carbon isotope study of biomarkers in the Maoming and the Jianghan Tertiary oil shale. *Chinese Science Bulletin*, 45, 90–6.

Yurewicz, D. A., Advocate, D. M., Lo, H. B. and Hernandez, E. A. (1998) Source rocks and oil families, Southwest Maracaibo Basin (Catatumbo Subbasin), Colombia. *American Association of Petroleum Geologists Bulletin*, 82, 1329–52.

Zachos, J. C., Lohmann, K. C., Walker, J. C. G. and Wise, S. W., Jr (1993) Abrupt climate change and transient climates during the Paleogene: a marine perspective. *Journal of Geology*, 101, 191–213.

Zachos, J. C., Pagani, M., Sloan, L., Thomas, E. and Billups, K. (2001) Trends, rhythms, and aberrations in global climate 65 Ma to Present. *Science*, 292, 686–93.

Zahnle, K. J. and Sleep, N. H. (1996) Impacts and the early evolution of life. In: *Comets and the Origin of Life* (P. J. Thomas, C. F. Chyba and C. P. McKay, eds.), Springer-Verlag, Berlin, pp. 175–208.

Zalessky, M. D. (1917) Sur le sapropelité marin de l'âge Silurien formé par une algue cyanophycée. *Ezhegodnik, Vsesokiluznoe Paleontologicheskoe Obshchestvo*, 1, 25–42.

Zander, J. M., Caspi, E., Pandey, G. N. and Mitra, C. R. (1969) The presence of tetrahymanol in *Oleandra wallichii*. *Phytochemistry*, 8, 2265–7.

Zanis, M. J., Soltis, D. E., Soltis, P. S., Mathews, S. and Donoghue, M. J. (2002) The root of the angiosperms revisited. *Proceedings of the National Academy of Science, USA*, 99, 6848–53.

Zaunbrecher, M. L. (1988) Hydrocarbon source potential of the Upper Roper Group, McArthur Basin, Northern Territory. B. Sc. thesis, University of Adelaide, Adelaide, Australia.

Zegers, T. E., White, S. H., de Wit, M. J. and Dann, J. (1998) Vaalbara, Earth's oldest assembled continent? A combined structural, geochronological, and palaeomagnetic test. *Terra Nova*, 10, 250–9.

Zekri, A. Y., Almehaideb, R. A. and Chaalal, O. (1999) Project of increasing oil recovery from UAE reservoirs using bacteria flooding. Presented at the 1999 Society of Petroleum Engineers Annual Technical Conference and Exhibition, October 3–6, 1999, Houston, TX.

Zemmels, I. and Walters, C. C. (1987) Variation of oil composition in vicinity of Arbuckle Mountains, Oklahoma. *American Association of Petroleum Geologists Bulletin*, 71, 998–9.

Zeng, X., Liu, S. and Ma, S. (1988) Biomarkers as source input indicators in source rocks of several terrestrial basins of China. In: *Geochemical Biomarkers* (T. F. Yen and J. M. Moldowan, eds.), Harwood Academic, New York, pp. 25–49.

Zengler, K., Richnow, H. H., Roselló-Mora, R., Michaelis, W. and Widdel, F. (1999) Methane formation from long-chain alkanes by anaerobic microorganisms. *Nature*, 401, 266–9.

Zhang, B. and Cech, T. R. (1997) Peptide bond formation by *in vitro* selected ribozymes. *Nature*, 390, 96–100.

Zhang, D., Huang, D. and Li, J. (1988) Biodegraded sequence of Karamay oils and semi-quantitative estimation of their biodegraded degrees in Junggar Basin, China. *Organic Geochemistry*, 13, 295–302.

Zhang, S. C., Hanson, A. D., Moldowan, J. M., *et al.* (2000) Paleozoic oil-source rock correlations in the Tarim Basin, NW China. *Organic Geochemistry*, 31, 273–86.

Zhang, C. L., Pancost, R. D., Sassen, R., Qian, Y. and Macko, S. A. (2003) Archaeal lipid biomarkers and isotopic evidence of anaerobic methane oxidation associated with gas hydrates in the Gulf of Mexico. *Organic Geochemistry*, 34, 827–36.

Zhi-Hua, H., Hui-Xiang, L., Rullkötter, J. and Mackenzie, A. S. (1986) Geochemical application of sterane and triterpane biological marker compounds in the Linyi Basin. *Organic Geochemistry*, 10, 433–9.

Zhou, W., Wang, R., Radke, M., *et al.* (2000) Retene pyrolyzates of algal and bacterial organic matter. *Organic Geochemistry*, 31, 757–62.

Ziegler, P. A. (1988) *Evolution of the Arctic-North Atlantic and the Western Tethys*. American Association of Petroleum Geologists, Tulsa, OK.

Zimmer, C. (1999) Ancient continent opens window on the early Earth. *Science*, 286, 2254–6.

Zimmerman, H. B., Boersma, A. and McCoy, F. W. (1987) Carbonaceous sediments and paleoenvironments of the Cretaceous South Atlantic Ocean. In: *Marine Petroleum Source Rocks* (J. Brooks and A. J. Fleet, eds.), Blackwell, Oxford, UK, pp. 271–86.

Zinniker, D. A. (2004) Formation, transformation, and utility of diterpenoid molecular fossils. Ph. D. thesis, Stanford University, Palo Alto, CA.

Zumberge, J. E. (1983) Tricyclic diterpane distributions in the correlation of Paleozoic crude oils from the Williston Basin. In: *Advances in Organic Geochemistry 1981* (M. Bjorøy, C. Albrecht, C. Cornford, *et al.* eds.), John Wiley & Sons, New York, pp. 738–45.

(1984) Source rocks of the La Luna (Upper Cretaceous) in the Middle Magdalena Valley, Colombia. In: *Geochemistry and Source Rock Potential of Carbonate Rocks* (J. G. Palacas, ed.), American Association of Petroleum Geologists, Tulsa, OK, p. 127–133.

(1987a) Prediction of source rock characteristics based on terpane biomarkers in crude oils: a multivariate statistical approach. *Geochimica et Cosmochimica Acta*, 51, 1625–37.

(1987b) Terpenoid biomarker distributions in low maturity crude oils. *Organic Geochemistry*, 11, 479–96.

Zumberge, J. E. and Ramos, S. (1996) Classification of crude oils based on genetic origin using multivariate modeling techniques. Presented at the 13th *Australian Geological Convention*, February 19–23, 1996, Canberra, Australia.

Zundel, M. and Rohmer, M. (1985) Prokaryotic triterpenoids 1. 3β-methylhopanoids from *Acetobacter* species and *Methylococcus capsulatus*. *European Journal of Biochemistry*, 150, 23–7.

Index

abietane 543, 600
abietene 547
abietic acid 552, 596
abiogenic methane *see* methane
abiotic origin of life 709
abiotic oxidation 653
abiotic synthesis 601
abnormal pressure *see* overpressure
Abo Formation, USA 813
Abu Durba seep, Gulf of Suez 915
accessory pigment *see* carotenoid
accommodation 837–8
accretion disk hypothesis 709
acenaphthene 644
acetate 646, 656, 657
acetogenic bacteria (acetogens) 656
acetyl-CoA (acetyl-coenzyme A) 650, 654, 711
acid gases (H_2S and CO_2) 649
acidic catalysis 630
acidic clay minerals 533
acidophilic
acid rain 746, 748
Acinetobacter 652
acritarchs 531, 588, 714, 716, 785
actinomycete 716
Actinpterygii 724
activating hydrocarbons 650
activation energy 814, 966
active source rock 976
acyclic isoprenoids *see* isoprenoids
adamantanes 699–700
Adiyaman region, Turkey 915, 917
Adriatic Basin 527, 535, 559, 613, 705, 977
Adriatic Sea 630
aerobic biodegradation 651, 652, 658
aerobic degradation rate 650
aerobic respiration 713
Agathis 550
Agbada Formation, Niger Delta 949–51
Agelas 495
age-related biomarkers 490–1, 493, 526, 541, 573, 588, 979, 980
Aglaophyton 727
Agrio Formation, Argentina 894–5

Akata Formation, Niger Delta 573, 949–50, 951
Akilia Island, Greenland 712
akinete 796
Akkas Formation, Iraq 791
Akkas well, Iraq 791
Alagamar Formation, Brazil 846, 849
Alagoas stage 835
Alaska National Wildlife Refuge (ANWR) 896
Alberta Basin 968
Alberta tar sands 658, 700
Alberta Trough, Canada 832
albertite 478
Albinia Formation, Australia 770
Albion-Scipio Field, Michigan 722
Alcaligenes 652
Alcanar Formation, Spain 947, 948
Alexander the Great 934
algae 524, 716, 725, 777
algaenan 520, 777, 780, 928
alginite 923, 928, 936
Aliambata Well, Indonesia 529, 544
alkadienes 517
alkaline facies 923
alkaline hydrolysis 539
alkalinity 533
alkane
  alkyl-substituted 493–6
  branched (*see* isoalkane)
  degradation 651, 653, 659, 666, 669
  macrocyclic 520–1
  normal (*see* normal alkane)
alkatriene 517
alkylated benzenes 665, 670–2
alkylated monocyclic alkanes 667–8
alkylation 582
alkylbenzenes 599, 653, 654, 670, 672
alkylbenzthiophenes 798
alkylbiphenyls 665, 672
alkylcyclohexanes 668, 669, 767
alkyldibenzothiophenes 485, 798
alkyldiphenylmethanes 665, 672
alk-1-enylbenzenes 599

alkyl esters 974
alkylnaphthalenes 599, 670
alkylphenols 670
2-alkylsteranes 532–3
3-alkylsteranes 532–3
alkyl-substitued alkanes 493–6
alkyltoluenes 600
Allen Hills meteorite (ALH84001), Antarctica 602, 980
Allgäu Formation, Germany 591
Allison Guyot, Mid-Pacific Mountains 921
alloxanthin (*see* carotenoid)
alluvial facies 924
Alpine Field, Alaska 826, 883
Alps, Europe 737, 741
Al Shomou Formation (*see* Athel silicilyte)
Altamont-Bluebell Field, Utah 479, 925, 927
Alum Shale (Alunskieffer), Sweden 540, 720, 774, 777, 778
Alunskieffer (*see* Alum Shale)
Amadeus Basin, Australia 770, 783
Amal well, Gulf of Suez 915
Amazon Basin, South America 732, 802
Amborella 739
Amino acids 711, 726
Ammonifex 497
ammonites 740
ammonoids 724, 732, 745
amphibians 724, 731, 744
Amposta Field, Spain 507, 520, 947
amyrin 547
amyrone 547
Anadarko Basin, Oklahoma 604, 700, 806, 808, 809, 810
anaerobic
  activation 654
  biodegradation 653–6, 658, 667–8
  degradation rate 650
  methane oxidation 507, 509, 511, 656–7
Ancestral Rocky Mountains 731, 809
Andes, South America 735
Andic cycle 894–5

angiosperms 491, 493, 547, 572–3, 577, 597, 617, 697, 699, 724, 729, 730, 736, 737, 738–9, 741, 942
Angola Basin, Africa 850, 935
Anguille Formation, Gabon Basin 850
anhydrite 487, 584, 946
animals 716–19, 744
 air-breathing 723
anoxia 499, 500, 561, 725, 744, 748, 749, 750, 774, 814, 892
anoxic *see* anoxia
anoxic carbonate 569
Antarctica 740, 742
Antares Field, Alaska 826, 827
anteisoalkanes 493
Antelope Shale, California 679
Antes Shale *see* Utica Shale
Anthocerophyta (Hornworts) 726
anthracene 834
anthrasteroids 586
anthraxolite 478
antioxidant 500
Antofagasta Province, Chile 688
Antrim Shale, USA 803
ANWR *see* Alaska National Wildlife Refuge
apatite 712
apatite fission track 686, 752
Apex Chert, Australia 712
API gravity 619, 631, 641, 658, 659, 660, 662, 663, 666, 706, 707
 see also American Petroleum Institute
Appalachian Basin, USA 803, 805
Appalachian Mountains, USA 731
Apsheron Peninsula, South Caspian 648
Apulian Block, Mediterranean 830, 831
Apure Basin, South America 907, 908
aquifer 702
Aquifex 497
Aquinet Ouernine (*see* Tanezzuft Shale)
Arab D, Saudi Arabia 860, 864
Arabian Basin 855, 859
Arabian Plate 855
Arabian Platform 859, 860, 861, 862
Ara Group, Oman 526, 771
Aramachay Formation, Peru 825
Arang Formation, Natuna Sea 941
Araripe Basin, South America 834
Araucariaceae 550, 551, 596, 600
arborinone 577
archaea 497, 498, 499, 511, 599, 647, 656
 lipids 503, 504–5, 506
archaeal acids 505–6
archaeal alcohols 505–6
archaeal lipids 928
archaebacteria *see* archaea
Archaefructus 572, 739

archaeocyathid 720, 747, 749
Archaeoglobus 648
Archaeopteris 728
archaeopyle 714
Archean Eon 711–14, 763
Archean petroleum systems 755–64
archenteron 717
archeols 510, 511, 957, 961
archosaurs 733
Ardennes 723
Ardjuna Basin, Indonesia 931
Ardmore Basin, Oklahoma 604, 700
Åre (Hitra) Formation, Norway 827–8
argon 749, 750
Arkoma Basin, USA 809
armored fish 721
Aroh well, Niger Delta 950
aromatic derivatives of the fernane/ arborane series 592–3
aromatic dinosteroids 490
aromatic hopanoids 580
aromatic hydrocarbon UCM *see* unresolved complex mixture
aromatic hydrocarbons (aromatics) 477, 594, 596
aromatic pseudo-reactions 975
aromatic steroids 665, 698–9, 798
aromatization 582, 586, 612
aromatization-isomerization (AI) diagram 968
aromatized C_{31} 8(14)-secohopanoid 588
Arrhenius constants (*see* rate constants)
Arrhenius equation 968
Arrhenius plot 967
Arthrobacter 681
Arthropleura 731
Arthur Creek Formation, Australia 774, 777
aryl isoprenoids (trimethylbenzenes) 485, 593, 780, 781, 798, 801, 811
Ashgill mass extinction 743, 749
Asmari Limestone, Iran 929
Aspen Shale, USA 900
asphalt
 Madagascar 680
 Malagasy 675, 685
asphaltenes 477, 479, 658
asphaltite 907
Asri Basin, Sumatra 939
assemblage A 779, 780, 783, 784, 788
assemblage B 779, 780, 788
Assiniboine Member, Favel Formation, Canada 903
Asteroxylon 727
Astrosclera 495
Asuka-881458 *see* carbonaceous chondrite

asymmetric carbon atom 556
Athabasca heavy oil, Canada 475, 645
Athabasca tar sand 544, 678
Athel silicilyte (Al Shomou Formation) 771, 772
atisane 490, 543, 551
atisene 550
Atlantic Ocean 736, 737, 738
atmospheric pressure chemical ionization (APCI) 603, 605
Atoka limestone, USA 732
atomic H/C 777
atomic H/C vs. O/C *see* van Krevelen diagram
atomic S/C 563
Austin Chalk, USA 882
Austral Basin, Argentina 894
authigenic carbonate 510
automated data inquiry for oil spills *see* oil spills
autoscale preprocessing 480
autotroph 711
Avalonia 722
Avilé Member, Argentina 895
Azerbaijan 932
Azile Formation, Gabon 849, 852

Bachaquero Field, Venezuela 911
Bach Ho Field, Vietnam 648, 939, 940
Bachu Uplift, China 785
Bacillus 649, 652
bacterial sulfate reduction (BSR) 815, 946
bacteriochlorophyll
 a and b 499
 d 605, 606
bacteriohopanepolyols 552, 588
bacteriohopanetetrol 552, 566
bacterivorous ciliates 576
Baghewala well, India 500, 530, 566, 567, 772
Bahamas Basin, Florida 879
Bahia Sul Basin, Brazil 847
Baikalian tectonic event 768
Bakken Shale, USA 477, 803–5
Bakr Field, Saudi Arabia 864, 865
Baku, South Caspian 934, 935
balanced-fill lake 817, 837, 838, 840, 844, 845, 846, 858, 923
Balcones Shale, Peru 930
Balikpapan Formation, Mahakam Delta 950–7
Balingian Delta, Borneo 942
Baltica 715, 719, 720, 722, 723, 730
Bambui Formation, Brazil 768
Banded Iron Formation (BIF) 712, 714
Banff Formation, Canada 803

bangiophyte 716
Bangladesh Basin 920, 922
Banuwati Shale, Sumatra 939
Baragwanathia 727
Baram Delta, Borneo 942
Barents Sea, Norway 483, 648
Barentsia 722
Barinas Basin, Peru 908
Barnett Shale, USA 813
Barney Creek Formation, Australia 765, 766, 767
Barreirinha Formation, South America 725, 802
Barreirinhas Basin, South America 837
barrels of oil equivalent (BOE) 756–7
Barrow Arch, Alaska 826
Barrow-Dampier Basin, Australia 874
Barrow-Prudhoe oil family 826
Barrow-Prudhoe oils 898
Barrow Sub-basin, Australia 563, 875
basement rocks 939
basin and range, USA 929
Baston Field, Texas 687
Bata Formation, Jordan 791
Bay of Campeche, Gulf of Mexico 878, 879
Baykit High, eastern Siberia 767
Bazhenov Formation, western Siberia 619, 681, 736, 888
Bazhenov-Neocomian(!) 755
Beatrice Oil, North Sea 477, 522, 523, 525, 526, 527, 528, 561, 583, 584, 794, 795, 884
Beaufort–Mackenzie Basin, Canada 580, 959–60
Bedo Formation, Africa 791
Be'eri Sulfur Mine, Israel 571, 572
Beggiatoa 562, 715, 945
Békés Basin, Hungary 947
Belait Formation, Brunei 943
belemnites 740, 745
Belle Fourche Formation, USA 901
Belluno Trough, Italy 830, 831
Belut Formation, Natuna Sea 941
Bemidji, Minnesota 669
Benakat Shale, Sumatra 939
Benin Formation, Nigeria 949
Ben Nevis Field, Newfoundland 888
Ben Nevis Formation, Jeanne d'Arc Basin 686
Benue Trough (Basin), Nigeria 579, 837
benzene 650, 658
  biodegradation 652, 654, 659
benzo(a)pyrene 602
benzocarbazole ratio 966
benzocarbazoles 966

benzo(e)pyrene 746
benzo[g,h,i]perylene 746
benzohopanes 588, 798
benzoic acids 702
benzoperylene 602, 750
benzopyrene 750
benzothiophenes 485, 562, 672
benzothiophenic acids 653
benzoyl-CoA (benzoyl-coenzyme A) 654
benzylsuccinate 654
benzylsuccinic acids 702
Bering Land Bridge 743
Beryl Complex, North Sea 884
Beta Field, California 563
betulinic acid 580
betulins 572, 580
beyerane 490, 543–51
beyerene 550
bicadinane maturity index (BMI) 626
  BMI-1 621, 622, 624, 627
  BMI-2 621, 624
  BMI-3 621–2, 624
bicadinanes 485, 547, 548, 549, 620, 621, 622, 624, 739
bicadinanes/hopane 549
bicadinane T/(bicadinane T + hopane) 549
bicyclic diterpanes 541–7
bicyclic sesquiterpanes 540–1, 542
bicyclic terpanes 668–70
BIF *see* Banded Iron Formation
Big Horn Basin, Wyoming 815
Bighorn Group, North America 805
biocide 649
biodegradation 535, 645–705
  aerobic 658
  bitumen 702
  branched alkanes 665, 666–7
  coal 702
  anaerobic 658, 669
  effects on petroleum composition 658–64
  kerogen 702
  methane 656–7
  oil seeps 704–5
  parameters 664–705
  quasi-sequential 661, 664
  scale 660, 661, 672, 701, 703
  sterane versus hopane 685
biofilm 982
biogenic CO_2 664
biological marker *see* biomarkers
biological oxygen demand (BOD) 725
biomarker biodegradation scale 661–2, 664, 666, 670, 674
biomarker kinetics 966

biomarker maturation index (BMAI) 629–30
biomarkers (biological markers, molecular or chemical fossils) 476
  age-related 490–3, 980
  extraterrestrial 980–2
  oldest indigenous 714
  Proterozoic rocks 765
  source-dependent 476
  taxon-specific 491
biomass 504–5, 725, 729, 730, 749, 764
biomineralization 716
biopolymer 517, 548
bioremediation 662
biosphere 647
biosurfactant 646
biotic *see* biogenic
biphenyl cyclization 975
biphenyls 593, 673, 816, 817
biphytane (bisphytane) 498, 504, 506, 507
biphytanediol 506
biphytanol 506
birds 736
Birdhead Formation, Australia 736
Birkhead Formation, Australia 873, 874, 895
bisabolene 582
25,30-bisnorhopane 675, 677, 678, 680, 689
28,30-bisnorhopane (BNH, 28,30-dinorhopane) 485, 561–3, 581, 611, 618–19, 678, 686, 945, 960
29,30-bisnorhop-17α-hopane 562, 691
24,28-bisnorlupane 581, 699
bisnorlupane/hopane 580
23,28-bisnorlupanes 485
28,30-bisnorlupanes 580
bisphytane *see* biphytane
Bitter Springs, Australia 716, 770
bitumen 477, 608
bituminites 816
bivalves 733
Biyang Basin, China 575
Blaa Mountain Group, Canada 825
Black Band Unit, UK 738
Black Rock Member, Falkland Islands 812
Black Sea 508, 519, 594, 657
black shale 770
black shale facies *see* Green River Formation
Black Warrior Basin, USA 809
Blake Nose, DSDP Site 390, North Atlantic 890
Blanca Lila Formation, Argentina 946

Blankenship Member, Alaska 883
Blueberry Debolt oils, Western Canada Basin 610
Blue Hill Shale, USA 901
Blue Mountain Shale, Canada 788
Buller Coal, New Zealand 573
BNH *see* 28, 30-Bisnorhopane
(BNH + TNH)/hopanes 618–19
boat conformation 564, 569
BOD *see* biological oxygen demand
BOE *see* barrels of oil equivalent
boghead coal 517, 597
Bohai Basin, China 538, 540, 772, 926, 928, 965
Bohemian Massif, Czech Republic 764
bolide 764
  impact 745–7, 749, 750
Bolivar Coastal Fields, Venezuela 695
Bombay Basin, India 741
Bonaparte Gulf Basin 809
Bonarelli Horizon, Europe 738
Bone Spring Formation, USA 813
Bonny Light oil 653
Boreal Ocean 899, 901
Borglum Formation, North Sea 885
Borsk (Pilsk) Formation, Russia 941–2
Boscan Field, Venezuela 603, 605
botryals 495
botryococcanes 484, 487, 490, 507, 516–18, 812, 930, 937
botryococcenes 516, 517, 812
Botryococcus 487, 493, 516–18, 520, 600, 721, 780, 784, 810, 812, 844, 853, 895, 926, 936, 939, 945, 948
  race A 496
  race B 812
  race L 519
  rubber 518
Boundary Creek Formation, Canada 959
Bowen Basin, Australia 702
brachiopods 724, 733, 747
brackish 488, 489, 838
Brainard Formation, USA 788
branched alkane 666–7
BraNobel 935
Brazilian oils 856
breccia 523, 527, 529, 574, 668, 830
Brent Formation, North Sea 885
Bridge Creek Limestone, USA 901, 903
Bright Angel Shale, Arizona 771
Brookian Sequence, Canada 959–60
Brooks Range, Alaska 812
Brora Beach, Scotland 523, 527, 529, 561, 574, 668
Brora Coal, Scotland 574
Browne Formation, Australia 770

Brown Limestone, Gulf of Suez 738, 915, 916
Brown Shale *see* Pematang Brown Shale
Brunei oils 624
Brushy Canyon Formation, USA 813, 814
Bryophyta (mosses) 724, 725, 726, 727
bryophyte *see* Bryophyta
bryozoans 720, 733
BSR *see* bacterial sulfate reduction or bottom-simulating reflector
Buah Formation, Oman 771, 772
Buckner Anhydrite, Gulf of Mexico 877
Bucomazi Formation, West Africa 575, 604, 841–2, 843, 844, 845, 853, 858
Bunte Breccia, Germany 830
Buracica Stage, 835
Burgan Formation, Arabian Platform 867
Burgan Rumaila High, Kuwait 863, 864
burial history chart 751, 753
butane 664
butanoyloxyfucoxanthin *see* carotenoid

C_{27}–C_{28}–C_{29} distribution *see* ternary diagram
C_{30} *ent*-isocopalane *see* tricyclohexaprenane
C_{30}-sterane index 527–30
C_{30} tetracyclic polyprenoids *see* tetracyclic polyprenoids
Caballos Formation, Colombia 908
Caballos Formation, South America 738
Cabinda, Angola 853
Cache Creek Formation, USA 733
cadalene (1, 6-dimethyl-4-isopropylnaphthalene) 582, 596, 597, 624, 981
cadinanes 490, 547–9
cadinene 548, 582
cadinol
calamenene 981
calamites *see* sphenopsid
Calcaires en Palquettes Formation, Jura, France 495
calcite 479, 533
  spurs 816
Caledonian Mountains 723
Cambay Basin, India 920
Cambay Shale, India 920
Cambrian explosion 718, 747
Cambrian paleo-plate reconstruction 719
Cambrian Period 719–20, 771
Cambrian source rocks 774–7

camphene 981
camphor 981
Campins Basin, Spain 945
Campos Basin, South America 835, 842–3, 846, 847, 852
Candeias Field, Brazil 842
Canning Basin, Australia 511
Canol Shale, Canada 725
canonical variable (CV) 499, 500, 569, 939
Canyon Formation, USA 813
Carapita Formation, Venezuela 942
Caravaca, Spain 746
carbocation 583
carbohydrates 725
carbonaceous chondrite 601, 711, 746, 750, 981
carbonate
  anhydrite 569
  clay ramp 861
  evaporite 555, 808
  oils 915
carbonate ramp intershelf system 861, 862
carbonate rock 487, 499, 613, 629
carbonate source rock *see* source rock
carbon cycle 724
carbon dioxide (CO_2) 919
Carboniferous coals 729
Carboniferous (Early) paleo-plate reconstruction 730
Carboniferous (Late) paleo-plate reconstruction 731
Carboniferous Period 730–2
Carboniferous source rocks 776, 807–12
carbon isotope ratio ($\delta^{13}C$) 510, 657, 749, 750
carbonium ion *see* carbocation
carbon preference index (CPI) 490, 493, 641–2, 781, 783
carboxylic acids 701
carcinogen 602
Cariaco Trench, Caribbean Sea 508, 518
Carlile Shale, USA 901
Carnarvon Basin, Australia 595, 597, 875
Carneros oil, California 585, 616
β-carotane (perhydro-β-carotene) 483, 485, 487, 521–4, 527, 593, 594, 796, 817, 818, 921
carotene 483, 522, 523, 593, 594, 726
carotenoid 521–4
Carpathian Mountains, Europe 947
Carpentaria Basin, Australia 905
Carrizo Field, Peru 930, 932, 933
Casablanca Field, Spain 947, 948
casbene 541

Caspian basins 742
Cassiporé Basin, Brazil 847
castaprenols 589
catabolism 661, 670
catagenesis 580, 608
Catalan oil shale, Spain 945–6
catalase 713
Catatumbo Basin, Venezuela 907
Cat Canyon Field, California 945, 946
catechol 650, 652
Catlin oil shale plant 931
Cavone Field, Italy 824
CDT (see Canyon Diablo Troilite)
Ceará Basin, South America 837, 846, 847, 849
cell-surface hydrophobicity 653
cellulose 725, 726
cembranoid diterpanes 538–40, 541, 542
cembrene 541, 542
Cenomanian-Turonian transgression and oceanic anoxic event (OAE 2) 890, 908, 909, 911
Central Atlantic Magmatic Province 748
Central Atlantic Ocean 734, 735
Central Basin Platform, USA 812
Central Brooks Range, Alaska 826
Central Colorado Trough, USA 810
Central Graben, North Sea 884
Centralian Superbasin, Australia 774
Centralian Supergroup, Australia 770
Central Saudi Arabian oils 631
centroid clustering 480
cephalopod 724, 736
Cerdanya Basin, Spain 945
Cerro Negro, Venezuela 668
certainty of correlation 751
Chacra Group, Peru 930
chain elongation 655
Chainman Shale, USA 732, 807
chair conformation 564, 569
Chaishiling Formation, China 871
chalcocite 815
chamaecydin 546
Charales 724, 726
Charco Bayo Field, Argentina 895, 896
charcoal 750
charge 475
charophytes 724, 725–6
Chattanooga Shale, USA 725, 803, 805
Chaunlinggou Formation, China 714
cheilanthanes see tricyclic terpanes
Chela Formation, Africa 841, 845
chemical fossil see biomarker
chemical ionization (CI) see ionization
chemoautotroph 562
chemocline 594, 749

chemometric analysis 479–82
Cherokee Shale, USA 732
Cherry Canyon Formation, USA 813
chert 712, 716, 727, 771, 796, 909, 917
Chiapas-Tabasco Basin 878
Chicxulub Impact Crater, Yucatan 738, 745, 747, 883
Chimney Rock Shale, Paradox Basin 810
Chira Formation, Peru 930, 932
Chloranthaceae 739
Chlorella 777
chlorite 479
chlorobactane 593
Chlorobactene 593
Chlorobiaceae (green sulfur bacteria) 496, 593, 603, 604, 605, 606, 797, 801
chlorococcalean algae 561
chlorophycean algae 716, 928
chlorophyll
  chlorophyll a 499, 606, 726
  chlorophyll b 726
  chlorophyll c 603
  chlorophyll d 603, 604
Chlorophyta 561, 725
chloroplast 716
5α-cholestanol 535
5α-cholest-5-en-3β-ol see cholesterol
cholest-5-en-3β-ol 525
cholesterol (5α-Cholest-5-en-3β-ol) 713
Chonta Formation, Peru 561, 738, 907, 908
C$_{35}$ hopane index see hopanes
chordate 717
Chorhat Sandstone, India 718
Chromatiaceae (purple sulfur bacteria) 593
chromatography see gas or liquid chromatography
chromium isotope ratio 745
chronostratigraphic unit 951
chrysenes 479
chrysophyte algae 527, 561
Chuar Group, Arizona 530, 540, 576, 771
cigarette smoke 602
Cimmeria 735
cladoxylate 728
cladoxylophyte 728
Clara Group, UK 720, 774
Clarkia Formation, Idaho 547, 579
clay 562, 629
Clearfork Formation, USA 813
Cleopatra 915
climate 837, 841
Clinograptus 722
Clinograptus Shale, Europe 777
cluster distance 481
Cnidaria 717

coal 551, 811, 837, 870, 950, 955
coalbed methane 902
coal/resin source rock 499, 500
coccolithophorid (coccolithophore) 526
co-elution 613
coenzyme Q see ubiquinone
Coetivy Island, Seychelles 873
Cold Lake heavy oil, Canada 645, 667
cold seep 497, 509, 510, 511
Colechaetales 726
Collingwood Member, Lindsey Formation, Canada 779, 785
collision-activated decomposition (CAD) 612
Colorado Shale, USA 900
Colpachucho Formation 801
Columbus Basin, Venezuela 907
Colville Basin, Alaska 812
Colville Delta, Alaska 826
comets 709
compound-specific isotope analysis (CSIA)
  correlation 476
  meteorites 981
  porphyrins 603
condensates 478
Condor Oil Shale, Hillsborough Basin, Australia 518
configurational isomerization see stereoisomerization
conformation 556, 564, 569
Congo Basin, Africa 841, 935
Congo Delta 850
congressane see diamantane
conifers (Gymnosperms) 539, 544, 547, 551, 577, 592, 596, 597, 600, 728, 729, 730, 731, 733, 736, 739
coniferophytes 578
conifer resins 547
conodont 744, 747
contamination 599
  bitumen and oil 964
  meteorite 602, 981–2
  rock 764
continental crust 709
conversion factors 762
Cook Inlet, Alaska 833–4
Cooksonia 724, 727
Cooper Basin, Australia 596, 702, 820, 821, 822, 874
coorongite 517
Copacabana Formation, South America 811
copper 814, 816
Coquinas Sequence 843
coral 539
cordaites see gymnosperms

Corg see total organic carbon
Cormorant Field, Australia 540
Cormorant Field, UK 886
Cornbrash Formation, UK 875
coronene 602, 746, 750
Corynebacterium 652
cosmic rays 745
cotyledon (seed leaf) 738
CPI (see carbon preference index)
cracking reactions 612
Crenarchaeota 503, 505
Cretaceous mass extinction see Cretaceous–Tertiary boundary
Cretaceous oceanic anoxic events 889–93
Cretaceous Period 736–8
Cretaceous paleo-plate reconstruction 737
Cretaceous source rocks 823, 907
Cretaceous–Tertiary (K–T) boundary 602, 740–1, 745, 746, 747, 749, 764, 883
Cretaceous-Tertiary paleo-plate reconstruction 737
C-ring monoaromatic steroids see monoaromatic steroids
crinoids 721
critical moment 753, 754, 789
crocetane 484, 502, 507–10, 511, 657
crocetene 509
crocodiles 736
crustal stretching (β) 967
β-cryptoxanthin see carotenoid
Crytograptus Shale, Europe 723
CSIA (see compound-specific isotope analysis)
C_{27} Ts see Ts
C_{29} Ts 552, 564, 565, 581, 617
C_{29} Ts/(C_{29} hopane + C_{29} Ts) 617
C_{30} Ts 581, 613
Cunaloo Member, Australia 875
Cupressaceae 547, 550, 551, 600
curcumene 981
Curimã Field, Brazil 846, 849
Curtis Sea, Utah 834
Curtisite 479
Curuá Group, Amazonas Basin 802
cuticle 724, 725
cutin 725, 727
Cuu Long Basin, Vietnam 938, 939
Cuyo Basin, Argentina 893
cyanobacteria 494, 569, 712–14, 725, 781, 784
cyanobacterial mats 493
cycads 730, 733
cyclicity of extinctions 747
cyclization 582

cycloalkane (naphthene, cycloparaffin) 650
cycloalkanoporphyrin (CAP) 604–5
cyclohexane 651, 668
cyclohexyl alkane 485
cycloparaffin see cycloalkane
cyclopentadecane 521
cyclothem 731, 736
p-cymene 582
Cymric Field, California 646, 679, 680, 682, 683, 684
Cytochrome 714
Cytophaga 652

Dacrydium 550
Dadas Formation, Turkey 791
Daekhurin Formation see Pilenga Formation
Dagong Field, China 628
Damar Field, Sumatra 517
Damborice Well, Czech Republic 538, 566, 586, 636, 637
Dameigou Formation, China 871, 872
dammar resin 548, 622
Daqing Field, China 904, 905
Dead Sea, Eastern Mediterranean 503, 512
Dead Sea asphalt 672, 917
Deadwood Sandstone, Williston Basin 789
dealkylation 582
Dean Formation, USA 813
de-A-steroidal tricyclic terpanes 696
Debolt Formation, Canada 807
decarboxylation 582, 700
Deccan Traps, India 748, 920
Decorah Formation, USA 540, 765, 780, 781, 783
Deep Sea Drilling Project (DSDP; later called Ocean Drilling Program, ODP) 889, 921
Deer-Boar(.) hypothetical petroleum system 753, 754, 961–3
degraded aromatic diterpanes 596
degradation index 909
dehydration 582
6,7-dehydroferruginol 546, 547
dehydrogenation 582
Delaware Basin, Texas 809, 810, 812, 813
Delaware Mountain Group, USA 813
Delle Phosphatic Member, Nevada 807
delta log R 755
demetallation 602
demethylated D-ring aromatized 8,14-secohopanes 693–5, 798

demethylated hopanes see 25-norhopanes, 28,30-bisnorhopane, 25, 28, 30-trisnorhopane
demethylated tricyclic terpanes 694, 695
dendrogram, HCA 480, 481
Dengying Formation, China 772
Denver Basin, Colorado 478, 810
deoxophylloerythroetioporphyrin (DPEP) 602
deoxyribonucleic acid (DNA) 711, 715, 726
Derdere Formation, Turkey 915
des-A-triene 593
des-E-hopane 559
16-desmethylbotryococcane 484, 516
10-desmethylhopanes see 25-norhopanes
desmethylhopanes 561
desmethylsteranes 527
desmethylsterols
Desulfosarcina 657
Desulfosarcinales 657
Desulfotomaculum 647
Desulfovibrio 648
deuterostomes 717
Devonian–Carboniferous boundary see Devonian–Mississpian boundary
Devonian mass extinction 743, 746
Devonian-Mississippian boundary 747, 797–8, 799, 803
Devonian paleo-plate reconstruction 723
Devonian Period 723–5
Devonian source rocks 776, 794–807
Dezful Embayment, Iran 741, 855, 867, 928, 929
Dhahaban Formation, Oman 526, 772, 775, 966
Dhruma Formation, Saudi Arabia 860, 864
Diacholestanes 537
dia/(dia + regular) C-ring monoaromatic steroids 584
diadinoxanthin see carotenoid
diagenesis 580, 603, 608
diahopanes (C_{29}*–C_{34}*) 564, 565, 696
17α-diahopane/18α-30-norneohopane (C_{30}*/C_{29} Ts) 563
diamantane (Congressane) 700
diaminopimelic acid (DAP) 720
diamond anvil 649
diamondoids 611
 biodegradation 699
diaromatic bicadinane ratios 624, 626
diaromatic carotenoid 801
diaromatic secobicadinane 620

diaromatic secobicadinane ratio (DSR) 622, 623, 627
diaromatic 8(14)-secohopane 589
diaromatic secotricadinane 620
diaromatic tricadinane ratio (DTR-1, DTR-2) 623, 627
diaryl isoprenoids 593
diasterane index *see* diasteranes/steranes
diasteranes (rearranged steranes) 485, 500, 662
   biodegradation 665, 675, 676
   C_{27} diasteranes 500
   C_{30} diasteranes 561
   20S/(20S + 20R) 631
diasteranes/steranes 483, 533–6, 617, 630–1, 856
diasterenes 533, 535, 630
diatomaceous 742, 942
Diatomaceous Zone, South Caspian 932, 934
diatomite 958
diatoms 491, 492, 512, 513–16, 526, 538, 607, 725, 859, 917, 943
dibenzo[a,h]fluorene 479
dibenzofuran 816
dibenzothiophenes 651, 672
Dicellograptus 722
Dicellograptus Shale, Scandanavia 777
dicot *see* angiosperm
Dicroidium 578, 733
Dictyonema Shale, Estonia 777
diesel 668
diether lipids 497, 503–5, 506
dihydro-*ar*-curcumene 582, 596, 670–1
dihydrophytol 499
Dillinger Ranch Field, Wyoming 814
dimer 547
dimethylalkane 493, 494
dimethylbiphenyls 672, 973
23,24-dimethylcholestane 485
17α, 18-dimethyl-des-E-hopane 572
2,6-Dimethylheptane 497
dimethylnaphthalene ratio (DNR, DMN, DBR, DPI) 625, 643–4, 671, 672, 974
dimethylnaphthalenes 643, 670, 973, 974
   1,6-dimethylnaphthalene 596, 624
   1,8-dimethylnaphthalene 644
12,17-dimethyloctacosane 497
13,16-dimethyloctacosane 494, 497
1,7-dimethylphenanthrene (DMP) 582, 596
dimethylxanthones 673
Dinaric Alps, Europe 947
Dingo Claystone, Australia 736, 821, 874, 875
Dingo Gas Field, Australia 770

dinoflagellates 491, 526, 531, 561, 588, 843, 859, 926
   bloom 483, 487
   cysts (hystrichospheres) 491, 588, 714, 919
28,30-dinorhopane (DNH) *see* 28, 30-bisnorhopane
dinorlupanes 697
24,25-dinorpentene 593
dinosaurs 736, 737, 740, 744, 745
dinostanol 531
dinosteranes 483, 485, 490, 491, 528, 531–2, 588, 859
dinosteranes/(dinosteranes + 3β-methyl-24-ethylcholestanes) 787
dinosteroid 491
dinosteroids/(dinosteroids + 3-methyl-24-ethylcholestanes) 787
dinosterol 483, 531, 588
dipentene *see* limonene
diphenylmethane 673
diphenylethane 673
diploid *see* sporphyte
diploptene 509, 552
diplopterol 552, 957, 960
Dipterocarpaceae 547, 620, 621, 624, 942
discriminant analysis 487, 983
disproportionation 582
diterpanes 543, 544
Diyab Formation, United Arab Emirates 863, 864, 866
DNA *see* deoxyribonucleic acid
Doba Basin, Chad 837
dodecane 654
dodecylsuccinic acids 654
Doig Formation, Canada 735, 825
dolichols 498
Dolni Lomna well, Czech Republic 538, 541, 566, 586, 636, 637
dolomite 575, 772, 783, 788, 794, 824
Domanik Basin, Russia 725
Domanik Shale, Russia 798, 799–800
Dong Ho, Vietnam 939
Dongying Depression, China 926
Dotternhausen Quarry, Germany 500, 829
Douala Basin, Cameroon 502, 851
Douglas Creek Formation, Uinta Basin 923
DPEP *see* deoxophylloerythroetioporphyrin
DPEP/etio 635
Drake, Colonel Edwin L. 934
Draupne Shale, North Sea 736, 827, 885
drepanophyte 727

drilling mud 479, 649, 671
drimane 540–1, 542, 552, 597, 670
Dry Piney Field, Wyoming 814
Duchesne River Formation, USA 479
Dukham Formation, United Arab Emirates 863, 864
Dunaliella 522
Dun Caan Shale, Scotland 970
Dunlin Formation, UK 885
Duri Field, Sumatra 518, 519, 520, 936
Duvernay Formation, Canada 725, 801, 802, 970
Duwi Member, Middle East 915
dysoxic *see* suboxic

Eagleford Formation, USA 881
Eagle Mills Formation, Gulf of Mexico 877
Early Proterozoic petroleum systems 764
Early Toarcian anoxic event 829–30
East African interior rift basins 853
East Berlin Formation, Connecticut 616
Eastern Drava Depression, Croatia 947
East Greenland 810–11
East Shetland Basin, UK 886
East Texas Basin 880
East Texas Field 881
east Texas oil boom 688
East-Zeit well, Gulf of Suez 915
EASY%R_o 611
echinoderms 717
ectothermic (cold-blooded) 733
Ediacara fauna 717, 718, 719, 747
Ediacara Hills, Australia 716
Eel River Basin, California 587, 618, 634
effective source rock 608
Egret Member, Rankin Formation, Newfoundland 686, 736, 887, 888–9
Egyházaskeszö Crater, Hungary 948
Eh 533, 604, 617, 630
eigenvector 480
   analysis *see* principal component analysis
Elko Formation, Nevada 526, 576, 929–30
Elk Point Basin, Canada 796–7
Ellesmerian(!) 825, 883
electron
   acceptor 646, 655
   donor (food) 646
   impact (EI) *see* ionization
Elenburger Formation, USA 813
Elika Formation, Iran 861
embryo 726
Emeishan-Panjal Volcanics 748
endosymbiotic hypothesis 715
endothermic 735

Ene Basin, South America 812, 813
Ene Formation, Peru 812, 813
enhanced oil recovery 647
Enteromorpha 513
Enterophysalis 779
Eocene–Oligocene boundary 747
Eocene paleo-plate reconstruction 740
eoembryophytic 727
eotracheophytic 724, 727
epicuticular wax 493
Epiya well, Nigeria 855
Equaluik oil, West Greenland 692
equilibrium reactions 610
Equisetum (horsetail) 728, 736
Eromanga Basin, Australia 596, 820, 821, 873, 874, 895, 905
Escabana Trough, Gorda Ridge, Pacific Ocean 964
Escherichia 720
Espírito Santo Basin, South America 835
ester decomposition 974, 975
estimated ultimate recovery (EUR) 736, 738, 955
Estonian oil shale 520
ethane 664
ether bond 504
ether-bound lipids 496
ether-bound methylated alkanes 496–7
ethylbiphenyls 672
24-ethylcholest-5-en-3β-ol 525
ethylnaphthalene isomer ratio (ENR) 643–4
10-ethyl-2,6,15,19-tetramethylicosane 511
etioporphyrin 602
ETR (extended tricyclic terpane ratio) *see* tricycylic terpanes
eudesmane 485, 540–1
eukarya *see* eukaryote
eukaryote 713–6
  oldest multicellular fossils 716
Eumeralla Formation, Australia 904–5
Euphorbiaceae 577
euphotic zone *see* photic zone
euphylloophytes 724
Euramerica *see* Laurussia
Eurasia 734, 735, 811
Europa 716
eustatic sea level rise 774
eutracheophyte 724, 727
euxinic 892
evaporite 744, 845
even-carbon preference *see* carbon preference index
events chart (timing-risk chart) 752, 754
Excello Shale, USA 732
exopolymer 713

exploratory data analysis (EDA) 480
Exshaw Formation, Alberta 477, 725, 803–5, 808
extended hopanes (*see* homohopanes)
extended tricyclic terpane ratio (ETR) *see* tricyclic terpanes
extraterrestrial biomarkers 980–2
Exuma Sound, Bahamas 713

Fahud Salt Basin, Oman 855, 859, 867
Fairport Chalk Member, USA 901
Falciferum zone 828, 829, 891
Falkland Plateau, South America 837
Farewell Formation, New Zealand 918
farnesane 606
farthest neighbor clustering 480
fatty acids
  C_4–C_{14} 701
fermentation 498
fernanes 577–8
fernenes 577–8, 592, 593
fernenol 577
Fernie Formation, Canada 830, 832
ferns 579, 724, 728, 729, 731, 733
Ferrar Province, Gondwana 829
ferrocenecarboxylic acid 670
ferruginol 546, 547
ficaprenols 589
fichtelite 543–7
field ionization (FI) *see* ionization
Filicopsida 728
Filletino oil, Italy 568
fingerprint 483
  m/z 191 551–5
  m/z 239 586
First White Speckled Shale, Canada 902
fish 724, 731
Fish River Shale, Canada 960
Fish Scales Zone, Canada 900
fission tracks *see* apatite fission tracks
flat spot 705
Flavobacterium 652
flood basalt 735, 747, 748, 749, 829, 889, 891, 919, 920
Florina Basin, Greece 551
flowering plants *see* angiosperms
flowers 738, 739
fluorenes 702
fluvial deltaic oils 549
fluvial-lacustrine lithofacies 838, 844, 846
Foinaven Complex, UK 574
foraminifera (forams) 725, 731, 732, 921, 922, 957, 960
forecasting efficiency 475, 476
formation water 649
Fortescue Group, Australia 714, 755

Forties Field, North Sea 884
Fort Kent Thermal Project, Alberta 667
Fort Worth Basin, Texas 809
Frasnian–Famennian boundary 743, 748, 801
free-base porphyrins 603
free radical scavengers 713
frequency factor 966
freshwater 488, 489
fritschiella 725
Frontier Shale, USA 738, 901
Fruitland Formation, USA 900, 902–4
frustule 492
fucoxanthin *see* carotenoid
Fulin Basin, China 522
fullerenes 745, 746, 749, 750, 764
fumarate 654–5
  addition 654–5
fungal hyphae 716
fungus 595, 645
fusulinids

Gabon Basin, Africa 835, 841, 847, 849
Gadvan Formation, Khuzestan 866
Galembo Member, Colombia 909
Galena Group, USA 783, 788
Gama well, Gulf of Suez 915
gametophyte 726
gamma-ray response 905
gamma-ray zone, Alaska (GRZ) 826, 885, 895–8
gammacerane 485, 554, 578, 591, 665, 697, 780, 801, 814, 818, 820, 885, 921, 923
gammacerane index 501, 575–7
gammacer-2-ene 575, 591
gammacer-3β-ol *see* tetrahymanol
Garau Formation, Lurestan 860, 865
Garden Banks well, Gulf of Mexico 878, 879
Garoupa Field, Brazil 843, 847
gas chromatography/atomic emission detection 670
gas chromatography/mass spectrometry *see* GCMS
gas chromatography/mass spectrometry/mass spectrometry *see* GCMS/MS
gas hydrate 657, 749, 919, 960
gas/liquid chromatography *see* gas chromatography
gas–oil ratio (GOR) 976
gas petroleum systems 756–7
gas–source rock correlation 752
gastropod 737
Gautier Formation, Venezuela 907
GC *see* gas chromatography

Index

GC/FID *see* gas chromatography
GC × GC *see* gas chromatography
GCMS 476
GCMS/MS 476, 612, 613, 617, 626, 629, 630
GDGT *see* gylcerol dialkyl glycerol tetraethers
Gebel Zeit, Gulf of Suez 915
Geiseltal lignites, Germany 547
gene 717
generation-accumulation efficiency (GAE) 789, 809
generation (conversion) parameters 608–9
generative basins 751
geochemical fossil *see* biomarker
geochemical log 840, 844, 846, 849, 853, 855, 872, 936, 950
geochromatography 629, 964–6
geographic extent 754
geoporphyrin *see* porphyrin
geopressure 649–50
Georgina Basin, Australia 770, 774, 777
geranylgeranyl diphosphate 539, 541
geranylgeranyl pyrophosphate (GGPP) 550
Ghaba Salt Basin, Oman 526, 771, 772
Ghadames Basin, North Africa 790, 791, 798
Ghakum Formation, Iran 861
Ghareb Formation, Jordan 917–18
Gharif Formation, Oman 772
Ghawar Field, Saudi Arabia 645
gibbane 550
gibberellins 550
Gigante Field, Colombia 913
gigantopterids 572, 739
gilsonite 478, 479, 522, 523, 577, 604
gilsonite dikes 479
ginkophyte (ginkos) 730, 733
Gippsland Basin, Australia 540, 547, 551, 580, 613, 670, 672, 753, 916–17, 918
GK well, South Sumatra Basin 626, 627
glaciation 716, 732, 740, 744, 747–8, 768, 777
glaciers 720, 731
Glamo oil, Yugoslavia 568
Glenwood Formation, USA 765, 783, 788
global oceanic anoxic event *see* oceanic anoxic event
global plume 770
global sea level 744, 890
global temperature 744
global warming 748, 749, 889–93, 919
Gloeocapso 779

Gloeocapsomorpha 492, 599, 721, 779–83, 784, 785, 788
glossopterid 730
Glossopteris 733
glycerol dialkyl glycerol tetraethers (GDGTs) 503, 504
Glyptostrobus 546, 547
gnetales 739
Golden Trend Field, USA 806, 808
Goldwyrn Formation, Canning Basin 765
Gomo Member, Candeias Formation, Brazil 840, 841, 842
Gondwana 715, 719, 720, 721, 722, 723, 730, 731, 732, 733, 734, 735, 737, 739, 747, 811, 872
GOR *see* gas–oil ratio
Gothic Shale, Paradox Basin 810
Gotland, Sweden 777, 778
Gotlandien Shale *see* Tanezzuft Shale
Gotnia Basin, Middle East 855, 859, 863, 864, 865
gracilicutes *see* Gram-negative bacteria
grahamite 478
grain size 646
Graminaeae (grasses) 577, 741
Gram-negative bacteria 652
Gram-positive bacteria 652
Grand Canyon, Arizona 576
Grande Coupure (Big Break) 740
Graneros Shale, USA 900, 901, 902
graphite 608, 712
graptolites 721, 748
graptolitic shale 722, 790–1
grasses *see* graminaeae
Great Basin, Nevada 807
Greater Green River Basin, USA 921
Greater Pangea 732, 733, 735, 825
Greenhorn Cyclothem 900–1, 902
Greenhorn Shale, USA 738, 901, 902, 903
greenhouse 723, 749, 889, 891
greenhouse gases 714
Greenland–Scotland Ridge 740
Green River(!) 926
Green River Formation, USA 479, 502, 522, 523, 555, 575, 741, 921–5, 927, 946, 978, 979
greenschist facies (memorphism) 764
Green sulfur bacteria *see Chlorobiaceae*
Green well, Iowa 783
greigite 980
Grenzbitumenzone Formation, Southern Alps 824
Grypania 714
GRZ *see* gamma-ray zone
Guadalupian Series, Permian Basin 813
Guaymas Basin, Gulf of California 964

Guelph-Salina Interval, Michigan Basin 794
Gulf of Mexico 875–83, 899
Gulf of Suez, Egypt 915, 916
Gulf Stream 742
Gulf of Thailand 939–41
Gullfaks Field, North Sea 674
Gumai Formation, South Sumatra Basin 627
Gunflint Banded Iron Formation, Ontario, Canada 714
Gurpi Formation, Khuzestan 867
Guttenberg Member, Decorah Formation 765, 780, 781, 783
gymnosperm *see* conifers
Gymnospermopsida 578

Hadean Eon 709–11
hafnium 711
halite 503, 771
halobacteria 483, 499
halophile 647
halophilic archaea 503
halophilic bacteria *see* halobacteria
Halsea 514
Haltenbanken Area, Norway 827–8
Hamelin Pool, Australia 713
Hamersley Group, Australia 714
Hamilton Dome, Wyoming 532, 566, 814, 815
Hangenberg Event 797
Hanifa Formation, Saudi Arabia 860, 864, 865
haploid *see* gametophyte
Hartford Basin, Connecticut 616, 824
Hartland Shale, USA 901, 902
Hassi Messaoud Field, Algeria 792
Haupdolomit Formation, Northern Alps 824
Haushi Sandstone, Oman 771
Hawtah Well, Saudi Arabia 631
HBI *see* highly branched isoprenoids
H/C *see* atomic H/C
HCA *see* hierarchical cluster analysis
head-to-head isoprenoids 484, 498
Heathfield Well, Australia 906
Heather Formation, North Sea 828, 885
Heath Formation, USA 809
Heath Formation, Peru 742, 930, 932, 933
Heath-Tyler(!) 809
heating rate 969, 970, 971, 975
heat of formation 566
Hebron Field, Newfoundland 686
helium 745, 746, 749, 750
Helmsdale Fault, Scotland 574

hematite 714
hepatophyta (liverworts) 726
herbaceous 733, 739, 814, 828, 845
Hercynides Mountains, Europe 731
Hermosa Group, USA 809–10
heterosporous 728
heterotroph 711
hexacyclic polyprenoids 591
hexadecane 656
hexahydrobenzohopanes (hexacyclic hopanoids) 485, 569, 570
hexanoyloxyfucoxanthin see carotenoid
HI see hydrogen index
Hibernia Field, Newfoundland 887, 888, 889
hierarchical cluster analysis (HCA) 480
higher-plant fingerprint (HPF) 597
higher-plant input 594, 596, 599
higher-plant input (HPI) parameter 596–7
higher-plant parameter (HPP) 596–7
higher plants 524, 596, 725, 728
highly branched isoprenoids (HBI) 484, 512–16
  C_{25} HBI 513, 514, 515, 516, 917, 945, 958
highly branched isoprenoid thiophenes (HBIT) 515
highly radioactive zone (HRZ) see gamma-ray zone
high-performance liquid chromatography (HPLC) 605
highstand systems tract (HST) 481, 482, 484, 862, 909, 910, 953, 955, 956
high-temperature gas chromatography/electron-impact mass spectrometry (HTGC/EIMS) 603
high-temperature gas chromatography/field ionization mass spectrometry (HTGC/FIMS) 603
Himalayan Mountains 742, 743
Himalayan orogenic collision 868
Hoggar Mountains, Algeria 834
Hollin Formation, Colombia 908
hominid 742
Homo 743
homodrimane 542, 670
homohopanes see hopanes
homohopanoic acids 568
homolog 476
homologous series see homolog
homolytic scission 632
C(14a)-homo-26-Nor-17a-hopanes 692
homopregnane 485, 674, 676
Hongyanchi Formation, China 818, 820

hopane isomerization [22S/(22S + 22R)] 613–14, 626, 627, 635, 683, 684, 966, 967, 968, 970
hopanes 552, 566, 567, 570, 572, 573, 581, 613, 662, 692, 797, 927
  biodegradation 665, 681, 682, 683
  C_{27} hopane II see Ts
  C_{29}/C_{30} 571
  C_{30} hopane 621, 663, 666
  C_{31}/C_{30} hopane 559, 569
  C_{35} hopanes 485, 814
  C_{35} homohopane index (C_{35} hopane index) 500, 573, 574, 684, 781, 952
  $C_{35}S/C_{34}S$ 571
  distributions 566, 568, 615, 687, 691
  $17\alpha,21\alpha(H)$-hopanes ($\alpha\alpha$-hopanes) 552, 553
  $17\alpha,21\beta(H)$-hopanes ($\alpha\beta$-hopanes) 555, 614
  $17\beta,21\alpha(H)$-hopanes ($\beta\alpha$-hopanes) see moretanes
  $17\beta,21\beta(H)$-hopanes ($\beta\beta$-hopanes) 614, 688
  22S and 22R 565, 613
  stability 614
C-10 hopanoic (hopanoid) acids 678, 689
hopanoic (hopanoid) acids 688, 702, 703
hop-17(21)-enes 614
hopenes 830
Δ2-hopenepolyol 571
Horneiophyton 727
Horn Valley Siltstone, Australia 779, 783
hornworts see Anthocerophyta
horses 741
horsetail see Equisetum
hot shale 790, 791
HPLC see high-performance liquid chromatography
Huab Formation, South Africa 812
Hue Shale, Alaska 825, 826, 896–8
Hue-Thompson(!) 896
Humble, Texas 688
huminite 953
hump see unresolved complex mixture
Huqf Formation, Oman 530, 771–2, 773
2-hydroarchaeol 511
hydrocarbon activation 654
hydrocarbon machine 751
hydrogen exchange 630
hydrogen index (HI) 567
hydrogen index vs. oxygen index see van Krevelen diagram, modified
hydrogen iodide (HI) 504, 506
hydrogen peroxide 713
hydrogen sulfide (H$_2$S) 604, 646, 649, 815
hydrolysis
  alkaline 539

hydrothermal silicification 755
hydrothermal vent 711
hydrous pyrolysis 615, 630, 966, 969, 970, 971, 975, 976, 977, 978, 979
hydroxyarcheols 510
hydroxyl radicals 713
hydroxytriterpenoids 547
hypercarnia 748
hypersaline 489, 499, 501, 503, 512, 575, 576, 615, 628, 629, 808, 810, 812, 814, 817, 820, 845, 850, 917, 923
  source rock 616, 641, 923
hyperthermophile 648
hyperthermophilic archaea 649
hypothetical(.) 752
hystrichospheres see dinoflagellate cysts

Iabe-Landana Source Rock, Angola 850, 853, 854, 936
Iapetus Ocean 719, 721, 722, 774
Ice Age 720, 748
Icebox Member, Williston Basin 765
ichnofossil see trace fossil
Icypachi Member, Peru 825
ichthyosaurs 737, 740
Ichtyol Shales, Swiss Alps 824
Illinois Basin, USA 788–9, 798
Illizi Basin, North Africa 798
illite 533
immature 608
Imperial Shale, Canada 725
independent petroliferous systems 751
India 737, 740
influx of sediment, water 837–8
inner ramp 861
Inoceramus Shale, South America 738
insects 724, 731, 733, 737
Interior Basin, Gabon 898
Internal Zone, Betic Cordillera, Spain 957
intershelf basins 861
Ionian Basin, Greece 830, 831
ionization
  chemical ionization (CI) 587, 603
  field ionization (FI) 587
ionizing chamber see source
iosene see phyllocladane
Iquiri Formation, Bolivia 801
Irati Formation, Brazil 812, 819–20
Iremeken Formation, eastern Siberia 768, 769
iridium 745, 746, 750
Irkutsk region, eastern Sibera 767
iron–nickel sulfides 711
iron-reducing bacteria 654, 655
isoalkane 493, 650
isoarborinol 577, 592, 593

isoarborinone 575
isocadalene (1,6-3-isopropylnaphthalene) 623, 624
isocadalene/(isocadalene + cadalene) 623–4, 626, 627
isochamaecydin 546
isodihydro-*ar*-curcumene 670–1
isoheptane ratio *see* K_1
isohexyl alkylaromatics 599–600, 601, 602
isomerization 582, 612
isoparaffin *see* isoalkane
isopentadecane (2-methylpentadecane) 497
isopimarane 543, 544, 545, 547
isoprenoid/*n*-alkane ratio 641
isoprenoids 764, 816
    acyclic 506–7, 517, 665
    acyclic ($>C_{20}$) 498–9
    C_{13}–C_{20} regular 502
    C_{21+} regular 503
    C_{25} regular 508
3-isopropylalkanes 496
24-isopropylcholestanes 490, 491, 530, 749, 771
24-isopropylcholestanes/*n*-propylcholestanes 787
6-isopropyl-1-isohexyl-2-methylnaphthalene (ip-iHMN) 596, 597
isorenieratane 485, 523, 591, 593–4
    ^{13}C-rich 593, 903
isorenieratene 485, 593, 595, 901, 903
Isthmus of Panama 742
Isua, Greenland 712
Izhara Formation, Qatar 860, 863

Jamison Formation 766
Japan Trench, Pacific Ocean 519
Jatobá Basin, South America 834
Jeanne d-Arc Basin, Canada 617, 685–6, 887, 888
Jenam Formation, Bangladesh 920
Jet Rock, Yorkshire, UK 830, 892
Jianghan Basin, China 531, 946
Jilh Formation, Abu Dhabi 861
Jingjingzigou (Jingjingzigou, Zingjingzigou) Formation, China 817, 818, 819, 820, 946
Jiquia Stage 835
Jubaila Formation, Saudi Arabia 860, 864
Junggar (Zhungeer) Basin, China 522, 674, 695, 696, 732, 817, 819, 820, 870, 871, 946
Jura-Bresse Basin, Europe 735
Jurassic-Cretaceous highstand 774
Jurassic (Early) paleo–plate reconstruction 734

Jurassic (Late) paleo–plate reconstruction 735
Jurassic Period 735–6
Jurassic source rocks 823
Jusepin Field, Maturin Basin, Venezuela 696

K_1 (Mango parameter) 667
Kaapvaal Craton, Australia 712, 763–4
Kaimiro Formation, New Zealand 919
Kalavryta Basin, Greece
Kalol Formation, India 920
Kalubik well, Alaska 883
Kamovsk Arch, Eastern Siberia 767
Kant-LaPlace hypothesis *see* accretion disk hypothesis
Kap Stewart Formation, Greenland 828
Kapuni coals, New Zealand 918–19
Karababa Formation, Turkey 915
Karabogaz Formation, Turkey 915
Karalian shungite, Russia 764
Karamay Black Hill Field, China 819
Karamay Field, China 817, 818
Karoo Basin, South Africa 812
Karoo-Ferra basalts 748, 919
Karoo Formation, South Africa and Madagascar 763, 830
Karoo Province, Gondwana 829
Karun Field, Iran 929
Kashaf Formation 861
Katia well, Italy 568, 976, 977
Kattegat, Denmark/Sweden 509
kaurane 490, 543–51, 820
kaurene 550
Kaveakak Field, Alaska 883
Kavearak Point Field, Alaska 883
Kavearak Well, Milne Field, Alaska 557
Kawaruppu, Hokkaido, Japan 750
Kazakhstania 723
Kazhdumi Formation, Khuzestan 738, 859, 860, 867, 868
Keel Field, Michigan Basin 786
Keg River Formation, Canada 796–7, 798
Kelamayi Field, China 522, 695, 696
kerogen assemblage A and B *see* assemblage A or B
kerogen 499, 500, 502, 539, 608, 713, 749, 764, 777, 819, 972
    see also type I, II, IIS, II/III, III, IV
Khazdumi Formation *see* Kazhdumi Formation
Khmer Trough, Gulf of Thailand 941
Khuzestan, Iran 855, 859, 863, 928
Khuzestan Basin *see* Dezful Embayment
Kimberlella 718
Kimmeridge Bay Field, UK 875

Kimmeridge Clay (Kimmeridge) Formation, North Sea 496, 525, 527, 583, 619, 736, 827, 884–8, 889
    Bitumen extracts 526, 528, 562, 584, 588, 594, 611, 619
    Oil 611
Kimmeridgian transgressive shales 884–9
kinetic fractionation 977
kinetic model for biomarker maturation 972
kinetic parameters 845, 966
Kingak-Blankenship Interval, Alaska 826, 883–4, 898
Kingak Formation, Alaska 557, 825, 826, 883, 885, 896, 897
Kiowa-Skull Creek cyclothem 900
Kisalföld Basin, Hungary 947
Kishenehn Formation, North America 579, 930, 934
Kissenda Formation, Gabon 844, 848, 898–9
K-nearest neighbor (KNN) 482, 955
KNN *see* K-nearest neighbor
known(!) 752
Knox Dolomite, USA 788
KOH *see* potassium hydroxide
Korea Bay Basin, Yellow Sea 635
Kössen Marl Formation, Hungary 947
Koviktinskoye Field, Eastern Siberia 767
Krebs tricarboxylic acid cycle 606
Kreyenhagen Formation, California water 701
Krishna-Goodavari Basin, India 741
Kugmallit Sequence, Canada 960
kukersite 765, 779, 783–4
Kuma Formation, South Caspian 932
Kuna-Lisburne Interval, Alaska 812
Kuparuk Field, Alaska 826, 883
Kupferschiefer Formation, Germany 501, 566, 577, 593, 605, 733, 811, 814–17
Kurra Chine Formation, Iran 861
Kutei Basin, Kalimantan 742, 950, 953
Kwagunt Formation, Utah 771
Kwanza Basin, Angola 835, 841, 854
Kyalla Member, Australia 766

LaClede Bed (Lower Laney Member), Wyoming 923, 924
labdane 543, 544, 550, 600
labyrinthodont *see* amphibian
lacustrine 487, 521, 560, 577, 837–45
    oils 561, 578, 856, 904
Laffan Formation, Southern Gulf Basin 860, 867

Lagoa Feia Formation, Brazil 842–3, 846, 847, 853
Lahat (Lemat) Formation, Sumatra 939
Lake Gosiute, USA 921
Lake Nyos, Cameroon 748
lakes 487
Lake Songliao, China 528
Lake Thetis, Australia 720
Lake Turkana, Kenya 853
Lake Uinta, USA 921
Lake Valencia, Venezuela 577
Lake Vostok, Antarctica 716
Lakhanda Group, Siberia 716
Lalla Rookh Formation, Australia 755
La Luna Formation, Venezuela 529, 549, 643, 738, 907–11, 942
La Luna-Misoa(!) 909–11
lamalginite 864
Lameignère Quarry, Orthez, France 590
laminae 713
lamination 488
Landana Formation see Iabe-Landana source rock
land plants 578, 721, 723, 724, 725–30
lanosterol 14α-demethylase 714
Laramide Uplift 921
Late Ashgill event see Ashgill mass extinction
Late Paleocene thermal maximum 919
Late Proterozoic (Neoproterozoic) petroleum systems 768–72
Late Proterozoic paleo-plate reconstruction 715
latex 547
laticifers 547
Latrobe Formation, Australia 738, 916–17, 918
Laurasia 735
Laurentia 715, 719, 720, 721, 722, 723, 725, 730
Laurineae 739
Laurussia 723, 730
Lava Basin, Greece 551
leaching see solubilization
lead ore 825
leaves see megaphylla
Leduc Reef, Canada 801
legumes 726
Leigh Creek Coalfield, Australia 578
Lemmaphyllum 579
lemurs 741
Leonard Shale, USA 733, 813
Leonard-Wolfcamp units, USA 812, 813
Lepidodendron 579, 729
lexane 523
Leyte oil, Philippines 622
LiAlH$_4$ see lithium aluminum hydride

Liaohoe Basin, China 634
Liaoning Province, China 739
Lias Shale, Europe 968
Libros Basin, Spain 945
lichens 726
light hydrocarbons C$_2$–C$_6$ 664
light oils 478
lignin 725, 727
lignite 547, 613, 926
limestone 495
limnic (lacustrine) coal 551
limonene 582
limonite 479
Lincoln Shale, USA 901, 902
Lindsey Formation, Canada 785
Linyi Basin, China 616, 968
lipids 725
Locker Shale, Australia 821–2, 875
liquid chromatography/mass spectrometry (LCMS) 603
lithium aluminum hydride (LiAlH$_4$) 504, 506
lithology 646
Little Plains Basin, Hungary 947
littoral facies 924
liverwort see non-vascular plant
Llanos Basin, South America 907
loadings plot 482
Lodgepole Formation, USA 803, 804
Logbaba Gas Field, Cameroon 851
Loma Chimico Shale, Costa Rica 516, 917, 919
Loma Montosa Formation, Argentina 894
Lombardy Basin, Italy 831
Lorca Basin, Spain 957–8
Los Angeles Basin, California 563
Los Molles Formation, Argentina 893
Los Monos Shale, South America 796
Louanne Salt, Gulf of Mexico 735, 877
Loufika outcrop, Congo 675, 691
Louise well, Mahakam Delta 950, 953
Lower Congo Basin, West Africa 841, 850
Lower Kittanning coal, USA 970
Lower Magdalena Basin, Colombia 907
lowstand systems tract 481, 482, 574, 749, 862, 909, 953, 955, 956
Lubna well, Czech Republic 538, 541, 566, 586, 636, 637
Lucaogou Formation, China 817, 818, 820
Lucina Formation, Gabon 898
Luman Tongue, Washakie Basin 921, 923, 924
lupane 574, 575, 580, 581, 582, 697, 960
lupanoid ketones 697

lup-20(29)-ene 580
lupeol 580
lurestan, Iran 855, 859, 863, 865, 928
lutein see carotenoid
lycopadiene 517, 520
Lycopodium (club moss) 579
lycopa-14(E), 18(E)-diene see lycopadiene
lycopane 498, 507, 518–20, 524
lycopene 519, 520, 524
lycophyte (club mosses) see lycopsids
lycopsids (lycopsida) 724, 727, 728, 729, 731, 733, 811

Maceió Formation, Brazil 846
Machareti Formation, South America 796
Mackenzie Delta, Canada 580, 742
Mackenzie Mountains, Canada 716
Mackeral Field, Australia 918
Macrocyclic alkanes 485, 520–1
Macroseeps 704
Madbi Formation, Yemen 863
Madison Group, USA 805
Madre de Dios Basin, Bolivia 801, 811, 813
Madrin Formation, east Siberia 768
Mae Sot Basin, Thailand 959
Magwa Field, Kuwait 863, 864
magnetite 980, 981
magnoliidae 739
Mahakam Delta, Borneo 481, 482, 613, 942, 953, 955
Mahogany Member see Green River Formation
Majaer Depression, China 785
Makassar Slope, Kalimantan 481, 482, 953
malabaricatrienes 556
Malay-Tho Chu Basin, Vietnam 938, 939
maleimides 485, 605–6
Malembo Formation, West Africa 853, 935, 936
Malm carbonates, Bavaria 501, 518
Malongo West Field, Congo Basin 843
Malossa Field, Italy 824
MA(I)/MA(I + II) see monoaromatic steroids
mammals 736, 737, 741
Mancos Shale, USA 901
Mandal Formation, North Sea 885
Mandan, North Dakota 668
Mangahewa Formation, New Zealand 919
manganese 655
Manicouagan impact crater, Quebec 746, 747
Maniguin Island, Visayan Sea, Philippines 518, 519

Manilla Embayment, Gulf of Mexico 877
Mannville-Grosmont reservoirs, Alberta, Canada 477
Maoming Shale, China 516, 518, 519, 926
Maqna Formation, Red Sea Basin 946
Maquoketa Shale, USA 779, 788–9
Maracaibo Basin, Venezuela 907, 911
Marañon Basin, South America 811, 812, 813, 907, 908
Marathon Orogeny 809
Maracaibo Basin, Venezuela 489
Marañon Basin, Peru 667, 825
Mara well, Canada 685, 686
Marco Polo 934
Mardin Group, Turkey 915
marine coal 551
marine oils 561, 856
marine organic matter 524, 527
marine reptiles 736, 740, 744, 745
Marmoul Field, Oman 771, 773
Marraat oil, west Greenland 581, 582
Marrat Formation, Saudi Arabia 860, 864
Mars 716, 980
Marrat Shale, Oman 736
marsupial *see* mammal
Martian bacteria 602, 980
Marun Field, United Arab Emirates 867, 868
mass extinction 559, 721, 724, 733, 737, 743–9, 797, 801, 919
MA-steroid *see* monoaromatic steroids
mature 608
Maturin Basin, Venezuela 696, 907, 942
Mattoon Formation, Illinois 551
Maui Field, New Zealand 544, 547, 919
Mauritanide Mountains, North America 731
maximum flooding surface (MFS) 790, 831, 894, 908, 909, 956
Maykop Series, South Caspian 931–5
Mazalij well, Saudi Arabia 631
McArthur Basin, Australia 719, 765, 767
McArthur Group, Australia 765–6, 767
McArthur River well, Cook Inlet 834
McKittrick Field, California 616
McMinn Formation, Australia 765, 766, 767
Meade Peak Member, USA 733, 814
Meganeura 731
megaphylla (true leaves) 728
Melania Formation, Gabon 844, 848, 898–9
Melke Formation, Norway 828
meiosis 726
Mereenie Field, Australia 783

Meren well, Niger Delta 951
Meride Limestones, Italy 822
mesohaline 501, 824
Mesoproterozoic petroleum systems *see* Middle Proterozoic petroleum systems
Mesostigma 725
Mesozoic source rocks 820–918
Mesozoic source rocks, Arabian Platform 823, 855–68
  jurassic 862–4
  Lower Cretaceous 865–7
  Middle Cretaceous 867
  Upper Cretaceous 867–8
Mesozoic source rocks, Gulf of Mexico 876
Mesozoic source rocks, South Atlantic margins 834–48, 858, 908
  lacustrine 837–45, 849
  marine carbonates 847
  marine shales 852
  transitional carbonates 845–7
Messel Shale, Germany 494, 496, 497, 504, 518, 577, 589, 590, 591, 593, 603, 605, 635, 926–8
metalloporphyrin *see* porphyrin
metastable reaction monitoring *see* MRM/GCMS
metaxylene *see* xylene
metazoa 713, 718, 720
methane 714, 763
  biodegradation 656–7, 664
  hydrate 829, 892, 919, 960–1
  methane monooxygenase 656
methane-oxidizing bacteria 572
methanogen 498, 503, 509, 512, 656
methanogenic archaea 507, 508, 509
methanogenic bacteria *see* methanogen
methanogenic biomass 504–5
Methanosarcina 508
Methanosarcinales 657
methanotroph 656, 688, 960
methanotrophic archaea 509
methanotrophic bacteria 571
methyladamantane/adamantane 700
methyladamantane/nC$_{11}$ 700
methylalkanes 764
methylbicadinanes 549, 621, 622
methylbiphenyls 672
methylcyclopentadecane 521
methyldiamantanes 700
methyldibenzothiophenes 672
methyldiphenylmethanes 672
2-methyldocosane 484
methylfluorenes 702, 973
2α-methylhopane index 571, 781
methylhopanes 569–71, 781

2-methylhopanes 485, 494, 571
2α-methylhopanes 569, 714
3-methylhopanes 569, 571
3β-methylhopanes 485, 571
methylnaphthalenes 643, 670
methylnaphthalene isomer ratios (MNR) 643–4
Methylococcus 531, 571
methylphenanthrene index (MPI-1) 642–3
  MPI-1 vs. R_o 642
methylphenanthrene ratio (PP-1) 974
methylphenanthrenes 582, 596, 642, 643, 672, 973, 974
methylsteranes 534, 674
2α-methylsteranes 763
4-methylsteranes 485, 487, 528, 530–2, 533, 561
4-methylsterols 531, 927
methylthioacetate 711
4α-methyltriaromatic steroid 591
methyltrimethyltridecylchroman (MTTC) 500–1
methylxanthones 673
Metula 695
Mexia-Talco Fault Zone, Texas 877
Miandoum Field, Chad 542
Michigan Basin, USA 794, 803, 975
microbial gas 509
microbial mat 494
  see also cyanobacterial mat
microfossils 712, 713, 714, 764, 765
microphylla (spines) 727
microseeps 704
microsphere 711
microtektites 746
microtubules 726
Mid-Atlantic Ridge 647
mid-chain monomethylalkanes (X-compounds) 484, 493, 767, 771, 772
Mid-Cretaceous anoxic event 749
Middle Green River Formation, Uinta Basin 923
Middle Magdalena Basin, Colombia 549, 907, 908, 909
Middle Proterozoic (Mesoproterozoic) petroleum systems 764–8
Midland Basin, USA 812, 813
mid-ramp 861
Midway Formation, Gulf of Mexico 882
migrated oil 477
Migration 616, 629
  distance 966
Mikkelsen Bay State well, Alaska 898
Milankovitch 824

Miller–Urey experiment 711
Milligans Formation, Bonaparte Gulf Basin 809
Milne Point Field, Alaska 883
Minagish Formation, Gotnia Basin 860, 866
Minas Field, Sumatra 517, 518, 519, 520, 936, 937
Minnelusa Shale, USA 811
Miocene paleo-plate reconstruction 741
Mississippian-Pennsylvanian Period *see* carboniferous period
Mississippi Canyon Well, Gulf of Mexico 881
mitosis 726
mixed oils 561, 685
MNR *see* methylnaphthalene ratio
Moho *see* Mohorovicic discontinuity
M-2 oil, Danish North Sea 553
Mokattam Shale *see* Tanezzuft Shale
Molasse Basin, Europe 699
molecular clock 717
molecular dynamics (MD) 966
molecular fossil *see* biological marker
molecular mechanics 556, 557, 565, 617, 629, 684, 695
molybdate 654
monazite 755
monoaromatic hopanoid parameter 634–5
monoaromatic polyprenoids 589–91
monoaromatic steroids (MA-steroids) 583, 584, 585, 631, 632
  MA(I)/MA(I + II) 632–3
  [TA/(MA + TA)] *see* steroid aromatization
monocot *see* angiosperm
monoethylalkane 900
monomethylalkane 493, 494–6, 767, 900
monooxygenases 650, 651, 652, 653
monotreme *see* mammal
Monte Prena, Italy 976, 977
Monterey Formation, California 505, 519, 540, 562, 568, 609, 615, 618, 679, 742, 944, 945
  bacterial mats 562
  biogenic silica 944
  lithofacies 563
  oils 475, 559, 562, 568, 603, 619, 628, 641, 646, 680, 706, 707, 708, 946, 977, 978
  Phosphatic Member 971, 978, 979
  Siliceous Member 971, 978
  water 701
montmorillonite 533, 535, 965
Montney Formation, Canada 825
Moravia, Czech Republic 538, 587
Moraxella 705

Moray Firth, UK 477, 523, 884, 885
Moreno Formation, California 509
moretanes 552, 614, 616, 621
  22S/(22S + 22R) 614
  moretanes/hopanes 614–15
mosasaurs 740
Moschopotamos Basin, Greece 551
Mosquito Creek Formation, Australia 755
Moura coals, Australia 702–4
Mount Cap Formation, Canada 780
Mount Holland Shale, UK 738
Mowry Shale, USA 738, 900, 901
MPI *see* methylphenanthrene index
MRM/GCMS 476, 528, 530, 612
Msimbati, Tanzania 688
MTTC *see* methyltrimethyltridecylchroman
mud volcano 497, 933
Mukluk well, Alaska 475
Mulhouse Basin, Alsace, France 604, 605
Multidimensional chromatography *see* gas chromatography
multiple ion detection (MID) *see* selected ion monitoring
multivariate statistics 479
Murchison meteorite 602, 981
murein 720
Muribeca Formation, Brazil 846
Murta Formation, Australia 874, 895
Murzuq Basin, Libya 791
Muskeg Member, Keg River Formation, Canada 796
Myanmar oils 622
Mycobacterium 652
m/z 191 *see* terpanes
m/z 217 *see* steranes

Nahr Umr Formation, Arabia 738
Nahrum Shale, Arabia 738
Najmah Formation, Kuwait 860, 863, 864, 866
Nam Con Son 938, 939
naphthalenes 651, 652, 672, 817
naphthene *see* cycloalkane
naphthenic acids 701
naphthenoaromatic acids 701
Naples Beach, California 562
Napo Formation, South America 738, 907, 911, 912
nappe 742
Narragansett Bay, Rhode Island
Narrows Gaben, Queensland, Australia 928
Narryer Gneiss, Australia 711
$NaSCH_3$ *see* sodium methane thiolate

Natih Formation, Oman 860, 861, 862, 867, 870
National Petroleum Reserve in Alaska (NPRA) 883, 896
Natuna Sea rift basins 939–41
Navay Formation, South America 907, 908
nC_{29} alkane 750
nearest neighbor clustering 480
negative correlation 477
Neiber Dome, Wyoming 585
Nemcicky well, Czech Republic 538
Nematothallus 726
Nemesis 747
Neogene Period 741–2
18α-neohopane
  C_{27} *see* Ts
  C_{29} *see* C_{29}Ts
  C_{30} *see* C_{30}Ts
Neoproterozoic Eon 715–19, 771
Neoproterozoic petroleum sytems *see* Late Proterozoic petroleum systems
neoxanthin *see* carotenoid
Nesson Anticline, Williston Basin 806
Neuquén Basin, Argentina 893, 894–5, 896
neurotoxin 919
New Idria, California 479
New Albany-Chesterian(!) 798
New Albany Shale, USA 609, 636, 725, 798
Newark Basin, North America 968
Newark Supergroup, North America 824
Newporte impact crater, Williston Basin 789
Niagaran pinnacle-reefs, Michigan Basin 794
nickel 663
nickel ion 604
nickel sulfide 604
Niger Delta, Nigeria 573, 618, 742, 837, 851, 875, 949, 951, 952
Niland Tongue, Wasatch Formation 921, 924
Niobrara cyclothem, USA 901
Niobrara Formation, USA 901, 903
nitrate
  -reducing anaerobes 653
  -reducing bacteria 654
nitrogen 641, 663
  fixation 726
Niuzhuang South Slope, China 965
Nkalagu Formation, Gabon 738
N'Komi Rift, Gabon 898
Nobel family 935
Nobel Prize 935

Nocardia 652, 653
Nocardioides 681
nomenclature (structural) terpanes 695
Nonesuch Shale, Michigan 610, 766
non-marine oil 528
non-vascular land plant *see* Bryophyta
norcholestanes 538, 540
   biodegradation 674
   21/(21 + 27) norcholestane ratio 631
   C24/(C24 + C27)
   21-norcholestanes 539
   24-norcholestanes 490, 492, 537–8, 859
   27-norcholestanes 560
Nordegg Formation, Canada 477, 832–3
24-nordiacholestane ratio [24/(24 + 27)] 541, 787
24-nordiacholestanes 490, 491, 492, 537–8
27-nordiacholestanes 560
Nördlinger Ries crater, Germany 539, 830
25-norhomohopane 20S 574, 675, 697
norhopanes
   α,β-norhopane 555, 662
   β,α-norhopane 689
   norhopane (C_{29} norhopane) 485, 675, 800
   15-norhopane ratio 684–5
   25-norhopanes (10-desmethylhopanes) 674, 675–85, 692, 705
   25, 28-norhopanes 692
   28-norhopanes 689, 692
   30-norhopane/hopane 569
   30-norhopanes 570, 571, 676, 680, 690
norisopimarane 543, 544
norkaurane 543
24-norlupane 580, 699
norlupanes 697
18α-30-norneohopane *see* C_{29}Ts
normal alkane (*n*-Alkane, *n*-Paraffin) 484, 493, 650, 668, 725, 750, 816
   biodegradation 651, 653, 654, 659, 664–5
   envelope 641, 659, 667, 669
Norphlet Formation, Gulf of Mexico 877
norphyllocladane 543
17-nortetracyclic diterpane 550
North Caspian Basin *see* Pricaspian Basin
North Equatorial Current 742
Northern Arabian Basin 855
Northern Permian Basin 814
North Falkland Basin, South America 834, 841
North Pacific Gyre 519
North Pole 712
North Sea
   oils 619

North Slope, Alaska 477, 827, 883, 897
Norwegian-Danish Basin 885
Northwest Peciko Well, Mahakam Delta 952, 953
25-nor-triene 593
novaculite 798, 806
NPRA *see* National Petroleum Reserve in Alaska
NSO compounds (resins) 477, 479, 658
Nubian Plate 861
Nubian Shelf 867
Nubian Shield 855, 861
nucleotide 711
nutrients 646
Nyalau Formation, Sarawak 942
nymphaeales (water lilies) 739

OAE *see* oceanic anoxic event
Ocal Section, Upper Magdalena Valley 910
oceanic anoxic event (OAE) 505, 719, 829, 889–93, 908, 919
Oceanic oil, California 616
octadecahydro-isorenieratene 594
odd-carbon preference *see* carbon preference index
odd-to-even predominance (OEP) 493, 635, 641–2, 783
OEP *see* odd-to-even predominance
Officer Basin, Australia 770, 780
Oficina Basin 942
Oil City Field, California 509
oil–oil correlation 477–9, 482, 976–9
oil sample *see* sample quality, selection, storage
Oil Shale Member, Dead Sea Graben 917
oil shales 945–6, 958
oil–source rock correlation 475–7, 752, 754, 976–9
oil spills
   biodegradation 662
oil-to-oil correlation *see* oil–oil correlation
oil-to-source rock correlation *see* oil–source rock correlation
oil window (oil-generative window) 608, 612
oil–water contact (OWC) 646
oil–water partitioning 966
okenone 593
Oklahoma Basin, USA 809
oldest free-flowing oil 710, 766
oldest giant oil field 710
oldest oil 710
Old Red continent *see* Laurussia
Old Red Sandstone, Scotland 794–6
$18\alpha/(18\alpha + 18\beta)$-oleananes 617–18

oleanane 485, 488, 489, 490, 491, 492, 529, 538, 547, 554, 573, 580, 600, 621, 665, 675, 684, 696, 697, 698, 699, 725, 730, 739, 947
oleanane/hopane *see* oleanane index
oleanane index [(Oleanane/(Oleanane + Hopane)] 527, 541, 549, 572–5, 617–18
18α-olean-12-ene 697, 699
olean-12-ene 697, 699
oleanenes 618
olefin 964
olefinic biomarkers 698
olnellids 747
Olpad Formation, India 920
Oman Salt Basin 771, 859
onoceranes 579–80, 930
Ontong-Java Plateau basalts, Pacific Ocean 919
Onverwacht Group, South Africa 712, 713
Ooraminna well, Australia 770
Oort Cloud 745, 747
optical activity 589
optical rotation 593
Oquirrh Basin, Nevada 807
Ora Field, Libya 914
Orca Basin, Gulf of Mexico 875
Orcadian Basin, Scotland 794, 983
Ordovician mass extinction 743
Ordovician paleo-plate reconstruction 721
Ordovician Period 721–2
Ordovician–Silurian boundary 746
Ordovician source rocks 776, 777–90, 983
organic acids 700–2
organic carbon *see* total organic carbon
organic matter
   depositional environment 487
   lacustrine 489, 560
   marine 489, 524, 527
   marine versus lacustrine 569
   marine versus terrigenous 488–90, 527
   mineralogy 487
   non-marine 524
   terrigenous 489, 524, 580
   type 487
organic productivity *see* productivity
organophosphorus insecticides 645
Orgueil meteorite *see* carbonaceous chondrite
Oriente Basin, South America 813, 907, 908, 911, 912
Orinoco Heavy Oil (Tar) Belt, Venezuela 645, 658, 942

ortho-substituted cyclization reacations 973
orthoxylene *see* xylene
Oseberg Field oil, Norwegian North Sea 692, 693
osmium isotope ratio 745
ostracodes 591
Ostracode Zone, Alberta 591
Otuk Formation 825, 883
Otway Basin, Australia 904–5, 906
Ouachita (Marathon) Mountains, USA 731, 779
Ouachita orogeny 809
Ouldburra Formation, Australia 780
outer ramp 861
overburden 754
overfilled lake 817, 818, 837, 838, 840, 844, 845, 846, 921
overmature *see* postmature
overpressure 479, 953
Overthrust Belt, Wyoming 616
Oxford Clay, UK 874–5
oxic 499, 567
oxic–anoxic interface 562
oxidation 816, 817
β-oxidation 650, 655
oxidative degradation 817
oxidation–reduction reaction *see* redox
oxygen
 free 713–15, 716, 747
oxygen isotopes 922
oxygen-minimum zone 790, 950
oxygen-producing cyanobacteria 569

Pabdeh Formation, Iran 741, 928–9
Pachuta Creek Field, Alabama 877
Pacific Ocean 735, 737
PAH *see* polycyclic aromatic hydrocarbons
PAL *see* Present atmospheric level
Palegreda Formation, Peru 930
Paleocene–Eocene boundary 740
paleoclimate 597
Paleogene Period 740–1
paleoheat flow 969
paleolatitude 524
Paleozoic Eon 771, 776
Palm Valley gas field, Australia 783
Palo Duro Basin, Texas 809
Panca well, Mahakam Delta 953, 955
Pando well, Bolivia 811
Pangea 731, 732, 734, 736, 740, 749, 824
Pannonian Basin, Europe 947–9
Pannotia 715, 719
Panthalassa 732
Panthalassic Ocean 715, 719, 721, 722, 730, 731, 732, 734, 735

Parachute Creek Member 923
Paradox Basin, USA 809–10
paraffin, normal *see* Alkane
paraffin dirt 687–9
n-paraffins *see* alkane
Paralic Basin, Europe 811
Paralic marine shale source rock 499, 500
Pará-Maranhão Basin, Brazil 847
Paraná basalts, South America 748
Paraná Basin, South America 812, 819–20
parasequence 925
paraxylene *see* xylene
Pardonet Formation, Canada 825
Parka 726
Pasco Field, Australia 875
Pasenhor 915
Patchawarra Formation, Australia 820
Pattani Basin, Gulf of Thailand 940
Pavlova 531
PCA *see* principal component analysis
Peace River heavy oil, Canada 645
Pearl River Mouth Basin, China 741
Pearsall Group, East Texas 880
peat 811
pebble–GRZ–Torok Interval, Alaska 826, 883, 895–898
pebble shale unit, Alaska 738, 826, 897
Pediastrum 853
Peedee Belemnite *see* PBD
Pelycosaurs 733
Pematang Brown Shale, Sumatra 518, 741, 935–39
Pennsylvanian coals, Southern Gas Basin 811
Pennsylvanian lacustrine rocks, East Greenland 810–811
Pennsylvanian-Permian Shales, North America 809
pentacyclic terpanes 662
pentacyclic triterpenoids 552
pentane 664
2,6,10,15,19-pentamethylicosane (PMI) 484, 498, 505, 507–10, 511, 512
pentakishomohopane 680
pepidoglycan *see* murein
perhydro-β-carotane *see* β-carotane
peridinin *see* carotenoid
Perintis well, Mahakam Delta 953, 956
periodic extinction 747
permeability 646
Permian coals 702
Permian Basin, USA 806, 812, 813
Permian extinction 743, 749

Permian paleo-place reconstruction 732
Permian Period 732–3
Permian source rocks 776, 812–20
Permian–Triassic boundary 746, 749, 764
Pertatatak Formation, Australia 770
Perth Basin, Australia 511, 548
Pertica 728
perylene 594–6
Petapahan Field, Sumatra 517
Peterborough Member, UK 874
petroleum 608
 system name
 system processes
 volumes (production and reserves) 755
petroleum systems 752, 756–7
 folio sheet 753, 754
 modeling *see* thermal modeling
 nomenclature 751–5
 non-US 762, 763
pH 533, 604, 617, 649
Phacoides oil, California 616
Phaeophyta 726
Phanerozoic Era 719–50
Phanerozoic petroleum systems 774–961
phenanthrene parameter (PP-1 and PP-1$_{modified}$) 643
phenanthrenes 643, 651, 670, 672, 817
phenolic hydrocarbons 650
phenylalkanes (linear alkylbenzenes) 598, 599, 600
Phoenix well, Alaska 826
Phosphatic Member *see* Monterey Formation
Phosphoria Formation, USA 628, 649, 686, 700, 733, 749, 814, 815, 972
photic zone
 anoxia 593, 901
photoinhibition 522
photosynthesis 713
 anoxygenic
 oxygenic 714
photosynthetic bacteria 576
photosystems I and II 713
phthalic acids 817
phyllocladane 485, 490, 543–51, 600, 820
phytane (Ph) 499, 507–8, 543, 591, 657, 666, 668, 981
phytane/nC_{18} 502, 641, 666, 674
phytanic acid 701
phytenic acid 499
phytol 499
phytyl side chain 499
Piceance Creek Basin, USA 479
picene 479
Piedra Redonda well, Peru 932
Pierre Shale, USA 902

Pilbara Craton, Australia 530, 712, 714, 755–63
Pilenga Formation, Russia 941–2
Pilsk Formation *see* Borsk Formation
pimarane 543, 544, 545, 550, 600
pimaric acid 546
Pimenteriras Formation, South America 725
Pimienta Formation, Gulf of Mexico 878
Pimienta-Tamabra(!) 878
Pinaceae 551
Pinda Formation, Angola 853, 935
Pindinga Formation, Nigeria 579
Pinnacle reefs, Western Canada Basin 475
Pipeline Shale, USA 813
Piper oil, North Sea 525, 526, 527, 528, 562, 583, 584, 611, 619
piston (seabottom) cores 704, 705
placental *see* mammal
placoderm 724
planetesimals 709
plate tectonics 712, 764, 836
Plattville Formation, USA 781, 783
play 751
Pleistocene paleo-plate reconstruction 742
Plenus Marl, Europe 738
plesiosaurs 740
Pleurosigma 515
PMI *see* 2,6,10,15, 19-Pentamethylicosane
PMP *see* porphyrin maturity parameter
pod of active source rock 753
Podocarpaceae 547, 550, 551, 600
podocarpane 543, 600
Podocarpus 550
Point Clairette Formation, Gabon Basin 850
Point Pleasant Formation, North America 779, 785
Point Thomson, Alaska 896
Poker Chip Shale, Canada 830
polar compound 658
pollen 573, 725, 739
pollination 739
pollution 507
polycadalene 623
polycadinenes 547, 548, 619, 620
polycyclic aromatic hydrocarbons (PAH, PNA) 479, 595, 600–2, 670–2
  biodegradation 665
  Cretaceous-Tertiary 750
  Martian 980
  pyrogenic (pyrosynthetic) 602
polyester 645
polyisoprenols 556
polymethylsiloxane 627

polymethylsqualanes 484, 487, 518, 937
polynuclear aromatic hydrocarbons *see* polycyclic aromatic hydrocarbons
polyprenoid sulfides 561
polysporangia 726, 727
polysulfides 520
Ponta do Tubarão Member, Brazil 846
Popigai impact crater, Russia 747
Porifera 491, 530, 717
porosity 646
porphyrins 602–5, 664, 928
  biodegradation 665, 700
  distributions 604
  V/(V + Ni) 603–4
porphyrin maturity parameter (PMP) 628, 635–6
Port Gentil Ocean Field, Gabon 852
Port Sussex Formation, Falkland Islands 812
Posidonienschiefer, Europe 830
Posidonia Shale, Europe 830, 831
positive correlation 477
postmature 608
Post-Neocomian interval, Alaska 826
potassium hydroxide 701
potential source rock 608, 976
Potiguar Basin, South America 834, 837, 846, 847
Po Valley, Italy 822
Powder River Basin, Wyoming 811
PP-1 or PP-1$_{modified}$ *see* phenanthrene parameter
prasinophyte algae 790, 802
prasinoxanthin *see* carotenoid
Precambrian organic matter 493
Precambrian oils and source rocks 495, 763, 771
Precambrian-Cambrian 710, 717, 719
Pre-Caspian Basin *see* Pricaspian Basin
prediction of physical properties 705, 706
pregnane 485, 674, 676
prehnite-pumpellyite facies 755
pre-rift facies 837
present atmospheric level of oxygen (PAL) 713, 714
preservation 725, 754, 798
pressure 649–50
Pricaspian Basin, Kazakhstan 807
primary migration 965
Prince William Sound, Alaska 705
principal component 480
principal component analysis (PCA, eigenvector analysis) 478, 480–2, 983
Prinos Field, Greece 554
pristane (Pr) 499, 501, 513, 657, 666, 668, 972, 973, 981

pristane formation index (PFI) 972, 975
pristane/nC_{17} 641, 659, 666, 705, 706
pristane/phytane, pristane/(pristane + phytane) 484, 499–502, 562, 575, 781
pristene 499
prist-1-ene 500, 501, 972
productivity 798
Progreso Basin, South America 907, 930
progymnosperm 728
prokaryotes (bacteria) 712, 713, 766
propane 664
24-n-propylcholestanes 485, 527, 529, 530, 537, 569
24-n-propylcholesterols 527
24-propyldiacholestane ratio 561
24-propyldiacholestanes 561
prospect 751
proteins 711, 725
proteobacteria 657, 716
Proterozoic Eon 714–19, 763
proto-continents 712
Proto-Mediterranean 741
Protosalviania 726
Proto-South Atlantic Ocean 735–6, 847
protostomes 717
Prototaxites 726
prototracheophytes 724, 727
Proto-Zagros 861
protozoa 576
Prudhoe Bay Field, Alaska 477, 563, 616, 812, 826, 883, 896
prymnesiophyte algae 531
pseudo-biomarker 501
Pseudofagus 579
pseudohomologous series 498, 514, 516, 524, 532, 552, 556, 564, 565, 615, 691
Pseudomonas 652, 653
psilophytes *see* vascular plant
psilophyton 728
psychrophilic 503, 505, 506
pteridosperms (seed ferns) 578, 596, 729, 733
pterosaur 745
PTV *see* injection
Pucará Group, Peru 825
Pueblo, Colorado 902
Puente Formation Schist-Conglomerate, California 563
Puertollano oil shale, Spain 693
Pujamana Member, Colombia 909
Purana basins, India 768
purple sulfur bacteria *see* Chromatiaceae
Putamayo Basin, South America 907
pycnocline 829, 875

Pyrenees 741
pyrite 815
pyrobitumen 764, 816
　　nodules 755, 763
Pyrodictium 647
Pyrolobus 647
pyrolysis/gas chromatography (py/GC) 828, 871, 906
pyrolysis/gas chromatography/mass spectrometry (py/GCMS) 479
pyrolyzate 664
pyrrhotite 980

Qaidam Basin, China 538, 544, 613, 871, 872, 946
Q compounds 591
Qingshankou Formation, China 738, 904
Q oils, Oman 526, 772, 774, 775, 966
Quaidam Basin, China 871
quantum structure–activity relationship (QSAR) 684
Quaternary Period 742–3
Queen Formation, USA 813
Quenguela Field, Angola 854
Querecual Formation, South America 942
quinones 655, 701
Quiriquire Field, Venezuela 696
Qunituco Formation, Argentina 893, 894
Qusaiba Member, Qalibah Formation, Saudi Arabia 631, 723, 791, 793

Rabi Kounga Field, Gabon 848
racemic 972, 973
radiogenic immobilization 755
radiolysis 698
Raghib well, Saudi Arabia 793
Ragusa Basin, Sicily 827
Rakopi Formation, New Zealand 919
Raney nickel 512, 520, 568, 594, 689, 690
rank *see* thermal maturity
Rann Formation, United Arab Emirates 791
Ras Fanar well, Gulf of Suez 915
Rastrites Shale, Sweden 723
Ratawai Shale, Iraq 867
rate constants 967, 968, 970, 971
Ravnefjeld Formation, Greenland 818
Ravni Kotari oil, Yugoslavia 554, 568
Raya Formation, South America 908
Ray Formation, South America 738
Raymond Field, Williston Basin 782
rearranged hopanes *see* diahopanes
rearranged sterane *see* diasteranes
rearrangement 582
recent sediments 504–5, 512, 518, 704
Recôncavo Basin, Brazil 840, 841

recovery *see* estimated ultimate recovery
Red Band fraction 498
red beds 714
Redondo Shale, Peru 930
redox potential (oxidoreduction potential) 499, 566, 603
Red River Formation, USA 765, 782, 789, 804
red tide 919
Redwater Reef Field, Canada 802
reflectance *see* vitrinite reflectance
regression 747, 749
regressive marine facies 837
Reindeer Supersequence, Canada 960
Renqui Field, China 772
reproduction 711, 716, 725
reptiles 732, 733
reservoir rock 475, 608
reservoir temperature 646, 647, 648, 659
resinite 542, 573
resorcinol 784
　　lipids 780
retene 544, 582, 596, 597, 602, 750
retention time 556
Retort Shale Member, USA 814
reversed tricarboxylic acid cycle 606
reverse methanogenesis *see* anaerobic methane oxidation
Rhaetogonyaulax 531
Rhenish Mountains 723
Rhine Basin, Germany 566
Rhizobium 726
rhizoid 727
Rhizolenia 514–15
Rhodococcus 652
Rhynie Chert, Scotland 727, 796
rhyniophyte 724, 727
ribonucleic acid (RNA) 711
ribosomal RNA (rRNA) 648, 657, 726
ribozyme 711
Ribesalbes Basin, Spain 945
RIC *see* reconstructed ion chromatogram
Richards Sequence, Canada 960
Rimbey-Meadowbrook Trend, Canada 966
rimuane 543, 544, 545
Rincon Formation, California 562, 944
ring-opening 582
Rinkabeena Formation, Australia 770
Rio da Serra Brazilian stage 834
Riphean shales, eastern Siberia 766–8
risk for oil biodegradation 650
Riva di Solto Shale, Italy 822
RNA *see* ribonucleic acid
RNA enzymes *see* ribozymes

RNA-world 711
R_0 *see* vitrinite reflectance
Roberts Mountains, Nevada 789
rock sample *see* sample quality, selection, storage
Rocky Mountains, USA 735
rodents 741
Rodda Formation, Australia 770
Rodinia 715
root nodule 726
Roper Groups Shale, Australia 765–6, 767
rosane 543, 544, 545
Rothschild family 935
Rotliegendes Group, Europe 814
Rough range oil, Australia 600
Rovesti Oil, Italy 568
Rozel Point, Utah 512, 515, 520
rRNA *see* ribosomal RNA
Rub al Khali Basin, Middle East 855
Rubiales Field, Llanos Basin, Colombia 555
Rubielos de Mora Basin, Spain 945
Rudeis Formation, Gulf of Suez 915
rudists 740, 745
Ruhr District, Germany 551
Rundle Group, Canada 807
Rundle oil shale, Australia 928
Rundle-type oil shale *see* lacosite
Rutaceae 577
ruthenium tetroxide (RuO_4) 568

Saanich Inlet, British Columbia 596
Saar District, Germany 551
Sabah Delta, Borneo 942
Sabhka 487, 584, 880
saddle dolomite 816
Sadlerochit Formation, Alaska 826
Sag River Formation, Alaska 826
Sag River oil, Alaska 556, 557
Saint-Cecile, Camargue, France 580
Sakhalin Island, Russia 695, 742, 941–2
Salada Member, Colombia 909
Salado Formation, New Mexico 649
Salina A-1 carbonates, USA 794
Salina Basin, Gulf of Mexico 878, 879
saline 488, 838, 839, 850
salinity 488–9, 501, 646, 649, 972
　　stratification 576, 814, 817
salt diapir 649
Salym Field, West Siberia 889
San Andres Formation, USA 813
San Cristobal Formation, Peru 930
Sandakan Delta, Borneo 942
sandstone breccia 527, 574, 668
Sangerhausen Basin, Germany 816
San Joaquin Basin, California 479, 742

San Jorge Basin, Argentina 894
San Juan Basin, New Mexico 834, 902, 904
Sanmiguelia 739
Santa Barbara Basin, California 715, 957, 960
Santa Barbara Channel, California 641
Santa Barbara–Ventura Basin, California 562, 945
Santa Cruz-Tarija Province, South America 796
Santa Maria Basin, California 510, 606, 619, 628, 641, 705, 706, 707, 945, 946, 977, 978
Santa Maria oil, Italy 570
Santa Monica Basin, California 576
Santos Basin, South America 835, 852
Santos well, Middle Magdalena Basin, Colombia 549
São Francisco Basin, Brazil 768
sapropel 741
Sarcinochrysidales 527
Sarcopterygii 724
sargasterol *see* fucosterol
Sargelu Formation, Iran 736, 863, 866
Sarriah Member, Safiq Formation, Oman 791
Sassendalen Group, North Atlantic 825
saturate/aromatic ratio 641
saturated hydrocarbons (saturates) 477
saturated hydrocarbon UCM *see* unresolved complex mixture
sauropods 736
Sava Depression, Croatia 947
Sayindere Formation, Turkey 915
Sayyala Field, Oman 774, 775
S/C *see* atomic S/C
Scales Formation, USA 788
scanning transmission X-ray microscopy (STXM) 727
scan run *see* scan analysis
Schei Point Group, Canada 825
Schistes Carton, Europe 830–2, 833
scores plot 481, 482
Scotian Shelf, Canada 968
scuticociliates 576
Sdom Formation, Israel 503, 512, 518
sea-bottom cores 704, 705
sea grass 489
sea level *see* global sea level
seal rock (cap rock) 754
8, 14-secohopanes 600, 692–3, 797
17, 21-secohopanes 552, 559, 572

secondary migration 965
secondary porosity 700
secondary processes 478
Second White Speckled (Second White Specks) Shale, Canada 901
sedimentation rate Pennsylvanian–Permian, USA 810
Sedlec well, Czech Republic 538
seed coat 739
seed ferns *see* pteridosperms
seed leaf *see* cotyledon
seed plants *see* vascular plants
seeds 729, 738
seep oil 537, 706
selected ion monitoring (SIM, multiple ion detection, MID) 221, 611
selected ion recording (SIR) *see* selected ion monitoring
Senonian Limestone, Dead Sea *see* Ghareb Formation
sequence stratigraphy 483, 803, 862, 909
Sergipe-Alagoas Basin, Brazil 835, 846, 847
Serpiano oil shale, Switzerland 603
serratanes 578–9, 691
Seychelles Microplate 872–4
Shahejie Formation, China 522, 579, 772, 926, 928
Sharawra Formation, Qatar 791
Shark Bay, Australia 713
sharks 731, 736
Sharon Springs Member, USA 902
Shengli Field, China 610, 647
Shilaif Formation, Southern Gulf Basin 860, 867, 869
Shinjo Basin, Japan 513
shocked quartz 745, 746
Shoemaker–Levy 9 comet 745
Shu'aiba Formation, Southern Gulf Basin 772, 860, 867
Shublik Formation, Alaska 556, 557, 735, 825, 827, 896, 897
Shublik–Otuk oils 883, 885
Shublik–Otuk source rocks 898
shungite *see* Karalian shungite
Shuntar Formation, East Siberia 768
Shuram Formation, Oman 771, 772
Shushufindi-Aguarico Field, Ecuador 912
Sibang Limestone, Gabon Basin 849
Siberia 715, 719, 723
Siberian Craton 767
Siberian Traps 748
Sichuan Basin, China 772
side-chain scission 632
siderite 714

Sierra Nevada Mountains, USA 735
Sigillaria 729
Siglia paraffin dirt, Chile 688
sigma bond *see* covalent bond
signal-to-noise 527
silica 714
silicalite *see* ZSM-5
silled basin 790
Silurian paleo-plate reconstruction 722
Silurian Period 722–3
Silurian–Devonian highstand 774
Silurian source rocks 776, 790–4
SIM *see* selected ion monitoring
SIMCA *see* soft independent modeling of class analogy
simonellite 596, 597, 834
Simpson Group, USA 765, 783, 812, 813
Simpson-Ellenburger(.), USA 789
simulated distillation *see* distillation
sinister *see* S configuration
Sirte Basin, Libya 914–15
Sirte Formation, Libya 738, 914–15
Sisi well, Mahakam Delta 952, 953, 955, 956
Sisquoc Formation, California 562
Skull Creek Shale, USA 738, 900
Slave Craton, Canada 711
Slavonija-Srijem Depression, Croatia 948
slime molds 716
Smackover Formation, Gulf of Mexico 735, 877
smoke 602
Smoking Hills Formation, Canada 959
Smoky Hill Member, USA 901, 903
snowball Earth 768
sodium methane thiolate 504
soft independent modeling of class analogy (SIMCA) 482
solanesol 589
solar flare 745
solar system 709
solid bitumens 478–9
solubilization 613, 964–5
solvent-split injection *see* injection
Somerset Island, Canada 716
Song Hong Basin, Vietnam 938, 939
Songliao Basin, China 904, 905
soot 602, 745, 750
sour crude *see* sulfur-rich oil
Sour Lake Field, Texas 687
source input and depositional environment 982–3
source potential index (SPI) 872
  Gulf of Mexico, Upper Jurassic 879
source-related biomarker parameters 476, 478, 483–90, 606

source rocks 475, 608, 647, 763
  carbonate 569, 571, 588, 616, 630, 641, 744, 771, 776, 851
  carbonate versus clastic 533
  carbonate versus shale 487–8
  character from oil composition 487–90
  evaporitic 588, 850
  gas-prone 752
  marine carbonate 536
  marine deltaic 490, 536
  marine shale 536
  marl 569
  oil-prone 752
South American basins 907
South Atlantic 835, 836, 858, 859
South Barrow wells, Alaska 812
South Belridge Field, California 686, 689, 691
Southern Gas Basin, Europe 811
Southern Gulf Basin (Rub Al Khali Basin) 859
South Florida Basin, USA 555
Southern Permian Basin, Europe 814
South Mekong Basin, Vietnam 940
South Sumatra Basin, Indonesia 573, 624
South Terekheveskaya Field, Russia 800
South Timbalier Field, Gulf of Mexico 647
space-filled projection *see* structural notation
Sparta Formation, Gulf of Mexico 882
Spechts Ferry Member, Decorah Formation 780, 781, 783
speculative(?) 752
Spekk Formation, Norway 827
spermatophyte 726
sphenophyte *see* sphenopsids
sphenopsids (sphenopsida) 724, 728, 729, 731, 733
Spindletop, Texas 687
split injection *see* injection
splitless injection *see* injection
sponges *see* porifera
sporangia 727
spore coloration index *see* thermal alteration index
spores 721, 724, 725, 726, 727
sporophyte 726
sporopollenin 725, 727
Sprayberry Formation, Texas 813
Sprendlinger Horst, Germany 926
squalane (2, 6, 10, 15, 19, 23-hexamethyltetracosane) 484, 498, 507, 512

squalene 498, 506, 507, 509, 511–12, 552, 713
squalene-hopene cyclase (squalene cyclase) 591
stable carbon isotope ratios
  C_{29} n-alkane 746
  carbonates 816, 817, 829, 890, 891
  forams 921, 922, 957
  oils 977, 978
  organic carbon 829, 892, 902
  wood 892
Standard Light Antarctic Precipitation *see* SLAP
Statfjord Formation, North Sea 828
steam flooding 689
sterane isomerization, $\beta\beta/(\beta\beta + \alpha\alpha)$ 628–629, 634, 662, 706, 970, 978
sterane isomerization, 20S/(20S + 20R) 566, 622, 625–8, 634, 674, 965, 966, 967, 968, 969
C_{30} sterane index 527–30
steranes 714, 763
  biodegradation 537, 665, 674–5
  C_{27} 662, 772
  C_{28}/C_{29} 490, 526–7
  C_{29} 607, 771, 818
  C_{29} $\alpha\alpha\alpha$ 20S 625
  C_{29} $\alpha\beta\beta$ 662
  C_{30} sterane index 574
  C_{30} steranes
  fingerprint 611
steranes/hopanes 524
steranes/hopanes (regular steranes/17α-hopanes) 524
steranes/triterpanes 524
$\Delta 2$-sterenes 533
sterenes 830
sterilization 648, 702
sterols 483, 533, 583, 714
stereoisomerization 611
steroid
  aromatization [TA/(MA + TA) 627, 631–2, 705, 706, 966, 967, 968, 970, 972
  ketones 653
stick projection *see* structural notation
stigmastanes (24-ethylcholestanes) 611, 768
Stirling Range Formation, Australia 718
Stockinford Shales, UK 774
stomata 725, 727
stonewart 725
STP *see* standard temperature and pressure
stratification (stratified water) 575, 576
stratified water 820, 845, 899

Stratigraphic-39 well, Saudi Arabia 631
Strawn Formation, USA 813
Strelly Pool Chert, Australia 712
Streppenosa Formation, Sicily 827
stretophyta 725
stromatolites 569, 712, 713, 714, 720
stromatoporid 530
strontium isotopes 749, 890
Sublett Basin, USA 814
sublittoral facies 924
suboxic 567, 721
success ratio 753, 822, 874
Sudbury Impact Crater, Ontario 747
Sudr Formation, Middle East 915
Suerte well, Middle Magdalena Basin, Colombia 549
sugiol 546, 547
Suket Shales, India 716
Sulaiy Formation, Iraq 860, 866
sulfate 657, 705
  reducing bacteria 509, 604, 653, 654, 655, 656, 705
sulfones 653
sulfur 488, 500, 573, 629, 641, 663, 707
  sulfur isotopes 749, 816, 817
sulfur-rich oils 563
Sumatra 931, 935–39
Sunda Basin, Sumatra 939
Sunfish well, Cook Inlet 834
Sunniland Field, Florida 880
Sunniland Formation, Florida 879–80
supercritical fluid extraction 981
supercritical fluid chromatography/mass spectrometry (SFC/MS) 603
supernova 745, 749
superoxide anions 713
superoxide dimutase 713
superplumes 768
supralittoral facies 924
surfactant 599
Surma Basin, Bangladesh 547
Svanbergfjellet Formation, Spitsbergen 716
Sverdrup Basin, Canada 825
Swanson River Field, Cook Inlet 554
sweet crude *see* low-sulfur oil
Sydney Basin, Australia 812
Sylvan Shale, USA 722, 777
symbiont 726
Synapsida 733
synrift facies 837
syntrophic consortia 509–11, 656, 657
syntrophic metabolism 656

Tabei Uplift, China 785, 787
table of hydrocarbon accumulations 752

tabulates 721
Tagrifet Formation 915
T-branched alkanes 493, 494–6
TAI *see* thermal alteration index
Talang Akar Formation, Sumatra 627, 931, 939, 976
Talara Basin, Peru 907, 930, 932, 933
Talc-Stevensitic Sequence 842
TA/(MA + TA) *see* monoaromatic steroid aromatization
Tambora Field, Mahakam Delta 951, 953, 954, 955
Tampen Spur, North Sea 648
Tampico-Misantla Basin, Gulf of Mexico 878
tandem mass spectrometry 605
see also GCMS/MS
Tanezzuft Shale, North Africa 723, 791, 792
Tanf Formation, Syria 791
Tanzanian paraffin dirt 688–9
Taoudenni Basin, Mauritania 720
Tapeats Sandstone, Arizona 771
Tarafaya Basin, Morocco 913–14
Tarakan Delta, Borneo 942
Taranaki Basin, New Zealand 544, 550, 597, 918–19
taraxastanes 580, 581
taraxerenes 618
tar ball 873
Tarim Basin, China 785, 787, 870, 871
Tarn Field, Alaska 826, 898
Tarraco oil, Spain 507, 520
Tarragona Basin, Spain 947, 948
tar sands 658
Tasmanite oil shale, Tasmania 591
Tasmanites 552, 556, 557, 796
Taxodiaceae 547, 551
taxodione acetate 546
Taxodium 546, 547
taxon-specific biomarkers 491, 492
Tazhong Uplift, China 785, 787
TBR *see* trimethylnaphthalene ratio
TCI *see* transmittance color index
TEHRGC *see* thermal extraction high-reolustion gas chromatogram
tektite 745
telalginite 779, 784, 785, 895
teleost fish 737
Tempisque Basin, Costa Rica 917
Tent Island Formation, Canada 960
Termit Basin, Africa 837
Terang-Sirasun Field, Indonesia 509
terminal addition 655

ternary diagram (C_{27}–C_{28}–C_{29}) 410–7.1
  diasteranes 536
  C-ring monoaromatic steroids 583–4, 978, 979
  steranes 476, 483, 485, 525–6, 536, 871, 978, 979
  triaromatic steroids 586–8
terpane m/z 191 fingerprint 551–5, 611, 697, 797, 880
terrane 720
Terra Nova Field, Newfoundland 888
terrigenous/aquatic ratios (TAR) 493
terrigenous oils 490, 939
terrigenous organic matter 499, 580
terrestrial impact craters 747
Tertiary source rocks 918–60
Tethys Sea (Ocean) 732, 733, 734, 735, 737, 740, 741, 901, 932, 947
tetracorals 721
tetracyclic diterpanes 549–51
tetracyclic polyprenoids (TPP) 487, 560–1
tetracyclic terpanes 552, 927
C_{24} tetracyclic terpane 485
C_{24} tetracyclic terpane ratios 559–60
tetrad spore 727
Tetraedon 927
tetraether lipids 497, 503–5, 506
5,6,7,8-tetrahydrocadalene 981
tetrahymanol (gammaceran-3β-ol) 575–6, 591
Tetrahymena 576
tetrakishomohopanes 808
1,2,3,4-tetramethylbenzene 801
2,6,10,14-tetramethylhexadecane *see* phytane
2,6,15,19-tetramethylicosane (TMI) 505, 511
1,2,5,6-tetramethylnaphthalene (TeMN) 582
1,3,6,7-tetramethylnaphthalene (1, 3, 6, 7-TeMN) 596, 597
tetramethylnaphthalene ratio (TeBR) 671, 672
tetramethylnaphthalenes 670
2,6,10,14-tetramethylpentadecane *see* pristane
tetramethylsqualane 518, 519
1,1,7,8-tetramethyl-1,2,3,4-tetrahydrophenanthrene 591
1,1,5,6-tetramethyltetralin 596
Thamama Group, Southern Gulf Basin 860, 867
therapod 737
therapsids 733, 735
thermal gradient 650

thermal maturity 535, 608–9
  criteria for biomarker parameters 609–10
thermal maturity-dependent biomarker ratios 478, 612
thermal stress parameters 608–9
thermoacidophilic archaea
Thermoanaerobacter 648
thermochemical sulfate reduction (TSR) 608, 815, 825
Thermodesulfobacterium 648
thermophile 647
thermophilic 599
  eubacteria 497
Thermodesulfobacterium 497
Thermoplasma 599
Thermopolis Member 900
thermospray 603
Thermotoga 648
thin-layer chromatography/flame ionization detection (TLC/FID) *see* iatroscan
Thiobacillus 562
Thiomargarita 562
Thiomicrospira 562
Thioploca 562
Thiothrix 562
Thompson B-F diagram *see* B-F diagram
thorite 755
thorium/uranium 719
thylakoid membrane 726
Tibet Plateau 871
TIC *see* reconstructed ion chromatogram
tillite 716
Timan-Pechora Basin, Russia 799, 800
timing-risk chart *see* events chart
Tithonian source rocks
TLC *see* thin-layer chromatography
TMI *see* 2,6,15,19-tetramethylicosane
Tm (C_{27} 17α-22,29,30-trisnorhopane or 17α-trisnorhopane) 555, 565, 581, 616, 617, 686–7
T_{max} 627
  vs. PMP 635
TNH *see* 25, 28, 30-trisnorhopane
TNH/(TNH + hopanes) 619
TNR *see* trimethylnaphthalene ratio
TNS *see* 25,28,30-trinorhopane
Toarcian Shale, Europe 532, 535, 584, 617, 628, 828–30, 831, 972
Toarcian Shale, Madagascar 831
tobacco 539
Toca Formation, Africa 841
TOC *see* total organic carbon
tocopherols 500, 501
Todilto Formation, New Mexico 834
toluene 653, 654, 658, 659, 828

TOM *see* total organic matter
Tomachi Formation, Bolivia 725, 801–2
Toolachee Formation, Australia 820, 821
Toolebuc Formation, Australia 738, 905–7
torbanites 495, 517, 518, 520, 521, 597, 812
Torok Shale, U.S.A. 738, 826, 895, 897
Torok-Nanushuk(.) 897
total acid number (TAN) 700–2
total ion chromatogram *see* reconstructed ion chromatogram
total organic carbon (TOC, C_{org})
  Agrio Formation 894–5
  Bakken Formation, North America 804, 805
  Bazhenov Formation, West Siberia 536, 540, 889
  Black Sea sediments 496
  Greenhorn Cyclothem 902
  Jet Rock 892
  Monterey Formation 945
  Paris Basin, Liassic 831
  Tyumen Formation, West Siberia 536
  Vaca Muerta 893
total petroleum system 751
TAPH *see* total polycyclic aromatic hydrocarbons
TPP *see* tetracyclic polyprenoids
trace fossils (ichnofossils) 717, 718
tracheid 727
tracheophytes 724, 726
Trafras Group, Brazil 768
training set 482
Trans-Brazilian Shear 834
transgression 748, 774, 777, 790, 864, 899, 905, 908
transgressive marine facies 837, 851
transgressive systems tract (TST) 481, 482, 484, 862, 909, 910, 956, 957
transpiration 725
trap 754
Traralgon Formation, Australia 917
Trenton–Black River Formation, Michigan 722, 783, 785, 786
Trenton Group, North America 765, 784–5
Trento Plateau, Italy 830
triaromatic dinosteroids 588
triaromatic (TA) steroids 586, 587, 632
  $C_{26}/C_{28}S$ vs. $C_{27}/C_{28}R$ 586
  C_{26} 20S/(20S + 20R) 634
  TA(I)/TA(I + II) 633–4
  20S/(20S + 20R) 634
Triassic Black Shale, Italy 593

Triassic carbonates
  Arabian Platform 824
  Swiss Alps 824
Triassic–Jurassic boundary 744, 746, 747
Triassic mass extinction 743
Triassic mudrocks, Barents Sea 483, 615, 822
Triassic paleo-plate reconstruction 734
Triassic Period 733–5
Triassic rift sequences, North America 824–5
Triassic source rocks 823, 825
tricadinanes 547, 548
tricyclic biphytane 506
tricyclic diterpanes 541–7, 552
tricyclics/17α-hopanes 556, 615–16
tricyclic terpanes (cheilanthanes) 485, 552, 555–8, 591, 615, 662, 694, 695, 696, 826, 927
  age 490
  extended tricyclic terpane ratio (ETR) 558–9, 749
  C_{22}/C_{21} 558
  C_{23} tricyclic terpane/hopane 556
  C_{24}/C_{23} 558
  C_{26}/C_{25} 487, 558, 559, 569
tricyclohexaprenane 555, 556
tricyclohexaprenol 552, 556, 557
tricyclooctaprenol 556
3,4,5-trihydroxybenzoic acid *see* gallic acid
trilete spore 727
trilobites 720, 721, 733, 745, 747
trimerophyte 727
trimethylalkanes 493
trimethylbiphenyls 672
4,23,24-trimethylcholesterols 588
trimethylchromans 485
2,6,10-trimethyldodecane (*see* farnesane)
2,6,10-trimethyl-7-(3-methylbutyl)-dodecane 502
trimethylnaphthalenes 597–9, 644, 670, 973, 974
1,2,5-trimethylnaphthalene (TMN) 582, 596, 597
1,2,7-trimethylnaphthalene 597
1,3,6-trimethylnaphthalene 596
trimethylnaphthalene depositional environment (TDE) ratios 597
trimethylnaphthalene ratio (TNR-1, TBR, TP-1) 644, 671, 672, 974
2,6,10-trimethylpentadecane *see* norpristane
trinitrotoluene (TNT) 645
17β-trinorhopane 688
triple quadrupole mass spectrometer 612

triple-sector mass spectrometer *see* triple quadrupole mass spectrometer
tripolblast bilateralians 718
Tripoli Unit, Spain 957–8
17α-trisnorhopane *see* Tm
17α-22,29,30-trisnorhopane *see* Tm
18α-trisnorhopane II *see* Ts
22,29,30-trisnor-17β-hopane 616
25,28,30-trisnorhopane (TNH) 485, 561–3, 618–19, 678, 686, 687, 694
18α-22,29,30-trisnorneohopane *see* Ts
Trombetas Formation, South America 723
Ts (C_{27}Ts or 18α-22,29,30-trisnorneohopane) 552, 553, 555, 565, 581, 616, 686–7
TSF *see* total scanning fluorescence
Ts/hopane 617
Ts/Tm *see* Ts/(Ts + Tm)
Ts/(Ts + Tm) 616–17, 618, 686, 687
tsunami 745
Tucano-Recôncavo Basin, South America 834
Tuna well, Australia 547, 624
Tunu well, Mahakam Delta 481, 952, 953, 955
turbidites 851
Turkmenistan 932
Turpan Basin, China 870, 871
Turpan-Hami Basin, China
Tuscaloosa Formation, USA 881
Tuwaiq/Hanifa-Arab Petroleum System 755
Tuwaiq Mountain Formation, Saudi Arabia 864, 865
Tuxedni Group, Alaska 833–4
two-dimensional GC *see* gas chromatography
two-dimensional NMR *see* nuclear magnetic resonance
Tynec well, Austria 538, 541, 566, 586
type I 725
type IS 946
type II 563, 725
type IIS 563, 879, 885
type II/III 563
type III 567, 642, 725, 730, 851
type curve *see* stable carbon isotope type-curve
Tyrannosaurus 737
Tyumen Formation, West Siberia 573, 574

Ucayali Basin, Peru 811, 812, 813, 825
UCM *see* unresolved complex mixture
Ugut well, West Siberia 889

Uinta Basin, Utah 479, 523, 675, 921, 926, 927
Uinta Formation, USA 479
ultralaminae 928
Umbrian Basin, Italy 830, 831
Umiat-Simpson Oils 898
Umir Formation, Colombia 549
underfilled lake 817, 837, 838, 839, 840, 844, 846, 858
underground leakage see oil spills
UNESCO see United Nations Educational Scientific and Cultural Organization
uniformitarianism 745
United Nations Educational, Scientific and Cultural Organization (UNESCO) 928
unresolved complex mixture (UCM) 659, 667, 668
Upanema Member, Brazil 846
Upper Magdalena Valley, Colombia 907, 908, 910, 911, 913
Upper Zakum Field, United Arab Emirates 864, 866
upwelling 492, 516, 721, 722, 789, 790, 798, 814, 825
 Peru 519, 930
Ural Mountains 723
Ural Trough 740
uraninite 714, 755
uranium 777, 905
urea adduction 494
ursanes 580, 697
urs-12-ene 697, 699
U-Shale Formation, Oman 771
US-Y molecular sieve see ultrastable-Y
Utica Shale, USA 722, 723, 777, 779, 785–8

Vaalbara Supercontinent 712
Vaca Muerta Formation, Argentina 893, 894, 896
vanadium 663
vanadyl ion 604
Van-Egan well, West Siberia 682, 683
Van Krevelen diagram
 modified 831, 870, 882, 934
vaporizing injection see injection
Vargner Ice Age 716
Várkeszö crater, Hungary 948
varve 834
vascular land plants see tracheophytes
vascular system (tracheids) 725
Vedreshev Formation, East Siberia 768
Velkerri Formation, Australia 765, 766, 767

Verkhne-Chonskaya Field, Eastern Siberia 767
Vermelha Formation, West Africa 853
Vermilion Field, Louisiana 965
Victorian brown coals (Traralgon Formation) 917
Vienna Basin, Austria 538, 947
Vienna PDB see VPDB
Vienna Standard Mean Ocean Water see VSMOW
Vietnam 938, 939
Viking Graben, North Sea 648, 814, 884
Villafortuna-Gaggiano Complex, Italy 824
Villahermosa Area, Gulf of Mexico 879
Ville Perdue Field, Paris Basin, France 833
Villeta Formation, Colombia 738, 907, 911, 913
Vindhyan Basin, India 768
Vinini Shale, USA 722, 777, 779, 789–90
violaxanthin see carotenoid
Visayan Basin, Philippines 624, 626
viscosity 659, 660, 662, 663
vitamin C 713
vitamin E 713
vitrinite 573, 727
vitrinite reflectance (R_o in oil, R_r random, or R_m mean) 609, 613, 618, 627, 642, 727, 753, 816, 822, 893, 894–5, 924
 versus biomarker ratios 612
 versus dimethylnaphthalene ratio (DNR) 643
 versus methylnaphthalene ratio (MNR) 643
 versus methylphenanthrene index (MPI) 642
volcanism 748
Voltziales 551, 730
Vredefort impact structure, Kaapvaal 747

Wabasca heavy oil, Canada 645
Walcott Member, Kwagunt Formation, Utah 771
Walvis Bay, West Africa 519, 595
Wancoocha Field, Australia 821
Warrawoona Group, Australia 712, 713, 755
Washakie Basin, USA 921
Wasia Group, Arabian Platform 867
Wastach Formation, USA 479, 921
Watchet, North Somerset, UK 531
water injection 649
water washing 658, 670, 702
weathering 725, 798
Weiyuan gas field, China 772

Wessex Basin, UK 648
West Belayim Field, Egypt 916
Western Canada Basin 477, 591, 645, 801, 807–9, 832–3
Western Interior Seaway 899–904
West Greenland 580
West Tevlin Field, Russia 888
whales 74
Whitby Mudstone Formation, Port Mulgrave area, Yorkshire, UK 892
Whitehill Formation, South Africa 812
White Pine, Michigan 610, 766
White Tiger Field, Vietnam see Bach Ho Field
Wilcox Formation, Gulf of Mexico 882
Wilkins Peak Member, USA 923, 924, 946
Williston Basin, North America 789, 796–7, 803, 806, 809
Wingayongo Seep, Tanzania 689, 690
Winnipeg Formation, North America 765, 783, 789, 805
Winnipegosis Formation, North America 796–7, 804, 808
Wiriagar Well, Indonesia 529, 544
wire-frame projection see structural notation
Witwatersrand Supergroup, South Africa 763
Womble Shale, USA 779
Woodbine Formation, East Texas 881
Woodford Shale, USA 725, 803, 805–6, 807, 808, 812, 813
Woodleigh Impact Crater, Australia 746, 747
wood tar see pitch
Wumishan Formation, China 772

Xanthomonas 652
xanthones 672–4
xanthophyll 522
X-compounds see mid-chain monomethylalkanes
Xiaomeigou Formation, China 871
Xinjiang Uygur, China 869–71, 872
xylenes 653, 659

Yalco Formation, Australia 765
Yanshan Fold Belt, China 772
Yeoman Formation, Williston Basin 765
Yunnan Province, China 764
Yurubchen-Tokhom Field, Eastern Siberia 710, 719, 767, 769

Zagros-Mesopotamia region 755, 763
Zagros Mountains, Iran 741, 859

Zakum Field, United Arab Emirates 867, 869
Zala Basin, Hungary 947
Zaonezhskaya Formation, Karalia, Russia 764
Zdanice well, Czech Republic 538, 541, 566, 586, 636, 637
zeaxanthin *see* carotenoid

Zechstein Basin, Germany
Zechstein Formation, Germany 814–17
Zeitz Formation, Germany 547
Zhungeer Basin, China *see* Junggar Basin
zigzag projection *see* structural notation

zinc ore 825
Zingjingzigou Formation *see* Jingjingzigou Formation
zircon 711
Zostera 489
zosterophyllophyte 727
ZSM-5 494
zygote 726

Printed in the United States
85163LV00003B